Modern Synthesis Processes and Reactivity of Fluorinated Compounds

Progress in Fluorine Science Series

Vol. 1 – "*Photonic and Electronic Properties of Fluoride Materials*", A. Tressaud, K. Poeppelmeir, Eds
Vol. 2 – "*New Fluorinated Carbons: Fundamentals and Applications*", O. Boltalina, T. Nakajima, Eds
Vol. 3 – "*Modern Synthesis Processes and Reactivity of Fluorinated Compounds*", H. Groult, F. Leroux, A. Tressaud, Eds

Modern Synthesis Processes and Reactivity of Fluorinated Compounds

Progress in Fluorine Science Series

Edited by

Henri Groult

Sorbonne Universités, UPMC, CNRS UMR 8234, Paris, France

Frédéric R. Leroux

University of Strasbourg, CNRS, UMR 7509, Strasbourg, France

Alain Tressaud

Emeritus Research Director, ICMCB-CNRS, University Bordeaux, Pessac, France

ELSEVIER

AMSTERDAM • BOSTON • HEIDELBERG • LONDON
NEW YORK • OXFORD • PARIS • SAN DIEGO
SAN FRANCISCO • SINGAPORE • SYDNEY • TOKYO
Academic Press is an imprint of Elsevier

Academic Press is an imprint of Elsevier
125 London Wall, London EC2Y 5AS, United Kingdom
525 B Street, Suite 1800, San Diego, CA 92101-4495, United States
50 Hampshire Street, 5th Floor, Cambridge, MA 02139, United States
The Boulevard, Langford Lane, Kidlington, Oxford OX5 1GB, United Kingdom

Library of Congress Cataloging-in-Publication Data
A catalog record for this book is available from the Library of Congress

British Library Cataloguing-in-Publication Data
A catalogue record for this book is available from the British Library

ISBN: 978-0-12-803740-9

For information on all Academic Press publications
visit our website at https://www.elsevier.com/

Publisher: John Fedor
Acquisition Editor: Katey Birtcher
Editorial Project Manager: Jill Cetel
Production Project Manager: Mohanapriyan Rajendran
Designer: Matthew Limbert

Typeset by TNQ Books and Journals

Contents

List of Contributors

F. Alloin University Grenoble Alpes, LEPMI, Grenoble, France; CNRS, LEPMI, Grenoble, France; Réseau sur le Stockage Electrochimique de l'Energie (RS2E), Amiens Cedex, France

B. Ameduri Institut Charles Gerhardt, UMR5253 (CNRS) E.N.S. Chimie Montpellier, Montpellier Cedex, France

C. Belhomme Areva, Pierrelatte Cedex, France

A.A. Berger Freie Universität, Berlin, Germany

T. Billard Institute of Chemistry and Biochemistry (ICBMS – UMR CNRS 5246), University Lyon 1, CNRS, Villeurbanne, France; CERMEP – In Vivo Imaging, Groupement Hospitalier Est, Lyon, France

D. Binet Sorbonne Universités, UPMC, CNRS UMR 8234, Paris, France

M. Braun Solvay Special Chem, Hannover, Germany

A. Carboni Institut de Chimie des Substances Naturelles, CNRS, Gif-sur-Yvette Cedex, France

H. Carreyre Université de Poitiers, CNRS UMR 7285 IC2MP, Poitiers Cedex 09, France

A. Colisson Areva, Pierrelatte Cedex, France

I. Crassous Areva, Pierrelatte Cedex, France

G. Dagousset Université de Versailles-Saint-Quentin, Versailles Cedex, France

D. Dambournet Sorbonne Universities, CNRS, UMR 8234, Paris, France

O. Danyliv University Grenoble Alpes, LEPMI, Grenoble, France; CNRS, LEPMI, Grenoble, France

D. Devilliers Sorbonne Universités, UPMC, CNRS UMR 8234, Paris, France

A.D. Dilman N.D. Zelinsky Institute of Organic Chemistry, Moscow, Russian Federation

J. Eicher Solvay Special Chem, Hannover, Germany

R. Fernandes Freie Universität, Berlin, Germany

H. Groult Sorbonne Universités, UPMC, CNRS UMR 8234, Paris, France

R. Hagiwara Kyoto University, Sakyo-ku, Kyoto, Japan

A. Higelin Freie Universität Berlin, Berlin, Germany

T. Honda Stony Brook University—State University of New York, Stony Brook, NY, United States

S. Huhmann Freie Universität, Berlin, Germany

N.V. Ignat'ev Merck KGaA, Darmstadt, Germany; Dr. Ignatiev – Chemistry Consultancy, Duisburg, Germany

C. Iojoiu University Grenoble Alpes, LEPMI, Grenoble, France; CNRS, LEPMI, Grenoble, France

M. Jiwei Ma Sorbonne Universities, CNRS, UMR 8234, Paris, France

H. Jönsson Lund University, Lund, Sweden

C.M. Julien Sorbonne Universités, UPMC, CNRS UMR 8234, Paris, France

E. Kemnitz Humboldt University of Berlin, Berlin, Germany

J.-H. Kim University of Fukui, Fukui, Japan

B. Koksch Freie Universität, Berlin, Germany

G. Kostov Bourgas As. Zlatarov University, Bourgas, Bulgaria

B.R. Langlois University of Lyon, Lyon, France; University Claude Bernard-Lyon 1, UMR 5246, Villeurbanne, France

F. Lantelme Sorbonne Universités, UPMC, CNRS UMR 8234, Paris, France

F.R. Leroux University of Strasbourg, CNRS, UMR 7509, Strasbourg, France

W. Li Sorbonne Universities, CNRS, UMR 8234, Paris, France

E. Magnier Université de Versailles-Saint-Quentin, Versailles Cedex, France

A. Mantoux Université Joseph Fourier, Saint Martin d'Hères, France

G. Masson Institut de Chimie des Substances Naturelles, CNRS, Gif-sur-Yvette Cedex, France

K. Matsumoto Kyoto University, Sakyo-ku, Kyoto, Japan

Z. Mazej Jožef Stefan Institute, Ljubljana, Slovenia

A. Mingot Université de Poitiers, CNRS UMR 7285 IC2MP, Poitiers Cedex 09, France

B. Morel Areva, Pierrelatte Cedex, France

J. Moschner Freie Universität, Berlin, Germany

M. Nappa Chemours Chemical Company, Wilmington, DE, United States

I. Ojima Stony Brook University—State University of New York, Stony Brook, NY, United States

S. Peng Chemours Chemical Company, Wilmington, DE, United States

G.K.S. Prakash University of Southern California, Los Angeles, CA, United States

F.-L. Qing Shanghai Institute of Organic Chemistry, Shanghai, China; Donghua University, Shanghai, China

S. Riedel Freie Universität Berlin, Berlin, Germany

N. Ryde Lund University, Lund, Sweden

G. Sandford Durham University, Durham, United Kingdom

G. Schäfer University of California, Berkeley, CA, United States

G. Scholz Humboldt University of Berlin, Berlin, Germany

X. Sun Chemours Chemical Company, Wilmington, DE, United States

M. Takashima University of Fukui, Fukui, Japan

S. Thibaudeau Université de Poitiers, CNRS UMR 7285 IC2MP, Poitiers Cedex 09, France

R. Thornbury University of California, Berkeley, CA, United States

F.D. Toste University of California, Berkeley, CA, United States

F. Toulgoat Institute of Chemistry and Biochemistry (ICBMS – UMR CNRS 5246), University Lyon 1, CNRS, Villeurbanne, France; CPE-Lyon, Villeurbanne, France

A. Tressaud Emeritus Research Director, ICMCB-CNRS, University of Bordeaux, Pessac, France

T. Umemoto Zhejiang Jiuzhou Pharmaceutical Co., Ltd., Taizhou, Zhejiang, China

S. Vukelić Freie Universität, Berlin, Germany

X.-H. Xu Shanghai Institute of Organic Chemistry, Shanghai, China

S. Yonezawa University of Fukui, Fukui, Japan

Z. Zhang University of Southern California, Los Angeles, CA, United States

Foreword

Most properties of fluorine and fluoride products derive from the dual character of this element: (1) so reactive, on the one hand, that it could neither be isolated as the F_2 molecule nor be found in nature for centuries and (2) yielding, on the other hand, very stable bonding with almost all other elements, especially with carbon. After the isolation of F_2 gas in 1886 by Henri Moissan, only in the 21st century was this gas found trapped on the Earth's surface in natural fluorite crystals. The cosmic synthesis of F_2 gas in stars or in stellar processes some 4.6 billion years ago will serve as a kind of introduction to the peculiar properties of fluorine chemistry nowadays.

This book is the third volume of the series "*Progress in Fluorine Science*" launched by Elsevier. The first two volumes concerned "*Photonic & Electronic Properties of Fluoride Materials*," and "*New Forms of Fluorinated Carbons*." This new volume deals with the synthesis and reactivity mechanisms of fluorine chemistry, illustrating the unique character of fluorinated products through examples taken from all classes of fluorides: organic, polymeric, and inorganic compounds, and focusing on the peculiarity of the synthetic routes. It is intended for a broad readership of professionals and researchers particularly interested in fluorine chemistry, ranging from Life Sciences to Material Sciences.

Fluorine has ceased to be considered as a laboratory curiosity since its discovery by Henri Moissan. At that time, nobody could imagine the importance of this discovery for the future life of humans. Indeed, despite being largely absent from natural products and biological processes, fluorine plays an increasingly important role in numerous areas of our daily life. Decades of chemical research have shown that fluorine atom and fluorine-containing motifs profoundly impact the structure, reactivity, and function of organic and inorganic molecules. Fluorine-containing compounds are nowadays synthesized in pharmaceutical, agrochemical, polymer, and electronic research on a routine basis. As an example, it is well established that fluorine atom(s) and/or fluoroalkyl group(s) can lead to many beneficial effects in biologically active molecules, such as enhanced metabolic stability, bioavailability, lipophilicity, and membrane permeability, as well as a strengthening of protein–ligand binding interactions. A logical consequence of these highly desirable properties is that more than 200 pharmaceuticals and 155 agrochemicals (among the 920 registered) containing at least one fluorine atom are currently on the market, which accounts for approximately 25% of the bioactive compounds. Material sciences are also deeply impacted by fluorine, as exemplified by the annual growth rate of the fluoropolymer field estimated at 6.5% through 2016.

Fluorine is an essential element in several industrial domains such as for the preparation of SF_6 (dielectric gas), NF_3 (cleaning agent in microelectronic), WF_6 chemical vapor deposition (CVD), etc.; it is also used in the nuclear industry for the preparation of UF_6 (uranium enrichment, 60% of the world production). More recently, nanofluorinated materials were synthesized, and much attention has been paid to their use in lithium batteries. Recently, F-ion batteries were proposed to possibly replace Li-ion battery.

In the 2000s, fluorine chemistry has greatly expanded with insightful contributions from research groups aiming at developing novel synthetic methods and reagents for the regio- and stereoselective introduction of fluorine or fluorine-containing groups into molecular scaffolds. Elegant stoichiometric and catalytic methods based on stable and economic fluoride and fluorine-containing precursors were devised. Classical reagents such as gaseous fluoroform or the Ruppert–Prakash reagent were tamed for efficient and practical fluorination processes. The discovery of new entities such as the electrophilic trifluoromethylation and trifluoromethylthiolation reagents or the interplay with new types of catalysis, such as dual catalysis and/or photocatalysis, led to highly efficient processes that opened new avenues for the academic and industrial communities. Indeed, fluorine chemistry field is still developing at a rapid pace.

In this volume, a particular attention has been paid to cover different aspects of fluorine chemistry by integrating contributions from industrial and organic synthetic chemists on topics such as innovative techniques like microfluidics, continuous flow, electro- and photochemical synthesis, drug discovery and medicinal chemistry, and fluoropolymers. In inorganic chemistry and material sciences, numerous original syntheses and applications have been also proposed, including sol–gel techniques, microwave-enhanced routes, photochemical syntheses in anhydrous HF, surface modifications using fluorinated reagents, synthesis of new fluorinated ionic liquids, stabilization of transition metals at high oxidation states, issues related to electrochemical synthesis of F_2 and stability in $KF \cdot 2HF$ medium, etc.

Therefore, the wide scope of entries occurring in this book, ranging from new syntheses, reactivity mechanisms, and outstanding physical and chemical properties to industrial applications, will illustrate how fluorides and fluorinated products may have strong positive impacts on various aspects of modern life, such as new drugs, health care, agrochemicals, refrigerants, catalytic processes, energy storage through batteries and fuel cells, etc. One of the goals of this book is to show the reader how innovation in synthesis and a better understanding of reaction mechanisms may provide efficient tools to incorporate fluorine in various aspects of modern academic and industrial research. Although benefits of fluoride molecules and materials for our society are evident in many fields, drawbacks caused to our environment by some fluorinated products should not be underestimated, as already pointed out in previous books of the "*Advances in Fluorine Science*" series, Elsevier (2006), and adapted solutions should be greatly welcome.

In this volume, researchers from public and private laboratories reveal to the reader an up-to-date overview on synthetic fluorine chemistry within 25 chapters, covering all aspects from their respective research. We wish to express our deepest gratitude to all the authors who contributed to this volume.

We sincerely hope that this volume will become a useful tool and inspire numerous researchers in academia and industry to further develop the importance and role of fluorine.

Henri Groult, Frédéric R. Leroux, and Alain Tressaud
Paris, Strasbourg and Bordeaux, April 2016

Chapter One

The Cosmic Origin of Fluorine: An Astronomer's View on Fluorine Synthesis

H. Jönsson, N. Ryde

Lund University, Lund, Sweden

Only the three very lightest elements were produced in the Big Bang, when our universe was created. All other elements have been, and continue to be, produced by stars of different types. The major formation sites of most elements and isotopes in the periodic table have been established over the years, but there are still a number of elements whose cosmic origin is elusive, fluorine being one of them.

Besides being interesting in its own right, the origin of the elements, once known, can be used to answer a plethora of astronomically interesting questions. The knowledge of what kind, or kinds, of star(s) particular key elements were formed in is necessary to be able to answer questions such as how galaxies formed, how our galaxy and its subcomponents formed, whether the very first stars were different from the stars being forming currently, and whether stars formed in different regions of our galaxy have different properties. Analyzing stars of different ages in a stellar system can therefore tell us, for example, how quickly a stellar population formed, and whether it formed from a common gas cloud in one place in space or whether it was built up by the merging of smaller, individually born, stellar populations. Indeed, one of the areas of most intense research in astronomy today is galaxy formation. One way of understanding this is through determining the abundances of different key elements in stars in different parts of our own galaxy, the Milky Way. From this, the formation and evolution of the different galactic subparts can be inferred and the puzzle of galaxy

Modern Synthesis Processes and Reactivity of Fluorinated Compounds. http://dx.doi.org/10.1016/B978-0-12-803740-9.00001-9

formation can be solved. It turns out that fluorine might have a part to play in understanding how the important innermost parts of our galaxy, the Bulge, formed and evolved. However, before fluorine can be used to trace galaxy formation, astronomers need to find out the more basic question of how fluorine is produced in the universe.

1. The Cosmic Origin of the Elements

All chemical elements, except the very lightest ones (i.e., hydrogen, helium, and lithium), that we come in contact with everyday, have been produced in different nuclear processes in different types of stars at different stages of evolution. The newly synthesized elements inside a star can either be locked up inside the star or stellar remnant or they can be expelled into the interstellar medium. In the latter case, a new generation of stars, with a different elemental composition than their parents, can be formed from such polluted interstellar gas clouds. This means that, in general, the younger a star is, the more heavy elements it contains. Astronomers call this concept chemical evolution.

The interior of a star, where the nuclear reactions take place, is hot, at least 10 million degrees, and gives off an electromagnetic blackbody spectrum. The cooler, outer parts of the star, called the stellar atmosphere, contain atoms, ions, and molecules that can get excited by absorbing the radiation from the inner parts of the star. By identifying these absorption lines, recorded in high-resolution stellar spectra, we can infer what elements are present in these outer layers of the star. If then the physical parameters of the star can be determined, such as its temperature, surface gravity, and age, stellar atmosphere models can be used to infer the atmospheric abundance of the elements for which absorption lines can be seen.

In most cases, the elements synthesized by the star in its nucleus do not pollute the atmosphere. Thus, for most types of "normal" stars, the abundances of elements in their atmospheres represent the primordial abundances, which means that the measured abundances represent the abundances of the elements in the interstellar medium at the time and place the star was formed.

Therefore, by choosing an ensemble of stars representing different epochs of the evolution of the Milky Way and of different stellar populations, the cosmic buildup and origin of an element can be deduced. Through these types of investigations the cosmic origin of many elements have been determined: for example, the astronomically common element iron is produced in supernova explosions types Ia and II, whereas oxygen is mainly produced in type II explosions. Fluorine, on the other hand, is one of the few elements whose origin in the cosmos is still debated.

2. The Cosmic Origin of Fluorine

Despite it being astrophysically interesting, astronomers have still not been able to precisely determine the cosmic origin of fluorine. The atomic nucleus of fluorine is fragile and is easily destroyed in nuclear processes in stellar interiors, which means that to produce fluorine that survives to the next generation of stars it has to be formed in a region of the star where the energy is high enough to initiate nuclear processes, but at the same time allow for the fluorine created to quickly be transferred to a region where it can survive. This sensitivity to the physical conditions of its production site means that measured fluorine abundances can set tight constraints on models of stellar evolution and also on the chemical evolution of certain stellar populations.

The cosmic origin of fluorine is uncertain mainly because of problems in detecting fluorine in stellar atmospheres using the most common approaches: there are no suitable spectral lines available for atomic or ionized fluorine in "normal" stars, and instead, fluorine abundance has to be inferred via the HF molecule, using spectral absorption lines around 2.3 μm. This in turn means that relatively cool stars need to be observed: the HF molecules dissociate in stars of rather moderate temperatures. For example, absorption lines of HF cannot be detected in the "normal," undisturbed atmosphere of the Sun, but only in the much cooler, magnetically induced sunspots. This means that to measure the fluorine abundance in a star, one has, in three ways, to step outside the comfort zone of most stellar spectroscopists: the moderately hot sun-like stars often used in stellar abundance determinations cannot be used. Spectral lines from a molecule as opposed to the more often used atomic or ionized spectral lines have to be used, and spectroscopic observations have to be carried out in the near-infrared (IR) region instead of the visual wavelength region often investigated.

2.1 *Theoretical Ideas*

Even if the cosmic origin of fluorine is puzzling, astronomers have narrowed down the possibilities to one or more of three different sources:

- Asymptotic Giant Branch stars, stars similar to the Sun but slightly heavier (2.5−4 times heavier) in a later evolutionary stage, produce fluorine via an intricate chain of nuclear reactions, including α, neutron, and proton captures, starting with ^{14}N. The synthesized fluorine is then saved from destruction by being transferred to the cooler, outer layers of the star via simultaneous thermally induced pulsations of the entire star. The production of fluorine by Asymptotic Giant Branch stars has been experimentally proved numerous times in different types of observations, for example, in the very first detection of fluorine in stellar atmospheres by Jorissen et al.[3]
- Core-collapse supernovae, also called type II supernovae, which are the explosions ending the lives of massive stars, are suggested to form fluorine via the exotic neutrino process: even though the cross-section for neutrinos to react with atomic nuclei is very small, the enormous neutrino flux created when the star is imploding to form a black hole has been theoretically suggested to be able to turn a significant fraction of the ^{20}Ne present in the outer layers of the collapsing star into fluorine by ejecting a proton or a neutron and a subsequent beta decay.[8]
- Wolf−Rayet stars are very massive stars (25−120 times the mass of the Sun) that will end their lives as a type II supernova, and before that experience a large mass loss due to a radiatively driven stellar wind. This means that they shed their outer envelopes, and any fluorine produced might be saved from destruction by this process.[5] Also here, fluorine is synthesized by nuclear reactions starting from ^{14}N, including α, neutron, and proton captures.

To conclude, it has been verified that the Asymptotic Giant Branch stars produce fluorine, but the question is whether any of the other processes would produce a significant amount of fluorine and/or whether they even could be the dominant source in the galactic chemical evolution of fluorine. Astronomers are still struggling to assess this point. By investigating stellar populations formed at different times it would be possible to determine the relative contribution of the different production channels, since these would dominate at different epochs. Stellar spectra of relevant objects are, however, still very challenging to record.

If it turns out that the Wolf–Rayet stars do produce significant amounts of fluorine, fluorine might be used as a proxy for the fraction of massive stars that has existed in a stellar population. For example, there exist several indications that the central parts of our galaxy, the Bulge, might have had a history of a larger fraction of massive stars compared with the solar neighborhood.[7] Comparing the chemical evolution of fluorine in the Bulge environment with that of the solar neighborhood would empirically test such a claim, given that the Wolf–Rayet stars do produce fluorine.

2.2 Observations

Over the years, there have been claims both for and against all three proposed production sites. Not many observations, however, have been made due to the difficulties in observing in the near-IR region. The spectral HF lines used in the analysis of stellar fluorine abundances are vibration–rotational lines around 2.3 μm. Recently, we also explored pure rotational lines of the HF molecule around 12 μm, observed with a mid-IR high-resolution spectrometer.[2] In that *Letter*, we also analyzed 25-year-old, near-IR spectra covering the 2.3-μm lines, recorded with the Fourier transform spectrometer at Kitt Peak National Observatory outside of Tucson in the United States. The analysis of the seven stars in the investigation indicates that the fluorine production by Asymptotic Giant Branch stars is enough to explain the abundances that we found. This conclusion was reached by comparing our fluorine, iron, and oxygen abundances and their ratios to those predicted by theoretical models of how the proportion of elements is expected to change over time as the cosmic cycle progresses.[4] Since all the stars in this investigation were of similar age, the next step would be to investigate both older and more recently formed stars. What we could conclude was that in the solar neighborhood and on the Earth, red giant stars seem to be able to produce most of the fluorine.

Since our publication, two articles on the creation of fluorine in the universe have been published with conclusions contradicting ours. A few interesting points have been raised, which open the question of the creation of fluorine in the universe again.

The first article[1] presents new models of how fluorine is expected to evolve if only produced in Asymptotic Giant Branch stars. These are very different from the models that we used to draw our conclusions.[4] These new models imply that to reach the fluorine abundances we measured in our article, one or both of the other two production sites would also be needed. These results demonstrate the uncertainties in current galactic chemical evolution models and call for a purely empirical and model-independent investigation of the cosmic origin of fluorine.

The second article[6] observed a larger sample of 51 stars, but still having an approach similar to ours. Due to the limited wavelength coverage of the spectrometer they used, they were able to determine fluorine abundances, but neither the iron nor the oxygen abundances for their stars. This limitation in wavelength coverage also meant that they were not able to determine the temperatures of the stars themselves, a parameter that is critical for the analysis, but were instead forced to take the temperatures, surface gravities, and iron abundances for their stars from different literature sources, unambiguously introducing systematic uncertainties in their sample. These uncertainties lead to a rather large spread in their fluorine evolutionary trend, but their results seem to indicate, nevertheless, that the neutrino process in type II supernovae is responsible for most of the fluorine found in the solar neighborhood.

2.3 What the Future Holds

To settle these issues, we have recently obtained, and are currently analyzing, a large set of fluorine measurements in stars. These observations were carried out during the summer of 2015 using the Phoenix spectrometer mounted on the 4-m Mayall telescope at Kitt Peak National Observatory and the FIES spectrometer at the Nordic Optical Telescope, on La Palma, Canary Islands. The near-IR Phoenix spectra are centered on the 2.3-μm HF line to allow for fluorine abundance determination. From the visual FIES spectra we are able to, homogeneously, determine the temperatures, surface gravities, and iron and oxygen abundances, which means that we can expect smaller uncertainties in our abundance ratios compared with earlier studies. Furthermore, the full optical spectral coverage will make it possible to determine abundances of several other elements with different known formation sites, allowing for a completely empirical, model-independent investigation. Preliminary results are very interesting and surprising, possibly indicating that we might know even less than we thought we did, regarding the elusive cosmic origin of the element that this entire book is dedicated to.

References

1. Abia, C.; Cunha, K.; Cristallo, S.; de Laverny, P. The Origin of Fluorine: Abundances in AGB Carbon Stars Revisited. *Astron. Astrophys.* **September 2015,** *581,* A88.
2. Jönsson, H.; Ryde, N.; Harper, G. M.; Richter, M. J.; Hinkle, K. H. Fluorine in the Solar Neighborhood: Is It All Produced in Asymptotic Giant Branch Stars? *Astrophys. J. Lett.* **July 2014,** *789* (2), L41.
3. Jorissen, A.; Smith, V. V.; Lambert, D. L. Fluorine in Red Giant Stars − Evidence for Nucleosynthesis. *Astron. Astrophys.* **July 1992,** *261,* 164−187.
4. Kobayashi, C.; Izutani, N.; Karakas, A.; Yoshida, T.; Yong, D.; Umeda, H. Evolution of Fluorine in the Galaxy With the ν-Process. *Astrophys. J. Lett.* **October 2011,** *739* (2), L57.
5. Meynet, G.; Arnould, M. Synthesis of ^{19}F in Wolf-Rayet Stars. *Astron. Astrophys.* **March 2000,** *355,* 176−180.

6. Pilachowski, C. A.; Pace, C. The Abundance of Fluorine in Normal G and K Stars of the Galactic Thin Disk. *Astron. J.* **September 2015,** *150* (3), 66.
7. Rich, R. M. The Galactic Bulge. *Planets* **2013,** 271.
8. Woosley, S. E.; Haxton, W. C. Supernova Neutrinos, Neutral Currents and the Origin of Fluorine. *Nature* **July 1988,** *334* (6177), 45–47. ISSN:0028-0836.

Chapter Two

The Fluorine Atom in Health Care and Agrochemical Applications: A Contribution to Life Science

M. Braun, J. Eicher

Solvay Special Chem, Hannover, Germany

Modern Synthesis Processes and Reactivity of Fluorinated Compounds. http://dx.doi.org/10.1016/B978-0-12-803740-9.00002-0

1. Introduction

More than 50% of all modern pharmaceutical and agrochemical compounds contain fluorine.[71]

Several unique properties introduced by fluorine, the most electronegative atom, e.g., into a drug, are quite well known. These are, for example:

- metabolic stability (substitution of the oxidizable C–H group by a C–F group),
- change in basicity (better membrane permeation, and as a result better bioavailability),
- higher lipophilicity, and
- higher polarity.

1.1 The History of Fluorine in Pharmaceutical Applications

Until 1957, no F-containing drug had been developed. A strong increase in the number of fluorinated drugs became noticeable in the 1980s. The introduction of trifluoromethyl groups in active pharmaceutical ingredients was the first step in the ongoing active trend toward more fluorine in life science.

The consequence was that after the phasing out of halon as a fire-extinguishing agent and of chlorofluorocarbons (CFCs) as refrigerants and foam-blowing agents in many countries, following the need for hydrofluorocarbons (HFCs) and hydrochlorofluorocarbons (HCFCs), the highest fluorine demand in industrial volumes was the one for the anesthetics isoflurane, desflurane, and sevoflurane (see Fig. 2.1).

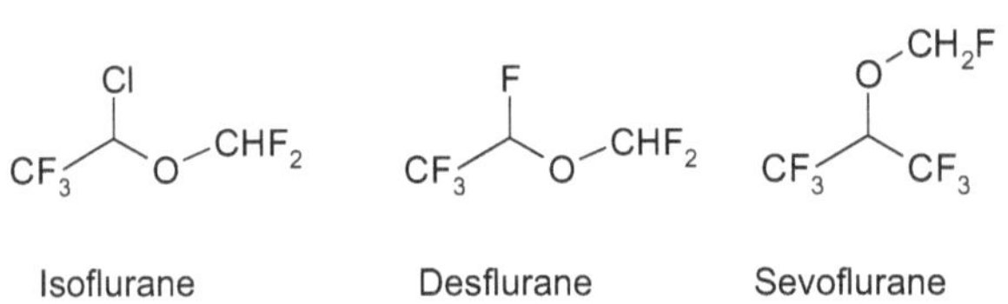

Figure 2.1 The most important anesthetics.[55]

Nevertheless, some CFCs remained as precursors for intermediates for life science products. In the early 1990s, in addition to the anesthetics, the antimalaria drug Lariam (Mefloquine, see Fig. 2.2), developed by Hoffman La Roche in Switzerland,[26,66,70] was almost the only drug that made it to the market.

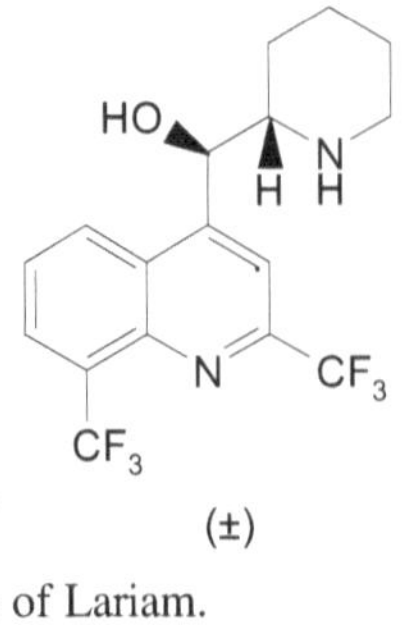

Figure 2.2 The molecular structure of Lariam.

Some of the early top-selling active pharmaceutical ingredients were Eli Lilly's antidepressant Prozac,[74] Pfizer's cholesterol-lowering drug Lipitor,[56] and Bayer's antibiotic drug Ciprobay (Fig. 2.3).[24] Since 1957, around 150 active ingredients have been commercialized. In today's development pipeline, around one-third of the investigated substances or processes contain fluorine or involve fluorinated organic compounds as a protecting group (e.g., CF_3-CO-), as a catalyst (triflic acid and derivatives[30,54,67]), or even as green solvents [CF_3-toluene, i.e., 1,1,1,3,3-pentafluorobutane, (HFC-365mfc)[61–64]].

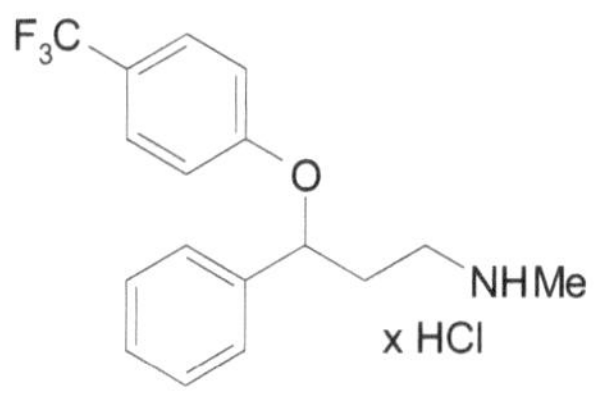

Fluoxetine (Prozac®/Sarafem®)

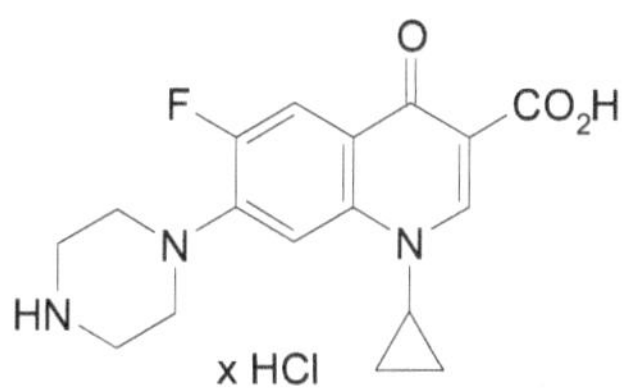

Ciprofloxacin (Ciprobay®)

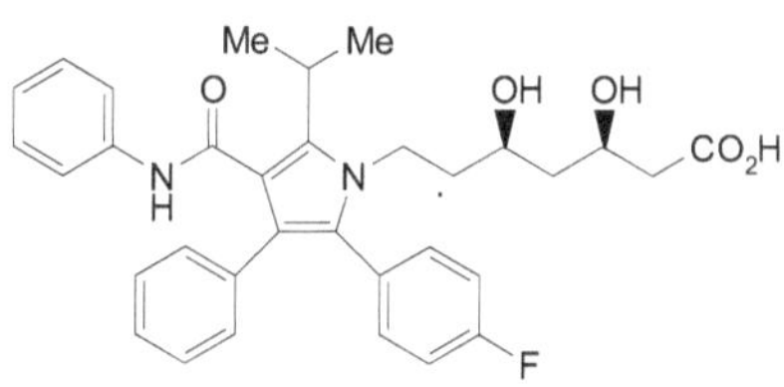

Atorvastatin (Lipitor®)

Figure 2.3 Early examples of commercial fluorinated active pharmaceutical ingredients.

1.2 The History of Fluorine in Agrochemical Applications

The trend to use more fluorine for agrochemicals also began later, probably due to pricing issues. Compared with the agrochemical industry the pharmaceutical industry was able to spend more money on funding the necessary R&D work including toxicity investigations.

For fluorine atoms in agrochemicals the development of new leads was and is still very empirical (trial and error); however, a very strong trend toward even more fluorine use in the big agrochemical industry's development pipeline is also noticeable.[14] The percentage of fluorinated versus nonfluorinated compounds (approximately 50%) is almost the same for fungicides, herbicides, and insecticides.

At the beginning of the "fluorine century," the focus in agrochemistry was on reasonably easily mastered technologies for the formation of the C−F bond. Key structures were fluorobenzene moieties as in Novalurone,[52] benzotrifluoride derivatives as in parafluron,[25] and CF_3 pyrethroid structures as in cyhalothrin (Fig. 2.4).[2,39,40,45,57]

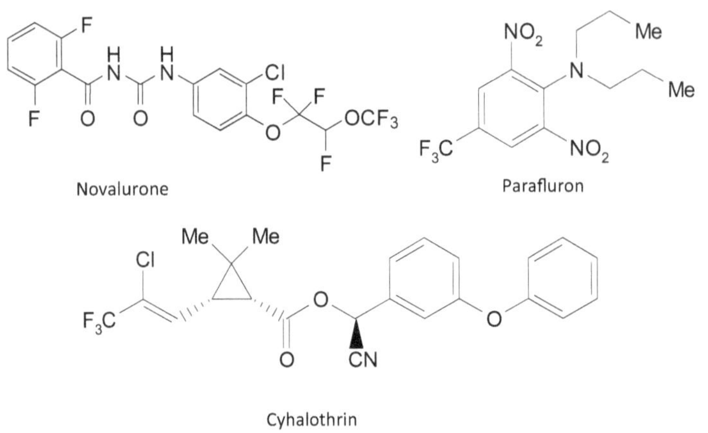

Figure 2.4 Early examples of commercial active agrochemical ingredients including fluorine atoms.

2. Methods to Introduce Fluorine Into Organic Molecules

For fluorobenzene derivatives the Balz–Schiemann reaction[23,53] was applied, using the respective phenyldiazonium tetrafluoroborate derivative, or simply the Halex reaction[19], using spray-dried KF as a fluorine source for halogen exchange reactions in polar aprotic solvents. Precipitated KF with a residual moisture content does not work efficiently in Halex reactions.

The skill of handling anhydrous HF became an obvious requirement to succeed in converting CCl_3 groups into CF_3 groups and to synthesize benzotrifluorides very fast.

The CF_3 group in the cyclopropyl moiety of cyhalothrin was and is still partly made by using the so-called CFC-113a (CF_3–CCl_3) as a starting material. CFC-113a is already the result of an $AlCl_3$-induced isomerization of CFC-113 (CF_2Cl–CCl_2F), which is obtained from perchloroethylene (PCE) and anhydrous HF. Due to the phasing out of CFC as a result of the Montreal protocol, the synthesis of the 2-chloro-1,1,1-trifluoromethyl propenyl cyclopropyl carboxylic acid moiety is more often based on HCFC-123 (CF_3–$CHCl_2$), which remains an intermediate for the synthesis of the refrigerant component HFC-125 (CF_3–CHF_2), and as a raw material for trifluoroacetyl chloride (TFAC), as described later.

2.1 *Fluorinating Agents*

There are many nucleophilic as well as electrophilic fluorinating agents in the market. The most firmly established ones are the nucleophilic Olah's reagent,[49] a mix of 30% pyridine and 70% HF; diethylaminosulfur trifluoride (DAST)[17]; and Deoxo-Fluor [bis-(2-methoxyethyl)aminosulfur trifluoride],[42] which are derivatives of SF_4 that are "safe and easier to handle." Frequently used electrophilic fluorinating agents are

Selectfluor,[10] *N*-fluorobenzenesulfonimide,[1] and *N*-fluoropyridinium tetrafluoroborate, or triflate,[38,72] besides many others (Fig. 2.5).

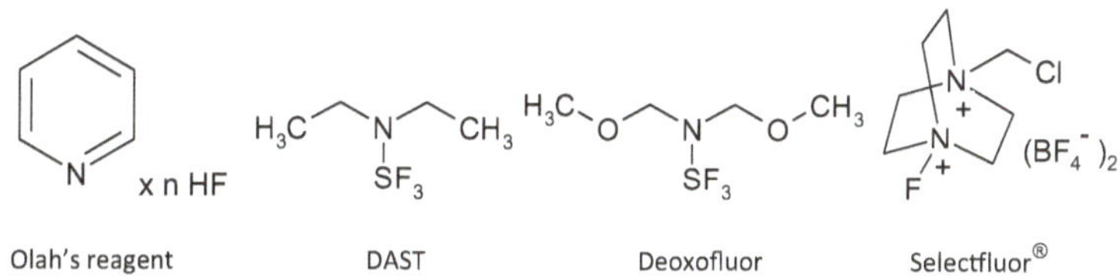

Figure 2.5 Some examples of fluorinating agents.

Except for the pyridine × *n*HF, which meets the criteria of "availability" in large quantities and "reasonable price," most other fluorinating agents are applied for lead screening purposes, but failed due to the already mentioned pricing and lack of availability at the industrial scale. At present, due to some emerging projects, SF_4 is becoming more attractive for larger scale applications.

2.2 Use of Anhydrous HF

The first applied method using anhydrous HF is its addition to a C−C double bond in a first step, followed by additional halogen−fluorine exchange reactions. This type of reaction is often catalyzed by Lewis acids and is among the best industrially feasible options for connecting a fluorine atom to carbon. With this technology, CF, CF_3, and CF_2 moieties are accessible, such as 1,1-difluoro-1,2,2-trichloroethane (HCFC-122) and 1,1,1-trifluoro-2,2-dichloroethane (HCFC-123).[6,15] They serve as intermediates for the so-called building block approach to obtain pharmaceutical and active agrochemical ingredients (Fig. 2.6).

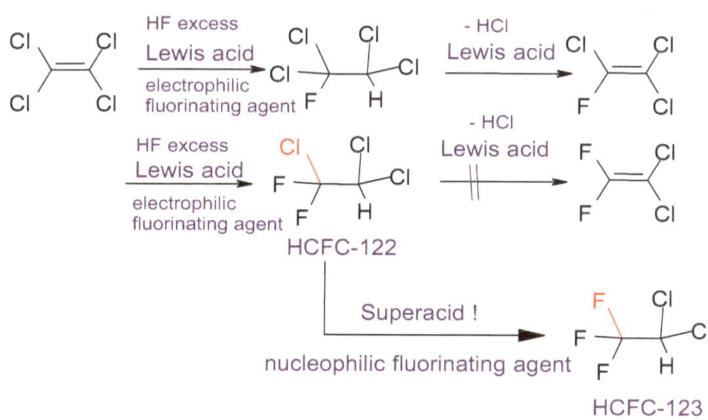

Figure 2.6 The synthesis of HCFC-122 and HCFC-123.

Another very interesting example of the use of anhydrous HF for different types of C−F bond formation is the synthesis of HFC-365mfc from pentachlorobutane

(Fig. 2.7). HFC-365mfc is a compound mainly used as a foam blowing agent, but it is also applied as a green solvent [365mfc: boiling point (BP) 40°C, close to that of CH_2Cl_2, BP, 39.7°C] in organic chemistry. The synthetic access route to HFC-365mfc follows a combination of two different mechanisms, which are:

1. the nucleophilic replacement of chlorine by fluorine atoms and
2. the addition of HF to double bonds formed by a Lewis acid–initiated HCl elimination.[4,50,73]

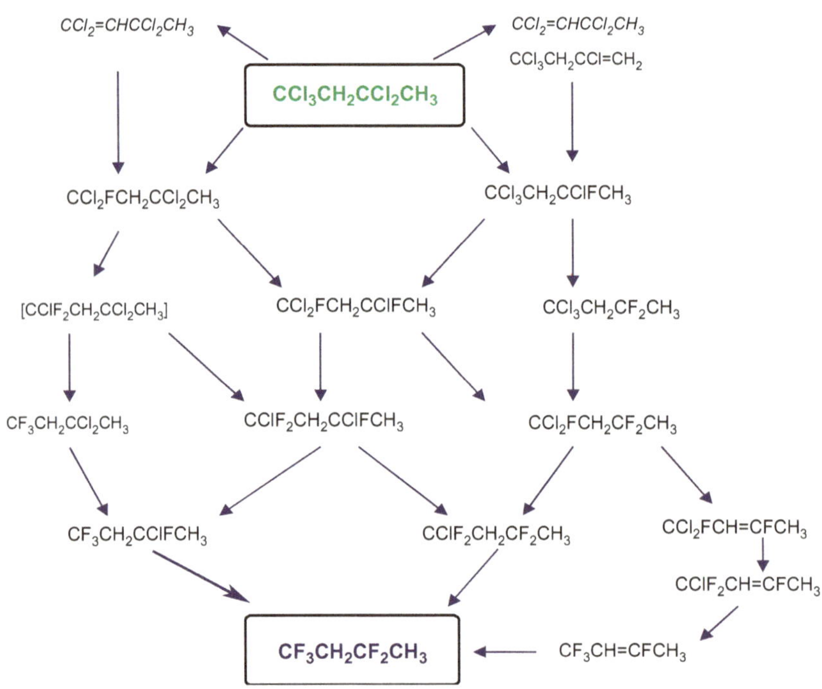

Figure 2.7 The synthesis of 1,1,1,3,3-pentafluorobutane, HFC-365mfc.[50]

Anhydrous HF is also applied as a solvent or coreagent in combination with sulfur tetrafluoride (SF_4) to form, e.g., a CF_3 group from an ester or acid function. Analogous reactions for the synthesis of CF_2 groups from aldehydes[12] and CF groups from alcohols were also published.[65]

The reaction described in Fig. 2.8, starting from the intermediate diethyl 2-(fluoromethoxy)malonate was developed for the investigation of an alternative route for the production of 1,1,1,3,3,3-hexafluoro-2-(fluoromethoxy)propane (Sevoflurane).[7]

As polar organic compounds are soluble in anhydrous HF, this opens up the possibility of using electrochemical synthesis to generate polyfluorinated or perfluorinated organic materials. The method of electrochemical fluorination at 4–8 V can even be applied to suspensions of insoluble organic compounds in HF if sufficient conductivity is induced by the addition of an electrolyte.[11]

A material applied in lithium ion battery technology more in 2010 is bis(pentafluoroethyl) phosphinic acid, which can also be obtained by electrochemical fluorination.[31]

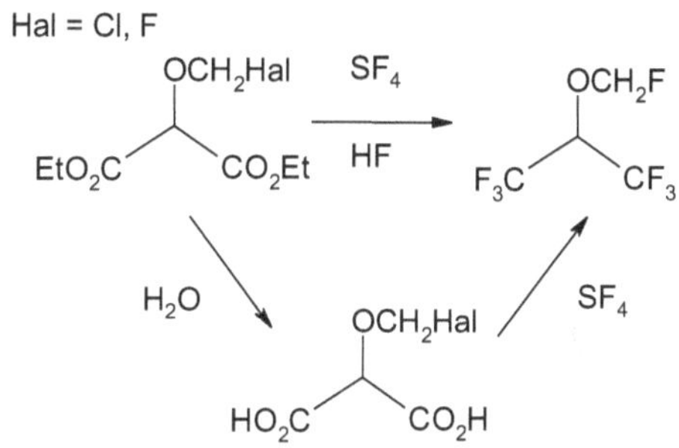

Figure 2.8 Fluorination step with SF_4 in anhydrous HF.

2.3 Indirect Fluorination: Use of a Building Block Containing a Fluorinated Group

There are four principal classes of lead structures that are used as fluorinated organic intermediates (building blocks) in life science.

These are:

1. the large and much more easily obtainable class of fluorobenzenes,
2. benzotrifluorides (not further discussed here),
3. the aliphatic CF_3, CF_2, and C−F compounds, and
4. chiral building blocks.

Almost all fluorine-containing aliphatic building blocks are downstream products of technologies described in the following two sections.

3. Use of Trifluoromethyl Group in Pharmaceuticals and Agrochemistry

As outlined previously, there are attractive alternatives to obtain CF_3 components relatively cheaply.

The two key compounds to start with are PCE and HCFC-123. The CF_3 building blocks can be synthesized via photooxidation of both substances.

3.1 Synthesis of Trifluoroacetyl Chloride and Trifluoroacetic Acid

The synthesis of TFAC from trifluorodichloroethane (HCFC-123) using chlorine-sensitized photooxidation[5] with O_2 is currently the most environment-friendly route. Only light is used as a catalyst, besides O_2 as a green oxidant. Only HCl is formed as a by-product. The HCl formed is also a commercial product. This technology could be used for the synthesis of a significant number of products containing aliphatic CF_3 groups that enter into the synthesis of many life science blockbusters. Numerous new projects are being developed (Fig. 2.9).

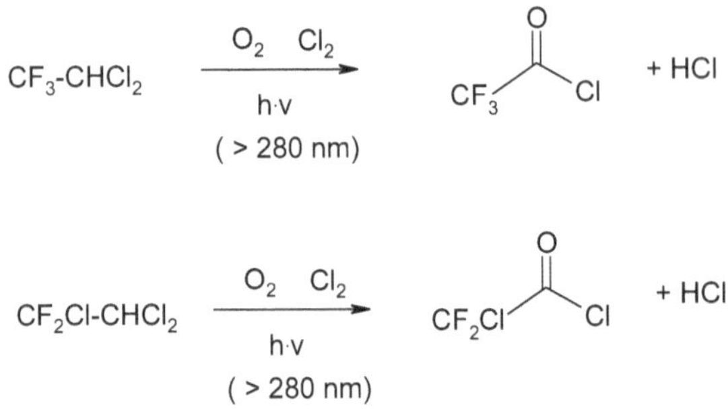

Figure 2.9 The synthesis of trifluoroacetylchloride (TFAC) and chlorodifluoroacetyl chloride (CDFAC).

As an alternative, starting with PCE is also an attractive option to obtain CF_3 compounds (Eqs. (2.1)–(2.3)). The first step is the photooxidation to trichloroacetylchloride, followed by fluorination with HF to obtain trifluoroacetylfluoride.

$$Cl_2C{=}CCl_2 + {}^1\!/_2\, O_2 \rightarrow Cl_3C{-}COCl \tag{2.1}$$

$$Cl_3C{-}COCl + 4HF \rightarrow CF_3COF + 4HCl \tag{2.2}$$

$$CF_3COF + H_2O \rightarrow CF_3COOH + HF \tag{2.3}$$

After hydrolysis the trifluoroacetic acid (TFA) obtained can serve as a starting material for further downstream chemistry.[3,21,22,51] Reactions such as esterification, decarboxylation, and sulfination are typical examples of this downstream chemistry. Fig. 2.10 gives an overview of the resulting CF_3 products.

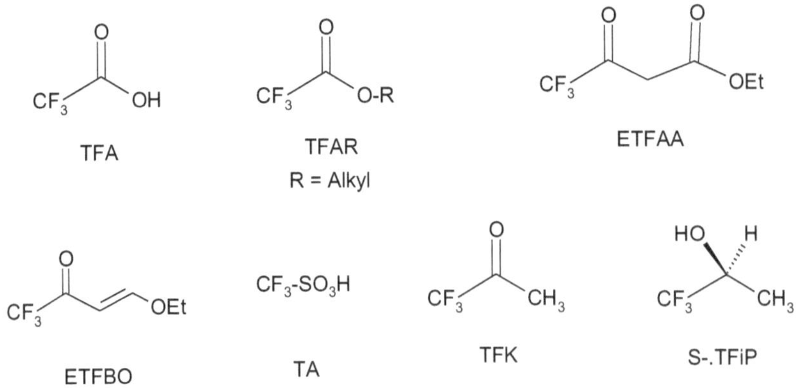

Figure 2.10 Examples of commercially available aliphatic CF_3 building blocks.

3.2 Some Active Agrochemical Ingredients Based on Aliphatic Trifluoromethylated Building Blocks

The CF_3 group connected to the thiodiazole in Bayer's herbicide flufenacet (Axiom) is made from TFA following the reaction sequence shown in Fig. 2.11.[20]

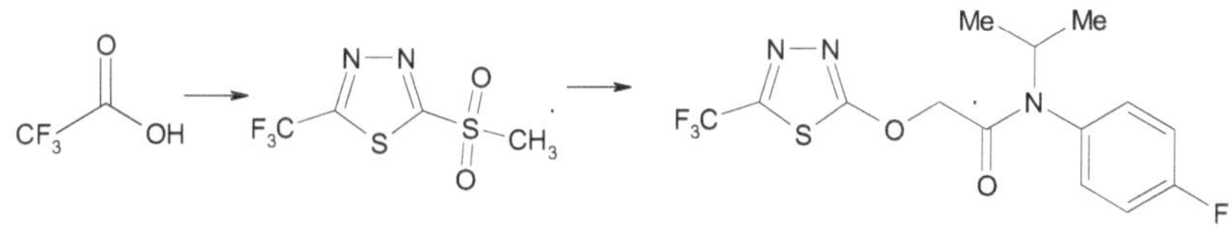

TFA Flufenacet

Figure 2.11 The synthesis of flufenacet. *TFA*, trifluoroacetic acid.

In Dow's Thiazopyr, the classical Hantzsch pyridine synthesis is used to cyclize two molecules of ethyl trifluoroacetoacetate to form a pyridine derivative with fluorinated carboxylic acid groups in the α-position.[16] Fig. 2.12 shows an overview starting from TFAC.

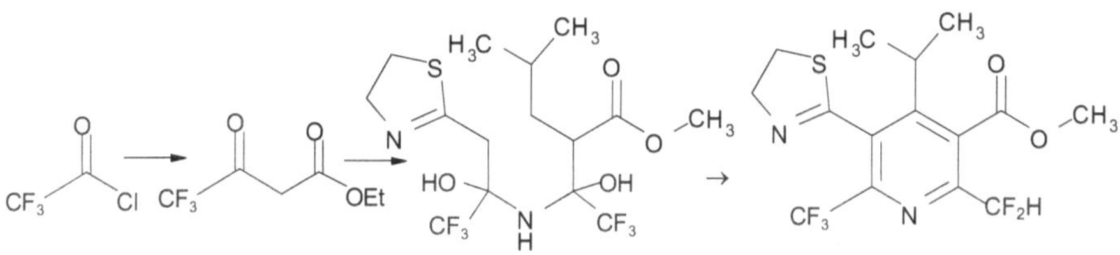

TFAC ETFAA Thiazopyr

Figure 2.12 Synthesis of Thiazopyr. *ETFAA*, ethyl trifluoroacetoacetate; *TFAC*, trifluoroacetyl chloride.

Among the building blocks that are used in industrial scale, 4-ethoxy-1,1,1-trifluoro-3-butene-2-one (ETFBO) is the newest. ETFBO offers an advantage especially in the synthesis of CF_3-containing heterocyclic compounds,[59] which also used to be synthesized by classical chlorination of the corresponding CH_3 hererocycle, followed by a reaction with HF. But that old classical way usually leads to side products as a consequence of the chlorination and fluorination of other functional groups present in the molecules, decreasing the economic viability of any given project in the end. Syngenta's bicyclopyrone[34,43] makes use of these special abilities of ETFBO (Fig. 2.13).

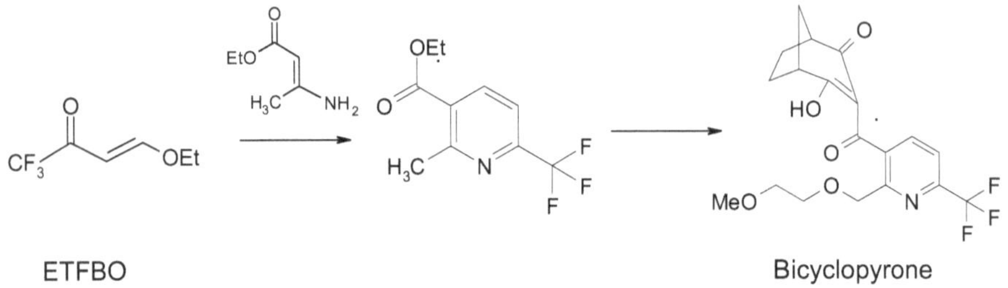

Figure 2.13 Example for 4-ethoxy-1,1,1-trifluoro-3-butene-2-one (ETFBO) as a starting material.

3.3 Some Active Pharmaceutical Ingredients Based on Aliphatic Trifluoromethylated Building Blocks

Sitagliptin is a drug developed by Merck[13,32,37] for the treatment of type 2 diabetes. The fluorinated starting material trifluoroacetic acid ethylester has been readily available for decades. In the last stage of the synthesis, amination to a crotonic acid derivative followed by Rh-catalyzed hydrogenation[75] yields the R-configuration at the amino group in sitagliptin (Fig. 2.14).

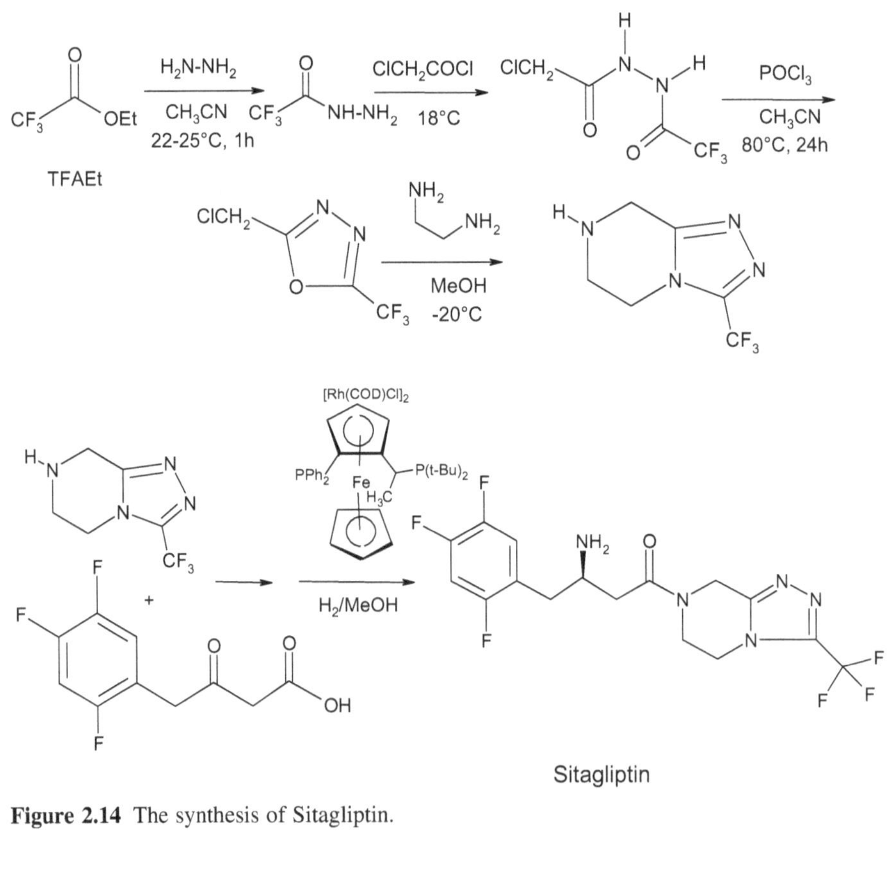

Figure 2.14 The synthesis of Sitagliptin.

As demonstrated in Fig. 2.15, the use of ETFBO opens up a very convenient route for the synthesis of 2-CF_3-5-aminopyridine, which is used for the preparation of Pradigastat, a new drug from Novartis also developed for the treatment of type 2 diabetes.[60,76]

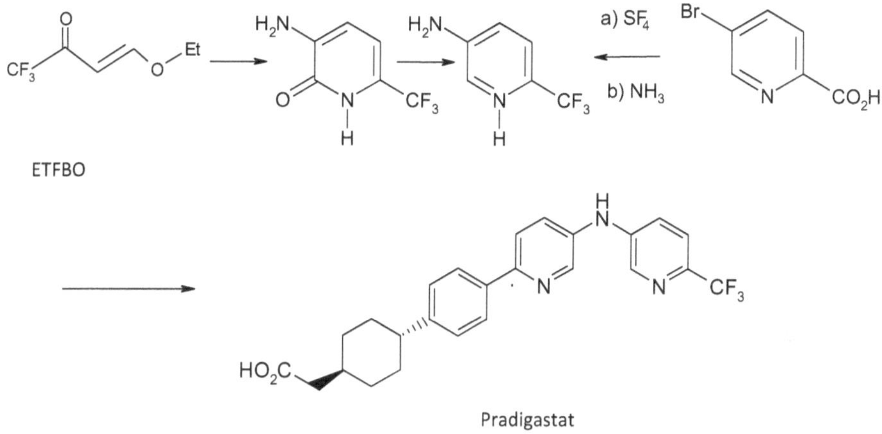

Figure 2.15 The synthesis of Pradigastat. *ETFBO*, 4-ethoxy-1,1,1-trifluoro-3-butene-2-one.

The intermediate 2-CF_3-5-aminopyridine shown can also be obtained from the corresponding pyridine-2-carboxylic acid via fluorination with SF_4, but with some drawbacks in connection with the availability of SF_4[60,76] (or DAST/Deoxo-Fluor) at the industrial scale, besides some additional waste management challenges.

4. Use of Difluoromethyl Group in Pharmaceuticals and Agrochemistry

The use of the CF_2 moiety containing active ingredients essentially began with the anesthetics isoflurane and desflurane, mentioned earlier. For their production, trifluoroethanol is reacted with difluorocarbene, which is obtained from chlorodifluoromethane (HCFC-22), to yield the corresponding difluoromethoxy trifluoromethylether. The generation of difluorocarbene from HCFC-22 is very well known, because monomeric tetrafluoroethylene (TFEe) is also produced by pyrolysis in that compound.[68]

Active ingredients launched early in this development are Nihon Nohyaku's herbicide Pyraflufen-ethyl[47]; the former Ciba Geigy's herbicide Primisulfuron-methyl,[46] which is now produced by several companies based in Asia; and last but not least, an antihypertensive drug developed by Latvian (Russia) called Riodipine (Fig. 2.16).[18,36]

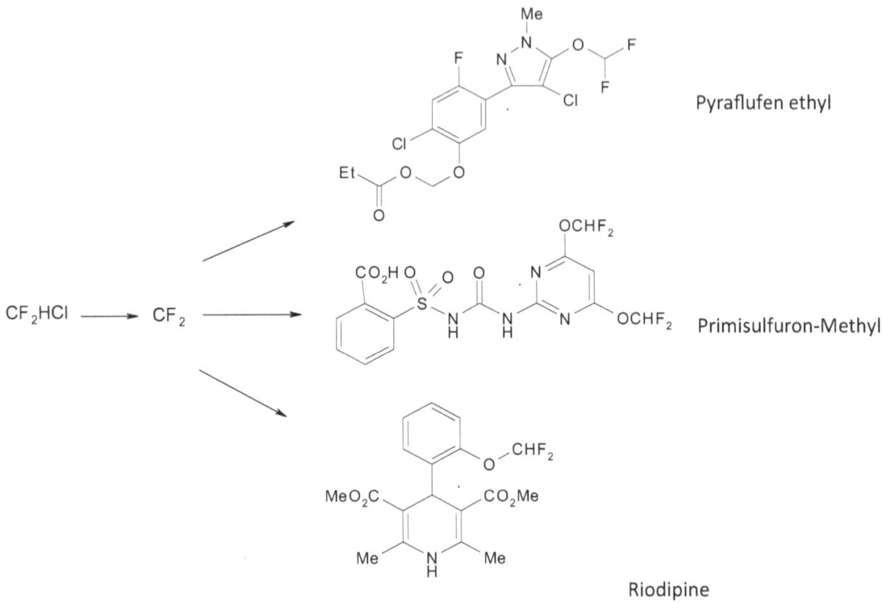

Figure 2.16 Active ingredients containing CF_2 groups.

4.1 Synthesis Strategy to Obtain CF_2 Groups in Active Ingredients

Besides the use of chlorodifluoromethane (CF_2ClH), where possible, there are not many inexpensive building blocks or fluorinating agents available, especially at the industrial scale. The remaining options to introduce a CF_2 group into an active ingredient are:

- Formation of a carbonyl group and fluorination by SF_4.
- Formation of a $-CCl_2-$ (or $-CHal_2$) group followed by a chlorine/fluorine exchange with anhydrous HF (if necessary catalyzed by a Lewis acid) or amine $\times$ *n*HF like DBN $\times$ 2,6-HF followed by regeneration.[33,69]
- Use of a CF_2-containing building block.

4.2 Examples of Active Ingredients Containing the CF_2 Group

Deramaxx is a veterinarian drug developed by Searle. At present, it is manufactured by Novartis Animal Health. The CF_2H group is introduced in a building block approach starting with ethyl difluoroacetate and a Claisen condensation, followed by reaction with the corresponding phenylhydrazine (Fig. 2.17).[33,69]

Another very successful drug containing a CF_2 group is Gemzar,[28] an anticancer drug developed by Eli Lilly. The synthetic route uses ethylbromodifluoro acetate,[27] which is reacted to the 2-desoxy-2,2-difluoro-1-oxoribose with 4-formyl-2,2-dimethyldioxolane. The 2-desoxy-2,2-difluoro-1-oxoribose is protected and then reacted with *N*-4-acetyl-*N*-4-O-bis(trimethylsilyl)cytosine, followed by the removal of the protective group[29] to obtain Gemzar (Fig. 2.18).

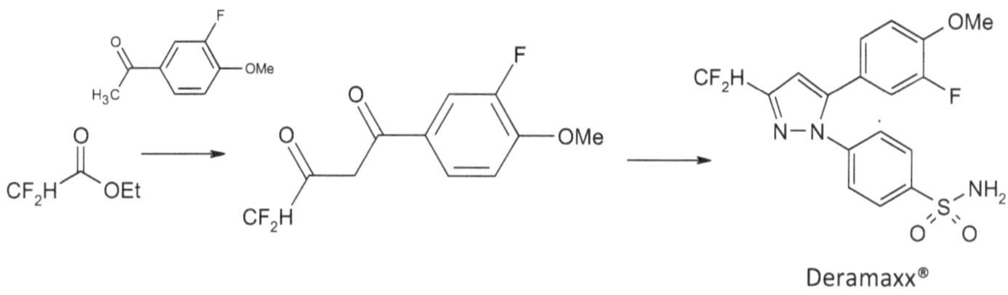

Figure 2.17 The synthesis of Deramaxx.

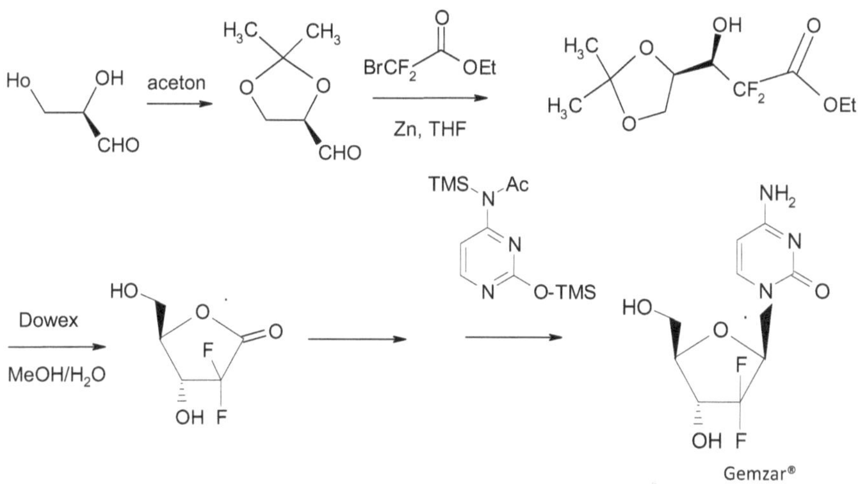

Figure 2.18 The synthetic route to Gemzar.

The new family of the so-called succinate dehydrogenase inhibitors (SDHI) fungicides is very important. This family consists of the fungicides Bixafen, Fluxapyroxad, Sedaxane, Benzovindiflupyr, and Isopyrazam (Fig. 2.19). These new active ingredients will significantly contribute to feeding the rapidly growing world population. Due to the importance of these compounds, their synthetic pathways are discussed in more detail later.

The 3-(difluoromethyl)-1-methyl-1H-pyrazole-4-carboxamide moiety is common to all structures shown in Fig. 2.20. It is obvious that a process that is environment friendly and as green as possible must be applied to prepare this key building block.

In a 2013 review published by Leroux et al.[44] different synthetic routes to this key intermediate were presented. The final anilide structure can also be made starting from the 3-(difluoromethyl)-1-methyl-1H-pyrazole-4-carboxlic acid (DFPA) or from the ethyl 3-(difluoromethyl)-1-methyl-1H-pyrazole-4-carboxylate (DFMMP).[9]

Thc main difference between the synthetic pathways is before the generation of ethyl difluoroacetate (DFAEt). After that it is a typical Claisen reaction leading to ethyl difluoroacetoacetate. The raw materials for the two different routes to DFAEt, trichloroethylene (TCE), and TFEe are industrially available in very large quantities. The

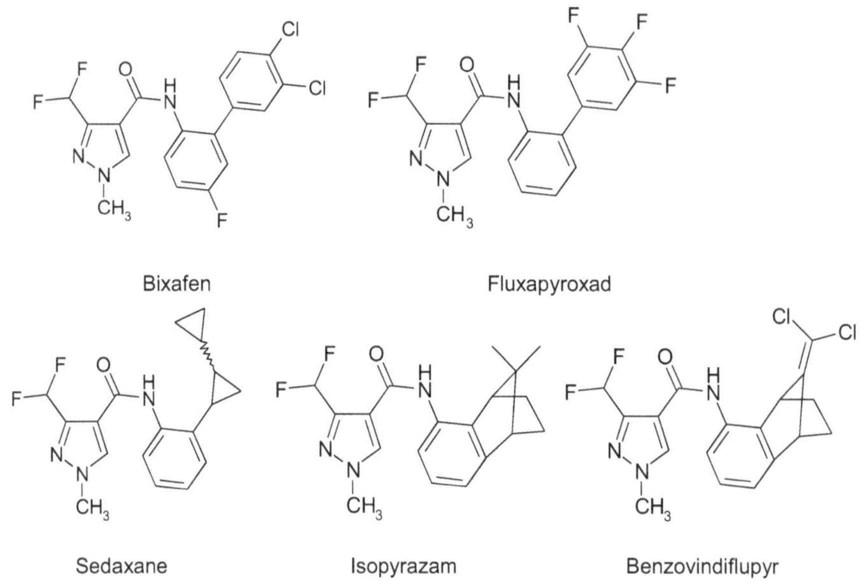

Figure 2.19 Succinate dehydrogenase inhibitors fungicides.

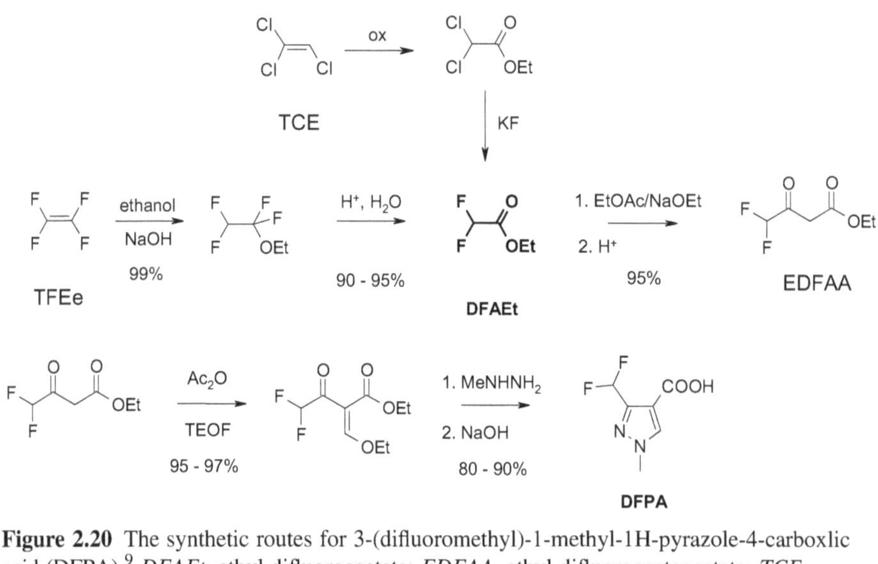

Figure 2.20 The synthetic routes for 3-(difluoromethyl)-1-methyl-1H-pyrazole-4-carboxlic acid (DFPA).[9] *DFAEt*, ethyl difluoroacetate; *EDFAA*, ethyl difluoroacetoacetate; *TCE*, trichloroethylene; *TFEe*, tetrafluoroethylene.

environmental issues related to this chemistry are more challenging. In the case of TCE as a starting material, the Halex reaction with KF requires excess KF and results in an aqueous KF/KCl solution contaminated with toxic material. Besides the complicated handling of TFEe, the TFEe route wastes two valuable fluorine atoms by forming HF as a by-product.

Avoiding the DFPA routes, Solvay developed a different route[9] to obtain the final anilide structure, which proved to avoid waste significantly. This different synthetic route begins with difluorotrichloroethane (see Fig. 2.9), which is oxidized to chlorodifluoroacetyl chloride in a first step.

The following steps involve scalable and high-yield chemistries (Fig. 2.21).

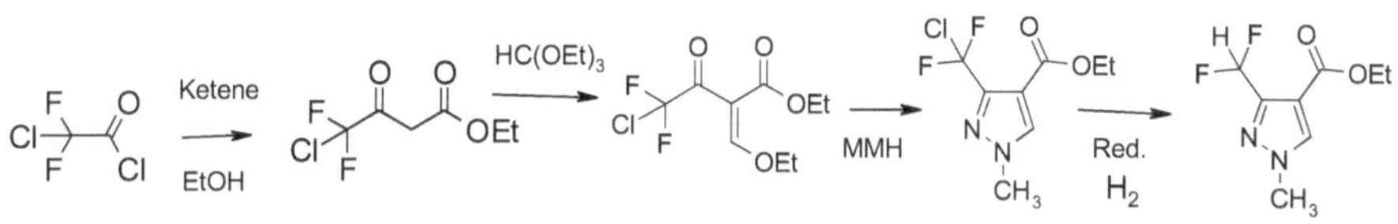

Figure 2.21 Synthesis of ethyl 3-(difluoromethyl)-1-methyl-1H-pyrazole-4-carboxylate (DFMMP).

It has been shown over all reaction steps that this route, which starts from HCFC-122, compared with the pathways starting from either TFEe or TCE, allows the amount of waste to be reduced from approximately 10 kg per kg of DFPA produced to less than 3 kg per kg of DFMMP produced, which is significant.

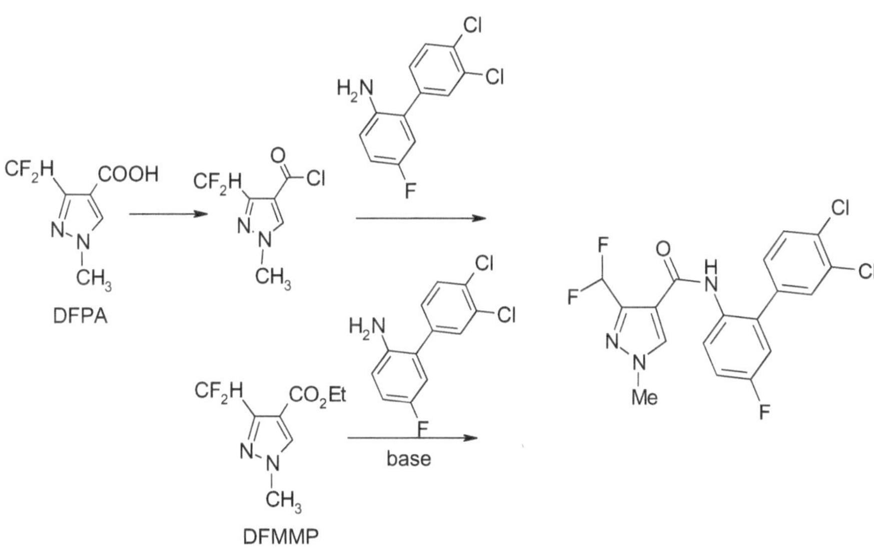

Figure 2.22 Synthetic routes to Bixafen from 3-(difluoromethyl)-1-methyl-1H-pyrazole-4-carboxlic acid (DFPA) and ethyl 3-(difluoromethyl)-1-methyl-1H-pyrazole-4-carboxylate (DFMMP).

An additional improvement in connection with the environmental aspects is the synthesis of the active ingredient directly from the related pyrazole ester, rather than from the pyrazole acid (see Fig. 2.22). Proceeding in this way saves the step of the preparation of the acid chloride from the ester with thionylchloride and the final purification step in water. In the case of Bixafen, this leads to an impressive further reduction of waste.[8,48]

4.3 Use of Difluoromethanesulfonyl Chloride

Difluoromethanesulfonyl chloride (DFSCl), also known as diflyl chloride, was developed by Solvay[58] and also serves as a building block with a high potential for CF_2-compounds (Fig. 2.23).

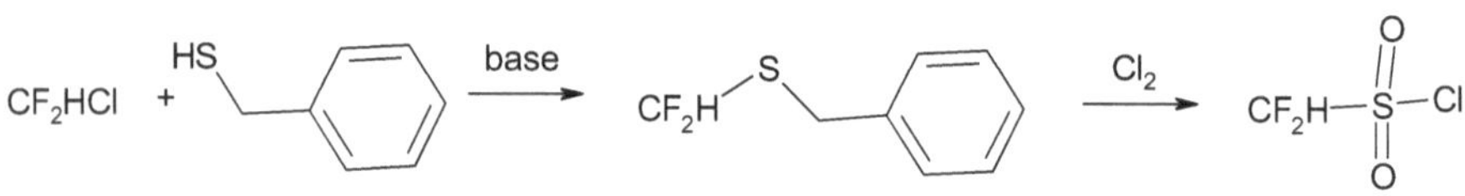

Figure 2.23 The synthesis of difluoromethanesulfonyl chloride (DFSCl) from difluorochloromethane (HCFC-22).

The main commercial application of DFSCl is as a raw material for sulfonanilide herbicides like Triafamone (Fig. 2.24).[35,41]

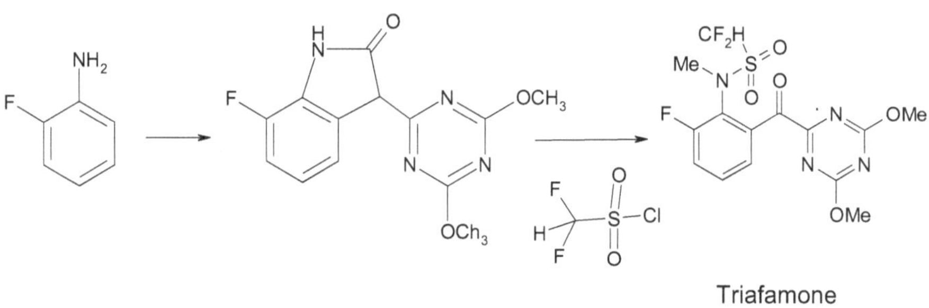

Figure 2.24 The Triafamone process route.

There are other potential blockbusters containing CF_2 groups currently under development, such as the chirally active ingredient 1,1-difluoromethanesulfonanilide[77] shown in Fig. 2.25.

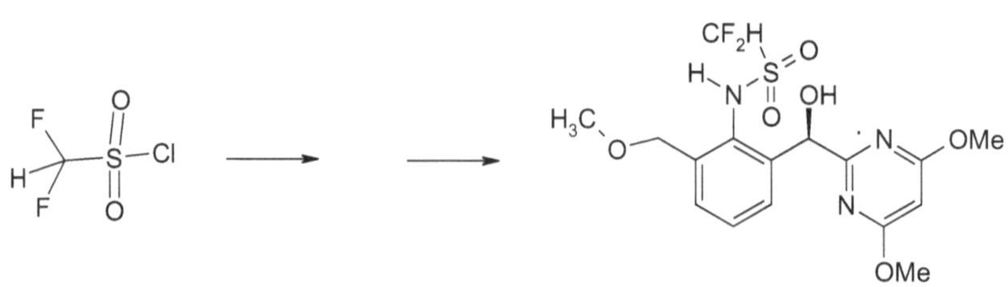

Figure 2.25 The use of difluoromethanesulfonyl chloride (DFSCl).

As the industrial availability of new CF_2 building blocks has in principal just begun, many new projects involving CF_2 groups may develop soon, especially as the generally positive contribution to the activity of a CF_2 group in a life science molecule has also been known for a long time.

References

1. Badoux, J.; Cahard, D. Electrophilic Fluorination with N-F- Reagents. *Org. React.* **2007,** *69,* 347.
2. Bentley, P. D.; et al. *Pestic. Sci.* **1980,** *11* (2), 156.
3. Besson, B.; Metz, F. WO 07128893, 2007.
4. Bertocchio, R.; Andre, A. EP 787707, 1997.
5. Braun, M.; Rudolph, W.; Eichholz, K. EP 659729, 1995.
6. Braun, M. WO 020675, 2003.
7. Braun, M. WO 018466, 2011.
8. Braun, M. WO 055864, 2012.
9. Braun, M.; Jaunzems, J. An Atom-Efficient Route to Ethyl 3-(difluoromethyl)-1-methyl-1H-pyrazole-4-carboxylate (DFMMP) − a Key Building Block for a Novel Fungicide Family. *Org. Process Res. Dev.* **2014,** *18* (8), 1055−1059.
10. Brunet, V. A.; O'Hagan, D. Catalytic Asymmetric Fluorination Comes of Age. *Angew. Chem. Int. Ed.* **2008,** *47* (7), 1179−1182.
11. Burdon, J.; Tatlow, J. C. The Electrochemical Process for the Synthesis of Fluoro-Organic Compounds. In *Advances in Fluorine Chemistry,* Vol. 1, Stacey, M., Tatlow, J. C., Sharpe, A. G., Eds.; Butterworths Scientific Publications: London, 1960; pp 129−165.
12. Burmakov, A. I.; Khassanein, S. M.; Kunshenko, B. V.; Alekseeva, L. A.; Yagupolskii, L. M. *Zh. Org. Khim.* **1986,** *22,* 1273.
13. Edmondson, S. D.; Fisher, M. H. WO004498, 2003.
14. Eicher, J.; Braun, M.; Jaunzems, J. The Importance of Fluorine in New Agrochemical Active Ingredients. In *2nd Indian International Symposium on Fluorine Chemistry, Hyderabad, India, Oral Presentation;* 2014.
15. Elsheikh, M. Y.; Tracy, J. M. US 6479718, 2002.
16. Fang, L. L.; Lawrence, S. EP 278944, 1988.
17. Fauq, A. H. N,N-Diethylaminosulfur Trifluoride. In *Encyclopedia of Reagents for Organic Synthesis;* Paquette, L., Ed.; J. Wiley & Sons: New York, 2004.
18. Fialkov, Y. A.; et al. *Ukr. Khim. Zh. Russ. Ed.* **1983,** *49,* 187−189.
19. Finger, G. C.; Kruse, C. W. Aromatic Fluorine Compounds. VII. Replacement of Aromatic −Cl and $-NO_2$ Groups by −F[1,2]. *J. Am. Chem. Soc.* **1956,** *78* (23), 6034−6037.
20. Förster, H. EP 348737, 1989.
21. Forat, G.; Mas, J. M.; Saint-Jalmes, L. FR 2732010, 1996.
22. Forat, G.; Mas, J. M.; Saint-Jalmes, L. FR 2732016, 1996.
23. Gattermann, L.; Wieland, H. *Die Praxis des organischen Chemikers,* Vol. 3; Auflage, de Gruyter: Berlin, 1982; p 617.
24. Grohe, K. DE 3142854, 1983.
25. Hansch, C. Biophysics Photosynthesis. *Biochim. Biophys. Acta* **1966,** *112,* 381−391.
26. Hardegger, B. EP 137375, 1984.
27. Hertel, L. W. EP 0122707, 1984.
28. Hertel, L. W.; et al. *J. Org. Chem.* **1988,** *53* (11), 2406−2409.

29. Hertel, L. W. US 4808614, 1989.
30. Howells, R. D.; Mc Cown, J. D. Trifluoromethanesulfonic Acid and Derivatives. *Chem. Rev.* **1977,** *77* (1), 69–92.
31. Ignatyev, N.; Bejan, D.; Willner, H. WO 012359, 2010.
32. Ikemoto, N.; Dreher, S. D. WO 0077901, 2004.
33. Isakson, P. C.; Telley, J. J. US 5756529, 1998.
34. Jackson, D. A.; Edmunds, A. US 232837, 2007.
35. Karig, G.; et al. WO 084854, 2012.
36. Kastron, V.; et al. DE 2900537, 1978.
37. Kim, D.; et al. MK-0431: A Potent, Orally Active DP-Iv Inhibitor for the Treatment of Type 2 Diabetes. In *228th ACS National Meeting (August 22–26, Philadelphia);* 2004. Abst MEDI 208.
38. Kiselyov, A. S. Chemistry of N-fluoropyridinium Salts. *Chem. Soc. Rev.* **2005,** *34,* 1031.
39. Klemmensen, D. WO 9942432, 1999.
40. Klemmensen, D. US 5986130, 1999.
41. Koichi, A.; et al. WO 024251, 2009.
42. Lal, G. S.; Pez, G. P.; Pesaresi, R. J.; Prozonic, F. M. Bis(2-methoxy-ethyl)aminosulfur Trifluoride: a New Broad-Spectrum Deoxofluorinating Agent with Enhanced Thermal Stability. *Chem. Comm.* **1999,** *2,* 215.
43. Lee, S. EP 338992, 1989.
44. Leroux, F. R.; et al. *J. Fluor. Chem..* **2013,** *152,* 2–11.
45. Lisseter, S. G.; et al. *J. Chromatogr..* **1991,** *539* (1), 207.
46. Meyer, W. EP 70804, 1982.
47. Miura, Y.; et al. EP 361114, 1989.
48. Moradi, W. A.; et al. US0128617, 2014.
49. Olah, G. A.; et al. *J. Fluor. Chem.* **1986,** *33,* 377–396.
50. Pennetreau, P.; Janssens, F. EP 699649, 1996.
51. Pevere, V.; Marx, E. WO 09852886, 1998.
52. Pietro, M. EP 0271923, 1987.
53. Roe, A. *Org. React.* **1949,** *5,* 193.
54. Roques, N.; et al. *Chim. Oggi* **2003,** 43–45.
55. Rossaint, W. Die Anästhesiologie. Allgemeine und spezielle Anästhesiologie. In *Schmerztherapie und Intensivmedizin;* Zwissler, Ed.; Springer: Berlin, 2008; pp 297–320, 2. Auflage.
56. Roth, B. D. US 4681893, 1987.
57. Ruzo, L. O.; et al. *J. Agric. Food Chem.* **1987,** *35* (6), 879.
58. Saint-Jalmes, L.; et al. US 7534906, 2009.
59. Schlosser, M. *Angew. Chem. Int. Ed.* **2006,** *45,* 5432–5446.
60. Serrano-Wu, M. H.; Kwak, Y. S. WO 126957, 2007.
61. Shibata, N.; et al. *Green Chem.* **2009,** *11,* 1733–1735.
62. Shibata, N.; et al. *Green Chem.* **2011,** *13,* 46.
63. Shibata, N.; et al. *Green Chem.* **2011,** *13,* 843.
64. Shibata, N.; et al. *Eur. J. Org. Chem.* **2012,** *8,* 1504–1508.
65. Smith, W. C. EP 823336, 1957.
66. Solange, A. M. EP 103259, 1983.
67. Subramanian, L. R. *Trifluoromethanesulfonic Acid. Encyclopedia of Reagents for Organic Synthesis;* John Wiley & Sons, 2006.
68. Sung, D. J.; Moon, D. J.; Lee, Y. J.; Hong, S. I. Catalytic Pyrolysis of Difluorochloromethane to Produce Tetrafluoroethylene. *Int. J. Chem. React. Eng.* **2004,** *2.*

69. Telley, J. J.; et al. US 6492411, 2002.
70. The Merck Index *An Encyclopaedia of Chemicals, Drugs and Biologicals,* Vol. 14; Auflage, 2006; p 1002.
71. Theodoridis, G. Fluorine-Containing Agrochemicals: An Overview of Recent Developments. In *Advances in Fluorine Science, Fluorine and the Environment, Agrochemicals, Archaeology, Green Chemistry & Water;* Tressaud, A., Ed.; ; Elsevier: Amsterdam, 2006, ; pp 121−176.
72. Umemoto, T.; El-Awa, A. *N-Fluoropyridinium Triflate. e-EROS Encyclopedia of Reagents for Organic Synthesis,* 2013. Online Publication.
73. Wilmet, V.; Janssens, F. WO 02012153, 2002.
74. Wong, D. T.; Horng, J. S.; Bymaster, F. P.; Hauser, K. L.; Molloy, B. B. A Selective Inhibitor of Serotonin Uptake: Lilly 110140, 3-(p-Trifluoromethylphenoxy)-n-methyl-3-phenylpropylamine. *Life Sci.* **1974,** *15* (3), 471−479.
75. Xiao, Y.; Armstrong, J. D. US 194977, 2006.
76. Yan, H.; et al. Pharmacokinetic and Pharmacodynamic Drug-Drug Interaction Assessment between Pradigastat and Digoxin or Warfarin. *J. Clin. Pharmacol.* **2014,** *54,* 800−808.
77. Yusuke, H.; et al. JP 005252, 2014.

Industrial Syntheses of Hydrohaloolefins and Related Products

3

M. Nappa, S. Peng, X. Sun
Chemours Chemical Company, Wilmington, DE, United States

Chapter Outline

Modern Synthesis Processes and Reactivity of Fluorinated Compounds. http://dx.doi.org/10.1016/B978-0-12-803740-9.00003-2

1. Introduction

Most climate scientists agree that global climate change is the result of the release of large amounts of CO_2 from the burning of fossil fuels with a contribution from long-lived man-made chemicals that accumulate in the upper atmosphere and absorb infrared radiation from the Earth.[1] Traditional fluorochemicals have high global warming potentials due to their infrared absorption cross-section, and many fluorochemicals are volatile and used in emissive applications. Hence, low-global-warming alternatives are constantly being sought after that have the appropriate characteristics and physical properties for end use applications such as refrigeration, air conditioning, foam expansion, clean agent fire suppression, specialty fluids, propellants, and semiconductor chip manufacture.[2] These replacement molecules also need to have zero ozone depletion potential (ODP), just as the hydrofluorocarbons (HFCs) that they are replacing. The chemical industry has been targeting hydrohaloolefins as replacements since the unsaturation reduces atmospheric life by providing a point of attack for hydroxyl and nitroxyl radicals in the atmosphere.[3] Over the years, research efforts in industry and academia have been focusing on identifying and developing synthetic schemes to hydrohaloolefin candidates that are economically viable and have low toxicity, low cost of manufacture, an appropriate level of flammability, and efficacy in use. Table 3.1 shows a list of olefins that will be discussed in this chapter along with some of their properties.

Traditional Swarts chemistry, which has been useful to make chlorofluorocarbons (CFCs) and even some hydrohalocarbons, is of limited utility in the synthesis of functionalized olefins but is sometimes used for the synthesis of intermediates to make these hydrohaloolefins. The products of the Swarts reaction are usually saturated and have some chlorine remaining in the molecule,[4] which can lead to a high-ozone-depletion potential, and therefore need to be converted into something of more value. Herein we review the efforts to develop selective synthetic methods to make hydrohaloolefins for commercial applications including both vapor phase and liquid-phase processes.

2. Hydrofluoroolefins

2.1 $CHF{=}CF_2$, 1,1,2-Trifluoroethylene or HFO-1123

Trifluoroethylene (HFO-1123) has been proposed as a replacement for room air-conditioning with a boiling point of −51°C.[5] By virtue of the double bond, the global

Table 3.1 HFOs and HCFOs

Name	Structure	BP (°C)	GWP
1123	$CHF{=}CF_2$	−51	0.3
E-1234ze	$CF_3CH{=}CHF$ (E)	−19	<1
Z-1234ze	$CF_3CH{=}CHF$ (Z)	−9	–
1234yf	$CF_3CF{=}CH_2$	−29.5	<1
Z-1225ye	$CF_3CF{=}CHF$ (Z)	−19	<1
E-1225ye	$CF_3CF{=}CHF$ (E)	−10	–
1243zf	$CF_3CH{=}CH_2$	−22	<1
E-1336mzz	$CF_3CH{=}CHCF_3$ (E)	7	–
Z-1336mzz	$CF_3CH{=}CHCF_3$ (Z)	33	2
E-1233zd	$CF_3CH{=}CHCl$ (E)	19	1
Z-1233zd	$CF_3CH{=}CHCl$ (Z)	38	<1
1233xf	$CF_3CCl{=}CH_2$	12	–
1214ya	$CF_3CF{=}Cl_2$	46	–

BP, boiling point; *GWP*, global warming potential; *HCFO*, hydrochlorofluoroolefin; *HFO*, hydrofluoroolefin.

warming potential (GWP) is low (0.3), but its reactivity is high. This molecule is known to undergo disproportionation to make CF_4 and HF liberating about 60 kcal/mol; it has also been reported to deflagrate under test conditions at 200 psig and 27°C, free of air and in the presence of a terpene inhibitor, and its reactivity and explosivity have been compared with those of tetrafluoroethylene (TFE).[6] Nonetheless, the routes to its manufacture are reviewed below.

An early report for the manufacture of 1123 is via the hydrogenation of CFC-113 (CCl_2FCF_2Cl) in a hot iron tube. By-product yields of 16% were obtained by passing vapors of 113 and H_2 over an iron reaction zone (empty tube) at 585°C and 3-s contact time with the main product being chlorotrifluoroethylene (CTFE).[7]

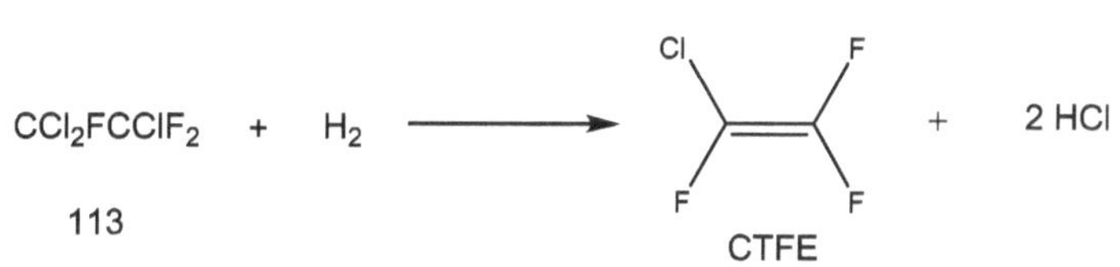

Shortly thereafter, a catalytic process was developed that showed a selectivity of 1123 of 42.5% from the hydrogenation of CTFE at 140–158°C over 5% Pd/C.[8]

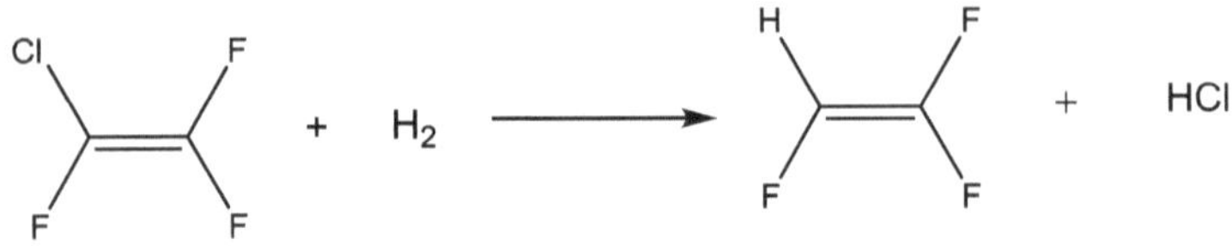

CTFE 1123

Much higher selectivities were observed when the reaction was carried out at higher temperature and short contact times. Diluting a 3.8% Pd/C with alumina (1:2) and hydrogenating CTFE at 300°C and 0.5-s contact time gives a selectivity of 94.3% 1123.[9] Silica impregnated with 0.8% Pd and 6.6% barium catalyzes the hydrogenation of CTFE (1:1 mol ratio) at 240°C and 43.5 psia to give 1123 in about 94% selectivity. Up to 97% selectivity is obtained by running deficient in hydrogen.[10] Hydrogenation of 113 using Pd containing Au, Te, Sb, Bi, and As adjuvants at 150°C gives a selectivity of only 65% with arsenic added to palladium.[11]

At about this time, HFC-134a (CH_2FCF_3) was being marketed as a replacement for CFC-12 (CCl_2CF_2) for mobile air-conditioning, so this molecule became available as a feedstock for the manufacture of other materials, and routes to 1123 began to appear based on 134a; 134a with an equimolar mount of N_2 was passed over 8% nickel of fluorinated alumina at 530°C and atmospheric pressure to give 31% conversion of 134a and a 94% selectivity to 1123.[12]

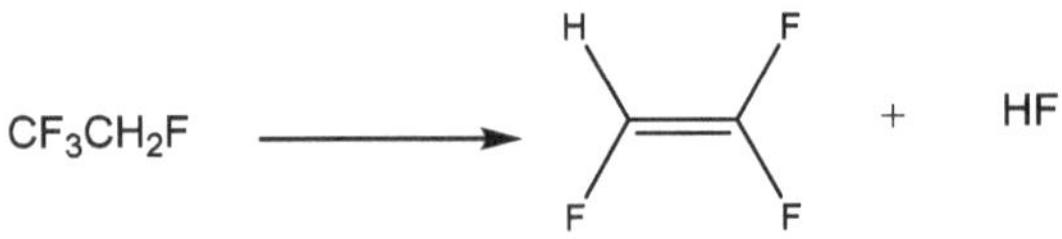

134a 1123

Alumina doped with iron was also effective at converting 134a at 26% to give 1123 in very high selectivity, 99.6%. The feed was predominantly N_2 (14.5:1), which drives the equilibrium to the olefin and HF from the saturated alkane.[13] The use of steam instead of nitrogen to drive the equilibrium in an uncatalyzed reactor was reported by the same laboratory, but the results were not as good, achieving a selectivity of only 75% at 1000°C.[14] The use of N_2 or CO_2 as a diluent and γ-alumina catalysts doped with Ni, Mg, or Zn are also effective in producing 1123 in high selectivity.[15]

The hydrogenation of TFE ($CF_2{=}CF_2$) to make 134 (CHF_2CHF_2) was coupled with the dehydrofluorination of 134 to make 1123. The heat generated from the hydrogenation is used with additional heat to drive the endothermic dehydrofluorination without catalyst, thereby reducing the utility costs.[16] 134 can also be dehydrofluorinated using a fluidized bed of potassium carbonate in 95% selectivity but only at 4% conversion, showing the difficulty in cracking 134.[17] Calcium oxide gave a higher conversion (15%) and higher selectivity (98%).[18]

2.2 E,Z-$CF_3CH{=}HF$, 1,3,3,3-Tetrafluoropropene or HFO-1234ze

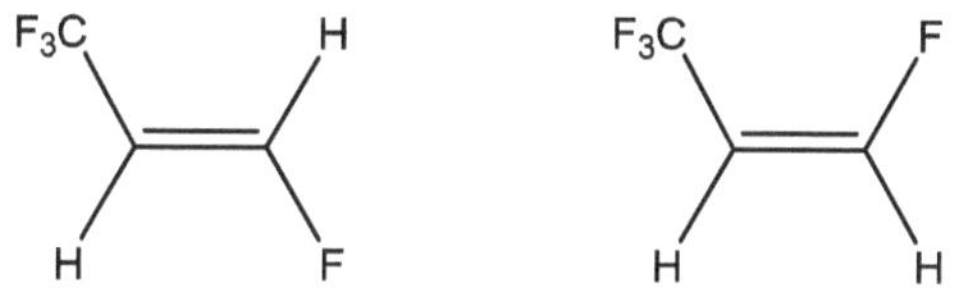

HFO-*E*-1234ze HFO-*Z*-1234ze

HFO-*E*-1234ze with a boiling point of −19°C has been proposed as a foam expansion agent for thermoplastic foam and as an aerosol propellant.[19] HFO-*Z*-1234ze has a boiling point of 9°C and can be used as a solvent, heat transfer fluid, and fluxing agent.

Hazeldine first reported the synthesis of HFO-*E*-1234ze in about 50% yield by the reaction of hydrochlorofluorocarbon-244fa (HCFC-244fa) (CF_3CH_2CHClF) with 10% alcoholic KOH.[20]

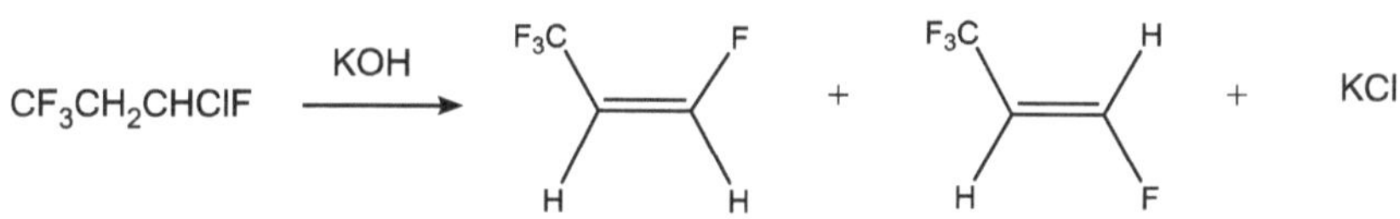

Z-1234ze *E*-1234ze

The yield was low most likely due to handling losses of the low-boiling product. Shortly thereafter, it was reported that HFC-245fa ($CF_3CH_2CHF_2$) can be made via the hydrogenation of HFO-1225zc ($CF_3CH{=}CF_2$), and then the 245fa was treated with KOH in dibutylether to give 1234ze in 70% yield.[21]

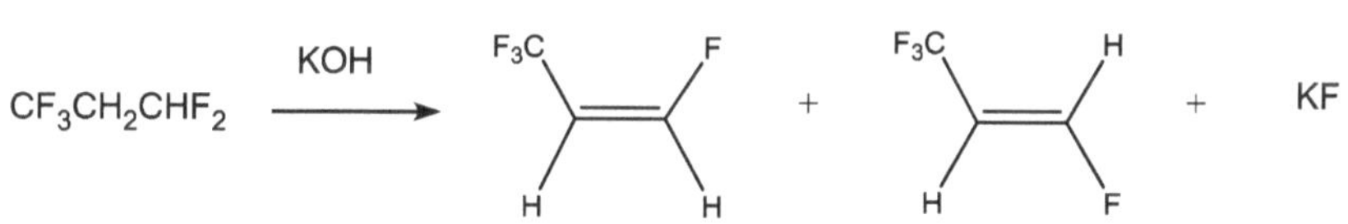

Z-1234ze *E*-1234ze

HF and HCC-240fa ($CCl_3CH_2CHCl_2$) were reacted over Cr_2O_3/C (activated PCB) at 400°C to yield 28.8% 1234ze with the remainder being 7.4% 245fa and 63.1% HCFO-1233zd ($CF_3CH{=}CHCl$).[22]

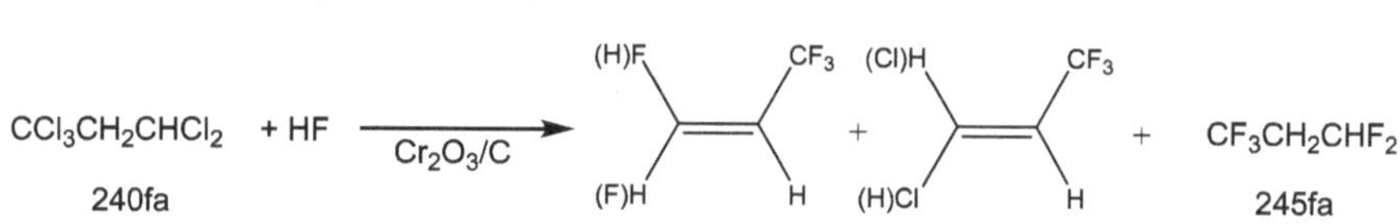

E/Z-1234ze *E/Z*-1233zd

It is not surprising that the sum of the olefins would predominate at high temperature because HX elimination is thermodynamically favored at these conditions. If HCFO-1233zd is reacted in place of 240fa, the yield of 1234ze increases to 73% under similar conditions.[23] At about the same time under very similar conditions, it was reported that HF reacts with 240fa at 270–350°C and a high HF ratio[20–29] to give mostly *E* and *Z*-1233zd (74–82% and 13–14%, respectively) and some *E* and *Z*-1234ze (1.6–6.5% and 0.4–3.2%, respectively). Very little 245fa (1.8–2.1%) survived these conditions due to equilibrium constraints. The highest yield of 1234ze was observed at 350°C.[24] Also at about the same time, HCO-1230za ($CHCl_2CH{=}CCl_2$) was reported to react with HF over a fluorinated chrome oxide catalyst to give up to 11.5% 1234ze when run at pressure (165 psig and in the presence of an oxygen cofeed—3% based on 1230za).

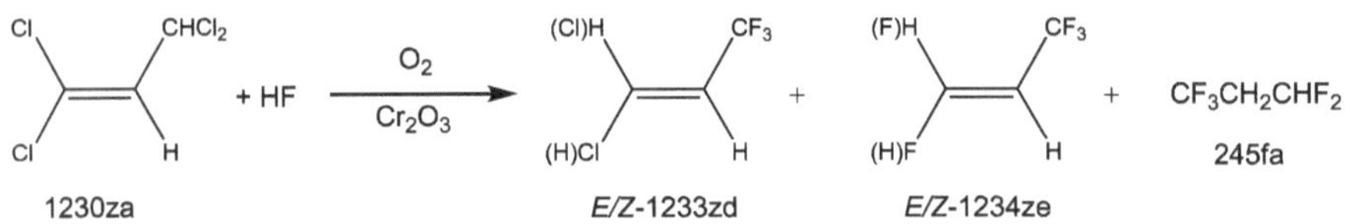

Not surprisingly, the main product was 1233zd.[25]

Crown ethers and other phase transfer catalysts (PTCs) were shown to be effective at catalyzing the reaction of hydroxide salts with 245fa in the absence of a solvent at room temperature to give 85% conversion and 67% overall yield to presumably a mixture of *E* and *Z*-1234ze; no conversion was observed in the absence of crown ether.[26] Seemingly in an attempt to avoid conditions normally used with PTCs, it was reported that 245fa can react in acetonitrile in the absence of a PTC to give 1234ze, although no actual yields and selectivity were provided.[27]

C1s and C2s were coupled to make precursors that were converted to 1234ze. HCFC-21 ($CHCl_2F$) and vinyl fluoride were coupled over a Cu catalyst to make HCFC-243fb ($CHClFCH_2CClF_2$), which when passed over a bed of Cr and Sn salts yields 1234ze in 40–65% isolated yields. Similarly, CHI_2F and vinylidene fluoride surprisingly yield the same level of 1234ze.[28]

Early on, as noted earlier, pentachloropropanes can be converted directly to the olefins by reaction of HF in the vapor phase due to favorable HX elimination at high temperature. These reactions can be decoupled, and 245fa can be converted directly in the vapor phase to make 1234ze in high selectivity, and the vapor-phase dehydrofluorination of 245fa to a mixture of *E* and *Z*-1234ze has been reported for a variety of catalysts. Acid-washed carbon (0.8, 1.1, and 1.8%), Ni/C (0.4, 0.6, 0.7, 1.0%), Ni/Cr_2O_3 (0.6, 0.7, and 1.0%), and Ni/alumina dehydrofluorinated 245fa to give conversions of 30–100% and selectivities of 43–100% at 475–515°C. The product of these reactions were mixtures of *E* and *Z*-1234ze.

Subsequently, integrated processes appeared wherein the undesired *Z* isomer was converted back to the *E* isomer in a separate step to recover its value.[29] In the first reactor, 245fa is converted (96%) over fluorinated chromium oxide at 350°C to give 80.6% selectivity to *E*-1234ze and 18.0% *Z*-1234ze. Similar results were obtained

Table 3.2 Conditions for 1234ze Isomerization

Catalyst	Reaction Temperature (°C)	Conversion, %*Z*-1234ze	Selectivity *E*-1234ze
Fluorinated Cr_2O_3	100	91.0	100.0
AlF_3	200	85.2	99.3
0.5% Co/acid-washed C	350	45.0	98.2

with AlF_3, 10% MgF_2/90% AlF_3, giving a lower conversion (80%) and higher selectivity of *Z*-1234ze at 525°C with Fe/C (23.4%). The organic product of this first reactor was purified by distillation and the *E*-1234ze separated. The high boilers, *E*-1234ze and 245fa, are fed into a low-temperature (depending on the catalyst) reactor to isomerize the *Z* to *E*-1234ze (Table 3.2).

The lower temperature is required due to the equilibrium constraints favoring the *Z* isomer more so at higher temperature, and the equilibrium effect was exploited to prepare the *Z* isomer.[30] The *E* isomer is passed over a catalyst at 200−500°C to convert it to a mixture of the *E* and *Z*, and then the *Z* is isolated by distillation. The *Z* isomer has a boiling point of 9°C, 25°C higher than that of the *E* isomer, and thus it is easily separated and purified by distillation. The boiling point of the *Z* isomer makes it a useful cleaning agent, flux, and heat transfer fluid.

The dehydrofluorination of 245fa is also plagued by side reactions, which reduces the selectivity of 1234ze. It is known that fluorinated chromium oxide can isomerize 1234ze to 1234yf, and some of this reaction may occur in the dehydrofluorination reaction to make 1234ze.[31] Potassium-doped aluminum oxide gave 99.9% selectivity to 1234ze from 245fa at 250−300°C with the conversion ranging from 27.3% to 56.3%.[32] The potassium dopant deactivates the electrophilic isomerization sites and makes the reaction more selective for 1234ze.

2.3 $CF_3CF{=}CH_2$, 2,3,3,3-Tetrafluoro-1-propene or HFO-1234yf

HFO-1234yf has a boiling point of −29.5°C and is a low-GWP near drop-in replacement for HFC-134a as a refrigerant for mobile air-conditioning applications,[33] and it is used as a component in blends for stationary applications.[34] Due to its adoption as a replacement for 134a, many routes have been developed for its manufacture based on a variety of feedstocks. The first that will be covered uses hexafluoropropylene since it is an item of commerce, and fluorinations are not required to make the final product, 1234yf, using this feedstock.

2.3.1 Hexafluoropropene Route

HFO-1234yf can be made from hexafluoropropene (HFP) via a 4-step process:

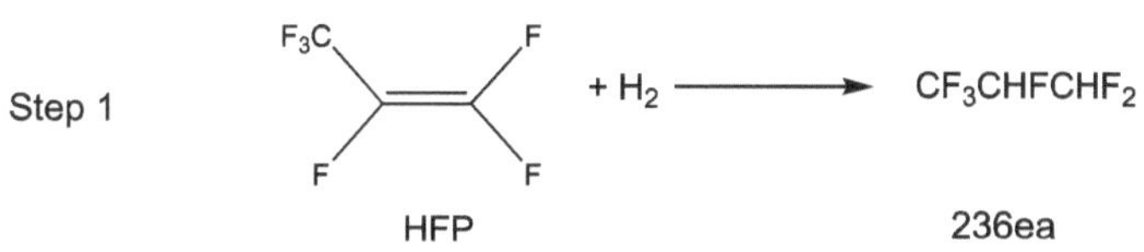

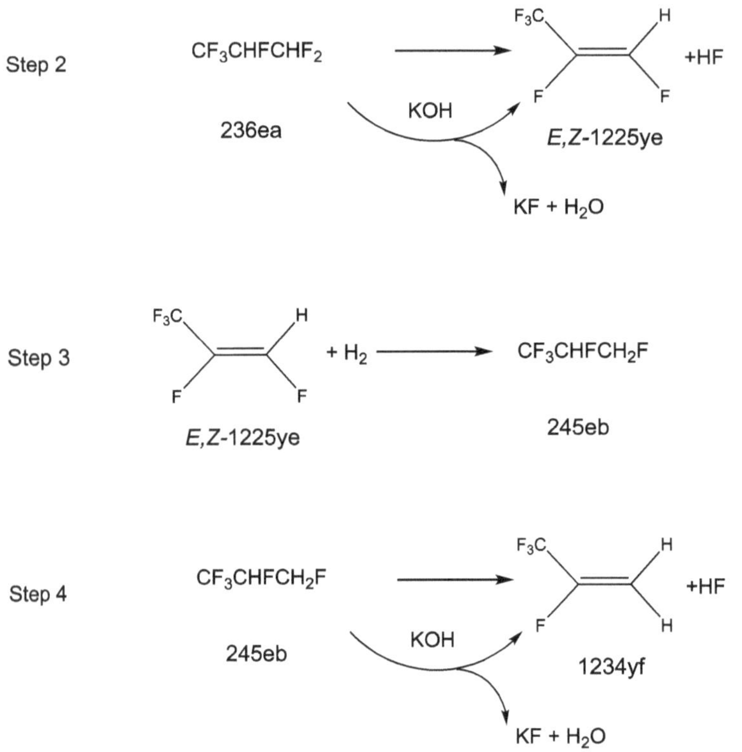

Steps 1 and 2 will be reviewed in the section for making HFO-1225ye. Similar to HFP, 1225ye was hydrogenated in the vapor phase to 245eb over Pd catalyst loaded on carbon or Al_2O_3 support.[35–43] The loading of Pd was typically between 0.1% and 10%. This reaction is also strongly exothermic, and the major side reaction was over-hydrogenation to 254eb. To control the reaction temperature and reduce overhydrogenation, several approaches were reported, such as using a multistage reactor,[36] diluting the catalyst with inert packing,[36] recycling 245eb as a heat carrier,[40] and using low catalyst loadings (0.001–0.2%). The catalyst support can also affect the performance of the catalyst. Pd loaded on α-Al_2O_3 was found to be more stable than Pd loaded on γ-Al_2O_3,[41] and the selectivity of 245eb achieved was 99%.

The conversion of 245eb to 1234yf was carried out in either the vapor or liquid phase. The noncatalytic thermal pyrolysis of 245eb only reached 6% conversion at 600°C,[44] so 245eb dehydrofluorination to 1234yf was carried out in the presence of catalysts such as Cr_2O_3, Al_2O_3, Cr/Co oxide, Zn/Cr oxide, $FeCl_3$/C, activated carbon, Pb/C, and MgO.[35,38,45–50] Transition metal oxide catalysts need to be activated by HF for conversion to an active catalyst. The 1234yf selectivity with fluorinated Cr_2O_3 and Al_2O_3 catalyst was higher than with other types.[35,45] Several side reactions also occurred in the vapor-phase 245eb dehydrofluorinations, leading to the production of 245cb ($CF_3CF_2CH_3$), 245fa ($CF_3CH_2CHF_2$), and 1234ze ($CF_3CH{=}CHF$).[48] 245cb Was the major by-product, which was as high as 30–60%, whereas 245fa and 1234ze were typically between 1 and 5%. Although 245cb can be converted to 1234yf, higher temperature and maybe a different catalyst are required. It was reported that doping of alkali metal to Cr_2O_3 catalyst can effectively reduce the formation of 245cb[48]; however, doping of alkali metals to metal oxides such as Cr_2O_3 and Al_2O_3 also reduces the activity of the catalyst.[48,50] Catalysts were also modified by adding components such as Ni, Co, Pd, Fe, Au, Cu, and Al for better 1234yf yield and catalyst life.[49,51] The 1234yf selectivity achieved was as high as 90%.

245eb was converted to 1234yf by reacting with caustic in the liquid phase. Unlike vapor-phase dehydrofluorinations, the reaction with caustic was much more selective to 1234yf. It did not form 245cb and 245fa and only formed very low amounts of 1234ze (<1%).[52] Solvents such as THF and PTCs such as Aliquat 336 were used to accelerate the reaction rate and allowed the reaction to occur at a lower temperature. The 245eb reaction with caustic was also run without a solvent and catalyst, but required a higher temperature (60°C).[47,53] This process was also carried out by passing 245eb gas through a KOH melt at temperatures above 160°C with a reported high conversion (83%) and high 1234yf selectivity (99%).[54] Acetate, formate, and monofluoroacetate (MFA)[53,55] were found in the spent KOH solution, which indicates the decomposition of 245eb and yield loss. Toxic MFA needed to be removed from the product waste stream, and thermal treatment of waste water at temperatures above 120°C was reported to effectively reduce the MFA level.[55]

Typically, KOH is used as caustic in this process, and the coproduct KF can be converted back to KOH by reacting with $Ca(OH)_2$ for reuse.[56]

2.3.2 Tetrafluoroethylene ($CF_2{=}CF_2$) Routes

Using TFE, there are two routes to make 1234yf; the first route uses CH_2F_2 in a four-step process. The first two steps make 1225ye as an intermediate, which has been discussed in Section 2.4.2. The last two steps convert 1225ye to 1234yf, which has been discussed in the section using HFP as a feedstock.

Step 1

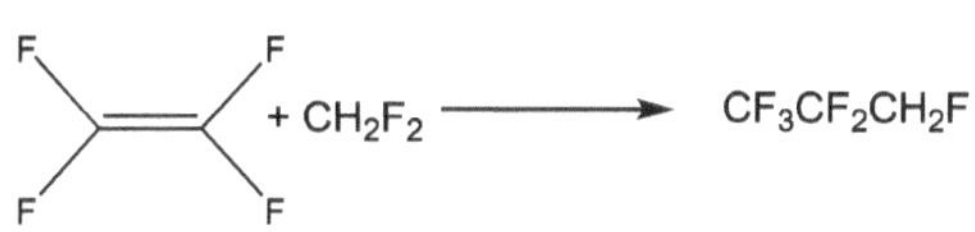

TFE 32 236cb

Step 2

236cb *E,Z*-1225ye

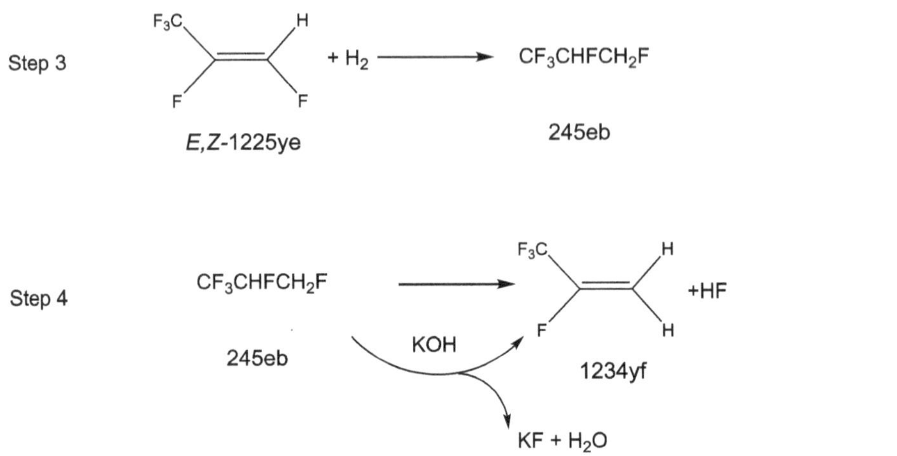

The second route from TFE is also a four-step process. Steps 1–3 will be discussed in the Section 4.4, and only step 4 will be discussed here.

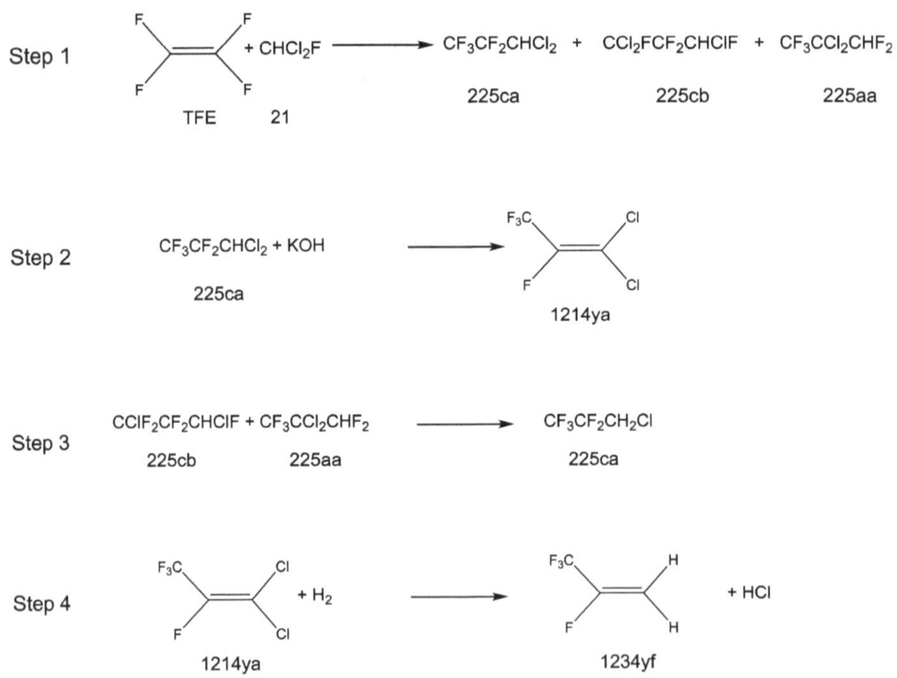

1214ya is converted to 1234yf by hydrogenation,[57–60] and the process can take place in vapor or liquid phase.[58] Catalysts used in this process were typically 0.1–2% Pd loaded on Al_2O_3, AlF_3, and activated carbon or silica gel in the temperature range between 50 and 200°C. The yield of 1234yf in vapor phase was greater than 90% but only around 75% in liquid-phase process. Several side reactions were observed in this process[61] such as hydrogenation of 1234yf to 254eb (CF_3CHFCH_3) or 1243zf ($CF_3CH{=}CH_2$) or hydrogenation of 1224yd ($CF_3CF{=}CHCl$) to 244eb (CF_3CHFCH_2Cl). The major yield loss is overhydrogenation of 1234yf to 254eb, which was lessened using Pd/C versus Pd/Al_2O_3.[61] Overhydrogenation to 245eb can be mitigated by using a low H_2/1214ya ratio (less than 0.5), keeping the catalyst bed cooler than 100°C, and running at atmospheric pressure. 1243zf is a troublesome by-product, since its boiling point is only 4°C lower than that of 1234yf, and hence its separation from 1234yf is very difficult. Careful control of the catalyst bed temperature to below 130°C was the key to reducing the 1243zf formation to less than 100 ppm.[62] It is also reported that the hydrogenation was improved by feeding partially liquefied 1214ya and 1224yd through a fixed bed Pd/C catalyst, which inhibited formation of HFO-1243zf.[63]

Instead of converting 225ca to 1214ya, 225ca was also converted to 245cb by hydrogenation with Pd/C catalyst. The reduction temperature was between 200 and 300°C, and both 225ca and 225cb hydrogenated at this condition.[64,65] 245cb selectivity was reported to be approximately 70%, and then 245cb was dehydrofluorinated to 1234yf.

$$\underset{225ca}{CF_3CF_2CHCl_2} + H_2 \longrightarrow \underset{245cb}{CF_3CF_2CH_3} + HCl$$

$$\underset{245cb}{CF_3CF_2CH_3} \longrightarrow \underset{1234yf}{CF_3CF{=}CH_2} + HF$$

Compared with the other 245 isomers such as 245eb and 245fa, the dehydrofluorination of 245cb is much more difficult. 245cb dehydrofluorination was carried out in the vapor phase over dehydrofluorination catalysts such as Cr_2O_3, Al_2O_3, and activated carbon.[66–68] There is one report of reacting in the liquid phase 245cb with 38% KOH at 50–60°C, and 80% conversion and 95% selectivity to 1234yf were achieved.[69] The activity of activated carbon was much lower than those of Cr_2O_3 and Al_2O_3 and the reaction required a high temperature (400–500°C). Even with Cr_2O_3 and Al_2O_3 catalysts, the reaction still required a temperature above 300°C and O_2 cofeed was needed to maintain catalyst activity. The typical selectivity of 1234yf was between 90% and 99%, and the selectivity was improved by cofeeding HF.[70]

2.3.3 Tetrafluoroethylene–Chloroform Route

The route using TFE and chloroform is similar to the HCFC-21/TFE route, but it is a three-step process as shown below.

Step 1 $$CHCl_3 + CF_2{=}CF_2 \longrightarrow \underset{224ca}{CClF_2CF_2CHCl_2}$$

Step 2 $CClF_2CF_2CHCl_2 + H_2 \longrightarrow CClF_2CF_2CH_3$

224ca 244cc

Step 3 $CClF_2CF_2CH_3 + HF \longrightarrow CF_3CF{=}CH_2 + HCl$

244cc 1234yf

Similar to the HCFC-21/TFE reaction, 224ca was synthesized by reaction of chloroform and TFE with a Lewis acid catalyst such as $AlCl_3$ or aluminum chlorofluoride (ACF)[71–73] in 85% yield. Then, 224ca was hydrogenated to 244cc in the liquid phase with 10% Pd/C as catalyst resulting in 60% yield.[74] The hydrofluorination with HF and dehydrofluorination of 244cc were done in the vapor phase over chromium oxide in one step to produce 1234yf in the temperature range of 280–400°C.[42,75] The 1234yf selectivity approached 95% with the major side reaction forming $CF_2ClCF{=}CH_2$ (1233yf).

2.3.4 *1230xa ($CCl_2{=}CClCH_2Cl$) Route*

Starting from 1230xa, there are two synthetic routes to 1234yf. Both routes share the same first step, but in the first route 1233xf is converted to 244bb in the second step.

Step 1 $CCl_2{=}CClCH_2Cl + 3\ HF \longrightarrow CF_3CCl{=}CH_2 + 3\ HCl$

1230xa 1233xf

Step 2 $CF_3CCl{=}CH_2 + HF \longrightarrow CF_3CClFCH_3$

1233xf 244bb

Step 3 $CF_3CClFCH_3 \longrightarrow CF_3CF{=}CH_2 + HCl$

244bb 1234yf

1230xa hydrofluorination to 1233xf has been discussed in the section for making 1233xf. As shown in the scheme above, 1233xf is converted to 244bb by reacting with HF[76–79] over Lewis acid catalysts such as $SbCl_5$, $TiCl_4$, SbF_5, $SbCl_xF_{5-x}$, $SnCl_4$, $SbCl_3$, TaF_5, and TiF_4. $SbCl_xF_{5-x}$ is the most commonly used catalyst due to the combination of high activity and low cost. Most of the reported processes are in liquid phase, but a vapor-phase process was also reported with $SbCl_5$/C as the catalyst. Although the selectivity of 244bb in the vapor phase reached 98%, the 1233xf conversion was only 92%. In the liquid-phase process, 244bb selectivity was between 80% and 99%, and the 1233xf conversion reached 99.9%. The catalyst at high activity also produced 245cb as by-product. When the catalyst deactivated, less 245cb was formed.[80] The $SbCl_xF_{5-x}$ catalyst deactivated over time, but can be regenerated with Cl_2, and cofeed of Cl_2 was also used to maintain catalyst activity. The underfluorinated impurities such as 1232xf in the feed 1233xf accelerated catalyst deactivation.[81] The liquid-phase process is normally preferred due to the high 1233xf conversion since the separation of 244bb from 1233xf is difficult due to their very close boiling points. It was reported that the addition of potassium bifluoride and sodium bifluoride[82] increased the 1233xf conversion and 244bb selectivity.

Researchers initially explored the catalytic conversion of 244bb to 1234yf in a vapor-phase process, and various catalysts such as activated carbon, Pd/C, Pt/C, MgF_2, Cr_2O_3, MgO, $FeCl_3$, $CsCl/MgF_2$, and KCl/C were reported to be effective.[83] Although Lewis acid catalysts such as Cr_2O_3 and $FeCl_3$ showed the highest activity, the selectivity to 1234yf only reached 85−90% with the formation of 1233xf.[83] Alkali metal halide supported on activated carbon also showed good activity and high selectivity to 1234yf. At 350°C, the reaction reached 47% conversion and 98% 1234yf selectivity, but catalyst life was an issue.[84]

The thermal pyrolysis of 244bb without a catalyst was found to be effective for high conversion and selectivity in the temperature range of 460−500°C.[85,86] The 1234yf selectivity reached greater than 98%; however, the selectivity of 1234yf still changed over time even without a catalyst present. This was explained by the fluorination and chlorination of the reactor wall, which generates an active catalyst for dehydrofluorination to make 1233xf.[87,88] Therefore, keeping HF and H_2O levels in the 244bb low was very critical to maintain high 1234yf selectivity. High 1233xf levels in 244bb were also found to reduce the 244bb conversion and increase the trifluoropropyne formation, and trifluoropropyne is expected to be highly toxic.[89]

244bb dehydrochlorination by KOH with a PTC was successful at 50°C with 64% conversion and 95% selectivity to 1234yf.[90]

In the second route using 1230xa as the feedstock, the second step converts the 1233xf directly to 1234yf and 245cb.

Step 1 $$\underset{\text{1230xa}}{CCl_2{=}CClCH_2Cl} + 3\ HF \longrightarrow \underset{\text{1233xf}}{CF_3CCl{=}CH_2} + 3\ HCl$$

Step 2 $$\underset{\text{1233xf}}{CF_3CCl{=}CH_2} + HF \longrightarrow \underset{\text{245cb}}{CF_3CF_2CH_3} + \underset{\text{1234yf}}{CF_3CF{=}CH_2}$$

There are many reports on the direct conversion of 1233xf to 1234yf, and the catalysts used were $SbCl_5/C$ and fluorinated Cr_2O_3.[91−95] The hydrofluorination of 1233xf is difficult and a temperature of over 300°C is required with fluorinated Cr_2O_3 catalyst and O_2 cofeed to maintain catalyst life.[92] Unlike the highly selective 1233xf to 244bb process followed by conversion of 244bb to 1234yf, the selectivity to 1234yf via the vapor-phase fluorination of 1233xf was only around 50–80%. The major by-product was 245cb, which ranged from 10% to 50% depending on the conditions. Many other by-products, such as 1233zd, 1243zf, 1234ze, and 245fa were also produced,[94] resulting in a 2–25% yield loss.[92]

There is an equilibrium between 245cb, 1234yf, and HF, and the equilibrium reaction favors 245cb at lower reaction temperatures. Hence, it was found that recycling 245cb back to the 1233xf fluorination step can suppress further formation of 245cb.[93] 245cb can be also be converted to 1234yf in a separate vapor-phase dehydrofluorination over activated Lewis acid catalysts such as fluorinated Al_2O_3 and Cr_2O_3.[66,68,96] Dehydrofluorination of 245cb is very difficult, and at 400°C with activated carbon, only 23% conversion was observed. Cr_2O_3 catalyst was much more active, and at 350°C, 61% conversion and 99% selectivity to 1234yf were achieved, but the catalyst deactivated rapidly.

2.3.5 240db ($CCl_3CClHCH_2Cl$) Route

240db can also be used as a chlorinated alkane feedstock in place of 1230xa, with the main difference being that four molecules of HCl are generated in the first step and a higher vaporization temperature is needed in the vapor-phase reaction.

Step 1 $$\underset{\text{240db}}{CCl_3CHClCH_2Cl} + 3\ HF \longrightarrow \underset{\text{1233xf}}{CF_3CCl{=}CH_2} + 4\ HCl$$

Step 2 $$\underset{\text{1233xf}}{CF_3CCl{=}CH_2} + HF \longrightarrow \underset{\text{244bb}}{CF_3CClFCH_3}$$

Step 2a $$\underset{\text{1233xf}}{CF_3CCl{=}CH_2} + HF \longrightarrow \underset{\text{245cb}}{CF_3CF_2CH_3} + \underset{\text{1234yf}}{CF_3CF{=}CH_2}$$

Step 3 $$\underset{\text{244bb}}{CF_3CClFCH_3} \longrightarrow \underset{\text{1234yf}}{CF_3CF{=}CH_2} + HCl$$

240db can be converted to 1234yf by the scheme shown above, and its conversion to 1233xf has been discussed in the section on using 1230xa as a feedstock.

2.3.6 1243zf ($CF_3CH{=}CH_2$) Route

1243zf has been used as a monomer to make elastomers, and its preparation is discussed here. It can also be used as a starting material to make 1234yf via a four-step process described below. Both the conversion of 1243zf to 1233xf and 1233xf to 1234yf have been discussed within.

Step 1 $$\underset{\text{1243zf}}{CF_3CH{=}CH_2} + Cl_2 \longrightarrow \underset{\text{243db}}{CF_3CHClCH_2Cl}$$

Step 2 $$\underset{\text{243db}}{CF_3CHClCH_2Cl} \longrightarrow \underset{\text{1233xf}}{CF_3CCl{=}CH_2} + HCl$$

Step 2a $$\underset{\text{243db}}{CF_3CHClCH_2Cl} + NaOH \longrightarrow \underset{\text{1233xf}}{CF_3CCl{=}CH_2} + NaCl + H_2O$$

Step 3 $$\underset{\text{1233xf}}{CF_3CCl{=}CH_2} + HF \longrightarrow \underset{\text{244bb}}{CF_3CClFCH_3}$$

Step 3a $$\underset{\text{1233xf}}{CF_3CCl{=}CH_2} + HF \longrightarrow \underset{\text{245cb}}{CF_3CF_2CH_3} + \underset{\text{1234yf}}{CF_3CF{=}CH_2}$$

Step 4 $$\underset{\text{244bb}}{CF_3CClFCH_3} \longrightarrow \underset{\text{1234yf}}{CF_3CF{=}CH_2} + HCl$$

2.3.7 114a (CF_2ClCF_2Cl) and Formaldehyde Route

114a, currently banned by the Montreal Protocol, was used as a refrigerant, and methods of its preparation have been known for many years; it can be converted to 1234yf via a three-step process.[97] The solvents in steps 1 and 3 are typical aprotic polar solvents such as dimethylformamide, dimethylacetamide, and pyridine and markedly influence product selectivity. The major side reaction in step 1 the is formation of 124 (CF_3CHClF) by proton transfer, and pyridine was found to give highest yield. Both steps 1 and 3 involve using activated zinc as a reactant, and it was found that an added zinc salt such as zinc acetate increased the zinc's activity and improved the product selectivity. Although the yield of each step is high, zinc recycling is potentially expensive.

Step 1 $CF_3CCl_2F + CH_2O + Zn \longrightarrow CF_3CClFCH_2OZnCl$

114a — 80-90% yield

Step 2 $CF_3CClFCH_2OZnCl + (CH_3CO)_2 \longrightarrow CF_3CClFCH_2OC(O)CH_3$

95% yield

Step 3 $CF_3CClFCH_2OC(O)CH_3 + Zn \longrightarrow CF_3CF{=}CH_2 + ZnClO(CO)CH_3)$

1234yf 98% yield

2.3.8 *Thermal Pyrolysis Processes*

Thermal pyrolysis of CFCs in the temperature range of 550°C–750°C was also explored[98,99] to make 1234yf; however, these high-temperature routes are usually plagued by low selectivity as shown in the following reactions:

$CH_3CF_2Cl + CF_3Cl \rightarrow CF_3CF{=}CH_2$, 17% selectivity to 1234yf[98]

$CF_3CHClF + CH_3Cl \rightarrow CF_3CF{=}CH_2$, 22% selectivity to 1234yf[98]

$CF_3CHClF + CH_4 \rightarrow CF_3CF{=}CH_2$, 23% selectivity to 1234yf[98]

$CF_2{=}CFCl + CH_3Cl \rightarrow CF_3CF{=}CH_2$, 87% selectivity to 1234yf[99a]

$CHClF_2 + CH_3Cl \rightarrow CF_3CF{=}CH_2$, ~10% selectivity to 1234yf[99a]

However, there is one report utilizing steam pyrolysis of TFE with chloromethane giving a selectivity of 95% using ZnF_2/C as a catalyst.[99b]

$$CF_2{=}CF_2 + CH_3Cl \rightarrow CF_3CF{=}CH_2 + HCl$$

2.4 *$CF_3CF{=}CHF$, 1,2,3,3,3-Pentafluoro-1-propene or HFO-1225ye*

1225ye was first reported in 1953 by Haszeldine,[100] and it was later suggested that it might have the appropriate physical properties to make it a good refrigerant.[101] Both HFP and TFE have been used as starting materials to make 1225ye as shown below.

2.4.1 Hexafluoropropene Route

It takes two steps to make 1225ye from HFP as shown below and this is the most studied route in the literature.

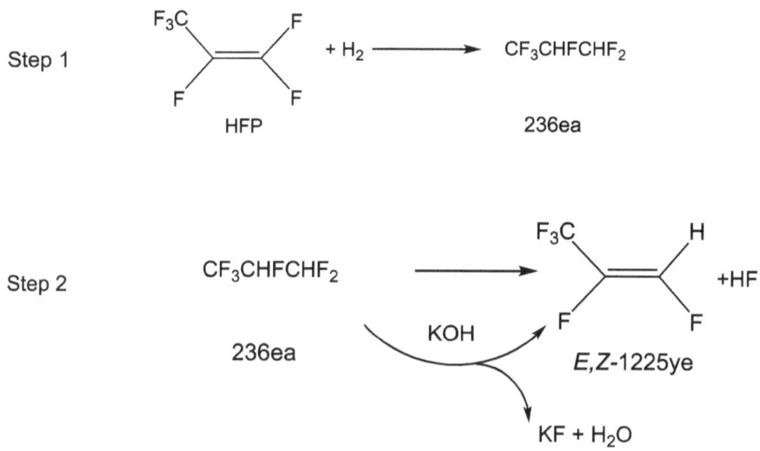

The hydrogenation of HFP was carried out in vapor phase with Pd loaded on various supports such as carbon, α-Al_2O_3, metal halide, and metal oxyhalides at temperatures between 50 and 200°C,[41,102–111] and high selectivities to 236ea were observed (85–99%).[105] The major by-product was the overhydrogenation product, 245eb ($CF_3CHFCHF_2$). The hydrogenation of HFP is highly exothermic, which can lead to poor temperature control, high levels of by-products, and safety concerns. The approach used by industry to control the heat included using multiple vapor-phase reaction stages,[110] diluting catalyst with inert packing,[110] and using an optimized low-catalyst loading.[105]

The hydrogenation of HFP was also carried out in the liquid phase in the presence of Pd catalysts supported on $BaSO_4$ and/or activated C at room temperature, and the selectivity for 236ea was reported to be in the range of 90–99%.[112]

The dehydrofluorination of 236ea was carried out in two ways, in liquid phase using caustic and in the vapor phase over a dehydrofluorination catalyst. The liquid-phase dehydrofluorination of 236ea was first reported with aqueous caustic in the presence of PTC and a solvent.[113] The KOH is typically used due to the higher solubility of KF versus NaF. The KF formed in this process can be converted back to KOH through $Ca(OH)_2$ for reuse.[114] Although this process achieved greater than 99% selectivity for 1225ye,[113] it also produced a high volume of waste. Toxic MFA was also found in the spent caustic solution, which comes from the reaction of 1225ye with caustic. The toxic MFA had to be removed from the aqueous waste stream; thermal decomposition treatment at temperature greater than 160°C was reported to be an effective way to reduce the MFA but the process consumes a large amount of energy.[55]

The dehydrofluorination of 236ea by caustic was also carried out by passing 236ea through 80% KOH at 170°C with 93% yield without using solvent and PTC.[53,115,116] The addition of KOH into 236ea was found to improve 1225ye yield in contrast to the reverse addition of 236ea into KOH. The yield was improved from 65% to 95% due to the reduced decomposition of 1225ye.[53]

The vapor-phase dehydrofluorination of 236ea was reported to be carried out with catalysts such as chromium fluoride, aluminum fluoride, oxyfluorides of chromium, oxyfluorides of aluminum, and activated carbon.[38,117–122] 1225ye selectivity from the vapor-phase reaction was generally between 90% and 96%. It was reported that the cofeed of H_2 and HF with 236ea further increased the 236ea conversion and 1225ye selectivity.[121] Although both liquid phase and vapor dehydrofluorinations of 236ea produced both *E*-1225ye and *Z*-1225ye, the vapor dehydrofluorination of 236ea produced about 10% *E*-1225ye and the liquid phase route produced only about 1% *E*-1225ye.[122,123]

Z-1225ye and *E*-1225ye boil at −18.5°C and −10°C respectively,[124,125] and the *Z* isomer is more thermodynamically stable than the *E* isomer. *E*-1225ye can be efficiently isomerized to *Z*-1225ye with several Lewis acid catalysts, such as SbF_5, ACF, as well as $SbClxF_5{-}x$, $SnClxF_4{-}x$, and $TiClxF_4{-}x$ loaded on a support.[126–128]

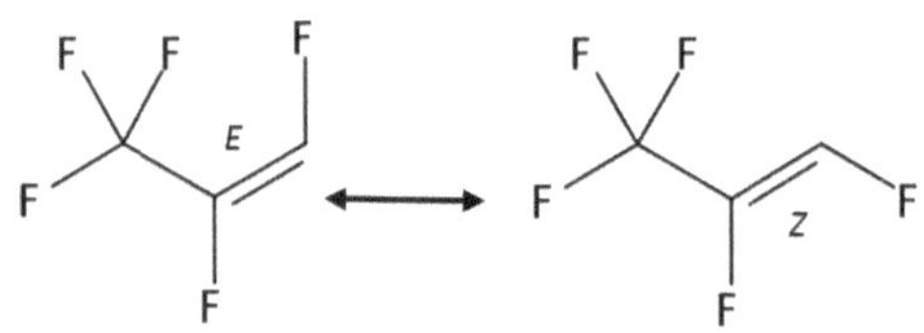

2.4.2 *Tetrafluoroethylene Routes*

Starting from TFE, there are several synthetic routes that lead to 1225ye, and the first starts with a Prins reaction between difluoromethane and TFE.

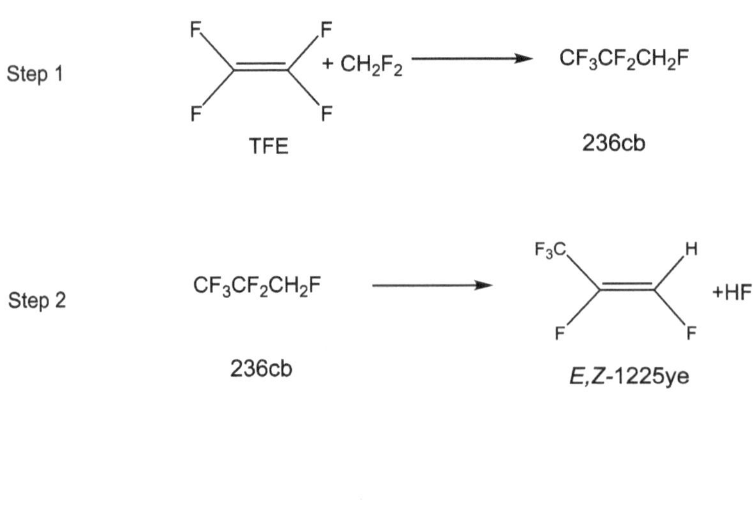

Difluoromethane (HFC-32) can be added to the double bond of TFE to form 1,1,1,2,2,3-pentafluopropane (236cb) in the presence of Lewis acids, such as SbF_5, ACF, and $AlCl_3$, as catalyst.[129–131] The selectivity of 236cb with SbF_5 catalyst reached greater than 99%, which is much higher than with ACF (67%).

Unlike 236ea, 236cb dehydrofluorination by caustic is very difficult. All the reported dehydrofluorinations of 236cb were done in vapor phase over a Lewis acid catalyst such as fluorinated chromium oxide and aluminum oxide in the temperature range of 300–450°C with about 92% selectivity to 1225ye.[119,120] Cofeed of H_2 with 236cb reduced catalyst deactivation.[132]

2.4.3 *Tetrafluoroethylene-CFC-12 (CCl_2F_2) Route*

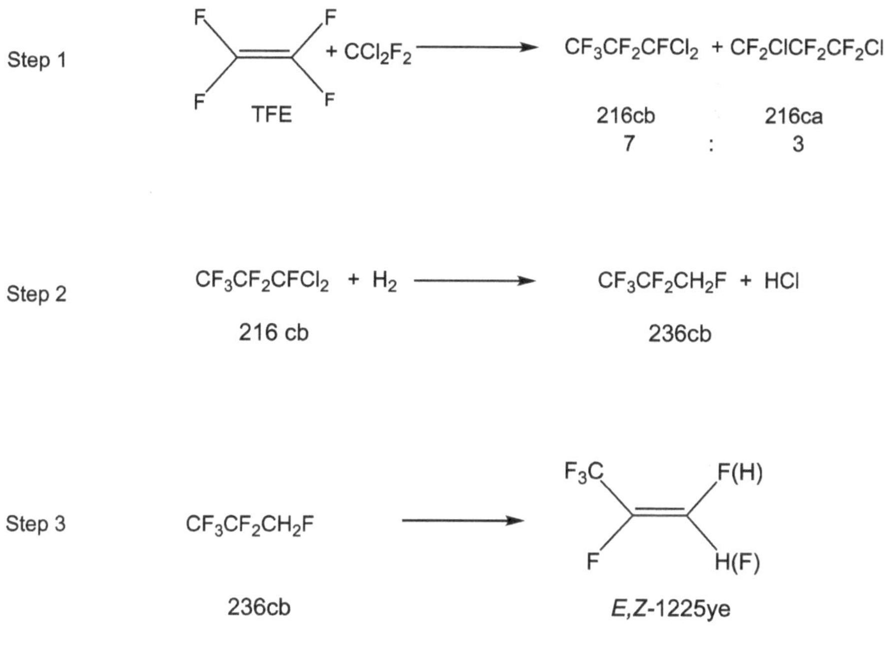

This route uses 236cb as a precursor to 1225ye. Similar to HFC-32, TFE inserts into CFC-12 using Lewis acid catalysts such as $AlCl_3$ and ACF between 30 and 80°C.[133] TFE insertion occurs into the C–F bond to make 216cb and the C–Cl bond to make 216ca. The selectivity of both 216 isomers together reached 92% with a typical molar ratio of 216cb to 216ca of about 7:3.[133] 216cb can be converted to 236cb by hydrogenation with Pd/C at 150°C with about 75% selectivity to 236cb and about 25% to 226cb ($CHF_2CF_2CH_2Cl$).[134]

2.5 $CF_3CH{=}CH_2$, 3,3,3-Trifluoropropene or HFO-1243zf

Trifluoropropene, which normally has a boiling point of −22°C, has uses as a propellant, refrigerant, and monomer for polymerizations.[135] An early reported synthesis of 1243zf is from the hydrofluorination of HCC-250fb.[136]

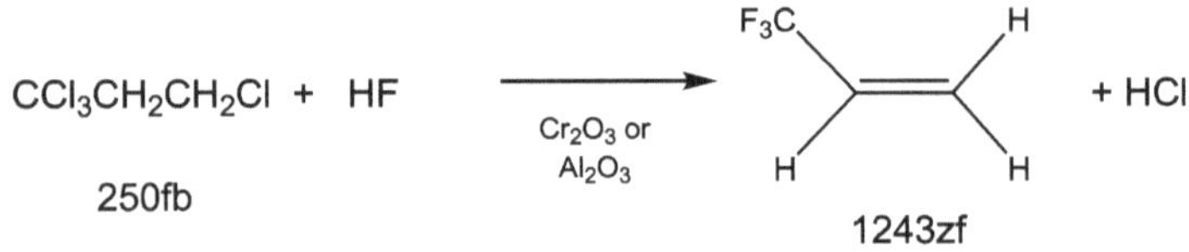

HCC-250fb was reacted with HF (11.4 mol ratio) in a vapor-phase reactor over a Cr/Al catalyst for 6 h at 350 °C at 3.4-s contact time and a conversion to 250fb of 94.2% and a selectivity of 97.6% was observed. Increasing the contact time to 4.9 s and the HF ratio to 16.7 improves the conversion to 99.8% and the selectivity to 99.5%. A cofeed of oxygen was used to sustain catalyst activity in the hydrofluorination of 250fb over a Cr/Al_2O_3 catalyst at 300°C and 4-s contact time, and after 300 h a selectivity of 96.5% was observed.[137]

A series of dehydrofluorinations and hydrogenations yielded 1243zf from HFC-245fa ($CF_3CH_2CHF_2$), and conversions and selectivities were high for each of the steps.[138]

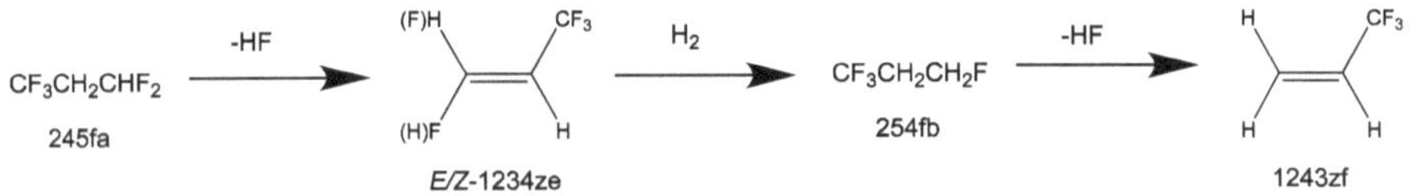

Another attempt to circumvent catalyst life issues was to run uncatalyzed using HCO-1240za ($CH_2ClCH{=}CCl_2$) as the organic feedstock.[139]

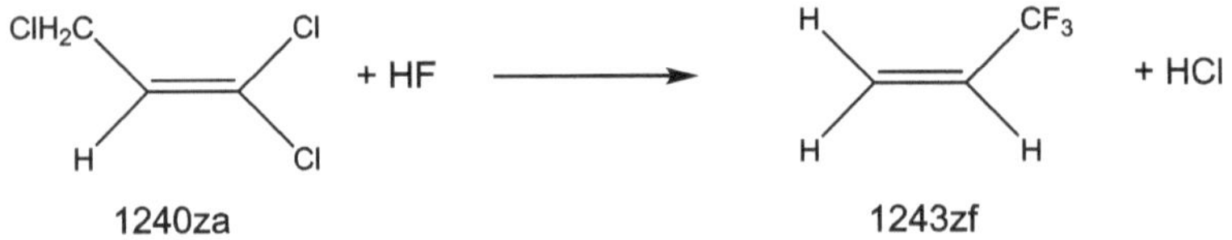

HCO-1240za and HF (150:1 HF/1240za) were reacted at 100° for 1 h up to 600 psig; the organic selectivity was greater than 99%, and there was no mention of oligomer formation. Under similar conditions, 250fb was reacted with HF at 100°C in an autoclave in the presence of iron, and the selectivity was greater than 99%. The same selectivity was observed using ZnO at 130°C.[140] It is not certain whether these additives are actually working as catalysts in these reactions or the reaction is primarily occurring uncatalyzed.

2.6 Z-$CF_3CH{=}CHCF_3$, 1,1,1,4,4,4-Hexafluoro-2-butene, Z-1336mzz

1,1,1,4,4,4-Hexafluorobut-2-ene (1336mzz) consists of *E* and *Z* isomers, and the *E* configuration is the thermodynamically favored isomer. *Z*-1336mzz is commercially manufactured for use as a replacement blowing agent for $CF_3CH_2CHF_2$ (245fa) and chiller fluid for CF_3CHCl_2 (123). *Z*-1336mzz has a GWP of 2, whereas that for 245fa is 1030 and for 123 is 76. In addition, 123 is a class II ozone-depleting substance with an ODP of 0.012, whereas *Z*-1336mzz has 0 ODP. *Z*-1336mzz is a nonflammable liquid and boils at 33°C. The excellent thermal and chemical stabilities of *Z*-1336mzz also make it an attractive candidate in applications like high-temperature heat pump and organic Rankine cycle for waste heat recovery. The routes to its preparation are discussed later.

An early report described a one-step synthesis employing 123 self-coupling to form *Z*-1336mzz and *E*-1336mzz in close to 1:1 ratio at 50% conversion of 123.[141] This liquid phase reaction was carried out at 25–60°C using stoichiometric copper powder and organic amines.

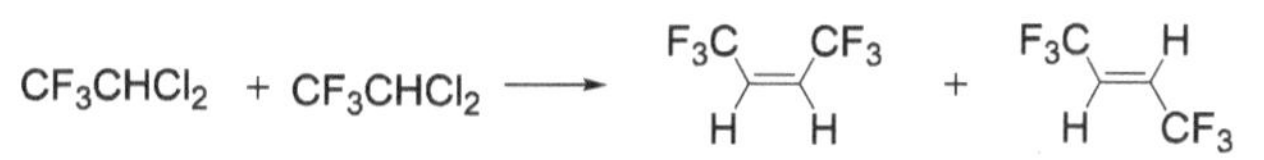

123 123 *Z*-1336mzz *E*-1336mzz

The preference for the nonthermodynamic *Z* isomer presents a significant synthetic challenge. Several multistep syntheses have been developed to generate *Z*-1336mzz exclusively. Several of them start with 1,1,1-trichloro-2,2,2-trifluoroethane (113a).

As shown below, 1,1,1-trichloro-2,2,2-trifluoroethane (113a) self-couples to form 2,3-dichloro-1,1,1,4,4,4-hexafluorobut-2-ene (1316mzz) in the presence of hydrogen and a nickel or ruthenium catalyst.[142] Subsequently, 1316mxx is hydrodechlorinated at elevated temperatures (200–400°C) with a copper/nickel/chromium catalyst on a calcium fluoride support to provide *Z*-1336mzz as the main product, and the by-products 2-chloro-1,1,1,4,4,4-hexafluorobut-2-ene (1326mxz) and hexfluoro-2-butyne.[143]

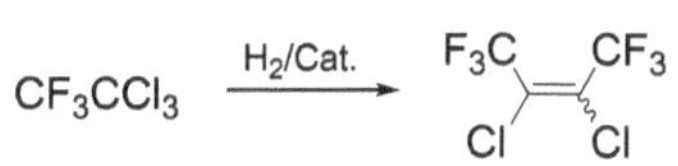

113a *E/Z*-1316mxx

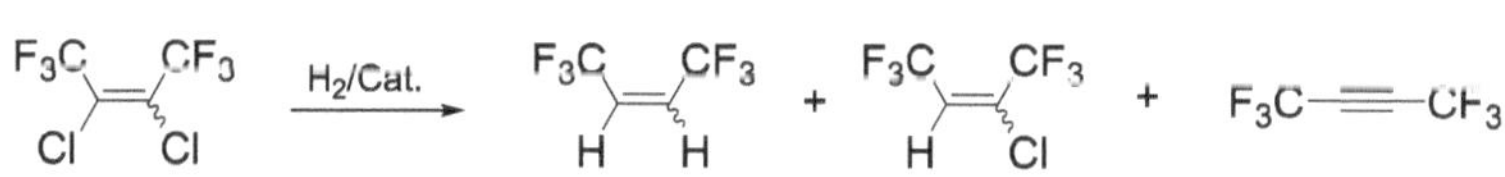

E/Z-1316mxx *E/Z*-1336mzz *E/Z*-1326mxz

The reaction of 113a and tetrachloroethene catalyzed by iron and tributylphosphate gives 1,1,1,2,2,3,3-heptachloro-4,4,4-trifluorobutane (313jaa). 313jaa is fluorinated by antimony pentachloride and hydrogen fluoride giving 1316mxx, followed by gas-phase hydrodechlorination in the presence of a copper catalyst to Z-1336mzz.[144]

$$CF_3CCl_3 + CCl_2{=}CCl_2 \longrightarrow CF_3CCl_2CCl_2CCl_3$$

113a 313jaa

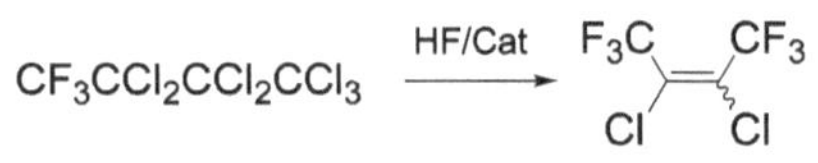

313jaa *E/Z*-1316mxx

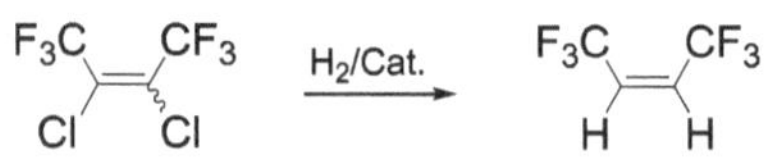

E/Z-1316mxx *E/Z*-1336mzz

Hydrogenation of acetylenes is one of the most well-known methods to prepare Z-olefins, and several processes are centered on the hydrogenation of hexafluorobutyne to selectively produce Z-1336mzz. Hydrogenation of hexafluorobutyne over Raney Ni leads to the production of Z-1336mzz successfully, but the reaction also forms over 10% over hydrogenated product such as $CF_3CH_2CH_2CF_3$ (356mff).[145] A variety of palladium catalysts poisoned by elements such as lead, bismuth, and barium are able to hydrogenate hexafluorobutyne to Z-1336mzz and avoid the formation of saturated by-product.[146]

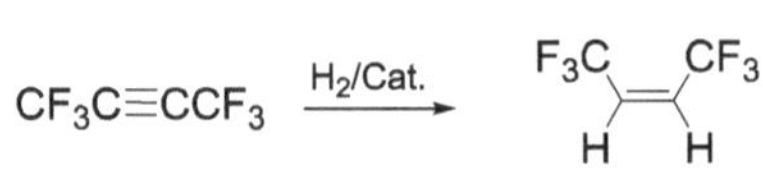

Z-1336mzz

The overall routes to Z-1336mzz via hexafluorobutyne hydrogenation are illustrated below. The first scheme describes a five-step synthesis. 1,2-Dichloro-3,3,3-trifluoroprop-1-ene ($CF_3CCl{=}CHCl$, 1223xd) inserts carbon tetrachloride with cuprous chloride catalyst to give 2,2,3,4,4,4-hexachloro-1,1,1-trifluorobutane

($CCl_3CHClCCl_2CF_3$, 323jxa). 323jxa is then fluorinated to 2,2,3-trichloro-1,1,1,4,4,4-hexafluorobutane ($CF_3CHClCl_2CF_3$, 326max) via an antimony catalyst and anhydrous hydrogen fluoride. Dehydrochlorination of 326max offers 1316mxx using aqueous potassium hydroxide, and the subsequent dechlorination of 1316mxx by zinc generates hexafluorobutyne.[146c]

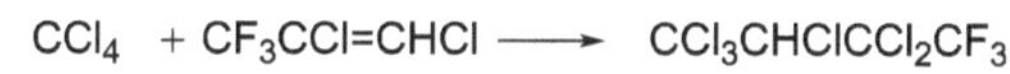

1223xd 323jxa

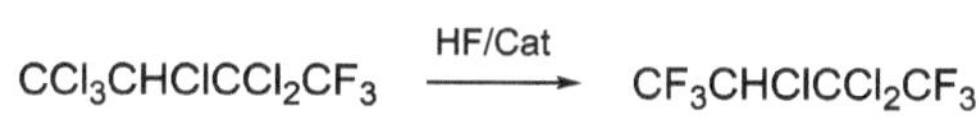

323jxa 326max

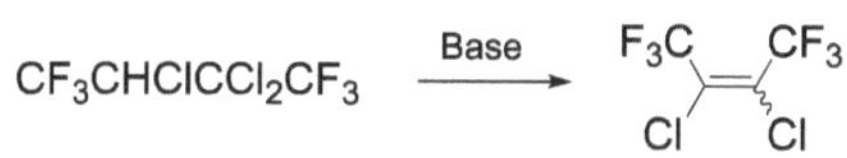

326max *E/Z*-1316mxx

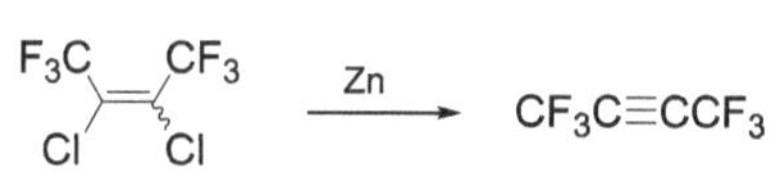

E/Z-1316mxx

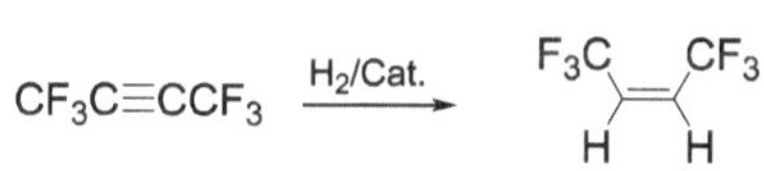

Z-1336mzz

In the second scheme, 1,1,1-trichloro-2,2,3,4,4,4-hexafluorobutane ($CCl_3CF_2CHFCF_3$, 326jcz) is produced from chloroform and hexafluoropropylene. Fluorination of 326jcz by antimony pentachloride and hydrogen fluoride gives 1,1,1,2,2,3,4,4,4-nonafluorobutane ($CF_3CF_2CHFCF_3$, 329mcz), and subsequent gas-phase dehydrohalogenation of 329mcz over a nickel/copper/chromium catalyst leads to hexafluorobutyne.[147]

$$CHCl_3 + CF_3CF{=}CF_2 \longrightarrow CCl_3CF_2CHFCF_3$$

326jcz

$$CCl_3CF_2CHFCF_3 \xrightarrow{HF/Cat} CF_3CF_2CHFCF_3$$

326jcz 329mcz

$$CF_3CF_2CHFCF_3 \xrightarrow{H_2/Cat.} CF_3C{\equiv}CCF_3$$

329mcz

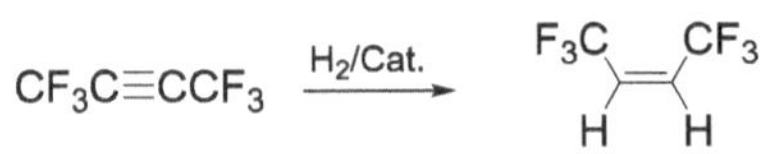

Z-1336mzz

Other methods have been reported for hexafluorobutyne preparation. For example, hexafluorobutadiene is isomerized to hexafluorobutyne at room temperature using AlOClF catalyst in 100% yield.[148] 1326mxz is dehydrochlorinated to hexafluorobutyne with sodium hydroxide in the presence of a PTC.[149]

E/Z-1326mxz

2.7 E-$CF_3CH{=}CHCF_3$, (E)-1,1,1,4,4,4-Hexafluorobut-2-ene, E-1336mzz

(*E*)-1,1,1,4,4,4-hexafluorobut-2-ene (*E*-1336mzz) is the thermodynamic isomer with a boiling point of 7°C. It is a candidate for applications such as electronic gas and foam blowing agent for thermoplastics. The routes to its manufacture are reviewed below.

Carbon tetrachloride is often used as the starting material in *E*-1336mzz synthesis. For instance, a four-step synthesis is reported by ethylene insertion of carbon tetrachloride to 1,1,1,3-tetrachloropropane ($CCl_3CH_2CH_2Cl$, 250fb), dehydrochlorination of 250fb to 3,3,3-trichloroprop-1-ene ($CCl_3CH{=}CH_2$, 1240zf) catalyzed by ferric chloride on carbon, reaction of 1240zf with carbon tetrachloride to 1,1,1,2,4,4,4-heptachlorobutane ($CCl_3CHClCH_2CCl_3$, 340jfd), and final fluorination of 340jfd by hydrogen fluoride over a chromium catalyst.[150]

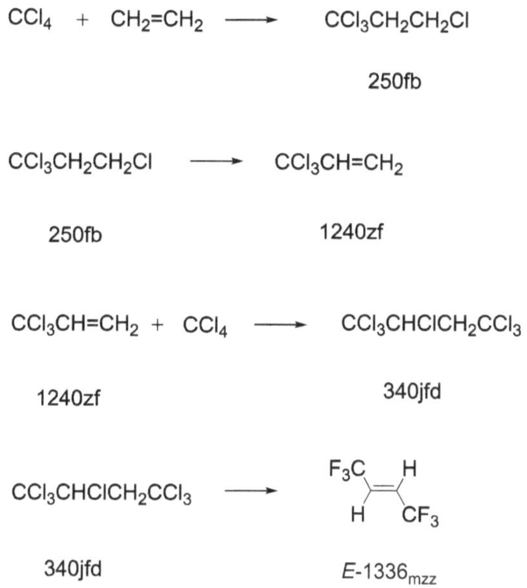

Besides the previously mentioned 123 coupling to form a *Z/E*-1336mzz mixture, 123 is also used in exclusive *E*-1336mzz synthesis. For instance, 123 reacts with vinyl chloride catalyzed by copper chloride and bipyridine in methanol at 120°C to form 2,4,4-trichloro-1,1,1-trifluorobutane ($CF_3CHClCH_2CHCl_2$, 353mdf), followed by chlorination of 353mdf under ultraviolet (UV) irradiation to provide 2,4,4,4-tetrachloro-1,1,1-trifluorobutane ($CF_3CHClCH2CCl_3$, 343jfd).[151] The final 343jfd fluorination over a chromium catalyst produces *E*-1336mzz in greater than 95% selectivity.[151,152]

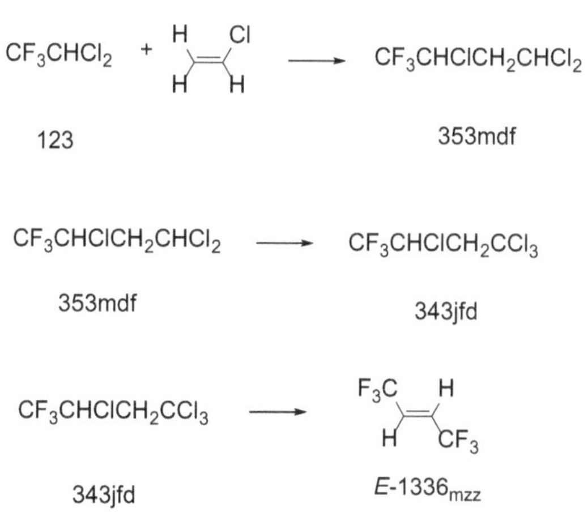

Vinylidene chloride can also be used instead of vinyl chloride,[151] and it shortens the synthesis to two steps.

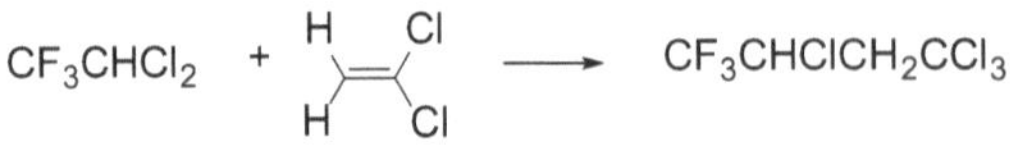

123 343jfd

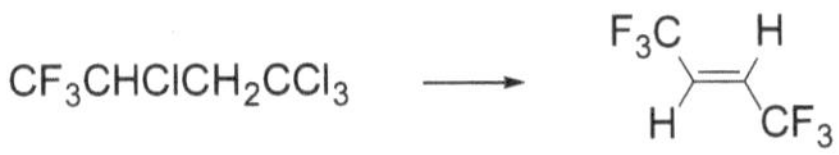

E-1336mzz

3. Other Hydrofluoroolefins

Other hydrofluoroolefins were also prepared for potential application as refrigerants, fire extinguishers, foam expansion agents, etc (Table 3.3). The detailed syntheses can be found in the publication by Yagupolskii et al.[153]

4. Hydrochlorofluoroolefins

4.1 E-CF_3CH=CHCl, E-3,3,3-Trifluoro-1-chloropropene, E-1233zd

(*E*)-1-chloro-3,3,3-trifluoroprop-1-ene (*E*-1233zd) is a hydrochlorofluoroolefin that has been developed as a foam blowing agent and a refrigerant for chillers. It boils at 19°C and has a GWP of 1. Although it contains chlorine and has ODP, its ODP is low (0.0003) due to its short atmospheric life time and it is reported to be nonflammable.

1233zd is produced as a mixture of *E* and *Z* isomers with the *E* isomer as the thermodynamically favored configuration (*E*/*Z* ratio >90/10). Isomerization is often used to improve the selectivity of the desired configuration. The scheme below shows the isomerization of *Z*-1233zd to *E*-1233zd; the reported catalysts are chromium, cobalt, alumina oxide, alumina fluoride, and ferric chloride on carbon.[154]

Z-1233zd *E*-1233zd

Table 3.3 Other Hydrofluorolefins

Structure	BP (°C)
$CHF_2CF{=}CH_2$	5–6
$CF_3CHF{=}CH_2$	24–25
$C_3F_7CF{=}CH_2$	31–32
$C_2F_5OCF{=}CH_2$	6–7
$CHF_2CF_2CF{=}CH_2$	27–28
$CF_3CHFCH{=}CHF$ (Z)	29–31
$CF_3CH_2CH{=}CHF$ (Z)	24–26
$C_3F_7CH{=}CHF$ (3/1 E/Z ratio)	36–38

BP, boiling point.

1233zd formation is often observed as an intermediate in the liquid-phase production of HFC foam blowing agent 245fa. As shown below, the fluorination of 1,1,1,3,3-pentachloropropane ($CF_3CH_2CHCl_2$, 240fa) in the presence of antimony catalyst gives rise to 245fa, 1233zd, 3-chloro-1,1,1,3-tetrafluoropropane (CF_3CH_2CHClF, 244fa), and 3,3-dichloro-1,1,1-trifluoropropane ($CF_3CH_2CHCl_2$, 243fa).[155]

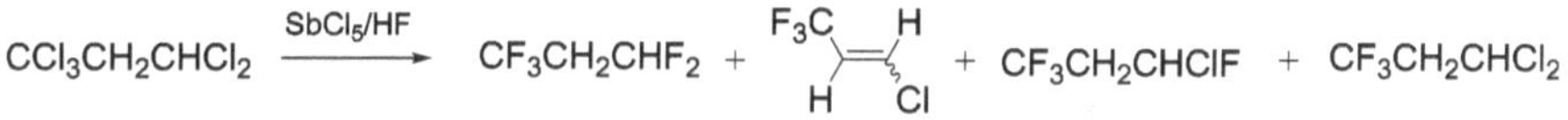

240fa   245fa   E/Z-1233zd   244fa   243fa

45~98%   0.5~13%   0.7~16%   0.1~25%

Under optimum reaction conditions 1233zd can be formed predominantly in 240fa fluorination, and this is the most widely reported route for 1233zd preparation. 240fa can be easily produced through the reaction of carbon tetrachloride and vinyl chloride.

Within this route, quite a few approaches have been reported. In a liquid-phase process, by switching the catalyst from SbF_5 to $TiCl_4$, the selectivity to 1233zd increased from 10% to 90% at 120°C.[156] The reaction can also be carried out successfully by noncatalytic fluorination with neat HF only at 110–200°C, leading to 1233zd in 80–90% selectivity.[157] Gas-phase fluorination has also been explored to improve 1233zd selectivity. By implementing a carbon-supported SbF_5 catalyst, the selectivity to 1233zd can be as high as 78.6%.[158] Furthermore, gas-phase fluorination over fluorinated chromium catalysts (made from Cr_2O_3 or $CrCl_3$ on alumina support) generates 1233zd in 80–95% selectivity at 300–350°C.[159] An oxygen/HF cofeed process is also reported to double the chromium catalyst life.[160]

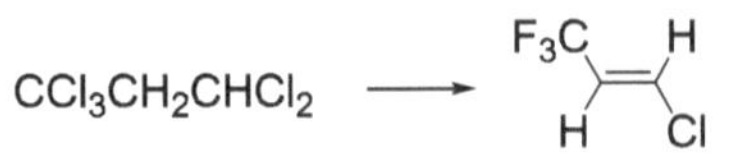

240fa *E/Z*-1233zd

Besides 240fa, other halogenated hydrocarbons are used as the raw material, as illustrated in the scheme below. 1,1,1,3-Tetrachloro-3-fluoropropane (241fb) made by reacting CCl_4 and vinyl fluoride is converted to 1233zd in greater than 60% yield above 280°C. The catalysts used in the fluorination include chromium and chromium/nickel.[161] A noncatalytic fluorination is used to make 1233zd at 82% selectivity at 400°C.[162] 1,1,3,3-Tetrachloropropane (250fa) is prepared from vinyl chloride and chloroform, followed by fluorination to 1233zd.[163]

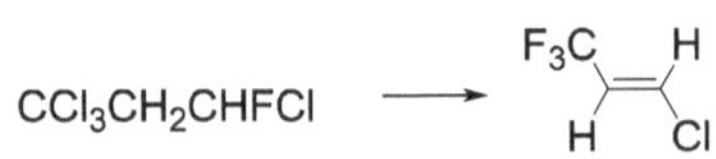

240fa *E*-1233zd

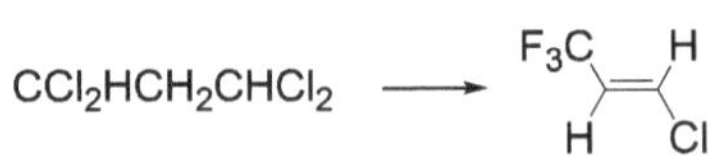

250fa *E*-1233zd

Halogenated olefins such as 1230za (1,1,3,3-tetrachloroprop-1-ene) and 1,3,3,3-tetrachloroprop-1-ene (1230zd) are also reported as raw materials in 1233zd preparation.[156,164] As described in the scheme below, 1230za is fluorinated to 1233zd by anhydrous HF.[164]

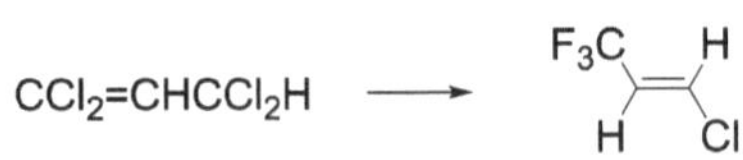

1230za *E*-1233zd

Dehydrohalogenation is another process reported frequently in 1233zd preparation. The common raw materials used in the process are 243fa and 244fa. The dehydrochlorination of 243fa is fairly straightforward via fluorinated chromium oxide or ferric chloride on carbon at 250–400°C,[154a] whereas conversion of 244fa to 1233zd is more challenging because it can simultaneously dehydrofluoridinate to 1233zd and dehydrochlorinate to 1,3,3,3-tetrafluoroprop-1-ene (1234ze). By manipulating the catalyst, either product can be obtained in good selectivity or exclusively in some cases.[165] For example, by using fluorinated chromium oxide, the selectivity to 1233zd is about 75% and to 1234ze is about 21%. A series of lithium-doped catalysts such as ferric chloride on carbon, alumina fluoride, and cerium fluoride give selectivities to 1233zd from 66 to 77% and to 1234ze from 5 to 22%. Graphite boosts the 1233zd selectivity to 95.5%.[165a] Another very interesting development is the observation that metal chlorides such as lithium chloride and potassium chloride on carbon supports give very high selectivity to 1234ze (>90%). Ferric chloride is an exception, favoring 1233zd, whereas the alkaline metal chloride–doped magnesium fluorides almost make 1234ze exclusively. Liquid-phase caustic dehydrohalogenation by KOH in the absence of a PTC also favors the formation of 1234ze (75% selectivity).

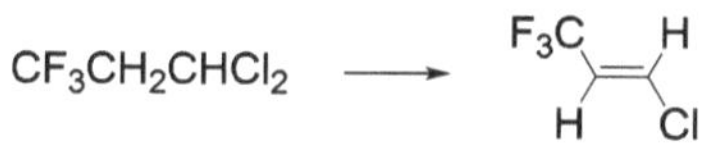

243fa E-1233zd

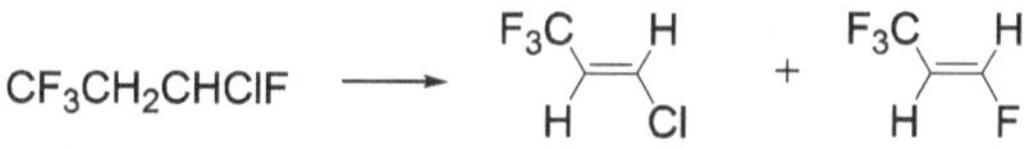

244fa E-1233zd E-1234ze

4.2 Z-CF_3CH=CHCl, E-3,3,3-Trifluoro-1-chloropropene, E-1233zd

(Z)-1-chloro-3,3,3-trifluoroprop-1-ene (Z-1233zd) is a stereoisomer of *E*-1233zd. It has a boiling point of 38°C and is considered as a solvent and chiller fluid.

Z-1233zd is made as a minor isomer (<20% of formed *E* isomer) in *E*-1233zd synthesis. To obtain it in a commercial quantity, the most commonly reported method is *E*-1233zd isomerization. This isomerization is achieved either by using 316 SS as the catalyst[166] or heating the substance at very high temperature (up to 450°C).[167]

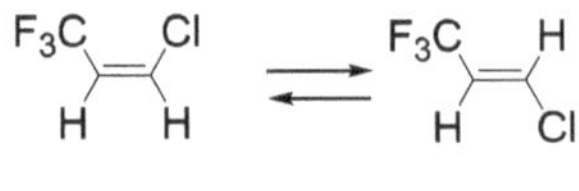

Z-1233zd *E*-1233zd

Other methods include hydrogenation of 1-chloro-3,3,3-trifluoroprop-1-yne over Lindlar catalyst[168] and hydrogen chloride addition to trifluoropropyne catalyzed by copper chloride and 1-butyl-3-methylimidazolium chloride.[169]

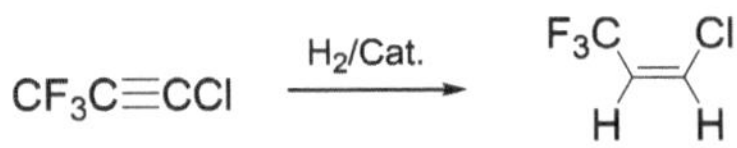

Z-1233zd

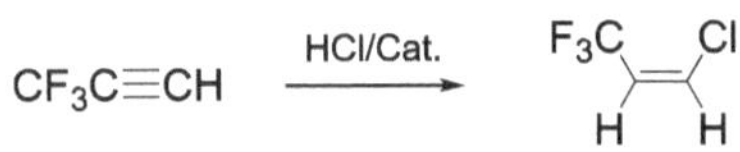

Z-1233zd

4.3 $CF_3CCl{=}CH_2$, 3,3,3-Trifluoro-2-chloro-1-propene, HFCO-1233xf

HFCO-1233xf is the most referenced precursor to make 1234yf. Its synthesis has been intensively studied, and four of its synthetic routes are summarized below.

4.3.1 1230xa ($CCl_2{=}CClCH_2Cl$) Route

The first route uses 1230xa as the feedstock, and this route has been the most extensively studied in the literature.

$$\underset{\text{1230xa}}{CCl_2{=}CClCH_2Cl} + 3\,HF \longrightarrow \underset{\text{1233xf}}{CF_3CCl{=}CH_2} + 3\,HCl$$

The majority of reported hydrofluorinations of 1230xa are done in vapor phase with Lewis acid catalysts, such as Cr_2O_3, Al_2O_3, Cr_2O_3/Al_2O_3, Cr_2O_3/AlF_3, Cr_2O_3/C, $CrCl_3/C$, and $CoCl_2/AlF_3$.[78,170–182] The selectivity of 1233xf in vapor-phase hydrofluorination of 1230xa is very high, and a value of 95% has been observed. However, the catalyst deactivated quickly, and the addition of stabilizers such as phenols, hydroquinones, and amines in 1230xa were reported to extend the catalyst life.[171] Cofeeding of oxygen was also reported to be effective in maintaining the catalyst stability[173]; however, added oxygen generally reduces selectivity due to the formation of chlorinated by-products. Other measures such as controlling the reaction hot spot in the catalyst bed,[175] moisture levels in the feeds,[176,178] using high-purity 1230xa,[177] modifying the catalyst with adjuvants,[181,182] and cofeeding of Cl_2[178] are all reported to be beneficial for catalyst life.

There are also reports on the liquid-phase hydrofluorination by HF over Lewis acids such as $SbCl_5$, $SbCl_3$, $TiCl_4$, $SnCl_4$, $NbCl_5$, $FeCl_3$, $TaCl_5$, and $MoCl_6$ with or without a solvent.[94,183] The liquid-phase hydrofluorination of 1230xa gave yields of less than 80%, much lower than reactions carried out in the vapor phase.

Nucleophilic substitution of 1230xa were carried out with fluoride salts such as alkali metal fluoride, alkali metal hydrogen fluoride, *tert*-amino hydrogen fluoride, fluorinated quaternary ammonium salts, or fluorinated quaternary ammonium hydrogen fluoride 1231xa with 85–98% selectivity, followed by hydrofluorination of 1231xa in the vapor phase to 1233xf, which reduces the need of stabilizer.[184]

4.3.2 2,3,3,3-Tetrachloropropene ($CCl_3CCl{=}CH_2$) 1230xf Route

HCCO-1230xf is a less thermally stable isomer of 1230xa, and it can be converted to 1233xf by vapor-phase hydrofluorination under similar conditions as 1230xa with ~92% selectivity to 1233xf.[170,185]

$$\underset{\text{1230xa}}{CCl_3CCl{=}CH_2 + 3\,HF} \longrightarrow \underset{\text{1233xf}}{CF_3CCl{=}CH_2 + 3\,HCl}$$

4.3.3 240db 1,1,1,2,3-Pentachloropropane Route

HCC-240db is a precursor to make 1230xa[185,186] and 1230xf; however, 240db itself can be directly hydrofluorinated to 1233xf with about 70–98% 1233xf selectivity[78,187–193] at conditions similar to 1230xa hydrofluorination.

$$\underset{\text{240db}}{CCl_3CHClCH_2Cl + 3\,HF} \longrightarrow \underset{\text{1233xf}}{CF_3CCl{=}CH_2 + 4\,HCl}$$

The liquid-phase hydrofluorination of 240 db[190,191] is less selective for making 1233xf than the vapor-phase processes.[188,192,193] Other pentachloropropanes such as 1,1,2,2,3-pentachloropropane (240aa) and 1,1,1,2,2-pentachloropropane(240ab) can be hydrofluorinated to 1233xf as well.[188,194,195]

4.3.4 3,3,3-Trifluoropropene ($CF_3CH{=}CH_2$) 1243zf Route

HFO-1243zf can be converted to 1233xf by vapor-phase chlorination and dehydrochlorination in one step with Cl_2 over an activated carbon catalyst[196]; however, the

1233xf selectivity is less than 88%. Higher 1233xf selectivity was achieved using the two-step process described below:

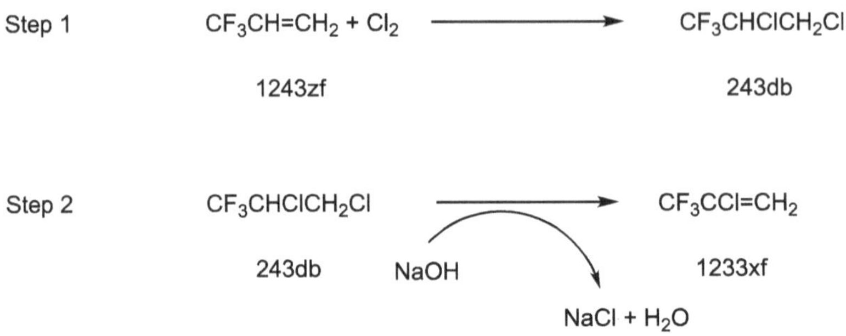

The chlorination of 1243zf was carried out via UV photochlorination[197] or thermally at 250°C in the vapor phase.[198] Although photochlorination achieved 99% selectivity, it is typically more expensive to scale-up. The vapor-phase chlorination 1243zf was also carried out using catalysts, such as activated carbon, alumina oxide, chromium oxide, oxides of a transition metals, and metal halides.[196,199–201] The catalytic process operates at milder temperatures such as 80–90°C and reported 98% selectivity, which is more selective than the thermal process (80–90% selectivity).

Liquid-phase chlorination of 1243zf was reported[201,202] without catalyst or with metal halide catalyst. Although the selectivity to 243db in liquid phase can reach 99%, which is higher than that in vapor phase, the liquid-phase process may be more difficult to control on a large scale.

There are many reports regarding the vapor-phase dehydrochlorination of 243db over catalysts such as activated carbon, alumina oxide, chromium oxide, and transition metal oxides.[199,203–206] The dehydrochlorination of 243db molecule can occur in two ways, making 1233xf ($CF_3CCl{=}CH_2$) or 1233zd ($CF_3CH{=}CHCl$) depending on which hydrogen is removed. The selectivity to 1233xf over transition metal oxides such as chromium oxide is around 50–92%, which is due to the loss of selectivity to 1233zd. However, if HF is cofed with 243db, 1233zd formation was suppressed and the 1233xf selectivity was improved to 92–95%.[204] 243db Dehydroclorination over catalysts such as activated carbon and alkaline metal halide loaded on carbon led to 98.5% selectivity for 1233xf.[205,206] The noncatalytic vapor-phase dehydrochlorination of 243db resulted in about 90% selectivity to 1233xf, lower than that of the catalytic processes.[207]

243db dehydrochlorination was also carried out in the liquid phase using aqueous caustic in the presence of a PTC[208] and it resulted in 90–97% yield of 1233xf. There is also a report of using CaO, CaO_2, or $Ca(OH)_2$ as caustic without a catalyst.[209]

4.4 $CF_3CF{=}Cl_2$, 2,3,3,3-Tetrachloro-1,1-dichloro-1-propene, HCFO-1214ya

HCFO-1214ya boils at 46°C,[210] and it has potential uses as solvent and as a heat transfer medium.[211–213] It is also an important precursor to make 1234yf and can be synthesized by the following routes.

4.4.1 TFE-HCFC-21 ($CHCl_2F$) Route

This route is a three-step process and has been commercialized in Japan by Asahi Glass. This process makes a 225 mixture, marketed as a solvent, which contains about 50% 225ca, which was converted to 1214ya. The other 225s in the mixture are 225cb and 225aa, which can be isomerized to 225ca.

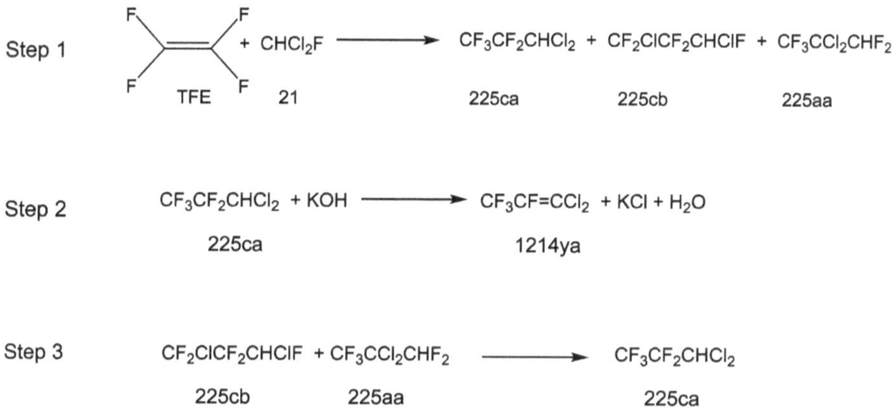

HCFC-21 reacts with TFE in the presence of Lewis acid catalysts, such as chlorides, chlorofluorides, bromides, and bromofluorides of Sb, Al, Nb, Ta, Ga, In, Zr, Hf, Ta, and Ga, at 10–120°C to make 225 isomers.[214–217]

This reaction gave 80% selectivity to 225 isomers, and the three primary isomers are 225ca, 225cb, and 225aa. Among them, about 50% was 225ca, 45% was 225cb, and about 2% was 225aa.[215] All 225 isomers have very similar boiling points (~51°C), and it very difficult to separate them in pure form. Different catalysts affected the selectivity of 225s and also the distribution of isomers.[215–217] Zirconium tetrachloride and hafnium tetrachloride had the highest catalytic activity and improved selectivity to HCFC-225ca and HCFC-225cb of greater than 96%.[216,217]

The main side reaction in step 1 above is the disproportion of HCFC-21 to HCFC-22 and chloroform, and the chloroform was very difficult to remove from the 225 isomers. Although the process was typically run lean in TFE for both safety and cost, this process was run with excess TFE to suppress chloroform formation. Lower reaction temperature also reduced chloroform formation.[216]

225ca was converted to 1214ya by reacting with aqueous KOH with a PTC such as tetrabutyl ammonium bromide at 45°C. Since 225cb is difficult to separate from the 225ca, the mixture of 225ca and 225cb was reacted with caustic KOH together; only the 225ca reacted with KOH and both 225cb and 225aa remained unreacted.[218,219] Since the boiling point of 225cb and 225aa are about 10°C higher than 1214ya, 1214ya was separated from the unreacted 225cb.[218]

225ca can also be dehydrofluorinated over vapor-phase catalysts[57,220] such as fluorinated Cr_2O_3 and Al_2O_3; however, it is not as selective as the liquid-phase process using caustic. Moreover, the vapor-phase process generated by-products such as 1214yb ($CF_2ClCF{=}CClF$), HCFC-224, and HCFC-226 isomers.

To improve the overall yield of this process, 225cb was isomerized to 225ca over Lewis acid catalysts such as metal halide and fluorinated metal oxide. When $AlCl_3$ is used, liquid-phase process was run at mild temperature (0–50°C). When fluorinated Al_2O_3 is used, a vapor-phase process was carried out at higher temperature (200–500°C).[218,219] 225aa was also formed during this isomerization process and can be recycled back to the isomerization reactor to recover its value.

4.4.2 Other Synthetic Routes to 1214ya

These routes are not as commercially attractive as those starting with TFE since they either use stoichiometric reagents such as Zn or expense hexafluoropropylene.

The dehalogenation of 1,1,1,2-tetrachloro-2,3,3,3-tetrafluoropropane by zinc is as follows:[221]

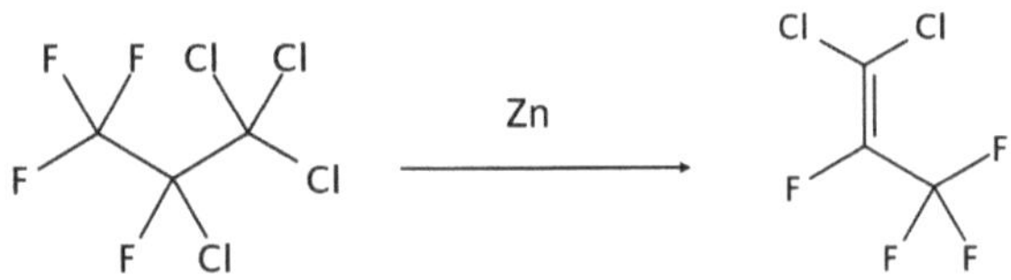

The disproportionation of 1-chloro-1,2,3,3,3-HFP over Cr_2O_3 catalyst proceeds as follows:[222]

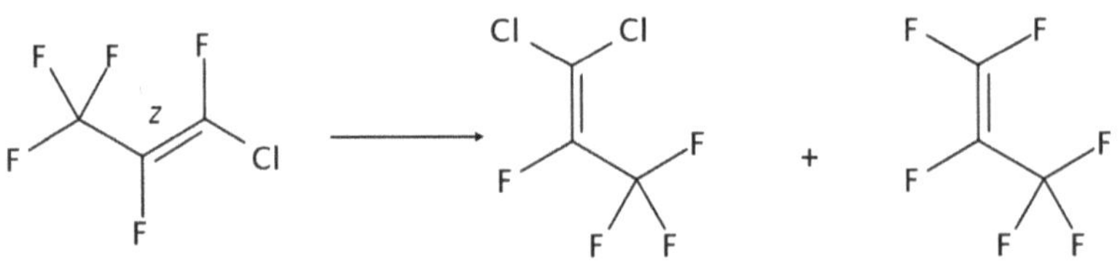

Hydrochlorination of HFP by HCl occurs as:[223]

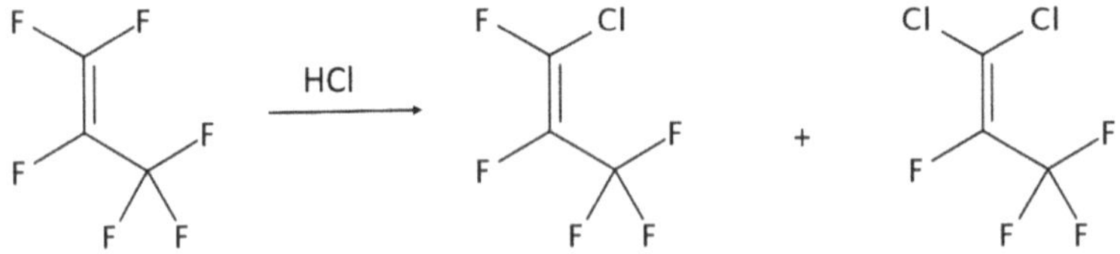

5. Conclusion

Interest in hydrohaloolefins has been growing markedly due to the increased global warming pressure on HFCs and the cessation of production of HCFCs as mandated by the Montreal Protocol. The number of carbon atoms used to make the olefins is on average more than in HFCs and HCFCs due to the need to adjust the physical property of the olefin to match the need in the marketplace; the more the number of carbons, the greater is the number of ways to connect the atoms to give molecules with the desired properties. This has led to many more chemicals that can be made and in many cases more involved and costly routes to make them. Moreover, some selections are stereoisomers that are thermodynamically unfavored, and this complicates the preparations as well. The number of literature references for the preparation of hydrohaloolefins is vast, and it was attempted to give a synopsis of the main routes to the most commercially significant "new" olefins that are now or soon likely to become commercially available. The task of finding a hydrofluoroolefin to replace the incumbents in the marketplace, while not daunting, is a challenge. The selected olefins need to meet all the important criteria such as having low toxicity, an appropriate level of flammability, and effectiveness in use; being environment friendly; and most importantly, having a process to make them that is economically viable.

Most of the routes discussed in this chapter fall into two categories, vapor phase and liquid phase. In general, the liquid-phase processes have better temperature control, run at lower temperature, use less costly reactors, and favor saturated molecules, whereas vapor-phase routes run at high temperature, favor olefins, and can have lower selectivity due to complex equilibria that can prevail at these temperatures giving a variety of products. This was certainly seen in the high-temperature pyrolysis reactions to make 1234yf. Intermediate temperatures for vapor-phase reactions give acceptable kinetics and selectivities for olefins and were commonly discussed within. The challenge is to find the right molecule and the right process having the appropriate number of individual steps that can be easily and safely be practiced on a large commercial scale.

References

1. Shepherd, J. G. *Phil. Trans. R. Soc. A* **2012,** 4166–4175.
2. Sievert, A. C.; Nappa, M. J.; Minor, B. H.; Leck, T. J.; Rao, V. N. M.; Swearingen, E. N.; Schmitz, C.; Mouli, N.; Perti, D. Patent WO 2007053697 to E. I. du Pont de Nemours and Co., 2007.
3. Wallington, T. J.; Sulbaek Andersen, M. P.; Nielsen, O. J. Atmospheric chemistry of short-chain haloolefins : photochemical ozone creation potentials (POCPs), global warming potentials (GWPs), and ozone depletion potentials (ODPs). *Chemosphere* **2015,** *129,* 135–141.
4. Swarts, F. *Bull. Acad. Roy. Belg* **1892,** *24,* 309–314.
5. Tanaka, T.; Okamoto, H.; Ueno, K.; Irisawa, J.; Otsuka, T.; Nogami, T.; Dobashi, R. *Development of a New Low-GWP Refrigerant Composed of HFO-1123 (Trifluoroethylene);* AICHE Meeting: Atlanta, GA, November 18, 2014.
6. Hulbert, J. D.; Feiring, A. E. *C&ENews* **1997,** *75* (51), 6.
7. Rucker, J.T.; Stormon, D.B. US 2,760,997 to Hooker Electochemical Company, 1956.
8. Miller, C. B.; Pearson, J. H.; Smith, L. B. US 2,802,887 to Allied Chemical & Dye Corporation, 1957.
9. Nakagawa, S. US 3,564,064 to Daikin Kogyo Katiushiki Kaisha, 1971.

10. Lurot, L.; Wilmet, V.; Pirotton, J. EP343707 A1 to Solvay & Cie, 1992.
11. Saiki, T.; Sumida, M.; Nakano, S.; Murakami, K. EP 471320 A1 to Japan Halon, 1992.
12. Hub, S.; Lantz, A.; Shirmann, J. -P. FR 2710054 to Elf Atochem, 1995.
13. Powell, R. L.; Sharratt, A. P. US 5,856,593 to Imperial Chemical Industries PLC, 1999.
14. Powell, R. L.; Sharatt, A. P. WO 9729065 to Imperial Chemical Industries PLC, 1997.
15. Zhu, Z.; Jia, W.; Wei, J. CN 103288589 A to Tonghi University, 2013.
16. Takeuchi, Y.; Furuta, S. JP2015117188 to Asahi Glass, 2015.
17. Nakamura, M.; Okamoto, H. WO2015115548 to Asahi Glass, 2015.
18. Nakamura, M.; Okamoto, H. WO2015115549 to Asahi Glass, 2015.
19. Singh, R.; Pham, H. T.; Wilson, D. P.; Thomas, R. H. WO2004037913 to Honeywell, 2004.
20. Hazeldine, R. N.; Steele, B. R. *J. Am. Chem. Soc.* **1953,** 1199.
21. Knunyants, I. L.; Krasuskaya, M. P.; Mysov, E. I. Izvestiya Akademii Nauk SSSR. *Seriya Khimicheskaya* **1960,** 1412.
22. Yoshikawa, S.; Tamai, R.; Hibino, Y. JP10007605 to Central Glass, 1998.
23. Yoshikawa, S.; Tamai, R.; Hibino, Y. JP10007604 to Central Glass, 1998.
24. Tung, H. S. US 5, 710, 352 to Allied Signal Inc., 1998.
25. Elsheikh, M. Y. US 5,811,603 to Elf Atochem North America, 1998.
26. Nair, H. K.; Van Der Puy, M.; Nalewajek, D.; Demmin, T. R.; Poss, A. J.; Bradley, D. E.; Shankland, I. R. US2003/0060670 A1 to Honeywell, 2003.
27. Merkel, D. C.; Singh, R. R.; Tung, H. S. US 7, 230146 B2 to Honeywell International, Inc., 2007.
28. Mukhopadhyay, S.; Nair, H. K.; Tung, H. S.; Van Der Puy, M. US 7,345,209 B2 to Honeywell International Inc., 2008.
29. Wang, H.; Tung, H. S.; Chie, Y.; Cerri, G.; Cottrell, S. A. US 7,485,760 B2 to Honeywell International Inc., 2009.
30. Okamoto, S.; Takada, N.; Sakyu, F.; Kitamoto, T.; Tonomura, T.; Kanai, M. US 20150112103 to Central Glass Company, Unlimited, 2015.
31. Petrov, V. A.; Nappa, M. J. US 7,420,094 to E. I. E. I. du Pont de Nemours and Co. de Nemours and Company, 2008.
32. Sun, X.; Nappa, M. J. US 2014/0350310 to E. I. E. I. du Pont de Nemours and Co. de Nemours and Company, 2014.
33. Minor, B.; Spatz, M. *International Refrigeration and Air Conditioning Conference. Paper 937,* 2008.
34. *Experimental Study of R-134a Alternative in a Supermarket Refrigeration System;* The Free Library. 2011 American Society of Heating, Refrigerating, and Air-Conditioning Engineers, Inc., October 29, 2015. http://www.thefreelibrary.com/Experimental+study+of+R-134a+alternative+in+a+supermarket...-a0272754897.
35. Rao, V. N. M.; Sievert, A. C.; Nappa, M. J. WO 2008030440 A2 to E. I. du Pont de Nemours and Co., 2008.
36. Van der Puy, M.; Cook, G. R.; Scheidle, P. H.; Uhrich, P. H.; Wang, K. D.; Wang, H.; Tung, H. S. US7560602 B2 to Honeywell International Inc., USA, 2009.
37. Low, R. E.; Sharratt, A. P. WO2009093047 A2 to Ineos Fluor Holdings Limited, 2009.
38. Smith, J. W.; McGuiness, C.; Sharratt, A. P.; Taylor, A. M. WO2009138764 A1 to Ineos Fluor Holdings Limited, UK, 2009.
39. Pigamo, A.; Devic, M.; Wendlinger, L. WO2010029240 A1 to Arkema, 2010.
40. Devic, M.; Doucet, N.; Wendlinger, L.; Cavallini, G. WO2010142878 A1 to Arkema, 2010.

41. Wang, H.; Tung, H. S.; Merkel, D. C. WO2011017138 A2 to Honeywell International Inc., USA, 2011.
42. Shiotani, Y.; Hanabusa, K.; Chaki, T.; Takahashi, K. WO2011122157 A1 to Daikin Industries, 2011.
43. Cheung, P.; Sievert, A. C.; Lamarca, C. WO2012158848 A2 to E. I. du Pont de Nemours and Co., 2012.
44. Nappa, M. J.; Rao, V. N. M.; Sievert, A. C. WO2008002499 A2 to E. I. du Pont de Nemours and Co., 2008.
45. Mukhopadhyay, S.; Nair, H. K.; Tung, H. S.; Van Der Puy, M.; Merkel, D. C.; Dubey, R. K.; Ma, J. J. US20070112230 to Honeywell International Inc., 2007.
46. Nappa, M. J. WO2007117391 A1 to E. I. du Pont de Nemours and Co., 2007.
47. Wang, H.; Tung, H. S.; Singh, R. R.; Shankland, I. EP2149543 A1 to Honeywell International Inc., 2010.
48. Nappa, M. J.; Jackson, A. WO2010141664 A1 to E. I. du Pont de Nemours and Co., 2010.
49. Elsheikh, M. Y.; Bonnet, P.; Keeley, O. C. N.; Chen, B. B. WO2011140013 A1 to Arkema, 2011.
50. Sharratt, A. P.; McGuinness, C. E.; Carolan, S. C. WO2013164618 A1 to Ineos Fluor Holdings Limited, 2013.
51. Han, C.; Zhang, Y.; Zhang, H.; Gao, Z.; Chen, Y. CN102267869 A to Shandong Dongyue Polymers, 2011.
52. Rao, V. N. M.; Sievert, A. C.; Nappa, M. J. WO2008030439 A2 to E. I. du Pont de Nemours and Co., 2008.
53. Cook, G. R.; Kopkalli, H.; Cottrell, S. A.; Chiu, Y.; Scheidle, P.; Merkel, D. C. US20110269999 A1 to Honeywell International Inc., USA, 2011.
54. Pigamo, A.; Devic, M.; Wendlinger, L. WO2010029239 to Arkema, 2010.
55. Nappa, M. J.; Cheung, P.; Krause, K. R.; Sisk, M. A. WO2010129844 A1 to E. I. du Pont de Nemours and Co., 2010.
56. Bossoutrot, J. -M.; Sedat, P. -M. WO2011010023 A1 to Arkema, 2011.
57. Rao, V. N. M.; Sievert, A. C. WO2008060614 A2 to E. I. du Pont de Nemours and Co., 2008.
58. Nose, M.; Suzuki, A.; Sugiyama, A. WO2010013576 A1 to Daikin Industries, 2010.
59. Okamoto, H.; Sasao, Y. WO2010074254 A1 to Asahi Glass Company, 2010.
60. Takagi, H.; Okamoto, H. WO2011162340 A1 to Asahi Glass Company, 2011.
61. Kawaguchi, S.; Okamoto, H.; Takeuchi, Y.; Takagi, H.; Watanabe, K.; Yanase, K. WO2011162338 A1 to Asahi Glass Company, 2011.
62. Kawaguchi, S.; Okamoto, H.; Takeuchi, Y.; Takagi, H.; Watanabe, K.; Yanase K. WO2011162337 A1 to Asahi Glass Company, 2011.
63. Nomura, S.; Okamoto, S.; Takagi, Y. JP2013180964 A to Asahi Glass Co., 2013.
64. Yoshikawa, S.; Takada, N.; Oshio, H. US5756869 A to Central Glass, 1998.
65. Zhang, Y.; Han, C.; Zhang, H.; Gao, Z.; Chen, Y. CN102295522 to Shandong Dongyue Polymer, 2011.
66. Miller, R. N.; Nappa, M. J.; Rao, V. N. M.; Sievert, A. C. US20060106263 A1 to E. I. du Pont de Nemours and Co., 2006.
67. Tung, H. S.; Mukhopadhyay, S.; Van der Puy, M.; Ma, J. J.; Merkel, D. C.; Bortz, C.; Light, B.; Phillips, S. D.; Fleming, K. M.; Ferguson, S. Compounds WO2007079435 A2 to Honeywell International Inc., 2007.
68. Takahashi, K.; Nose, M.; Chaki, T. WO2009066580 A1 to Daikin Industries, 2009.
69. Bektesevic, S.; Tung, H. S.; Wang, H.; Merkel, D. C.; Johnson, R. C. US20110270000 A1 to Honeywell International Inc., 2011.

70. Komatsu, Y.; Karube, D. WO2011099604 to Daikin Industries, 2011.
71. Joyce, R. M., Jr. US2462402 to E. I. du Pont de Nemours and Co., 1949.
72. Coffman, D. D.; Cramer, R.; Rigby, G. W. *J. Am. Chem. Soc.* **1949,** *71,* 979−980.
73. Morikawa, S.; Samejima, S.; Ohnishi, K.; Okamoto, H.; Ohmori, T.; Tanuma, T. WO9101287 A1 to Asahi Glass Co. Ltd., 1991.
74. Nose, M.; Karube, D.; Sugiyama, A.; Shibanuma, T.; Chaki, T. WO2010013796 A1, to WO2010013795, 2010.
75. Karube, D.; Chaki, T.; Yuko, S.; Sugiyama, A.; Nose, M.; Shibanuma, T. WO2010013795 A1 to WO2010013795, 2010.
76. Tung, H. S.; Mukhopahyay, S.; Van der Puy, M.; Ma, J. J.; Bortz, C.; Light, B.; Phillips, S. D.; Dubey, R.; Merkel, D. C. WO2007079431 A2 to Honeywell International Inc., 2007.
77. Merkel, D. C.; Johnson, R. C.; Tung, H. US20090182179 A1 to Honeywell International Inc., 2009.
78. Merkel, D. C.; Tung, H. S.; Pokrovski, K. A.; Bektesevic, S.; Johnson, R. C.; Wang, H. EP2103587 A2 to Honeywell International Inc., USA, 2009.
79. Nappa, M. J.; Jackson, A.; Merkel, D. C. WO2013071024 A1 to E. I. du Pont de Nemours and Co., 2013.
80. Wang, H.; Merkel, D. C.; Tung, H. S.; Kopkalli, H. US20140275646 A1 to Honeywell International Inc., 2014.
81. Johnson, R. C.; Merkel, D. C. WO2012009114 A2 A1 to Honeywell International Inc., 2012.
82. Chen, X.; Wang, S.; Zhou, Q.; Zhou, L.; Lei, J.; He, H. CN103449962 A to Juhua Technology Center, 2013.
83. Wang, H.; Tung, H. S. WO2009009421 A1 to Honeywell International Inc., 2009.
84. Sun, X.; Nappa, M. J. US20120215037 A1 to E. I. du Pont de Nemours and Co., 2012.
85. Nappa, M. J. US20120232317 to E. I. du Pont de Nemours and Co., 2012.
86. Zhang, W.; Lv, J.; Zeng, J.; Wang, B.; Hao, Z.; Feng, L.; Yuan, J.; Shi, L. CN101913988 B to Xi'An Modern Chemistry Research Institute, 2013.
87. Wang, H.; Tung, H. S.; Merkel, D. C. WO2013055722 A1 to Honeywell International Inc., 2013.
88. Sun, X.; Nappa, M. J.; Wang, H.; Kopkalli, H. WO2013106305 A1 to E. I. du Pont de Nemours and Co., 2013.
89. Wang, H.; Tung, H. S.; Kopkalli, H.; Chiu, Y.; Cerri, G.; Ball, J.; Da Prato, P.L.; Sun, X.; Nappa, M. J. WO2013119919 A1 to Honeywell International Inc., 2013.
90. Bektesevic, S.; Tung, H. S.; Wang, H.; Merkel, D. C.; Johnson, R. C. WO2011139646 A2 Honeywell International Inc., 2011.
91. Phillips, S.; Fleming, K.; Dubey, R.; Bortz, C.; Mukhopadhyay, S. WO2009026526 A1 to Honeywell International Inc., 2009.
92. Suzuki, A.; Nose, M.; Yamashita, T. WO2010123154 A2 to Daikin Industries, 2010.
93. Pigamo, A.; Deur-Bert, D.; Wendlinger, L. WO2011077191 A1 to Arkema, 2010.
94. Elsheikh, M. Y.; Bonnet, P.; Chen, B. B. US 20120078020 A1 to Arkema Inc., 2012.
95. Takahashi, K.; Motoyama, H.; Chaki, T.; Karube, D.; Kishimoto, M. WO2014025065 A1 to Daikin Industries, 2014.
96. Mukhopadhyay, S.; Tung, H. S.; Van Der Puy, M.; Merkel, D. C.; Ma, J. J.; Bortz, C. L.; Light, B. A.; Phillips, S. D.; Fleming, K. M.; Ferguson, S. A. US 20070197841 A1 to Honeywell International Inc., USA, 2007.
97. Nappa, M. J.; Sun, X.; Yagupolskii, L. M.; Filatov, Andrey A.; Boiko, V. N.; Yagupolskii, Y. L. WO2009067571 A1 to E. I. du Pont de Nemours and Co., 2009.

98. Mukhopadhyay, S.; Nair, H. K.; Tung, H. S.; Van Der Puy, M.; Johnson, R. C.; Merkel, D. C. US20070112227 A1 to Honeywell International Inc., 2007.
99. a. Van Der Puy, M. US20090253946 A1 to Honeywell International Inc., 2009.b. Han, C.; Zhang, Y.; Zhang, H.; Chen, Y. CN102442880 to Shandong Dongyue Polymer Materials, 2012.
100. Haszeldine, R. N.; Steele, B. R. *J. Chem. Soc.* **1953,** *1592.*
101. Brown, S. J.; Zilio, C.; Cavalini, A. *Int. J. Refrig.* **2010,** *33,* 235.
102. Kim, C. S.; Yoo, K.; Ahn, B. S.; Lee, H. J.; Ha, J. M.; Kim, H.G., US20150045589 to Korea Institute of Science and Technology, S. Korea, 2015.
103. Kim, C. S.; Yoo, K. S. *Res. Chem. Intermed.* **2014,** *40* (7), 2471.
104. Vanderpuy, M.; Cook, G. R.; Scheidle, P. H.; Ulrich, K. D.; Wang, H.; Tung, H. S.; Shankland, I. WO2008057794 A1 to Honeywell International Inc., 2008.
105. Cheung, P.; Sievert, A. C.; Lamarca, C. WO2012158848 A2 to E. I. du Pont de Nemours and Company, 2012.
106. Han, C.; Zhang, Y.; Zhang, H.; Gao, Z.; Chen, Y. CN102267869 A to Shandong Dongyue Polymer Material Co., Ltd., Peop. Rep. China, 2011.
107. Shiotani, Y.; Hanabusa, K.; Chaki, T.; Takahashi, K. WO2011122157 A1 by to Daikin Industries, Ltd., Japan, 2011.
108. Wang, H.; Tung, H. S. WO2011028415 A2 to Honeywell International Inc., USA, 2011.
109. Lv, J.; Zhang, C.; Zhang, W.; Hao, Z.; He, F.; Kou, L.; Wang, B.; Pang, G. CN 101628849 A to Xi'an Modern Chemistry Research Institute, P. R. China, 2010.
110. Van Der Puy, M.; Cook, G. R.; Scheidle, P. H.; Uhrich, K. D.; Wang, H.; Tung, H. S. US7560602 B2 to Honeywell International Inc., USA, 2009.
111. Yang, S.; Tung, H. S. WO2009064939 A2 to Honeywell International Inc., USA, 2009.
112. Aoyama, H.; Yamamoto, A.; Shibata, N. JP08165256 A to Daikin Ind Ltd, Japan, 1996.
113. Rao, V. N. M.; Sievert, A. C.; Nappa, M. J. WO2008030439 A2 to E. I. du Pont de Nemours and Company, 2008.
114. Avril, K.; Collier, B. WO2011010025 A1 to Arkema France, Fr., 2011.
115. Pigamo, A.; Devic, M.; Wendlinger, L. WO2010029240 A1 by to Arkema France, Fr., 2010.
116. Bossoutrot, J. M.; Sedat, P. M. WO2011010023 A1 by to Arkema France, Fr., 2011.
117. Rao, V. N. M.; Munirpallam, S. A. US6031141 A to E. I. Du Pont de Nemours & Co., USA, 2000.
118. Nappa, M. J. WO2007117391 A1 to E. I. Du Pont de Nemours & Co., USA, 2007.
119. Sievert, A. C.; Nappa, M. J.; Mahler, B. A. WO 2008030441 A1 to E. I. Du Pont de Nemours & Co., USA, 2008.
120. Pigamo, A.; Devic, M.; Wendlinger, L. WO 2009153493 A1 to Arkema France, Fr., 2009.
121. Hua, Y. CN104311385 A, 2015.
122. Uenveren, E.; Kemnitz, E.; Ruediger, S.; Dimitrov, A.; Eicher, J. WO 2009010472 A1 to Solvay Fluor GmbH, Germany and Humboldt-Universitaet Zu Berlin, 2009.
123. Rao, V. N. M.; Sievert, A. C.; Nappa, M. J. WO 2008030439 A2 to to E. I. Du Pont de Nemours & Co., USA, 2008.
124. Haszeldine, R. N. *J. Chem. Soc.* **1953,** *1592.*
125. Synquest Laboratories, 13201 Rachael Blvd, Alachua, FL 32615.
126. Burton, D. J.; Spawn, T. D.; Heinze, P. L.; Bailey, A. R.; Shin-ya, S. *J. Fluor. Chem.* **1989,** *44,* 167.
127. Nappa, M. J.; Toton, D. J.; Rao, V. N. M. WO2008008351 A2 to E. I. Du Pont de Nemours & Co., USA, 2008.

128. Nappa, M. J. WO2009026080 A1 to E. I. Du Pont de Nemours & Co., USA, 2009.
129. Belen'kii, G. G.; Petrov, V. A.; Resnick, P. R. *J. Fluor. Chem.* **2001,** *108* (1), 15–20.
130. Sun, X.; Petrov, V. A.; Nappa, M. J.; Rao, V. N. M.; Swearingen, E. N. WO2008008252 A1 to E. I. Du Pont de Nemours & Co., USA, 2008.
131. Sun, X.; Nappa, M. J.; Rao, V. N. M.; Petrov, V. A.; Swearingen, E. N.; Delyagina, N. I.; Cherstkov, V. F. WO 2008013779 A2 to E. I. Du Pont de Nemours & Co., USA, 2008.
132. Devic, M.; Pigamo, A.; Wendlinger, L. WO 2009153492 A1 to Arkema France, Fr., 2009.
133. Sievert, A. C.; Nappa, M. J. WO 9516655 A1 to E. I. Du Pont de Nemours & Co., USA, 1995.
134. Sievert, A. C., WO 2008030444 A2 to E. I. Du Pont de Nemours & Co., USA, 2008.
135. Butler, A. J. US 3,723,318 to Dow Corning Corporation, 1973.
136. Ruh, R. P.; Stowe, R. A.; Davis, R. A. US 2,889,379 to Dow Chemical Corporation, 1959.
137. Jimumaa, M. F.; Marupasu, D. F.; Sumisu, J. P. 59108726 A U. E., to General Electric, 1984.
138. Van Der Puy, M., US 5986151 to AlliedSignal Inc., 1999.
139. Elsheikh, M. Y.; Chen, B. US 6211421 to Atofina Chemicals, Inc., 2001.
140. Zhou, Q.; Ye, Z.; Wu, Z.; Wang, S.; Tang, Y.; Chen, K. CN 101306978A to Juhua Group Corporation, 2008.
141. Aoyama, H. WO9412454 to Daikin, 1994.
142. a. Takada, N.; Yoshikawa, S. JP07053420 to Central Glass, 1995.b. Johnson, R. W.; Moeller, W. C.; Van Der Puy, M.; Kaiser, M.; Wang, L. g products. WO9505353 to AlliedSignal Inc., USA, 1995.c. Duzick, T. C.; Rao, V. N. M.; Subramanian, M. A. WO9719751 to E. I. Du Pont de Nemours & Co., USA, 1997.
143. Nappa, M. J.; Swearingen, E. N. WO2009006358 to E. I. Du Pont de Nemours & Co., USA, 2009.
144. Poss, A. J.; Van Der Puy, M.; Singh, R. R.; Nalewajek, D.; Nair, H. K. WO20110288350 to Honeywell International Inc., USA, 2011.
145. Haszeldine, R. N. *J. Chem. Soc.* **1952,** 2504–2513.
146. a. Swearingen, E. N. WO2009142642 to E. I. Du Pont de Nemours & Co., USA, 2009.
b. Jia, X.; Zhou, X.; Quan, H.; Tamura, M.; Sekiya, A. *J. Fluor. Chem.* **2011,** *132* (12), 1188–1193.
c. Poss, A. J.; Nalewajek, D.; Nair, H. K.; Van Der Puy, M. WO2011146820 to Honeywell International Inc., USA, 2011.
147. Poss, A. J.; Nalewajek, D.; Van Der Puy, M.; Nair, H. K. WO20110288349 to Honeywell International Inc., USA, 2011.
148. Quan, H. -D.; Tamura, M.; Mizukado, J.; Sekiya, A. WO2011158790 to National Institute of Advanced Industrial Science and Technology, Japan, 2011.
149. Peng, S.; Nappa, M. J. WO2014052695 to E. I. Du Pont de Nemours & Co., 2014.
150. Tung, H. S.; Wang, H. WO20110237843 to Honeywell International Inc., USA, 2011.
151. Lv, J.; Zhao, B.; Mao, W.; Wang, B.; Zeng, J.; Zhang, W.; Hao, Z.; Tang, X.; Ma, H.; Li, F. CN104370690 to Xi'an Modern Chemistry Research Institute, Peop. Rep. China, 2015.
152. Hedrick, V.; Chien, J.; Boggs, J.; Jackson, A.; Edwards, E. B.; Brandstadter, S. WO2007059468 to E. I. Du Pont de Nemours & Co., USA, 2007.
153. Yagupolskii, Y. L.; Pavlenko, N. V.; Shelyazhenko, S. V.; Filatov, A. A.; Kremlev, M. M.; Mushta, A. I.; Gerus, I. I.; Peng, S.; Petrov, V. A.; Nappa, M. *J. Fluor. Chem.* **2015**. Ahead of Print.

154. a. Wang, H.; Tung, H. S. WO20120271069 to Honeywell International Inc., USA, 2012.b. Okamoto, S.; Sakyu, F. WO2013077189 to Central Glass Company, Limited, Japan, 2013.c. Elsheikh, M. Y.; Bonnet, P.; Wismer, J. A.; Seshadri, S. R. WO20140228600 to Arkema Inc., USA, 2014.
155. Tamai, R.; Yoshikawa, S.; Sakyu, F.; Hibino, Y. EP0729932 to Central Glass Company, Limited, Japan, 1996.
156. Tung, H. S.; Ulrich, K. D.; Merkel, D. C. US6844475 to Honeywell International Inc., USA, 2005.
157. a. Nakata, T.; Shibata, N.; Shibanuma, T. JPH11180908 to Daikin Industries, Ltd., Japan, 1999.b. Pokrovski, K. A.; Merkel, D. C.; Tung, H. S. WO20120059199 to Honeywell International Inc., USA, 2012.
158. Quan, H. -D.; Yang, H. -E.; Tamura, M.; Sekiya, A. *J. Fluor. Chem.* **2007,** *128* (3), 190−195.
159. a. Nakada, T.; Aoyama, H.; Yamamoto, A. WO9724307 to Daikin Industries, Ltd., 1997. b. Goto, Y.; Tamai, R.; Hibino, Y. JPH09194404 to Central Glass Company, Limited, Japan, 1997.
160. Merkel, D. C.; Pokrovski, K. A.; Tung, H. S. WO20110224465 to Honeywell International Inc., USA, 2011.
161. Nose, M.; Yamashita, T.; Sugiyama, A. WO2010101198 to Daikin Industries, Ltd., Japan, 2010.
162. Nose, M. WO2012020992 to Daikin Industries, Ltd., Japan, 2012.
163. Nakada, T.; Yamamoto, A.; Shibata, N.; Shibanuma, T. WO9840335 to Daikin Industries, Ltd., Japan, 1998.
164. Chen, B.; Bolmer, M. S.; Elsheikh, M. Y. US6166274 to Arkema, 2000.
165. a. Wang, H.; Tung, H. S. US20090417 to Honeywell International Inc., USA, 2009.b. Merkel, D. C.; Johnson, R.; Tung, H. S.; Wang, H.; Pokrovski, K. A. WO20110245549 to Honeywell International Inc., USA, 2011.
166. Merkel, D. C.; Pokrovski, K. A. WO2014149807 to Honeywell International Inc., USA, 2014.
167. Pokrovski, K. A.; Merkel, D. C.; Tung, H. S. WO20140288336 to Honeywell International Inc., USA, 2014.
168. Nair, H. K.; Poss, A. J.; Singh, R. R.; Van Der Puy, M.; Hulse, R. J. WO20120215041 to Honeywell International Inc., USA, 2012.
169. Zhai, Y.; Poss, A. J.; Singh, R. R. WO20130150633 to Honeywell International Inc., USA, 2013.
170. Mukhopadhyay, S.; Tung, H. S.; Light, B. A.; Phillips, S. D.; Ma, J. J.; Bortz, C. L.; Van Der Puy, M.; Merkel, D. C.; Dubey, R. K. US20070197842 A1 to Honeywell International Inc., USA, 2007.
171. Merkel, D. C.; Tung, H. S. WO2009015317 A1 to Honeywell International Inc., USA, 2009.
172. Kopkalli, H.; Chiu, Y.; Cottrell, S.; Tung, H. S.; Ulrich, K. D. WO 2011056441 A2 to Honeywell International Inc., USA, 2011.
173. Merkel, D. C.; Pokrovski, K. A.; Tung, H. S. WO 2011126679 A2 to Honeywell International Inc., USA, 2011.
174. Cottrell, S. A.; Chiu, Y.; Tung, H. S.; Uhrich, K. D.; Scheidle, P. WO2012099776 A1 to Honeywell International Inc., USA, 2012.
175. Bektesevic, S.; Tung, H. S.; Wang, H. WO2013049743 A1 by to to Honeywell International Inc., USA, 2013.

176. Wang, H.; Bektesevic, S.; Tung, H. S. WO2013049744 A2 to Honeywell International Inc., USA, 2013.
177. Bektesevic, S.; Merkel, D. C.; Nappa, M. J.; Sun, X.; Tung, H. S.; Wang, H. WO2013055894 A1, 2013.
178. Komatsu, Y.; Kishimoto, M. WO 2013065617 A1 to Daikin Industries, Ltd., Japan, 2013.
179. Elsheikh, M. Y.; Bonnet, P.; Chen, B. B. WO2013074324 A1 to Arkema Inc., 2013.
180. Takahashi, K.; Motoyama, H.; Chaki, T.; Karube, D.; Kishimoto, M. WO2014025065 A1 to Daikin Industries, Ltd., Japan, 2014.
181. Mao, W.; Wang, B.; Ma, Y.; Zhang, W.; Du, Y.; Qin, Y.; Kang, J.; Lu, J. *Catal. Commun.* **2014,** *49,* 73.
182. Wang, F.; Fan, J. L.; Zhao, Y.; Zhang, W. X.; Liang, Y.; Lu, J. Q.; Luo, M. F.; Wang, Y. J. *J. Fluor. Chem.* **2014,** *166,* 78.
183. Pigamo, A. WO2011110889 A1 to Arkema France, Fr., 2011.
184. Lv, J.; Zeng, J.; Zhang, W.; Du, Y.; Hao, Z.; Wang, B.; Mao, W.; Qin, Y. CN102617274 A to Xi'an Modern Chemistry Research Institute, P. R. China, 2012.
185. Nose, M.; Suzuki, A.; Yamashita, T.; Katsukawa, K. WO2010123148 A1, to Daikin Industries, Ltd., Japan, 2010.
186. Woodard, S. S. EP131560 A1 to Monsanto Co., USA, 1985.
187. Zhou, X.; Jia, X.; Bao, P. CN102199071 A to Beijing Yuji Science and Technology Development Co., Ltd., P. R. China, 2011.
188. Pigamo, A.; Wendlinger, L.; Bonnet, P. WO2011135395 A1 to Arkema France, Fr., 2011.
189. Wendlinger, L.; Bonnet, P.; Pigamo, A.; Doucet, N. WO2012052797 A1 to Arkema France, Fr., 2012.
190. Bonnet, P.; Pigamo, A.; Wendlinger, L.; Doucet, N. WO2012052798 A1 to Arkema France, Fr., 2012.
191. Takahashi, K.; Komatsu, Y.; Ueda, A. WO 2012057367 A1 to Daikin Industries, Ltd., Japan, 2012.
192. Wendlinger, L.; Pigamo, A.; Bonnet, P. WO 2012066375 A1 to Arkema France, Fr., 2012.
193. Merkel, D. C.; Tung, H. S.; Pokrovski, K. A.; Wang, H.; Bektesevic, S. WO 2012094477 A2 to Honeywell International Inc., USA, 2012.
194. Lv, J.; Wang, B.; Zeng, J.; Li, F.; Liu, B.; Du, Y.; Qin, Y.; Zhao, F.; Feng, L. CN101913986 A to Xi'an Modern Chemistry Research Institute, P. R. China, 2010.
195. Wang, B.; Lv, J.; Zhang, W.; Zeng, J.; Ma, Y.; Du, Y.; Yuan, J.; Xu, Q.; Feng, L. CN102001910 A to Xi'an Modern Chemistry Research Institute, P. R. China, 2011.
196. Smith, J. W.; McGunness, C.; Sharratt, A. P. WO 2009125200 to Ineos Fluor Holdings Limited, UK, 2009.
197. Wu, X.; Zhou, Q.; Wang, S.; Su, l.; He, H.; Chen, X.; Gong, H. CN102503766 (A) to Juhua Group, P. R. China, 2012.
198. Wu, X.; Chen, X.; Zhou, Q.; Wang, S.; Wang, X.; Zhou, H.; Xu, D.; Su, L.; Gong, H.; Chen, K. CN102320917 A to Juhua Group, P. R. China, 2012.
199. Smith, J.; McGuiness, C.; Sharratt, A. P. WO2009125199 A2 to Ineos Fluor Holdings Limited, UK, 2009.
200. Wu, X.; Chen, X.; Zhou, Q.; Wang, S.; Wang, X.; Zhou, H.; Xu, D.; Su, L.; Gong, H.; Chen, K. CN102336630 A to Juhua Group, P. R. China, 2012.
201. Sun, X.; Nappa, M. J. WO 2015095497 to E.I. Du pont de Nemours and Company, USA, 2015.
202. Pigamo, A.; Devic, M.; Wendlinger, L. WO2010029419 A1 to Arkema France, Fr., 2010.
203. Rao, M. V. N.; Sievert, A. C. WO 2008054781 A1 to E.I. Du pont de Nemours and Company, USA, 2008.

204. Nappa, M. J. US20120215035 A1 to E.I. Du pont de Nemours and Company, USA, 2012.
205. Sun, X.; Nappa, M. J. US20120215036 A1 to E.I. Du pont de Nemours and Company, USA, 2012.
206. Sun, X.; Nappa, M. J. WO 2012115957 A1 to E.I. Du pont de Nemours and Company, USA, 2012.
207. Nose, M.; Suzuki, A. JP2011121942 A to Daikin Industries, Ltd., Japan, 2011.
208. Yamashita, T.; Nose, M.; Katsukawa, K. WO2011102538 A2 to Daikin Industries, Ltd., Japan, 2011.
209. Zhou, Q.; Zhou, L.; Chen, X.; Wang, S.; Lei, J.; He, H., CN103449958 A to Juhua Group, P. R. China, 2013.
210. Agahigian, H.; Woolf, C. DE1139831 to Allied Chemical Corp, 1962.
211. Tsusaki, M.; Hashimoto, M.; Mitsuoka, H. JP2013224383 A to Asahi Glass Co., Ltd., Japan, 2013.
212. Tsuzaki, M.; Hashimoto, M.; Mitsuoka, H. WO2013161723 A1 to Asahi Glass Co., Ltd., Japan, 2013.
213. Nishiguchi, A.; Okamoto, S.; Sakyu, F. JP2014005418 A to Central Glass Co., Ltd., Japan, 2014.
214. Joyce, R. M., Jr. US2462402 to E. I. du Pont de Nemours & Co, 1949.
215. Ohnishi, K.; Okamoto, H.; Tanuma, T.; Yanase, K.; Kawasaki, T.; Takei, R. WO 9108183 A1 to Asahi Glass Co., Ltd., Japan, 1991.
216. Aoyama, H.; Yasuhara, T.; Kono, S.; Koyama, S.; Ueda, S. EP492527 A1 to Daikin Industries, Ltd., Japan, 1992.
217. Tanuma, T.; Okamoto, H.; Ohnishi, K.; Morikawa, S. *Appl. Catal. A General* **2008,** *348,* 236.
218. Okamoto, H.; Sasao, Y. WO 2010074254A1 to Asahi Glass Company, Limited, Japan, 2010.
219. Okamoto, H. WO 2010082662A1 to Asahi Glass Company, Limited, Japan, 2010.
220. Rao, V. N. M.; Sievert, A. C. WO2008060612 A2 to E. I. du Pont de Nemours & Co., 2008.
221. Paleta, O.; Kvicala, J.; Gunter, J.; Dedek, V. *Bull. la Soc. Chim. Fr.* **1986,** *920.*
222. Coulson, D. R. *J. Fluor. Chem.* **1990,** *50,* 77–87.
223. Rao, V. N. M.; Nappa, M. J.; Sievert, A. C.; Knapp, J. P. WO2008079265 A1 to E. I. du Pont de Nemours & Co., 2008.

Electrochemical Fluorination: A Powerful Tool for the Preparation of Organofluorine Compounds

4

N.V. Ignat'ev [1,2]
[1]Merck KGaA, Darmstadt, Germany; [2]Dr. Ignatiev – Chemistry Consultancy, Duisburg, Germany

Chapter Outline

1. Introduction and Historical Overview

The electrochemical approach stimulated very much the development of fluorine chemistry. Fluorine itself was born in the electrochemical process when Henri Moissan in 1886 subjected HF containing traces of KF to electrolysis.[1] Until today it is the only way for the industrial production of commercial quantities of F_2 by means of electrolysis of KF·2HF (40.8 wt% of HF).[2,3] The typical operating temperatures are 85–100°C to keep the electrolyte in a molten state. During the electrolysis the HF is consumed and F_2 is formed. To balance the electrolyte concentration, fresh HF is added to the cell continuously or in intervals.

Modern Synthesis Processes and Reactivity of Fluorinated Compounds. http://dx.doi.org/10.1016/B978-0-12-803740-9.00004-4

A useful approach to electrochemical syntheses of organofluorine chemicals is the anodic fluorination of organic compounds on a Pt electrode in organic solvents in the presence of fluoride anion[4] or in liquid electrolytes $(Et)_4NF \cdot mHF$ (m = 3–5)/CH_3CN.[5–16] Some additional information on this topic can be found in review articles.[17–23]

Electrochemistry can be applied not only for introduction of fluorine into organic molecules, but also for the further derivatization of organofluorine compounds. Probably the most prominent example is the Kolbe electrolysis of carboxylic acids.[24,25] In the case of the electrolysis of fluorinated or perfluorinated carboxylic acids, dimerization of anodically generated radicals results in the formation of fluorinated alkanes (Scheme 4.1).[26–28]

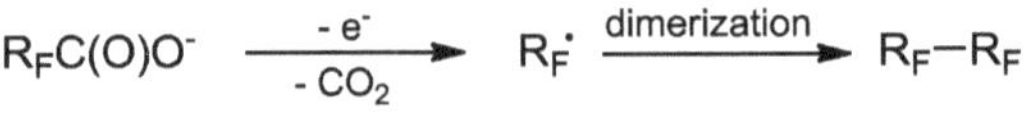

Scheme 4.1 Kolbe electrolysis of fluorinated carboxylic acids.

Electrochemical oxidation of the dianion of sulfodifluoroacetic acid is accompanied not by desulfonation, but by decarboxylation resulting in the formation of the sulfodifluoromethyl radical (Scheme 4.2).[29] Sulfodifluoroacetic acid, $HOSO_2CF_2C(O)OH$, can be easily obtained from readily available tetrafluoroethane-β-sultone.[30]

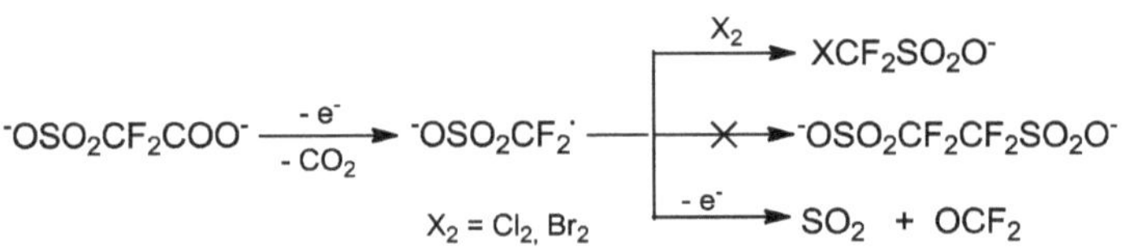

Scheme 4.2 Electrochemical oxidation of the dianion of sulfodifluoroacetic acid, $HOSO_2CF_2C(O)OH$.

Surprisingly, the Kolbe dimerization product was not detected by ^{19}F nuclear magnetic resonance (NMR) spectroscopy in the reaction mixture after electrolysis. If the electrolysis of the dianion of sulfodifluoroacetic acid was carried out in the presence of halogens (Cl_2 or Br_2) the formation of the corresponding halodifluoromethanesulfonic acids was detected (Scheme 4.2).[29]

The trifluoromethyl radical, $CF_3^{\cdot}$, electrochemically generated by anodic oxidation of trifluoroacetate in the presence of unsaturated compounds undergoes addition to double[31] or triple bonds[32] or it reacts with active methylene compounds, for instance, with β-ketoesters and 1,3-diketones, affording trifluoromethylated substances.[33] The system $CF_3COOH/CF_3COONa/CH_3CN/H_2O$ was applied for the electrochemical trifluoromethylation of 3-sulfolene.[34]

Generation of perfluoroalkyl radicals by means of electrochemical reduction of perfluoroalkyl halides, R_FX (X = I, Br; Scheme 4.3), was intensively studied in the 1990s.[35,36]

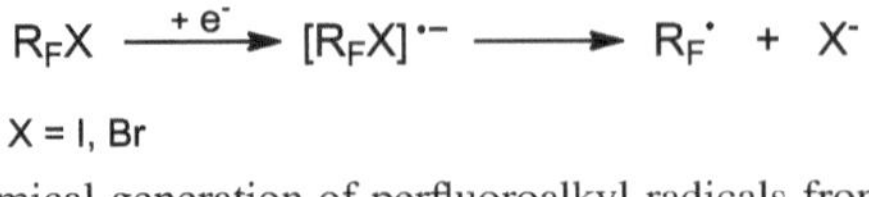

Scheme 4.3 Electrochemical generation of perfluoroalkyl radicals from R_FX (X = I, Br).

Electrochemically generated electrophilic perfluoroalkyl radicals are able to react with various nucleophiles, for instance, C nucleophiles,[36,37] oxygen-containing nucleophiles,[37,38] and aliphatic and aromatic thiols[37,39] (Scheme 4.4). Low consumption of electricity (much less than 1F/mol) indicates that the reaction proceeds via chain radical-anion $S_{RN}1$ mechanism.[37,39]

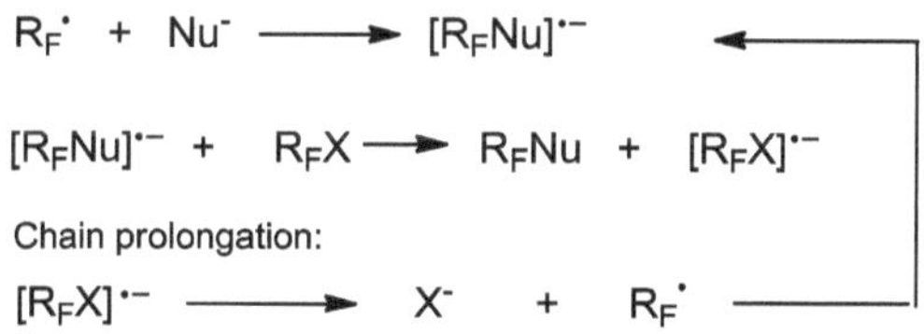

Scheme 4.4 Chain radical-anion $S_{RN}1$ mechanism of the interaction of electrochemically generated perfluoroalkyl radicals with nucleophiles (Nu).

Electrochemical radical-chain trifluoromethylation of thiophenols with bromotrifluoromethane can be carried out in the presence of SO_2, which plays the role of a mediator in this process.[40] Reaction of polyfluoroalkyl iodides of the type $H(CF_2CF_2)_nCH_2I$ (n = 1, 2, 3) with *p*-chlorothiophenol proceeds similarly and results in $H(CF_2CF_2)_nCH_2SC_6H_4Cl$. To complete this reaction, less than 1 F/mol of electricity is required.[41] Alternating current can be successfully applied instead of constant current to initiate efficiently the radical ion reactions of perfluoroalkyl halides with nucleophiles.[42] This approach makes it possible to simplify the setup required for electrochemical syntheses.

Due to their electrophilic nature, the electrochemically generated perfluoroalkyl radicals are able to react with weak nucleophiles, for example, with the methoxysulfinate $CH_3OSO_2^-$ (Scheme 4.5).[43] This reaction results in the formation of perfluoroalkylsulfinate anions, $R_FSO_2^-$, which can be oxidized with Cl_2 to the corresponding perfluoroalkylsulfonyl chlorides, R_FSO_2Cl ($R_F = CF_3$, C_3F_7, C_8F_{17}).

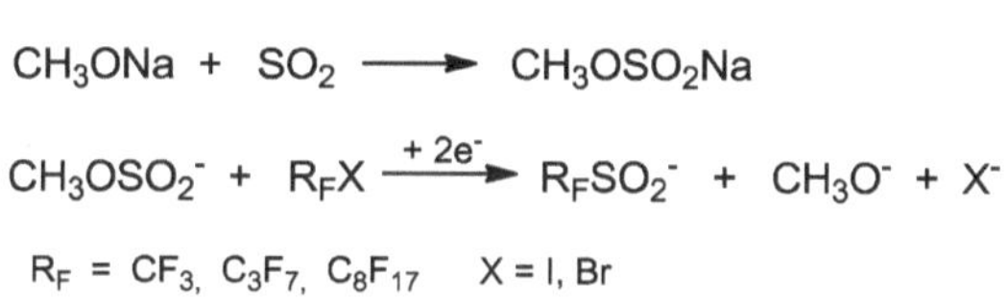

Scheme 4.5 Electrochemical synthesis of perfluoroalkylsulfinates.

Electrochemically generated perfluoroalkyl radicals react not only with charged nucleophiles but also with electron-rich systems. For instance, perfluoroalkyl iodides R_FI ($R_F = C_4F_9$, C_6F_{13}) react with alkynes in an electrolytic cell giving an addition product. Carbon fiber was used as cathode material and an aqueous solution of KCl as the electrolyte.[44] Electrochemically induced addition of C_4F_9I to ethylenic and acetylenic substrates was described by Dapremont et al.[45] Electron-rich heterocycles react with electrochemically generated perfluoroalkyl radicals forming the corresponding perfluoroalkyl-substituted heterocyclic compounds.[46–48]

The electrochemical cross-coupling of bromotrifluoromethane with aromatic or heteroaromatic iodides and bromides was successfully carried out with a sacrificial copper anode.[49] The intermediate formation of CF_3Cu was postulated. A similar approach was used by Prakash et al. for the preparation of the very versatile reagent $(CH_3)_3SiCF_3$

(Ruppert–Prakash reagent) by means of electrolysis of CF_3Br in an undivided cell equipped with a sacrificial aluminum anode.[50] Later, a similar methodology was applied to the electrosynthesis of various clorofluoroalkyl silanes in the cell having a stainless-steel cathode and a sacrificial Zn anode.[51] Addition of fluoroalkyltrimethyl silanes to carbonyl compounds can be initiated electrochemically.[52] Presumably, the generation of fluoroalkyl anions R_F^- proceeds with the participation of cathodically formed anion radicals of carbonyl compound, which act as a redox mediator in this process. The R_F^- anion undergoes addition to carbonyl group. Only 0.02–0.33 F/mol of electricity is required to complete this reaction. Electrochemically generated anion radicals of nitrobenzene can be also used to reduce fluoroalkyltrimethyl silanes. However, the major products in this case were the coupled substances R_F–R_F.[52] Electrochemical reduction of bis(trifluoromethyl)mercury resulted in the intermediate formation of the trifluoromethyl anion, CF_3^-, which was trapped by reactions with benzaldehyde, Me_3SiCl, and bromobenzonitrile.[52] For a review of the electrochemical syntheses of fluorinated organosilicon compounds using sacrificial anode, see Ref. 53.

Preparative electrochemical reduction of perfluoroalkyl halides (C_3F_7I, $C_6F_{13}I$, $C_6F_{13}Br$, $C_8F_{17}Br$, and $CF_3OC_2F_4I$) in polar organic solvents in the presence of oxygen leads to 62–76% yield of the corresponding carboxylic acids having one CF_2 group less than the starting perfluoroalkyl halides.[54] According to cyclic voltammetry (CVA), oxygen acts as a redox shuttle and as a reagent at the same time in this reaction. Trifluoroacetic acid is formed in 65% yield by electrochemical reduction of CF_3CCl_3 in the presence of oxygen.[54]

Similar to nonfluorinated acyl halides, irreversible electrochemical reduction of perfluoroacyl halides, i.e., $C_4F_9C(O)I$, $C_4F_9C(O)Br$, $C_6F_{13}C(O)Br$, $C_6F_{13}C(O)Cl$, and $C_8F_{17}C(O)F$, results in the formation of the corresponding acyl radicals, which undergo fast dimerization to α-diketones.[55] The reduction potentials $-E_p$ and $-E_{1/2}$ of perfluoroacyl halides are decreased in the following order: $-C(O)F > -C(O)Cl > -C(O)Br > -C(O)I$.

Perfluoroalkyl radicals can be generated not only by the anodic oxidation of perfluoroalkyl carboxylic acids or by the reduction of perfluoroalkyl halides but also by electrochemical reduction of perfluoroalkyl iodonium salts, for instance $[R_F{-}I{-}C_6H_5]^+ X^-$ ($R_F = C_3F_7$, $i{-}C_3F_7$, C_6F_{13}; $X^- = BF_4^-$, $CF_3SO_3^-$, $CF_3C(O)O^-$),[56,57] or trifluoromethyl sulfonim salts.[58] Perfluoroalkyl iodonium salts are reduced much more easily when compared with the corresponding perfluoroalkyl halides. This enables the electrosynthesis at low cathodic potentials. The preparative electrochemical reduction of heptafluoropropyl iodonium tetrafluoroborate in the presence of sodium sulfinates results in the formation of the corresponding sulfones, $RSO_2C_3F_7$.[56,57] Low consumption of electricity (0.2 F/mol) confirms the chain radical-anion mechanism.

Interaction of electrochemically generated fluoroalkyl radicals with "small" molecules (CO_2, SO_2, CS_2) leads to the formation of carboxylic, sulfinic, or dithiocarboxylic acids.[59]

The reactivity of perfluoroalkyl halides toward various nucleophiles has been studied by means of CVA.[60,61]

Electrochemical syntheses of diethyl fluoromethanephosphonate and fluoroalkylphosphines were published previously.[62,63]

The aforementioned examples demonstrate the power of the electrochemical approach to the syntheses of organofluorine compounds. Some more examples can be found in reviews.[23,64–71]

However, according to Lund[72] "The most important industrial application of organic electrolysis introduced in the period 1940-1960 was the electrochemical fluorination (ECF) of organic substrates, and the electrochemical manufacture of fluorinated compounds has since continued to be an important industrial process." Joseph H. Simons has discovered the electrochemical fluorination (ECF) of organic substances in liquid hydrogen fluoride (known as Simons process) in the early 1940s. However, this invention was classified due to the relationship to the Manhattan project.[73,74] The first publication on this topic appeared only in 1949.[75–80] In 1948, the Minnesota Mining & Manufacturing Company (3M Company) patented the invention of J.H. Simons.[81] However, as W.H. Pearlson wrote,[82] expenditures at 3M on the Simons process had already begun as early as 1946, when several laboratory ECF cells were put into operation and construction of a pilot plant cell begun.[82] Nowadays, many fluorochemicals are produced by the Simons process or are derived from ECF products. Takashi Abe, who greatly contributed to the development of this fantastic process, called the ECF "a locomotive for the development of fluorine chemistry."[83,84] The ECF (Simons process) was and is a vital process for the production of high-performance materials, such as artificial blood substitutes,[85,86] surfactants, materials for polymer chemistry, reagents, fluorochemicals for the electronic industry, and fluorinated materials for energy conversion. For many years the ECF process was the process of choice to produce long-chain perfluorinated substances, like perfluorooctane sulfonic acid (PFOS) and perfluorooctanoic acid (PFOA). These chemicals and their derivatives were used as fluorosurfactants[87] in chromium electroplating; for the production of fluoropolymers; as water-, soil, and oil repellents for carpets, paper, and textiles[87]; and as fire-fighting foams ("Light Water").[88] However, for more than a decade, the long-chain perfluorinated substances are more and more under criticism,[89] since they are highly persistent, bioaccumulative, and have been detected globally in the environment.[90–93] Thus, they are recognized as global contaminants of high concern.[94] Due to these reasons, the use of PFOS, PFOA, and their derivatives is strongly restricted (limited), and many companies over the world have already stopped the production of these chemicals.

Since its discovery, the application of the Simons process, in particular the ECF of carboxylic acid derivatives and aliphatic and heterocyclic amines, has been reviewed several times.[4,17,20,83,95–112]

In some reviews, the Simons process (ECF) and electrochemical partial fluorination of organic compounds on Pt anode in organic or ionic liquids (ILs) solvents in the presence of a fluoride source were presented together, in spite of the dissimilarities in technique, their mechanisms, and in resulting products derived from these two different electrochemical processes.

The present review deals with the ECF in anhydrous HF (aHF) (Simons process) of short-chain organic molecules having C–S, C–P, and S–N bonds and covers the literature published during the past 20 years. For a comparison, selected earlier publications will also be referred.

2. Technical Aspects of the Electrochemical Fluorination (Simons Process)

2.1 Cell Construction and General Requirements for the Electrochemical Fluorination Process

Electrochemistry is probably the most sustainable way to utilize solar energy for preparation of chemical products. The following are the advantages of the electrochemical processes:

- The electrochemical cell has a rather simple construction and operates typically under atmospheric pressure in a moderate temperature range.
- Electron (current) is a cheap reagent.
- The process can be run as a single-stage, single-pot operation.
- The parameters of the electrochemical process, e.g., voltage and current, can be easily controlled and regulated.
- The electrochemical process can be easily automatized, controlled, and operated by computer.
- An electrochemical plant can consist of several small units providing flexibility in the production's capacity.

More information about the industrial electroorganic chemistry can be found in Ref. 113.

ECF (Simons process) has all of the advantages mentioned earlier, but it is a dangerous process. First, aHF [boiling point (BP) 20°C; very toxic substance] is used as a solvent and reagent in the ECF process. Second, the ECF of oxygen containing compounds, i.e., alcohols, ethers, and carboxylic acid derivatives, can result in the formation of oxygen difluoride, OF_2 (a strong oxidizer) as by-product. Formation of stoechiometric amount of hydrogen at the cathode has to be taken into account. For example, an ECF cell operating at 2000 A (relatively small cell) can theoretically produce 37.3 mol (approx. 0.84 m^3) of gaseous H_2 within 1 h, which together with OF_2 or F_2 or even simply with air can result in an explosive mixture. Therefore, only skilled and well-trained personnel should work with the ECF process. Anyway, the ECF cell should be placed in a separate room equipped with corresponding suitable ventilation. The power supply equipment, control instruments, and operating system should be placed separately, outside of the ECF chamber.

Since the invention of the ECF process by Simons, the construction of the electrochemical cell has not changed much. Fig. 4.1 shows the Simons original draft of the electrochemical cell and setup for the ECF process.[81]

Nickel is the only metal suitable for manufacturing anode plates for the ECF process.[102,114,115] All attempts to carry out the ECF by the Simons procedure using another metal as anode failed. Iron or nickel can be used as cathode material. The

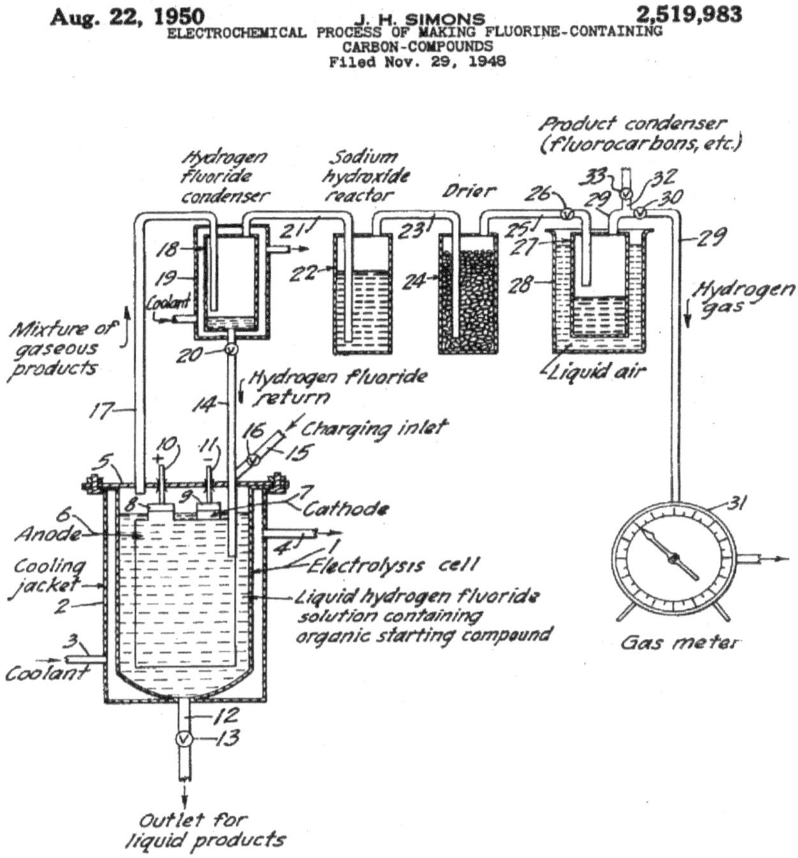

Figure 4.1 Original Simons drawing from the US patent 2,519,983.
Reproduced with the permission from 3M Company.

author personally prefers to use a nickel (anode)—nickel (cathode) electrode array to have the possibility to change the polarity of the electrodes and to avoid the problems with the corrosion of the iron cathode. Fig. 4.2 shows the laboratory cylindrical cell, which has been used by author for many years to carry out ECFs of various organic substances by the Simons process. The laboratory setup to carry out the ECF in aHF is shown in Fig. 4.3.

Nickel quality is crucial for successful ECF. The Ni must have a quality of minimum 99.6 (Nickel 201) and it should not contain much impurities, in particular Si. The Ni plates for manufacturing of the electrodes are approximately 2 mm thick depending on the size of the cell. The distance between the electrodes is crucial to ensure a safe and efficient process. On the one side, the distance should not be too large to avoid high resistance causing energy losses. On the other side, it should not be too small to avoid a shortcut by electrode deformation or by small metal pieces that are, for example, released from the reflux condenser as a result of corrosion. Smooth

Figure 4.2 Cylindrical laboratory cell for electrochemical fluorination in anhydrous HF (according to the Simons process).

electrolyte circulation in the cell should be ensured by optimal distance between the electrodes, which is typically in the range 2–10 mm depending on the size of the cell. The laboratory cell presented in Fig. 4.2 has 3 mm space between the electrodes. The distance between the electrodes is fixed by polytetrafluoroethylene (PTFE) spacers, which also prevent the direct electrical contact (shortcut) of the electrodes. During the electrolysis the electrode array should be completely covered with electrolyte to avoid corrosion at the border between the liquid and gaseous phases. Preferably, the constant current power supply should have an automatically regulated limitation of current and voltage to avoid overvoltage at the clips of the cell. The necessary control instruments are shown in Fig. 4.3: ammeter (A), voltmeter (V_1), and a current counter (A·min). All these instruments are very important to control and regulate the ECF process. Differences in the values of V_1 and V_2 (typically $V_1 < V_2$) shows that energy is consumed for the electrochemical process. The case $V_1 = V_2$ indicates that the electrodes are blocked with a polymeric film or other deposits preventing the ECF of the substrate. The current counter (A•min) allows the observation of the progress of the ECF process, since the amount of the electricity required for the complete fluorination of the starting material loaded into the cell can be calculated by the Faraday law taking into account that for the conversion of a C–H bond into a C–F bond two electrons are required. After passing 100% or less electricity, the cell should be charged again with starting material. Electrolysis of an "empty" cell can be dangerous due to the possibility of formation of F_2 if the voltage rises above 6.5–7 V. Therefore, the use of constant current power supplier with automatic regulation (limitation) of the voltage is advantageous. The current yield can be high, but it cannot reach 100% in the electrochemical processes. It means that part of the energy from the power supplier is not used for the formation of new chemical bond but is transformed into heat. The

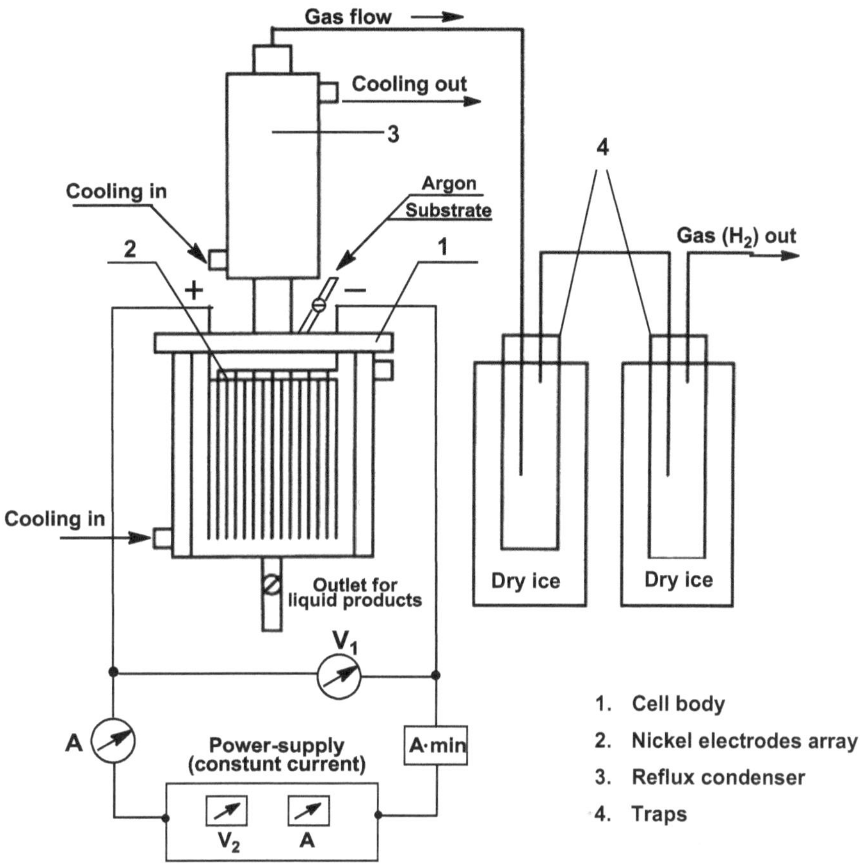

Figure 4.3 Laboratory setup for electrochemical fluorination experiments in anhydrous HF.

boiling point of HF, which is the solvent and the source of the fluorine for the C−F bond formation in the ECF process, is 20°C. Thus, to keep the electrolyte in the liquid state, effective cooling of the electrochemical cell is required. Typically, the temperature of the electrolyte in the ECF cell is maintained between 0 and 15°C depending on the nature and solubility of the starting material. To avoid loss of HF from the cell together with the stream of gaseous reaction products and H_2, an effective reflux condenser is required. Usually the temperature of the reflux condenser is well below 0°C. Liquid perfluorinated products having boiling points above 50°C can be discharged from the bottom of the ECF cell. More volatile perfluorinated products are usually released from the cell through the reflux condenser and collected in the traps cooled with the dry ice (Fig. 4.3). Cooling with liquid nitrogen is not recommended because of a possible collection of liquid oxygen difluoride, F_2O, in the traps (**danger of severe explosion in the mixture with H_2!**). F_2O can be eliminated from gaseous stream by passing it through an additional trap filled with a suitable reducing material able to react with oxygen difluoride, for instance, with an aqueous solution of sodium

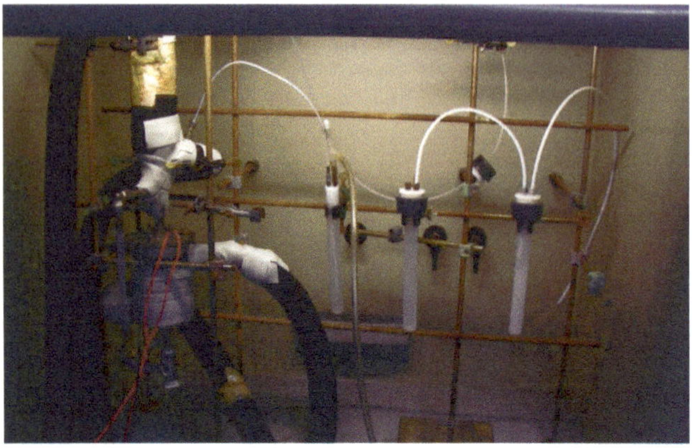

Figure 4.4 Laboratory ECF cell with traps made from perfluoroalkoxy-polymer (the cooling system is not shown).

sulfite. In this case, the additional hydraulic resistance and refreshing of the sodium sulfite solution in proper time has to be taken into account. Obviously, the perfluorinated products that react with water or Na_2SO_3 will be lost. For industrial production of fluorochemicals by the Simons process, special absorbers for the removal of HF and F_2O are often used. In the laboratory practice, the author has used the setup presented in Fig. 4.3 and shown in Fig. 4.4.

Additional technical information regarding ECF cell construction and setup for carrying out the ECF in aHF (Simons process) can be found in the original articles and reviews.[4,95,97,99,100,102,116]

2.2 Requirements for Starting Materials for the Electrochemical Fluorination (Simons) Process

The success of ECF in aHF strongly depends on the right choice of the starting material. For example, ECF of acetic acid results in the formation of perfluoroacyl fluoride, CF_3COF, in 17% yield. However, ECF of acetic anhydride, $(CH_3CO)_2O$, or acetyl fluoride, CH_3COF, gives CF_3COF in 32% and 80% yields, respectively.[114]

In general, the starting material for the Simons process should have the following properties:

- It should be soluble in aHF in the concentration range 2–10 wt%.
- It must contain a polar (basic) group, which can undergo protonation in aHF to form an electroconductive solution (electrolyte).
- The solution in aHF should be stable at least for some hours at temperatures below 20°C.
- The ECF of small molecules having short carbon chains usually results in higher yields of the respective perfluorinated products.
- The shorter the way (the less numbers of C—H bonds should be fluorinated) to the final substance, the better the yield in the perfluorinated product.

- The partially fluorinated starting materials (which are still soluble in aHF) provide better yields of perfluorinated products when compared with the nonfluorinated starting compounds. Some examples can be found in the literature.[117–125]
- The perfluorinated product should exhibit low solubility in aHF to avoid further ECF under C–C or C–heteroatom bond scission (fluorination).

Typically, organic molecules containing heteroatoms O, S, N, and P possess good solubility in aHF. Protonation of a basic center gives the corresponding onium salts (**A**) with polyhydrogenfluoride counteranion (Scheme 4.6).

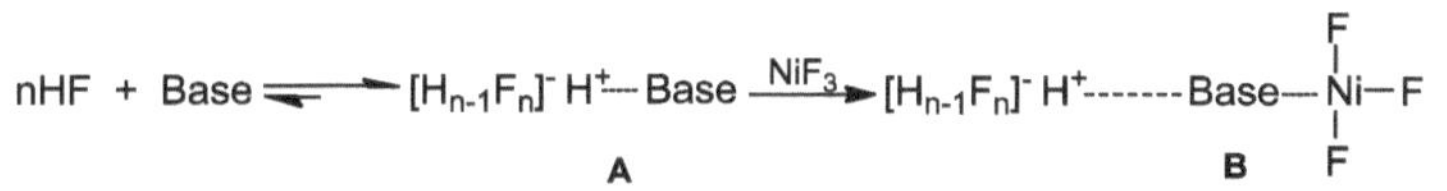

Scheme 4.6 Protonation of an organic substrate and its coordination on the surface of the $NiF_3(NiF_4)$/Ni anode.

The reaction of organic substrates (bases) with HF is an equilibrium process. In aHF, the equilibrium is shifted to the side of complex **A**. However, NiF_3, electrochemically formed on the surface of the Ni anode (see Section 3), acts as a strong Lewis acid facilitating the adsorption of the organic substrate on the surface of the Ni anode. Due to the interaction with NiF_3, the coordination of organic substrates with proton becomes much weaker in the vicinity of the Ni anode. Nevertheless, organic molecules are oriented at the surface of the anode in the way that the basic side is kept away from the surface of the Ni anode. Probably this is why ECF starts at the end of a hydrocarbon chain. This hypothesis explains the formation of partially fluorinated compounds having $-CF_2-CH_2-N$ and $-CF_2-CH_2-P$ fragments as the major impurities by the ECF of trialkylamines[123] and trialkylphosphines.[126]

2.3 Practical Recommendation for the Fluorination of Organic Compounds by the Electrochemical Fluorination Process

The following recommendations are based on the personal experience of the author gained during the practical work using ECF (Simons process) over the past 30 years. These recommendations are not instructions, but only useful suggestions.

The nickel electrode array is the heart of the electrochemical cell. As already mentioned, the electrodes should be made from good-quality nickel and assembled in a single array (see Fig. 4.2) using PTFE spacers. Good welding connections to the current collector, separately for anode and cathode plates, are essential to avoid additional resistance during the electrolysis. The new or used nickel electrode package should be cleaned off any deposit on the surface of the nickel plates. Mechanical cleaning with abrasive materials containing silicon dioxide is not recommended. The author favors chemical cleaning by immersing the electrode package into aqua regia for some minutes. Subsequently, the electrodes should be washed with plenty of deionized water and carefully dried overnight. The clean and dry set of Ni electrodes should be placed into the cell body strictly vertical, to ensure that gas bubbles (H_2) generated at the cathode plates cannot move to the anode plates. The absence of electrical shortcut between the cathode and anode plates and between the electrode package and the cell body should be ensured, as well.

After connecting all parts together (see Fig. 4.4) the complete ECF system should be tested with argon for (1) free outlet of gaseous products at the top of the reflux condenser and (2) the absence of leakages in the cell body and reflux condenser. Penetration of cooling liquid (typically ethanol) from the cell or reflux cooler jacket into the electrolyte can cause a very dangerous situation. The electrolyte circulation in the cell is important not only for keeping the temperature constant in the electrolysis area but also to reduce the concentration polarization phenomena.[110] The hydrogen bubbles developed at the cathode provide sufficient electrolyte mixing in the laboratory cells. However, for industrial cell an effective electrolyte circulation can be achieved by a special cell design (two-compartment cell) or by the use of the so-called air-lift system.[4]

The quality of the aHF has a strong influence on the ECF process. According to Conte et al.[110] the presence of anions other than F^-, e.g., Cl^- and HO^-, which are favorably discharged on the anode, reduces the yield of the desired perfluorinated product. Some examples for the ECFs of chlorine-containing compounds are given in Section 5.1.

NiF_3 doped with NiF_4 is the actual fluorinating reagent in the Simons process (see Section 3). In contrast to other metal fluorides, NiF_3 and NiF_4 are not soluble in aHF.[115] NiF_3 is electrochemically formed on the surface of the nickel anode during the so-called preelectrolysis. After filling the cell with aHF and setting up the temperature of the cell body and reflux condenser (see Section 2.1), the Ni electrodes are polarized with a constant current (cell voltage $V_1 = 5-5.5$ V; Fig. 4.3) until the current drops and remains unchanged for 1−2 h. Typically, this procedure takes 6−24 h depending on the water content of the aHF. After that, the cell can be charged with the organic substrate and the electrochemical fluorination process can start.

Depending on the aggregate state, liquid or solid, the starting material can be charged into electrochemical cell either in pure form or as a solution in aHF. For strong bases, like amines or phosphines, the preliminary preparation of a solution with approx. 30−40 wt% concentration accomplished by slow addition of the starting material to stirred, cold (well below 0°C) aHF placed into a perfluoroalkoxy-polymer flask is strongly recommended. The cell can be charged with starting material periodically, after consumption of a certain amount of electricity (A·h), or continuously to have an almost constant concentration of the substrate. The method of substrate loading into the cell depends on the properties of the starting material and the final product. It is important that the perfluorinated product leaves the electrolysis area (space between the cathode and anode) as soon as it is completely fluorinated to avoid further degradation. This can be achieved by optimization of the temperature profile, electrolyte circulation, current density, and proper loading of the starting material into the cell just in time. Typical parameters for the ECF process are cell voltage, 4.5−6.0 V; current densities, 0.5−3.0 (for industrial cells) A/dm^2; cell temperature, 0−15°C; and initial substrate concentration, 4−10 wt%. For every product the optimal process parameters are very individual. Typically 1−2 days are required to achieve the optimal parameters and best current and product yields. At these optimal conditions the ECF process can be performed continuously for a weeks or even months.

The perfluorinated product can be collected from the bottom of the cell, in the traps, or in a collector after separation from gaseous steam. The isolation method depends on the physical properties of the perfluorinated product, namely: boiling point, vapor pressure, solubility in aHF, electrolyte temperature, circulation, and cell construction. Usually, the raw product contains some amount of partially fluorinated substances, which can be separated by fractional distillation.

3. Mechanism of Electrochemical Fluorination (Simons Process)

The mechanism of ECF of organic substances in aHF was under discussion since the discovery by Simons more than 70 years ago. Many scientists contributed to this topic offering various models for the Simons process. In general, all these models can be separated into the following mechanisms of ECF:

- Electrochemical generation of atomic or molecular fluorine on the surface of the Ni anode. This mechanism was considered by Simons himself as a possible radical pathway to perfluorinated substances.[100,127]
- Participation of high-valent (III and IV) nickel fluorides in the course of the ECF according to the Simons process was considered in the early 1950s[116,128] and in later publications.[99]
- Rüdiger et al. postulated a weak complex of fluorine and Ni fluorides ("active fluorine") as a key reagent in the ECF process in aHF.[129]
- A carbocationic mechanism was proposed by Meinert et al.[130] According to this mechanism the C−H bond in organic molecules adsorbed on the Ni anode is polarized by the interaction with aHF. A subsequent two-electron electrochemical oxidation and addition of F^- anion to the intermediately generated carbocation results in the formation of the C−F bond.
- A related mechanism was proposed earlier by Gambaretto.[131] This EC_bEC_N mechanism is based on the assumption that organic substances undergo direct electrochemical oxidation (step E) during the ECF (Simons process) to the corresponding radical cation, which is deprotonated (step C_b) to give a radical. One-electron oxidation of this radical (electrochemical step E) leads to the formation of the carbocationic intermediate that adds a F^- anion (step C_N). According to this mechanism, the ECF of organic substrates should not take place without electricity, because the first reaction step in this mechanism is the electrochemical oxidation.

In course of our study, we have obtained strong experimental evidences that fluorination of organic compound, added to the cell after preelectrolysis, took place even when the ECF cell was disconnected from the current supplier.[124] This means that the EC_bEC_N mechanism of the Simons process is not realistic because the predicted initial direct electrochemical oxidation of the substrate (limiting step) in the ECF process is obviously not required. A similar observation was reported later.[132]

We have studied the electrochemical behavior of a Ni disk electrode (Ni-wire pressed into PTFE; disk surface area 3.14 mm^2) in aHF by means of linear sweep

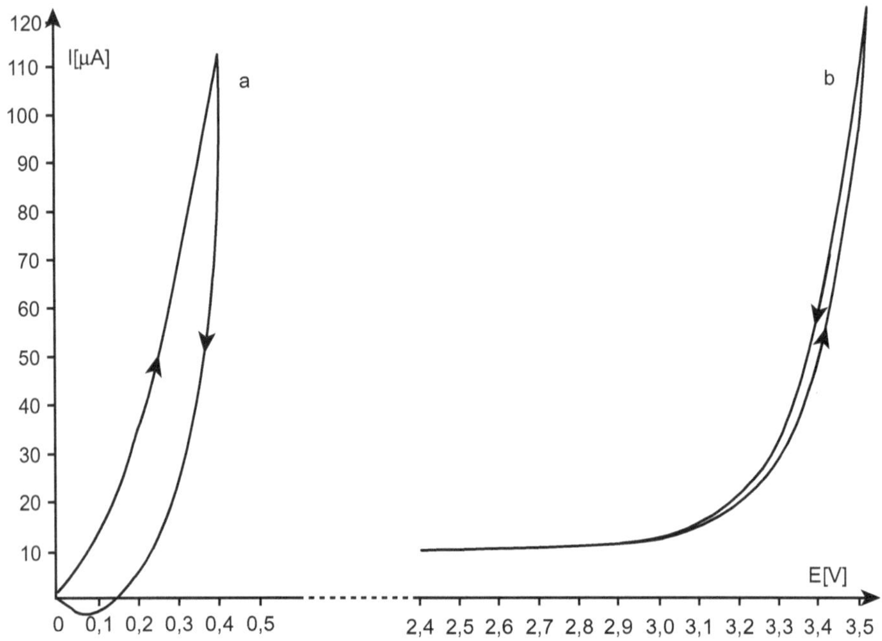

Figure 4.5 Linear sweep voltammogram of a Ni electrode in aHF. Peak A: freshly polished electrode; peak B: Ni electrode that was previously anodically polarized in aHF for 10 min with a constant current (current density c. 300 mA/cm^2).
Reprinted from Dimitrov A.; St Rüdiger, Ignat'ev N. V.; Datsenko S. D. J. Fluorine Chem. **1990,** 50, 197–205 with permission from Elsevier.

voltammetry and found that the freshly polished Ni electrode was electrochemically oxidized in aHF at low anodic potentials of about +0.1 V referring to a Ni plate cathode (nonpolarizable electrode; surface area 15 cm^2) (Fig. 4.5A).[129] Presumably the peak A in the voltammogram corresponds to electrochemical oxidation of Ni^0 in aHF to NiF_2, which is not soluble in aHF[115] and forms a film on the surface of the Ni anode (Scheme 4.7).

$$Ni^0 \xrightarrow[aHF]{-2e^-} NiF_2 + 2H^+$$

Scheme 4.7 Electrochemical oxidation of Ni^0 in aHF.

However, if the Ni electrode has been previously anodically polarized for 10 min in aHF (current density c. 300 mA/cm^2), no current response is observed at potentials below +3.1 V; at higher potentials, the current raises significantly (Fig. 4.5B).

Presumably, at high anodic potentials, further oxidation of NiF_2 to NiF_3 and NiF_4 takes place resulting in the formation of an electroconductive composite consisting of NiF_3 doped with NiF_4. The anodic behavior of Ni in HF was the research topic for many scientists.[133–136] G.G. Totir et al. have demonstrated that a mirrorlike Ni film is oxidized on the surface to give NiF_2 upon exposure to gaseous aHF after contact with the laboratory atmosphere.[137]

In 1995, Zemva et al. presented chemical syntheses of NiF_3 and NiF_4.[138] The authors described NiF_3 as a black solid stable at 20°C. However, NiF_4 (tan powder) is not stable above −55°C, liberating F_2 and converting into black NiF_3. Both NiF_3 and NiF_4 are practically insoluble in aHF. R−NiF_3 was found to be a strong fluorinator converting perfluoropropene into perfluoropropane, C_3F_8, at 20°C in aHF[138]; cyclic partially fluorinated ether into perfluorinated one; and adamantane derivative into corresponding substituted perfluoroadamantane.[139] These examples demonstrate the fluorinating power of NiF_3.

Apparently, the black deposit that can be found on the surface of the Ni anode after opening of a Simons cell in an inert atmosphere (see Fig. 4.6) is NiF_3.

Obviously, the NiF_3 deposit on the anode plates is protected from contact with atmosphere by a film of the perfluorinated substance, which covers the Ni electrodes (Fig. 4.6). However, after evaporation of this protecting film, the color of the deposit is changed within short time (some minutes) to yellow-green, which is typical for NiF_2.[140]

Figure 4.6 Black deposit on the surface of the Ni anode after opening of a Simons cell in an inert atmosphere.
Reprinted from Ignat'ev N. V.; Welz-Biermann U.; Heider U.; Kucheryna A.; von Ahsen S.; Habel W.; Sartori P.; Willner H. J. Fluorine Chem. **2003,** 124, 21−37 with permission from Elsevier.

On the basis of all experimental data described earlier in this chapter, we have proposed the $NiF_2/NiF_3(NiF_4)$-mediated mechanism of ECF in aHF (Simons process) in 1995.[109,124,141] According to this mechanism, the formation of a NiF_3 film (presumably doped with NiF_4) on the surface of the Ni anode takes place during preelectrolysis in aHF (Scheme 4.8).

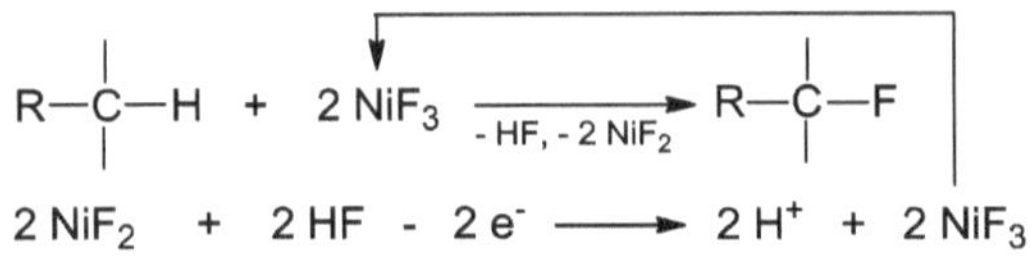

Scheme 4.8 The $NiF_2/NiF_3(NiF_4)$-mediated mechanism of electrochemical fluorination in anhydrous HF (Simons process).[141]

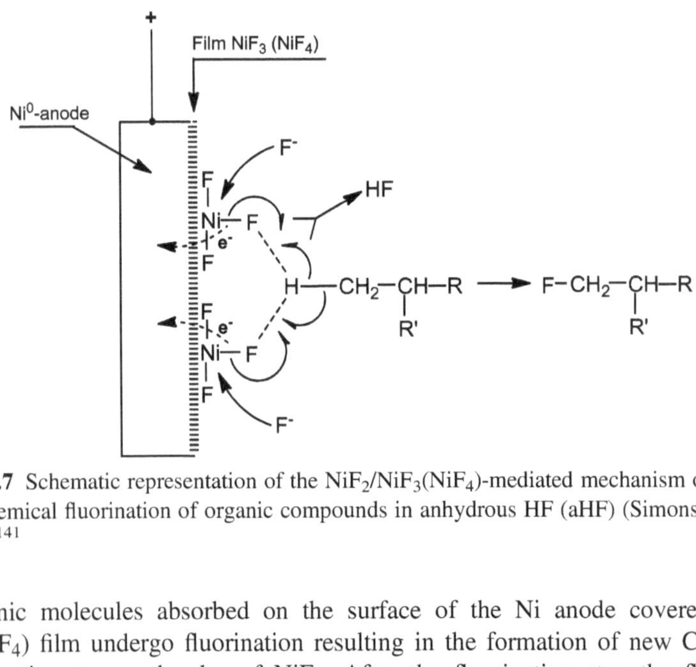

Figure 4.7 Schematic representation of the $NiF_2/NiF_3(NiF_4)$-mediated mechanism of electrochemical fluorination of organic compounds in anhydrous HF (aHF) (Simons process).[141]

Organic molecules absorbed on the surface of the Ni anode covered with a $NiF_3(NiF_4)$ film undergo fluorination resulting in the formation of new C–F bond and liberating two molecules of NiF_2. After the fluorination step the fluorinated molecules are desorbed into the electrolyte, whereas the NiF_2 remains on the surface of the Ni anode and is electrochemically reoxidized back to NiF_3 (Fig. 4.7).[109,141]

In summary, the ECF of organic compounds in aHF (Simons process) proceeds as follow[141]:

- The ECF of organic compounds in aHF (Simons process) proceeds via a $NiF_2/NiF_3(NiF_4)$-mediated mechanism.
- Nickel fluorides, that are electrochemically formed on the surface of the nickel anode, act as mediators in the Simons process for the transfer of the oxidation potential from the anode and fluorine atoms from HF to the substrate.
- Protons are discharged at the cathode liberating hydrogen, $2H^+ + 2e \rightarrow H_2$.

- ECF in HF is a step by step process. During a single elemental stage of the process, only one hydrogen atom (or another equivalent atom) is replaced, and for this replacement two electrons and one molecule of HF are required.
- Partially fluorinated compounds are intermediates on the way from nonfluorinated starting material to perfluorinated product.
- Organic molecules are adsorbed and orientated on the surface of the anode in the way that the part of the molecule that bears the maximum electron density is as close as possible to the nickel fluoride layer on the surface of the anode.
- The fluorination preferably starts at the most hydrogenated position of the molecule.
- Replacement of every hydrogen atom is always in competition with the possible cleavage of the neighboring bonds, thus reducing the yield of the desired perfluorinated product due to degradation of the starting material.
- The less hydrogen atoms that have to be replaced by fluorine during the perfluorination of the starting material, the higher will be the yield of the final product.
- The ECF in aHF is a radical process. Radicals, which are formed on the surface of the nickel anode via oxidation of starting materials or intermediates by nickel fluorides, can undergo the following transformations: fluorination (addition of a fluorine atom), oxidation to carbocationic intermediate, isomerization, or cyclization (see Section 4).

By now, the $NiF_2/NiF_3(NiF_4)$-mediated mechanism of the ECF of organic compounds in aHF (Simons process) is well recognized and accepted by many scientists.[17,20,100,108,111,121,132,142,143] Many observations of the ECF of organic compounds in aHF (Simons process) may be explained based on the proposed mechanism. This mechanism is a useful tool for the optimization and application of the Simons process in the laboratory and for the industrial production of fluorinated materials.

4. Carbon Chain Isomerization During Electrochemical Fluorination (Simons Process)

Carbon chain isomerization during the ECF in aHF is well documented in the literature.[4,83,99,119,144,145] The formation of cyclic products or the conversion of branched starting compounds into products that have a linear structure during the ECF process results in a complex mixture of the fluorinated substances. In particular, this phenomenon takes place by the ECF of long-chain derivatives or compounds having *iso*-alkyl groups in the structure. The composition of the fluorinated and perfluorinated products produced by ECF of either type of starting material is very specific and can be used to identify the origin of these products from the ECF process like a "fingerprint".[90,91,93] Cyclic compounds formed by the ECF of aliphatic carboxylic acid derivatives,[145–148] aliphatic primary alcohols and aldehydes,[149] or substituted alkylamines[150,151] are sometimes the main products and can be isolated in reasonable yields. If the formation of cyclic or other isomeric products should be suppressed, the application of partially fluorinated compounds as starting materials for the ECF process is favored.[119,120,124,125,152–155]

Some hypotheses about the possible mechanism of the isomerization process during the ECF were discussed in the literature. Plashkin et al.[156] have rationalized the isomerization of a branched alkyl chain to a linear one by migration of the methyl (alkyl) group to the neighbor carbon atom, for instance (Scheme 4.9).

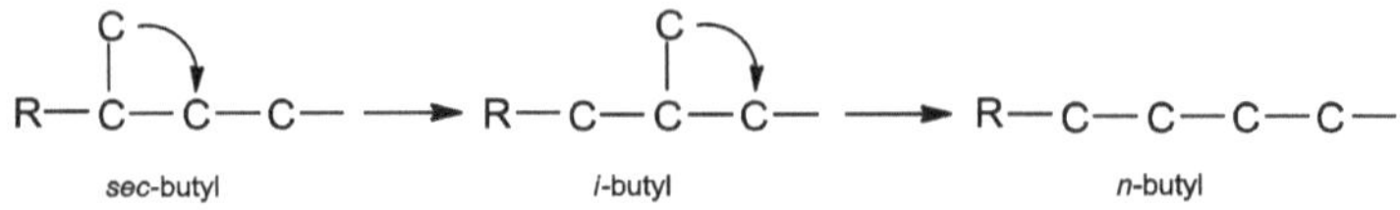

Scheme 4.9 Rationalization of the isomerization of a branched alkyl chain by migration of the methyl group according to Plashkin et al.[156]

This simple approach does not explain the isomerization of a linear alkyl group into a branched one observed by the ECF of various compounds.[119,131,157,158]

A carbocationic mechanism of the isomerization in the Simons process was proposed by G.B. Gambareto et al.[131] based on the hypothesis of the four-step (EC_bEC_N) mechanism. As demonstrated in Section 3, the Simons process does not proceed via a direct electrochemical oxidation of the substrate on the Ni anode, but by a $NiF_2/NiF_3(NiF_4)$-mediated mechanism.[104,124,141]

According to Scheme 4.8 and Fig. 4.7, the ECF proceeds in two reaction steps at the NiF_3 surface under formation of an intermediate radical species.[141] In some cases, in particular in the late fluorination stage, these radical species are stable enough[159–168] to migrate from the surface of the Ni anode without addition of the fluorine atom.

If the geometry of the organic molecule is suitable for the adsorption on the surface of the Ni anode by two sites, for example, in the case of the *iso*-butyl structure (see Fig. 4.8) the formation of a biradicals is possible. In this case, the interaction of both radical centers leads to the formation of cyclic product, having, for example, a cyclopropane structure (Fig. 4.8).[140]

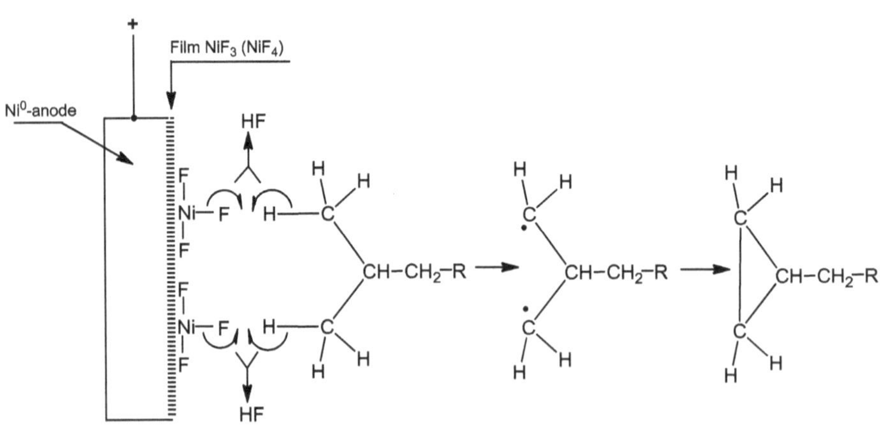

Figure 4.8 Generation of biradicals and formation of a cyclopropane structure in the course of the electrochemical fluorination process.[140]

Žemva et al.[138] have investigated the structure of NiF_3 and determined the Ni–Ni distance in the solid state to be approximately 3.5 Å. This distance fits with the 1,3-coordination of *iso*-alkane derivatives on the surface of the Ni anode, covered with NiF_3 (see Fig. 4.8). Results of quantum chemical calculations demonstrate that the cyclization of β-biradicals into cyclopropane derivatives is a thermodynamically favored process.[140]

NiF_3 is a strong oxidizer.[138,139] The formation of a cyclopropane derivative is also possible via a carbocationic mechanism by one-site, two-electron oxidation of a methyl group (Fig. 4.9).

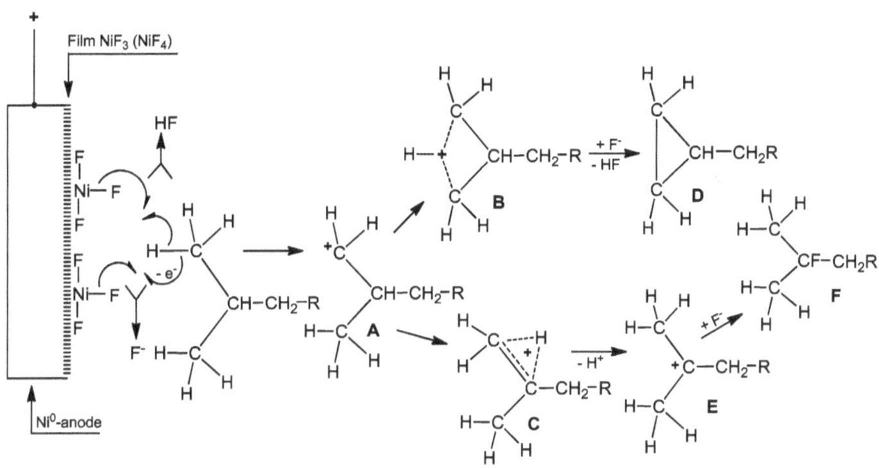

Figure 4.9 Formation of a cyclopropane intermediate via a carbocationic mechanism.

Oxidation of the substrate having an *iso*-butyl structure by two units of NiF_3 (Fig. 4.9) results in the formation of the carbocation **A**. The HF in the Simons cell is rigorously anhydrous and shows H° degree $= -15.0$.[169,170] In this superacidic medium (aHF) the carbocation **A** can be stabilized via the intermediate formation of a "nonclassical" carbocation **B**, which under the action of F^- anion undergoes conversion to cyclopropane **D**. Carbocation **B** is a typical carbenium ion as proposed by Olah.[169] The carbocation **A** can also undergo transformation into carbocation **E** via hydrogen migration. Addition of fluoride ion to carbocation **E** leads to the formation of the fluorinated product **F** (R = SO_2F; Scheme 4.11), which was found in the reaction mixture in significant quantity.

Both the radical (Fig. 4.8) and the carbocationic (Fig. 4.9) routes, which result in the ring structure **D**, are reasonable, but the radical route seems to be more favorable by facilitating single-electron oxidation (radical formation) when compared with a two-electron oxidation (cation formation, Fig. 4.9), assuming that oxidation of each methyl group (biradical formation, Fig. 4.8) proceeds independently. It is also possible that in the early stage of the fluorination (the molecule is still enriched with hydrogen atoms) the carbocationic way predominates, but after introduction of some fluorine atoms into the molecule, the radical way becomes more favorable. The subsequent

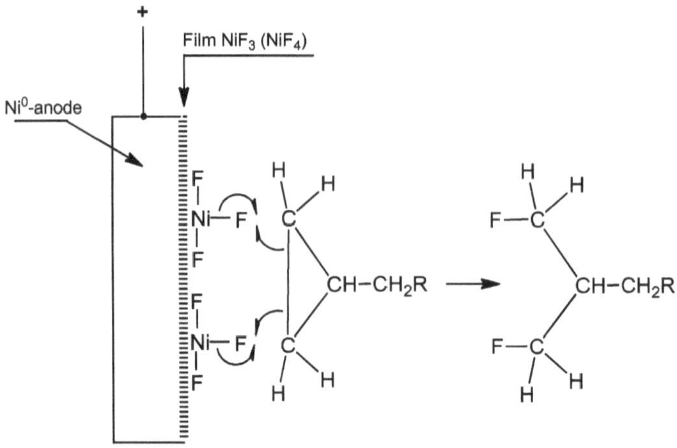

Figure 4.10 Fluorination of the cyclopropane derivate after adsorption on the anode covered with NiF_3.

ring-opening reactions (Fig. 4.10) in the cyclopropane structure during the ECF process result in the formation of an isomeric mixture of fluorinated products (Scheme 4.10).

The cyclopropane structure during the ECF process can be fluorinated (open) in two possible ways (**1**) and (**2**) leading to linear and *iso*-structure correspondingly (Scheme 4.10).

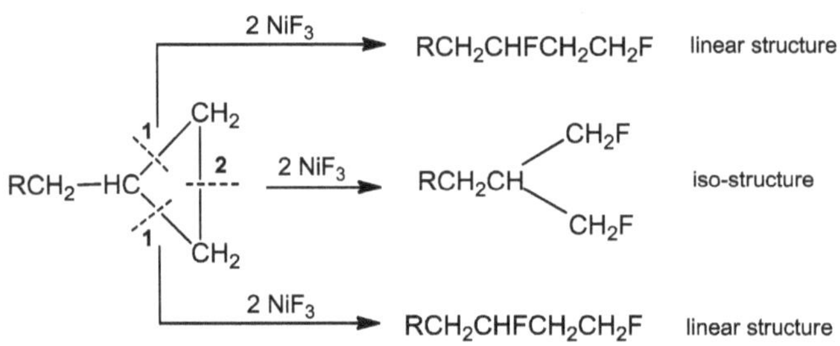

Scheme 4.10 Reaction paths (**1**) and (**2**) for cyclopropane ring fluorination.

This mechanism of the carbon chain isomerization via intermediate cyclopropane ring formation was briefly presented in the article devoted to the ECF of trialkylphosphines[126] and at the 16th International Symposium on Fluorine Chemistry.[171]

To support the proposed mechanism, the ECF of various organic compounds was examined in detail, e.g., *iso*-butylsulfonyl fluoride, *iso*-propylsulfonyl fluoride, *iso*-butyryl chloride, and cyclopropane carboxylic acid fluoride.[140] The ECF of *iso*-butyl sulfonyl fluoride results, as expected, in a mixture of *n*-perfluorobutylsulfonyl fluoride and *iso*-perfluorobutylsulfonyl fluoride together with small quantities of *sec*- and *tert*-perfluorobutylsulfonyl fluorides (Scheme 4.11).

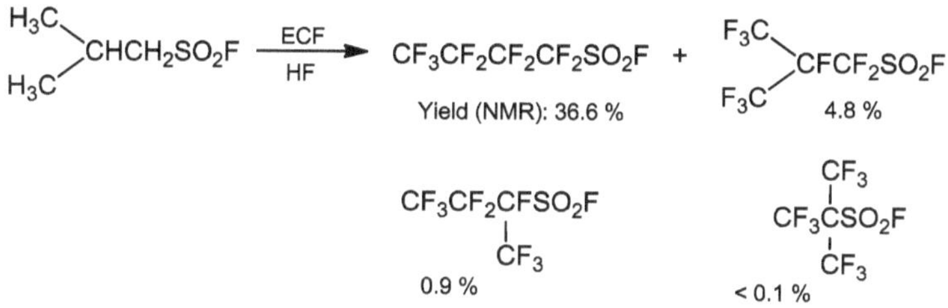

Scheme 4.11 Product distribution pattern from the ECF of *iso*-butyl sulfonyl fluoride.[140] Yields were determined by means of *NMR*, nuclear magnetic resonance.

Scheme 4.11 indicates that the reaction mixture is enriched with *n*-perfluorobutylsulfonyl fluoride. According to Scheme 4.10, opening of every cyclopropane ring leads to the formation of double the quantity of linear perfluorobutylsulfonyl fluoride than branched perfluorobutylsulfonyl fluoride. Theoretically, if every fluorination step is accompanied by the formation of a cyclopropane structure, the relation between the linear structure and *iso*-structure should be maximal 26: 1. Taking into account the parallel one-site fluorination via the radical (Fig. 4.7, $R' = CH_3$) or the carbocationic mechanism (Fig. 4.9), this theoretical value should be reduced and the *iso*-structure formation should be observed in a large quantity. Analysis of the product shows that the ratio of linear to *iso*-structure is 7.6: 1. This is in good agreement with the proposed model (Fig. 4.10; Scheme 4.10).[140] *Sec*- and *tert*-perfluorobutylsulfonyl fluorides were also found in the reaction mixture (Scheme 4.11). This indicates that ring closure reaction during the ECF process proceeds not only between methyl groups but also to some extent between the CH_2 group in the α-position and the methyl group (1, 3—ring closure) of starting *iso*-butylsulfonyl fluoride. Remarkably, the formation of perfluorocyclopropane derivatives including perfluorocyclopropane itself was observed by the ECF of pivaloyl chloride.[172] This observation strongly supports the cyclopropane intermediate mechanism of carbon chain isomerization during the ECF (Simons process).

The ECF of the *iso*-propylsulfonyl fluoride results in the formation of two types of perfluoropropylsulfonyl fluorides exhibiting branched and linear structures (Scheme 4.12) together with some amounts of partially fluorinated compounds bearing one or two hydrogen atoms and the products resulting from the decomposition of the starting material.[140]

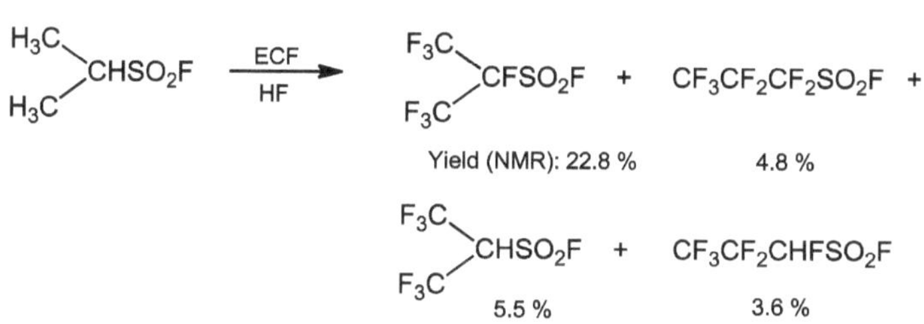

Scheme 4.12 Product distribution pattern of the ECF of *iso*-propyl sulfonyl fluoride.[140] Yields were determined by means of *NMR*, nuclear magnetic resonance.

The ratio between the molecules with branched and linear structures is about 3: 1 (see Scheme 4.12). This ratio is not in agreement with prediction derived from the cyclopropane ring contraction model (Figs. 4.8 and 4.9). The *iso*-structure is somehow preserved by the bulky $-SO_2F$ group attached to the *iso*-position. This steric hindrance makes the fluorination in α-position more difficult (Scheme 4.12) and disfavors the cyclopropane ring contraction via the biradical formation (Fig. 4.8). Nevertheless, presence of the linear perfluorinated and partially fluorinated substances (Scheme 4.12) in the reaction mixture after the ECF of the *iso*-propylsulfonyl fluoride indicates that the cyclopropane ring contraction and the following fluorination still take place to some extent.

Gambaretto et al. has reported the same observation by the ECF of *iso*-butyryl chloride (Scheme 4.13).[131]

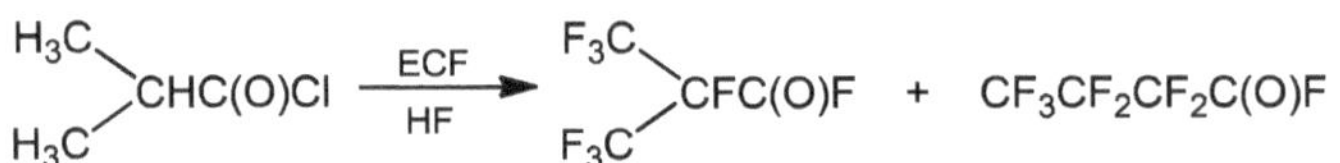

1.8 : 1

Scheme 4.13 Product distribution pattern from the electrochemical fluorination (ECF) of *iso*-butyryl chloride.[131]

The reported ratio between the *iso*- and linear structure is about 1.8: 1,[131] which varies from the prediction made by the cyclopropane ring contraction model. Again, a bulky $-C(O)F$ group in the *iso*-position, analogous to the $-SO_2F$ group, hinders the isomerization process via the intermediate cyclopropane ring contraction.

A similar result was reported by Velayutham et al. from ECF experiments with the *iso*-butyryl chloride.[172] The authors have found that the ratio between perfluoro-*iso*-butyric acid and perfluoro-*n*-butyric acid (obtained by the hydrolysis of the corresponding perfluorobutyryl fluoride) is about 1: 1.

Remarkably, ECF of the cyclopropane carboxylic acid fluoride as the starting material gave the result that is in accordance with the cyclopropane ring contraction model (Scheme 4.14). The cyclopropane carboxylic acid fluoride is stable in HF solution, but as it can be expected, it undergoes ECF in the Simons cell.[140] Contrary to the ECF of *iso*-butyryl chloride[131,172] (Scheme 4.13), *n*-heptafluorobutyryl fluoride and *iso*-heptafluorobutyryl fluoride are formed by the ECF of the cyclopropane carboxylic acid fluoride in a molar ratio of 2.3: 1. This is in good agreement with the prediction from the cyclopropane ring contraction model (Scheme 4.14).[140] Small quantities of perfluoroethane, perfluoropropane, and perfluorocyclobutane were found in the reaction mixture. This means that the decomposition process had a minor influence on the product distribution pattern in this case.

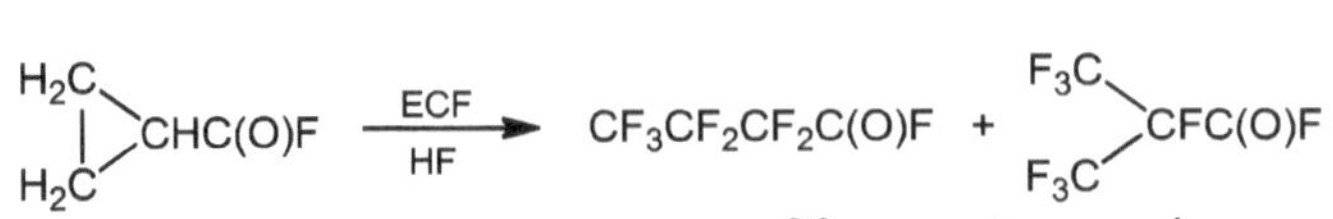

2.3 : 1

Scheme 4.14 Products distribution pattern of the electrochemical fluorination (ECF) of the cyclopropane carboxylic acid fluoride.

The intermediate cyclopropane ring contraction model of carbon chain isomerization during the Simons fluorination process is applicable to the analysis of the literature data.

In 1974, Plashkin et al. reported the isomerization of heterocyclic *N-iso*-alkylamines during the ECF in aHF.[156] The migration of the methyl group (Scheme 4.9) was proposed to explain the isomerization of *iso*-alkyl group into a linear one. The analysis of the reaction mixture after ECF of *N-iso*-butylpyrrolidine confirmed the formation of *n*-perfluorobutyl and *iso*-perfluorobutyl octafluoropyrrolidine in a molar ratio 2: 1. This is in harmony with the intermediate cyclopropane ring contraction model. A similar conclusion can be drawn from the analysis of the reaction mixture after ECF of *N-iso*-butylpiperidine.[156] The molar ratio of the sum of products having an *n*-structure and the sum of product with an *iso*-structure plus the product, which is formed by splitting off the methyl group, is again 2: 1 (Scheme 4.15).[140]

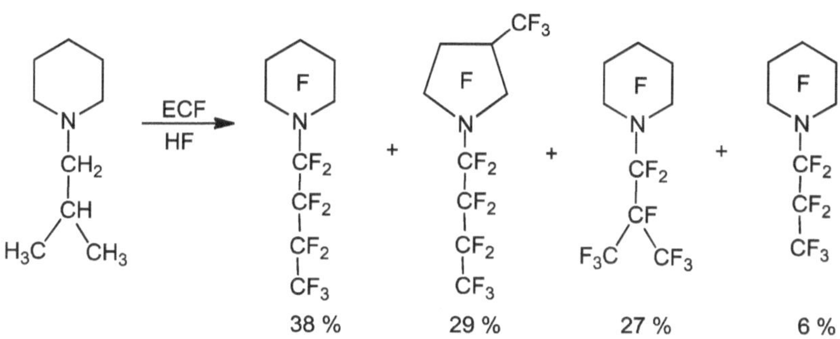

Scheme 4.15 Composition of the reaction mixture obtained from the electrochemical fluorination (ECF) of *N-iso*-butylpiperidine.[156]

The picture is different in the case of the ECF of *N-iso*-propylpiperidine.[156] The content of the isomerization products, namely, perfluoro-*N*-(*n*-propyl)piperidine and perfluoro-*N*-(*n*-propyl)-3-(trifluoromethyl)pyrrolidine, is less than the content that can be predicted from the intermediate cyclopropane ring contraction model. It seems that the *N*-heterocycle bonded to the *iso*-position of the alkyl chain somehow hinders the adsorption of the starting molecule at the Ni anode in a way that would be necessary to enable the cyclopropane ring contraction−fluorination via a biradical mechanism (Figs. 4.8 and 4.10). A similar observation was reported for the ECF of $(CH_3)_2CHC(O)Cl$[172] and $(CH_3)_2CHSO_2F$.[140] Presumably, in all these cases the adsorption of the starting molecules on the surface of the Ni anode facilitates one-site fluorination with NiF_3 (Fig. 4.7). This preserves, to some extent, the structure of starting compound and leads to larger quantity of *iso*-products in the reaction mixture after ECF of $(CH_3)_2CH{-}X$ (X = C(O)Cl, SO_2F, and *N*-heterocycle) than can be predicted from the intermediate cyclopropane ring contraction model.

The influence of nitrogen on the efficiency of five- and six-membered ring formation during the ECF process is known.[157,173,174] Abe et al.[174] have proposed the incorporation of the nitrogen atom into aliphatic chain to prevent the intramolecular cyclization during the Simons process. More information about cyclic products formed during the ECF of *N*-containing carboxylic acid derivatives, amino alcohols, and oxolane derivatives can be found in previous publications.[151,157,174–179] In general, a ring closure reaction during the Simons process proceeds via attack of electrophilic fluorinated radicals on carbonyl or hydroxyl oxygen. Formation of cyclic products by the ECF of *N,N,N′,N′*-tetramethylethylenediamine and *N,N,N′,N′*-tetramethylsulfondiamides can be rationalized through intramolecular recombination of carbon–carbon-centered or carbon–nitrogen-centered biradicals.[142,180] In many cases the cyclization process is accompanied by carbon–carbon, carbon–nitrogen, and carbon–sulfur bond scission and carbon chain isomerization.

It is well known that a large quantity of cyclic products are formed by the production of PFOA, $C_7F_{15}COOH$, with the Simons ECF method.[119,181,182] According to Napoli et al.[119] approximately 60% of the starting material, $C_7H_{15}C(O)F$, is converted into perfluorooxolanes and perfluorooxanes bearing perfluoro-*n*-alkyl and perfluoro-*iso*-alkyl groups in position 2 (Scheme 4.16). The content of perfluoro-2-alkyl-oxolanes in the mixture after the ECF process is particularly high and corresponds to 42 wt%.[119] Three main oxolane derivatives were identified in the reaction mixture (Scheme 4.16).

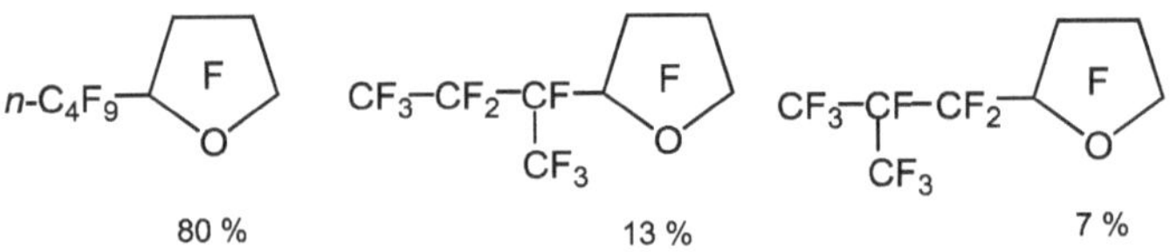

Scheme 4.16 Perfluorinated oxolanes found in the reaction mixture after electrochemical fluorination of $C_7H_{15}C(O)F$.

Formation of the perfluorooxolane with *n*-butyl substituent at position 2 mirrors the intramolecular cyclization during the Simons process via attack of an electrophilic carbon-centered radical on oxygen.[119] Products with *s*-butyl and *iso*-butyl substituents (Scheme 4.16) are formed similarly but with subsequent isomerization of *n*-butyl into *s*-butyl or *iso*-butyl groups, probably via intermediate formation of cyclopropane derivatives as follows (Scheme 4.17).

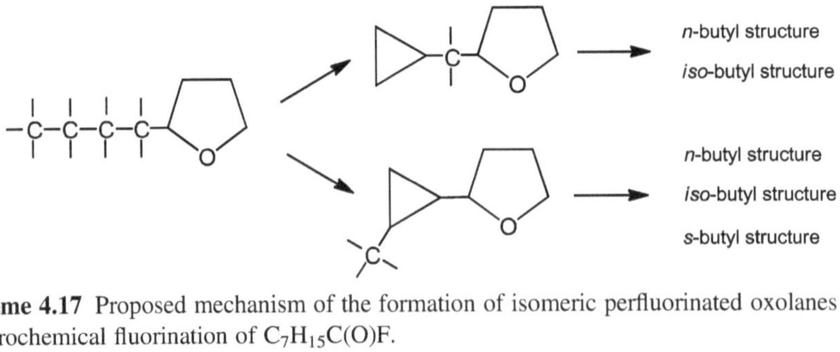

Scheme 4.17 Proposed mechanism of the formation of isomeric perfluorinated oxolanes by electrochemical fluorination of $C_7H_{15}C(O)F$.

The formation of a *tert*-Bu structure is not possible in this way, and therefore products of this type were not found in the reaction mixture.[119] However, the carbon chain isomerization prior to cyclization cannot be excluded during the ECF of long-chain carboxylic acid fluorides.[182]

These examples indicate that not only isomerization of branched alkyl chain but also isomerization of linear alkyl chain into branched one is possible during the ECF in aHF (Simons process). Several examples were reported in the literature. Obviously, for the cyclopropane ring contraction of at least three carbon atoms are required. However, cyclopropane ring formation from *n*-propyl substituent is very unfavorable. Therefore, only traces of $(CF_3)_2CHC(O)ONa$ (formed by the hydrolysis of the corresponding acyl fluoride with aqueous NaOH) were observed in the reaction mixture after ECF of *n*-butyryl chloride.[172] A small quantity of perfluoro-*iso*-propylsulfonyl fluoride, $(CF_3)_2CFSO_2F$ (4.4% calculating on the amount of the *n*-$C_3F_7SO_2F$) was found in the reaction mixture after ECF of *n*-$C_3H_7SO_2F$.[183]

The extent of isomerization of linear alkyl chain into branched one during the ECF process strongly depends on the chain length. The isomerization of *n*-butyl groups into branched structures occurs to some extent by the ECF of tri(*n*-butyl)phosphine.[126] The perfluorinated products containing branched alkyl chain are typical side products in the production of PFOS by means of ECF of octanesulfonyl fluoride, *n*-$C_8H_{17}SO_2F$.[91,93] Validation of the isomerization process through the intermediate cyclopropane ring contraction in the octane chain is presented in Scheme 4.18.

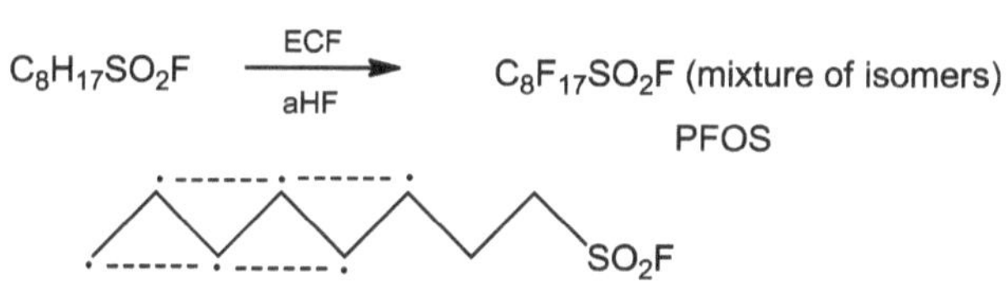

Scheme 4.18 Isomerization in the octane chain during the electrochemical fluorination (ECF) of *n*-$C_8H_{17}SO_2F$ in *aHF*, anhydrous HF.

The intermediate cyclopropane ring contraction model can also be applied for the ring isomerization during the Simons process. For example, ECF of *N*-(ω-chloroalkyl)-hexahydroazepines yields a mixture of the corresponding fluoro-(*N*-alkyl-hexahydroazepines) together with isomeric fluoro-(1-alkyl-methylpiperidines) (Scheme 4.19).[184]

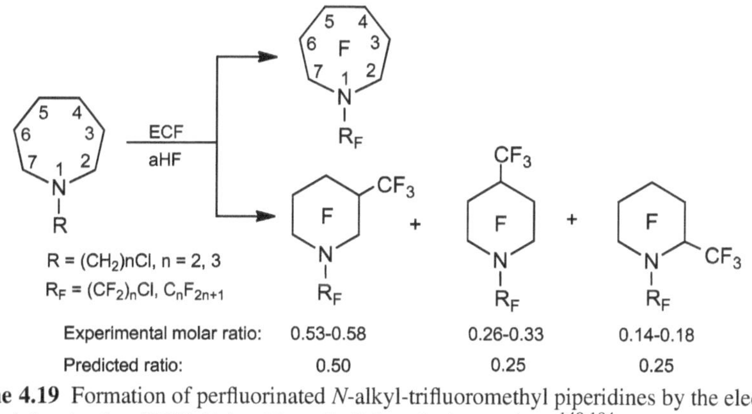

Scheme 4.19 Formation of perfluorinated *N*-alkyl-trifluoromethyl piperidines by the electrochemical fluorination (ECF) *N*-(ω-chloroalkyl)-hexahydroazepines.[140,184] *aHF* means anhydrous HF.

The formation of biradicals is possible by the simultaneous oxidation of two C—H bonds with NiF_3 on the surface of the anode at positions 2 and 4, 3 and 5, 4 and 6, and 5 and 7 of the hexahydroazepine ring. Ring closure reactions at these positions with subsequent C—C bond fluorination (Fig. 4.10) in the cyclopropane derivatives led to the formation of mixed perfluoro-(1-alkyl-methylpiperidines) with a predicted molar ratio for 3-CF_3:4-CF_3:2-CF_3 of 0.50:0.25:0.25. Considering the possible influence of the nitrogen atom on the cyclopropane ring formation, the yield of the *N*-(fluoroalkyl)-2-CF_3-perfluoropiperidine may be lower. Consequently, the experimental product distribution of 0.55: 0.29: 0.16 (see Scheme 4.19) is in good agreement with the intermediate cyclopropane ring formation model.[140]

The contraction of a six-membered ring, for example, during the ECF of *N*-substituted piperidines, leads to the formation of perfluoro-3-CF_3-pyrrolidine derivatives.[130,158,176] Obviously, the cyclopropane ring closure reaction at positions 2 and 4 and 4 and 6 of the piperidine ring is hindered by nitrogen. The only possibility is the ring formation at positions 3 and 5 (Scheme 4.20). In this case, the subsequent cyclopropane ring-opening reactions by the action of NiF_3 during the ECF process can give either 3-CF_3-pyrrolidine derivatives or compounds with the structure of the starting substances. This is in harmony with the experimental results.[130,158,176]

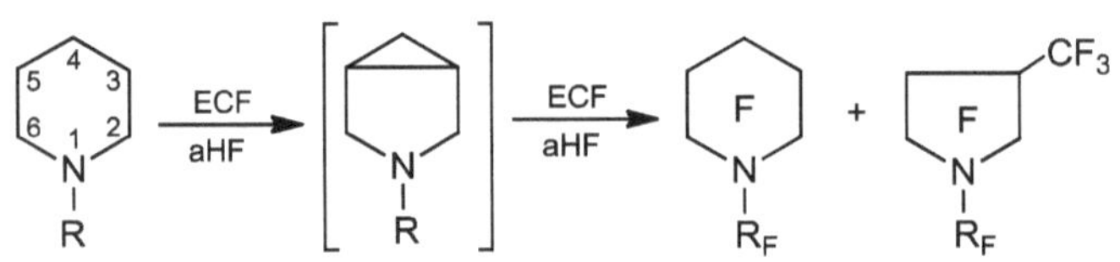

Scheme 4.20 Ring contraction during the electrochemical fluorination (ECF) of *N*-substituted piperidines in *aHF*, anhydrous HF.

In summary, the following conclusion can be drawn from the experimental results reported in the literature:

- Carbon chain isomerization is always accompanied by the ECF (Simons process) of carbon–hydrogen bonds if the starting material has a carbon chain longer than C_3 or a branched alkyl chain.
- The isomerization of branched to linear alkyl chain proceeds much more extensively than the isomerization of linear alkyls into branched ones.
- The isomerization of *iso*-alkyls into *n*-alkyls can be reduced by the introduction of bulky groups to *iso*-position.
- The intermediate cyclopropane ring contraction model provides a plausible explanation for the carbon chain isomerization during the ECF process. This model can predict the pathway of the isomerization from the molecular geometry of the starting compound.
- The ring contraction during the ECF of cyclic compounds is a common process that can be elucidated on the basis of the intermediate cyclopropane ring formation model.

5. Electrochemical Fluorination of the Elementorganic Compounds

5.1 Electrochemical Fluorination of the Compounds With C–S Bond and Follow-Up Chemistry

Trifluoromethanesulfonic acid, CF_3SO_2OH (triflic acid), was one of the first fluorochemicals produced by means of the ECF technology (Simons process). The 3M Company introduced this chemical to the market under the trade name Fluorad.[95] Trifluoromethanesulfonic acid has many applications in organic chemistry (e.g., as catalyst)[185–187] and as starting material for the preparation of various derivatives (triflates), for instance, alkyl[186,188,189] and silyl esters.[186,189,190] Nowadays, triflic acid is produced in multiton quantities by several companies throughout the world. Some of them are still using an electrochemical technology, which provides the flexibility to adjust the production capacities to the market demand.

ECF of methanesulfonyl chloride, CH_3SO_2Cl, or methanesulfonyl fluoride, CH_3SO_2F, is the key step in the production of triflic acid by the Simons process.[191] The conversion of CH_3SO_2Cl into CF_3SO_2F via the ECF process proceeds smoothly and results in the formation of trifluoromethanesulfonyl fluoride in good chemical (80–87%) and current (79–82%) yields.[116,192] Gramstadt and Haszeldine have reported a slightly better yield of CF_3SO_2F (91%) by the ECF of CH_3SO_2F instead of CH_3SO_2Cl. This again confirms the fact that the shorter the way to perfluorinated molecules, the better the yield of the final product. In any case, the ECF of small molecules CH_3SO_2Cl, CH_3SO_2F, and $CH_3C(O)F$ (formed in situ from $CH_3C(O)Cl$ by the reaction with aHF) provides the best yield of perfluorinated substances, CF_3SO_2F and $CF_3C(O)F$ produced by the Simons process. The trifluoromethanesulfonyl fluoride can be easily hydrolyzed by aqueous sodium or barium hydroxide to the corresponding salts.[116,192,193] Trifluoromethanesulfonic acid, CF_3SO_2OH, can be recovered by the action of concentrated sulfuric acid on sodium or barium salts.

Surprisingly, the ECF of a very similar compound, ethanesulfonyl chloride, $C_2H_5SO_2Cl$, results in the formation of a very complex mixture of chlorofluoroethanes together with chlorofluoroethane sulfonyl fluorides and chlorides: $CF_3CF_2SO_2F$ (39 mol %), $ClCF_2CF_2SO_2F$ (3.8 mol %), $CF_3CFClSO_2F$ (1.5 mol %), $CF_3CF_2SO_2Cl$ (1 mol %), and CF_3CHFSO_2F (0.2 mol %).[194] It was shown that the chloride ion, which is formed by cathodic reduction of $C_2H_5SO_2Cl$ in the ECF cell, triggered the chain process: chlorination–reduction–chlorination, which proceeds in parallel with the Simons process and causes a severe decomposition of the starting material. According to CVA measurements, the $C_2H_5SO_2Cl$ is irreversibly reduced on a glassy carbon electrode at -1.34 V versus saturated calomel electrode (SCE) (Scheme 4.21).[194] Subsequent electrochemical oxidation of the chloride anion by NiF_3 or electrochemically on the Ni anode leads to the generation of chlorine radicals, which react with the starting material $CH_3CH_2SO_2Cl$ producing $ClCH_2CH_2SO_2Cl$.

Cathodic process:

$$CH_3CH_2SO_2Cl + 2e^- \longrightarrow CH_3CH_2SO_2^- + Cl^-$$

Anodic process:

$$Cl^- - e^- \longrightarrow Cl^\bullet$$

Scheme 4.21 Generation of chlorine radicals in the electrochemical cell during the electrochemical fluorination of the $CH_3CH_2SO_2Cl$.

β-Chloroethanesulfonyl chloride, $CH_2ClCH_2SO_2Cl$, is reduced at lower cathodic potentials (−0.99 vs. SCE)[194] liberating the chloride anion, which extends the cyclic process: chlorination− reduction−chlorination and reduces the yield of $CF_3CF_2SO_2F$ by ECF of the $CH_3CH_2SO_2Cl$. Sartori et al. have reported the same problem, heavy destruction of starting material, by the ECF of the *α*-chloroethanesulfonyl chloride, $CH_3CHClSO_2Cl$,[195] and chloromethylsulfonyl chloride, CH_2ClSO_2Cl.[196] Absence or limited concentration of chloride anions in the composition of the electrolyte in Simons cell is one of the requirements for the successful ECF of organic compounds in aHF.[110]

In contrast to $CH_3CH_2SO_2Cl$, the ECF of $C_2H_5SO_2F$ gave practically pure $C_2F_5SO_2F$ in 82% yield under the same conditions.[194] Obviously, the large differences in the yield of $C_2F_5SO_2F$, obtained by means of the Simons process from $C_2H_5SO_2Cl$ or $C_2H_5SO_2F$ as starting materials, is determined by the chlorination process, which takes place during the ECF of ethanesulfonyl chloride. The difference in the yield of CF_3SO_2F is much less by the ECF of CH_3SO_2Cl and CH_3SO_2F (of about 10%). The pathway to the perfluorinated molecule in the case of CH_3SO_2Cl is much shorter and the probability of chlorination of the methyl group during the ECF of methanesulfonyl chloride is much lower.

In 1986, Klink et al. have reported the synthesis of difluoromethanedisulfonyl fluoride, $CF_2(SO_2F)_2$, in 75−82% yield by the ECF of methanedisulfonyl fluoride, $CH_2(SO_2F)_2$[197] Perfluoroethane, 1,3-propane, and 1,4-butane disulfonyl fluoride have been prepared by the ECF of the corresponding 1,n-alkane-bis(sulfonylfluorides) (n = 1−4) (Scheme 4.22).[103,183,198−200]

$$FSO_2\text{-}(CH_2)_n\text{-}SO_2F \xrightarrow[\text{aHF}]{\text{ECF}} FSO_2\text{-}(CF_2)_n\text{-}SO_2F$$

n = 1 - 4

Scheme 4.22 Synthesis of perfluoroalkane disulfonyl fluorides by the electrochemical fluorination (ECF) process in *aHF*, anhydrous HF.

Similar to the preparation of triflic acid, pefluoroalkane disulfonyl fluorides can be converted into the corresponding pefluorinated disulfonic acids,[103,201] cyclic anhydrides,[103] imides,[202−204] and perfluorinated ionomers for applications in rechargeable lithium batteries.[205,206] The electrochemical properties of cyclic Li imides were studied.[207,208]

Bis(trifluoromethylsulfonyl)imide, $(CF_3SO_2)_2NH$, and its salts are available from various companies or can be prepared in the laboratory from trifluoromethylsulfonyl fluoride by previously reported protocols.[209,210] The acid $(CF_3SO_2)_2NH$ and alkali metal salts $(CF_3SO_2)_2NM$ (M = Li, Na, K)[211] are very convenient starting materials for the preparation of very popular hydrophobic ILs with the bis(trifluoromethylsulfonyl)imide anion (Scheme 4.23).[212] Synthesis and properties of the C-H acids $(CF_3SO_2)_2CH_2$, $(CF_3SO_2)_3CH$ and their derivatives (Scheme 4.23) were reviewed.[213]

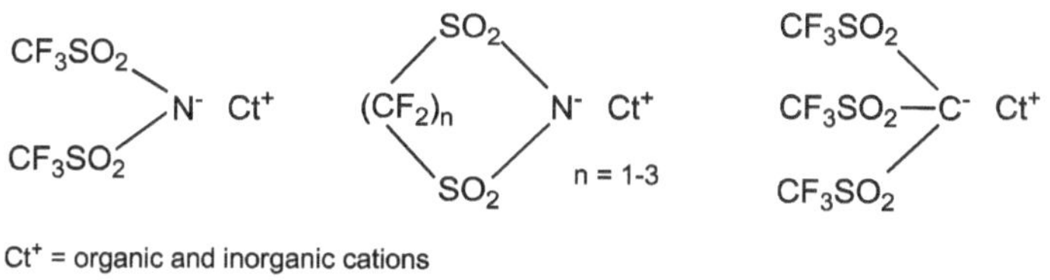

Scheme 4.23 Weakly coordinating anions derived from the electrochemical fluorination products, CF_3SO_2F and $FSO_2(CF_2)_nSO_2F$.

Triflic acid is the starting material for the preparation of various derivatives, e.g., metal salts, triflic anhydride, and silyl and alkyl esters.[189] Methyl triflate, $CF_3SO_2OCH_3$, is a very powerful methylating reagent (**Caution! Strong methylating reagents are toxic**). Several methods to prepare methyl triflate have been reported in the literature, e.g., by the reaction of silver triflate with methyl iodide[116] or triflic acid anhydride with methanol.[214] Both methods are not the best ones to produce this chemical. Presumably, commercial quantities of methyl triflate are produced by heating the triflic acid with dimethyl sulfate (**Caution! Toxic compound**) (Scheme 4.24).[215] Disadvantages of this protocol are (1) not perfect yield of methyl triflate, (2) the use of toxic starting material, and (3) the waste of sulfuric acid contaminated with dimethyl sulfate.

$$2\ CF_3SO_2OH + (CH_3O)_2SO_2 \xrightarrow{\Delta} 2\ CF_3SO_2OCH_3 + H_2SO_4$$

Yield: 81 %

Scheme 4.24 Synthesis of methyl triflate by reaction of the triflic acid with dimethyl sulfate.

A more convenient and environment-friendly protocol to produce methyl esters of trifluoromethane and pentafluoroethane sulfonic acids was developed recently (Scheme 4.25).[216,217]

$$(CF_3SO_2)_2O + (CH_3O)_2CO \xrightarrow{CF_3SO_2OH\ (cat.)} 2\ CF_3SO_2OCH_3 + CO_2$$

Yield: 99 %

$$(C_2F_5SO_2)_2O + (CH_3O)_2CO \xrightarrow{C_2F_5SO_2OH\ (cat.)} 2\ C_2F_5SO_2OCH_3 + CO_2$$

Yield: 84 %

Scheme 4.25 Synthesis of methyl esters of triflic and pentafluoroethane sulfonic acids by reaction with dimethyl carbonate.

Nontoxic and cheap dimethyl carbonate serves as a source of the methyl group in this technology. Triflic anhydride can be replaced by cheaper triflic acid in solvent-free, three-component synthesis of methyl and ethyl triflate (Scheme 4.26).[217,218]

$$CF_3SO_2OH + (CH_3O)_2CO + C_6H_5C(O)Cl \xrightarrow{\Delta} CF_3SO_2OCH_3 + C_6H_5C(O)OCH_3 + HCl + CO_2$$

Yield: 93 %

$$CF_3SO_2OH + (C_2H_5O)_2CO + C_6H_5C(O)Cl \xrightarrow{\Delta} CF_3SO_2OC_2H_5 + C_6H_5C(O)OC_2H_5 + HCl + CO_2$$

Yield: 92 %

Scheme 4.26 Synthesis of methyl and ethyl triflate by reaction of triflic acid with dimethyl or diethyl carbonate.

Methyl and ethyl triflate are very powerful alkylators. They can alkylate not only organic bases (Scheme 4.27)[217,219] but also very quickly react with chloride or bromide anions liberating chloroform or bromoform, respectively (Scheme 4.28).[220]

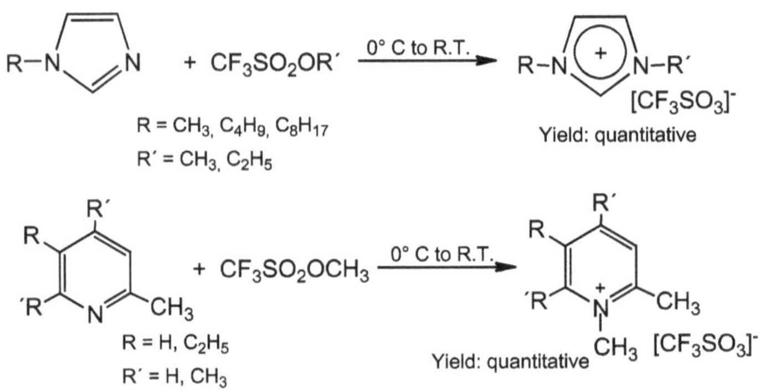

Scheme 4.27 Alkylation of organic bases with methyl and ethyl triflate.[217]

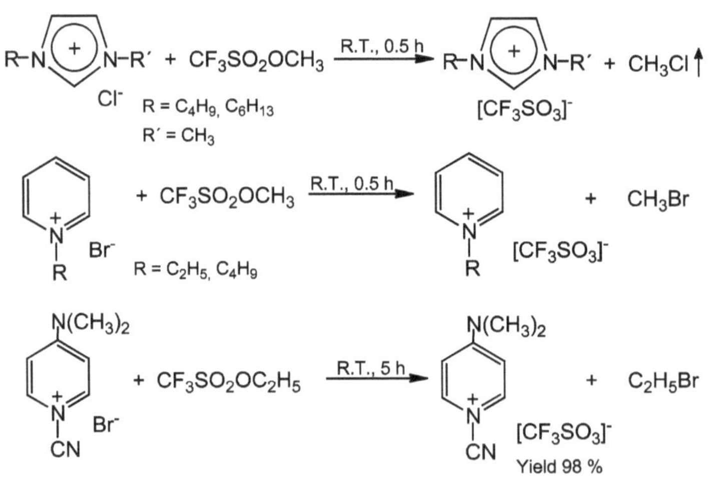

Scheme 4.28 Conversion of organic chlorides or bromides into triflates ionic liquids.[220] *RT*, room temperature.

This protocol provides a very convenient access to unsymmetrical ILs (Scheme 4.28).[220]

The examples mentioned earlier demonstrate the utility of the parent compounds, like CF_3SO_2F, produced by means of ECF (Simons process) technology.

5.2 Electrochemical Fluorination of the Compounds with S–N Bond and Follow-Up Chemistry

First reports about the successful ECF of sulphonamides were published in 1992.[124,221,222] All previous attempts to carry out the ECF in aHF of sulfonamides had failed. For instance, the ECF of H_2NSO_2F results in the gaseous decomposition products, SO_2F_2, NF_3, N_2, and traces of N_2F_2.[223] Similar results were observed by the ECF of alkanesulfonamides, RSO_2NH_2 ($R = CH_3$, C_2H_5, n-C_3H_7, i-C_3H_7).[224] Only the corresponding perfluoroalkysulfonyl fluorides, NF_3, and other gaseous products were identified in the reaction mixture after ECF, which indicates quantitative scission of the S–N bond.

The ECF of the $CF_3SO_2NH_2$ also resulted in S–N bond scission (Scheme 4.29).

$$CF_3SO_2NH_2 \xrightarrow[\text{aHF}]{\text{ECF}} CF_3SO_2F \quad + \quad NF_3$$

Scheme 4.29 Electrochemical fluorination (ECF) of the $CF_3SO_2NH_2$ in *aHF*, anhydrous HF.

Presumably, good solubility of $CF_3SO_2NF_2$ in aHF enables the S–N bond scission and completes fluorination of $CF_3SO_2NH_2$ to CF_3SO_2F and NF_3.

However, semifluorinated *N,N*-dimethylsulfonamides of the formula $(CH_3)_2NSO_2R$ ($R = F$, CF_3, C_4F_9) undergo successful fluorination in aHF to the corresponding perfluorinated *N,N*-bis(trifluoromethyl)sulfonamides (Scheme 4.30).[103,124,221,222] During the ECF process not only perfluorinated but also partially fluorinated sulfonamides of the general formula $(CF_{3\text{-}x}H_x)(CF_{3\text{-}y}H_y)NSO_2R$ ($R = F$, CF_3; x, $y = 1, 2$) are formed. The perfluorinated *N,N*-bis(trifluoromethyl)sulfonamide (major product) can be separated from partially fluorinated compounds by fractional distillation.[104,124,222]

$$R_FSO_2N(CH_3)_2 \xrightarrow[\text{aHF}]{\text{ECF}} R_FSO_2N(CF_3)_2$$

$$R_F = F,\ CF_3,\ C_4F_9$$

Scheme 4.30 Electrochemical fluorination (ECF) of the $R_FSO_2N(CH_3)_2$ ($R_F = F$, CF_3, C_4F_9) in *aHF*, anhydrous HF.

In contrast to these results, the ECF of *N,N*-dimethylmethanesulfonamide, $CH_3SO_2N(CH_3)_2$, gave $CF_3SO_2N(CF_3)_2$ in a very low yield (Scheme 4.31).[124]

$$CH_3SO_2N(CH_3)_2 \xrightarrow[\text{aHF}]{\text{ECF}} \text{decomposition}$$

Scheme 4.31 Electrochemical fluorination (ECF) of the $CH_3SO_2N(CH_3)_2$ in *aHF*, anhydrous HF.

Presumably, the perfluoroalkyl group bonded to the sulfur atom protects the S–N bond against cleavage during the Simons process. Probably, the perfluoroalkyl tail forces the molecule of *N,N*-dimethylperfluoroalkanesulfonamide to be attached to the surface of the Ni anode in a way that facilitates the fluorination of the $N(CH_3)_2$ group first.[124] These examples demonstrate how different the outcome of the ECF of very similar organic substances containing a S—N bond can be. The success of the ECF process strongly depends on the nature of the groups bonded to the sulfur atom.

In contrast to the ECF of $R_FSO_2N(CH_3)_2$, the ECF of *N,N*-dimethylacyl amides of the formula $R_FC(O)N(CH_3)_2$ ($R_F = C_2F_5$ and C_3F_7) resulted in the formation of the corresponding perfluoroacyl fluorides, $R_FC(O)F$ as major products. The $R_FC(O)N(CF_3)_2$ were the minor products in the mixture after ECF of the $R_FC(O)N(CH_3)_2$. Traces of $(CF_3)_2NC(O)F$ were also detected in the reaction mixtures by NMR spectroscopy.[125]

It is interesting to note that gas-phase fluorination of $(CH_3)_2NSO_2F$ with CoF_3 at 200—250°C yields partially fluorinated compound $(CHF_2)_2NSO_2F$ as the main product.[222] This nicely demonstrates that $NiF_3(NiF_4)$ electrochemically formed in the Simons cell at 0°C is a stronger fluorinator than CoF_3 at 250°C.

In the course of ECF of *N,N,N′,N′*-tetramethylsulfondiamides of formula $(CH_3)_2NSO_2(CF_2)_nSO_2N(CH_3)_2$ ($n = 0, 1$) not only the perfluorinated and partially fluorinated compounds with the structure of starting molecules but also minor quantities of cyclic products were detected in the reaction mixture by ^{19}F NMR spectroscopy (Scheme 4.32).[180]

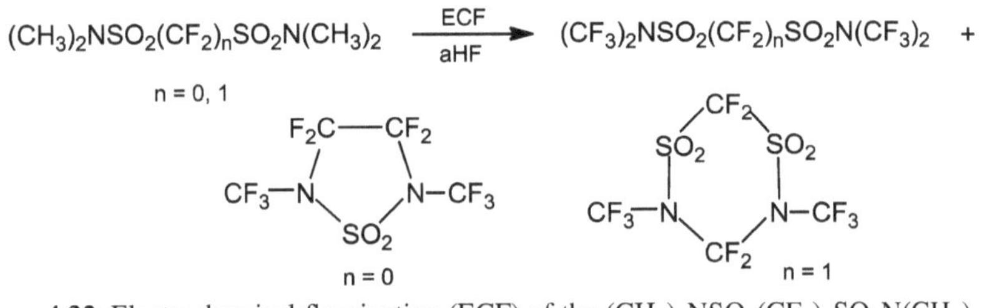

Scheme 4.32 Electrochemical fluorination (ECF) of the $(CH_3)_2NSO_2(CF_2)_nSO_2N(CH_3)_2$ ($n = 0, 1$) in *aHF*, anhydrous HF.

N,N-bis(trifluoromethyl)frifluoromethanelsulfonamide (BP 55—56°C) is a convenient reagent for the generation of the $(CF_3)_2N^-$ anion (Scheme 4.33).[225–227]

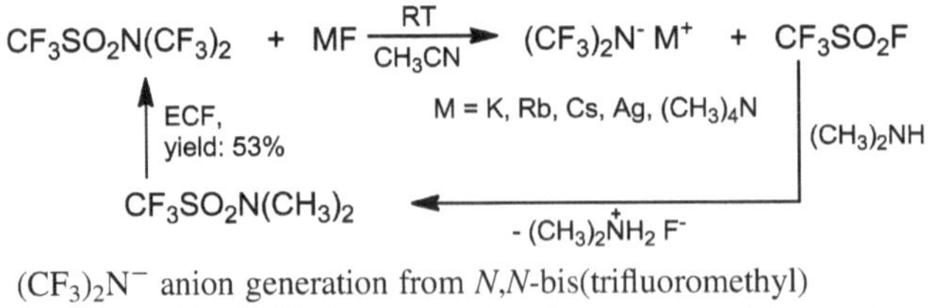

Scheme 4.33 $(CF_3)_2N^-$ anion generation from *N,N*-bis(trifluoromethyl) frifluoromethanesulfonamide. *ECF*, electrochemical fluorination; RT, room temperature.

The side product, CF_3SO_2F, can be trapped with dimethyl amine to produce $CF_3SO_2N(CH_3)_2$—the starting material for the ECF process. In summary, this technology provides a convenient route to convert inexpensive dimethylamine into $(CF_3)_2N^-$ salts (Scheme 4.33). Organic salts of the $(CF_3)_2N^-$ anion were prepared by metathesis of tetrafluoroborates with $Rb[N(CF_3)_2]$.[225] The $(CF_3)_2N^-$ anion generated either in situ from $CF_3SO_2N(CH_3)_2$ or its salts prepared in advance reacts as a nucleophile with various organic substrates affording one-pot, one-step synthesis of aliphatic compounds with $(CF_3)_2N$ group (Scheme 4.34).[225,227,229]

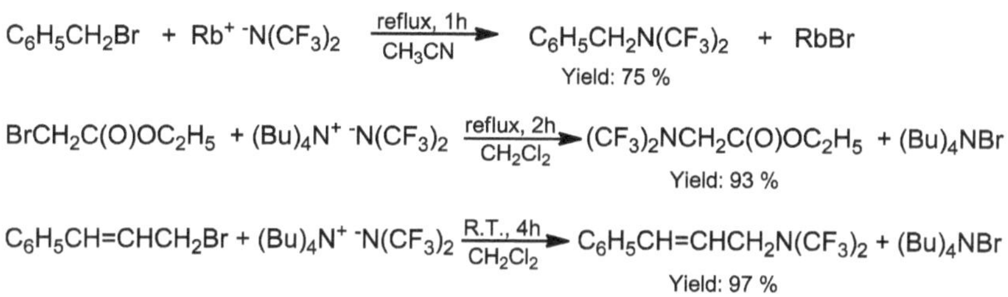

Scheme 4.34 Introduction of the $(CF_3)_2N$ group into aliphatic compounds via S_N1 nucleophilic substitution reactions.

However, attempts to introduce the $(CF_3)_2N$ group into aromatic ring via nucleophilic substitution reactions were unsuccessful. For example, in the course of the reaction of 2,4-dinitrochlorobenzene with $(CF_3)_2N^-$ salts, there was not substitution of Cl against $(CF_3)_2N$ group, but replacement of chlorine by fluorine.[228,229]

However, the synthesis of aromatic $(CF_3)_2N$ compounds starting from the corresponding diazonium salts[230] was successful. Decomposition of these salts in the presence of a suitable Cu(I) catalyst resulted in the formation of the corresponding $(CF_3)_2N$-aromatic compounds in reasonable to good yields (Scheme 4.35).[229]

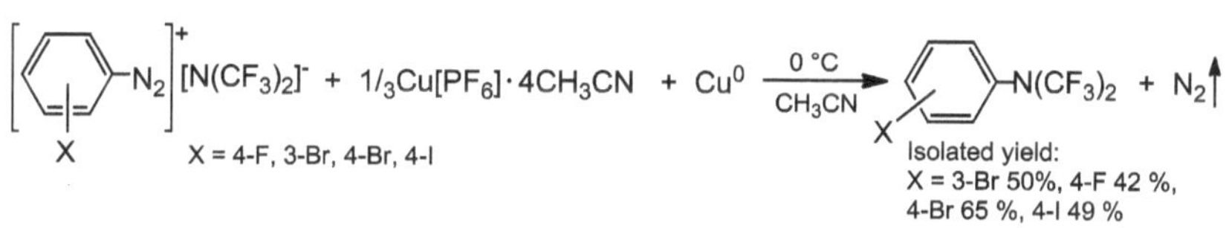

Scheme 4.35 Synthesis of Ar-$N(CF_3)_2$ compounds via diazonium salts.

$[Ar(Ar')I][N(CF_3)_2]$ ($Ar/Ar' = C_6H_5/4\text{-}FC_6H_4$, $3\text{-}FC_6H_4/4\text{-}FC_6H_4$, $3\text{-}FC_6H_4/4\text{-}FC_6H_4$) salts were prepared similarly to diazonium salts by anion metatheses. These iodonium salts can be stabilized in solution by the solvent CH_3CN or in the solid state by coordination of 1,10-phenanthroline. *N*-base stabilization avoided fluoride transfer from the labile anion to the electrophilic cation. Fluoride transfer reduces the potential reactivity of the $[Ar(Ar')I]$ $[N(CF_3)_2]$ salts to form arylbis(trifluoromethyl) anilines.[231]

The electrochemistry of $(CF_3)_2N$ compounds and electrochemical synthesis of Ar-$N(CF_3)_2$ were reviewed earlier.[66]

In conclusion, the preparation of $CF_3SO_2N(CF_3)_2$ via the Simons process tremendously accelerated the development of the $(CF_3)_2N$ chemistry.

5.3 Electrochemical Fluorination of the Compounds With C–P Bond and Follow-Up Chemistry

The first example on ECF of organophosphorus compounds in aHF was reported by Yagupolskii et al.[232a,232b] The authors claimed the preparation of tris(perfluoroalkyl) phosphine oxides by the ECF (Simons process) of the corresponding trialkylphosphine oxides (Scheme 4.36). Later on, it was found that tris(perfluoroalkyl)phosphine oxides are not the primary products of the ECF of trialkylphosphine oxides, but tris(perfluoroalkyl)difluorophosphoranes, $(R_F)_3PF_2$. The difluorophosphoranes were hydrolyzed with air moisture to the corresponding phosphine oxides liberating HF during workup (Scheme 4.36).[233]

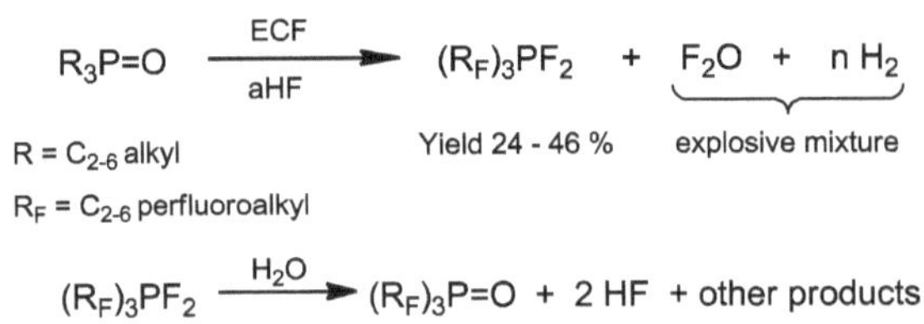

Scheme 4.36 Preparation of tris(perfluoroalkyl)difluorophosphoranes, $(R_F)_3PF_2$ by electrochemical fluorination (ECF) of trialkylphosphine oxides in *aHF*, anhydrous HF.

The main disadvantages of this protocol are the low yield of tris(perfluoroalkyl)diflorophosphoranes and the formation in parallel of equimolar amount of the very toxic gas, oxygen difluoride, which in the mixture with hydrogen can cause severe explosions.

In course of our study, we have shown that the ECF of trialkylphosphines in aHF proceeds smoothly and leads to the formation of tris(perfluoroalkyl)difluorophosphoranes in high yields (Scheme 4.37).[126,234] Trialkylphosphines, in particular with short alkyl chains, are pyrophoric materials (**Caution! Can ignite in contact with air**). However, by dissolving in aHF (**Caution! Very exothermic reaction**) they form salts that are suitable for the ECF process. Trialkylphosphines do not contain oxygen. This excludes the formation of OF_2 during the ECF process.

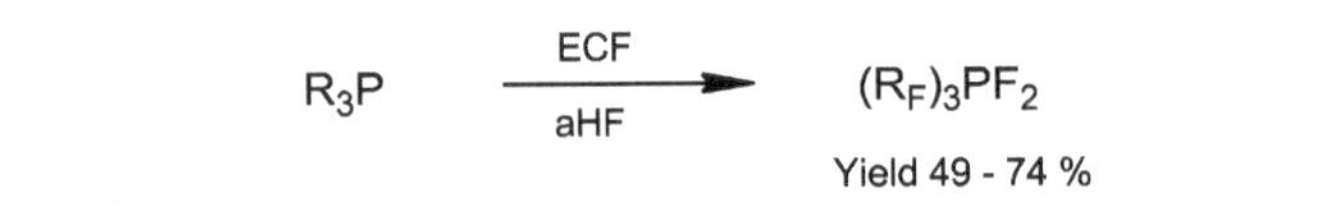

Scheme 4.37 Preparation of the tris(perfluoroalkyl)difluorophosphoranes, $(R_F)_3PF_2$, by electrochemical fluorination (ECF) of trialkylphosphines in *aHF*, anhydrous HF.

Especially, the ECF of the triethylphosphine, $(C_2H_5)_3P$, gives high yield of perfluorinated product, tris(pentafluoroethyl)difluorophosphorane, $(C_2F_5)_3PF_2$. The raw product that is obtained from the ECF cell contains about 95–97% of perfluorinated material. The main impurities are phosphoranes having CF_3CHF and CF_3CH_2 groups bonded to phosphorus. The $(C_2F_5)_3PF_2$ can be obtained in the quality 99.5+ % by fractional distillation of the crude tris(pentafluoroethyl)difluorophosphorane.

The ECF of 1,2-bis(diethylphosphino)ethane, $(C_2H_5)_2PCH_2CH_2P(CH_2CH_3)_2$, proceeds similar to the ECF of trialkylphosphines, but it is accompanied by the formation of a larger number of partially fluorinated compounds and C–P bond scission results in the formation of $(C_2F_5)_3PF_2$ as a side product (Scheme 4.38).[235]

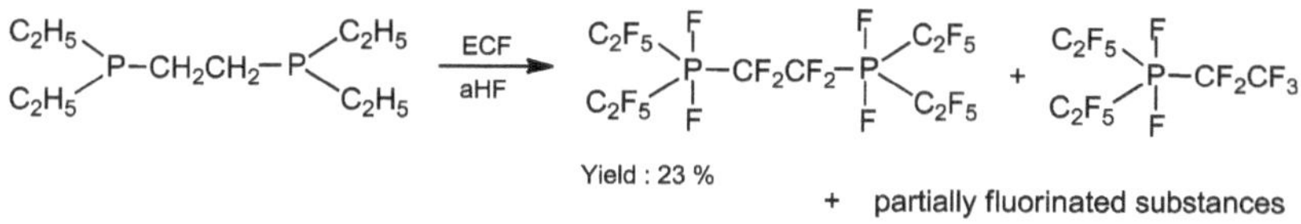

Scheme 4.38 The electrochemical fluorination of 1,2-bis(diethylphosphino)ethane, $(C_2H_5)_2PCH_2CH_2P(CH_2CH_3)_2$ in *aHF*, anhydrous HF; *ECF*, electrochemical fluorination.

The ECF of dialkylphosphines, for instance di(*iso*-propyl)chlorophosphine, is accompanied with strong decomposition of the starting material presumably due to relatively good solubility of fluorinated products in aHF. The yield of perfluorinated di(*iso*-propyl)trifluorophosphorane is only 8.5% (Scheme 4.39). Scission of C–P bond during the ECF leads to the formation of the *iso*-heptafluoropropyl-tetrafluorophosphorane, *iso*-$C_2F_7PF_4$, and perfluoropropane.[236] The presence of partially fluorinated molecules with one and two hydrogen in the *iso*-position in the reaction mixture indicates the steric hindrance of the *iso*-C–H bond against fluorination, similar to the aforementioned observation during the ECF of *iso*-propylsulfonyl fluoride.

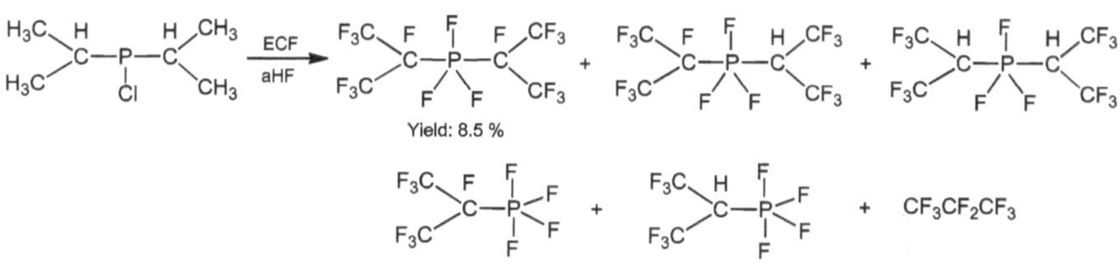

Scheme 4.39 Product distribution pattern by the electrochemical fluorination (ECF) of di(*iso*-propyl)chlorophosphine.[236] *aHF*, anhydrous HF.

Fluorination at phosphorus in di(*iso*-propyl)chlorophosphine with SbF_3, prior to the ECF, increases the yield of perfluorinated substance to 30% (Scheme 4.40). This confirms again that the use of partially fluorinated substances as starting materials for Simons process gives advantages in product yields. Nevertheless, some quantities of partially fluorinated compounds with hydrogen in *iso*-position were also detected in the reaction mixture after the ECF of di(*iso*-propyl) trifluorophosphorane.

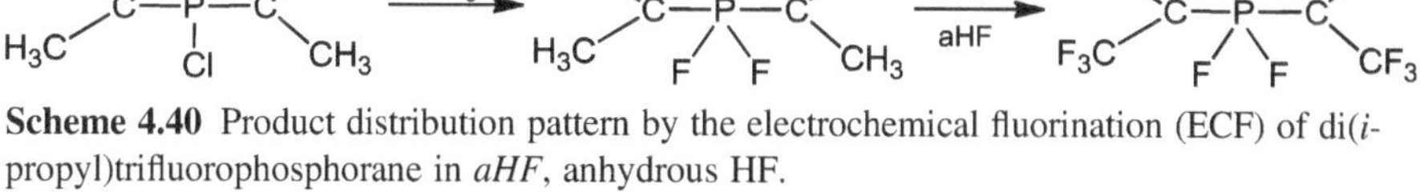

Scheme 4.40 Product distribution pattern by the electrochemical fluorination (ECF) of di(*i*-propyl)trifluorophosphorane in *aHF*, anhydrous HF.

There seems to be a general tendency toward the formation of partially fluorinated compounds by the ECF of *iso*-alkyl phosphanes due to the steric hindrance in the replacement of the hydrogen atom in the *iso*-position.

Tris(perfluoroalkyl)difluorophosphoranes, $(R_F)_3PF_2$, are very reactive compounds. They can be easily converted into various derivatives, which are interesting for practical applications (Scheme 4.41)[237,238] (**Caution! Perfluoroalkylfluorophosphoranes are toxic to inhale and liberate HF by contact with water and moisture; appropriate protections for skin and eyes are required**).

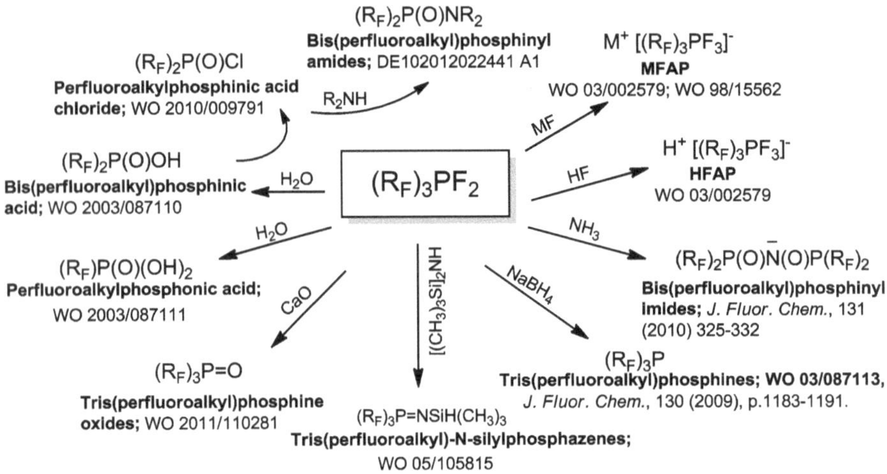

Scheme 4.41 Conversion of tris(perfluoroalkyl)difluorophosphoranes, $(R_F)_3PF_2$, to various derivatives.
Reprinted from Ignat'ev N. V.; Bader J.; Koppe K.; Hoge B.; Willner H. J. Fluorine Chem. **2015**, 171, 36–45 with permission from Elsevier.

Tris(perfluoroalkyl)difluorophosphoranes are strong Lewis acids forming stable complexes with donor molecules (Lewis bases), for instance, with 4-*N*,*N*-dimethylaminopyridine,[237,239] which are convenient starting materials for preparation of various tris(perfluoroalkyl)trifluorophosphates (FAP salts).[237] Quantum chemical calculations predict the high Lewis acidity of tris(perfluoroalkyl)difluorophosphoranes.[237,240] $(R_F)_3PF_2$ are able to abstract a fluoride anion from BF_4^- or PF_6^- anions forming the corresponding FAP salts (see Scheme 4.42).[241]

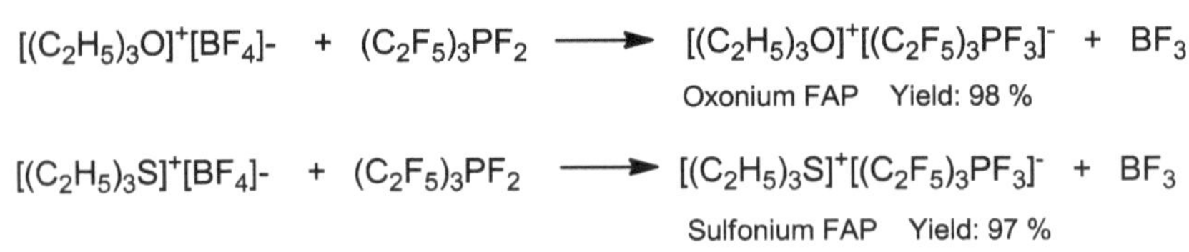

Scheme 4.42 Syntheses of oxonium and sulfonium salts with tris(perfluoroalkyl) trifluorophosphates (FAP) anion.

Due to the extraordinary Lewis acidity of tris(perfluoroalkyl)difluorophosphoranes, they can serve as active catalysts in organic synthesis. Applications of $(C_2F_5)_3PF_2$ as a

Figure 4.11 Tris(pentafluoroethyl)difluorophosphorane, $(C_2F_5)_3PF_2$, on supporting material (charcoal).

Reprinted from Ignat'ev N. V.; Bader J.; Koppe K.; Hoge B.; Willner H. J. Fluorine Chem. **2015,** 171, 36–45 with permission from Elsevier.

catalyst in Diels–Alder reaction, Michael addition, cyanosilylation of aldehydes and ketones, and Strecker reaction were published.[237,242,243] Tris(perfluoroalkyl)difluorophosphoranes can be used not only as liquid catalysts but also as heterogeneous catalyst after impregnation on suitable supporting materials, for instance, on charcoal (Fig. 4.11).[237]

The high fluoride affinity of fluoroalkylfluorophosphoranes was reported earlier.[244] For instance, mono- and tris(perfluoroalkyl)difluorophosphoranes readily react with NaF, KF, or CsF.[244,245] The same approach has been used for the preparation of lithium tris(pentafluoroethyl)trifluorophosphate, $Li[(C_2F_5)_3PF_3]$ (LiFAP).[236,246] This salt can be obtained straightforwardly as a solvate via interaction of tris(pentafluoroethyl)difluorophosphorane with LiF in polar organic solvents or organic solvent mixtures (Scheme 4.43). The reaction between bis(pentafluoroethyl)trifluorophosphorane, $(C_2F_5)_2PF_3$, and LiF proceeds similarly, but due to the high symmetry that enables the $F\cdots Li^+\cdots F$ coordination, the $Li[(C_2F_5)_2PF_4]$ can be obtained in solvate-free form as a solid material (Scheme 4.43).[238]

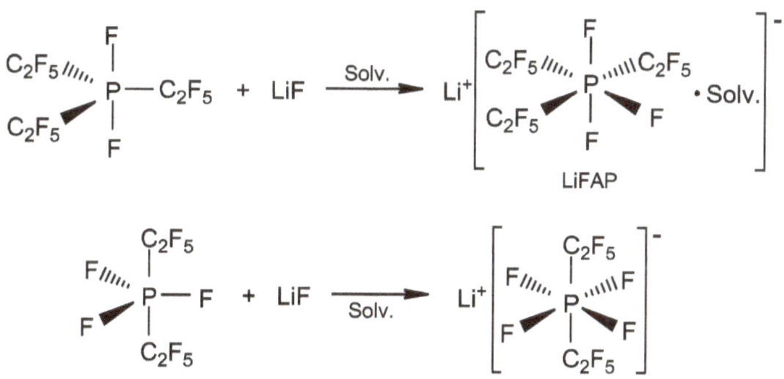

Scheme 4.43 Preparation of $Li[(C_2F_5)_3PF_3]\cdot$Solvent (LiFAP) and $Li[(C_2F_5)_2PF_4]$ by interaction of pentafluoroethylfluorophosphoranes with LiF.

Table 4.1 Conductivity of $Li[(C_2F_5)_2PF_4]$ in the mixture of ethylene carbonate and dimethyl carbonate (EC:DMC; 1:1 w/w) at Different Temperatures

Electrolyte: 1 mol/l in EC:DMC (1:1, w/w)								
Temperature,°C: Conductivity, mS/cm:	−40	−20	−10	0	20	40	60	80
$Li[(C_2F_5)_2PF_4$	0.15	2.7	4.1	5.6	9.4	13.7	18.5	23.5
$LiPF_6$		3.2	4.7	6.4	10.5	15.2	19.8	

LiFAP and $Li[(C_2F_5)_2PF_4]$ are of interest for applications in Li-ion batteries as new conducting salt to replace the hydrolytically unstable and thermally less stable $LiPF_6$[246] The $Li[(C_2F_5)_2PF_4]$, dissolved in an organic solvent mixture of ethylene carbonate and dimethyl carbonate (EC:DMC; 1:1 w/w), shows a very similar conductivity as the benchmark lithium hexafluorophosphate (Table 4.1). However, it is hydrolytically much more stable than $LiPF_6$. The hydrolysis rate at room temperature of Li $[(C_2F_5)_2PF_4]$ is about 1% (^{31}P NMR measurements) within 48 h after addition of 20% of water to the electrolyte EC:DMC (1:1 w/w).[238]

Hence, two pentafluoroethyl groups bonded to phosphorus are enough to retain the high hydrolytic stability as reported earlier for LiFAP.[246] The reduced molecular mass of the $[(C_2F_5)_2PF_4]$ anion when compared with the $[(C_2F_5)_3PF_3]$ anion improves the conductivity of $Li[(C_2F_5)_2PF_4]$-based electrolytes when compared with that of LiFAP electrolyte.[246] In contrast to LiFAP, $Li[(C_2F_5)_2PF_4]$ forms only one isomer, which has almost ideal D_{4h} symmetry as confirmed by ^{31}P NMR and ^{19}F NMR measurements.[238]

Tris(perfluoroalkyl)difluorophosphoranes, $(R_F)_3PF_2$, can be converted into the corresponding tris(perfluoroalkyl)phosphine oxides, $(R_F)_3P{=}O$, for example, by reaction with CaO[247]; hydrolyzed with water to bis(perfluoroalkyl)phosphinic, $(R_F)_2P(O)OH$, or perfluoroalkylphosphonic acids, $R_FP(O)(OH)_2$[248]; or reduced to tris(perfluoroalkyl) phosphines, $(R_F)_3P$[249] (Scheme 4.41). $(R_F)_2P(O)OH$ and $(R_F)P(O)(OH)_2$ are strong Brønsted acids, which are of interest as catalysts[248] and as components of proton-conducting membranes for fuel cell application.[250,251] Lanthanide salts of the bis(perfluoroalkyl)phosphinic acid are active catalysts for application in organic syntheses.[252]

Tris(perfluoroalkyl)difluorophosphoranes are moisture-sensitive compounds, but due to their high fluoride affinity they do not undergo hydrolysis by interaction with aqueous HF, although they form the corresponding tris(perfluoroalkyl)trifluorophosphoric acids, for example, $[H_3O][(C_2F_5)_3PF_3]\cdot 4H_2O$ (HFAP), by the addition of fluoride anion to phosphorus.[253] HFAP is formed as a mixture of two possible isomers: meridional (C_s symmetry) and facial (C_{3v} symmetry).[238] According to the ^{19}F NMR spectra, HFAP with meridional structure is preferably formed in this reaction. HFAP and its alkali metal salts are convenient starting materials for the synthesis of ILs with the FAP anion (Scheme 4.44).[253,254]

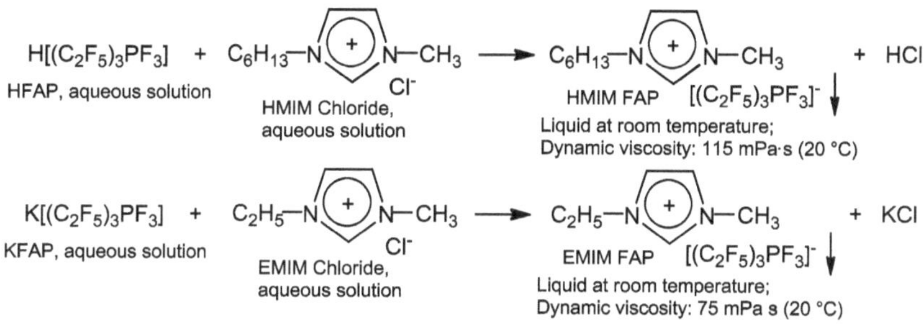

Scheme 4.44 Synthesis of tris(perfluoroalkyl)trifluorophosphates (FAP) ionic liquids using $[H_3O][(C_2F_5)_3PF_3]\cdot 4H_2O$, HFAP or $K[(C_2F_5)_3PF_3]$, KFAP as sources of the FAP anion.[253,254] *EMIM*, 1-ethyl-3-methylimidazolium; *HMIM*, 1-hexyl-3-methylimidazolium.

FAP-based ILs possess high hydrolytic, thermal, and electrochemical stability.[254] The properties of ILs with FAP anions strongly depend on the structure and molecular mass of the perfluoroalkylfluorophosphate anion.[238] The 1-ethyl-3-methylimidazolium (EMIM) salt with the symmetrical $[(C_2F_5)_2PF_4]^-$ anion is a solid. It has a melting point of 62–64°C. In contrast, the EMIM ILs with the asymmetrical $[(C_2F_5)_3PF_3]^-$ and the $[(C_2F_5)_1PF_5]^-$ anions are liquids at room temperature.[238]

Tris(perfluoroalkyl)phosphine oxides, $(R_F)_3P{=}O$, provide entry to a new alkylation technology using alcohols as alkylators. Reaction of organic bases with alcohols in the presence of $(R_F)_3P{=}O$ yields quaternary salts with the bis(perfluoroalkyl)phosphinate anion (Scheme 4.45)[255] This new reaction results in the formation of hydrophilic (water soluble) ILs in good yields and the C_2F_5H (F 125) is the only side product of this process. This new protocol permits the alkylation of organic compounds with alcohols—cheap and in many cases nontoxic source of alkyl groups (Scheme 4.45).

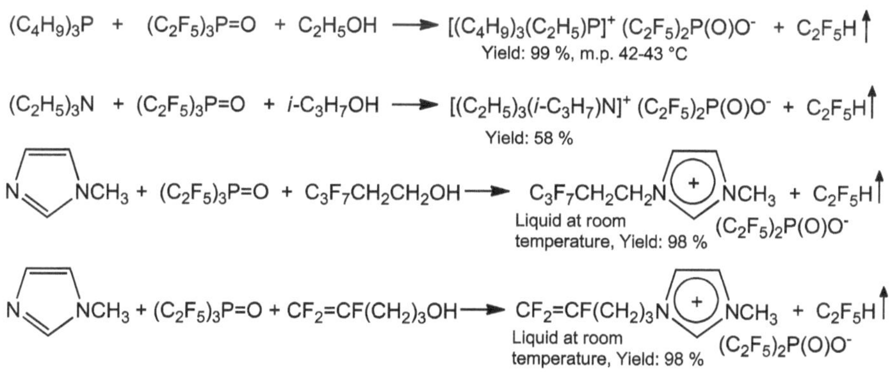

Scheme 4.45 Alkylation of organic bases with alcohols in the presence of tris(pentafluoroethyl) phosphine oxides, $(C_2F_5)_3P{=}O$.

Alternatively, ILs with bis(perfluoroalkyl)phosphinate anions can be obtained by alkylation of organic bases with esters of bis(perfluoroalkyl)phosphinic acids. For example, various esters of bis(pentafluoroethyl)phosphinc acid were prepared by reaction of the $(C_2F_5)_3P{=}O$ with alcohols using KF as a catalyst (Scheme 4.46).[256]

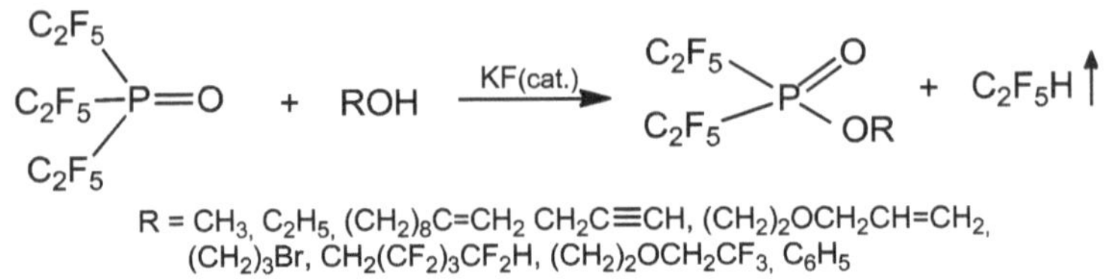

Scheme 4.46 Synthesis of alkyl esters of bis(pentafluoroethyl)phosphinic acid.[256]

This protocol provides a convenient access to various alkyl bis(perfluoroalkyl) phosphinates (Scheme 4.46). This method is not limited to short-chain alkyl phosphinates but also includes the preparation of alkyl phosphinates with functional groups such as alkenyl, alkynyl, halogenalkyl, or alkoxy groups and can also be extended to aromatic compounds.[256] Bis(perfluoralkyl)phosphinic acid esters are strong alkylating agents and have to be handled with care. The methyl ester of bis(pentafluoroethyl)phosphinic acid was also prepared by the reaction of dimethylsulfate with $(C_2F_5)_2P(O)OH$.[257]

Treatment of bis(perfluoroalkyl)phosphinates or perfluoroalkylphosphonates with aHF leads to the formation of the corresponding bis(perfluoroalkyl)tetrafluorophosphates or perfluoroalkylpentafluorophosphates (Scheme 4.47).[258]

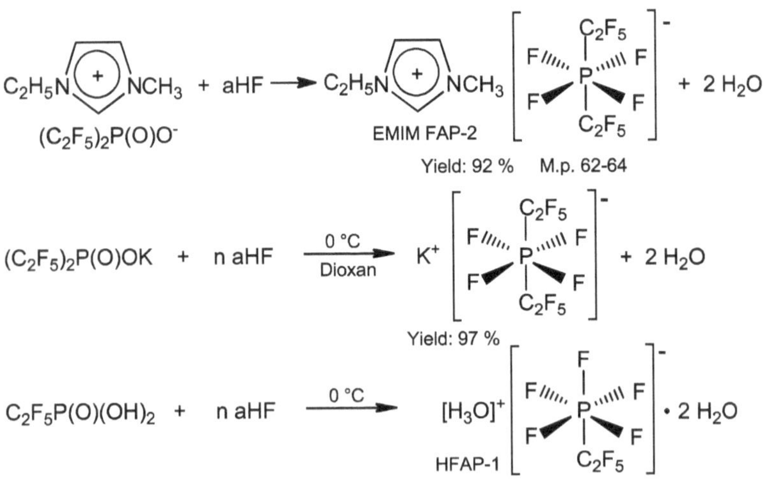

Scheme 4.47 Converting of bis(pentafluoroethyl)phosphinates into corresponding phosphates by treatment with anhydrous HF (aHF).

The reaction of SbF_5 or $SbCl_5$ with alkali metal or organic salts with the $[(C_2F_5)_2PF_4]^-$ or the $[(C_2F_5)_1PF_5]^-$ anions results in liberation of volatile bis(pentafluoroethyl)trifluorophosphorane, $(C_2F_5)_2PF_3$ or pentafluoroethyltetrafluorophosphorane, $C_2F_5PF_4$ (Scheme 4.48).[258]

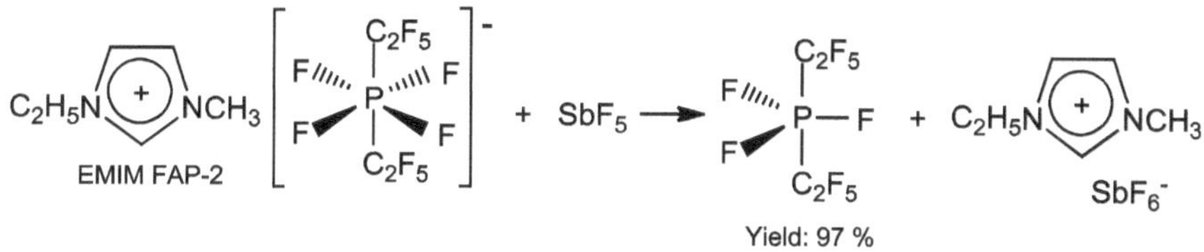

Scheme 4.48 Synthesis of bis(pentafluoroethyl)trifluorophosphorane, $(C_2F_5)_2PF_3$, by interaction of 1-ethyl-3-methylimidazolium (EMIM) salt with SbF_5.

Tris(pentafluoroethyl)phosphine oxides react with ammonia, hydrazines, as well as primary aliphatic and aromatic amines yielding the corresponding bis(pentafluoroethyl)phosphinyl amides or hydrazides in good to excellent yield (Scheme 4.49).[237,259] The lower yield observed in the reaction with aniline is attributed to the reduced nucleophilicity of aromatic amines. Presumably, the reaction proceeds via intermediate complex **A**, which is not stable under the reaction conditions and is converted into the final product, bis(pentafluoroethyl)phosphinyl amide or hydrazide (Scheme 4.49). The reactions proceed at low temperature (−30 to 20°C). The hydrolytic stability of bis(pentafluoroethyl)phosphinyl amides depends on the substituent R. Amides with longer alkyl chain are more stable against hydrolysis when compared with $(C_2F_5)_2P(O)NH_2$, which by dissolving in wet solvents forms the ammonium salt $(C_2F_5)_2P(O)O^-\ NH_4^+$.

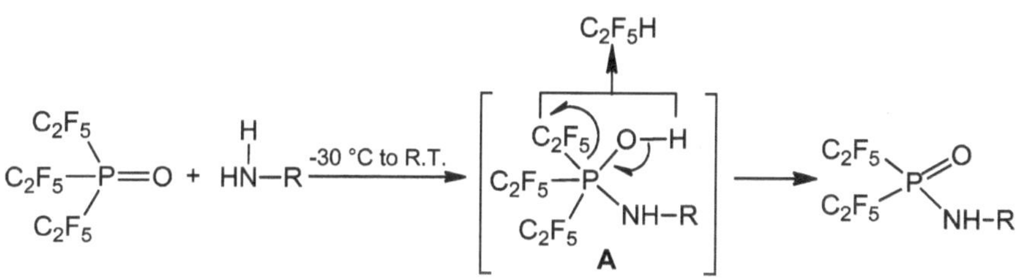

R = H, C_6H_5, C_6H_4-4-CH_3, $CH_2C_6H_5$, $CH_2C_6H_4$-4-CH_3, $CH_2C_6H_4$-4-OCH_3, $CH_2CH_2C_6H_5$, C_4H_9, $CH_2CH(C_2H_5)(CH_2)_3CH_3$, $HNCH_3$, $N(CH_3)_2$, HNC_6H_5, $HNCH_2[C_6H_{2\text{-}3,5}(t\text{-}C_4H_9)_{2\text{-}4}\text{-}OH]$

Scheme 4.49 Preparation of bis(pentafluoroethyl)phosphinyl amides and hydrazides. *RT*, room temperature.

Tris(perfluoroalkyl)phosphine oxides undergo hydrolysis to the corresponding bis(perfluoroalkyl)phosphinic acids.[248] The reaction of bis(perfluoroalkyl)phosphinic acids with $PhPCl_4$ or PCl_5 yields bis(perfluoroalkyl)phosphinyl chlorides.[248,260,261] Anhydrides of bis(perfluoroalkyl)phosphinic acids were prepared by interaction of the corresponding acids with P_2O_5[262] The anhydrides $(R_F)_2P(O)OP(O)(R_F)_2$ are convenient starting materials for the preparation of various derivatives of bis(perfluoroalkyl)phosphinic acids, for instance, silyl esters, $(R_F)_2P(O)OSi(CH_3)_3$[262] Bis(perfluoroalkyl)phosphinic acid fluorides can be prepared from the corresponding bis(perfluoroalkyl)trifluorophosphoranes by interaction with hexamethyldisiloxane, $[(CH_3)_3Si]_2O$.[263] The reaction of tris(perfluoroalkyl)difluorophosphoranes with hexamethyldisilazane, $[(CH_3)_3Si]_2NH$, results in the formation of the corresponding tris(perfluoroalkyl)-*N*-trimethylsilylphosphazenes, $(R_F)_3P{=}N{-}Si(CH_3)_3$ (Scheme 4.41).[264]

Bis(perfluoroalkyl)phosphinic acid chlorides are more reactive toward secondary amines than tris(perfluoroalkyl)phosphine oxides providing an access to the corresponding amides (Scheme 4.50).[237,263]

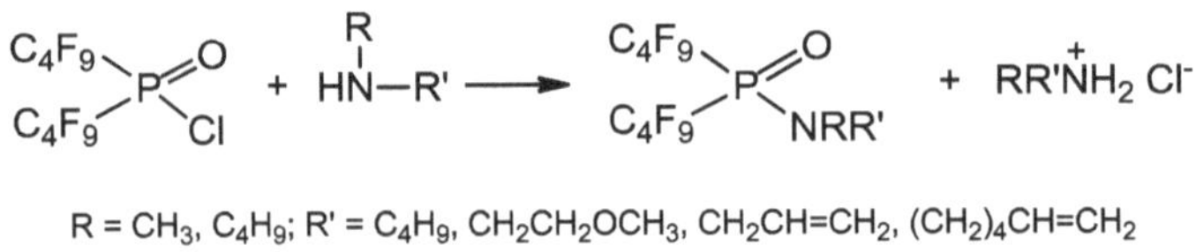

Scheme 4.50 Synthesis of secondary phosphinyl amides from $(C_4F_9)_2P(O)Cl$.

Secondary bis(perfluoroalkyl)phosphinyl amides are more hydrolytically stable than the primary ones.

The reaction of bis(perfluoroalkyl)phosphinic chloride, $(R_F)_2P(O)Cl$, and bis(perfluoroalkyl)phosphinyl amide, $(R_F)_2P(O)NH_2$, yields the corresponding bis(di(perfluoroalkyl)phosphinyl)imide $[(R_F)_2P(O)]_2NH$.[265–267] Bis(di(pentafluoroethyl)phosphinyl)imide salts can be prepared in better yields in a one-pot synthesis by the interaction of bis(pentafluoroethyl)phosphinic acid chloride, $(C_2F_5)_2P(O)Cl$, with ammonia (Scheme 4.51).[267]

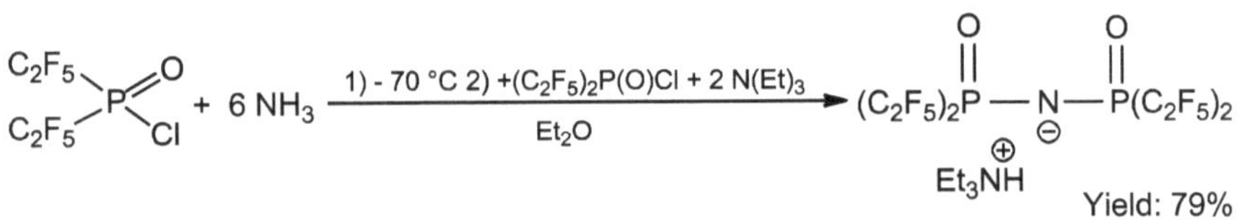

Scheme 4.51 Synthesis of the triethylammonium salt of the bis(di(pentafluoroethyl)phosphinyl) imide, $[(C_2F_5)_2P(O)]_2NH$.

The free acid, bis(di(pentafluoroethyl)phosphinyl)imide $[(C_2F_5)_2P(O)]_2NH$ (H-FPI) was obtained by treatment of the triethylammonium salt with concentrated sulfuric acid.[267] Bis(di(pentafluoroethyl)phosphinyl)imide, $[(C_2F_5)_2P(O)]_2NH$, is the phosphorus analog of the well-known bis(trifluoromethylsulfonyl)imide, $(CF_3SO_2)_2NH$ (H-TFSI; DesMarteau imide).[209] $[(C_2F_5)_2P(O)]_2NH$ is hydrolytically stable in aqueous solution for several weeks. It is a strong Brønsted acid with a p*K*a value of 1.9.[237,267] This value is well comparable to the p*K*a value of 1.7 of $(CF_3SO_2)_2NH$.[209] Due to the strong acidic properties of $[(C_2F_5)_2P(O)]_2NH$ (H-FPI), it is of interest as a Brønsted acid catalyst.[248] Rare-earth metal salts of FPI can be used as Lewis acid catalysts.[268] Many organic salts of the FPI anion are ILs; some of them are liquid at room temperature.[237,248,269]

ILs with the $[(C_2F_5)_2P(O)]_2N^-$ (FPI) anion are analogs to the well-known ILs with the bis(trifluorometylsulfonyl)imide anion, $(CF_3SO_2)_2N^-$, (TFSI anion). FPI ILs were prepared by metathesis in water using $[(C_2F_5)_2P(O)]_2NM$ (M = H, Na or K) as starting materials.[266,269] The $[(C_2F_5)_2P(O)]_2N^-$ anion is much more bulky than the $(CF_3SO_2)_2N^-$ anion and more hydrophobic.[270] Probably, the FPI anion is the most bulky, hydrolytically stable, low-coordinating anion known, so far.[271] FPI ILs are hydrolytically indifferent and have a large electrochemical window.[270] They exhibit thermal stabilities of up to 300°C (measured by thermogravimetry).[269]

Tris(perfluoroalkyl)phosphines, $(R_F)_3P$, were obtained by a solvent-free reduction of the corresponding tris(perfluoroalkyl)difluorophosphoranes with $NaBH_4$[249] By the interaction with strong bases, for instance, BuLi, tris(pentafluoroethyl)phosphine splits off the C_2F_5 group, which can be trapped with various electrophiles.[249]

The presented examples demonstrate the utility and versatility of perfluoralkylfluorophosphoranes chemistry. Nowadays, tris(pentafluoroethyl)difluorophosphorane, $(C_2F_5)_3PF_2$, is produced in multiton quantity by ECF (Simons process). Some of its derivatives have already found practical applications and are produced in commercial quantities.

Acknowledgments

I would like to thank Prof. P. Sartori and Prof. H. Willner for valuable suggestions and support during my research work at the University of Duisburg and at the Merck KGaA Company (Darmstadt, Germany). I also thank Prof. M. Finze (University of Würzburg, Germany) for taking care about the further research in the electrochemical fluorination of organic substrates in aHF. I thank Dr. J. Bader and Prof. B. Hoge (University of Bielefeld, Germany) for their enthusiastic and fruitful research in the field of perfluoralkylfluorophosphoranes chemistry.

References

1. a. Flahaut, J.; Viel, C. The Life and Scientific Work of Henri Moissan. In *Fluorine. The First Hundred Years (1886–1986);* Banks, R. E., Sharp, D. W. A., Tatlow, J. C., Eds.; Elsevier Sequoia S.A.: Lausanne, New York, 1986; pp 27–42.
   b. Banks, E. *J. Fluor. Chem.* **1986,** *33,* 3–26.
2. Ellis, J. F.; May, G. F. Modern Fluorine Generation. In *Fluorine. The First Hundred Years (1886–1986);* Banks, R. E., Sharp, D. W. A., Tatlow, J. C., Eds.; Elsevier Sequoia S.A.: Lausanne, New York, 1986; pp 133–146.
3. Hough, P. T. Fluorine Production and Use – an Overview. In *Electrochemistry in the Preparation of Fluorine and its Compounds;* Childs, W. V., Fuchigami, T., Eds.; *Proceedings*; The Electrochemical Society, Inc.: Pennington, NJ, 1997, Vols. 97–15; pp 113–141.
4. Rozhkov, I. N. Anodic Fluorination. In *Organic Electrochemistry – an Introduction and a Guide;* Baizer, M. M., Lund, H., Eds., 2nd ed.; Marcel Dekker: New York, 1983 (chapter 24).
5. Momota, K. Anodic Fluorination of Toluene and Fluorotoluenes. In *Electrochemistry in the Preparation of Fluorine and its Compounds;* Childs, W. V., Fuchigami, T., Eds.; *Proceedings*; The Electrochemical Society, Inc.: Pennington, NJ, 1997, Vols. 97–15; pp 74–82.
6. Konno, A.; Fuchigami, T. Regioselective Anodic Difluorination of Sulphides Using Novel Fluorine Source $Et_4NF \cdot 4HF$. In *Electrochemistry in the Preparation of Fluorine and its Compounds;* Childs, W. V., Fuchigami, T., Eds.; *Proceedings*; The Electrochemical Society, Inc.: Pennington, NJ, 1997, Vols. 97–15; pp 101–105.

7. Fuchgami, T.; Imagi, S.; Atobe, M. *Fundamentals and Applications of Organic Electrochemistry: Synthesis, Materials, Devices;* Wiley, 2014; pp 104−115, 156−161.
8. Shaaban, M. R.; Inagi, S.; Fuchigami, T. *Electrochim. Acta* **2009,** *54,* 2635−2639.
9. Inagi, S.; Sawamura, T.; Fuchigami, T. *Electrochem. Commun.* **2008,** *10,* 1158−1160.
10. Fuchigami, T. Electrolytic Partial Fluorination of Heteroatom Compounds. Highly Selective Direct and Indirect Anodic Monofluorination of Heterocyclic Compounds. In *Electrochemistry in the Preparation of Fluorine and its Compounds;* Childs, W. V., Fuchigami, T., Eds.; *Proceedings*; The Electrochemical Society, Inc.: Pennington, NJ, 1997, Vols. 97−15; pp 65−73.
11. Fuchgami, T. Electrochemical Partial Fluorination. In *Organic Electrochemistry;* Lund, H., Hammerich, O., Eds., 4th ed.; Marcel Dekker: New York, Basel, 2001; pp 1035−1050.
12. Baroux, P.; Tardivel, R.; Simonet, J. Electrofluorination of Olefins Substituted by an Electron-Withdrawing Group. In *Electrochemistry in the Preparation of Fluorine and its Compounds;* Childs, W. V., Fuchigami, T., Eds.; *Proceedings*; The Electrochemical Society, Inc.: Pennington, NJ, 1997, Vols. 97−15; pp 91−100.
13. Yoneda, N. *J. Fluor. Chem.* **2000,** *105,* 205−207.
14. Fukuhara, T.; Sawaguchi, M.; Yoneda, N. *Electrochem. Commun.* **2000,** *2,* 259−261.
15. Yoneda, N. Electrochemical Fluorination of Carbonyl Compounds. Fluorination of Cycloalkylidieneacetates Accompanying with or without Ring Expansion. In *Electrochemistry in the Preparation of Fluorine and Its Compounds;* Childs, W. V., Fuchigami, T., Eds.; *Proceedings*; The Electrochemical Society, Inc.: Pennington, NJ, 1997, Vols. 97−15; pp 83−90.
16. Sawaguchi, M.; Fukuhara, T.; Hara, S.; Yoneda, N. Preparation of Organofluorine Compounds by the Anodic Oxidative Fluorination of Iodine Compounds. In *Reactive Intermediates in Organic and Biological Electrochemistry;* Peters, D. C., Schaefer, H. J., Workentin, M. S., Yoshida, J., Eds.; *Proceedings*; The Electrochemical Society, Inc.: Pennington, NJ, 2001, Vols. 2001−14; pp 25−28.
17. Dinoiu, V. *Rev. Roum. Chim.* **2007,** *52,* 453−466.
18. Noel, M.; Suryanarayanan, V.; Chellammal, S. *J. Fluor. Chem.* **1997,** *83,* 31−40.
19. Fuchigami, T.; Tajima, T. *J. Fluor. Chem.* **2005,** *126,* 181−187.
20. Noel, M.; Suryanarayanan, V. *J. Appl. Electrochem.* **2004,** *34,* 357−369.
21. Fuchigami, T. *J. Fluor. Chem.* **2007,** *128,* 311−316.
22. Yoshida, J. -I.; Kataoka, K.; Horcajada, R.; Nagaki, A. *Chem. Rev.* **2008,** *108,* 2265−2299.
23. Francke, R.; Little, R. D. *Chem. Soc. Rev.* **2014,** *43,* 2492−2521.
24. Kolbe, H. *Justus Liebigs Ann. Chem.* **1849,** *69,* 257−294.
25. Tori, S.; Tanaka, H. Carboxylic Acids. In *Organic Electrochemistry;* Lund, H., Hammerich, O., Eds., 4th ed.; Marcel Dekker: New York, Basel, 2001; pp 499−543.
26. Weiper-Idelmann, A.; Kahmen, M.; Schäfer, H. J.; Gockeln, M. *Acta Chem. Scand.* **1998,** *52,* 672−682.
27. Schäfer, H. J. Kolbe-Reactions. In *Comprehensive Organic Synthesis,* Vol. 3, Trost, B. M., Fleming, I., Eds.; Pergamon Press, 1991; pp 633−658.
28. Schäfer, H. J. Electrolytic Oxidative Coupling. In *Organic Electrochemistry;* Lund, H., Hammerich, O., Eds., 4th ed.; Marcel Dekker: New York, Basel, 2001; pp 883−967.
29. Smertenko, E. A.; Volkov, V. D.; Datsenko, S. D.; Ignat'ev, N. V. *Theor. Exper. Chem.* **1999,** *35,* 231−233.
30. Sikol'skii, G. A.; Knunyants, I. L. *Izv. Akad. Nauk UdSSR, Ser. Khim.* **1967,** *6,* 1289−1294.

31. Schäfer, H. J. Electrolytic Oxidative Coupling. In *Organic Electrochemistry;* Lund, H., Hammerich, O., Eds., 4th ed.; Marcel Dekker: New York, Basel, 2001; pp 941−951.
32. Dmowski, W.; Biernacki, A. *J. Fluor. Chem.* **1996,** *78,* 193−194.
33. Uneyama, K.; Ueda, K. *Chem. Lett.* **1988,** 853−854.
34. Dmowski, W.; Koslowski, T. *J. Fluor. Chem.* **1998,** *87,* 179−183.
35. Andrieux, C. P.; Gelis, L.; Medebielle, M.; Pinson, J.; Saveant, J.-M. *J. Am. Chem. Soc.* **1990,** *112,* 3509−3520.
36. Ignat'ev, N. V.; Datsenko, S. D.; Pazenok, S. V.; Yagupolskii, L. M. *Zh. Org. Khim. (Russ.)* **1990,** *26,* 1740−1746.
37. Medebielle, M.; Pinson, J.; Saveant, J. -M. *J. Am. Chem. Soc.* **1991,** *113,* 6872−6879.
38. Datsenko, S. D.; Ignat'ev, N. V.; Yagupolskii, L. M. *Sov. Electrochem.* **1991,** *27,* 1016−1019.
39. Ignat'ev, N. V.; Datsenko, S. D.; Yagupolskii, L. M. *Zh. Org. Khim. (Russ.)* **1991,** *27,* 905−910.
40. Koshechko, V. G.; Kiprianova, L. A.; Fileleeva, L. I.; Rozhkova, Z. Z. *J. Fluor. Chem.* **1995,** *70,* 277−278.
41. Datsenko, S. D.; Ignat'ev, N. V.; Yagupolskii, L. M. *Sov. Electrochem.* **1991,** *27,* 1484−1486.
42. Datsenko, S. D.; Ignat'ev, N. V.; Brighinev, A. B. *Electrochem. (Russ.)* **1995,** *31,* 7375−7379.
43. Ignat'ev, N.; Datsenko, S. *J. Fluor. Chem.* **1992,** *58,* 270.
44. Calas, P.; Moreau, P.; Commeyras, A. *J. Chem. Soc., Chem. Commun.* **1982,** 433−434.
45. Dapremont, C.; Calas, P.; Commeyras, A. *J. Fluor. Chem.* **1992,** *56,* 249−258.
46. Medebielle, M.; Pinson, J.; Saveant, J. -M. *Tetrahedron Lett.* **1990,** *31,* 1279−1282.
47. Medebielle, M.; Pinson, J.; Saveant, J. -M. *Tetrahedron Lett.* **1992,** *33,* 7351−7354.
48. Medebielle, M.; Oturan, M. A.; Pinson, J.; Saveant, J.-M. *J. Org. Chem.* **1996,** *61,* 1331−1340.
49. Paratian, J. M.; Sibille, S.; Perichon, J. *J. Chem. Soc., Chem. Commun.* **1992,** 53−54.
50. Prakash, G. K. S.; Deffieux, D.; Yudin, A. K.; Olah, G. A. *Synlett* **1994,** 1057−1058.
51. Martynow, B. I.; Stepanov, A. A. *J. Fluor. Chem.* **1997,** *85,* 127−128.
52. Stepanov, A. A. *J. Fluor. Chem.* **2002,** *114,* 225−228.
53. Bordeau, M.; Biran, C.; Serein-Spirau, F.; Leger-Lambert, M. -P.; Deffieux, D.; Moreau, C.; Dunogues, J. Electrochemical Access to Fluorinated Organosilicon Synthons by the Sacrificial Anode Process. In *Electrochemistry in the Preparation of Fluorine and its Compounds;* Childs, W. V., Fuchigami, T., Eds.; *Proceedings*; The Electrochemical Society, Inc.: Pennington, NJ, 1997, Vols. 97−15; pp 201−212.
54. Ignat'ev, N. V.; Datsenko, S. D.; Yagupolskii, L. M. *Sov. Electrochem.* **1992,** *28,* 405−408.
55. Smertenko, E. A.; Datsenko, S. D.; Ignat'ev, N. V.; Deev, L. E.; Bil'dinov, I. K.; Podsevalov, P. D. *Russ. J. Electrochem.* **1994,** *30,* 1172−1174.
56. Ignat'ev, N. V.; Nechitailo, L. A.; Mironova, A. A.; Maletina, I. I.; Orda, V. V. *Sov. Electrochem.* **1992,** *28,* 408−412.
57. Ignat'ev, N. V.; Nechitailo, L. A.; Orda, V. V.; Maletina, I. I.; Mironova, A. A.; Yagupolskii, L. M. *J. Fluor. Chem.* **1991,** *54,* 402.
58. Mizuta, S.; Verhoog, S.; Wang, X.; Shibata, N.; Gouverneur, V.; Medebielle, M. *J. Fluor. Chem.* **2013,** *155,* 124−131.
59. Pokhodenko, V. D.; Koshechko, V. G.; Titov, V. E.; Sednev, D. V. *Theor. Experim. Chem.* **1992,** *28,* 85−89.
60. Ignat'ev, N. V.; Datsenko, S. D. *Russ. J. Electrochem.* **1995,** *31,* 1235−1239.

61. Datsenko, S.; Ignat'ev, N. *J. Fluor. Chem.* **1992,** *58,* 269.
62. Tomilov, A. P.; Martynov, B. I.; Smirnov, Y. D.; Semchenko, F. M. *J. Fluor. Chem.* **1997,** *82,* 39–41.
63. Mikhaylov, D. Y.; Gryaznova, T. V.; Dudkina, Y. B.; Polyancev, F. M.; Latypov, S. K.; Sinyashin, O. G.; Budnikova, Y. H. *J. Fluor. Chem.* **2013,** *153,* 178–182.
64. Uneuyama, K. *J. Fluor. Chem.* **2000,** *105,* 209–210.
65. Ignat'ev, N.; Datsenko, S.; Smertenko, E.; Yagupolskii, L. Electrochemical Initiation of Radical-Ion Reactions of Organo-Fluorine Compounds. In *Electrochemistry in the Preparation of Fluorine and its Compounds;* Childs, W. V., Fuchigami, T., Eds.; *Proceedings*; The Electrochemical Society, Inc.: Pennington, NJ, 1997, Vols. 97–15; pp 166–189.
66. Ignat'ev, N. V.; Hilarius, V.; Heider, U.; Kühner, A.; Schmidt, M.; Kucherina, A.; Sartori, P. Electrochemistry of $-N(CF_3)_2$ Compounds. In *Reactive Intermediates in Organic and Biological Electrochemistry;* Peters, D. C., Schaefer, H. J., Workentin, M. S., Yoshida, J., Eds.; *Proceedings*; The Electrochemical Society, Inc.: Pennington, NJ, 2001, Vols. 2001–14; pp 113–116.
67. Ignat'ev, N.; Datsenko, S.; Smertenko, E. Electrochemical Generation of Fluorinated Active Species – a New Approach to the Synthesis of Organo-fluorine Compounds. In *Novel Trends in Electroorganic Synthesis;* Tori, S., Ed.; Springer-Verlag: Tokyo, 1998; pp 287–292.
68. Ignat'ev, N.; Datsenko, S.; Smertenko, E. Electrochemical Introduction of Fluorinated Substituents into Organic Molecules. In *Novel Trends in Electroorganic Synthesis;* Tori, S., Ed.; Springer-Verlag: Tokyo, 1998; pp 305–308.
69. Fuchigami, T. Electrolytic Reactions. In *Electrochemical Aspects of Ionic Liquids;* Ohno, H., Ed.; John Wiley & Sons, Inc.: Hoboken, NJ, 2005; pp 89–109.
70. Sterlin, S. R.; Grinberg, V. A. *Russ. J. Electrochem.* **2010,** *46,* 843–870.
71. Fuchigami, T.; Imagi, S. Fluorinations. In *Organic Electrochemistry;* Hammerich, O., Speiser, B., Eds.; Taylor & Francis, 2015 (chapter 14).
72. Lund, H. A Century of Organic Electrochemistry. *J. Electrochem. Soc.* **2002,** *149,* S21–S33.
73. Goldwhite, H. The Manhattan Project. In *Fluorine. The First Hundred Years (1886–1986);* Banks, R. E., Sharp, D. W. A., Tatlow, J. C., Eds.; Elsevier Sequoia: Lausanne, NY, 1986; pp 109–131.
74. Simons, J. H. *J. Fluor. Chem.* **1986,** *32,* 7–24.
75. Simons, J. H. *J. Electrochem. Soc.* **1949,** *95,* 47–52.
76. Simons, J. H.; Francis, H. T.; Hogg, J. A. *J. Electrochem. Soc.* **1949,** *95,* 53–55.
77. Simons, J. H.; Harland, W. J. *J. Electrochem. Soc.* **1949,** *95,* 55–59.
78. Simons, J. H.; Pearlson, W. H.; Brice, T. J.; Wilson, W. A.; Dresdner, R. D. *J. Electrochem. Soc.* **1949,** *95,* 59–64.
79. Simons, J. H.; Dresdner, R. D. *J. Electrochem. Soc.* **1949,** *95,* 64–67.
80. Dresdner, R. D. *J. Fluor. Chem.* **1986,** *32,* 25–28.
81. Simons, J. H. US 2,519,983, Minnesota Mining & Manufacturing Company.
82. Pealson, W. H. *J. Fluor. Chem.* **1986,** *32,* 29–40.
83. Abe, T. *J. Fluor. Chem.* **2000,** *105,* 181–183.
84. Abe, T. Electrochemical Fluorination as a Locomotive for the Development of Fluorine Chemistry. In *Fluorine Chemistry at the Millennium. Fascinated by Fluorine;* Banks, R. E., Ed.; Elsevier: Amsterdam, NY, Tokyo, 2000; pp 272–275.
85. Le Blanc, M.; Riess, J. G. Artificial Blood Substituents Based on Perfluorochemicals. In *Preparation, Properties and Industrial Applications of Organofluorine Compounds;* Banks, R. E., Ed.; Ellis Horwood, Ltd: Chichester, England, 1982; pp 83–138.

86. Lowe, K. C. Properties and Biomedical Applications of Perfluorochemicals and Their Emulsions. In *Organofluorine Chemistry. Principles and Commercial Applications;* Banks, R. E., Smart, B. E., Tatlow, J. C., Eds.; Plenum Press: NY, 1994; pp 555–577.
87. Rao, N. S.; Baker, B. E. Textile Finishes and Fluorosurfactants. In *Organofluorine Chemistry. Principles and Commercial Applications;* Banks, R. E., Smart, B. E., Tatlow, J. C., Eds.; Plenum Press: NY, 1994; pp 321–338.
88. Alsmeyer, Y. W.; Childs, W. V.; Flynn, R. M.; Moore, G. G.; Smeltzer, J. C. Electrochemical Fluorination and its Applications. In *Organofluorine Chemistry. Principles and Commercial Applications;* Banks, R. E., Smart, B. E., Tatlow, J. C., Eds.; Plenum Press: NY, 1994; pp 130.
89. *Chem. Eng. News* **July 13, 2015,** 27–29.
90. Wang, Z.; Cousins, I. T.; Scheringer, M.; Buck, R. C.; Hungerbühler, K. *Environ. Int.* **2014,** *70,* 62–75.
91. Hekster, F. M.; Laane, R. W. P. M.; de Voogt, P. *Rev. Environ. Contam. Toxicol.* **2003,** *179,* 99–121.
92. Jackson, D. A.; Mabury, S. A. *Environ. Sci. Technol.* **2013,** *47,* 382–389.
93. Yamashita, N.; Yeung, L. W. Y.; Taniysu, S.; Kwok, K. Y.; Petrick, G.; Gamo, T.; Guruge, K. S.; Lam, P. K. -S.; Loganathan, B. G. Global Distribution of PFOS and Related Chemicals. In *Global Contamination Trends of Persistent Organic Chemicals;* Loganathan, B. G., Lam, P. K. -S., Eds.; CRC Press, 2011; pp 593–624.
94. Scheringer, M.; Trier, X.; Cousins, I. T.; de Voogt, P.; Fletcher, T.; Wang, Z.; Webster, T. F. *Chemosphere* **2014,** *114,* 337–339.
95. Abe, T.; Nagase, S. Electrochemical Fluorination (Simons Process) as a Route to Perfluorinated Organic Compounds of Industrial Interest. In *Preparation, Properties and Industrial Applications of Organofluorine Compounds;* Banks, R. E., Ed.; Ellis Horwood, Ltd: Chichester, England, 1982; pp 19–43.
96. Watanabe, N. *J. Fluor. Chem.* **1983,** *22,* 205–230.
97. Banks, R. E.; Tatlow, J. C. A Guide to Modern Organofluorine Chemistry. In *Fluorine. The First Hundred Years (1886–1986);* Banks, R. E., Sharp, D. W. A., Tatlow, J. C., Eds.; Elsevier Sequoia: Lausanne, NY, 1986; pp 269–272.
98. Sartori *Bull. Electrochem.* **1990,** *6,* 471–482.
99. Alsmeyer, Y. W.; Childs, W. V.; Flynn, R. M.; Moore, G. G.; Smeltzer, J. C. Electrochemical Fluorination and its Applications. In *Organofluorine Chemistry. Principles and Commercial Applications;* Banks, R. E., Smart, B. E., Tatlow, J. C., Eds.; Plenum Press: NY, 1994; pp 121–143.
100. Drakesmith, F. D. Electrofluorination of Organic Compounds. In *Topics in Current Chemistry,* Vol. 193; Springer Verlag Berlin Heidelberg, 1997; pp 197–242.
101. Sartori, P.; Ignat'ev, N.; Jüschke, R. Recent Development in Electrochemical Fluorination (Simons Process). In *Electrochemistry in the Preparation of Fluorine and its Compounds;* Childs, W. V., Fuchigami, T., Eds.; *Proceedings*; The Electrochemical Society, Inc.: Pennington, NJ, 1997, Vols. 97–15; pp 13–21.
102. Pohmer, K.; Bulan, A. Electrochemical Introduction of Fluorine. In *Houben-Weyl, Science of Synthesis;* 1998 Vol. E10a; pp 305–320.
103. Sartori, P.; Ignat'ev, N.; Juenger, C.; Jüschke, R.; Rieland, P. *J. Solid State Electrochem.* **1998,** *2,* 110–116.
104. Ignat'ev, N.; Sartori, P. *J. Fluor. Chem.* **2000,** *101,* 203–207.
105. Berenblit, V. V. *Fluorine Notes* **1999,** *1,* 1–8.
106. Berenblit, V. V. *Fluorine Notes* **1999,** *2,* 1–7.
107. Berenblit, V. V. *Fluorine Notes* **1999,** *3,* 1–6.

108. Jayaraman, K.; Noel, M. *Bull. Electrochem.* **2001,** *17,* 227–238.
109. Ignat'ev, N. V.; Welz-Biermann, U.; Weiden, M.; Heider, U.; Kucheryna, A.; Sartori, P.; Willner, H. Mechanistic and Synthetic Aspects of the Simons Process. Electrochemical Synthesis of New Organo-phosphorus Compounds. In *Mechanistic and Synthetic Aspects of Organic and Biological Electrochemistry;* Peters, D. G., Simonet, J., Tanaka, H., Eds.; *Proceedings*; The Electrochemical Society, Inc.: Pennington, NJ, 2003, Vols. 2003–12; pp 81–84.
110. Conte, L.; Gambaretto, G. P. *J. Fluor. Chem.* **2004,** *125,* 139–144.
111. Misra, S.; Jethva, A. D.; Gohil, M. S.; Elsheimer, S.; Ramachandraiah, G.; Ghosh, P. K. *Rev. Chem. Eng.* **2003,** *19,* 357–386.
112. Kirsch, P. *Modern Fluoroorganic Chemistry, Synthesis, Reactivity, Applications;* Wiley-VCH Verlag GmbH: Weinheim, Germany, 2013; pp 34–35.
113. Puetter, H. Industrial Electroorganic Chemistry. In *Organic Electrochemistry;* Lund, H., Hammerich, O., Eds., 4th ed.; Marcel Dekker: New York, Basel, 2001; pp 1259–1307.
114. Abe, T.; Nagase, S. Electrochemical Fluorination (Simons Process) as a Route to Perfluorinated Organic Compounds of Industrial Interest. In *Preparation, Properties and Industrial Applications of Organofluorine Compounds;* Banks, R. E., Ed.; Ellis Horwood, Ltd: Chichester, England, 1982; p 20.
115. Hackerman, N.; Snavely, E. S.; Fiel, L. D. *Corros. Sci.* **1967,** *7,* 39–50.
116. Gramstad, T.; Haszeldine, R. N. *J. Chem. Soc.* **1956,** 173–180.
117. Ono, T.; Inoue, Y.; Fukaya, C.; Arakawa, Y.; Naito, Y.; Yokoyama, K.; Yamanouchi, K.; Kobayashi, Y. *J. Fluor. Chem.* **1985,** *27,* 333–346.
118. Napoli, M.; Conte, L.; Gambaretto, G. P.; Carlini, F. M. *J. Fluor. Chem.* **1989,** *45,* 213–224.
119. Napoli, M.; Scipioni, A.; Gambaretto, G. P.; Carlini, F. M.; Bertola, M. *J. Fluor. Chem.* **1994,** *67,* 261–264.
120. Moldavsky, D. D.; Kaurova, G. I.; Bispen, T. A.; Furin, G. G. *J. Fluor. Chem.* **1993,** *63,* 193–201.
121. Sartori, P.; Velayutham, D.; Ignat'ev, N.; Noel, M. *J. Fluor. Chem.* **1998,** *87,* 31–36.
122. Chambers, R. D.; Fuss, R. W.; Jones, M.; Sartori, P.; Swales, A. P.; Herkelmann, R. *J. Fluor. Chem.* **1990,** *49,* 409–419.
123. Dimitrov, A.; Radeck, W.; Rüdiger, S.; Platonov, V. E. *J. Fluor. Chem.* **1991,** *52,* 317–331.
124. Sartori, N.; Ignat'ev, N.; Datsenko, S. *J. Fluor. Chem.* **1995,** *75,* 157–161.
125. Ignat'ev, N. V.; Schmidt, M.; Heider, U.; Kucherina, A.; Sartori, P.; Helmy, F. M. *J. Fluor. Chem.* **2002,** *113,* 201–205.
126. Ignat'ev, N.; Sartori, P. *J. Fluor. Chem.* **2000,** *103,* 57–61.
127. Simons J. H. In *Fluorine Chemistry*, Vol. 1, Academic Press: NY, 1950.
128. Haszeldine, R. N.; Nyman, F. *J. Chem. Soc.* **1956,** 2684–2689.
129. Dimitrov, A.; Rüdiger, St; Ignat'ev, N. V.; Datsenko, S. D. *J. Fluor. Chem.* **1990,** *50,* 197–205.
130. Meinert, H.; Fackler, R.; Mader, J.; Reuter, P.; Röhlke, W. *J. Fluor. Chem.* **1992,** *59,* 351–365.
131. Gambaretto, G. P.; Napoli, M.; Conte, L.; Scipioni, A.; Armelli, R. *J. Fluor. Chem.* **1985,** *27,* 149–155.
132. Rangarajan, T. M.; Sathyamoorthi, S.; Velayutham, D.; Noel, M.; Singh, R. P.; Brahma, R. *J. Fluor. Chem.* **2011,** *132,* 107–113.
133. Noel, M. *Bull. Electrochem.* **1999,** *15,* 466–471.

134. Watanabe, N.; Haruta, M. *Electrochim. Acta* **1980,** *25,* 461–464.
135. Hackerman, N.; Snavely, E. S.; Fiel, L. D. *Electrochim. Acta* **1967,** *12,* 535–551.
136. Stein, L.; Neil, J. M.; Alms, G. R. *Inorg. Chem.* **1969,** *8,* 2472–2476.
137. Totir, G. G.; Chottiner, G. S.; Gross, C. L.; Ves Childs, W.; Scherson, D. A. *J. Electrochem. Soc.* **2000,** *147,* 4212–4216.
138. Zemva, B.; Lutar, K.; Chacon, L.; Fele-Beuermann, M.; Allman, J.; Shen, C.; Bartlett, N. *J. Am. Chem. Soc.* **1995,** *117,* 10025–10034.
139. Bartlett, N.; Chambers, R. D.; Roche, A. J.; Spink, R. C. H.; Chacon, L.; Whalen, J. M. *Chem. Commun.* **1996,** 1049–1050.
140. Ignat'ev, N. V.; Welz-Biermann, U.; Heider, U.; Kucheryna, A.; von Ahsen, S.; Habel, W.; Sartori, P.; Willner, H. *J. Fluor. Chem.* **2003,** *124,* 21–37.
141. Sartori, P.; Ignat'ev, N. *J. Fluor. Chem.* **1998,** *87,* 157–162.
142. Sartori, P.; Velayutham, D.; Ignat'ev, N.; Noel, M. *J. Fluor. Chem.* **1997,** *83,* 1–8.
143. Suriyanarayanan, N.; Noel, M. *J. Solid State Electrochem.* **2008,** *12,* 1453–1460.
144. Barbour A. K. In *Organofluorine Chemicals and Their Industrial Application;* Banks, R. E., Ed.; Ellis Horwood: Chichester, 1979; p. 56.
145. Abe, T.; Kodaira, K.; Baba, H.; Nagase, S. *J. Fluor. Chem.* **1978,** *12,* 1–25.
146. Abe, T.; Baba, H.; Hayashi, E.; Nagase, S. *J. Fluor. Chem.* **1983,** *23,* 123–146.
147. Prokop, H. W.; Zhou, H. -J.; Xu, S. -Q.; Wu, C. -H.; Chuey, S. R.; Liu, C. C. *J. Fluor. Chem.* **1989,** *43,* 277–290.
148. Drakesmith, F. G.; Hughes, D. A. *J. Appl. Electrochem.* **1979,** *9,* 685–697.
149. Abe, T.; Nagase, S.; Baba, H. *Bull. Chem. Soc. Japan* **1976,** *49,* 1888–1892.
150. Abe, T.; Soloshonok, I.; Baba, H.; Sekiya, A. *J. Fluor. Chem.* **1999,** *99,* 51–57.
151. Abe, T.; Fukaya, H.; Hayashi, E.; Ono, T.; Nishida, M.; Soloshonok, I.; Okuhara, K. *J. Fluor. Chem.* **1999,** *99,* 229–237.
152. a. U.K. Pat. 1 007 288, Minnesota Mining and Manufacturing Co., 1965. b. Chem. Abs. 64 (1966) 3363h.
153. Napoli, M.; Gonte, L.; Gambaretto, G. P.; Carlini, F. M. *J. Fluor. Chem.* **1989,** *45,* 213–224.
154. Gard, G. L.; Hamel, N. N.; Mohtasham, J.; Waterfeld, A.; Mews, R. *J. Fluor. Chem.* **1991,** *55,* 313–321.
155. Totir, G. G.; Chottiner, G. S.; Gross, C. L.; Scherson, D. A. *J. Electrochem. Soc.* **2001,** *148,* 262–266.
156. Plashkin, V. S.; Pushkina, L. N.; Sokolov, S. V. *Zh. Org. Khim. (Russ.)* **1974,** *10,* 1215–1224.
157. Abe, T.; Hayashi, E.; Fukaya, H.; Baba, H. *J. Fluor. Chem.* **1990,** *50,* 173–196.
158. Ono, T.; Inoue, Yo; Arakawa, Yo; Naito, Yo; Fukayra, C.; Yamanouchi, K.; Yokoyama, K. *J. Fluor. Chem.* **1989,** *43,* 67–85.
159. Gervits, L. L.; Kadyrov, A. A.; Makarov, K. N.; Tumanskii, B. L.; Dimitrov, A.; Gross, U.; Rüdiger, S. *J. Fluor. Chem.* **1991,** *54,* 304.
160. Dolbier, W. R. *Chem. Rev.* **1996,** *96,* 1557–1584.
161. Gross, U.; Rüdiger, S.; Dimitrov, A. *J. Fluor. Chem.* **1996,** *76,* 139–144.
162. Cherstkov, V. F.; Tumanskii, B. L.; Delyagina, N. I.; Sterlin, S. R.; German, L. S. *J. Fluor. Chem.* **1996,** *79,* 93–95.
163. Bartberger, M. D.; Dolbier, W. R. *Tetrahedron* **1997,** *53,* 9857–9880.
164. Avetisyan, E. A.; Cherstkov, V. F.; Tumanskii, B. L.; Sterlin, S. R. *Izv. Akad. Nauk, Ser. Khim.* **1998,** 2511–2515.

165. Ono, T.; Fukaya, H.; Nishida, M.; Terasawa, N.; Abe, T. *Chem. Commun.* **1996,** 1579–1580.
166. Ono, T.; Fukaya, H.; Hayashi, E.; Saida, H.; Abe, T.; Henderson, P. B.; Fernandez, R. E.; Scherer, K. V. *J. Fluor. Chem.* **1999,** *97,* 173–182.
167. Ono, T.; Ohta, K. *J. Fluor. Chem.* **2014,** *167,* 198–202.
168. Allayarov, S. R.; Gordon, D. A.; Henderson, P. B.; Fernandez, R. E.; Jackson, V. E.; Dixon, D. A. *J. Fluor. Chem.* **2015,** *180,* 240–247.
169. Olah, G. A. *J. Org. Chem.* **2001,** *66,* 5943–5957.
170. Olah, G. A.; Surya Prakash, G. K.; Sommer, J. *Superacids;* John Wiley & Sons: NY, 1985; p 4934.
171. Ignat'ev, N.; Sartori, P. *16th International Symposium on Fluorine Chemistry;* University of Durham: UK, July 16–21, 2000 (Abstract C-12).
172. Velayutham, D.; Jayaraman, K.; Kulangiappar, K.; Ilayaraja, N.; Babu, Y. R.; Rao, P. S.; Reddy, S. N.; Babu, K. V.; Noel, M. *J. Fluor. Chem.* **2006,** *127,* 1111–1118.
173. Abe, T.; Hayashi, E.; Fukaya, H.; Hayakawa, Y.; Baba, H. *J. Fluor. Chem.* **1992,** *57,* 101–111.
174. Abe, T.; Baba, H.; Soloshonok, I. *J. Fluor. Chem.* **2001,** *108,* 215–228.
175. Abe, T.; Hayashi, E.; Baba, H.; Fukaya, H. *J. Fluor. Chem.* **1990,** *48,* 257–279.
176. Abe, T.; Fukaya, H.; Hayashi, E.; Hayakawa, Y.; Nishida, M.; Baba, H. *J. Fluor. Chem.* **1994,** *66,* 193–202.
177. Abe, T.; Fukaya, H.; Ono, T.; Hayashi, E.; Soloshonok, I.; Okuhara, K. *J. Fluor. Chem.* **1998,** *87,* 193–201.
178. Abe, T.; Baba, H.; Okuhara, K.; Fukaya, H. *J. Fluor. Chem.* **2001,** *111,* 115–128.
179. Abe, T.; Tamura, M.; Sekiya, A. *J. Fluor. Chem.* **2005,** *126,* 325–332.
180. Sartori, P.; Ignat'ev, N.; Jüschke, R. *J. Fluor. Chem.* **1997,** *81,* 115–121.
181. Ryabinin, N. A.; Kolenko, I. P.; Lundin, B. N.; Butina, I. B. *Trudi Inst. Khim., Ural Div. UdSSR* **1968,** *15,* 67–74.
182. Ilayaraja, N.; Manivel, A.; Velayutham, D.; Noel, M. *J. Appl. Electrochem.* **2008,** *38,* 175–186.
183. Hollitzer, E.; Sartori, P. *J. Fluor. Chem.* **1987,** *35,* 329–341.
184. Abe, T.; Pandey, S. K.; Baba, H. *J. Fluor. Chem.* **2000,** *105,* 149–157.
185. Howells, R. D.; Mc Cown, J. D. *Chem. Rev.* **1977,** *77,* 69–92.
186. Banks, R. E., Smart, B. E., Tatlow, J. C., Eds. *Organofluorine Chemistry. Principles and Commercial Applications;* Plenum Press: NY, 1994.
187. Rakita, P. E. *Chimica Oggi, Chem. Today* **March/April 2004,** 48–50.
188. Burdon, J.; McLoughlin, V. C. R. *Tetrahedron* **1965,** *21,* 1–4.
189. Katsuhara, Y.; Aramaki, M.; Ishii, A.; Kume, T.; Kawashima, C.; Mitsumoto, S. *J. Fluor. Chem.* **2006,** *127,* 8–17.
190. Noyori, R.; Murata, S.; Suzuki, M. *Tetrahedron* **1981,** *37,* 3899–3910.
191. Brice T. J.; Trott P. W. US 2 732 398 (1956), Chem. Abs. 50 (1956) P13982.
192. Shopen V.; Sapozhnikov M.; Lvov V.; Varfalameev L.; Turnaev M.; Cherviachkov Yu.; Yurochkin V.; Ignat'ev N. In *Eleventh European Symposium on Fluorine Chemistry, 17–22 September 1995, Bled, Slovenia,* Abstracts, p. 90.
193. Burdon, J.; Farazmafid, I.; Stacey, M.; Tatlow, J. T. *J. Chem. Soc.* **1957,** 2574–2578.
194. Ignat'ev, N.; Kucherina, A.; Sartori, P. *Acta Chem. Scand.* **1999,** *53,* 1110–1116.
195. Sartori, P.; Habel, W. *J. Fluor. Chem.* **1981,** *18,* 131–141.
196. Sartori, P.; Habel, W. *J. Fluor. Chem.* **1980,** *16,* 265–276.
197. Klink, F. W.; Wasser, D. J.; Liu, C. C. *J. Fluor. Chem.* **1986,** *32,* 89–101.
198. Herkelmann, R.; Sartori, P. *J. Fluor. Chem.* **1989,** *44,* 299–307.

199. Sartori, P.; Jüschke, R. *J. Fluor. Chem.* **1994,** *69,* 157–162.
200. Jüschke, R.; Velayutham, D.; Sartori, P. *J. Fluor. Chem.* **1997,** *83,* 145–149.
201. Rieland P. Dissertation, Universität-Gesamthochschule-Duisburg, 1993.
202. Pohl L.; Hilarius V.; Sartori P.; Jüschke R. WO 9731909 A1, 1997.
203. DesMarteau, D. D.; Zhu, S.; Pennington, W. T.; Gotoh, Y.; Witz, M.; Zuberi, S.; Singh, S. *J. Fluor. Chem.* **1989,** *45,* 24.
204. Conte, L.; Gambaretto, G. P.; Caporiccio, G.; Alessandrini, F.; Passerini, S. *J. Fluor. Chem.* **2004,** *125,* 243–252.
205. DesMarteau, D. D. *J. Fluor. Chem.* **1995,** *72,* 203–208.
206. Geiculescu, O. E.; Xie, Y.; Rajagopal, R.; Creager, S. E.; DesMarteau, D. D. *J. Fluor. Chem.* **2004,** *125,* 1179–1185.
207. Murmann, P.; Niehoff, P.; Schmitz, R.; Nowak, S.; Gores, H.; Ignatiev, N.; Sartori, P.; Winter, M.; Schmitz, R. *Electrochim. Acta* **2013,** *114,* 658–666.
208. Murmann, P.; Streipert, B.; Kloepsch, R.; Ignatiev, N.; Sartori, P.; Winter, M.; Cekic-Laskovic, I. *Phys. Chem. Chem. Phys.* **2015,** *17,* 9352–9358.
209. Foropoulos, J.; DesMarteau, D. D. *Inorg. Chem.* **1984,** *23,* 3720–3724.
210. DesMarteau, D. D.; Witz, M. *J. Fluor. Chem.* **1991,** *52,* 7–12.
211. Xue, L.; Padgett, C. W.; DesMarteau, D. D.; Pennington, W. T. *Solid State Sci.* **2002,** *4,* 1535–1545.
212. Wasserscheid, P.; Welton, T. In *Ionic Liquids in Synthesis,* 2nd Eds.; Wiley-VCH: Weinheim, 2008. vols. 1 and 2.
213. Yanai, H.; Taguchi, T. *J. Fluor. Chem.* **2015,** *174,* 108–119.
214. Booth, B. L.; Haszeldine, R. N.; Laali, K. *J. Chem. Soc., Perkin Trans. I* **1980,** *12,* 2887–2894.
215. Beard, C. D.; Baum, K.; Grakauskas, V. *J. Org. Chem.* **1973,** *38,* 3673–3677.
216. Ignatyev N.; Schmidt M.; Heider U.; Sartori P.; Kucheryna A. WO 2002/098844, Merck Patent GmbH, Darmstadt, Germany.
217. Ignat'ev, N. V.; Barthen, P.; Kucheryna, A.; Willner, H.; Sartori, P. *Molecules* **2012,** *17,* 5319–5338.
218. Ignatyev N.; Schmidt M.; Heider U.; Sartori P.; Kucheryna A. WO 2003/053918, Merck Patent GmbH, Darmstadt, Germany.
219. Barthen, P.; Frank, W.; Ignatiev, N. *Ionics* **2015,** *21,* 149–159.
220. Ignatyev N.; Welz-Biermann U.; Kuchryna A.; Willner H. WO 2006/063656 A1, Merck Patent GmbH, Darmstadt, Germany.
221. Ignat'ev, N.; Datsenko, S.; Yagupolskii, L.; Rüdiger, St; Dimitrov, A.; Radeck, W. *J. Fluor. Chem.* **1992,** *58,* 146.
222. Ignat'ev, N. V.; Datsenko, S. D.; Yagupolskii, L. M.; Dimitrov, A.; Radeck, W.; Rüdiger, St. *J. Fluor. Chem.* **1995,** *74,* 181–184.
223. Sartori, P.; Lattasch, K. -D. *J. Fluor. Chem.* **1992,** *57,* 113–119.
224. Sartori, P.; Jünger, C. *J. Fluor. Chem.* **1996,** *79,* 71–75.
225. Heider U.; Schmidt M.; Sartori P.; Ignatyev N.; Kucheryna A. EP 1081129 A2, 2003, Merck Patent GmbH, Darmstadt, Germany.
226. Hilarius V.; Buchholz H.; Ignatiev N.; Sartori P.; Kucherina A.; Datsenko S. WO 2000/046180, Merck Patent GmbH, Darmstadt, Germany.
227. Ignatyev N.; Welz-Biermann U.; Schmidt M.; Willner H.; Kucheryna A. WO 2004/054991, Merck Patent GmbH, Darmstadt, Germany.
228. Kucheryna A. Dissertation, Bergischen Universität Wuppertal, 2006.
229. Hirschberg, M. E.; Ignat'ev, N. V.; Wenda, A.; Frohn, H.-J.; Willner, H. *J. Fluor. Chem.* **2012,** *135,* 176–182.

230. Hirschberg, M. E.; Ignat'ev, N. V.; Wenda, A.; Frohn, H.-J.; Willner, H. *J. Fluor. Chem.* **2012,** *135,* 183–186.
231. Hirschberg, M. E.; Wenda, A.; Frohn, H. -J.; Ignat'ev, N. V. *J. Fluor. Chem.* **2012,** *138,* 24–27.
232. a. Yagupolskii L. M.; Zavatskii V. N.; Semenii V. Y.; Bildinov K. N.; Serebrov P. V.; Goncharenko A. A.; Kirsanov A. A.; Lyapunov M. I.; Feshchenko N. G. Institute of Organic Chemistry Ukrainian Academy of Sciences, Kiev, *Chem. Abs.* **1978,** *89,* 43766j, p. 666. b.; *Chem. Abs.* **1979,** *90,* 87670y, p. 645.
233. Semenii, V. Ya.; Stepanov, V. A.; Ignat'ev, N. V.; Furin, G. G.; Yagupolskii, L. M. *Zh. Obshch. Chim.* **1985,** *55,* 2415–2417 (English Translation).
234. Sartori P.; Ignat'ev N. DE 19846636, 1998.
235. Schmidt M.; Kühner A.; Ignatyev N.; Sartori P. DE 10109756 A1, 2002, Merck Patent GmbH.
236. Sartori P.; Ignatiev N. WO 98/15562, 1998.
237. Ignat'ev, N. V.; Bader, J.; Koppe, K.; Hoge, B.; Willner, H. *J. Fluor. Chem.* **2015,** *171,* 36–45.
238. Ignat'ev, N. V.; Willner, H.; Sartori, P. *J. Fluor. Chem.* **2009,** *130,* 1183–1191.
239. Bader, J.; Ignat'ev, N.; Hoge, B. *Inorg. Chem.* **2014,** *53,* 7547–7553.
240. Steinhauer, S.; Bader, J.; Stammler, H. -G.; Ignat'ev, N.; Hoge, B. *Angew. Chem. Int. Ed.* **2014,** *53,* 5206–5209.
241. Ignatyev N.; Bissky G.; Willner H. WO 2007/087949, Merck Patent GmbH, Darmstadt, Germany.
242. Bittner, B.; Koppe, K.; Bilir, V.; Frank, W.; Willner, H.; Ignat'ev, N. *J. Fluor. Chem.* **2015,** *169,* 50–60.
243. Bittner, B.; Koppe, K.; Frank, W.; Ignat'ev, N. *J. Fluor. Chem.* **2016,** *182,* 22–27.
244. Chan, S. S.; Willis, C. J. *Can. J. Chem.* **1968,** *46,* 1237–1248.
245. Pavlenko, N. V.; Yagupolskii, L. M. *Zh. Obshch. Khim.* **1989,** *59,* 528–534.
246. Schmidt, M.; Heider, U.; Kuehner, A.; Oesten, R.; Junguitz, M.; Ignat'ev, N.; Sartori, P. *J. Power Sources* **2001,** *97–98,* 557–560.
247. Ignatyev N.; Wiebe W.; Willner H. DE 10 2010 011 034 A1, WO 2011/110281 A1, Merck Patent GmbH, Darmstadt, Germany.
248. Ignat'ev, N. V.; Bejan, D.; Willner, H. *Chimica Oggi, Chem. Today* **2011,** *29* (5), 32–34.
249. Welz-Biermann U.; Ignatyev N.; Weiden M.; Schmidt M.; Heider U.; Miller A.; Willner H.; Sartori P. WO 2003/087113, Merck Patent GmbH, Darmstadt, Germany.
250. Ignat'ev N. V.; et al. In *4th International Conference on Polymer Batteries and Fuel Cells, August 2–6, Yokohama, Japan*, 2009 (Abstract).
251. Huebner G.; Huth A. DE 102006054737, Volkswagen AG, Wolfsburg, Germany.
252. Ignatyev N.; Meyer M.; Barthen P.; Frank W. WO 2015/000543 A1, Merck Patent GmbH, Darmstadt, Germany.
253. Ignatyev N.; Schmidt M.; Kühner A.; Hilarius V.; Heider U.; Kucheryna A.; Sartori P.; Willner H. WO2003/002579, Merck Patent GmbH, Darmstadt, Germany.
254. Ignat'ev, N. V.; Welz-Biermann, U.; Kucheryna, A.; Bissky, G.; Willner, H. *J. Fluor. Chem.* **2005,** *126,* 1150–1159.
255. Ignatyev N.; Welz-Biermann U.; Willner H.; Kucheryna A. WO 2005/049555, Merck Patent GmbH, Darmstadt, Germany.
256. Ignatyev N.; Schulte M.; Jablonka C. A.; Koppe K.; Frank W. WO 2013/1274931A1, Merck Patent GmbH, Darmstadt, Germany.
257. Ignatyev N.; Welz-Biermann U.; Kucheryna A.; Willner H. WO 2006/063655, Merck Patent GmbH, Darmstadt, Germany.

258. Ignatyev N.; Welz-Biermann U.; Schmidt M.; Kucheryna A.; Willner H. WO 2005/049628 A1, Merck Patent GmbH, Darmstadt, Germany.
259. Bejan, D.; Dinoiu, V.; Ignat'ev, N.; Bernhardt, E.; Willner, H. *Phosphorus Sulfur Silicon Relat. Elem.* **2014,** *189,* 1–14.
260. Kovaleva, T. V.; Martynyukh, E. G.; et al. *J. Gen. Chem. USSR* **1989,** *59,* 2245–2248.
261. Ignatyev N.; Aust E. F.; Bejan D.; Willner H. WO 2010/009791 A1, Merck Patent GmbH, Darmstadt, Germany.
262. Ignatyev N.; Wiebe W.; Willner H. WO 2013/013766 A1, Merck Patent GmbH, Darmstadt, Germany.
263. Ignatyev N.; Friedrich R.; Schellenberger S.; Koppe K.; Frank W. DE 10 2012 022441 A1 (2014), Merck Patent GmbH, Darmstadt, Germany.
264. Ignatyev N.; Welz-Biermann U.; Bissky G.; Willner H.; Kucheryna A. WO 2005/105815, Merck Patent GmbH, Darmstadt, Germany.
265. Pavlenko, N. V.; Matyushecheva, G. I.; Semenii, V. Y.; Yagupol'skii, L. M. *J. Gen. Chem. USSR* **1985,** *55,* 1586–1590.
266. Ignatiev N.; Weltz-Biermann U.; Heckmeier M.; Bissky G.; Willner H. WO 2006/128563, Merck Patent GmbH, Darmstadt, Germany.
267. Bejan, D.; Willner, H.; Ignatiev, N.; Lehmann, C. W. *Inorg. Chem.* **2008,** *47,* 9085–9089.
268. Bejan D. Dissertation, University of Wuppertal, Germany, 2010.
269. Bejan, D.; Ignat'ev, N. V.; Willner, H. *J. Fluor. Chem.* **2010,** *131,* 325–332.
270. Ignat'ev, N. V.; Finze, M.; Sprenger, J. A. P.; Kerpen, C.; Bernhardt, E.; Willner, H. *J. Fluor. Chem.* **2015,** *177,* 46–54.
271. Beichel, W.; Yu, Y.; Dlubek, G.; Krause-Rehberg, R.; Pionteck, J.; Pfefferkorn, D.; Bulut, S.; Bejan, D.; Friedrich, C.; Krossing, I. *Phys. Chem. Chem. Phys.* **2013,** *15,* 8821–8830.

Once Upon a Time Was the Langlois' Reagent: a "Sleeping Beauty"

5

B.R. Langlois [1,2]
[1]University of Lyon, Lyon, France; [2]University Claude Bernard-Lyon 1, UMR 5246, Villeurbanne, France

Chapter Outline

1. Introduction

Because of its unique properties, fluorine has become a "magic" element, and its introduction in various molecules is now considered as an essential tool to strongly modify several properties of a lot of compounds in different domains, from polymers, materials, and surfactants to pharmaceutical and agrochemical ingredients. Fluoro compounds are especially useful in the latter two areas since fluorine-containing substituents can deeply alter the chemical, physical, and biological properties of the compounds bearing them, which consequently gain a better metabolic stability and bioavailability.[1] Fluorinated moieties act first by their electronic properties, which can be continuously regulated from F [the most electronegative element by induction (Pauling scale: 3.98)[2a] but electron donating by conjugation] to CF_3, OCF_3, and SCF_3 (the strongest inductively electron-withdrawing groups). They also act by their hydrophobicity (related in some sense to the electronegativity of fluorine), which can be modulated to a rather large extent, as indicated by the Hansch–Leo parameters Π_R[3] (Table 5.1). Their hydrophobicity governs the bioavailability of the substrates bearing them.

Among the fluorinated substituents, CF_3, with an electronegativity of 3.5 (higher than that of Cl: 3.0),[2b] constitutes a good compromise for improving the hydrophobicity, and consequently, the bioavailability of substrates bearing it, without inducing too high an accumulation of these substrates in lipidic systems such as

Modern Synthesis Processes and Reactivity of Fluorinated Compounds. http://dx.doi.org/10.1016/B978-0-12-803740-9.00005-6

Table 5.1 Hansch–Leo Parameters of Fluorinated Moieties and Hydrogenated Analogs (Logarithmic Scale)

Moiety	Π_R (X = H)	Π_R (X = F)
X	0.00	+0.14
CX_3	+0.56	+0.88
OCX_3	−0.02	+1.04
SCX_3	+0.61	+1.44
SO_2CX_3	−1.63	+0.55
$NHSO_2CX_3$	−1.18	+0.92

$\Pi_R > 0$, hydrophobic substituent; $\Pi_R < 0$, hydrophilic substituent.

the nervous system. That is why, during the past 50 years, an incredibly large number of trifluoromethylated compounds have been proposed (and developed in many cases) as pharmaceuticals, anesthetics, "blood substitutes," and agrochemicals.[4] It must be pointed out that, in a large fraction of these bioactive ingredients, the CF_3 group is linked to an aromatic or heteroaromatic nucleus.

From the 1960s to the 1980s, trifluoromethylated aromatics (or heteroaromatics) were essentially obtained by a two-step synthesis from the corresponding methylated precursors, after radical chlorination and subsequent fluorination with SbF_3 (Swarts' reagent),[5] anhydrous hydrogen fluoride,[6,2b] or tamed HF-containing mixtures, such as Et3N-3HF[7] or HF/pyridine (Olah's reagent).[8] It must be pointed out that such a technique is quite exclusively restricted to aromatic substrates. Thus, to overcome this limitation, it rapidly appeared that it would be far better to have in hands specific reagents dedicated to the transfer of the CF_3 group onto a larger panel of organic skeletons. For this purpose, three tools can be considered: generators (or equivalents) of the very unstable trifluoromethyl anion ($^-CF_3$), equivalents of the putative trifluoromethyl cation ("$^+CF_3$"), or generators of the reasonably stable trifluoromethyl radical ($\bullet CF_3$).[9]

The first successes were obtained with nucleophilic trifluoromethylating reagents ($^-CF_3$ equivalents), first through the thermal decarboxylation of alkaline trifluoroacetates in the presence of cuprous iodide (a method that is, nevertheless, quite efficient only for the substitution of aryl or heteroaryl iodides),[10] later improved by the action of the ternary system CF_3CO_2Me/CsCl/CuI on aromatic halides[11] and the mild trifluoromethylation of carbonyl substrates with β-(trimethylsilyoxy)trifluoroacetamides/F^-.[12] Then appeared the very popular Ruppert–Prakash reagent (CF_3SiMe_3),[13] first used to add $^-CF_3$ to unsaturated electrophiles such as carbonyl compounds (as did Langlois et al. with fluoroform),[14] the scope of which has been extended, year after year, to the substitutive carbon trifluoromethylation of nonactivated aliphatic or aromatic compounds, as well as the sulfur trifluoromethylation of thiocyanates leading to trifluoromethyl sulfides.[15]

Then appeared the electrophilic trifluoromethylating reagents ($^+CF_3$ equivalents), essentially Umemoto's trifluoromethylsulfonium salts[16] and Togni's trifluoromethyliodonium salts,[17] (may be inspired from the pioneering work of L.M. Yagupolskii with perfluoroalkyliodonium salts,[18] but extensively extended to a very broad panel of substrates).

As far as radical trifluoromethylation is concerned, many generations of the CF_3 radical have been reported during the prehistory of fluorine chemistry, through photolysis of toxic precursors [$Hg(CF_3)_2$, $Te(CF_3)_2$] or putatively carcinogenic ones [$CF_3N = NCF_3$, $CF_3N(NO)SO_2R$], single-electron reduction of highly reactive compounds [$CF_3C(O)CF_3$, $(CF_3CO_2)_2$], or single-electron oxidation of trifluoroacetic acid at a very high potential. Trifluoromethyl iodide (CF_3I) was considered at that time as the less unsatisfactory reagent for radical trifluoromethylation, despite its toxicity. Moreover, radical trifluoromethylation was considered for a long time to be insufficiently regioselective to be used for targeted syntheses, until the appearance of chemical libraries for which the simultaneous production of several isomers is beneficial, provided that these isomers could be separated. For these reasons, radical trifluoromethylation attracted less attention during the late 1990s and the first decade of the 21st century.

2. Origin of the Trifluoromethanesulfinate Salts and Their First Uses

Nevertheless, it should be mentioned that great interest was accorded, from the mid-1980s, to trifluoromethyl bromide (CF_3Br), which is not toxic and was at that time commercially available, as a fire extinguisher, until the late 1990s, when its production was stopped because of its significant ozone-depleting effect. CF_3Br can indeed be not only transformed into stabilized forms of the anion $^-CF_3$ but can also be reduced, through a single-electron transfer, into Br^- and $\bullet CF_3$, an electrophilic radical that can be trapped by nucleophilic substrates. In the early studies—those of C. Wakselman and M. Tordeux[19a] and L.M. Yagupolskii et al.[19b]—strong and soft nucleophiles (thiolates[20a,b] or strongly delocalized carbanions[20c]) were used both to reduce CF_3Br (in N,N-dimethylformamide and under a slight pressure) and then trap the resulting $\bullet CF_3$ radical.

Later, a collaboration was established between C. Wakselman's group and B. Langlois' group (from the Rhône-Poulenc Co. at that time) to find new single-electron reducers of CF_3Br, different from the nucleophilic substrates used to trap the generated $\bullet CF_3$.[19a] It appeared that generators of the stable radical anion SO_2^- (Zn/SO_2, HCO_2Na/SO_2, $Na_2S_2O_4$) were the best candidates and were able to reduce CF_3Br in the presence of substrates such as organic disulfides (leading to trifluoromethyl sulfides)[21] or electron-rich aromatics,[22] as shown in Scheme 5.1. Interestingly, it was observed that, in the absence of any substrate, $\bullet CF_3$ was trapped by the persistent radical anion SO_2^- to provide a trifluoromethanesulfinate ("triflinate") anion.[23] This stable salt, after oxidation with hydrogen peroxide and acidification with sulfuric acid then oleum, opened a new and purely chemical route to triflic acid, which was

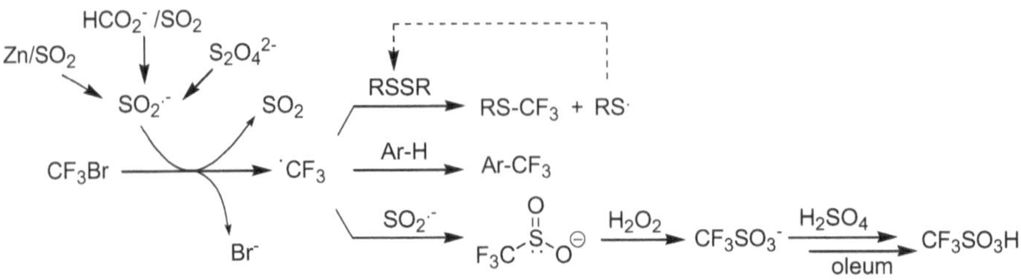

Scheme 5.1 Generation and use of •CF_3 from CF_3Br and SO_2^-.

previously obtained by Simmons' electrochemical fluorination only. This synthesis has been industrially developed by Rhône-Poulenc Co.[24]

However, it was observed that only hydrogen peroxide was able to cleanly transform sodium triflinate (CF_3SO_2Na) into sodium triflate (CF_3SO_3Na). None of the other oxidizers that were tested (*tert*-butyl hydroperoxide and peroxide, cumyl hydroperoxide and peroxide, N_2O_4, cerium ammonium nitrate, sodium persulfate, and sodium hypochlorite) were able to transfer an oxygen atom onto CF_3SO_2Na. All of them led, at room temperature, to the same reaction: the cleavage of the carbon−sulfur bond, the formation of sodium hydrogen sulfite and sulfate, as well as the evolution of sulfur dioxide and other gaseous products. Thus, it was suspected that they could act as single-electron oxidizers and transform the anion $CF_3SO_2^-$ into the corresponding radical $CF_3SO_2^{\cdot}$, which is known to be very unstable[25] and to collapse into SO_2 and the trifluoromethyl radical •CF_3. tert-Butyl hydroperoxide appeared to be the most efficient oxidizer to get such a reaction (Scheme 5.2).

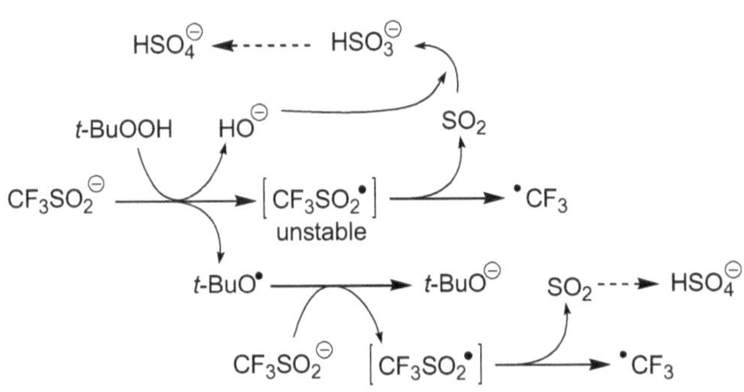

Scheme 5.2 Single-electron oxidation of CF_3SO_2Na by ^tBuOOH.

To verify the involvement of •CF_3 during such an oxidation process, disulfides, known to trap efficiently this radical,[21] were added to the medium. Indeed, trifluoromethyl sulfides were produced (Scheme 5.3),[26] even from cystine and homocystine without any loss of enantioselectivity.[26c] Usually, no added catalyst was needed to get good results.[27] Nevertheless, when the substrate contained electron-withdrawing moieties, the yield was improved by some $MoO_2(acac)_2$, known to facilitate the production of ^tBuO• radicals.[28] Thus, it could be deduced that these radicals oxidize the triflinate anion more efficiently than the hydroperoxide does.

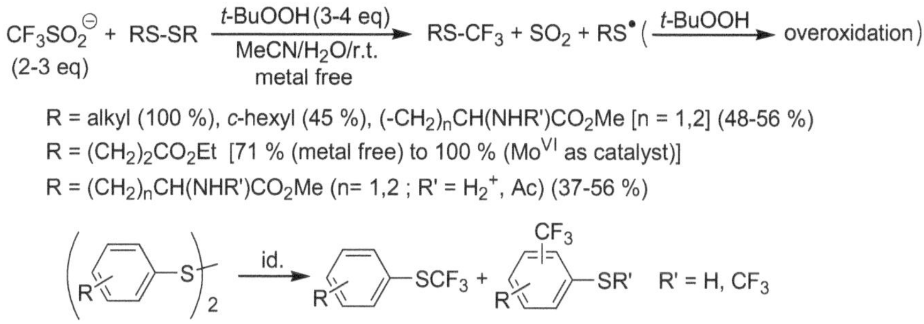

Scheme 5.3 Oxidative sulfur trifluoromethylation with CF_3SO_2Na.

As it was observed that S and C trifluoromethylation coexisted during the reaction of aryl disulfides, the behavior of electron-rich aromatics was also examined. Indeed, CF_3SO_2Na and *tert*-butyl hydroperoxide were able to produce trifluoromethylated aromatics under very mild conditions, especially in the presence of Cu(II) catalysts (Scheme 5.4).[29] This system was also able to trifluoromethylate pyrrole derivatives.

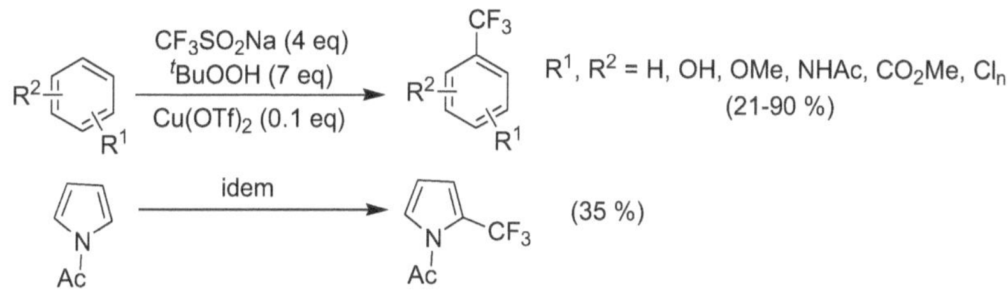

Scheme 5.4 Oxidative trifluoromethylation of aromatics.

The behavior of 1,3-(dimethoxy)benzene, a very enriched substrate, indicated that this reaction can be carried out without any metallic catalyst, that Fe(III) improves the yield but without modifying the products ratio to any significant extent, and that Cu(II) is the most effective catalyst for enhancing the global yield and, in addition, minimizes the amount of bis-trifluoromethylated products (Scheme 5.5). The last of these effects is probably due to the better efficiency of Cu(II) to oxidize the intermediate trifluoromethylated cyclohexadienyl radical into the corresponding cation, as postulated in

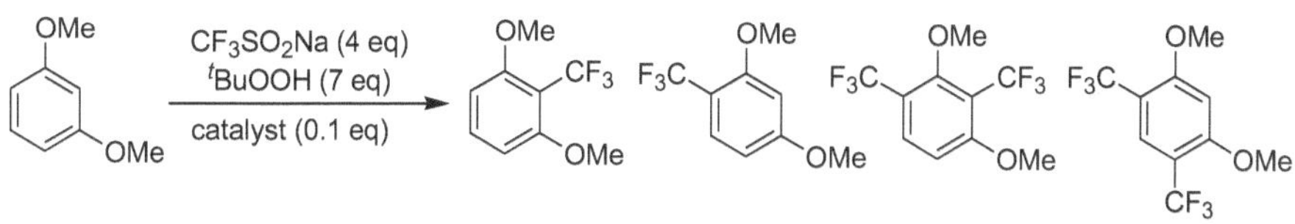

Cata :				
none (yield: 58 %)	9	25	9	14
$Fe(NO_3)_3$ (yield: 67 %)	11	31	10	15
$Cu(OTf)_2$ (yield: 90 %)	24	60	4	3

Scheme 5.5 Trifluoromethylation of (1,3-dimethoxy)benzene.

Scheme 5.6. Despite the fact that Cu(II) exhibits a lower oxidation potential than *tert*-BuOOH, its better efficiency may be related to the fact that Cu(II) transfers an electron in an inner-sphere process, whereas *tert*-BuOOH does so in an outer-sphere process. Such hypotheses led the authors to qualify this reaction as a "pseudoelectrophilic" trifluoromethylation.

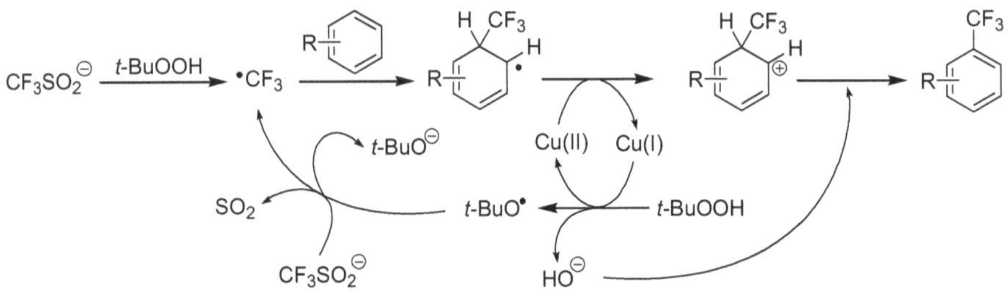

Scheme 5.6 Postulated mechanism for the aromatic trifluoromethylation.

Our later experiments,[30] as well as those of Smertenko et al.,[31] concerning electrochemical oxidation of potassium triflinate in the presence of aromatic substrates corroborated this mechanism: when exactly 2 F/mole of substrate were used, no bis-trifluoromethylated product was produced.

As it could be expected that such a mechanism, implying the oxidation of an intermediate trifluoromethylated radical, would be applied to other electron-rich unsaturated substrates, enol esters were submitted to the oxidative trifluoromethylation by sodium triflinate/*tert*-butyl hydroperoxide.[32] As expected, α-(trifluoromethyl)ketones were obtained under very mild conditions (Scheme 5.7).[33] This result was interpreted in a similar way as in Scheme 5.6. Interestingly, phenyl vinyl sulfide was also cleanly trifluoromethylated, provided that a nucleophile, such as methanol, was added to the medium. The role of methanol is to reduce the oxidation power of Cu(II) (and thus, avoid oxidation of the sulfide) and to trap the intermediate carbocation, which is too stabilized to evolve through deprotonation; in this way, a thioketal of 3,3,3-trifluoropropionaldehyde was obtained (Scheme 5.7).[34]

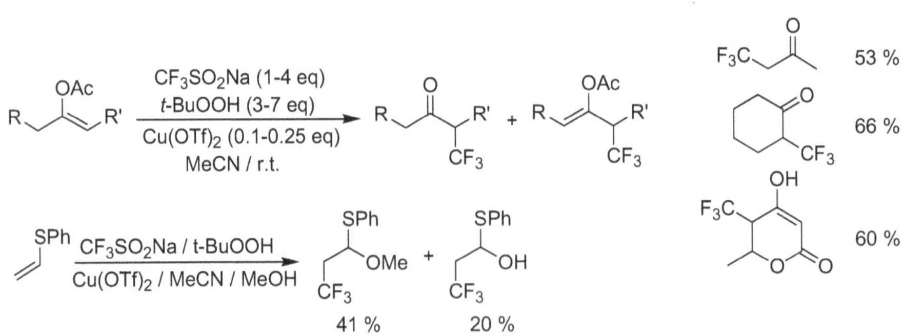

Scheme 5.7 Oxidative trifluoromethylation of enol esters and vinyl sulfides.

As for the trifluoromethylation of less enriched unsaturated substrates, electrochemical oxidation of CF_3SO_2Na was a better tool: with 1.2−1.5 F/mole of substrate, that is, under conditions that prevented the oxidation of the intermediate trifluoromethylated radical and allowed its disproportionation, a mixture of saturated and unsaturated products were obtained.[30]

However, Langlois' seminal results, which opened a promising methodology for the radical trifluoromethylation, were largely ignored for almost 15 years, perhaps because of the lack of availability of alkaline triflinates after the banishment of bromotrifluoromethane, although other environmentally benign sources of such salts have been reported during this period (Scheme 5.8).[35−37]

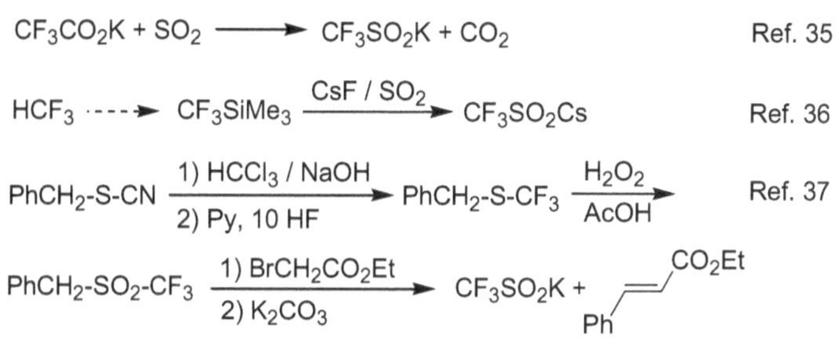

Scheme 5.8 Alternative syntheses of alkaline triflinates.

3. Rebirth of the "Langlois' Reagent"

Surprisingly, in mid-2011, Baran et al. published the trifluoromethylation of various nitrogen-containing heterocycles with sodium triflinate and *tert*-butyl hydroperoxide. They not only cited Langlois' work from the 1990s but also, as perfect gentlemen, named CF_3SO_2Na the "Langlois' reagent" ("an old reagent with a new use" sic).[38] Indeed, the novelty was that they applied this reaction to a large panel of nitrogen-containing heterocycles (although they omitted to cite Langlois' result on *N*-acetyl pyrrole) and observed its high tolerance toward various substituents as well as the ease with which it can be carried out. More importantly, they demonstrated that it is possible to trifluoromethylate sophisticated compounds in the last stages of a multistep synthesis. This greatly facilitates the preparation of libraries of potentially bioactive products. Concerning the technical aspect, Baran et al. carried out the reaction in a biphasic water/organic system, observed the influence of stirring on the yield and that of the organic solvent on the regioselectivity of trifluoromethylation, and claimed that no metallic catalyst was needed [a fact that Langlois et al. already reported while mentioning that Cu(II) improved the yield].

Since Baran's work was commented in *Nature*[39] and in different widely diffused journals [C&N News (December 2011), Royal Chemical Society News (August 2011), and the Scripps Institute advertisements], it quickly became popular. It was rapidly followed by M. Sanford et al. who published a substitutive trifluoromethylation of aryl boronic acids with the "Langlois' reagent."[40] Very soon thereafter, the use of CF_3SO_2M (M = Na, K), associated with various oxidants, blossomed and gave rise to a broad family of radical or pseudoelectrophilic trifluoromethylations applied to a large variety of substrates.[41] No less than 50 original articles appeared between 2012 and 2015!

All these publications are summarized in the following schemes (Schemes 5.9–5.13), each of them being devoted to a type of substrate. For each article is given the main author, the year of publication, the claimed oxidant, and the reference. As it

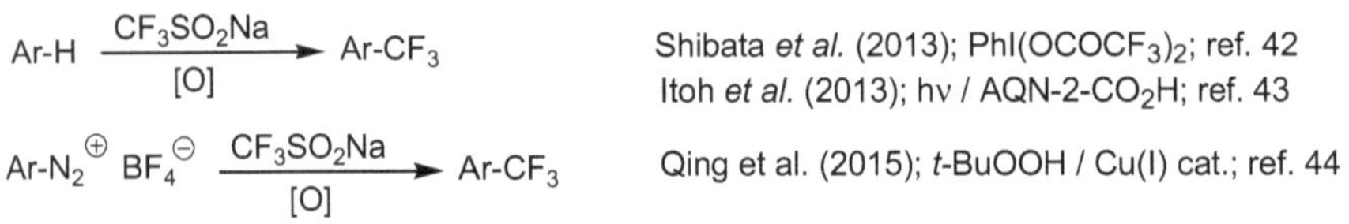

Scheme 5.9 Oxidative trifluoromethylation of aromatics.

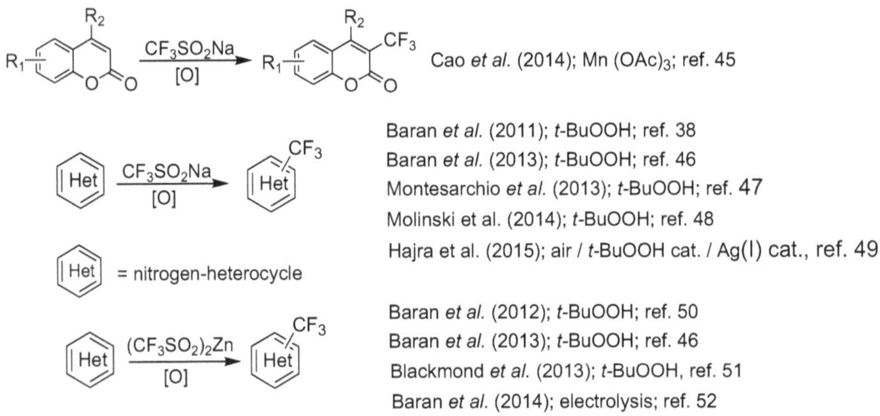

Scheme 5.10 Oxidative trifluoromethylation of heterocycles.

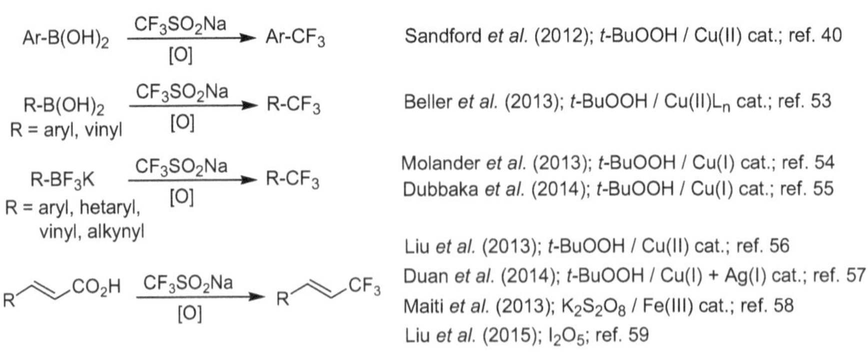

Scheme 5.11 Substitutive trifluoromethylation of boronic acids, trifluoroborates, and acrylic acids.

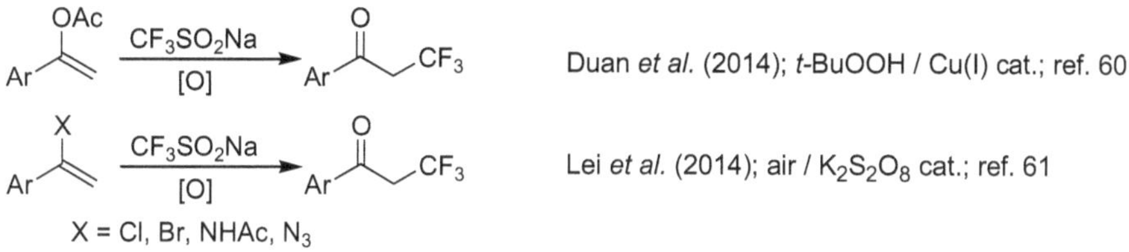

Scheme 5.12 Oxidative trifluoromethylation of functionalized alkenes.[62]

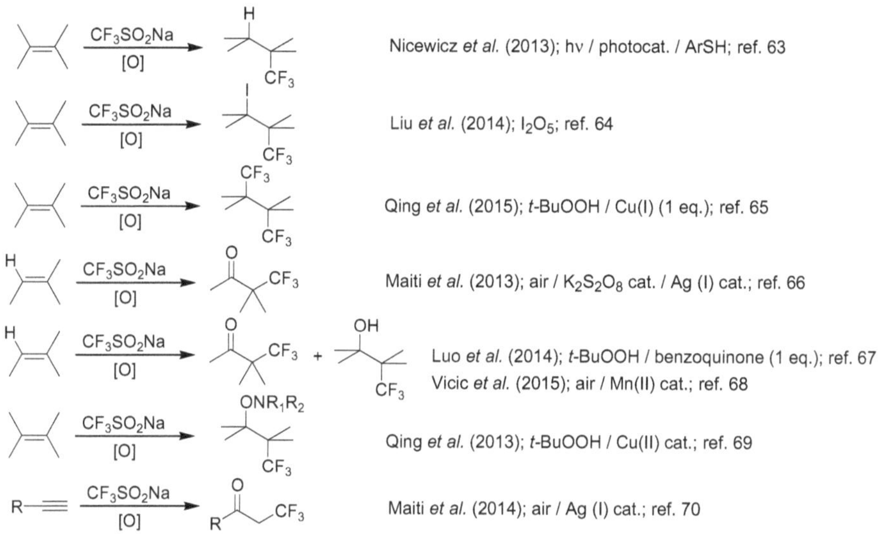

Scheme 5.13 Conjugated trifluoromethylation of nonfunctionalized alkenes and alkynes.

can be seen, a large panel of substrates have been used to prepare numerous and valuable products, but the principle of these reactions remains constant. Far more interesting is the variety of oxidant that have been experimented with, in a constant progression toward cheaper and environment-friendly reagents (vide infra).

Radical trifluoromethylation with triflinate salts was also frequently used to induce cyclizations leading to potentially bioactive compounds, based on nitrogen- or oxygen-containing heterocycles. For this purpose, all the described oxidants have been

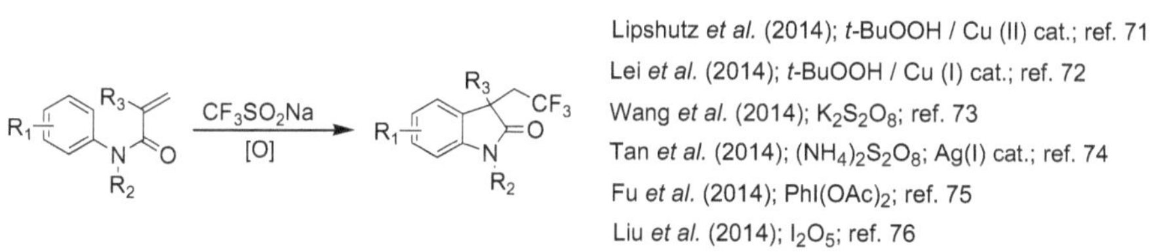

Scheme 5.14 Cyclizing trifluoromethylations leading to oxindoles.

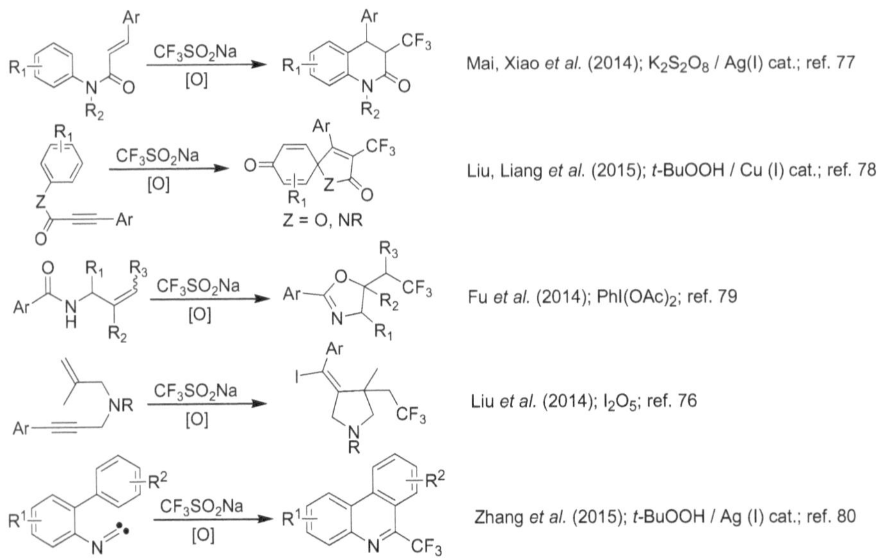

Scheme 5.15 Cyclizing trifluoromethylations leading to various heterocycles.

employed. Trifluoromethylated oxindoles constituted the main target (Scheme 5.14), although other five-membered or six-membered cycles were prepared (Scheme 5.15).

Apart from all the possibilities afforded by triflinate salts to carry out various trifluoromethylations, an analogous breakthrough must be emphasized: the use of zinc difluoromethanesulfinate $(HCF_2SO_2)_2Zn$, reported for the first time by Baran et al., as well as Blackmond et al., to carry out radical difluoromethylation upon heterocycles, especially those containing nitrogen (Scheme 5.16).[46,50–52,81] Later, this

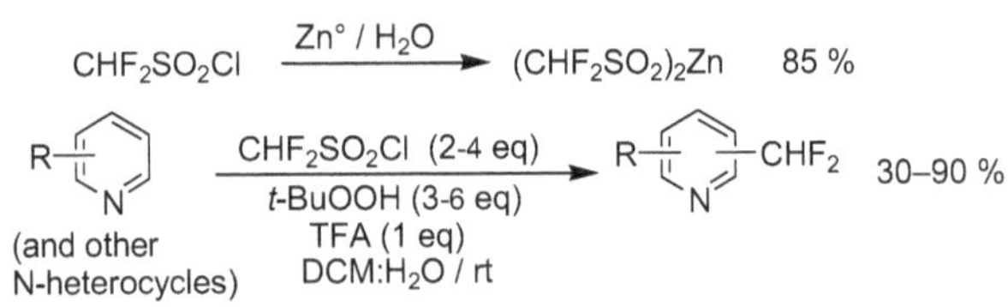

Scheme 5.16 Preparation of $(HCF_2SO_2)_2Zn$ and its use for radical difluormethylation.

difluoromethylation was applied to acrylic acids[56] and for the synthesis of difluoromethylated oxindoles.[74]

Apart from the fact that $(HCF_2SO_2)_2Zn$ (zinc "diflinate") can be easily prepared from difluoromethanesulfonyl chloride and is now commercially available,[81] the interest in such a reaction is the fact that the difluoromethyl moiety has become more and more popular as a replacement for or complement to the trifluoromethyl one, for, at least, three reasons. The first is that, although interestingly stable, it is less stable than CF_3. The second one is that, while remaining a highly hydrophobic group, its hydrophobicity is lower than that of CF_3. Thus, for these two reasons, difluoromethylated compounds are less persistent in situ than their trifluoromethylated analogs, for example, in the soil (for agrochemicals) or in lipidic systems (for pharmaceuticals), which gives it a serious advantage. The third reason is that •CHF_2 being less electrophilic than •CF_3, the regioselectivity of attack of these two radicals can differ, as it was nicely illustrated by Baran

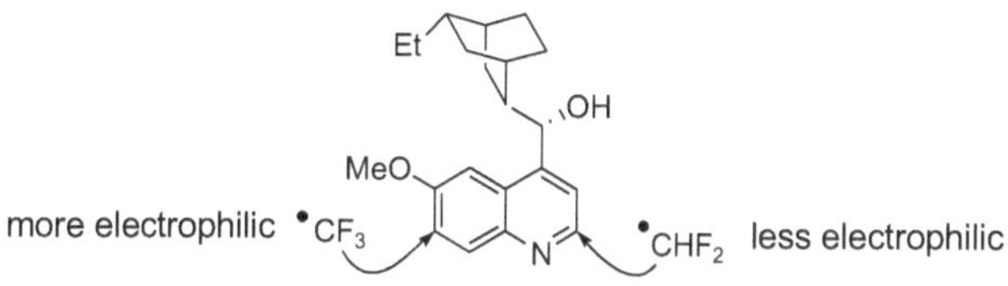

Scheme 5.17 Difference in regioselectivities for •CF_3 and •CHF_2.

et al. on dihydroquinine and varenicline (Scheme 5.17).[81] The interest in such a finding could allow us to name $(HCF_2SO_2)_2Zn$ as the "Baran–Blackmond reagent"!

4. Conclusions

In conclusion, radical trifluoromethylation with trifluoromethanesulfinate salts under oxidative conditions, first discovered by Langlois' et al. in the early 1990s, remained relatively little known until the very beginning of the 2010s decade when Baran et al., as well as Blackmond et al. brought it back to life, named sodium triflinate the "Langlois' reagent," and opened new perspectives for mild, easy, robust, and cheap syntheses of trifluoromethylated products, even sophisticated ones that could be of interest for bioactive specialties. Simultaneously, Baran et al., as well as Blackmond et al., also developed a similar method to graft difluoromethyl moieties on organic substrates from zinc difluoromethanesulfinate, which could have a bright future.[82]

Surprisingly, the rebirth of the Langlois' reagent received a wide echo in the chemical community and induced an important blossom of publications. Although the principle remained the one postulated by Langlois et al. in 1991–92, important improvements were brought to the methodology, especially concerning the nature of the single-electron oxidant. Although until now, *tert*-butyl hydroperoxide, usually associated with copper or silver salts as catalysts, remains the most employed oxidant, more stable and cheaper reagents such as $K_2S_2O_8$ and I_2O_5 are emerging. However, the breakthrough came from the possibility, published in 2015, to use air as oxidizer and various salts ($K_2S_2O_8$) or metallic cations, especially Mn(II), as catalysts.[61,66,68,70]

Acknowledgments

The author warmly acknowledges Dr. Thierry Billard and Dr. Fabien Toulgoat for helping him very efficiently during the elaboration of this chapter and carefully checking all the articles published between 2010 and 2015.

References

1. a. Kirsch, P. *Modern Fluoroorganic Chemistry: Synthesis, Reactivity, Applications;* Wiley-VCH: Weinheim, 2004.
   b. Bégué, J. P.; Bonnet-Delpon, D. *Bioorganic and Medicinal Chemistry of Fluorine;* John Wiley & Sons: Hoboken, 2008.
   c. Ojima, I. *Fluorine in Medicinal Chemistry and Chemical Biology;* Wiley-Blackwell: Oxford, 2009.
   d. Uneyama, K. *Organofluorine Chemistry;* Blackwell Publishing: Oxford, 2006.
   e. Hiyama, T. *Organofluorine Compounds. Chemistry and Applications;* Springer: Berlin, 2000. Soloshonok, V. A. *Fluorine-Containing Synthons*, American Chemical Society (ACS Symposium Series 911): Washington DC, 2005.
   f. Ametamey, S. M.; Honer, M.; Schubiger, P. A. *Chem. Rev.* **2008,** *108,* 1501–1516.
2. a. Sen, K. D.; Jorgensen, C. K. *Electronegativity;* Springer-Verlag: New York, 1987.
   b. McClinton, M. A.; McClinton, D. A. *Tetrahedron* **1992,** *48,* 6555.
3. Hansch, C.; Leo, A. *Substituent Constants for Correlation Analysis in Chemistry & Biology;* Wiley: New York, 1979.
4. a. Schlosser, M. *Angew. Chem. Int. Ed.* **2006,** *45,* 5432.
   b. Müller, K.; Faeh, C.; Diederich, F. *Science* **2007,** *317,* 1881.
   c. Hagmann, W. K. *J. Med. Chem.* **2008,** *51,* 4359.
   d. Kirk, K. L. *Org. Process Res. Dev.* **2008,** *12,* 305.
   e. Purser, S.; Moore, P. R.; Swallow, S.; Gouverneur, V. *Chem. Soc. Rev.* **2008,** *37,* 320.
   f. Tomashenko, O. A.; Grushin, V. V. *Chem. Rev.* **2011,** *111,* 4475.
5. a. Swarts, F. *Bull. Acad. Roy. Med. Belg.* **1892,** *24,* 309.
   b. Swarts, F. *Bull. Class Sci. Acad. Roy. Belg.* **1898,** *35,* 375.
6. a. Osswald, P.; Muller, F.; Steinhaüser, F. Ger. Patent 575,593 (1933) (to I.G. Farben). *Chem. Abstr.* **1933,** *27,* 4813.
   b. Holt, L. C.; Mattison, E. L. U.S. Patent 2,005,712 (1935) (to Kinetic Chemicals Co.). *Chem. Abstr.* **1935,** *29,* 5123.
   c. McBee, E. T.; Haas, H. B.; Hodnett, E. M. *Ind. Eng. Chem.* **1947,** *39,* 389.
   d. McBee, E. T.; Pierce, O. R.; Bolt, R. O. *Ind. Eng. Chem.* **1947,** *39,* 391.
   e. Sheppard, W. A.; Sharts, C. M. *Organic Fluorine Chemistry;* Benjamin: New York, 1969.
   f. Banks, R. E.; Smart, B. E.; Tatlow, J. C. In *Organofluorine Chemistry: Principles and Commercial Applications;* Plenum Press: New York, 1994.
7. Haufe, G. *J. Prakt. Chem.* **1996,** *338,* 99–113.
8. a. Bergstrom, C. G.; Nicholson, R. T.; Dodson, R. M. *J. Org. Chem.* **1963,** *28,* 2633.
   b. Olah, G. A.; Nojima, M. *Synthesis* **1973,** 779.
   c. Olah, G. A.; Shih, J. G.; Prakash, G. K. S. In *Fluorine, the First Hundred Years;* Banks, R. E., Sharp, D. W. A., Tatlow, J. C., Eds.; 1986; p 377.
9. For recent reviews on trifluoromethylation, see:
   Langlois, B. R.; Billard, T. *Synthesis* **2003,** 185–194.
   b. Studer, A. *Angew. Chem. Int. Ed.* **2012,** *51,* 8950–8958.
   c. Liu, H.; Gu, Z.; Jiang, X. *Adv. Synth. Catal.* **2013,** *355,* 617–626.
   d. Egami, H.; Sodeoka, M. *Angew. Chem. Int. Ed.* **2014,** *53,* 8294–8308.
   e. Zhang, C. *Adv. Synth. Catal.* **2014,** *356,* 2895–2906.

f. Zhang, C. P.; Chen, Q. Y.; Guo, Y.; Xiao, J. C.; Gu, Y. C. *Chem. Soc. Rev.* **2012,** *41,* 4536–4559.
g. Ma, J. A.; Cahard, D. *J. Fluorine Chem.* **2007,** *128,* 975–996.
h. Landelle, G.; Panossian, A.; Pazenok, S.; Vors, J. P.; Leroux, F. R. *Beilstein J. Org. Chem.* **2013,** *9,* 2476–2536.
i. Liu, X.; Xu, C.; Wang, M.; Liu, Q. *Chem. Rev.* **2015,** *115,* 683–730.
j. Macé, Y.; Magnier, E. *Eur. J. Org. Chem.* **2012,** 2479–2494.
10. a. Matsui, K.; Tobita, E.; Ando, M.; Kondo, K. *Chem. Lett.* **1981,** 1719.
b. Carr, G. E.; Chambers, R. D.; Holmes, T. F.; Parker, D. G. *J. Chem. Soc. Perkin Trans. I* **1988,** 921.
11. Langlois, B. R.; Roques, N. *J. Fluorine Chem.* **2007,** *128,* 1318.
12. Joubert, J.; Christophe, C.; Roussel, S.; Billard, T.; Langlois, B. R. *Angew. Chem. Int. Ed.* **2003,** *42,* 3133.
13. a. Prakash, G. K. S.; Krishnamurti, R.; Olah, G. A. *J. Am. Chem. Soc.* **1989,** *111,* 393.
b. Prakash, G. K. S.; Yudin, A. K. *Chem. Rev.* **1997,** *97,* 757–786.
c. Prakash, G. K. S.; Jog, P. V.; Batamack, P. T. D.; Olah, G. A. *Science* **December 7, 2012,** *338,* 1324–1327.
14. Large, S.; Roques, N.; Langlois, B. R. *J. Org. Chem.* **2000,** *65,* 8848–8856.
15. a. Billard, T.; Large, S.; Langlois, B. R. *Tetrahedron Lett.* **1997,** *38,* 65–68.
b. Bouchu, M. N.; Large, S.; Steng, M.; Langlois, B. R.; Praly, J. P. *J. Carbohydrate Res.* **1998,** *314* (1–2), 37–45.
c. Granger, C. E.; Félix, C. P.; Parrot-Lopez, H. P.; Langlois, B. R. *Tetrahedron Lett.* **2000,** *41,* 9257–9260.
16. a. Umemoto, T.; Ishihara, S. *Tetrahedron Lett.* **1990,** *31,* 3579.
b. ibid *J. Am. Chem. Soc.* **1993,** *115,* 2156.
c. Umemoto, T.; Adachi, K. *J. Org. Chem.* **1994,** *59,* 5692.
17. Eisenberger, R.; Gischig, S.; Togni, A. *Chem. Eur. J.* **2006,** *12,* 2579.
18. Yagupolskii, L. M.; Maletina, I. I.; Kondratenko, N. V.; Orda, V. V. *Synthesis* **1978,** 835.
19. a. Wakselman, C.; Tordeux, M. In *The Roots of Organic Development;* Desmurs, J. R., Ratton, S., Eds.; *Industrial Chemistry Library*; Elsevier: Amsterdam, 1996, Vol. 8.
b. Ignatev, N. V.; Boiko, V. N.; Yagupolskii, L. M. *Zh. Org. Khim.* **1985,** *21,* 653.
20. a. Wakselman, C.; Tordeux, M. *J. Chem. Soc. Chem. Commun.* **1984,** 793.
b. Wakselman, C.; Tordeux, M. *J. Org. Chem.* **1985,** *50,* 4047.
c. Bey, P.; Vevert, J. P. *Tetrahedron Lett.* **1978,** *14,* 1215.
21. a. Wakselman, C.; Tordeux, M.; Clavel, J. L.; Langlois, B. *J. Chem. Soc. Chem. Commun.* **1991,** 993–994.
b. Clavel, J. L.; Langlois, B.; Wakselman, C.; Tordeux, M. *Phosphorus Sulfur Silicon* **1991,** *59,* 3691–3694.
c. Clavel, J. L.; Langlois, B.; Nantermet, R.; Tordeux, M.; Wakselman, C. *J. Chem. Soc. Perkin Trans. I* **1992,** 3371–3375.
d. Clavel, J.L.; Langlois, B.; Nantermet, R.; Tordeux, M.; Wakselman, C. Eur. Patent 374,061 (**1990**) (to Rhône-Poulenc Co.). *Chem. Abstr., 114,* 5483.
22. a. Tordeux, M.; Langlois, B.; Wakselman, C. *J. Chem. Soc. Perkin Trans. I* **1990,** 2293–2299.
b. Tordeux, M.; Wakselman, C.; Langlois, B. Eur. Patent 298,803 (**1989**) (to Rhône-Poulenc Co.). *Chem. Abstr., 111,* 173730.

23. Tordeux, M.; Langlois, B.; Wakselman, C. *J. Org. Chem.* **1989,** *54,* 2452–2453.
24. a. Tordeux, M.; Langlois, B.; Wakselman, C. Eur. Patent 278,822 (**1988**) (to Rhône-Poulenc Co.). *Chem. Abstr., 110,* 94514.b. Tordeux, M.; Langlois, B.; Wakselman, C. French Patent 2,593,808 (**1987**) (to Rhône-Poulenc Co.). *Chem. Abstr., 108,* 166975.c. Forat, G.; Langlois, B.; Lorieau, P.; Poppei, J. Eur. Patent 396,458 (**1990**) (to Rhône-Poulenc Co.). *Chem. Abstr., 114,* 121482.
25. Gonbeau, D.; Guimon, M. F.; Duplantier, S.; Ollivier, J.; Pfister-Guillouzo, G. *Chem. Phys.* **1989,** *135,* 85–89.
26. a. Clavel, J. L.; Langlois, B.; Laurent, E.; Roidot, N. *Phosphorus Sulfur Silicon* **1991,** *59,* 169–172.
    b. Langlois, B.; Montègre, D.; Roidot, N. *J. Fluorine Chem.* **1994,** *68,* 63–66.
    c. Clavel, J.L.; Langlois, B.; Laurent, E.; Roidot, N. Eur. Patent 458,684 (**1991**) & U.S. 5,565,689 (**1996**) (to Rhône-Poulenc Co.). *Chem. Abstr., 116,* 127821.
27. Since sodium triflinate was produced in a glass-lined apparatus by a catalyst-free process, it was supposed to be free of any metallic impurity. Thus, we assumed that it was directly oxidized by the hydroperoxide.
28. a. Sharpless, K. B.; Verhoeven, T. R. *Aldrichim Acta* **1979,** *12,* 63.
    b. Kochi, J. K. *Free Radicals;* Wiley: New York, 1973; p 591.
29. Langlois, B.; Laurent, E.; Roidot, N. *Tetrahedron Lett.* **1991,** *32,* 7525–7528.
30. Tommasino, J. B.; Brondex, A.; Médebielle, M.; Thomalla, M.; Langlois, B. R. *Synlett* **2002,** *10,* 1697–1699.
31. Smertenko, E. A.; Datsenko, S. D.; Ignatev, N. V. *Russ. J. Electrochem.* **1998,** *34,* 46.
32. Langlois, B.; Laurent, E.; Roidot, N. *Tetrahedron Lett.* **1992,** *33,* 1291–1294.
33. Enol ethers were not appropriate substrates since they were too sensitive to *tert*-butyl hydroperoxide.
34. Langlois, B.; Billard, T.; Guérin, S.; Large, S.; Roidot-Pérol, N. *Phosphorus Sulfur Silicon* **1999,** *153–154,* 323–324.
35. Forat, G.; Mas, J.M.; Saint-Jalmes, L. Eur. Patent 735,023 (**1996**) (to Rhône-Poulenc Co.).
36. Singh, R. P.; Shreeve, J. M. *Chem. Commun.* **2002,** 1818–1819.
37. Langlois, B.; Billard, T.; Mulatier, J. C.; Yézéguélian, C. *J. Fluorine Chem.* **2007,** *128,* 851–856.
38. a. Ji, Y.; Brueckl, T.; Baxter, R. D.; Fujiwara, Y.; Seiple, I. B.; Su, S.; Blackmond, D. G.; Baran, P. S. *Proc. Natl. Acad. Sci. U.S.A.* **2011,** *108,* 14411.
    b. Brückl, T.; Baxter, R. D.; Ishihara, Y.; Baran, P. S. *Acc. Chem. Res.* **2012,** *45,* 826–839.
39. Parsons, A. T.; Buchwald, S. L. *Nature* **December 08 , 2011,** *480* (7376), 184–185.
40. Ye, Y.; Künzi, S. A.; Sandford, M. S. *Org. Lett.* **2012,** *14,* 4979–4981.
41. As far as *t*-BuOOH is concerned as oxidant, it must be noticed that, except Baran et al. and some medicinal chemists, most of the scientists used a metallic catalyst, essentially Cu(II).
42. Yang, Y. D.; Iwamoto, K.; Tokunaga, E.; Shibata, N. *Chem. Commun.* **2013,** *49,* 5510–5512.
43. Cui, L.; Matusaki, Y.; Tada, N.; Miura, T.; Uno, B.; Itoh, A. *Adv. Synth. Catal.* **2013,** *355,* 2203–2207.
44. Zhang, K.; Xu, X. H.; Qing, F. L. *J. Org. Chem.* **2015,** *80,* 7658–7665.
45. Cao, X. H.; Pan, X.; Zhou, P. J.; Zou, J. P.; Taiwo Asekun, O. *Chem. Commun.* **2014,** *50,* 3359–3362.
46. O'Hara, F.; Blackmond, D. G.; Baran, P. S. *J. Am. Chem. Soc.* **2013,** *135,* 12122–12134.

47. Musumeci, D.; Irace, C.; Santamaria, R.; Montesarchio, D. *Med. Chem. Commun.* **2013,** *4,* 1405–1410.
48. Paige Stout, E.; Choi, M. Y.; Castro, J. E.; Molinski, T. F. *J. Med. Chem.* **2014,** *57,* 5085–5093.
49. Monir, K.; Kumar Bagdi, A.; Ghosh, M.; Hajra, A. *J. Org. Chem.* **2015,** *80,* 1332–1337.
50. Fujiwara, Y.; Dixon, J. A.; O'Hara, F.; Daa Funder, E.; Dixon, D. D.; Rodriguez, R. A.; Baxter, R. D.; Herlé, B.; Sach, N.; Collins, M. R.; Ishihara, Y.; Baran, P. S. *Nature* **December 6, 2012,** *492,* 95–99.
51. Baxter, R. D.; Blackmond, D. G. *Tetrahedron* **2013,** *69,* 5604–5608.
52. O'Brien, A. G.; Maruyama, A.; Inokuma, Y.; Fujita, M.; Baran, P. S.; Blackmond, D. G. *Angew. Chem. Int. Ed.* **2014,** *53,* 11868–11871.
53. Li, Y.; Wu, L.; Neumann, H.; Beller, M. *Chem. Commun.* **2013,** *49,* 2628–2630.
54. Presset, M.; Oehlrich, D.; Rombouts, F.; Molander, G. A. *J. Org. Chem.* **2013,** *78,* 12837–12843.
55. a. Dubbaka, S. R.; Salla, M.; Bolisetti, R.; Nizalapur, S. *RSC Adv.* **2014,** *4,* 6496–6499. b. Dubbaka, S. R.; Nizalapur, S.; Atthunuri, A. R.; Salla, M.; Mathew, T. *Tetrahedron* **2014,** *70,* 2118–2121.
56. Li, Z.; Cui, Z.; Liu, Z. Q. *Org. Lett.* **2013,** *15,* 406–409.
57. Yin, J.; Li, Y.; Zhang, R.; Jin, K.; Duan, C. *Synthesis* **2014,** *46,* 607–612.
58. Patra, T.; Deb, A.; Manna, S.; Sharma, U.; Maiti, D. *Eur. J. Org. Chem.* **2013,** 5247–5250.
59. Shang, X. J.; Li, Z.; Liu, Z. Q. *Tetrahedron Lett.* **2015,** *56,* 233–235.
60. Lu, Y.; Li, Y.; Zhang, R.; Jin, K.; Duan, C. *J. Fluorine Chem.* **2014,** *161,* 128–133.
61. Lu, Q.; Liu, C.; Huang, Z.; Ma, Y.; Zhang, J.; Lei, A. *Chem. Commun.* **2014,** *50,* 14101–14104.
62. The work described in reference 60 is almost a pure pliagiarism of Langlois' method (Ref. 32) since the only difference is the exclusive use of α-(hetero)aryl-enol acetates as substrates instead of aliphatic or cycloaliphatic enol acetates.
63. Wilger, D. J.; Gesmundo, N. J.; Nicewicz, D. A. *Chem. Sci.* **2013,** *4,* 3160–3165.
64. Hang, Z.; Li, Z.; Liu, Z. Q. *Org. Lett.* **2014,** *16,* 3648–3651.
65. Yang, B.; Xu, X. H.; Qing, F. L. *Org. Lett.* **2015,** *17,* 1906–1909.
66. Deb, A.; Manna, S.; Modak, A.; Patra, T.; Maity, S.; Maiti, D. *Angew. Chem. Int. Ed.* **2013,** *52,* 9747–9750.
67. Luo, H. Q.; Zhang, Z. P.; Dong, W.; Luo, X. Z. *Synlett* **2014,** *25,* 1307–1311.
68. Yang, Y.; Liu, Y.; Jiang, Y.; Zhang, Y.; Vicic, D. A. *J. Org. Chem.* **2015,** *80,* 6639–6648.
69. Jiang, X. Y.; Qing, F. L. *Angew. Chem. Int. Ed.* **2013,** *52,* 14177–14180.
70. Maji, A.; Hazra, A.; Maiti, D. *Org. Lett.* **2014,** *16,* 4524–4527.
71. Yang, F.; Klumphu, P.; Liang, Y. M.; Lipshutz, B. H. *Chem. Commun.* **2014,** *50,* 936–938.
72. Lu, Q.; Liu, C.; Peng, P.; Liu, Z.; Fu, L.; Huang, J.; Lei, A. *Asian J. Org. Chem.* **2014,** *3,* 273–276.
73. Wei, W.; Wen, J.; Yang, D.; Liu, X.; Guo, M.; Dong, R.; Wang, H. *J. Org. Chem.* **2014,** *79,* 4225–4230.
74. Liu, J.; Zhuang, S.; Gui, Q.; Chen, X.; Yang, Z.; Tan, Z. *Eur. J. Org. Chem.* **2014,** 3196–3202.
75. Shi, L.; Yang, X.; Wang, Y.; Yang, H.; Fu, H. *Adv. Synth. Catal.* **2014,** *356,* 1021–1028.
76. Zhang, L.; Li, Z.; Liu, Z. Q. *Org. Lett.* **2014,** *16,* 3688–3691.
77. Mai, W. P.; Wang, J. T.; Yang, L. R.; Yuan, J. W.; Xiao, Y. M.; Mao, P.; Qu, L. B. *Org. Lett.* **2014,** *16,* 204–207.

78. Hua, H. L.; He, Y. T.; Qiu, Y. F.; Li, Y. X.; Song, B.; Gao, P.; Song, X. R.; Guo, D. H.; Liu, X. Y.; Liang, Y. M. *Chem. Eur. J.* **2015,** *21,* 1468−1473.
79. Yu, J.; Yang, H.; Fu, H. *Adv. Synth. Catal.* **2014,** *356,* 3669−3675.
80. Liu, Y. R.; Tu, H. Y.; Zhang, X. G. *Synthesis* **2015,** *47.* A-G.
81. Fujiwara, Y.; Dixon, J. A.; Rodriguez, R. A.; Baxter, R. D.; Dixon, D. D.; Collins, M. R.; Blackmond, D. G.; Baran, P. S. *J. Am. Chem. Soc.* **2012,** *134,* 1494−1497.
82. a. Rewcastle, G. W.; et al. *J. Med. Chem.* **2011,** *54,* 7105−7126.
b. Meanwell, N. A. *J. Med. Chem.* **2011,** *54,* 2529−2591. and references therein.
c. Furuya, T.; Kuttruff, C.; Ritter, T. *Curr. Opin. Drug Disc. Dev.* **2008,** *11,* 803−819.

Toward CF_3S Group: From Trifluoromethylation of Sulfides to Direct Trifluoromethylthiolation

6

F. Toulgoat[1,2], T. Billard[1,3]
[1]Institute of Chemistry and Biochemistry (ICBMS – UMR CNRS 5246), University Lyon 1, CNRS, Villeurbanne, France; [2]CPE-Lyon, Villeurbanne, France; [3]CERMEP – In Vivo Imaging, Groupement Hospitalier Est, Lyon, France

Chapter Outline

Modern Synthesis Processes and Reactivity of Fluorinated Compounds. http://dx.doi.org/10.1016/B978-0-12-803740-9.00006-8

1. Introduction

The introduction of fluorine onto organic molecules is today well known to modify their properties.[1–7] Moreover, the demand of compounds with specific characteristics continues to increase and contributes to the quest of new fluorinated groups with specific properties. Association of fluorinated moieties with heteroatoms is one of the possibilities of obtaining original substituents with interesting specificities. In this field, trifluoromethanesulfenyl group (CF_3S) is of particular interest. Indeed, this substituent possesses high Hansch parameters ($\pi_R = 1.44$), which confers to molecules an increased lipophilicity.[8] Such a property is of great interest in various fields and in particular in life sciences by favoring the membrane's permeation and, consequently, the bioavailability of substrates.[9–13]

Such a specific property has contributed to the high interest in developing various efficient strategies and methods to synthesize CF_3S compounds.[14–17]

In front of trifluoromethyl sulfides, different retrosynthetic analyses can be proposed (Scheme 6.1). On the one hand, the sulfur and fluorine atoms of the trifluoromethanesulfenyl group (CF_3S) can be introduced at different stages of the synthesis. In such methods, which can be considered as indirect, trifluoromethyl sulfides were obtained starting from methyl sulfides, thiols, disulfides, or functionalized sulfur components (chloro sulfides, thiocyanates, thiosulfonates). On the other hand, a more direct approach consists of introducing the trifluoromethanesulfenyl group (CF_3S) in a single step.

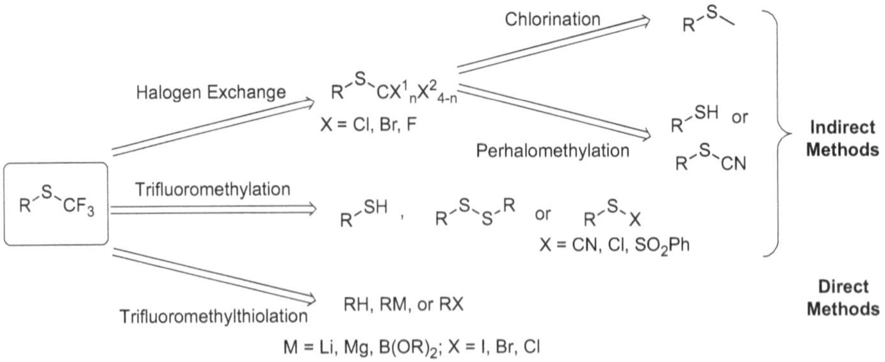

Scheme 6.1 Retrosynthetic analysis of trifluoromethyl sulfides.

2. Indirect Methods

As depicted in Scheme 6.1, several indirect methods were developed to synthesize trifluoromethyl sulfides. They can be classified into two parts: halogen exchange on perhalomethyl sulfides and trifluoromethylation of sulfur-containing products.

2.1 Preparation Involving Halogen Exchange Reaction

Syntheses of trifluoromethyl sulfides based on a halogen exchange require the preparation of the intermediate perhalomethyl sulfides. Thus, a two-step process was developed in the early 20th century and still represents one of the most important route to aryl trifluoromethyl sulfides at the industrial scale (Scheme 6.2).[14,15] This strategy was considered more convenient for large-scale preparation of aryl trifluoromethyl sulfides than known strategies in 1998.[18] Actually the first step consists in a photochemical chlorination of aryl methyl sulfides. Then, a halogen exchange (known as Swarts reaction) was applied to the previously obtained aryl trichloromethyl sulfides using mainly SbF_3 or HF as the fluorinating reagent.[19–28] The more easy-to-handle at laboratory scale complex Et_3N-3HF was also used in this second transformation.[18,29] However, harsh conditions are usually required during the halogen exchange reaction, which limits the scope of the reaction.

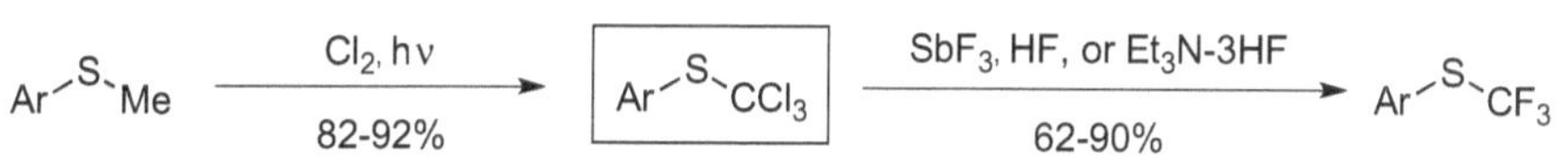

Scheme 6.2 Aryl trifluoromethyl sulfides synthesis by chlorination–halogen exchange process.

In the 1980's, it was found that the chlorination step can be replaced by a perhalomethylation step with the creation of the sulfur–carbon bond (Scheme 6.3). Indeed, by analogy to the trifluoromethylation of thiolates with bromotrifluoromethane (see Section 2.2.1.2), bromodifluoromethylation was developed using dibromodifluoromethane or bromochlorodifluoromethane (Scheme 6.3).[18,30] The obtained bromodifluoromethane sulfides react in halogen exchange reactions to furnish alkyl-, aryl-, or heteroaryl trifluoromethyl sulfides. Of note, a large panel of reaction conditions is reported for the halogen exchange step. It may depend on the fluorinated reagent and/or the substrate. In some cases, the reaction proceeds at room temperature.

Scheme 6.3 Synthesis of trifluoromethyl sulfides by bromodifluoromethylation–halogen exchange sequence. *DMF*, *N,N*-Dimethylformamide.

Later on, a similar route was also applied to the synthesis of few alkyl trifluoromethyl sulfides (Scheme 6.4). The first step differs from the previous one as bromodifluoromethylation of thiolates is replaced by trichloromethylation of alkyl thiocyanates.[31,32] Then, the halogen exchange can be performed using the complex Et_3N-3HF or pyridine-10HF at room temperature (known as Olah reagent).[32,33] It is worth noting that the replacement of the original chlorination step offers access to the alkyl trifluoromethyl sulfides.

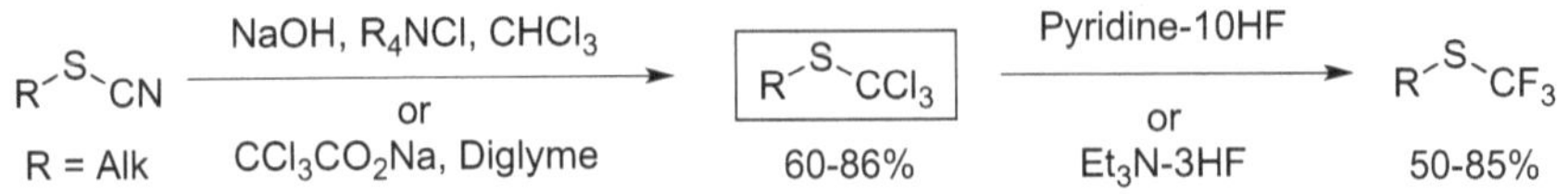

Scheme 6.4 Synthesis of alkyl trifluoromethyl sulfides by trichloromethylation–halogen exchange process.

2.2 *Trifluoromethylation of Sulfur Compounds*

The second class of indirect preparation of trifluoromethyl sulfides is the trifluoromethylation of sulfur-containing compounds. Indeed, trifluoromethylation has received considerable attention for many years[34,35] and was obviously applied to sulfur components. Such reaction can be classified depending on the reactivity of the trifluoromethylated reagent, i.e., whether this one takes part in a radical or SET (single electron transfer) mechanism, in a nucleophilic or electrophilic process.

2.2.1 *Trifluoromethylation of Thiols or Disulfides Involving Radical or SET Mechanisms*

Several reagents, in various conditions, were used to perform the trifluoromethylation of thiols or disulfides. Among them, the reactivities of trifluoroiodomethane and bromotrifluoromethane were intensively studied. In addition, other reagents (CF_2Br_2, CF_2BrCl, $CF_3N(NO)SO_2Ph$, CF_3SO_2Na, CF_3SO_2SR, CF_3COSR) proved to be efficient to prepare trifluoromethyl sulfides.

2.2.1.1 Trifluoroiodomethane as Trifluoromethylating Reagent

Historically, trifluoroiodomethane was the first reagent considered to prepare aryl or alkyl trifluoromethane sulfides from disulfides.[36,37] Such reactions were performed under photochemical activation (Scheme 6.5).[36,37] Considering the two reported examples of alkyl disulfide trifluoromethylation, this method appeared to take very long time (more than 20 days of reaction) due to proton abstraction by trifluoromethyl radical.[36] In 2011, metal-dissolved reduction conditions (Na/NH_3), instead of photochemical activation, were applied to trifluoroiodomethane (Scheme 6.5) to form in situ the trifluoromethyl radical and prepare few grams of trifluoromethylthiolated amino acids from the corresponding disulfides.[38]

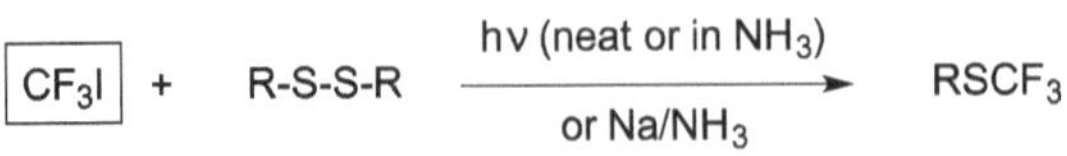

Scheme 6.5 Trifluoromethylation of disulfides with CF_3I.

As disulfides, thiols react with trifluoroiodomethane (Scheme 6.6a) in liquid ammonia under photochemical irradiation to afford the corresponding trifluoromethyl sulfides.[39–43] Similar yields are obtained in phase transfer conditions under photochemical activation (Scheme 6.6b).[44] In 2014, a ruthenium photocatalyst, activated by ultraviolet (UV) irradiation, proved to be very efficient in this transformation (Scheme 6.6c).[45] Indeed, aryl, heteroaryl trifluoromethyl sulfides were obtained with very good yields in mild conditions. Although the reaction of trifluoroiodomethane with alkylthiols was found to be tedious as formation of disulfides was often observed,[38,45] three examples of alkyl trifluoromethyl sulfides were reported with fairly good yields.[45] Moreover, performing this transformation in a continuous flow process allows reducing the large excess of CF_3I required during batch reaction.

A chemical activation can also be envisaged to avoid photochemistry processes. Thus, methylviologen[46,47] or thiolates are used as catalyst or initiator (Scheme 6.6d and e).[18,48,49] In the latter case, thiolates function as electron donors to initiate the reaction via the formation of radical anion $CF_3I^{\bullet-}$ (observed by electron spin resonance spectroscopy), which collapses into trifluoromethyl radical (S_RN1 mechanism).[50] This radical anion can also be generated, based on a methodology previously developed with bromotrifluoromethane[51,52] and then iodoperfluoroalkyl,[53] from reduction of trifluoroiodomethane by the sodium formate–sulfite system (Scheme 6.6f).[54] It was applied to the trifluoromethylation of 2-mercaptoethanol as unique example.[54] Of note, the synthesis of *p*-chlorobenzene trifluoromethyl sulfide was described using the liquid complex trifluoroiodomethane–tetramethylguanidine instead of gaseous trifluoroiodomethane.[55]

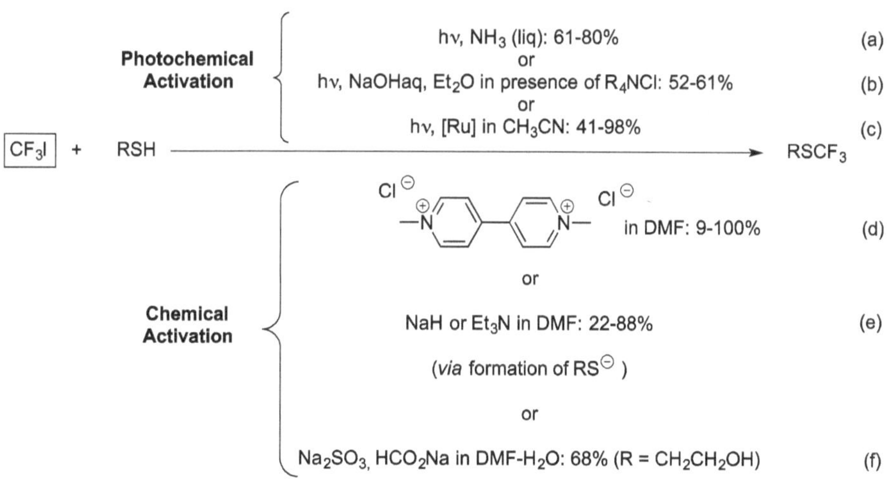

Scheme 6.6 Trifluoromethylation of thiols with CF_3I.

2.2.1.2 Bromotrifluoromethane as Trifluoromethylating Reagent

As trifluoroiodomethane, its bromoanalog (CF_3Br) reacts with arylthiolates to furnish the corresponding aryl trifluoromethyl sulfides via a S_RN1 mechanism with the formation of the radical anion $CF_3Br^{\bullet-}$ as intermediate (Scheme 6.7).[50,56,57] Similarly, SO_2 can be chemically[51,52,58–60] or electrochemically[61] reduced to form the corresponding radical anion, which acts as a reducing agent of bromotrifluoromethane (Scheme 6.7). In such conditions, aryl trifluoromethyl sulfides were formed starting from the corresponding thiols[59,61] or disulfides.[51,58] Moreover, three examples of alkyl trifluoromethyl sulfides were reported using the system Rongalite–sulfur dioxide.[58] It should be noted that due to the low solubility of bromotrifluoromethane in solvent, these reactions have to be performed under pressure.

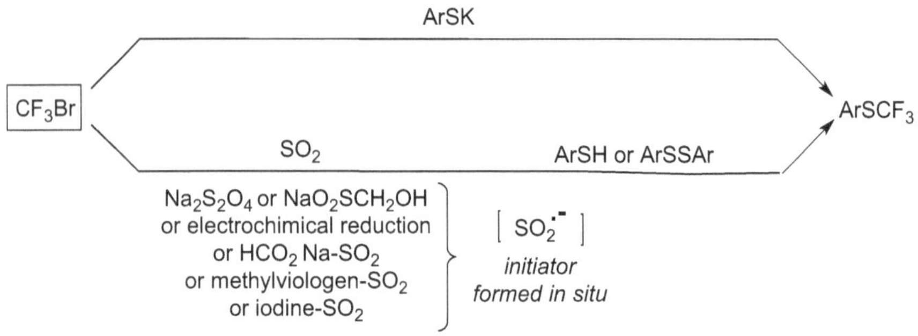

Scheme 6.7 Trifluoromethylation of thiols with CF_3Br.

2.2.1.3 Miscellaneous Trifluoromethylating Reagent

During reactions involving radical anion of SO_2 and CF_3Br (Scheme 6.7), trifluoromethanesulfinate appeared to be a by-product. Actually, this reaction corresponds to one of the described synthetic route of this compound (see chapter on Langlois' reagent). Interestingly, this latter proved to be a precursor of trifluoromethyl radical under oxidative conditions (see chapter on Langlois' reagent). In such conditions, the trifluoromethyl radical can be trapped by disulfides to afford trifluoromethyl sulfides (Scheme 6.8).[62,63] Thus, several alkyl trifluoromethyl sulfides were obtained thanks to this methodology.

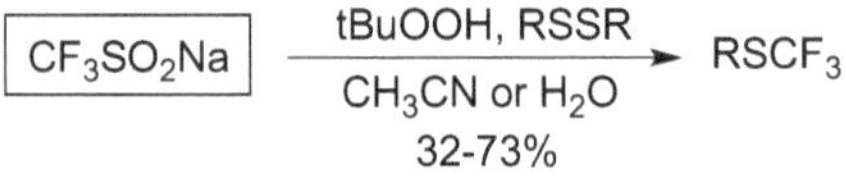

Scheme 6.8 Trifluoromethylation of disulfides with Langlois' reagent.

In addition to CF_nX_{4-n} (X = Cl or Br), three other reagents are efficient in radical trifluoromethylation of disulfides (Scheme 6.9).[32,64–66]

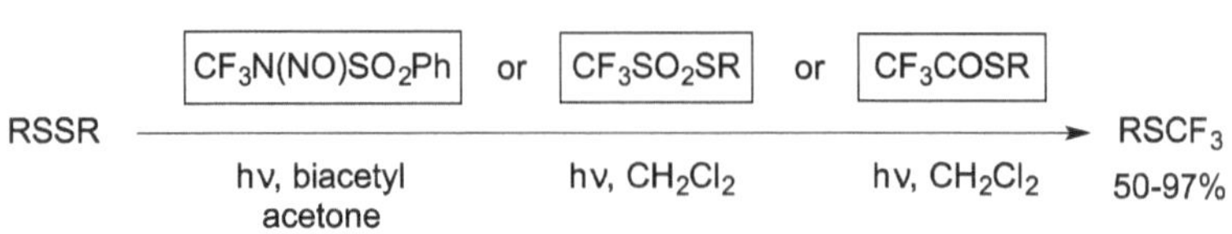

Scheme 6.9 Trifluoromethylation of disulfides.

It must be emphasized that the use of halons, which is subjected to restriction due to their deleterious effect upon stratospheric ozone, could be avoided for the preparation of CF_3SO_2Na, CF_3SO_2SR, as well as CF_3COSR.[65] Indeed, trifluorothioacetates can be prepared from trifluoroacetic anhydride and thiols,[65] and are precursors of trifluoromethanesulfinate,[32] which itself is a precursor of trifluoromethanethiosulfonates (Scheme 6.10).[65] The use of *N*-trifluoromethyl-*N*-nitrosobenzensulfonamide is probably hampered due to the involvement of the highly toxic gaseous trifluoronitrosomethane during its synthesis.[67]

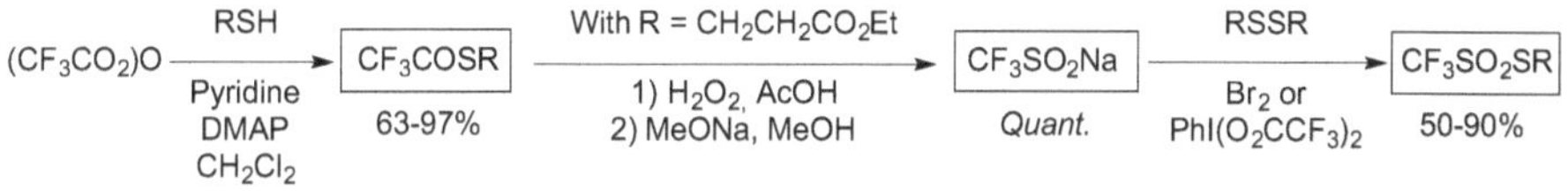

Scheme 6.10 Preparation of trifluoromethylating reagents.

2.2.2 *Nucleophilic Trifluoromethylation*

To avoid radical conditions, nucleophilic trifluoromethylation of sulfur compounds has been developed. Such approach has been readily facilitated by the development of stable reagent "generator" of CF_3^- anion (see Chapter 11 on nucleophilic fluoroalkylation).

2.2.2.1 Bromotrifluoromethane and Trifluoroiodomethane as Trifluoromethylating Reagents

Even if CF_3Br[68] and CF_3I[69,70] have been largely used in radical or SET trifluoromethylations (see Section 2.2.1), they can also generate CF_3^- anion and, then, perform nucleophilic trifluoromethylation of thiocyanates or disulfides. Thus, in the presence of zinc, in pyridine, CF_3Br can trifluoromethylate thiocyanates in low yields (Scheme 6.11a). Concerning CF_3I, a double electron transfer, induced by the tetrakis(dimethylamino)ethylene (TDAE), generates the trifluoromethyl anion, stabilized by the dication arising from TDAE, which can react with disulfides (Scheme 6.11b). In this last case, both parts of disulfides are used in the presence of a large excess of CF_3I: the released thiolate during the nucleophilic attack carrying out a S_RN1 reaction with CF_3I, as previously described.

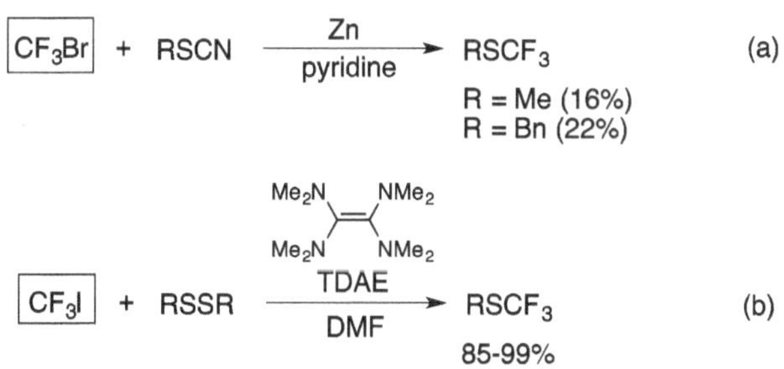

Scheme 6.11 Trifluoromethylation of thiocyanates and disulfides with CF_3X (X = Br, I).

2.2.2.2 Trifluoromethyltrimethylsilane (Ruppert–Prakash Reagent)

The Ruppert–Prakash reagent (CF_3SiMe_3) is certainly the most known and used generator of CF_3^- anion.[71] This efficient reagent has been used with a large panel of various sulfur compounds to obtain easily trifluoromethyl sulfides (Scheme 6.12). In this way, disulfides,[72] thiocyanates,[73] sulfenyl chlorides,[74] thiones,[75] and dithioesters[76,77] have been trifluoromethylated.

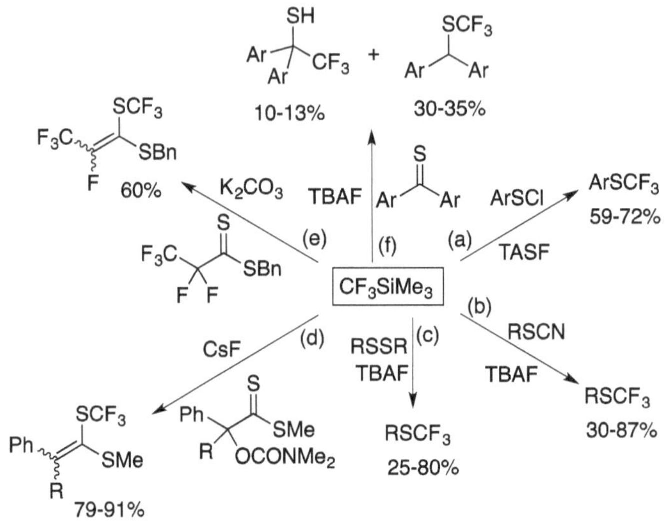

Scheme 6.12 Trifluoromethylation of sulfur compounds with CF_3SiMe_3.

The strategy starting from thiocyanates appears as the most efficient, because not only the starting material is easy to synthesize but also, contrary to disulfides, all the sulfur moiety is used. Furthermore, the fluoride can be catalytic, the released cyanide anion being able to desilylate CF_3SiMe_3. Moreover, this method has been applied onto some carbohydrates with yields in the range 39–68% and high selectivities.[78] It has also been applied to the synthesis of trifluoromethylthio ferrocenes.[79] Confirming this highest interest, this strategy has been efficiently adapted in one-pot tandem or cascade processes where thiocyanates are preformed in situ, without further purifications, from the corresponding aryl diazoniums salts (preformed or generated in situ from amines), electron-rich aromatic compounds, or alkyl halides (Scheme 6.13a–d).[80–83]

In a similar strategy, trifluoromethylation of S-alkyl or S-aryl sulfothiolates ($RSSO_3Na$), formed in situ with $Na_2S_2O_3$, has been elaborated to achieve with good yields alkyl or aryl trifluoromethyl sulfides. This system requires copper catalysis to favor the trifluoromethylation of the intermediate sulfothiolates but also to catalyze the formation of S-aryl sulfothiolates (Scheme 6.13e).[84]

2.2.2.3 Fluoroform and Fluoral Hemiaminals

In an atom economy point of view, fluoroform is the best reagent. Its ability to deliver by deprotonation a CF_3^- anion, able to react with various electrophiles, has been largely

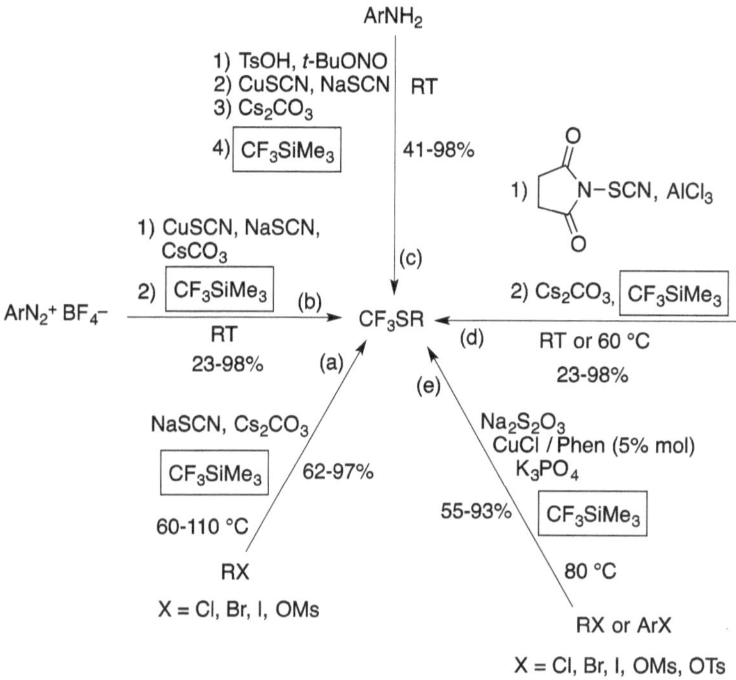

Scheme 6.13 One-pot process thiocyanation or sulfothiolation/trifluoromethylation.

demonstrated.[85] In particular, deprotonated HCF_3 can react with disulfides, sulfenyl chlorides, and thiosulphonates (Scheme 6.14a).[86–88]

In the presence of formamides, fluoroform can provide fluorinated hemiaminals by attacking the carbonyl function. Such compounds can be also obtained by the condensation of corresponding amines with trifluoroacetaldehyde (fluoral).[89] These stable hemiaminals (free or silylated forms) can expel a CF_3^- anion under bases or fluorides action. In the presence of disulfides, the corresponding trifluoromethyl sulfides can be then obtained (Scheme 6.14c).[90]

2.2.2.4 Trifluoromethanesulfinamides and Trifluoromethylsulfones

Trifluoromethanesulfinamides can be easily synthesized from triflinate ($CF_3SO_2^-$, Langlois' reagent) and corresponding amines.[91] These compounds can undergo a nucleophilic attack of tertiobutylate anion onto their electrophilic sulfur atom, triggering a CF_3^- anion release. This anion can then react with disulfides (Scheme 6.14g).[92] In a similar manner, trifluoromethyl phenyl sulfone also generates a trifluoromethyl anion under tBuOK activation. In the presence of phenyldisulfide, the corresponding trifluoromethylthioether is formed (Scheme 6.14f).[93]

2.2.2.5 Miscellaneous Reagents

Other nucleophilic trifluoromethylating reagents have been described as reacting with disulfides to provide trifluoromethylthioethers (Scheme 6.14). Thus,

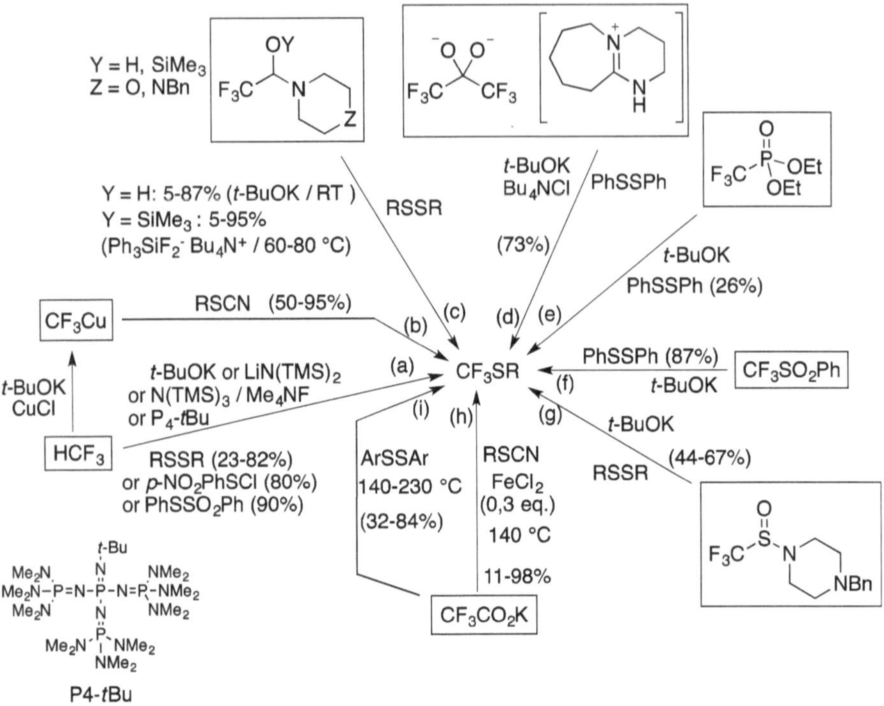

Scheme 6.14 Nucleophilic trifluoromethylation of disulfides with various reagents.

temperature-mediated decarboxylation of potassium trifluoroacetate is efficient with aryl disulfides (Scheme 6.14i).[94,95] This decarboxylation has been adapted to trifluoromethylate thiocyanates, under iron catalysis (Scheme 6.14h).[96] Commercially available amidinate salt of hexafluoroacetone hydrate is also able to trifluoromethylate phenyldisulfide in the presence of tBuOK and Bu_4NCl (Scheme 6.14d).[97] Finally, diethyl trifluoromethylphosphonate under tertiobutylate activation can also trifluoromethylate phenyldisulfide with modest yield (Scheme 6.14e).[98] In situ–formed trifluoromethylcopper from fluoroform was found to react with thiocyanates to provide the corresponding trifluoromethyl sulfides (Scheme 6.14b).[99]

2.2.3 Electrophilic Trifluoromethylation

In addition to radical and nucleophilic trifluoromethylation, electrophilic trifluoromethylation has also been developed.[100,101] Actually, two classes of electrophilic reagents have been considered to prepare trifluoromethyl sulfides starting from thiols: derivatives of aryl trifluoromethyl sulfonium salts (or selenophenium salts) and hypervalent trifluoromethyl-iodine(III) reagents.

2.2.3.1 Aryl Trifluoromethyl Sulfonium Salts (Umemoto's Reagents)

Aryl trifluoromethyl sulfonium salts (or selenophenium salts) have been involved in few thiol trifluoromethylations (Scheme 6.15).[100,102–104] Although some are

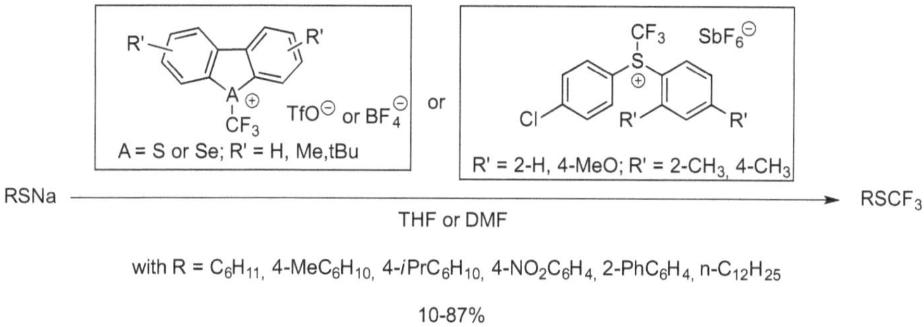

Scheme 6.15 Trifluoromethylation of sodium thiolates with trifluoromethyl sulfonium salts (or selenophenium salts). *DMF*, *N,N*-Dimethylformamide; *THF*, Tetrahydrofuran.

commercially available, such reagents do not appear so useful in the considered context as they are often prepared from trifluoromethyl sulfides.[100–103,105]

2.2.3.2 Hypervalent Trifluoromethyl-Iodine(III) Reagents (Togni's Reagents)

Two new hypervalent trifluoromethyl-iodine(III) reagents were prepared and isolated by Togni et al. (Scheme 6.16).[106] Of note, the trifluoromethyl part is introduced using the Ruppert–Prakash reagent (CF_3SiMe_3).[106,107] Among the panel of reactions performed with these reagents,[107] trifluoromethylation of thiols has also been achieved (Scheme 6.16).[108] The scope of the reaction is quite large: aryl,[108] heteroaryl,[108] alkyl,[108,109] carbohydrate,[108] and amino acid derivatives,[108] as well as peptides[110,111] and thiols were trifluoromethylated in quite good yields. Although these reagents are commonly considered as electrophilic trifluoromethyl reagent, further investigations suggest the involvement of a radical mechanism.[107,112]

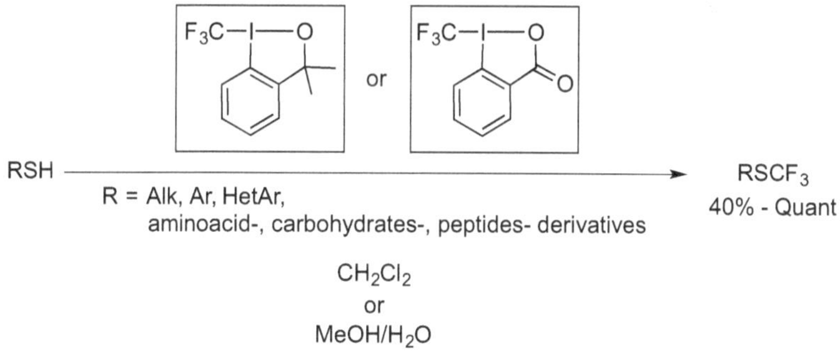

Scheme 6.16 Trifluoromethylation with Togni's reagent.

3. Direct Preparation: Trifluoromethylthiolation

Despite the efficiency of some of the previous methods, they require the preliminary synthesis of sulfur compounds. This preliminary step could constitute a major drawback. Consequently, the direct introduction of CF_3S group onto organic substrates represents a valuable alternative. Furthermore, this strategy is more convenient from a retrosynthetic point of view, as the direct disconnection of the CF_3S group could be envisaged anywhere onto the target and at any time of the synthesis.

3.1 Nucleophilic Trifluoromethylthiolation

Nucleophilic trifluoromethylthiolation had been considered more than 50 years ago, but new reactions have been developed from 2010, thanks to the breakthrough in organometallic catalysis in fluorine chemistry.[113–115] Nevertheless, all reactions based on nucleophilic trifluoromethylthiolation have to overcome the low stability of the trifluoromethyl thiol or its corresponding anion.[116] Indeed, potassium trifluoromethylthiolate can liberate potassium fluoride and gaseous difluorophosgene (Scheme 6.17).[117–121] Therefore, several strategies have been developed and nucleophilic trifluoromethylthiolations have been performed mainly using metallic trifluoromethylthiolate (Ag, Cu), ammonium trifluoromethylthiolate, and also sulfur compounds formed in situ.

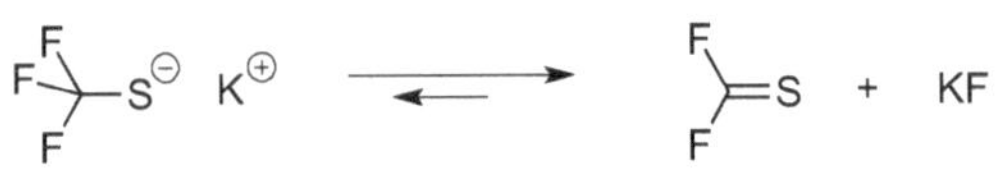

Scheme 6.17 Equilibrium between potassium trifluoromethylthiolate and difluorophosgene–potassium fluoride.

3.1.1 Trifluoromethylthiosilver(I)

Trifluoromethylthiosilver(I) (CF_3SAg) was involved in several nucleophilic trifluoromethylthiolation processes. It allows to convert methyl iodide,[122] propargylic halides,[123] tropylium bromide,[124] α-haloketones,[125,126] and electron-deficient benzyl halides.[126] It should be emphasized that the addition of iodide salt was necessary in some cases to form in situ a more active species.[126] The requirement of such additives has been previously underscored during CF_3SAg-based aryl trifluoromethyl sulfides synthesis.[127] Recently, alkyl trifluoromethyl sulfides have been prepared from the corresponding alcohol instead of alkyl halides (Scheme 6.18).[128] Actually, the alcohol function reacts on difluorophosgene formed by the decomposition of CF_3SAg and is converted into a carbonofluoridothioate, which acts as a good leaving group. Moreover, CF_3SAg was involved in the trifluoromethylthiolation of diazoesters

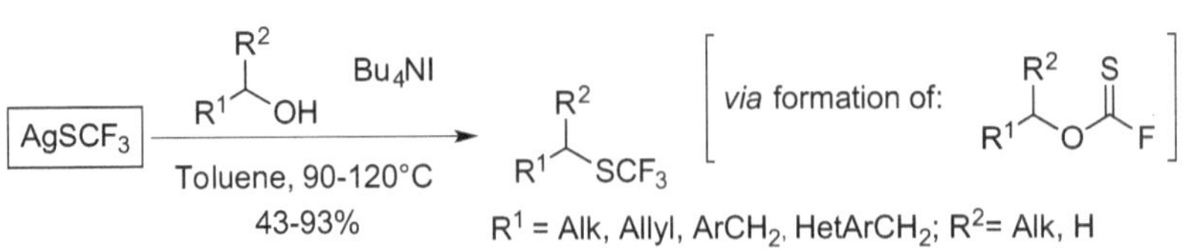

Scheme 6.18 Trifluoromethylthiolation of alcohols with CF_3SAg.

(Scheme 6.23).[129–131] However, CF_3SAg is here only a precursor as it is converted in situ into CF_3SCu, which is the trifluoromethylthiolating reagent.

Until 2000, aryldiazonium[132] or electron-poor aryl halides (in the presence of potassium iodide),[127] were the only substrates involved in CF_3SAg-based syntheses of aryl trifluoromethyl sulfides (Scheme 6.19a and b). However, the scope has been extended thanks to the advances in metal-catalyzed reactions.[113–115] In 2011, a Pd-catalyzed trifluoromethylthiolation of aryl and heteroaryl bromide with good yields was developed (Scheme 6.19c).[133] This reaction was performed in presence of not only biaryl ligands but also an iodide additive. This additive was found to be crucial in this transformation. Similar to previous study,[127] the formation of an anionic "ate" complex was proposed.[126,133] CF_3SAg can also be involved in Cu-catalyzed trifluoromethylthiolations (Scheme 6.19d).[134] Compared with the Pd-catalyzed reaction developed earlier, no additives are necessary, but substrates have to bear an ortho-directing group such as pyridyl, ester, ketone, amide, imine, or oximes. In 2015, a Cu-catalyzed reaction of diaryl-λ^3-iodanes has been reported (Scheme 6.19e).[135,136] Preliminary mechanism experiments revealed that a trifluoromethylthiocopper complex is formed and can be considered as the active trifluoromethylthiolating reagent. This method presents a good functional group tolerance. Thus, several aryl trifluoromethyl sulfides are obtained in good yields. In addition to coupling reactions, C(sp2)-H trifluoromethylthiolation using CF_3SAg was reported (Scheme 6.19f).[137]

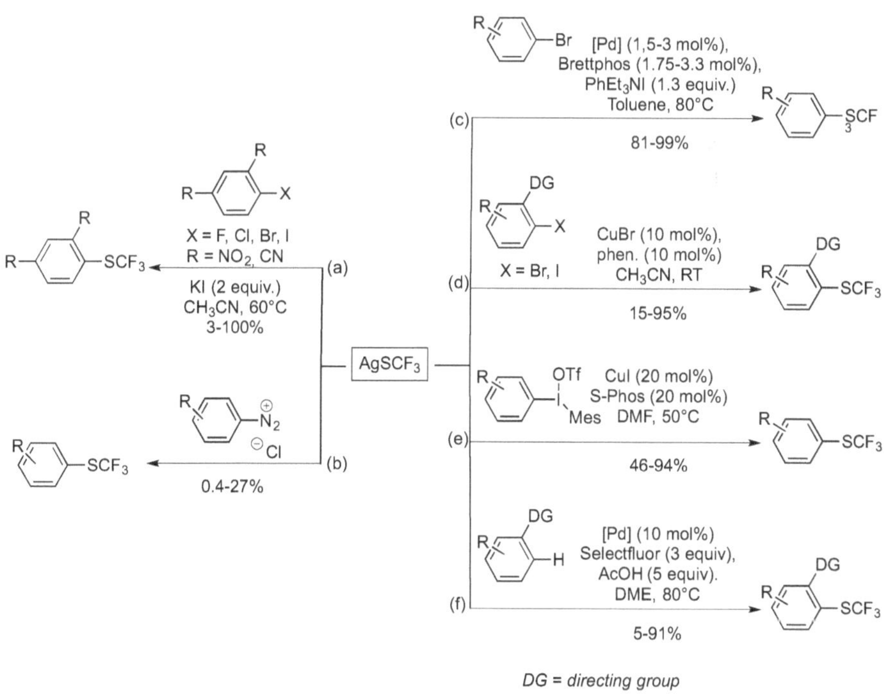

Scheme 6.19 Aryl trifluoromethyl sulfides synthesis from CF_3SAg.

This Pd-catalyzed transformation gives access to aryl trifluoromethyl sulfides ortho-substituted with nitrogen heterocycles or 8-aminoquinoline. Use of Selectfluor as oxidant as well as use of AcOH as additive were found to be essential in this reaction.

To access aryl trifluoromethyl sulfides from CF_3SAg, an alternative to the functionalization of aromatic components is based on alkyne activation by silver (Scheme 6.20).[138,139] In such an alternative, construction of the aromatic ring and incorporation of the trifluoromethylthio group occur in a single step. In the presence of silver salts, a thermal hexadehydro-Diels–Alder reaction of triyne leads to a silver-coordinated aryne, which can be trapped by fluorinated nucleophiles, especially trifluoromethylthiolate (Scheme 6.20a).[138] During this reaction, CF_3SAg plays both roles: alkyne activation and nucleophile. Starting with a similar alkyne activation, 2-alkynylbenzaldoxime can easily be converted into 1-(trifluoromethythio)isoquinolines in the presence of silver triflate as catalyst, a base, and *p*-methoxybenzenesulfonyl chloride (Scheme 6.20b).[139]

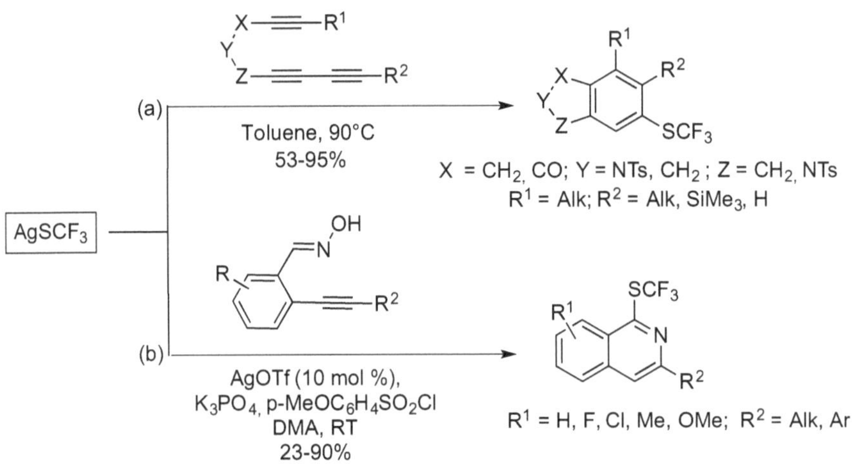

Scheme 6.20 Silver-catalyzed alkyne activation, aromatic ring formation, and trifluoromethylthiolation processes.

Of note, CF_3SAg can be involved not only in nucleophilic but also in radical or electrophilic (as precursor of the electrophile) processes (see Sections 3.2.7 and 3.3.2).

3.1.2 Trifluoromethylthiocopper(I)

Trifluoromethylthiocopper (CF_3SCu) was considered as a reagent in several nucleophilic trifluoromethylthiolation reactions. In contrast to its mercury[140,141] or silver analogs, this copper reagent allows performing the trifluoromethylthiolation of aryl iodides without any additives (Scheme 6.21).[142–145] It also gives better results during reactions involving diazonium salts (Scheme 6.21).[146] However, the same limitation was observed in these two types of reaction: electron-rich aryl halides give low yield. Of note, the trifluoromethylation of aryl iodides still proceeds when the copper reagent is formed in situ starting from bis(trifluoromethylthio)mercury(II).[147–149]

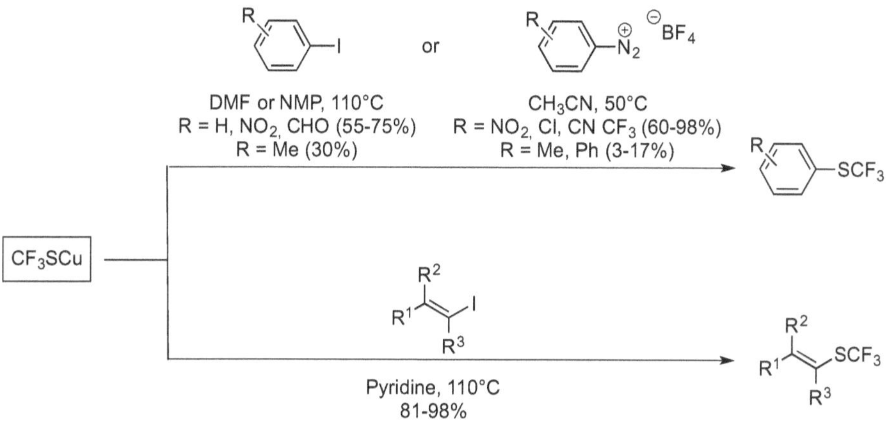

Scheme 6.21 Trifluoromethylthiolation on C(sp2) using CF_3SCu as reagent.

This trifluoromethylthiolation was extended to aryl bromides using a KI-CuI-alumina system followed by the addition of an alumina-supported CF_3SCu reagent to the in situ–formed iodoaryl.[150] Reactivity of vinyl iodides was also investigated[151,152]: trifluoromethylthiolation of electron-poor as well electron-rich vinyl iodides in pyridine was achieved in very good yields (Scheme 6.21).[152]

Significant advances involving trifluoromethylthiocopper were achieved thanks to its stabilization using bipyridine as the chelating ligand (Scheme 6.22).[153] Such complexes are air-stable[153] in contrast to unstabilized CF_3SCu.[154] More interestingly, aryl and heteroaryl trifluoromethysulfides are prepared from the corresponding aryl and heteroaryl halides in good yields and, in contrast to unstabilized CF_3SCu,[142,143] regardless of the electron-donating or electron-withdrawing aromatic substituents

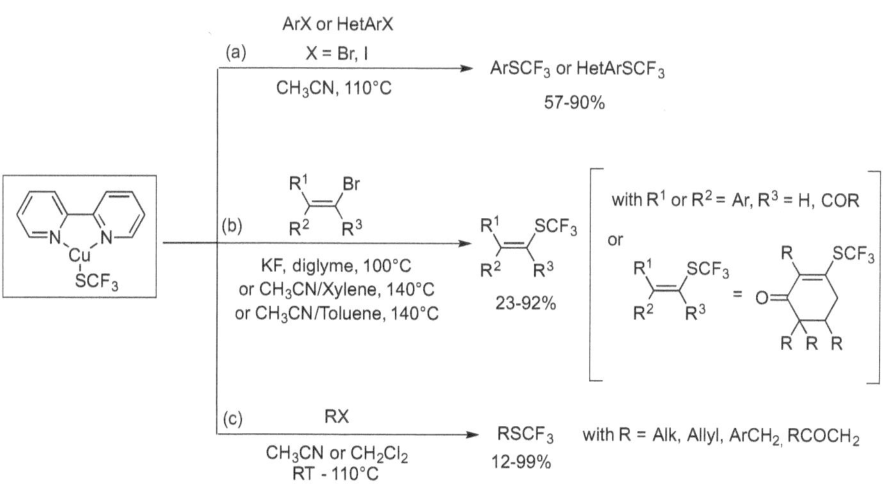

Scheme 6.22 Reactivity of $CF_3SCu(bpy)$ complexes.

(Scheme 6.22).[153] These complexes are able to convert alkyl,[155] allyl,[156] benzyl,[157] and vinyl[158–160] halides and also α-bromoketones[161] into the corresponding trifluoromethyl sulfides (Scheme 6.22). Of note, a trifluoromethylthiocopper complex stabilized by two triphenylphosphines instead of the bipyridine was reported to react with allylic bromides.[162]

However, the reactivity of CF_3SCu is still being investigated: reaction with diazoesters was reported and provides alkyl trifluoromethyl sulfides (Scheme 6.23a).[163] In parallel, similar transformations were reported to occur with an in situ–formed CF_3SCu obtained by the reaction of the silver salt CF_3SAg with CuI or CuCl.[129–131] Moreover, alcohols were converted to the corresponding alkyl trifluoromethyl sulfides using a $CF_3SCu–BF_3.Et_2O$ system (Scheme 6.23b).[164] These latter methods are thus complementary to trifluoromethylthiolation of alkyl halides.

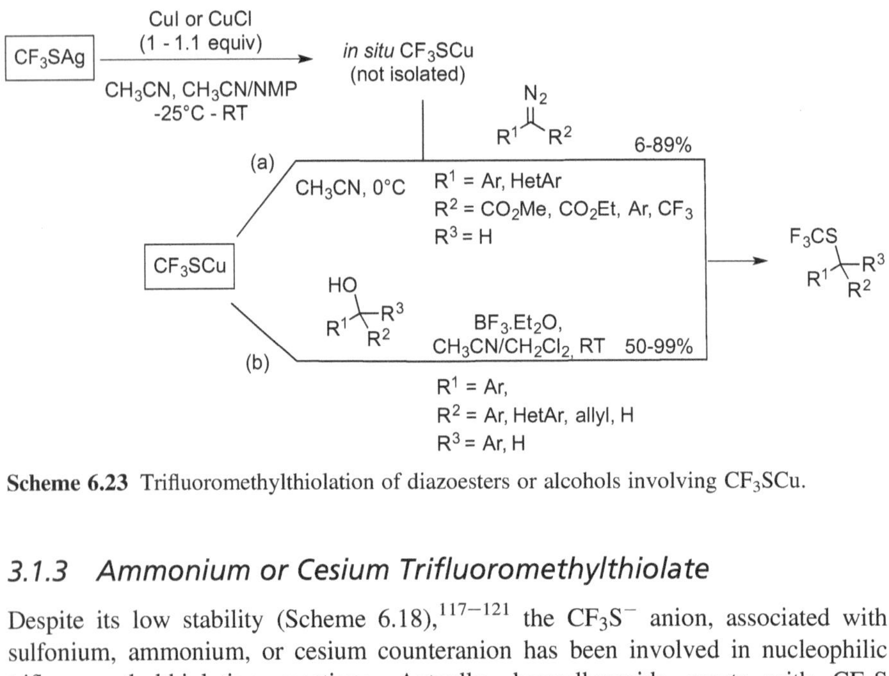

Scheme 6.23 Trifluoromethylthiolation of diazoesters or alcohols involving CF_3SCu.

3.1.3 *Ammonium or Cesium Trifluoromethylthiolate*

Despite its low stability (Scheme 6.18),[117–121] the CF_3S^- anion, associated with sulfonium, ammonium, or cesium counteranion has been involved in nucleophilic trifluoromethylthiolation reactions. Actually, benzylbromide reacts with $CF_3S[S(NMe_2)_3]$[165] and $(CF_3S)_2TDAE$[166] to provide the corresponding trifluoromethyl sulfide in 82% and 95% yield, respectively. Using cesium or tetramethylammonium trifluoromethylthiolate, few examples of nucleophilic trifluoromethylthiolations on iodoethane, benzylbromide, allylbromide, and acyl chlorides were reported.[167,168] Moreover, two nucleophilic aromatic substitutions with these cesium, tetramethylammonium,[167] or TDAE[166] salts on tetrafluoropyridine and 1-chloro-2,4-dinitrobenzene were described.

Of note, cesium or ammonium salts can be prepared by the reaction of Ruppert–Prakash reagent with elemental sulfur in the presence of a fluoride source (Scheme 6.24).[167] Moreover, theses salts are precursors of trifluoromethylthiomercury,

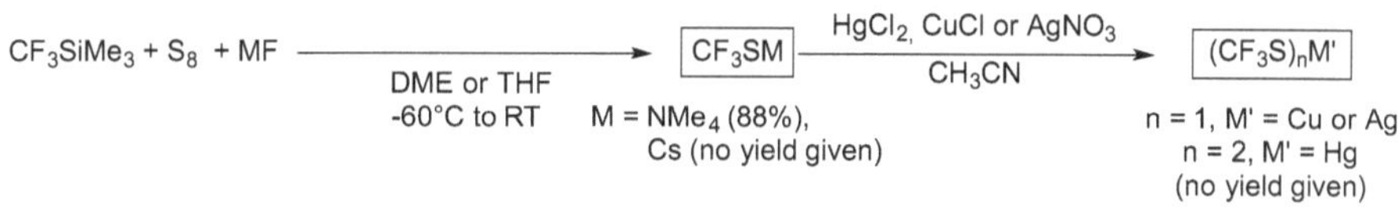

Scheme 6.24 Preparation of trifluoromethylthiolate salts from Ruppert–Prakash reagent.

trifluoromethylthiosilver, or trifluoromethylthiocopper. This route avoids use of bis(trifluoromethyl) disulfide or carbon disulfide, which were initially involved in the preparation of these metallic trifluoromethylthiolates.[122,140,147–149,169] Thus, Ruppert–Prakash reagent represents a precursor of several nucleophilic and also radical trifluoromethylthiolation reagents.[167]

As it can be expected, without any metal, no trifluoromethylthiolation of aryl halides occurs.[167] Nevertheless, a nickel-catalyzed trifluoromethylation of aryl iodides or bromides was reported by Vicic et al. in 2012 (Scheme 6.25a).[170] As they found that aryl chlorides were unreactive in those conditions, they developed an oxidative coupling of tetramethylammonium trifluoromethylthiolate and aryl boronic acid (Scheme 6.25b), the latter being prepared from the corresponding aryl chlorides.[171] In 2015, the clue was found by Schoenebeck et al: to convert aryl chlorides into aryl trifluoromethyl sulfides in good yields via a nickel-catalyzed trifluoromethylatiolation coupling, wide-bite-angle phosphines such as xantphos, 1,1′-ferrocenediyl-bis (diphenylphosphine), or BINAP, are necessary (Scheme 6.25c).[172] Moreover, addition

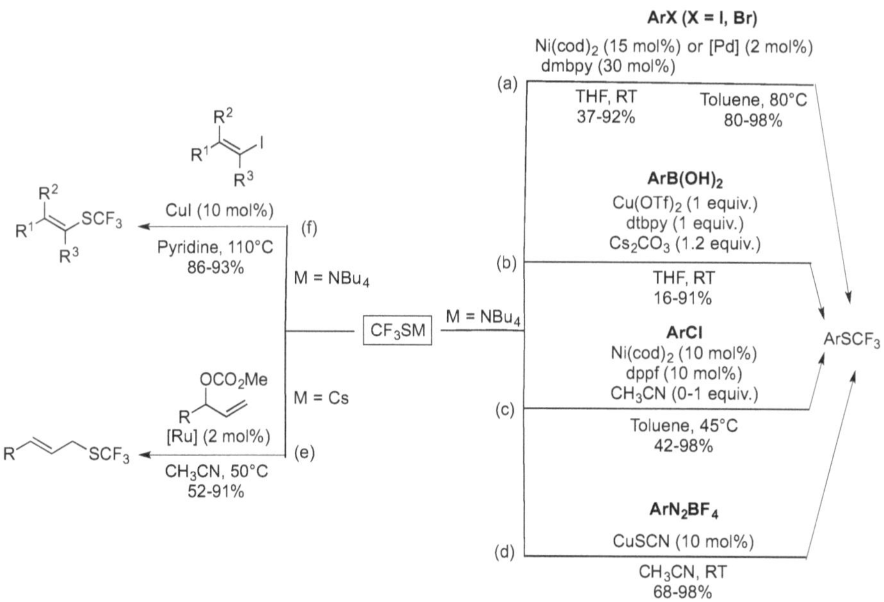

Scheme 6.25 Catalyzed reactions using tetramethylammonium or cesium trifluoromethylthiolate [cod, 1,5-cyclooctadiene; dmbpy, 4,4-dimethoxybipyridine; dtbpy, 4,4′-di-tert-butyl-2,2′-bipyridine; dppf, 1,1′-Ferrocenediyl-bis(diphenylphosphine)]. *RT*, Room temperature.

of a nitrile in the reaction leads to in situ formation of a more reactive nickel catalyst and the yields are increased.[172] In parallel, the same group developed trifluoromethylthiolation of aryl iodides or bromides catalyzed by an easy-to-recover dinuclear Pd complex (Scheme 6.25a).[173] Moreover, a series of aryl trifluoromethyl sulfides have been prepared thanks to Sandmeyer-type reaction using tetramethylammonium trifluoromethylthiolate and copper thiocyanate as catalysts (Scheme 6.25d).[174] Compared with the reaction of diazonium salts with trifluoromethylthiocopper (Scheme 6.21),[146] and although a trifluoromethylthiocopper is proposed as an intermediate in the present mechanism, it should be noticed that the reaction is not limited to diazonium-bearing electron-withdrawing substituents.[174] In addition to trifluoromethylation of aromatic compounds, Cu-catalyzed trifluoromethylthiolation of vinyl iodides was reported using tetramethylammonium trifluoromethylthiolate (Scheme 6.25f), similar to the one based on CF_3SCu.[152] In 2014, catalyzed allylic trifluoromethylthiolation received attention: a Ru-catalyzed reaction of allylic carbonates with cesium trifluoromethylthiolate was developed (Scheme 6.25e).[175] This reaction was found to be regioselective: irrespective of the structure of starting allylic carbonate (linear or branched), the obtained allylic trifluoromethyl sulfides is linear.

3.1.4 Difluorothiophosgene

An alternative to avoid isolation of the less stable trifluoromethylthiolate anion involves generating and using it directly without any isolation. The first strategy was based on the existing equilibrium between difluorothiophosgene and the trifluoromethylthiolate anion (Scheme 6.17).[117–121] It was applied to nucleophilic aromatic substitution and also extended to the addition of fluoride to other difluorothiophosgene analogs (Scheme 6.26). Such SNAr (nucleophilic aromatic substitution) trifluoromethylthiolation was limited to electron-deficient substrates.[117–121]

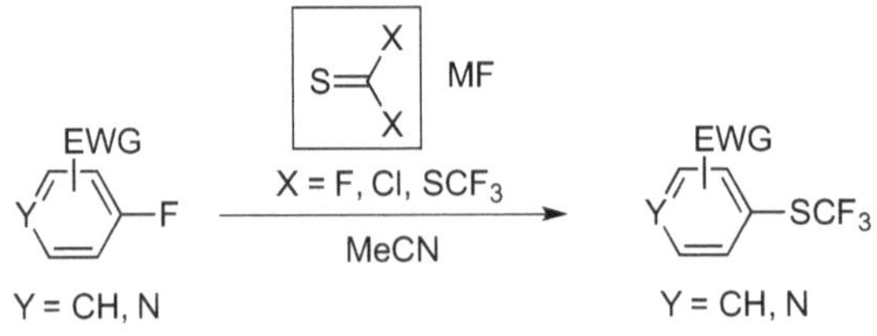

Scheme 6.26 Trifluoromethylthiolation based on thiophosgene derivatives.

3.1.5 CF_3^- Anion/Elemental Sulfur

Another strategy was based on the use of elemental sulfur, which can generate in situ a thiolate intermediate. Indeed, elemental sulfur can react with the trifluoromethylating reagent to generate the trifluoromethylthiolate anion (Scheme 6.24),[167] or with a copper complex and then a nucleophilic species to form the corresponding copper thiolate.[176] Actually, both pathways were proposed to occur. In the presence of copper iodide and elemental sulfur, methylfluorosulfonyldifluoroacetate reacts with aryl iodides to provide aryl trifluoromethyl sulfides (Scheme 6.27e).[177] Trifluoromethylthiocopper, formed by the addition of trifluoromethyl anion to sulfur, was proposed as a key intermediate. A system based on Ruppert–Prakash reagent as the trifluoromethyl donor, instead of methylfluorosulfonyldifluoroacetate, and elemental sulfur was applied to metal-free trifluoromethylthiolation of terminal alkynes (Scheme 6.27c).[178] Although the mechanism is still not completely understood, trifluoromethylthiolate was proposed to be an intermediate in that reaction. Moreover, this system CF_3SiMe_3/S_8 was also considered as a precursor of trifluoromethylthiolate during copper-catalyzed allylic or propargylic trifluoromethylthiolation,[179,180] during copper-catalyzed substitution of α-bromoketones[181,182] or during trifluoromethylthiolation of Morita–Baylis–Hillman carbonates (Scheme 6.27a, b, and d).[183] A copper-catalyzed coupling of aryl boronic using this system CF_3SiMe_3/S_8 was also developed

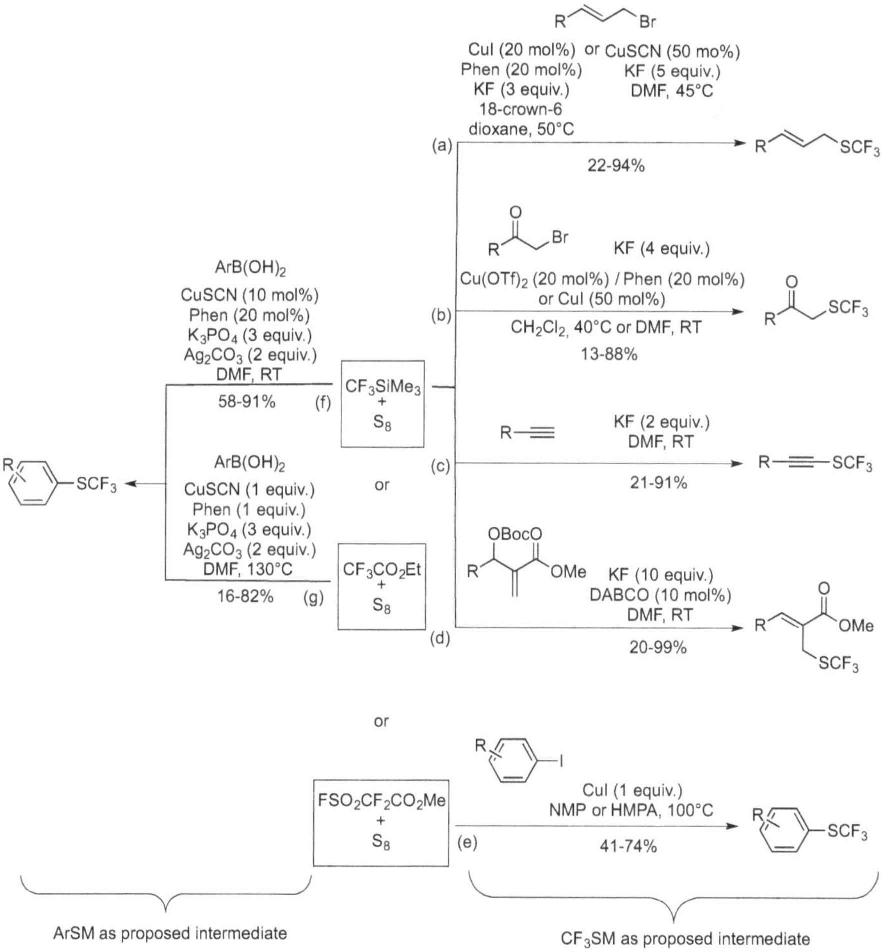

Scheme 6.27 Trifluoromethylthiolation based on a system CF_3SiMe_3 (or analogs)/S_8. *HMPA*, Hexamethylphosphoramide; *NMP*, N-Methylpyrrolidinone; *RT*, room temperature.

(Scheme 6.27f).[176] Yields of aryl trifluoromethyl sulfides are quite good regardless of the substituents on the aromatic ring. In this case, a comparison of the reactivity of CF_3SCu and ArSCu suggests that copper thiophenolate is the key intermediate here. A similar system was developed by using sodium trifluoroacetate instead of Ruppert–Prakash reagent as the trifluoromethyl donor, but requires a stoechiometric amount of copper (Scheme 6.27g).[184]

3.1.6 Bis(trifluoromethyl)disulfide and O-Octadecyl-S-trifluorothiolcarbonate

Due to the high electronegativity of the fluorine atoms, the trifluoromethanesulfenyl group can be considered as a leaving group. Thus, the addition of nucleophiles to trifluoromethylated compounds can liberate the expected anion, which can be directly used. This way was initially reported using toxic bis(trifluoromethyl)disulfide[185] as CF_3S^- donor in an Arbuzov-type reaction of the disulfide with a phosphite or a diamidoalkylphosphite.[186,187] As the latter are obtained from the corresponding alcohols, the whole process can be regarded as a two-step trifluoromethylthiolation of alcohols. In 2013, the trifluoromethylthiolate anion was generated by the addition of a nucleophile (pyrrolidine, DABCO (1,4-Diazabicyclo[2.2.2]octane), or dimethylamide liberated in situ from gramines) to the stable crystalline *S*-trifluorothiolcarbonate (Scheme 6.28).[183,188] This method was applied to trifluoromethylthiolation of gramines,[188]

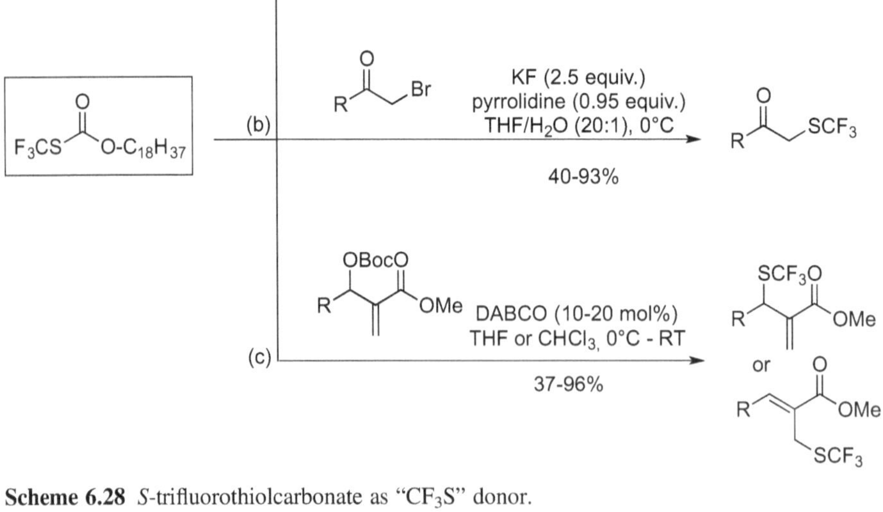

Scheme 6.28 *S*-trifluorothiolcarbonate as "CF_3S" donor.

α-bromoketones,[188] or Morita–Baylis–Hillman carbonates (Scheme 6.28).[183,189] Conditions plays a crucial role as the kinetic products or thermodynamic products can be isolated depending on the solvent or reaction time.[183,189] Of note, trifluoromethylthiolation of Morita–Baylis–Hillman carbonates occurs faster with *S*-trifluorothiolcarbonate than with the system CF_3SiMe_3-S_8-KF.[183] Actually, a dual role was attributed to DABCO: it activates the *S*-trifluorothiolcarbonate, and also generates a more active trifluoromethylthiolating species.[189] Indeed, no reaction was observed when *S*-trifluorothiolcarbonate was replaced by ammonium trifluoromethylthiolate alone.

3.1.7 From an Electrophilic Reagent: Umpolung Reactivity of Trifluoromethanesulfenamides

Although described as efficient electrophilic trifluoromethylthiolating reagents (see Section 3.2.2), trifluoromethanesulfenamides can also perform nucleophilic trifluoromethylthiolations. In the presence of Bu_4NI, the transient formation of CF_3SI has been postulated. This compound, under iodide activation can generate CF_3S^- species, which can be, then, engaged in nucleophilic substitutions (Scheme 6.29).[190]

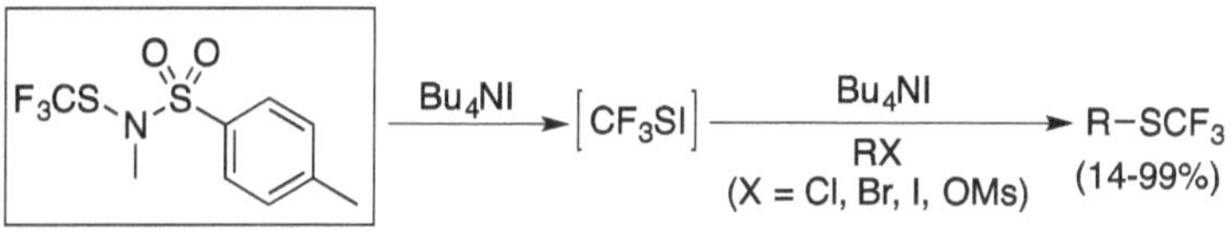

Scheme 6.29 Umpolung trifluoromethylthiolation with trifluoromethanesulfenamide.

3.2 Electrophilic Trifluoromethylthiolation

For a long time, electrophilic trifluoromethylthiolation has been the prerogative of CF_3SCl.[14–16] However, in these last years some new efficient reagents have been developed to circumvent the high toxicity of CF_3SCl.[185]

3.2.1 Trifluoromethanesulfenyl Chloride and bis-Trifluoromethyl Disulfide

Despite the high toxicity of gaseous trifluoromethanesulfenyl chloride, this reagent has been, for a long time, the only possibility to efficiently carry out direct electrophilic trifluoromethylthiolations. CF_3SCl can add onto olefins[191] and perform aromatic electrophilic substitutions with arenes[192–194] or heteroarenes.[193,195–201] It also reacts with Grignard reagents[148,195,202,203] and with enolates[204–208] or enamines[209] (Scheme 6.30a–d). It should be noticed that due to the high electrophilicity of CF_3SCl, some by-products of bis-trifluoromethanesulfenylation are often observed. In 2015, in situ formation of CF_3SCl starting from stable trifluoromethylthiolating reagent (Munavalli's reagent, see Section 3.2.3) to trifluoromethylthiolate N-heteroarenes has been reported (Scheme 6.30e).[210]

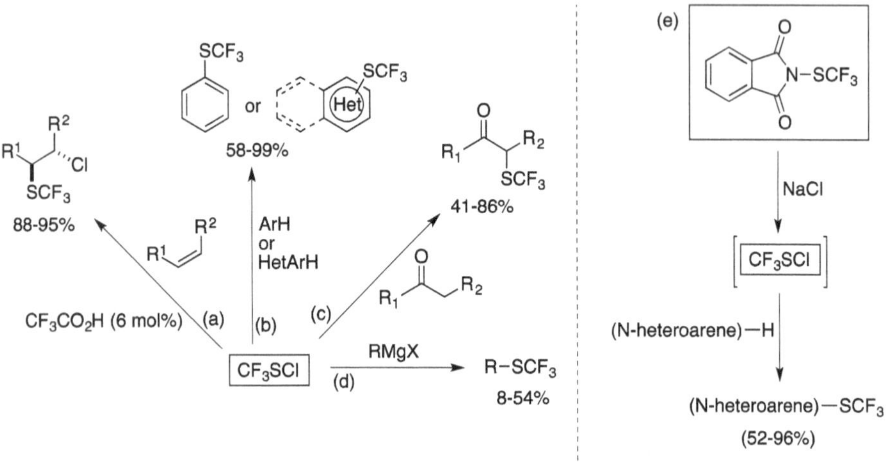

Scheme 6.30 Trifluoromethylthiolation with trifluoromethanesulfenyl.

In the mid-20th century, some attempts to replace the toxic gas CF_3SCl by CF_3SSCF_3 had been unsuccessful[202]; however, in 2012 C(sp^2)−H bond trifluoromethanesulfenylation has been successfully performed by using CF_3SSCF_3 (Scheme 6.30).[211] However, bis-trifluoromethyl disulfide is a volatile liquid (boiling point = 34°C) as toxic as trifluoromethanesulfenyl chloride,[185] which can restrict its use.

The in situ generation of CF_3SSCF_3, from CF_3SO_2Na, has been, also, described to perform electrophilic trifluomethylthiolations of some indoles, two pyrroles, and two enamines (Scheme 6.31).[212]

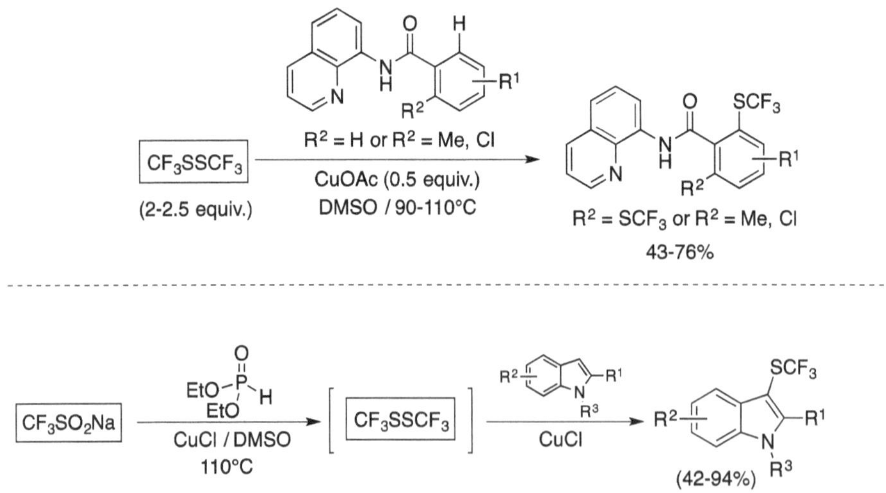

Scheme 6.31 Trifluoromethylthiolation with CF_3SSCF_3.

3.2.2 *Trifluoromethanesulfenamides (Billard's Reagents)*

Since 2009, trifluoromethanesulfenamide reagents have been described as efficient equivalents of "CF_3S^+" species.[213] These compounds are easily synthesized from readily commercially available CF_3SiMe_3, DAST ((Diethylamino)sulfur trifluoride), and amines in large scale (up to 100 g).[214] To this day, three reagents have been efficiently involved in various electrophilic trifluoromethylthiolations (Scheme 6.32). Aromatic and heteroaromatic trifluoromethylthiolations have been performed through electrophilic substitutions (Scheme 6.32a),[215–218] nucleophilic attacks with organometallic species (Scheme 6.32b–d, and h),[218–223] or cross-coupling reactions with aryl boronic acids (Scheme 6.32l).[224] Organometallic attacks can be extended to alkyls or alkynes, whereas coupling with boronic acids is also possible with vinylic reagents.

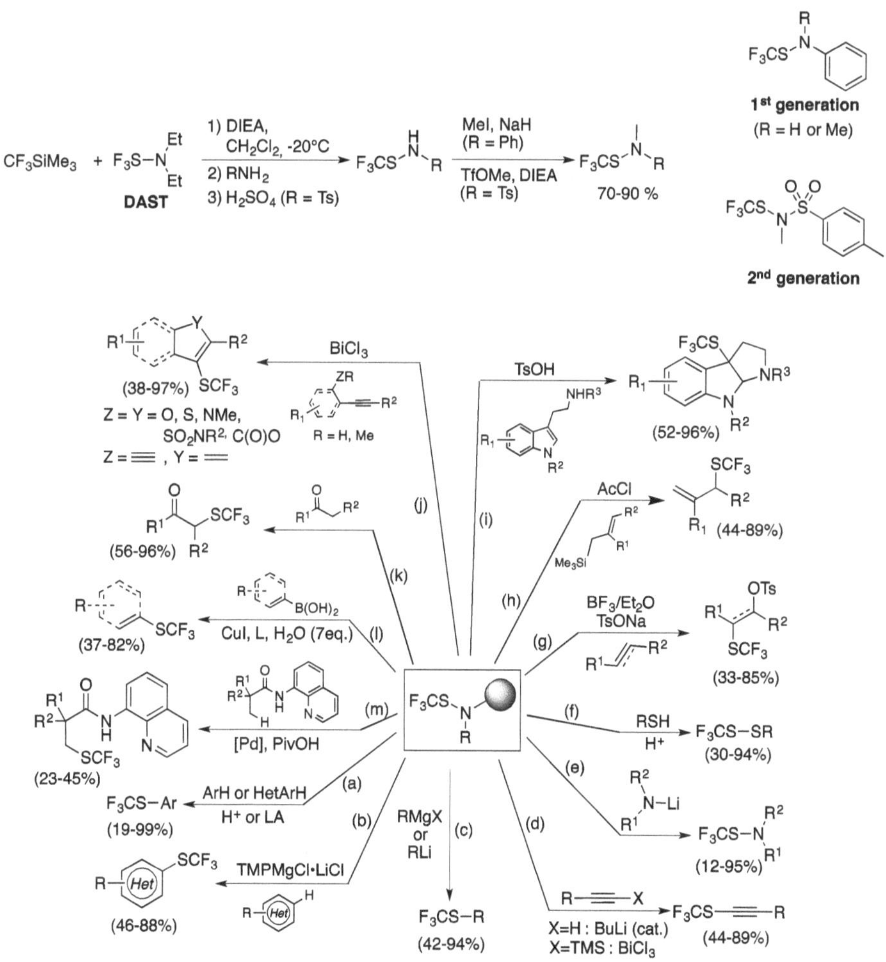

Scheme 6.32 Synthesis and reactivity of trifluoromethanesulfenamides.

Electrophilic additions onto alkenes and alkynes lead to functionalized compounds (Scheme 6.32g).[213,218] Cyclization reactions afford access to various heterocycles (Scheme 6.32j).[225–231] Reactions with enols or enolates open the way to various valuable building blocks (Scheme 6.32k).[232–234] Trifluoromethylthiolation of $C(sp^3)$—H bonds has also been successful but with modest yields (Scheme 6.32m).[235] Finally, amines and thiols can also react with these reagents to give, respectively, other trifluoromethanesulfenamides (Scheme 6.32e)[218,236] and trifluoromethyl disulfides (Scheme 6.32f).[237] In general, the second-generation trifluoromethanesulfenamides, bearing a tosyl group on nitrogen atom, appear to be more reactive.[218]

Of note, trifluoromethanesulfenamides can be also involved in nucleophilic processes (see Section 3.1.7).

3.2.3 N-Trifluoromethylthiophtalimide (Munavalli's Reagent) and N-Trifluoromethylsuccinimide

In 2000, *N*-trifluoromethylthiophtalimide was reported to react with two enamines to obtain the corresponding trifluoromethylthioketones. However, the synthesis of this reagent from CF_3SCl constituted a major drawback. Nevertheless, in 2013, this reagent reappeared to perform catalytic asymmetric trifluoromethanesulfenylation of cyclic ketoesters[238](Scheme 6.33j). The renewal of interest for this reagent has led to the development of a more convenient synthesis starting not from a toxic reagent but from trifluoromethylthiolate anions CF_3SAg[239] or CF_3SCu[240] (Scheme 6.33b and c). Finally, this reagent can also react with amines and thiols to give, respectively, other trifluoromethanesulfenamides and trifluoromethyl disulfides (Scheme 6.33g and h)[241]; with aryl and vinyl boronic acids through cross-coupling reactions (Scheme 6.33e)[239,240]; and with terminal alkynes (Scheme 6.33f)[240] and oxindoles (Scheme 6.33i).[242] As with previous trifluoromethanesulfenamides, some trifluoromethylthiolations of $C(sp^3)$—H bonds have been also realized but always with modest yields (Scheme 6.33d).[235]

A derivative of the *N*-trifluoromethylthiophtalimide, namely, *N*-trifluoromethylthiosuccinimide, was also synthesized, from CF_3SCl, in 1996,[243] but used as trifluoromethylthiolating reagent only in 2014. Thus, aryl C—H bond trifluoromethylthiolation has been performed. It is noteworthy that an alternative synthesis from CF_3SAg has been also developed (Scheme 6.34).[244] The asymmetric synthesis of trifluoromethylated dithioketals has also been described, but the reaction time is 6 days.[245] The in situ generation of this reagent has been described to trifluoromethylthiolate terminal alkynes.[246]

Of note, *N*-trifluoromethylthiophtalimide can be also involved in radical processes (see Section 3.3.3).

3.2.4 Trifluoromethylsulfenates (Shen's Reagent)

First described as a hypervalent iodine derivative, a "CF_3S-Togni's reagent,"[247] the structure has been later corrected as being a trifluoromethanesulfenate.[248] These reagents appear to be efficient with a broad scope of substrates (Scheme 6.35).[249]

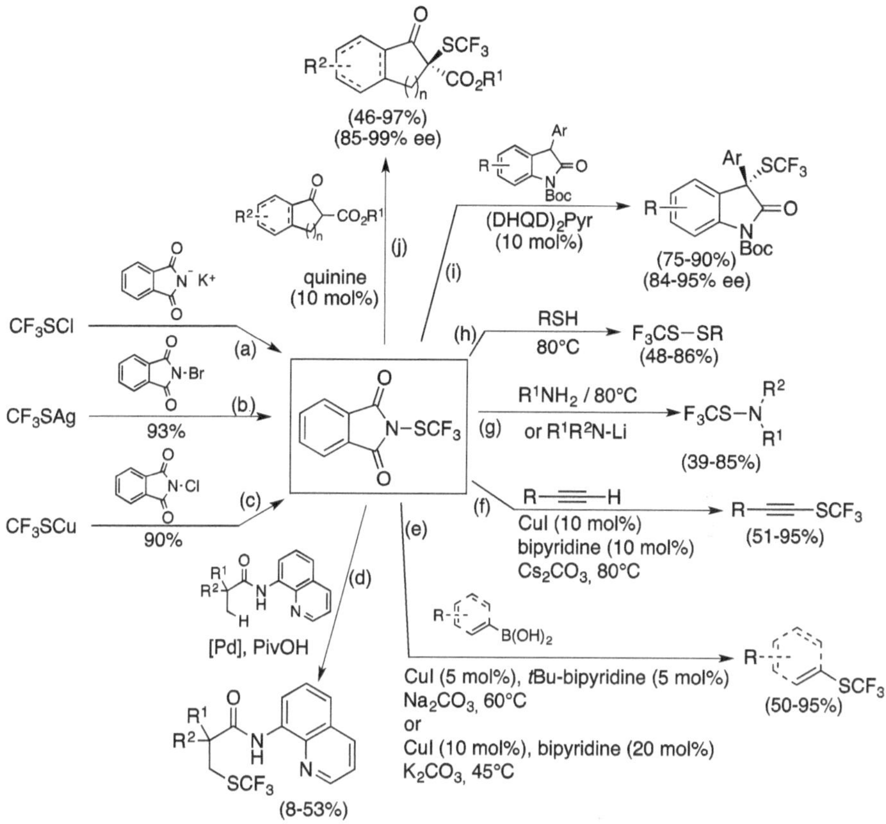

Scheme 6.33 Synthesis and reactivity of *N*-trifluoromethylthiophtalimide.

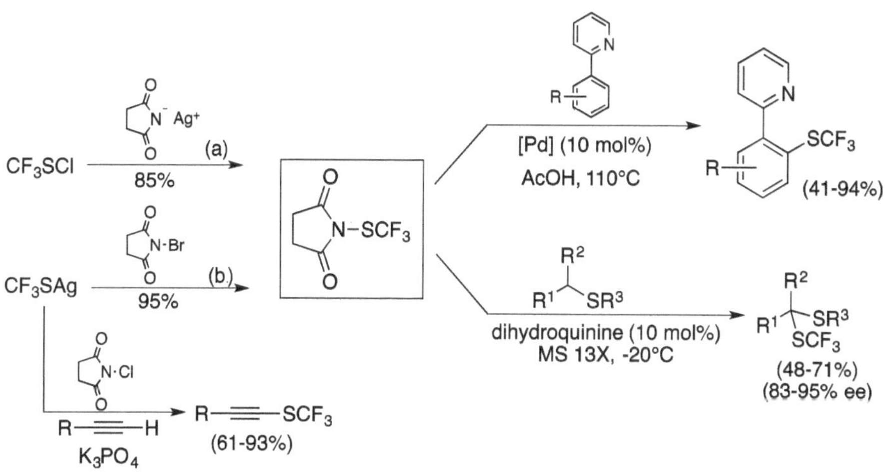

Scheme 6.34 Synthesis and reactivity of *N*-trifluoromethylthiosuccinimide.

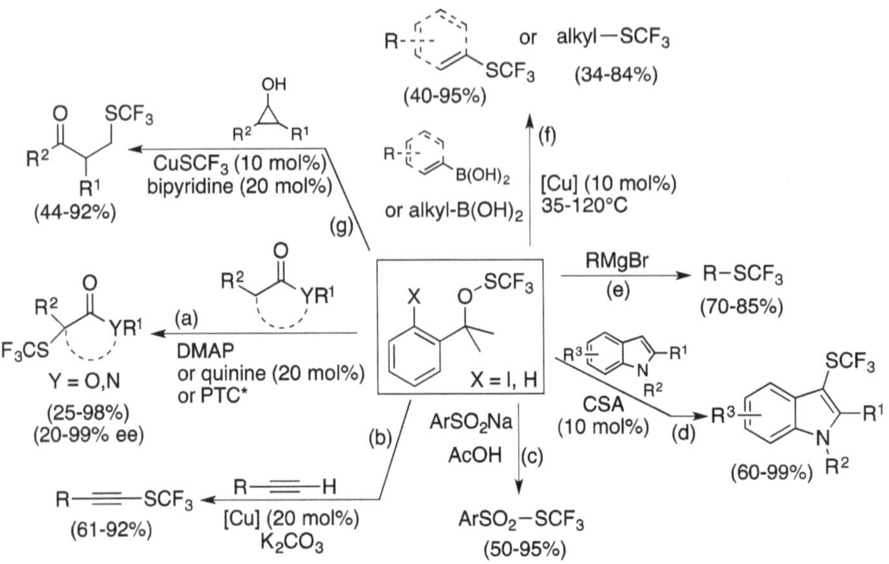

Scheme 6.35 Reactivity of trifluoromethanesulfenates.

Thus, trifluoromethylthiolations of indoles,[250,251] Grignard reagents,[251] and terminal alkynes[247,251] have been performed. Cross-coupling reactions have been successful with aryl, vinyl, and even alkyl boronic acids.[247,251,252] Racemic and asymmetric trifluoromethylthiolations of ketoesters and oxindoles have been also achieved.[247,251,253–255] By reaction with sodium sulfinates, various trifluoromethylthiolated sulfonates have been obtained.[251] Cyclopropanols have been opened with this reagent to afford β-SCF_3-carbonyl compounds.[256]

Of note, trifluoromethanesulfenate can be also involved in radical processes (see Section 3.3.3).

3.2.5 N-Trifluoromethylthiosaccharin (Shen's Reagent)

Inspired by the *N*-trifluoromethylphtalimide, another reagent, more reactive, arising from saccharin has been described. This compound is again synthesized starting from CF_3SAg.[257] The reactivity of this reagent covers the same panel of substrates as the other reagents. However, because of its higher reactivity, it covers a broader field of application close to trifluoromethanesulfenamides.[249] Aromatic and heteroaromatic trifluoromethylthiolations have been performed via Friedel–Crafts reactions (Scheme 6.36a)[257,258] or C–H activation[259] (Scheme 6.36i). Terminal alkynes can also be functionalized through a copper-catalyzed reaction (Scheme 6.36b).[257] α-Trifluoromethylation of various carbonyl compounds is also efficient with this reagent (Scheme 6.36h).[257] Heteronucleophiles (alcohols, thiols, and amines) also react (Scheme 6.36e–g).[257] In particular, starting from alcohols, other trifluoromethanesulfenates could be obtained (Scheme 6.36e). Finally, electrophilic

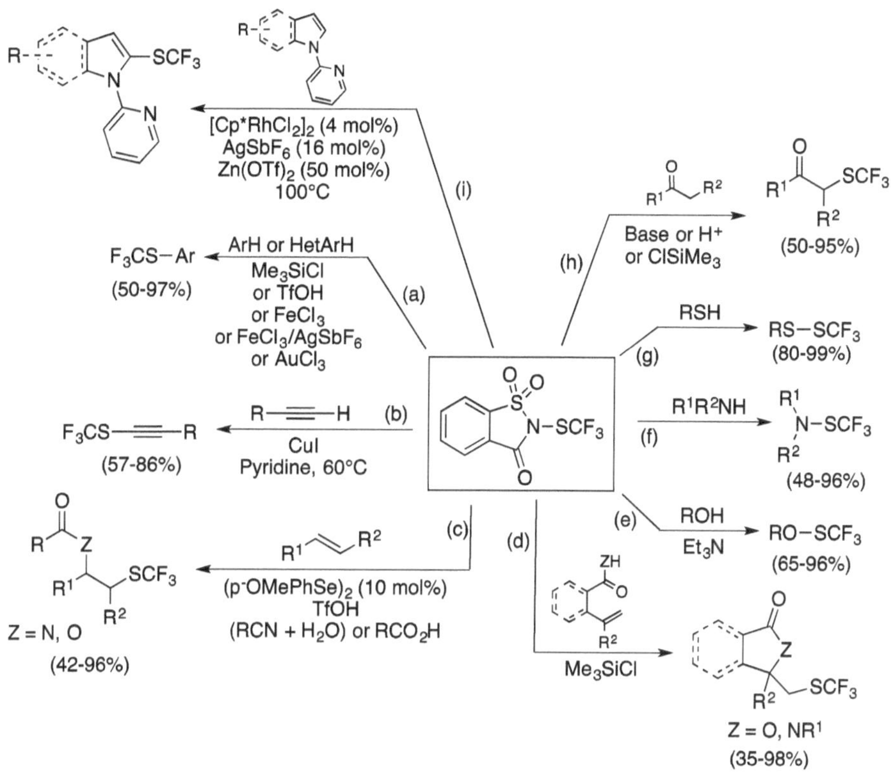

Scheme 6.36 Reactivity of *N*-trifluoromethylthiosaccharin.

additions onto double bonds have been carried out to obtain functionalized products (Scheme 6.36c and d).[260,261] By using this strategy, lactones and lactames have been synthesized (Scheme 6.36d).[260]

3.2.6 *Trifluoromethanesulfonyl Hypervalent Iodonium Ylide (Shibata's Reagent)*

To avoid the use of DAST or CF_3SM to synthesize electrophilic reagent, a trifluoromethanesulfonyl hypervalent iodonium ylide reagent has been described.[262] Indeed, this compound can be obtained starting from sulfones. To this day, the reactivity of this reagent appears less extended than other reagents (Scheme 6.37). Enamines, silyl enol ethers, and two ketoesters led to α-trifluoromethylthiocarbonyl compounds or their masked forms.[262] Friedel–Crafts reactions are limited to indoles[262] and pyrroles.[263] Few cross-coupling reactions with aryl and vinyl boronic acids have also been realized.[264] Allylsilanes give corresponding allylic trifluoromethylthiolated products.[265] Finally, amines can also react to provide trifluoromethanesulfenamides.[266]

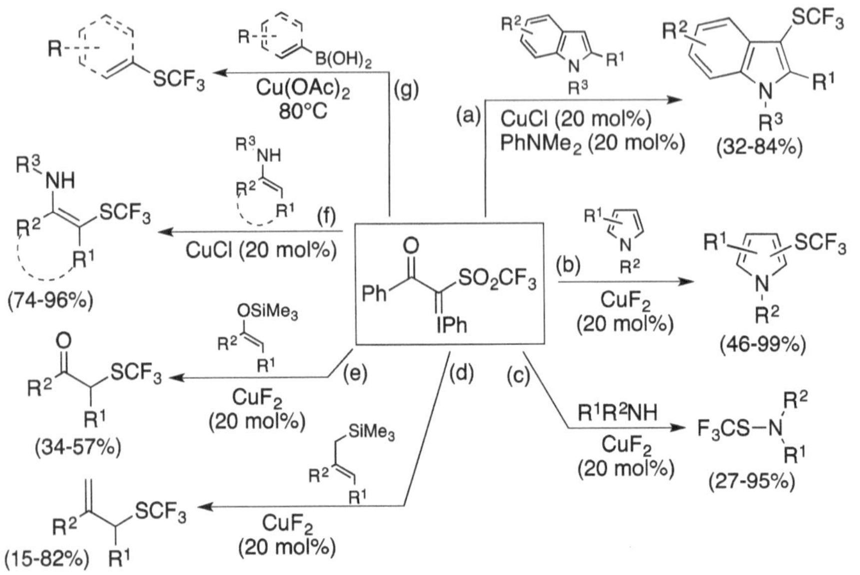

Scheme 6.37 Reactivity of trifluoromethanesulfonyl hypervalent iodonium ylide.

3.2.7 From CF_3SAg

In situ generation of CF_3S^+ species, which can react with nucleophiles, has been described by mixing CF_3SAg with trichloroisocyanuric acid.[267] Even if the effective active SCF_3 intermediate has not been identified, this strategy has been applied to the asymmetric trifluoromethylthiolation of oxindoles[267] and the synthesis of 3-((trifluoromethyl)thio)-4H-chromen-4-one[268] (Scheme 6.38). This strategy looks like the in situ formation of *N*-trifluoromethylthiosuccinimide by mixing also CF_3SAg and a chlorinating/oxidant reagent, namely, *N*-chlorosuccinimide[246] (see Section 3.2.3).

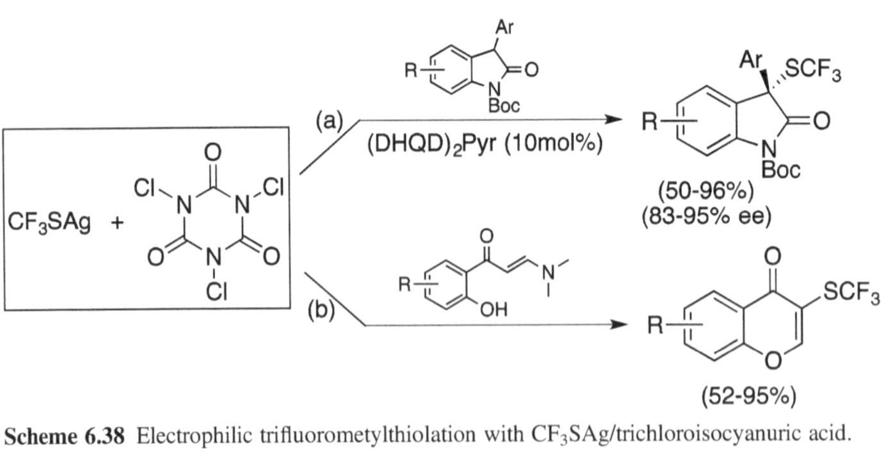

Scheme 6.38 Electrophilic trifluorometylthiolation with CF_3SAg/trichloroisocyanuric acid. *ee*, Enantiomeric excess.

3.3 Radical Trifluoromethylthiolation

Historically, the first reagents considered during direct radical trifluoromethylthiolations were trifluoromethanethiol[269] and trifluoromethanesulfenyl chloride.[270] A usual reagent considered in nucleophilic trifluoromethylthiolation, trifluoromethanethiosilver, and two usual electrophilic reagents were, also, involved in radical pathways.

3.3.1 Historical Reagents: Trifluoromethanethiol and Trifluoromethanesulfenyl Chloride

Initially, reactions of trifluoromethanethiol, CF_3SH,[269] and trifluoromethanesulfenyl chloride, CF_3SCl,[270] with alkenes under UV irradiation, X-ray activation, or in the presence of radical initiators were investigated. The expected trifluoromethyl sulfides were obtained in both cases in various yields (Scheme 6.39a). Shortly afterward, radical trifluoromethylthiolation was extended to unfunctionalized alkanes,[271] ketenes,[272] acrylates,[272] and trimethylsilyl enol ethers[273] with CF_3SCl and to methylpropiolate with CF_3SH.[272] However, the synthetic utility of such methods remains low: a mixture of products was obtained in most cases (Scheme 6.39).

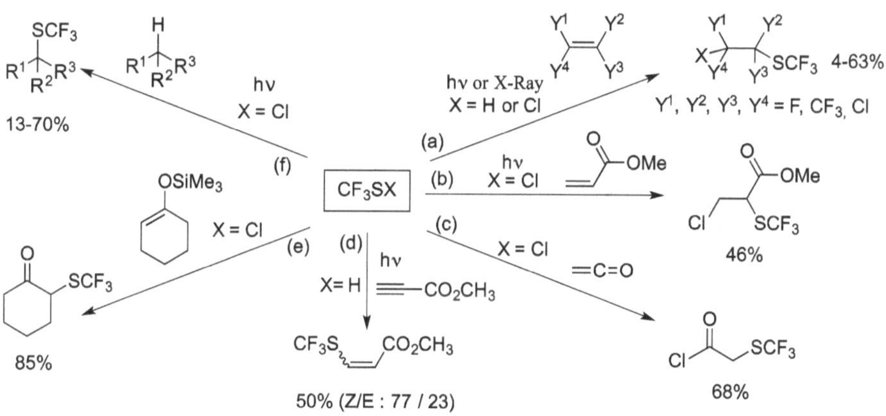

Scheme 6.39 Radical trifluoromethylthiolation involving CF_3SCl or CF_3SH.

3.3.2 Trifluoromethylthiosilver(I)

Initially developed for nucleophilic trifluoromethylthiolation (see Section 3.1.1), trifluoromethythiosilver was also involved in radical pathways: the trifluoromethythiolate anion can be oxidized, usually by $K_2S_2O_8$, to provide the corresponding $CF_3S^{\cdot}$ radical (Scheme 6.40). This system CF_3SAg/oxidant was involved in copper-mediated trifluoromethylthiolation on terminal alkenes (Scheme 6.40b)[274] or on benzylic position (Scheme 6.40a).[275] More recently, a copper-free $C(sp^3)$-H trifluoromethylthiolation was reported to occur using the system CF_3SAg/oxidant (Scheme 6.40c).[276,277] Thanks to the formation of new radical after the first addition of CF_3S

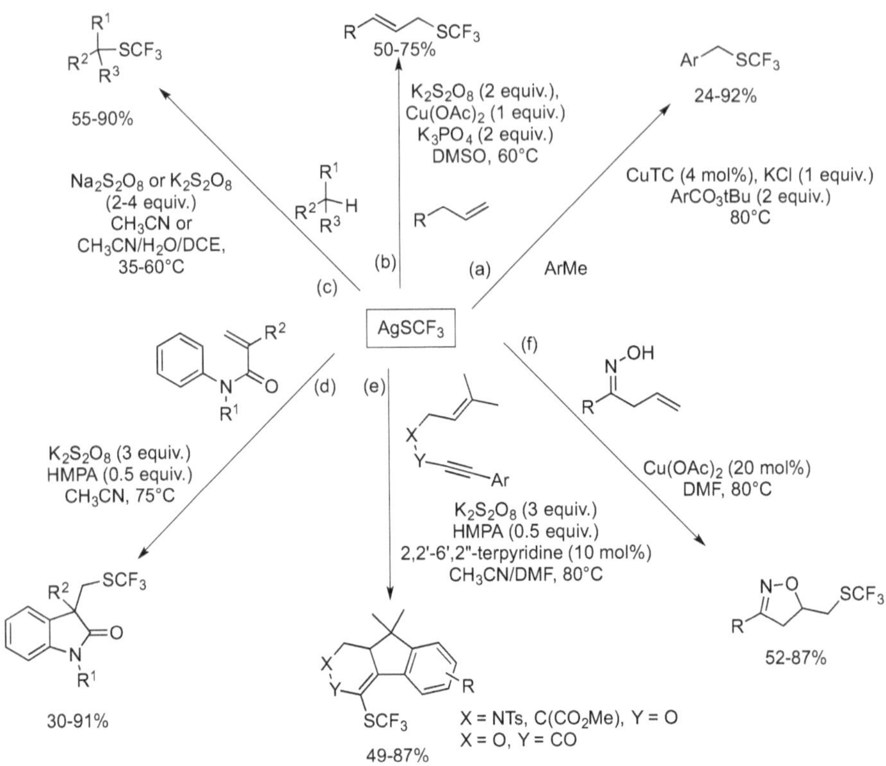

Scheme 6.40 Radical trifluoromethylthiolation using $AgSCF_3$.

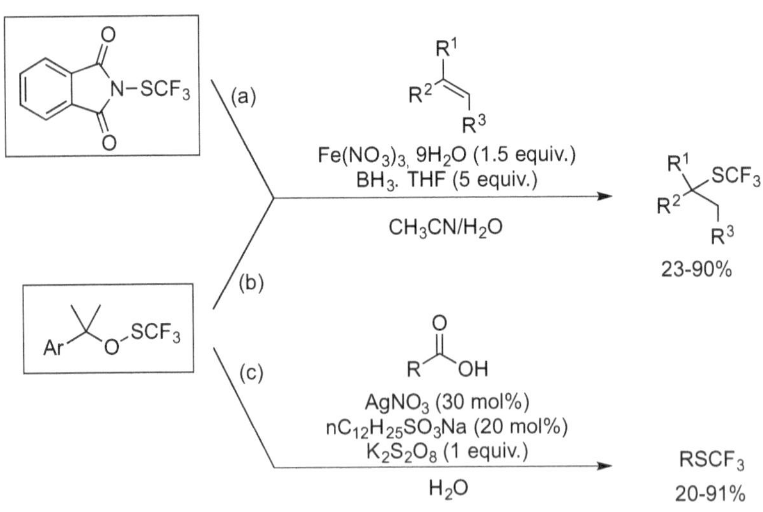

Scheme 6.41 Radical trifluoromethylthiolation using Munavalli's or Shen's reagents.

on a multiple bond, cascade processes including radical trifluoromethylthiolation–cyclization were developed to offer access to sophisticated scaffolds (Scheme 6.40e).[278–282]

3.3.3 Electrophilic Reagents in Radical Pathways

In addition, two reagents usually considered electrophilic were involved in radical trifluoromethylations of alkenes[283] or of carboxylic acids[284] to provide alkyl trifluoromethyl sulfides (Scheme 6.41).

4. Conclusions

Some methods to introduce the trifluoromethanesulfenyl group (CF_3S) onto organic molecules have been reported since the 1950's. However, some of the historical methods suffered from limitations such as low substrate scope, low reaction selectivity, and/or toxicity of the involved reagents. That's why, despite the promising properties of CF_3S molecules and their potential application in various fields, this class of compounds remained underdeveloped. Nevertheless, the continuously increasing demand of compounds with specific properties has contributed to a renewed interest in the synthesis of trifluoromethyl sulfides. From 2009 to 2015, a huge effort (growth rate of publications of around 40% over 6 years[285]) has been made to develop new efficient processes to synthesize CF_3S molecules. At present, previous limitations have been surpassed. A large panel of methods with easy-to-handle reagents is now available. Different modern methodologies of organic chemistry especially transition metal catalysis have been applied to perform trifluoromethylthiolations and to extend the substrate scope. Moreover, introduction of the CF_3S group at the late stage of synthesis can now be envisaged thanks to the mild reaction conditions and functional group tolerance of several methods. With such methods in chemists' hands, it could be expected that, similar to trifluoromethylated compounds in the late 1990's, applications of CF_3S molecules will benefit from all these innovative achievements in synthesis and thus will be developed in the near future.

References

1. Dolbier, W. R., Jr. *J. Fluor. Chem.* **2005,** *126,* 157–163.
2. Wang, J.; Sánchez-Roselló, M.; Aceña, J. L.; del Pozo, C.; Sorochinsky, A. E.; Fustero, S.; Soloshonok, V. A.; Liu, H. *Chem. Rev.* **2014,** *114,* 2432–2506.
3. Smart, B. E. *J. Fluor. Chem.* **2001,** *109,* 3–11.
4. Hird, M. *Chem. Soc. Rev.* **2007,** *36,* 2070–2095.
5. Purser, S.; Moore, P. R.; Swallow, S.; Gouverneur, V. *Chem. Soc. Rev.* **2008,** *37,* 320–330.
6. Gillis, E. P.; Eastman, K. J.; Hill, M. D.; Donnelly, D. J.; Meanwell, N. A. *J. Med. Chem.* **2015,** *58,* 8315–8359.

7. Muller, K.; Faeh, C.; Diederich, F. *Science* **2007,** *317,* 1881–1886.
8. Hansch, C.; Leo, A.; Taft, R. W. *Chem. Rev.* **1991,** *91,* 165–195.
9. Avdeef, A. *Absorption and Drug Development: Solubility, Permeability, and Charge State;* Wiley: Hoboken, 2012.
10. Clark, D. E. *Drug Discov. Today* **2003,** *8,* 927–933.
11. Kelder, J.; Grootenhuis, P. D. J.; Bayada, D. M.; Delbressine, L. P. C.; Ploemen, J.-P. *Pharm. Res.* **1999,** *16,* 1514–1519.
12. Norinder, U.; Haeberlein, M. *Adv. Drug Del. Rev.* **2002,** *54,* 291–313.
13. Smith, D. A.; Allerton, C.; Kubinyi, H.; Walker, H.; Walker, D. K.; Mannhold, R.; Folkers, G. *Pharmacokinetics and Metabolism in Drug Design;* Wiley: Weinheim, 2012.
14. Boiko, V. N. *Beilstein J. Org. Chem.* **2010,** *6,* 880–921.
15. Xu, X.-H.; Matsuzaki, K.; Shibata, N. *Chem. Rev.* **2015,** *115,* 731–764.
16. Toulgoat, F.; Alazet, S.; Billard, T. *Eur. J. Org. Chem.* **2014,** 2415–2428.
17. Tlili, A.; Billard, T. *Angew. Chem. Int. Ed.* **2013,** *52,* 6818–6819.
18. Umemoto, T.; Ishihara, S. *J. Fluor. Chem.* **1998,** *92,* 181–187.
19. Farbenind, I.G. Patent D.R.P. 682,971, 1936.
20. Gregory, W.A. U.S. Patent 2,763,692, 1956.
21. Gregory, W.G. U.S. Patent. 2,776,992, 1957.
22. Nodiff, E. A.; Lipsehutz, S.; Craig, P. N.; Gordon, M. *J. Org. Chem.* **1960,** *25,* 60–65.
23. Yagupolskii, L. M.; Gruz, B. E. *Zh. Obshch. Khim.* **1961,** *31,* 1315–1320.
24. Buchanan, J.B.; Gregory, W.A. U.S. Patent 3,061,645, 1962.
25. Yagupolskii, L. M.; Orda, V. V. *Zh. Obshch. Khim.* **1964,** *34,* 1979–1984.
26. Yagupolskii, L. M.; Gandelsman, L. Z.; Trushanina, L. I. *Ukr. Khim. Zh* **1965,** *31,* 1301–1305.
27. Yagupolskii, L. M.; Boiko, V. N. *Zh. Obshch. Khim.* **1969,** *39,* 195–198.
28. Tohyama, Y.; Kanematsu, K. *Appl, P. C. T. I. WO* **2009,** *8* (287).
29. Kremsner, J. M.; Rack, M.; Pilger, C.; Oliver Kappe, C. *Tetrahedron Lett.* **2009,** *50,* 3665–3668.
30. Suda, M.; Hino, C. *Tetrahedron Lett.* **1981,** *22,* 1997–2000.
31. Makosza, M.; Fedorynski, M. *Synthesis* **1974,** *1974,* 274–276.
32. Langlois, B. R.; Billard, T.; Mulatier, J.-C.; Yezeguelian, C. *J. Fluor. Chem.* **2007,** *128,* 851–856.
33. Saint-Jalmes, L. *J. Fluor. Chem.* **2006,** *127,* 85–90.
34. Alonso, C.; Martínez de Marigorta, E.; Rubiales, G.; Palacios, F. *Chem. Rev* **2015,** *115,* 1847–1935.
35. Ma, J. A.; Cahard, D. *J. Fluor. Chem.* **2007,** *128,* 975–996.
36. Haszeldine, R. N.; Rigby, R. B.; Tipping, A. E. *J. Chem. Soc. Perkin I* **1972,** 2180–2182.
37. Gandel'sman, L. Z.; Boiko, V. N. *Ukr. Khim. Zh. Russ. Ed.)* **1977,** *43,* 1224–1225.
38. Yasui, H.; Yamamoto, T.; Tokunaga, E.; Shibata, N. *J. Fluor. Chem.* **2011,** *132,* 186–189.
39. Soloshonok, V.; Kukhar, V.; Pustovit, Y.; Nazaretian, V. *Synlett* **1992,** *1992,* 657–658.
40. Ebner, S.; Jaun, B.; Goenrich, M.; Thauer, R. K.; Harmer, J. *J. Am. Chem. Soc.* **2010,** *132,* 567–575.
41. Goenrich, M.; Mahlert, F.; Duin, E.; Bauer, C.; Jaun, B.; Thauer, R. *J. Biol. Inorg. Chem.* **2004,** *9,* 691–705.
42. Gregory, W. A.; Brittelli, D. R.; Wang, C. L. J.; Kezar, H. S.; Carlson, R. K.; Park, C. H.; Corless, P. F.; Miller, S. J.; Rajagopalan, P. *J. Med. Chem.* **1990,** *33,* 2569–2578.
43. Houston, M. E.; Honek, J. F. *J. Chem. Soc. Chem. Commun.* **1989,** 761–762.
44. Popov, V. I.; Boiko, V. N.; Yagupolskii, L. M. *J. Fluor. Chem.* **1982,** *21,* 365–369.

45. Straathof, N. J. W.; Tegelbeckers, B. J. P.; Hessel, V.; Wang, X.; Noel, T. *Chem. Sci.* **2014,** *5,* 4768–4773.
46. Koshechko, V. G.; Kiprianova, L. A.; Fileleeva, L. I. *Tetrahedron Lett.* **1992,** *33,* 6677–6678.
47. Park, C.-M.; Bruncko, M.; Adickes, J.; Bauch, J.; Ding, H.; Kunzer, A.; Marsh, K. C.; Nimmer, P.; Shoemaker, A. R.; Song, X.; Tahir, S. K.; Tse, C.; Wang, X.; Wendt, M. D.; Yang, X.; Zhang, H.; Fesik, S. W.; Rosenberg, S. H.; Elmore, S. W. *J. Med. Chem.* **2008,** *51,* 6902–6915.
48. Boiko, V. N.; Shchupak, G. M. *J. Fluor. Chem.* **1994,** *69,* 207–212.
49. Harsányi, A.; Dorkó, É.; Csapó, Á.; Bakó, T.; Peltz, C.; Rábai, J. *J. Fluor. Chem.* **2011,** *132,* 1241–1246.
50. Wakselman, C.; Tordeux, M. *J. Chem. Soc. Chem. Commun.* **1984,** 793–794.
51. Wakselman, C.; Tordeux, M.; Clavel, J.-L.; Langlois, B. *J. Chem. Soc. Chem. Commun.* **1991,** 993–994.
52. Clavel, J. L.; Langlois, B.; Wakselman, C.; Tordeux, M. *Phosphorus Sulfur Silicon Relat. Elem.* **1991,** *59,* 129–132.
53. Magnier, E.; Tordeux, M.; Goumont, R.; Magder, K.; Wakselman, C. *J. Fluor. Chem.* **2003,** *124,* 55–59.
54. Sokolenko, L. V.; Maletina, I. I.; Yagupolskii, L. M.; Yagupolskii, Y. L. *Synlett* **2010,** *2010,* 2075–2078.
55. Sladojevich, F.; McNeill, E.; Börgel, J.; Zheng, S.-L.; Ritter, T. *Angew. Chem. Int. Ed.* **2015,** *54,* 3712–3716.
56. Wakselman, C.; Tordeux, M. *J. Org. Chem.* **1985,** *50,* 4047–4051.
57. Goumont, R.; Faucher, N.; Moutiers, G.; Tordeux, M.; Wakselman, C. *Synthesis* **1997,** *1997,* 691–695.
58. Anselmi, E.; Blazejewski, J.-C.; Tordeux, M.; Wakselman, C. *J. Fluor. Chem.* **2000,** *105,* 41–44.
59. Koshechko, V. G.; Kiprianova, L. A.; Fileleeva, L. I.; Tsanov, K. G. *J. Fluor. Chem.* **1999,** *96,* 163–166.
60. Magnier, E.; Wakselman, C. *Synthesis* **2003,** *2003,* 0565–0569.
61. Koshechko, V. G.; Kiprianova, L. A.; Fileleeva, L. I.; Rozhkova, Z. Z. *J. Fluor. Chem.* **1995,** *70,* 277–278.
62. Clavel, J.-L.; Langlois, B.; Laurent, E.; Roidot, N. *Phosphorus, Sulfur Silicon Relat. Elem.* **1991,** *59,* 169–172.
63. Langlois, B.; Montègre, D.; Roidot, N. *J. Fluor. Chem.* **1994,** *68,* 63–66.
64. Umemoto, T.; Miyano, O. *Tetrahedron Lett.* **1982,** *23,* 3929–3930.
65. Billard, T.; Roques, N.; Langlois, B. R. *J. Org. Chem.* **1999,** *64,* 3813–3820.
66. Billard, T.; Roques, N.; Langlois, B. R. *Tetrahedron Lett.* **2000,** *41,* 3069–3072.
67. Urben, P. *Bretherick's Handbook of Reactive Chemical Hazards;* Elsevier Science: Oxford, 2007.
68. Tordeux, M.; Francese, C.; Walkselman, C. *J. Fluor. Chem.* **1989,** *43,* 27–34.
69. Pooput, C.; Medebielle, M.; Dolbier, W. R. *Org. Lett.* **2004,** *6,* 301–303.
70. Pooput, C.; Dolbier, W. R.; Médebielle, M. *J. Org. Chem.* **2006,** *71,* 3564–3568.
71. Liu, X.; Xu, C.; Wang, M.; Liu, Q. *Chem. Rev.* **2015,** *115,* 683–730.
72. Billard, T.; Langlois, B. R. *Tetrahedron Lett.* **1996,** *37,* 6865–6868.
73. Billard, T.; Large, S.; Langlois, B. R. *Tetrahedron Lett.* **1997,** *38,* 65–68.
74. Movchun, V. N.; Kolomeitsev, A. A.; Yagupolskii, Y. L. *J. Fluor. Chem.* **1995,** *70,* 255–257.

75. Large-Radix, S.; Billard, T.; Langlois, B. R. *J. Fluor. Chem.* **2003,** *124,* 147–149.
76. Timoshenko, V. M.; Portella, C. *J. Fluor. Chem.* **2009,** *130,* 586–590.
77. Gouault-Bironneau, S.; Timoshenko, V. M.; Grellepois, F.; Portella, C. *J. Fluor. Chem.* **2012,** *134,* 164–171.
78. Bouchu, M.-N.; Large, S.; Steng, M.; Langlois, B.; Praly, J.-P. *Carbohydr. Res.* **1998,** *314,* 37–45.
79. Hess, J.; Konatschnig, S.; Morard, S.; Pierroz, V.; Ferrari, S.; Spingler, B.; Gasser, G. *Inorg. Chem.* **2014,** *53,* 3662–3667.
80. Danoun, G.; Bayarmagnai, B.; Gruenberg, M. F.; Goossen, L. J. *Chem. Sci.* **2014,** *5,* 1312–1316.
81. Jouvin, K.; Matheis, C.; Goossen, L. J. *Chem. Eur. J.* **2015,** *21,* 14324–14327.
82. Matheis, C.; Wang, M.; Krause, T.; Goossen, L. J. *Synlett* **2015,** *26,* 1628–1632.
83. Bayarmagnai, B.; Matheis, C.; Risto, E.; Goossen, L. J. *Adv. Synth. Catal.* **2014,** *356,* 2343–2348.
84. Zhong, W.; Liu, X. *Tetrahedron Lett.* **2014,** *55,* 4909–4911.
85. Prakash, G. K. S.; Jog, P. V.; Batamack, P. T. D.; Olah, G. A. *Science* **2012,** *338,* 1324–1327.
86. Russell, J.; Roques, N. *Tetrahedron* **1998,** *54,* 13771–13782.
87. Kawai, H.; Yuan, Z.; Tokunaga, E.; Shibata, N. *Org. Biomol. Chem.* **2013,** *11,* 1446–1450.
88. Large, S.; Roques, N.; Langlois, B. R. *J. Org. Chem.* **2000,** *65,* 8848–8856.
89. Langlois, B. R.; Billard, T. *Synthesis* **2003,** 185–194.
90. Blond, G.; Billard, T.; Langlois, B. R. *Tetrahedron Lett.* **2001,** *42,* 2473–2475.
91. Billard, T.; Greiner, A.; Langlois, B. R. *Tetrahedron* **1999,** *55,* 7243–7250.
92. Inschauspe, D.; Sortais, J.-B.; Billard, T.; Langlois, B. R. *Synlett* **2003,** 233–235.
93. Prakash, G. K. S.; Hu, J.; Olah, G. A. *Org. Lett.* **2003,** *5,* 3253–3256.
94. Quiclet-Sire, B.; Saicic, R. N.; Zard, S. Z. *Tetrahedron Lett.* **1996,** *37,* 9057–9058.
95. Roques, N. *J. Fluor. Chem.* **2001,** *107,* 311–314.
96. Exner, B.; Bayarmagnai, B.; Jia, F.; Goossen, L. J. *Chem. Eur. J.* **2015,** *21,* 17220–17223.
97. Riofski, M. V.; Hart, A. D.; Colby, D. A. *Org. Lett.* **2013,** *15,* 208–211.
98. Cherkupally, P.; Beier, P. *Tetrahedron Lett.* **2010,** *51,* 252–255.
99. Potash, S.; Rozen, S. *J. Fluor. Chem.* **2014,** *168,* 173–176.
100. Yagupolskii, L. M.; Kondratenko, N. V.; Timofeeva, G. N. *Zh. Org. Khim.* **1984,** *20,* 115–118.
101. Umemoto, T. *Chem. Rev.* **1996,** *96,* 1757–1778.
102. Teruo, U.; Sumi, I. *Tetrahedron Lett.* **1990,** *31,* 3579–3582.
103. Umemoto, T.; Ishihara, S. *J. Am. Chem. Soc.* **1993,** *115,* 2156–2164.
104. Carcenac, Y.; Tordeux, M.; Wakselman, C.; Diter, P. *New J. Chem.* **2006,** *30,* 447–457.
105. Shibata, N.; Matsnev, A.; Cahard, D. *Beilstein J. Org. Chem.* **2010,** *6,* 65.
106. Eisenberger, P.; Gischig, S.; Togni, A. *Chem. Eur. J.* **2006,** *12,* 2579–2586.
107. Charpentier, J.; Früh, N.; Togni, A. *Chem. Rev.* **2015,** *115,* 650–682.
108. Kieltsch, I.; Eisenberger, P.; Togni, A. *Angew. Chem. Int. Ed.* **2007,** *46,* 754–757.
109. Ausín, C.; Kauffman, J. S.; Duff, R. J.; Shivaprasad, S.; Beaucage, S. L. *Tetrahedron* **2010,** *66,* 68–79.
110. Capone, S.; Kieltsch, I.; Flögel, O.; Lelais, G.; Togni, A.; Seebach, D. *Helv. Chim. Acta* **2008,** *91,* 2035–2056.
111. Charkoudian, L. K.; Liu, C. W.; Capone, S.; Kapur, S.; Cane, D. E.; Togni, A.; Seebach, D.; Khosla, C. *Protein Sci.* **2011,** *20,* 1244–1255.

112. Sala, O.; Lüthi, H. P.; Togni, A.; Iannuzzi, M.; Hutter, J. *J. Comput. Chem.* **2015,** *36,* 785–794.
113. Furuya, T.; Kamlet, A. S.; Ritter, T. *Nature* **2011,** *473,* 470–477.
114. Liang, T.; Neumann, C. N.; Ritter, T. *Angew. Chem. Int. Ed.* **2013,** *52,* 8214–8264.
115. Landelle, G.; Panossian, A.; Pazenok, S.; Vors, J.-P.; Leroux, F. R. *Beilstein J. Org. Chem.* **2013,** *9,* 2476–2536.
116. Haszeldine, R. N.; Kidd, J. M. *J. Chem. Soc. (Resumed)* **1955,** 3871–3880.
117. Dmowski, W.; Haas, A. *J. Chem. Soc. Perkin Trans.* **1987,** *1,* 2119–2124.
118. Dmowski, W.; Haas, A. *J. Chem. Soc., Perkin Trans.* **1988,** *1,* 1179–1181.
119. Adams, D. J.; Clark, J. H.; Heath, P. A.; Hansen, L. B.; Sanders, V. C.; Tavener, S. J. *J. Fluor. Chem.* **2000,** *101,* 187–191.
120. Clark, J. H.; Tavener, S. J. *J. Fluor. Chem.* **1997,** *85,* 169–172.
121. Tavener, S. J.; Adams, D. J.; Clark, J. H. *J. Fluor. Chem.* **1999,** *95,* 171–176.
122. Emeléus, H. J.; MacDuffie, D. E. *J. Chem. Soc.* **1961,** 2597.
123. Hanack, M.; Massa, F. W. *Tetrahedron Lett.* **1981,** *22,* 557–558.
124. Hanack, M.; Kühnle, A. *Tetrahedron Lett.* **1981,** *22,* 3047–3048.
125. Yagupolskii, L. M.; Smirnova, O. D. *Zh. Org. Khim.* **1972,** *8,* 1990–1991.
126. Jiang, M.; Zhu, F.; Xiang, H.; Xu, X.; Deng, L.; Yang, C. *Org. Biomol. Chem.* **2015,** *13,* 6935–6939.
127. Adams, D. J.; Clark, J. H. *J. Org. Chem.* **2000,** *65,* 1456–1460.
128. Liu, J.-B.; Xu, X.-H.; Chen, Z.-H.; Qing, F.-L. *Angew. Chem. Int. Ed.* **2015,** *54,* 897–900.
129. Hu, M.; Rong, J.; Miao, W.; Ni, C.; Han, Y.; Hu, J. *Org. Lett.* **2014,** *16,* 2030–2033.
130. Wang, X.; Zhou, Y.; Ji, G.; Wu, G.; Li, M.; Zhang, Y.; Wang, J. *Eur. J. Org. Chem.* **2014,** *2014,* 3093–3096.
131. Emer, E.; Twilton, J.; Tredwell, M.; Calderwood, S.; Collier, T. L.; Liégault, B.; Taillefer, M.; Gouverneur, V. *Org. Lett.* **2014,** *16,* 6004–6007.
132. Kondratenko, N. V.; Sambur, V. P. *Ukr. Khim. Zh. Russ. Ed.)* **1975,** *41,* 516–519.
133. Teverovskiy, G.; Surry, D. S.; Buchwald, S. L. *Angew. Chem. Int. Ed.* **2011,** *50,* 7312–7314.
134. Xu, J.; Mu, X.; Chen, P.; Ye, J.; Liu, G. *Org. Lett.* **2014,** *16,* 3942–3945.
135. Saravanan, P.; Anbarasan, P. *Adv. Synth. Catal.* **2015,** *357,* 3521–3528.
136. Xie, F.; Zhang, Z.; Yu, X.; Tang, G.; Li, X. *Angew. Chem. Int. Ed.* **2015,** *54,* 7405–7409.
137. Yin, W.; Wang, Z.; Huang, Y. *Adv. Synth. Catal.* **2014,** *356,* 2998–3006.
138. Wang, K.-P.; Yun, S. Y.; Mamidipalli, P.; Lee, D. *Chem. Sci.* **2013,** *4,* 3205–3211.
139. Xiao, Q.; Sheng, J.; Ding, Q.; Wu, J. *Eur. J. Org. Chem.* **2014,** *2014,* 217–221.
140. Man, E. H.; Coffman, D. D.; Muetterties, E. L. *J. Am. Chem. Soc.* **1959,** *81,* 3575–3577.
141. Harris, J. F. *J. Org. Chem.* **1967,** *32,* 2063–2074.
142. Yagupolskii, L. M.; Kondratenko, N. V.; Sambur, V. P. *Synthesis* **1975,** *1975,* 721–723.
143. Kondratenko, N. V.; Kolomeytsev, A. A.; Popov, V. I.; Yagupolskii, L. M. *Synthesis* **1985,** *1985,* 667–669.
144. Munavalli, S.; Hassner, A.; Rossman, D. I.; Singh, S.; Rohrbaugh, D. K.; Ferguson, C. P. *J. Fluor. Chem.* **1995,** *73,* 7–11.
145. Kirsch, P.; Lenges, M.; Kühne, D.; Wanczek, K.-P. *Eur. J. Org. Chem.* **2005,** *2005,* 797–802.
146. Adams, D. J.; Goddard, A.; Clark, J. H.; Macquarrie, D. J. *Chem. Commun.* **2000,** 987–988.
147. Remy, D. C.; Rittle, K. E.; Hunt, C. A.; Freedman, M. B. *J. Org. Chem.* **1976,** *41,* 1644–1646.

148. Haas, A.; Lieb, M. *J. Heterocycl. Chem.* **1986,** *23,* 1079−1084.
149. Rhode, C.; Lemke, J.; Lieb, M.; Metzler-Nolte, N. *Synthesis* **2009,** *2009,* 2015−2018.
150. Clark, J. H.; Jones, C. W.; Kybett, A. P.; McClinton, M. A.; Miller, J. M.; Bishop, D.; Blade, R. J. *J. Fluor. Chem.* **1990,** *48,* 249−253.
151. Haas, A.; Krächter, H.-U. *Chem. Ber.* **1988,** *121,* 1833−1840.
152. Rueping, M.; Tolstoluzhsky, N.; Nikolaienko, P. *Chem. Eur. J.* **2013,** *19,* 14043−14046.
153. Weng, Z.; He, W.; Chen, C.; Lee, R.; Tan, D.; Lai, Z.; Kong, D.; Yuan, Y.; Huang, K.-W. *Angew. Chem. Int. Ed.* **2013,** *52,* 1548−1552.
154. Clark, J. H.; Smith, H. *J. Fluor. Chem.* **1993,** *61,* 223−231.
155. Lin, Q.; Chen, L.; Huang, Y.; Rong, M.; Yuan, Y.; Weng, Z. *Org. Biomol. Chem.* **2014,** *12,* 5500−5508.
156. Tan, J.; Zhang, G.; Ou, Y.; Yuan, Y.; Weng, Z. *Chin. J. Chem.* **2013,** *31,* 921−926.
157. Kong, D.; Jiang, Z.; Xin, S.; Bai, Z.; Yuan, Y.; Weng, Z. *Tetrahedron* **2013,** *69,* 6046−6050.
158. Huang, Y.; Ding, J.; Wu, C.; Zheng, H.; Weng, Z. *J. Org. Chem.* **2015,** *80,* 2912−2917.
159. Zhu, P.; He, X.; Chen, X.; You, Y.; Yuan, Y.; Weng, Z. *Tetrahedron* **2014,** *70,* 672−677.
160. Hou, C.; Lin, X.; Huang, Y.; Chen, Z.; Weng, Z. *Synthesis* **2015,** *47,* 969−975.
161. Huang, Y.; He, X.; Li, H.; Weng, Z. *Eur. J. Org. Chem.* **2014,** *2014,* 7324−7328.
162. Wang, Z.; Tu, Q.; Weng, Z. *J. Organomet. Chem.* **2014,** *751,* 830−834.
163. Lefebvre, Q.; Fava, E.; Nikolaienko, P.; Rueping, M. *Chem. Commun.* **2014,** *50,* 6617−6619.
164. Nikolaienko, P.; Pluta, R.; Rueping, M. *Chem. Eur. J.* **2014,** *20,* 9867−9870.
165. Farnham, W. B.; Midddleton, W. J. (du Pont de Nemours, E. I. and Co., USA), EP164124A2, *Chem. Abstr.* **1985**, *105,* 114600.
166. Kolomeitsev, A.; Medebielle, M.; Kirsch, P.; Lork, E.; Roschenthaler, G.-V. *J. Chem. Soc. Perkin Trans.* **2000,** *1,* 2183−2185.
167. Tyrra, W.; Naumann, D.; Hoge, B.; Yagupolskii, Y. L. *J. Fluor. Chem.* **2003,** *119,* 101−107.
168. Kremlev, M. M.; Tyrra, W.; Naumann, D.; Yagupolskii, Y. L. *Tetrahedron Lett.* **2004,** *45,* 6101−6104.
169. Brandt, G. A. R.; Emeleus, H. J.; Haszeldine, R. N. *J. Chem. Soc. (Resumed)* **1952,** 2198−2205.
170. Zhang, C.-P.; Vicic, D. A. *J. Am. Chem. Soc.* **2012,** *134,* 183−185.
171. Zhang, C.-P.; Vicic, D. A. *Chem. Asian J.* **2012,** *7,* 1756−1758.
172. Yin, G.; Kalvet, I.; Englert, U.; Schoenebeck, F. *J. Am. Chem. Soc.* **2015,** *137,* 4164−4172.
173. Yin, G.; Kalvet, I.; Schoenebeck, F. *Angew. Chem. Int. Ed.* **2015,** *54,* 6809−6813.
174. Matheis, C.; Wagner, V.; Goossen, L. J. *Chem. Eur. J.* **2016,** *22,* 79−82.
175. Ye, K.-Y.; Zhang, X.; Dai, L.-X.; You, S.-L. *J. Org. Chem.* **2014,** *79,* 12106−12110.
176. Chen, C.; Xie, Y.; Chu, L.; Wang, R.-W.; Zhang, X.; Qing, F.-L. *Angew. Chem. Int. Ed.* **2012,** *51,* 2492−2495.
177. Chen, Q.-Y.; Duan, J.-X. *J. Chem. Soc. Chem. Commun.* **1993,** 918−919.
178. Chen, C.; Chu, L.; Qing, F.-L. *J. Am. Chem. Soc.* **2012,** *134,* 12454−12457.
179. Rong, M.; Li, D.; Huang, R.; Huang, Y.; Han, X.; Weng, Z. *Eur. J. Org. Chem.* **2014,** *2014,* 5010−5016.
180. Li, J.; Wang, P.; Xie, F.-F.; Yang, X.-G.; Song, X.-N.; Chen, W.-D.; Ren, J.; Zeng, B.-B. *Eur. J. Org. Chem.* **2015,** *2015,* 3568−3571.
181. Huang, Y.; He, X.; Lin, X.; Rong, M.; Weng, Z. *Org. Lett.* **2014,** *16,* 3284−3287.
182. Li, J.; Xie, F.-F.; Wang, P.; Wu, Q.-Y.; Chen, W.-D.; Ren, J.; Zeng, B.-B. *Tetrahedron* **2015,** *71,* 5520−5524.

183. Dai, X.; Cahard, D. *Synlett* **2015,** *26,* 40–44.
184. Zhai, L.; Li, Y.; Yin, J.; Jin, K.; Zhang, R.; Fu, X.; Duan, C. *Tetrahedron* **2013,** *69,* 10262–10266.
185. Stump, E. C. *Chem. Eng. News* **1967,** *45,* 44.
186. Haas, A.; Lopusiński, A. *Chem. Ber.* **1981,** *114,* 3176–3178.
187. Kolomeitsev, A. A.; Chabanenko, K. Y.; Röschenthaler, G. V.; Yagupolskii, Y. L. *Synthesis* **1994,** *1994,* 145–146.
188. Li, S.-G.; Zard, S. Z. *Org. Lett.* **2013,** *15,* 5898–5901.
189. Yang, H.-B.; Fan, X.; Wei, Y.; Shi, M. *Org. Chem. Front.* **2015,** *2,* 1088–1093.
190. Glenadel, Q.; Bordy, M.; Alazet, S.; Tlili, A.; Billard, T. *Asian J. Org. Chem.* **2016,** *5,* 428–433.
191. Haas, A.; Lieb, M.; Zhang, Y. *J. Fluor. Chem.* **1985,** *30,* 203–210.
192. Andreades, S.; Harris, J. F.; Sheppard, W. A. *J. Org. Chem.* **1964,** *29,* 898–900.
193. Haas, A.; Hellwig, V. *J. Fluor. Chem.* **1975,** *6,* 521–532.
194. Scribner, R. M. *J. Org. Chem.* **1966,** *31,* 3671–3682.
195. Haas, A.; Niemann, U. *Chem. Ber.* **1977,** *110,* 67–77.
196. Gerstenberger, M. R. C.; Haas, A.; Liebig, F. *J. Fluor. Chem.* **1982,** *19,* 461–474.
197. Belanger, P. C.; Atkinson, J. G.; Rooney, C. S.; Britcher, S. F.; Remy, D. C. *J. Org. Chem.* **1983,** *48,* 3234–3241.
198. Mirek, J.; Haas, A. *J. Fluor. Chem.* **1981,** *19,* 67–70.
199. Gerstenberger, M. R. C.; Haas, A. *J. Fluor. Chem.* **1983,** *23,* 525–540.
200. Haas, A.; Hellwig, V. *Chem. Ber.* **1976,** *109,* 2475–2484.
201. Croft, T. S.; McBrady, J. J. *J. Heterocycl. Chem.* **1975,** *12,* 845–849.
202. Sheppard, W. A. *J. Org. Chem.* **1964,** *29,* 895–898.
203. Haas, A.; Niemann, U. *J. Fluor. Chem.* **1978,** *11,* 509–518.
204. Bayreuther, H.; Haas, A. *Chem. Ber.* **1973,** *106,* 1418–1422.
205. Bauer, M.; Haas, A.; Muth, H. *J. Fluor. Chem.* **1980,** *16,* 129–136.
206. Kolasa, A. *J. Fluor. Chem.* **1987,** *36,* 29–40.
207. Popov, V. I.; Haas, A.; Lieb, M. *J. Fluor. Chem.* **1990,** *47,* 131–136.
208. Zankowska-Jasinska, W.; Zaleska, B.; Haas, A. *J. Fluor. Chem.* **1984,** *24,* 363–368.
209. Bogdanowicz-Szwed, K.; Kawałek, B.; Lieb, M. *J. Fluor. Chem.* **1987,** *35,* 317–327.
210. Honeker, R.; Ernst, J. B.; Glorius, F. *Chem. Eur. J.* **2015,** *21,* 8047–8051.
211. Tran, L. D.; Popov, I.; Daugulis, O. *J. Am. Chem. Soc.* **2012,** *134,* 18237–18240.
212. Jiang, L.; Qian, J.; Yi, W.; Lu, G.; Cai, C.; Zhang, W. *Angew. Chem. Int. Ed.* **2015,** *54,* 14965–14969.
213. Ferry, A.; Billard, T.; Langlois, B. R.; Bacqué, E. *Angew. Chem. Int. Ed.* **2009,** *48,* 8551–8555.
214. Ferry, A.; Billard, T.; Langlois, B. R.; Bacque, E. *J. Org. Chem.* **2008,** *73,* 9362–9365.
215. Alazet, S.; Billard, T. *Synlett* **2015,** *26,* 76–78.
216. Ferry, A.; Billard, T.; Bacqué, E.; Langlois, B. R. *J. Fluor. Chem.* **2012,** *134,* 160–163.
217. Jereb, M.; Gosak, K. *Org. Biomol. Chem.* **2015,** *13,* 3103–3115.
218. Glenadel, Q.; Alazet, S.; Billard, T. *J. Fluor. Chem.* **2015,** *179,* 89–95.
219. Alazet, S.; Zimmer, L.; Billard, T. *J. Fluor. Chem.* **2015,** *171,* 78–81.
220. Alazet, S.; Zimmer, L.; Billard, T. *Angew. Chem. Int. Ed.* **2013,** *52,* 10814–10817.
221. Baert, F.; Colomb, J.; Billard, T. *Angew. Chem. Int. Ed.* **2012,** *51,* 10382–10385.
222. Liu, J.; Chu, L.; Qing, F.-L. *Org. Lett.* **2013,** *15,* 894–897.
223. Sheng, J.; Wu, J. *Org. Biomol. Chem.* **2014,** *12,* 7629–7633.
224. Glenadel, Q.; Alazet, S.; Tlili, A.; Billard, T. *Chem. Eur. J.* **2015,** *21,* 14694–14698.
225. Yang, Y.; Jiang, X.; Qing, F.-L. *J. Org. Chem.* **2012,** *77,* 7538–7547.

226. Chen, D.-Q.; Gao, P.; Zhou, P.-X.; Song, X.-R.; Qiu, Y.-F.; Liu, X.-Y.; Liang, Y.-M. *Chem. Commun.* **2015,** *51,* 6637–6639.
227. Sheng, J.; Fan, C.; Wu, J. *Chem. Commun.* **2014,** *50,* 5494–5496.
228. Sheng, J.; Li, S.; Wu, J. *Chem. Commun.* **2014,** *50,* 578–580.
229. Xiao, Q.; Sheng, J.; Chen, Z.; Wu, J. *Chem. Commun.* **2013,** *49,* 8647–8649.
230. Li, Y.; Li, G.; Ding, Q. *Eur. J. Org. Chem.* **2014,** *2014,* 5017–5022.
231. Xiao, Q.; Zhu, H.; Li, G.; Chen, Z. *Adv. Synth. Catal.* **2014,** *356,* 3809–3815.
232. Alazet, S.; Ismalaj, E.; Glenadel, Q.; Le Bars, D.; Billard, T. *Eur. J. Org. Chem.* **2015,** 4607–4610.
233. Wu, W.; Zhang, X.; Liang, F.; Cao, S. *Org. Biomol. Chem.* **2015,** *13,* 6992–6999.
234. Alazet, S.; Zimmer, L.; Billard, T. *Chem. Eur. J.* **2014,** *20,* 8589–8593.
235. Xiong, H.-Y.; Besset, T.; Cahard, D.; Pannecoucke, X. *J. Org. Chem.* **2015,** *80,* 4204–4212.
236. Alazet, S.; Ollivier, K.; Billard, T. *Beilstein J. Org. Chem.* **2013,** *9,* 2354–2357.
237. Jereb, M.; Dolenc, D. *RSC Adv.* **2015,** *5,* 58292–58306.
238. Bootwicha, T.; Liu, X.; Pluta, R.; Atodiresei, I.; Rueping, M. *Angew. Chem. Int. Ed.* **2013,** *52,* 12856–12859.
239. Kang, K.; Xu, C.; Shen, Q. *Org. Chem. Front.* **2014,** *1,* 294–297.
240. Pluta, R.; Nikolaienko, P.; Rueping, M. *Angew. Chem. Int. Ed.* **2014,** *53,* 1650–1653.
241. Pluta, R.; Rueping, M. *Chem. Eur. J.* **2014,** *20,* 17315–17318.
242. Rueping, M.; Liu, X.; Bootwicha, T.; Pluta, R.; Merkens, C. *Chem. Commun.* **2014,** *50,* 2508–2511.
243. Haas, A.; Möller, G. *Chem. Ber.* **1996,** *129,* 1383–1388.
244. Xu, C.; Shen, Q. *Org. Lett.* **2014,** *16,* 2046–2049.
245. Liao, K.; Zhou, F.; Yu, J.-S.; Gao, W.-M.; Zhou, J. *Chem. Commun.* **2015,** *51,* 16255–16258.
246. Zhu, S.-Q.; Xu, X.-H.; Qing, F.-L. *Eur. J. Org. Chem.* **2014,** *2014,* 4453–4456.
247. Shao, X.; Wang, X.; Yang, T.; Lu, L.; Shen, Q. *Angew. Chem. Int. Ed.* **2013,** *52,* 3457–3460.
248. Vinogradova, E. V.; Müller, P.; Buchwald, S. L. *Angew. Chem. Int. Ed.* **2014,** *53,* 3125–3128.
249. Shao, X.; Xu, C.; Lu, L.; Shen, Q. *Acc. Chem. Res.* **2015,** *48,* 1227–1236.
250. Ma, B.; Shao, X.; Shen, Q. *J. Fluor. Chem.* **2015,** *171,* 73–77.
251. Shao, X.; Xu, C.; Lu, L.; Shen, Q. *J. Org. Chem.* **2015,** *80,* 3012–3021.
252. Shao, X.; Liu, T.; Lu, L.; Shen, Q. *Org. Lett.* **2014,** *16,* 4738–4741.
253. Yang, T.; Shen, Q.; Lu, L. *Chin. J. Chem.* **2014,** *32,* 678–680.
254. Wang, X.; Yang, T.; Cheng, X.; Shen, Q. *Angew. Chem. Int. Ed.* **2013,** *52,* 12860–12864.
255. Deng, Q.-H.; Rettenmeier, C.; Wadepohl, H.; Gade, L. H. *Chem. Eur. J.* **2014,** *20,* 93–97.
256. Li, Y.; Ye, Z.; Bellman, T. M.; Chi, T.; Dai, M. *Org. Lett.* **2015,** *17,* 2186–2189.
257. Xu, C.; Ma, B.; Shen, Q. *Angew. Chem. Int. Ed.* **2014,** *53,* 9316–9320.
258. Wang, Q.; Qi, Z.; Xie, F.; Li, X. *Adv. Synth. Catal.* **2015,** *357,* 355–360.
259. Wang, Q.; Xie, F.; Li, X. *J. Org. Chem.* **2015,** *80,* 8361–8366.
260. Xu, C.; Shen, Q. *Org. Lett.* **2015,** *17,* 4561–4563.
261. Luo, J.; Zhu, Z.; Liu, Y.; Zhao, X. *Org. Lett.* **2015,** *17,* 3620–3623.
262. Yang, Y.-D.; Azuma, A.; Tokunaga, E.; Yamasaki, M.; Shiro, M.; Shibata, N. *J. Am. Chem. Soc.* **2013,** *135,* 8782–8785.
263. Huang, Z.; Yang, Y.-D.; Tokunaga, E.; Shibata, N. *Org. Lett.* **2015,** *17,* 1094–1097.
264. Arimori, S.; Takada, M.; Shibata, N. *Dalton Trans.* **2015,** *44,* 19456–19459.
265. Arimori, S.; Takada, M.; Shibata, N. *Org. Lett.* **2015,** *17,* 1063–1065.

266. Huang, Z.; Yang, Y.-D.; Tokunaga, E.; Shibata, N. *Asian J. Org. Chem.* **2015,** *4,* 525–527.
267. Zhu, X.-L.; Xu, J.-H.; Cheng, D.-J.; Zhao, L.-J.; Liu, X.-Y.; Tan, B. *Org. Lett.* **2014,** *16,* 2192–2195.
268. Xiang, H.; Yang, C. *Org. Lett.* **2014,** *16,* 5686–5689.
269. Harris, J. F.; Stacey, F. W. *J. Am. Chem. Soc.* **1961,** *83,* 840–845.
270. Harris, J. F. *J. Am. Chem. Soc.* **1962,** *84,* 3148–3153.
271. Harris, J. F. *J. Org. Chem.* **1966,** *31,* 931–935.
272. Harris, J. F. *J. Org. Chem.* **1972,** *37,* 1340–1346.
273. Munavalli, S.; Rohrbaugh, D. K.; Rossman, D. I.; Wagner, W. G.; Durst, H. D. *Phosphorus Sulfur Silicon Relat. Elem.* **2002,** *177,* 1021–1031.
274. Zhang, K.; Liu, J.-B.; Qing, F.-L. *Chem. Commun.* **2014,** *50,* 14157–14160.
275. Chen, C.; Xu, X.-H.; Yang, B.; Qing, F.-L. *Org. Lett.* **2014,** *16,* 3372–3375.
276. Guo, S.; Zhang, X.; Tang, P. *Angew. Chem. Int. Ed.* **2015,** *54,* 4065–4069.
277. Wu, H.; Xiao, Z.; Wu, J.; Guo, Y.; Xiao, J.-C.; Liu, C.; Chen, Q.-Y. *Angew. Chem. Int. Ed.* **2015,** *54,* 4070–4074.
278. Yin, F.; Wang, X.-S. *Org. Lett.* **2014,** *16,* 1128–1131.
279. Zhu, L.; Wang, G.; Guo, Q.; Xu, Z.; Zhang, D.; Wang, R. *Org. Lett.* **2014,** *16,* 5390–5393.
280. Qiu, Y.-F.; Zhu, X.-Y.; Li, Y.-X.; He, Y.-T.; Yang, F.; Wang, J.; Hua, H.-L.; Zheng, L.; Wang, L.-C.; Liu, X.-Y.; Liang, Y.-M. *Org. Lett.* **2015,** *17,* 3694–3697.
281. Fuentes, N.; Kong, W.; Fernández-Sánchez, L.; Merino, E.; Nevado, C. *J. Am. Chem. Soc.* **2015,** *137,* 964–973.
282. Zeng, Y.-F.; Tan, D.-H.; Chen, Y.; Lv, W.-X.; Liu, X.-G.; Li, Q.; Wang, H. *Org. Chem. Front.* **2015,** *2,* 1511–1515.
283. Yang, T.; Lu, L.; Shen, Q. *Chem. Commun.* **2015,** *51,* 5479–5481.
284. Hu, F.; Shao, X.; Zhu, D.; Lu, L.; Shen, Q. *Angew. Chem. Int. Ed.* **2014,** *53,* 6105–6109.
285. Alazet, S.; Billard, T.; Toulgoat, F. Determined from Scifinder Data between January 2000 and August 2015.

Nucleophilic Di- and Trifluoromethylation of C=O and C=N Bonds

7

A.D. Dilman
N.D. Zelinsky Institute of Organic Chemistry, Moscow, Russian Federation

Chapter Outline

1. Introduction

The importance of trifluoromethyl- and difluoromethyl-substituted compounds for drug discovery and related fields[1–4] has stimulated the development of methods for direct introduction of CF_3 and CHF_2 groups. Given the availability of carbonyl compounds and azomethines, which successfully interact with wide variety of nucleophiles, there has been a constant interest in the development of reactions allowing to effect direct addition of fluorinated nucleophiles to C=O and C=N bonds. Indeed, these reactions would provide straightforward access to valuable alcohols and amines in a single step starting from conventional organic compounds such as aldehydes, ketones, and imines. However, application of fluorinated lithium and Grignard reagents as nucleophiles, the method of choice in nonfluorinated chemistry, is not feasible, and other organometallic reagents have to be considered.

Concerning fluorinated reagents, the choice of metal is crucial, since the nature of metal determines the stability and reactivity of the reagent. Trifluoromethyl carbanion has long been considered elusive by virtue of unsuccessful attempts to prepare lithium and magnesium reagents even at very low temperatures. In fact, the instability of these

Modern Synthesis Processes and Reactivity of Fluorinated Compounds. http://dx.doi.org/10.1016/B978-0-12-803740-9.00007-X

organometallics is associated with metal Lewis acidity, which fosters the decomposition of M-CF_3 to metal fluoride and difluorocarbene. Only in 2014, Prakash demonstrated that free trifluoromethyl carbanion can be generated in solution as a long-lived species at −78°C if it is devoid of Lewis acidic counterion (metal cation complexed with a crown ether).[5] On the other hand, a number of relatively less polar organometallics such as derivatives of zinc, cadmium, and copper have been described, but their nucleophilic reactivity is moderate, and they are more suitable for transition-metal-mediated substitution reactions.

Nucleophilic trifluoromethylation of CO and CN bonds has been intensively investigated over the past 23 years, and this topic was the subject of several reviews.[6–14] In this regard, the present chapter highlights the most important concepts and approaches in the field. At the same time, nucleophilic difluoromethylation has been less studied,[15] in part, because of limited availability of corresponding reagents.

2. Trifluoromethylation Reactions

2.1 Silicon Reagents

Trimethyl(trifluoromethyl)silane (the Ruppert−Prakash reagent, **1**) is one of the most widely used fluorinated chemicals, which is applied to effect nucleophilic trifluoromethylation reactions.[8–11] First described for the trifluoromethylation of carbonyl compounds in 1989,[16,17] this method rapidly expanded for many other electrophiles, and now constitutes a reliable approach for the synthesis of CF_3-containing molecules.

Reaction of silane **1** with aldehydes proceeds in the presence of Lewis basic activators (Scheme 7.1). Silyl ethers **2** are the primary products, which can be readily desilylated affording trifluoromethylated alcohols **3**.

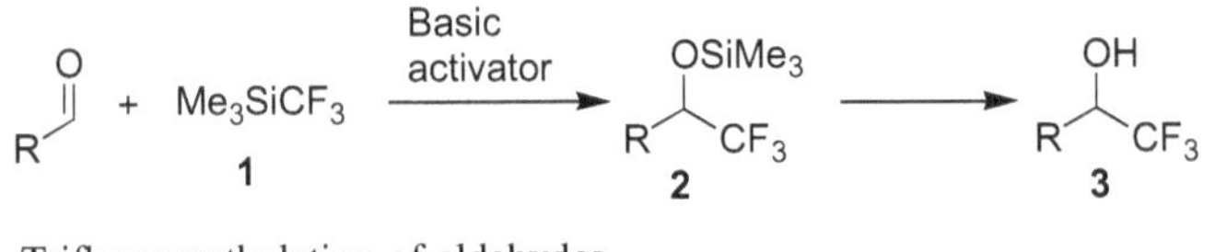

Scheme 7.1 Trifluoromethylation of aldehydes.

The role of Lewis base is usually considered to initiate the reaction (Scheme 7.2). Upon interaction of a base with the silane pentacoordinate species **4** is generated. The generation of such intermediates was firmly established by low-temperature nuclear magnetic resonance measurements, as well as X-ray analysis.[18,19] In the structure of **4**, the most electron-withdrawing groups CF_3 and Y occupy apical positions of the trigonal bipyramid. Then, the intermediate **4** transfers the trifluoromethyl group from the silicon atom to the carbonyl group with the formation of alkoxide anion **5**. The latter itself can serve as an efficient activator with respect to silane **1** thereby allowing for a chain process. Although this general scheme is widely accepted, the precise mechanism for the transfer of the trifluoromethyl group from intermediate **4** onto C=O bond still remains unclear, and no studies and virtually no discussions about

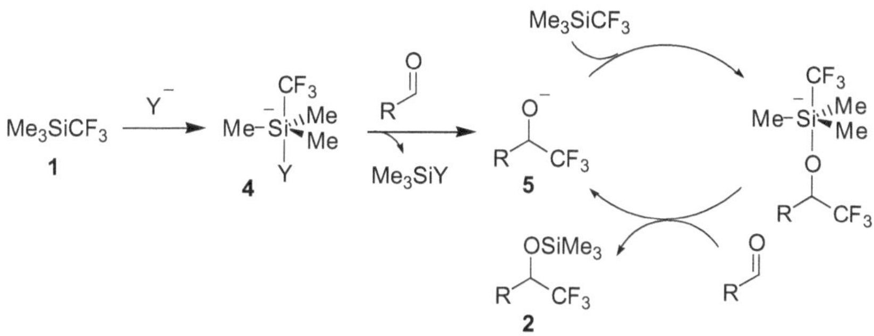

Scheme 7.2 Mechanism of trifluoromethylation.

this particular step have been reported. Two limiting mechanisms can be proposed for this step. First mechanism is the generation of free trifluoromethyl carbanion with its subsequent trapping by carbonyl electrophile. An alternative route involves concerted transfer of the CF_3 fragment from silicon to carbon.

As activators, silaphilic Lewis bases are most effective (Scheme 7.3). Typically, fluoride ion in various soluble forms (TBAF, TBAT) is considered to be the strongest activator. However, tetrabutylammonium acetate, which can work in various solvents, was found to be less active but highly efficient.[20] More available alkaline acetates can also be employed provided that dimethylformamide (DMF) is used as solvent. The role of DMF is likely to coordinate metal ion rendering acetate anion more reactive. In DMF even simple potassium carbonate can serve an activator. It was proposed that carbonate anion can form hexacoordinate siliconate species by binding of two negatively charged oxygen atoms to the silicon atom of reagent **1**.[21]

The chain process, shown on Scheme 7.2 is very efficient, and virtually any system, which can initiate the primary addition event, would work as a "catalyst." For example, a wide variety of basic activators were reported to promote the trifluoromethylation of aromatic aldehydes involving *N*-heterocyclic carbene,[22] bulky phosphines,[23,24] nucleophilic solvents, and activators (dimethylsulfoxide,[25] trimethylamine *N*-oxide,[26] ferrate salt[27]).

Basic activator:

F^-	AcO^-	CO_3^{2-}
Bu_4NF (TBAF)	Bu_4NOAc	K_2CO_2/DMF
$Bu_4N\ Ph_3SiF_2$ (TBAT)	AcOM (M = Li, Na, K)/DMF	

Scheme 7.3 Lewis bases.

The trifluoromethylation of carbonyl compounds has a very high scope. Indeed, aldehydes and ketones, involving enolizable ones, can be effectively used. α,β-Unsaturated substrates react exclusively at the carbonyl group, that is, in 1,2-fashion. Esters can also be trifluoromethylated using silane **1**[20,28] (Scheme 7.4). In contrast to nonfluorinated nucleophiles, which in reactions with esters typically provide alcohols due to double addition, for silane **1** only trifluoromethylketones **8** are

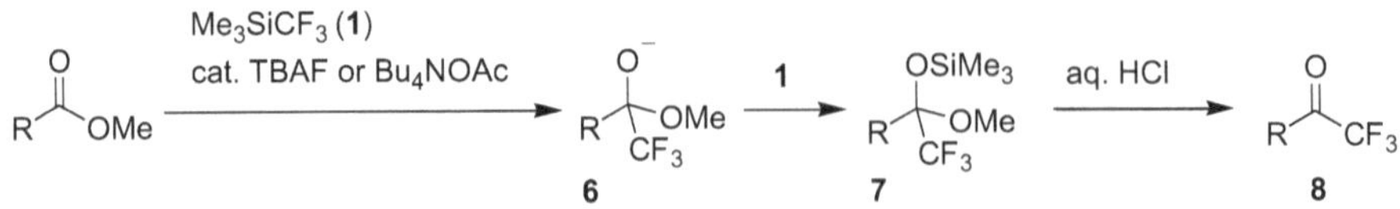

Scheme 7.4 Trifluoromethylation of esters.

formed. In the latter case, tetrahedral intermediate **6** has sufficient lifetime, apparently because of the electron-withdrawing CF_3 substituent, to get silylated by the silicon reagent to maintain the chain propagation. The formed ketals **7** are the primary products, which upon acidic treatment can be converted to ketones **8**.

Heterocycles **9** bearing lactone and imine fragments were trifluoromethylated exclusively at the carbonyl group furnishing lactols **10**, which were reduced to give *anti*-alcohols **11** (Scheme 7.5). Interestingly, prior hydrogenation of C=N bond followed by trifluoromethylation/reduction sequence provides *syn*-diastereoisomers. Finally, chiral auxiliary can be removed from nitrogen furnishing isomeric 1,2-aminoalcohols.[29]

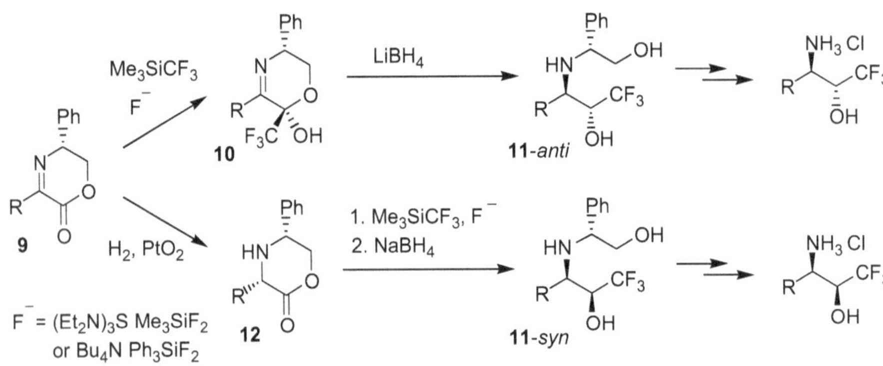

Scheme 7.5 Synthesis of amino alcohols.

Various types of activated carboxylic acid derivatives such anhydrides, haloanhydrides, or pentafluorophenyl esters react with Me_3SiCF_3 furnishing bis(trifluoromethylated) alcohols **13** (Scheme 7.6). However, these reactions require stoichiometric amounts of fluoride activator (up to two equivalents), since the first addition step leading to ketones is not catalytic (if X is not F), while at the second addition step, the tetrahedral alkoxide intermediate is poorly nucleophilic to effect the chain process.[30]

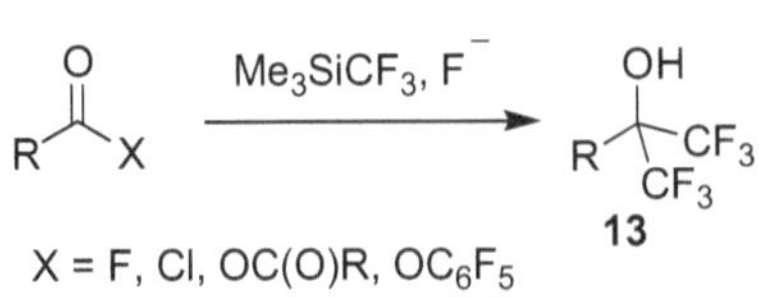

Scheme 7.6 Trifluoromethylation of activated carboxylic acid derivatives.

Amides are typically unreactive toward the Ruppert–Prakash reagent. However, Weinreb amides **14** were trifluoromethylated applying cesium fluoride as a basic activator (Scheme 7.7). Primary addition products were desylilated and hydrolyzed affording trifluoroketones.[31] Imides undergo trifluoromethylation at one carbonyl group.[32] For bicyclic substrates **15**, only one diasteroisomer is formed corresponding to *trans*-orientation of the trifluoromethyl group with respect to the fused cycle (Scheme 7.7).

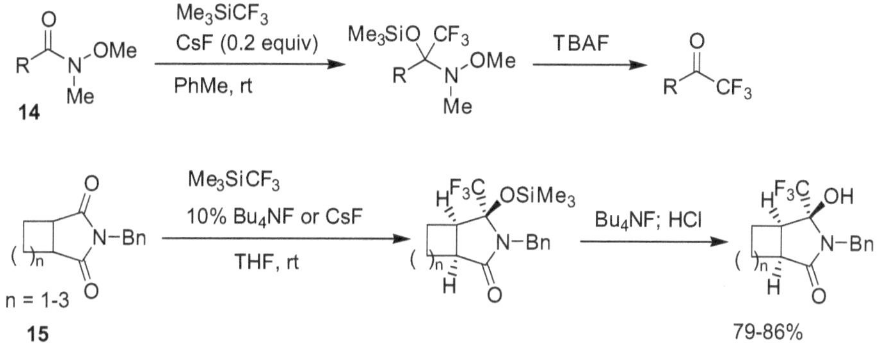

Scheme 7.7 Trifluoromethylation of amides and imides.

Reaction of acylsilanes **16** with Me_3SiCF_3 at 0°C leads to difluorosubstituted silyl enol ethers **17** (Scheme 7.8). The mechanism involves generation of alkoxide intermediate **18**, which does not have time to activate the silicon reagent, but rapidly undergoes Brook rearrangement followed by fluoride elimination.[33,34] However, performing the reaction at low temperature allows isolation of the trifluoromethylation

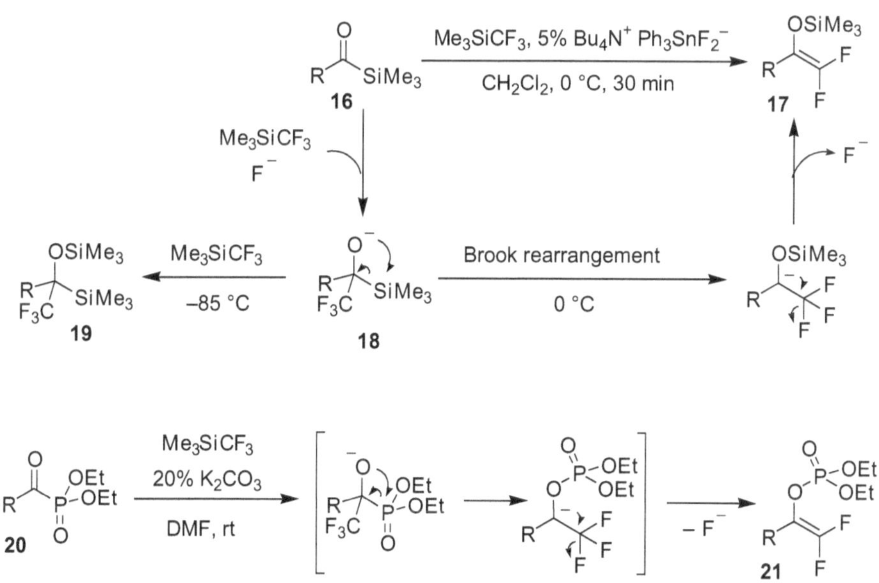

Scheme 7.8 Trifluoromethylation of acylsilanes and acylphosphonates.

product **19**. Enol ethers **17** may couple with various electrophiles such as aldehydes, imines, and Michael acceptors affording a series of *gem*-difluorinated products.[35,36] Analogously to acylsilanes, acylphosphonates **20** react with Me_3SiCF_3 in DMF in the presence of potassium carbonate affording enolphosphates **21**.[37]

Since trifluoromethylation of carbonyl compounds creates a stereogenic center, it is desirable to perform this reaction enantioselectively. Furthermore, the fragment of α-CF_3-substituted alcohol (or its derivative) is present in several drugs. In earlier reports good enantioselectivities were observed, but reactions were highly substrate specific.[38,39] Only in 2007, the first general protocol for catalytic asymmetric trifluoromethylation of ketones was disclosed[40] (Scheme 7.9, upper equation). The reaction was catalyzed by tetramethylammonium fluoride and chiral dicationic salt derived from cinchona alkaloids. However, the reaction requires high loading of the catalyst (10 mol%). Later studies demonstrated the applicability of this method for the trifluoromethylation of alkynyl ketones (ee up to 96%),[41] whereas for aldehydes, moderate selectivities (50–70% ee) were observed.[42] Another method involves application of catalytic amounts (2 mol%) of a combination of copper fluoride complexed with *N*-heterocyclic carbene together with chiral quaternary ammonium salt affording products with good asymmetric induction[43,44] (Scheme 7.9, bottom equation). Despite these successful examples, there is still need for more effective and more general reactions for catalytic asymmetric trifluoromethylation. Apparently, the major problem associated with realization of enantioselective catalysis in this case is the facile background chain reaction that starts just after the primary addition of CF_3 carbanion at the carbonyl group.

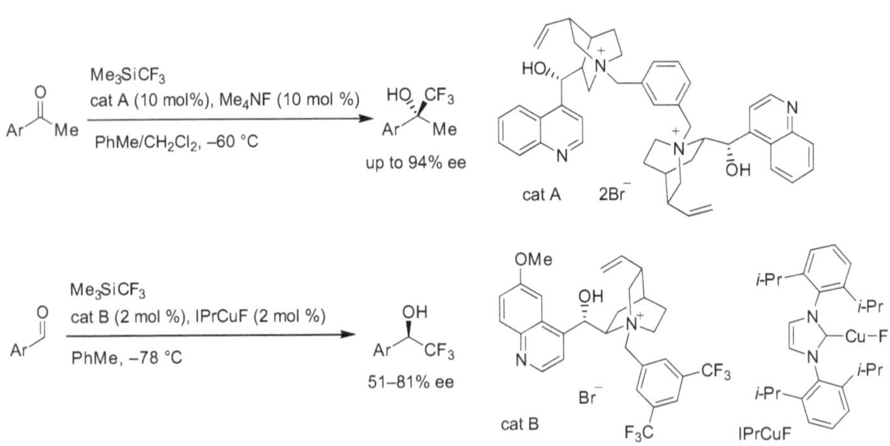

Scheme 7.9 Catalytic asymmetric trifluoromethylation.

Conditions for trifluoromethylation reactions of C=N bond depends on the substrate.[11] *N*-tosyl and *N*-sulfinylimines (**22** and **24**, respectively) react with the Ruppert–Prakash reagent under typical conditions of fluoride activation leading to products **23** and **25**[45,46] (Scheme 7.10). However, stoichiometric quantities of fluoride may be required since after addition of CF_3 carbanion to C=N bond, the resulting

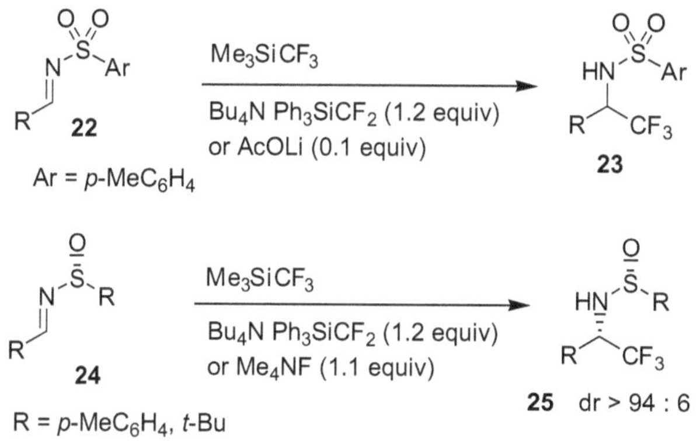

Scheme 7.10 Trifluoromethylation of *N*-sulfonyl- and *N*-sulfinylimines.

anionic species is insufficiently nucleophilic to activate the silane. For sulfinyl imines **24**, high diastereoselectivities may be achieved.[47,48] Importantly, corresponding sulfinamides ($RS(O)NH_2$) are readily available in enantiomerically pure form, which means that the use of chiral imines **24** and the subsequent removal of sulfinyl group from products **25** can lead to chiral trifluoromethylated primary amines.

N-phenylimines are notably less reactive compared to *N*-tosylimines,[49] whereas *N*-alkylimines are completely unreactive, apparently because of low electrophilicity of C=N bond. A different approach was proposed to trifluoromethylate these substrates[50] (Scheme 7.11). The reaction is performed using potassium hydrodifluoride in combination with strong protic acid such as trifluoroacetic acid or triflic acid taken in stoichiometric amounts. The proposed mechanism involves protonation of imine to give iminium ion, whereas hydrodifluoride anion activates the silicon reagent to generate pentacoordinate species. Transfer of the trifluoromethyl group from silicon to iminium electrophile may proceed in a concerted fashion without generation of free CF_3 carbanion. This method has a wide scope and is applicable to *N*-alkyl, *N*-

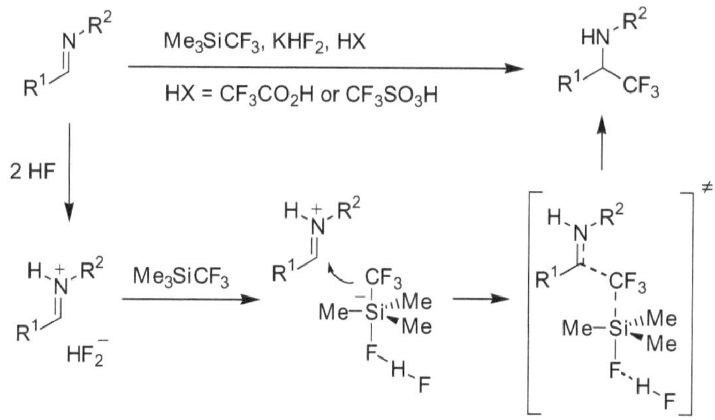

Scheme 7.11 Trifluoromethylation under acidic conditions.

benzyl, and *N*-arylimines.[50–52] Enamines may also be trifluoromethylated using KHF_2/acid system.[53]

Various iminium ions **26** were successfully trifluoromethylated using silane **1** and a stoichiometric amount of fluoride or acetate activator affording tertiary amines **27** (Scheme 7.12). Iminium ions **26** can be obtained by alkylation of imines or from aldehydes and *N*-silylamines.[54] Alkylation of *N*-benzylpyrrolidone with methyltriflate also provides cationic species **28**, which is successfully trifluoromethylated to semiaminal product **29**. The methoxy group in compound **29** can be substituted by various silicon nucleophiles under Lewis acidic conditions.[55] Trifluoromethylation of pyridinium ions **31** was performed using potassium fluoride as activator and fluorotriphenylstannane serving as a carrier of fluoride ion. Oxidation of addition products by ceric ammonium nitrate effected aromatization of the azine ring along with removal of the substituent from nitrogen. Overall, the transformation of pyridines **30** to products **32** corresponds to substitution of hydrogen by the trifluoromethyl group.[56]

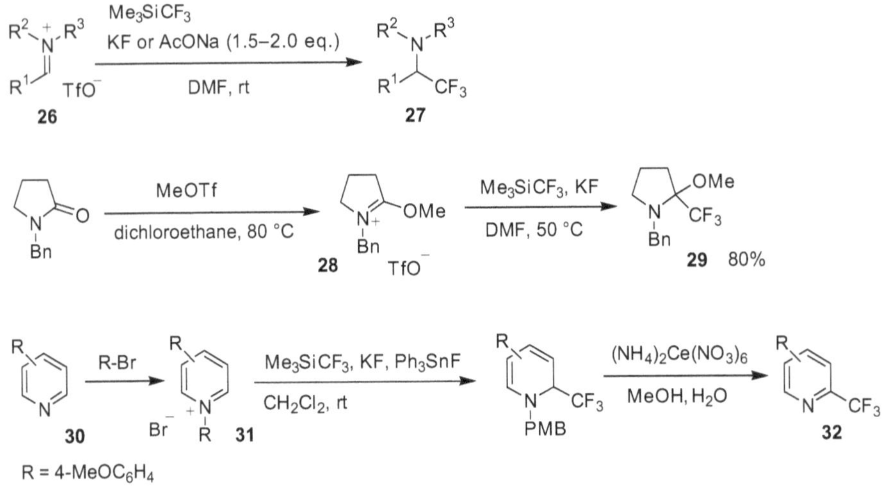

Scheme 7.12 Trifluoromethylation of iminium ions.

Isoquinolinium ions generated in situ by different means serve as good electrophiles for nucleophilic trifluoromethylation. For example, reaction of alkynyl substituted imines **33** with the Ruppert-Prakash reagent was performed using catalytic amount of a silver salt in combination with stoichiometric amount of acetic acid[57] (Scheme 7.13). It is likely that, at first, silver cation activates the triple bond toward nucleophilic attack by imine nitrogen affording isoquinolinium ion **34** together with acetate counterion followed by trifluoromethylation event. Reaction of tetrahydroisoquinolines **35** with Me_3SiCF_3 is performed in the presence of benzoylperoxide furnishing a mixture of products. In this case, iminium ion is formed by the oxidation of benzylic hydrogen, with the benzoate anion formed serving as activator with respect to the silicon reagent.[58] Generation of iminium ions from tetrahydroisoquinolines **35** under visible light irradiation using an organic dye followed by reaction with Me_3SiCF_3/KF system was also reported.[59,60]

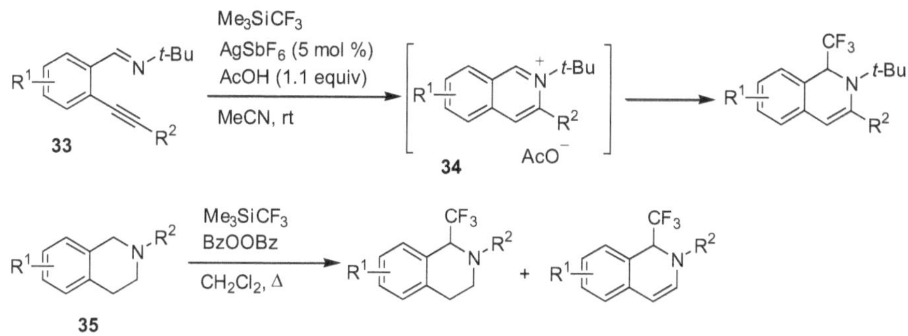

Scheme 7.13 Trifluoromethylation of isoquinolinium ions.

Hydrazones possess relatively unreactive C=N bond, which can be activated by internal Lewis acidic complexation (Scheme 7.14). For example, treatment of *N*-benzoyl hydrazones **36** with boron trifluoride and allylic silane generates five-membered boryl complexes **37**. These five-membered complexes readily react with the Ruppert–Prakash reagent and sodium acetate affording after hydrolytic workup trifluoromethylation products.[61] Similarly, hydrazones **38** can be transformed into six-membered complexes **39** followed by nucleophilic trifluoromethylation.[62] A similar Lewis acid chelation approach was realized for trifluoromethylation of salicyl aldimines[63] and imines obtained from aldehydes and 8-aminoquinoline.[64] Oxidative trifluoromethylation of phenanthrolines also likely involves activation of aromatic ring by the formation of a chelate complex.[65]

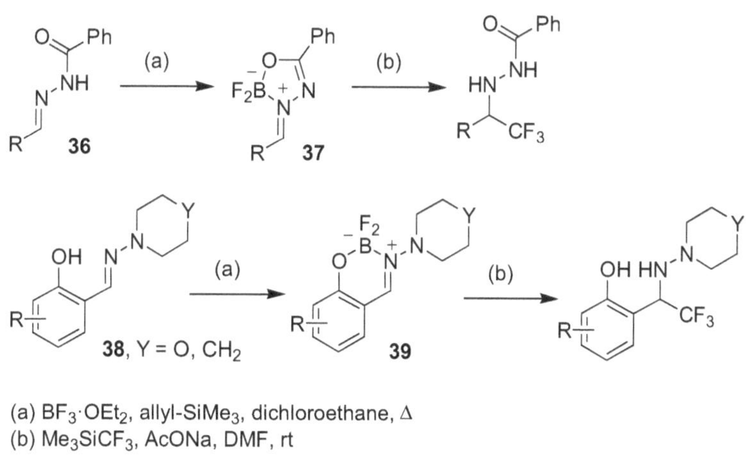

Scheme 7.14 Reactions of difluoroboryl complexes.

Asymmetric trifluoromethylation of C=N bond still is a challenging problem. The only successful example of such process was reported by Shibata for the addition to azomethine imines **40**[66,67] (Scheme 7.15). Cinchona alkaloid–derived quaternary ammonium salt was employed as a catalyst at 10 mol % amount. However, subsequent transformation of products **41** into primary amines requires harsh conditions.

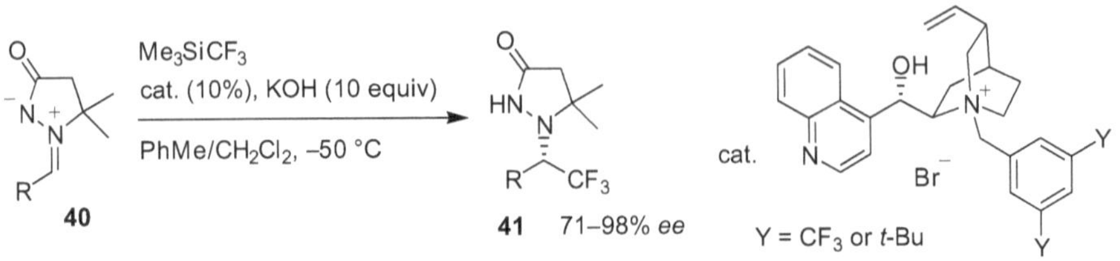

Scheme 7.15 Trifluoromethylation of azomethine imines.

2.2 *Derivatives of Trifluoroacetic Acid and Trifluoroacetaldehyde*

Trifluoroacetic acid is one of the cheapest fluorinated building blocks, and its application for nucleophilic trifluoromethylation is very attractive.[68] Decarboxylation of trifluoroacetate anion to generate trfluoromethyl carbanion can indeed proceed, but requires high temperatures, usually above 150°C. There were several reports describing nucleophilic trifluoromethylation of aldehydes using trifluoroacetate salts (Scheme 7.16). In these reactions, excess of trifluoroacetate (4 equiv) was employed. The addition of stoichiometric amount of copper iodide allowed to improve the yields, presumably, due to stabilization of the intermediate trifluoromethyl carbanion.[69,70]

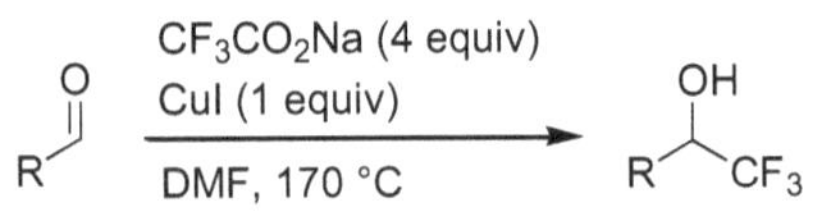

Scheme 7.16 Trifluoromethylation using sodium trifluoroacetate.

The *N*-trifluoroacetyl-*O*-trimethylsilyl derivative of ephedrine **42** was used for trifluoromethylation of aldehydes in the presence of catalytic amounts of fluoride (Scheme 7.17). Fluoride generates alkoxide, which attacks at the carbonyl group to give tetrahedral species **43**. Subsequent expulsion of CF_3 carbanion from **43** is driven by the formation of stable carbamate moiety. The alkoxide formed after the addition of the CF_3 carbanion to aldehyde then leads the chain process. Despite

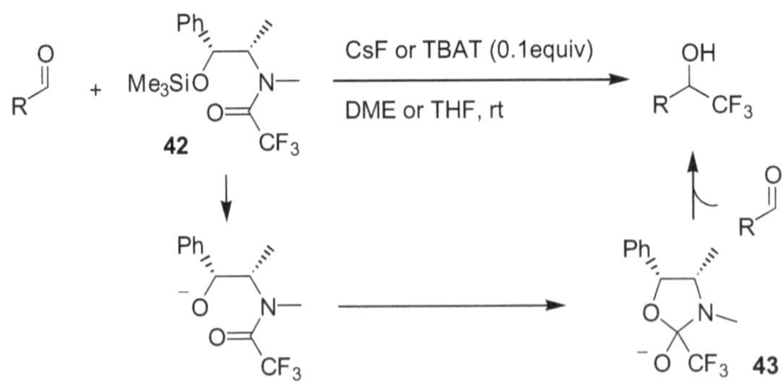

Scheme 7.17 Reactions of *N*-trifluoroacetyl-*O*-trimethylsilylephedrine.

the fact that the starting reagent **42** is chiral, the final products were formed with low enantioselectivities.[71,72]

Reaction of aldehydes with trifluoroacetaldehyde hydrate **44** is performed in the presence of excess of potassium *tert*-butoxide[73] (Scheme 7.18). It was proposed that *tert*-butoxide abstracts two acidic protons followed by transfer of the CF_3 group from dianionic intermediate onto aldehyde. However, due to the high acidity of the reaction medium, enolizable aldehydes are not suitable for this process. A similar mechanism is operative for the reaction of aldehydes with a salt of hexafluoroacetone hydrate and DBU (1,8-diazabicycloundec-7-ene) (**45**). Salt **45** is air-stable nonhygroscopic crystalline solid with good solubility in polar solvents. The optimal results were observed when potassium *tert*-butoxide was used in combination with tetrabutylammonium chloride in DMF as a solvent.[74]

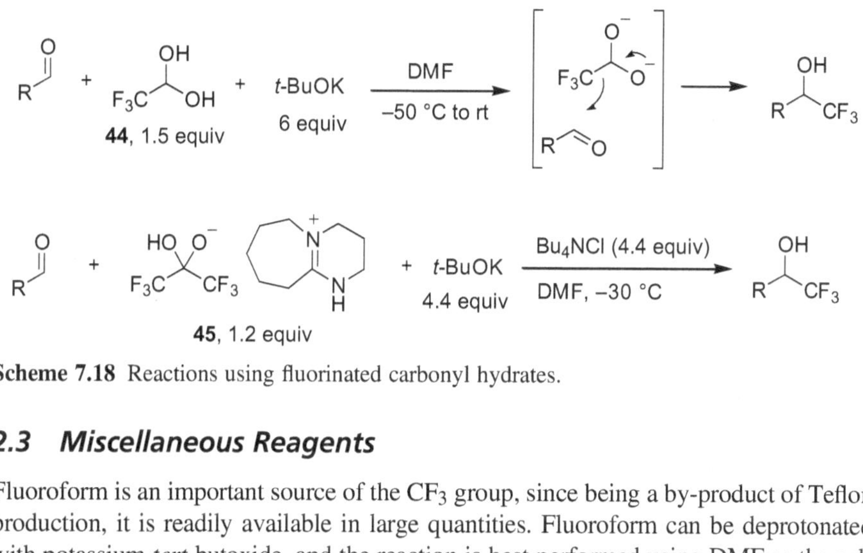

Scheme 7.18 Reactions using fluorinated carbonyl hydrates.

2.3 Miscellaneous Reagents

Fluoroform is an important source of the CF_3 group, since being a by-product of Teflon production, it is readily available in large quantities. Fluoroform can be deprotonated with potassium *tert*-butoxide, and the reaction is best performed using DMF as the solvent[68] (Scheme 7.19). The role of DMF is to trap trifluoromethyl carbanion, which is unstable in the presence of potassium counterion, to generate more stable adduct **46**. The latter species itself can serve as a source of the trifluoromethyl fragment in reaction with aldehydes.[75–78] Trifluoromethylation of benzophenones with fluoroform was

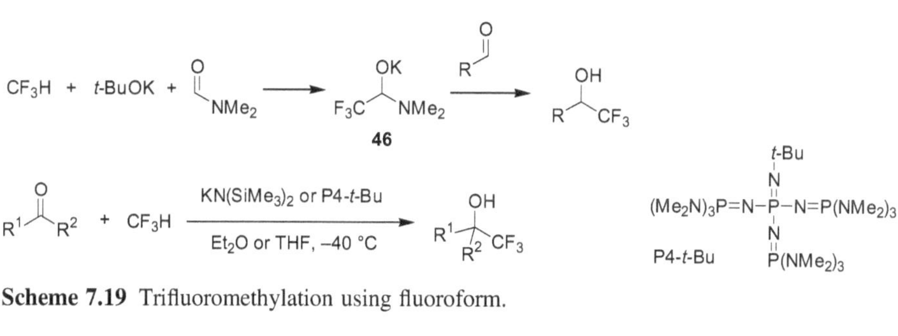

Scheme 7.19 Trifluoromethylation using fluoroform.

performed in ether but employing potassium hexamethyldisilazide as a base. This particular base was found optimal for deprotonation of fluoroform to trap trifluoromethyl carbanion with various electrophiles.[79] Phosphazene base P4-*t*-Bu can also be used to effect trifluoromethylation of ketones and aldehydes using tetrahydrofuran as solvent.[80,81]

Combination of tetrakis(dimethylamino)ethylene (TDAE, **47**) with trifluoromethyl iodide generates an orange-red charge transfer complex, which is stable at 0°C. This complex reacts with nonenolizable carbonyl compounds, *N*-tosylimines and *N*-sulfinylimines[12] (Scheme 7.20). The reaction is best performed in DMF. For aldehydes, irradiation with a sun lamp was important to achieve good yields of products.

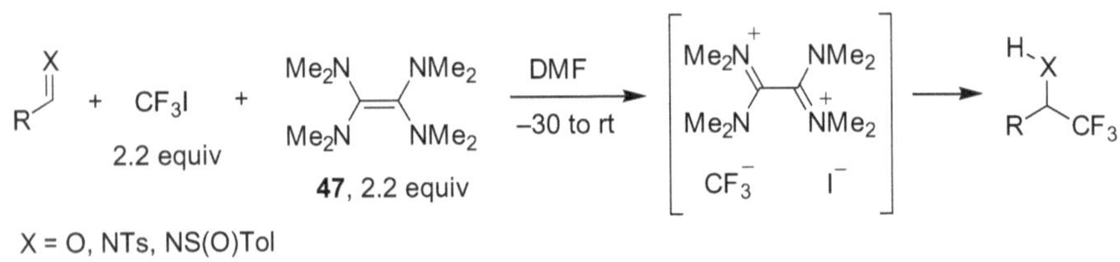

Scheme 7.20 Trifluoromethylation using trifluoroiodomethane.

Trifluoromethyl-substituted sulfone **48**[82,83] or phosphonate **49**[84] were used for trifluoromethylation of carbonyl compounds and imines (Scheme 7.21). These reagents have to be activated by alkoxide anion to generate five-coordinate intermediates, which serve as sources of the CF_3 carbanion.

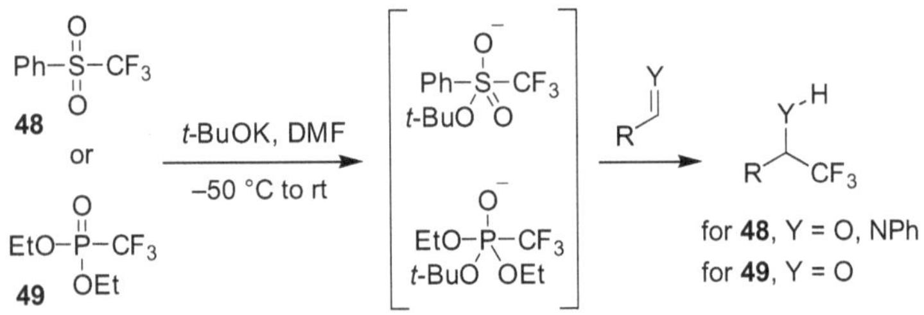

Scheme 7.21 Trifluoromethylation using sulfur and phosphorus reagents.

Borate salt **50** reacts with aldehydes and *N*-tosylimines at slightly elevated temperatures in DMF without any additional activator[85] (Scheme 7.22). Salt **50**, which is a stable crystalline solid, is obtained from trimethylborate, the Ruppert–Prakash reagent, and potassium fluoride. Although this salt is moderately basic, the highest yields were observed for nonenolizable aldehydes and ketones.

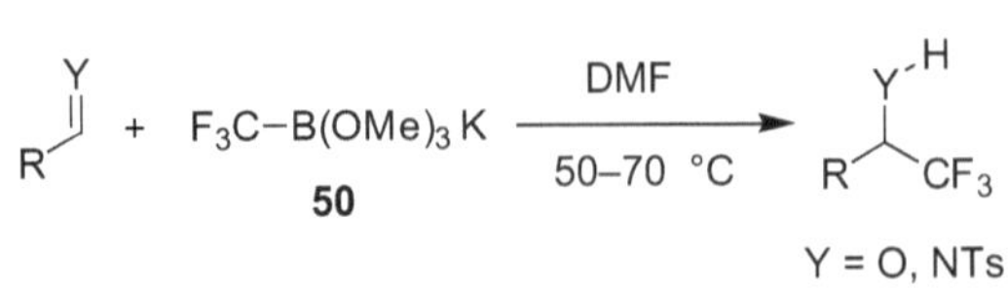

Scheme 7.22 Trifluoromethylation using borate reagent.

3. Difluoromethylation Reactions

3.1 Silicon Reagents

(Difluoromethyl)trimethylsilane (**51**) has long been hardly accessible until 2011, when Igoumnov reported the synthesis of **51** by reduction of the Ruppert−Prakash reagent with sodium borohydride in diglyme[86] (Scheme 7.23). Although the mechanism of this reaction is not clear, this method renders silane **51** readily available, and now this reagent is offered by several companies even on kilogram scale.

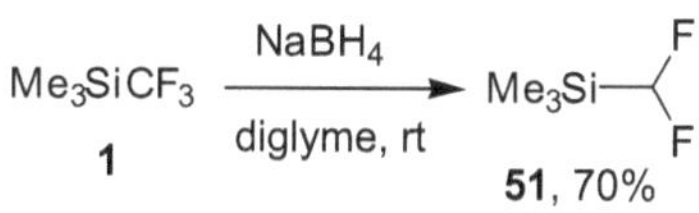

Scheme 7.23 Synthesis of (difluoromethyl) trimethylsilane.

Compared to silane **1**, reagent **51** is more difficult to activate to effect transfer of the difluoromethyl group. Optimal conditions involve use of catalytic amounts of cesium fluoride (10 mol%) in DMF[87,88] (Scheme 7.24). Interestingly, even enolizable substrates provided products in reasonable yields. Phosphazene base P4-*t*-Bu was also reported to be a competent mediator of reaction of aromatic aldehydes with silane **51**.[89] Chiral *N*-sulfinyl imines were also difluoromethylated by combination of silane **51** and potassium *tert*-butoxide furnishing products with good diastereoselectivities.[87]

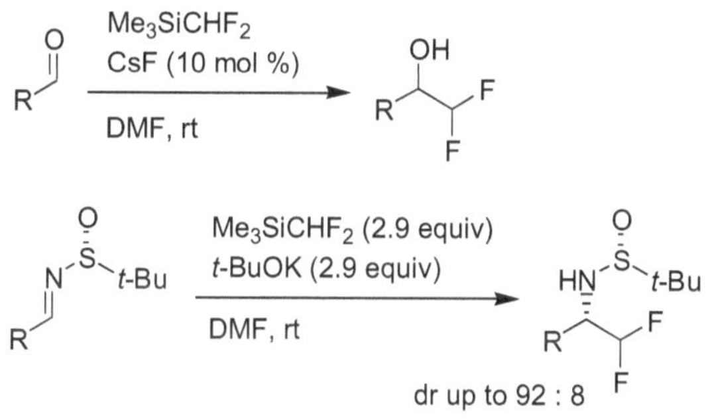

Scheme 7.24 Reactions using (difluoromethyl) trimethylsilane.

Silicon reagents **52**−**54** bearing sulfinyl,[90,91] sulfonyl,[92,93] or phosphoryl group[94] were used as equivalents of difluoromethyl carbanion (Scheme 7.25). The reactivity of these reagents is similar to that of the Ruppert−Prakash reagent, and coupling with aldehydes proceeds readily in the presence of fluoride activator leading to products **55**. To access difluoromethylated alcohols, the heteroatom group has to be removed. The sulfinyl fragment can be substituted by hydrogen either under free radical conditions using tributyltin hydride or by oxidation into sulfonyl followed by reduction of the C−S bond by magnesium or sodium. The phosphoryl group is removed under basic conditions.[95] Reagents **52** and **53** were also successfully used for reactions with *N*-alkyl imines mediated by potassium hydrodifluoride/trifluoroacetic acid.[93]

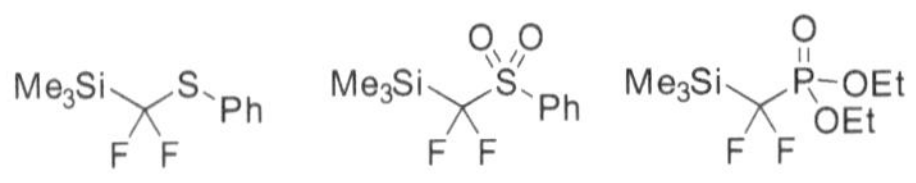

52 53 54

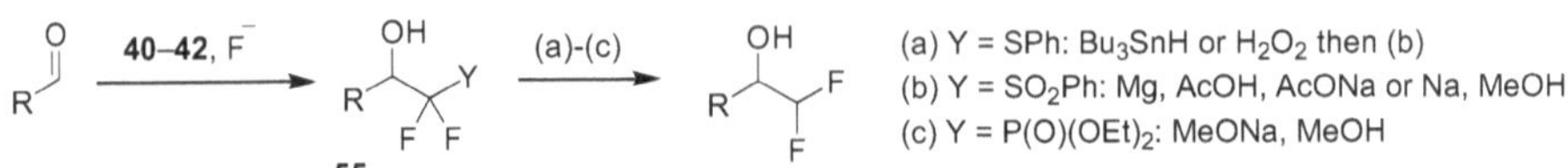

55

Scheme 7.25 Difluromethylation using sulfur and phosphorus reagents.

3.2 Miscellaneous Reagents

The ability of sulfur and phosphorus atoms to stabilize adjacent negative charge allows for the relatively straightforward generation of carbanionic species **56** or organometallic intermediates **57** (Scheme 7.26). Deprotonation of difluoromethyl-substituted starting compound or bromine/metal exchange (or reduction) of CF_2Br group provides intermediates **56** or **57**. Their reactions with C=O electrophiles afford products **55**, which can be further transformed into difluorinated alcohols (vide supra).

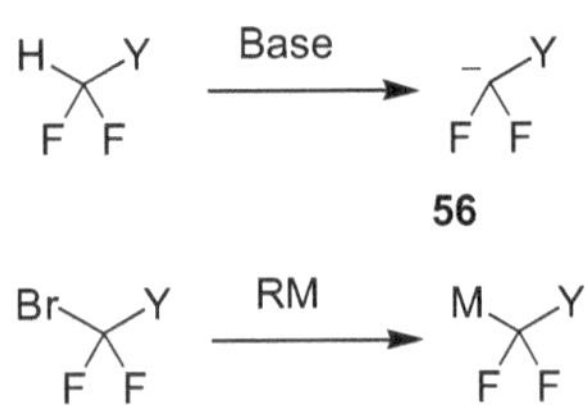

Y = SPh, SO_2Ph, $P(O)(OEt)_2$

Scheme 7.26 Generation of difluorosubstituted carbanionic species.

For thio-substituted species, the reactions are usually performed by the addition of a base to a mixture of aldehyde and difluoromethylphenyl sulfone **58** (Scheme 7.27). Aqueous base is sufficient to generate the carbanionic species, which is trapped by aldehyde, although these conditions are suitable only for aromatic or branched aliphatic aldehydes.[96] An asymmetric version of this processes was evaluated using chiral quaternary ammonium salt and a solid alkali hydroxide base with enantiomeric excess up to 47%.[97] Reaction in the presence of LiHMDS (lithium hexamethyldisilazide) as base has wider scope,[98] whereas combination of tris(trimethylsilylamine) and a catalytic amount of tetramethylammonium fluoride works well for enolizable aldehydes.[99] In the latter case, fluoride ion generates amide anion following chain process, wherein alkoxide anion formed after addition of the carbanion to the aldehyde maintains the chain propagation. The reaction of bromodifluoromethylphenyl sulfone CF_2BrSO_2Ph with aldehydes mediated by TDAE also gives addition products.[100] Reaction of carbonyl compounds with difluoromethylphenylsulfide **59** was performed using potassium *tert*-butoxide, solid potassium hydroxide,[101] or phosphazene base P4-*t*-Bu.[102]

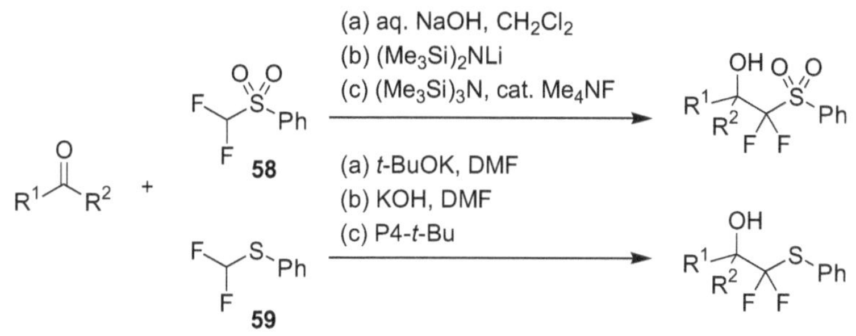

Scheme 7.27 Synthesis of difluorinated sulfones and sulfides.

Organolithium reagent **60** can be easily generated by the deprotonation of the corresponding difluorophosphonate with lithium diisopropylamide (Scheme 7.28). The reagent is stable at −78°C, and reacts with a wide variety of electrophiles including carbonyl compounds and imines.[103–105]

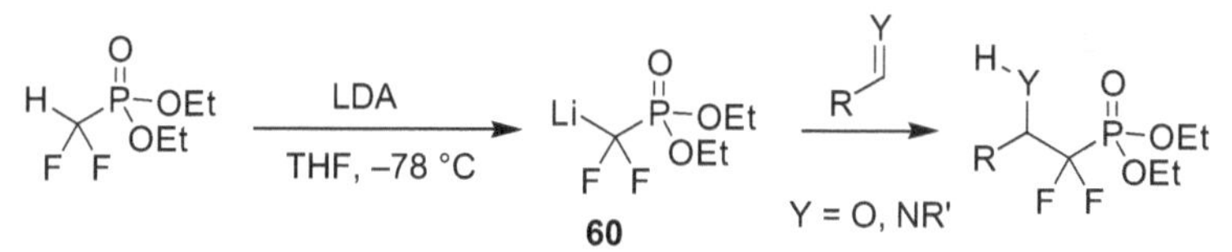

Scheme 7.28 Reactions using diethyl difluorophosphonate.

Difluoromethylene phosphabetaine **61** readily available from triphenylphosphine and potassium bromodifluoroacetate can serve as equivalent of difluoromethyl carbanion in reaction with aldehydes and azomethines[106] (Scheme 7.29). Phosphabetaine **61** is an air-stable solid poorly soluble in organic solvent.[107] Upon mild heating it

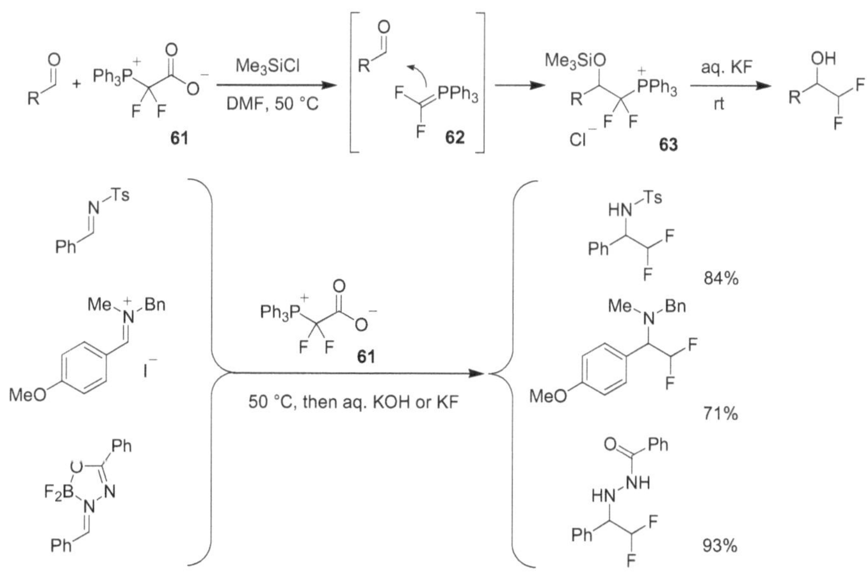

Scheme 7.29 Difluromethylation using phosphabetaine reagent.

undergoes decarboxylation to generate difluorinated phosphorus ylide **62**, which is trapped by aldehyde and chlorotrimethylsilane with the formation of phosphonium salt **63**. The presence of chlorosilane is essential, since otherwise products of the Wittig reaction are produced. When this salt is treated with aqueous potassium fluoride, smooth hydrolysis of carbon–phosphorus bond occurs leading to difluorinated alcohols. The reaction has a wide scope, and is applicable to both aromatic and enolizable aliphatic alcohols. The reagent was employed for difluoromethylation of *N*-tosylimines, iminium ions, and difluoroboryl complexes derived from *N*-benzoyl hydrazone.

4. Conclusion

Significant progress has been achieved in the trifluoromethylation and difluoromethylation of carbonyl compounds and azomethines. Reactions of aldehydes and ketones have a particularly wide scope. Among the variety of synthetic equivalents of trifluoromethyl carbanion, methods involving the Ruppert–Prakash reagent strongly dominate. A recent discovery of convenient procedure for the preparation of (difluoromethyl)trimethylsilane has sparked the interest in nucleophilic difluoromethylation process, and novel modes of its applications may follow in the near future. Another significant problem, which has to be tackled, is performing catalytic asymmetric tri- and difluoromethylation reactions.

Acknowledgments

This work was supported by the Ministry of Science (project MD-3256.2015.3) and Russian Foundation for Basic Research (projects 14-03-00293, 15-33-20133).

References

1. Wang, J.; Sanchez-Roselló, M.; Aceña, J. L.; del Pozo, C.; Sorochinsky, A. E.; Fustero, S.; Soloshonok, V. A.; Liu, H. *Chem. Rev.* **2014,** *114,* 2432–2506.
2. Müller, K.; Faeh, C.; Diederich, F. *Science* **2007,** *317,* 1881–1886.
3. Purser, S.; Moore, P. R.; Swallow, S.; Gouverneur, V. *Chem. Soc. Rev.* **2008,** *37,* 320–330.
4. O'Hagan, D. *J. Fluor. Chem.* **2010,** *131,* 1071–1081.
5. Prakash, G. K. S.; Wang, F.; Zhang, Z.; Haiges, R.; Rahm, M.; Christe, K. O.; Mathew, T.; Olah, G. A. *Angew. Chem. Int. Ed.* **2014,** *53,* 11575–11578.
6. Prakash, G. K. S.; Yudin, A. K. *Chem. Rev.* **1997,** *97,* 757–786.
7. Liu, X.; Xu, C.; Wang, M.; Liu, Q. *Chem. Rev.* **2015,** *115,* 683–730.
8. Prakash, G. K. S.; Mandal, M. *J. Fluor. Chem.* **2001,** *112,* 123–131.
9. Singh, R. P.; Shreeve, J. M. *Tetrahedron* **2000,** *56,* 7613–7632.
10. Rubiales, G.; Alonso, C.; de Marigorta, E. M.; Palacios, F. *ARKIVOC* **2014,** 362–405.
11. Dilman, A. D.; Levin, V. V. *Eur. J. Org. Chem.* **2011,** 831–841.
12. Medebielle, M.; Dolbier, W. R., Jr. *J. Fluor. Chem.* **2008,** *129,* 930–942.

13. Massicot, F.; Monnier-Benoit, N.; Deka, N.; Plantier-Royon, R.; Portella, C. *J. Org. Chem.* **2007,** *72,* 1174–1180.
14. Boechat, N.; Bastos, M. M. *Curr. Org. Synth.* **2010,** *7,* 403–413.
15. Hu, J.; Zhang, W.; Wang, F. *Chem. Commun.* **2009,** 7465–7478.
16. Prakash, G. K. S.; Krishnamurti, R.; Olah, G. A. *J. Am. Chem. Soc.* **1989,** *111,* 393–395.
17. Krishnamurti, R.; Bellew, D. R.; Prakash, G. K. S. *J. Org. Chem.* **1991,** *56,* 984–989.
18. Kolomeitsev, A.; Bissky, G.; Lork, E.; Movchun, V.; Rusanov, E.; Kirsch, P.; Röschenthaler, G. -V. *Chem. Commun.* **1999,** 1017–1018.
19. Maggiarosa, N.; Tyrra, W.; Naumann, D.; Kirij, N. V.; Yagupolskii, Y. L. *Angew. Chem. Int. Ed.* **1999,** *38,* 2252–2253.
20. Kawano, Y.; Kaneko, N.; Mukaiyama, T. *Bull. Chem. Soc. Jpn.* **2006,** *79,* 1133–1145.
21. Prakash, G. K. S.; Panja, C.; Vaghoo, H.; Surampudi, V.; Kultyshev, R.; Mandal, M.; Rasul, G.; Mathew, T.; Olah, G. A. *J. Org. Chem.* **2006,** *71,* 6806–6813.
22. Song, J. J.; Tan, Z.; Reeves, J. T.; Gallou, F.; Yee, N. K.; Senanayake, C. H. *Org. Lett.* **2005,** *7,* 2193–2196.
23. Mizuta, S.; Shibata, N.; Sato, T.; Fujimoto, H.; Nakamura, S.; Toru, T. *Synlett* **2006,** 267–270.
24. Matsukawa, S.; Saijo, M. *Tetrahedron Lett.* **2008,** *49,* 4655–4657.
25. Iwanami, K.; Oriyama, T. *Synlett* **2006,** 112–114.
26. Prakash, G. K. S.; Mandal, M.; Panja, C.; Mathew, T.; Olah, G. A. *J. Fluor. Chem.* **2003,** *123,* 61–63.
27. Klein, J. E. M. N.; Rommel, S.; Plietker, B. *Organometallics* **2014,** *33,* 5802–5810.
28. Wiedemann, J.; Heiner, T.; Mloston, G.; Prakash, G. K. S.; Olah, G. A. *Angew. Chem. Int. Ed.* **1998,** *37,* 820–821.
29. Fustero, S.; Albert, L.; Acena, J. L.; Sanz-Cervera, J. F.; Asensio, A. *Org. Lett.* **2008,** *10,* 605–608.
30. Babadzhanova, L. A.; Kirij, N. V.; Yagupolskii, Y. L.; Tyrra, W.; Naumann, D. *Tetrahedron* **2005,** *61,* 1813–1819.
31. Rudzinski, D. M.; Kelly, C. B.; Leadbeater, N. E. *Chem. Commun.* **2012,** *48,* 9610–9612.
32. Hoffmann-Röder, A.; Seiler, P.; Diederich, F. *Org. Biomol. Chem.* **2004,** *2,* 2267–2269.
33. Brigaud, T.; Doussot, P.; Portella, C. *J. Chem. Soc. Chem. Commun.* **1994,** 2117–2118.
34. Lefebvre, O.; Brigaud, T.; Portella, C. *J. Org. Chem.* **2001,** *66,* 4348–4351.
35. Portella, C.; Brigaud, T.; Lefebvre, O.; Plantier-Royon, R. *J. Fluor. Chem.* **2000,** *101,* 193–198.
36. Jonet, S.; Cherouvrier, F.; Brigaud, T.; Portella, C. *Eur. J. Org. Chem.* **2005,** 4304–4312.
37. Demir, A. S.; Eymur, S. *J. Org. Chem.* **2007,** *72,* 8527–8530.
38. Iseki, K.; Nagai, T.; Kobayashi, Y. *Tetrahedron Lett.* **1994,** *35,* 3137–3138.
39. Caron, S.; Do, N. M.; Arpin, P.; Larivee, A. *Synthesis* **2003,** 1693–1698.
40. Mizuta, S.; Shibata, N.; Akiti, S.; Fujimoto, H.; Nakamura, S.; Toru, T. *Org. Lett.* **2007,** *9,* 3707–3710.
41. Kawai, H.; Tachi, K.; Tokunaga, E.; Shiro, M.; Shibata, N. *Org. Lett.* **2010,** *12,* 5104–5107.
42. Kawai, H.; Mizuta, S.; Tokunaga, E.; Shibata, N. *J. Fluor. Chem.* **2013,** *152,* 46–50.
43. Wu, S.; Zeng, W.; Wang, Q.; Chen, F. -X. *Org. Biomol. Chem.* **2012,** *10,* 9334–9337.
44. Wu, S.; Guo, J.; Sohail, M.; Cao, C.; Chen, F. -X. *J. Fluor. Chem.* **2013,** *148,* 19–29.
45. Prakash, G. K. S.; Mandal, M.; Olah, G. A. *Synlett* **2001,** 77–78.
46. Kawano, Y.; Fujisawa, H.; Mukaiyama, T. *Chem. Lett.* **2005,** *34,* 422–423.
47. Prakash, G. K. S.; Mandal, M.; Olah, G. A. *Angew. Chem. Int. Ed.* **2001,** *40,* 589–590.
48. Prakash, G. K. S.; Mandal, M.; Olah, G. A. *Org. Lett.* **2001,** *3,* 2847–2850.

49. Kirij, N. V.; Babadzhanova, L. A.; Movchun, V. N.; Yagupolskii, Y. L.; Tyrra, W.; Naumann, D.; Fischer, H. T. M.; Scherer, H. *J. Fluor. Chem.* **2008,** *129,* 14–21.
50. Levin, V. V.; Dilman, A. D.; Belyakov, P. A.; Struchkova, M. I.; Tartakovsky, V. A. *Eur. J. Org. Chem.* **2008,** 5226–5230.
51. Shevchenko, N. E.; Vlasov, K.; Nenajdenko, V. G.; Röschenthaler, G. -V. *Tetrahedron* **2011,** *67,* 69–74.
52. Radchenko, D. S.; Michurin, O. M.; Chernykh, A. V.; Lukin, O.; Mykhailiuk, P. K. *Tetrahedron Lett.* **2013,** *54,* 1897–1898.
53. Gritsenko, R. T.; Levin, V. V.; Dilman, A. D.; Belyakov, P. A.; Struchkova, M. I.; Tartakovsky, V. A. *Tetrahedron Lett.* **2009,** *50,* 2994–2997.
54. Levin, V. V.; Kozlov, M. A.; Song, Y. -H.; Dilman, A. D.; Belyakov, P. A.; Struchkova, M. I.; Tartakovsky, V. A. *Tetrahedron Lett.* **2008,** *49,* 3108–3111.
55. Levin, V. V.; Kozlov, M. A.; Dilman, A. D.; Belyakov, P. A.; Struchkova, M. I.; Tartakovsky, V. A. *Russ. Chem. Bull. Int. Ed.* **2009,** *58,* 484–486.
56. Loska, R.; Majcher, M.; Makosza, M. *J. Org. Chem.* **2007,** *72,* 5574–5580.
57. Wang, X.; Qiu, G.; Zhang, L.; Wu, J. *Tetrahedron Lett.* **2014,** *55,* 962–964.
58. Chu, L.; Qing, F. -L. *Chem. Commun.* **2010,** *46,* 6285–6287.
59. Fu, W.; Guo, W.; Zou, G.; Xu, C. *J. Fluor. Chem.* **2012,** *140,* 88–94.
60. Pan, Y.; Wang, S.; Kee, C. W.; Dubuisson, E.; Yang, Y.; Loh, K. P.; Tan, C. -H. *Green Chem.* **2011,** *13,* 3341–3344.
61. Dilman, A. D.; Arkhipov, D. E.; Levin, V. V.; Belyakov, P. A.; Korlyukov, A. A.; Struchkova, M. I.; Tartakovsky, V. A. *J. Org. Chem.* **2008,** *73,* 5643–5646.
62. Dilman, A. D.; Levin, V. V.; Belyakov, P. A.; Korlyukov, A. A.; Struchkova, M. I.; Tartakovsky, V. A. *Mendeleev Commun.* **2009,** *19,* 141–143.
63. Dilman, A. D.; Arkhipov, D. E.; Levin, V. V.; Belyakov, P. A.; Korlyukov, A. A.; Struchkova, M. I.; Tartakovsky, V. A. *J. Org. Chem.* **2007,** *72,* 8604–8607.
64. Zhang, Y. -Q.; Liu, J. -D.; Xu, H. *Org. Biomol. Chem.* **2013,** *11,* 6242–6245.
65. Zhu, C. -L.; Zhang, Y. -Q.; Yuan, Y. -A.; Xu, H. *Synlett* **2015,** *26,* 345–349.
66. Kawai, H.; Kusuda, A.; Nakamura, S.; Shiro, M.; Shibata, N. *Angew. Chem. Int. Ed.* **2009,** *48,* 6324–6327.
67. Okusu, S.; Kawai, H.; Xu, X. -H.; Tokunaga, E.; Shibata, N. *J. Fluor. Chem.* **2012,** *143,* 216–219.
68. Langlois, B. R.; Billard, T. *In Fluorine-containing Synthons,* vol. 911; American Chemical Society, 2005; pp 57–86.
69. Chang, Y.; Cai, C. *Tetrahedron Lett.* **2005,** *46,* 3161–3164.
70. Chang, Y.; Cai, C. *J. Fluor. Chem.* **2005,** *126,* 937–940.
71. Joubert, J.; Roussel, S.; Christophe, C.; Billard, T.; Langlois, B. R.; Vidal, T. *Angew. Chem. Int. Ed.* **2003,** *42,* 3133–3136.
72. Roussel, S.; Billard, T.; Langlois, B. R.; Saint-James, L. *Chem. Eur. J.* **2005,** *11,* 939–944.
73. Prakash, G. K. S.; Zhang, Z.; Wang, F.; Munoz, S.; Olah, G. A. *J. Org. Chem.* **2013,** *78,* 3300–3305.
74. Riofski, M. V.; Hart, A. D.; Colby, D. A. *Org. Lett.* **2013,** *15,* 208–211.
75. Shono, T.; Ishifune, M.; Okada, T.; Kashimura, S. *J. Org. Chem.* **1991,** *56,* 2–4.
76. Russell, J.; Roques, N. *Tetrahedron* **1998,** *54,* 13771–13782.
77. Folleas, B.; Marek, I.; Normant, J. -F.; Saint-Jalmes, L. *Tetrahedron* **2000,** *56,* 275–283.
78. Large, S.; Roques, N.; Langlois, B. R. *J. Org. Chem.* **2000,** *65,* 8848–8856.
79. Prakash, G. K. S.; Jog, P. V.; Batamack, P. T. D.; Olah, G. A. *Science* **2012,** *338,* 1324–1327.

80. Kawai, H.; Yuan, Z.; Tokunaga, E.; Shibata, N. *Org. Biomol. Chem.* **2013,** *11,* 1446–1450.
81. Zhang, Y.; Fujiu, M.; Serizawa, H.; Mikami, K. *J. Fluor. Chem.* **2013,** *156,* 367–371.
82. Prakash, G. K. S.; Hu, J.; Olah, G. A. *Org. Lett.* **2003,** *5,* 3253–3256.
83. Prakash, G. K. S.; Wang, Y.; Mogi, R.; Hu, J.; Mathew, T.; Olah, G. A. *Org. Lett.* **2010,** *12,* 2932–2935.
84. Cherkupally, P.; Beier, P. *Tetrahedron Lett.* **2010,** *51,* 252–255.
85. Levin, V. V.; Dilman, A. D.; Belyakov, P. A.; Struchkova, M. I.; Tartakovsky, V. A. *Tetrahedron Lett.* **2011,** *52,* 281–284.
86. Tyutyunov, A. A.; Boyko, V. E.; Igoumnov, S. M. *Fluorine Notes* **2011,** *74,* 1–3.
87. Zhao, Y.; Huang, W.; Zheng, J.; Hu, J. *Org. Lett.* **2011,** *13,* 5342–5345.
88. Tyutyunov, A. A.; Boyko, V. E.; Igoumnov, S. M. *Fluorine Notes* **2011,** *78,* 1–5.
89. Du, G. -F.; Wang, Y.; Gu, C. -Z.; Dai, B.; He, L. *RSC Adv.* **2015,** *5,* 35421–35424.
90. Prakash, G. K. S.; Hu, J.; Wang, Y.; Olah, G. A. *J. Fluor. Chem.* **2005,** *126,* 529–534.
91. Soorukram, D.; Kuhakarn, C.; Reutrakul, V.; Pohmakotr, M. *Synlett* **2014,** *25,* 2558–2573.
92. Ni, C.; Hu, J. *Tetrahedron Lett.* **2005,** *46,* 8273–8277.
93. Kosobokov, M. D.; Dilman, A. D.; Struchkova, M. I.; Belyakov, P. A.; Hu, J. *J. Org. Chem.* **2012,** *77,* 2080–2086.
94. Alexandrova, A. V.; Beier, P. *J. Fluor. Chem.* **2009,** *130,* 493–500.
95. Beier, P.; Alexandrova, A. V.; Zibinsky, M.; Prakash, G. K. S. *Tetrahedron* **2008,** *64,* 10977–10985.
96. Stahly, G. P. *J. Fluor. Chem.* **1989,** *43,* 53–66.
97. Ni, C.; Wang, F.; Hu, J. *Beilstein J. Org. Chem.* **2008,** *4,* 21.
98. Prakash, G. K. S.; Hu, J.; Wang, Y.; Olah, G. A. *Eur. J. Org. Chem.* **2005,** 2218–2223.
99. Hu, M.; Gao, B.; Ni, C.; Zhang, L.; Hu, J. *J. Fluor. Chem.* **2013,** *155,* 52–58.
100. Prakash, G. K. S.; Wang, Y.; Hu, J.; Olah, G. A. *J. Fluor. Chem.* **2005,** *126,* 1361–1367.
101. Hu, M.; Wang, F.; Zhao, Y.; He, Z.; Zhang, W.; Hu, J. *J. Fluor. Chem.* **2012,** *135,* 45–58.
102. Punirun, T.; Soorukram, D.; Kuhakarn, C.; Reutrakul, V.; Pohmakotr, M. *Eur. J. Org. Chem.* **2014,** 4162–4169.
103. Obayashi, M.; Ito, E.; Matsui, K.; Kondo, K. *Tetrahedron Lett.* **1982,** *23,* 2323–2326.
104. Cherkupally, P.; Beier, P. *J. Fluor. Chem.* **2012,** *141,* 76–82.
105. Chunikhin, K. S.; Kadyrov, A. A.; Pasternak, P. V.; Chkanikov, N. D. *Russ. Chem. Rev.* **2010,** *79,* 371–396.
106. Levin, V. V.; Trifonov, A. L.; Zemtsov, A. A.; Struchkova, M. I.; Arkhipov, D. E.; Dilman, A. D. *Org. Lett.* **2014,** *16,* 6256–6259.
107. Zheng, J.; Cai, J.; Lin, J. -H.; Guo, Y.; Xiao, J. -C. *Chem. Commun.* **2013,** *49,* 7513–7515.

Oxidative Trifluoromethylation and Trifluoromethylthiolation

8

F.-L. Qing[1,2], X.-H. Xu[1]
[1]Shanghai Institute of Organic Chemistry, Shanghai, China; [2]Donghua University, Shanghai, China

Chapter Outline

1. Introduction

It is well known that the incorporation of fluorine-containing groups into an organic molecule often drastically perturbs the chemical, physical, and biological properties of the parent compound.[1] Thus, fluorinated organic compounds are receiving increased attention in the medicinal, pharmaceutical, agricultural, and material sciences.[2] However, only a dozen fluorinated organic compounds have been identified in nature. It is not surprising that numerous efforts have been devoted to the development of new methodologies for the preparation of fluorinated organic compounds.[3]

Normally, methods for the incorporation of fluorine-containing groups into organic molecules are considered as nucleophilic, electrophilic, or free radical processes. Traditionally, nucleophilic process involves the electrophile and the fluorine-containing anion (Scheme 8.1A), whereas electrophilic process involves the nucleophile and the fluorine-containing cation (Scheme 8.1B). Inspired by the Glaser—Hay coupling[4] and Chan—Evans—Lam reaction,[5] where two nucleophiles underwent cross-coupling in the presence of oxidants,[6] we questioned whether it might be possible to prepare fluorinated organic compounds by the oxidative methodology, the reaction of a nucleophile and a nucleophilic fluorine-containing agent in the presence of an exogenous oxidant (Scheme 8.1C).

Modern Synthesis Processes and Reactivity of Fluorinated Compounds. http://dx.doi.org/10.1016/B978-0-12-803740-9.00008-1

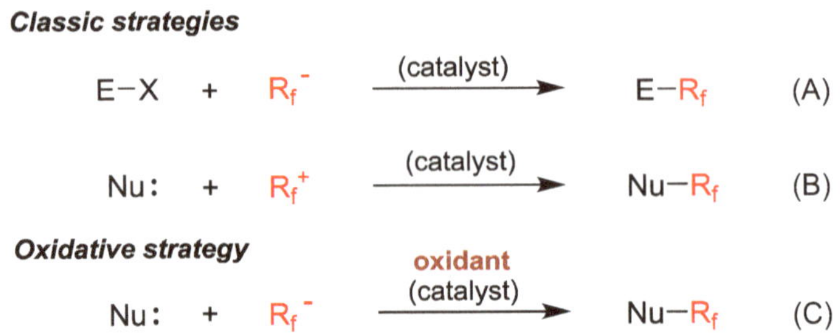

Scheme 8.1 Different strategies for the preparation of fluorinated compounds.

In 2010, our group initially reported the construction of C(sp, sp^2, sp^3)–CF_3 bonds[7–9] by oxidative trifluoromethylation, the reaction of nucleophilic substrates and nucleophilic trifluoromethylation reagents in the presence of oxidants. Given the variability of nucleophiles and the tunability of exogenous oxidants, more and more oxidative trifluoromethylation reactions have been developed since then.[10] Furthermore, this oxidative strategy has been extended to the preparation of other fluorinated organic compounds, such as trifluoromethylthiolated and perfluoroalkylated compounds.

In this chapter, we will present the development of methods for the preparation of fluorinated organic compounds by oxidative strategies, mainly oxidative trifluoromethylation and trifluoromethylthiolation from 2010 to 2015. It is noteworthy that only the reactions involving nucleophilic fluorine-containing reagents are included in this chapter. The reactions of nucleophiles with electrophilic or radical fluorine-containing reagents in the presence of oxidants (sometimes the electrophilic reagents themselves act as oxidants) belong to a broader definition of oxidative strategy and are not included in this chapter.

2. Oxidative Trifluoromethylation

2.1 Oxidative Trifluoromethylation for the Formation of C(sp)–CF_3 Bond

Trifluoromethylated acetylenes have been widely used as versatile building blocks in medicinal, agrochemical, and material science. In addition, they are also excellent precursors for the synthesis of other trifluoromethylated building blocks. Despite their apparent simplicity, the preparation of these building blocks is far from being trivial. This quest for direct methods for the trifluoromethylation of alkynes recently resulted in the development of innovative and more practical routes to trifluoromethylated acetylenes. In 2010, our group reported the copper-mediated oxidative trifluoromethylation of alkynes with Ruppert–Prakash reagent ($TMSCF_3$) (Scheme 8.2).[7] This work was the first example of the synthesis of trifluoromethylated alkynes from the direct trifluoromethylation of alkynes.

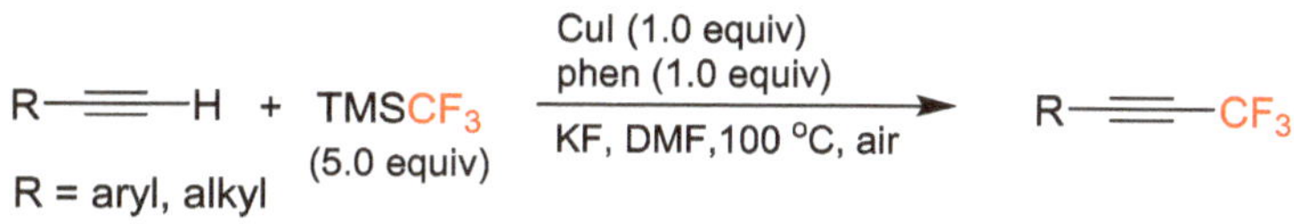

Scheme 8.2 Oxidative trifluoromethylation of terminal alkynes. *DMF*, dimethyl formamide.

Although this reaction represents the first example of oxidative cross-coupling of a nucleophile and $TMSCF_3$, it had several limitations including the use of stoichiometric amounts of both the copper salt and the ligand, an excess of $TMSCF_3$, and high temperatures. To address these limitations, we developed two improved procedures in 2012. The first one is the copper-catalyzed protocol, which is achieved by adding both the alkynes and a portion of $TMSCF_3$ slowly using a syringe pump to the reaction mixture.[11] The second one proceeds at room temperature by using (phen)$CuCF_3$ generated in situ from CuCl/phen/*t*-BuOK/$TMSCF_3$.[12] After 2 years, Blanchard and Evano reported two complementary methods for the synthesis of trifluoromethylated alkynes.[13] One is based on the readily available and bench-stable copper acetylides (Scheme 8.3A), whereas the other is based on (tetramethyl ethylenediamine) TMEDA-chelated trifluoromethylcopper complex (Scheme 8.3B). These two processes are very simple to set up under very mild conditions.

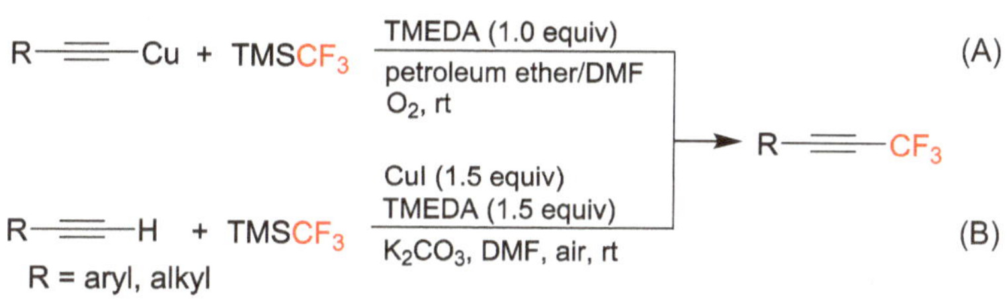

Scheme 8.3 Practical oxidative trifluoromethylation of terminal alkynes. *TMEDA*, tetramethyl ethylenediamine.

Mikami et al.[14] and Hu et al.,[15] respectively, reported the treatment of $PhCOCF_3$ and $PhSOCF_3$ with CuCl and *t*-BuOK to generate "$CuCF_3$," which then underwent oxidative cross-coupling with terminal alkynes for the preparation of trifluoromethylated alkynes (Scheme 8.4). However, both of these methods need the generation of trifluoromethyl copper prior to its reaction with the alkyne.

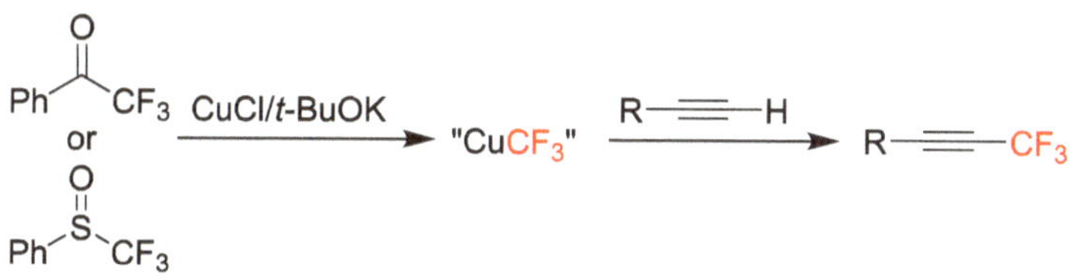

Scheme 8.4 Oxidative trifluoromethylation of terminal alkynes with $PhCOCF_3$ or $PhSOCF_3$.

Xiao and coworkers developed a Cu-promoted oxidative trifluoromethylation of terminal alkynes with difluoromethylene phosphobetaine, a known difluorocarbene reagent (Scheme 8.5).[16]

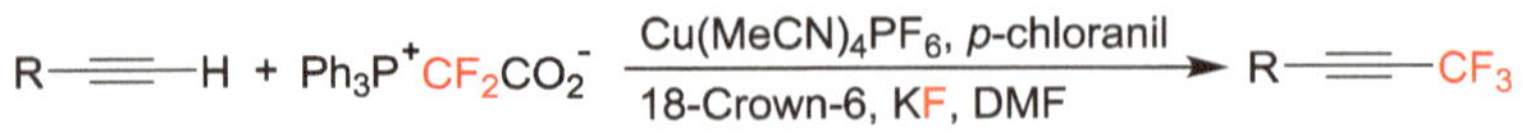

Scheme 8.5 Oxidative trifluoromethylation of terminal alkynes with difluoromethylene phosphobetaine.

2.2 *Oxidative Trifluoromethylation for the Formation of C(sp^2)–CF$_3$ Bond*

Aryl boronic acid derivatives are compatible with a broad range of common functional groups, commercially available and stable to air and moisture. Thus, aryl boronic acid derivatives have been widely used in organic synthesis, such as the oxidative coupling with nucleophiles, which is named Chan–Evans–Lam coupling reactions.[5] A wide range of heteroatom nucleophiles, including amines, amides, nitrogen heterocycles, alcohols, and phenols, were used in this oxidative coupling reaction. In 2010, our group successfully extended the substrate scope to $TMSCF_3$.[8] The copper-mediated cross-coupling of aryl or alkenyl boronic acids with $TMSCF_3$ using Ag_2CO_3 as the oxidant afforded the corresponding trifluoromethylated products in moderate to excellent yields (Scheme 8.6A). Soon after our work on the oxidative trifluoromethylation of boronic acids, Buchwald and coworkers also reported a similar but more economical protocol, in which dry O_2 was used as the oxidant (Scheme 8.6B).[17]

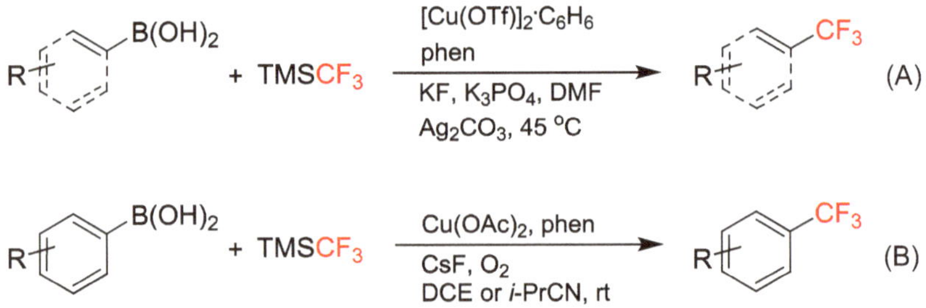

Scheme 8.6 Oxidative trifluoromethylation of aryl and alkenyl boronic acids. *DCE*, dichloroethane; *DMF*, dimethyl formamide.

In 2012, Gooβen[18] and Hartwig[19] successively developed two user-friendly oxidative trifluoromethylation of aryl boronates with shelf-stable $K[CF_3B(OMe)_3]$ (Scheme 8.7A) or preprepared thermally stable (phen)$CuCF_3$ (Scheme 8.7B). Both these trifluoromethylating reagents are solid and easy to handle.

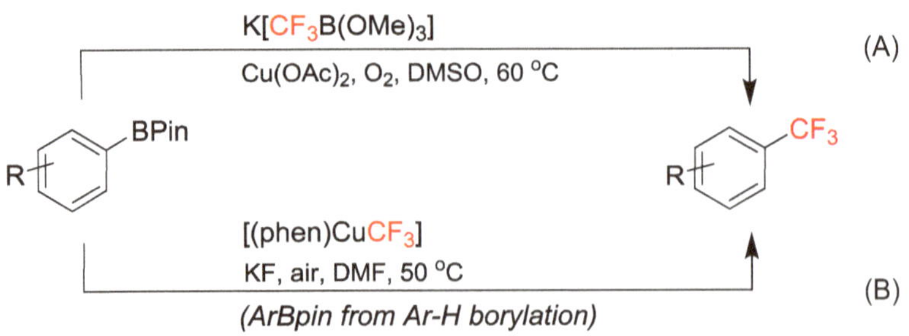

Scheme 8.7 Oxidative trifluoromethylation of aryl boronates. *DMF*, dimethyl formamide; *DMSO*, dimethyl sulphoxide.

In 2012, Grushin and coworkers disclosed a simple and low-cost oxidative trifluoromethylation of aryl boronic acids with fluoroform-derived $CuCF_3$ reagent in an atmosphere of air (Scheme 8.8A).[20] It was noteworthy that this oxidative trifluoromethylation reaction proceeded smoothly at room temperature in the absence of any extra ligands/additives. This method was applied for the radiosynthesis of pharmacologically relevant [^{18}F]trifluoromethylarenes and heteroarenes via the [^{18}F]$CuCF_3$ intermediate (Scheme 8.8B).[21]

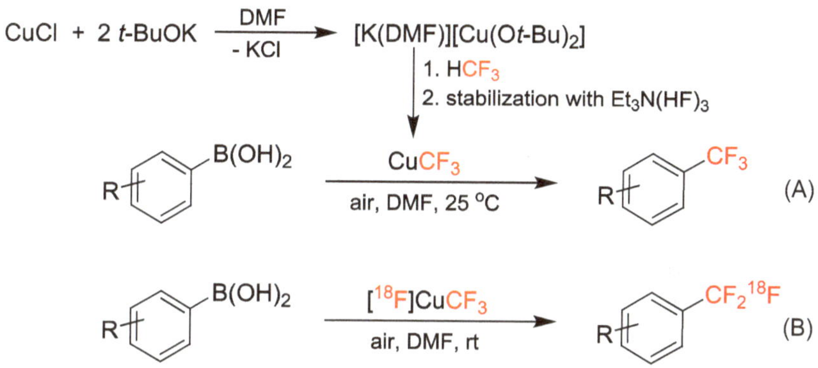

Scheme 8.8 Oxidative trifluoromethylation of aryl boronic acids with fluoroform-derived $CuCF_3$.

Compared with oxidative trifluoromethylation of aryl boronic acids, the direct trifluoromethylation of aryl C–H bonds is more straightforward and more economical because it does not require the prefunctionalized starting materials. In 2011, Sanford and coworkers reported a Ag-mediated oxidative C–H trifluoromethylation of aromatic substrates with $TMSCF_3$ under mild conditions (Scheme 8.9A).[22a] They proposed that the reaction proceeded via a $AgCF_3$ intermediate and that CF_3 radical might be involved in further transformation. Two years later, Greaney and coworkers reported a Ag-catalyzed trifluoromethylation system for electron-rich aromatic and heteroaromatic substrates (Scheme 8.9B).[22b] In the same year, our group disclosed a $PhI(OAc)_2$-mediated oxidative trifluoromethylation of arenes with $TMSCF_3$ under metal-free conditions (Scheme 8.9C).[22c] This protocol precludes the need of substrate prefunctionalization and metal catalysts. The addition of catalytic amount of benzoquinonine (BQ) is necessary to achieve high yield, although the role of BQ remains to be elucidated.

Another important type of C–H trifluoromethylation is regioselective trifluoromethylation of aromatic substrates with directing groups. This protocol provides a more precise installation of the trifluoromethyl group than radical trifluoromethylation. For example, Hafner and Bräse described a highly selective *ortho*-trifluoromethylation of aromatic triazenes (Scheme 8.10A).[23] Triazenes are useful equivalents of protected diazonium salts and therefore can be easily transformed into various functional groups. In 2014, Dai and Yu reported a direct copper(II)-promoted C–H *ortho*-trifluoromethylation of benzamides, which were prepared from the condensation of the corresponding benzoic acids and 2-(2-oxazolyl)aniline (Scheme 8.10B).[24] This amide-directing group is also synthetically useful and could be easily transformed into other functional groups.

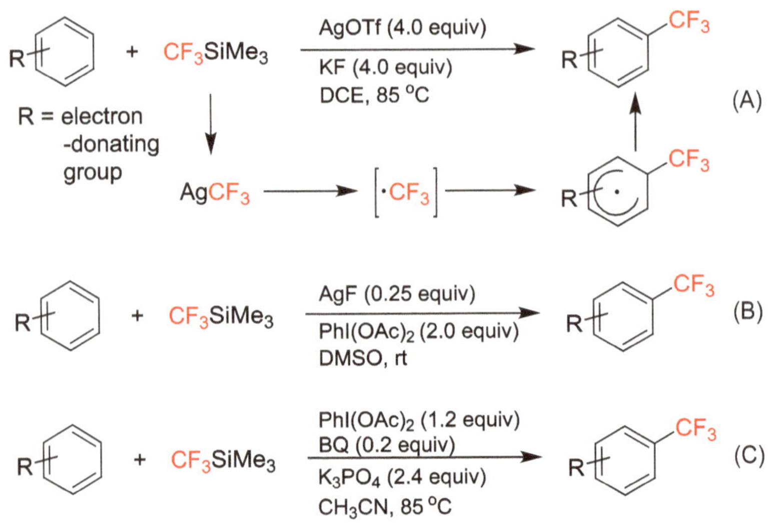

Scheme 8.9 Oxidative C–H trifluoromethylation of electron-rich aromatic and heteroaromatic substrates.

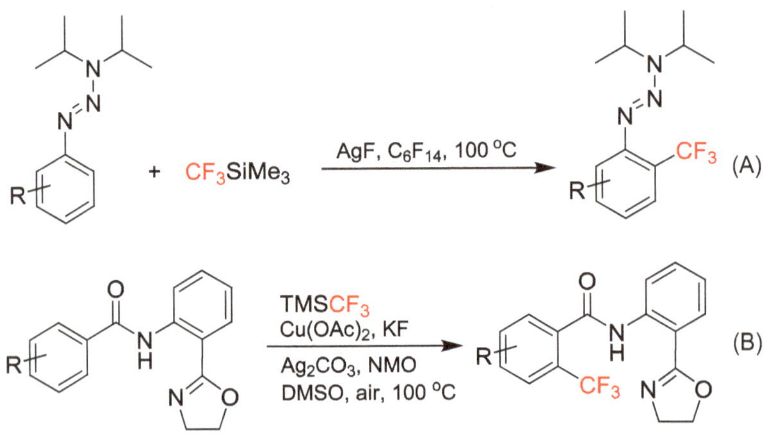

Scheme 8.10 Directing group-assisted oxidative C–H trifluoromethylation. *NMO*, 4-methylmorpholine N-oxide.

Inspired by our previous work of oxidative trifluoromethylation of terminal alkynes with $TMSCF_3$,[7] we envisioned that the copper-mediated direct C–H oxidative trifluoromethylation strategy might be applicable to heteroarenes, which bear similar acidic C–H bonds as those of terminal alkynes. In 2012, we accomplished the copper-catalyzed oxidative trifluoromethylation of heteroarenes and highly

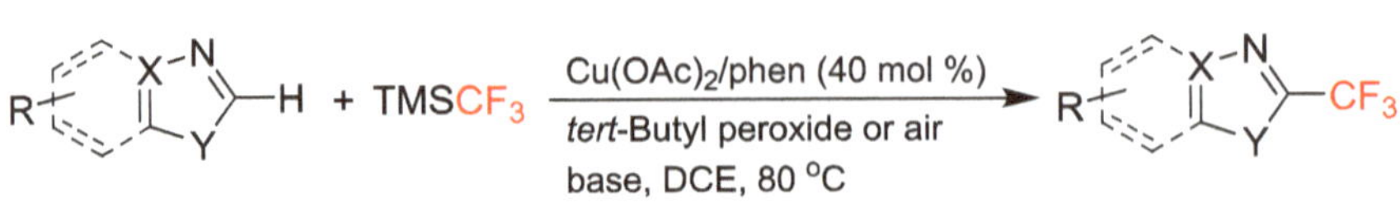

Scheme 8.11 Oxidative trifluoromethylation of heteroarenes.

electron-deficient arenes with $TMSCF_3$ through C−H activation (Scheme 8.11).[25] Various functionalized heteroarenes including 1,3,4-oxadiazole, benzo[*d*]oxazoles, benzo[*d*]imidazoles, imidazoles, and benzo[*d*]thiazoles, as well as electron-deficient polyfluoroarenes were regioselectively trifluoromethylated, affording the corresponding trifluoromethylated heteroarenes and arenes in moderate to excellent yields.

Liu and coworkers explored a Pd-catalyzed oxidative trifluoromethylation of indoles at room temperature, in which $PhI(OAc)_2$ was used as an oxidant (Scheme 8.12A).[26] In this reaction, a $Pd^{II/IV}$ pathway is proposed to lead to the trifluoromethylated indoles. Wang and Liu reported a transition-metal-free oxidative trifluoromethylation of indoles (Scheme 8.12B).[27] They proposed that this transformation proceeded through the formation of PhI^+CF_3, which was followed by nucleophilic attack of the nucleophilic C3 atom of indoles, reduction elimination, and rearomatization.

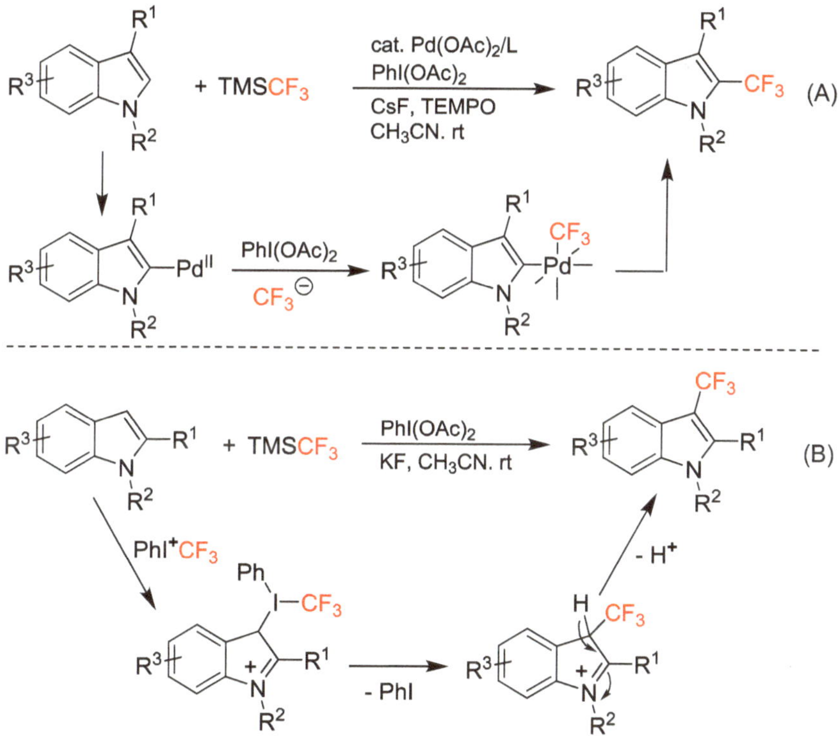

Scheme 8.12 Oxidative trifluoromethylation of indoles.

The transition-metal-free oxidative trifluoromethylation could also be applied in the synthesis of other trifluoromethylated compounds. Zhou and coworkers used the method that we developed (Scheme 8.9C) for the synthesis of 6-(trifluoromethyl) phenanthridines through oxidative cyclization of 2-isocyanobiphenyls with $TMSCF_3$ (Scheme 8.13).[28] The CF_3 radical was probably involved in this transformation.

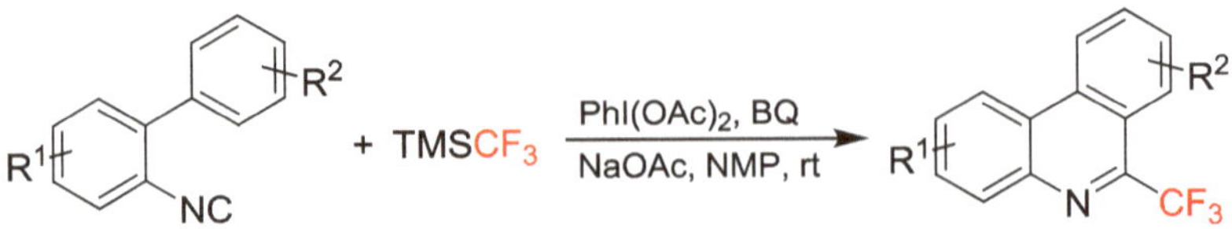

Scheme 8.13 Oxidative cyclization of 2-isocyanobiphenyls. *BQ*, benzoquinonine.

A transition-metal-free, regioselective C-5 trifluoromethylation of 2,3-dihydropyridin-4(1*H*)-ones (cyclic enaminones) with $TMSCF_3$ for the preparation of medicinally important 3-trifluoromethylpiperidine derivatives was developed by Georg and coworkers (Scheme 8.14A).[29] A radical mechanism is also proposed for this transformation. Yu and coworkers reported an efficient Cu^{II}-catalyzed trifluoromethylation of α-oxoketene dithioacetals (Scheme 8.14B).[30] It was noteworthy that both these works represented the rare examples of direct trifluoromethylation of internal olefinic C–H bonds.

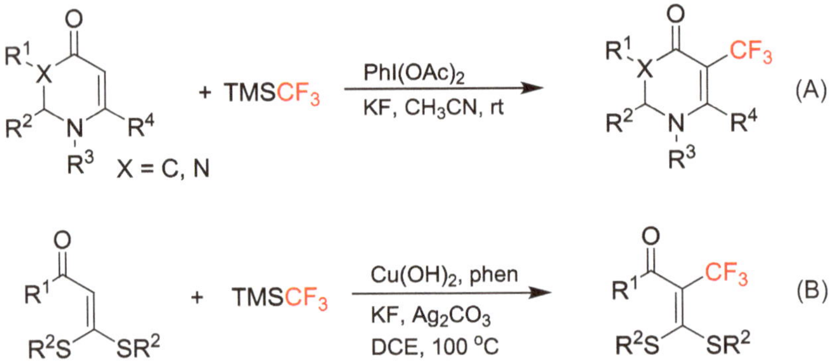

Scheme 8.14 Oxidative trifluoromethylation of activated alkenes.

2.3 Oxidative Trifluoromethylation for the Formation of C(sp^3)–CF$_3$ Bond

The oxidation of sp^3 C–H bond adjacent to nitrogen atom in tertiary amine, followed by attack of a nucleophile, is a powerful strategy for the synthesis of functionalized amines. As tetrahydroisoquinoline derivatives are important structural features of pharmaceutical and natural products, our group focused on the direct oxidative trifluoromethylation of tetrahydroisoquinoline derivatives. As shown in Scheme 8.15, various 1-trifluoromethylated tetrahydroisoquinolines were prepared through benzoyl peroxide (BPO)–promoted oxidative trifluromethylation.[9] In most cases, overoxidation was observed, producing the corresponding 1,2-dihydro analogs, which could be easily converted to the tetrahydroisoquinoline derivatives via hydrogenation in the presence of Pd/C.

The similar oxidative trifluoromethylation of tetrahydroisoquinoline derivatives were also reported by other several groups. In 2011, Mitsudera and Li developed a Cu-catalyzed trifluoromethylation via oxidative sp^3 C–H activation at the

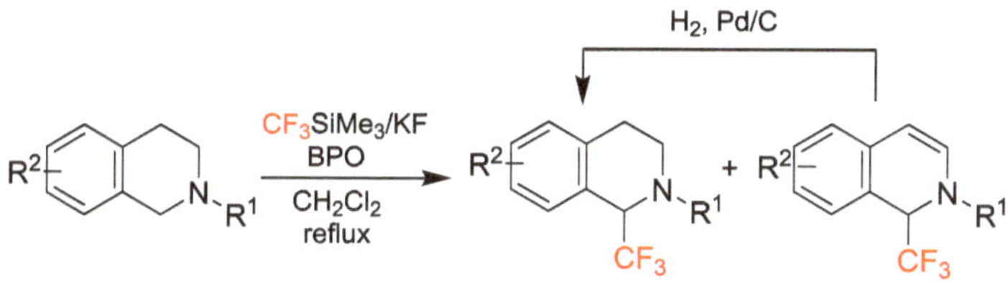

Scheme 8.15 Benzoyl peroxide (BPO)–promoted oxidative trifluoromethylation of tetrahydroisoquinolines.

α-position of nitrogen in tetrahydroisoquinoline derivatives using 2,3-dichloro-5,6-dicyanobenzoquinone as the oxidant (Scheme 8.16A).[31] Soon later, Tan[32a] and Fu,[32b] respectively, reported oxidative trifluoromethylation of tetrahydroisoquinolines promoted by Rose Bengal and visible light (Scheme 8.16B). In both protocols, no overoxidized by-product was detected.

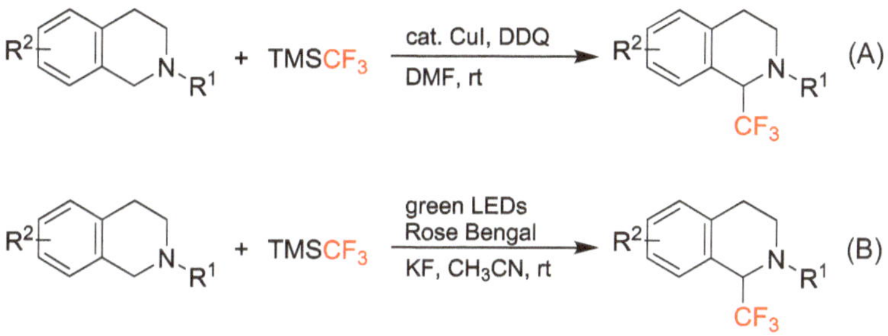

Scheme 8.16 Other oxidative trifluoromethylation of tetrahydroisoquinolines. *DDQ*, 2,3-dichloro-5,6-dicyanobenzoquinone; *LED*, light-emitting diode.

Compared with oxidative trifluoromethylation at the activated sp^3-hybridized carbon, the analogous reaction at the unactivated sp^3-hybridized carbon is more difficult. In 2012, Fu and coworkers successfully extended our oxidative trifluoromethylation of aryl boronic acids[8] to alkyl boronic acids (Scheme 8.17).[33] They developed a general approach for the construction of $C(sp^3)-CF_3$ bonds and provided a rare example for the Cu-promoted cross-coupling reaction of alkyl boronic acids.

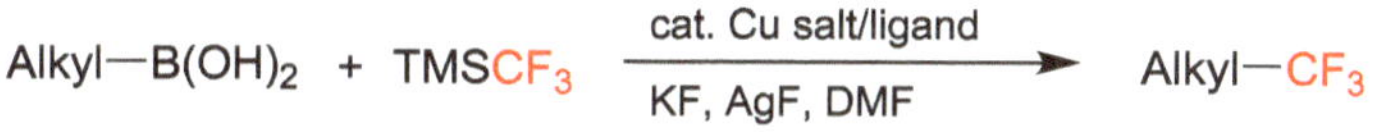

Scheme 8.17 Oxidative trifluoromethylation of alkyl boronic acids.

In 2012, our group developed a Cu-catalyzed oxidative trifluoromethylation of terminal alkenes (Scheme 8.18A).[34] This method allows access to a variety of trifluoromethylated allylic compounds. Preliminary mechanism investigations precluded the involvement of a CF_3 radical or an allylic radical in this process. We surmised that this transformation occurred via $PhI(OAc)_2$-induced oxidation of terminal alkene for the formation of radical cation intermediate as a key intermediate. One year later, our group developed a silver-catalyzed hydrotrifluoromethylation of unactivated alkenes using $TMSCF_3$ in the presence of $PdI(OAc)_2$ (Scheme 8.18B).[35] This reaction is interesting in terms of the fact that selective formation of the hydrotrifluoromethylated product by radical trifluoromethylation of alkenes remains challenging because of the competitive radical atom transfer processes and deprotonative trifluoromethylation. The success of this reaction rests on the use of a silver(I) salt and 1,4-cyclohexadiene as the H donor to inhibit the competitive reactions.

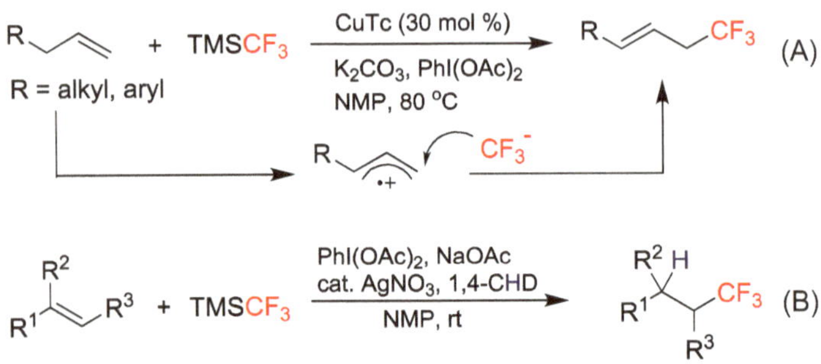

Scheme 8.18 Oxidative trifluoromethylation of alkenes. *CHD*, 1,4-cyclohexadiene; *NMP, N-methyl-2-pyrrolidone*.

The difunctionalization of alkenes provides an efficient strategy to construct diverse chemical structures. This strategy was also applied in the preparation of trifluoromethylated alkanes. In 2012, Liu and coworkers explored a palladium-catalyzed intramolecular oxidative aryltrifluoromethylation reaction of activated alkenes (Scheme 8.19A).[36] This reaction allows for an efficient synthesis of a variety of CF_3-containing oxindoles. Preliminary mechanistic study indicated that a Pd^{IV} intermediate was involved in this transformation. Two years later, Tan and Liu reported a similar reaction under metal-free conditions.[37] The reaction of alkenes and $TMSCF_3$ using $PhI(OAc)_2$ as the oxidant furnished a diverse collection of synthetically valuable trifluoromethylated heterocycles (Scheme 8.19B).

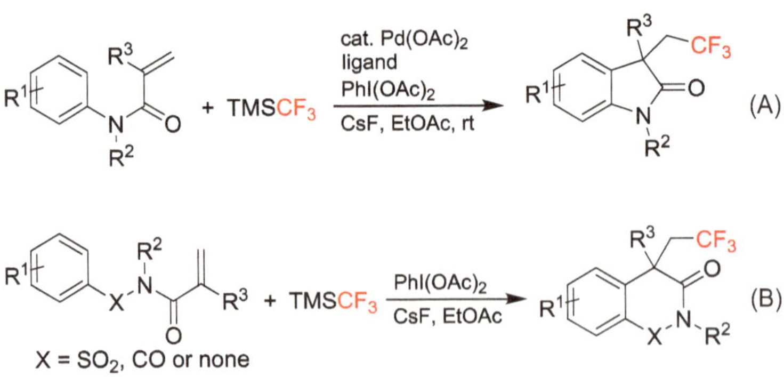

Scheme 8.19 Oxidative aryltrifluoromethylation of alkenes.

The first example of a copper(I)-catalyzed intramolecular aminotrifluoromethylation of unactivated alkenes with $TMSCF_3$ was reported by Tan and Liu.[38] A broad range of electronically and structurally varied substrates underwent this transformation with the concurrent construction of a five- or six-membered ring and a C−CF_3 bond for the preparation of different trifluoromethylated azaheterocycles (Scheme 8.20).

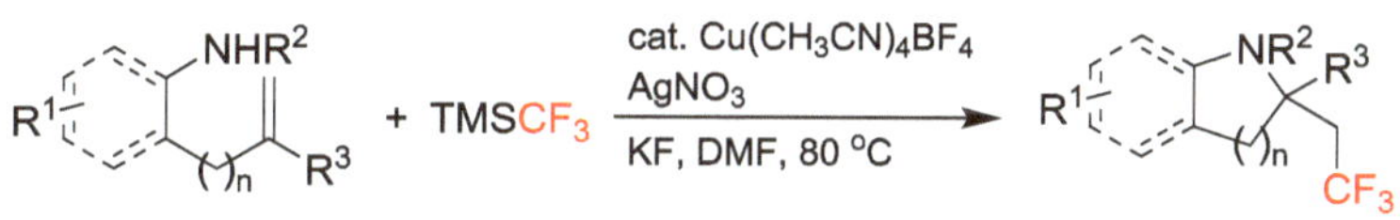

Scheme 8.20 Oxidative aminotrifluoromethylation of unactivated alkenes.

Our group developed a silver-mediated oxidative fluorotrifluoromethylation of unactivated alkenes using Selectfluor as the fluorine source and $TMSCF_3$ as the trifluoromethyl source (Scheme 8.21).[39] This reaction provided a mild and facile method for the simultaneous introduction of a fluorine atom and a trifluoromethyl group into organic molecules. The silver salt turns out to be crucial for this transformation.

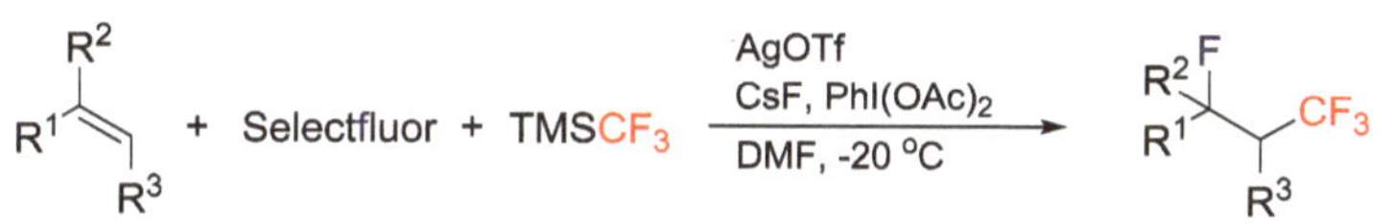

Scheme 8.21 Oxidative fluorotrifluoromethylation of unactivated alkenes.

Vinyl azides have served as versatile materials for the synthesis of various nitrogen-containing molecules, especially azaheterocycles. Chiba and coworkers described a $PhI(OAc)_2$-mediated radical trifluoromethylation of vinyl azides with $TMSCF_3$ to generate α-trifluoromethyl azines, which could be transformed to valuable fluorine-containing molecules such as α-trifluoromethylketones, β-trifluoromethyl amines, and 5-fluoropyrazoles (Scheme 8.22A).[40] They also reported an oxidative radical trifluoromethylation of biaryl vinyl azides to produce trifluoroethylated aza-polycyclic aromatic hydrocarbons (Scheme 8.22B).[41]

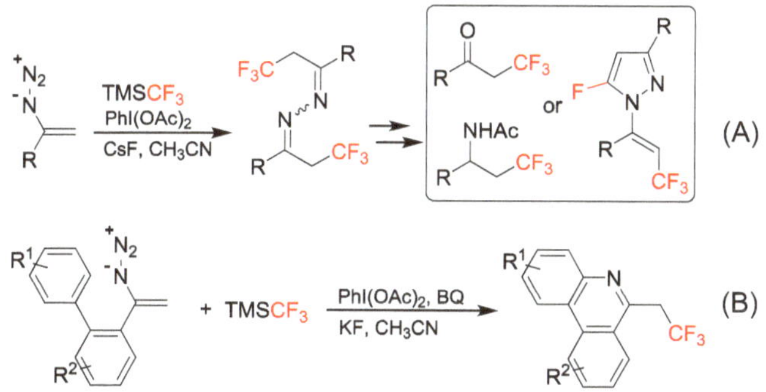

Scheme 8.22 Oxidative trifluoromethylation of vinyl azides.

2.4 Oxidative Trifluoromethylation for the Formation of X(P, O, N)–CF_3 Bond

The oxidative trifluoromethylation reactions could be applied to not only C-based nucleophiles but also heteroatom-based nucleophiles. We found that several H-phosphonates were trifluoromethylated with $TMSCF_3$ in the presence of catalytic amounts of Cu(II) salt and phen under an air atmosphere (Scheme 8.23).[42]

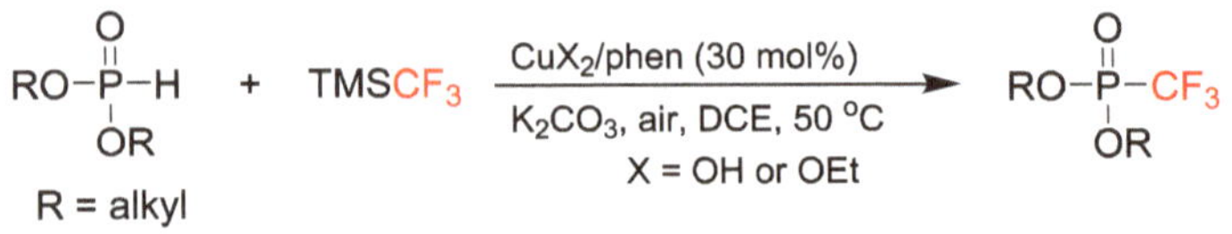

Scheme 8.23 Oxidative trifluoromethylation of H-phosphonates.

Our group have accomplished the silver-mediated oxidative *O*-trifluoromethylation of alcohol (Scheme 8.24A)[43] and phenols (Scheme 8.24B)[44] with $TMSCF_3$ for the preparation of alkyl and aryl trifluoromethyl ethers, which are difficult to prepare by common synthetic methods. Both these reactions are challenging because some side reactions easily happen. In the case of phenols, the *C*-trifluoromethylation of phenols is the major competitive reaction, whereas alcohols might be oxidized into aldehydes or ketones in the presence of oxidants. We assumed that these competitive reactions could be eliminated or reduced by choosing the appropriate metal salt, ligand, and oxidant. After careful screening of different metal salts, AgOTf was found to be optimal. The use of 2-fluoropyridine as the ligand was important to achieve good yields. Selectfluor was selected as the oxidant in the former reaction, whereas the latter needed both Selectfluor and NSFI as the oxidants. It was noteworthy that the solvents also played an important role in these transformations. If the substrates contained the electron-rich aryl group, it was necessary to add 2,6-di-*tert*-butylphenol to reduce the competitive *C*-trifluoromethylation of electron-rich phenyl ring with electrophilic CF_3 radical, which was easily generated from the combination of $TMSCF_3$/KF/AgOTf. Finally, preliminary mechanism experiments demonstrated that $Ag^{III}(CF_3)$ intermediates were involved in these transformations.

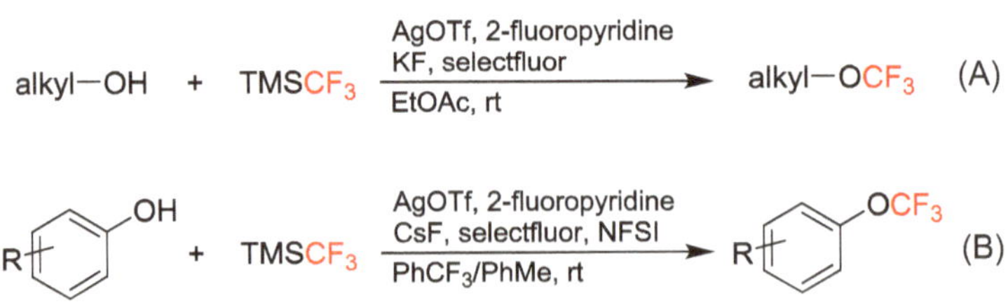

Scheme 8.24 Oxidative trifluoromethylation of alcohols and phenols.

Cheng and Bolm reported the silver-mediated oxidative *N*-trifluoromethylation of sulfoximines to give the previously unknown products (Scheme 8.25).[45]

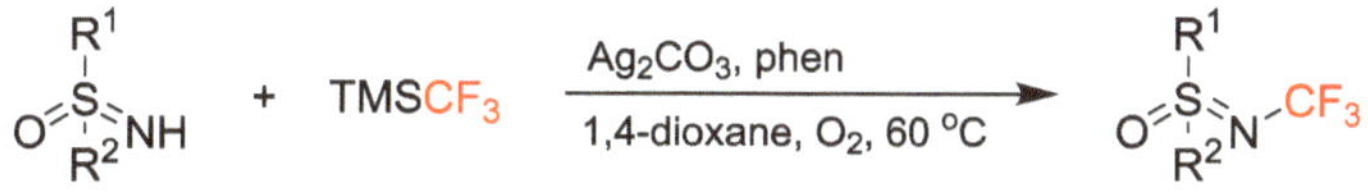

Scheme 8.25 Oxidative trifluoromethylation of sulfoximines.

3. Oxidative Trifluoromethylthiolation

The trifluoromethanesulfenyl group (SCF_3) is one of the most important fluorine-containing groups. Its unique properties, such as its strong electron-withdrawing effects and extremely high lipophilicity, have received increased attention from the pharmaceutical and agrochemical industries. Inspired by our work on oxidative trifluoromethylation of terminal alkynes,[7] we sought to develop analogous oxidative trifluoromethylthiolation of terminal alkynes. In 2012, our group reported an efficient metal-free oxidative trifluoromethylthiolation of terminal alkynes with $TMSCF_3$ and elemental sulfur at room temperature (Scheme 8.26A).[46] Further investigation revealed that in situ–generated SCF_3 anion species might be the active species in this transformation. This method represented a straightforward approach to a broad range of alkynyl trifluoromethyl sulfides. However, an excess amount of $TMSCF_3$ was required. Two years later, our group developed another type of oxidative trifluoromethylthiolation of terminal alkynes with $AgSCF_3$ for the preparation of alkynyl trifluoromethyl sulfides (Scheme 8.26B),[47] which avoided the use of excess amount of $TMSCF_3$. During this reaction, an electrophilic active intermediate was formed in situ and then was involved in the silver-mediated trifluoromethylthiolation reaction.

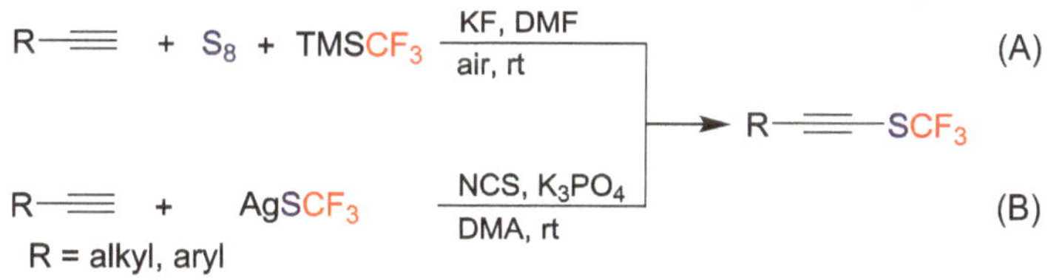

Scheme 8.26 Oxidative trifluoromethylthiolation of terminal alkynes.

The first example of the Cu-catalyzed oxidative trifluoromethylthiolation of aryl boronic acids with $TMSCF_3$ and elemental sulfur at room temperature was developed by our group (Scheme 8.27A).[48] An organocopper disulfide complex was proposed as the key intermediate in the catalytic reaction. Soon later, Li and Duan developed a similar reaction with the inexpensive CF_3CO_2Na as the trifluoromethylating reagent (Scheme 8.27B).[49] However, their method required stoichiometric amounts Cu salt and ligand as well as high reaction temperature. Zhang and Vicic reported a complementary oxidative trifluoromethylthiolation of aryl boronic acids with Me_4NSCF_3 as the SCF_3 source (Scheme 8.27C).[50]

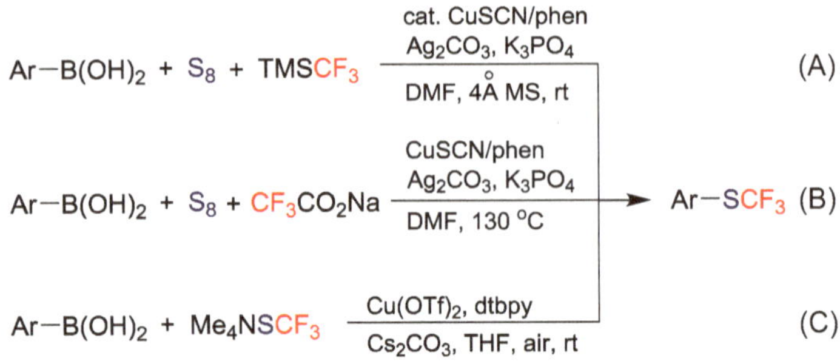

Scheme 8.27 Oxidative trifluoromethylthiolation of aryl boronic acids.

Direct C−H trifluoromethylthiolation reactions are highly valuable synthetic tools for the late-stage modification of drug candidates. Huang and coworkers reported a palladium-catalyzed *ortho*-selective monotrifluoromethylthiolation reaction of arenes (Scheme 8.28).[51] This reaction proceeded through a key ligand exchange pathway using readily available $AgSCF_3$. Acetic acid was found to be crucial to minimize oxidative dimerization of the starting materials and to facilitate the transfer of SCF_3 anion.

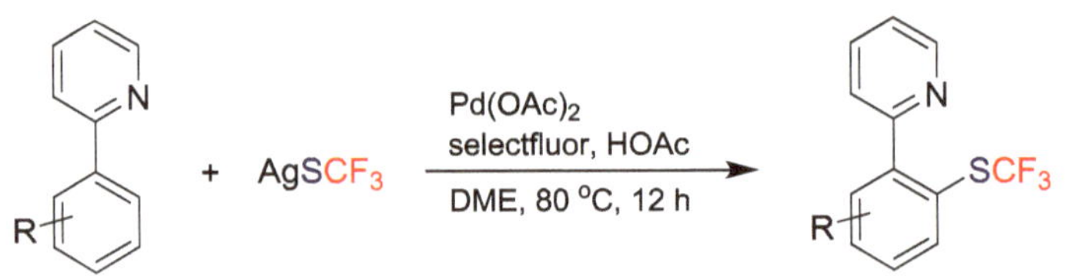

Scheme 8.28 Oxidative C−H trifluoromethylthiolation of aromatic compounds.

Quinone motifs are prevalent in a number of naturally occurring biologically active compounds. Our group developed an efficient and practical approach to previously unknown trifluoromethylthiolated quinones by Cu-mediated oxidative C−H trifluoromethylthiolation of quinones with $AgSCF_3$ (Scheme 8.29).[52] This reaction proceeded through a radical process.

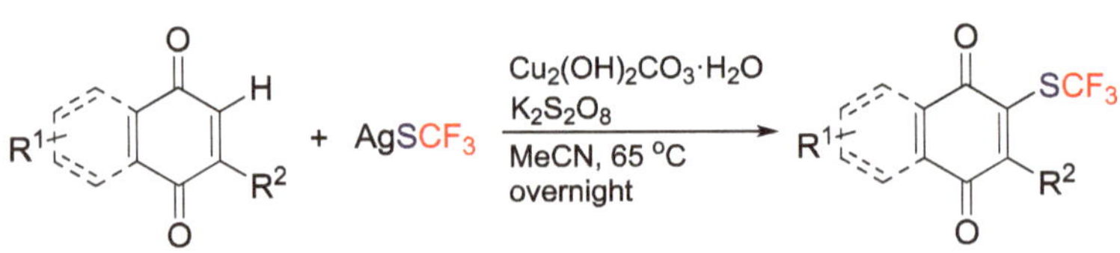

Scheme 8.29 Oxidative C−H trifluoromethylthiolation of quinones.

Oxidative C−H trifluoromethylthiolation/cyclization reactions provided a useful strategy for the preparation of SCF_3-substituted cyclic compounds. Yang and Xiang presented a facile and general synthetic route to a range of 3-(trifluoromethylthio)-chromones via the in situ generation of electrophilic trifluoromethylthio species from trichloroisocyanuric acid (TCCA) and $AgSCF_3$ (Scheme 8.30A).[53] Wang and coworkers demonstrated a mild and convenient oxidative radical cyclization of aryl alkynoate esters for the synthesis of 3-trifluoromethylthiolated coumarins using $AgSCF_3$ as the SCF_3 radical source (Scheme 8.30B).[54] A rare example of $AgSCF_3$-mediated radical cascade cyclization/trifluoromethylthiolation of 1,6-enynes triggered by a C—C triple bond was reported by Liang and coworkers (Scheme 8.30C).[55] This protocol provided a novel opportunity to construct a valuable trifluoromethylthio-substituted polycyclic fluorene system through the formations of one C—SCF_3 bond and two C—C bonds in a single step.

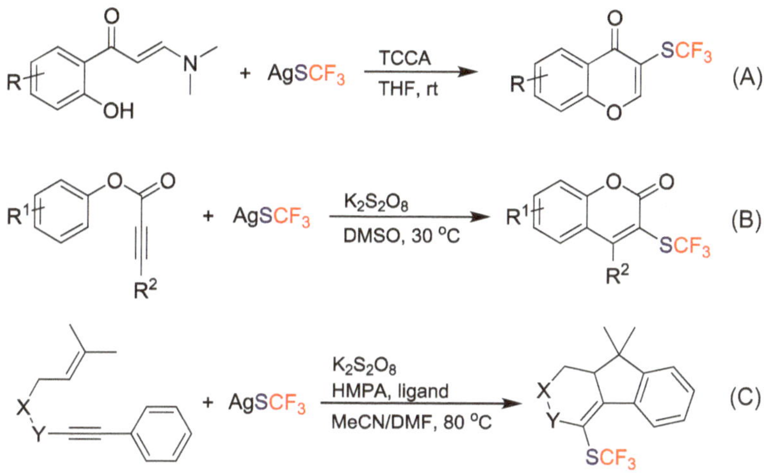

Scheme 8.30 Oxidative trifluoromethylthiolation/cyclization reactions. *DMSO*, dimethyl sulphoxide; *HMPA*, hexamethylphosphoramide; *THF*, tetrahydrofuran.

In 2014, our group developed a copper-catalyzed trifluoromethylthiolation of benzylic sp^3 C—H bonds via nondirected oxidative C—H activation using $AgSCF_3$ (Scheme 8.31A).[56] This work presents the first example of transition-metal-catalyzed direct formation of $C(sp^3)$—SCF_3 bonds, and it is the most straightforward method for the preparation of benzyl trifluoromethyl sulfides, compared with previous methods. In the same year, our group applied the oxidative trifluoromethylthiolation strategy for the synthesis of allyl trifluoromethyl sulfides.[57] The reaction of terminal alkenes and $AgSCF_3$ in the presence of $Cu(OAc)_2$ and $K_2S_2O_8$ gave a series of trifluoromethylthiolated allylic compounds (Scheme 8.31B). The major advantage of this protocol is avoiding the use of functionalized allylic substrates. In 2015, two research groups (Tang[58] and Chen/Liu[59]) independently reported the oxidative trifluoromethylthiolation of unactivated aliphatic C—H bonds by $AgSCF_3/K_2S_2O_8$ (Scheme 8.31C). Preliminary investigations showed that a radical reaction pathway was involved, in which $K_2S_2O_8$ acted not only as an oxidant but also as a good H atom abstractor.

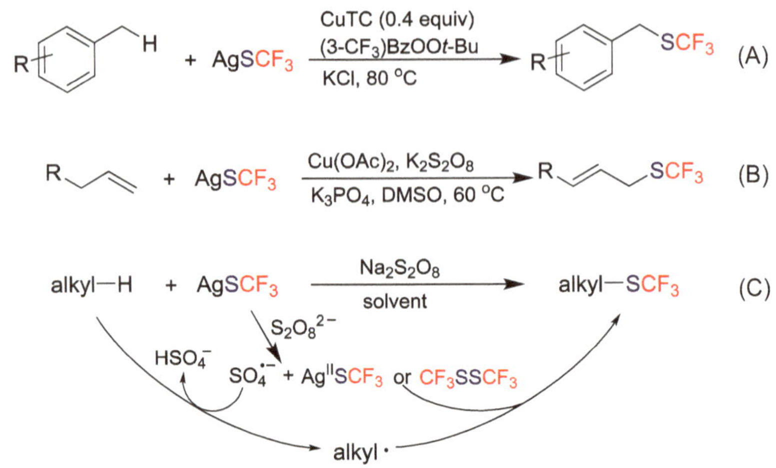

Scheme 8.31 Oxidative trifluoromethylthiolation for the preparation of benzyl, allyl, and alkyl trifluoromethyl sulfides.

In 2014, Wang and Yin reported the first example of silver-mediated oxidative aryl-trifluoromethylthiolation of activated alkenes to produce valuable trifluoromethylthio-containing oxindoles (Scheme 8.32A).[60] Mechanistic investigations indicated that this novel transformation proceeded through a unique SCF_3 radical addition path. One year later, this protocol was applied for the synthesis of trifluoromethylated heterocyclic scaffolds by Nevado and coworkers (Scheme 8.32B).[61] The reaction of *N*-[(2-ethynyl)arylsulfonyl]acrylamides and $AgSCF_3$ in the presence of $Na_2S_2O_8$ gave SCF_3-containing indolo[2,1-*a*]isoquinolin-6(5*H*)-ones through a multistep radical reaction cascade. The process involves the one-pot formation of four new bonds, a formal 1,4-aryl migration, and desulfonylation of the starting material.

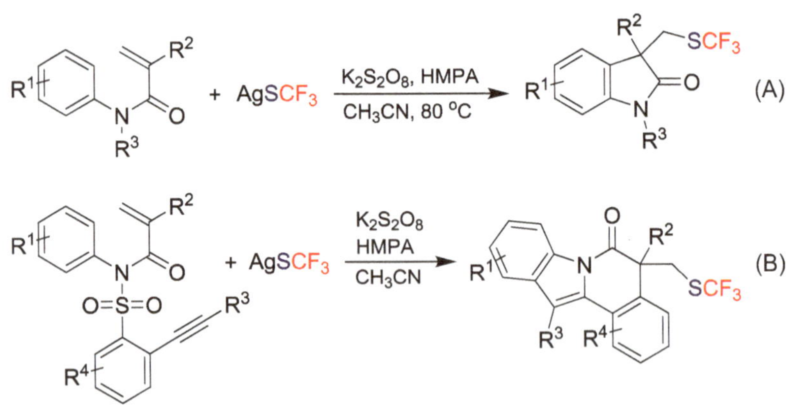

Scheme 8.32 Oxidative aryltrifluoromethylthiolation.

A copper-catalyzed intermolecular oxytrifluoromethylthiolation of unactivated alkenes was reported by Xu and Wang (Scheme 8.33).[62] This methodology allows rapid access to a variety of isoxazolines bearing a SCF_3 substituent. They proposed that this difunctionalization proceeded through a radical process.

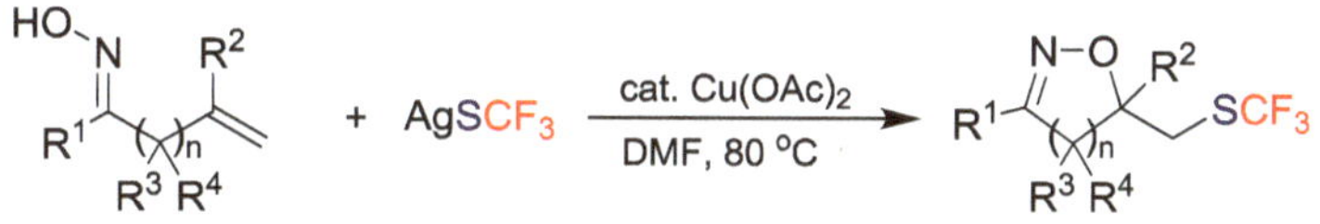

Scheme 8.33 Oxidative oxytrifluoromethylthiolation.

In 2014, Liu and Tan developed an organocatalytic asymmetric trifluoromethylthiolation of oxindoles via the in situ generation of active electrophilic trifluoromethylthio species involving TCCA and $AgSCF_3$ (Scheme 8.34).[63]

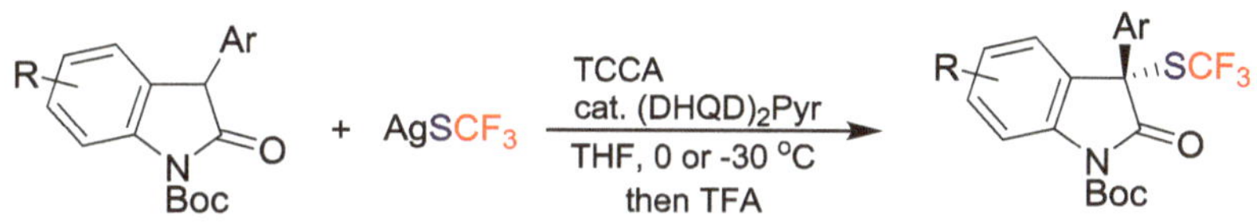

Scheme 8.34 Enantioselective trifluoromethylthiolation of oxindoles. *TFA*, trifluoroacetic acid.

4. Other Oxidative Fluoroalkylations

The difluoromethyl group is considered isosteric and isopolar to a hydroxyl and could act as a lipophilic hydrogen bond donor. Inspired by our work on oxidative trifluoromethylation of terminal alkynes,[7] we anticipated that a similar copper-mediated oxidative C–H difluoromethylation of terminal alkynes with $TMSCF_2H$ might be possible. However, this transformation is more challenging than trifluoromethylation of alkynes, because the Si–CF_2H bond is more inert than the Si–CF_3 bond and difluoromethyl copper complexes are less stable than trifluoromethyl copper complexes. Despite these challenges, our group achieved a copper-mediated oxidative difluoromethylation of terminal alkynes with $TMSCF_2H$ using 9,10-phenanthraquinone as the oxidant (Scheme 8.35).[64] This reaction provided a direct and efficient method for the preparation of difluoromethylated alkynes.

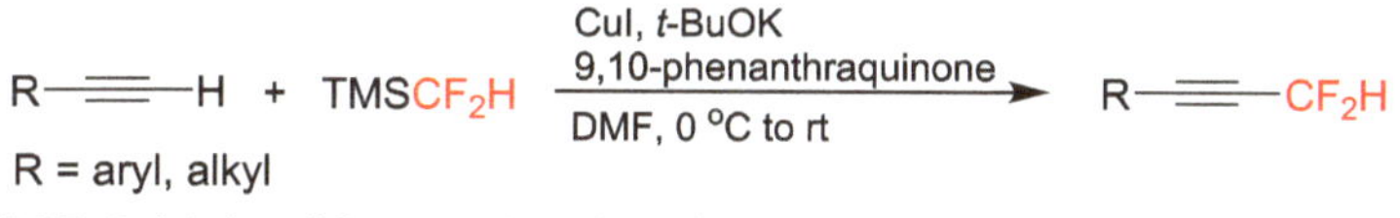

Scheme 8.35 Oxidative difluoromethylation of terminal alkynes.

Difluorophosphonates have gained special attention because of their potential as analogs of biologically important phosphate esters and utilization in the design and development of new drugs. Consequently, it is highly important to develop new methods for the preparation of these compounds. Our group has successfully applied the copper-mediated oxidative coupling strategy to the preparation of α,α-difluoropropargylphosphonates (Scheme 8.36A)[65] and (aryl)difluoromethylphosphonates (Scheme 8.36B).[66]

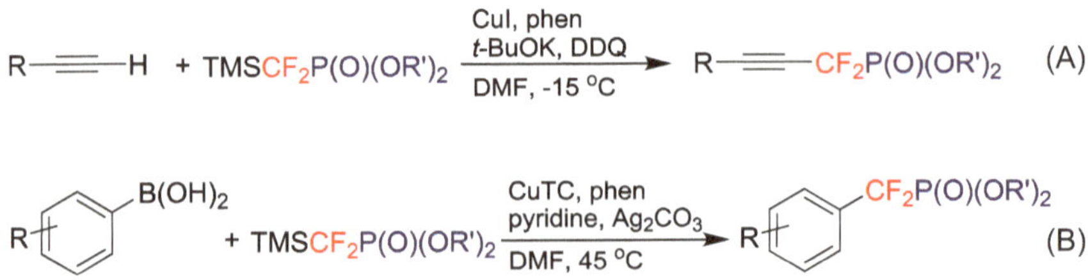

Scheme 8.36 Oxidative difluoromethylenation.

Among the functionalized difluoromethylenated moieties, the ethoxycarbonyldifluoromethyl (CF_2CO_2Et) moiety is extremely appealing due to the huge possibility of transformation. In 2014, Hao and Wan reported a regioselective hydrodifluoromethylenation of unactivated alkenes (Scheme 8.37).[67] This reaction affords a practical instance of hydro-(ethoxycarbonyl)difluoromethylation of the vicinal C=C bond for construction of $C(sp^3)-CF_2$ bond.

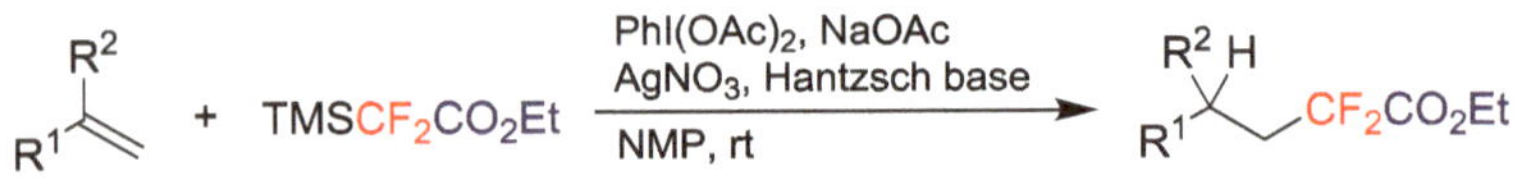

Scheme 8.37 Oxidative hydrodifluoromethylenation of alkenes.

Compared with trifluoromethylation reactions, pentafluoroethylation reactions are less developed, despite the fact that in certain cases C_2F_5 derivatives exhibit properties that are superior to those of their CF_3 counterparts. In 2013, Grushin and Lishchynskyi disclosed that the ligandless CuC_2F_5 reagent in situ prepared from the direct cupration of economical C_2F_5H underwent oxidative coupling with aryl boronic acid to provide the corresponding pentafluoroethylated product (Scheme 8.38A).[68] The intrinsic drawback of this method is that the gaseous C_2F_5H is hard to handle in common academic laboratories. One year later, Mikami and coworkers developed a complementary synthesis of CuC_2F_5 from operationally simple ethyl pentafluoropropionate (Scheme 8.38B).[69] This CuC_2F_5 reagent was also successfully applied to the oxidative coupling with aryl boronic acids.

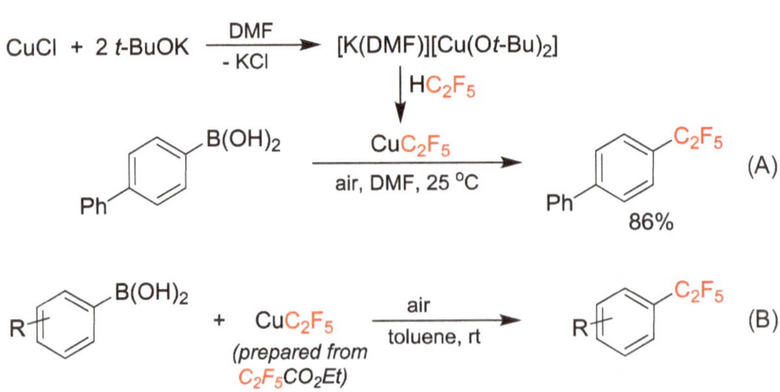

Scheme 8.38 Oxidative pentafluoroethylation of aryl boronic acid.

The first oxidative heptafluoroisopropylation of aryl boronic acids was reported by Wu and Gong (Scheme 8.39).[70] They prepared heptafluoroisopropyl silver reagent by the nucleophilic addition of silver fluoride to hexafluoropropene, a fire-fighting and readily available agent. This method provides a new way to synthesize a variety of synthetically useful heptafluoroisopropylated arenes.

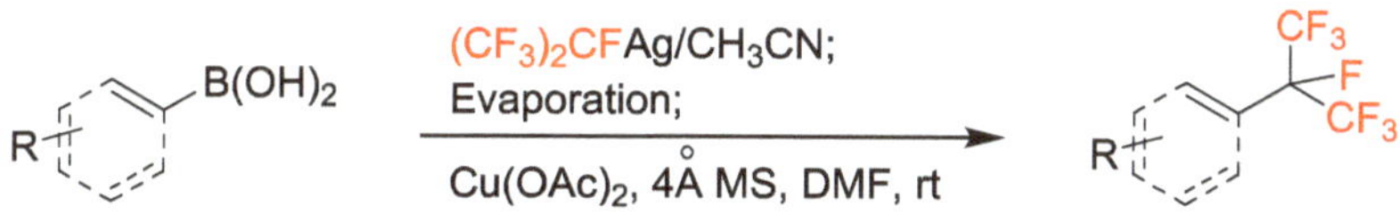

Scheme 8.39 Oxidative heptfluoroisopropylation of aryl boronic acids.

In 2012, Hartwig and coworkers demonstrated that the oxidative coupling of [(phen)CuR_f] and arylboronate esters provided a general strategy for the preparation of perfluoroalkylated arenes (Scheme 8.40A).[19] These copper reagents are thermally stable, easily handled solids. Another type of oxidative perfluoroalkylation reaction was disclosed by Bräse and coworkers.[71] The reaction of aromatic triazenes and $TMSR_f$ in the presence of AgF provided various *ortho*-perfluoroalkylated triazenes (Scheme 8.40B).

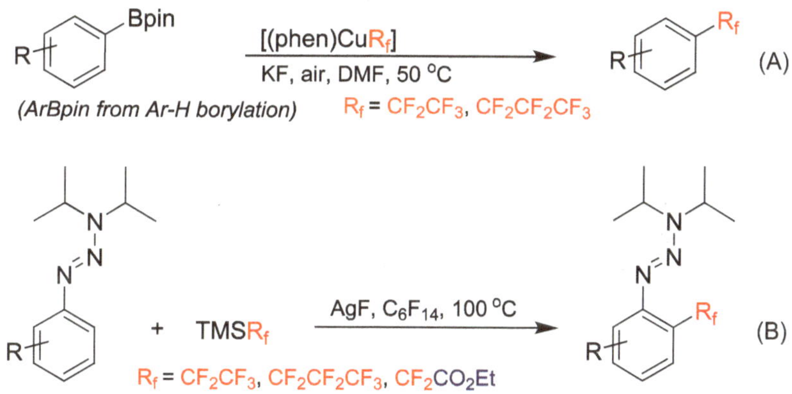

Scheme 8.40 Oxidative perfluoroalkylation.

5. Conclusion

Since we proposed the concept of oxidative trifluoromethylation in 2010, oxidative trifluoromethylation reactions have been widely used in the preparation of various CF_3-containing compounds. This oxidative strategy has also been extended to the preparation of other fluorinated organic compounds, such as trifluoromethylthiolated, difluoromethylated, and perfluoroalkylated compounds. Because of the importance of fluorinated compounds in pharmaceuticals and agrochemicals, we believe these oxidative reactions would find wide application in the synthesis of biologically interesting fluorinated organic compounds.

References

1. a. Uneyama, K. *Modern Fluoroorganic Chemistry;* Wiley-VCH: Weinheim, 2004.
   b. Bégué, J. -P.; Bonnet-Delpon, D. *Bioorganic and Medicinal Chemistry of Fluorine;* John Wiley & Sons: Hoboken, 2008.
   c. Ojima, I. *Fluorine in Medicinal Chemistry and Chemical Biology;* Wiley-Blackwell: Oxford, 2009.
   d. Gouverneur, V.; Müller, K. *Fluorine in Pharmaceutical and Medicinal Chemistry: From Biophysical Aspects to Clinical Applications;* Imperial College Press: London, 2012.
2. a. Müller, K.; Faeh, C.; Diederich, F. *Science* **2007,** *317,* 1881.
   b. Purser, S.; Moore, P. R.; Swallow, S.; Gouverneur, V. *Chem. Soc. Rev.* **2008,** *37,* 320.
   c. Meanwell, N. A. *J. Med. Chem.* **2011,** *54,* 2529.
   d. Wang, J.; Sánchez-Roselló, M.; Aceña, J. L.; del Pozo, C.; Sorochinsky, A. E.; Fustero, S.; Soloshonok, V. A.; Liu, H. *Chem. Rev.* **2014,** *114,* 2432.
   e. Gillis, E. P.; Eastman, K. J.; Hill, M. D.; Donnelly, D. J.; Meanwell, N. A. *J. Med. Chem.* **2015,** *58,* 8315.
3. a. Liang, T.; Neumann, C. N.; Ritter, T. *Angew. Chem. Int. Ed.* **2013,** *52,* 8214.
   b. Charpentier, J.; Fruh, N.; Togni, A. *Chem. Rev.* **2015,** *115,* 650.
   c. Liu, X.; Xu, C.; Wang, M.; Liu, Q. *Chem. Rev.* **2015,** *115,* 683.
   d. Xu, X. H.; Matsuzaki, K.; Shibata, N. *Chem. Rev.* **2015,** *115,* 731.
   e. Ni, C.; Hu, M.; Hu, J. *Chem. Rev.* **2015,** *115,* 765.
   f. Yang, X.; Wu, T.; Phipps, R. J.; Toste, F. D. *Chem. Rev.* **2015,** *115,* 826.
   g. Alonso, C.; Martinez de Marigorta, E.; Rubiales, G.; Palacis, F. *Chem. Rev.* **2015,** *115,* 1847.
4. a. Glaser, C. *Ber. Dtsch. Chem. Ges.* **1869,** *2,* 422.
   b. Hay, A. S. *J. Org. Chem.* **1962,** *27,* 3320.
5. a. Chan, D. M. T.; Monaco, K. L.; Wang, R.-P.; Winters, M. P. *Tetrahedron Lett.* **1998,** *39,* 2933.
   b. Evans, D. A.; Katz, J. L.; West, T. R. *Tetrahedron Lett.* **1998,** *39,* 2937.
   c. Lam, P. Y. S.; Clark, C. G.; Saubern, S.; Adams, J.; Winters, M. P.; Chan, D. M. T.; Combs, A. *Tetrahedron Lett.* **1998,** *39,* 2941.
6. a. Liu, C.; Zhang, H.; Shi, W.; Lei, A. *Chem. Rev.* **2011,** *111,* 1780.
   b. Yeung, C. S.; Dong, V. M. *Chem. Rev.* **2011,** *111,* 1215.
   c. Kumar, R.; Van der Eychen, E. V. *Chem. Soc. Rev.* **2013,** *42,* 1121.
7. Chu, L.; Qing, F. -L. *J. Am. Chem. Soc.* **2010,** *132,* 7262.
8. Chu, L.; Qing, F. -L. *Org. Lett.* **2010,** *12,* 5060.
9. Chu, L.; Qing, F. -L. *Chem. Commun.* **2010,** *46,* 6285.
10. Chu, L.; Qing, F. -L. *Acc. Chem. Res.* **2014,** *47,* 1513.
11. Jiang, X. L.; Chu, L.; Qing, F. -L. *J. Org. Chem.* **2012,** *77,* 1251.
12. Zhang, K.; Qiu, X. -L.; Huang, Y.; Qing, F. -L. *Eur. J. Org. Chem.* **2012,** 58.
13. Tresse, C.; Guissart, C.; Schweizer, S.; Bouhoute, Y.; Chany, A. -C.; Goddard, M. -L.; Blanchard, N.; Evano, G. *Adv. Synth. Catal.* **2014,** *356,* 2051.
14. Serizawa, H.; Aikawa, K. *Mikami. K. Chem. Eur. J.* **2013,** *19,* 17692.
15. Li, X.; Zhao, J.; Zhang, L.; Hu, M.; Wang, L.; Hu, J. *Org. Lett.* **2015,** *17,* 298.
16. Deng, X.; Lin, J.; Zheng, J.; Xiao, J. *Chin. J. Chem.* **2014,** *32,* 689.
17. Senecal, T. D.; Parsons, A. T.; Buchwald, S. L. *J. Org. Chem.* **2011,** *76,* 1174.
18. Khan, B. A.; Buba, A. E.; Gooßen, L. J. *Chem. Eur. J.* **2012,** *18,* 1577.
19. Litvinas, N. D.; Fier, P. S.; Hartwig, J. F. *Angew. Chem. Int. Ed.* **2012,** *51,* 536.

20. Novák, P.; Lishchynskyi, A.; Grushin, V. V. *Angew. Chem. Int. Ed.* **2012,** *51,* 7767.
21. a. Ivashkin, P.; Lemonnier, G.; Cousin, J.; Grégoire, V.; Labar, D.; Jubault, P.; Pannecoucke, X. *Chem. Eur. J.* **2014,** *20,* 9514.
b. Born, D.; Sewing, C.; Herscheid, J. D. M.; Windhorst, A. D.; Orru, R. V. A.; Vugts, D. J. *Angew. Chem. Int. Ed.* **2014,** *53,* 11046.
22. a. Ye, Y.; Lee, S. H.; Sanford, M. S. *Org. Lett.* **2011,** *13,* 5464.
b. Seo, S.; Taylor, J. B.; Greaney, M. F. *Chem. Commun.* **2013,** *49,* 6385.
c. Wu, X.; Chu, L.; Qing, F. -L. *Tetrahedron Lett.* **2013,** *54,* 249.
23. Hafner, A.; Bräse, S. *Angew. Chem. Int. Ed.* **2012,** *51,* 3713.
24. Shang, M.; Sun, S. -Z.; Wang, H. -L.; Laforteza, B. N.; Dai, H. -X.; Yu, J. -Q. *Angew. Chem. Int. Ed.* **2014,** *53,* 10439.
25. Chu, L.; Qing, F. -L. *J. Am. Chem. Soc.* **2012,** *134,* 1298.
26. Mu, X.; Chen, S.; Zhen, X.; Liu, G. *Chem. Eur. J.* **2011,** *17,* 6039.
27. Xu, C.; Liu, J.; Ming, W.; Liu, Y.; Liu, J.; Wang, M.; Liu, Q. *Chem. Eur. J.* **2013,** *19,* 9104.
28. Wang, Q.; Dong, X.; Xiao, T.; Zhou, L. *Org. Lett.* **2013,** *15,* 4846.
29. Yu, Y. -Y.; Ranade, A. R.; Georg, G. I. *Adv. Synth. Catal.* **2014,** *356,* 3510.
30. Mao, Z.; Huang, F.; Yu, H.; Chen, J.; Yu, Z.; Xu, Z. *Chem. Eur. J.* **2014,** *20,* 3439.
31. Mitsudera, H.; Li, C. -J. *Tetrahedron Lett.* **2011,** *52,* 1898.
32. a. Pan, Y.; Wang, S.; Kee, C. W.; Dubuisson, E.; Yang, Y.; Loh, K. P.; Tan, C. -H. *Green Chem.* **2011,** *13,* 3341.
b. Fu, W.; Guo, W.; Zou, G.; Xu, C. *J. Fluor. Chem.* **2012,** *140,* 88.
33. Xu, J.; Xiao, B.; Xie, C. -Q.; Luo, D. -F.; Liu, L.; Fu, Y. *Angew. Chem. Int. Ed.* **2012,** *51,* 12551.
34. Chu, L.; Qing, F. -L. *Org. Lett.* **2012,** *14,* 2106.
35. Wu, X.; Chu, L.; Qing, F. -L. *Angew. Chem. Int. Ed.* **2013,** *52,* 2198.
36. Mu, X.; Wu, T.; Wang, H. -Y.; Guo, Y. -L.; Liu, G. *J. Am. Chem. Soc.* **2012,** *134,* 878.
37. Li, L.; Deng, M.; Zheng, S. -C.; Xiong, Y. -P.; Tan, B.; Liu, X. -Y. *Org. Lett.* **2014,** *16,* 504.
38. Lin, J. -S.; Liu, X. -G.; Zhu, X. -L.; Tan, B.; Liu, X. -Y. *J. Org. Chem.* **2014,** *79,* 7084.
39. Yu, W.; Xu, X. -H.; Qing, F. -L. *Adv. Synth. Catal.* **2015,** *357,* 2039.
40. Wang, Y. -F.; Lonca, G. H.; Chiba, S. *Angew. Chem. Int. Ed.* **2014,** *53,* 1067.
41. Wang, Y. -F.; Lonca, G. H.; Runigo, M. L.; Chiba, S. *Org. Lett.* **2014,** *16,* 4272.
42. Chu, L.; Qing, F. -L. *Synthesis* **2012,** 1521.
43. Liu, J.-B.; Xu, X. -H.; Qing, F. -L. *Org. Lett.* **2015,** *17,* 5048.
44. Liu, J. -B.; Chen, C.; Chu, L.; Chen, Z. -H.; Xu, X. -H.; Qing, F. -L. *Angew. Chem. Int. Ed.* **2015,** *54,* 11839.
45. Teng, F.; Cheng, J.; Bolm, C. *Org. Lett.* **2015,** *17,* 3166.
46. Chen, C.; Chu, L.; Qing, F. -L. *J. Am. Chem. Soc.* **2012,** *134,* 12454.
47. Zhu, S. -Q.; Xu, X. -H.; Qing, F. -L. *Eur. J. Org. Chem.* **2014,** 4453.
48. Chen, C.; Xie, Y.; Chu, L.; Wang, R. -W.; Zhang, X.; Qing, F. -L. *Angew. Chem. Int. Ed.* **2012,** *51,* 2492.
49. Zhai, L.; Li, Y.; Yin, J.; Jin, K.; Zhang, R.; Fu, X.; Duan, C. *Tetrahedron* **2013,** *69,* 10262.
50. Zhang, C. -P.; Vicic, D. A. *Chem. Asian. J.* **2012,** *7,* 1756.
51. Yin, W.; Wang, Z.; Huang, Y. *Adv. Synth. Catal.* **2014,** *356,* 2998.
52. Li, C.; Zhang, K.; Xu, X. -H.; Qing, F. -L. *Tertahedron Lett.* **2015,** *56,* 6273.
53. Xiang, H.; Yang, C. *Org. Lett.* **2014,** *16,* 5686.
54. Zeng, Y. -F.; Tan, D. -H.; Chen, Y.; Lv, W. -X.; Liu, X. -G.; Li, Q.; Wang, H. *Org. Chem. Front.* **2015,** *2,* 1511.
55. Qiu, Y. -F.; Zhu, X. -Y.; Li, Y. -X.; He, Y. -T.; Yang, F.; Wang, J.; Hua, H. -L.; Zheng, L.; Wang, L. -C.; Liu, X. -Y.; Liang, Y. -M. *Org. Lett.* **2015,** *17,* 3694.

56. Chen, C.; Xu, X. -H.; Yang, B.; Qing, F. -L. *Org. Lett.* **2014,** *16,* 3372.
57. Zhang, K.; Liu, J. -B.; Qing, F. -L. *Chem. Commun.* **2014,** *50,* 14157.
58. Guo, S.; Zhang, X.; Tang, P. *Angew. Chem. Int. Ed.* **2015,** *54,* 4065.
59. Wu, H.; Xiao, Z.; Wu, J.; Guo, Y.; Xiao, J. -C.; Liu, C.; Chen, Q. -Y. *Angew. Chem. Int. Ed.* **2015,** *54,* 4070.
60. Yin, F.; Wang, X. -S. *Org. Lett.* **2014,** *16,* 1128.
61. Fuentes, N.; Kong, W.; Fernández-Sánchez, L.; Merino, E.; Nevado, C. *J. Am. Chem. Soc.* **2015,** *137,* 964.
62. Zhu, L.; Wang, G.; Guo, Q.; Xu, Z.; Zhang, D.; Wang, R. *Org. Lett.* **2014,** *16,* 5390.
63. Zhu, X. -L.; Xu, J. -H.; Cheng, D. -J.; Zhao, L. -J.; Liu, X. -Y.; Tan, B. *Org. Lett.* **2014,** *16,* 2192.
64. Zhu, S. Q.; Xu, X. -H.; Qing, F. -L. *Org. Chem. Front.* **2015,** *2,* 1022.
65. Jiang, X.; Chu, L.; Qing, F. -L. *Org. Lett.* **2012,** *14,* 2870.
66. Jiang, X.; Chu, L.; Qing, F. -L. *New J. Chem.* **2013,** *37,* 1736.
67. Ma, G.; Wan, W.; Li, J.; Hu, Q.; Jiang, H.; Zhu, S.; Wang, J.; Hao, J. *Chem. Commun.* **2014,** *50,* 9749.
68. Lishchynskyi, A.; Grushin, V. V. *J. Am. Chem. Soc.* **2013,** *135,* 12584.
69. Serizawa, H.; Aikawa, K.; Mikami, K. *Org. Lett.* **2014,** *16,* 3456.
70. Li, Y.; Wang, X.; Guo, Y.; Zhu, Z.; Wu, Y.; Gong, Y. *Chem. Commun.* **2016,** *52,* 796.
71. Hafner, A.; Bihlmeier, A.; Nieger, M.; Klopper, W.; Bräse, S. *J. Org. Chem.* **2013,** *78,* 7938.

Catalytic Enantioselective Fluorination

9

R. Thornbury, G. Schäfer, F.D. Toste
University of California, Berkeley, CA, United States

Chapter Outline

1. Introduction

The introduction of fluorine into small organic compounds has become a major research topic in organic chemistry over the past two decades. The growing interest in fluorination chemistry can be mainly attributed to the positive effects that fluorine substituents exert in pharmaceuticals and agrochemicals.[48,62] Among these positive attributes are the increased metabolic stability of many fluorinated compounds compared with their non-fluorinated analogs, and the possibility of modulating the acidity or basicity of a given functional group by the introduction of a fluorine in close proximity. Therefore, it comes as no surprise that 35% of agrochemicals and 20% of pharmaceuticals currently on the market contain fluorine atoms.[47,83]

Modern Synthesis Processes and Reactivity of Fluorinated Compounds. http://dx.doi.org/10.1016/B978-0-12-803740-9.00009-3

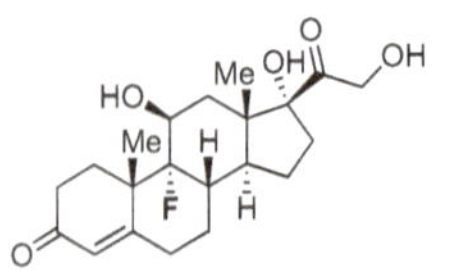

Fludrocortisone

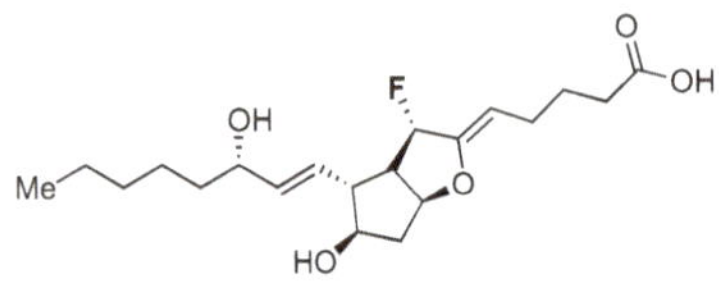

7-F-PGI_2

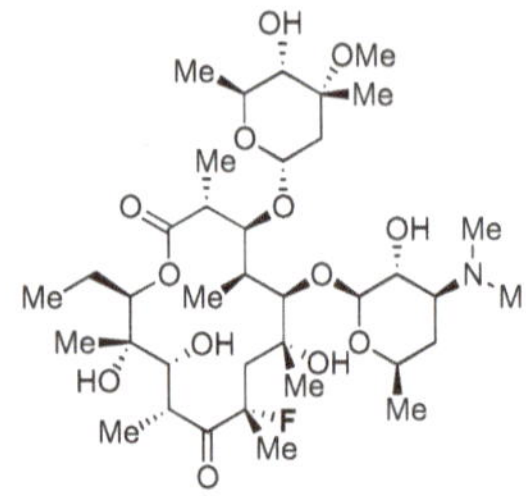

Fluorithromycin (Ritro)

In contrast, in nature, only 21 biosynthesized, fluorine-containing molecules are known and only a single enzyme (5′-fluoro-5′-deoxyadenosine synthase) that can introduce fluorine into organic molecules has been identified.[13,63] This scarce occurrence of natural, fluorine-containing compounds is a consequence of the high oxidation potential of fluorine, making the existence of fluoroperoxidases problematic. Additionally, the high solvation energy of the fluoride ion in aqueous media results in a tightly bound hydration shell, dramatically lowering the nucleophilicity of fluoride.

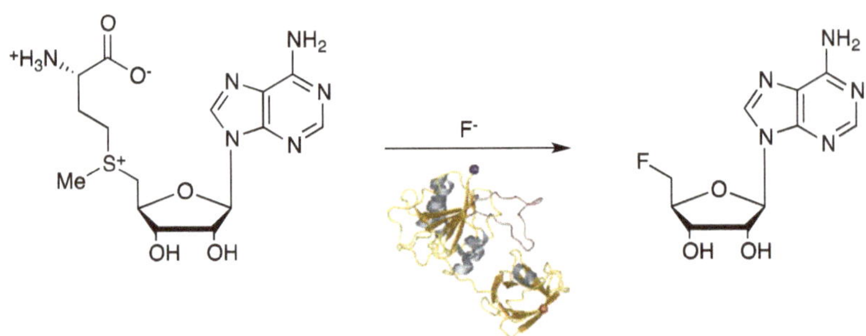

The lack of chiral-pool organofluorine compounds makes the development of asymmetric fluorination methods of particular interest and importance.[57] Although there has been a tremendous progress in fluorination chemistry, the development of enantioselective reactions has been a much more challenging task. This challenge can, in part, be understood in comparison with other enantioselective halogenation reactions, e.g., the enantioselective iodolactonization. This halocyclization reaction proceeds via the initial formation of a cyclic iodonium intermediate, followed by the catalyst-controlled and enantiodetermining attack of the carboxylate. In contrast, although chlorine, bromine, and iodine readily react with unactivated alkenes via these cyclic halonium ions, the corresponding fluoronium ion is significantly more difficult to access.[32,77] Due to the low reactivity of unactivated alkenes toward electrophilic fluorinating reagents, significant effort has been focused on the electrophilic fluorination of activated alkene derivatives, such as enols or enamines. More recently, nucleophilic fluoride sources, in combination with transition metal catalysts, have been used in epoxide openings and allylic substitution reactions. Furthermore, Toste and coworkers have introduced a novel enantioselective fluorination concept that relies on anion exchange processes of ionic fluorinating reagents with chiral anions leading to an increased solubility of these reagents in a new, chiral environment. This approach has proven to be very fruitful in the development of synthetic routes for enantioenriched, fluorine-containing compounds.

In this chapter, we provide a comprehensive overview of the progress in the field of *catalytic enantioselective fluorination* through September 2015. Due to our desire to solely focus on catalytic methods, we will not include diastereoselective processes, stoichiometric reactions, or asymmetric functionalization of fluorine-containing molecules. To enhance readability, the chapter will be separated by topic into electrophilic and nucleophilic fluorination methods. Electrophilic enantioselective fluorinations have been intensively studied, and therefore will be further subdivided into six subsections: fluorination of metal enolates, tertiary amine catalysis, enamine catalysis, miscellaneous organocatalysts, chiral anion phase transfer catalysis, and metal-catalyzed fluorinations not involving enolates. Nucleophilic fluorination will be divided into two subsections: ring opening of strained heterocycles and metal-catalyzed allylic substitution.

2. Electrophilic Fluorination Methods

2.1 Fluorination of Metal Enolates

The emergence of novel enantioselective fluorination protocols in the past two decades can be directly linked to the introduction of easily handled and bench-stable electrophilic fluorinating reagents such as *N*-fluorobenzenesulfonamide (NFSI), *N*-fluoropyridinium salts, and 1-(chloromethyl)-4-fluoro-1,4-diazoniabicyclo[2.2.2]octane ditetrafluoroborate (Selectfluor). The reason for the successful implementation of these mild reagents in asymmetric fluorinations lies in their relatively slow reaction with a given nucleophile (e.g., an enolate), opening the possibility for catalyst-accelerated process and, therefore, reaction control by a chiral catalyst.

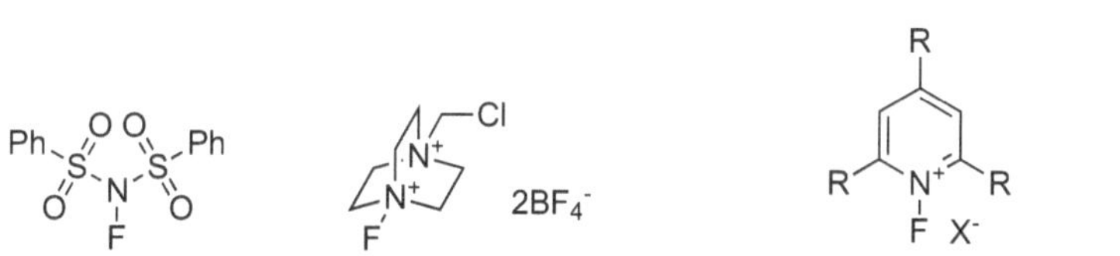

NFSI
Honeywell
200-1000 €/kg

Selectfluor
Air Products, Sinofluoro
800 €/kg

***N*-Fluoropyridinium salts**
R = H, X = OTf: **NFPY·OTf**
R = H, X = BF_4: **NFPY·BF$_4$**
R = Me, X = OTf: **NFCO·OTf**
Tosoh, ca. 350 €/kg

In the early stages of enantioselective fluorination, metal enolates proved to be the substrates of choice. The transformation of a given carbonyl compound into its activated enolate by bidentate coordination to a metal center bearing chiral ligands, resulted in the formation of a rigid, chiral environment that was key to obtaining enantiomerically enriched, fluorinated products. One of the first breakthroughs was reported by Differding and Lang.[12] They described the enantioselective fluorination of a β-ketoester metal enolate using stoichiometric amount of a chiral electrophilic fluorinating reagent derived from camphor; however, it was not until 2000 that the first catalytic fluorination of enolates was reported by Hintermann and Togni[5,33]. The use of a (α, α, α', α'-tetraaryl-1,3-dioxolane-4,5-dimethanol) TADDOL-based Ti-Lewis acid catalyst in combination with Selectfluor allowed for the formation of fluorinated β-ketoesters in high yields and up to 90% ee (enantiomeric excess).

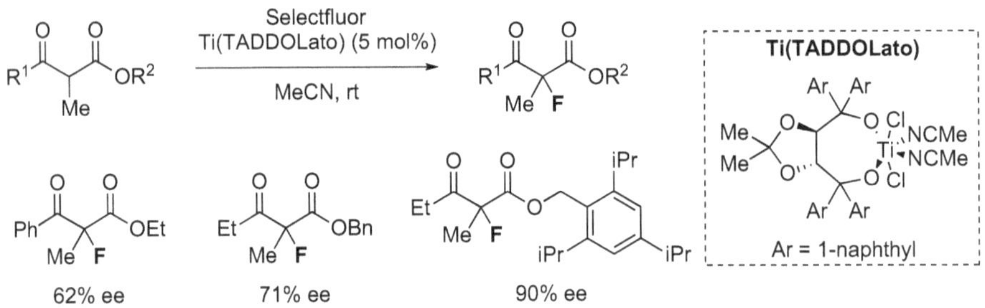

The same catalytic system was also compatible with the enantioselective heterodihalogenation of β-ketoesters. Employing Selectfluor and NCS (*N*-chlorosuccinimide) as halogenating reagents afforded interesting α-fluoro-α-chloro-β-ketoesters, albeit with only asymmetric induction.[18]

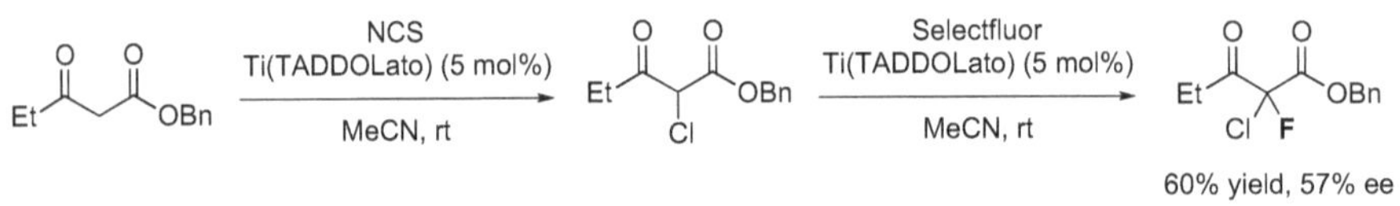

A more general enantioselective fluorination protocol was disclosed by the group of Sodeoka.[24] Using a cationic palladium–BINAP ((1,1′-binaphthalene-2,2′-diyl)bis(diarylphosphine)) complex, a wide range of β-ketoesters, including cyclic and acyclic substrates, were fluorinated with excellent enantioselectivity and in high yield.

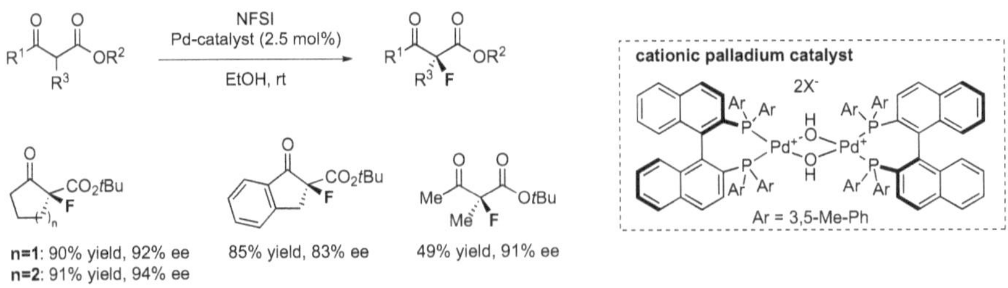

The authors proposed a mechanism in which the β-ketoester is coordinated to the cationic palladium catalyst in a bidentate manner. Through this coordination, the acidity of the α-proton is significantly increased, allowing for the facile generation of the metal enolate and subsequent fluorination. The stereochemical outcome of the reaction was rationalized by proposing a square planar Pd−enolate complex in which the steric interactions between the aryl groups of the ligand and the *tert*-butyl group of the ester are minimized. This arrangement makes the *Si* face inaccessible, forcing the NFSI to approach the enolate from the less hindered *Re* face.[28]

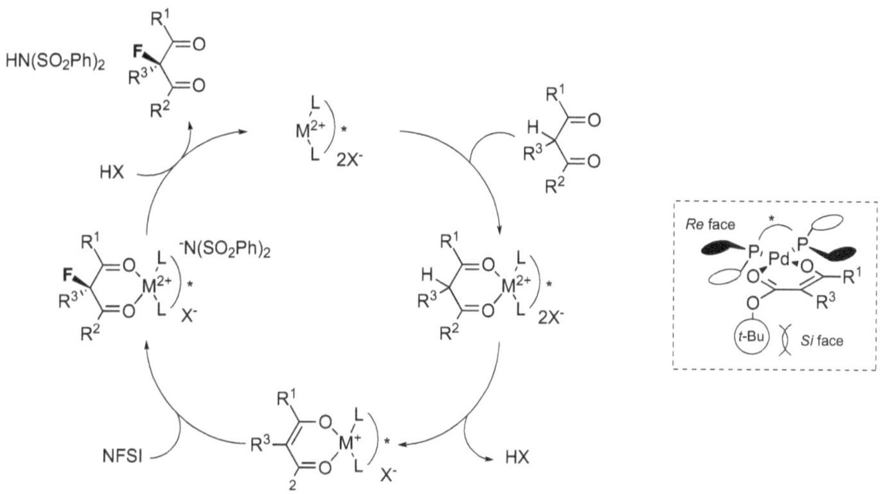

The same palladium catalyst could also be immobilized onto ionic liquids and be reused up to 10 times in the same reaction with similar enantioselectivity.[25] Sodeoka also reported the first Pd-catalyzed enantioselective fluorination of oxindoles[26] and the enantioselective fluorination of more exotic substrate classes, including β-ketophosphonates[27] and α-aryl-α-cyanophosphonates.[60]

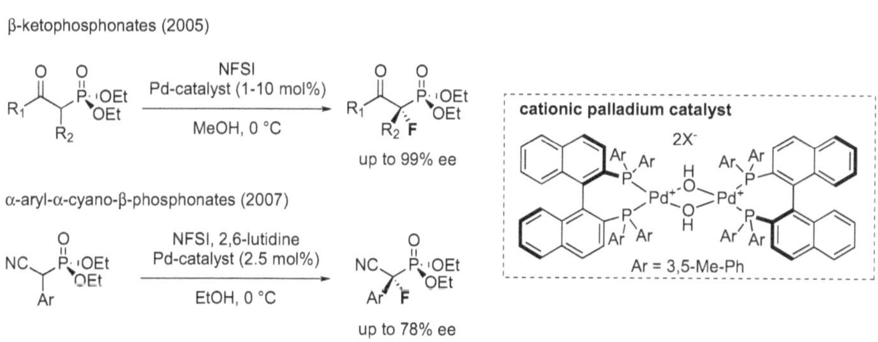

The major limitation of the earlier reports on Pd-catalyzed enantioselective fluorination was that less acidic substrates, such as esters, could not be successfully fluorinated. In order to address this problem, Sodeoka and coworkers[80] introduced a novel approach by using a $NiCl_2$—BINAP/Et_3SiOTf/2,6-lutidine triad that allowed for the fluorination of ester equivalents. 2,6-Lutidine and triethylsilyl triflate played a crucial role in the success of this reaction by promoting the enolization of the substrate and by generating a dicationic nickel triflate complex. The fluorinated α-acetic acid derivatives were obtained in good yields and enantioselectivity, and the products were transformed into the corresponding carboxylic acids without erosion of enantiomeric excess.

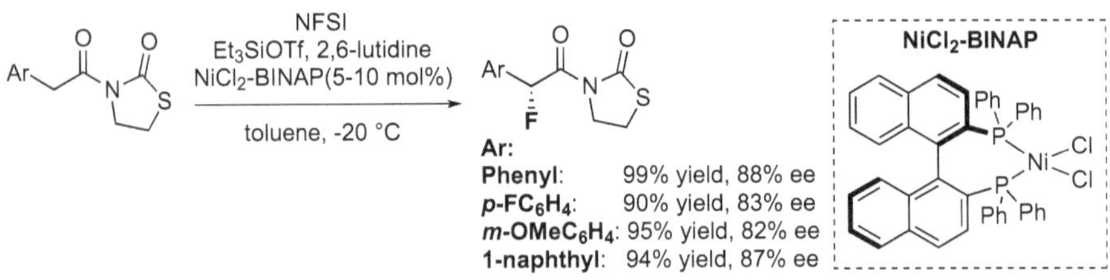

Sodeoka's[79] group also developed a catalytic, enantioselective monofluorination of α-ketoesters using a palladium π-hydroxo complex. To facilitate purification and avoid potential racemization, the fluorinated products were reduced in situ to the β-fluoro-α-hydroxy esters. Depending on the reducing agent employed, the *anti*-fluorohydrin (with DIBAL or enzymatically) or *syn*-diastereomer (with L-selectride) were obtained selectively.

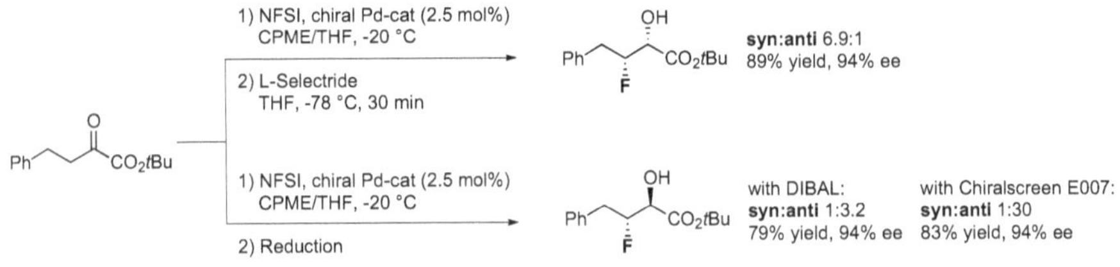

Bis(oxazline)-ligand (Box)—containing metal complexes have been proved to be privileged catalysts in many asymmetric transformations. The first use of a Box—metal complex in the area of enantioselective fluorination was reported by Ma and Cahard.[56] The enantioselective fluorination of β-ketoesters was achieved using only 1.0 mol% of a chiral bis(oxazoline) copper catalyst. The addition of hexafluoroisopropanol (HFIP) was crucial for achieving good enantioselectivities; however, only a single cyclopentanone-derived substrate was obtained in synthetically useful enantiomeric excess.

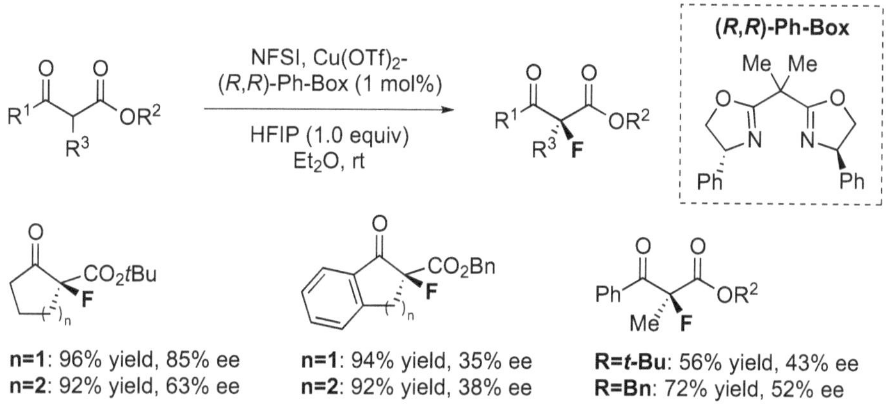

The same year, Shibata et al.[71] expanded on the findings of Ma and introduced an improved catalyst system based on nickel–Box and copper–Box complexes. The use of methyl-*tert*-butylether dramatically improved the enantioselectivity of the copper-catalyzed reaction, whereas in the nickel-catalyzed case, dichloromethane was the solvent of choice. Interestingly, the sense of enantioinduction for the respective Ni and Cu catalysts bearing the same (S,S)-bis(oxazoline)-Ph ligand was not identical, which allowed for the selective synthesis of both enantiomers of the fluorinated β-ketoester from a single ligand. Based on evidence obtained throughout the work, this intriguing outcome was not the result of a solvent effect, but most likely due to a change in metal geometry and therefore also in the transition state of the enantiodetermining step.

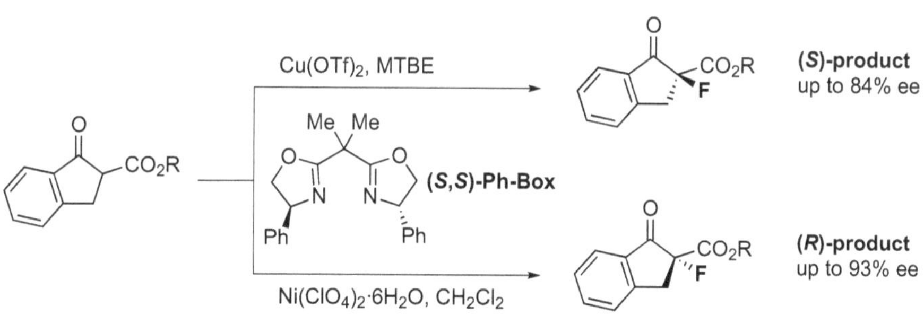

One year later, Shibata et al.[72] introduced a more general and effective set of conditions that allowed for the enantioselective fluorination of 3-substituted oxindoles (93–96% ee), cyclic β-ketoesters (93–99% ee), and challenging acyclic β-ketoesters (83% ee). The remarkable power of the Ni/DBFOX (dibenzofuran-bis-oxazoline)-Ph catalyst to distinguish between the enantiotopic faces of a given enolate was showcased with the first catalytic, enantioselective preparation of the potassium ion channel modulator MaxiPost. Furthermore, the authors observed a positive, nonlinear effect of chiral amplification: the use of a DBFOX-Ph ligand with only 45% enantiopurity led to the formation of the fluorinated product in 99% ee. In previous reports by the same group on chiral amplifications, this effect was attributed to the precipitation of an S_4-symmetric meso 2:1 complex between DBFOX-Ph and Ni(II) and/or the associative formation of a 1:1 heterochiral complex by the aid of hydrogen bonds based on aqua ligands to produce stable meso oligomers.

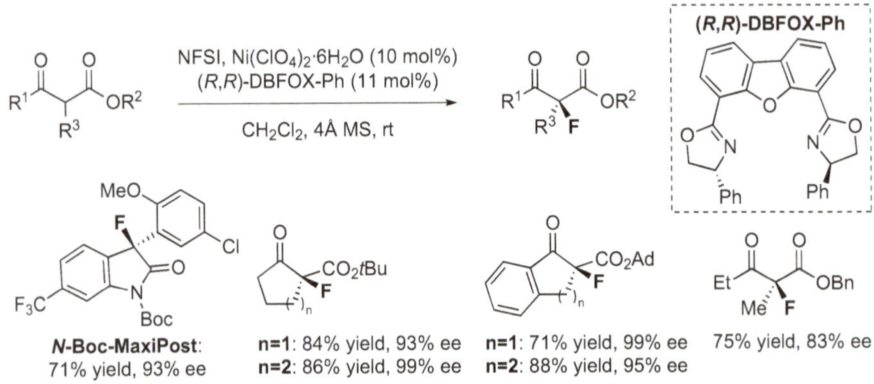

The comprehensive work of Shibata's group also included the catalytic, enantioselective fluorination of malonates[67] and ester equivalents[37] employing Box–nickel and Box–zinc complexes. Tridentate Box ligands, in combination with $Ni(ClO_4)_2$, were used by Iwasa[73] and later also by Gade[11] for the highly enantioselective fluorination of β-ketoesters and oxindoles.

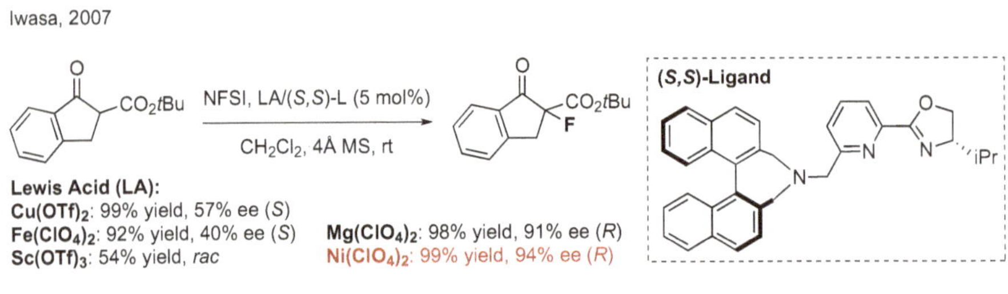

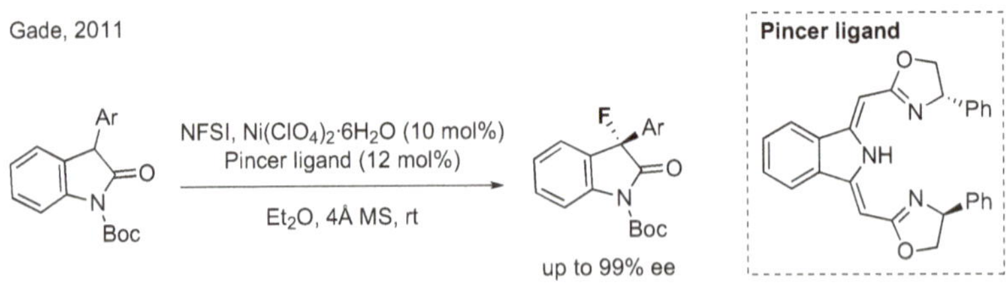

Finally, Kesavan and coworkers[3] described a novel tartrate-derived Box–Cu(II) complex that was an active catalyst in the fluorination of aliphatic cyclic and acyclic β-ketoesters; however, the corresponding products were obtained with only moderate enantioselectivities. This copper-catalyzed approach was refined by Iwasa.[61] They used an improved catalyst system—derived from $Cu(OTf)_2$ and a chiral spiro-oxazoline ligand that enabled the fluorination to proceed in a highly enantioselective manner.

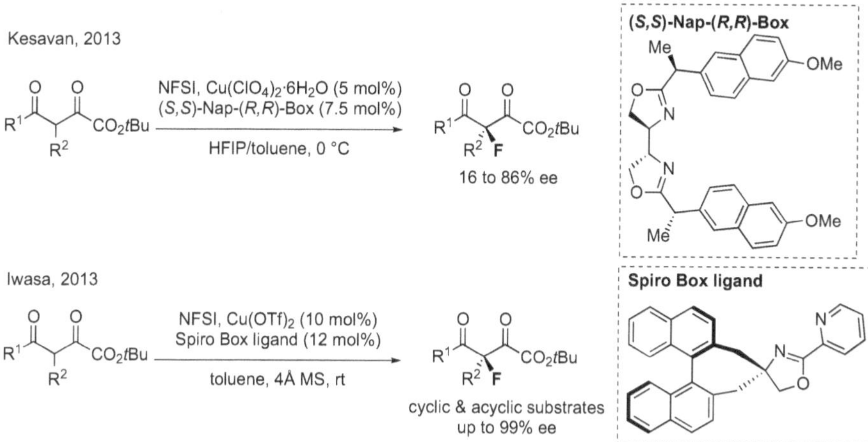

Metal complexes containing ligands that are based on the TADDOL, BINAP, or Box core structures were not the only successful catalysts used in the enantioselective fluorination of metal enolates. Inanaga and coworkers[78] synthesized a rare earth perfluorinated binaphthyl phosphate, Sc[(*R*)-F_8BNP]$_3$, and applied it in the enantioselective fluorination of β-ketoesters. This report by Inanaga was remarkable not only because of the exotic catalyst but also because it is one of the rare examples of successful implementation of an *N*-fluoropyridinium salt (NFPY-OTf) as an electrophilic fluorine source in an enantioselective fluorination reaction. A different scandium catalyst was used by Feng and coworkers[54] for the enantioselective fluorination of oxindoles, providing 3-alkyl-3-fluoro-2-oxindoles with high enantioselectivity.

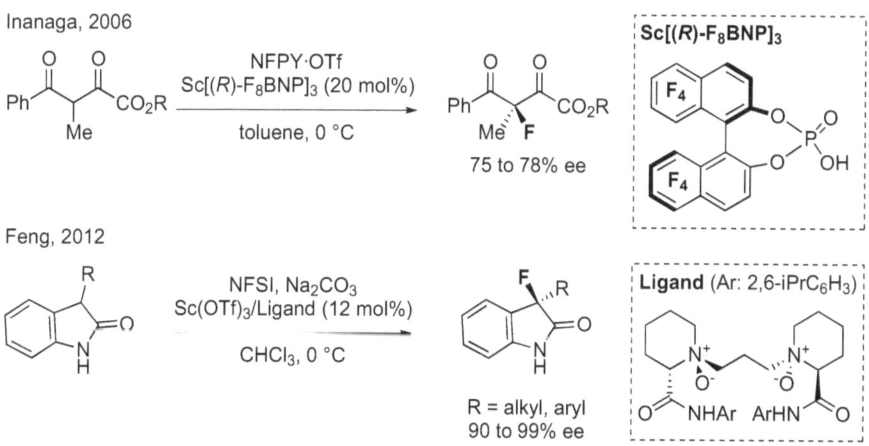

Mezzetti and Togni popularized the use of ruthenium complexes containing chiral tetradentate *PNNP* ligands in a wide range of atom transfer reactions. These complexes, upon activation with a stoichiometric amount of $(Et_3O)PF_6$, were also successfully applied in the enantioselective fluorination of 1,3-dicarbonyl compounds.[1] Oxygen donors, in particular diethyl ether as cosolvent, increased the activity of the catalyst and led to improved enantioselectivities in certain cases; an observation that underlined the importance of proton transfer processes promoted by these oxygen donors.

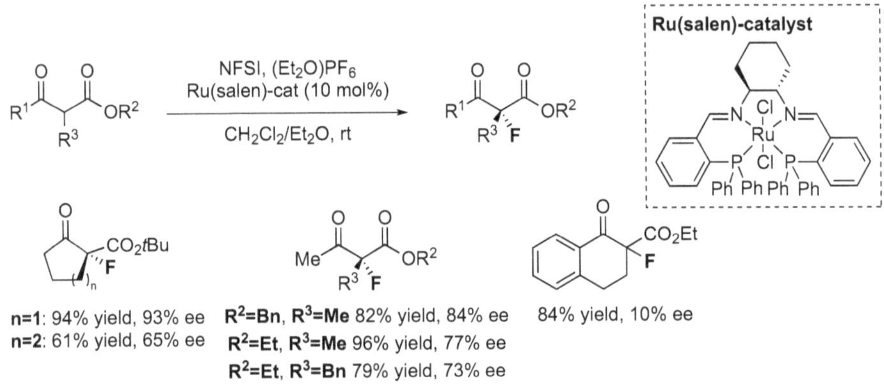

A cobalt-catalyzed enantioselective fluorination was reported by Itoh et al.[46] Using Jacobsen's salen ligand and $Co(acac)_2$, several acyclic and cyclic β-ketoesters were fluorinated in good yields and enantioselectivity. Combinations of the salen ligand with other metals was also extensively studied, culminating in the 2014 report of Xu et al.[22] on the iron-catalyzed α-fluorination of β-ketoesters and *N*-Boc-oxindoles. The corresponding products were obtained in high yields and good to excellent enantiomeric excess.

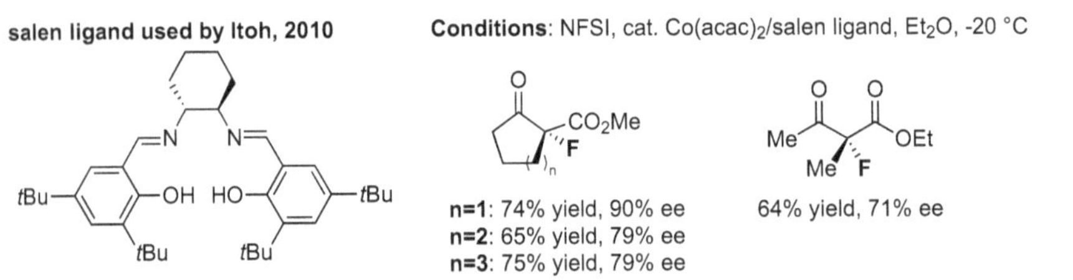

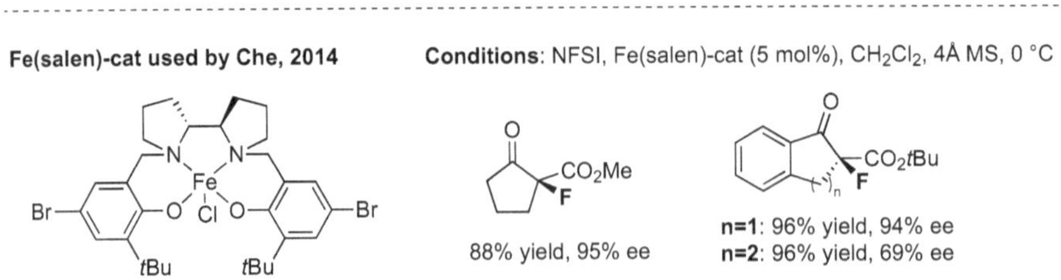

2.2 Tertiary Amine Catalysis

As described in the beginning of the chapter, the reactivity of most electrophilic fluorinating reagents is centered around the polarized N-F bond. Therefore, the preparation of electrophilic chiral N-fluoro reagents presented an opportunity for enantioselective fluorination. One of the first breakthroughs was disclosed by Differding and Lang.[12] They described the enantioselective fluorination of a β-ketoester enolate using a stoichiometric electrophilic fluorinating reagent derived from camphor. Soon afterward, Shibata et al.[70] and Cahard et al.[8] independently reported on the isolation of stable *N*-fluoroammonium salts of cinchona alkaloids, prepared by the reaction of Selectfluor with the tertiary amine group of the alkaloid. These reagents were extremely effective in the enantioselective fluorination of preformed enolate equivalents (e.g., metal enolates or silyl enol ethers). These reactions are not catalytic and will therefore not be discussed in greater detail; however, they present interesting reactivity and can be seen as important groundwork for the development of catalytic reactions.

The challenge of developing a tertiary amine–catalyzed, enantioselective fluorination reaction lies in the fact that most preformed enolate equivalents are reactive toward electrophilic *N*-fluoro reagents even in the absence of a catalyst, and therefore a subject of significant racemic background reactions. The transition to a catalytic method was first achieved by Shibata and coworkers,[19] employing a combination of a cinchona alkaloid and Selectfluor. The substrates, *O*-acetylated enol ethers, were unreactive toward Selectfluor at ambient temperature allowing for the in situ formation of the active catalytic species—a chiral fluorinated cinchona alkaloid. The enantioinduction was still relatively low, but the demonstration of feasibility of a catalytic process was a significant step.

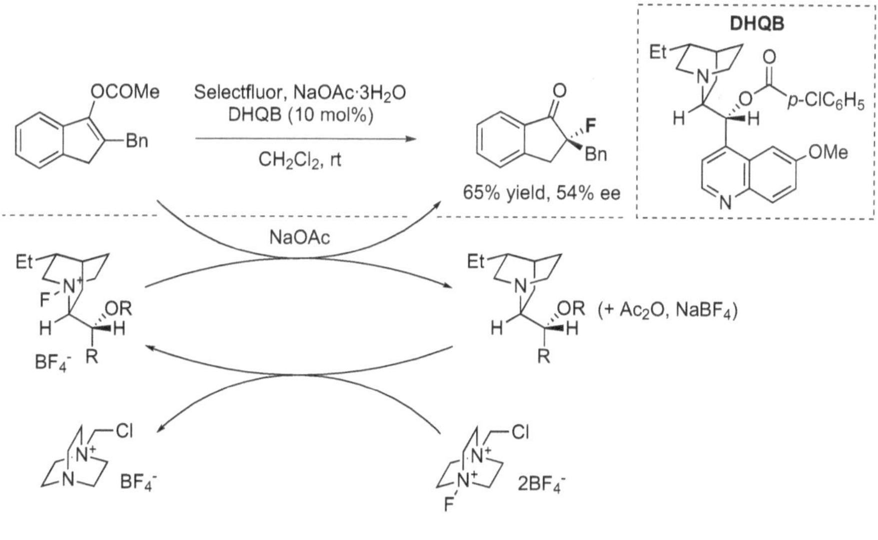

Two years after their seminal report, Shibata and coworkers[38] improved their own method and described the first highly enantioselective catalytic process, using NFSI as the electrophilic fluorinating reagent, a bis-cinchona alkaloid catalyst and an excess of base. This modified set of conditions allowed for the effective fluorodesylation of silyl enol ethers and allyl silanes and provided access to enantioenriched, fluorine-substituted quaternary carbon centers. The authors proposed a mechanism involving initial formation of an *N*-fluoroammonium-KCO_3 intermediate, in which the counter-anion is responsible for the desylation of the substrate, followed by the catalyst-controlled enantioselective fluorine transfer to the enolate.

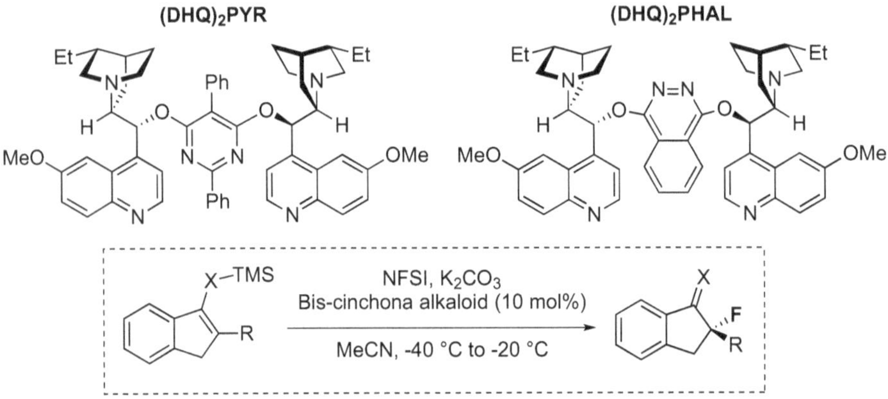

for X = CH_2, with (DHQ)$_2$PYR
R = Bn: 75% yield, 94% ee
R = *p*-Cl-Bn: 81% yield, 94% ee
R = *o*-Me-Bn: 58% yield, 93% ee
R = Me 73% yield, 72% ee
R = H 58% yield, 51% ee

for X = O, with (DHQ)$_2$PHAL
R = Bn: 82% yield, 82% ee
R = *p*-Cl-Bn: 74% yield, 86% ee
R = *p*-Me-Bn: 79% yield, 86% ee
R = Et 95% yield, 67% ee

In the following years, the groups of Gouverneur[55] and He[88] used conditions very similar to those reported by Shibata for two distinct catalytic enantioselective fluorination reactions. First, Gouverneur et al.[55] reported the enantioselective fluorocyclization of protected indoles. The reaction allows for the simultaneous installation of two chiral centers, including the fluorine-substituted quaternary benzylic stereogenic center. In cases in which the nucleophilic tether was attached onto the nitrogen of an unsubstituted indole, the corresponding difluorinated, tricyclic product was obtained.

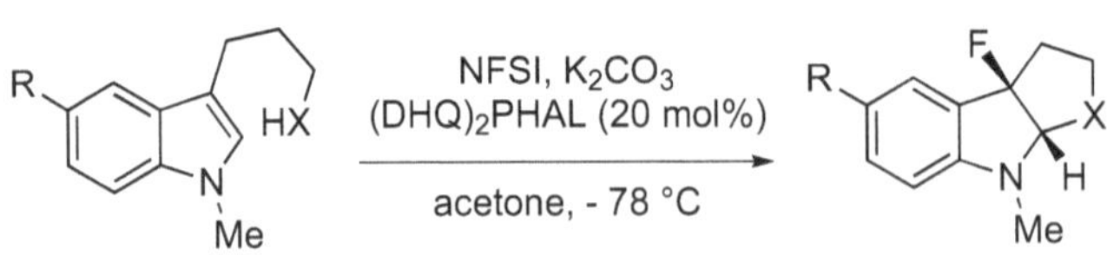

for X = O
R = H: 72% yield, 66% ee
R = OMe: 65% yield, 74% ee
R = Ph: 61% yield, 61% ee
R = Mes 55% yield, 84% ee

for X = NTs
R = H: 59% yield, 64% ee
R = OMe: 51% yield, 70% ee
R = Ph: 70% yield, 70% ee
R = Mes 80% yield, 84% ee

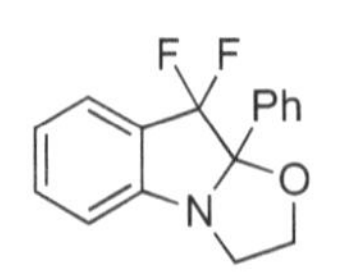

50% yield, 60% ee

He and coworkers[88] used the $(DHQD)_2PHAL$ (dihydroquinidine 1,4-phthalazinediyl diether) catalyst for the enantioselective fluorination of substituted oxindoles. This report also describes a rare investigation of how structural and electronic changes in the electrophilic fluorinating reagent (NFSI analogs) influence the stereochemical outcome of the transformation. They found that the bis-4-*tert* butyl-NFSI analog enhanced the enantioselectivity, albeit at the expense of diminished reaction rate.

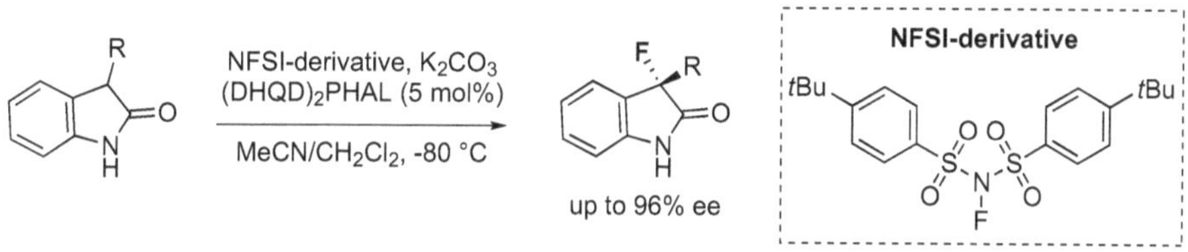

A different cinchona alkaloid–catalyzed fluorination was reported by Tu's group.[9] They discovered an asymmetric fluorination/semipinacol rearrangement reaction sequence of 2-oxa allylic alcohols. The corresponding β-fluoroketone products were obtained in moderate to good enantioselectivity.

2.3 Enamine Catalysis

2.3.1 Linear Aldehydes

Since the pioneering studies of Hajos and Parrish[23] as well as Eder et al.[15] on the proline-catalyzed asymmetric aldol reaction, enamine catalysis has emerged as one of the cornerstones in the field of asymmetric organocatalysis. The application of this concept to enantioselective fluorination reactions has been intensively investigated by many different research groups. Compared with the tertiary amine–catalyzed approach, the enamine approach does not rely on the in situ formation of a chiral fluorinating reagent, but on the condensation of the chiral amine catalyst with a carbonyl substrate to form a chiral enamine intermediate.

In 2005, Enders and Hüttl[16] reported the first direct organocatalytic α-fluorination of aldehydes and ketones using a proline-derived catalyst; however, the asymmetric inductions were low (ee $\leq$ 36%), and in some cases the enantiopurities of the products were not reported.

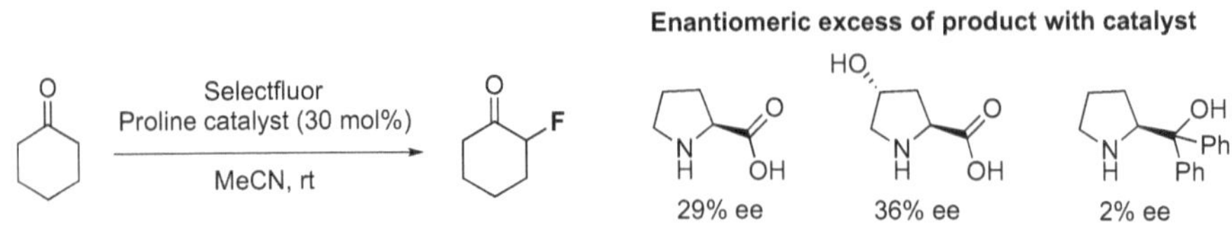

In the same year, the enantioselective α-fluorination of aldehydes was disclosed by Jorgensen and coworkers.[59] Using their privileged trimethylsilylated prolinol derivative (Jorgensen–Hayashi catalyst), the fluorination proceeded effectively at room temperature with only 1.0 mol% catalyst loading and provided the β-fluoroalcohols (in situ reduction of the unstable α-fluorinated aldehydes) with high enantioselectivity.

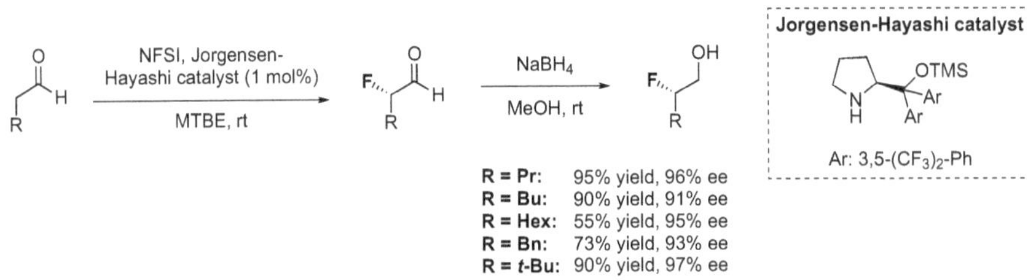

The authors rationalized the stereochemical outcome of the reaction by the formation of the more stable *E*-enamine, in which the *Re* face of the double bond is blocked by the bulky 3,5-bis-(trifluoromethyl)phenyl groups, and as a consequence electrophilic fluorination selectively occurred from the *Si* face.

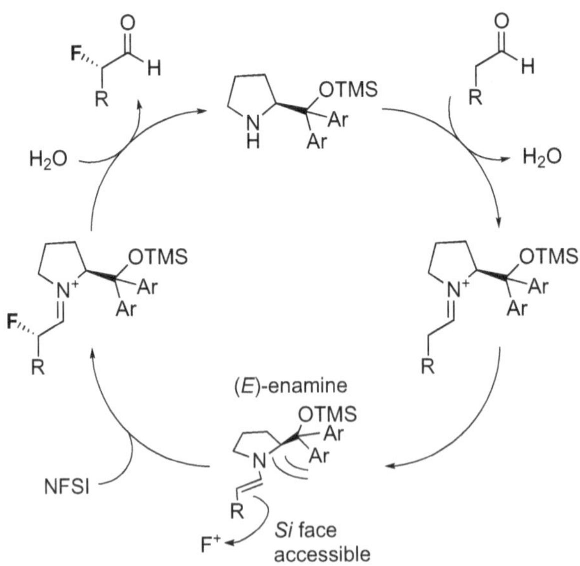

Shortly after the report of Jorgensen, Beeson and Macmillan[4] reported the use of an imidazolidinone catalyst for the fluorination of linear aldehydes. Remarkably, a wide range of functional groups, including olefins, esters, and carbamates, were compatible with the reaction conditions and the corresponding functionalized β-fluoroalcohols were obtained with outstanding enantioselectivity. An interesting solvent effect was observed as the addition of 10% isopropanol dramatically improved the enantioinduction and efficiency of the catalyst; however, the origin of this effect was not rationalized.

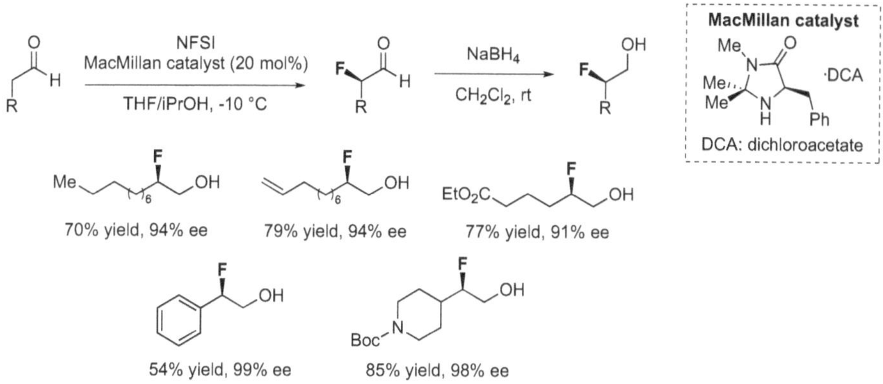

Aside from the in situ reduction to the β-fluoroalcohol, other reports focused on complementary conversions of the intermediate fluorinated aldehyde. The Jorgensen group[39] reported the one-pot, enantioselective fluorination/Seyferth–Gilbert homologation of linear aldehydes to obtain enantiomerically enriched propargyl fluorides in up to 99% ee. Pushing this methodology even further, the authors applied click chemistry to synthesize optically active triazoles without isolation of any intermediates. In addition, the Wittig reactions proved fully compatible with the organocatalytic, enantioselective fluorination process. A one-pot reaction setup was designed, and allyl fluorides were isolated in good yields and with high enantioselectivity.

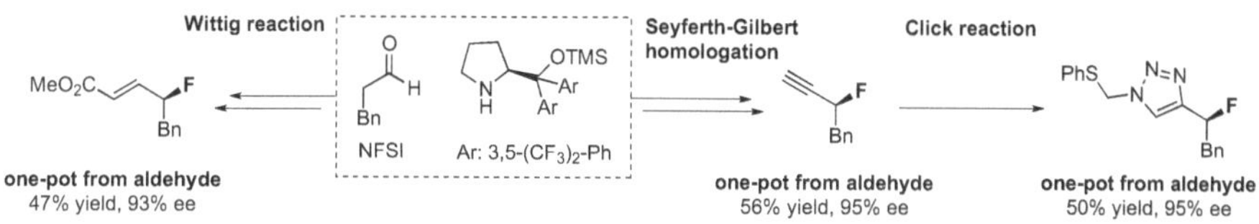

In 2009, Fadeyi and Lindsley[17] published the one-pot catalytic, enantioselective synthesis of chiral β-fluoroamines, further highlighting the versatility and importance of the intermediate α-fluoroaldehyde building block. Both, high yields and enantioselectivity were obtained using MacMillans imidazolidinone catalyst and isopropanol as cosolvent, followed by direct reductive amination with different secondary amines and $NaBH(OAc)_3$.

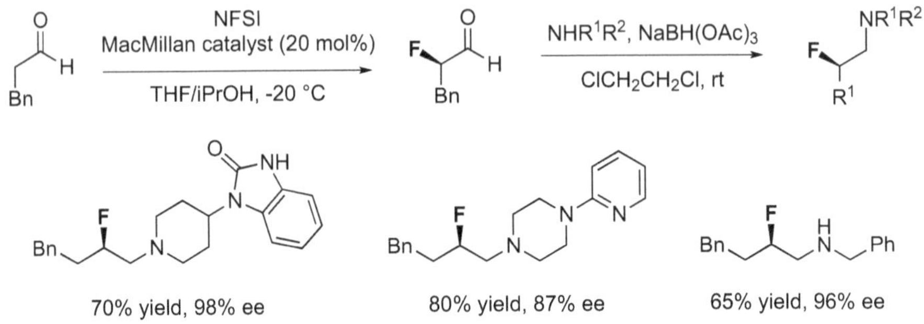

2.3.2 *Branched Aldehydes*

In the seminal report of Jorgensen and coworkers[59] on the enantioselective fluorination of linear aldehydes, the authors also showed preliminary results on the challenging fluorination of α-branched substrates. Using a sterically less encumbered catalyst and higher reaction temperature, the formation of the quaternary stereocenter proceeded in good yield and with moderate (but still noteworthy) stereoinduction (48% ee). Barbas et al.[76] also tackled the challenge of finding an effective method for the enantioselective fluorination of branched aldehydes. An intensive catalyst screening revealed the best results where obtained using a tetrazole-substituted secondary amine or a (Triisopropylsilyl) TIPS-protected prolinol derivative as the catalyst; however, high catalyst loadings were necessary and the enantioselectivities remained relatively low.

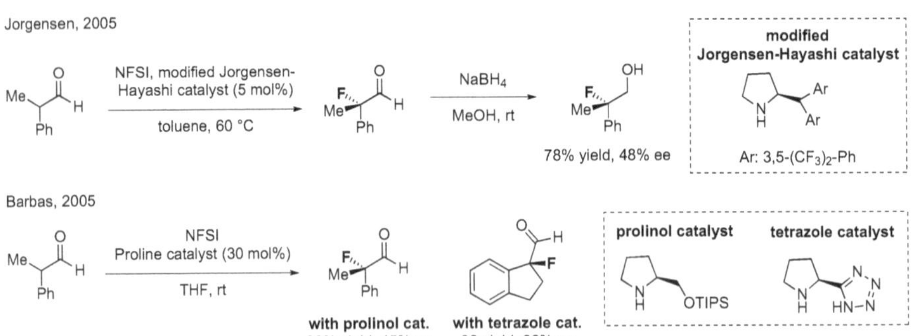

In a different approach by Jorgensen's group,[7] an atropochiral 8-amino-2-naphthol derivative was used as an organocatalyst enabling the asymmetric fluorination of branched aldehydes. The enantiocontrol of the catalyst was found to be the highest for branched, benzylic aldehydes, and the products were obtained in up to 90% ee, albeit in low yields due to partial decomposition of the products during column chromatography. In contrast, aldehydes bearing two aliphatic α-substituents proved to be very poor substrates, highlighting the difficulties of finding a general method for the fluorination of branched aldehydes using enamine catalysis—the problem remains unresolved.

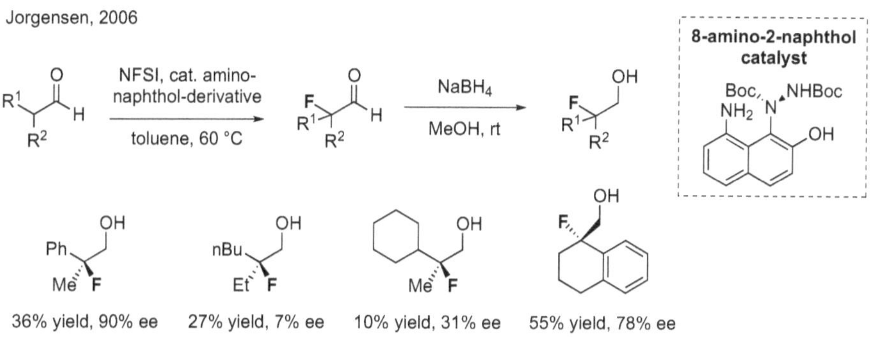

2.3.3 Ketones

The first example of asymmetric fluorination of ketones using enamine catalysis was reported by MacMillan's group.[49] An extensive high-throughput screening of a library of chiral amines identified a cinchona alkaloid–derived catalyst as ideal. Using their optimized system, consisting of a combination of the trichloroacetate salt of the catalyst, NFSI, and a base, the fluorination of a variety of carbo- and heterocyclic substrates proceeded with high yields and outstanding enantioselectivity. Remarkably, both diastereomers of cyclic ketones with a preexisting stereocenter were selectively obtained by the right choice of catalyst. Furthermore, the protocol allowed for the fluorination of complex carbonyl systems, such as the steroidal derivatives *allo*-pregnanedione and cholestanone. A detailed computational study of the origin of enantioselectivity in this reaction was reported 3 years later.[50] Similar to the cinchona alkaloid–catalyzed fluorination of enolate equivalents, the initial fluorination of the quinuclidine moiety of the catalyst with NFSI was proposed. The primary amine moiety of the catalyst then condenses with the ketone and initiates the enantiodetermining, intramolecular fluorine transfer to the formed enamine. Calculation of the lowest energy transition state revealed that the enantiocontrol is governed by the preferred chairlike conformation in a seven-membered ring transition state and the strong preference of the bulky quinoline substituent of the catalyst to occupy a pseudoequatorial position.

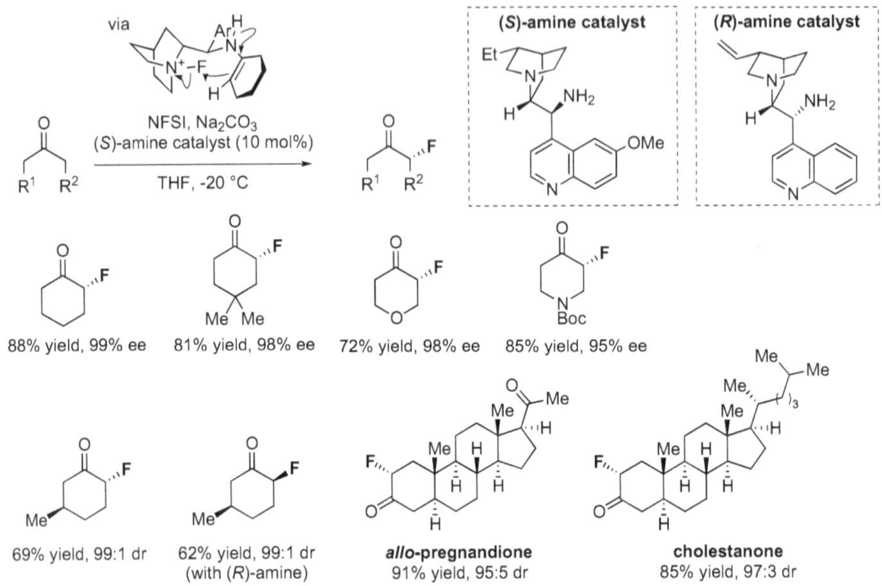

2.3.4 Other Substrates

The formal enantioselective addition of HF across the double bond of α,β-unsaturated aldehydes was achieved by MacMillan and coworkers.[35] They used a sequence of transfer hydrogenation (employing Hantzsch ester) and electrophilic fluorination (employing NFSI) to convert the enals into the corresponding α-fluoroaldehydes. The authors realized that the diastereomeric outcome of the reaction could be altered by controlling the iminium and enamine steps with separate catalysts. As an example, (*E*)-3-phenylbut-2-enal was selectively transformed into the formal *anti*- or *syn*-HF additions products using the same chiral imidazolidinone catalyst in the transfer hydrogenation step, followed by choice of enantiomers of the MacMillan catalyst for the diastereoselectively controlling the fluorination step. Remarkably, both diastereomers were obtained with essentially perfect enantioselectivity and in good yields.

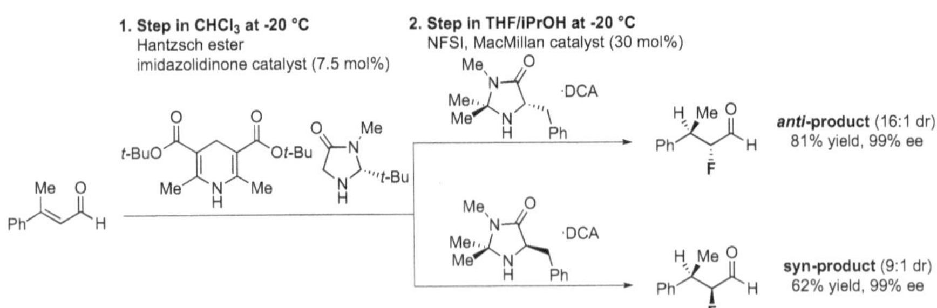

A further proof of the efficiency of the enamine/iminium catalysis cascade was illustrated by Appayee and Brenner-Moyer.[2] They synthesized optically active α-fluoro-β-amino aldehydes from simple enals using the Jorgensen-Hayashi catalyst.

The mild reaction conditions allowed for broad functional group tolerance and fluorinated products bearing olefins, ethers and nitriles were obtained with excellent enantioselectivity.

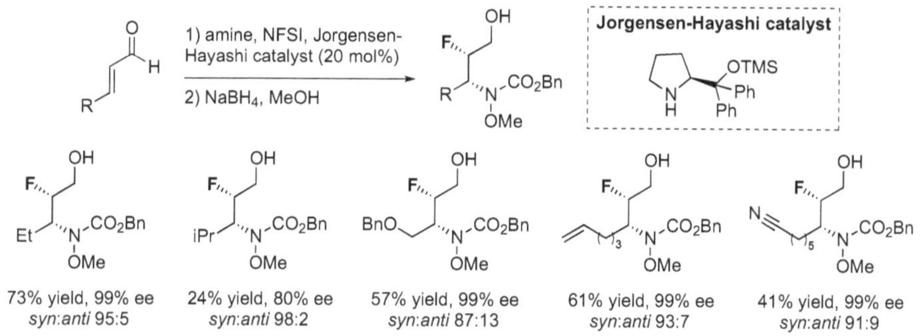

2.4 *Miscellaneous Organocatalysis Catalysts*

Since the discovery of Ugai et al.[82] that thiazolium salts can replace cyanide as a catalyst in benzoin condensations, the unique capability of *N*-heterocyclic carbenes (NHCs) to induce an Umpolung aldehyde reactivity has found widespread application in organic synthesis. Sun and coworkers[90] hypothesized that electrophilic enals bearing a leaving group in the γ-position were viable precursors for the generation of nucleophilic, vinylogous NHC enolates that could subsequently be trapped by an electrophilic fluorinating reagent. When the authors used Bode's triazolium NHC catalyst in combination with NFSI, the generated dienolate was fluorinated selectively at the α-position to give β,γ-unsaturated α-fluoroesters in good yields and with high enantioselectivity. The mild reaction conditions left sensitive functional groups such as aldehydes, ketones, unprotected alcohols, and nitriles completely untouched, making this NHC-catalyzed reaction highly valuable. However, as with every methodology, this reaction is not without limitations. Substituents on the α-position of the enal retarded the reaction drastically (<10% conversion) and led to diminished enantioselectivity (55% ee).

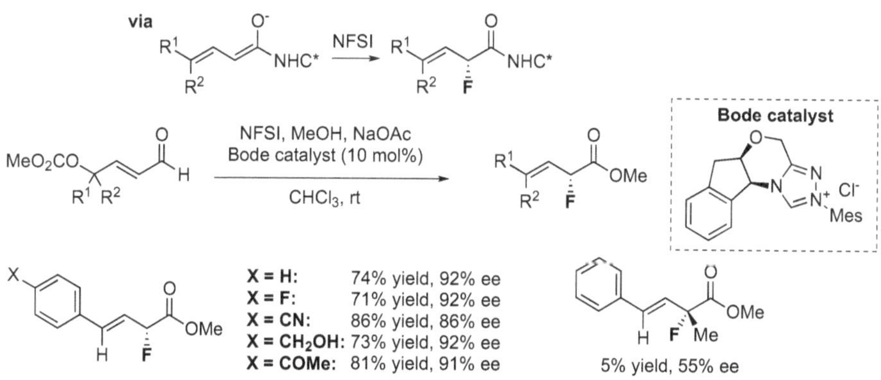

In 2015, the groups of Sun[14] and Wang[53] independently reported the enantioselective fluorination of simple aldehydes using NHC catalysis. Notably, the intermediate acyl azolium species was generated by using NFSI as an oxidant and as the source of "F^+" in the subsequent fluorination of the formed enolate. This remarkable finding paved the way for the NHC-catalyzed synthesis of a wide range of α-fluorinated esters, thioesters, and amides with excellent enantioselectivities and from simple starting materials. Similar to the first report of Sun et al., the substrate scope was very broad, and products bearing ethers, esters, basic amines, olefins, and alkynes were obtained. In Sun et al.'s study the commercially available Bode catalyst was used, and the reaction worked well with simple alcohols such as methanol or benzyl alcohol (necessary for ester formation and catalyst turnover). In contrast, Wang and coworkers employed a modified version of the catalyst bearing an unusual tribromophenyl group, and their reaction only worked with more hindered alcohols, as only trace product was obtained when methanol or ethanol was used. The authors speculated that sterically less hindered alcohols prefer to react directly with the initially formed acyl azolium species before fluorination could occur.

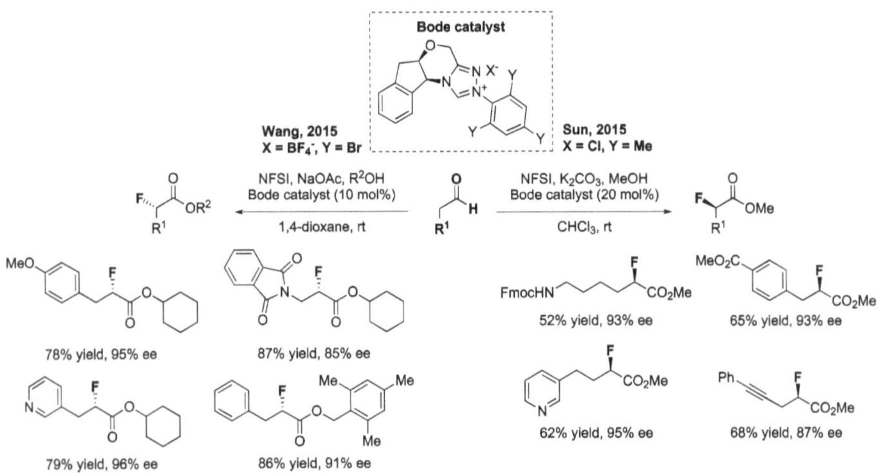

A thiourea-catalyzed approach for the enantioselective fluorination of β-ketoesters was disclosed by Niu and coworkers.[85] An initial screening of substrates with chiral bifunctional thiourea catalysts revealed that the choice of the ester component in the starting material was crucial for achieving high enantioselectivity, with methyl and benzyl esters providing the best results. With this information in hand, the authors turned their attention to the expansion of the substrate scope. A wide range of 5-membered, cyclic β-ketoesters were efficiently fluorinated to afford the desired products in high yield and good enantioselectivity. However, the use of tetralone derivatives and acyclic substrates led to generally significantly lower enantioselectivities.

The authors proposed an enantiodetermining transition state in which the NFSI is bound via hydrogen bonds to the urea moiety of the catalyst and therefore held in close proximity to the β-ketoester, which interacts with the basic nitrogen of the catalyst.

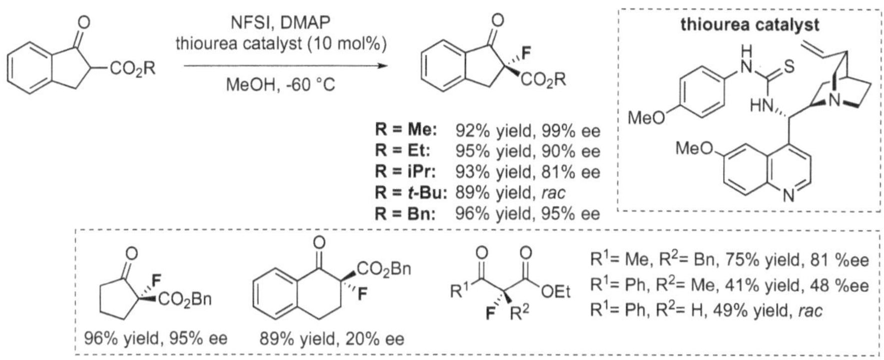

Fu and coworkers[51] used their chiral DMAP catalyst for the synthesis of challenging, tertiary α-fluoroesters from alkyl aryl ketenes. Mechanistically, the catalyst reacts with the ketene to give a chiral enolate that undergoes electrophilic fluorination with NFSI. The formed α-fluoroacetylpyridinium ion is then attacked by the pentafluorophenoxide additive to release the catalyst and to form the desired product. Tertiary α-fluoroesters were obtained in nearly quantitative yields and with high levels of enantioselectivity.

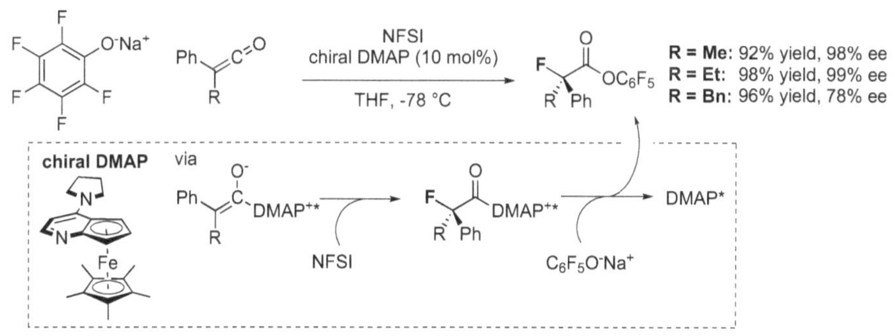

2.5 Chiral Anion Phase Transfer Catalysis

As previously mentioned, the development of new fluorinating reagents had a profound impact on the development of catalytic asymmetric fluorination reactions, providing opportunities for novel strategies to be designed. One such strategy was chiral anion phase transfer (CAPT) catalysis. CAPT catalysis largely remained an undeveloped field in organic synthesis, despite the success of its cationic counterpart.[6,58,64] Selectfluor, a doubly cationic reagent, is sparingly soluble in most organic solvents, making it an ideal reagent to employ in CAPT catalysis. In their seminal report, Toste and coworkers developed a fluorocyclization reaction with cyclic enol ethers using Selectfluor under CAPT conditions.[68] They envisaged that a chiral phosphate anion would participate in salt metathesis with Selectfluor, affording a soluble chiral ion pair, that could then react in an enantioselective fashion with olefinic substrates, followed by intramolecular trapping by a nucleophile.

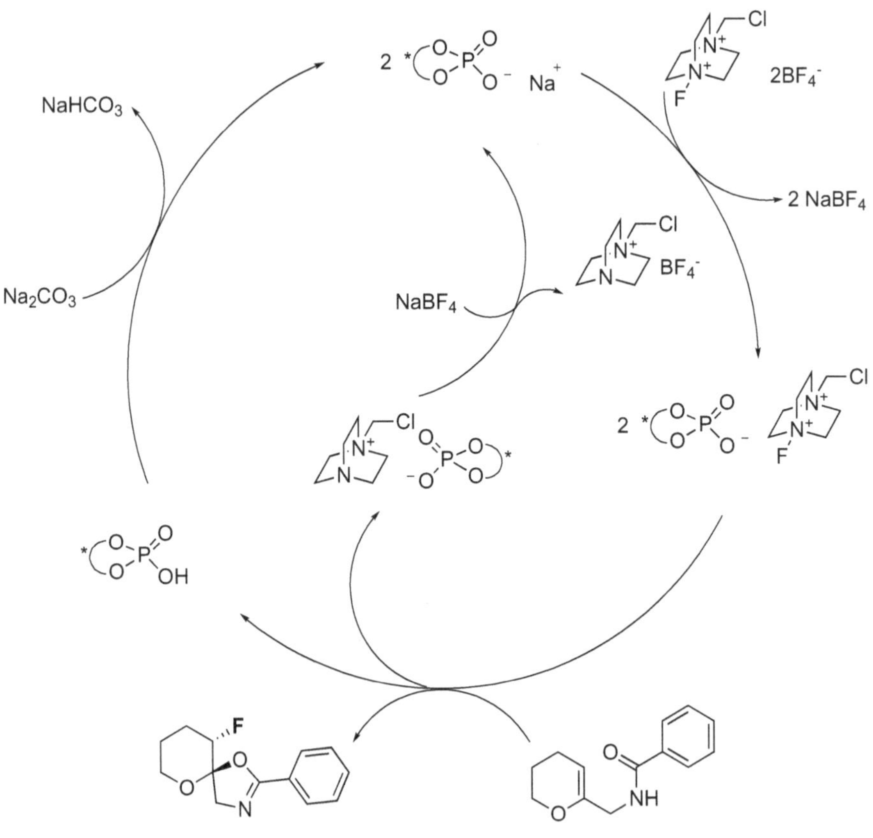

Due to the solubility properties of Selectfluor, a racemic background reaction would be mitigated by use of a nonpolar solvent. Employing the highly lipophilic phosphate catalyst C_8-TRIP (Triisopropylphenyl), in combination with fluorobenzene (C_6H_5F) as a solvent, the fluorocylized products were obtained in high enantiomeric excess and excellent diastereoselectivity. The deprotonated phosphate catalyst is bifunctional, serving both as a counteranion and a hydrogen bond acceptor. Hydrogen bonding interactions between the phosphate and the substrate were proposed as key interactions for the organization of the transition state, and available interactions would prove to be important in future CAPT fluorination reactions.

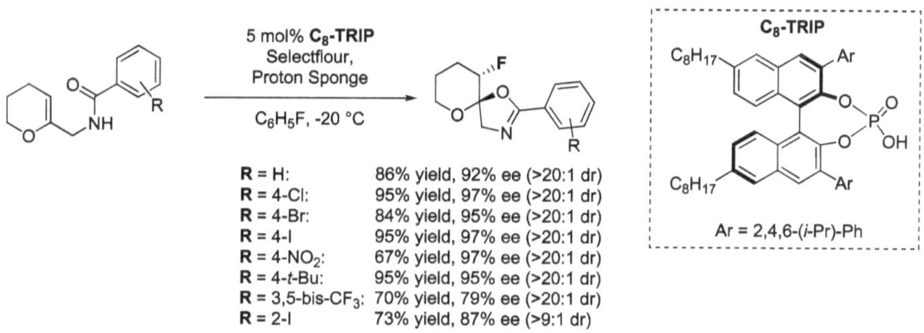

A major benefit of the phase transfer strategy was the tempered reactivity of the strongly oxidizing Selectfluor under heterogeneous conditions. When benzothiophene substrates with sensitive functional groups were treated with Selectfluor under homogeneous conditions (MeCN as solvent), a complex reaction mixture containing only traces of the desired fluorocyclized product was obtained. However, under phase transfer conditions, the products were obtained in good yield and excellent enantioselectivity.

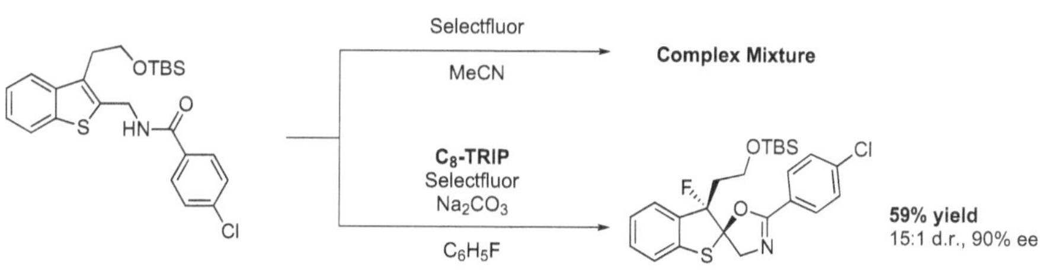

In addition to benzothiophene substrates, fluorocylization reactions of chromenes and dihydronapthalenes also proceeded with high enantio- and diastereoselectivity. An example utilizing an acyclic olefin was also reported, which afforded the product in good yield, but modest enantioselectivity. These less reactive olefins represent a challenging substrate class for future improvements of this methodology.

The CAPT platform was also found to be amenable to other olefin difunctionalization reactions. The 1,4-aminofluorination[74] of conjugated dienes, again utilizing an amide as an intramolecular nucleophile, was described. For some substrates, the best results were obtained when utilizing more electrophilic derivatives of Selectfluor.

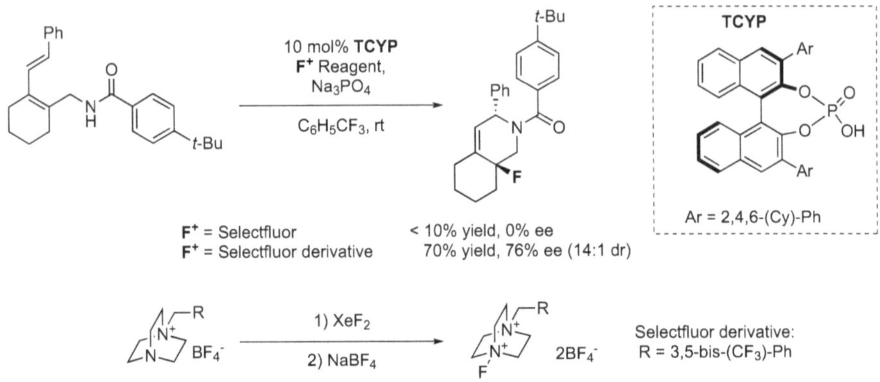

A highly enantio- and diasteroselective 1,2-oxyfluorination of (*Z*)-enamides was achieved by Toste and coworkers,[34] utilizing water and alcohols as nucleophiles. It was posited that both steps of this transformation, olefin fluorination and oxygen nucleophile addition to the *N*-acyliminium intermediate, were influenced by the phosphate catalyst. Therefore, formation of the second stereocenter was complicated by matched or mismatched pairing of catalyst-controlled selectivity and the inherent diastereoselectivity imparted by the first stereocenter. Consequently, a rather limited number of phosphate catalyst structures yielded satisfactory results, and the key to the success of these reactions was the use of the doubly axially chiral phosphate catalyst (*R*,*R*)-PhDAP (phenyl-substituted doubly axially chiral phosphate).

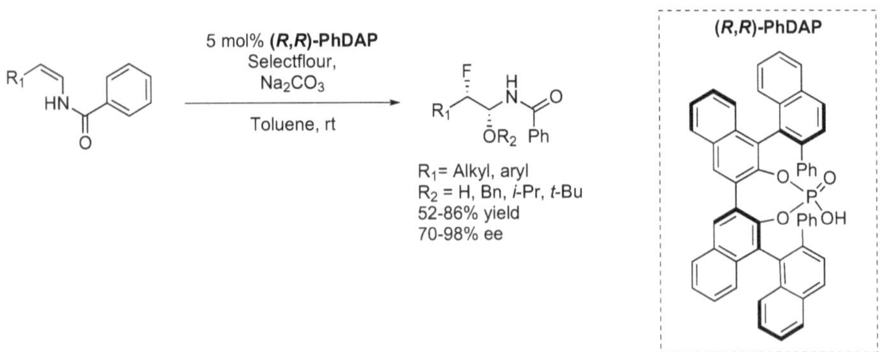

The enantioselective fluorination of *N*-benzoylenamines to afford the corresponding α-fluoroimine products was also achieved by Toste and coworkers [65]. Fluorine containing tertiary and quaternary stereocenters were accessed using this method, including unique fluoro, chloro and fluoro, bromo quaternary stereocenters, derived from chloro- and bromo-substituted enamines. When a racemic substrate that contained an existing stereocenter was employed, both the *syn* and *anti* diastereomers were obtained with high enantiomeric excess. This implied that this process was highly catalyst controlled, overriding the inherit substrate-controlled diastereoselectivity.

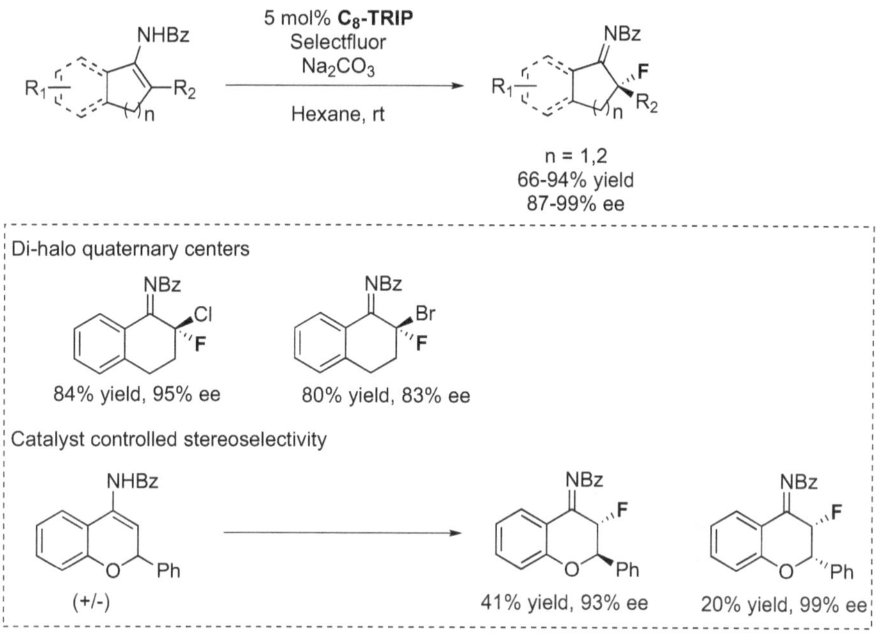

A model for rationalizing the observed absolute stereochemistry was proposed based on previous models for (1,1′-Bi-2-naphthol) BINOL-derived phosphoric acid–catalyzed reactions.[75] This model also illustrated the bifunctional nature of these catalysts: In addition to serving as a counteranion for Selectfluor, the catalysts also activated and organized the substrate in the transition state via hydrogen bonding interactions with the phosphoryl oxygen.

Building on the bifunctional nature of chiral phosphate catalysts, an enantioselective fluoro-dearomitization of phenols was developed.[66] For a variety of 2,3-disubstituted and 2,3,4-trisubstituted phenols, *ortho*-fluoro-dearomitization products were obtained in both high yield and enantiomeric excess. For substrates with 2-, 2,4-, and 2,5-substitution, fluoro-dearomitization products were isolated as [4 + 2] cycloaddition dimers. The *para*-fluoro-dearomatization was achieved for a few substrates, demonstrating the impressive ability of the chiral phosphates to control the enantioinduction even four carbon atoms away from the phenol oxygen.

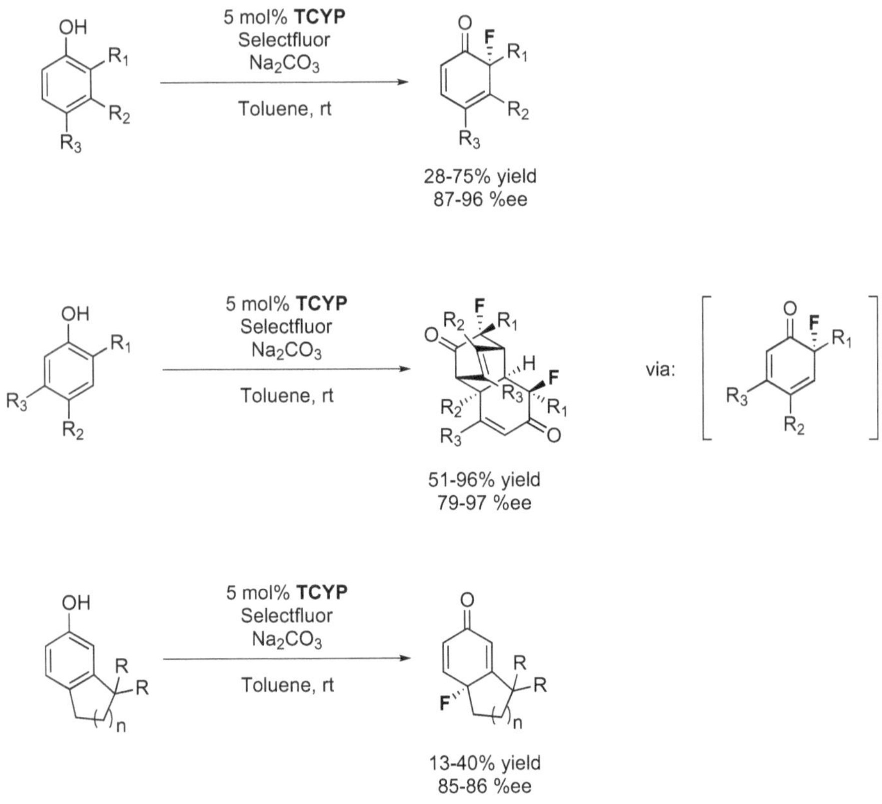

Toste and coworkers[84] also reported the enantioselective fluorination of alkenes utilizing STRIP (spiro-Triisopropylphenyl) as a catalyst. Key to the success of these reactions was the incorporation of an allylic amide–directing group capable of participating in hydrogen bonding interactions with the phosphate anion; when an *N*-methyl variant of the substrate was utilized, low conversion was observed. On the basis of mechanistic experiments, it was proposed that fluorination and deprotonation occur in a concerted process, with assistance of the phosphate and the directing group, to generate the corresponding allylic fluorides. The 2-hydroxyphenol substituent also proved to be a suitable directing group. Interestingly, in this case, selective fluorination of the alkene instead of fluoro-dearomatization of the phenol was observed.

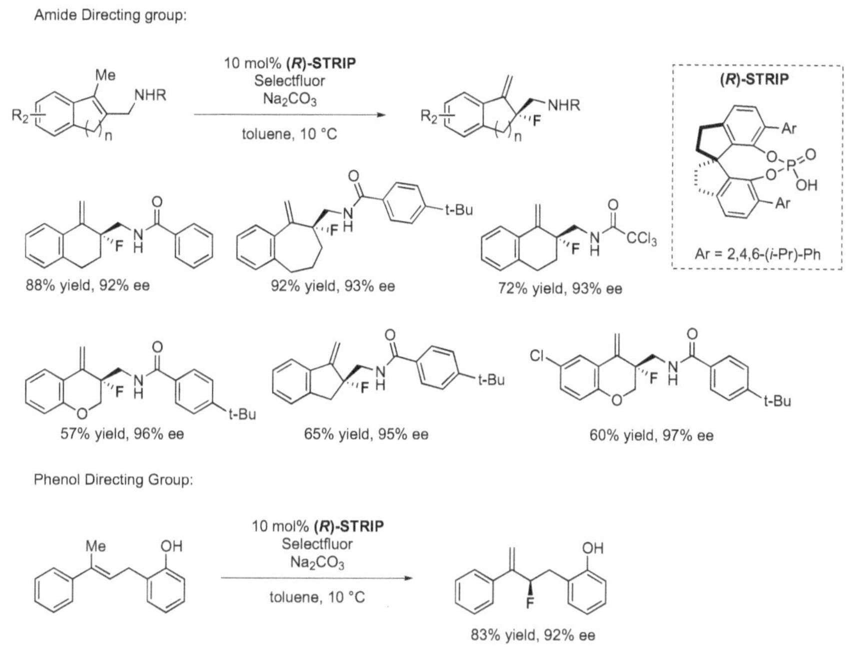

The fluorination of allylic alcohols was realized by applying an in situ–directing group strategy.[92] Having previously observed a distance dependence between the directing group and the alkene, it was hypothesized that a simple allylic alcohol would not serve as an appropriate directing group. Thus, based on a report by Falck and coworkers,[52] boronic acids were chosen as an additive, which generated in situ a monoboronic ester–directing group. Utilizing this strategy, a wide range of acyclic allylic fluorides were synthesized in high yield and with excellent enantioselectivity.

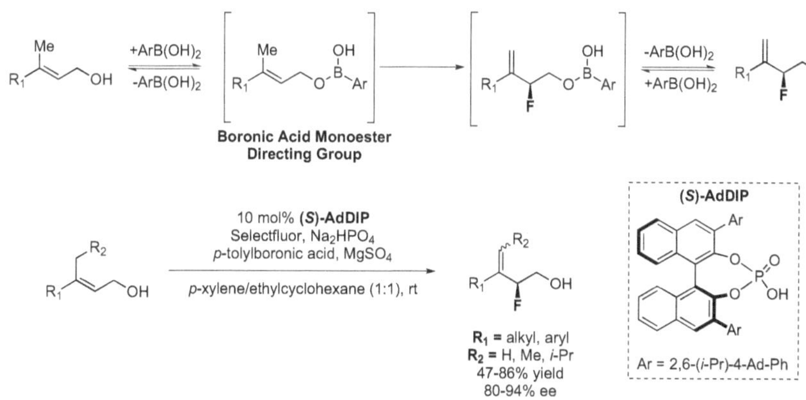

A rare CAPT example of carbofluorination of alkenes via a fluorination-induced Wagner–Meerwein rearrangement was realized by Alexakis and coworkers.[69] This stands in contrast to most CAPT alkene difunctionalizations involving fluorination. By using a highly lipophilic catalyst, the fluorinated ring expansion products were obtained in high yield, high enantioselectivity, and good to excellent diastereoselectivity.

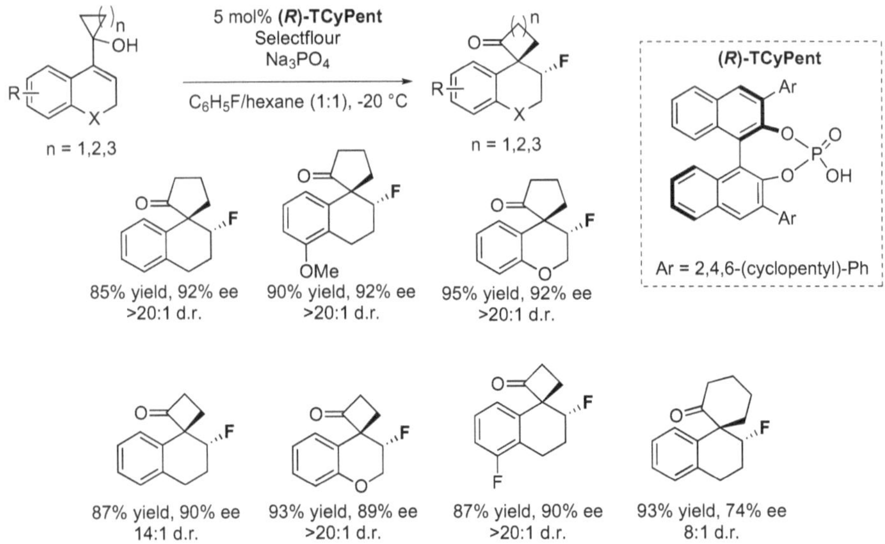

In 2014, Toste and coworkers[86] reported the dual catalytic fluorination of α-branched ketones. Under CAPT conditions, the fluorination of enamines provided only traces of the desired fluorinated ketones. However, the addition of a cocatalytic amount of a primary amine for the in situ generation an enamine dramatically increased the yield. Moreover, the use of a chiral amine in conjunction with a chiral phase transfer catalyst led to significantly increased enantioselectivity in the matched case. Merging both CAPT and enamine catalysis, the enantioselective fluorination of α-aryl, alkenyl, and alkynyl cyclohexanones was achieved in high yield (with respect to Selectfluor) and excellent enantioselectivity. Although the substrate scope was limited to α-branched cyclohexanones, this report provided a rare example of dual catalysis in the context of enantioselective fluorination and afforded α-quaternary fluorine-containing stereocenters derived from ketone substrates, which are not accessible by enamine catalysis alone (Section 2.3.3).

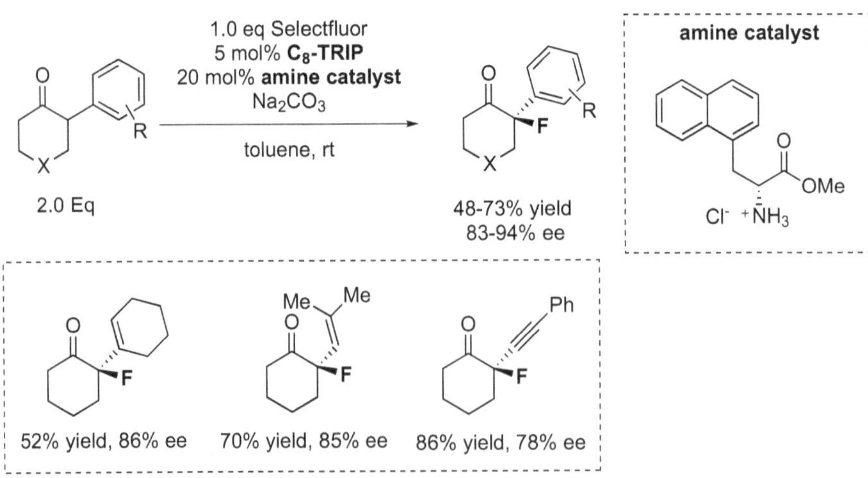

2.6 Metal-Catalyzed Fluorinations Not Involving Enolates

In the late 2000's, reports of C-F bond formation from high-valent transition metal intermediates began to emerge in the literature, primarily focusing on sp^2-C-F bond formation. Sanford and coworkers[36] reported the fluorination of aryl C-H bonds catalyzed by palladium in the presence of oxidizing N-fluoropyridinum salts. They postulated that C-F bond formation occurs via reductive elimination from a Pd(IV) fluoride. Two years later, Furuya and Ritter[20] reported the isolation of Pd(IV) fluoride organometallic complexes and observation of C-F reductive elimination from these complexes.

A few years after these seminal studies, catalytic asymmetric sp^3-C-F bond-forming reactions via high-valent transition metal intermediates were reported. Gagne's group[10] disclosed the platinum-catalyzed polyene cyclization that was terminated by C-F bond formation. On the basis of their previous mechanistic studies,[89] it was proposed that the cationic platinum (II) species, supported by (*S*)-xylyl-phanephos, initiated polyene cyclization to form platinum (II) alkyl intermediate. This intermediate was subsequently oxidized by xenon difluoride to a platinum (IV) intermediate that underwent reductive elimination to form the fluorinated polycyclic products. Moderate to good enantioselectivities were obtained. Notably, the C-F bond-forming step was found to be stereoretentive, implicating an inner sphere reductive elimination.

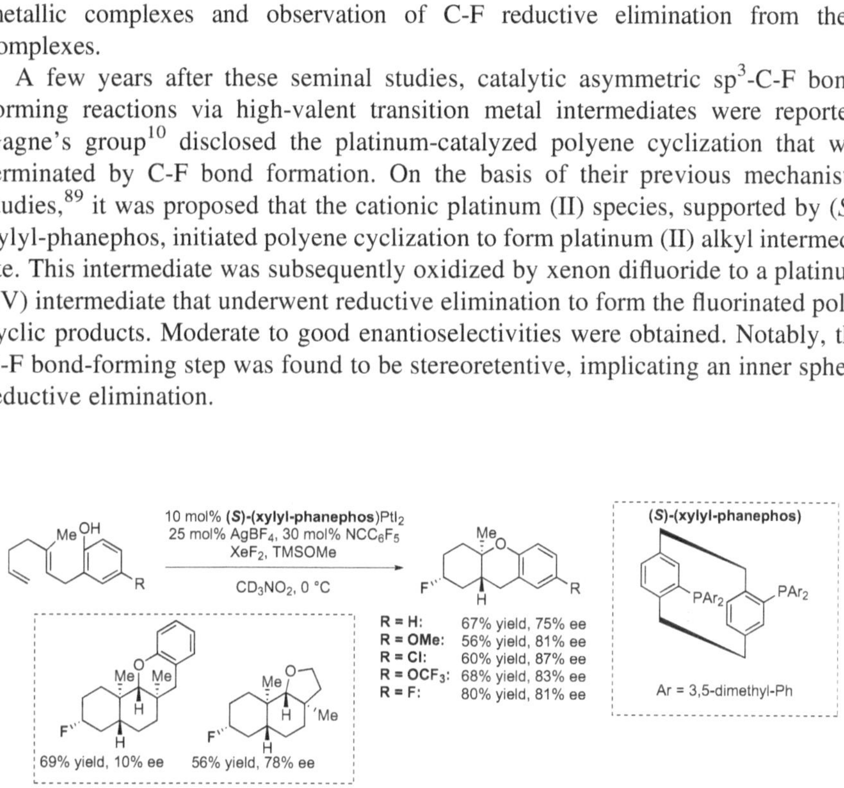

In 2014, Toste and coworkers[81] reported the palladium-catalyzed 1,2-fluoroarylation of styrenes via a three-component coupling of Selectfluor, boronic acids, and *ortho*-amide–substituted styrenes. Utilizing a chiral pyridine–oxazoline ligand, good yields and high enantiomeric excess was obtained for a variety of different amide-directing groups and boronic acid coupling partners.

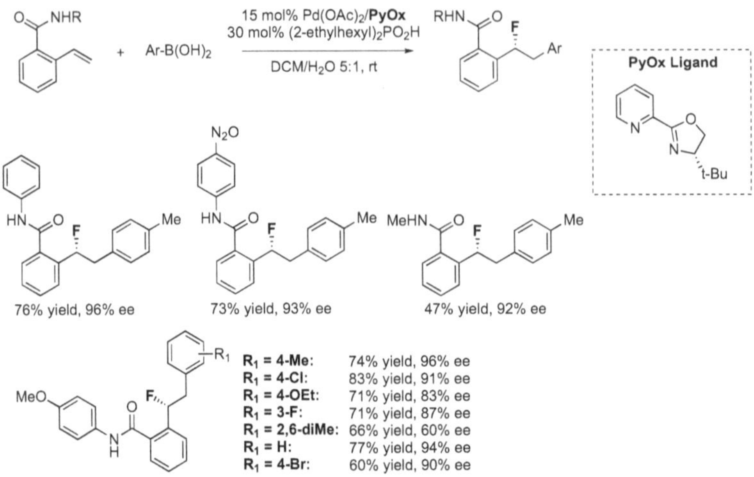

A mechanism in which an alkylpalladium(II) species was formed after transmetallation of the boronic acid, followed by insertion of the styrene substrate into the palladium aryl bond was proposed. The *ortho*-amide substituent stabilizes the Heck-type intermediate, preventing β-hydride elimination and olefin dissociation from occurring and allowing for oxidation by Selectflour to a palladium (IV) intermediate. Subsequent C-F reductive elimination regenerated the catalyst and formed the 1,2-functionalized product.

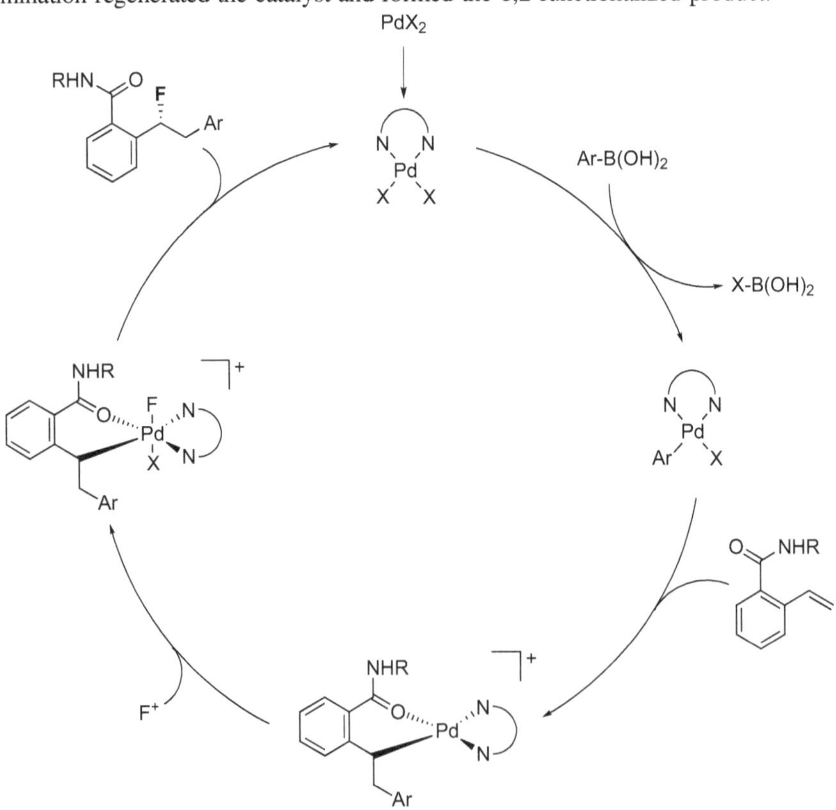

One year later,[31] this methodology was extended to the 1,1-fluoroarylation of protected allylic amines. The fluorinated products were obtained in excellent enantiomeric excess when a chiral *bis*-oxazoline ligand was employed. In this case, aryl-palladation of the olefin was followed by β-hydride elimination and reinsertion to form a benzyl palladium species, which was then oxidized by Selectfluor to form the benzylic fluorides. The lack of 1,2-fluoroarylation product suggested that C-F bond formation was slow relative to β-hydride elimination, and consequently the regioselectivity was controlled by the thermodynamic stability of palladium alkyl intermediates.

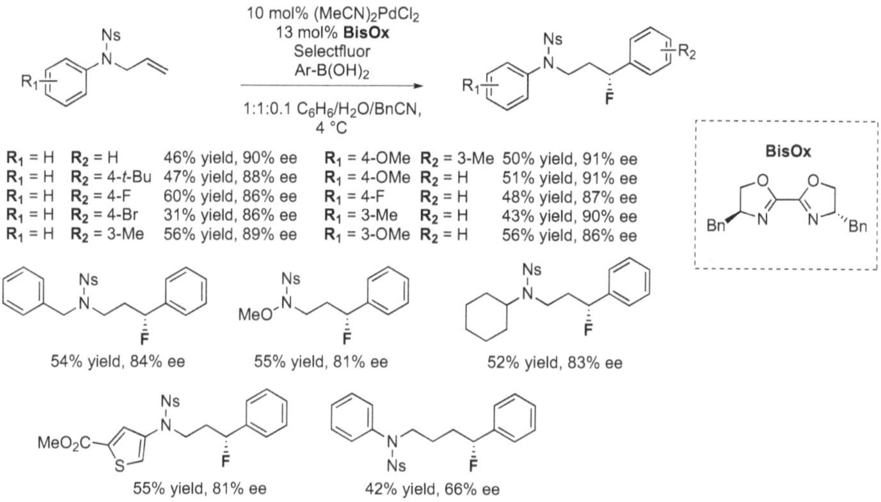

3. Nucleophilic Fluorination

3.1 Ring Opening of Strained Heterocycles

Compared with electrophilic fluorination reactions, significantly fewer enantioselective nucleophilic fluorination methodologies have been developed. This can be mainly attributed to the weak nucleophilicity and high basicity of the fluoride anion. However, from 2010 to 2015, formidable methodologies began to emerge for enantioselective nucleophilic fluorination, enabled by judicious choice of the fluoride source and unique speciation of the fluoride ions. The earliest examples of enantioselective nucleophilic fluorination were reported by Haufe et al.[29] and Haufe and Bruns,[30] using chromium salen complexes as Lewis acid catalysts for the ring opening of *meso*-epoxides. These reactions, however, afforded the products with only modest enantioenrichment (up to 66% ee). Kalow and Doyle[40] extended this methodology utilizing chiral amines and cobalt salen complexes as cocatalysts. High levels of enantioselectivity were achieved for a variety of cyclic *meso*-epoxide substrates under mild conditions as well as in the kinetic resolution of terminal epoxides via fluoride-promoted ring opening. A combination of benzoyl fluoride and HFIP was employed as a "slow release hydrogen fluoride source." In the case of *meso*-epoxides, both the chiral amine and the chiral Co(salen) catalysts were required to achieve the highest levels of enantioselectivity.

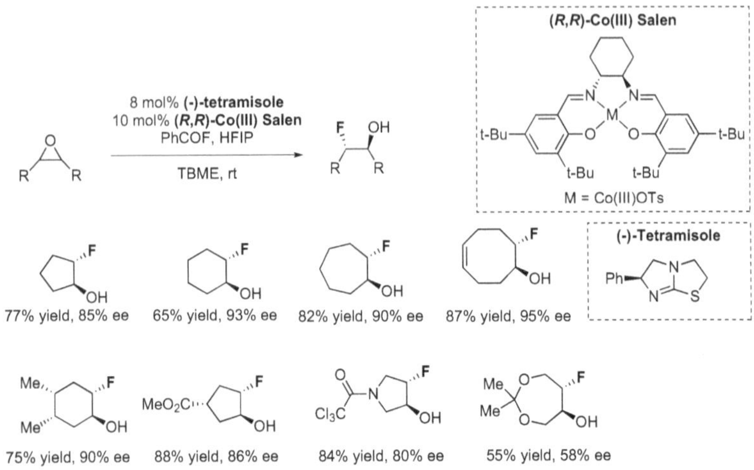

Elucidating the role of the amine cocatalyst in this reaction, along with other mechanistic aspects, was the subject of careful mechanistic studies by Kalow and Doyle.[41] After analysis of a variety of mechanistic probes, including substituent effects, nonlinear effects, and reaction kinetics with linked bimetallic catalysts, the authors were able to gain evidence for a bimetallic ring-opening step that was rate limiting. Dissociation of resting-state dimeric species was promoted by the amine additive, followed by Lewis acid activation of the epoxide, and delivery of the fluoride to the substrate by a (salen)CoF(HF) species. Notably the linked, bimetallic catalyst systems led to significantly improved reaction rate, scope, and enantioselectivity, and enabled decreased catalyst loading. Information gained from these mechanistic studies guided Doyle and coworkers[21] in their synthesis of chiral [^{18}F](salen)CoF catalysts and was applied to the enantioselective synthesis of [^{18}F]fluorohydrins.

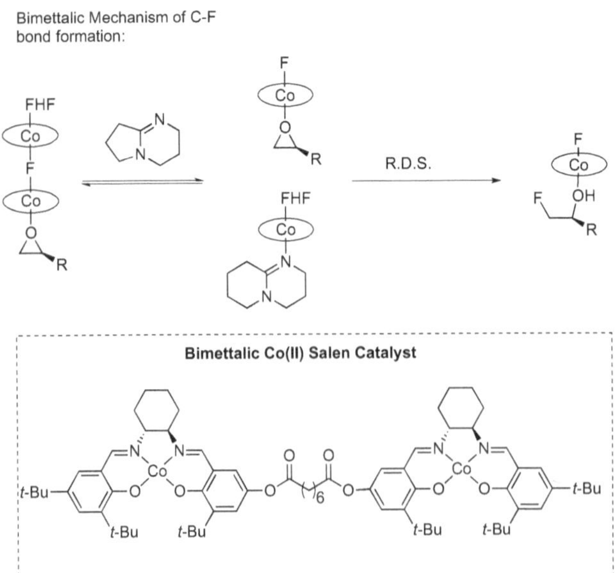

Kalow and Doyle [42] were able to extend their Co(salen) methodology to the enantioselective ring opening of protected aziridines. The choice of protecting group on the aziridine and an achiral titanium cocatalyst were crucial to the success of this reaction. Modest to high enantioselectivites were obtained for selected substrates. Mechanistic studies suggested that activation of the aziridine was achieved by the achiral titanium catalyst and that the fluoride was delivered by the Co(salen) catalyst. Preliminary studies with chiral titanium complexes revealed matched/mismatched effects on the enantioselectivity of the reaction, further supporting the proposal of cooperative catalysis.

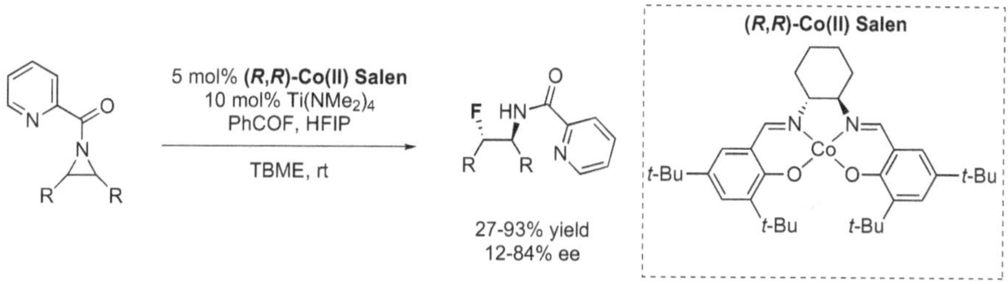

Lautens and coworkers[91] have reported the enantioselective ring opening of oxabicycloalkenes. Using a rhodium(I)/Josiphos catalyst system, and triethylamine trihydrofluoride as a fluoride source, high enantioselectivities were obtained for a variety of benzofused oxabicycloalkene substrates. Although the mechanistic details of this reaction were not clear at the time of the report, including the speciation of the fluoride, the authors suggested the possibility of an S_N2-type mechanism based on the stereochemical outcome of the reaction.

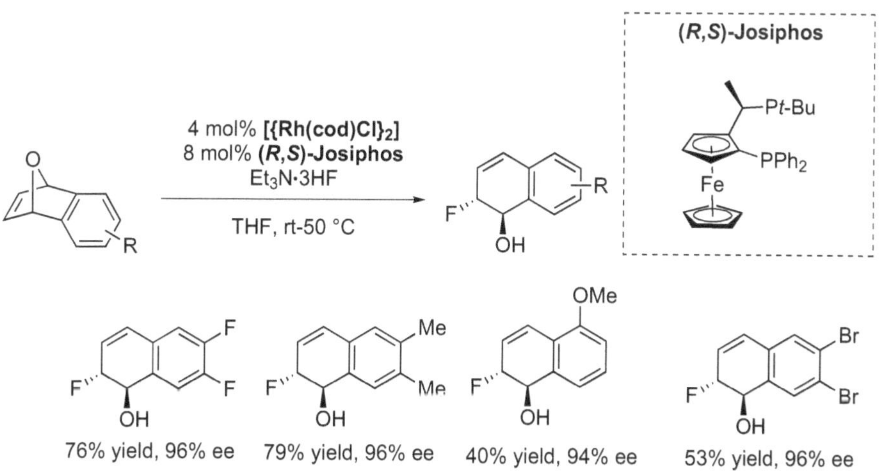

3.2 Metal-Catalyzed Allylic Substitution

A variety of strategies have been developed for enantioselective fluorinations involving transition metal–catalyzed asymmetric allylic substitution. These strategies follow the general mechanistic framework of transition metal allyl formation via oxidative addition, followed by C-F reductive elimination promoted by a fluorine nucleophile. Katcher and Doyle[43] developed the palladium-catalyzed enantioselective fluorination of racemic allylic chlorides. Utilizing a Trost bisphosphine ligand, and silver fluoride as the fluoride source, the allylic fluoride products were obtained in high yields and enantiomeric excess. Notably, the fluoride was delivered to the same diastereotopic face as the allylic chloride. The authors rationalized this outcome by a double inversion of configuration, namely, via an S_N2-type oxidative addition, followed by outer-sphere S_N2-type reductive elimination. Future mechanistic studies on acyclic systems further supported the hypothesis of outer-sphere C-F reductive elimination, albeit by a slightly different mechanism than anticipated.[44]

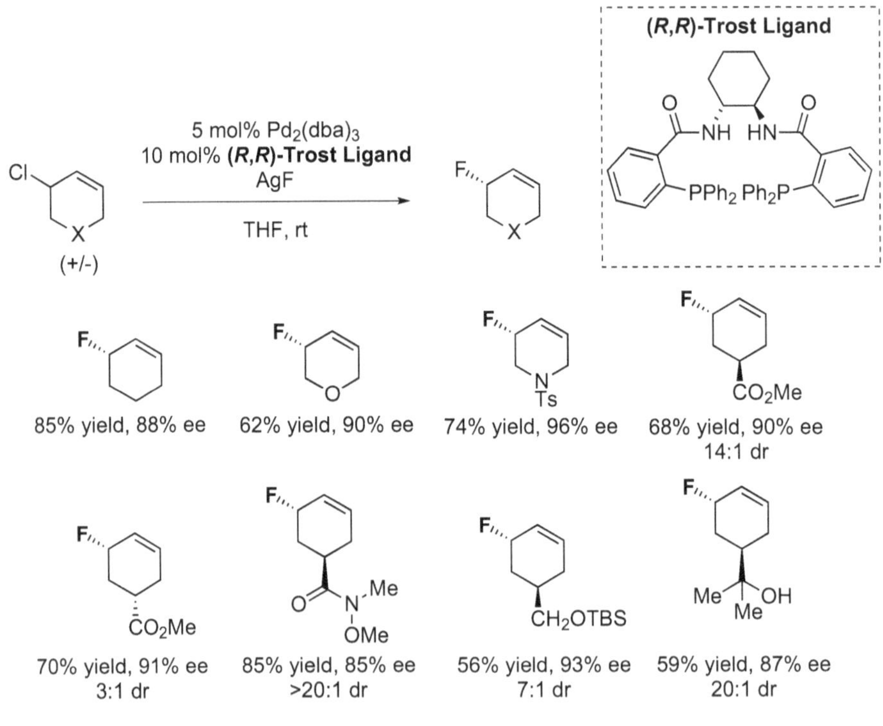

This methodology was extended by Doyle's group[45] to acyclic primary allylic chloride substrates to afford branched allylic fluoride products with excellent yield, regioselectivity, and high enantioselectivity. A combined experimental and computational study led the authors to propose that C-F bond formation occurred via an outer-sphere attack of fluoride on a palladium allyl intermediate.[44] Palladium fluoride serving as the active fluorinating agent helped to rationalize the surprising chemeoselectivity observed in these reactions, namely, the tolerance of alcohols and silyl ethers in the cyclic substrates. Additionally, the unique reactivity of allylic chlorides, when compared with that of allylic acetates and carbonates, was rationalized by the need of the oxidative addition products to undergo counterion exchange with AgF, which is facile with halides.

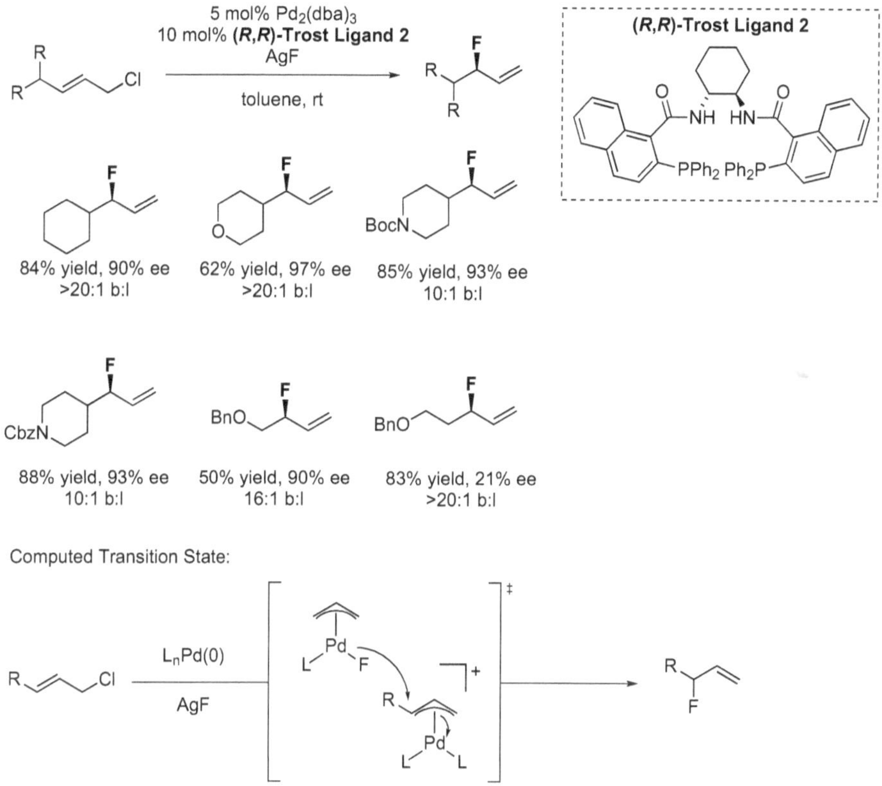

Nguyen and coworkers[87] reported the enantioselective fluorination of racemic, allylic trichloroacetimidates catalyzed by iridium in a dynamic kinetic asymmetric transformation process. Utilizing triethylamine trihydofluoride as a fluoride source and a chiral iridium/diene complex, high yields and enantiomeric excess were obtained for a variety of acyclic substrates. The palladium-catalyzed allylic substitution systems provided the best enantioselectivities for products with branching in α-position to the allylic fluoride, whereas the iridium system provided complementarity selectivity, as the products with linear substitution are well tolerated.

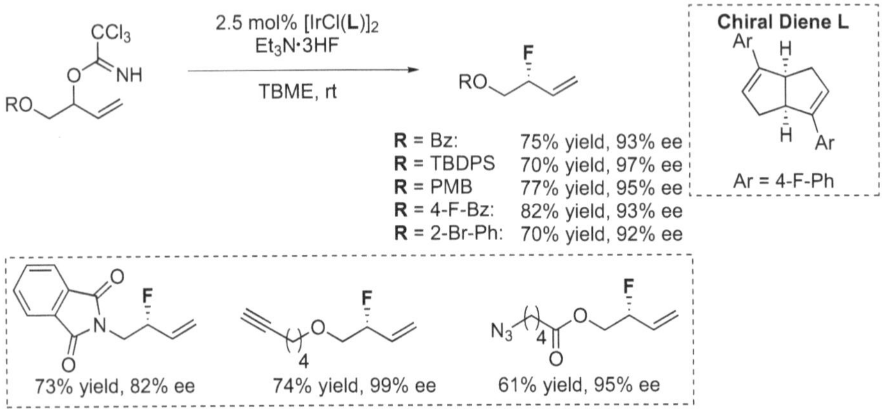

4. Conclusions and Outlook

The field of enantioselective fluorination has made tremendous advances in the past 15 years. Nearly all the examples described in this chapter have been reported during this period. The development of new reagents, such as Selectfluor and NFSI, played a crucial role in the increase of reports on enantioselective electrophilic fluorination, and even allowed for completely new areas of catalysis, such as CAPT, to emerge. Comparatively, enantioselective nucleophilic fluorination remains largely underdeveloped to this day; however, this area will likely continue to grow and contribute to the development of novel enantioselective fluorination methods. Transition metal–catalyzed fluorination reactions are also newcomers in the field of enantioselective fluorination, and many opportunities exist in this area of research. The importance of fluorinated organic molecules in pharmaceutical, agrochemical, and advanced materials industries will continue to propel the development of methods for the efficient and site-specific incorporation of fluorine into organic structures.

References

1. Althaus, M.; Becker, C.; Togni, A.; Mezzetti, A. Ruthenium-catalyzed Asymmetric Electrophilic Fluorination of 1,3-Dicarbonyl Compounds. *Organometallics* **2007,** *26,* 5902–5911.
2. Appayee, C.; Brenner-Moyer, S. E. Organocatalytic Enantioselective Olefin Aminofluorination. *Org. Lett.* **2010,** *12,* 3356–3359.
3. Balaraman, K.; Vasanthan, R.; Kesavan, V. Enantioselective Fluorination of β-ketoesters Using Tartrate Derived Bidentate Bioxazoline-Cu(II) Complexes. *Tetrahedron Asymmetry* **2013,** *24,* 919–924.
4. Beeson, T. D.; Macmillan, D. W. Enantioselective Organocatalytic Alpha-fluorination of Aldehydes. *J. Am. Chem. Soc.* **2005,** *127,* 8826–8828.
5. Bertogg, A.; Hintermann, L.; Huber, D. P.; Perseghini, M.; Sanna, M.; Togni, A. Substrate Range of the Titanium TADDOLate Catalyzed Asymmetric Fluorination of Activated Carbonyl Compounds. *Helv. Chim. Acta* **2012,** *95,* 353–403.

6. Brak, K.; Jacabosen, E. N. Asymmetric Ion-pairing Catalysis. *Angew. Chem. Int. Ed.* **2013,** *52,* 534–561.
7. Brandes, S.; Niess, B.; Bella, M.; Prieto, A.; Overgaard, J.; Jorgensen, K. A. Non-biaryl Atropisomers in Organocatalysis. *Chem. Eur. J.* **2006,** *12,* 6039–6052.
8. Cahard, D.; Audouard, C.; Plaquevent, J. C.; Roques, N. Design, Synthesis, and Evaluation of a Novel Class of Enantioselective Electrophilic Fluorinating Agents: N-fluoro Ammonium Salts of Cinchona Alkaloids (F-CA-BF(4)). *Org. Lett.* **2000,** *2,* 3699–3701.
9. Chen, Z. M.; Yang, B. M.; Chen, Z. H.; Zhang, Q. W.; Wang, M.; Tu, Y. Q. Organocatalytic Asymmetric Fluorination/Semipinacol Rearrangement: An Efficient Approach to Chiral Beta-fluoroketones. *Chem. Eur. J.* **2012,** *18,* 12950–12954.
10. Cochrane, N. G.; Nguyen, H.; Gagne, M. R. Catalytic Enantioselective Cyclization and C3-fluorination of Polyenes. *J. Am. Chem. Soc.* **2013,** *135,* 628–631.
11. Deng, Q. H.; Wadepohl, H.; Gade, L. H. The Synthesis of a New Class of Chiral Pincer Ligands and Their Applications in Enantioselective Catalytic Fluorinations and the Nozaki-Hiyama-Kishi Reaction. *Chem. Eur. J.* **2011,** *17,* 14922–14928.
12. Differding, E.; Lang, R. W. New Fluorinating Reagents – I. The First Enantioselective Fluorination Reaction. *Tetrahedron Lett.* **1988,** *29,* 6087–6090.
13. Dong, C.; Huang, F.; Deng, H.; Schaffrath, C.; Spencer, J. B.; O'Hagan, D.; Naismith, J. H. Crystal Structure and Mechanism of a Bacterial Fluorinating Enzyme. *Nature* **2004,** *427,* 561–565.
14. Dong, X.; Yang, W.; Hu, W.; Sun, J. N-Heterocyclic Carbene Catalyzed Enantioselective Alpha-fluorination of Aliphatic Aldehydes and Alpha-chloro Aldehydes: Synthesis of Alpha-fluoro Esters, Amides, and Thioesters. *Angew. Chem. Int. Ed.* **2015,** *54,* 660–663.
15. Eder, U.; Sauer, G.; Wiechert, R. New Type of Asymmetric Cyclization to Optically Active Steroid CD Partial Structures. *Angew. Chem. Int. Ed.* **1971,** *10,* 496–497.
16. Enders, D.; Hüttl, M. R. M. Direct Organocatalytic A-Fluorination of Aldehydes and Ketones. *Synlett* **2005,** 0991–0993.
17. Fadeyi, O. O.; Lindsley, C. W. Rapid, General Access to Chiral Beta-fluoroamines and Beta,beta-difluoroamines via Organocatalysis. *Org. Lett.* **2009,** *11,* 943–946.
18. Frantz, R.; Hintermann, L.; Perseghini, M.; Broggini, D.; Togni, A. Titanium-catalyzed Stereoselective Geminal Heterodihalogenation of Beta-ketoesters. *Org. Lett.* **2003,** *5,* 1709–1712.
19. Fukuzumi, Y.; Shibata, N.; Sugiura, M.; Nakamura, S.; Toru, T. Enantioselective Fluorination Mediated by Cinchona Alkaloids/Selectfluor Combinations: A Catalytic Approach. *J. Fluor. Chem.* **2006,** *127,* 548–551.
20. Furuya, T.; Ritter, T. Carbon-fluorine Reductive Elimination From a High Valent Palladium Fluoride. *J. Am. Chem. Soc.* **2008,** *130,* 10060–10061.
21. Graham, T. J. A.; Lambert, R. F.; Ploessl, K.; Kung, H. F.; Doyle, A. G. Enantioselective Radiosynthesis of Positron Emission Tomography (PET) Tracers Containing [^{18}F]Fluorohydrins. *J. Am. Chem. Soc.* **2014,** *136,* 5291–5294.
22. Gu, X.; Zhang, Y.; Xu, Z. J.; Che, C. M. Iron(III)-salan Complexes Catalysed Highly Enantioselective Fluorination and Hydroxylation of Beta-keto Esters and N-boc Oxindoles. *Chem. Commun.* **2014,** *50,* 7870–7873.
23. Hajos, Z. G.; Parrish, D. R. Asymmetric Synthesis of Bicyclic Intermediates of Natural Product Chemistry. *J. Org. Chem.* **1974,** *39,* 1615–1621.
24. Hamashima, Y.; Yagi, K.; Takano, H.; Tamas, L.; Sodeoka, M. An Efficient Enantioselective Fluorination of Various Beta-ketoesters Catalyzed by Chiral Palladium Complexes. *J. Am. Chem. Soc.* **2002,** *124,* 14530–14531.

25. Hamashima, Y.; Takano, H.; Hotta, D.; Sodeoka, M. Immobilization and Reuse of Pd Complexes in Ionic Liquid: Efficient Catalytic Asymmetric Fluorination and Michael Reactions with Beta-ketoesters. *Org. Lett.* **2003,** *5,* 3225–3228.
26. Hamashima, Y.; Suzuki, T.; Takano, H.; Shimura, Y.; Sodeoka, M. Catalytic Enantioselective Fluorination of Oxindoles. *J. Am. Chem. Soc.* **2005a,** *127,* 10164–10165.
27. Hamashima, Y.; Suzuki, T.; Shimura, Y.; Shimizu, T.; Umebayashi, N.; Tamura, T.; Sasamoto, N.; Sodeoka, M. An Efficient Catalytic Enantioselective Fluorination of β-Ketophosphonates Using Chiral Palladium Complexes. *Tetrahedron Lett.* **2005b,** *46,* 1447–1450.
28. Hamashima, Y.; Suzuki, T.; Takano, H.; Shimura, Y.; Tsuchiya, Y.; Moriya, K.-I.; Goto, T.; Sodeoka, M. Highly Enantioselective Fluorination Reactions of β-ketoesters and β-ketophosphonates Catalyzed by Chiral Palladium Complexes. *Tetrahedron* **2006,** *62,* 7168–7179.
29. Haufe, G.; Bruns, S.; Runge, M. Enantioselective Ring-opening of Epoxides by HF-reagents. Asymmetric Synthesis of Fluoro Lactones. *J. Fluor. Chem.* **2001,** *112,* 55–61.
30. Haufe, G.; Bruns, S. (Salen)chromium Complex Mediated Asymmetric Ring Opening of *Meso*-epoxides with Different Fluoride Sources. *Adv. Synth. Catal.* **2002,** *344,* 165–171.
31. He, Y.; Yang, Z.; Thornbury, R. T.; Toste, F. D. Palladium-catalyzed Enantioselective 1,1-fluoroarylation of Aminoalkenes. *J. Am. Chem. Soc.* **2015,** *137,* 12207–12210.
32. Hennecke, U. A New Approach Towards the Asymmetric Fluorination of Alkenes Using Anionic Phase-transfer Catalysts. *Angew. Chem. Int. Ed.* **2012,** *51,* 4532–4534.
33. Hintermann, L.; Togni, A. Catalytic Enantioselective Fluorination of β-Ketoesters. *Angew. Chem. Int. Ed.* **2000,** *39,* 4359–4362.
34. Honjo, T.; Phipps, R. J.; Rauniyar, V.; Toste, F. D. A Doubly Axially Chiral Phosphoric Acid Catalyst for Asymmetric Tandem Oxyfluorination of Enamides. *Angew. Chem. Int. Ed.* **2012,** *51,* 9684–9688.
35. Huang, Y.; Walji, A. M.; Larsen, C. H.; Macmillan, D. W. Enantioselective Organo-cascade Catalysis. *J. Am. Chem. Soc.* **2005,** *127,* 15051–15053.
36. Hull, K. L.; Anani, W. Q.; Sanford, M. S. Palladium-catalyzed Fluorination of Carbon-hydrogen Bonds. *J. Am. Chem. Soc.* **2006,** *128,* 7134–7135.
37. Ishimaru, T.; Shibata, N.; Reddy, D. S.; Horikawa, T.; Nakamura, S.; Toru, T. DBFOX-Ph/metal Complexes: Evaluation as Catalysts for Enantioselective Fluorination of 3-(2-arylacetyl)-2-thiazolidinones. *Beilstein J. Org. Chem.* **2008a,** *4,* 16.
38. Ishimaru, T.; Shibata, N.; Horikawa, T.; Yasuda, N.; Nakamura, S.; Toru, T.; Shiro, M. Cinchona Alkaloid Catalyzed Enantioselective Fluorination of Allyl Silanes, Silyl Enol Ethers, and Oxindoles. *Angew. Chem. Int. Ed.* **2008b,** *47,* 4157–4161.
39. Jiang, H.; Falcicchio, A.; Jensen, K. L.; Paixao, M. W.; Bertelsen, S.; Jorgensen, K. A. Target-directed Organocatalysis: A Direct Asymmetric Catalytic Approach to Chiral Propargylic and Allylic Fluorides. *J. Am. Chem. Soc.* **2009,** *131,* 7153–7157.
40. Kalow, J. A.; Doyle, A. G. Enantioselective Ring Opening of Epoxides by Fluoride Anion Promoted by Dual-catalyst System. *J. Am. Chem. Soc.* **2010,** *132,* 3268–3269.
41. Kalow, J. A.; Doyle, A. G. Mechanistic Investigations of Cooperative Catalysis in the Enantioselective Fluorination of Epoxides. *J. Am. Chem. Soc.* **2011,** *133,* 16001–16012.
42. Kalow, J. A.; Doyle, A. G. Enantioselective Fluoride Ring Opening of Aziradines Enabled by Co-operative Lewis Acid Catalysis. *Tetrahedron* **2013,** *69,* 5702–5709.
43. Katcher, M. H.; Doyle, A. G. Palladium-catalyzed Asymmetric Synthesis of Allylic Fluorides. *J. Am. Chem. Soc.* **2010,** *132,* 17402–17404.
44. Katcher, M. H.; Norrby, P. -O.; Doyle, A. G. Mechanistic Investigations of Palladium-catalyzed Allylic Fluorination. *Organometallics* **2014,** *33,* 2121–2133.

45. Katcher, M. H.; Sha, A.; Doyle, A. G. Palladium-catalyzed Regio- and Enantioselective Fluorination of Acyclic Allylic Halides. *J. Am. Chem. Soc.* **2011,** *133,* 15902–15905.
46. Kawatsura, M.; Hayashi, S.; Komatsu, Y.; Hayase, S.; Itoh, T. Enantioselective α-Fluorination and Chlorination of β-Ketoesters by Cobalt Catalyst. *Chem. Lett.* **2010,** *39,* 466–467.
47. Kirk, K. L. Fluorination in Medicinal Chemistry: Methods, Strategies, and Recent Developments. *Org. Process Res. Dev.* **2008,** *12,* 305–321.
48. Kirsch, P. *Modern Fluoroorganic Chemistry;* Wiley-VCH: Weinheim, 2013.
49. Kwiatkowski, P.; Beeson, T. D.; Conrad, J. C.; Macmillan, D. W. Enantioselective Organocatalytic Alpha-fluorination of Cyclic Ketones. *J. Am. Chem. Soc.* **2011,** *133,* 1738–1741.
50. Lam, Y. H.; Houk, K. N. How Cinchona Alkaloid-derived Primary Amines Control Asymmetric Electrophilic Fluorination of Cyclic Ketones. *J. Am. Chem. Soc.* **2014,** *136,* 9556–9559.
51. Lee, S. Y.; Neufeind, S.; Fu, G. C. Enantioselective Nucleophile-catalyzed Synthesis of Tertiary Alkyl Fluorides via the Alpha-fluorination of Ketenes: Synthetic and Mechanistic Studies. *J. Am. Chem. Soc.* **2014,** *136,* 8899–8902.
52. Li, D. R.; Murugan, A.; Falck, J. R. Enantioselective, Organocatalytic Oxy-michael Addition to γ/δ-hydroxy-α,β-enones: Boronate Amine Complexes as Chiral Hydroxide Synthons. *J. Am. Chem. Soc.* **2008,** *130,* 46–48.
53. Li, F.; Wu, Z.; Wang, J. Oxidative Enantioselective Alpha-fluorination of Aliphatic Aldehydes Enabled by N-heterocyclic Carbene Catalysis. *Angew. Chem. Int. Ed.* **2015,** *54,* 656–659.
54. Li, J.; Cai, Y.; Chen, W.; Liu, X.; Lin, L.; Feng, X. Highly Enantioselective Fluorination of Unprotected 3-substituted Oxindoles: One-step Synthesis of BMS 204352 (MaxiPost). *J. Org. Chem.* **2012,** *77,* 9148–9155.
55. Lozano, O.; Blessley, G.; Martinez Del Campo, T.; Thompson, A. L.; Giuffredi, G. T.; Bettati, M.; Walker, M.; Borman, R.; Gouverneur, V. Organocatalyzed Enantioselective Fluorocyclizations. *Angew. Chem. Int. Ed.* **2011,** *50,* 8105–8109.
56. Ma, J. -A.; Cahard, D. Copper(II) triflate-bis(oxazoline)-catalysed enantioselective electrophilic fluorination of β-ketoesters. *Tetrahedron Asymmetry* **2004,** *15,* 1007–1011.
57. Ma, J. A.; Cahard, D. Update 1 of: Asymmetric Fluorination, Trifluoromethylation, and Perfluoroalkylation Reactions. *Chem. Rev.* **2008,** *108,* PR1–43.
58. Mahlau, M.; List, B. Asymmetric Counteranion-directed Catalysis: Concept, Definition, and Applications. *Angew. Chem. Int. Ed.* **2013,** *52,* 518–533.
59. Marigo, M.; Fielenbach, D.; Braunton, A.; Kjaersgaard, A.; Jorgensen, K. A. Enantioselective Formation of Stereogenic Carbon-fluorine Centers by a Simple Catalytic Method. *Angew. Chem. Int. Ed.* **2005,** *44,* 3703–3706.
60. Moriya, K. -I.; Hamashima, Y.; Sodeoka, M. Pd(II)-Catalyzed Asymmetric Fluoronation of α-Cyanophosphonates with the Aid of 2,6-Lutidine. *Synlett* **2007,** 1139–1142.
61. Narayama, A.; Shibatomi, K.; Soga, Y.; Muto, T.; Iwasa, S. Copper(II)-Catalyzed Enantioselective Fluorination of β-Keto Esters Using Chiral Spiro Oxazoline Ligands. *Synlett* **2013,** 375–378.
62. O'Hagan, D.; Harper, D. B. The Fluorinated Natural Products. *Nat. Prod. Rep.* **1994,** *11,* 123–133.
63. O'Hagan, D.; Schaffrath, C.; Cobb, S. L.; Hamilton, J. T.; Murphy, C. D. Biochemistry: Biosynthesis of an Organofluorine Molecule. *Nature* **2002,** *416,* 279–280.
64. Phipps, R. J.; Hamilton, G. L.; Toste, F. D. The Progression of Chiral Anions from Concepts to Applications in Asymmetric Catalysis. *Nat. Chem.* **2012a,** *4,* 603–614.

65. Phipps, R. J.; Hiramatsu, K.; Toste, F. D. Asymmetric Fluorination of Enamides: Access to A-fluoroimines Using an Anionic Phase-transfer Catalyst. *J. Am. Chem. Soc.* **2012b,** *134,* 8376–8379.
66. Phipps, R. J.; Toste, F. D. Chiral Anion Phase-transfer Catalysis Applied to the Direct Enantioselective Fluorinative Dearomatization of Phenols. *J. Am. Chem. Soc.* **2013,** *135,* 1268–1271.
67. Reddy, D. S.; Shibata, N.; Nagai, J.; Nakamura, S.; Toru, T.; Kanemasa, S. Desymmetrization-like Catalytic Enantioselective Fluorination of Malonates and its Application to Pharmaceutically Attractive Molecules. *Angew. Chem. Int. Ed.* **2008,** *47,* 164–168.
68. Rauniyar, V.; Lackner, A. D.; Hamilton, G. L.; Toste, F. D. Asymmetric Electrophilic Fluorination Using an Anionic Chiral Phase-transfer Catalyst. *Science* **2011,** *334,* 1681–1683.
69. Romanov-Michailidis, F.; Guenée, L.; Alexakis, A. Enantioselective Organocatalytic Fluorination-induced Wagner-Meerwein Rearrangement. *Angew. Chem. Int. Ed.* **2013,** *52,* 9266–9270.
70. Shibata, N.; Suzuki, E.; Takeuchi, Y. A Fundamentally New Approach to Enantioselective Fluorination Based on Cinchona Alkaloid Derivatives/Selectfluor Combination. *J. Am. Chem. Soc.* **2000,** *122,* 10728–10729.
71. Shibata, N.; Takehisa, I.; Tatsuya, N.; Junji, K.; Takeshi, Y. First Enantio-flexible Fluorination Reaction Using Metal-bis(oxazoline) Complexes. *Synlett* **2004,** 1703–1706.
72. Shibata, N.; Kohno, J.; Takai, K.; Ishimaru, T.; Nakamura, S.; Toru, T.; Kanemasa, S. Highly Enantioselective Catalytic Fluorination and Chlorination Reactions of Carbonyl Compounds Capable of Two-point Binding. *Angew. Chem. Int. Ed.* **2005,** *44,* 4204–4207.
73. Shibatomi, K.; Tsuzuki, Y.; Nakata, S. -I.; Sumikawa, Y.; Iwasa, S. Synthesis of Axially Chiral Cyclic Amine-substituted 2-(Oxazolinyl)pyridine Ligands for Catalytic Asymmetric Fluorination of β-Keto Esters. *Synlett* **2007,** 551–554.
74. Shunatona, H. P.; Fruh, N.; Wang, Y. -M.; Rauniyar, V.; Toste, F. D. Enantioselective Fluoroamination: 1,4-addition to Conjugated Dienes Using Anionic Phase-transfer Catalysis. *Angew. Chem. Int. Ed.* **2013,** *52,* 7724–7727.
75. Simon, L.; Goodman, J. M. A Model for the Enantioselectivity of Imine Reactions Catalyzed by BINOL-phosphoric Acid Catalysts. *J. Org. Chem.* **2012,** *76,* 1775–1788.
76. Steiner, D. D.; Mase, N.; Barbas, C. F., III Direct Asymmetric Alpha-fluorination of Aldehydes. *Angew. Chem. Int. Ed.* **2005,** *44,* 3706–3710.
77. Struble, M. D.; Scerba, M. T.; Siegler, M.; Lectka, T. Evidence for Symmetrical Fluoronium Ion in Solution. *Science* **2013,** *340,* 57–60.
78. Suzuki, S.; Furunob, H.; Yokoyamac, Y.; Inanaga, J. Asymmetric Fluorination of β-keto Esters Catalyzed by Chiral Rare Earth Perfluorinated Organophosphates. *Tetrahedron Asymmetry* **2006,** *17,* 504–507.
79. Suzuki, S.; Kitamura, Y.; Lectard, S.; Hamashima, Y.; Sodeoka, M. Catalytic Asymmetric Mono-fluorination of Alpha-keto Esters: Synthesis of Optically Active Beta-fluoro-alpha-hydroxy and Beta-fluoro-alpha-amino Acid Derivatives. *Angew. Chem. Int. Ed.* **2012,** *51,* 4581–4585.
80. Suzuki, T.; Hamashima, Y.; Sodeoka, M. Asymmetric Fluorination of Alpha-aryl Acetic Acid Derivatives with the Catalytic System NiCl2-Binap/R3SiOTf/2,6-lutidine. *Angew. Chem. Int. Ed.* **2007,** *46,* 5435–5439.
81. Talbot, E. P. A.; Fernandes, T.de A.; McKenna, J. M.; Toste, F. D. Asymmetric Palladium-catalyzed Directed Intermolecular Fluoroarylation of Styrenes. *J. Am. Chem. Soc.* **2014,** *136,* 4101–4104.
82. Ugai, T.; Tanaka, S.; Dokawa, S. *J. Pharm. Soc. Jpn.* **1943,** *63,* 269–300.

83. Wang, J.; Sanchez-Rosello, M.; Acena, J. L.; del Pozo, C.; Sorochinsky, A. E.; Fustero, S.; Soloshonok, V. A.; Liu, H. Fluorine in Pharmaceutical Industry: Fluorine-containing Drugs Introduced to the Market in the Last Decade (2001–2011). *Chem. Rev.* **2013,** *114,* 2432–2506.
84. Wu, J.; Wang, Y. -M.; Drljevic, A.; Rauniyar, V.; Phipps, R. J.; Toste, F. D. A Combination of Directing Groups and Chiral Anion Phase-transfer Catalysis for Enantioselective Fluorination of Alkenes. *Proc. Nat. Acad. Sci. U.S.A.* **2013,** *110,* 13729–13733.
85. Xu, J.; Hu, Y.; Huang, D.; Wang, K. -H.; Xu, C.; Niu, T. Thiourea-catalyzed Enantioselective Fluorination of β-Keto Esters. *Adv. Synth. Catal.* **2012,** *354,* 515–526.
86. Yang, X.; Phipps, R. J.; Toste, F. D. Asymmetric Fluorination of α-branched Cyclohexenones Enabled by a Combination of Chiral Anion Phase-transfer Catalysis and Enamine Catalysis Using Protected Amino Acids. *J. Am. Chem. Soc.* **2014,** *136,* 5225–5228.
87. Zhang, Q.; Stockdale, D. P.; Mixdorf, J. C.; Topczewski, J. J.; Nguyen, H. M. Iridium-catalyzed Enantioselective Fluorination of Racemic, Secondary Allylic Trichloroacetimidates. *J. Am. Chem. Soc.* **2015,** *137,* 11912–11915.
88. Zhang, Y.; Yang, X. -J.; Xie, T.; Chen, G. -L.; Zhu, W. -H.; Zhang, X. -Q.; Yang, X. -Y.; Wu, X. -Y.; He, X. -P.; He, H. -M. Comparative Studies on the Enantioselective Fluorination of Oxindoles with Structurally Modified N-fluorobenzenesulfonimides. *Tetrahedron* **2013,** *69,* 4933–4937.
89. Zhao, S. -B.; Becker, J. J.; Gagne, M. R. Steric Crowding Makes Challenging Csp^3-F Reductive Eliminations Feasible. *Organometallics* **2011,** *30,* 3926–3929.
90. Zhao, Y. M.; Cheung, M. S.; Lin, Z.; Sun, J. Enantioselective Synthesis of Beta,gamma-unsaturated Alpha-fluoroesters Catalyzed by N-heterocyclic Carbenes. *Angew. Chem. Int. Ed.* **2012,** *51,* 10359–10363.
91. Zhu, J.; Ysui, G. C.; Lautens, M. Rhodium-catalyzed Enantioselective Nucleophilic Fluorination: Ring Opening of Oxabicyclic Alkenes. *Angew. Chem. Int. Ed.* **2012,** *51,* 12353–12356.
92. Zi, W.; Wang, Y. -M.; Toste, F. D. An In-situ Directing Group Strategy for Chiral Anion Phase-transfer Fluorination of Allylic Alcohols. *J. Am. Chem. Soc.* **2014,** *136,* 12864–12867.

Development of Electrophilic Trifluoromethylating Reagents

10

T. Umemoto
Zhejiang Jiuzhou Pharmaceutical Co., Ltd., Taizhou, Zhejiang, China

Chapter Outline

1. Introduction

Since the CF_3 group has unique properties such as high electronegativity, stability, and lipophilicity, the introduction of the CF_3 group into an organic molecule could bring about a remarkable change in the properties of the original molecule. This is the reason why the CF_3 group has been drawing much interest of researchers in modifying lead compounds to develop new effective medicines and agrochemicals.[1] Some typical examples developed so far are shown in Fig. 10.1.

Electrophilic trifluoromethylation is one of the fundamental methods for trifluoromethylation, the others being nucleophilic and free radical. However, since it is very hard to generate CF_3^+ in solution due to the strongly electron-withdrawing effect of the CF_3 group, electrophilic trifluoromethylation is extremely difficult compared with electrophilic alkylation in hydrocarbon chemistry, which is easily accomplished by treating a methyl halide with a nucleophile ($CH_3X + Nu^- \rightarrow CH_3Nu + X^-$).[2a] Accordingly, studies on making useful electrophilic trifluoromethylating agents have long been continued, as described in this chapter. There are several reviews for electrophilic trifluoromethylating reagents and their reactions.[2]

Modern Synthesis Processes and Reactivity of Fluorinated Compounds. http://dx.doi.org/10.1016/B978-0-12-803740-9.00010-X

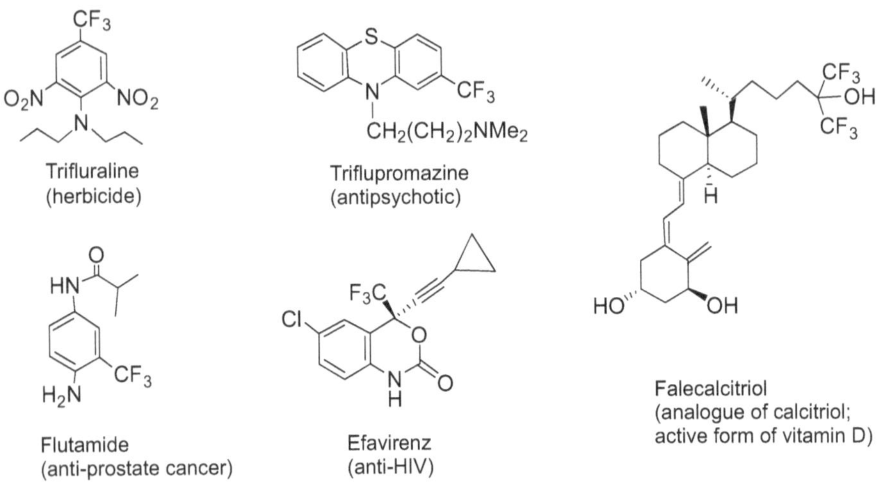

Figure 10.1 Some examples of trifluoromethyl-containing medicines and agrochemicals. *HIV*, human immunodeficiency virus.

2. Historical Background

Electrophilic perfluoroalkylating reagents having two or more carbons (C_nF_{2n+1}, $n \geq 2$) were developed around 1980. The reagents were (perfluoroalkyl)aryliodonium chlorides **4** by Yagupolskii[3] and triflates **5** by Umemoto,[4,2a] which were prepared from perfluoroalkyl iodides (Rf-I) as shown in Scheme 10.1. The latter (perfluoroalkyl)phenyliodonium trifluoromethanesulfonates **5** (FITS reagents) were stable crystals, but were very reactive because of strong activation by TfO group. FITS reacted with many kinds of nucleophilic organic substrates under mild conditions to give various perfluoroalkylated compounds.[2a]

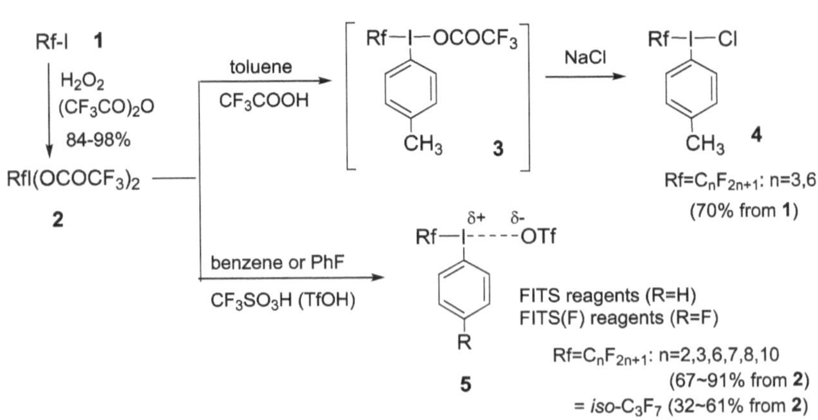

Scheme 10.1 Synthesis of electrophilic perfluoroalkylating reagents **4** and **5**.

However, both Yagupolskii and Umemoto failed to synthesize the corresponding (trifluoromethyl)aryliodonium salts from CF_3I, probably due to the instability of the intermediates or final products.[4b,5]

Another approach reported by Umemoto in 1982[6] and 1986[7] was an attempt to make trifluoromethyl azo compounds **8** by reaction of CF_3NO with NH_2OH followed by treatment with an arene- or perfluoroalkane-sulfonyl halide (Scheme 10.2). He expected to generate CF_3^+ with evolution of N_2 from **8** by heterolytic bond cleavage.[8] However, the isolated products were *N*-CF_3-*N*-nitroso derivatives **9a–e**.

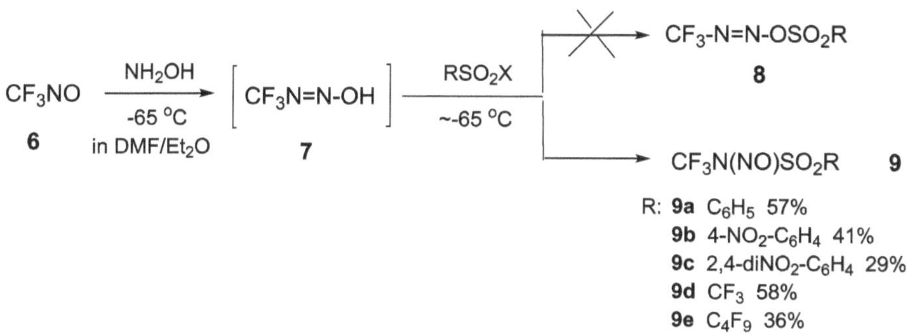

Scheme 10.2 Synthesis of *N*-trifluoromethyl-*N*-nitroso sulfonamides.

It was found that **9a** reacted photochemically with activated aromatics, disulfides, and others and **9d** did so both photo- and thermo-chemically to produce trifluoromethylated compounds. The mechanism suggested was that CF_3 radicals were generated via isomerization to the diazo compound **8** followed by homolytic cleavage (Scheme 10.3). Thus, these trifluoromethylations were considered radical reactions.

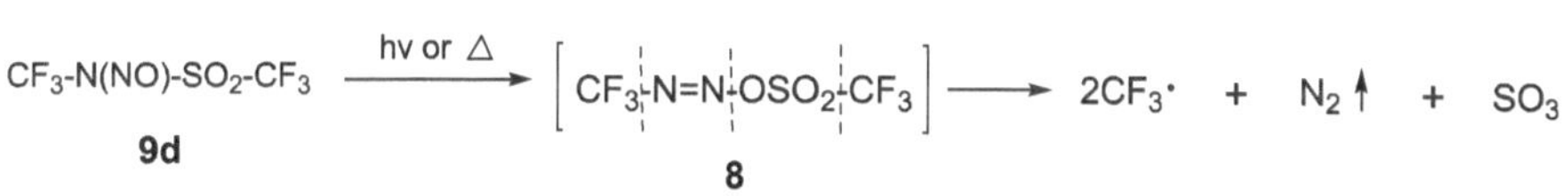

Scheme 10.3 Mechanism of trifluoromethylation with **9d**.

3. Electrophilic Trifluoromethylating Reagents

3.1 Sulfonium, Selenonium, and Telluronium Salts

In 1984, Yagupolskii et al. synthesized stable (trifluoromethyl)diarylsulfonium salts **12a,b**[9] (Scheme 10.4). Sulfoxide **10** was deoxofluorinated with $SF_3^+SbF_6^-$ and then treated with SbF_5 to give intermediate **11**, which was then reacted with *m*-xylene and anisole to produce **12a** and **12b**, respectively. Since **11** was extremely hygroscopic, the procedure was carried out in completely dry atmosphere.

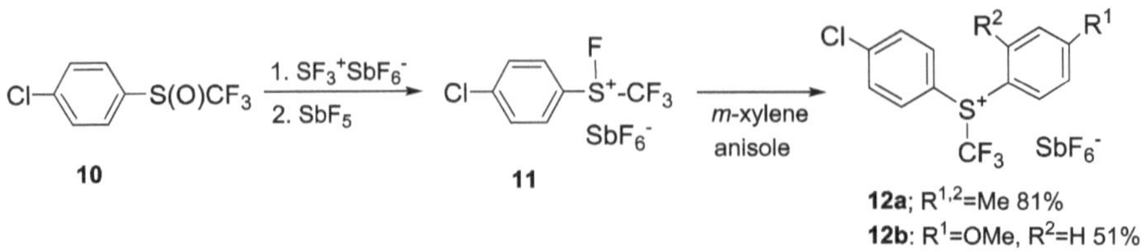

Scheme 10.4 Synthesis of *S*-CF_3-diarylsulfonium salts.

12a,b reacted with *p*-nitrobenzenethiolate **13** in *N,N*-dimethylformamide (DMF) at room temperature to produce CF_3 sulfide **14** in 65% yield (Scheme 10.5). This was the first report that trifluoromethylsulfonium salts acted as electrophilic trifluoromethylating reagents, but only for an arenethiolate. **12a,b** did not react with an activated aromatic substrate, *N,N*-dimethylaniline, even at temperatures up to 70°C.

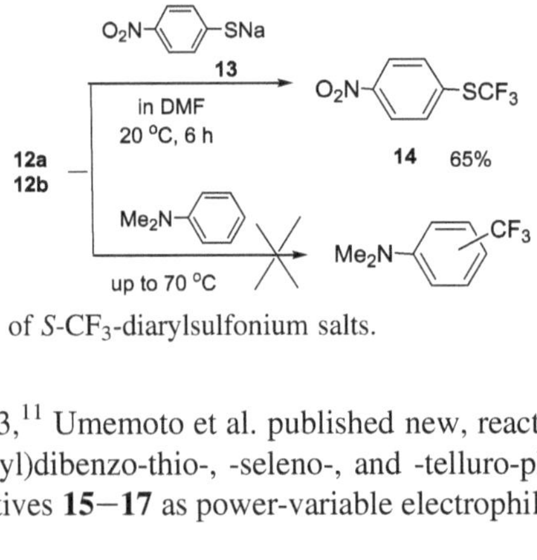

Scheme 10.5 Reactions of *S*-CF_3-diarylsulfonium salts.

In 1990[10] and 1993,[11] Umemoto et al. published new, reactive, and stable *S*-, *Se*-, and *Te*-(trifluoromethyl)dibenzo-thio-, -seleno-, and -telluro-phenium salts and their alkyl and nitro derivatives **15–17** as power-variable electrophilic trifluoromethylating reagents (Fig. 10.2).

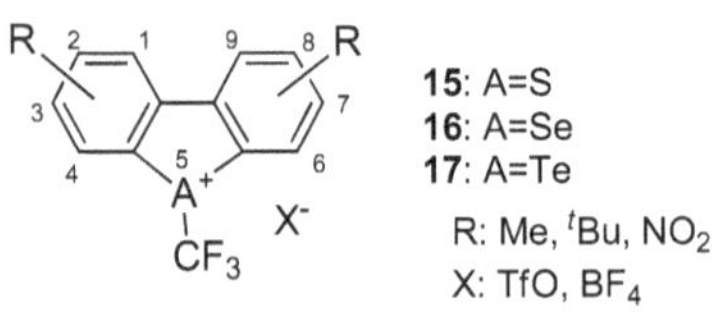

Figure 10.2 *S*-, *Se*-, and *Te*-CF_3-dibenzoheterocyclic onium salts.

Thiophenium salts **15** were synthesized by oxidation of sulfide **18** with *m*-chloroperbenzoic acid (MCPBA) followed by easy cyclization of the resulting sulfoxides **19** with triflic anhydride (Tf_2O) in high yields (Scheme 10.6). **15** were also prepared from **18** by a one-pot reaction, direct fluorination using F_2/N_2 (1/9 v/v) in the presence of HX (TfOH, HBF_4 or $BF_3 \cdot Et_2O$), or stepwise treatment of the direct fluorination followed by treatment with HX. *S*-(Perfluoro-ethyl, -propyl, -butyl, and -octyl)dibenzothiophenium triflates were synthesized in a similar manner.[11]

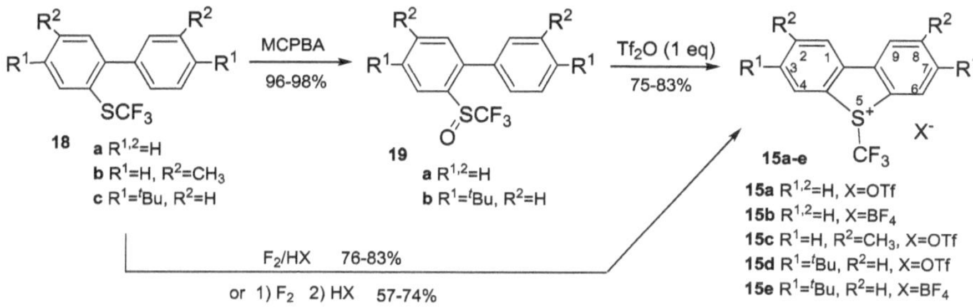

Scheme 10.6 Synthesis of *S*-CF_3-dibenzothiophenium salts.

The mononitration of **15a** was carried out by treatment of nitronium triflate, which was in situ prepared from nitric acid and Tf_2O in nitromethane solvent (Scheme 10.7). The dinitration of **15a** was completed in high yield by using excess of nitronium triflate without nitromethane.

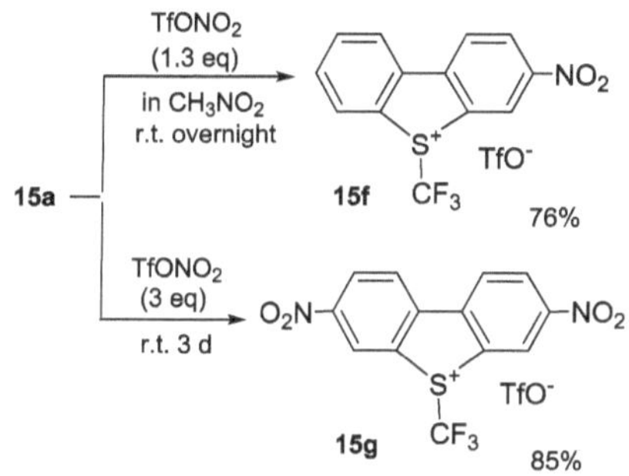

Scheme 10.7 Nitration of *S*-CF_3-dibenzothiophenium salts.

Selenophenium salts **16a** (A = Se, R = H, X = TfO), **16b** (A = Se, R = 2,8-diMe, X = TfO), and dinitro salt **16c** (A = Se, *R* = 3,7-diNO_2, X = TfO) were synthesized in the same manner as thiophenium salts **15**. Salt **16a** was dinitrated much faster than **15a** to give **16c**.

Tellurophenium salts **17** were synthesized in a different manner (Scheme 10.8). 2-(CF_3Te)biphenyl **20** underwent quick intramolecular cyclization by treatment with Tf_2O in the presence of dimethyl sulfoxide (DMSO) to give **17a** in 84% yield. **17a** was also synthesized in 63% yield by treatment of **20** with bromine followed by heating with TfOH. Dinitration of **17a** was easily carried out to give **17d** in 76% yield. BF_4 salt **17c** was synthesized in 73% yield by conversion to bromide **17b** followed by treatment with $AgBF_4$.

Another heterocyclic salt **22** was synthesized by the intramolecular cyclization of **21**. However, the reaction was very slow and the yield was very low (Scheme 10.9).

An unsubstituted acyclic salt, *S*-(trifluoromethyl)diphenylsulfonium triflate (**24**), was synthesized by condensation of sulfoxide **23** with benzene by action of Tf_2O (Scheme 10.10).[11] The reaction was slow.

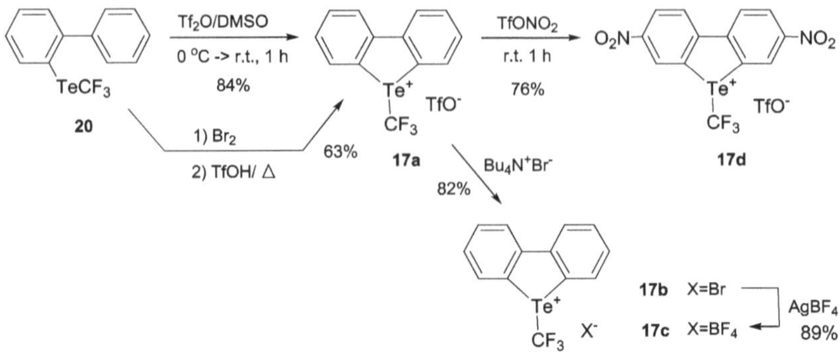

Scheme 10.8 Synthesis of *Te*-CF_3-dibenzotellurophenium salts.

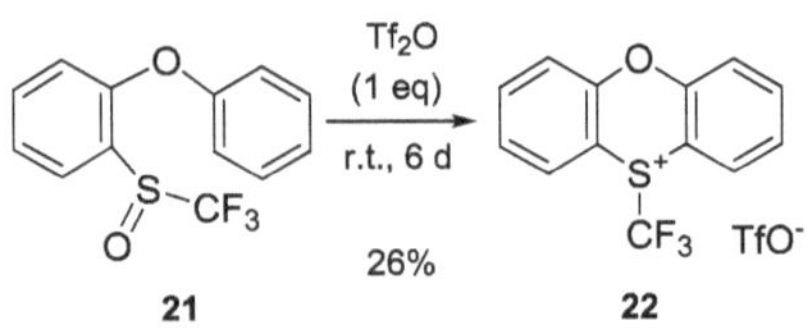

Scheme 10.9 Synthesis of *S*-CF_3-phenoxathiinium triflate.

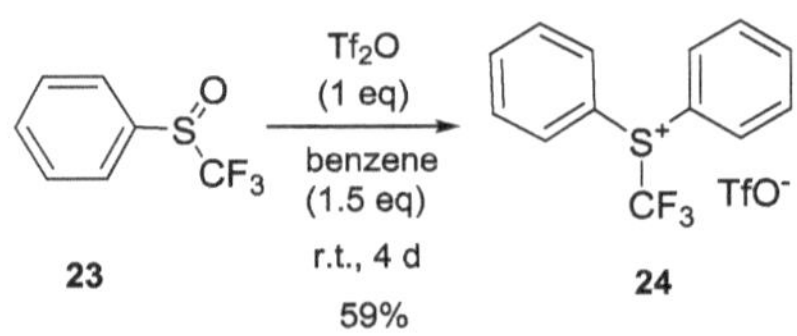

Scheme 10.10 Synthesis of *S*-CF_3-diphenylsulfonium triflate.

It was shown that the power (reactivity) of trifluoromethylating reagent increased in the order **17a** (Te) < **24** (S, acyclic) < **15c** (S, 2,8-diMe) < **15d** (S, 3,7-di^{t}Bu) < **16a** (Se) ≤ **17d** (Te, 3,7-diNO_2) < **15a** (S) < **15f** (S, 3-NO_2) < **16c** (Se, 3,7-diNO_2) < **15g** (S, 3,7-diNO_2). The power order in each of the structures, heteroatoms, and substituents was as follows: acyclic structure < cyclic structure; Te < Se < S; and 2,8-dimethyl < 3,7-di-*tert*-butyl < H < 3-nitro < 3,7-dinitro. Thus, the power depended on the structure and electronegativity of the heteroatom and ring substituent. Importantly, the effect of NO_2 group overcame the effect of heteroatoms. The substituent effect was clearly reflected in the ^{19}F chemical shift of CF_3 groups. In the S series, the chemical shift appeared in the order 2,8-diMe (−53.8 ppm) < 3,7-di^{t}Bu (−53.1) < unsubstituted (−52.6) < 3-nitro (−50.5) < 3,7-dinitro (−48.4). Thus, the power depended on the electron density of the CF_3 group. The higher reactivity of the cyclic structure than the acyclic structure was clearly demonstrated by kinetic studies and explained by its greatly enhanced entropy ΔS.[12]

The (trifluoromethyl)dibenzoheterocyclic salts **15–17** first made it possible to trifluoromethylate a wide range of nucleophilic substrates.[11] The suitable CF_3-dibenzoheterocyclic salt reagent was chosen for the substrate. In general, a less reactive substrate was trifluoromethylated by a powerful reagent, whereas a reactive substrate was satisfactorily done with a less powerful reagent. Actually, trifluoromethylations of many different substrates were carried out by using three different salts, **15a**, **15g**, and **16a** (power order **16a** < **15a** < **15g**) (Scheme 10.11 and Table 10.1). These trifluoromethylations were accompanied by quantitative formation of dibenzoheterocycle **25**.

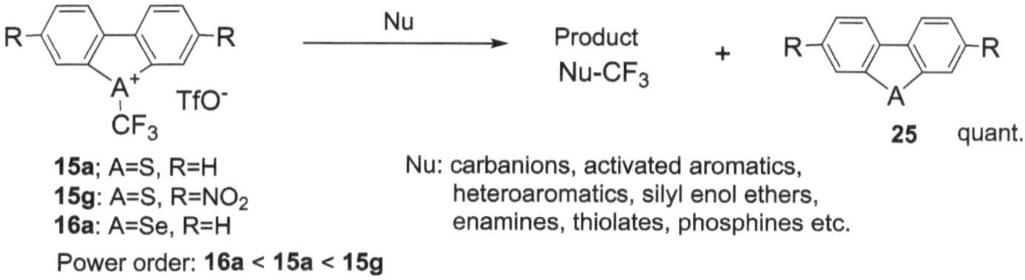

Scheme 10.11 Trifluoromethylation with **15a, 15g,** and **16a**.

As seen in Table 10.1, **15a** was suitable for salts of active methylene compounds, silyl enol ethers, enamines, heteroaromatics such as pyrrole, and sodium iodide and nitrate.[11,13] The most powerful **15g** was suitable for trifluoromethylations of aromatics such as aniline, dihydroquinone and naphthol, and phosphines, whereas the least powerful **16a** was suitable for thiolates and metal acetylides.[11] Although the yields were a little low, dialkyl *S*-CF_3 salt **15c** or **15d** could be replaced with Se-CF_3 salt **16a**.[11]

Table 10.1 **Trifluoromethylation of various substrates with 15a, 15g, and 16a**

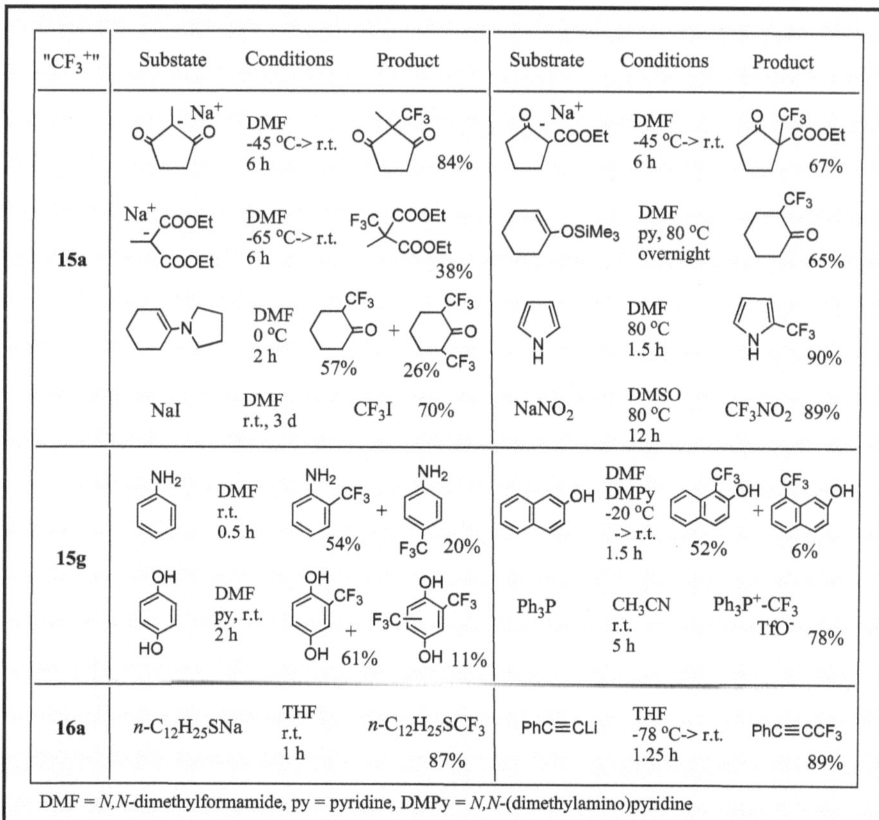

"CF_3^{+}"	Substate	Conditions	Product	Substrate	Conditions	Product
15a	O, O, Na+	DMF -45 °C-> r.t. 6 h	O, O, CF_3 84%	O, Na+, COOEt	DMF -45 °C-> r.t. 6 h	O, CF_3, COOEt 67%
	Na+ COOEt, COOEt	DMF -65 °C-> r.t. 6 h	F_3C COOEt, COOEt 38%	-OSiMe3	DMF py, 80 °C overnight	CF_3, =O 65%
	N	DMF 0 °C 2 h	CF_3, =O 57% + CF_3, =O, CF_3 26%	N H	DMF 80 °C 1.5 h	N H, CF_3 90%
	NaI	DMF r.t., 3 d	CF_3I 70%	$NaNO_2$	DMSO 80 °C 12 h	CF_3NO_2 89%
15g	NH_2	DMF r.t. 0.5 h	NH_2, CF_3 54% + NH_2, F_3C 20%	OH	DMF DMPy -20 °C -> r.t. 1.5 h	CF_3, OH 52% + CF_3, OH 6%
	OH, HO	DMF py, r.t. 2 h	OH, CF_3, OH 61% + OH, CF_3, F_3C, OH 11%	Ph_3P	CH_3CN r.t. 5 h	Ph_3P^+-CF_3 TfO^- 78%
16a	n-$C_{12}H_{25}SNa$	THF r.t. 1 h	n-$C_{12}H_{25}SCF_3$ 87%	PhC≡CLi	THF -78 °C-> r.t. 1.25 h	PhC≡CCF_3 89%

DMF = *N,N*-dimethylformamide, py = pyridine, DMPy = *N,N*-(dimethylamino)pyridine

Metal enolates of ketones could not be trifluoromethylated in satisfactory yields with any of the salts. For this case, the complexes of potassium enolates with 2-phenyl-1,3,2-benzodioxaborole as a Lewis acid were in situ prepared and then treated with **15a** to give the desired α-trifluoromethyl ketones in high yields.[14] The first enantioselective α-trifluoromethylation of a ketone made by this methodology provided a 45%ee yield (chemical yield 54%).[14]

In these trifluoromethylations, dibenzoheterocycle **25** was formed as another product, which was soluble in an organic layer. Therefore, the separation of CF_3 products from **25** was difficult. To solve it, *S*-, *Se*-, and *Te*- (trifluoromethyl)dibenzo-thio-, -seleno-, and -telluro-phenium-3-sulfonates, **26a-c**, **27a**, and **28a,** were synthesized (Scheme 10.12).[15] *S*-(Perfluoro-ethyl, -butyl, and -octyl)dibenzothiophenium-3-sulfonates were also synthesized.[15]

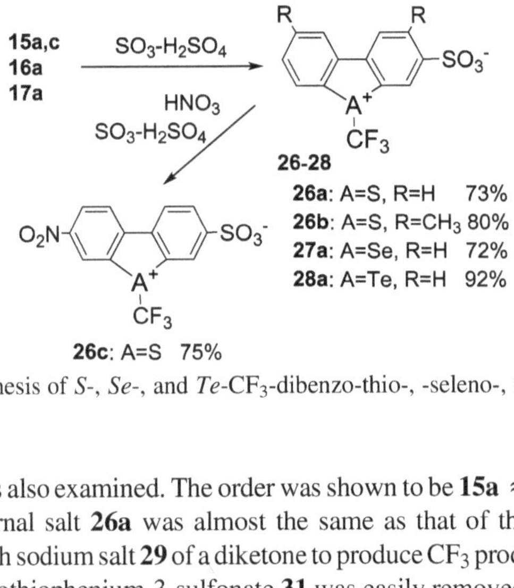

Scheme 10.12 Synthesis of *S*-, *Se*-, and *Te*-CF_3-dibenzo-thio-, -seleno-, and -telluro-phenium-sulfonates.

Their power was also examined. The order was shown to be **15a** ≈ **26a** < **26c** < **15f**. The power of internal salt **26a** was almost the same as that of the external salt **15a**. Salt **26a** reacted with sodium salt **29** of a diketone to produce CF_3 product **30** in 86% yield, and sodium dibenzothiophenium-3-sulfonate **31** was easily removed by washing the reaction mixture with water (Scheme 10.13).

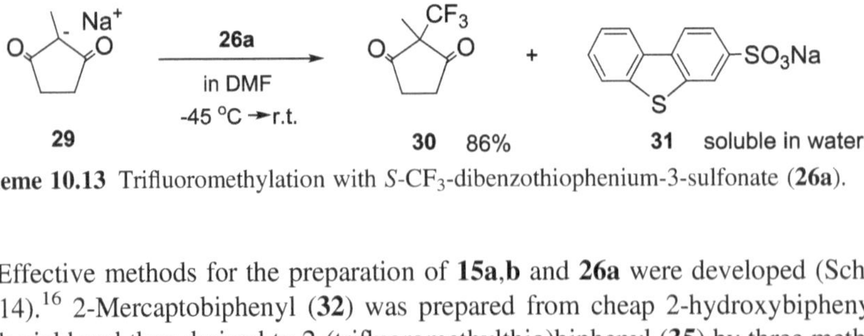

Scheme 10.13 Trifluoromethylation with *S*-CF_3-dibenzothiophenium-3-sulfonate (**26a**).

Effective methods for the preparation of **15a,b** and **26a** were developed (Scheme 10.14).[16] 2-Mercaptobiphenyl (**32**) was prepared from cheap 2-hydroxybiphenyl in high yield and then derived to 2-(trifluoromethylthio)biphenyl (**35**) by three methods. **35** was treated with H_2O_2 in acetic acid, and the resulting sulfoxide **36** was reacted with fuming sulfuric acid, followed by treatment with NaOTf and $NaBF_4$ to give **15a** and **15b**, respectively, in high yields. **36** was treated with excess fuming sulfuric acid to directly give the internal salt **26a** in high yield. These reactions were carried out at a large scale (a half kg).

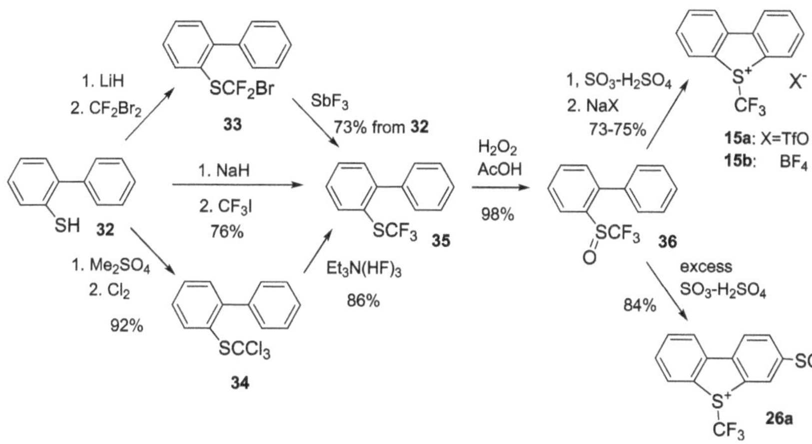

Scheme 10.14 Effective methods for preparation of CF_3 reagents, **15a,b** and **26a**.

S-(Trifluoromethyl)dibenzothiophenium salts **15a,b** called later as Umemoto's reagents have been commercialized and distributed worldwide since the early 2000s by chemical reagent companies. Since then, many new reactions of **15a,b** have been reported,[2e] which include iodide anion-catalyzed trifluoromethylation of active methylene compounds and others,[17] Pd(II)-catalyzed *o*-trifluoromethylation of aromatics,[18] Cu-catalyzed trifluoromethylation of aryl boronic acids[19]; CuCl-catalyzed trifluoromethylation of terminal alkynes[20]; oxy-, amino-, and keto-trifluoromethylation of alkenes catalyzed by photoredox catalyst under visible light[21]; Ru-photocatalyzed hydrotrifluoromethylation of alkenes[22]; Cu-promoted Sandmeyer trifluoromethylation of anilines[23]; direct aromatic trifluoromethylation via an electron donor−acceptor complex[24]; one-step synthesis of 4-trifluoromethyl-2,3-dihydropyroliums from amino alkynes[25]; one-step preparation of 3-trifluoromethylindoles from 2-alkynylanilines[26]; and one-step preparation of 2-trifluoromethyl-dihydronaphthalene derivatives.[27] Highly enantioselective trifluoromethylation of β-ketoesters was also reported.[28] Newly designed high-valent CF_3-Ni(IV) catalysts were prepared through the reaction with Ni(II) precursors.[29]

Umemoto suggested that the reaction mechanism for the trifluoromethylation with the *S*- and *Se*-CF_3 reagents could vary from CF_3 free radical to CF_3^+ cation species, depending on the reactivity of nucleophiles, the trifluoromethylating power of CF_3 reagents, and reaction conditions.[30] Umemoto also discussed the possible insertion mechanism to the doubly positive $S^{\delta+}$-$C^{\delta+}$ bond of the S-CF_3 group accompanied by a one-electron or two-electron exchange, which was completely different from the S_N2 mechanism.[12] Magnier suggested a single electron-transfer pathway on the basis of the radical trapping experiment of trifluoromethylation of a silyl enol ether with Umemoto reagent **15a** and analogous reagents.[31] Shibata discussed the reaction mechanism based on the calculations.[32] Many metal-mediated trifluoromethylations with Umemoto reagents were explained by the reductive mechanism via a single electron-transfer from metal.[2i]

In 1998, Shreeve et al. published acyclic *S*-(trifluoromethyl)diarylsulfonium salts **37a–e** having electron-withdrawing substituent(s) as another set of power-variable reagents (Scheme 10.15).[33] **37a–c** were prepared by the reaction of **23** with a large excess amount of each of Tf_2O and an aromatic compound. Salts **37d,e** were prepared by nitration of **37a,b**.

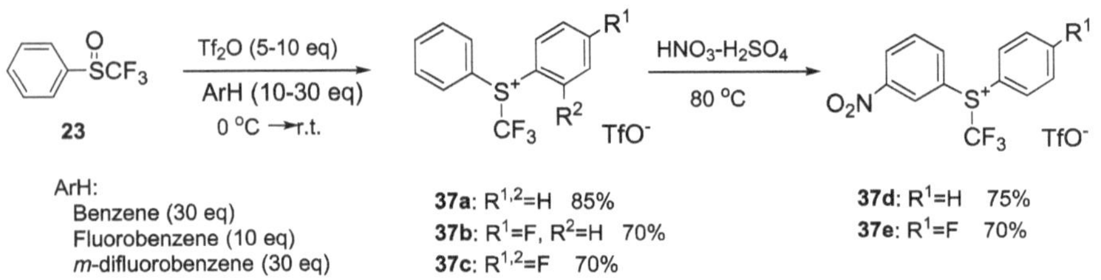

Scheme 10.15 Synthesis of *S*-CF_3-diarylsulfonium triflates **37**.

The trifluoromethylation experiments of dihydrobenzoquinone, pyrrole, and aniline showed that the power increased in the order of **37a** ≈ **37b** < **37c** < **37d** ≈ **37e**. Thus, two fluorine atoms or a nitro group definitely increased the reactivity. The reaction pattern was the same as for the *S*-CF_3-dibenzothiophenium salts.

In 2005, Adachi et al. disclosed 1-oxo-1-(trifluoromethyl)-1λ^6-benzo[d]isothiazol-3-one **39** and *S*-(trifluoromethyl)-*N*-triflyl- and -benzoyl-*S*-phenylsulfoximide **41** and **42** as electrophilic trifluoromethylating reagents in their patent application,[34] although their structure was assigned by ^{1}H and ^{19}F nuclear magnetic resonance (NMR) data alone (Schemes 10.16 and 10.17). **39** and **42** reacted with lithium arylacetylides and sodium arenethiolates to give the corresponding CF_3 products in 47–73% yields. It was also reported that **39** and **41** reacted with phenylmagnesium bromide in tetrahydrofuran (THF)/hexamethylphosphoramide (1/3) at −30°C to room temperature to give benzotrifluoride in 24% and 15% yield, respectively, whereas Umemoto reagent **15a** gave benzotrifluoride in only 3% yield under the same reaction conditions.

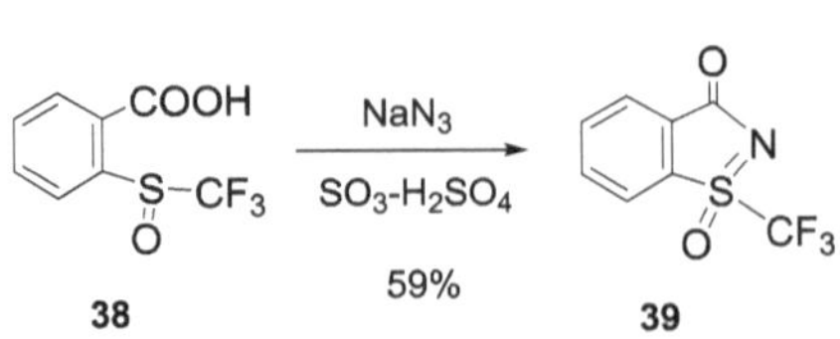

Scheme 10.16 Synthesis of 1-oxo-1-CF_3-1λ^6-benzo[d]isothiazol-3-one (**39**).

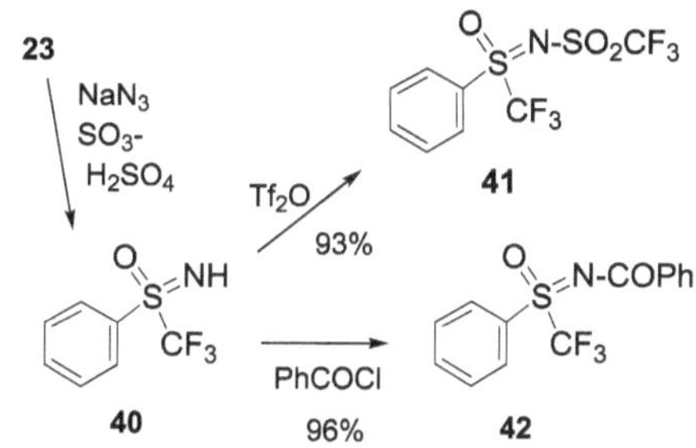

Scheme 10.17 Synthesis of *S*-CF_3-phenylsulfoximides **41** and **42**.

In 2006, Magnier et al. reported the one-pot preparation of *S*-(trifluoromethyl) diarylsulfonium triflates **24** and **43a–d** from the reaction of an arene, a trifluoromethanesulfinate salt, and Tf_2O (Scheme 10.18).[35] The emergence of the one-pot process aroused usefulness of acyclic **24** among chemists, and new reactions of **24** were developed, which were trifluoromethylations of aryl iodides,[36] aryl boronic acids,[37] benzyl bromides,[38] styrenes,[39] and alkynes[40] in the presence of copper, cupper iodide, rongalite, or a photocatalyst.

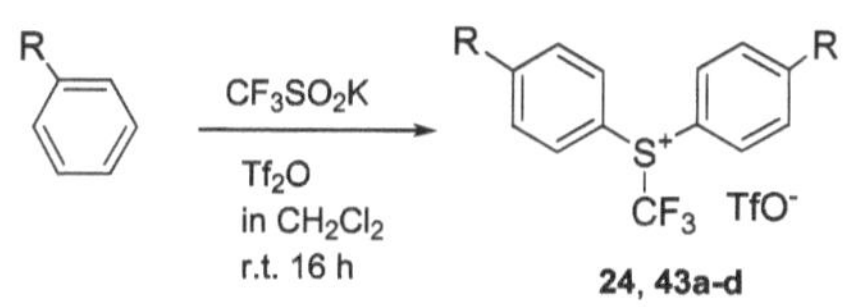

R: H (**24**, 60%), CH_3 (**43a***, 73%), F (**43b**, 46%)
Cl (**43c**, 50%), OCF_3 (**43d**, 15%)
*Isomeric ratio 85:15

Scheme 10.18 One-pot preparation of *S*-CF_3-diarylsulfonium triflates.

In 2009, they extended it to the preparation of *S*-(trifluoromethyl)dibenzothiophenium triflate **15a** and its methyl and nitro derivatives **44a–d** (Scheme 10.19).[41] However, the yields were very low. Their examination showed that methylated salts **44a–d** trifluoromethylated aniline in better yields than unsubstituted Umemoto reagent **15a**.

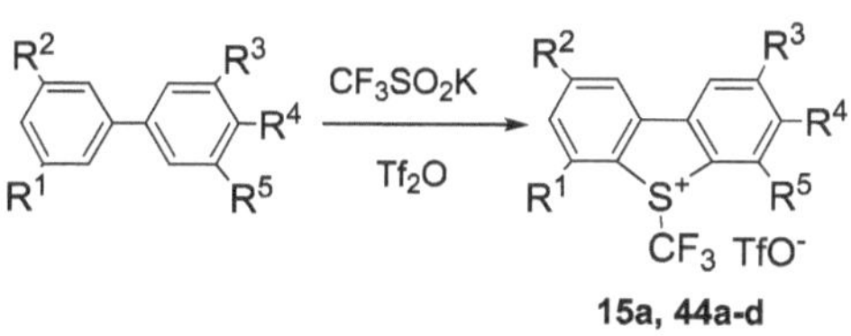

15a $R^{1\text{-}5}$=H 12%
44a $R^{1,3,5}$=H,$R^{2,3}$=CH_3/$R^{1,3,4}$=H,$R^{2,5}$=CH_3 15% (7/3)
44b $R^{1,2}$=CH_3, $R^{3,4,5}$=H 36%
44c $R^{1,2,3,5}$=CH_3, R^4=H 46%
44d $R^{1,2}$=CH_3, $R^{3,5}$=H, R^4=NO_2 27%

Scheme 10.19 One-pot preparation of Umemoto reagent **15a** and its multi-methylated analogs.

In 2014, Laali et al. synthesized *S*-(trifluoromethyl)dibenzothiophenium triflate **45** possessing an electron-withdrawing pentafluorosulfanyl group in 55% yield by the same one-pot synthesis (Scheme 10.20).[42] They surveyed the reactivity of **45** toward reactive aromatics. However, the distinguished effect of SF_5 group was not observed.

45

Scheme 10.20 One-pot preparation of SF_5-substituted *S*-CF_3-dibenzothiophenium triflate.

The one-pot method could not apply for direct synthesis of (trifluoromethyl)diarylsulfonium salts possessing strong electron-withdrawing group(s) such as NO_2. In 2008, Yagupolskii et al. published another method from diaryldifluorosulfurane **47** by the action of (trifluoromethyl)trimethylsilane ($TMSCF_3$)/F^- followed by treatment with $BF_3 \cdot Et_2O$ to produce **48a–e**[43] (Scheme 10.21). *S*-(4-Nitrophenyl)-*S*-phenyl-*S*-(trifluoromethyl)sulfonium triflate (**49**) could be prepared in 42.5% yield by the reaction of 4-nitrophenyl trifluoromethyl sulfoxide with a large excess of benzene (51 eq) and Tf_2O (5 eq).[43]

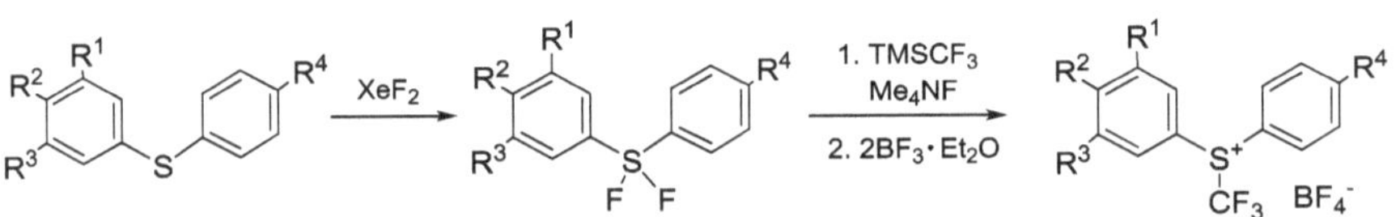

46 **47** **48a-e**

48a: $R^{1,2,3,4}$=H 75%
48b: $R^{1,3}$=H, $R^{2,4}$=F 61%
48c: $R^{1,3}$=CF_3, $R^{2,4}$=H 42%
48d: $R^{1,3,4}$=H, R^2=NO_2 40%
48e: $R^{1,3}$=H, $R^{2,4}$=NO_2 35%

Scheme 10.21 Synthesis of *S*-CF_3-diarylsulfoinium salts **48a–e**.

The power order was **48a** (H) < **48d** (NO_2) < **48e** (diNO_2) from the comparison in the reaction with NaI, giving CF_3I. *N*-Methylpyrrole, *N,N*-dimethyaniline, tetraethylthiourea, and Michler's thioketone were trifluoromethylated with **48d** or triflate **49** in 72–90% yields. The thiourea and thioketone produced *S*-trifluoromethylated salts.

In 2008, Shibata et al. reported *N,N*-dimethyl-*S*-phenyl-*S*-(trifluoromethyl)sulfoximinium tetrafluoroborate (**52**) (Scheme 10.22).[44] As triflate **51** was viscously oil, it was transformed to **52**, which was a crystalline compound. **52** trifluoromethylated β-ketoesters and dicyanoalkylidenes in the presence of 1,8-diazobicyclo[5.4.0]undec-7-ene (DBU) or a phosphazene base to give the *C*-CF_3 products in fair to high yields. Later, Magnier et al. reported that triflate **51** reacted with lithium arylacetylides to give CF_3 alkynes ($ArC \equiv CCF_3$) in 12–53% yields.[45]

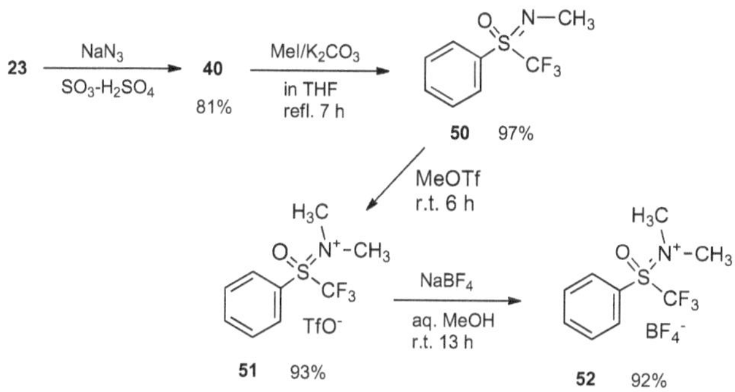

Scheme 10.22 Synthesis *S*-CF_3-phenylsulfoximinium salt **52**.

In 2009, Magnier et al. reported a safe process for the preparation of sulfoximine **40**, which avoided the use of NaN_3 in hot oleum (Scheme 10.23).[46] Sulfoxide **23** was treated with acetonitrile in the presence of Tf_2O, followed by oxidation with $KMnO_4$ to produce **40** in high yield.

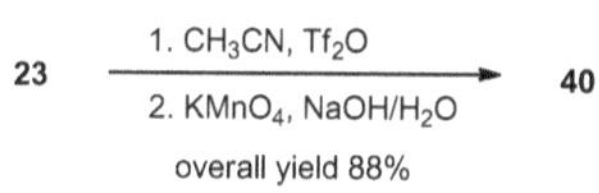

Scheme 10.23 Safe preparation of *S*-CF_3-phenylsulfoximine **40**.

In 2010, Shibata et al. reported another type of cyclic salts, *S*-(trifluoromethyl) benzothiophenium triflates **54a–j**, which were synthesized by intramolecular cyclization of **53** with TfOH (Scheme 10.24).[47] They were useful reagents for trifluoromethylation of β-ketoesters and dicyanoalkylidenes. It was later shown that **54** undertook reverse reaction in DMF at room temperature to get back to the starting materials **53**.[48]

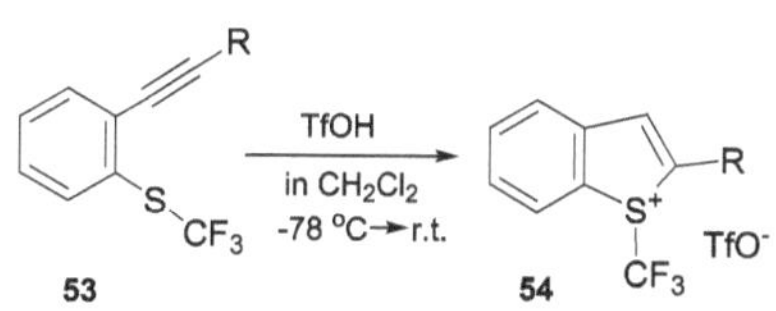

	R			R	
54a	Ph	79%	**54g**	2,4-diF-C_6H_3	72%
54b	*p*-Me-C_6H_4	94%	**54h**	*p*-I-C_6H_4	78%
54c	*m*-Br-C_6H_4	87%	**54i**	cyclopropyl	80%
54d	*p*-Br-C_6H_4	82%	**54j**	*tert*-butyl	64%
54e	*p*-MeO-C_6H_4	73%	**54k**	CH_2Ph	0%
54f	*o*-Cl-C_6H_4	64%			

Scheme 10.24 Synthesis of *S*-CF_3-benzothiophenium triflates.

They also synthesized optically active salt **57** having (+)-camphor substituent (Scheme 10.25). However, no asymmetric trifluoromethylation occurred in the reaction with a β-ketoester.[47]

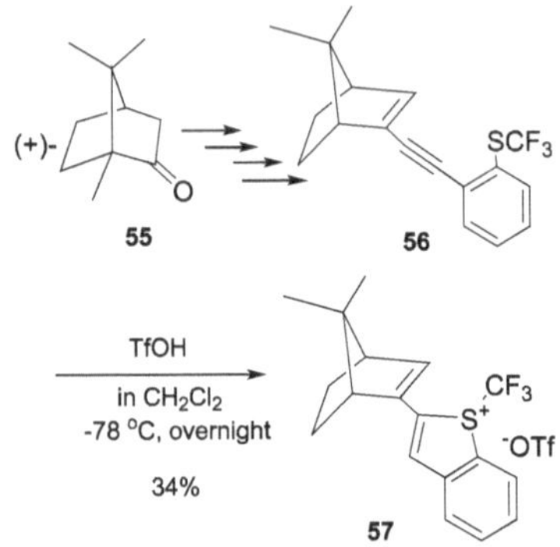

Scheme 10.25 Synthesis of optically active *S*-CF_3-benzothiophenium triflate **57**.

Most lately, Lu and Shen synthesized trifluoromethyl sulfonium ylide **59** as an electrophilic trifluoromethylating reagent (Scheme 10.26).[49] **59** was not air and moisture sensitive and did not decompose for 3 months. **59** reacted with various cyclic β-ketoesters in DMF at 100°C in the presence of K_2CO_3 to give the corresponding α-CF_3-β-ketoesters in 48−75% yields. Treatment with aryl iodides in the presence of Cu afforded the CF_3 arenes in high yields.

Scheme 10.26 Synthesis of *S*-CF_3 sulfonium ylide **59** and its reactions.

Shibata et al. reported another trifluoromethylating agent **61** as a similar strategy, which was prepared by diazotization of **60** with TsN_3 in the presence of DBU.[50] **61** reacted with nucleophiles such as ketoesters, dicyanoalkylidenes, and trimethylsilyl enol ethers in the presence of $Rh_2(OAc)_4$ catalyst in acetonitrile under reflux, giving the corresponding trifluoromethyl compounds in 23−80% yield (Scheme 10.27). Sulfonium ylide **62** was suggested as a reactive intermediate.

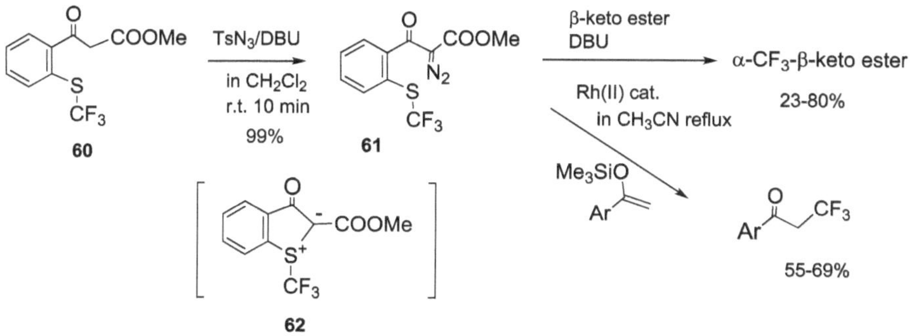

Scheme 10.27 Synthesis of CF_3S diazoketoester **61** and its reactions.

3.2 Oxonium Salts

In 1996 and 2007, Umemoto et al. published the synthesis and reactivity of the first trifluoromethyl oxonium salts, *O*-(trifluoromethyl)dibenzofuranium salts **64**, which were thermally unstable at $>-70°C$ (Scheme 10.28).[2a,30] They were in situ prepared from diazonium salts **63** by photoreaction at very low temperature ($-90 \sim -100°C$) and identified by ^{19}F and ^1H NMR analysis at $-70°C$.

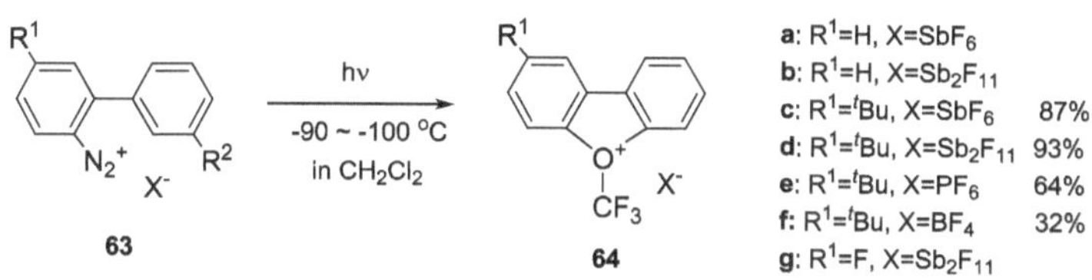

Scheme 10.28 In situ synthesis of *O*-CF_3-dibenzofuranium salts.

It was revealed that the stability depended on the nucleophilicity of the counteranion X^-. It increased in the order of $BF_4^- < PF_6^- < SbF_6^- < Sb_2F_{11}^-$. The most nonnucleophilic $Sb_2F_{11}^-$ gave the highest yield and stability. The stability also depended on the electronic nature of the ring substituent R: F < H < ^{t}Bu. Electron-donating *tert*-butyl group gave the highest stability of the substituents examined. Salt **64d** ($X = Sb_2F_{11}$, R = ^{t}Bu) had the longest half-life of 415 min at $-60°C$. Salt **64g** ($X = BF_4$, R = F) had the shortest half-life of 13 min at $-60°C$.

These salts were extremely reactive. They decomposed to give CF_4 and a dibenzofuran. Surprisingly, CF_4 was formed even when X^- was $Sb_2F_{11}^-$, which has been known to be a really nonnucleophilic anion. It was thus suggested that real CF_3^+ was generated as a transit species in these reactions. It was a sharp contrast to the fact that *O*-methyldibenzofuranium salt was stable up to 80°C. This meant that CF_3^+ was generated much easier than CH_3^+, which had been suggested by the calculations.[51]

O-Trifluoromethyl oxonium salt **64c** acted as a source of real CF_3^+. **64c** was in situ generated by photo reaction of **63c** at the low temperature and then treated with *N*- and

O-nucleophilies such as amines, alcohols, and phenols to give the corresponding *N*- and *O*-CF_3 products in fair to high yields (Scheme 10.29). Tertiary amines and pyridines were trifluoromethylated to give *N*-CF_3 quaternary ammonium and pyridinium salts in fair to high yields.

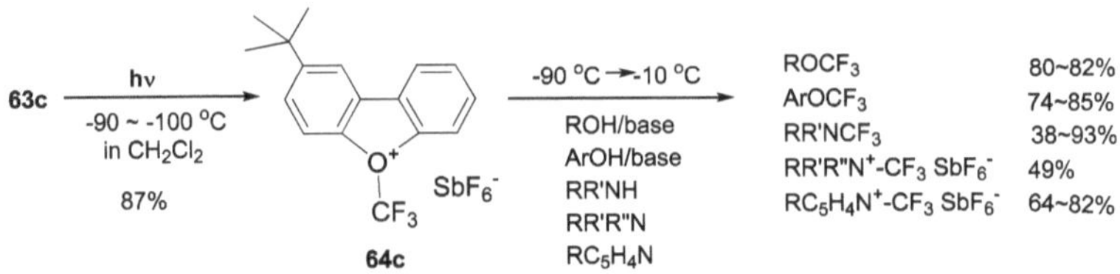

Scheme 10.29 *O*- and *N*-trifluoromethylations with *O*-CF_3-dibenzofuranium salt **64c**.

Some of the *O*- and *N*-trifluoromethylations were carried out by the thermal decomposition method of diazonium salt **63a**, in which a mixture of **63a** and an *O*- or *N*-nucleophile was heated in dichloromethane under reflux for 3 h.[2a,30] Recently, the synthesis of *N*-(trifluoromethoxy)pyridinium hexafluoroantimonates containing an electron-withdrawing substituent(s) on the pyridine ring from pyridine *N*-oxides by this thermal method was disclosed.[52]

3.3 Iodine(III) Compounds

Long after the attempt by Yagupolskii and Umemoto, Togni et al. succeeded in synthesizing cyclic type of CF_3-iodine(III) compounds, 1-(trifluoromethyl)-1,2-benziodoxol-3(1H)-one (**69**) and 1-(trifluoromethyl)-1,3-dihydro-3,3-dimethyl-1,2-benziodoxole (**74**), by the Umpolung methodology using $TMSCF_3$ in 2006 (Schemes 10.30 and 10.31).[53] $TMSCF_3$ was found to be useful for nucleophilic trifluoromethylation in 1989 by Prakash et al.[54]

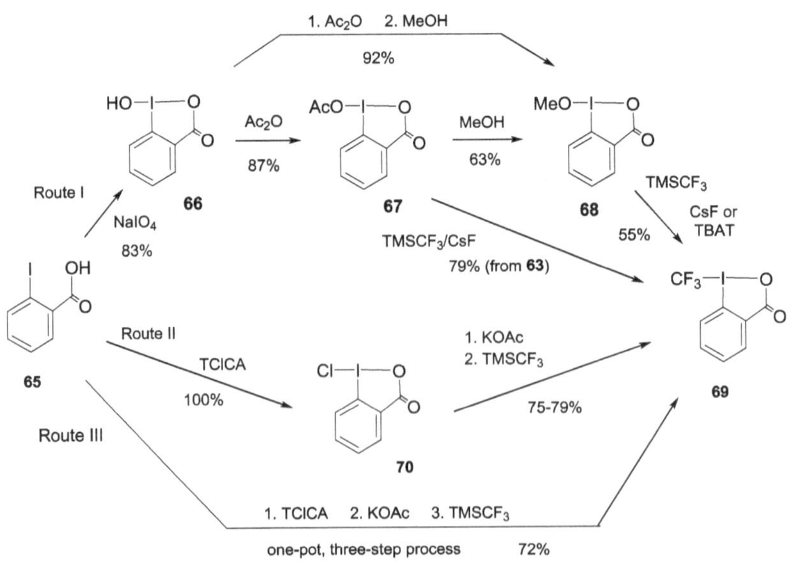

Scheme 10.30 Synthesis of 1-CF_3-1,2-benziodoxol-3(H)-one (**69**).

At first, **69** was synthesized by ligand exchange of **68** with $TMSCF_3$ in the presence of tetra-*n*-butylammonium difluorotriphenylsilicate (TBAT) or CsF (route I, Scheme 10.30). Reassessment of the synthetic route showed that chloride **70** was a suitable intermediate (route II) and finally provided a one-pot, three-step process (route III) for the preparation of **69.**[55]

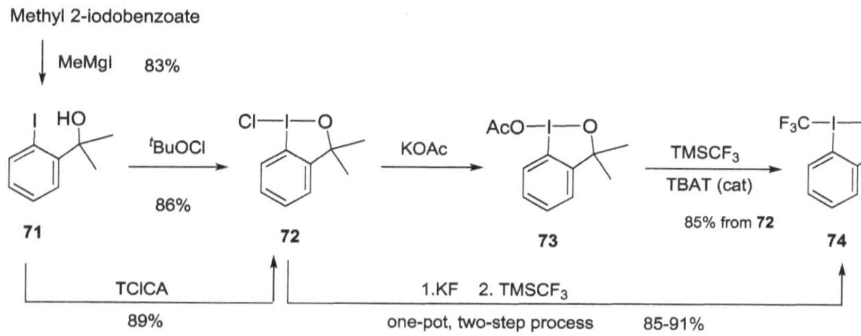

Scheme 10.31 Synthesis of 1-CF_3-1,3-dihydro-3,3-dimethyl-1,2-benziodoxole (**74**).

74 was synthesized from **71** by oxidation with ^{t}BuOCl followed by treatment with KOAc and then $TMSCF_3$[56] (Scheme 10.31). For large-scale application, trichlorocyanuric acid (TCICA) was used in place of ^{t}BuOCl and the chlorine atom of **72** was replaced with a fluorine atom. A one-pot, two-step process of **72** to **74** was finally developed.[55]

Many $CF_3I(III)$ compounds **75**–**78** with different side chains were synthesized and evaluated. Even though a certain trend was indicated in their reactivity, no strong correlation could be established among them.[57] Nitro derivative **79** was also synthesized.[58] **79** showed high reactivity to *p*-toluenesulfonic acid. However, the desired CF_3 product, *p*-$TsOCF_3$, was not formed. Another limitation of **79** was its poor solubility in common organic solvents [57,2h] (Fig. 10.3).

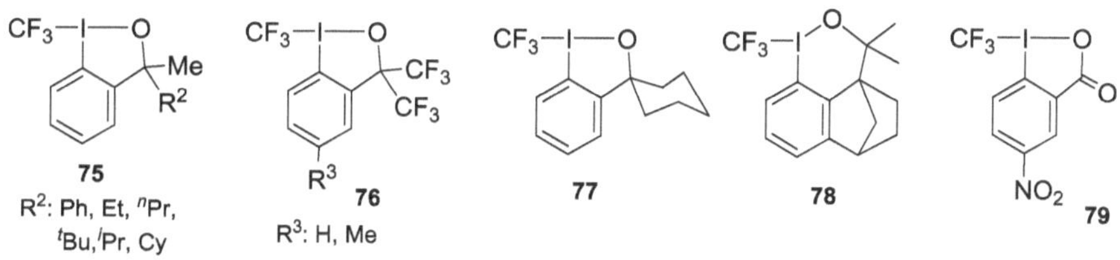

Figure 10.3 1-CF_3-1,2-benziodoxol analogs.

Since **69** and **74**, known as Togni's reagents, could be prepared by short steps and they have been sold commercially, many new reactions with them have been developed by many research groups.[2h] As seen in Scheme 10.32, Togni reagents were reactive and versatile reagents for trifluoromethylations of various substrates such as carbonyl compounds,[59,60] aromatics,[61] heteroaromatics,[61] silyl enol ethers,[62] enamides,[63] aryl and vinyl boronic acids,[64–66] alkynes,[67] alkenes,[68–70] as well as *O*-, *S*-, and *P*-nucleophiles such as alcohols,[71] sulfonates,[72,73] and phosphines[74] and *N*-nucleophiles.[75] The starting material **65** or **71** was restored from the trifluoromethylations.

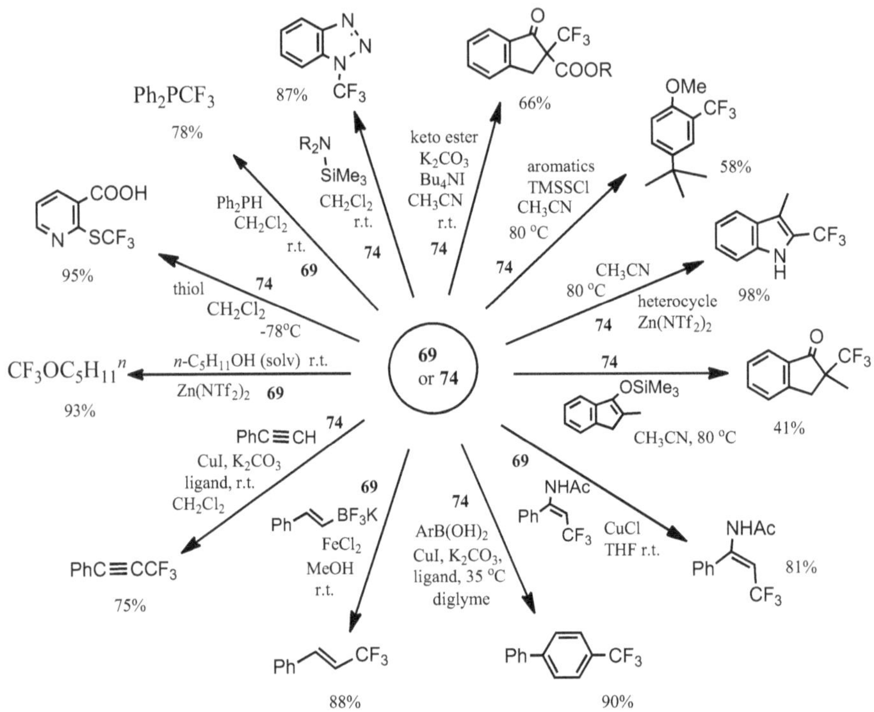

Scheme 10.32 Trifluoromethylation of various substrates with **69** and **74**.

In addition, alkoxy-,[76] azido-,[77] cyano-,[78] oxy-,[79] and amino-trifluoromethylations[80] including concomitant cyclization and trifluoromethylations with other C-C bond formation and rearrangement reactions[2h] were developed. Highly enantioselective trifluoromethylations of aldehydes[81] and β-ketoesters[28b] were successful. Lewis acids, Brϕnsted acids, copper catalysts, organocatalysts, and photocatalysts were successfully used for many trifluoromethylations with **69** and **74**[2h]. However, it was cautioned that **74** caused a rapid exothermic decomposition above its melting point (78°C)[56,82] and **69** was dangerously explosive.[82]

Electrochemical data for the first reduced reaction of different CF_3 reagents were reported[48]: **15a** (or **15b**), −0.63 V; **54i**, −0.72 V; **54a**, −0.49 V; **54e**, −0.49 V; **54l**, −0.47 V; **69**, −1.10 V; **74**, −1.82 V (V vs. Ag/Ag^+ in CH_3CN) (**54l**, *R* = 2,4-dinitrophenyl in **54**). Thus the oxidation potential of the trifluoromethylating reagents increased in the order **74** << **69** << **54i** < **15a** (**15b**) < **54e** = **54a** < **54l**. The oxidation potentials of $CF_3I(III)$ compounds were considerably lower than those of the CF_3 sulfonium salts. In the series **54**, the electronic nature of substituent at 2-position was reflected in the potentials: **54i** (cyclopropyl) << **54e** (4-methoxyphenyl) = **54a** (phenyl) < **54l** (2,4-dinitrophenyl). Another group's data[21a,40] were as follows (V vs. Cp_2Fe in CH_3CN): **74** (−1.49 V) < **69** (−1.34 V) < **24** (−1.11 V) << **15b** (−0.75 V). Acyclic sulfonium salt **24** had higher potential than Togni reagents **69** and **74**, but lower than Umemoto reagent **15b**.

In 2013, Wang and Liu reported the trifluoromethylation of nucleophilic substrates with $PhI(OAc)_2$ and $TMSCF_3$ in the presence of KF (Scheme 10.33).[83] In their studies, they detected a peak corresponding to $[PhICF_3]^+$ species **82** by mass spectrometric analysis of the reaction mixture in CH_3CN. It was thus proposed that acyclic (trifluoromethyl)phenyliodonium acetate (**81**) was formed as an intermediate, which was the target compound attempted by Yagupolskii and Umemoto around 35 years ago. Some other trifluoromethylation reactions with $PhI(OAc)_2/TMSCF_3$/base were explained by a similar mechanism.[84]

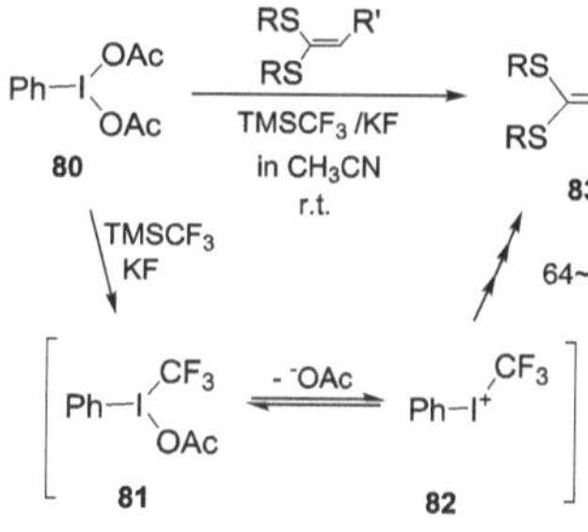

Scheme 10.33 Trifluoromethylation with $PhI(OAc)_2/TMSCF_3/KF$.

4. Conclusion[85]

Many electrophilic trifluoromethylating reagents have been developed so far. Many new trifluoromethylations with them have been reported by many newcomers to the fluorine chemistry during the past decade. Accordingly, the field of electrophilic trifluoromethylation, which had been considered to be left undeveloped, has markedly been advanced. The electrophilic trifluoromethylating reagents have a deep and complex chemistry because their reactions may include electron transfer or reduction/oxidation processes. There is still a lot of room to be investigated. Since this chapter could not satisfactorily cover the reactions by the trifluoromethylating reagents because of limited space, readers are advised to read original papers and reviews cited.

References

1. a. Kirsch, P. *Modern Fluoroorganic Chemistry;* Wiley-VCH: Weinheim, 2004.
   b. Bégué, J. -P.; Bonnet-Delpon, D. *Bioorganic and Medicinal Chemistry of Fluorine;* Wiley: Hoboken, NJ, 2008.
   c. Ojima, I. *Fluorine in Medicinal Chemistry and Chemical Biology;* Wiley-Blackwell: Chichester, U.K., 2009.
2. Reviews for Electrophilic Trifluoromethylating Agents and Their Reactions:
   a. Umemoto, T. *Chem. Rev.* **1996,** *96,* 1757–1777.
   b. Ma, J. -A.; Cahard, D. J. *J. Fluor. Chem.* **2007,** *128,* 975–996.
   c. Shibata, N.; Matsnev, A.; Cahard, D. *Beilstein J. Org. Chem.* **2010,** *6* (65).
   d. Macé, Y.; Magnier, E. *Eur. J. Org. Chem.* **2012,** 2479–2494.
   e. Zhang, C. *Org. Biomol. Chem.* **2014,** *12,* 6580–6589.
   f. Barata-Vallejo, S.; Lantaño, B.; Postigo, A. *Chem. Eur. J.* **2014,** *20,* 65806–65829.
   g. Koike, T.; Akita, M. *J. Fluor. Chem.* **2014,** *167,* 30–36.
   h. Charpertier, J.; Früh, N.; Togni, A. *Chem. Rev.* **2015,** *115,* 650–682.
   i. Wang, S. -M.; Han, J. -B.; Zhang, C. -P.; Qin, H. -L.; Xiao, J. -C. *Tetrahedron* **2015,** *71,* 7949–7976.
3. Yagupolskii, L. M.; Maletina, I. I.; Kondratenko, N. V.; Orda, V. V. *Synthesis* **1978,** 835–837.
4. a. Umemoto, T.; Kuriu, Y.; Shuyama, H.; Miyano, O.; Nakayama, S. -I. *J. Fluor. Chem.* **1982,** *20,* 695–698.
   b. Umemoto, T.; Kuriu, Y.; Shuyama, H.; Miyano, O.; Nakayama, S. -I. *J. Fluor. Chem.* **1986,** *31,* 37–56.
5. Yagupolskii, L. M. *J. Fluor. Chem.* **1987,** *36,* 1–28.
6. Umemoto, T.; Miyano, O. *Tetrahedron Lett.* **1982,** *23,* 3929–3930.
7. Umemoto, T.; Ando, A. *Bull. Chem. Soc. Jpn.* **1986,** *59,* 447–452.
8. Umemoto, T. *J. Fluor. Chem.* **2014,** *167,* 3–15.
9. Yagupolskii, L. M.; Kondratenko, N. V.; Timofeeva, G. N. *J. Org. Chem. USSR* **1984,** *20,* 103–106.
10. Umemoto, T.; Ishihara, S. *Tetrahedron Lett.* **1990,** *31,* 3579–3582.
11. Umemoto, T.; Ishihara, S. *J. Am. Chem. Soc.* **1993,** *115,* 2156–2164.
12. Ono, T.; Umemoto, T. *J. Fluor. Chem.* **1996,** *80,* 163–166.
13. Muralidharan, K.; Chakraborty, R.; Shreeve, J. M. *J. Fluor. Chem.* **2004,** *125,* 1967–1968.
14. Umemoto, T.; Adachi, K. *J. Org. Chem.* **1994,** *59,* 5692–5699.
15. Umemoto, T.; Ishihara, S.; Adachi, K. *J. Fluor. Chem.* **1995,** *74,* 77–82.
16. Umemoto, T.; Ishihara, S. *J. Fluor. Chem.* **1999,** *98,* 75–81.
17. Ma, J. -A.; Cahard, D. *J. Org. Chem.* **2003,** *68,* 8726–8729.
18. a. Wang, X.; Truesdale, L.; Yu, J. -O. *J. Am. Chem. Soc.* **2010,** *132,* 3648–3649.
    b. Zhang, X. -G.; Dai, H. -X.; Wasa, M.; Yu, J. -Q. *J. Am. Chem. Soc.* **2012,** *134,* 11948–11951.
    c. Zhang, L. -S.; Chen, K.; Chen, G.; Li, B. -J.; Luo, S.; Guo, Q. -Y.; Wei, J. -B.; Shi, Z. -J. *Org. Chem.* **2013,** *15,* 10–13.
19. Xu, J.; Luo, D. -F.; Xiao, B.; Liu, Z. -J.; Gong, T. -J.; Fu, Y.; Liu, L. *Chem. Commun.* **2011,** *47,* 4300–4302.
20. Luo, D. -F.; Xu, J.; Fu, Y.; Guo, Q. X. *Tetrahedron Lett.* **2012,** *53,* 2769–2772.
21. a. Yasu, Y.; Koike, T.; Akita, M. *Angew. Chem. Ind. Ed.* **2012,** *51,* 9567–9571.
    b. Yasu, Y.; Koike, T.; Akita, M. *Org. Lett.* **2013,** *15,* 2136–2139.
    c. Yasu, Y.; Arai, Y.; Tomita, R.; Koike, T.; Akita, M. *Org. Lett.* **2014,** *16,* 780–783.

d. Tomita, R.; Yasu, Y.; Koike, T.; Akita, M. *Angew. Chem. Int. Ed.* **2014,** *53,* 7144–7148.
e. Dagousset, G.; Carboni, A.; Magnier, E.; Masson, G. *Org. Lett.* **2014,** *16,* 4340–4343.
f. Noto, N.; Miyazawa, K.; Koike, T.; Akita, M. *Org. Lett.* **2015,** *17,* 3710–3713.
g. Wei, Q.; Chen, J. -R.; Hu, X. -Q.; Yang, X. -C.; Lu, B.; Xiao, W. -J. *Org. Lett.* **2015,** *17,* 4464–4467.

22. Mizuta, S.; Verhoog, S.; Engel, K. M.; Khotavivattana, T.; O'Dwill, M.; Wheelhouse, K.; Rascias, G.; Médebielle, M.; Gouverneur, V. *J. Am. Chem. Soc.* **2013,** *135,* 2505–2508.
23. Dai, J. J.; Fang, C.; Xiao, B.; Yi, J.; Xu, J.; Liu, Z. J.; Lu, X.; Liu, L.; Fu, Y. *J. Am. Chem. Soc.* **2013,** *135,* 8436–8439.
24. Cheng, Y.; Yuan, X.; Ma, J.; Yu, S. *Chem. Eur. J.* **2015,** *21,* 1–6.
25. Ge, G. -C.; Huang, X. -J.; Ding, C. -H.; Wan, S. -L.; Dai, L. -X. *Chem. Commun.* **2014,** *50,* 3048–3051.
26. Ge, G.; Huang, X.; Ding, C.; Li, H.; Wan, S.; Hou, X. *Chim. J. Chem.* **2014,** *32,* 727–733.
27. Xu, J.; Wang, Y. -L.; Gong, T. -J.; Xiao, B.; Fu, Y. *Chem. Commun.* **2014,** *50,* 12915–12918.
28. a. Noritake, S.; Shibata, N.; Nomura, Y.; Huang, Y.; Matsnev, A.; Nakamura, S.; Toru, T.; Cahard, D. J. *Org. Biomol. Chem.* **2009,** *7,* 3599–3604.
b. Deng, Q. -H.; Wadepohl, H.; Gade, L. H. *J. Am. Chem. Soc.* **2012,** *134,* 10769–10772.
29. Camasso, N. M.; Sanford, M. S. *Science* **2015,** *347* (6227), 1203–1204.
30. Umemoto, T.; Adachi, K.; Ishihara, S. *J. Org. Chem.* **2007,** *72,* 6905–6917.
31. Macé, Y.; Pradet, C.; Popkin, M.; Blazejewski, J. C.; Magnier, E. *Tetrahedron Lett.* **2010,** *51,* 5388–5391.
32. Yang, Y. -D.; Lu, X.; Liu, G.; Tokunaga, E.; Tsuzuki, S.; Shibata, N. *ChemistryOpen* **2012,** *1,* 221–226.
33. Yang, J. -J.; Kirchmeier, R. I.; Shreeve, J. M. *J. Org. Chem.* **1998,** *63,* 2656–2660.
34. Adachi, K.; Ishihara, S. Japan Tokyo Kokai, 2005-145917
35. Magnier, E.; Blazejewski, J. -C.; Tordeux, M.; Wakselman, C. *Angew. Chem. Int. Ed.* **2006,** *45,* 1279–1282.
36. Zhang, C. P.; Wang, Z. L.; Chen, Q. Y.; Zhang, C. T.; Gu, Y. C.; Xiao, J. C. *Angew. Chem. Int. Ed.* **2011,** *50,* 1896–1900.
37. Zhang, C. P.; Cai, J.; Zhou, C. B.; Wang, X. P.; Zheng, X.; Gu, Y. C.; Xiao, J. C. *Chem. Commun.* **2011,** *47,* 9516–9518.
38. Kawai, H.; Furukawa, T.; Nomura, Y.; Tokunaga, E.; Shibata, N. *Org. Lett.* **2011,** *13,* 3596–3599.
39. Zhang, C. P.; Wang, Z. L.; Chen, Q. Y.; Zhang, C. T.; Gu, Y. C.; Xiao, J. C. *Chem. Commun.* **2011,** *47,* 6632–6634.
40. Tomita, R.; Koike, T.; Akita, M. *Angew. Chem. Int. Ed.* **2015,** *54,* 12923–12927.
41. Macé, Y.; Raymondeau, B.; Pradet, C.; Blazejewski, J. -C.; Magnier, E. *Eur. J. Org. Chem.* **2009,** 1390–1397.
42. Okazaki, T.; Laali, K. K.; Reddy, A. S. *J. Fluor. Chem.* **2014,** *165,* 91–95.
43. Yagupolskii, L. M.; Matsnev, A. V.; Orlova, R. K.; Deryabkin, B. G.; Yagupolskii, Y. L. *J. Fluor. Chem.* **2008,** *129,* 131–136.
44. Noritake, S.; Shibata, N.; Nakamura, S.; Toru, T. *Eur. J. Org. Chem.* **2008,** 3465–3468.
45. Urban, C.; Cadoret, F.; Blazejewski, J. -C.; Magnier, E. *Eur. J. Org. Chem.* **2011,** 4862–4867.
46. Macé, Y.; Urban, C.; Pradet, C.; Marrot, J.; Blazejewski, J. -C.; Magnier, E. *Eur. J. Org. Chem.* **2009,** 3150–3153.
47. Matsnev, A.; Noritake, S.; Nomura, Y.; Tokunaga, E.; Nakamura, S.; Shibata, N. *Angew. Chem. Int. Ed.* **2010,** *49,* 572–576.

48. Mizuta, S.; Verhoog, S.; Wang, X.; Shibata, N.; Gouverneur, V.; Médebielle, M. *J. Fluor. Chem.* **2013,** *155,* 124–131.
49. Liu, Y.; Shao, X.; Zhang, P.; Lu, L.; Shen, Q. *Org. Lett.* **2015,** *17,* 2752–2755.
50. Saidalimu, I.; Tokunaga, E.; Shibata, N. *ACS Catal.* **2015,** *5,* 4668–4672.
51. Raynolds, C. H. *J. Chem. Soc. Chem. Commun.* **1991,** 975–976.
52. Umemoto, T.; Zhu, M.; Hu, J. *Chinese Patent Application.* CN 105017143A, 2015.
53. Eisenberger, P.; Gischig, S.; Togni, A. *Chem. Eur. J.* **2006,** *12,* 2579–2586.
54. Prakash, G. K. S.; Yudin, A. K. *Chem. Rev.* **1997,** *97,* 757–786.
55. Matoušek, V.; Pietrasiak, E.; Schwenk, R.; Togni, A. *J. Org. Chem.* **2013,** *78,* 6763–6768.
56. Eisenberger, P.; Kieltsch, I.; Koller, R.; Stanek, K.; Togni, A. *Org. Synth.* **2011,** *88,* 168–180.
57. Niedermann, K.; Welch, J. M.; Koller, R.; Cvengroš, J.; Santschi, N.; Battaglia, P.; Togni, A. *Tetrahedron* **2010,** *66,* 5753–5761.
58. Santschi, N.; Sarott, R. C.; Otth, E.; Kissner, R.; Togni, A. *Beilstein J. Org. Chem.* **2014,** *10,* 1–6.
59. Keiltsch, I.; Eisenbeger, P.; Togni, A. *Angew. Chem. Int. Ed.* **2007,** *46,* 754–757.
60. Matousek, V.; Togni, A.; Bizet, V.; Cahard, D. *Org. Lett.* **2011,** *13,* 5762–5765.
61. Wiehn, M. S.; Vinogradova, E. V.; Togni, A. *J. Fluor. Chem.* **2010,** *131,* 951–957.
62. Kieltsch, I.; Eisenberger, P.; Stanek, K.; Togni, A. *Chimia* **2008,** *62,* 260–263.
63. Feng, C.; Loh, T. -P. *Chem. Sci.* **2012,** *3,* 3458–3462.
64. Liu, T.; Shen, Q. *Org. Lett.* **2011,** *13,* 2342–2345.
65. Parsons, A. T.; Senecal, T. D.; Buchwald, S. L. *Angew. Chem. Int. Ed.* **2012,** *51,* 2947–2950.
66. Yasu, Y.; Koike, T.; Akita, M. *Chem. Commun.* **2013,** *49,* 2037–2039.
67. Weng, Z.; Li, H.; He, W.; Yao, L. -F.; Tan, J.; Chen, J.; Yuan, Y.; Huang, K. -W. *Tetrahedron* **2012,** *68,* 2527–2531.
68. Parsons, A. T.; Buchwald, S. L. *Angew. Chem. Int. Ed.* **2011,** *50,* 9120–9123.
69. Wang, X.; Ye, Y.; Zhang, S.; Feng, J.; Xu, Y.; Zhang, Y.; Wang, J. *J. Am. Chem. Soc.* **2011,** *133,* 16410–16413.
70. Wang, X. -P.; Lin, J. -H.; Zhang, C. -P.; Xiao, J. -C.; Zheng, X. *Beilstein J. Org. Chem.* **2013,** *9,* 2635–2640.
71. Koller, R.; Stanek, K.; Stolz, D.; Aardoom, R.; Niedermann, K.; Togni, A. *Angew. Chem. Int. Ed.* **2009,** *48,* 4332–4336.
72. Koller, R.; Huchet, Q.; Battaglia, P.; Welch, J. M.; Togni, A. *Chem. Commun.* **2009,** 5993–5995.
73. Kieltsch, I.; Eisenberger, P.; Togni, A. *Angew. Chem., Ind. Ed.* **2007,** *46,* 754–757.
74. Eisenberger, P.; Kieltsch, I.; Armanino, N.; Togni, A. *Chem. Commun.* **2008,** 1575–1577.
75. Niedermann, K.; Fruh, N.; Senn, R.; Czarniecki, B.; Verel, R.; Togni, A. *Angew. Chem. Int. Ed.* **2012,** *51,* 6511–6515.
76. Egami, H.; Shimizu, R.; Sodeoka, M. *Tetrahedron Lett.* **2012,** *53,* 5503–5506.
77. Wang, F.; Qi, X.; Liang, Z.; Chen, P.; Liu, G. *Angew. Chem. Int. Ed.* **2014,** *53,* 1881–1886.
78. He, Y. -T.; Li, L. -H.; Yang, Y. -F.; Zhou, Z. -Z.; Hua, H. -L.; Liu, X. -Y.; Liang, Y. -M. *Org. Lett.* **2014,** *16,* 270–273.
79. a. Li, Y.; Studer, A. *Angew. Chem. Int. Ed.* **2012,** *51,* 8221–8224.
b. Zhu, R.; Buchwald, S. L. *J. Am. Chem. Soc.* **2012,** *134,* 12462–12465.
c. Zhu, R.; Buchwald, S. L. *Angew. Chem. Int. Ed.* **2013,** *52,* 12655–12658.
d. Yu, Q.; Ma, S. *Chem. Eur. J.* **2013,** *19,* 13304–13308.
e. He, Y. -T.; Li, L. -H.; Yang, Y. -F.; Wang, Y. -Q.; Luo, J. -Y.; Liu, X. -Y.; Liang, Y. -M. *Chem. Commun.* **2013,** *49,* 5687–5689.

80. a. Lin, J. -S.; Xiong, Y. -P.; Ma, C. -L.; Zhao, L. -J.; Tan, B.; Liu, X. -Y. *Chem. Eur. J.* **2014,** *20,* 1332–1340.
b. Egami, H.; Kawamura, S.; Miyazaki, A.; Sodeoka, M. *Angew. Chem. Int. Ed.* **2013,** *52,* 7841–7844.
c. Ji, G.; Wang, X.; Zhang, S.; Xu, Y.; Ye, Y.; Li, M.; Zhang, Y.; Wang, J. *Chem. Commun.* **2014,** *50,* 4361–4363.
81. Allen, A. E.; MacMillan, D. W. C. *J. Am. Chem. Soc.* **2010,** *132,* 4986–4987.
82. Fiederling, N.; Haller, J.; Schramm, H. *Org. Process Res. Dev.* **2013,** *17,* 318–319.
83. Xu, C.; Liu, J.; Ming, W.; Liu, Y.; Liu, J.; Wang, M.; Liu, Q. *Chem. Eur. J.* **2013,** *19,* 9104–9109.
84. a. Wang, Q.; Dong, X.; Xiao, T.; Zhou, L. *Org. Lett.* **2013,** *15,* 4846–4849.
b. Li, L.; Deng, M.; Zheng, S. -C.; Xiong, Y. -P.; Tan, B.; Liu, X. -Y. *Org. Lett.* **2014,** *16,* 504–507.
c. Fu, W.; Xu, F.; Fu, Y.; Xu, C.; Li, S.; Zou, D. *Eur. J. Org. Chem.* **2014,** 709–712.
85. After this chapter was completed, a significant paper appeared; Li, M.; Xue, X. -S.; Guo, J.; Wang, Y.; Cheng, J. -P.. *J. Org. Chem.* **2016,** *81,* 3119–3126. The paper provided an energetic guide for estimating the trifluoromethyl cation (CF_3^+)-donating abilities (TC^+DA) of the electrophilic trifluoromethylating reagents. The TC^+DA values clearly revealed the relative reactivity of various types of electrophilic trifluoromethylating agents developed so far.

New Nucleophilic Fluoroalkylations

11

G.K.S. Prakash, Z. Zhang
University of Southern California, Los Angeles, CA, United States

Chapter Outline

1. Introduction

Fluoroorganic compounds have been playing important roles in the fields of chemistry, materials, and pharmaceutical sciences due to their unique chemical, physical, and biological properties.[1] The increasing demand of these fluorine-containing molecules and the lack of naturally occurring organofluorine compounds[2] in turn have been

Modern Synthesis Processes and Reactivity of Fluorinated Compounds. http://dx.doi.org/10.1016/B978-0-12-803740-9.00011-1

stimulating the development of novel methodologies toward the construction of interesting fluoroorganics. In 1835, Dumas and Péligot reported the synthesis of the first organofluorine molecule—fluoromethane, from potassium fluoride and dimethyl sulfate.[3] During the subsequent century, an enormous effort has been dedicated to investigate fluorination reactions utilizing various inorganic fluorides (such as anhydrous HF, SbF_3, KF, BF_3, tetrafluoroborate salts, F_2, and CoF_3) as fluorinating reagents.[4] In the past half century, considerable amount of more practical fluorinating/fluoroalkylating reagents and protocols have been developed and improved to prepare fluoroorganics via nucleophilic, electrophilic, carbene, ylide, electrochemical, and radical pathways.[5] To date, the effort has been mostly devoted to the development of versatile user-friendly reagents and new synthetic protocols under mild conditions with high efficacies, facilitating the late-stage fluorine incorporation.[6]

Despite the fact that fluoroalkyl moieties can be obtained by direct fluorinations with various fluorinating reagents such as SF_4 derivatives and fluorides, employment of fluoroalkyl synthons prevails over the former because of higher efficiency and greater functional group compatibility. In the field of fluoroalkylation chemistry, a number of key reaction intermediates are involved mechanistically, including α-fluorocarbanions, α-fluorocarbocations, α-fluorocarbenes, α-fluorinated radicals, and β-fluorocarbanions (Scheme 11.1). Although both free radical and electrophilic fluoroalkylations are useful methods to prepare fluoroorganics, nucleophilic fluoroalkylation has become one of the most important protocols for the preparation of fluoroorganic compounds.[1,7] Nucleophilic fluoroalkylation typically involves the transfer of a free fluorinated carbanion $\left(R_f^-\right)$ or equivalent (a species that has similar reactivity character to a fluorinated carbanion such as pentacoordinated silicon species), or a fluoroalkyl metal species (R_fM) with nucleophilic behavior to an electrophile. Over the past four decades, a number of nucleophilic fluoroalkylating reagents and protocols have been extensively investigated toward a variety of electrophiles, including aldehydes, ketones, enones, alkynes, imines, nitrones, indanones, esters, lactones, cyclic anhydrides, oxazolidinones, amides, imides, azirines, nitroso compounds, thiocyanates, selenocyanates, alkyl triflates, alkyl halides, aryl halides, epoxides, cyclic sulfates, and sulfamidates.[8] For the sake of brevity, in this chapter the majority of our attention is focused on the innovative achievements in the development of novel nucleophilic fluoroalkylation protocols over the past 10 years (since 2006). For a more general and exhaustive survey of fluorination and fluoroalkylation methods, one can refer to previously published, excellent reviews.[9]

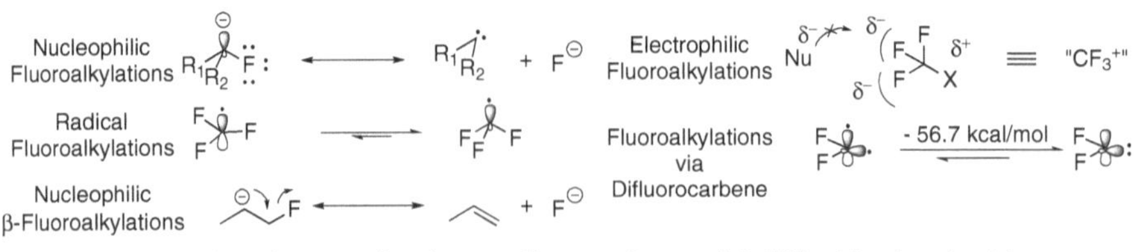

Scheme 11.1 Various key reaction intermediates and potential difficulties involved in fluoroalkylations.

2. Nucleophilic Trifluoroalkylation Reactions

Perfluoroalkyl groups, including the trifluoromethyl group, play important roles in agrochemical, pharmaceutical, and materials sciences due to their unique properties, including increased lipophilicity as well as thermal and metabolic stabilities.[1d,1e,9a,9b,10] For quite a long time, the attempts of direct nucleophilic trifluoromethylation using trifluoromethyl alkali/alkaline earth metal species have failed.[11] It was widely assumed that the naked trifluoromethyl anion was unstable and the decomposition to a fluoride anion and a stabilized difluorocarbene species occurred spontaneously. Two popular assumptions are the negative charge–lone pair (fluorine *2p* orbital) repulsion and the strong metal–fluoride interactions if a metal ion is present. However, an elaborate study of the first bulk synthesis of CF_3^- anion along with computational studies accomplished in 2014 by Prakash and coworkers suggested that the intermediacy of CF_3^- in nucleophilic trifluoromethylation was evident.[12] It has been shown that the formation of CF_3^- from difluorocarbene and fluoride is thermodynamically more favored than the reverse reaction. Furthermore, the isolation and characterization of α-fluorobis(phenylsulfonyl)methanide anion[13] clearly indicated that the lability of C–F bond is questionable, thus the hypothesis of negative charge–lone pair repulsion should be reconsidered (Scheme 11.2).

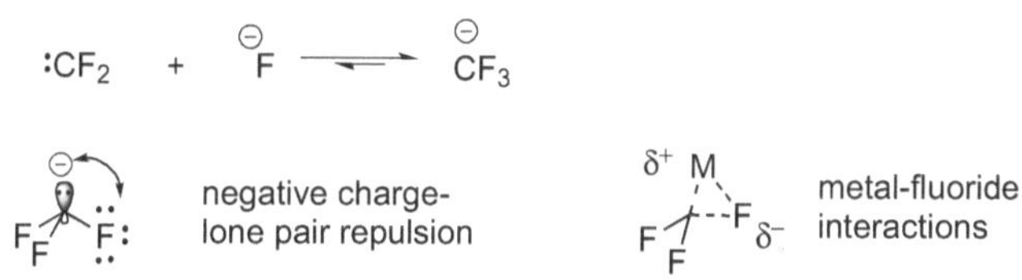

Scheme 11.2 Previous hypotheses for the instability of CF_3^- anion.

In spite of the persistency of CF_3^- in tetrahydrofuran (THF), a general, mild, and robust trifluoromethylating protocol had not been established until in 1989, when Prakash et al. made the breakthrough by employing (trifluoromethyl)trimethylsilane ($TMSCF_3$, originally synthesized by Ruppert, now known as Ruppert–Prakash reagent) for the nucleophilic trifluoromethylation of carbonyl compounds with high efficacy.[14] The lability of the $Si–CF_3$ bond facilitates the transfer of the CF_3^- anion under mild conditions to a wide range of functional groups, including aldehydes, ketones, esters, imines, nitriles, nitrones, alkyl halides, and also metallic substrates.[15] Subsequently, the utilization of inexpensive fluoroform (CF_3H) was first demonstrated by Shono et al.[16] and further developed by Périchon et al.,[17] Normant et al.,[18] Langlois et al.,[19] and others.[20] Later, the development of sulfur-based nucleophilic trifluoromethylating reagents such as trifluoromethyl phenyl sulfone ($PhSO_2CF_3$) extended the library of versatile nucleophilic trifluoromethylating reagent toward carbonyl compounds and imines.[21] In this section, developments of novel reagents and protocols of nucleophilic trifluoroalkylations since 2006 will be discussed.

2.1 Nucleophilic Metal-Free Trifluoromethylation Reactions

2.1.1 (Trifluoromethyl)Trimethylsilane as a Nucleophilic Trifluoromethyl Source

$TMSCF_3$ has been widely utilized in nucleophilic trifluoromethylation reactions of a wide range of electrophilic functionalities under mild conditions. The carbonyl moiety is one of the most widely studied functional groups in nucleophilic fluoroalkylations, and the development of selective trifluoromethylation in the presence of other reactive functional groups has prospered in the past two decades. Despite the existence of an amide functionality in the molecule, a highly diastereoselective trifluoromethylation on the ketone carbonyl site in a tartaric acid–derived ketoamide **1** has been reported by Portella et al.[22] It has been shown that the diastereoselectivities are generally higher for aromatic/heteroaromatic substrates than for aliphatic substrates. This one-pot procedure has been subsequently applied to the synthesis of all four enantiopure stereoisomers of 2-(trifluoromethyl) tetrahydronaphthalene-1,2-diols and their 2-O-allyl derivatives **3** possessing potential antiinflammatory bioactivities.[23] In the presence of an additional imine group, optically pure lactones **4** containing an imine group also underwent diastereoselective trifluoromethylation exclusively on the carbonyl functionality followed by stereoselective reduction with $LiBH_4$ to afford *anti*-amino diols **6** as the major isomers with good selectivities.[24] On the other hand, the *syn*-amino diols **8** were synthesized by pretreating the starting substrates **4** to form **7** with the amino group protected by a benzyl group. With the same protocol, optically pure fluorinated quaternary piperidines were also prepared from a chiral iminolactone.[25] Furthermore, selective reaction with the carbonyl group of acyclic imino ketones **9** afforded β-amino-α-trifluoromethyl alcohol derivatives **10** in high yields via a one-pot nucleophilic trifluoromethylation, followed by subsequent desilylation and selective reduction of the C=N bond with $NaBH_4$ or $LiAlH_4$.[26] It is worth mentioning that by the use of 1,1,1,3,3-pentafluorobutane (Solkane 365mfc) as a novel environmentally benign solvent, Shibata and coworkers[27] have obtained high diastereoselective trifluoromethylation of oxazolidine-5-one **11** with $TMSCF_3$ in excellent yield. Furthermore, α,β-unsaturated ketones are good substrates to construct trifluoromethylated allylic alcohols. Kobayashi and coworkers have reported the selective trifluoromethylation of the ketone carbonyl group of **13** in the presence of another ester functionality in a highly diastereoselective fashion with moderate yields.[28] Later, Levin et al. have reported the preparation of trifluoromethylated allylic alcohol ethers **16** from acylated Baylis–Hillman adducts **15**.[29] It has been found that no Michael addition products were observed under these reaction conditions. Starting from propargyl ketone (**17**), trifluoromethylated polyalcohols **18** were prepared efficiently from **17** and $TMSCF_3$ in moderate yields.[30] In 2013, Pohmakotr et al. have demonstrated the selective trifluoromethylation of only one carbonyl group of anhydride **19** with $TMSCF_3$ to afford γ-trifluoromethylated-γ-butyrolactone **21** upon subsequent treatment with Grignard reagents (see Scheme 11.3).[31]

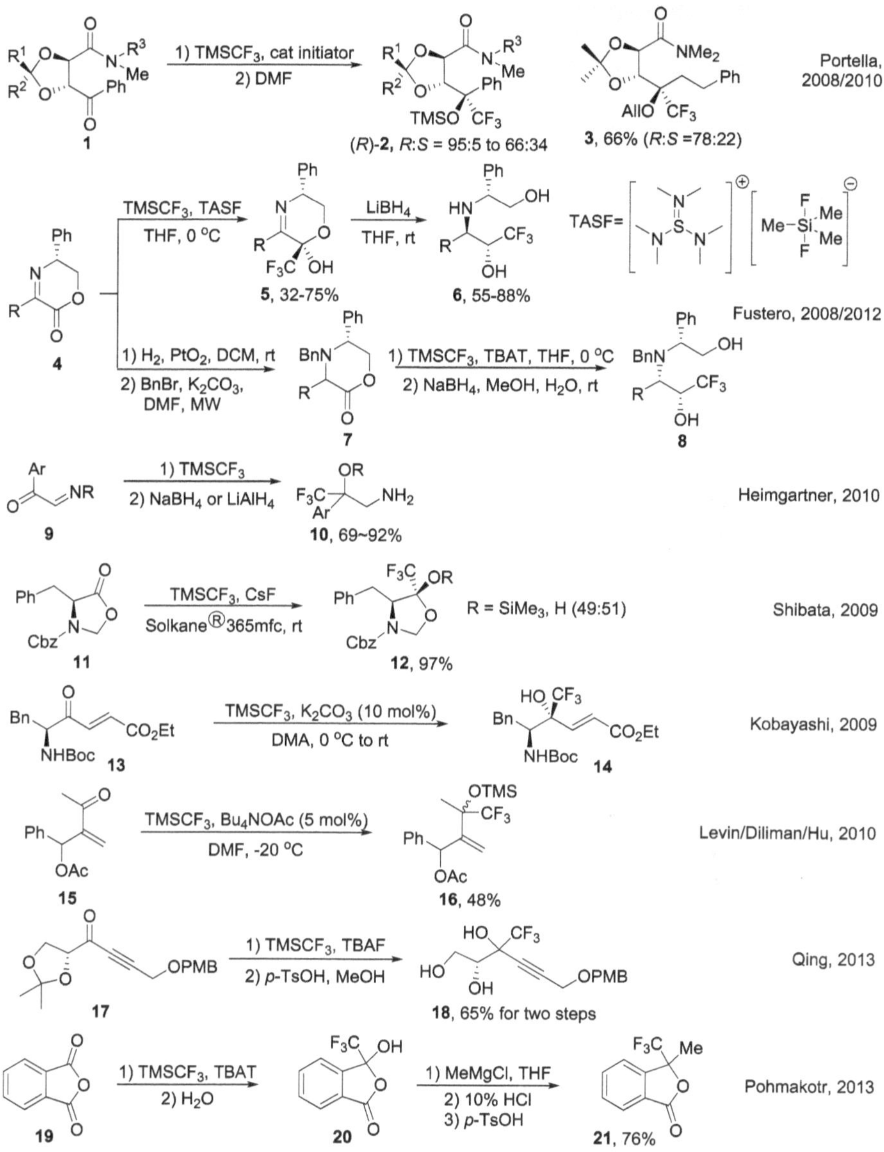

Scheme 11.3 Nucleophilic trifluoromethylation of selective carbonyl groups in the presence of other reactive functionalities using $TMSCF_3$.

The catalytic enantioselective nucleophilic trifluoromethylation of nonchiral carbonyl compounds is one of the most important methods for the preparation of optically active trifluoromethylated alcohols.[32] Cinchona alkaloid–derived catalysts are privileged chiral scaffolds in asymmetric catalysis[33] and have been employed in the catalytic enantioselective nucleophilic trifluoromethylations. The first enantioselective trifluoromethylation of carbonyl compounds was developed in 1994 by Kobayash et al.[34] and Yudin,[35] independently using $TMSCF_3$ and cinchona alkaloid–derived chiral ammonium fluorides as phase transfer catalysts. Although this system provided the optically active trifluoroethanol in excellent yields (87–99%), the enantioselectivity was poor (enantiomeric excess (*ee*) 15–51%). In 2007, Feng et al. used 1,1′-bi-2-naphthol (BINOL)/sodium salt **23** as a cocatalyst for the cinchona alkaloid derivative **22** to furnish trifluoromethylated alcohols with higher enantiomeric excess.[36] The disodium (*R*)-binaphtholate intermediate was proposed to serve as Lewis base to activate $TMSCF_3$. Later, the same group employed sodium hydride (NaH) as a Lewis base additive in the trifluoromethylation of aromatic ketones with **24** as a catalyst in high yields and good enantioselectivities.[37] Acting as both a catalyst and Lewis base initiator, chiral cinchonidine-derived quaternary ammonium phenoxide **27** was applied in the trifluoromethylation of ketones and α-ketoeseters in high yields and good enantioselectivities.[38] Under the initiation of tetramethylammonium fluoride (TMAF), a novel cinchona alkaloid–derived catalyst **25** was used for the enantioselective trifluoromethylation of acyclic and cyclic aryl ketones. The trifluoromethylated alcohols with a tetrasubstituted carbon atom were obtained in excellent yields with enantioselectivities up to 94%.[39] Aside from cinchona alkaloid derivatives, Shibata et al. employed chiral crown ether **26** as the catalyst for the trifluoromethylation of carbonyl compounds with modest enantioselectivity.[40] Crown ether **26** was easily synthesized from 3-iodobinaphthol and used together with catalytic amount of potassium phenoxide as a Lewis base. In 2013, the same group described the catalytic enantioselective trifluoromethylation of aromatic aldehydes using sterically demanding cinchona alkaloid–derived phase transfer catalyst **28** and TMAF as an initiator.[41] This methodology provided medicinally important α-trifluoromethyl alcohols in high chemical yields with moderate enantioselectivities (see Scheme 11.4).

Aside from carbonyl compounds, imines readily undergo nucleophilic trifluoromethylations to form a variety of trifluoromethylated amines. Dilman and coworkers have developed the nucleophilic trifluoromethylation of a series of aromatic imines **29** promoted by hydrofluoric acid generated in situ from KHF_2 and trifluoroacetic acid.[42] This protocol resulted in a new chemoselective trifluoromethylation since the C=N bond was found to be more reactive than the carbonyl group of **31** under developed reaction conditions. In contrast, the selective trifluoromethylation of only the carbonyl group was realized through the conventional reaction condition with tetrabutyl ammonium fluoride as an initiator. Later, a new organocatalytic protocol employing a phase-transfer system of tetra-*N*-butylammonium bromide (TBAB) and sodium phenoxide for the trifluoromethylation of imines **34** or amines **35** has been reported by Bernardi and coworkers.[43] The general applicability of this new reaction was demonstrated by the applications toward other silicon nucleophiles, such as $TMSC_6F_5$ and $TMSCH_2CN$. Sugar-derived cyclized nitrones **38** also reacted with $TMSCF_3$ to afford trifluoromethylated polyhydroxypyrrolidine derivatives **39**, which were further converted to trifluoromethylated analogs of the natural product 6-deoxy-2,5-dihydroxymethyl-3,4-dihydroxypyrrolidine **40** via cleavage of the N–O bond and removal of protecting groups.[44] The introduction of a trifluoromethyl group may significantly improve the bioavailability of polyhydroxypyrrolidines. In 2014,

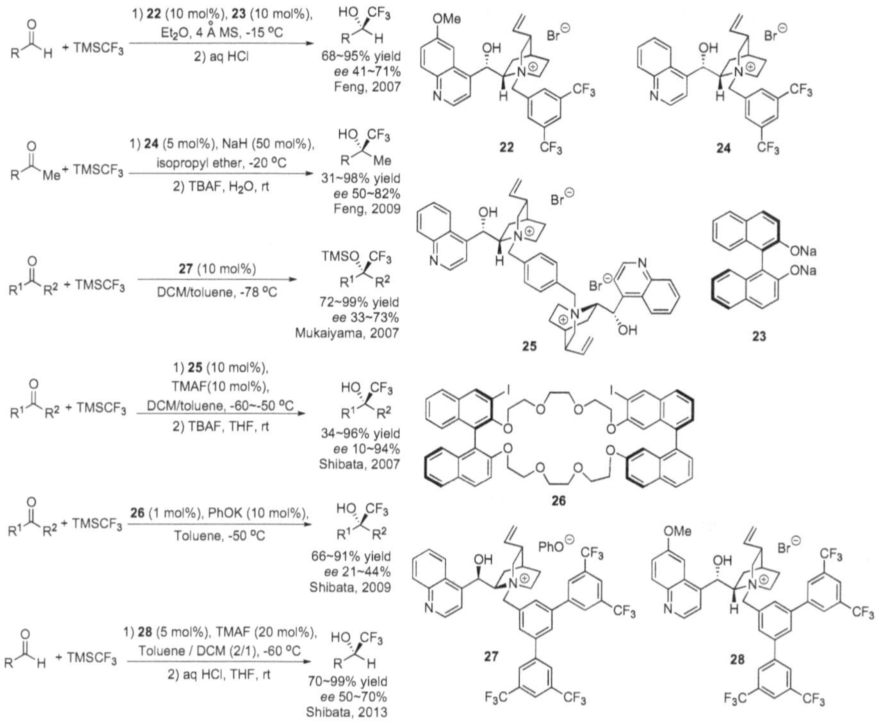

Scheme 11.4 Enantioselective nucleophilic trifluoromethylation of carbonyl compounds using $TMSCF_3$.

Larionov et al. developed a novel methodology to prepare 2-perfluoroalkyl/aryl-substituted nitrogenous heterocycles **42** from heterocyclic *N*-oxides **41** and perfluoroalkylsilanes (including $TMSCF_3$) in good to excellent yields.[45] This transformation was used to access fluorinated heterocyclic scaffolds of practical synthetic and medicinal significance, as demonstrated by the synthesis of two new fluorine-containing analogs of the antimicrobial drug, chlorquinaldol **43b** and **43c** (see Scheme 11.5).

A highly stereoselective tandem nucleophilic trifluoromethylation/intramolecular aza-Michael reaction of sulfinimine derivatives **44** gave trifluoromethylated 1,3-disubstituted isoindolines **45** in good yields and diastereoselectivities.[46] First, the catalytic enantioselective trifluoromethylation of imine equivalents was accomplished by Shibata et al. using cinchonium salt **24** or **47** as a catalyst and azomethine imines **46** as starting substrates to afford trifluoromethylated amines **49** in good yields and enantioselectivities.[47] Later, the same group applied the environmentally benign solvent Solkane 365mfc to this reaction with a novel catalyst, iododerivative of cinchoninium salt **48** to give high yields and enantioselectivities (see Scheme 11.6).[48] The trifluoromethylated adducts were readily converted into pharmaceutically important trifluoromethylated amines.

Trifluoromethylations of electrophilic alkenes provide efficient synthesis of a variety of β-functionalized trifluoromethylated compounds (see Scheme 11.7). Dilman and coworkers have elaborated a convenient method for the trifluoromethylation of

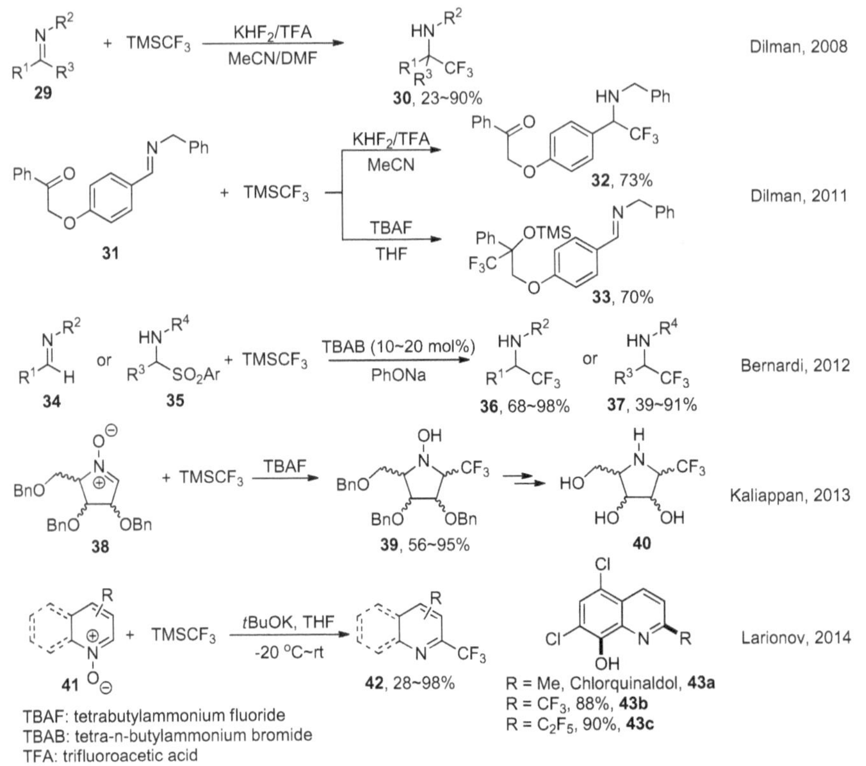

Scheme 11.5 Nucleophilic trifluoromethylation of imine derivatives using $TMSCF_3$.

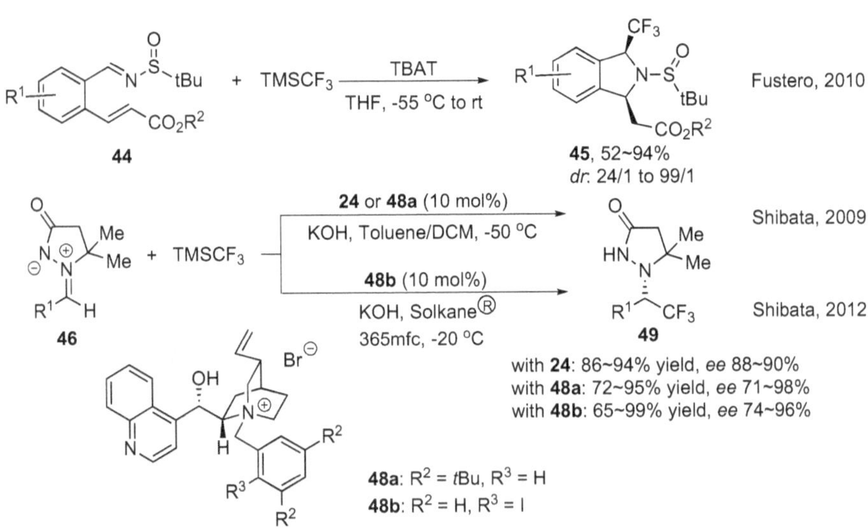

Scheme 11.6 Diastereo- and enantioselective nucleophilic trifluoromethylation of imine derivatives using $TMSCF_3$.

a range of arylidenemalononitriles **50** with $TMSCF_3$ using sodium acetate as a Lewis base in high yields.[49] It is interesting to note that in the case of a substrate possessing two sites for possible nucleophilic attack, only one product was obtained, in which the position adjacent to the malononitrile moiety was trifluoromethylated. β-Trifluoromethylated alcohols **54** and esters **55** were prepared from arylidene Meldrum's acids **52** and $TMSCF_3$ followed by subsequent transformations via a one-pot reaction in good overall yields.[50] In addition to C=C functionality, a nucleophilic *S*-trifluoromethylation of dithioesters bearing a nucleofugal group at α-position has been developed by Portella et al.[51] While the majority of S–CF_3 bond constructions are via electrophilic trifluoromethylation pathway, dithioesters **56** readily reacted with $TMSCF_3$ under the initiation of fluoride to afford unprecedented S–CF_3 ketene dithioacetals via a tandem S-trifluoromethylation–β-elimination process.

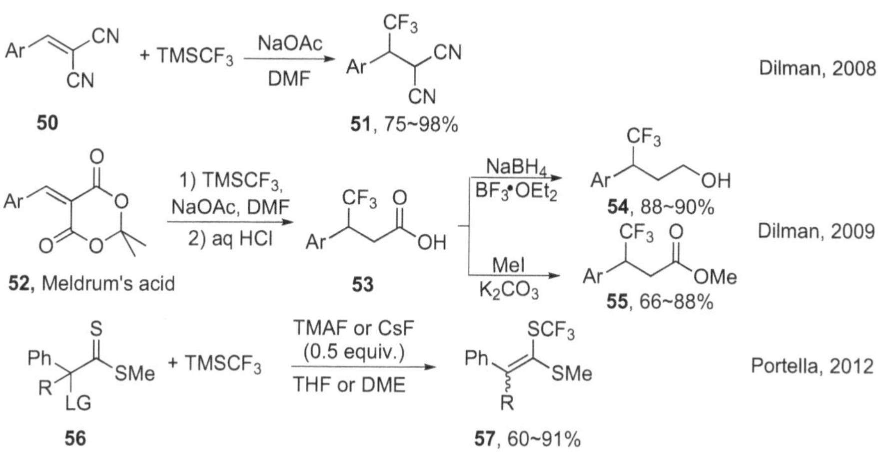

Scheme 11.7 Nucleophilic trifluoromethylations of alkenes and dithioesters.

2.1.2 CF_3H as a Trifluoromethyl Source

Fluoroform is a large-volume by-product in the manufacture of Teflon, refrigerants, fire extinguishing agents, and foams. It is also a potent and stable greenhouse gas that has an estimated global warming potential 11,700 times greater than that of CO_2.[52] Despite its earlier utilization as the trifluoromethyl source by Shono,[16] Normant et al.,[18] Langlois et al.,[19] and others,[20] its application in direct nucleophilic trifluoromethylation has received little attention due to its low boiling point (−83°C), high pK_a (25–28 in water), and strong basicity of the CF_3^- anion. Many previously reported trifluoromethylations using CF_3H suggested the necessity and importance of dimethylformamide (DMF) as a CF_3^- reservoir[17,18]; nevertheless, Prakash and coworkers have reported the direct nucleophilic trifluoromethylations of silicon, boron, sulfur, and carbon-based electrophiles using stoichiometric amount of fluoroform in common solvents other than DMF (see Scheme 11.8).[53] Synthesis of the Ruppert–Prakash reagent was easily achieved from fluoroform instead of ozone-depleting CF_3Br as the starting material. In 2013, Shibata et al. has reported the direct trifluoromethylation of aromatic aldehydes with CF_3H using *t*Bu-P_4 **59** organic base in good to excellent yields.[54] Unfortunately, the reaction system was not applicable for the trifluoromethylation of aliphatic aldehydes. Similarly, Mikami and collaborators have applied the organic base protocol to both aromatic and aliphatic carbonyl, ester, acid halide, epoxide, and carbon dioxide substrates to afford the corresponding products in good overall yields even under organocatalytic conditions.[55] Furthermore, [^{18}F]-trifluoromethane has been successfully prepared from difluoroiodomethane **60** and [^{18}F]fluoride/kryptofix 2.2.2 in a fast and efficient fashion at room temperature.[56] A variety of [^{18}F]-trifluoromethyl-containing carbinols **62** were synthesized from [^{18}F]-trifluoromethane **61** with *t*BuOK in moderate to high yields. The easy access to radioactive [^{18}F]-containing derivatives allowed its application in positron emission tomography imaging (Scheme 11.8). It is worth mentioning that chloroform, a haloform, was employed by Shibata et al. in 2015 in the enantioselective trichloromethylation of Morita–Baylis–Hillman (MBH)-type allylic fluoride analogs, in which $TMSCF_3$ played a very interesting role.[57] Trifluoromethanide anion generated in situ from $TMSCF_3$ deprotonated chloroform to generate CCl_3^-.

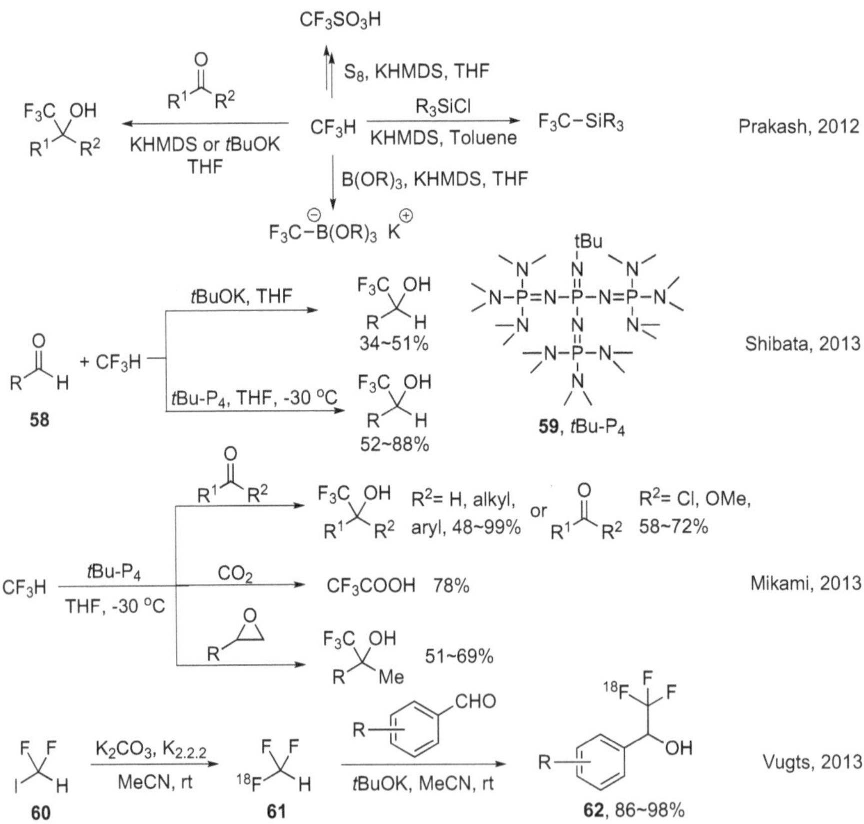

Scheme 11.8 Nucleophilic trifluoromethylations using CF_3H.

2.1.3 Sulfur-Derived Trifluoromethylating Reagents

The sulfur-derived fluorinating and fluoroalkylating reagents have rapidly grown in the 2000s and are applicable to various electrophiles including carbonyl compounds, imine derivatives, aziridines, epoxides, and olefins (see Scheme 11.9).[9b] The introduction of sulfur-based auxiliary group not only facilitates the generation of nucleophilic fluorinated carbanion species and increases the stability of the carbanion intermediates but also allows further transformation via reductive desulfonylation or elimination. For example, trifluoromethyl phenyl sulfoxide and sulfone **64**, being the most widely used sulfur-based nucleophilic trifluoromethylating reagents, can readily release trifluoromethyl anion upon treatment with a nucleophile such as alkoxide owing to the strong S−O bond formation and the high polarity of the C−S bond. In addition to carbonyl compounds, diastereoselective preparation of trifluoromethyl amines can be achieved from trifluoromethyl phenyl sulfone in good yields.[58] It has been shown that trifluoromethyl sulfinamides were also obtained in high diastereoselectivity from their corresponding sulfinimine precursors via this protocol. As a complementary route, an unprecedented reductive nucleophilic trifluoromethylation of aldehydes using phenyl trifluoromethyl sulfone has been developed by Hu and coworkers. This new procedure allowed efficient trifluoromethylation of nonenolizable or enolizable aldehydes **66** to afford trifluoromethylated alcohols **67** in good overall yields.[59] In 2013, Sakavuyi and Petersen have reported a mild process for the conjugate addition of the trifluoromethanide anion to activated Michael acceptors such as Meldrum's acid derivatives **68** and arylidenemalononitriles **71** via a similar reductive trifluoromethylation fashion.[60] This new methodology has been applied to the preparation of befloxatone **73**, which is a reversible and selective monoamine oxidase A inhibitor. Although *S*-trifluoromethyl sulfoximines are generally utilized as electrophilic trifluoromethylating reagents, in 2014, Hu et al. found that aldehyde **74** underwent direct nucleophilic trifluoromethylation with *N*-tosyl sulfoximine **75** in the presence of sodium phenolate as a nucleophile (Scheme 11.9).[61]

2.1.4 Trifluoromethylations Using Phosphonate-, Hexafluoroacetone-, Trifluoroacetaldehyde, and Zinc-Based Reagents

A new alkoxide-induced nucleophilic trifluoromethylation of a variety of carbonyl compounds has been developed by Beier et al. using diethyl trifluoromethylphosphonate **77** in moderate to good yields.[62] Although phosphonate **77** does not surpass $TMSCF_3$ in terms of availability and substrate range applicability, it expands the array of available trifluoromethylating reagents with a new phosphorus-containing member. Later, Colby et al. synthesized an air-stable amidinate salt of hexafluoroacetone hydrate **78** as a novel nucleophilic trifluoromethylating reagent toward a range of aromatic carbonyl compounds.[63] Subsequently, Prakash and coworkers developed an efficient and more atom-economical method employing trifluoroacetaldehyde hydrate **79** for the trifluoromethylation of carbonyl compounds.[64] This method has not only provided a facile synthetic access toward α-trifluoromethyl alcohols but also allowed maximum utilization of the CF_3 moiety in the precursor, in an atom-economical way (Scheme 11.10). In 2015,

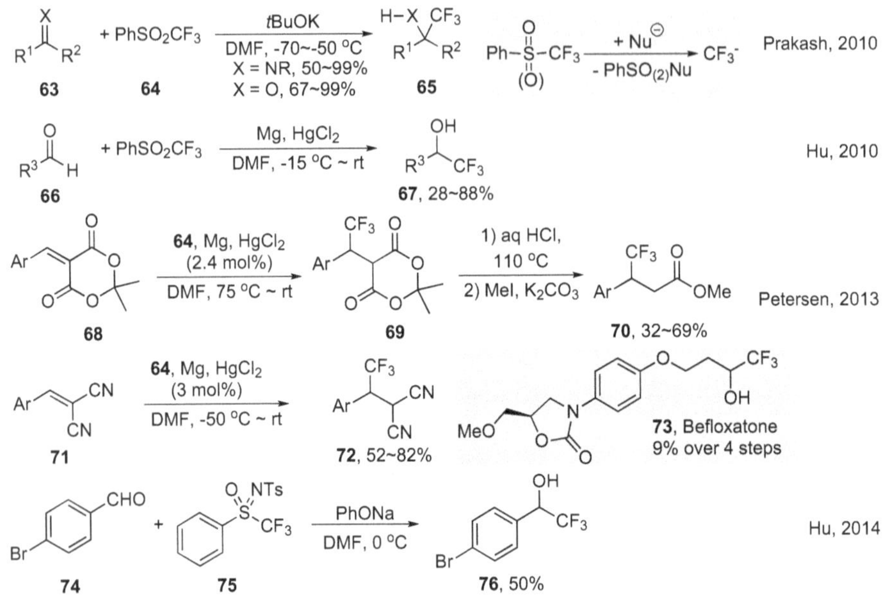

Scheme 11.9 Nucleophilic trifluoromethylations with sulfur-derived reagents.

Mikami and coworkers developed an air-stable zinc-based trifluoromethylating reagent $Zn(CF_3)_2$[*N*,*N*,*N′*,*N′*-tetramethylethylenediamine (TMEDA)] **80b** prepared from CF_3I, $ZnEt_2$, and TMEDA, which was utilized as either a trifluoromethyl anion source with ammonium salt as an initiator or a difluorocarbene source for the *gem*-difluoroolefination of carbonyl compounds in the presence of phosphine (Scheme 11.10).[65]

2.2 Metal-Assisted Nucleophilic Trifluoromethylation Reactions

2.2.1 Copper-Mediated Trifluoromethylations

Copper has been the most efficient and widely utilized metal in metal-mediated trifluoromethylation reactions. In 1991, Fuchikami and coworkers reported the cross-coupling reactions between alkenyl and aryl halides and $CuCF_3$ complex, which was in situ prepared from $TMSCF_3$ in the presence of KF and copper(I) iodide at 80°C in DMF.[66] Other copper(I) species, including cyanide, bromide, and chloride were found to be equally effective; however, copper(I) species did not work catalytically under developed reaction conditions. (Trifluoromethyl)copper complex can be prepared via various methods, among which the transfer of CF_3 group to a variety of electrophiles with nucleophilic CF_3 sources will be discussed in this section.

Ruppert–Prakash reagent, $TMSCF_3$ and other trifluoromethyltrialkylsilanes have been found to be suitable in the copper-mediated trifluoromethylations toward aryl halides (C_{sp2}–CF_3 bond) since 1991 with great developments taking place in the 2000s (Scheme 11.11). In 2008, Vicic et al. reported the synthesis and characterization of first N-heterocyclic carbene (NHC)-stabilized CF_3Cu intermediate **81**, which readily reacts

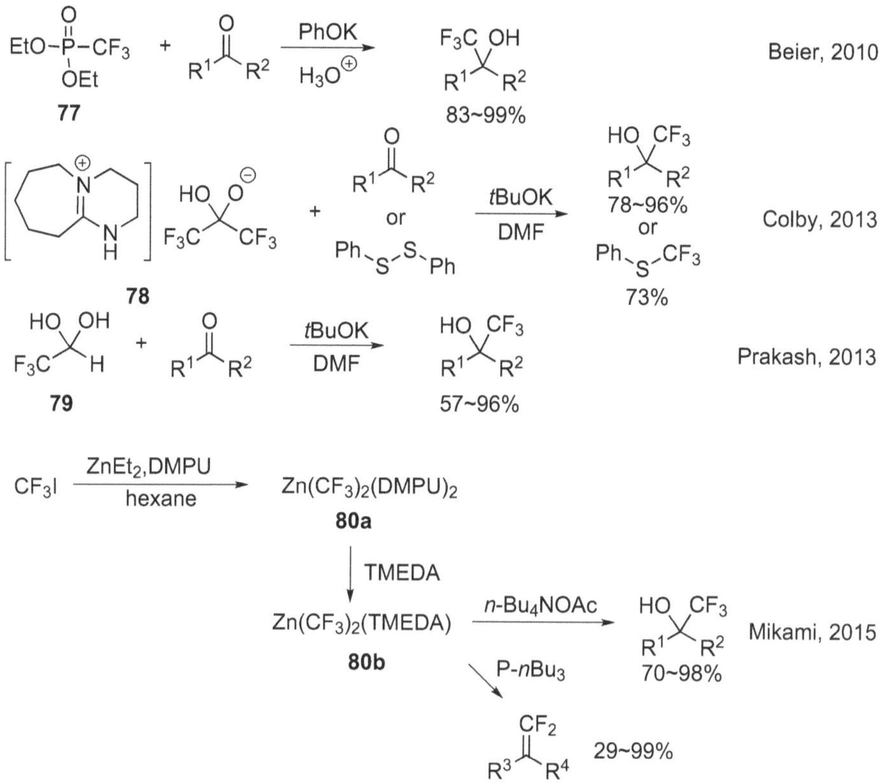

Scheme 11.10 Nucleophilic trifluoromethylations with phosphonate-, hexafluoroacetone-, and trifluoroacetaldehyde-derived and zinc-based reagents.

with aryl halides to give quantitative yields of trifluoromethylated products.[67] In 2009, Amii and coworkers reported the catalytic trifluoromethylation of aryl iodides with trifluorotriethylsilane as a nucleophilic CF_3 source and 1,10-phenanthroline as the ligand.[68] Later, Hartwig et al.[69] and Grushin et al.[70] developed similar trifluoromethylations employing 1,10-phenanthroline (**82**) and triphenylphosphine (**83**) as ligands, respectively. The reactions were conducted with either the isolated (trifluoromethyl)copper(I) complex or generated in situ from copper halides and $TMSCF_3$. In 2010, Qing et al. reported the first copper-mediated oxidative trifluoromethylation of terminal alkynes using **84** with air as the oxidant.[71] This protocol was later improved by the same group using only catalytic amount of copper.[72] Subsequently, they have applied this novel oxidative trifluoromethylation protocol to heteroarenes **85** via C–H activation.[73] Both air and di-*t*-butyl peroxide were used as oxidants for 1,3,4-oxadiazoles, but only di-*t*-butyl peroxide worked for 1,3-azoles and perfluoroarenes, whereas only silver carbonate worked for the C2-trifluoromethylation of indoles. In 2012, Nishibayashi et al. demonstrated a regioselective reaction between allylic bromide **86** and $TMSCF_3$ with CuTc (Tc = thiophene-2-carboxylate) as a catalyst.[74] Later, the same group applied this new methodology to the trifluoromethylation of propargyl halides **87** in high yields under mild conditions.[75] In

2014, Yu and coworkers have reported a copper(II) mediated *ortho*-trifluoromethylation of (hetero)aromatic amides **88**, bearing an oxazolyl-directing group using $TMSCF_3$ via C—H activation within 30 min (Scheme 11.11).[76] Mechanistic study supported a C—H activation pathway rather than an electrophilic aromatic substitution or a radical pathway.

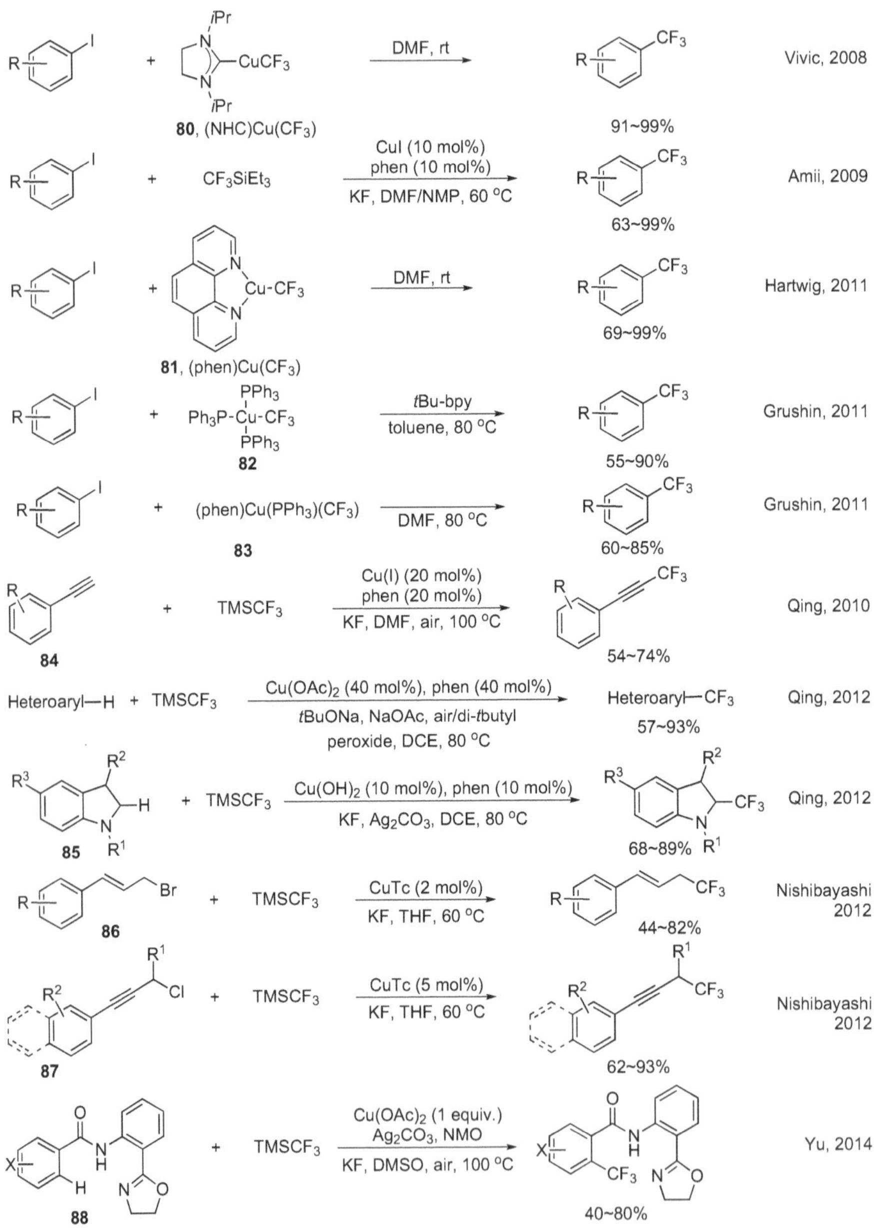

Scheme 11.11 Copper-mediated nucleophilic trifluoromethylations using (trifluoromethyl)trialkylsilanes and related reagents.

Catalytic enantioselective copper-mediated trifluoromethylation of aromatic aldehydes has been receiving increasing attention using cinchona alkaloid-derived catalysts. In 2013, Chen and coworkers developed the enantioselective protocol using quinidine-derived quaternary ammonium salt **22** as a catalyst and (IPr)CuF ([IPr = 1,3-*bis*(2′,6′-di*iso*propylphenyl)imidazol-2-ylidene]) as the copper source (Scheme 11.12).[77] This reaction proceeded via the key intermediate [(IPr)CuCF_3], which readily transferred the trifluoromethyl group to a wide range of aromatic aldehydes in high levels of enantiomeric excess.

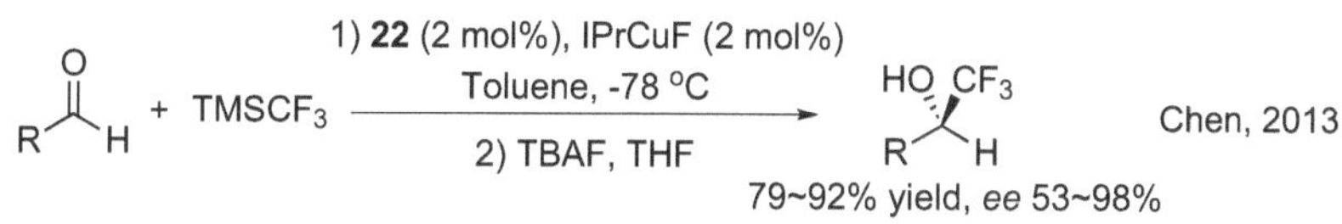

Scheme 11.12 Enantioselective copper-catalyzed nucleophilic trifluoromethylations using $TMSCF_3$.

Grushin and coworkers have reported a direct cupration of fluoroform from readily available and inexpensive CuCl and *t*BuOK without addition of any ligands in nearly quantitative yield (Scheme 11.13).[78] The generated $CuCF_3$ species were efficiently stabilized with triethylamine trihydrofluoride to furnish $CuCF_3$ reagents that did not decompose significantly for at least 1−2 days at room temperature. These reagents were used to transfer the trifluoromethyl group to another metal/element and for the trifluoromethylations of organic electrophiles such as (hetero)aryl and alkyl halides. Later, the same group applied this new protocol for the trifluoromethylation of a variety of α-haloketones with the carbonyl group remaining intact for further transformations (Scheme 11.13).[79] The products were obtained smoothly at room temperature in moderate to high yields with unprecedented functional group tolerance.

In 2015, Mikami and coworkers reported the aromatic perfluoroalkylation mediated by CuI with novel bis(trifluoroalkyl)zinc reagent **89**, which was prepared from trifluoroalkyl iodide as the trifluoromethyl source, 1,3-dimethyl-3,4,5,6-tetrahydro-2-pyrimidinone (DMPU), and diethylzinc.[80] The air-stable and easy-to-handle reagent can be utilized in the absence of a ligand or an activator, applicable not only for the trifluoromethylation but also the perfluoroalkylation of aromatic and vinylic iodide substrates (Scheme 11.13).

Trifluoroacetaldehyde derivatives and trifluoroacetates are also good nucleophilic CF_3 sources to prepare key intermediate $CuCF_3$ in situ (Scheme 11.14). Amii and coworkers have employed trifluoromethylated hemiaminal **90** derived from trifluoroacetaldehyde as novel trifluoromethylating reagent with a high tolerance of functionalities under mild conditions.[81] Alkyl and metal trifluoroacetates have been used as inexpensive and efficient CF_3 source for the in situ generation of CF_3Cu by Langlois and Roques,[82] Beller et al.,[83] Duan et al.,[84] Chen and Buchwald,[85] and others. In the presence of cesium fluoride or cesium chloride, methyl trifluoroacetate **91** constituted a valuable trifluoromethylating agent in the Cu(I)-mediated trifluoromethylation of (hetero)aromatic iodides and bromides. It has been declared that methyl trifluoroacetate is easier to access than methyl chlorodifluoroacetate, previously used by Chen et al.[86] and McNeil and Burton[87] as a trifluoromethylating agent, considering that chlorodifluoroacetic derivatives are usually obtained as by-products in the manufacture of

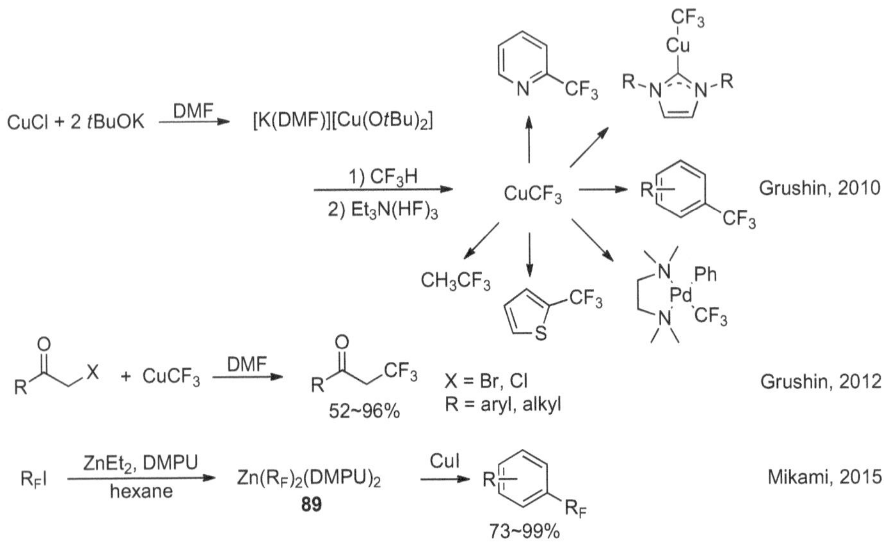

Scheme 11.13 Copper-mediated trifluoromethylations using CF_3H and CF_3I.

trifluoroacetic derivatives on an industrial scale. From sodium trifluoroacetate **92a**, a practical Cu-catalyzed ligand-free decarboxylative trifluoromethylation of aryl iodides has been developed by Duan et al. A variety of trifluoromethyl-substituted aromatics were obtained in moderate to excellent yields with great tolerance of functional group under relatively mild conditions. Buchwald and coworkers have developed a novel operationally simple method by applying flow chemistry to the copper-mediated aromatic trifluoromethylation. A broad spectrum of trifluoromethylated (hetero)aromatic compounds were prepared from CF_3CO_2K **92b** and corresponding (hetero)aryl iodides with this protocol requiring only minutes of residence time (Scheme 11.14). However, high temperature was often necessary for these transformations.

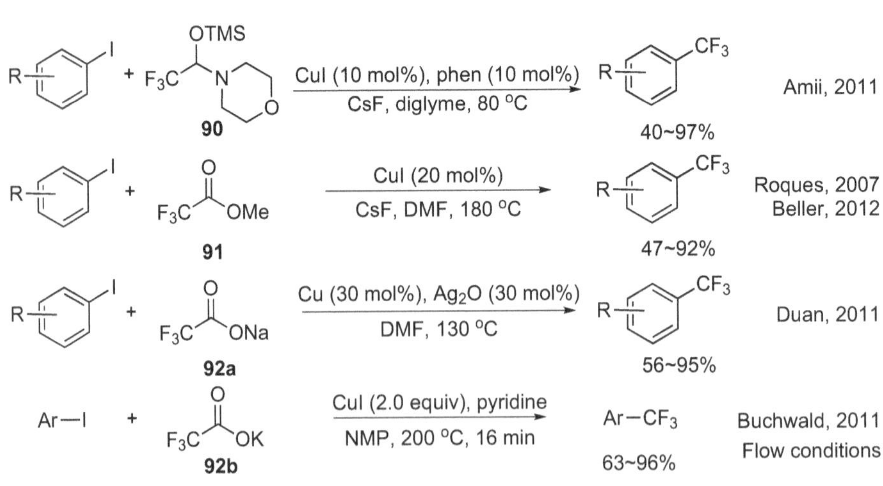

Scheme 11.14 Copper-mediated trifluoromethylations using trifluoroacetaldehyde and acetate derivatives.

In 2011, Goossen et al. have prepared potassium (trifluoromethyl)trialkoxyborates **93** as a storage-stable trifluoromethylating reagent (Scheme 11.15).[88] Both electron-deficient and electron-rich aryl/heteroaryl halides were readily trifluoromethylated in high yields under mild and base-free conditions with CuI/1,10-phenanthroline complex as a catalyst. Later, trifluoromethyl ketone $PhCOCF_3$ was demonstrated as low-cost trifluoromethyl source for the generation of CF_3Cu in nearly quantitative yields.[89] With this protocol, terminal alkynes, arylboronic acids, and aryl iodides were trifluoromethylated to provide the corresponding products in good to high yields. Apart from direct transfer of the trifluoromethyl group, the CF_3Cu complex was also prepared from the in situ−generated trifluoromethyl group from difluorocarbene and fluoride. Altman et al. have developed a two-step copper-catalyzed decarboxylative protocol for converting allylic alcohols to trifluoroalkenes from NaO_2CCF_2Br and KF (Scheme 11.15).[90] Conjugated and nonconjugated substrates bearing a variety of functional groups afforded α-substituted trifluoromethylated products in moderate to good yields with excellent diastereoselectivities (Scheme 11.15). In 2015, Hu et al. developed the copper-mediated trifluoromethylation of aryl iodides and bromides with phenyl trifluoromethyl sulfoxide as the trifluoromethyl source to generate $CuCF_3$.[91] It is worth noting that $PhSOCF_3$ showed higher reactivity than $PhSO_2CF_3$ in the formation of $CuCF_3$ species, which has been found to be remarkably efficient in the trifluoromethylation of aryl iodides in the absence of additional ligands, while some activated aryl bromides can also be trifluoromethylated in moderate to good yields. Furthermore, the oxidative cross-coupling of terminal alkynes or arylboronic acids with the "$CuCF_3$" species has afforded the corresponding products in good yields (Scheme 11.15).

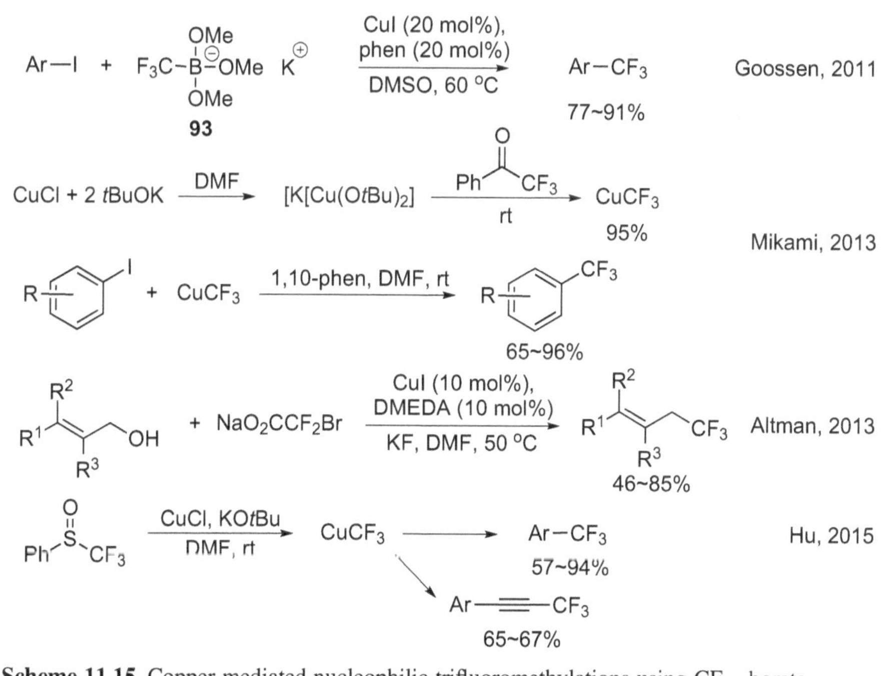

Scheme 11.15 Copper-mediated nucleophilic trifluoromethylations using CF_3−borate, CF_3−ketone, CF_3−sulfoxide, and difluorocarbene source.

2.2.2 Palladium-Catalyzed Trifluoromethylations

In 2010, Buchwald and coworkers reported the first palladium-catalyzed trifluoromethylation of aryl chlorides using trifluoromethyltriethylsilane as a nucleophilic CF_3 source (Scheme 11.16).[92] The bulky and electron-rich phoshphine ligand (BrettPhos, **94**) was the key to the success of this reaction. Both electron-deficient and electron-rich aryl/heteroaryl chlorides underwent trifluoromethylations in good yields. The method tolerated a wide scope of functional groups including esters, amides, ethers, acetals, nitriles, and tertiary amines, thus showing applicability to late-stage modifications of advanced intermediates. Later, the same group applied this methodology to the trifluoromethylation of vinyl sulfonates using a more hindered phosphine ligand (*t*BuXPhos, **96**).[93] Various cyclohexenyl triflates and nonaflates were trifluoromethylated in moderate yields, providing a direct method to access trifluoromethylated alkenes from vinyl electrophiles. Following the pioneering work by Buchwald, Samant et al. reported a similar palladium-catalyzed trifluoromethylation of aryl bromides using reverse micelles as the reaction medium.[94] The micellar conditions resulted in good yields and were applicable to bromoaromatics with ketone, aldehyde, hydroxyl, and amine functionalities. To realize milder reaction conditions, Liu et al. have developed a novel protocol of palladium-catalyzed oxidative trifluoromethylation of indoles **97** at room temperature with $TMSCF_3$.[95] $PhI(OAc)_2$ was found to be the best oxidant, and bidental nitrogen-containing ligand **98** largely increased the reaction yields. A wide scope of indoles were successfully transformed into the corresponding trifluoromethylated indoles in good yields. Later, the same group applied this methodology to the intramolecular oxidative trifluoromethylation of activated alkenes **99** to afford trifluoromethylated oxindoles **100** in good yields.[96] This reaction allows an easier access to construct a variety of bioactive molecules containing trifluoromethylated oxindole moieties. In addition to synthetic methodologies, elaborate mechanistic investigation has also been conducted. Sanford and coworkers reported a rational design of the generation of aryl$-CF_3$ via reductive elimination and the change of the oxidation state at the Pd center.[97] By changing the metal center of **101** from Pd(II) to Pd(IV), the reductive elimination toward the formation of aryl$-CF_3$ bond was proposed to accelerate. It has been found that the reaction temperature was largely decreased to room temperature from 80°C by switching the ligand from di-*t*-butylbipyridine to TMEDA (Scheme 11.16).

2.2.3 Silver-Mediated Trifluoromethylations

Despite the high affinity of silver(I) salts to halides, only a few examples of nucleophilic trifluoromethylation employing (trifluoromethyl)silver have been reported. In 2007, Yagupolskii and coworkers demonstrated the use of (trifluoromethyl)silver in the selective synthesis of trifluoromethyl ketones of aromatics, heteroaromatics, and olefins starting from the corresponding acyl chlorides (Scheme 11.17).[98] Later, Sanford et al. have reported the silver-mediated C$-$H trifluoromethylation of aromatic substrates using $TMSCF_3$ in good to excellent yields.[99] In 2013, Hu et al. demonstrated the use of (trifluoromethyl)silver in the one-step trifluoromethylation$-$iodination of

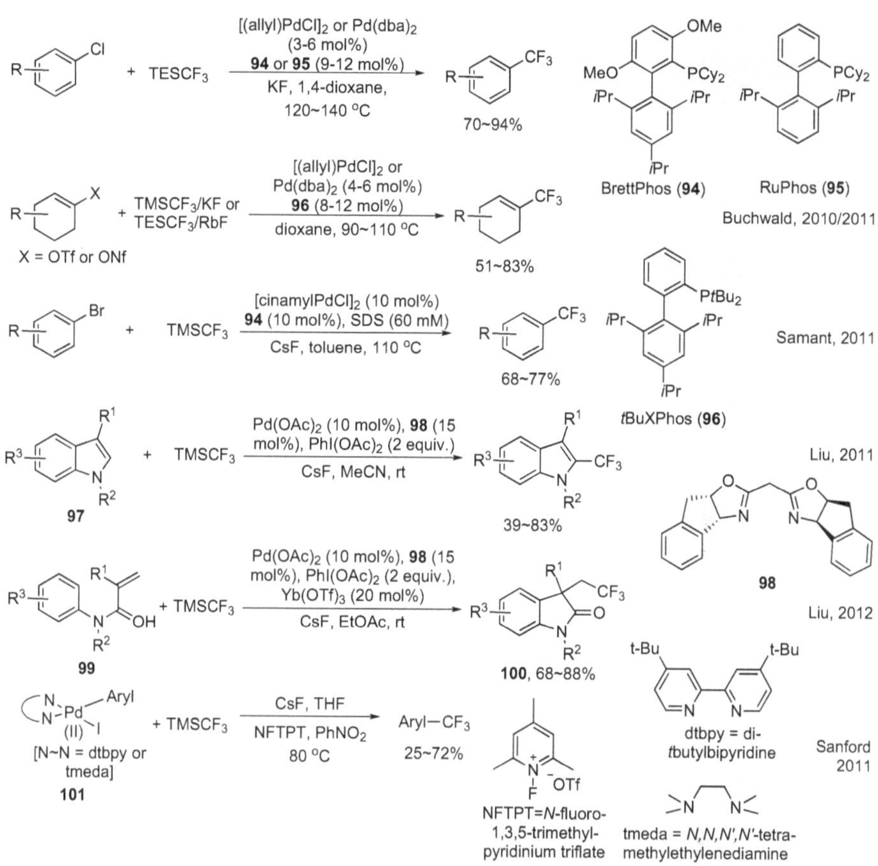

Scheme 11.16 Palladium-catalyzed nucleophilic trifluoromethylations.

arynes in situ generated from **102**.[100] While no reaction was observed between benzynes and (trifluoromethyl)zinc or (trifluoromethyl)copper species, (trifluoromethyl) silver readily added to the in situ–generated benzyne followed by iodination with 1-iodophenylacetylene to give *o*-trifluoromethyliodoarenes in modest to excellent yields. This method provides a simple and efficient route to various *o*-trifluoromethylated iodoarenes **103** that are otherwise difficult to synthesize.[99,101] With trifluoromethylated arynes as synthetic targets, Lee et al. have developed a new approach starting from nonaromatic building blocks, in which indoline and isoindoline derivatives **105** were synthesized from *bis*-1,3-diynes **104**.[102] Although the extra alkyne moiety on the *bis*-1,3-diynes is necessary to facilitate the hexadehydro-Diels–Alder reaction and thus compromises the general substrate scope, it does enable further elaboration to useful functional groups via simple manipulations such as semihydrogenation and ozonolysis. Another application of $AgCF_3$ was examined by Wang and coworkers, who have demonstrated the silver-mediated aromatic trifluoromethylation of arenediazonium salts **106** and successfully converted aromatic amino group into the trifluoromethyl group.[103] A wide range of anilines are suitable for this reaction at low temperature

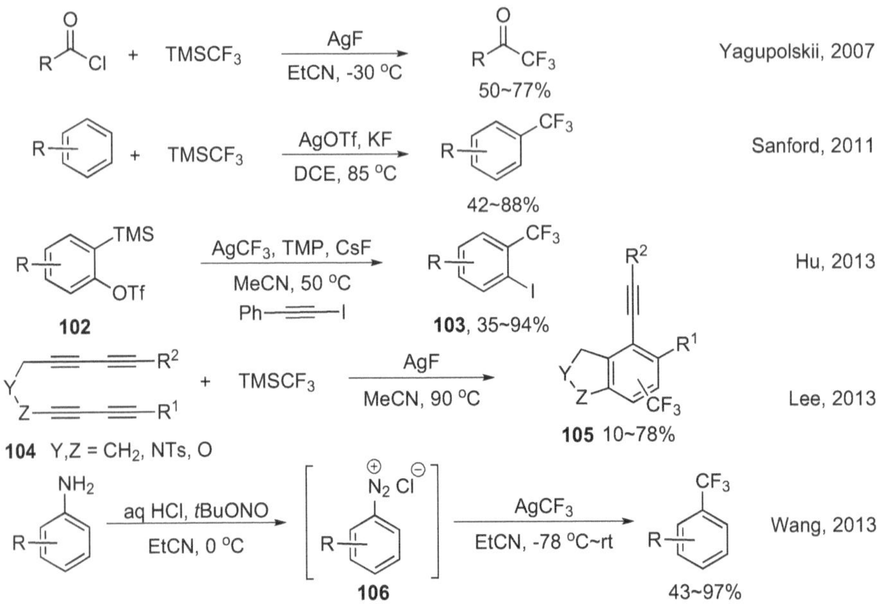

Scheme 11.17 Silver-mediated trifluoromethylations.

with great functional group tolerance (Scheme 11.17). The amino group was readily protected and deprotected, rendering this route suitable for late-stage introduction of a CF_3 group onto an aromatic motif.

2.3 Other Perfluoroalkylation Reactions

Unlike the labile trifluoromethyl alkali/alkaline earth metal species, perfluoroalkyllithium and magnesium bromide were used as nucleophilic perfluoroalkylating reagents back in the 1950s.[104] Nucleophilic perfluoroalkylation of aryl halides was first accomplished by Thrower in 1969, who prepared perfluoroalkylarenes from iodoperfluoroalkanes and copper.[105] Subsequently, many other nucleophilic perfluoroalkylating reagents have been developed, including perfluoroalkyl calcium, tin, zinc, and trimethylsilyl species.[106] In 2006, Dolbier et al. developed the novel tetrakis(dimethylamino)ethylene/perfluoroalkyl halide system, which was applicable for the nucleophilic perfluoroalkylation of aldehydes, ketones, and imines (Scheme 11.18).[107] Later, Prakash and coworkers demonstrated the alkoxide-induced pentafluoroethylation of carbonyl compounds and imines using sulfur-based reagent $PhSO_2C_2F_5$.[58] In 2011, Hartwig et al. reported the improved perfluoroalkylation of alkenyl and aryl halides using isolated perfluoroalkyl copper complexes [(phen)Cu($CF_2CF_2CF_3$)] **108** with phenanthroline as ligand,[69] which afforded both electron-rich and electron-poor perfluoroalkylated (hetero)arenes in excellent yields. Later, Dougulis et al. reported the reaction between aryl iodides and perfluoroalkanes with DMPU as solvent, zinc *bis*-2,2,6,6-tetramethylpiperidide (TMP_2Zn) as a base, and CuCl/phen as the catalyst (Scheme 11.18).[108]

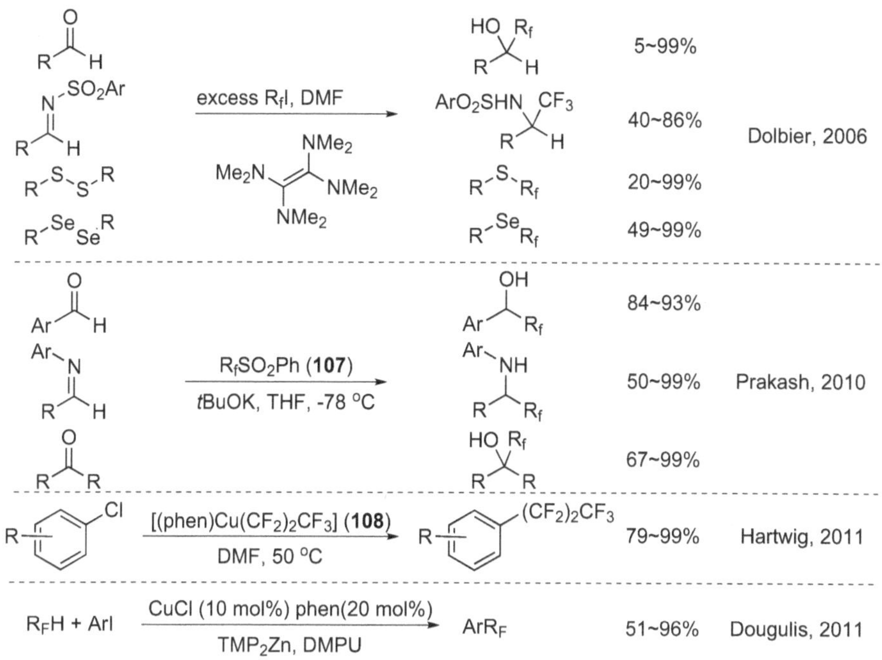

Scheme 11.18 Other nucleophilic perfluoroalkylations.

3. Difluoroalkylation Reactions

The difluoromethylene group ($-CF_2-$) is known to be isosteric and isopolar with the ethereal oxygen, which has potential applications in bioactive fluorinated analogs of oxygenated compounds such as sugars.[9e] On the other hand, the difluoromethyl group ($-CF_2H$) has been found to be a steric and electronic mimic of a hydroxyl group but a more lipophilic weak hydrogen bond donor. Over the past two decades, difluoromethyl- and difluoromethylene-containing organic compounds have been attracting increasing attention due to their diverse applications.[1d] However, compared with the well-investigated trifluoromethylations, the available reagents for direct difluoromethylation are largely limited to electrophilic-type reactions with lack of general applicability.[109] In this section, we will focus on the development of novel nucleophilic difluoroalkylation methods and reagents.

3.1 Nucleophilic Metal-Free Difluoromethylation and Difluoromethylenation Reactions

Aside from the silicon-based reagent, the phosphorous-based difluoromethylating reagent diethyl difluoromethylphosphonate **109** has been used for the nucleophilic introduction of difluoromethylphosphonate moiety into various electrophiles including carbonyl compounds, imines, carboxylic acid esters, as well as chlorides, unsaturated compounds, and triflates (Scheme 11.19). In 2008, Beier and coworkers applied this reagent to transfer difluoromethyl and difluoromethylene group to carbonyl compounds, affording difluoromethylated alcohols and 2,2-difluoro-1,3-diols, respectively.[110] Nevertheless, the drawback of this protocol is the use of strong bases, which limited the substrate scope to only nonenolizable carbonyl compounds and the requirement of isolation of intermediate phosphonates **110** before reacting with the second electrophile. To tackle these obstacles, **109** was modified by silylation to generate **111**, which readily transferred the CF_2 group to aldehydes and ketones with a catalytic amount of initiator in a one-pot reaction.[111] This methodology was compatible with nonenolizable and in some cases enolizable carbonyl compounds to afford 2,2-difluoro-1,3-diols in moderate to good overall yields. In addition, Prakash et al. have successfully accomplished the efficient synthesis of bis(difluoromethylene) triphosphoric acid from phosphonate **109**.[112] The triphosphoric acid has been used to synthesize interesting fluorinated DNA and RNA analogs, which can be used as probes to elucidate subtle differences in the catalytic activity of different enzymes. On the other hand, unlike the widely used Ruppert–Prakash reagent as trifluoromethylating reagent, $TMSCF_2H$ **112** has been widely believed to be an inefficient difluoromethylating agent, which requires harsh reaction conditions due to difficulty in achieving facile $Si-CF_2H$ bond cleavage. It was not until 2011 when Hu et al. made progress in the long-sought-after direct nucleophilic difluoromethylation.[113] By selecting a proper Lewis base activator and solvent, $TMSCF_2H$ efficiently transferred the difluoromethyl group to various aldehydes, ketones, and imines giving the corresponding products in good to excellent yields at room temperature or even at low temperatures. In 2014, Larionov and coworkers employed $TMSCF_2H$ in the regioselective direct fluoroalkylation reaction with heterocyclic *N*-oxides to afford 2-difluoroalkyl-substituted nitrogenous heterocycles in good to excellent yields.[45] As a member of the silicon-based difluoroalkylating reagent family, difluoro(trimethylsilyl)acetonitrile **113** efficiently transferred the cyanodifluoromethyl group to aldehydes, *N*-tosylimines, *N*-alkylimines, and enamines, whereas the nitrile group was further reduced to an amine group or transformed to an amide functionality (Scheme 11.19).[114]

As key reaction intermediates in nucleophilic difluoromethylations, difluoromethyl carbanion species tend to decompose via α-elimination of fluoride, which leads to significant competitive reactions diminishing the synthetic applicability of the intermediates. Therefore, various auxiliary groups possessing electron-withdrawing and electron-releasing substituents have been employed to modulate the stability and reactivity of these carbanion species. Another advantage of using auxiliary groups is the potential for further transformations of many auxiliaries. Aside from the previously mentioned silicon- and phosphorous-based auxiliary groups, sulfur-based functionalities have become one of the most widely used auxiliary groups. Therefore, α-difluorinated sulfides, sulfoxides, sulfones, and sulfoximines constitute a major domain for the nucleophilic introduction of a CF_2H or CF_2 group.[9b] The facile reductive cleavage of $S-CF_2$ bond and further transformations of the sulfur-based moieties add to their superior synthetic advantages.

Sulfide-based reagent $PhSCF_2TMS$ **114** has been used in difluoromethylation of a wide range of substrates since its first preparation in 2003[21a] and its first application in the difluoromethylation of aldehydes in 2005.[115] In addition to simple aldehydes and

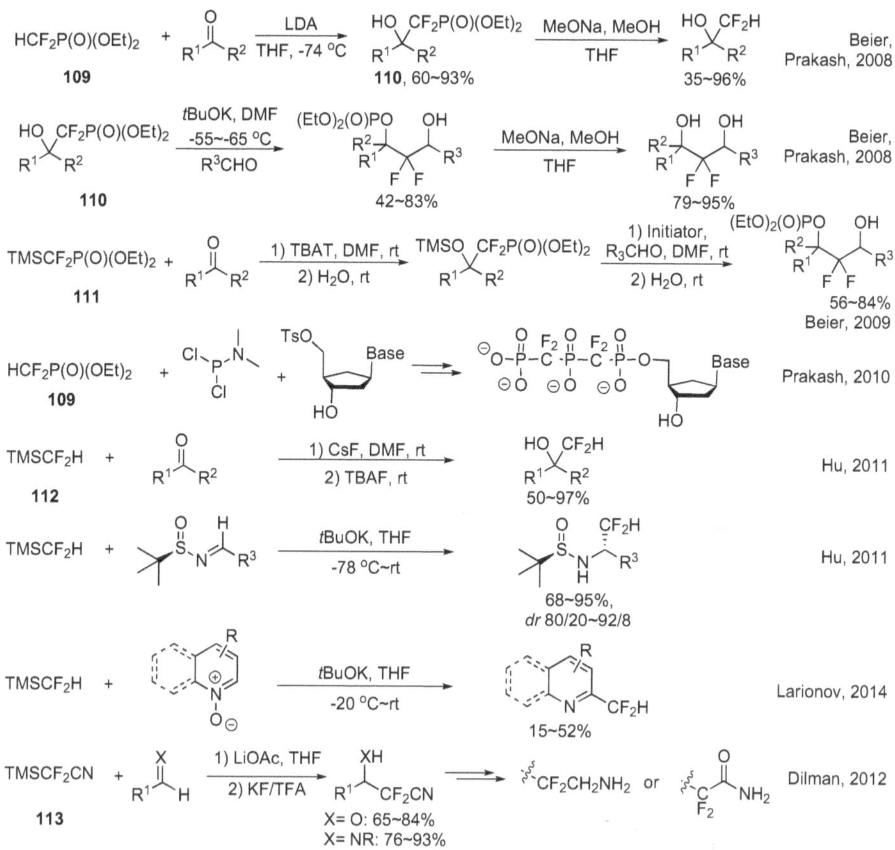

Scheme 11.19 Nucleophilic difluoromethylations using silicon and phosphorous-based reagents.

ketones, polycarbonyl compounds such as α- or γ-ketoesters,[116] cyclic imides,[117] and succinic anhydrides[118] also undergo difluoro(phenylthio)methylation with $PhSCF_2TMS$ in high yields (Scheme 11.20). Subsequent desulfenylation with Bu_3SnH in the presence of 2,2′-azobisisobutyronitrile or oxidative desulfonylation procedure gave the corresponding *gem*-difluoromethylated alcohols **116** in good yields. Aside from carbonyl compounds, imines and enamines are also good electrophiles for the nucleophilic difluoromethylations. Hu et al. have reported the diastereoselective reaction between (*R*)-(*N*-*t*-butylsulfinyl)imines and $PhSCF_2TMS$ to afford the corresponding products **117** in good yields and high diastereoselectivity.[119] The obtained $PhSCF_2$-containing sulfinamides were further transformed into chiral 2,4-*trans*-disubstituted 3,3-difluoropyrrolidines through an intramolecular radical cyclization pathway. *N*-alkylated imines and *N,N*-dialkylated enamines, which are unreactive toward fluoroalkylsilane reagents under conventional conditions,[42b] undergo reactions after activation by a Brønsted acid or an *N*-alkylating reagent to form the iminium salts. For instance, imines **118** and enamines **119** react with $PhSCF_2TMS$ under the promotion of hydrofluoric acid, which was generated in situ from potassium hydrodifluoride and trifluoroacetic acid. The difluoromethylated tertiary amine products **120** and **121** were thus obtained in good yields from **118** and **119**, respectively.[120] Alkyl halides such as primary alkyl bromides or alkyl iodides also undergo nucleophilic difluoromethylation with $PhSCF_2TMS$ under initiation of CsF in the presence of 15-crown-5 ether as a critical additive.[121] The thus formed $PhSCF_2$-containing products are readily transformed into the CF_2H-containing compounds via a free radical desulfurization method. As alternatives to **114**, a series of sulfur-, selenium- and tellurium-containing (phenylchalcogen) difluoromethylating reagents $PhSCF_2H$ (**123a**), $PhSeCF_2H$ (**123b**), and $PhTeCF_2H$ (**123c**), respectively, have been prepared and applied to similar reactions with aldehydes, ketones, and activated imines.[122] Reagents **123a** and **123b** were found to have better reactivity than **123c** toward carbonyl compounds and imines in the presence of *t*BuOK. In 2014, Pohmakotr et al. employed phosphazene organic base in place of *t*BuOK in the (difluoromethyl)phenylthiolation of nonenolizable carbonyl compounds, cyclic imides, and acid anhydrides (Scheme 11.20).[123] The method efficiently provides α-*gem*-difluoromethylated adducts in good yields despite the limitation of substrate scope to only nonenolizable aldehydes.

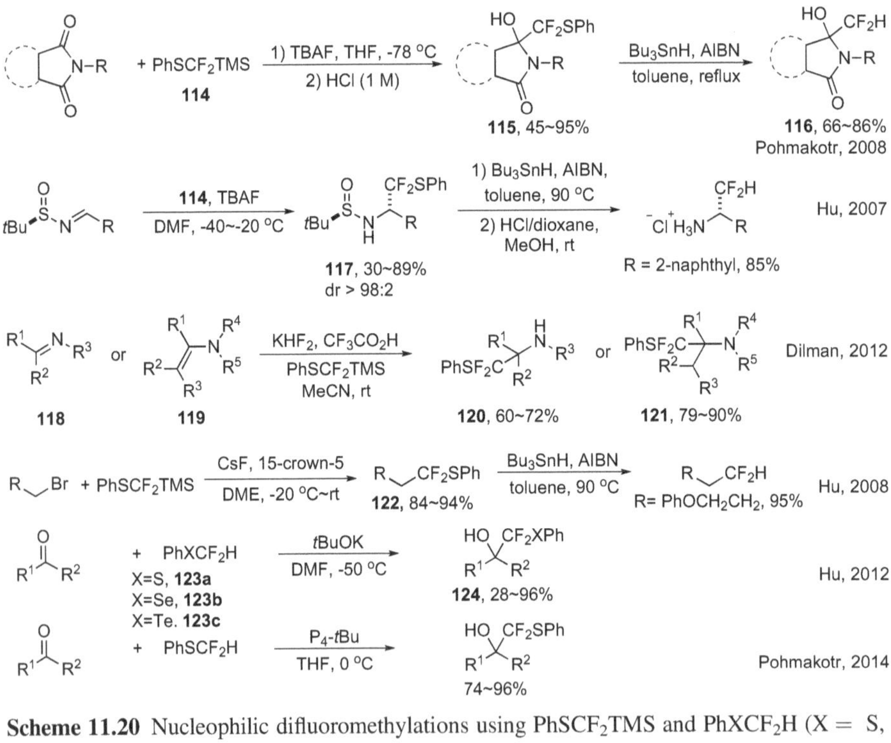

Scheme 11.20 Nucleophilic difluoromethylations using $PhSCF_2TMS$ and $PhXCF_2H$ (X = S, Se, Te).

Compared with difluoro(phenylthio)methylation, difluoro(phenylsulfonyl)methylation is superior not only because the removal of phenylsulfonyl group is easier with a metal reductant but also because the phenylsulfonyl group is a better auxiliary group to facilitate the introduction of CF_2 moiety in nucleophilic difluoro(phenylsulfonyl) methylation.[9c] Three major reagents have been used for the nucleophilic difluoro(phenylsulfonyl)methylation: difluoromethyl phenyl sulfone ($PhSO_2CF_2H$ **126**),[124] [difluoro(phenylsulfonyl)methyl]trimethylsilane ($PhSO_2CF_2TMS$ **131**),[125] and bromodifluoromethyl phenyl sulfone ($PhSO_2CF_2Br$) (Scheme 11.21).[115] In 2007, Hu et al. reported the nucleophilic difluoromethylation of 1,2-cyclic sulfates and sulfamidates **125** to afford β-difluoromethylated or β-difluoromethylenated alcohols and amines **127** in good yields.[126] In 2013, Hu and coworkers developed a new base system using the combination of stoichiometric amount of $N(TMS)_3$ and substoichiometric quantity of Me_4NF for the difluoro(phenylsulfonyl)methylation of both aromatic and aliphatic aldehydes with $PhSO_2CF_2H$.[127] It is worth noting that the steady-state in situ generation of low concentration of the base $(TMS)_2N^-$ suppressed the enolization of aliphatic aldehydes. Moreover, when using an alkoxide as base such as *t*BuOK, $PhSO_2CF_2H$ reacted with two molecules of aryl aldehydes to give 2,2-difluorinated *anti*-1,3-diols with high diastereoselectivity.[128] Aside from cyclic sulfates, sulfamidates, and carbonyl compounds, many other electrophiles including activated imines,[46,129] alkyl halides,[124c] esters,[130] and halogens[131] are also suitable for the reaction with $PhSO_2CF_2H$. On the other hand, enantiomerically pure *N*-(*t*-butylsulfinyl)imines have been employed by Hu and coworkers in the highly diastereoselective difluoro(phenylsulfonyl)methylation reaction to afford enantioenriched α-difluoromethyl amines **128** (Scheme 11.21) with the use of **126**.[129a–c] A variety of both chiral aldimines and ketimines were difluoromethylated under current conditions, of which the latter could not be difluoromethylated using other reagents such as $TMSCF_2H$, $PhSCF_2TMS$,[119] or $PhSCF_2H$.[122] With proper cinchona alkaloid–derived ammonium salt **129** as a catalyst, the first enantioselective nucleophilic difluoromethylation of aldehydes using $PhSO_2CF_2H$ has been successfully accomplished with up to 64% ee.[132] Aside from the widely used $PhSO_2CF_2H$, compound $PhSO_2CF_2TMS$ has been used as an alternative along with catalytic amount of initiator instead of stoichiometric quantity of a strong Brønsted base. For example, the diastereoselective difluoromethylation of α-amino aldehydes **130** with $PhSO_2CF_2TMS$ **131** provides higher yields and diastereoselectivities than obtained when using the $PhSO_2CF_2H$/lithium hexamethyldisilazide (LiHMDS) system.[133] Other examples showing the superiority of $PhSO_2CF_2TMS$ over $PhSO_2CF_2H$ include the difluoro(phenylsulfonyl)methylation of nonactivated imines,[120] enamines,[99] *N*,*N*-acetals,[134] as well as alkyl, propargyl, and alkynyl halides.[121a,121b,135]

Within the family of sulfone-based difluoromethylating reagents, other potential reagents were prepared in 2010 by altering the phenyl group by a series of heterocyclic aromatic functionalities (Scheme 11.22).[136] Reagent **134** reacted with primary alkyl halides and triflates to afford the corresponding difluoro(sulfonato)methylated products **135** in good yields, which were further transformed into stable α-difluorinated sulfinate and sulfonate salts **135/136** via dearylation.[137] Subsequently, Hu et al. developed a formal nucleophilic iodo- and bromodifluoromethylation for carbonyl compounds. To tackle the high tendency of the decomposition of CF_2Br^- and CF_2I^-, a strategy to halogenate the in situ–generated sulfinate salt was proposed to change the reaction pathway from traditional olefination to alkylation. With this new method, a wide range of aldehydes and ketones were transformed to the corresponding iodo- and bromodifluoromethylated products **140** in high yields.[138]

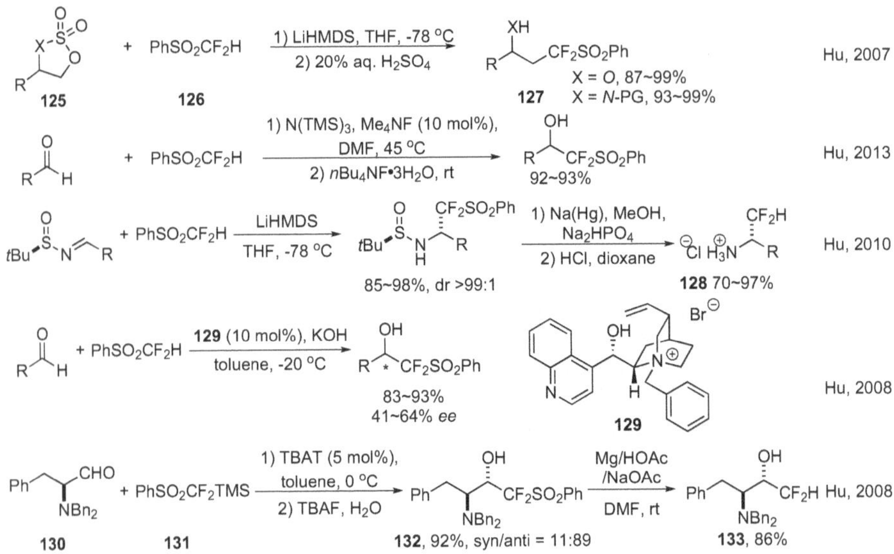

Scheme 11.21 Nucleophilic difluoromethylations using $PhSO_2CF_2H$ and $PhSO_2CF_2TMS$.

Later, Prakash et al. examined the reaction between chiral *N*-*t*-butanesulfinyl imines and **134** for the synthesis of novel optically active α, α-difluoro-β-amino sulfinic and sulfonic acids **142/143**.[139] The reaction proceeds in a highly stereoselective fashion and the difluorinated β-amino sulfinates **142** are suitable for constructing difluorinated peptidosulfonamides. As a member of sulfone-based CF_2 synthon, **144** was successfully prepared and it reacted with one or two molecules of aromatic aldehydes to afford difluoromethylene-containing alcohols **145** or 1,3-(2,2,difluoromethylene) diols in good yields (Scheme 11.22).[140]

S-difluoromethyl sulfoximines are generally used as robust electrophilic difluoromethylating reagents. However, Hu et al. have found that the reactivity of the difluoromethyl sulfoximines was tuned from electrophilic to nucleophilic by changing the *N*-substituent from an electron-withdrawing group (EWG) (*p*-toluenesulfonyl, tosyl) to an electron-donating group (*t*-butyldimethylsilyl, TBS).[141] With the enantiopure TBS-sulfoximine **146**, a variety of aromatic ketones undergo nucleophilic diastereoselective addition followed by reductive desulfonimidoylation to afford enantioenriched difluoromethyl tertiary alcohols **148** in good yields (Scheme 11.23). In 2014, the same group reported a highly regioselective ring-opening nucleophilic difluoromethylation of epoxides **149** with racemic TBS-sulfoximine **146**, which is also amenable to four- or five-membered oxacycles and sterically hindered 1,2-disubstituted oxiranes.[142] The ring-opening products were easily transformed to β-difluoromethyl, γ-difluoromethyl, and β-difluoromethylenyl alcohols via desulfonimidoylation (Scheme 11.23). It was noted that the preorganization of the sulfoximine and epoxides with $BF_3\cdot Et_2O$ was crucial to accomplish this reaction.

In 2015, Mikami and coworkers employed fluoroform as an ideal difluoromethylating reagent and developed the difluoromethylation on sp^3 carbon of various nitrile compounds to provide the α-difluoromethylated nitrile products with an all-carbon

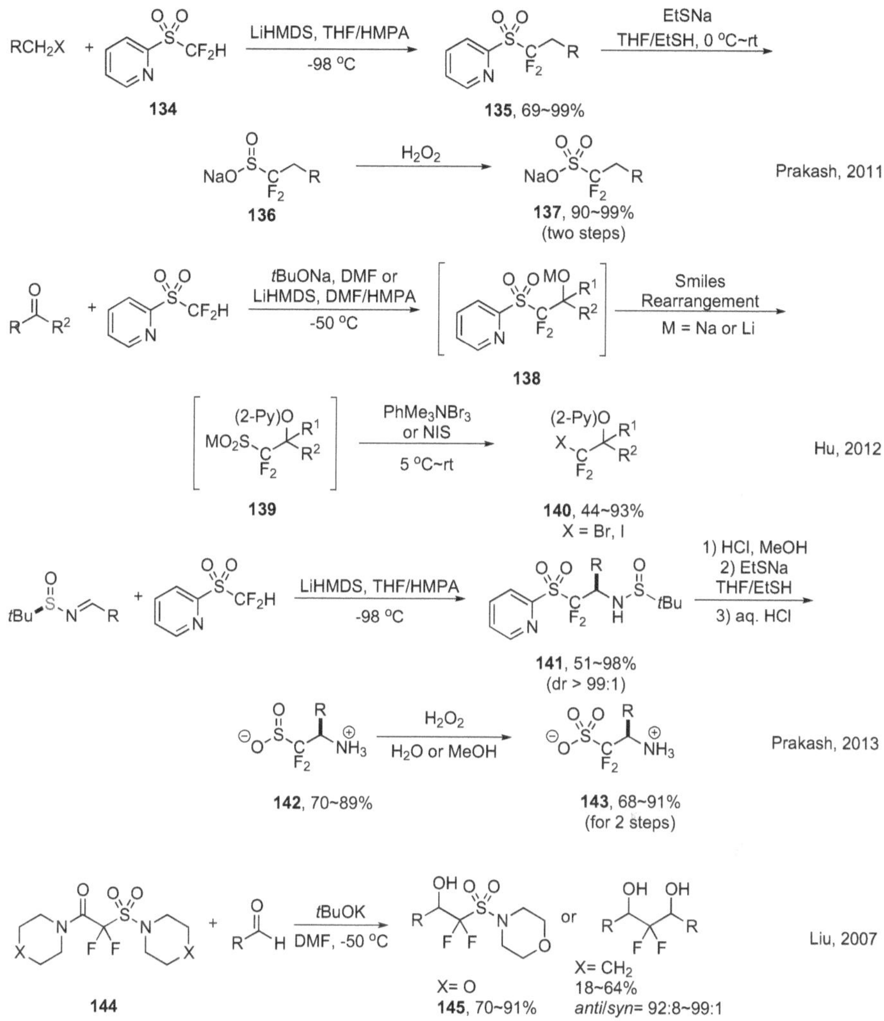

Scheme 11.22 Nucleophilic difluoromethylations using difluoromethylpyridylsulfone **134** and diamide **144**.

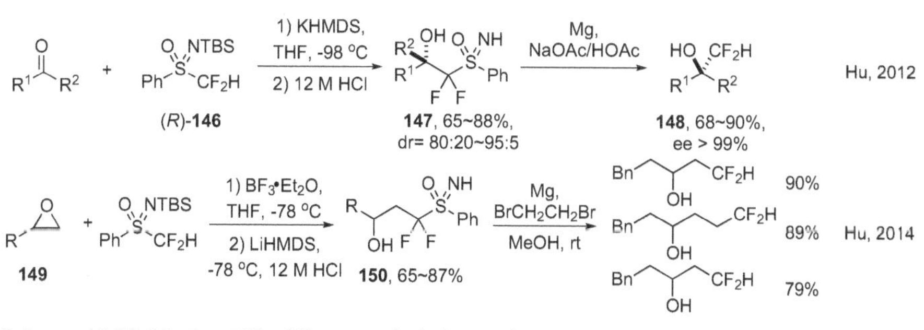

Scheme 11.23 Nucleophilic difluoromethylations using difluoromethylphenylsulfoximine **146**.

quaternary center in moderate to high yields.[143] The Ruppert–Prakash reagent, which is normally used as a robust trifluoromethylating reagent, was also applicable to this reaction affording the α-siladifluoromethylated nitrile products. In the same year, Shibata et al. reported the difluoromethylation of terminal alkynes using fluoroform as a difluorocarbene source.[144] A variety of terminal alkynes were efficiently difluoromethylated to afford the corresponding difluoromethylalkynes in moderate to good yields with great functionality tolerance. The further modification of the products was demonstrated by a one-step transformation to the difluoromethylated isoxazoles and 1,2,3-triazoles (Scheme 11.24).

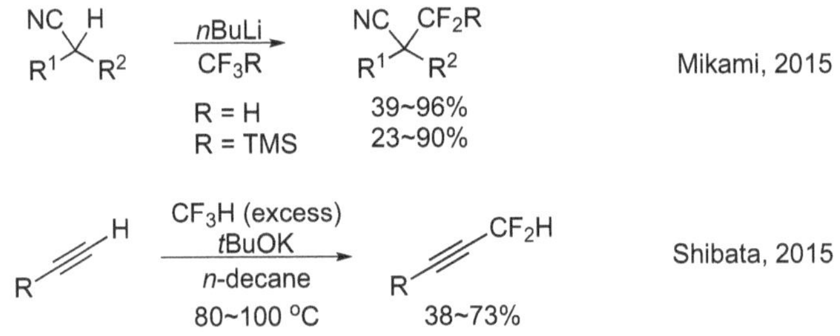

Scheme 11.24 Difluoromethylation using CF_3H and $TMSCF_3$.

3.2 *Transition Metal-Assisted Nucleophilic Difluoromethylation Reactions*

Similar to metal-mediated nucleophilic trifluoromethylation, copper has been the most widely used metal either in stoichiometric or catalytic amounts in nucleophilic difluoromethylations. In 2011, Amii and coworkers reported a CuI-catalyzed three-step reaction between aryl iodide and α-silyldifluoroacetate **151**, followed by hydrolysis and decarboxylation, to obtain difluoromethylarenes in modest overall yields (Scheme 11.25).[145] Meanwhile, Hu et al. took advantage of the sulfone-based reagent **131** and accomplished a copper-mediated difluoro(phenylsulfonyl)methylation of propargyl chlorides and alkynyl halides under the promotion of CuI, giving $PhSO_2CF_2$-substituted allenes **153** and alkynes **152**, respectively. Subsequent reductive desulfonylation of the difluoromethylphenylsulfonylalkyne with $Mg/HgCl_2$ afforded the difluoromethylated alkyne compound in good yield.[146] Besides silyl-based reagents, tributyl(difluoromethyl)stannane **154** has also been used by Prakash et al. in the efficient difluoromethylation of electron-rich and electron-poor iodo/heteroarenes and β-styryl halides in good yields (Scheme 11.25).[147] Despite the previously developed methodologies, metal-mediated protocols using more readily available, less toxic, and more atom-economical reagent $TMSCF_2H$ had remained a challenge for a long time. In 2012, Hartwig et al. successfully achieved a one-step copper-mediated difluoromethylation using $TMSCF_2H$ to form difluoromethylarenes and difluoromethylated alkenes in high yields.[148] However, the substrate scope is limited to electron-rich and electron-neutral iodoarenes with the use of a significant excess of $TMSCF_2H$. Furthermore, this protocol is not compatible with aldehydes and ketones owing to the competitive nucleophilic addition at the carbonyl center. Later, Qing and coworkers extended this method to electron-deficient aryl iodides, as well as heteroaryl and β-styryl halides under mild reaction conditions. They were able to reduce the amount of $TMSCF_2H$ and reaction temperature to room temperature by employing phenanthroline as the ligand and *t*BuOK as the base.[149] In 2014, Goossen and co-workers used CuSCN as the copper source for a Sandmeyer-type difluoromethylation of (hetero)arenediazonium salts, which were either used after isolation or generated in situ from various aromatic amines.[150] Aside from copper, Shen and coworkers developed a Pd/Ag bimetallic catalyst system that cooperatively catalyzed the direct difluoromethylation of aryl bromides and iodides using $TMSCF_2H$ under mild conditions (Scheme 11.25).[151] This protocol has great compatibility with a wide range of functional groups, including ester, amide, cyclopropyl, bromide, and heteroaryl subunits such as pyrrole, benzothiazole, carbazole, and pyridine whereas phenol and ketone functionalities required the prior installation of protecting groups.

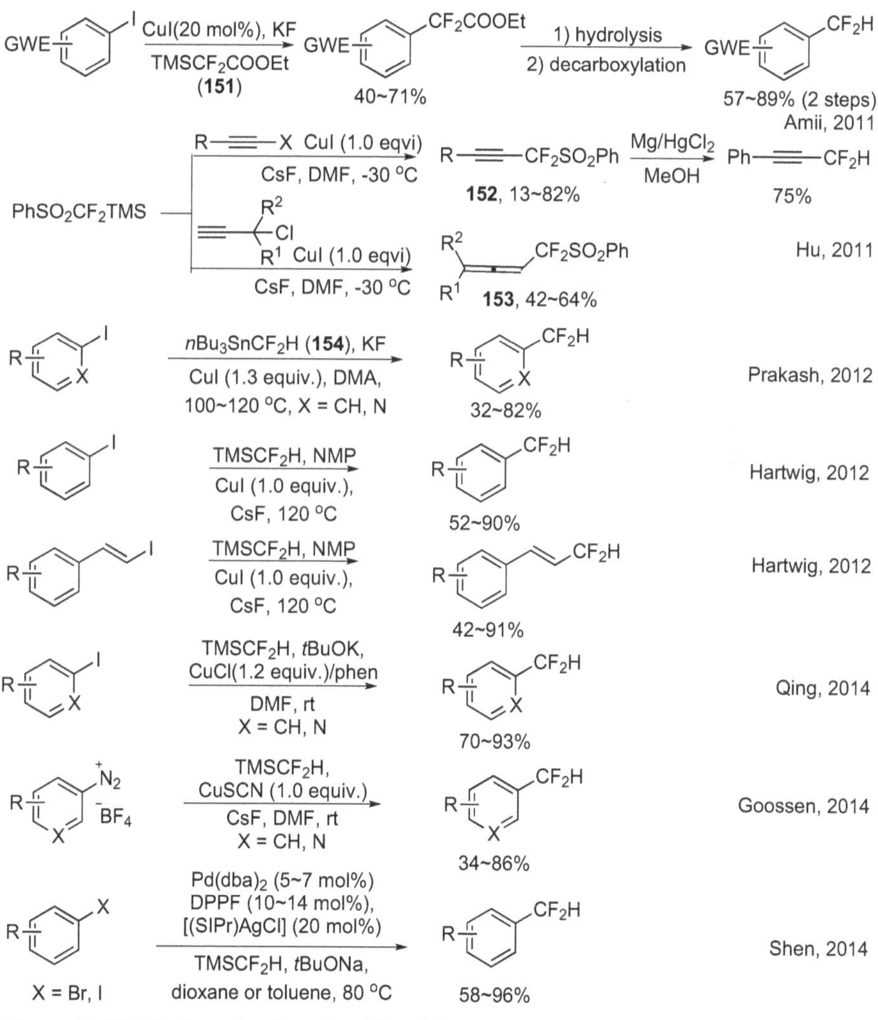

Scheme 11.25 Metal-mediated nucleophilic difluoromethylations of aromatic and aliphatic halides.

3.3 Difluoroolefination Reaction

As described previously, difluoromethyl phenyl sulfoxide readily reacts with primary alkyl iodide,[120] enolizable and nonenolizable aldehydes, and ketones[152] to afford the corresponding products in good to excellent yields (Scheme 11.26). The products were either converted to difluoromethylated alkanes and alcohols through reductive desulfonylation, or to *gem*-difluoroalkenes via elimination reaction such as thermal elimination or dehydrosulfonylation. Similarly, difluoro(phenylthio)methyl compounds were first oxidized to sulfoxides followed by flash vacuum pyrolysis to furnish difluoroolefins.[153] Later, Hu and coworkers developed the one-pot Julia-Kocienski-type *gem*-difluoroolefination reaction of diaryl ketones and enolizable aliphatic aldehydes with difluoromethyl 2-pyridyl sulfone **134** providing difluoroalkenes in good yields.[154] The reaction was proposed to proceed through an acid-promoted Smiles rearrangement of the carbinol intermediate, which is otherwise difficult to achieve through a conventional Julia–Kocienski olefination protocol due to the retroaldol-type decomposition of the key intermediate.[136] The use of milder base generated in situ from CsF/$(TMS)_3N$ not only diminished the undesired enolization of aliphatic aldehydes but also provided a powerful synthetic method for chemoselective *gem*-difluoroolefination of polycarbonyl compounds. Aside from these protocols, the difluorinated sulfoximine products generated from **156** affords difluoroalkenes via base-induced dehydrosulfonimidoylation (Scheme 11.26).[142] In 2015, Hu et al. and Wang et al. simultaneously reported the transition-metal-free *gem*-difluoroolefination using diazo compounds. Wang and coworkers employed $TMSCF_2Br$ as the difluorocarbene source and TBAB as the promoter affording the 1,1-difluoroolefins under mild conditions with great functionality tolerance.[155] Hu and coworkers exploited both $TMSCF_2Br$ and $TMSCF_3$ as the difluorocarbene source.[156] Initiated by either NaI or $n$$Bu_4NBr$, the in situ–generated difluorocarbene underwent the direct C-difluoromethylenation of diazo compounds followed by β-elimination of N_2 (Scheme 11.26).

4. Nucleophilic Monofluoroalkylation Reactions

The monofluoromethyl group is of great importance in isostere-based drug design as well.[157] Direct fluorinations for the selective construction of a single C–F bond have been well investigated. Nevertheless, the employment of monofluorinated synthons constitutes a significant alternative pathway. In this section, novel nucleophilic monofluoroalkylating reagents and methods will be discussed.

4.1 Nucleophilic Metal-Free Assisted Monofluoromethylation and Monofluoromethylenation Reactions

Mechanistically, unlike the extremely labile trifluoromethyl and difluoromethyl carbanions, many α-monofluorocarbanions possess reasonable thermal stability if their anionic centers are adjacent to EWGs. In fact, in 2009, Prakash and coworkers accomplished the first isolation and characterization of an α-fluorinated methanide anion by

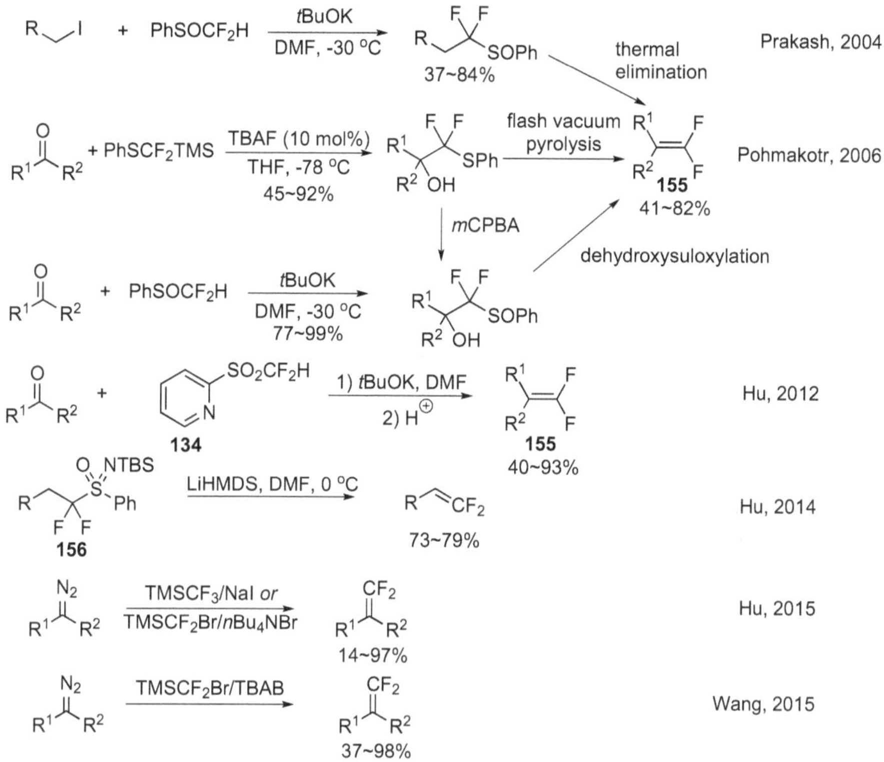

Scheme 11.26 Syntheses of *gem*-difluoroolefins.

single crystal X-ray diffraction.[13] Therefore, EWG-substituted monofluorocarbanions generally can be generated in situ, and they readily undergo reactions with electrophilic substrates. Many nucleophilic monofluoroalkylating reagents involving one or two carbonyl or nitro groups as the EWGs have been developed in the past half century and reviewed previously.[9b] Another important type of reagents based on sulfoxide was developed to efficiently react with aldehydes, aryl, and alkyl halides back in the 1980s.[158] On the other hand, sulfone- and sulfoximine-based monofluoromethylating reagents have also received increasing attention (Scheme 11.27). Although first synthesized in 1968[159] and first used as a nucleophilic fluoro(phenylsulfonyl)methylating reagent in 1985,[160] the synthetic potential of $PhSO_2CH_2F$ **157** as an efficient monofluoromethylating reagent was not fully recognized until 2006. Hu et al. have reported a diastereoselective monofluoromethylation of enantiopure *N-t*-butylsulfinylaldimines using $PhSO_2CH_2F$ to provide **158** and **159** with high diastereoselectivities and good yields.[161] Later, the same group has applied this methodology to both α-amino aldimines[129b] and ketimines.[162] Compared with $PhSO_2CH_2F$, the introduction of a second EWG further facilitates the generation and stabilization of monofluorocarbanions. Therefore, some reactions that are not feasible using $PhSO_2CH_2F$ readily occur with α-EWG-substituted monofluoromethyl sulfones under mild conditions.[8] For example, fluorobis(phenylsulfonyl)methane (FBSM, **162**) is easily deprotonated under

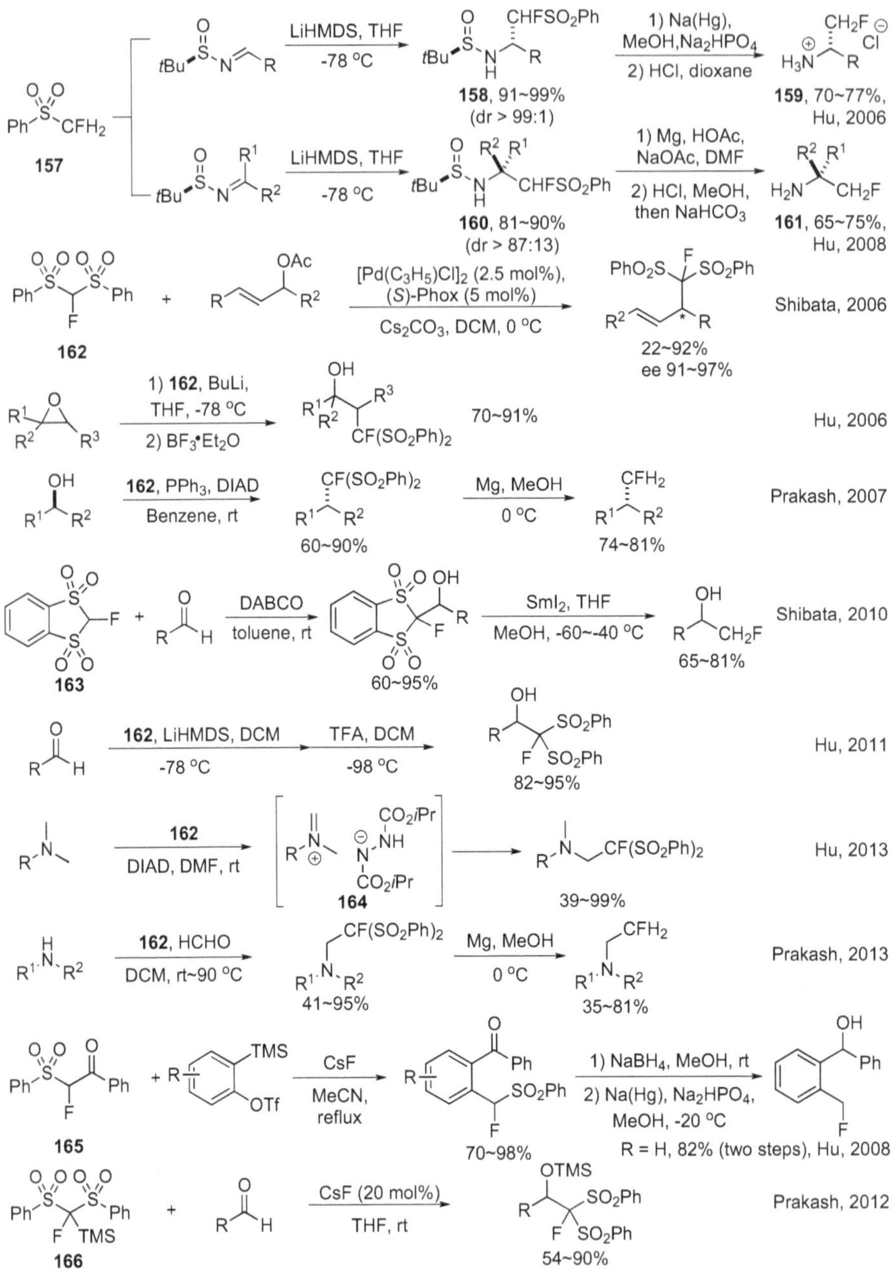

Scheme 11.27 Sulfone-based nucleophilic monofluoromethylations and monofluoromethylenations.

mild conditions, and its application as a monofluoromethylating reagent toward epoxides and allyl acetates was first reported by Hu et al.[163] and Shibata et al.[164] independently. Subsequently, Prakash et al. applied FBSM in enantioselective nucleophilic substitution reaction (Mitsunobu reaction) with primary, secondary, allylic, benzylic, and alicyclic alcohols in moderate to good yields (Scheme 11.27).[165] Starting from chiral alcohols, excellent enantiospecificity is observed with full inversion of the configuration. Later, the same group extended the substitution reaction to alkyl halides with high efficacy.[166] In 2010, the monofluoromethylation of aldehydes using FBSM was attempted by Shibata and coworkers without success. Alternatively, they developed 2-fluoro-1,3-benzodithiole-1,1,3,3-tetraoxide **163** as a new monofluoromethylating reagent toward aldehydes at room temperature (Scheme 11.27).[167] However, Hu et al. tackled this problem and successfully accomplished the monofluoromethylation of aldehydes with FBSM by using LiHMDS as a base.[168] The employment of strong Li–O coordination at low temperatures played a key role in the successful nucleophilic addition of FBSM to aldehydes. The same group has also applied FBSM to the dehydrogenative coupling reaction with aliphatic tertiary amines to afford β-fluorinated amines **164** using diisopropyl azodicarboxylate as catalyst.[169] Another protocol to prepare β-fluorinated amines has been developed by Prakash and coworkers via a three-component Mannich reaction between formaldehyde, amine, and FBSM (Scheme 11.27).[170] The reaction is highly efficient toward both primary and secondary amines under mild conditions, and no base is required in many cases. Replacing one phenylsulfonyl group with an acyl group, monofluoro(phenylsulfonyl)(phenylacyl)methane **165** reacted with in situ–generated arynes efficiently.[8] Further reduction with $NaBH_4$ followed by reductive desulfonylation with Na(Hg) afforded the corresponding aromatic monofluoromethylated alcohol in good yields. In addition, Prakash et al. synthesized [fluorobis(phenylsulfonyl)methyl]trimethylsilane as a new monofluoroalkylating reagent, which reacted with aldehydes upon initiation by catalytic amount of CsF.[171] The reaction conditions were very mild with **166** functioning as both a pronucleophile and a Lewis acid, which allowed the one-step addition of the FBSM anion toward various aldehydes via a self-quenching mechanism.

The improved preparation of FBSM since it was first developed by Shibata and Hu in 2006 has contributed to the facile use of FBSM as a robust monofluoroalkylating reagent. The conventional preparation of the reagent is through electrophilic fluorination[163,164] and/or electrochemical fluorination (ECF) processes.[172] Later, Hu et al. reported a superior synthetic route via the sulfoxidation of fluoromethyl phenyl sulfone followed by oxidation, which avoids the tedious purification procedure involved in the conventional protocol.[130a] Prakash and coworkers designed a practical one-step synthesis of FBSM with fluoromethyl phenyl sulfone and less expensive phenyl sulfonyl fluoride, affording FBSM with high efficacy and purity (Scheme 11.28).[173]

The catalytic enantioselective monofluoromethylation has also received increasing interest. Chiral α-fluoromethylated amides and amines have been prepared from FBSM via Mannich-type reactions. Shiabta et al. reported the first enantioselective monofluoromethylation of the in situ–generated imines using FBSM in good yields and high enantioselectivities with *N*-benzylquinidinium chloride **167** as the phase transfer catalyst (Schemes 11.29A and B).[174] The α-fluorobisphenylsulfonylmethylated

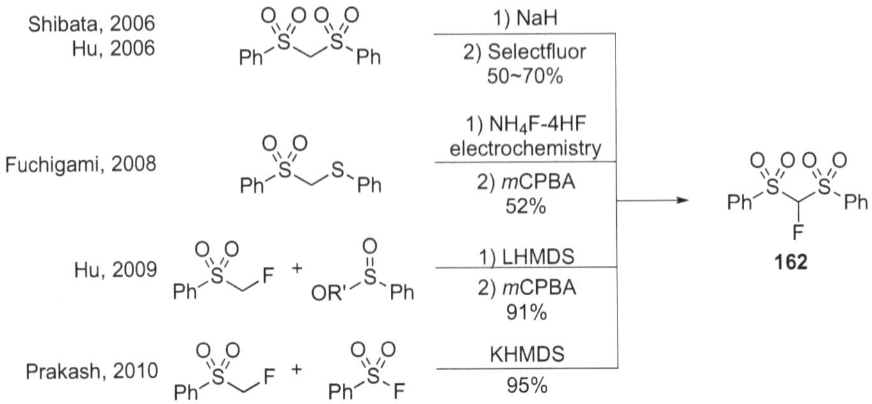

Scheme 11.28 Protocols for the synthesis of fluorobis(phenylsulfonyl)methane.

amines were converted to α-monofluoromethylamines by reductive desulfonylation in good yields. In 2013, the same group applied this methodology for the monofluoromethylation of arylsulfonyl C2-arylindoles using the benzylcinchoninium salt bearing a sterically demanding substituent **179** as the catalyst.[175] The reaction proceeded via the in situ–generated vinylogous imino intermediates by desulfonylation and afforded the chiral monofluoromethylated 2-arylindoles in high yields and high enantioselectivities. Aside from imines, Michael acceptors such as α,β-unsaturated carbonyl compounds react with FBSM to afford β-monofluoromethylated carbonyl compounds after reductive desulfonylation. In 2008, the first enantioselective conjugate addition between FBSM and aryl vinyl ketones was reported by Shibata et al. using Cs_2CO_3 as the base and a sterically hindered quinidinium salt **168** as catalyst.[176] Almost simultaneously, Prakash et al. developed the stereoselective Michael addition of α-fluoro-α-nitro(phenylsulfonyl) methane **170** to chalcone derivatives via a hydrogen-bonding activating mechanism using the cinchona alkaloid-derived thiourea catalyst **171**.[177] Subsequently, Kim et al. examined alkyl vinyl ketones as the Michael acceptors using epi-9-amino-9-dexoyquinine as the chiral bifunctional organocatalyst with good efficiency and enantioselectivity.[178] Simultaneously, α,β-enals have also been explored as Michael acceptors with TMS and TBS ethers of diarylprolinols as catalysts.[179] In 2014, Rios and coworkers reported the progress of the enantioselective synthesis of fluoromethylated indanes **173** and chromanols **174** from FBSM and enals via a double Michael reaction or Michael-hemiacetal formation.[180] In 2011, Tan et al. reported the organocatalyzed reaction between FBSM and MBH carbonates **175** with hydroquinidine(anthraquinone-1,4-diyl) diether ($(DHQD)_2AQN$) as a catalyst to furnish enantiopure β-methyl-γ-monofluoromethyl-substituted alcohols **177** in good yields and extremely high diastereoselectivities (Schemes 11.29A and B).[181] Meanwhile, Shibata and coworkers have reported very similar chemistry independently, and it has been found that the addition of Lewis acid such as $FeCl_2$ and $Ti(OiPr)_4$ improves the yield and enantioselectivity.[182] Starting from saturated aldehydes, enantioenriched β-monofluoromethylated aldehydes **178** was achieved via C–H activation in the presence of 2-iodoxybenzoic acid as oxidant and diarylprolinol derivative **172** as organocatalyst.[183] With subsequent reduction by $NaBH_4$, enantiopure γ-monofluoromethylated alcohols was obtained in modest yields and good enantioselectivities. In 2013, FBDT **163** was applied to the enantioselective monofluoromethylation of aldehydes in good yields and varying enantioselectivities using a bifunctional thiourea **180** as a catalyst (Schemes 11.29A and B).[184]

Similar to *S*-difluoromethyl sulfoximines, *S*-monofluoromethyl sulfoximines are used as nucleophilic monofluoromethylating reagents as well. In contrast to their nonfluorinated analogs, which are widely used as nucleophiles, monofluoromethylated sulfoximines have received much less attention.[185] In 2010, Hu et al. developed the synthesis of monofluorinated epoxides **183** from α-fluorosulfoximines **182** and ketones.[186] Although the monofluorinated epoxides were obtained in moderate to excellent yields, the purification was found to be difficult for the products, which lacked a strong electron-withdrawing substituent such as NO_2 on the aryl group and thus the substrate scope was limited. However, the reactive nature of the epoxides was applied in the one-pot synthesis of α-monofluorinated carbonyl compounds **184** directly from electron-neutral and even electron-rich aryl and alkyl ketones followed by treatment with catalytic amount of TiF_4 or pyridine-9HF reagent (Olah's reagent) (Scheme 11.30). In 2014, the same group has also observed the 1,2- and 1,5-fluorine migrations in cyclopropyl-substituted fluoroepoxides without addition of any external fluorine source.[187] On the other hand, enantiopure monofluoromethylated sulfoximines have been employed in asymmetric monofluoromethylation. In 2012, Hu et al. reported the first highly enantio- and diastereoselective monofluorocyclopropanations of α,β-unsaturated carbonyl compounds **186** using the first homochiral monofluoromethylenating reagent (*R*)-**185**.[141b,188] The reaction was proposed to undergo Michael addition/elimination to afford the corresponding optically enriched cyclopropanes bearing a fluorinated tertiary carbon center. Furthermore, enantiopure α-methyl-α-fluorinated sulfoximine **188** were prepared from (*R*)-**185** with MeI as a methylating reagent and used to prepare enantiomerically enriched cyclopropanes **189** containing fluorinated tertiary carbon centers (Scheme 11.30).

4.2 Transition Metal–Mediated Monofluoromethylation and Monofluoromethylenation Reactions

Alkali/alkaline earth metal–monofluoromethyl complexes such as fluoromethyllithium and fluoromethyl Grignard reagent are shown to have low thermal stability.[189] Transition metals such as copper, zinc, palladium, and iridium have been demonstrated to efficiently mediate or catalyze nucleophilic monofluoroalkylation reactions. To realize the metal-assisted monofluoromethylation of aryl halides, Inoue patented the copper(I)-catalyzed coupling reaction between aryl/heteroaryl iodides and the novel reagent, *t*-butyl 2-fluoro-2-(phenylsulfonyl)acetate (**190**) (Scheme 11.31).[190] The reaction is conducted using *t*BuOK as a base with catalytic quantities of CuI in the presence of picolinic acid as ligand at 70–100°C. After hydrolysis-decarboxylation of **191**, fluoro(phenylsulfonyl)methylated arenes **192** were obtained as the major

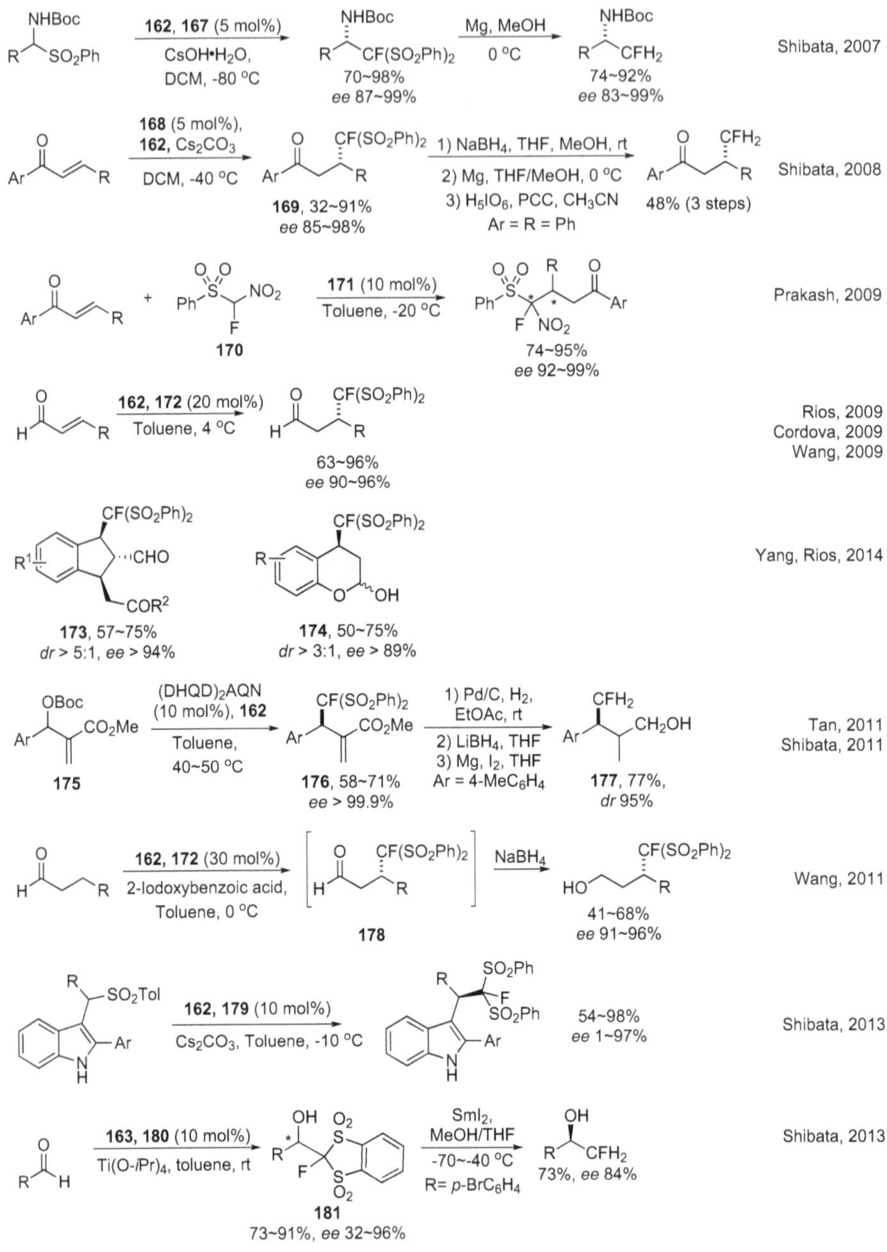

Scheme 11.29A Sulfone-based enantioselective nucleophilic monofluoromethylations.

products in 13–91% yields. Later, Hu and coworkers developed fluoroiodomethyl 2-pyridyl sulfone (**193**) as an efficient monofluoromethylating reagent that in situ generates the reactive species, [fluoro(2-pyridylsulfonyl)-methyl]copper via zinc–iodide exchange followed by transmetalation with CuI.[191] A wide range of aryl iodides

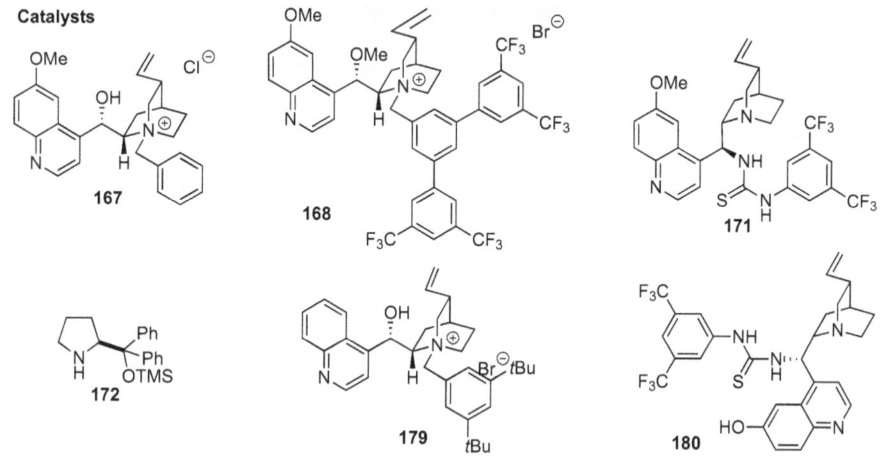

Scheme 11.29B Enantioselective catalysts.

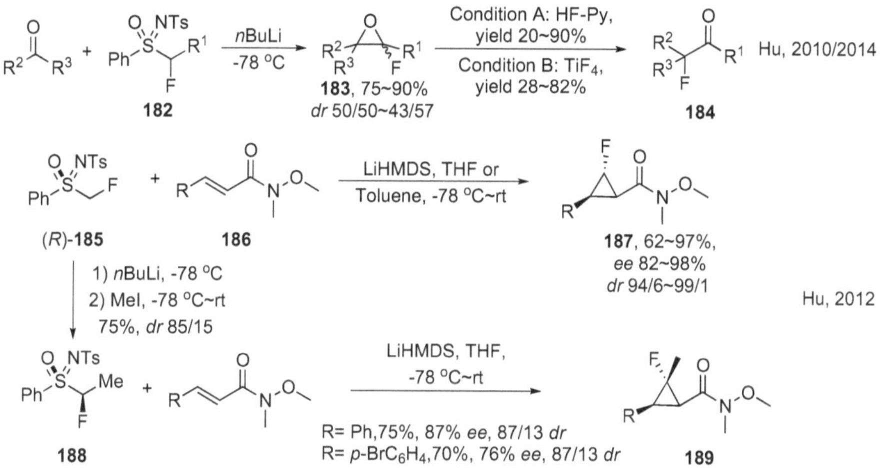

Scheme 11.30 Sulfoximine-based racemic and enantioselective monofluoromethylations.

readily reacted with **193** under the action of either stoichiometric or catalytic quantities of CuI to afford the desired monofluoroalkylated products **194** with great functionality tolerance (Scheme 11.31). During the investigation of Hurtley-type reaction between active fluorinated methylene and aryl halides, it was surprising to find FBSM (**162**) and α-fluoro-β-ketosulfone **165** was not reactive. Instead, by using reagent α-fluoro-β-keto-2-pyridylsulfone **195**, the same group was able to prepare various fluoromethylated arenes **197** through a copper-catalyzed debenzoylative fluoroalkylation of aryl iodides in the presence of $NaHCO_3$ followed by reductive desulfonylation (Scheme 11.31).[192] The (2-pryridyl)sulfonyl group was proposed to promote the copper-catalyzed cross-coupling and was removed easily via a Bu_3SnH-mediated desulfonylation.

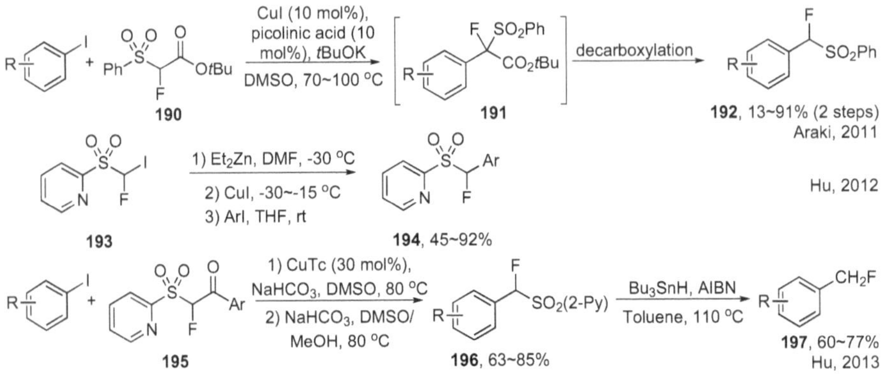

Scheme 11.31 Copper-mediated nucleophilic monofluoroalkylation of aryl halides.

Application of FBSM in catalytic enantioselective monofluoromethylation has been first investigated by Shibata and coworkers, who used FBSM in palladium-catalyzed asymmetric allylic monofluoromethylation of 1,3-disubstituted allyl acetates.[164] The obtained products **199** were further transformed to monofluoromethylated molecules of biological interest via desulfonylation and oxidation (Scheme 11.32). Later You and coworkers reported the iridium-catalyzed asymmetric reaction of monosubstituted allyl carbonates **200** with FBSM (**162**) in the presence of a chiral phosphoramidite ligand **201**.[193] The reaction is highly regiospecific to give the branched products **202** bearing a terminal alkene predominantly with high enantioselectivities (Scheme 11.32). The products were further transformed to the fluorobis(phenylsulfonyl)methylated ibuprofen family in a highly efficient manner. By switching to palladium catalyst and using 1,2-bis(diphenylphosphino)ethane (DPPE) as a ligand, a variety of terminal monofluoromethylated compounds were synthesized from the same precursors with high yields and regioselectivities.[193]

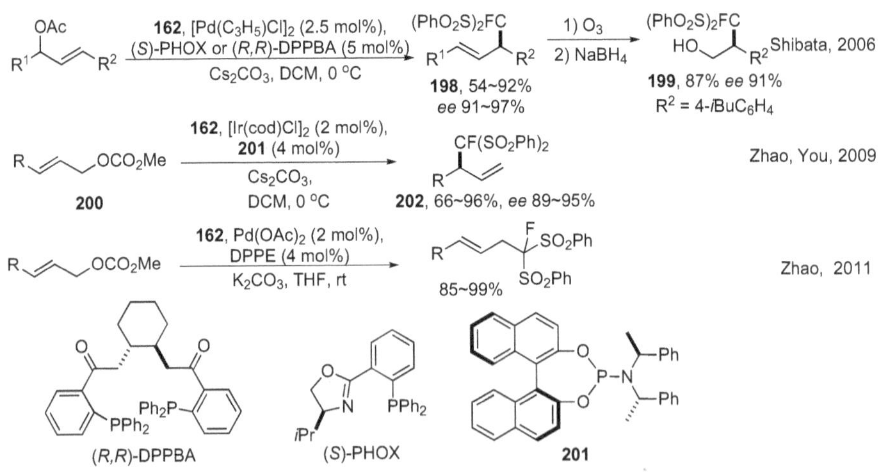

Scheme 11.32 Enantioselective allylic monofluoroalkylation catalyzed by palladium and iridium.

4.3 Monofluoroolefination Reaction

A variety of α-fluorinated olefins have been prepared from α-monofluorinated sulfones and carbonyl compounds via nucleophilic addition/Smiles rearrangement (Julia-Kocienski-type reaction). The pioneering work was accomplished by Lequeux and Pazenok in 2003, in which 2-(1-fluoroethyl)sulfonyl-1,3-benzothiazoles **203** were prepared and used as monofluoroolefinating reagents toward various carbonyl compounds.[194] Since 2006, Lequeux and Zajc have made great contributions to Julia-Kocienski-type monofluoroolefination with 1,3-benzothiazol-2-yl (BT) sulfones. From various α-monofluorinated BT sulfones, a wide range of functionalized monofluoroolefins were synthesized, including stilbene- and styrenelike fluoroolefins,[195] α-fluoroacrylates,[196] α-fluoroacrylonitriles,[197] α-fluorovinyl Weinreb amides and ketones,[198] 2-fluoro-1,3-enynes,[199] α-fluorovinyl triazoles,[200] and 2-fluoroallkyl amines (Scheme 11.33).[201] It has been shown that the dominance of *E*- or *Z*-isomer could be fine-tuned by altering the reaction conditions. Most of the reactions have limited application in the case of ketones, which can be attributed to the steric hindrance in the initial addition step. Aside from BT sulfones, other reagents have also been developed including 3,5-bis(trifluoromethyl)-phenyl,[202] 1-phenyl-1H-tetrazol-5-yl (**204a**),[203] 1-*t*-butyl-1H-tetrazol-5-yl (**204b**),[204] and 2-pyrimidinyl (**205**) sulfones.[205] These sulfone-based reagents exhibit similar stereocontrol ability as the BT sulfone reagents. In addition to Julia-Kocienski-type reaction, monofluoroolefins were also synthesized from α-fluoro-β-silylsulfones and α-fluoro-β-ketosulfones.[206] Prakash and coworkers developed an efficient one-pot methodology for the nucleophilic stereospecific fluoromethylation of benzyl halides using **162** and **165** in the presence of cesium carbonate to provide γ-fluorovinyl compounds **206** including γ-fluorostyrylsulfones, γ-fluorocinnamates, and γ-fluorochalcones.[166] In 2011, Jørgenson et al. reported a catalytic enantioselective nucleophilic fluorovinylation with α-fluoro-β-keto-(BT sulfones) **207** and a chiral primary amine **210** as catalyst.[207] Both cyclic and acyclic α,β-unsaturated ketones underwent the reaction to afford the monofluorovinylated products **209** in high yields and excellent enantioselectivities.

Similar to sulfone-based reagents, *S*-monofluoromethyl sulfoximines have been used to furnish monofluoroolefins as well. In 2009, Hu and coworkers reported the first highly stereoselective synthesis of monofluoroalkenes from racemic α-fluorosulfoximines **211** and nitrones **212**.[208] The reaction was proposed to take place by initial nucleophilic addition followed by 1,2-elimination to afford the products in good yields and excellent *Z*/*E* ratios. In 2014, the easily available ethyl bromofluoroacetate **213** was employed in the highly diastereoselective reaction with nitrones to afford monofluoroalkenoates and monofluorinated isoxazolidinones depending on the nature of *N*-substituents (Scheme 11.34).[209] The different behavior of *N*-phenyl- vs *N*-*t*-butyl-substituted nitrones was attributed to electronic origin. Moreover, 2-hydroxydifluorophosphonates **216**, which is the difluoromethylated product from diethyl difluoromethylphosphonate and carbonyl substrates, has been shown to undergo H–F elimination with the treatment of a proper base to afford monofluoroenol phosphonates **217** with high *Z*/*E* ratios (Scheme 11.34).[210]

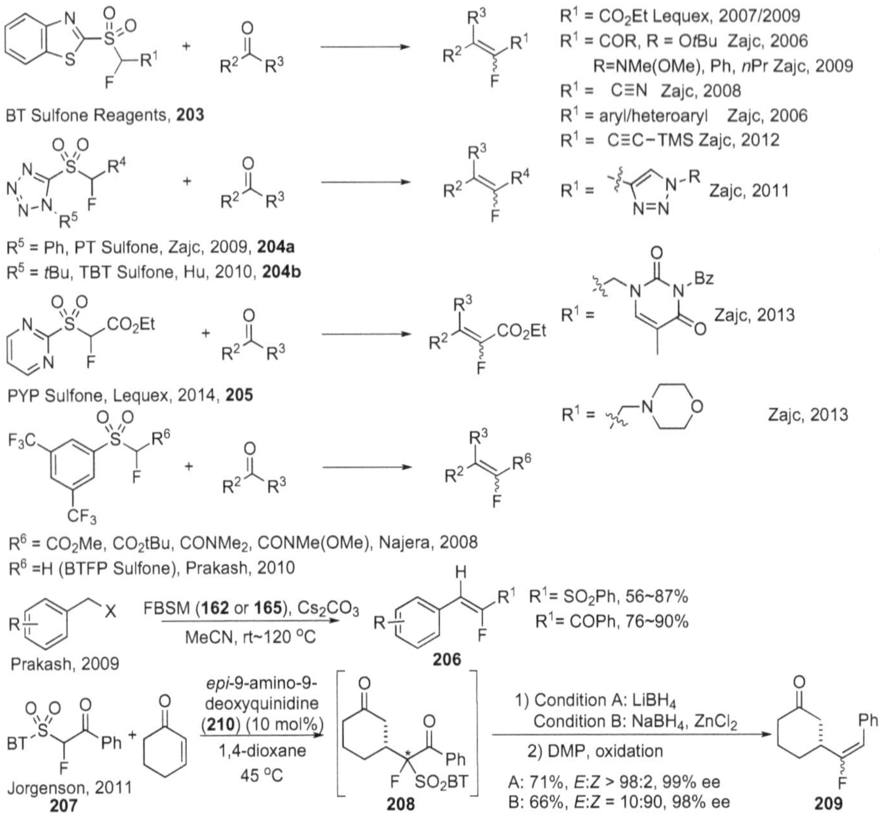

Scheme 11.33 Sulfone-based nucleophilic monofluoroolefinations.

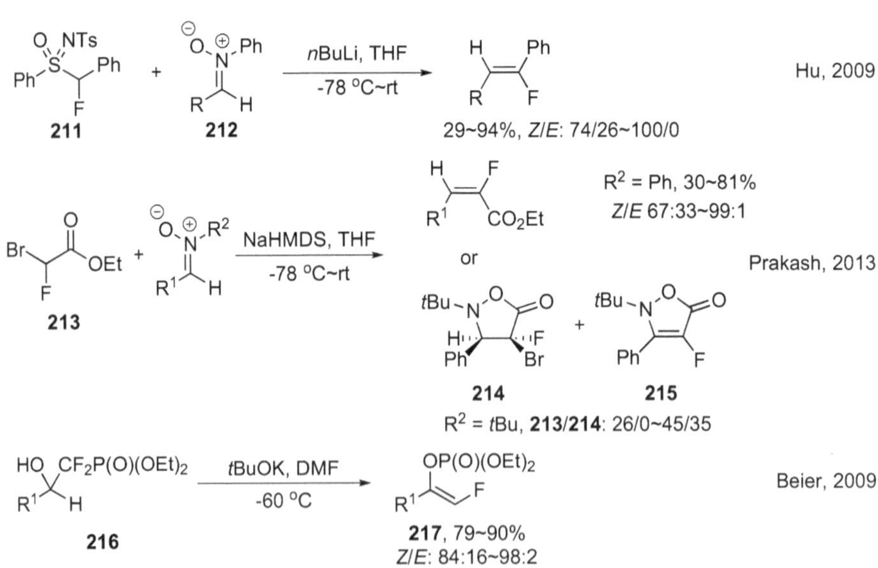

Scheme 11.34 Nucleophilic monofluoroolefinations using sulfoximine-, acetate-, and phosphonate-based reagents.

References

1. a. Hiyama, T. *Organofluorine Compounds: Chemistry and Applications;* Springer: New York, 2000.
   b. Chambers, R. D. *Fluorine in Organic Chemistry;* Blackwell: Oxford, 2004.
   c. Uneyama, K. *Organofluorine Chemistry;* Blackwell: Oxford, 2006.
   d. Bégué, J. -P.; Bonnet-Delpon, D. *Bioorganic and Medicinal Chemistry of Fluorine;* Wiley: Hoboken, New Jersey, 2008.
   e. Ojima, I. *Fluorine in Medicinal Chemistry and Chemical Biology;* Wiley-Blackwell: Chichester, U. K, 2009.
   f. Gouverneur, V.; Müller, K. *Fluorine in Pharmaceutical and Medicinal Chemistry: From Biophysical Aspects to Clinical Applications;* Imperial College Press: London, 2012.
   g. Kirsch, P. *Modern Fluoroorganic Chemistry: Synthesis, Reactivity, Applications,* 2nd ed.; Wiley-VCH: Weinheim, Germany, 2013.
2. a. Gribble, G. W. *Chem. Soc. Rev.* **1999,** *28,* 335–346.
   b. O'Hagan, D.; Schaffrath, C.; Cobb, S. L.; Hamilton, J. T. G.; Murphy, C. D. *Nature* **2002,** *416,* 279–280.
3. a. Dumas, J.; Peligot, E. *Ann. Pharm.* **1835,** *15,* 246.
   b. Dumas, J.; Peligot, E. *Ann. Chim. Phys.* **1836,** *61,* 193.
4. a. Hudlicky, M. *Chemistry of Organic Fluorine Compounds,* 2nd ed.; Ellis Horwood: New York, 1976.
   b. Banks, R. E.; Sharp, D. W. A.; Tatlow, J. C. *Fluorine: The First Hundred Years (1886–1986);* Elsevier: New York, 1986.
   c. Hudlicky, M.; Pavlath, A. E. *Chemistry of Organic Fluorine Compounds II. A Critical Review;* ACS: Washington, DC, 1995.
   d. Banks, R. E. *Fluorine Chemistry at the Millennium. Fascinated by Fluorine;* Elsevier: Amsterdam, 2000.
5. Smart, B. E. *Chem. Rev.* **1996,** *96,* 1555–1824.
6. Campbell, M. G.; Ritter, T. *Org. Process Res. Dev.* **2014,** *18,* 474–480. and references therein.
7. a. Prakash, G. K. S.; Hu, J. New Nucleophilic Fluoroalkylation Chemistry. In *Fluorine-containing Synthons;* Soloshonok, V. A., Ed.; ACS: Washington/DC, 2005. and references therein.
   b. Prakash, G. K. S.; Hu, J. *Acc. Chem. Res.* **2007,** *40,* 921–930. and references therein.
8. Ni, C.; Zhang, L.; Hu, J. *J. Org. Chem.* **2008,** *73,* 5699–5713. and references therein.
9. a. Prakash, G. K. S.; Wang, F. *Chemica Oggi* **2012,** *30,* 30–36.
   b. Prakash, G. K. S.; Wang, F. Flourishing Frontiers in Organofluorine Chemistry. In *Organic Chemistry Breakthroughs and Perspectives;* Ding, K., Dai, L.-X., Eds.; Wiley-VCH: Weinheim, 2012.
   c. Landelle, G.; Panossian, A.; Pazenok, S.; Vors, J. P.; Leroux, F. R. *Beilstein J. Org. Chem.* **2013,** *9,* 2476–2536.
   d. Liang, T.; Neumann, C. N.; Ritter, T. *Angew. Chem. Int. Ed.* **2013,** *52,* 8214–8264.
   e. Ni, C.; Hu, M.; Hu, J. *Chem. Rev.* **2015,** *115,* 765–825.
   f. Liu, X.; Xu, C.; Wang, M.; Liu, Q. *Chem. Rev.* **2015,** *115,* 683–730.
   g. Yang, X.; Wu, T.; Phipps, R. J.; Toste, F. D. *Chem. Rev.* **2015,** *115,* 826–870.
   h. Huang, Y.; Yang, X.; Chen, Z.; Verpoort, F.; Shibata, N. *Chem. Eur. J.* **2015,** *21,* 8664–8684.
   i. Catalán, S.; Munoz, S. B.; Fustero, S. *Chimia* **2014,** *68* (6), 382–409.

10. a. Kirk, K. L. *Org. Process Res. Dev.* **2008,** *12,* 305–321.
b. Müller, K.; Faeh, C.; Diederich, F. *Science* **2007,** *317,* 1881–1886.
c. Hagmann, W. K. *J. Med. Chem.* **2008,** *51,* 4359–4369.
d. Wang, J.; Sanchez-Rosello, M.; Aceña, J. L.; del Pozo, C.; Sorochinsky, A. E.; Fustero, S.; Soloshonok, V. A.; Liu, H. *Chem. Rev.* **2014,** *114,* 2432–2506.
11. a. Emeléus, H. J.; Haszeldine, R. N. *J. Chem. Soc.* **1949,** 2948–2952.
b. Haszeldine, R. N. *Nature* **1951,** *167,* 139–140.
c. Pierce, O. R.; McBee, E. T.; Judd, G. F. *J. Am. Chem. Soc.* **1954,** *76,* 474–478.
d. Haszeldine, R. N. *J. Chem. Soc.* **1954,** 1273–1279.
12. Prakash, G. K. S.; Wang, F.; Zhang, Z.; Haiges, R.; Rahm, M.; Christe, K. O.; Mathew, T.; Olah, G. A. *Angew. Chem. Int. Ed.* **2014,** *53,* 11575–11578.
13. Prakash, G. K. S.; Wang, F.; Shao, N.; Mathew, T.; Rasul, G.; Haiges, R.; Stewart, T.; Olah, G. A. *Angew. Chem. Int. Ed.* **2009,** *48,* 5358–5362.
14. a. Prakash, G. K. S.; Krishnamurti, R.; Olah, G. A. *J. Am. Chem. Soc.* **1989,** *111,* 393–395.
b. Ruppert, I.; Schlich, K.; Volbach, W. *Tetrahedron Lett.* **1984,** *25,* 2195–2198.
15. For review articles on the application of the Ruppert–Prakash reagent, see:
a. Prakash, G. K. S.; Yudin, A. K. *Chem. Rev.* **1997,** *97,* 757–786.
b. Singh, R. P.; Shreeve, J. M. *Tetrahedron* **2000,** *56,* 7613–7632.
c. Prakash, G. K. S.; Mandal, M. *J. Fluor. Chem.* **2001,** *112,* 123–131.
d. Shibata, N.; Mizuta, S.; Kawai, H. *Tetrahedron Asymmetry* **2008,** *19,* 2633–2644.
16. Shono, T.; Ishifune, M.; Okada, T.; Kashimura, S. *J. Org. Chem.* **1991,** *56,* 2–4.
17. Barhdadi, R.; Troupel, M.; Périchon, J. *J. Chem. Soc. Chem. Commun.* **1998,** 1251–1252.
18. a. Folléas, B.; Marek, I.; Normant, J. F.; Jalmes, L. S. *Tetrahedron Lett.* **1998,** *39,* 2973–2976.
b. Folléas, B.; Marek, I.; Normant, J. F.; Jalmes, L. S. *Tetrahedron* **2000,** *56,* 275–283.
19. Large, S.; Roques, N.; Langlois, B. R. *J. Org. Chem.* **2000,** *65,* 8848–8856.
20. Russel, J.; Roques, N. *Tetrahedron* **1998,** *54,* 13771–13782.
21. a. Prakash, G. K. S.; Hu, J.; Olah, G. A. *J. Org. Chem.* **2003,** *68,* 4457–4463.
b. Prakash, G. K. S.; Hu, J.; Olah, G. A. *Org. Lett.* **2003,** *5,* 3253–3256.
c. Prakash, G. K. S.; Hu, J. *Acc. Chem. Res.* **2007,** *40,* 921–930.
22. Nonnenmacher, J.; Massicot, F.; Grellepois, F.; Portella, C. *J. Org. Chem.* **2008,** *73,* 7990–7995.
23. Massicot, F.; Nonnenmacher, J.; Grellepois, F.; Portella, C. *Eur. J. Org. Chem.* **2010,** 275–279.
24. Fustero, S.; Albert, L.; Aceña, J. L.; Sanz-Cervera, J. F.; Asensio, A. *Org. Lett.* **2008,** *10,* 605–608.
25. Fustero, S.; Albert, L.; Mateu, N.; Chiva, G.; Miró, J.; González, J.; Aceña, J. L. *Chem. Eur. J.* **2012,** *18,* 3753–3764.
26. a. Obijalska, E.; Mloston, G.; Linden, A.; Heimgartner, H. *Helv. Chim. Acta* **2010,** *93,* 1725–1736.
b. Mloston, G.; Obijalska, E.; Heimgartner, H. *J. Fluor. Chem.* **2011,** *132,* 951–955.
27. Kusuda, A.; Kawai, H.; Nakamura, S.; Shibata, N. *Green Chem.* **2009,** *11,* 1733–1735.
28. Kobayashi, K.; Narumi, T.; Oishi, Sh; Ohno, H.; Fujii, N. *J. Org. Chem.* **2009,** *74,* 4626–4629.
29. Zemtsov, A. A.; Levin, V. V.; Dilman, A. D.; Struchkova, M. I.; Belyakov, P. A.; Tartakovsky, V. A.; Hu, J. *Eur. J. Org. Chem.* **2010,** 6779–6785.
30. Cheng, J. L.; Chu, L. L.; Qing, F. L. *J. Fluor. Chem.* **2013,** *152,* 70–76.
31. Masusai, C.; Soorukram, D.; Kuhakarn, C.; Tuchinda, P.; Reutrakul, V.; Pohmakotr, M. *J. Fluor. Chem.* **2013,** *154,* 37–42.

32. Zhen, Y.; Ma, J. A. *Adv. Synth. Catal.* **2010,** *352,* 2745–2750.
33. a. Yoon, T. P.; Jacobsen, E. N. *Science* **2003,** *299,* 1691–1693.
b. Kolb, H. C.; VanNieuwenhze, M. S.; Sharpless, K. B. *Chem. Rev.* **1994,** *94,* 2483–2547.
c. Dolling, U. H.; Davis, P.; Grabowski, E. J. J. *J. Am. Chem. Soc.* **1984,** *106,* 446–447.
d. Chen, Y.; Tian, S. K.; Deng, L. *J. Am. Chem. Soc.* **2000,** *122,* 9542–9543.
e. Vakulya, B.; Varga, S.; Csámpai, A.; Soós, T. *Org. Lett.* **2005,** *7,* 1967–1969.
f. Iwabuchi, Y.; Nakatani, M.; Yokoyama, N.; Hatakeyama, S. *J. Am. Chem. Soc.* **1999,** *121,* 10219–10220.
g. Roger, L. M. A.; Rouden, J.; Lecomte, L.; Lasne, M. C. *Tetrahedron Lett.* **2003,** *44,* 3047–3050.
h. McCooey, S. H.; Connon, S. J. *Org. Lett.* **2007,** *9,* 599–602.
i. Balázsik, K.; Martinek, T. A.; Bucsi, I.; Szõllõsi, G.; Fogassy, G.; Bartók, M.; Olah, G. A. *J. Mol. Catal. A Chem.* **2007,** *272,* 265–274.
34. Iseki, K.; Nagai, T.; Kobayashi, Y. *Tetrahedron Lett.* **1994,** *35,* 3137–3138.
35. Yudin, A. K. From Nucleophilic Perfluoroalkylations to Electrophilic Intermediates (Ph.D. thesis); University of Southern California, 1996.
36. Zhao, H.; Qin, B.; Liu, X.; Feng, X. *Tetrahedron* **2007,** *63,* 6822–6826.
37. Hu, X.; Wang, J.; Li, W.; Lin, L.; Liu, X.; Feng, X. *Tetrahedron Lett.* **2009,** *50,* 4378–4380.
38. Nagao, H.; Kawano, Y.; Mukaiyama, T. *Bull. Chem. Soc. Jpn.* **2007,** *80,* 2406–2412.
39. Mizuta, S.; Shibata, N.; Akiti, S.; Fujimoto, H.; Nakamura, S.; Toru, T. *Org. Lett.* **2007,** *9,* 3707–3710.
40. Kawai, H.; Kusuda, A.; Mizuta, S.; Nakamura, S.; Funahashi, Y.; Masuda, H.; Shibata, N. *J. Fluor. Chem.* **2009,** *130,* 762–765.
41. Kawai, H.; Mizuta, S.; Tokunaga, E.; Shibata, N. *J. Fluor. Chem.* **2013,** *152,* 46–50.
42. a. Levin, V. V.; Dilman, A. D.; Belyakov, P. A.; Struchkova, M. I.; Tartakovsky, V. A. *Eur. J. Org. Chem.* **2008,** 5226–5230.
b. Dilman, A. D.; Levin, V. V. *Eur. J. Org. Chem.* **2011,** 831–841.
43. Bernardi, L.; Indrigo, E.; Pollicino, S.; Ricci, A. *Chem. Commun.* **2012,** *48,* 1428–1430.
44. Khangarot, R. K.; Kaliappan, K. P. *Eur. J. Org. Chem.* **2013,** 2692–2698.
45. Stephens, D. E.; Chavez, G.; Valdes, M.; Dovalina, M.; Arman, H. D.; Larionov, O. V. *Org. Biomol. Chem.* **2014,** *12,* 6190–6199.
46. Fustero, S.; Moscardó, J.; Sánchez-Roselló, M.; Rodríguez, E.; Barrio, P. *Org. Lett.* **2010,** *12,* 5494–5497.
47. Kawai, H.; Kusuda, A.; Nakamura, S.; Shiro, M.; Shibata, N. *Angew. Chem. Int. Ed.* **2009,** *48,* 6324–6327.
48. Okusu, S.; Kawai, H.; Xu, X. H.; Tokunaga, E.; Shibata, N. *J. Fluor. Chem.* **2012,** *143,* 216–219.
49. Dilman, A. D.; Levin, V. V.; Belyakov, P. A.; Struchkova, M. I.; Tartakovsky, V. A. *Tetrahedron Lett.* **2008,** *49,* 4352–4354.
50. Zemtsov, A. A.; Levin, V. V.; Dilman, A. D.; Struchkova, M. I.; Belyakov, P. A.; Tartakovsky, V. A. *Tetrahedron Lett.* **2009,** *50,* 2998–3000.
51. Gouault-Bironneau, S.; Timoshenko, V. M.; Grellepois, F.; Portella, C. *J. Fluor. Chem.* **2012,** *134,* 164–171.
52. Han, W.; Li, Y.; Tang, H.; Liu, H. *J. Fluor. Chem.* **2012,** *140,* 7–16.
53. Prakash, G. K. S.; Jog, P. V.; Batamak, P. T. D.; Olah, G. A. *Science* **2012,** *338,* 1324–1327.
54. Kawai, H.; Yuan, Z.; Tokunaga, E.; Shibata, N. *Org. Biomol. Chem.* **2013,** *11,* 1446–1450.

55. Zhang, Y.; Fujiu, M.; Serizawa, H.; Mikami, K. *J. Fluor. Chem.* **2013,** *156,* 367–371.
56. Van der Born, D.; Herscheid, J. D. M.; Orru, R. V. A.; Vugts, D. J. *Chem. Commun.* **2013,** *49,* 4018–4020.
57. Nishimine, T.; Taira, H.; Tokunaga, E.; Shiro, M.; Shibata, N. *Angew. Chem. Int. Ed.* **2015,** *54,* 1–6.
58. Prakash, G. K. S.; Wang, Y.; Mogi, R.; Hu, J.; Mathew, T.; Olah, G. A. *Org. Lett.* **2010,** *12,* 2932–2935.
59. Zhao, Y.; Zhu, J.; Ni, C.; Hu, J. *Synthesis* **2010,** 1899–1904.
60. Sakavuyi, K.; Petersen, K. S. *Tetrahedron Lett.* **2013,** *54,* 6129–6132.
61. Shen, X.; Zhou, M.; Ni, C.; Zhang, W.; Hu, J. *Chem. Sci.* **2014,** *5,* 117–122.
62. Cherkupally, P.; Beier, P. *Tetrahedron Lett.* **2010,** *51,* 252–255.
63. Rioski, M. V.; Hart, A. D.; Colby, D. A. *Org. Lett.* **2013,** *15,* 208–211.
64. Prakash, G. K. S.; Zhang, Z.; Wang, F.; Munõz, S.; Olah, G. A. *J. Org. Chem.* **2013,** *78,* 3300–3305.
65. Aikawa, K.; Toya, W.; Nakamura, Y.; Mikami, K. *Org. Lett.* **2015,** *17,* 4996–4999.
66. Urata, H.; Fuchikami, T. *Tetrahedron Lett.* **1991,** *32,* 91–94.
67. Dubinina, G. G.; Furutachi, H.; Vicic, D. A. *J. Am. Chem. Soc.* **2008,** *130,* 8600–8601.
68. Oishi, M.; Kondo, H.; Amii, H. *Chem. Commun.* **2009,** 1909–1911.
69. Morimoto, H.; Tsubogo, T.; Litvinas, N. D.; Hartwig, J. F. *Angew. Chem. Int. Ed.* **2011,** *50,* 3793–3798.
70. Tomashenko, O. A.; Escudero-Adan, E. C.; Belmonte, M. M.; Grushin, V. V. *Angew. Chem. Int. Ed.* **2011,** *50,* 7655–7659.
71. Chu, L. L.; Qing, F. L. *J. Am. Chem. Soc.* **2010,** *132,* 7262–7263.
72. Jiang, X.; Chu, L. L.; Qing, F. -L. *J. Org. Chem.* **2012,** *77,* 1251–1257.
73. Chu, L. L.; Qing, F.-L. *J. Am. Chem. Soc.* **2012,** *134,* 1298–1304.
74. Miyake, Y.; Ota, S.; Nishibayashi, Y. *Chem. Eur. J.* **2012,** *18,* 13255–13258.
75. Miyake, Y.; Ota, S.; Shibata, M.; Nakajima, K.; Nishibayashi, Y. *Chem. Commun.* **2013,** *49,* 7809–7811.
76. Shang, M.; Sun, S. Z.; Wang, H. L.; Laforteza, B. N.; Dai, H. X.; Yu, J. Q. *Angew. Chem. Int. Ed.* **2014,** *53,* 10439–10442.
77. Wu, S.; Guo, J.; Sohail, M.; Cao, Ch; Chen, F. -X. *J. Fluor. Chem.* **2013,** *148,* 19–29.
78. Zanardi, A.; Novikov, M. A.; Martin, E.; Benet-Buchholz, J.; Grushin, V. V. *J. Am. Chem. Soc.* **2011,** *133,* 20901–20913.
79. Novak, P.; Lishchynskyi, A.; Grushin, V. V. *J. Am. Chem. Soc.* **2012,** *134,* 16167–16170.
80. Aikawa, K.; Toya, W.; Nakamura, Y.; Mikami, K. *Chem. Eur. J.* **2015,** *21,* 96–100.
81. Kondo, H.; Oishi, M.; Fujikawa, K.; Amii, H. *Adv. Synth. Catal.* **2011,** *353,* 1247–1252.
82. Langlois, B. R.; Roques, N. *J. Fluor. Chem.* **2007,** *128,* 1318–1325.
83. Schareina, T.; Wu, X. F.; Zapf, A.; Cotté, A.; Gotta, M.; Beller, M. *Top. Catal.* **2012,** *55,* 426–431.
84. Duan, C.; Li, Y.; Chen, T.; Wang, H.; Zhang, R.; Jin, K.; Wang, X. *Synlett* **2011,** 1713–1716.
85. Chen, M.; Buchwald, S. L. *Angew. Chem. Int. Ed.* **2013,** *52,* 11628–11631.
86. Su, D. B.; Duan, J. X.; Chen, Q. Y. *Tetrahedron Lett.* **1991,** *32,* 7689–7690.
87. McNeil, J. G.; Burton, D. J. *J. Fluor. Chem.* **1991,** *55,* 225–227.
88. Knauber, T.; Arikan, F.; Röschenthaler, G. V.; Gooßen, L. J. *Chem. Eur. J.* **2011,** *17,* 2689–2697.
89. Serizawa, H.; Aikawa, K.; Mikami, K. *Chem. Eur. J.* **2013,** *19,* 17692–17697.
90. Ambler, B. R.; Altman, R. A. *Org. Lett.* **2013,** *15,* 5578–5581.
91. Li, L.; Zhao, J.; Zhang, L.; Hu, M.; Wang, L.; Hu, J. *Org. Lett.* **2015,** *17,* 298–301.

92. Cho, E. J.; Senecal, T. D.; Kinzel, T.; Zhang, Y.; Watson, D. A.; Buchwald, S. L. *Science* **2010,** *328,* 1679–1681.
93. Cho, E. J.; Buchwald, S. L. *Org. Lett.* **2011,** *13,* 6552–6555.
94. Samant, B. S.; Kabalka, G. W. *Chem. Commun.* **2011,** *47,* 7236–7238.
95. Mu, X.; Chen, S.; Zhen, X.; Liu, G. *Chem. Eur. J.* **2011,** *17,* 6039–6042.
96. Mu, X.; Wu, T.; Wang, H. Y.; Guo, Y. L.; Liu, G. *J. Am. Chem. Soc.* **2012,** *134,* 878–881.
97. Ball, N. D.; Gary, J. B.; Ye, Y.; Sanford, M. S. *J. Am. Chem. Soc.* **2011,** *133,* 7577–7584.
98. Kremlev, M. M.; Mushta, A. I.; Tyrra, W.; Naumann, D.; Fischer, H. T. M.; Yagupolskii, Y. L. *J. Fluor. Chem.* **2007,** *128,* 1385–1389.
99. Ye, Y.; Lee, S. H.; Sanford, M. S. *Org. Lett.* **2011,** *13,* 5464–5467.
100. Zeng, Y.; Zhang, L.; Zhao, Y.; Ni, C.; Zhao, J.; Hu, J. *J. Am. Chem. Soc.* **2013,** *135,* 2955–2958.
101. a. Jones, R. G. *J. Am. Chem. Soc.* **1947,** *69,* 2346–2350.
     b. Uchiyama, M.; Naka, H.; Matsumoto, Y.; Ohwada, T. *J. Am. Chem. Soc.* **2004,** *126,* 10526–10527.
102. Wang, K. P.; Yun, S. Y.; Mamidipalli, P.; Lee, D. *Chem. Sci.* **2013,** *4,* 3205–3211.
103. Wang, X.; Xu, Y.; Mo, F. Y.; Ji, G. J.; Qiu, D.; Feng, J. J.; Ye, Y. X.; Zhang, S. N.; Zhang, Y.; Wang, J. B. *J. Am. Chem. Soc.* **2013,** *135,* 10330–10333.
104. a. Pierce, O. R.; McBee, E. T.; Judd, G. F. *J. Am. Chem. Soc.* **1954,** *76,* 474–478.
     b. Johncock, P. *J. Organomet. Chem.* **1969,** *19,* 257–265.
     c. Gassman, P. G.; Oreilly, N. J. *Tetrahedron Lett.* **1985,** *26,* 5243–5246.
     d. Gassman, P. G.; Oreilly, N. J. *J. Org. Chem.* **1987,** *52,* 2481–2490.
     e. McBee, E. T.; Roberts, C. W.; Meiners, A. F. *J. Am. Chem. Soc.* **1957,** *79,* 335–337.
     f. McBee, E. T.; Braendlin, H. P.; Battershell, R. D. *J. Org. Chem.* **1963,** *28,* 1131–1133.
     g. Denson, D. D.; Smith, C. F.; Tamborsk, C. *J. Fluor. Chem.* **1974,** *3,* 247–258.
     h. Smith, C. F.; Soloski, E. J.; Tamborsk, C. *J. Fluor. Chem.* **1974,** *4,* 35–45.
     i. Thoai, N. *J. Fluor. Chem.* **1975,** *5,* 115–125.
     j. Von Werner, K.; Gisser, A. *J. Fluor. Chem.* **1977,** *10,* 387–394.
105. McLoughlin, V. C. R.; Thrower, J. *Tetrahedron* **1969,** *25,* 5921–5940.
106. a. Santini, G.; Leblanc, M.; Riess, J. G. *J. Chem. Soc. Chem. Commun.* **1975,** 678–679.
     b. Santini, G.; Leblanc, M.; Riess, J. G. *J. Organomet. Chem.* **1977,** *140,* 1–9.
     c. Kitazume, T.; Ishikawa, N. *Chem. Lett.* **1981,** 1337–1338.
     d. Krishnamurti, R.; Bellew, D. R.; Prakash, G. K. S. *J. Org. Chem.* **1991,** *56,* 984–989.
     e. OReilly, N. J.; Maruta, M.; Ishikawa, N. *Chem. Lett.* **1984,** 517–520.
     f. Kitazume, T.; Ishikawa, N. *J. Am. Chem. Soc.* **1985,** *107,* 5186–5191.
107. Pooput, C.; Dolbier, W. R.; Medebielle, M. *J. Org. Chem.* **2006,** *71,* 3564–3568.
108. Popov, I.; Lindeman, S.; Daugulis, O. *J. Am. Chem. Soc.* **2011,** *133,* 9286–9289.
109. Hu, J.; Ni, C. Difluoro- and Fluoromethylation. In *Science of Synthesis: C-1 Building Blocks in Organic Synthesis,* Vol. 2, van Leeuwen, P. W. N. M., Ed.; Thieme: Stuttgart, 2014; pp 409–457.
110. Beier, P.; Alexandrova, A. V.; Zibinsky, M.; Prakash, G. K. S. *Tetrahedron* **2008,** *64,* 10977–10985.
111. Alexandrova, A. V.; Beier, P. *J. Fluor. Chem.* **2009,** *130,* 493–500.
112. Prakash, G. K. S.; Zibinsky, M.; Upton, T. G.; Kashemirov, B. A.; McKenna, C. E.; Oertell, K.; Goodman, M. F.; Batra, V. K.; Pedersen, L. C.; Beard, W. A.; et al. *Proc. Natl. Acad. Sci. U.S.A.* **2010,** *107,* 15693–15698.
113. Zhao, Y.; Huang, W.; Zheng, J.; Hu, J. *Org. Lett.* **2011,** *13,* 5342–5345.
114. Kosobokov, M. D.; Dilman, A. D.; Levin, V. V.; Struchkova, M. I. *J. Org. Chem.* **2012,** *77,* 5850–5855.

115. Prakash, G. K. S.; Wang, Y.; Hu, J.; Olah, G. A. *J. Fluor. Chem.* **2005,** *126,* 1361–1367.
116. Pohmakotr, M.; Panichakul, D.; Tuchinda, P.; Reutrakul, V. *Tetrahedron* **2007,** *63,* 9429–9436.
117. Bootwicha, T.; Panichakul, D.; Kuhakarn, C.; Prabpai, S.; Kongsaeree, P.; Tuchinda, P.; Reutrakul, V.; Pohmakotr, M. *J. Org. Chem.* **2009,** *74,* 3798–3805.
118. Pharikronburee, V.; Punirun, T.; Soorukram, D.; Kuhakarn, C.; Tuchinda, P.; Reutrakul, V.; Pohmakotr, M. *Org. Biomol. Chem.* **2013,** *11,* 2022–2033.
119. Li, Y.; Hu, J. *Angew. Chem. Int. Ed.* **2007,** *46,* 2489–2492.
120. Kosobokov, M. D.; Dilman, A. D.; Struchkova, M. I.; Belyakov, P. A.; Hu, J. *J. Org. Chem.* **2012,** *77,* 2080–2086.
121. a. Li, Y.; Hu, J. *J. Fluor. Chem.* **2008,** *129,* 382–385.
b. Levin, V. V.; Elkin, P. K.; Struchkova, M. I.; Dilman, A. D. *J. Fluor. Chem.* **2013,** *154,* 43–46.
122. Hu, M.; Wang, F.; Zhao, Y.; He, Z.; Zhang, W.; Hu, J. *J. Fluor. Chem.* **2012,** *135,* 45–58.
123. Punirun, T.; Soorukram, D.; Kuhakarn, C.; Reutrakul, V.; Pohmakotr, M. *Eur. J. Org. Chem.* **2014,** 4162–4169.
124. a. Stahly, G. P. *J. Fluor. Chem.* **1989,** *43,* 53–66.
b. Serafinowski, P. J.; Barnes, C. J. *Synthesis* **1997,** 225–231.
c. Prakash, G. K. S.; Hu, J.; Wang, Y.; Olah, G. A. *Org. Lett.* **2004,** *6,* 4315–4317.
d. Ye, J. D.; Liao, X.; Piccirilli, J. A. *J. Org. Chem.* **2005,** *70,* 7902–7910.
e. Prakash, G. K. S.; Hu, J.; Wang, Y.; Olah, G. A. *Angew. Chem. Int. Ed.* **2004,** *43,* 5203–5206.
125. Ni, C.; Hu, J. *Tetrahedron Lett.* **2005,** *46,* 8273–8277.
126. Ni, Y.; Liu, J.; Zhang, L.; Hu, J. *Angew. Chem. Int. Ed.* **2007,** *46,* 786–789.
127. Hu, M.; Gao, B.; Ni, C.; Zhang, L.; Hu, J. *J. Fluor. Chem.* **2013,** *155,* 52–58.
128. Prakash, G. K. S.; Hu, J.; Mathew, T.; Olah, G. A. *Angew. Chem. Int. Ed.* **2003,** *42,* 5216–5219.
129. a. Li, Y.; Hu, J. *Angew. Chem. Int. Ed.* **2005,** *44,* 5882–5886.
b. Liu, J.; Li, Y.; Hu, J. *J. Org. Chem.* **2007,** *72,* 3119–3121.
c. Liu, J.; Hu, J. *Chem. Eur. J.* **2010,** *16,* 11443–11454.
d. Fustero, S.; Ibañez, I.; Barrio, P.; Maestro, M. A.; Catalan, S. *Org. Lett.* **2013,** *15,* 832–840.
130. a. Ni, C.; Zhang, L.; Hu, J. *J. Org. Chem.* **2009,** *74,* 3767–3771.
b. Sanz-Marco, A.; García-Ortiz, A.; Blay, G.; Fernandez, I.; Pedro, J. R. *Chem. Eur. J.* **2014,** *20,* 668–672.
131. a. Zheng, J.; Li, Y.; Zhang, L.; Hu, J.; Meuzelaar, G. J.; Federsel, H. J. *Chem. Commun.* **2007,** 5149–5151.
b. Borys, K. M.; Korzynski, M. D.; Ochal, Z. *Tetrahedron Lett.* **2012,** *53,* 6606–6610.
132. Ni, C.; Wang, F.; Hu, J. *Beilstein J. Org. Chem.* **2008,** *4* (21).
133. Liu, J.; Ni, C.; Wang, F.; Hu, J. *Tetrahedron Lett.* **2008,** *49,* 1605–1608.
134. Huang, W.; Ni, C.; Zhao, Y.; Gao, B.; Hu, J. *J. Fluor. Chem.* **2012,** *143,* 161–166.
135. Zhu, L.; Li, Y.; Zhao, Y.; Hu, J. *Tetrahedron Lett.* **2010,** *51,* 6150–6152.
136. Zhao, Y.; Huang, W.; Zhu, L.; Hu, J. *Org. Lett.* **2010,** *12,* 1444–1447.
137. Prakash, G. K. S.; Ni, C.; Wang, F.; Hu, J.; Olah, G. A. *Angew. Chem. Int. Ed.* **2011,** *50,* 2559–2563.
138. Zhao, Y.; Gao, B.; Hu, J. *J. Am. Chem. Soc.* **2012,** *134,* 5790–5793.
139. Prakash, G. K. S.; Ni, C.; Wang, F.; Zhang, Z.; Haiges, R.; Olah, G. A. *Angew. Chem. Int. Ed.* **2013,** *52,* 10835–10839.
140. Li, J. L.; Liu, J. T. *Tetrahedron* **2007,** *63,* 898–903.

141. a. Zhang, W.; Wang, F.; Hu, J. *Org. Lett.* **2009,** *11,* 2109–2112.
b. Shen, X.; Zhang, W.; Ni, C.; Gu, Y.; Hu, J. *J. Am. Chem. Soc.* **2012,** *134,* 16999–17002.
142. a. Shen, X.; Liu, Q.; Luo, T.; Hu, J. *Chem. Eur. J.* **2014,** *20,* 6795–6800.
b. Shen, X.; Liu, Q.; Ni, C.; Hu, J. *Chin. J. Chem.* **2014,** *32,* 703–708.
143. Aikawa, K.; Maruyama, K.; Honda, K.; Mikimi, K. *Org. Lett.* **2015,** *17,* 4882–4885.
144. Okusu, S.; Tokunaga, E.; Shibata, N. *Org. Lett.* **2015,** *17,* 3802–3805.
145. Fujikawa, K.; Fujioka, Y.; Kobayashi, A.; Amii, H. *Org. Lett.* **2011,** *13,* 5560–5563.
146. Zhu, J.; Wang, F.; Huang, W.; Zhao, Y.; Ye, W.; Hu, J. *Synlett* **2011,** *7,* 899–902.
147. Prakash, G. K. S.; Ganesh, S. K.; Jones, J. P.; Kulkarni, A.; Masood, K.; Swabeck, J. K.; Olah, G. A. *Angew. Chem. Int. Ed.* **2012,** *51,* 12090–12094.
148. Fier, P. S.; Hartwig, J. F. *J. Am. Chem. Soc.* **2012,** *134,* 5524–5527.
149. Jiang, X. L.; Chen, Z. H.; Xu, X. H.; Qing, F. L. *Org. Chem. Front.* **2014,** *1,* 774–776.
150. Matheis, C.; Jouvin, K.; Goossen, L. J. *Org. Lett.* **2014,** *16,* 5984–5987.
151. Gu, Y.; Leng, X.; Shen, Q. *Nat. Comm.* **2014,** *7* (5), 5405.
152. Zhu, L.; Li, Y.; Ni, C.; Hu, J.; Beier, P.; Prakash, G. K. S.; Olah, G. A. *J. Fluor. Chem.* **2007,** *128,* 1241–1247.
153. Pohmakotr, M.; Boonkitpattarakul, K.; Ieawsuwan, W.; Jarussophon, S.; Duangdee, N.; Tuchinda, P.; Reutrakul, V. *Tetrahedron* **2006,** *62,* 5973–5985.
154. Gao, B.; Zhao, Y.; Hu, M.; Ni, C.; Hu, J. *Chem. Eur. J.* **2014,** *20,* 7803–7810. and references therein.
155. Zhang, Z.; Yu, W.; Wu, C.; Wang, C.; Zhang, Y.; Wang, J. *Angew. Chem. Int. Ed.* **2015,** *54.* http://dx.doi.org/10.1002/anie.201509711.
156. Hu, M.; Ni, C.; Li, C.; Han, Y.; Hu, J. *J. Am. Chem. Soc.* **2015,** *137,* 14496–14501.
157. a. Kollonitsch, J.; Patchett, A. A.; Marburg, S.; Maycock, A. L.; Perkins, L. M.; Doldouras, G. A.; Duggan, D. E.; Aster, S. D. *Nature* **1978,** *274,* 906–908.
b. Grunewald, G. L.; Caldwell, T. M.; Li, Q.; Slavica, M.; Criscione, K. R.; Borchardt, R. T.; Wang, W. *J. Med. Chem.* **1999,** *42,* 3588–3601.
c. Grunewald, G. L.; Seim, M. R.; Regier, R. C.; Martin, J. L.; Gee, L.; Drinkwater, N.; Criscione, K. R. *J. Med. Chem.* **2006,** *49,* 5424–5433.
d. Rothman, S. C.; Johnston, J. B.; Lee, S.; Walker, J. R.; Poulter, C. D. *J. Am. Chem. Soc.* **2008,** *130,* 4906–4907.
158. a. Satoh, T.; Onda, K.; Itoh, N.; Yamakawa, K. *Tetrahedron Lett.* **1991,** *32,* 5599–5600.
b. Satoh, T.; Takano, K. *Tetrahedron* **1996,** *52,* 2349–2358.
c. Reutrakul, V.; Rukachaisirikul, V. *Tetrahedron Lett.* **1983,** *24,* 725–728.
d. Nájera, C.; Yus, M.; Hässig, R.; Seebach, D. *Helv. Chim. Acta* **1984,** *67,* 1100–1103.
159. Yagupol'skii, L. M.; Aleksandrov, A. M. *Zh. Obshch. Khim.* **1968,** *38,* 1503–1509.
160. Inbasekaran, M.; Peet, N.; McCarthy, J. R.; LeTourneau, M. E. *J. Chem. Soc. Chem. Commun.* **1985,** 678–679.
161. Li, Y.; Ni, C.; Liu, J.; Zhang, L.; Zheng, J.; Zhu, L.; Hu, J. *Org. Lett.* **2006,** *8,* 1693–1696.
162. Liu, J.; Zhang, L.; Hu, J. *Org. Lett.* **2008,** *10,* 5377–5380.
163. Ni, C.; Li, Y.; Hu, J. *J. Org. Chem.* **2006,** *71,* 6829–6833.
164. Fukuzumi, T.; Shibata, N.; Sugiura, M.; Yasui, H.; Nakamura, S.; Toru, T. *Angew. Chem. Int. Ed.* **2006,** *45,* 4973–4977.
165. Prakash, G. K. S.; Chacko, S.; Alconcel, S.; Stewart, T.; Mathew, T.; Olah, G. A. *Angew. Chem. Int. Ed.* **2007,** *46,* 4933–4936.
166. Prakash, G. K. S.; Chacko, S.; Vaghoo, H.; Shao, N.; Gurung, L.; Mathew, T.; Olah, G. A. *Org. Lett.* **2009,** *11,* 1127–1130.
167. Furukawa, T.; Goto, Y.; Kawazoe, J.; Tokunaga, E.; Nakamura, S.; Yang, Y.; Du, H.; Kakehi, A.; Shiro, M.; Shibata, N. *Angew. Chem. Int. Ed.* **2010,** *49,* 1642–1647.

168. Shen, X.; Zhang, L.; Zhao, Y.; Zhu, L.; Li, G.; Hu, J. *Angew. Chem. Int. Ed.* **2011,** *50,* 2588–2592.
169. Huang, W.; Ni, C.; Zhao, Y.; Hu, J. *New J. Chem.* **2013,** *37,* 1684–1687.
170. Prakash, G. K. S.; Gurung, L.; Jog, P. V.; Tanaka, S.; Thomas, T. E.; Ganesh, N.; Haiges, R.; Mathew, T.; Olah, G. A. *Chem. Eur. J.* **2013,** *19,* 3579–3583.
171. Prakash, G. K. S.; Shao, N.; Zhang, Z.; Ni, C.; Wang, F.; Haiges, R.; Olah, G. A. *J. Fluor. Chem.* **2012,** *133,* 27–32.
172. Nagura, H.; Fuchigami, T. *Synlett* **2008,** 1714–1718.
173. a. Prakash, G. K. S.; Wang, F.; Ni, C.; Thomas, T. J.; Olah, G. A. *J. Fluor. Chem.* **2010,** *131,* 1007–1012.
b. Prakash, G. K. S.; Shao, N.; Wang, F.; Ni, C. *Org. Synth.* **2014,** *90,* 130–144.
174. Mizuta, S.; Shibata, N.; Goto, Y.; Furukawa, T.; Nakamura, S.; Toru, T. *J. Am. Chem. Soc.* **2007,** *129,* 6394–6395.
175. Matsuzaki, K.; Furukawa, T.; Tokunaga, E.; Matsumoto, T.; Shiro, M.; Shibata, N. *Org. Lett.* **2013,** *15,* 3282–3285.
176. Furukawa, T.; Shibata, N.; Mizuta, S.; Nakamura, S.; Toru, T.; Shiro, M. *Angew. Chem. Int. Ed.* **2008,** *47,* 8051–8054.
177. Prakash, G. K. S.; Wang, F.; Stewart, T.; Mathew, T.; Olah, G. A. *Proc. Natl. Acad. Sci. U.S.A.* **2009,** *106,* 4090–4094.
178. Moon, H. W.; Cho, M. J.; Kim, D. Y. *Tetrahedron Lett.* **2009,** *50,* 4896–4898.
179. a. Alba, A. N.; Companyo, X.; Moyano, A.; Rios, R. *Chem. Eur. J.* **2009,** *15,* 7035–7038.
b. Ullah, F.; Zhao, G. L.; deiana, L.; Zhu, M.; Dziedzic, P.; Ibrahem, I.; Hammar, P.; Sun, J.; Cordova, A. *Chem. Eur. J.* **2009,** *15,* 10013–10017.
c. Zhang, S.; Zhang, Y.; Ji, Y.; Li, H.; Wang, W. *Chem. Commun.* **2009,** 4886–4888.
180. Kim, Y. S.; Kim, S. M.; Wang, B.; Companyo, X.; Li, J.; Moyano, A.; Im, S.; Tosner, Z.; Yang, J. W.; Riosa, R. *Adv. Synth. Catal.* **2014,** *356,* 437–446.
181. Yang, E.; Wei, X.; Pan, Y.; Lee, R.; Zhu, B.; Liu, H.; Yan, L.; Huang, K.-W.; Jiang, Z.; Tan, C.-H. *Chem. Eur. J.* **2011,** *17,* 8066–8070.
182. Furukawa, T.; Kawazoe, J.; Zhang, W.; Nishimine, T.; Tokunaga, E.; Matsumoto, T.; Shiro, M.; Shibata, N. *Angew. Chem. Int. Ed.* **2011,** *50,* 9684–9688.
183. Zhang, S. L.; Xie, H. X.; Zhu, J.; Li, H.; Zhang, X. S.; Li, J.; Wang, W. *Nat. Comm.* **2011,** *2,* 211.
184. Ma, H.; Matsuzaki, K.; Yang, Y.-D.; Tokunaga, E.; Nakane, D.; Ozawa, T.; Masuda, H.; Shibata, N. *Chem. Commun.* **2013,** *49,* 11206–11208.
185. a. Boys, M. L.; Collington, E. W.; Finch, H.; Swanson, S.; Whitehead, J. F. *Tetrahedron Lett.* **1988,** *29,* 3365–3368.
b. Kahraman, M.; Sinishtaj, S.; Dolan, P. M.; Kensler, T. W.; Peleg, S.; Saha, U.; Chuang, S. S.; Bernstein, G.; Korczak, B.; Posner, G. H. *J. Med. Chem.* **2004,** *47,* 6854–6863.
186. Zhang, W.; Hu, J. *Adv. Synth. Catal.* **2010,** *352,* 2799–2804.
187. Luo, T.; Zhang, R.; Zhang, W.; Shen, X.; Umemoto, T.; Hu, J. *Org. Lett.* **2014,** *16,* 888–891.
188. Shen, X.; Zhang, W.; Zhang, L.; Luo, T.; Wan, X.; Gu, Y.; Hu, J. *Angew. Chem. Int. Ed.* **2012,** *51,* 6966–6970.
189. Kail, D. C.; Krizkova, P. M.; Wieczorek, A.; Hammerschmidt, F. *Chem. Eur. J.* **2014,** *20,* 4086–4091.
190. Inoue, M.; Araki, K. Jpn. Kokai Tokyo Koho **2011**, 2011162521.
191. Zhao, Y.; Gao, B.; Ni, C.; Hu, J. *Org. Lett.* **2012,** *14,* 6080.
192. Zhao, Y.; Ni, C.; Jiang, F.; Gao, B.; Shen, X.; Hu, J. *ACS Catal.* **2013,** *3,* 631–634.

193. a. Liu, W. B.; Zheng, S. C.; He, H.; Zhao, X. M.; Dai, L. -X.; You, S. L. *Chem. Commun.* **2009,** 6604–6606.
b. Zhao, X.; Liu, D.; Zheng, S.; Gao, N. *Tetrahedron Lett.* **2011,** *52,* 665–667.
194. Chevrie, D.; Lequeux, T.; Demoute, J. P.; Pazenok, S. *Tetrahedron Lett.* **2003,** *44,* 8127–8130.
195. del Solar, M.; Ghosh, A. K.; Zajc, B. *J. Org. Chem.* **2008,** *73,* 8206–8211.
196. a. Pfund, E.; Lebargy, C.; Rouden, J.; Lequeux, T. *J. Org. Chem.* **2007,** *72,* 7871–7877.
b. Calata, C.; Catel, J. M.; Pfund, E.; Lequeux, T. *Tetrahedron* **2009,** *65,* 3967–3973.
c. Zajc, B.; Kake, S. *Org. Lett.* **2006,** *8,* 4457–4460.
197. Ghosh, A. K.; Zajc, B. *Org. Lett.* **2006,** *8,* 1553–1556.
198. Ghosh, A. K.; Banerjee, S.; Sinha, S.; Kang, S. B.; Zajc, B. *J. Org.Chem.* **2009,** *74,* 3689–3697.
199. Kumar, R.; Zajc, B. *J. Org. Chem.* **2012,** *77,* 8417–8427.
200. Kumar, R.; Pradhan, P.; Zajc, B. *Chem. Commun.* **2011,** *47,* 3891–3893.
201. a. Prunier, A.; Calata, C.; Legros, J.; Maddaluno, J.; Pfund, E.; Lequeux, T. *J. Org. Chem.* **2013,** *78,* 8083–8097.
b. Calata, C.; Pfund, E.; Lequeux, T. *Tetrahedron* **2011,** *67,* 1398–1405.
c. Habib, S.; Larnaud, F.; Pfund, E.; Lequeux, T.; Fenet, B.; Goekjian, P. G.; Gueyrard, D. *Eur. J. Org. Chem.* **2013,** 1872–1875.
202. a. Alonso, D. A.; Fuensanta, M.; Gomez-Bengoa, E.; Najera, C. *Adv. Synth. Catal.* **2008,** *350,* 1823–1829.
b. Prakash, G. K. S.; Shakhmin, A.; Zibinsky, M.; Ledneczki, I.; Chacko, S.; Olah, G. A. *J. Fluor. Chem.* **2010,** *131,* 1192–1197.
203. Ghosh, A. K.; Zajc, B. *J. Org. Chem.* **2009,** *74,* 8531–8540.
204. Zhu, L.; Ni, C.; Zhao, Y.; Hu, J. *Tetrahedron* **2010,** *66,* 5089–5100.
205. Larnaud, F.; Pfund, E.; Linclau, B.; Lequeux, T. *Tetrahedron* **2014,** *70,* 5632–5639.
206. Fujii, A.; Usuki, Y.; Iio, H.; Tokoryama, T. *Synlett* **1994,** 725–726.
207. Jacobsen, C. B.; Nielsen, M.; Worgull, D.; Zweifel, T.; Fisker, E.; Jørgensen, K. A. *J. Am. Chem. Soc.* **2011,** *133,* 7398–7404.
208. Zhang, W.; Huang, W.; Hu, J. *Angew. Chem. Int. Ed.* **2009,** *48,* 9858–9861.
209. Prakash, G. K. S.; Zhang, Z.; Wang, F.; Iuliucci, M.; Ni, C.; Rahm, M.; Haiges, R.; Olah, G. A. *Chem. Eur. J.* **2014,** *20,* 831–838.
210. Beier, P.; Pohl, R.; Alexandrova, A. V. *Synthesis* **2009,** *6,* 0957–0962.

Continuous Flow Selective Direct Fluorination Using Fluorine Gas

12

G. Sandford
Durham University, Durham, United Kingdom

Chapter Outline

1. Introduction

A crucial stage in the synthesis of all fluorinated organic compounds is the construction of the carbon–fluorine bond whether at the late stage of a synthetic strategy or for the preparation of a fluorinated "building block" that is used as a substrate for transformation into more structurally sophisticated systems.[1,8,22] In particular, syntheses of fluorinated derivatives for the pharmaceutical industry, from the discovery phase to the manufacturing scale are, in many cases, very challenging given the structural complexity of many advanced pharmaceutical intermediates (API) and marketed drugs. Consequently, over the years, a range of fluorinating reagents have been developed to meet the varied synthetic requirements of the life science industry[1] and many reagents are commercially available from the usual specialty and bulk chemical suppliers.

The choice of fluorinating agent to use for any particular transformation depends on a number of factors including functional group tolerance, synthetic process, appropriate substrate, availability, facilities, and expertise. In particular, the reaction scale plays a role in choosing a particular fluorinating agent. For example, medicinal chemists requiring 10–20 mg of a fluorinated molecule for biological screening assays need easily handled, nontoxic, shelf-stable, and noncorrosive reagents that are

Modern Synthesis Processes and Reactivity of Fluorinated Compounds. http://dx.doi.org/10.1016/B978-0-12-803740-9.00012-3

amenable for use in automated parallel synthesis and purification systems. For such applications, the cost of a reagent is not a consideration and purification is usually performed automatically by column chromatography using preloaded silica gel cartridges. In contrast, larger scale syntheses to produce multi-kilogram quantities of fluorinated product for advanced clinical trials and manufacturing, the cost of API production, and the ease of purification, usually by recrystallization, become very important factors in the choice of the fluorinating agent and synthetic strategy used. In general, therefore, reagents used for the discovery and early clinical trial phases of a drug development campaign are often very different from those used on the manufacturing scale, principally due to reagent cost, ease of API production, and purification.[15]

All fluorinating reagents are derived from anhydrous hydrogen fluoride (aHF), which is manufactured on a very large scale from mined fluorspar.[16] Consequently, fluorinating agents that are sufficiently inexpensive for use on the manufacturing scale are, in general, aHF, KF, and F_2, which are easily accessible in one or two steps from fluorspar raw material. Systems such as Selectfluor, diethylaminosulfur trifluoride (DAST), Deoxo-Fluor and other electrophilic fluorinating agents of the N-F class are, in general, too expensive for manufacturing procedures, although, of course, some exceptions exist.[2]

Anhydrous HF is the most widely used fluorinating agent in manufacturing including extensive application for the synthesis of fluoro- and trifluoromethyl aromatic derivatives, which appear in the majority of fluorinated drug structures, despite the extreme hazards associated with using this very volatile (bp 19°C), highly corrosive and toxic reagent. KF is used for some halogen exchange processes for the synthesis of some appropriately activated fluoroheterocyclic systems, for example, although the low solubility and harsh reaction conditions required for useful yield preclude the general applicability of this fluoride ion source. In contrast, the use of fluorine gas for the manufacture of APIs and downstream pharmaceutical products has received little attention, perhaps due to the lack of suitable synthetic methodology available, until recent years. The general perception that fluorine is too reactive for effective use as a selective fluorinating agent has also hampered the development of F_2 as a reagent for organic synthesis despite the fact that F_2 is synthesized by electrolysis of aHF and is, therefore, relatively inexpensive. Indeed, 5-fluorouracil (oncology) has been manufactured for over 50 years by direct fluorination of uracil in aqueous acetic acid and, in 2001, a process for the synthesis of the 5-fluoropyrimidine subunit of the antifungal agent fluconazole (V-Fend, Pfizer) using F_2 began operation.[4] These two successes point to the opportunities for using F_2 for the synthesis of fluorinated APIs on the manufacturing scale.

In this chapter, we discuss how direct fluorination reactions using F_2 can be used for the synthesis of fluorinated derivatives of interest to life science programs by continuous flow procedures that are, in principle, suitable for scale-up. There is a considerable effort worldwide directed towards developing a vast array of continuous flow procedures, and reactor design is an especially important area.[14,23] There are many advantages associated with using continuous flow reactors for chemical synthesis at both the laboratory and manufacturing stages, and those discussed widely include high throughput; use of very small quantities of material when appropriate; reduced waste streams; low manufacturing, operation, and maintenance costs; low power consumption; increased precision and accuracy; and disposability. Continuous flow processes may also lead to increased performance of a system due to optimization of contact between reagents because of very rapid mixing in such devices. The concept of "scale-out" by running many reactors in parallel is also very appealing where bench operation would exactly mirror the manufacturing situation, thus reducing the resources required for good manufacturing practice accreditation. Additionally, there are special advantages to developing continuous flow reactors for potentially hazardous processes, where the inventory of reactants is sufficiently small in any one reactor section, such that levels of control and safety could be achieved, which would not be possible by conventional means.

This chapter is not a comprehensive review of all processes involving fluorine gas in continuous flow reactor processes but highlights some of the opportunities available and derived from research reported by the Durham group.[17,19,21]

2. Continuous Flow Processes Using Fluorine Gas

When performing a synthetic procedure using a gaseous reagent, careful consideration of the experimental technique is required due to inherent associated handling issues. Batch processes, in which gaseous reagents are passed through the rapidly stirred liquid phase, have been used very successfully for many years. However, gas mass transfer into the liquid phase, gas—liquid mixing via a small interfacial area, difficulties in heating or cooling a multiphase process, and handling waste gas streams efficiently have led to the development of complementary continuous flow technology for both the laboratory and manufacturing stages. The passage of gas and liquid substrate streams through a small-volume reactor channel by a continuous flow process offers far easier reaction control, increased safety due to the relatively small amounts of gas and liquid reagents in contact at any given time, more effective reaction temperature control leading to increased reaction selectivity, and better phase mixing, which allows more controlled gas mass transfer.[18]

Several types of flow reactor have been used for gas—liquid reactions including falling film, gas-permeable tube-in-tube membrane systems, and "mesh-" modified micro- or mesoscale systems, and these have been reviewed in 2015.[18]

2.1 Durham Continuous Flow Direct Fluorination Equipment

At Durham University (United Kingdom), several continuous flow reactor systems have been developed during a research program that began in 1995, and the most frequently used single- and multichannel systems used in Durham for fluorine gas reactions are described briefly in the following sections.[5,7,9] Full details of the equipment and experimental procedures may be obtained from the appropriate cited references and not repeated here.

2.1.1 Single-Channel Flow Reactors for Direct Fluorination

The first reactor designed for reactions involving fluorine gas was reported in 1999.[5,7] The reactor (Fig. 12.1) is, essentially, a nickel metal block that has a 0.5-mm-diameter reaction channel etched into its surface. A transparent polychlorotrifluoroethylene (PCTFE) plate and a bolted stainless-steel sheet complete the reactor channel. Fluorine gas and other reagent streams are introduced into the reactor channel via stainless-steel tubing, and the product outflow is collected. The temperature of the reaction block may be controlled by an external, circulating heating/cooling cryostat. Gas and liquid reagents are introduced into the reactor channel at prescribed rates by commercially available gas mass flow controller and either a syringe or high-performance liquid chromatography pump addition, respectively. In this reactor design, the gas and liquid flow concurrently down the reactor channel by laminar or "pipe" flow, as viewed through the PCTFE window, which enables excellent phase mixing and gas transport into the liquid phase.

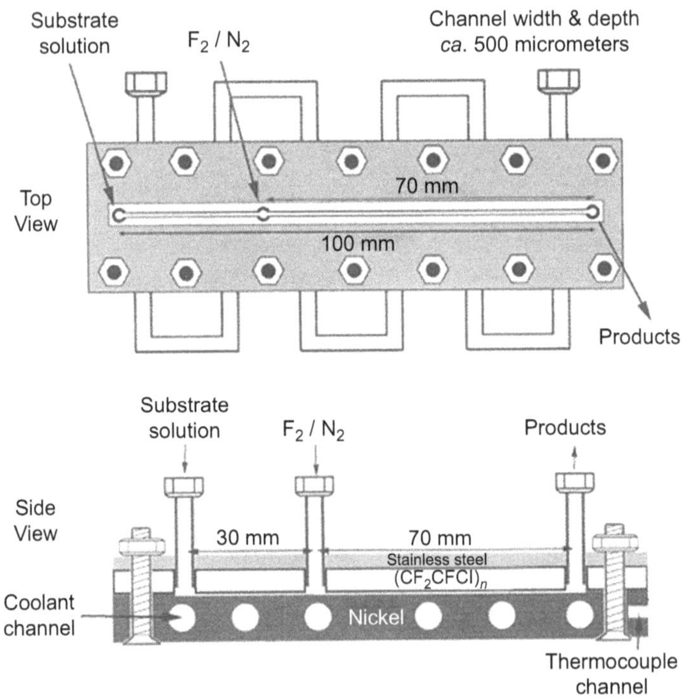

Figure 12.1 Durham single-channel laminar flow reactor.

A simpler single-channel continuous flow system is composed of a 0.1-mm-internal-diameter stainless-steel tube fitted with a T-piece union that enables introduction of both gas and liquid feedstocks. The tube can be coiled to allow the entire apparatus to be submerged in a cooling or heating fluid as appropriate depending upon the reaction conditions.

2.1.2 Multi-Channel Flow Reactors for Direct Fluorination

For larger scale syntheses, use of single-channel systems that have a much longer path length may allow higher throughput of material but, scaling out, which uses a reactor that presents a number of channels in parallel, offers a complementary approach. Multichannel reactors (Fig. 12.2) for direct fluorination have been developed at Durham[9] and allow the synthesis of 100 g quantities of the product material in an 8-h reaction using a 9-channel parallel reactor.

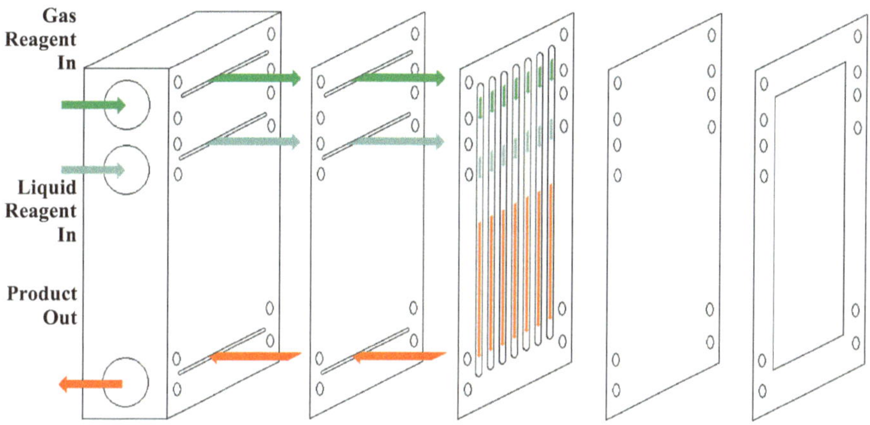

Figure 12.2 Multichannel continuous laminar flow reactor.

The reactor, described in detail elsewhere, basically consists of a steel base block that incorporates two substrate reservoirs within the body of the block and a stainless-steel reactor channel plate of typically 9, 18, or 30 channels that are 0.5 mm wide and 0.5 mm deep. Transparent PTCFE and stainless-steel window plates compress the reactor system together. Continuous operation of various fluorination processes over long reaction time periods of 150 h have been performed with no loss of yield or conversion.

The use of fluorine gas in any of the continuous flow systems described earlier, of course, requires experience and expertise in handling fluorine gas mixtures, careful passivation of all metal surfaces and reaction channels, and correct choice of appropriate resistant valves and mass flow meters before operation can begin.

3. Selective Direct Fluorination by Continuous Flow Processes

Various selective direct fluorination reactions using fluorine gas are presented in the following subsections using the laminar flow systems described earlier to illustrate the opportunities available using this usually undervalued and underused reagent.[19]

Given the low expense of fluorine gas and low capital expenditure required to purchase or construct a continuous gas flow reactor system and subsequent scale-up opportunities, all reactions are, in principle, appropriate for scale-up for applications within the life science industries.

3.1 Fluorination of Aromatic Systems

Reactions of aromatic derivatives with fluorine gas in acidic media such as formic or sulfuric acids give rise to products consistent with an electrophilic aromatic substitution process (Scheme 12.1).[6] As with other halogenation reactions, mixtures of fluorinated aromatic products are formed due to activation of several sites of the aromatic substrate and, in many cases, polyfluorination. Substrates that bear an electron-donating group *para* to an electron-withdrawing group, however, give rise to mainly monofluorinated derivatives. For example, 4-methoxynitrobenzene gives high yields of 3-fluoro-4-methoxynitrobenzene along with some difluorinated by-products. Reaction control may be mediated by the continuous flow reactor system, giving rise to less difluorinated by-product compared with the more usual batch processes, because the monofluorinated product flows out of the reaction channel preventing further fluorination.

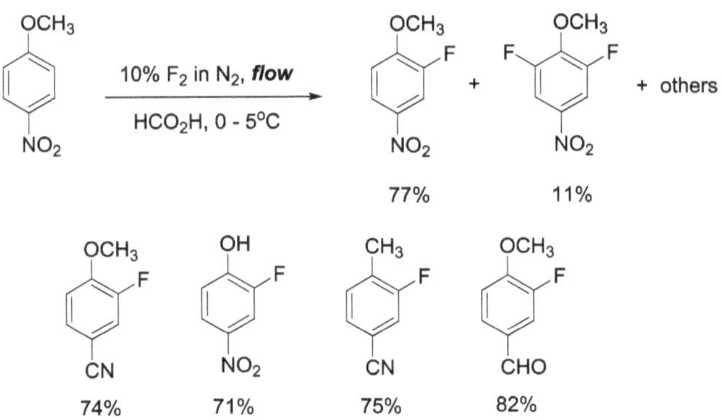

Scheme 12.1 Fluorination of 1,4-disubstituted aromatic derivatives.

The larger gas–liquid interface possible in a small-diameter continuous flow system allows even very unreactive aromatic systems to react efficiently with fluorine. A series of 1,3-dinitrobenzene substrates gave the corresponding 5-fluorinated products upon reaction with fluorine in formic acid in a flow process (Scheme 12.2).[12] Related batch fluorination processes require the use of at least 5–10 equivalents of fluorine to be passed through a rapidly stirred solution of the dinitrobenzene substrate due to slow reaction and loss of unreacted fluorine gas to the attached scrubbing system.

Benzaldehyde derivatives react with fluorine gas to give the corresponding fluoroaromatic or aryl carboxylic acid fluoride products depending on the nature of the substrate (Scheme 12.3).[13] Aromatic substrates bearing electron-donating groups such as methoxy give fluoroaromatic derivatives as the major products because, in

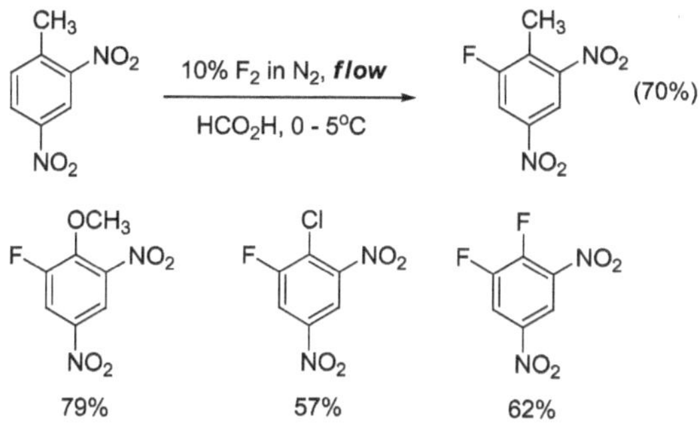

Scheme 12.2 Fluorination of 1,3-disubstituted aromatic derivatives.

these cases, the aromatic ring is sufficiently nucleophilic to undergo electrophilic substitution. In contrast, benzaldehyde derivatives bearing electron-withdrawing groups such as trifluoromethyl are deactivated toward electrophilic attack, and displacement of the aldehydic proton to give the carboxylic acid fluoride occurs. The acid fluoride can be isolated as the corresponding ester by the addition of an appropriate alcohol into the crude reaction mixture. Although fluorination of the arene ring is consistent with an electrophilic substitution process, the mechanism for the fluorination of the aldehyde group is less clear, but the selectivity of the process also suggests that an electrophilic process is more likely especially under such favorable acidic reaction conditions (Scheme 12.3).

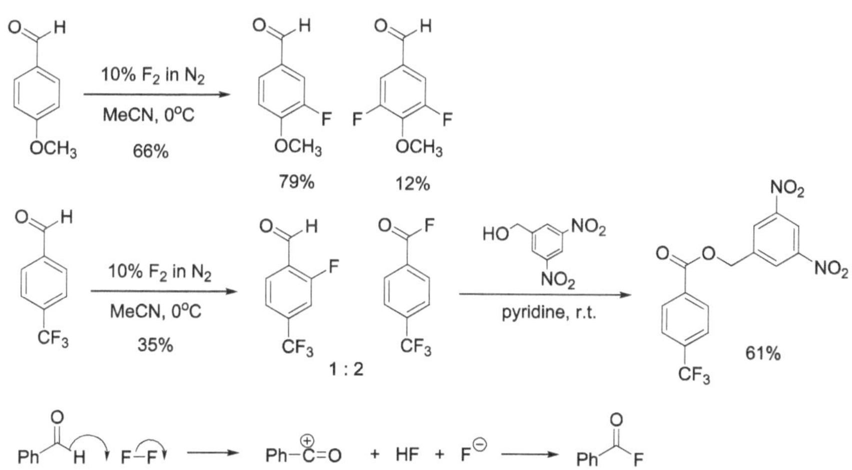

Scheme 12.3 Fluorination of benzaldehyde derivatives.

3.2 *Fluorination of 1,3-Dicarbonyl Systems*

Continuous flow fluorination of ranges of 1,3-diketone, ketoester, and diester substrates have been studied extensively[10,11] given the importance of fluorinated dicarbonyl derivatives as building blocks for the pharmaceutical industry. Ethyl acetoacetate mainly gives the desired 2-fluoro-derivative upon continuous flow fluorination in formic acid at 0–5°C. Some difluorinated products are formed, but these can be readily separated from the monofluorinated system by simple aqueous washing of the crude product material (Scheme 12.4).

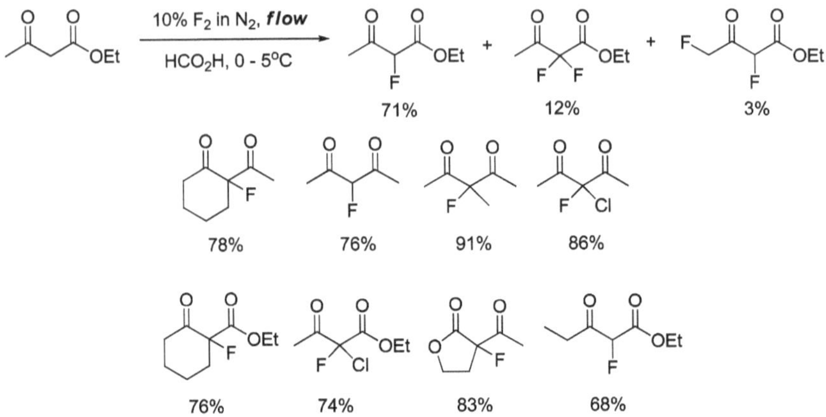

Scheme 12.4 Fluorination of 1,3-diketone and 1,3-ketoester derivatives.

Meldrum's acid gives a mixture of mono- and difluorinated systems on reaction with fluorine in a flow system, and these products are most easily separated by addition to ethanol and distillation of the corresponding ethyl diesters (Scheme 12.5).[11]

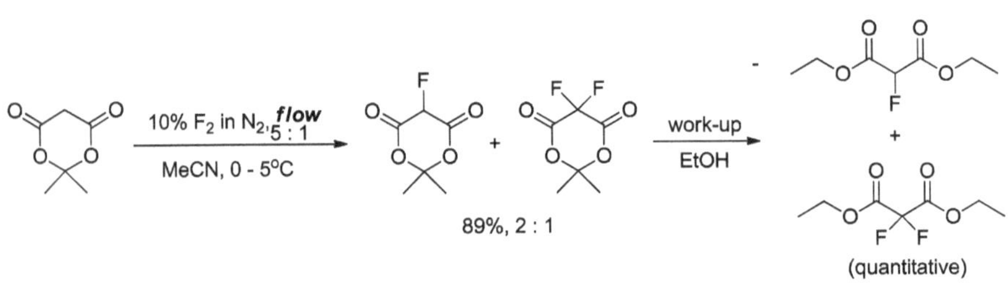

Scheme 12.5 Fluorination of Meldrum's acid.

Sequential gas–liquid/liquid–liquid flow processes allow the preparation of fluoropyrazoles in a two-step, one continuous flow process.[3] The single-channel reactor system described earlier can be adapted to include three inputs of fluorine: 1,3-dicarbonyl substrate and hydrazine solution, which enables sequential fluorination of the dicarbonyl substrate, followed by cyclization to the pyrazole product by established condensation reactions (Fig. 12.3). For example, pentane-2,4-dione can be transformed in one process to 4-fluoro-3,5-dimethyl pyrazole in high yield using a three-input flow reactor system (Scheme 12.6).

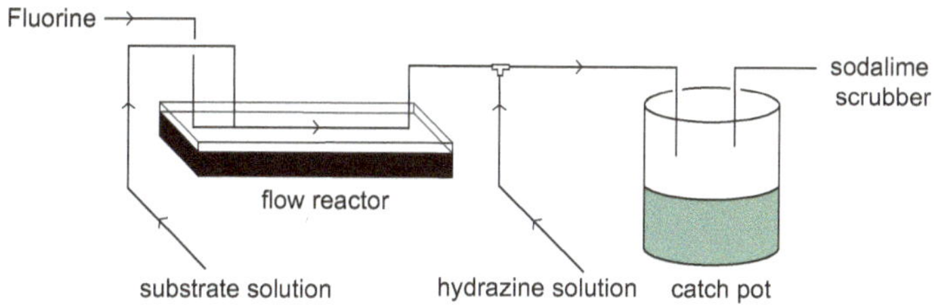

Figure 12.3 Flow reactor for sequential gas–liquid/liquid–liquid flow processes.

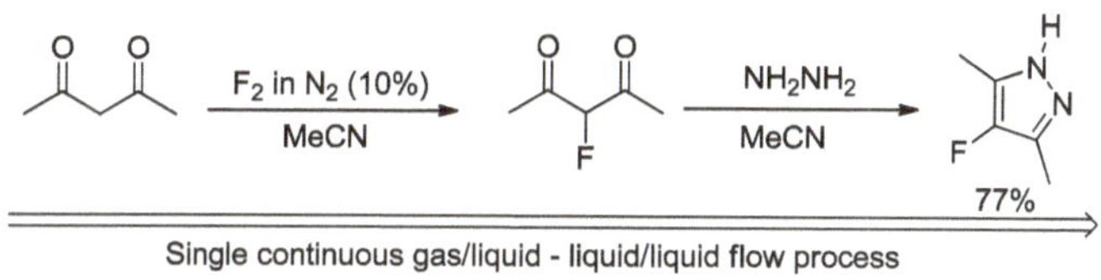

Scheme 12.6 Synthesis of 4-fluoro-3,5-dimethylpyrazole.

3.3 Fluorination of Aniline Derivatives

Synthesis of aryl–NF_2 systems can be achieved by reaction of the appropriate aniline derivatives with fluorine.[20] Passing a sixfold excess of fluorine and perfluoro-2-nitroaniline or a related perfluroaniline system concurrently through a continuous flow reactor gave good yields of the corresponding difluoroamine product (Scheme 12.7). Perfluorinated aniline derivatives were used as substrates since no competing fluorination of the aryl ring is possible.

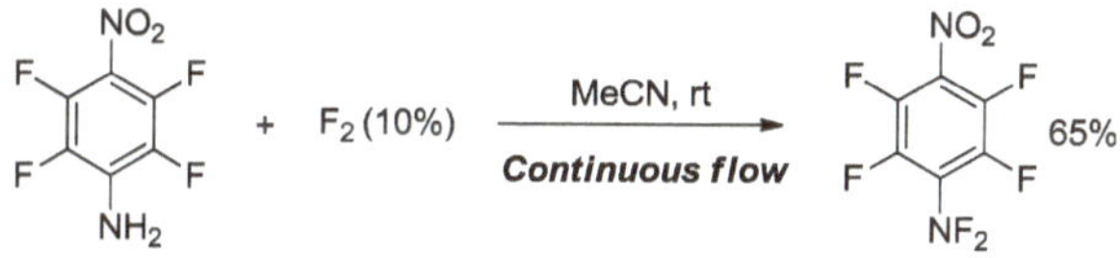

Scheme 12.7 Synthesis of Ar-NF_2 derivatives.

4. Conclusions

In summary, easily constructed single- and multichannel continuous flow reactors may be used for selective direct fluorination of a range of aromatic, 1,3-dicarbonyl and amine substrates using fluorine gas. These processes could, in principle, be scaled up if required using either longer path length single-channel reactors or multichannel systems. The use of fluorine gas can be particularly beneficial for large-scale synthesis due to the low cost of this reagent, and process development chemists should consider the use of continuous flow processes using fluorine gas in development and manufacturing phases of new fluorinated pharmaceuticals and agrochemicals.

References

1. Baasner, B.; Hagemann, H.; Tatlow, J. C. In *Houben-Weyl Organofluorine Compounds*, Vol. E10a; Thieme: Stuttgart, 2000.
2. Banks, R. E.; Smart, B. E.; Tatlow, J. C. *Organofluorine Chemistry. Principles and Commercial Applications;* Plenum: New York, 1994.
3. Breen, J. R.; Sandford, G.; Yufit, D. S.; Howard, J. A. K.; Fray, J.; Patel, B. Continuous Gas/Liquid – Liquid/Liquid Flow Synthesis of 4-Fluoropyrazole Derivatives by Selective Direct Fluorination. *Beilstein J. Org. Chem.* **2011,** *7,* 1048–1054.
4. Butters, M.; Ebbs, J.; Green, S. P.; MacRae, J.; Morland, M. C.; Murtiashaw, C. W.; Pettman, A. J. Process Development of Voriconazole: A Novel Broad-Spectrum Triazole Antifungal Agent. *Org. Process Res. Dev.* **2001,** *5,* 28–36.
5. Chambers, R. D.; Spink, R. C. H. Microreactors for Elemental Fluorine. *Chem. Commun.* **1999,** 883–884.
6. Chambers, R. D.; Hutchinson, J.; Sparrowhawk, M. E.; Sandford, G.; Moilliet, J. S.; Thomson, J. Fluorination of 1,4-Disubstituted Aromatic Compounds. *J. Fluor. Chem.* **2000,** *102,* 169–173.
7. Chambers, R. D.; Holling, D.; Sandford, G.; Spink, R. C. H. Gas-Liquid Thin Film Microreactors for Selective Direct Fluorination. *Lab Chip* **2001,** *1,* 132–137.
8. Chambers, R. D. *Fluorine in Organic Chemistry;* Blackwell: Oxford, 2004.
9. Chambers, R. D.; Fox, M. A.; Holling, D.; Nakano, T.; Okazoe, T.; Sandford, G. Versatile Thin-Film Gas-Liquid Multi-Channel Microreactors for Effective Scale-Out. *Lab Chip* **2005,** *5,* 191–198.
10. Chambers, R. D.; Fox, M. A.; Sandford, G. Selective Direct Fluorination of 1,3-Ketoesters and 1,3-Diketones Using Microreactor Technology. *Lab Chip* **2005,** *5,* 1132–1139.
11. Chambers, R. D.; Fox, M. A.; Holling, D.; Nakano, T.; Okazoe, T.; Sandford, G. Versatile Gas-Liquid Microreactors for Industry. *Chem. Eng. Technol.* **2005,** *28,* 344–352.
12. Chambers, R. D.; Fox, M. A.; Sandford, G.; Trmcic, J.; Goeta, A. Direct Fluorination of Deactivated Aromatic Systems Using Microreactor Techniques. *J. Fluor. Chem.* **2007,** *28,* 29–33.
13. Chambers, R. D.; Sandford, G.; Trmcic, J.; Okazoe, T. Direct Fluorination of Benzaldehyde Derivatives. *Org. Process Res. Dev.* **2008,** *12,* 339–344.
14. Ehrfeld, W.; Hessel, V.; Lowe, H. *Microreactors. New Technology for Modern Chemistry;* Wiley-VCH: New York, 2000.
15. Gadamasetti, K. G. *Process Chemistry in the Pharmaceutical Industry;* Marcel Dekker: New York, 1999.
16. Harsanyi, A.; Sandford, G. Organofluorine Chemistry: Applications, Sources and Sustainability. *Green Chem.* **2015,** *17,* 2081–2086.
17. Hutchinson, J.; Sandford, G. Elemental Fluorine in Organic Chemistry. *Top. Curr. Chem.* **1997,** *193,* 1–43.
18. Mallia, C. J.; Baxendale, I. R. The Use of Gases in Flow Synthesis. *Org. Process Res. Dev.* **2015,** *20*. ASAP article.
19. McPake, C. B.; Sandford, G. Selective Continuous Flow Processes Using Fluorine Gas. *Org. Process Res. Dev.* **2012,** *16,* 844–851.
20. McPake, C. B.; Murray, C. B.; Sandford, G. Continuous Flow Synthesis of Difluoroamine Systems by Direct Fluorination. *Aust. J. Chem.* **2013,** *66,* 145–150.
21. Sandford, G. Elemental Fluorine in Organic Synthesis (1997–2006). *J. Fluor. Chem.* **2007,** *128,* 90–104.
22. Uneyama, K. *Organofluorine Chemistry;* Blackwell: Oxford, 2006.
23. Wiles, C.; Watts, P. *Microreaction Technology in Organic Synthesis;* CRC Press: Boca Raton, 2011.

Synthesis of Fluorinated Nitrogen-Containing Compounds Through Superelectrophilic Activation in Superacid HF/SbF_5*

13

S. Thibaudeau, H. Carreyre, A. Mingot
Université de Poitiers, CNRS UMR 7285 IC2MP, Poitiers Cedex 09, France

Chapter Outline

* This work is dedicated to Pr Jean-Claude Jacquesy with great consideration.

Modern Synthesis Processes and Reactivity of Fluorinated Compounds. http://dx.doi.org/10.1016/B978-0-12-803740-9.00013-5

1. Introduction

The term "superacid" was first mentioned in the literature by Hall and Conant[74] for acids that were capable of protonating weak bases such as carbonyl compounds. These solutions of sulfuric acid or perchloric acid in glacial acetic acids were called "superacid solutions." Not followed up, their proposal, however, disappeared from the literature until the 1960s when Olah's studies focused on the use of highly acidic nonaqueous systems for studying long-lived carbocations.[159] This impetus culminated in 1994, when Pr. Olah was awarded a Nobel Prize in Chemistry "for his contribution to carbocation chemistry."[155] In 1971, Gillespie, whose pioneering work on the inorganic chemistry of superacids was also of immense value to the field, proposed an arbitrary but since widely accepted definition of superacids.[68] A superacid is a system more acidic than 100% sulfuric acid with H_0 values lower than -12.[69] Very well-known Brønsted superacids are HSO_3F, perfluorinated sulfonic acids, carboranes,[172] *bis*-(trifluoromethylsulfonyl)imide and its related derivative[66] *tris*-(trifluoromethylsulfonyl)methane,[202] or hexafluoroisopropoxysulfuric acid,[16] mainly designed by introducing strong electron-withdrawing substituents, following the Yagupolskii principle.[209,99,210,165] Concomitant to the development of Brønsted superacid, interest in Lewis superacid grew over the recent past. In their first superacid monograph,[148] Olah and coauthors arbitrarily defined Lewis superacids as those that are stronger than anhydrous aluminum chloride, the most used Friedel–Crafts catalyst. Based on the use of fluoride ion affinity,[118] the term Lewis superacid was also defined as a molecular Lewis acid, which is stronger than monomeric SbF_5 in the gas phase.[128] An approach using increasingly lower nucleophilic noncoordinating anions,[102] allowed significant recent progress in this field,[8,175] and ongoing efforts in this direction are awaited.[185] By combining Brønsted superacids with Lewis superacids, binary or ternary strong superacid systems, 10^7 to 10^{10} times stronger than 100% sulfuric acid, can be obtained. The exploitation of these superacids resulted in a significant new field of chemistry, named "superacid chemistry."

Under these highly acidic conditions, for example, in hydrofluorhydric/antimony pentafluoride mixtures (HF/SbF_5), functionalized organic substrates are mono- (or poly-) protonated (superelectrophilic activation), leading to original reactivity. The HF/SbF_5 superacid media had its glorious days in the 1970's–1980s when Olah's group exploited its ability to generate long-lived carbocations.[159] At the same time, Jacquesy intensively exploited superacid chemistry to perform oxidation, carbonylation, arylation, and isomerization of organic compounds going from simple models to elaborated synthetic and natural products, allowing the direct synthesis of analogs of bioactive compounds.[81–84,86,87] As mentioned by Olah in his autobiography,[155] "*Nature is able to perform its own transformations in ways* that chemists have only begun to understand. At enzymatic sites many significant transformations that take place, in a generalized sense, are also acid catalyzed (for example electron-deficient metal ion-catalyzed processes). Because the unique geometry at enzyme sites can make multidentate interactions possible, these superelectrophilic chemical activations may have their equivalent in some enzymatic process." Such a parallel was exemplified by Berkessel and Thauer[17] and reinforced when alkaloids (dihydrovincadifformine[18] and yohimbine[60]) and natural phenol[31]–hydroxylated metabolites were found to be directly synthesized under HF/SbF_5 conditions, in the presence of an oxidizing agent. Thus, the abundance of innovative development, directly related to HF/SbF_5 superacid use, triggered a strong interest in studying HF/SbF_5 solution composition.

1.1 HF/SbF_5 Superacid Solutions

Anhydrous hydrogen fluoride is an important chemical. It is notably widely used in the industry for the production of fluoroaromatic and trifluoromethylated aromatics with the highest volume fluorination reactions currently carried out.[177,178] Being also the origin of all fluorinated agents known to date,[77] its structure has been logically explored. Diffraction data from liquid hydrogen fluoride was shown to be consistent with the presence of short, bent, strongly hydrogen-bonded chains, the chains appearing to be curtailed in length at seven HF molecules per chain.[121] Addition of antimony pentafluoride to anhydrous liquid HF causes the dissociation of the linear polymeric chain and the subsequent formation of poorly solvated protons. By using infrared spectroscopy investigation, Bonnet and Marschepa[19] showed that over a large concentration range (0–40 mol% of SbF_5) in the HF/SbF_5 system, the HF-solvated $H_3F_2^+$ is the predominant species (Eq. (13.1)). The fluoronium ion H_2F^+ is only observed in concentrated solutions in SbF_5 (40–100 mol%) with highly polymerized and weakly basic counterions (Eq. (13.2)).

$$3\,HF + SbF_5 \rightleftharpoons H_3F_2{}^+ + SbF_6{}^- \tag{13.1}$$

$$2\,HF + 2\,SbF_5 \rightleftharpoons H_2F^+ + SbF_6{}^-\ (Sb_2F_{11}{}^-) \tag{13.2}$$

On the basis of ^{19}F and ^{1}H nuclear magnetic resonance (NMR) spectroscopic studies, the HF/SbF_5 composition was further investigated.[45] ^{19}F NMR studies showed a fast proton exchange between the different cations $H_{n+1}F_n{}^+$, the proton being always partly solvated by HF molecules. The anionic HF/SbF_5 composition study revealed that for SbF_5 concentrations lower than 10 mol%, $SbF_6{}^-$ is practically the only anionic species present. In slow equilibrium on the NMR time scale, in the concentration range of 10–22 mol% SbF_5 in HF, the anions are essentially $SbF_6{}^-$ and $Sb_2F_{11}{}^-$. For higher SbF_5 concentrations, the formation of polymeric ions $Sb_3F_{16}{}^-$ and $Sb_4F_{21}{}^-$ occurs (Fig. 13.1).

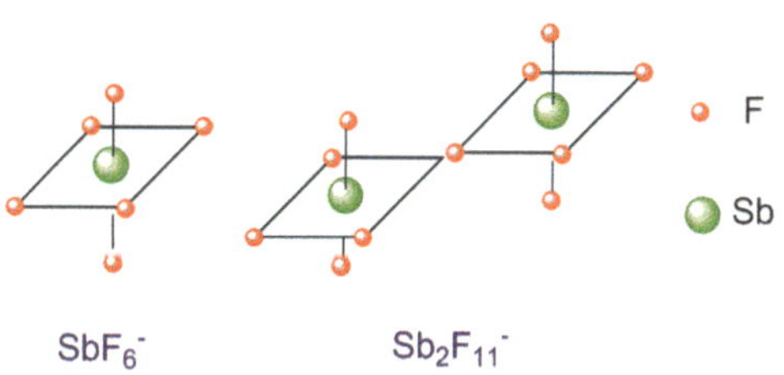

Figure 13.1 $SbF_6{}^-$ and $Sb_2F_{11}{}^-$, the main anionic species in the HF/SbF_5 system.

The HF/SbF_5 system has also been investigated by theoretical methods such as the density functional theory (DFT) theory[63] and ab initio molecular dynamics,[92,93] and this has allowed the estimation of the relative concentration of the most energetically favorable species present in HF/SbF_5 and very fast diffusion of the excess proton along the hydrogen-bonded HF chains.

In addition, following the excellent contribution of Mootz and Bartman[126] who characterized by crystal structure analysis the fluoronium ions H_2F^+ and $H_3F_2{}^+$, in 2015, Bour et al.[21] obtained single crystals composed of $[\{(IPr \cdot InBr_3)(Ag \cdot (CH_2Cl_2))\}_2][SbF_6]_2$ and two identical $HSbF_6$ units. These results are in accordance with similar data obtained with HF/AsF_5 systems after decomposition of trimethylsilyl *N,N*-dimethylcarbamate.[11] The HF autoprotolysis constant being of the order 10^{-12} at 0°C, the SbF_5-induced reaction causes a dramatic steep rise in acidity with the H_0 value going from −15.1 for neat HF to reach values lower than −21 for measured HF/SbF_5 mixture,[70] thus favoring reactions with no equivalent in classical conditions.

1.2 Superelectrophilic Activation in Superacid

As mentioned the previous section, the extreme acidity of the media allows for polyprotonation of functionalized substrates and for converting unreactive species to reactive ones. These generated polycationic species, named superelectrophiles, can be defined as electrophiles that are capable of further interaction (coordination, solvation) with strong Brønsted or Lewis acids.[150] Through the further complexation (solvation) of the monocationic electrophile by Brønsted or Lewis acid,[157] neighboring group participation (stabilization) with the electrophilic center is decreased, and the resulting electrophiles show increased electron deficiency and reactivity toward nucleophilic partners.[187] In other terms, one could be consider that superelectrophiles are electrophilic species that are activated through protonation or complexation allowing their further reaction with poor nucleophiles.

Boosted by this original reactivity, superelectrophiles and their reactions are an emerging and fast growing field.[158] Superelectrophiles can be categorized into two groups: the distonic (distant) and the gitonic (close) superelectrophiles (Table 13.1). Geminal, vicinal, and 1,3-dications are gitonic, whereas any superelectrophile in which the positive charge centers are separated by one or two more carbons or heteroatoms are distonic. The exploitation of superelectrophiles in contemporary chemistry is now well documented and has been reviewed in 2013.[131]

Since 2006, following this pioneering and fantastic solid fund of knowledge on superelectrophilic activation in superacid HF/SbF_5, this activation method has been further exploited for the design of new methodologies toward the direct synthesis of fluorinated nitrogen-containing compounds.

1.3 Fluorinated Nitrogen-Containing Compounds: A Superfamily for Medicinal Chemistry SAR Studies

Enhancement of therapeutic efficacy and improvement of pharmacological properties of a drug, the unavoidable dream for any medicinal chemist, can sometimes be rationally predicted through the use of fluorine chemistry.[15] This well-established strategy

Table 13.1 Classes and Examples of Superelectrophiles

Gitonic Superelectrophiles			Distonic Superelectophiles
Geminal	**Vicinal**	**1,3-Dications**	
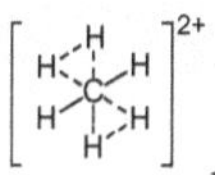 Olah et al.[156]	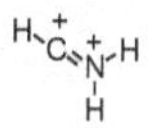 Sato et al.[179]	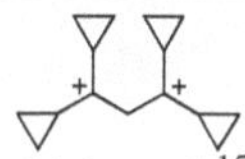 Olah et al.[154]	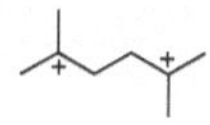 Olah et al.[147]
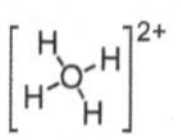 Reddy et al.[171]	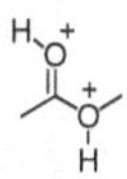 Olah et al.[151]	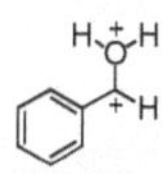 Olah et al.[152]	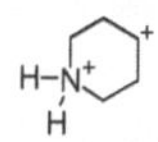 Klumpp et al.[95]
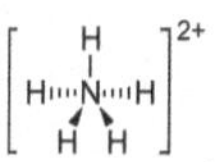 Olah et al.[153] and Rasul et al.[169]	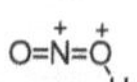 Olah et al.[149] and Weiske et al.[207]	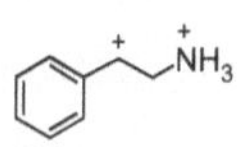 Klumpp et al.[96]	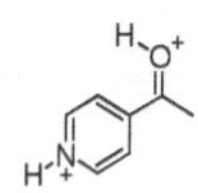 Klumpp et al.[94]

in the design of new drugs is tremendously used,[206] as shown by the number and diversity of fluorine-containing compounds currently in phase II–III clinical trials.[214] This success can be attributed to the many effects that the fluorine atom(s) and fluorinated moieties can have on the physicochemical properties of a molecule,[136] with evident consequences on its biological properties.[139] According to these benefits, fluorine is now “the second-favorite heteroatom” after nitrogen in drug design.[138] Fluorine atom is commonly used to tune amine basicity[127] as shown for example in 2014 by Jansen et al.[89] with consequences on penetration and bioavailability. In addition, the fluorine conformational effect, used in organocatalysis,[215] emerges today as a new strategy for medicinal chemistry molecular design.[80,71] As a consequence, considerable efforts have been put in the development of efficient synthetic approaches to vicinal fluorinated amines. The ring opening of aziridines,[51,90] the deoxyfluorination of aminoalcohols,[9,2] the reductive amination of α-fluorinated carbonyl-containing compounds,[117,64] and nucleophilic addition to fluorinated imines[163] are the frequently used methods for the synthesis of β-fluorinated amines.[67] Recently, elegant strategies have also been added to this synthetic arsenal, going from Julia modified reaction–based strategy,[27] CH activation/fluorination,[113,114] to multicomponent approach.[167] In this context, synthetic pathways to further extend the scope of fluorinated amines containing building blocks is of strong interest.

2. Hydrofluorination of C–C Double Bond in Superacid HF/SbF_5

2.1 Hydrofluorination of Unsaturated Amines

Hydrofluorination of unsaturated substrates can be considered as a straightforward route to monofluorinated products. However, radical hydrofluorination, compatible with a large variety of substrates, has never been applied to unprotected amines,[13,181] and the Pd-catalyzed hydrofluorination, developed in 2014, cannot be applied to amines.[62] In addition, the application of the well-known hydrofluorination using Olah's reagent to unsaturated amines is limited to a few examples in the literature, and reported yields are modest.[23,55–57,203] This difficulty can be directly related to the electronic deactivation of the unsaturation after protonation of the amino group, and neither gold catalysis[1,129] nor the HF-based fluorination reagent, designed in 2014, overcomes this obstacle.[140]

Starting from unsaturated amines, when the double bond is too deactivated to react in "classical" conditions, the superacidity of the media allows a subsequent protonation in superacid HF/SbF_5 to form an ammonium–carbenium dication. Interestingly, when the "noncoordinating" properties of antimony-complexed anions were fantastically exploited by Olah for studying long-lived carbocations, ammonium–carbenium superelectrophilic activation shattered the myth of noncoordinating anions[176] (Scheme 13.1). The superelectrophilic character of the cation **A** is strongly enhanced by the electron-withdrawing effect of the proximal protonated function, allowing its trapping by SbF_6^- or $Sb_2F_{11}^-$ weakly coordinating anions.[102]

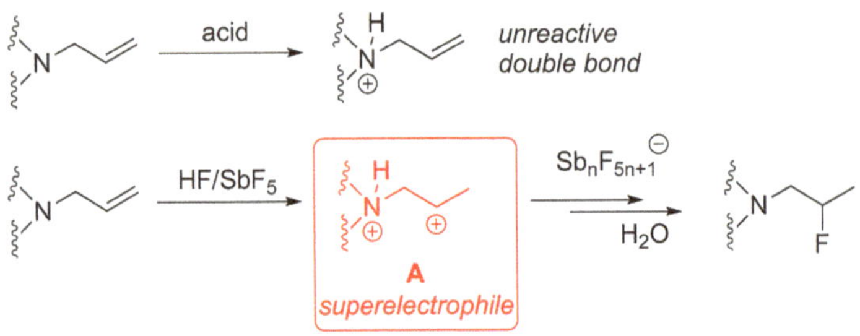

Scheme 13.1 Ammonium–carbenium superelectrophilic activation in HF/SbF_5 and hydrofluorination of unsaturated amines.

Superacid-catalyzed hydrofluorination of unsaturated amines **1** showed a broad scope of application, with the efficient synthesis of a variety of fluorinated amines **2** (Fig. 13.2).[200,108] As shown by the formation of products **2l-2n** in reasonable yields, substituted alkenes can also be fluorinated under similar conditions.

2.2 Hydrofluorination of Unsaturated Chlorinated Amines

The superacid-catalyzed hydrofluorination was then extended to the direct synthesis of *gem*-chlorofluorinated products **4** starting from chlorinated amines **3** (Fig. 13.3).[111] At −20°C, a variety of *gem*-chlorofluorinated amines, amides, imides, and sulfonamides could be generated in good yields.

Interestingly, starting from the same substrates, probably through carbon–chlorine bond activation/elimination/fluorination process after formation of the *gem*-chlorofluorinated intermediates, the corresponding *gem*-difluorinated products

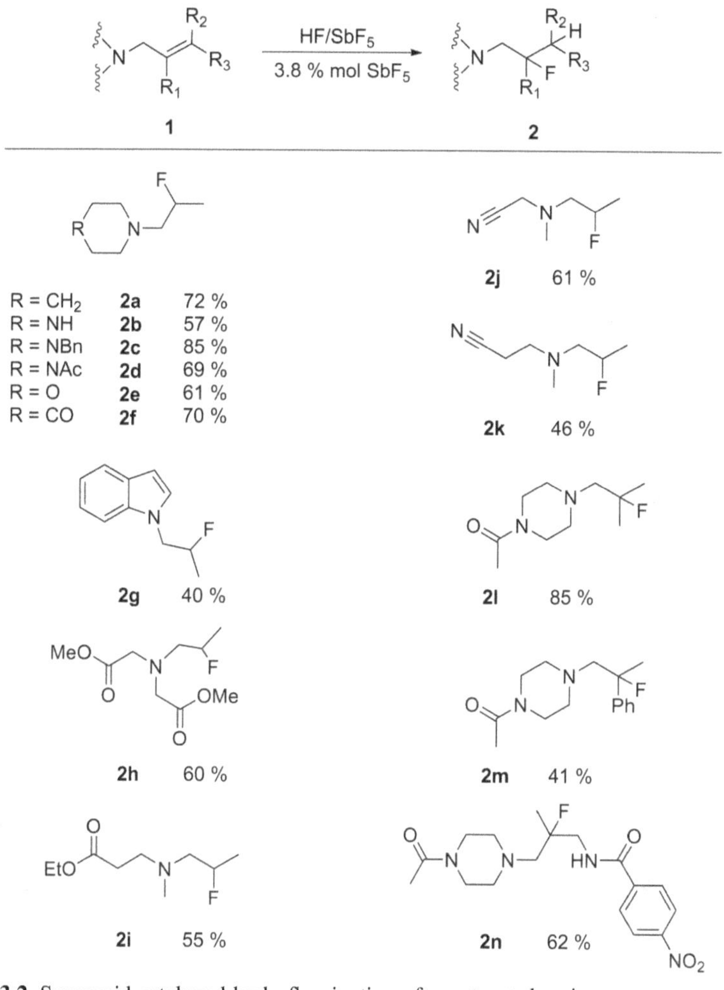

Figure 13.2 Superacid-catalyzed hydrofluorination of unsaturated amines.

5 could be generated under more acidic conditions at 0°C (%mol SbF_5 = 13.6 mol%) (Fig. 13.4).

The ability to generate *gem*-chlorofluorinated and *gem*-difluorinated products **4** and **5** from chlorinated olefins **3** was attributed to the superelectrophilic character of generated ammonium-α-chloronium dications **B** under HF/SbF_5 superacid conditions (Scheme 13.2).

As mentioned earlier, the hydrofluorination of halogen-substituted amide **3e** and imide **3f** allowed for the synthesis of the corresponding *gem*-chlorofluorinated and *gem*-difluorinated products. These results were particularly intriguing as when starting from analogous nonhalogenated substrates **1o** and **1p**, only hydroxylated products **6o** and **6p** could be generated under the same conditions (Scheme 13.3).

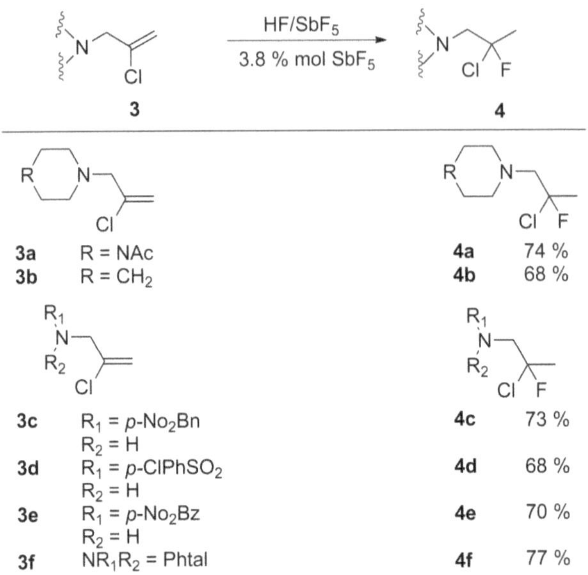

Figure 13.3 Synthesis of *gem*-chlorofluorinated nitrogen-containing compounds in superacid HF/SbF_5.

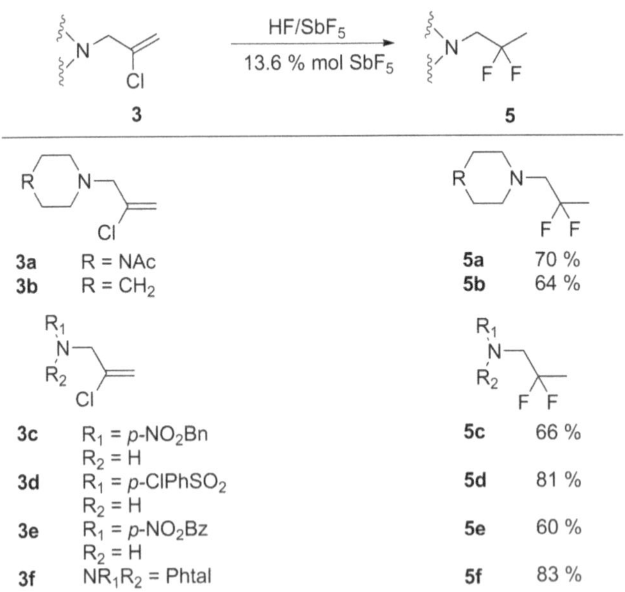

Figure 13.4 Synthesis of *gem*-difluorinated nitrogen-containing compounds in superacid HF/SbF_5.

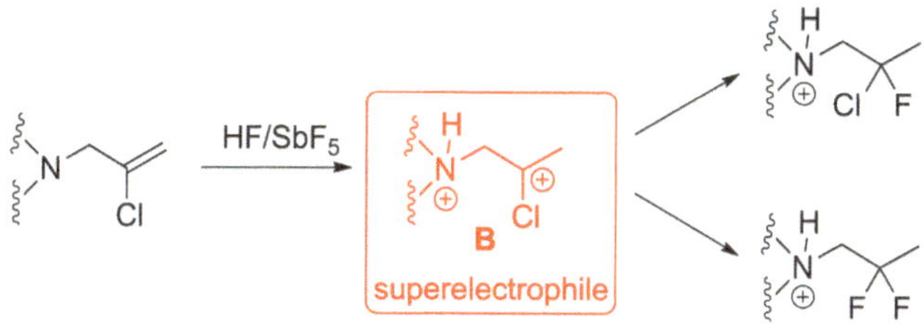

Scheme 13.2 Ammonium-α-chloronium superelectrophilic activation in HF/SbF_5.

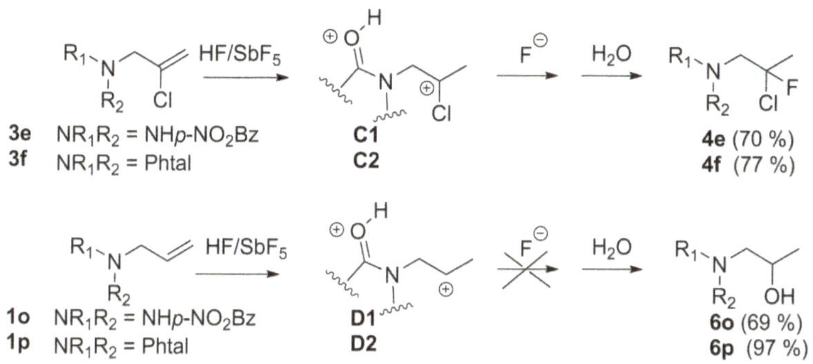

Scheme 13.3 Reaction of amides and imides in superacidic HF/SbF_5 (3.8% mol SbF_5).

In α-chloronium ions, the chlorine atom becomes both a π- and σ-donor substituent.[38] As a consequence, it could have been expected that the stabilizing effect of the chlorine atom in intermediates **C1** and **C2**, which must reduce the electrophilic character of the generated cation (compared to **D1** and **D2**), should therefore prevent them from fluorination. However, it is well accepted that amides and imides exist as neutral and protonated forms under strongly acidic conditions.[12,44,144,162,166] As a consequence, a fast equilibrium between protonated and nonprotonated forms of amides and imides can be envisaged, followed by the participation of the carbonyl group to the stabilization of the generated carbenium ions.[199] To confirm the involvement of the resulting oxazolinium ions in the mechanism, the starting materials were dissolved in HF/SbF_5 and studied by ^{13}C NMR experiments at $-20°C$. For example, a clean ^{13}C NMR spectrum consistent with the formation of an oxazolinium ion intermediate **E1** was obtained from substrate **1o** (Fig. 13.5). This study, which allowed to trap a variety of long-lived oxazolinium ion intermediates under superacid HF/SbF_5 conditions at low temperatures, confirmed the intramolecular partial stabilization of the generated carbenium ion intermediate by the carbonyl of amide and imide functions in HF/SbF_5.

On this basis, the reactivity of amides and imides was explained with the mechanism depicted in Scheme 13.4. After diprotonation in superacid, dications **C** and **D** can be generated, being in equilibrium with the monocationic species **C′** and **D′**. Then, intramolecular trapping of the cation by the carbonyl group allows for the subsequent formation of oxazolinium ions **E** and **E′**. When the inductive effect of electron-withdrawing chlorine atom in ion **E** enhances the electrophilic character of

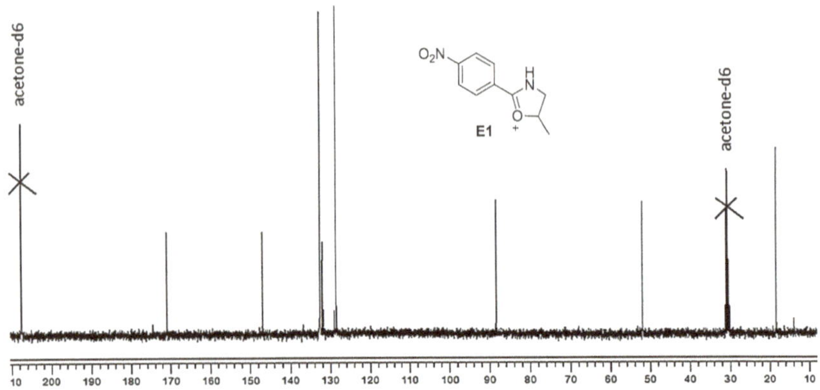

Figure 13.5 In situ ^{13}C nuclear magnetic resonance spectrum (100 MHz, CD_3COCD_3) of oxazolinium ion intermediate **E1**, resulting from substrate **1o** reaction in HF/SbF_5 at −20°C.

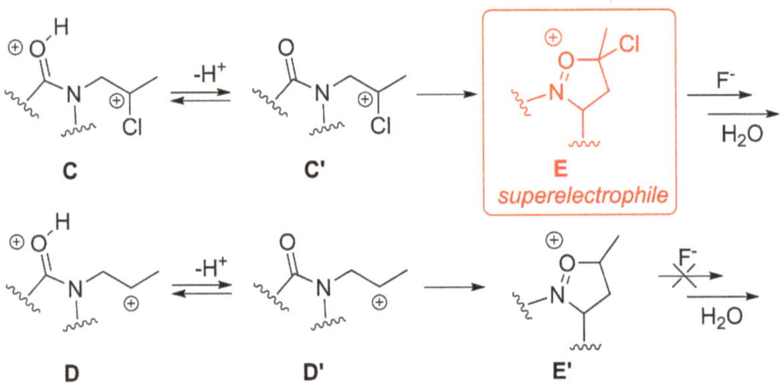

Scheme 13.4 Behavior of (chlorinated) amides and imides in superacid HF/SbF_5.

the cation allowing its fluorination, the analogous nonchlorinated ion **E′** is not electrophilic enough to be fluorinated and leads after hydrolysis to the corresponding hydroxylated products. As already showed with the trifluoromethyl substituent,[135] the inductive electron-withdrawing effect of the chlorine atom helps the oxazolinium to cross the superelectrophilic barrier, allowing the trapping of the generated ion with a poor nucleophile such as solvated fluorine in the polymeric form $Sb_nF_{5n+1}^-$.

After chlorine–fluorine exchange and hydrofluorination under similar conditions, 1,1-dichloroalkenes were also converted to the corresponding terminal trifluorinated compounds.[29]

2.3 Application to a Difluorinated Anticancer Agent Synthesis

In superacid HF/SbF_5, the involvement of superelectrophilic ammonium-α-chloronium dications of type **B** was previously postulated by Jacquesy et al. to rationalize the formation of the difluorinated anticancer agent vinflunine (Javlor) from vinorelbine after reaction in $HF/SbF_5/CCl_4$ (Scheme 13.5).[65,87]

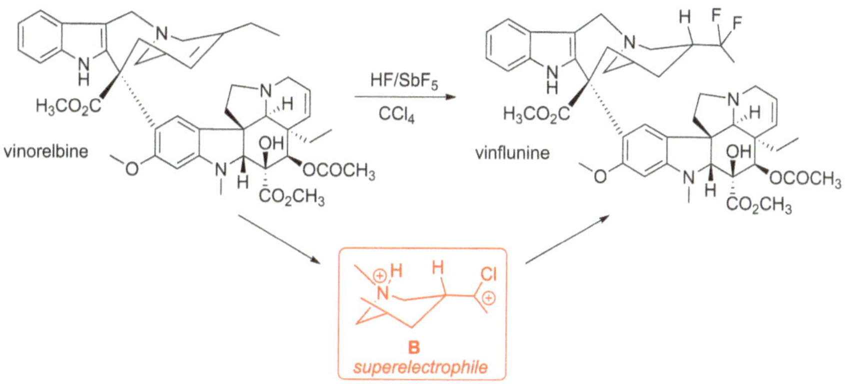

Scheme 13.5 Synthesis of vinflunine in superacid $HF/SbF_5/CCl_4$.

It was postulated that after protonation in superacid, vinorelbine must lead to a superelectrophilic ammonium–carbenium dication **A** (Scheme 13.6). After chlorination in the presence of CCl_4, a chlorinated intermediate (**I**) could be generated in addition to the resulting chlorocarbocation CCl_3^+. CCl_3^+ in HF/SbF_5 can act as a hydride-abstracting agent,[46,72,173,183,216] allowing the ionic electrophilic activation of alkanes. Through similar process, superelectrophile **B** must be generated, allowing a subsequent fluorination, halogen exchange to deliver vinflunine. A similar process was also beneficially applied to *Cinchona* alkaloids.[47]

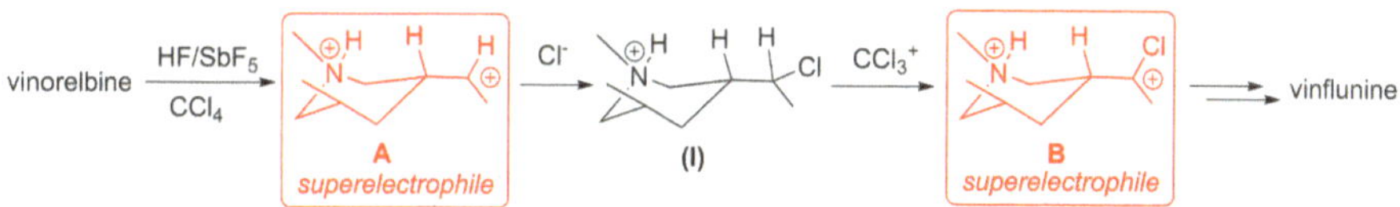

Scheme 13.6 Superelectrophilic activation of *Vinca* alkaloids in superacid.

2.4 Anti-Markovnikov Hydrofluorination of Unsaturated Amine

In due course of this work on the superacid-catalyzed hydrofluorination in superacid HF/SbF_5, the intriguing formation of a γ-fluorinated product **7d** beside the desired product **2d** from substrate **1d** led to a thorough mechanistic investigation (Scheme 13.7).[39]

The Markovnikov product **2d** and the anti-Markovnikov product **7d** could, respectively, come from the nucleophilic trapping of a secondary carbocation **A** (β-position) and of an intriguing primary carbocation **G** (ω-position),[98] as well as from of a nonclassical intermediate **B** (μ-hydrido-bridged structure).[7,170,184] By submitting fluorinated products to HF/SbF_5 conditions, the fluorinated products **2d** and **7d** were

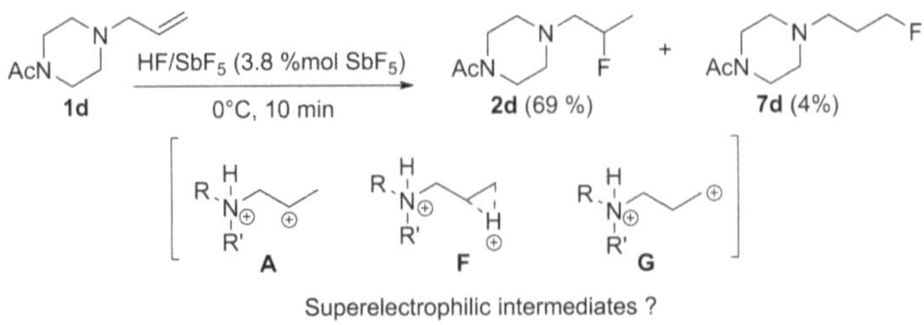

Scheme 13.7 Hydrofluorination of unsaturated amines in superacid HF/SbF_5 and hypothetical intermediates, precursors of the Markovnikov **2d**, and the anti-Markovnikov **7d** products.

shown to be under thermodynamic equilibrium through C—F bond activation/carbocation formation/fluorination process under the reaction conditions.[145,32] As a consequence, intramolecular trapping of the generated cationic intermediate through stable C—C bond formation was envisioned. As postulated, starting from *N*-allylic aniline derivatives, intramolecular trapping of the generated carbocation with aromatics allowed for the synthesis of indolines **8** and tetrahydroquinolines **9**, which were found to be stable under reaction conditions. A dramatic variation of behavior was observed between anilines when substituted with electron-donating groups and electron-withdrawing groups. For example, aniline **1q** led to a mixture of products **8q** and **9q** in a 1.6/1 ratio whereas, nitro-substituted substrate **1r** led only to the formation of the tetrahydroquinoline product **9r** (Scheme 13.8).

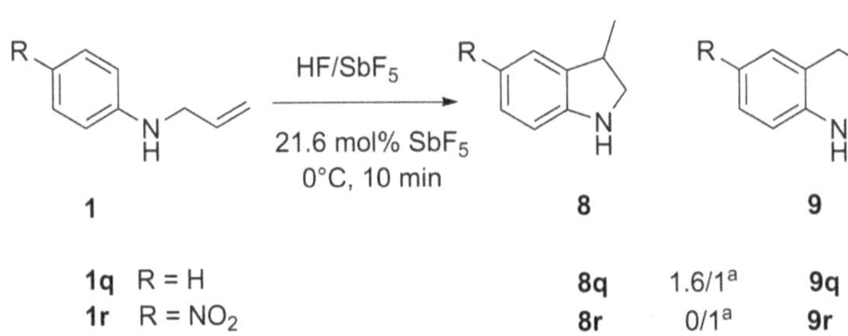

a: molar ratio determined by analysis of the ^{1}H NMR spectrum of the crude reaction

Scheme 13.8 Reactivity of *N*-allylanilines in superacid HF/SbF_5.

The dramatic behavioral differences between *N*-allylaniline **1q** and *N*-allyl-4-nitroaniline **1r** in superacid suggested that the protonation of the substrates needed to be evaluated through DFT calculation (M06-2X level). This study revealed that for "nondeactivated" aniline derivatives (anilines are partially deactivated in superacid due to nitrogen protonation), if polyprotonation occurs, it leads to the formation of dicationic ammonium—arenium species **H**. The species obtained after protonation on the nitrogen atom and on C—C double bond site has also stationary points on the potential energy surface, but 7.7 kcal/mol higher in energy than **H** (Fig. 13.6). Similar calculations from 4-nitroaniline derivative **1r** predicted the first protonation of the nitro

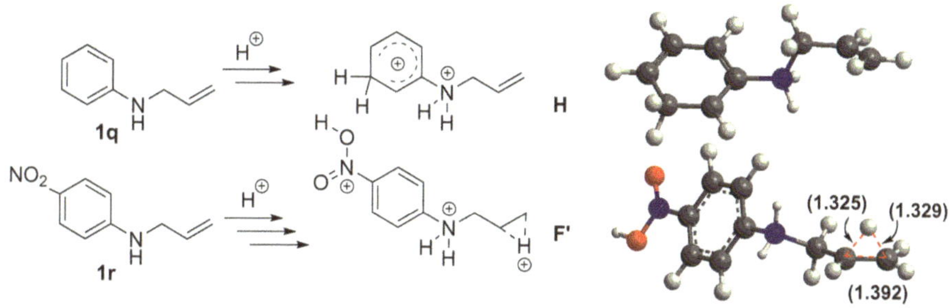

Figure 13.6 Proposed superelectrophile precursors as predicted from density functional theory calculations (distances in Angström).

group and a second protonation of the amino group. Interestingly, the calculated energies indicated that a further protonation, on the double bond, would lead to a triprotonated carbenium ion of type **A**. This reaction was predicted to be energetically more favorable (by more than 40 kcal/mol) than the protonation of the aromatic ring, so that, for deactivated substrates, the tetrahydroquinoline products would not arise from the protonation of the aromatic ring. Such a third protonation would produce a secondary carbenium ion of type **A** (Scheme 13.7) but a second stationary point (energy minimum) was located close in energy, at +1.1 kcal/mol. This optimized **F**′ structure (Fig. 13.6) involves a three-center two-electron bond intermediate and corresponds to a symmetrically μ-hydrido-bridged species (μ – H···C distances: 1.33 Å, activated C=C bond length: 1.39 Å).

As a consequence, the selectivity of the formation of the 6-membered ring products, from deactivated substrates, was attributed to the relative stabilities of the Wheland intermediates, **I** and **I**′, resulting from the cyclization via trapping of the intermediates **A** and/or **F**′ (Fig. 13.7). In this case, the difference of energy between **I** and more-strained **I**′ intermediates accounted for the exclusive formation of the tetrahydroquinoline product.

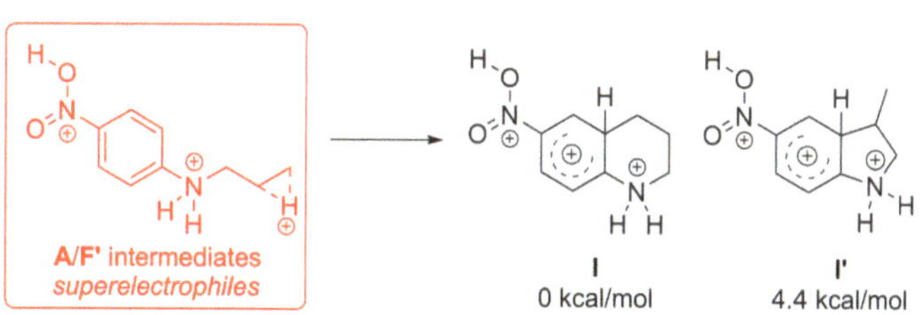

Figure 13.7 Schematic nucleophilic attack of intermediates **A** and/or **F**′ to form five- and six-membered rings and calculated relative free energies of the resulting Wheland intermediates **I** and **I**′.[75]

The diprotonation of the 4-nitroaniline moiety and the absence of aromatic ring protonation in case of substrate **1r** were then confirmed by in situ low-temperature NMR analysis. When *N*-methyl-4-nitroaniline was dissolved in superacid HF/SbF_5 and characterized by low-temperature NMR spectroscopy (−20°C), immaculate ^{1}H and ^{13}C spectra were obtained, showing the sole formation of a diprotonated species **J** (Fig. 13.8).

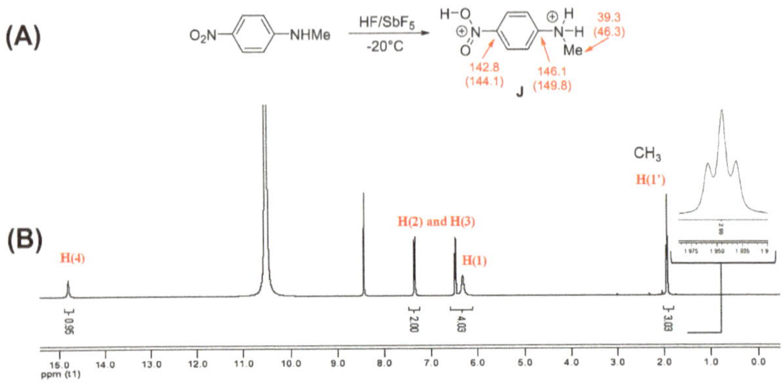

Figure 13.8 (A) Formation of dication **J** from *N*-methyl-4-nitroaniline after protonation in superacid HF/SbF_5 at −20°C and selected observed ^{13}C chemical shifts by in situ ^{13}C NMR (compared with theoretical values in parts per million). (B) In situ ^{1}H nuclear magnetic resonance (NMR) spectrum of dication **J** at −20°C.

This anti-Markovnikov process was further exploited for the synthesis of elaborated nitrogen-containing polycyclic compounds.[41] For example, starting from substrate **1r**, through a tandem cyclization/allylation/cyclization process, the highly challenging nitro-substituted julolidine **11r** was synthesized in an overall 65% yield (Scheme 13.9).

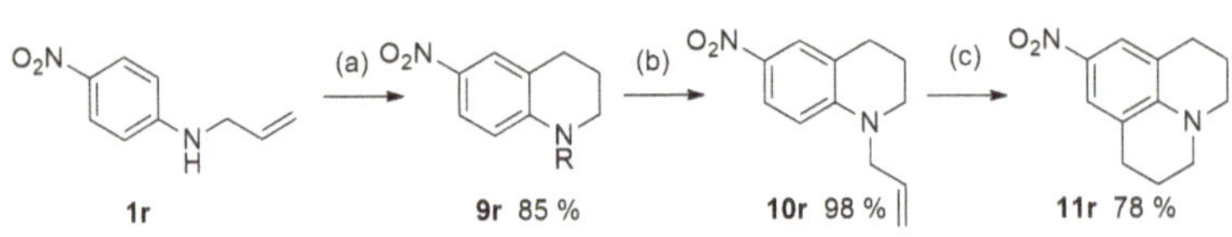

Conditions: (a) HF/SbF_5 (SbF_5, 21.6% mol); (b) NaH, allylbromide; (c) HF/SbF_5 (SbF_5, 21.6% mol).

Scheme 13.9 Synthesis of nitro-substituted julolidine **11r**.

3. Hydrofluorination of C–C Triple Bond in Superacid HF/SbF_5

3.1 Hydrofluorination of Propargylic Amines

Henne and Zimmerschied[79] and Olah et al.[146] successively described the *gem*-difluorination of alkynes after hydrofluorination in fluorhydric acid and HF/pyridine, the method being noncompatible with aminoalkynes, for the same reasons as mentioned earlier. More recently, Klumpp demonstrated that through superelectrophilic activation under superacid CF_3SO_3H conditions, similar aminoalkynes could be arylated.[97] Following a similar strategy and successive formation of dicationic superelectrophiles **K** and **M**, propargylic amines **12** could be efficiently converted to *gem*-difluorinated amines **13** (Fig. 13.9).[30]

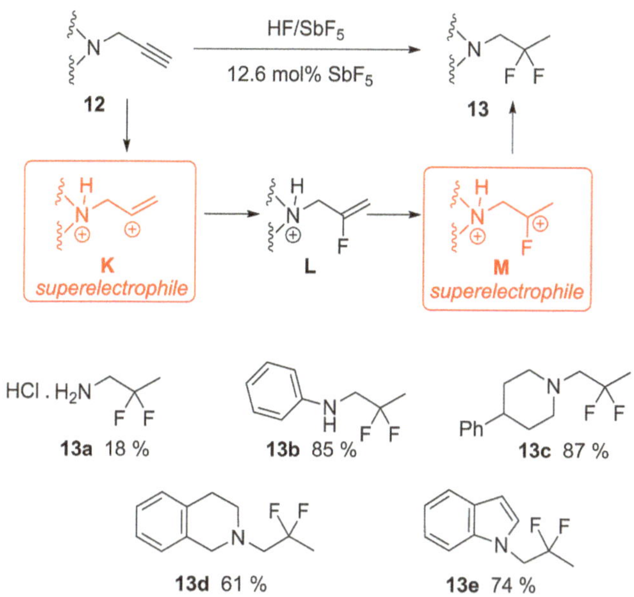

Figure 13.9 Synthesis of *gem*-difluorinated amines from propargylic amines in superacid HF/SbF_5.

3.2 Hydrofluorination of Ynamides

3.2.1 Synthesis of Fluoroenamides

Over the past decade, due to their unique structural properties,[59,212] the chemistry of urea has undergone nothing short of a renaissance with remarkable application in many fields of chemistry (organocalysts,[130] anion chemosensors,[14,28] artificial membranes[106,133]). In addition, the recent development of urea-based anticancer and anti-infective molecules[107,61] unambiguously demonstrated their strong potential as key structural elements and/or pharmacophores in medicinal chemistry, not only for anti-diabetic properties.[100,48] In addition, bioisosteric substitution of functional group has been a common strategy in structure-activity relationship (SAR) studies since more than 20 years.[37,132,161] Considering the well-known bioisosterism between fluoroolefins and amides[43] and the poverty of reported urea bioisosters in SAR studies,[25,78,134] in collaboration with Pr. Evano, our group has developed the stereoselective hydrofluorination of ynamides **14** in HF as an entry to α-fluoroenamides **15**, potent urea bioisosters (Scheme 13.10).[40]

In neat HF conditions, after protonation, highly reactive keteniminum intermediates **N** can be further fluorinated to stereoselectively generate fluoroenamides **15**. After

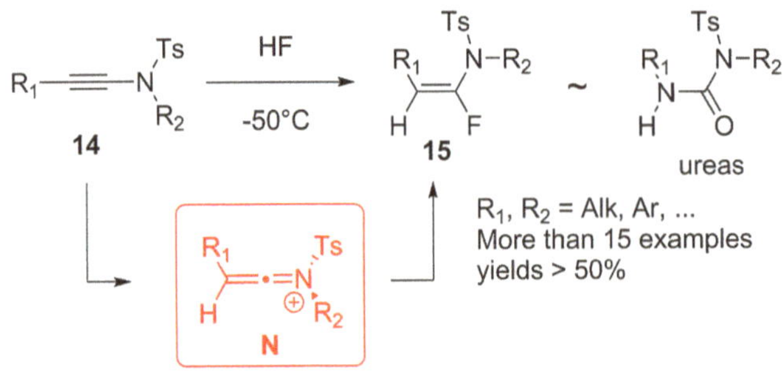

Scheme 13.10 Hydrofluorination of ynamides in HF.

showing that similar reaction also occurs in HF/pyridine,[124] rationalization of the observed steroselectivity was suggested, relying on the structure of the sterically favorable incoming approach of the fluorinating complex.

3.2.2 Fluoroenamides: Ureido Derivative Isosters?

Following this work, theoretical calculations were carried out on selected and representative substituted ureas and the postulated isosters, exploring geometric parameters, natural bond orbital charges, free energies of solvation, and electrostatic potential. In the light of the results issued from this systematical study, the following general remarks were made: in terms of steric demand fluorenamine and ureido units are similar. Fluoroenamine imparts the proper polarity to mimic that of a urea (with lowest magnitude of charge separation). Fluoroenamine electronically mimics urea with analogous continuous charge distribution and electrostatic potential profiles (Fig. 13.10).

Finally, the question remains regarding the overall effectiveness of the fluorenamines as ureido bioisosters. It could be considered that there are a lot of effects that determine or influence the effect of a group to act as a bioisoster, and the ideal evaluation would come from bioactivity comparison of fluoroenamine-containing bioactive derivatives.

3.2.3 Extension to Cascade Reaction

Interestingly, when the targeted fluoroenamide **15a** was generated in 63% yield from ynamide **14a** after reaction in HF/pyridine conditions, the tetracyclic compound **18a** was also generated in HF as a single diastereoisomer in 79% yield (Scheme 13.11).[198]

Based on a series of experiments with labeled substrates and reactants, to rationalize this original cationic polycyclization of ynamides, the following mechanism was proposed (Scheme 13.12). It subsequently involves keteniminum ion **N** formation and [1,5]-sigmatropic hydrogen shift to form **O** (in resonance with the *bis*-allylic carbocation **P**). The first cycle would then be formed by 4π conrotatory electrocyclization yielding diastereoselectively **Q** (in the manner of Nazarov reaction). The final cyclization with

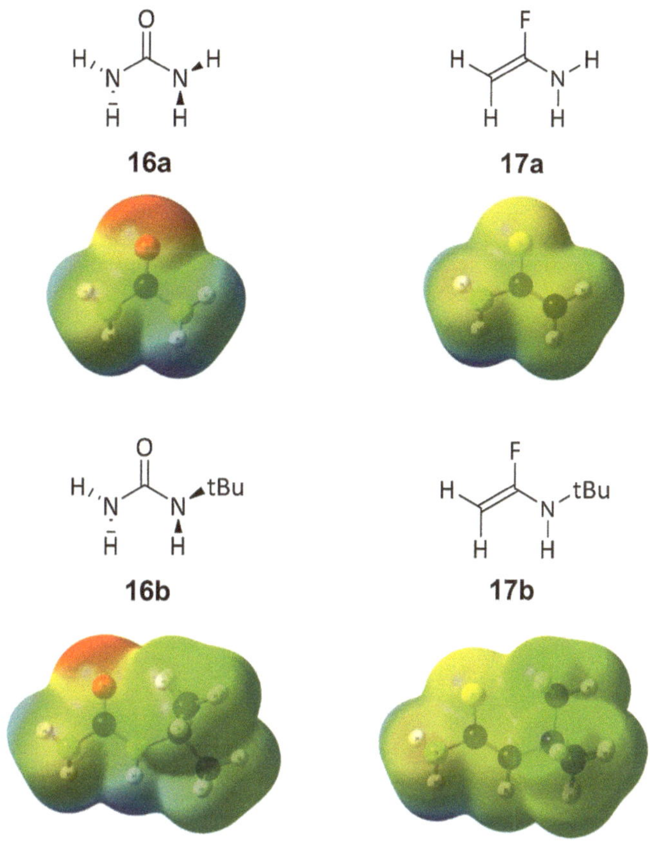

Figure 13.10 Electrostatic potential of the ureas **16a** and **16b** and their fluoroenamine isosters **17a** and **17b** (color alternation from red to blue describes the shift of the electronically rich to deficient circumstance).

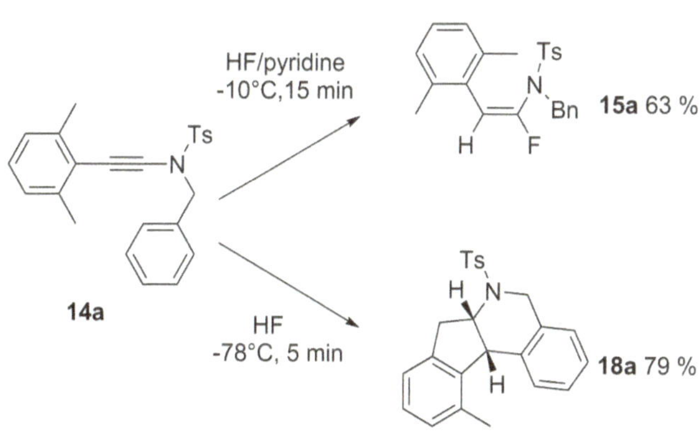

Scheme 13.11 Ynamide **14a** behavior in HF/pyridine and HF conditions.

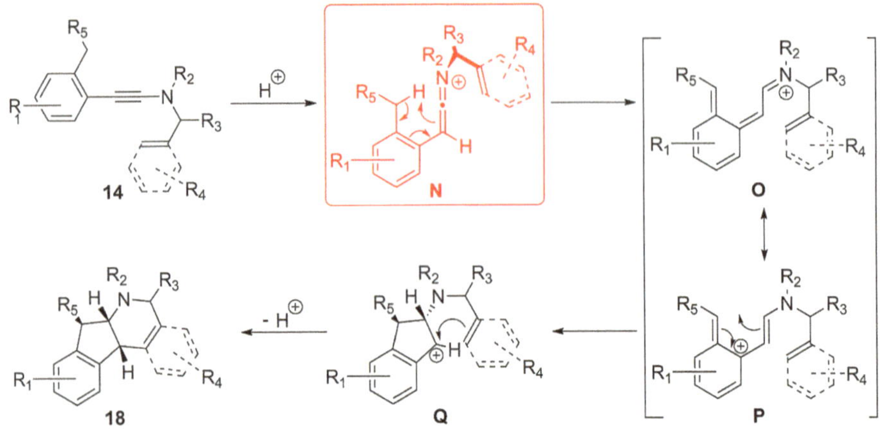

Scheme 13.12 Proposed mechanism for the cationic polycyclization of ynamides **14** to polycycle **18**.

the benzylic cation **Q** and the arene finally accounted for the formation of the tetracycle as well as its *cis* ring junction.

Having showed that this acid catalyzed polycyclization could be conducted in the presence of a catalytic amount of triflimide (Tf_2NH 20 mol%), the reaction was generalized to a variety of ynamides **14** (Fig. 13.11).

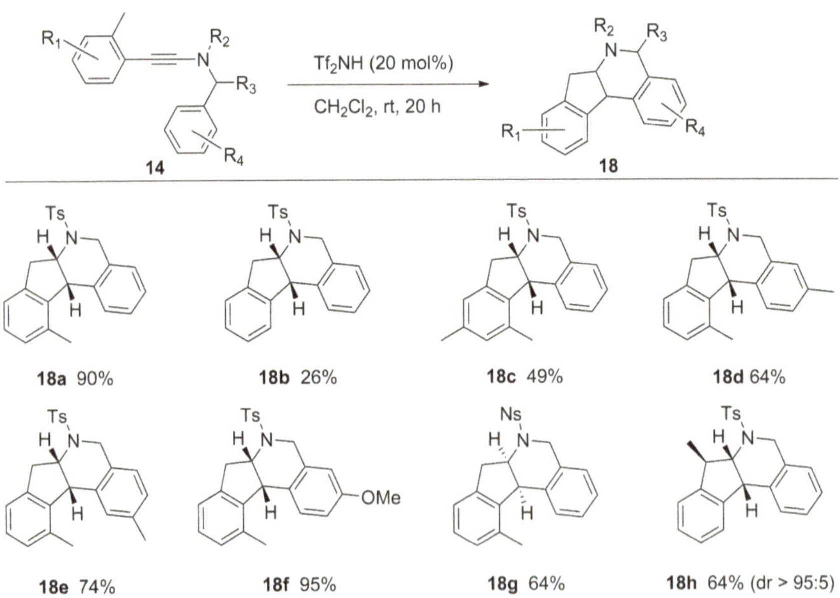

Figure 13.11 Examples of nitrogen-containing chiral tetracycles **18** generated from ynamides **14** in the presence of a catalytic amount of Tf_2NH.

4. Cyclization and Cyclization/Fluorination in Superacid HF/SbF_5

4.1 Selective Hydrofluorination or Cyclization of N-Allylic Derivatives in HF/SbF_5

As mentioned earlier, starting from *N*-allylic aniline derivatives (**1q** and **1r**, for example, Scheme 13.8), by modifying superacid conditions, hydrofluorination could be avoided allowing the intramolecular trapping of the generated dicationic intermediate, and thus even with deactivated nucleophile. At high SbF_5 concentration (% mol SbF_5 $\geq$13.6 mol%), C−F bond can be activated, thus displacing the equilibrium toward the formation of the dicationic superelectrophile **A**, prone to be quenched in an intramolecular way (Scheme 13.13).

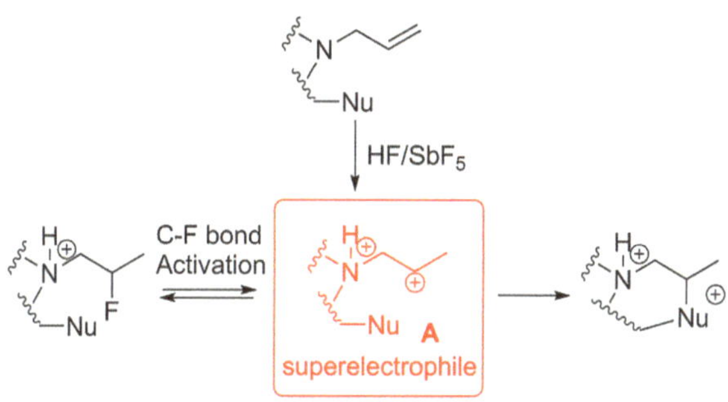

Scheme 13.13 C-F bond activation and intramolecular trapping of ammonium−carbenium superelectrophile in HF/SbF_5.

Since the discovery of antibiotic and antibacterial properties of sulfa drugs,[24] sulfonamides and sultams became very popular in SAR studies in medicinal chemistry research with a wide range of applications.[168,182,188,190,195,201] As a consequence, diversifying the synthetic arsenal for their preparation found strong interest.[116]

Exploiting the stability of fluorinated products in HF/SbF_5 conditions, when SbF_5 molar ratio is lower than 10 mol%, and the ability to activate C-F bond at higher acidity with a concomitant decrease of the nucleophilicity of fluoride source (see introduction), benzofused sultams **20** or fluorinated sulfonamides **21** could be selectively synthesized from *N*-allyl benzenesulfonamides **19** (Scheme 13.14).[109] Through this method, a large variety of products **20** and **21** incorporating alkyl, halogen, amide, ester, ketone, and other functions were generated in good yields.

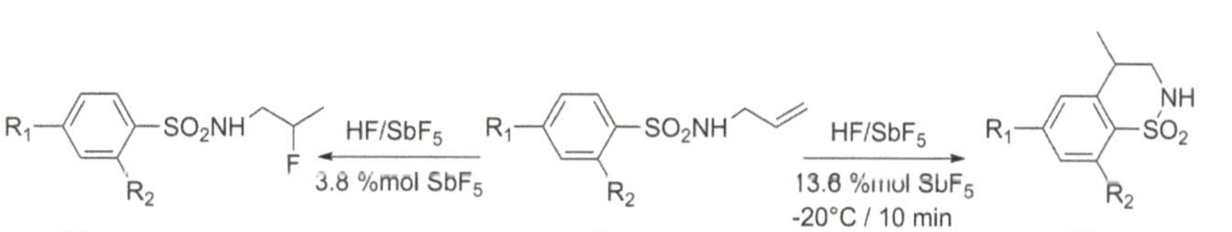

21
more than 12 examples
yields ≥ 42%

19

20
more than 20 examples
yields ≥ 47%

Scheme 13.14 Selective hydrofluorination and cyclization of *N*-allyl benzenesulfonamides.

Aminobenzenesulfonamide derivatives being the target of choice for many therapeutic applications through the inhibition of matrix metalloprotease,[36] carbonic anhydrases (CAs),[189,208] human immunodeficiency virus proteases,[186] and pyruvate kinase,[22] for example, the ability to incorporate these methods in elaborated synthetic plan needed to be evaluated. Starting from halogen-substituted derivatives, through a combination of superacid-catalyzed cyclization followed by a superacid-catalyzed hydrofluorination and Cu-I-catalyzed coupling reaction with amines, a full set of fluorinated 4-aminobenzofused sultams and fluorinated 4-aminobenzenesulfonamides could be generated.[110] For example, starting from brominated *N*-allylbenzene sulfonamide **19a**, through a divergent and multistep synthesis, elaborated fluorinated 4-morpholino-benzenesulfonamide **23a** and fluorinated 4-morpholino-benzofuzed sultam **26a** were synthesized in overall good yields, confirming the synthetic potential of the method (Scheme 13.15).

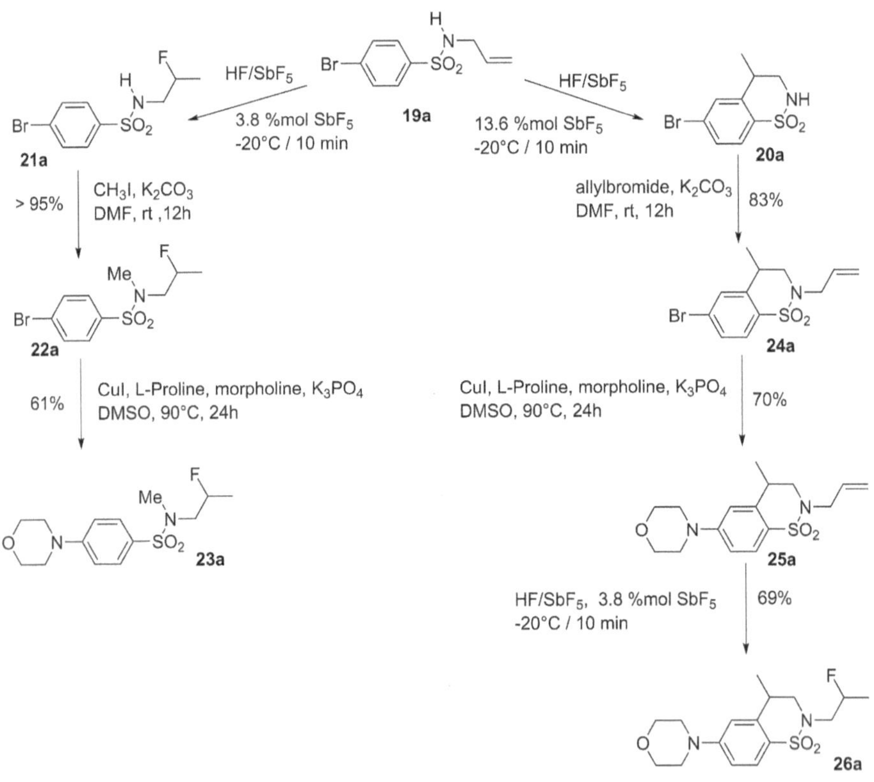

Scheme 13.15 Superacid/copper-mediated multistep divergent synthesis of 4-aminobenzenesulfamides.

4.2 Cyclization/Fluorination of N-Allylic Derivatives in HF/SbF_5

As shown for the benzofuzed sultam synthesis, the superelectrophilic character of ammonium–carbenium dications allows for their intramolecular trapping with poor nucleophiles. This was further exploited using double bond as a nucleophile. The cyclization/fluorination of variously substituted *N*-dienes **27** was shown to be an especially versatile entry to high-valued fluorinated piperidines **28** (Fig. 13.12).[204]

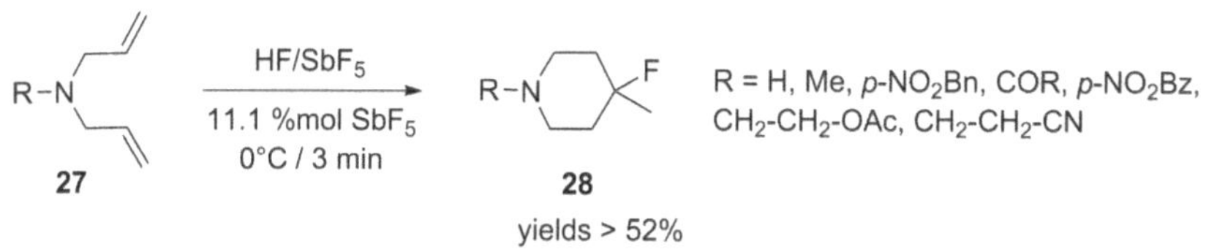

Figure 13.12 Cyclization/fluorination of *N*-dienes in HF/SbF_5.

In order to explore the mechanism of this reaction, low-temperature experiments were conducted, and after reaction of amine **27a** in HF/SbF_5 at −78°C, 21% of 3-fluoropiperidine **29a** was generated, beside its 4-fluoro analog **28a** (Scheme 13.16A). In addition, **27a** was shown to generate only **28a** after reaction in the superacid conditions at 0°C.

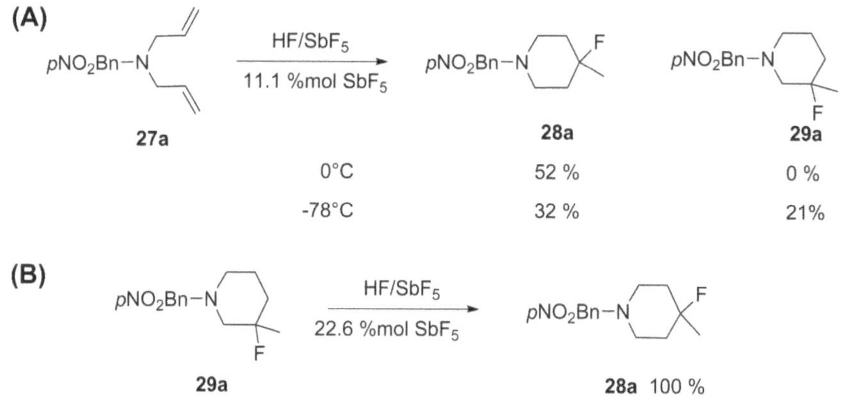

Scheme 13.16 (A) Formation of 3- and 4-fluoropiperidines in HF/SbF_5. (B) Formation of 4-flurorpiperidines from 3-fluoropiperidines in HF/SbF_5.

Mild conditions were obviously needed to obtain selectively the 3-fluoropiperidine, which indicated that this isomer could be an intermediate in the reaction course. The quantitative formation of compound **28a** starting from the 3-fluoropiperidine **29a** confirmed this hypothesis (Scheme 13.16B). After protonation of the nitrogen atom, the strong acidity of the medium allows for the formation of the ammonium–carbenium dication **A** (Scheme 13.17). The proximity of the charges strongly activates its electrophilic character, allowing nucleophilic substitution with poor nucleophile

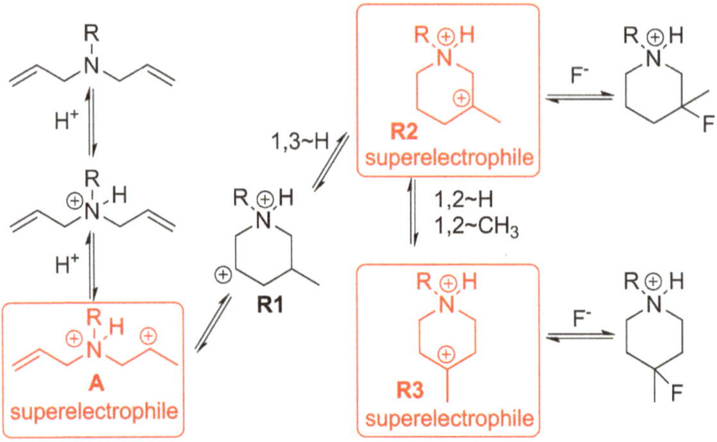

Scheme 13.17 Proposed mechanism for the cyclization/fluorination reaction.

partners. In this case, the nonprotonated double bond can act as a nucleophile, despite its electronic deactivation by the strong withdrawing effect of the proximal ammonium ion, and undergoes intramolecular cyclization to form **R1**. The formation of the 3-fluoropiperidine **29a** in mild conditions confirmed the possible 1,3-hydride shift, leading to inductively more stabilized intermediate **R2** precursor of 3-fluoropiperidine. In standard conditions, the more stabilized intermediate **R3** (optimal charges repulsion and inductive stabilization) must be generated after 1,2-hydride and methyl shifts and then trapped by fluoride ion to give the precursor of the 4-fluoropiperidine.

By exploiting the superelectrophilic character of ammonium-α-chloronium dications, the cyclization/fluorination was then extended to the direct synthesis of difluorinated piperidine. For example, starting from chlorinated *N*-diene **27b**, the difluorinated piperidine **30b** was directly synthesized in 57% yield (Scheme 13.18).[205]

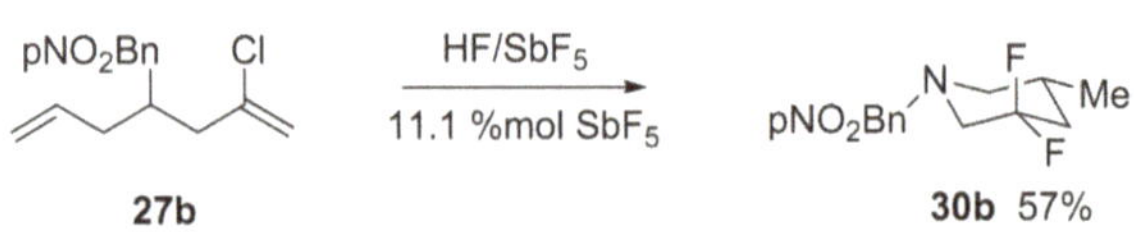

Scheme 13.18 Direct synthesis of difluorinated piperidine through cyclization/fluorination reaction.

5. Halofluorination of Unsaturated Nitrogen-Containing Substrates in Superacid HF/SbF_5

One of the simplest fluorinated amine building blocks would be the halofluorinated ones, and their synthesis could involve the halofluorination of unsaturated substrates.[49] Although electrophilic activation of alkenes via halenium ion attack has been extensively used for the formation of C–C, C–O, C–N, and C–X bonds,[10,34,35,88,91,194,213] its application to the halofluorination of allylic amine derivatives is limited to a few examples.[6,33,54,56,115,164,180,193,203] The successful use of HF/base/NXS (*N*-halosuccinimide) method for the halofluorination of carbon−carbon double bond is a lot less efficient when applied to allylic amine derivatives. The main issue in the application of the chlorofluorination to unsaturated amines is the dramatic decrease of nucleophilicity of the double bond after protonation of the amino group, which prevents any reaction with halonium source. In addition, if an alternative could be found to "force the reaction," after chlorenium ion attack, the regiochemical outcome of the chlorofluorination must be related to the nature of the generated chloronium ion intermediate. Olah's seminal work on cyclic halonium ions (haliranium ions) demonstrated that all the explored iodonium and bromonium ions are cyclic as well as many chloronium ions.[141−143] However, several chloriranium ions could not be observed due to their low stability in the experimental conditions used. In addition, theoretical calculations were also conducted showing that the substituents, and so their electronic effects, strongly influence the equilibrium that could exist between the chloromethyl carbenium, the α-chloronium ion, and the cyclic chloronium ions.[137,160,174] Reported stereoselective chloramination[26] and chlorolactonization reactions[211] also clearly demonstrated that their respective stereochemical and regiochemical outcomes are directly related to the respective generation of a cyclic chloronium ion and of a chloromethyl carbenium ion. These opposite data further emphasize the fact that the chloronium ion shapes are strongly affected by the nature of the reactive substrates. In this context, tandem superelectrophilic activation of unsaturated nitrogen-containing compounds and *N*-chlorosuccinimide (NCS) was found to be a valuable alternative to overcome the classically encountered difficulties for the halofluorination of unsaturated amines.

First, through diprotonation in superacid, the classical chlorenium source NCS can be activated. Dissolved in superacid HF/SbF_5 solution, the NCS protonation was studied by ^1H and ^{13}C NMR experiments at −20°C. The presence of one singlet at 3.30 ppm in the ^1H NMR spectrum confirmed that the two methylene units of the species generated from NCS in superacid are equivalent. In addition, the singlet at 9.40 ppm could be attributed to the protonation of both carbonyl groups of the imide function. In addition, in the ^{13}C spectrum, the observed signal for the two carbonyl groups, which must be equivalent, is strongly deshielded to 200.4 ppm, confirming the protonation of the two carbonyl groups (Fig. 13.13); these results are in accordance with previously reported data obtained from theoretical calculations.[166]

As a consequence, the electrophilicity of NCS must be exalted by protonation (formation of superelectrophile **S**) allowing to deliver the chlorenium ion to the electron-poor alkene coming from the protonation of unsaturated amines **1** (Scheme 13.19). The resulting ammonium−cyclic chloronium dication **T** showed a superelectrophilic character, allowing its subsequent fluorination to deliver β-fluoro-γ-chlorinated amines (Scheme 13.19). This reaction was found to be compatible with a significant set of unsaturated nitrogen-containing derivatives.[105]

Exploiting the ability of halogen substituent to orientate the fluoride ion attack on halogen-substituted ammonium−cyclic chloronium superelectrophiles **U** and **U′**, the chlorofluorination reaction in HF/SbF_5 was extended to olefins **3a**, **31**, and **32** substituted in internal or terminal position of the unsaturation with halogen atoms for the synthesis of versatile original high-valued polyhalogenated building blocks **33** (Fig. 13.14).

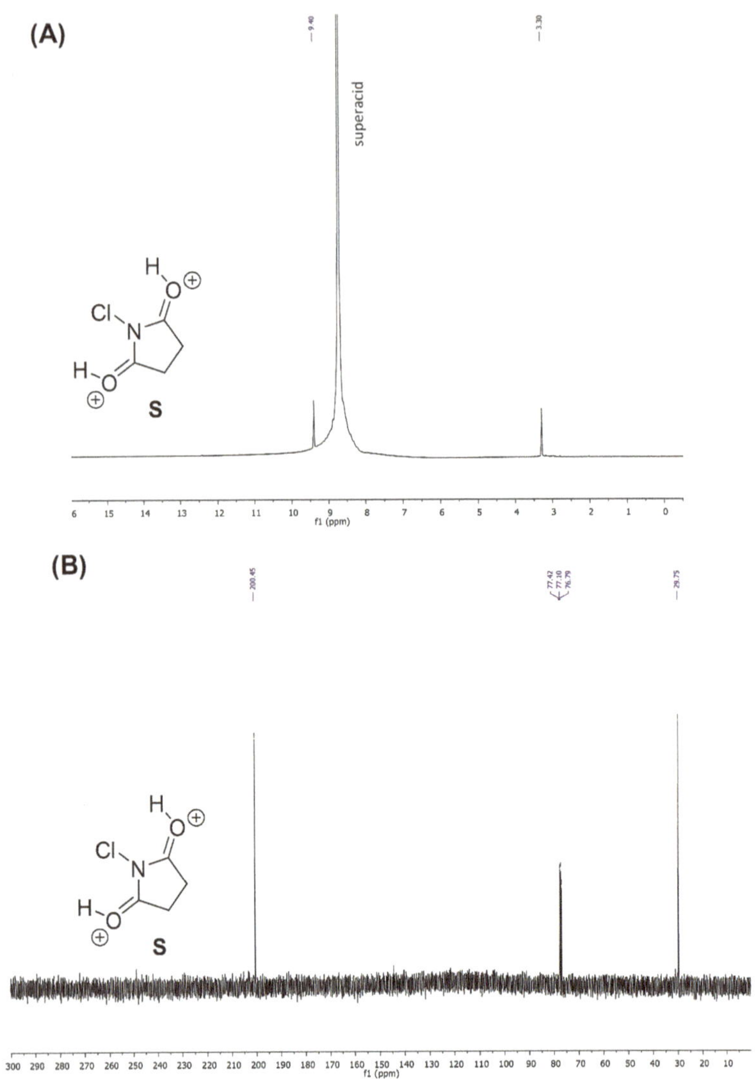

Figure 13.13 Results from low-temperature (−20°C) NMR studies of *N*-chlorosuccinimide dissolved in HF/SbF_5. (A) ^1H NMR spectrum. (B) ^{13}C NMR spectrum ($CDCl_3$ used as external standard).

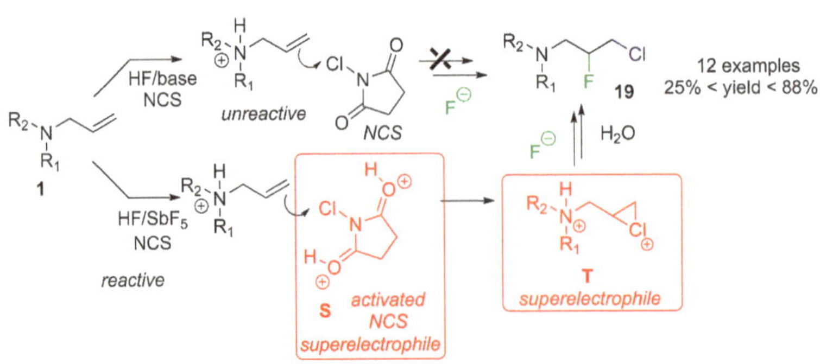

Scheme 13.19 HF/SbF_5/*N*-chlorosuccinimide tandem superelectrophilic activation for the chlorofluorination of allylic amines.

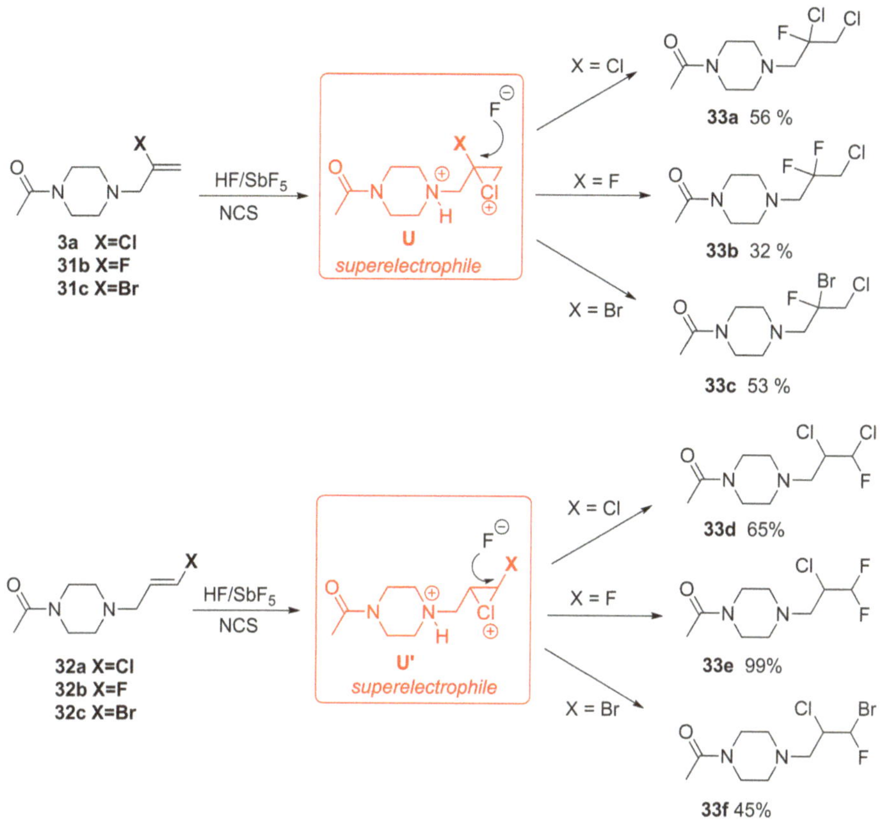

Figure 13.14 Chlorofluorination of halogenated amines in superacid HF/SbF_5 through halogen-directed regioselective fluorination of halogen-substituted ammonium–halonium superelectrophiles **U** and **U**′.

6. Exploitation of Superacid-Catalyzed Hydrofluorination of Unsaturated Amines in Superacid HF/SbF_5 for Bioorganic Studies

The most famous application of superacid-catalyzed fluorination is the synthesis of vinflunine from navelbine, performed by Jacquesy et al.[85] The original difluorinated semisynthetic *Vinca* alkaloid, acting as microtubule inhibitor, is prescribed as a treatment for metastatic bladder cancer under the trade name Javlor since 2009. More recently, the ability to generate a large variety of fluorinated analogs of nitrogen-containing derivatives in a divergent way was exploited to design human CA inhibitors for the treatment of hypoxic tumors.

CAs (EC 4.2.1.1) are widespread metalloenzymes that catalyze the reversible hydratation of carbon dioxide to bicarbonate and protons.[101,58] Human CAs (hCAs) exist in 16 isoforms, belonging to the so-called α-class and differ widely in their cellular localization (cytosol, mitochondria, cell membrane).[20,50,76,103,125,188,189,191,192] A recent application of the human carbonic anhydrase inhibitors (CAIs) regards their utility in the treatment of hypoxic tumors, through the selective inhibition of transmembrane hCA IX and XII, which are highly overexpressed in tumors. Sulfonamides and their bioisosteres (sulfamates, sulfamides, etc.) constitute the most investigated inhibitors of these enzymes ([3,4,197,53]; In their deprotonated forms, sulfonamides bind the Zn^{2+} ion of the active site, disrupting the catalytic process,[188]). However, most of the side effects of the sulfonamide-based CA inhibitors[5,52] are due to the inhibition of the cytosolic catalytically highly active isoform CA II, which is abundant in many tissues (organs) and involved in numerous physiological functions (pH homeostasis, secretion of electrolytes, transport of anions, etc.).[5] In addition, modifications of the aromatic ring include generally a strong variation in the hCA inhibition profile (modification of the selectivity), probably due to variation of interactions between the molecule and the hydrophobic and hydrophilic half parts of the isoforms (Fig. 13.15).

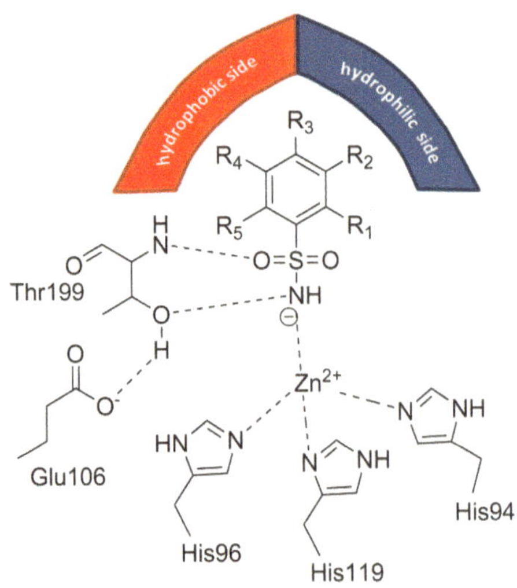

Figure 13.15 Binding of sulfonamide inhibitors to carbonic anhydrase. Key interactions between a generic benzenesulfonamide inhibitor and the hCA II active site.

As few hCA IX and XII isoform-specific inhibitors are known yet in the benzenesulfonamides series, the quest for new hCA inhibitors has given rise to the need for new chemotypes in this family.[120] To design inhibitors that would preferentially inhibit transmembrane isoforms over all other isoforms,[73] new selective inhibitors of tumor-associated CAs have been discovered, which act as non–zinc-binding inhibitors. In this series, (thio)coumarins and recently sulfocoumarins were shown to be hydrolyzed by esterase CA activity, and then anchored in the enzyme without direct interactions within the Zn^{2+} ion, into the so-called coumarin-binding site.[196] These data clearly demonstrated that not only zinc-binding inhibitors may exist but also that targeting different modes of inhibition for benzenesulfonamide inhibitors may be envisioned.

In this context, targeting secondary and tertiary benzenesulfonamides was explored, and a series of fluorinated superacid-catalyzed benzenesulfonamides were found to be selective tumor-associated isoforms validating this approach. Among all the tested derivatives ($\approx$ 100 compounds)[104,112,113,42] an especially interesting feature can be highlighted (Table 13.2).

Table 13.2 hCA Inhibition Data of Benzenesulfonamides 23

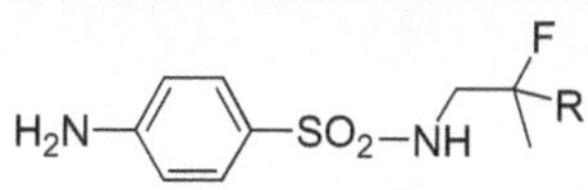

23

Substrate		Ki (nm)[a]			
		hCA I	hCA II	hCA IX	hCA XII
R = Me	23b	174	75	83	39.7
R = Cl	23c	5179	1400	2471	1521
R = F	23d	151	85.3	47.0	79.4

[a]Errors in the range of ±5% of the reported data from three different assays.

Fluorinated 4-aminobenzenesulfonamide **23** showed interesting inhibition properties at the nanomolar level against hCA isoforms. The effect of the fluorine atom in β-position (poor micromolar inhibition for nonfluorinated analogs against hCA I, II, IX, and XII) could have been attributed to a decrease of pK_a of the sulfonamide correlated with an increase of efficiency. However, there is a strong opposite effect of the addition of a chlorine atom (inhibition at the micromolar level for **23c**) compared with the addition of a second fluorine atom in the β-position (similar inhibition properties between **23b** and **23d**). This clearly ruled out this hypothesis and suggested a peculiar interactivity of the fluorine atom(s) within the enzyme active site, reinforcing the potential of this strategy in the quest of selective novel benzenesulfonamide-type inhibitors.

Targeting the so-called coumarin-binding site, with the ability to design benzenesulfonamide inhibitors adorned with precious fluorine-containing bullet point, a series of fluorinated tertiary benzenesulfonamides **34** were synthesized (Fig. 13.16).[122,123]

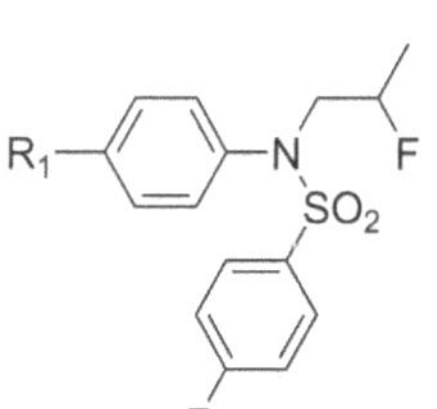

34

> 15 examples
Ki (hCA IX) = 4.2 - 81.2 nM
Selectivity hCA II / hCA IX > 1000

Figure 13.16 Fluorinated tertiary benzenesulfonamides acting as selective hCA IX nanomolar inhibitors.

Almost all the designed compounds in this fluorinated series were shown to be highly selective tumor-associated CAIs, without inhibiting the widespread off-target isoform hCA II, revealing a new mode of tumor-associated CA inhibition by sulfonamides.

7. Conclusion

The term "HF/SbF_5 superacid media" often scares and sometimes whets the curiosity. Fantastically exploited by Pr. Olah and others for studying long-lived carbocationic species and for designing new synthetic protocols, elegantly explored as an original media for elaborated bioactive products direct modifications by Jacquesy et al., it is now shown to be an especially versatile and powerful synthetic tool in the design of nitrogen-containing fluorinated (bioactive) compounds. These reviewed data, in addition to the ability to use superelectrophilic activation for CH bond fluorination (*to be published*) or for carbohydrate activation[119] let us envisage further advances in this field with original applications for this chemistry in the near future!

References

1. Akana, J. A.; Bhattacharrya, K. X.; Müller, P.; Sadighi, J. P. Reversible C-F bond formation and the Au-catalysed hydrofluorination of alkynes. *J. Am. Chem. Soc.* **2007,** *129,* 7736–7737.
2. Al-Maharik, N.; O'Hagan, D. Organofluorine Chemistry: Deoxyfluorination Reagents for C–F Bond Synthesis. *Aldrichim. Acta* **2011,** *44,* 65–75.
3. Alterio, V.; Di Fiore, A.; D'Ambrosio, K.; Supuran, C. T.; De Simone, G. X-Ray Crystallography of Carbonic Anhydrase Inhibitors and Its Importance in Drug Design. In *Drug Design of Zinc-Enzyme Inhibitors: Functional, Structural, and Disease Applications;* Supuran, C. T., Winum, J. Y., Eds.; Wiley: Hoboken (NJ), 2009a; pp 73–138.
4. Alterio, V.; Hilvo, M.; Di Fiore, A.; Supuran, C. T.; Pan, P.; Parkkila, S.; et al. Crystal Structure of the Catalytic Domain of the Tumor-associated Human Carbonic Anhydrase IX. *Proc. Natl. Acad. Sci. U.S.A* **2009b,** *106,* 16233–16238.
5. Alterio, V.; Hilvo, M.; Di Fiore, A.; Supuran, C. T.; De Simone, G. Multiple Binding Modes of Inhibitors to Carbonic Anhydrases: How to Design Specific Drugs Targeting 15 Different Isoforms? *Chem. Rev.* **2012,** *112,* 4421–4468.
6. Alvernhe, G.; Laurent, A.; Haufe, G. Triethylamine Tris-hydrofluoride [$(C_2H_5)_3N \cdot 3HF$]: A Highly Versatile Source of Fluoride Ion for the Halofluorination of Alkenes. *Synlett* **1987,** *6,* 562–564.
7. Andrei, H. S.; Solcà, N.; Dopfer, O. IR Spectrum of the Ethyl Cation: Evidence for the Nonclassical Structure. *Angew. Chem. Int. Ed.* **2008,** *47,* 395–397.
8. Antoniotti, S.; Dalla, V.; Duñach, E. Metal Triflimidates: Better than Metal Triflates as Catalysts in Organic Synthesis-the Effect of a High Delocalized Counteranion. *Angew. Chem. Int. Ed.* **2010,** *49,* 7860–7888.
9. Anxionnat, B.; Robert, B.; George, P.; Ricci, G.; Perrin, M. A.; Pardo, D. G.; et al. Ring Expansion of Cyclic B-amino Alcohols Induced by Diethylaminosulfur Trifluoride: Synthesis of Cyclic Amines with a Tertiary Fluorine at C3. *J. Org. Chem.* **2012,** *77,* 6087–6099.

10. Ashtekar, K. D.; Marzijarani, N. S.; Jaganathan, A.; Holmes, D.; Jackson, J. E.; Borhan, B. A New Tool to Guide Halofunctionalization Reactions: the Halenium Affinity (HalA) Scale. *J. Am. Chem. Soc.* **2014,** *136,* 13355–13362.
11. Axhausen, J.; Lux, K.; Kornath, A. The Existence of Hexafluoroarsenic(V) Acid. *Angew. Chem. Int. Ed.* **2014,** *53,* 3720–3721.
12. Bagno, A.; Bujnicki, B.; Bertrand, S.; Comuzzi, C.; Dorigo, F.; Janvier, P.; et al. Site of Protonation of Carboxylic and Non-carboxylic Amides in the Gas Phase and in Water. *Chem. Eur. J.* **1999,** *5,* 523–536.
13. Barker, T. J.; Boger, D. L. Fe(III)/$NaBH_4$ Mediated Free Radical Hydrofluorination of Unactivated Alkenes. *J. Am. Chem. Soc.* **2012,** *134,* 13588–13591.
14. Beer, P. D.; Gale, P. A. Anion Recognition and Sensing: The State of the Art and Future Perspectives. *Angew. Chem. Int. Ed.* **2001,** *40,* 486–516.
15. Bégué, J. P.; Bonnet-Delpon, D. *Chimie bioorganique et médicinale du fluor;* EDP CNRS Editions: Paris, 2005.
16. Beichel, W.; Panzer, J. M. U.; Hätty, J.; Ye, X.; Himmel, D.; Krossing, I. Straightfoward Synthesis of the Brønsted Acid $hfipOSO_3H$ and its Application for the Synthesis of Protic Ionic Liquids. *Angew. Chem. Int. Ed.* **2014,** *53,* 6637–6640.
17. Berkessel, A.; Thauer, R. K. On the Mechanism of Catalysis by a Metal-free Hydrogenase from *Metrhogenic archaea*: Enzymatic Transformation of H_2 without a Metal and Its Analogy to the Chemistry of Alkanes in Superacids. *Angew. Chem. Int. Ed.* **1995,** *34,* 2247–2250.
18. Berrier, C.; Jacquesy, J. C.; Jouannetaud, M. P.; Vidal, Y. Direct Bromination or Hydroxylation at C-11 in Vincadifformine and Tabersonine Derivatives in Superacids. *Tetrahedron* **1990,** *46,* 827–832.
19. Bonnet, B.; Marschepa, G. Ionic Species in HF-SbF_5 Solutions. An Infrared Spectroscopy Investigation. *Inorg. Chem.* **1980,** *19,* 785–788.
20. Borras, J.; Scozzafava, A.; Menabuoni, L.; Mincione, F.; Briganti, F.; Mincione, G.; et al. Carbonic Anhydrase Inhibitors: Synthesis of Water-Soluble, Topically Effective Intraocular Pressure Lowering Aromatic/Heterocyclic Sulfonamides Containing 8-quinoline-sulfonyl Moieties: Is the Tail More Important than the Ring? *Bioorg. Med. Chem.* **1999,** *7,* 2397–2406.
21. Bour, C.; Guillot, R.; Gandon, V. First Evidence for the Existence of Hexafluoroantimonic(V) Acid. *Chem. Eur. J.* **2015,** *21,* 6066–6069.
22. Boxer, M. B.; Jiang, J. K.; Vander Heiden, M. G.; Shen, M.; Skoumbourdis, A. P.; Southall, N.; et al. Evaluation of Substituted N,N′-diarylsulfonamides as Activators of the Tumor Cell Specific M2 Isoform of Pyruvate Kinase. *J. Med. Chem.* **2010,** *53,* 1048–1055.
23. Brabandt, W. V.; Verniest, G.; Smaele, D. D.; Duvey, G.; De Kimpe, N. Synthesis of 3-fluoroazetidines. *J. Org. Chem.* **2006,** *71,* 7100–7102.
24. Bryskier, A. *Antimicrobial Agents: Antibacterials and Antifugals;* ASM: Washington, 2005.
25. Brown, A.; Dave, E.; Wallace, O.; Ralph, M. Identification of a Urea Bioisostere of a Triazole Oxytocin Antagonist. *Bioorg. Med. Chem. Lett.* **2010,** *20,* 1851–1853.
26. Cai, Y.; Liu, X.; Chen, J.; Lin, L.; Feng, X. Catalytic Asymmetric Chloroamination Reaction of α,β-Unsaturated γ-Keto Esters and Chalcones. *J. Am. Chem. Soc.* **2011,** *133,* 5636–5639.
27. Calata, C.; Pfund, E.; Lequeux, T. Toward the Synthesis of Benzothiazolyl Fluoroaminosulfones. *J. Org. Chem.* **2009,** *74,* 9399–9405.

28. Caltagirone, C.; Gale, P. A. Anion Receptor Chemistry: Highlights from 2007. *Chem. Soc. Rev.* **2009,** *38,* 520–563.
29. Cantet, A. C.; Gesson, J. P.; Renoux, B.; Jouannetaud, M. P. A New Synthesis of Trifluorinated Compounds *via* 1,1-dichloro-1-alkenes in Superacid. *Tetrahedron Lett.* **2007,** *48,* 5255–5260.
30. Cantet, A. C.; Carreyre, H.; Gesson, J. P.; Jouannetaud, M. P.; Renoux, B. *Gem*-difluorination of Aminoalkynes *via* Highly Reactive Dicationic Species in Superacid HF/SbF_5: Application to the Efficient Synthesis of Difluorinated *Cinchona* Alkaloid Derivatives. *J. Org. Chem.* **2008,** *73,* 2875–2878.
31. Carreyre, H.; Coustard, J. M.; Carré, G.; Vandebrouck, C.; Bescond, J.; Ouédraogo, M. Natural Product Hybrid and its Superacid Synthesized Analogues: Dodoneine and its Derivatives Show Selective Inhibition of Carbonic Anhydrase Isoforms I, III, XIII and XIV. *Bioorg. Med. Chem.* **2013,** *21,* 3790–3794.
32. Champagne, P. A.; Benhassine, Y.; Desroches, J.; Paquin, J. F. Friedel-Crafts Reaction of Benzyl Fluorides: Selective Activation of C-F Bonds as Enabled by Hydrogen Bonding. *Angew. Chem. Int. Ed.* **2014,** *53,* 1–6.
33. Cheidi, I.; Chaabouni, M. M.; Baklouti, A. Bromofluoration des alcools allyliques par NBS/Et_3N, 3HF: une voie simple d'accès aux épifluorhydrines. *Tetrahedron Lett.* **1989,** *30,* 3167–3170.
34. Chen, G.; Ma, S. Enantioselective Halocyclization Reactions for the Synthesis of Chiral Cyclic Compounds. *Angew. Chem. Int. Ed.* **2010,** *49,* 8306–8308.
35. Chen, J.; Zhou, L. Recent Progress in the Asymmetric Intermolecular Halogenation of Alkenes. *Synthesis* **2014,** *46,* 586–595.
36. Cheng, X. C.; Wang, Q.; Fang, H.; Xu, W. F. Role of Sulfonamide Group in Matrix Metalloproteinase Inhibitors. *Curr. Med. Chem.* **2008,** *15,* 368–373.
37. Choudhary, A.; Raines, R. T. An Evaluation of Peptide-bond Isosteres. *Chem. Bio. Chem.* **2011,** *12,* 1801–1807.
38. Christe, K. O.; Zhang, X.; Bau, R.; Hegge, J.; Olah, G. A.; Prakash, G. K. S.; et al. Relative Abilities of Fluorine and Chlorine to Stabilize Carbenium Ions: Crystal Structures of Two Fluoro-Substituted Carbocations and $As_2F_{11}^-$. *J. Am. Chem. Soc.* **2000,** *122,* 481–487.
39. Compain, G.; Martin-Mingot, A.; Frapper, G.; Bachmann, C.; Jouannetaud, M. P.; Thibaudeau, S. Anti-Markovnikov Addition on *N*-allylic Derivatives. *Chem. Commun.* **2012a,** *48,* 5877–5879.
40. Compain, G.; Jouvin, K.; Martin-Mingot, A.; Evano, G.; Marrot, J.; Thibaudeau, S. Stereoselective Hydrofluorination of Ynamides: A Straightforward Synthesis of Novel α-fluoroenamides. *Chem. Commun.* **2012b,** *48,* 5196–5198.
41. Compain, G.; Bonneau, C.; Martin-Mingot, A.; Thibaudeau, S. Selective Anti-Markovnikov Cyclization and Hydrofluorination Reaction in Superacid HF/SbF_5: A Tool in the Design of Nitrogen-containing (Fluorinated) Polycyclic Systems. *J. Org. Chem.* **2013a,** *78,* 4463–4472.
42. Compain, G.; Martin-Mingot, A.; Maresca, A.; Thibaudeau, S.; Supuran, C. T. Superacid Synthesis of Halogen *N*-substituted-4-aminobenzenesulfonamides: New Selective Tumor-associated Carbonic Anhydrase Inhibitors. *Bioorg. Med. Chem.* **2013b,** *21,* 1555–1563.
43 Couve-Bonnaire, S.; Cahard, D.; Pannecoucke, X. Chiral Dipeptide Mimics Possessing a Fluoroolefin Moiety: A Relevant Tool for Conformational and Medicinal Studies. *Org. Biomol. Chem.* **2007,** *5,* 1151–1157.
44. Cox, R. A.; Druet, L. M.; Klausner, A. E.; Modro, T. A.; Wan, P.; Yates, K. Protonation Acidity Constants for Some Benzamides, Acetamides, and Lactams. *Can. J. Chem.* **1981,** *59,* 1568–1573.

45. Culmann, J. C.; Fauconet, M.; Jost, R.; Sommer, J. Fluoroanions and Cations in the HF-SbF_5 Superacid System. *New J. Chem.* **1999,** *23,* 863–867.
46. Culmann, J. C.; Simon, M.; Sommer, J. Selective Carbonylation of Propane in Superacid Media *via* Hydride Abstraction by the Chlorocarbocations: CCl_3^+ and $CHCl_2^+$. *J. Chem. Soc. Chem. Commun.* **1990,** 1098–1100.
47. Debarge, S.; Thibaudeau, S.; Violeau, B.; Martin-Mingot, A.; Jouannetaud, M. P.; Jacquesy, J. C.; et al. Rearrangement of *Gem*-difluorination of Quinine and 9-epiquinine and Their Acetates in Superacid. *Tetrahedron* **2005,** *61,* 2065–2073.
48. Del Prato, S.; Pulizzi, N. The Place of Sulfonylureas in the Therapy for Type 2 Diabetes Mellitus. *Metabolism* **2006,** *55,* S20–S27.
49. Denmark, S. E.; Kuester, W. E.; Burk, M. T. Catalytic, Asymmetric Halofunctionalization of Alkenes-a Critical Perspective. *Angew. Chem. Int. Ed.* **2012,** *51,* 10938–10953.
50. De Simone, G.; Di Fiore, A.; Supuran, C. T. *Curr. Pharm. Des.* **2008,** *14,* 655–660.
51. D'Hooghe, M.; Catak, S.; Stankovic, S.; Waroquier, M.; Kim, Y.; Ha, H. J.; et al. Systematic Study of Halide-induced Ring Opening of 2-substituted Aziridinium Salts and Theoretical Rationalization of the Reaction Pathways. *Eur. J. Org. Chem.* **2010,** *25,* 4920–4931.
52. Di Fiore, A.; Truppo, E.; Supuran, C. T.; Alterio, V.; Dathan, N.; Bootorabi, F.; et al. Crystal structure of the C183S/C217S Mutant of Human CA VII in Complex with Acetazolamide. *Bioorg. Med. Chem. Lett.* **2010,** *20,* 5023–5026.
53. Di Fiore, A.; Maresca, A.; Alterio, V.; Supuran, C. T.; De Simone, G. Carbonic Anhydrase Inhibitors: X-ray Crystallographic Studies for the Binding of *N*-substituted Benzene-sulfonamides to Human Isoform II. *Chem. Commun.* **2011,** *47,* 11636–11638.
54. Dolenc, D.; Šket, B. Chlorofluorination of Alkenes and Alkynes Using a New Reagent *N*-chlorosaccharin-HF/pyridine System. *Synlett* **1995,** *4,* 327–328.
55. Dolensky, B.; Kirk, K. L. New Building Blocks for Fluorinated Imidazole Derivatives: Preparation of beta-fluoro- and beta,beta-difluorohistamine. *J. Org. Chem.* **2001,** *66,* 4687–4691.
56. Dolensky, B.; Kirk, K. L. New Building Blocks for Fluorinated Bioimidazole Derivatives II: Preparation of *β*-fluorourocanic Acids. *J. Org. Chem.* **2002,** *67,* 3468–3473.
57. Dolensky, B.; Narayanan, J.; Kirk, K. L. Preparation of β-fluoro- and β,β-difluoro-histidinols. *J. Fluor. Chem.* **2003,** *123,* 95–99.
58. Domsic, J. F.; Avvaru, B. S.; Kim, C. U.; Sol, M.; Gruner, S. M.; Agbandje-McKenna, M.; Silverman, D. N.; et al. Entrapment of Carbon Dioxide in the Active Site of Carbonic Anhydrase II. *J. Biol. Chem.* **2008,** *283,* 30766–30771.
59. Doyle, A. G.; Jacobsen, E. N. Small-molecule H-bond Donors in Asymmetric Catalysis. *Chem. Rev.* **2007,** *107,* 5713–5743.
60. Duflos, A.; Redoules, F.; Fahy, J.; Jacquesy, J. C.; Jouannetaud, M. P. Hydroxylation of Yohimbine in Superacidic Media: One-step Access to Human Metabolites 10 and 11-hydroxyyohimbine. *J. Nat. Prod.* **2001,** *64,* 193–195.
61. Dumas, J. Protein Kinase Inhibitors from the Urea Class. *Curr. Opin. Drug. Discov. Dev.* **2002,** *5,* 718–727.
62 Emer, E.; Pfeifer, L.; Brown, J. M.; Gouverneur, V. *Cis*-specific Hydrofluorination of Alkenylarenes under Palladium Catalysis through an Ionic Pathway. *Angew. Chem. Int. Ed.* **2014,** *53,* 4181–4185.
63. Esteves, P. M.; Ramirez-Solis, A.; Mota, C. J. A. The Nature of Superacid Electrophilic Species in HF/SbF_5: A Density Functional Theory Study. *J. Am. Chem. Soc.* **2002,** *124,* 2672–2677.

64. Fadeyi, O. O.; Lindsey, C. W. Rapid, General Access to Chiral β-fluoroamines and β,β-difluoroamines *via* Organocatalysis. *Org. Lett.* **2009,** *11,* 943–946.
65. Fahy, J.; Duflos, A.; Ribet, J. P.; Jacquesy, J. C.; Berrier, C.; Jouannetaud, M. P.; et al. *Vinca* alkaloids in Superacidic Media: A Method for Creating a New Family of Antitumor Derivatives. *J. Am. Chem. Soc.* **1997,** *119,* 8576–8577.
66. Forropoulos, J., Jr.; Desmarteau, D. D. Synthesis, Properties and Reactions of bis((trifluoromethyl)sulfonyl)-imide, $(CF_3SO_2)_2NH$. *Inorg. Chem.* **1984,** *23,* 3720–3723.
67. Fustero, S.; Sanz-Cervera, J. F.; Aceña, J. L.; Sánchez-Roselló, M. Nitrogen-Containing Organofluorine Derivatives: An Overview. *Synlett* **2009,** *4,* 525–549.
68. Gillespie, R. J.; Peel, T. E. Superacid Systems. *Adv. Phys. Org. Chem.* **1971,** *9,* 1–24.
69. Gillespie, R. J.; Peel, T. E. The Hammett Acidity Function for Some Superacid Systems. II. The Systems H_2SO_4-HSO_3F, KSO_3F-HSO_3F, HSO_3F-SO_3, HSO_3F-AsF_5, HSO_3F-SbF_5 and HSO_3F-SbF_5-SO_3. *J. Am. Chem. Soc.* **1973,** *95,* 5173–5178.
70. Gillespie, R. J.; Liang, J. Superacid Solutions in Hydrogen Fluoride. *J. Am. Chem. Soc.* **1988,** *110,* 6053–6057.
71. Gillis, E. P.; Eastma, K. J.; Hill, M. D.; Donnelly, D. J.; Meanwell, N. A. Applications of Fluorine in Medicinal Chemistry. *J. Med. Chem.* **2015,** *58,* 8315–8359.
72. Giovanelli, E.; Leroux, S.; Moisan, L.; Carreyre, H.; Thuéry, P.; Buisson, D. A.; et al. On the Elucidation of the Mechanism of *Vinca* Alkaloid Fluorination in Superacidic Medium. *Org. Lett.* **2011,** *12,* 4116–4119.
73. Guler, O. O.; De Simone, G.; Supuran, C. T. Drug Design Studies of the Novel Antitumor Targets Carbonic Anhydrases IX and XII. *Curr. Med. Chem.* **2010,** *17,* 1516–1526.
74. Hall, N. F.; Conant, J. B. A Study of Superacid Solutions. I. The Use of the Chloranil Electrode in Glacial Acetic Acid and the Strength of Certain Weak Bases. *J. Am. Chem. Soc.* **1927,** 3047–3061.
75. Hammond, G. S. A Correlation of Reaction Rates. *J. Am. Chem. Soc.* **1955,** *77,* 334–338.
76. Harju, A. K.; Bootorabi, E.; Kuushahti, M.; Supuran, C. T.; Parkkila, S. Carbonic Anhydrase III: A Neglected Isozyme Is Stepping into the Limelight. *J. Enz. Inhib. Med. Chem.* **2013,** *28,* 231–239.
77. Harsanyi, A.; Sandford, G. Organofluorine Chemistry: Applications, Sources and Sustainability. *Green Chem.* **2015,** *17,* 2081–2086.
78. Hawryluk, N. A.; Merit, J. E.; Lebsack, A. D.; Branstetter, B. J.; Hack, M. D.; Swanson, N.; et al. Discovery and Synthesis of 6,7,8,9-tetrahydro-5H-pyrimido-[4,5-d] azepines as Novel TRPV1 Antagonists. *Bioorg. Med. Chem. Lett.* **2010,** *20,* 7137–7141.
79. Henne, A. L.; Zimmerschied, W. J. Fluorinated Acids. *J. Am. Chem. Soc.* **1947,** *69,* 281–283.
80. Hu, X. G.; Hunter, L. Stereoselectively Fluorinated *N*-heterocycles: a Brief Survey. *Beilstein J. Org. Chem.* **2013,** *9,* 2696–2708.
81. Jacquesy, J. C.; Berrier, C.; Gesson, J. P.; Jouannetaud, M. P. New Syntheses in Super Acids. *Ann. Chim.* **1984,** *9,* 641–648.
82. Jacquesy, J. C.; Berrier, C.; Gesson, J. P.; Jouannetaud, M. P. Electrophilic Hydroxylation of Aromatics in Superacids. *Bull. Soc. Chim. Fr.* **1994,** *131,* 658–664.
83. Jacquesy, J. C. Natural Products Chemistry in Superacids. In *Stable Carbocation in Chemistry;* Prakash, G. K. S., Schleyer, P.von R., Eds.; Wiley and Sons: New-York, 1997; pp 549–574.
84. Jacquesy, J. C.; Jouannetaud, M. P.; Martin, A. Functionalization of Nonactivated Bonds in Superacidic Media. *Bull. Soc. Chim. Fr.* **1997,** *134,* 425–438.

85. Jacquesy, J. C.; Fahy, J. Cancer: Superacid Generation of New Antitumor Agents. In *Biomedical Chemistry;* Torrence, P. F., Ed.; Wiley and Sons: New-York, 2000; pp 227–246.
86. Jacquesy, J. C. Organic Synthesis in Superacids. In *Carbocation Chemistry;* Olah, G. A., Prakash, G. K. S., Eds.; John Wiley-Interscience: New-York, 2004; pp 359–376.
87. Jacquesy, J. C. Reactivity of *Vinca* Alkaloids in Superacid: An Access to Vinflunine, A Novel Anticancer Agent. *J. Fluor. Chem.* **2006,** *127,* 1484–1487.
88. Jaganathan, A.; Garzan, A.; Whitehead, D. C.; Staples, R. J.; Borhan, B. A Catalytic Asymmetric Chlorocyclization of Unsaturated Amides. *Angew. Chem. Int. Ed.* **2011,** *50,* 2593–2596.
89. Jansen, K.; Heirbaut, L.; Verkerk, R.; Cheng, J. D.; Joossens, J.; Cos, P.; et al. Extended Structure-activity Relationship and Pharmacokinetic Investigation of (4-quinolinoyl) glycyl-2-cyanopyrrolidine Inhibitors of Fibroblast Activation Protein (FAP). *J. Med. Chem.* **2014,** *57,* 3053–3074.
90. Kalow, J. A.; Schmitt, D. E.; Doyle, A. G. Synthesis of β-fluoroamines by Lewis Base Catalyzed Hydrofluorination of Aziridines. *J. Org. Chem.* **2012,** *77,* 4177–4183.
91 Karpfen, A. Theoretical Characterization of the Trends in Halogen Bonding. In *Halogen Bonding: Fundamentals and Applications,* Vol. 126, Metrangolo, P., Resnati, G., Eds.; Springer-Verlag: Berlin, 2008; pp 1–15.
92. Kim, D.; Klein, M. L. Liquid Hydrogen Fluoride with an Excess Proton: *Ab Initio* Molecular Dynamics Study of a Superacid. *J. Am. Chem. Soc.* **1999,** *121,* 11251–11252.
93. Kim, D.; Klein, M. L. *Ab Initio* Molecular Dynamics Study of the Superacid System SbF_5/HF Solution. *J. Phys. Chem. B* **2000,** *104,* 10074–10079.
94. Klumpp, D. A.; Garza, M.; Sanchez, G. V.; Lau, S.; De Leon, S. J. Electrophilic Activation of Acetyl-Substituted Heteroaromatic Compounds. *J. Org. Chem.* **2000,** *65,* 8997–9000.
95. Klumpp, D. A.; Beauchamp, P. S.; Sanchez, G. V., Jr.; Aguirre, S.; De Leon, S. Superacid-catalyzed Preparation of Aryl-substituted Piperidines *via* Dicationic Electrophiles. *Tetrahedron Lett.* **2001a,** *42,* 5821–5823.
96. Klumpp, D. A.; Aguirre, S. L.; Sanchez, G. V.; De Leon, S. J. Reactions of Amino Alcohols in Superacid: the Direct Observation of Dicationic Intermediates and Their Application in Synthesis. *Org. Lett.* **2001b,** *3,* 2781–2784.
97. K.lumpp, D. A.; Rendy, R.; Zhang, Y.; McElrea, A.; Gomez, A.; Dang, H. Reactive Dications: the Superacid-catalyzed Reactions of Alkynes Bearing Adjacent *N*-heterocycles or Amine Groups. *J. Org. Chem.* **2004,** *69,* 8108–8110.
98. Klumpp, D. A. Superelectrophiles: Charge-Charge Repulsive Effects. *Chem. Eur. J.* **2008,** *14,* 2004–2015.
99. Koppel, I. A.; Burk, P.; Koppel, I.; Leito, I. Generalized Principle of Designing Neutral Superstrong Brønsted Acids. *J. Am. Chem. Soc.* **2002,** *124,* 5594–5600.
100. Krentz, A. J.; Bailey, C. J. Oral Antidiabetic Agents: Current Role in Type 2 Diabetes Mellitus. *Drugs* **2005,** *65,* 385–411.
101. Krishnamurthy, V. M.; Kaufman, G. K.; Urbach, A. R.; Gitlin, I.; Gudiksen, K. L.; Weibel, D. B.; et al. Carbonic Anhydrase as a Model for Biophysical and Physical-Organic Studies of Proteins and Protein-Ligand Binding. *Chem. Rev.* **2008,** *108,* 946–1051.
102. Krossing, I.; Raabe, I. Non Coordinating Anions-Fact or Fiction ? A Survey of Likely Candidates. *Angew. Chem. Int. Ed.* **2004,** *43,* 2066–2090.
103. Krungkrai, J.; Supuran, C. T. The Alpha-carbonic Anhydrase from the Malaria Parasite and its Inhibition. *Curr. Pharm. Des.* **2008,** *14,* 631–640.

104. Le Darz, A.; Mingot, A.; Bouazza, F.; Castelli, U.; Karam, O.; Tanc, M.; et al. Fluorinated Pyrrolidines and Piperidines Incorporating Tertiary Benzesulfonamide Moieties Are Selective Carbonic Anhydrase II Selective Inhibitors. *J. Enz. Inhib. Med. Chem.* **2014,** *30,* 737–745.
105. Le Darz, A.; Castelli, U.; Mokhtari, N.; Martin-Mingot, A.; Marrot, J.; Bouazza, F.; et al. Tandem Superelectrophilic Activation for the Regioselective Chlorofluorination of Recalcitrant Allylic Amines. *Tetrahedron* **2016,** *72,* 674–689.
106 Le Duc, Y.; Michau, M.; Gilles, A.; Gence, V.; Legrand, Y. M.; Van der Lee, A.; et al. Imidazole-Quartet Water and Proton Dipolar Channels. *Angew. Chem. Int. Ed.* **2011,** *50,* 11366–11372.
107. Li, H. Q.; Lv, P. C.; Yan, T.; Zhu, H. L. Urea Derivatives as Anticancer Agents. *Anticancer Agents Med. Chem.* **2009,** *9,* 471–480.
108. Liu, F.; Martin-Mingot, A.; Jouannetaud, M. P.; Karam, O.; Thibaudeau, S. Substitution Effect on the Hydrofluorination Reaction of Unsaturated Amines in Superacid HF/SbF_5. *Org. Biomol. Chem.* **2009,** *7,* 4789–4797.
109. Liu, F.; Martin-Mingot, A.; Jouannetaud, M. P.; Zunino, F.; Thibaudeau, S. Superelectrophilic Activation in Superacid HF/SbF_5 and Synthesis of Benzofused Sultams. *Org. Lett.* **2010a,** *12,* 868–871.
110. Liu, F.; Musadji, N. Y.; Lecornué, F.; Jouannetaud, M. P.; Thibaudeau, S. Superacid Mediated Reactions to 4-aminobenzofused Sultams and Fluorinated 4-aminobenzene Sulfonamides Synthesis. *Tetrahedron* **2010b,** *66,* 7112–7118.
111. Liu, F.; Martin-Mingot, A.; Jouannetaud, M. P.; Bachmann, C.; Frapper, G.; Zunino, F.; et al. Selective Synthesis of *gem*-chlorofluorinated Nitrogen-Containing Derivatives after Superelectrophilic Activation in Superacid HF/SbF_5. *J. Org. Chem.* **2011,** *76,* 1460–1463.
112. Liu, F.; Martin-Mingot, A.; Lecornué, F.; Jouannetaud, M. P.; Maresca, A.; Thibaudeau, S.; Supuran, C. T. Carbonic Anhydrases Inhibitory Effects of New Benzenesulfonamides Synthesized by Using Superacid Chemistry. *J. Enz. Inhib. Med. Chem.* **2012a,** *27,* 886–891.
113. Liu, W.; Huang, X.; Cheng, M. J.; Nielsen, R. J.; Goddard, W. A., III; Groves, J. T. Oxidative Aliphatic C-H Fluorination with Fluoride Ion Catalyzed by a Manganese Porphyrin. *Science* **2012b,** *337,* 1322–1325.
114. Liu, W.; Groves, J. T. Manganese-catalyzed Oxidative Benzylic C–H Fluorination by Fluoride Ions. *Angew. Chem. Int. Ed.* **2013,** *52,* 6024–6027.
115. Lubke, M.; Skupin, R.; Haufe, G. Regioselectivity of Bromofluorination of Functionalized 1-alkenes. *J. Fluor. Chem.* **2000,** *102,* 125–135.
116. Majumdar, K. C.; Mondal, S. Recent Developments in the Synthesis of Fused Sultams. *Chem. Rev.* **2011,** *111,* 7749–7773.
117. Malamakai, R. M.; Hess, W. R.; Davis, T. A. Highly Efficient Diastereoselective Reduction of α-fluoroimines. *Org. Lett.* **2010,** *12,* 2186–2189.
118. Mallouk, T. E.; Rosenthal, G. L.; Müller, G.; Brusasco, R.; Bartlett, N. Fluoride Ions Affinities of GeF_4 and BF_3 from Thermodynamic and Structural Data for $(SF_3)_2GeF_6$, ClO_2GeF_5 and ClO_2BF_4. *Inorg. Chem.* **1984,** *23,* 3167–3173.
119. Martin, A.; Arda, A.; Désiré, J.; Martin-Mingot, A.; Probst, N.; Sinaÿ, P.; et al. Catching Elusive Glycosyl Cations in a Condensed Phase with HF/SbF_5 Superacid. *Nat. Chem.* **2016,** *8,* 186–191.
120. McDonald, P. C.; Winum, J. Y.; Supuran, C. T.; Dedhar, S. Recent Developments in Targeting Carbonic Anhydrase IX for cancer Therapeutics. *Oncotarget* **2012,** *3,* 84–97.
121. McLain, S. E.; Benmore, C. J.; Siewenie, J. E.; Urbiqui, J.; Turner, J. F. C. On the Structure of Liquid Hydrogen Fluoride. *Angew. Chem. Int. Ed.* **2004,** *43,* 1951–1955.

122. Métayer, B.; Mingot, A.; Vullo, D.; Supuran, C. T.; Thibaudeau, S. New Superacid Synthesized (Fluorinated) Tertiary Benzenesulfonamides Acting as Selective hCA IX Inhibitors: toward a New Mode of Carbonic Anhydrase Anhibition by Sulfonamides. *Chem. Commun.* **2013a,** *49,* 6015–6017.
123. Métayer, B.; Martin-Mingot, A.; Vullo, D.; Supuran, C. T.; Thibaudeau, S. Superacid Synthesized Tertiary Benzenesulfonamides and Benzofuzed Sultams Act as Selective hCA IX Inhibitors: Toward Understanding a New Mode of Inhibition by Tertiary Sulfonamides. *Org. Biomol. Chem.* **2013b,** *11,* 7540–7549.
124. Métayer, B.; Compain, G.; Jouvin, K.; Martin-Mingot, A.; Bachmann, C.; Marrot, J.; et al. Chemo- and Stereoselective Synthesis of Enamides from Ynamides in HF/Pyridine: Second-generation Approach to Potent Ureas Bioisosters. *J. Org. Chem.* **2015,** *80,* 3397–3410.
125. Mincione, F.; Scozzafava, A.; Supuran, C. T. Antiglaucoma Carbonic Anhydrase Inhibitors as Ophthalomologic Drugs. In *Drug Design of Zinc-Enzyme Inhibitors: Functional, Structural, and Disease Applications;* Supuran, C. T., Winum, J. Y., Eds.; Wiley: Hoboken (NJ), 2009; pp 139–154.
126. Mootz, D.; Bartmann, K. The Fluoronium Ions H_2F^+ and $H_3F_2{}^+$: Characterization by Crystal Structure Analysis. *Angew. Chem. Int. Ed.* **1988,** *27,* 391–392.
127. Morgenthaler, M.; Schweizer, E.; Hoffmann-Röder, F. B.; Rainer, E. M.; Jaeschke, G.; Wagner, B.; et al. Predicting and Tuning Physicochemical Properties in Lead Optimization: Amine Basicities. *Chem. Med. Chem.* **2007,** *2,* 1100–1115.
128. Müller, L. O.; Himmel, D.; Stauffer, J.; Steinfeld, G.; Slattery, J.; Santiso-Quiñones, G.; et al. Simple Access to the Non-oxidizing Lewis Superacid $PhF \rightarrow Al(OR^F)_3$ ($R^F = C(CF_3)_3$). *Angew. Chem. Int. Ed.* **2008,** *47,* 7659–7663.
129. Nahra, F.; Patrick, S. R.; Bello, D.; Brill, M.; Obled, A.; Cordes, D. B.; et al. Hydrofluorination of Alkynes Catalyzed by Gold Bifluorides. *Chem. Cat. Chem.* **2015,** *7,* 240–244.
130. Nakanishi, W. Related Organocatalysts (2): Urea Derivatives. In *Superbases for Organic Synthesis: Guanidines, Amidines, Phosphazenes and Related Organocatalysts;* Ishikawa, T., Ed.; John Wiley and Sons: New York, 2009; pp 273–293.
131. Naredla, R. R. N.; Klumpp, D. A. Contemporary Carbocation Chemistry: Applications in Organic Synthesis. *Chem. Rev.* **2013,** *113,* 6905–6948.
132. Narumi, T.; Hayashi, R.; Tomita, K.; Kobayashi, K.; Tanahara, N.; Ohno, H.; et al. Synthesis and Biological Evaluation of Selective CXCR4 Antagonists Containing Alkene Dipeptide Isosteres. *Org. Biomol. Chem.* **2010,** *8,* 616–621.
133. Nasr, G.; Macron, T.; Gilles, A.; Petit, E.; Barboiu, M. Systems Membranes – Combining the Supramolecular and Dynamic Covalent Polymers for Gas-Selective Dynameric Membranes. *Chem. Commun.* **2012,** *48,* 7398–7400.
134. Natarajan, A.; Guo, Y.; Arthanari, H.; Wagner, G.; Halperin, J. A.; Chorev, M. Synthetic Studies toward aryl-(4-aryl-4H-[1,2,4]triazole-3-yl)-amine from 1,3-diarylthiourea as Urea Mimetics. *J. Org. Chem.* **2005,** *70,* 6362–6368.
135. O'Connor, M. J.; Boblak, K. N.; Topinka, M. J.; Kindelin, P. J.; Briski, J. M.; Zheng, C.; et al. Superelectrophiles and the Effects of Trifluoromethyl Substituents. *J. Am. Chem. Soc.* **2010,** *132,* 3266–3267.
136. O'Hagan, D. Understanding Organofluorine Chemistry. An Introduction to the C-F Bond. *Chem. Soc. Rev.* **2008,** *37,* 308–319.
137. Ohta, B. K.; Hough, R. E.; Schubert, J. W. Evidence for β-chlorocarbenium and β-bromocarbenium Ions. *Org. Lett.* **2007,** *9,* 2317–2320.

138. Ojima, I. *Fluorine in Medicinal Chemistry and Chemical Biology;* Wiley-Blackwell: Chichester, United-Kingdom, 2009.
139. Ojima, I. Exploration of Fluorine Chemistry at the Multidisciplinary Interface of Chemistry and Biology. *J. Org. Chem.* **2013,** *78,* 6358–6383.
140. Okoromoba, O. E.; Han, J.; Hammond, G. B.; Xu, B. Designer HF-based Fluorination Reagent: Highly Regioselective Synthesis of Fluoroalkenes and *gem*-difluoromethylene Compounds from Alkynes. *J. Am. Chem. Soc.* **2014,** *136,* 14381–14384.
141. Olah, G. A.; Bollinger, J. M. Stable Carbonium Ions. XLVIII. Halonium Ion Formation *via* Neighboring Halogen Participation. Tetramethylethylene Halonium Ions. *J. Am. Chem. Soc.* **1967,** *89,* 4744–4752.
142. Olah, G. A.; Bollinger, J. M. Stable Carbonium Ions. LVII. Halonium Ion Formation *via* Neighboring Halogen Participation. Trimethyl- and 1,1-dimethylethylenehalonium Ion. *J. Am. Chem. Soc.* **1968,** *90,* 947–953.
143. Olah, G. A.; Bollinger, J. M.; Brinich, J. Stable Carbonium Ions. LXII. Halonium Ion Formation *via* Neighboring Halogen Participation: Ethylenehalonium, Propylenehalonium, and 1,2-dimethylethylenehalonium Ions. *J. Am. Chem. Soc.* **1968,** *90,* 2587–2594.
144. Olah, G. A.; Schlosberg, R. H. Stable Carbonium Ions. LXXI. Protonated Imides. *J. Am. Chem. Soc.* **1968,** *90,* 6464–6467.
145. Olah, G. A.; Mo, Y. K.; Halpern, Y. Stable Carbocations. CXXXI. Intermolecular Fluorine Exchange of the Methylfluorocarbenium and Dimethylfluorocarbenium Ions in HF-SbF_5-SO_2ClF Solution. *J. Org. Chem.* **1972,** *8,* 1169–1174.
146. Olah, G. A.; Nojima, M.; Kerekes, I. Synthetic Methods and Reactions II. Hydrofluorination of Alkenes, Cyclopropane and Alkynes with Poly-hydrogen Fluoride/Pyridine (Trialkylamine)reagents. *Synthesis* **1973,** 779–780.
147. Olah, G. A.; Grant, J. L.; Spear, R. J.; Bollinger, J. M.; Serianz, A.; Sipos, G. Stable Carbocations. 192. Acyclic Carbodications. *J. Am. Chem. Soc.* **1976,** *98,* 2501–2507.
148. Olah, G. A.; Prakash, G. K. S.; Sommer, J. *Superacids;* Wiley Intersciences: New-York, 1985.
149. Olah, G. A.; Rasul, G.; Aniszfeld, R.; Prakash, G. K. S. Considered Onium Ions. Part 40. Protonitronium Dication, (NO_2H^{2+}. *J. Am. Chem. Soc.* **1992,** *114,* 5608–5609.
150. Olah, G. A. Superelectrophiles. *Angew. Chem. Int. Ed.* **1993,** *32,* 767–788.
151. Olah, G. A.; Hartz, N.; Rasul, G.; Burrichter, A.; Prakash, G. K. S. Chemistry in Superacids. Part 18. Ester Cleavage in Superacid Media Involving Diprotonated Gitonic Carboxonium Dications. *J. Am. Chem. Soc.* **1995a,** *117,* 6421–6427.
152. Olah, G. A.; Rasul, G.; York, C.; Prakash, G. K. S. Superacid-catalyzed Condensation of Benzaldehyde with Benzene. Study of Protonated Benzaldehydes and the Role of Superelectrophilic Activation. *J. Am. Chem. Soc.* **1995b,** *117,* 11211–11214.
153. Olah, G. A.; Burrichter, A.; Rasul, G.; Prakash, G. K. S. The Protiotetramethylammonium Dication $(CH_3)_3NCH_4^{2+}$: Hydrogen/deuterium Exchange and Calculational Studies. Search for the Parent Protioammonium Dication NH_5^{2+}. *J. Am. Chem. Soc.* **1997,** *119,* 4594–4598.
154. Olah, G. A.; Reddy, V. P.; Rasul, G.; Prakash, G. K. S. Search for Long-lived 1,3-carbodications and Preparation of the Persistent 1,1,3,3-tetracyclopropyl-1,3-propanediyl Dication. *J. Am. Chem. Soc.* **1999,** *121,* 9994–9998.
155. Olah, G. A. *A Life of Magic Chemistry: Autobiographical Reflexions of a Nobel Prize Winner. 2001;* John Wiley and Sons: New-York, 2001.
156. Olah, G. A.; Prakash, G. K. S.; Rasul, G. Gitonic and Distonic Alkanonium Dications (Diprotonated Alkane Dications $C_nH_{2n+4}^{2+}$, $n = 1-4$). *J. Org. Chem.* **2001,** *66,* 2907–2910.

157. Olah, G. A.; Klumpp, D. A. Superelectrophiles Solvation. *Acc. Chem. Res.* **2004,** *37,* 211–220.
158. Olah, G. A.; Klumpp, D. A. *Superelectrophiles and Their Chemistry;* Wiley Interscience: Hoboken, 2008.
159. Olah, G. A.; Prakash, G. K. S.; Molnar, A.; Sommer, J. *Superacids,* 2nd ed.; Wiley Intersciences: New-York, 2009.
160. Olah, G. A.; Prakash, G. K. S.; Rasul, G. Study of the Fluoro- and Chlorodimethylbutyl Cations. *Proc. Natl. Acad. Sci.* **2013,** *21,* 8427–8430.
161. Patani, G. A.; LaVoie, E. J. Bioisosterism: A Rational Approach in Drug Design. *Chem. Rev.* **1996,** *96,* 3147–3176.
162. Perrin, C. L.; Lollo, C. P.; Johnston, E. R. NMR Site-to-site Rate Constants and the Mechanisms of Acid-catalyzed Proton Exchange in Secondary Amines. *J. Am. Chem. Soc.* **1984,** *106,* 2749–2753.
163. Pierry, C.; Zoute, L.; Jubault, P.; Pfund, E.; Lequeux, T.; Cahard, D.; et al. Diastereocontrolled Addition of Organometallic Reagents to *S*-chiral *N*-(*tert*-butanesulfinyl)-α-fluoroenimines. *Tetrahedron Lett.* **2009,** *50,* 264–266.
164. Podichetty, A. K.; Wagner, S.; Schröer, S.; Fast, A.; Schäfers, M.; Schober, O.; et al. Fluorinated Isatin Derivatives. Part 2. New *N*-substituted 5-pyrrolidinylsulfonyl Isatins as Potential Tools for Molecular Imaging of Caspases in Apoptosis. *J. Med. Chem.* **2009,** *52,* 3484–3495.
165. Posternak, A. G.; Garlyauskayte, R. Y.; Polovinko, V. V.; Yagupolski, L. M.; Yagupolski, Y. L. New Kind of Organic Superacids. Bis(perfluoroalkylsulfonylimino) trifluoromethanesulfonic Acids and Their Trimethylsilyl Esters. *Org. Biomol. Chem.* **2009,** *7,* 1642–1645.
166. Prakash, G. K. S.; Mathew, T.; Hoole, D.; Esteves, P. M.; Wang, Q.; Rasul, G.; et al. N-halosuccinimide/BF_3–H_2O, Efficient Electrophilic Halogenating Systems for Aromatics. *J. Am. Chem. Soc.* **2004,** *126,* 15770–15776.
167. Prakash, G. K. S.; Gurung, L.; Jog, P. V.; Tanaka, S.; Thomas, T. E.; Ganesh, N.; et al. Direct Synthesis of Diverse β-fluoroethylamines by a Multicomponent Protocol. *Chem. Eur. J.* **2013,** *19,* 3579–3583.
168. Rabasseda, T.; Hopkins, S. J. Ampiroxicam: A Prodrug of Piroxicam Devoid of Gastrointestinal Toxicity. *Drugs Today* **1994,** *30,* 557–563.
169. Rasul, G.; Prakash, G. K. S.; Olah, G. A. XH_5^{2+} Dications and XH_6^{3+} Trications (X = N, P, and As). *J. Am. Chem. Soc.* **1997,** *119,* 12984–12985.
170. Rasul, G.; Prakash, G. S. K.; Olah, G. A. MP2, CCSD(T) and Density Functional Theory Study of the 2-butyl Cation: New Insight into the Methyl- and Hydrogen-Bridged Structures. *J. Phys. Chem. A* **2015,** *119,* 5762–5769.
171. Reddy, V. P.; Sinn, E.; Olah, G. A.; Prakash, G. K. S.; Rasul, G. Theoretical Investigation of Superelectrophilic Fluorooxonium Dications FOH_3^{2+} and $F_2OH_2^{2+}$: Comparison with Parent H_4O^{2+} Dication. *J. Phys. Chem. A* **2004,** *108,* 4036–4039.
172. Reed, C. A.; Fackler, N. L. P.; Kim, K. C.; Stasko, D.; Evans, D. R. Isolation of Protonated Arenes (Wheland Intermediates) with BAr^F and Carborane Anions. A Novel Crystalline Superacid. *J. Am. Chem. Soc.* **1999,** *121,* 6314–6315.
173. Reynolds, C. H. Trihalomethyl Cations: Relative Stability of CX_3^+ (X = F, Cl, Br). *J. Chem. Soc. Chem. Commun.* **1991,** 975–976.
174. Re.ynolds, C. H. Structure and Relative Stability of Halogenated Carbocations: the $C_2H_4X^+$ and $C_4H_8X^+$ (X = Fluoro, Chloro, Bromo) Cations. *J. Am. Chem. Soc.* **1992,** *114,* 8676–8682.

175. Rohde, M.; Müller, L. O.; Himmel, D.; Scherer, H.; Krossing, I. A Janus-Headed Lewis Superacid: Simple Access to, and First Application of $Me_3Si\text{-}F\text{-}Al(OR^F)_3$. *Chem. Eur. J.* **2014,** *20,* 1218−1222.
176. Rosenthal, M. R. The Myth of the Non-coordinating Anion. *J. Chem. Educ.* **1973,** *50,* 331−335.
177. Sandford, G. Trihalides. In *Comprehensive Organic Functional Group Transformations 2;* Katritzky, A. R., Rees, C. J., Eds.; Elsevier: Amsterdam, 2004; pp 1−22.
178. Sandford, G. Fluoroarenes. In *Sciences of Synthesis. Houben-weyl Methods of Molecular Transformations,* Vol. 31a, Ramsden, C. A., Ed.; Thieme: Stuttgart, 2007; pp 21−78. compounds with two carbon-heteroatom bonds.
179. Sato, Y.; Yato, M.; Ohwada, T.; Saito, S.; Shudo, K. Involvement of Dicationic Species as the Reactive Intermediates in Gattermann, Houben-Hoesch, and Friedel-Crafts Reactions of Nonactivated Benzenes. *J. Am. Chem. Soc.* **1995,** *117,* 3037−3043.
180. Shellhamer, D. F.; Horney, M. J.; Pettus, P. J.; Pettus, T. L.; Stringer, J. M.; Heasley, V. L.; et al. Generation of Interhalogen Fluorides under Mild Conditions: A Comparison of Sluggish and Reactive Interhalogen Fluorides. *J. Org. Chem.* **1999,** *64,* 1094−1098.
181. Shigehisa, H.; Nishi, E.; Fujisawa, M.; Hiroya, K. Cobalt-Catalysed Hydrofluorination of Unactivated Olefins: A Radical Approach of Fluorine Transfer. *Org. Lett.* **2013,** *15,* 5158−5161.
182. Silver, L. H. Clinical Efficacy and Safety of Brinzolamide (Azopt), a New Topical Carbonic Anhydrase Inhibitor for Primary Open-angle Glaucoma and Ocular Hypertension. Brinzolamide Primary Therapy Study Group. *Am. J. Ophtalmol.* **1998,** *126,* 400−408.
183. Sommer, J.; Bukala, J. Selective Electrophilic Activation of Alkanes. *Acc. Chem. Rev.* **1993,** *26,* 370−376.
184. Sorensen, T. S. Investigation on μ-hybridbridged Carbocations in Superacids: the Effect of Ring Size on Transannular Participation. In *Stable Carbocation Chemistry;* Prakash, G. S. K., Schleyer, P.von R., Eds.; Wiley and Sons: New-York, 1997; pp 75−136.
185. Srivastava, A. K.; Misra, N. Hydrogenated Superhalogens Behave as Superacids. *Polyhedron* **2015,** *102,* 711−714.
186. Stranix, B. R.; Lavallée, J. F.; Sévigny, G.; Yelle, J.; Perron, V.; LeBerre, N.; et al. Lysine Sulfonamides as Novel HIV-protease Inhibitors: Nepsilon-acyl Aromatic Alpha-amino Acids. *Bioorg. Chem. Lett.* **2006,** *16,* 3459−3462.
187. Streidl, N.; Denegri, B.; Kronja, O.; Mayr, H. A Practical Guide for Estimating Rates of Heterolysis Reactions. *Acc. Chem. Res.* **2010,** *43,* 1537−1549.
188. Supuran, C. T. Carbonic Anhydrases: Novel Therapeutic Applications for Inhibitors and Activators. *Nat. Rev. Drug Discov.* **2008a,** *7,* 168−181.
189. Supuran, C. T. Diuretics: from Classical Carbonic Anhydrase Inhibitors to Novel Applications of the Sulfonamides. *Curr. Pharm. Des.* **2008b,** *14,* 641−648.
190. Supuran, C. T.; Di Fiore, A.; De Simone, G. Carbonic Anhydrase Inhibitors as Emerging Drugs for the Treatment of Obesity. *Expert Opin. Emerg. Drugs* **2008,** *13,* 383−392.
191. Supuran, C. T. Carbonic Anhydrase Inhibitors. *Bioorg. Med. Chem. Lett.* **2010a,** *20,* 3467−3474.
192. Supuran, C. T. Carbonic Anhydrase Inhibition/activation: Trip of a Scientist Around the World in the Search of Novel Chemotypes and Drug Targets. *Curr. Pharm. Des.* **2010b,** *16,* 3233−3245.
193. Takeuchi, Y.; Yamada, A.; Suzuki, T.; Koizumi, T. Synthetic Studies towards Prline Amide Isosters, Potentially Useful Molecules for Biological Investigations. *Tetrahedron* **1996,** *52,* 225−232.

194. Tan, C. K.; Yeung, Y. Y. Recent Advances in Stereoselective Bromofunctionalization of Alkenes Using *N*-bromoamide Reagents. *Chem. Commun.* **2013,** *49,* 7985–7996.
195. Tanimukai, H.; Inui, M.; Hariguchi, S.; Kaneko, Z. Antiepileptic Property of Inhibitors of Carbonic Anhydrase. *Biochem. Pharmacol.* **1965,** *14,* 961–970.
196. Tars, K.; Vullo, D.; Kazaks, A.; Leitans, J.; Lends, A.; Grandane, A.; et al. Sulfocoumarins (1,2-benzothiine-2,2-dioxides): A Class of Potent and Isoform-selective Inhibitors of Tumor-associated Carbonic Anhydrases. *J. Med. Chem.* **2013,** *56,* 293–300.
197. Temperini, C.; Cecchi, A.; Scozzafava, A.; Supuran, C. T. Carbonic Anhydrase Inhibitors. Sulfonamide Diuretics Revisited—old Leads for New Applications? *Org. Biomol. Chem.* **2008,** *6,* 2499–2506.
198. Theunissen, C.; Métayer, B.; Henry, N.; Compain, G.; Marrot, J.; Martin-Mingot, A.; et al. Keteniminum Ion-initiated Cascade Cationic Polycyclization. *J. Am. Chem. Soc.* **2014,** *136,* 12528–12531.
199. Thibaudeau, S.; Martin-Mingot, A.; Jouannetaud, M. P.; Jacquesy, J. C. Functionalization of Non-activated C–H Bonds in Ketones and Imines with HF/SbF5/CCl4. *Tetrahedron* **2002,** *58,* 6643–6649.
200. Thibaudeau, S.; Martin-Mingot, A.; Jouannetaud, M. P. J.; Karam, O.; Zunino, F. A Novel and Facile Route to β-fluoroamines by Hydrofluorination Using Superacid HF-SbF_5. *Chem. Commun.* **2007,** 3198–3200.
201. Thiry, A.; Dogné, J.-M.; Supuran, C. T.; Masereel, B. Anticonvulsant Sulfonamides/Sulfamates/Sulfamides with Carbonic Anhydrase Inhibitory Activity: Drug Design and Mechanism of Action. *Curr. Pharm. Des.* **2008,** *14,* 661–671.
202. Turowsky, L.; Seppelt, K. Tris((trifluoromethyl)sulfonyl)methane, $HC(SO_2CF_3)_3$. *Inorg. Chem.* **1988,** *27,* 2135–2137.
203. Van Hende, E.; Verniest, G.; Deroose, F.; Thuring, J. W.; Macdonald, G.; De Kimpe, N. Synthesis of 1-Boc-3-fluoroazetidine-3-carboxylic Acid. *J. Org. Chem.* **2009,** *74,* 2250–2253.
204. Vardelle, E.; Gamba-Sanchez, D.; Martin-Mingot, A.; Jouannetaud, M. P.; Thibaudeau, S.; Marrot, J. Cyclisation/fluorination of Nitrogen Containing Dienes in Superacid HF/SbF5: A New Route to 3 and 4-fluoropiperidines. *Chem. Commun.* **2008,** 1473–1475.
205. Vardelle, E.; Martin-Mingot, A.; Jouannetaud, M. P.; Bachmann, C.; Marrot, J.; Thibaudeau, S. Substitution Effect on Cylisation/fluorination Reaction of *N*-dienes in Superacid. *J. Org. Chem.* **2009,** *74,* 6025–6034.
206. Wang, J.; Sánchez-Roselló, M.; Aceña, J. L.; del Pozo, C.; Sorochinsky, A. E.; Fustero, S.; et al. Fluorine in Pharmaceutical Industry: Fluorine-containing Drugs Introduced to the Market in the Last Decade (2001–2011). *Chem. Rev.* **2014,** *114,* 2432–2506.
207. Weiske, T.; Koch, W.; Schwarz, H. Experimental Evidence for the Existence of the Protonitronium Dication ($HONO_2^{+}$) in the Gas Phase and *Ab Initio* Molecular Orbital Calculations of its Potential Energy Surface. *J. Am. Chem. Soc.* **1993,** *115,* 6312–6316.
208. Winum, J. Y.; Rami, M.; Scozzafava, A.; Montero, J. L.; Supuran, C. Carbonic Anhydrase IX: A New Druggable Target for the Design of Antitumor Agents. *Med. Res. Rev.* **2008,** *28,* 445–463.
209. Yagupolskii, L. M. *Aromatic and Heterocyclic Compounds with Fluorine-Containing Substituents;* Nauokova Dumka: Kiev, 1988.
210. Yagupolskii, L. M.; Petrik, V. N.; Kondratenko, N. V.; Sooväli, L.; Kaljurand, I.; Leito, I.; et al. The Immense Acidifying Effect of the Supersubstituent $=NSO_2CF_3$ on the Acidity of Amides and Amidines of Benzoic Acids in Acetonitrile. *J. Chem. Soc. Perkin. Trans.* **2002,** *2,* 1950–1955.

211. Yousefi, R.; Ashtekar, K. D.; Whitehead, D. C.; Jackson, J. E.; Borhnan, B. Dissecting the Stereocontrol Elements of a Catalytic Asymmetric Chlorolactonization: *Syn* Addition Obviates Bridging Chloronium. *J. Am. Chem. Soc.* **2013,** *135,* 14524–14527.
212. Zhang, Z.; Schreiner, P. R. (Thio)urea Organocatalysis—What Can Be Learnt from Anion Recognition? *Chem. Soc. Rev.* **2009,** *38,* 1187–1198.
213. Zheng, S.; Schienebeck, C. M.; Zhang, W.; Wang, H. Y.; et al. Cinchona Alkaloids as Organocatalysts in Enantioselective Halofunctionalization of Alkenes and Alkynes. *Asian. J. Org. Chem.* **2014,** *3,* 366–376.
214. Zhou, Y.; Wang, J.; Gu, Z.; Wang, S.; Zhu, W.; Aceña, J. L.; et al. Next Generation of Fluorine-containing Pharmaceuticals, Compounds Currently in Phase II-III Clinical Trials of Major Pharmaceutical Companies: New Structural Trends and Therapeutic Areas. *Chem. Rev.* **2016,** *116,* 422–518.
215. Zimmer, L. E.; Sparr, C.; Gilmour, R. Fluorine Conformational Effets in Organocatalysis: an Emerging Strategy for Molecular Design. *Angew. Chem. Int. Ed.* **2011,** *50,* 11860–11871.
216. Zunino, F.; Liu, F.; Berrier, C.; Martin-Mingot, A.; Thibaudeau, S.; Jouannetaud, M. P.; et al. *Gem*-difluorination in Superacid: the Dramatic Role of Halonium Ions. *J. Fluor. Chem.* **2008,** *129,* 775–780.

Visible Light–Induced (Per)fluoroalkylation by Photoredox Catalysis

14

G. Dagousset[1], A. Carboni[2], G. Masson[2], E. Magnier[1]
[1]Université de Versailles-Saint-Quentin, Versailles Cedex, France; [2]Institut de Chimie des Substances Naturelles, CNRS, Gif-sur-Yvette Cedex, France

Chapter Outline

1. Introduction

Visible light photoredox catalysis has received increasing attention since the seminal reports of the groups of Yoon[1] and MacMillan[2] in 2008. This powerful tool involves the use of photocatalysts (PCs), which are able to perform single electron transfer (SET) processes with organic substrates under visible light irradiation. The revival of this synthetic methodology has a particular impact in organofluorine chemistry. The (per)fluoroalkyl radicals are indeed reactive species that are very frequently encountered in the literature. As a result, an increasing number of reports using photoredox catalysis have been

Modern Synthesis Processes and Reactivity of Fluorinated Compounds. http://dx.doi.org/10.1016/B978-0-12-803740-9.00014-7

recently published, in particular for the synthesis of fluorine-containing molecules. They are summarized in this chapter, from seminal works until September 2015. Only the reactions involving (per)fluoroalkyl radical intermediates will be covered. Therefore, photoredox-catalyzed decarboxylative fluorination[3–5] or trifluoromethylation[6] of carboxylic acids, hydrofluorination of alkenes,[7] and α-fluoroalkylation of amines[8–10] are outside the scope of this chapter. Furthermore, PC-free visible light–promoted trifluoromethylation involving the in situ generation of electron donor–acceptor complexes will not be reported herein.[11,12] Additionally, particular attention will be given to photoredox-catalyzed trifluoromethylation procedures.

The general pathway of photoredox catalysis is briefly explained in the following discussion and illustrated in Scheme 14.1.

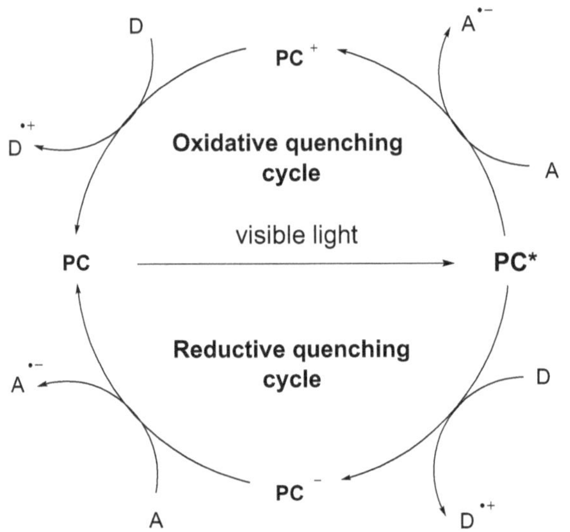

Scheme 14.1 General behavior of a photoredox catalyst PC. *PC*, photocatalyst.

A PC (see Fig. 14.1) can absorb a photon in the visible region, leading to a long-lived excited state (PC*) via a metal-to-ligand charge transfer. The resulting excited photocatalyst (PC*) is more oxidizing and more reducing than the ground-state PC and can then act as a one-electron oxidant (reductive quenching cycle) or a one-electron reductant (oxidative quenching cycle). In the reductive quenching cycle, a strong reductant PC^{-} is generated, which may then donate an electron to some acceptor A to return to the ground-state species PC, thus completing the photocatalytic cycle. Alternatively, in the oxidative quenching cycle, PC* performs a SET with an acceptor A (in the case of the present chapter, the source of (per)fluoroalkyl radical, see Fig. 14.2) to give a strong oxidant PC^{+}, which can then react with some donor D to afford the corresponding radical cation D^{+} (Scheme 14.1).

It is also worth noting that photoredox catalysis requires very mild and ecofriendly reaction conditions, compared with the classical radical reactions, which often need high-energy ultraviolet light equipment or highly reactive and toxic radical initiators. Indeed, most photoredox-catalyzed reactions are generally carried out at room temperature, with low loadings of PC (around 1 mol%). Moreover, as most organic compounds do not absorb in the visible region, side reactions resulting from the photoexcitation of the substrates are usually minimized, and a large variety of functional groups can therefore be tolerated in photoredox processes.

Among the three essential partners (PC, radical precursor, light), the influence of the source of light has not been extensively studied. It is noteworthy to state that visible light irradiation is typically performed with inexpensive and easy-to-handle fluorescent light bulbs or blue light-emitting diode (LED) lamps (often used as rubbers).

Most of the commonly employed PCs are the well-known ruthenium (II) polypyridyl complexes **1a–c** such as $[Ru(bpy)_3]^{2+}$ (bpy = 2,2′-bipyridine) **1a** and **b**, the iridium (III) cyclometalated derivatives **1d–f** such as *fac*-Ir(ppy)$_3$ (ppy = 2-phenylpyridine) **1d**, or organic dyes such as eosin Y **1k**. For the sake of clarity, all the PCs **1a–1n** used in this book chapter will be referenced as depicted in Fig. 14.1 in which BiOBr nanosheets **1i** and mesoporous graphitic carbon nitride, mpg-CN **1o**, are not included.

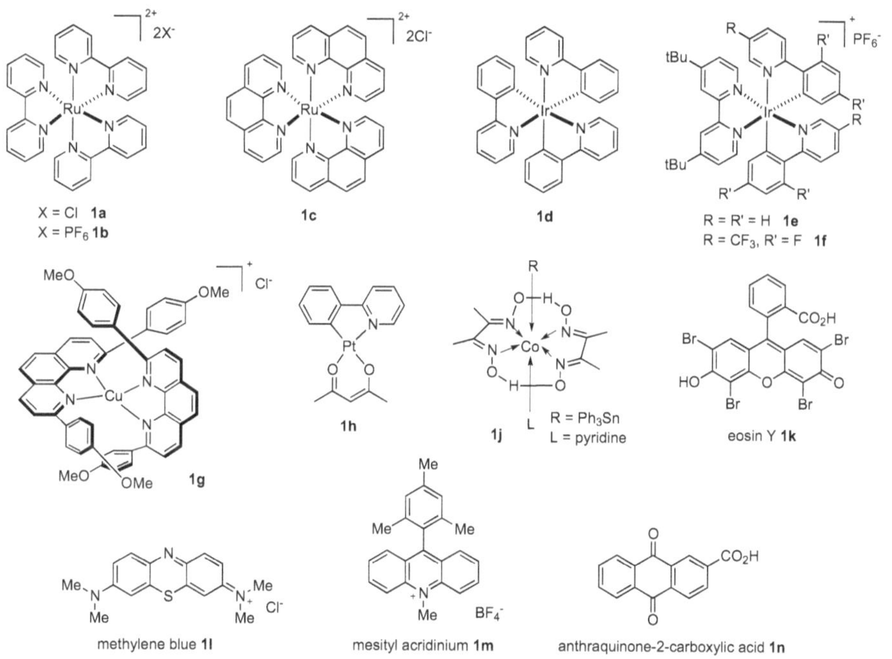

Figure 14.1 Structures of photocatalysts employed in this chapter.

A large variety of (per)fluoroalkyl radical sources are readily available to the organic chemist and quite exhaustively listed in Fig. 14.2. The most widely employed reagents **2a–2i**, used to generate a CF_3 radical, are listed in Fig. 14.2A. Apart from the Langlois reagent **2h**, which may be oxidized by PCs to lead to a CF_3 radical, all these reagents (iodotrifluoromethane **2a**, Umemoto reagents **2b–c**,[13] Yagupol'skii reagent **2d**,[14] Togni reagents **2e–f**,[15] triflyl chloride **2g**, and trifluoroacetic anhydride (TFAA)/pyridine *N*-oxide adduct **2i**) are electrophilic CF_3 reagents that can undergo a one-electron reduction to afford a CF_3 radical.

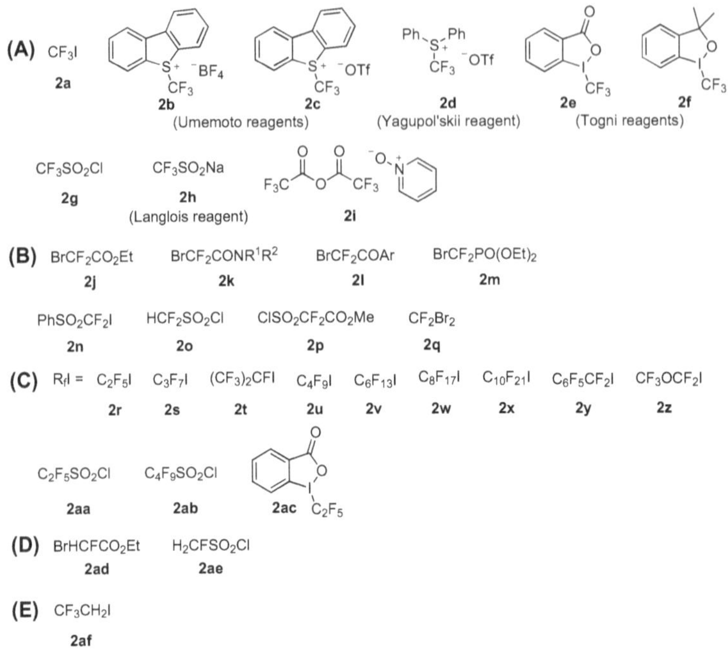

Figure 14.2 (A) Trifluoromethyl radical sources. (B) Difluoromethyl radical sources. (C) perfluoroalkyl radical sources. (D) Monofluoromethyl radical sources. (E) Trifluoroethyl radical sources.

The redox properties of a PC depend on its structure. Therefore, as a large structural diversity of PCs is available to the organic chemist (see Fig. 14.1), a wide range of redox potentials can be accessed. The choice of the PC is often made based on both the source of the radical and the substrate. The two first partners (PC and radical source) have to be tuned and adapted to each other. Thanks to photoredox catalysis, an important number of radicals can then be produced from various organic substrates.

Due to the growing interest in photocatalyzed (per)fluoroalkylation reactions, several reviews have been already published in this field.[16–20] This chapter will be divided into two main parts according to the type of reactions used to introduce a (per)fluoroalkyl moiety. First, photoredox-induced (per)fluoroalkylations via the monofunctionalization of various substrates will be thoroughly described. Then, the second part will focus on photoredox-catalyzed (per)fluoroalkylations caused by the difunctionalization of unsaturated systems with concomitant introduction of a (per)fluoroalkyl moiety and another functional group.

2. (Per)fluoroalkylation Reactions Using Photoredox-Induced Monofunctionalization

2.1 (Per)fluoroalkylation of Enol and Enamine Derivatives

The first photocatalyzed (per)fluoroalkylation reaction, combined with an enamine-type organocatalysis, was reported by MacMillan et al. in 2009.[21] The enantioselective α-trifluoromethylation of aldehydes could be achieved using PC **1e(Ir)**, CF_3I **2a** as the $\cdot CF_3$ source, and chiral imidazolidinone **A** as the organocatalyst to give rise to a set of alcohols after in situ reduction process (Scheme 14.2).

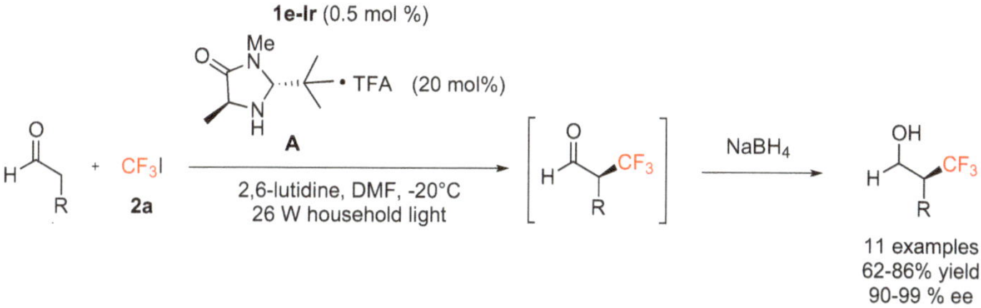

Scheme 14.2 Enantioselective α-trifluoromethylation of aldehydes.

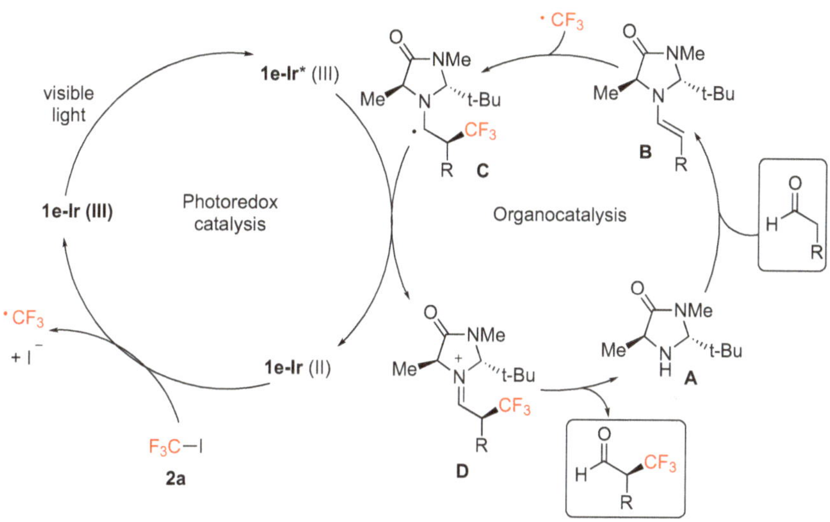

Scheme 14.3 Proposed mechanism for the enantioselective α-trifluoromethylation of aldehydes.

The proposed mechanism of such transformation is shown in Scheme 14.3. First, PC **1e(Ir)** is excited to PC* **1e(Ir)*** under visible light irradiation. Then, this excited state undergoes a reductive quenching with a sacrificial amount (0.5 mol%) of enamine **B** to give **1e(Ir)** (II). This strong reductant ($E = -1.51$ V vs. saturated calomel electrode (SCE) in CH_3CN) gives an electron to **2a** ($E = -1.22$ V vs. SCE in dimethylformamide (DMF)), affording the desired electrophilic CF_3 radical and regenerating the ground-state PC **1e(Ir)**. Regarding the organocatalytic cycle, condensation of an aldehyde substrate with imidazolidinone **A** delivers the corresponding electron-rich enamine **B**. Then, the key step of the transformation consists of the stereoselective addition of the CF_3 radical controlled by the chirality of **B**. The resulting α-amino radical **C** is easily oxidized into iminium ion **D** by **1e(Ir)***. Hydrolysis of **D** finally regenerates the organocatalyst **A** and furnishes the expected enantioenriched α-trifluoromethylated aldehyde.

Remarkably, this protocol was also successfully extended to a variety of perfluoroalkyl iodides (**2j**, **2q–u**, and **2y** and **z**), affording the desired α-perfluoroalkylated aldehydes in 67–85% yield with high enantioselectivities (96–99% ee). Similar studies were reported by Zeitler et al. with the use of eosin Y **1k** as the photoredox catalyst[22] (Scheme 14.4).

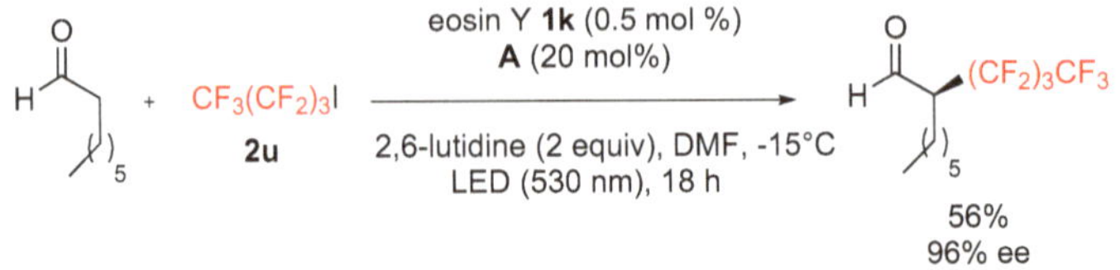

Scheme 14.4 Enantioselective organophotoredox-catalyzed α-perfluoroalkylation of aldehydes.

A few years later, this α-perfluoroalkylation process was efficiently applied to enol silane derivatives, giving access to α-perfluoroalkylated ketones, esters, and amides as depicted in Scheme 14.5.[23]

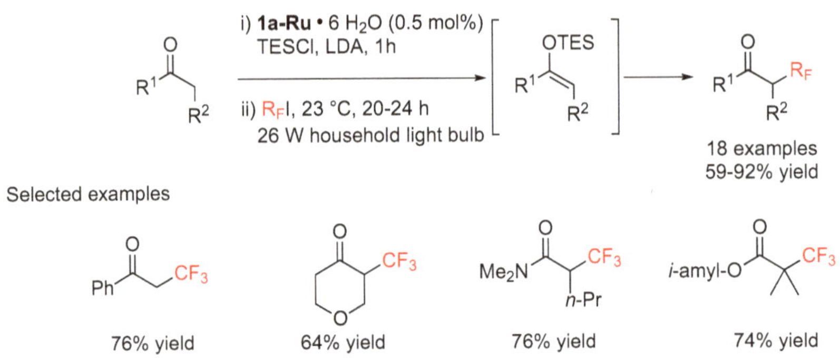

Scheme 14.5 One-pot photoredox-promoted α-perfluoroalkylation of enol silane derivatives.

One of the main issues of CF_3I **2a** or C_2F_5I **2r** is to measure the exact amount of these reagents, as they are gases at room temperature under atmospheric pressure. An elegant way to circumvent this issue was reported by Ritter et al. in 2015, by making halogen-bonded adducts of these perfluoroalkylating reagents with tetramethylguanidine or dimethylsulfoxide (DMSO). These two complexes are easy-to-handle liquids under classical conditions (1 atm, 23°C).[24] The silyl enol ether, derived from methyl phenyl ketone, was trifluoromethylated with 2DMSO.CF_3I into the corresponding α-CF_3 derivative under photoredox catalysis (Scheme 14.6).

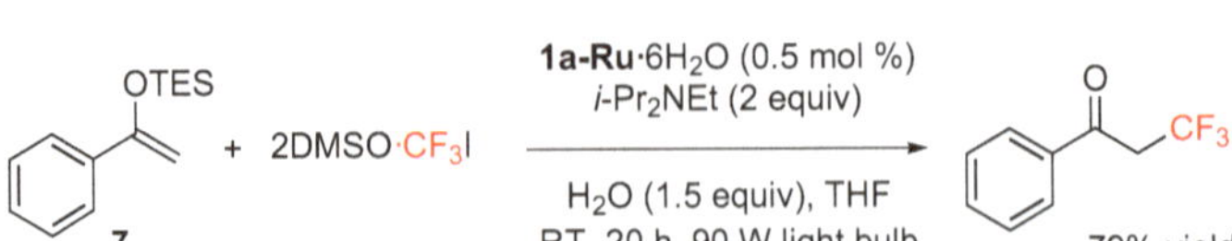

Scheme 14.6 Use of CF_3I·DMSO adduct in photocatalyzed trifluoromethylation reactions.

Using continuous-flow techniques, a similar synthesis of α-trifluoromethylated ketones was achieved with eosin Y **1k** as the PC and trifluoromethanesulfonyl chloride **2g** as the CF_3 source. A wide range of aromatic, heteroaromatic, and aliphatic ketones are compatible with these conditions. Thanks to this protocol, the reaction could be performed on a larger scale (1.5 mmol) and with a decreased reaction time (30 min instead of 24 h, Scheme 14.7).[25]

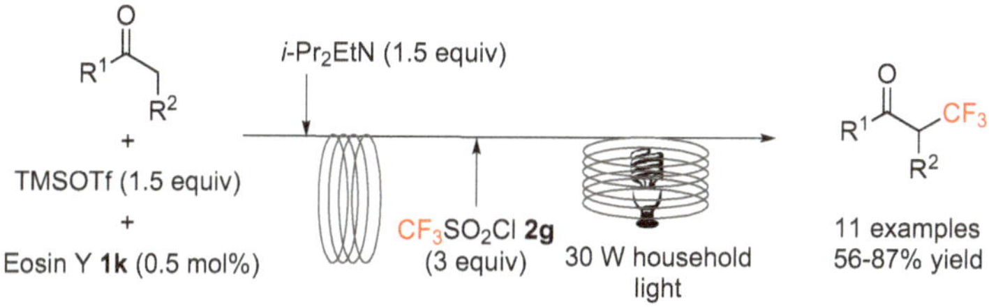

Scheme 14.7 Continuous-flow photocatalyzed preparation of α-trifluoromethylated ketones.

The trifluoromethylation and carbomethoxydifluoromethylation of α-aryl and α-alkyl enol acetates, respectively, were also successfully catalyzed under visible light irradiation by iridium complexes **1e(Ir)** and **1f(Ir)**.[26,27] Surprisingly, in the case of α-unsubstituted enol acetates, the chloroacetoxy difluoromethylated adduct is formed instead of the expected aldehyde (Scheme 14.8).

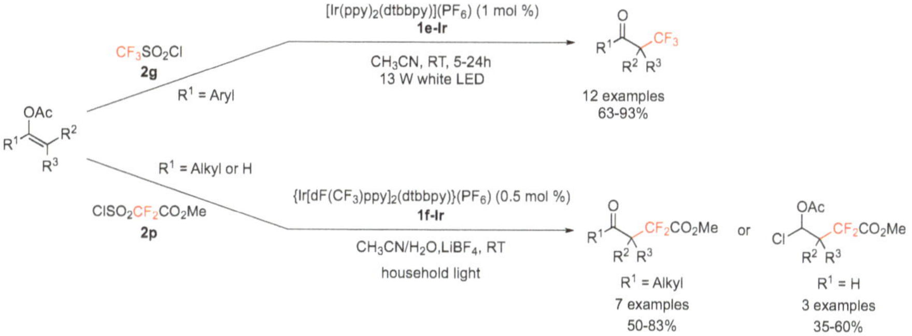

Scheme 14.8 α-Fluoroalkylation of enol acetates under visible light photoredox catalysis.

The first example of photoredox-catalyzed trifluoromethylation of enamides was developed by Yu et al. in 2012.[28] In this report, treatment of two enamides with the compound **2g**, in the presence of 0.5 mol% of iridium PC **1e(Ir)**, allowed the synthesis of the corresponding trifluoromethylated adducts in good yields (Scheme 14.9).

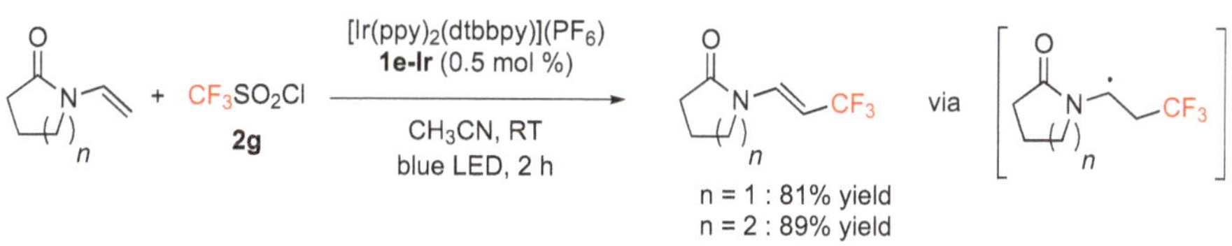

Scheme 14.9 Trifluoromethylation of enamides by means of photoredox catalysis.

In the same article, the authors also reported the difluoromethylation of three enamides with **2j** in the same reaction conditions.

2.2 (Per)fluoroalkylation of Arenes and Heteroarenes

Trifluoromethylated arenes and heteroarenes constitute an important class of pharmacophores in medicinal chemistry. To access such organic structures, visible light photoredox catalysis has recently emerged as a powerful method available to incorporate at a late stage a CF_3 moiety onto an aromatic or heteroaromatic system. Such reaction was first reported by MacMillan in 2011 (Scheme 14.10).[29] Treatment of various unactivated arenes, as well as five-membered and six-membered heteroarenes, with **2g** in the presence of a catalytic amount of PC $Ru(phen)_3Cl_2$ **1c(Ru)** gave rise to the corresponding trifluoromethylated adducts in good to excellent yields.

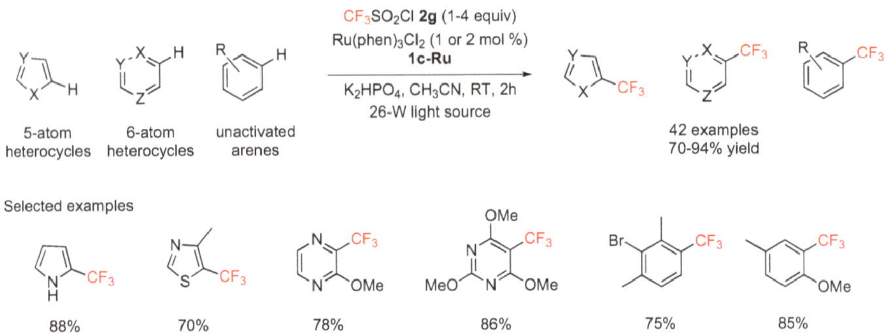

Scheme 14.10 Trifluoromethylation of arenes and heteroarenes.

The methodology was later also applied to indoles by using slightly modified conditions (photoredox catalyst **1a(Ru)** or **1h(Pt)** with **2a** as CF_3 source),[30,31] which also proved efficient for free anilines.[32] The latter was a remarkable result because of the high propensity of the NH group to be oxidized (Scheme 14.11).

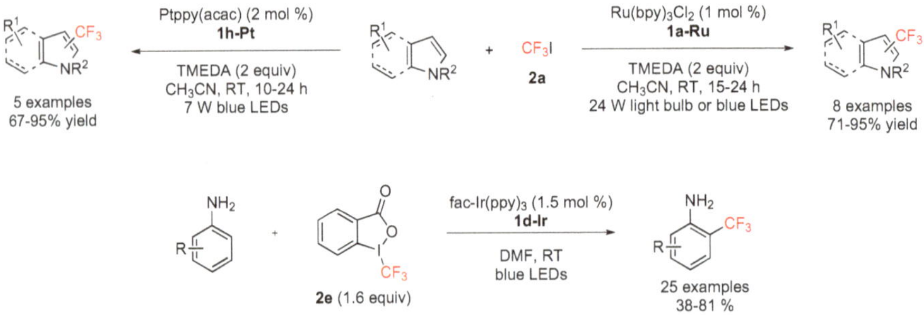

Scheme 14.11 Trifluoromethylation of anilines and heteroarenes.

If other approaches were also successfully developed by using organophotocatalysts such as anthraquinone-2-carboxylic acid **1n**[33] and methylene blue **1l**[34] or continuous-flow techniques,[35] one of the most significant improvements in this field was made by Stephenson et al. in 2015.[36] In this report, the radical CF_3 source was

TFAA, which is inexpensive and easy to handle compared with the others. The innovative strategy is based on the use of pyridine *N*-oxide as an electron-poor redox relay. This reagent can combine with TFAA to form adduct **2i** ($E = -1.10$ V vs. SCE in CH_3CN), which can be reduced after decarboxylation into $\cdot CF_3$ by the excited state **1a(Ru)*** of $Ru(bpy)_3{}^{2+}$ via an oxidative quenching cycle (as shown by Stern–Volmer quenching experiments). Following this procedure, a variety of aryl and heteroaryl substrates were trifluoromethylated in moderate yields (Scheme 14.12). Some compounds have been prepared on multigram scale by batch or flow synthesis.

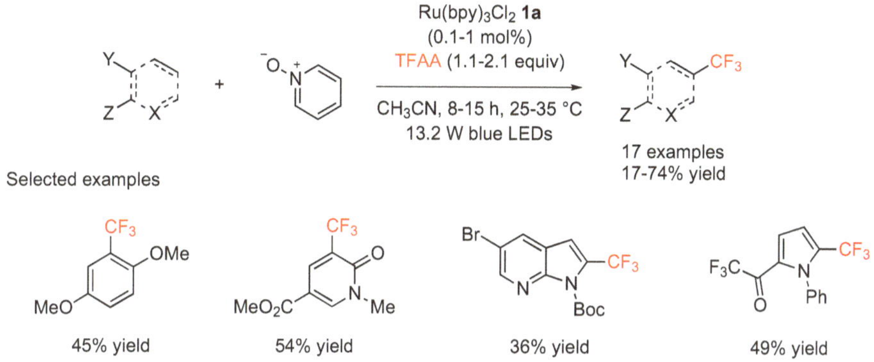

Scheme 14.12 Trifluoromethylation of arenes and heteroarenes using TFAA as the CF_3 source.

It is also worth noting that arenes and heteroarenes were suitable substrates for visible light–promoted introduction of various difluoromethyl groups such as CF_2CO_2Et, $CF_2CONR^1R^2$, and CF_2SO_2Ph, by using the corresponding difluoromethyl radical sources $BrCF_2CO_2Et$ **2j**,[37,38] $BrCF_2CONR^1R^2$ **2k**[39] and $PhSO_2CF_2I$ **2n**,[40] respectively (Scheme 14.13).

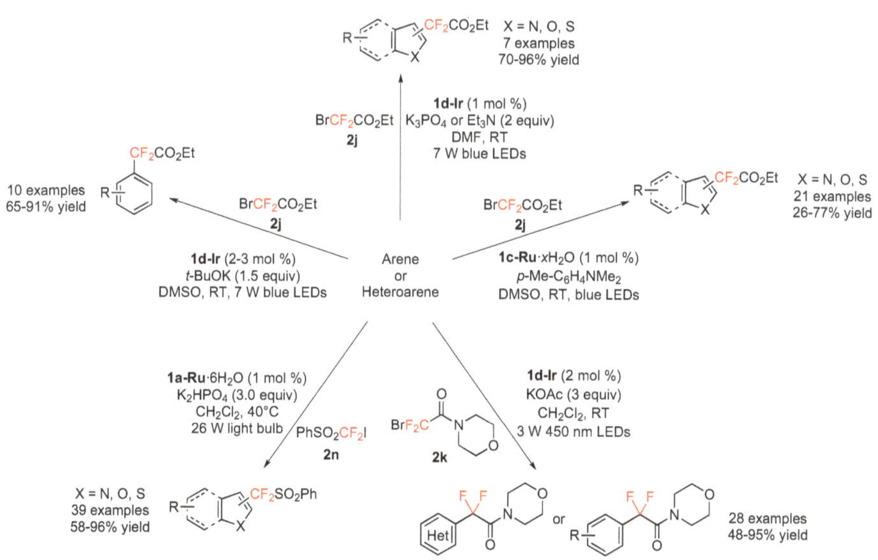

Scheme 14.13 Visible light–mediated difluoromethylation of arenes and heteroarenes.

An uncommon PC, a mesoporous graphitic carbon nitride (mpg-CN **1o**), was successfully used by Blechert et al. for the perfluoroalkylation of a family of heterocycles.[41] The recyclable catalyst **1o** was employed with **2g**, **2u**, and **2ac** to give rise to the corresponding trifluoromethyl and nonafluorobutyl heterocycles.

All the aforementioned reactions are direct (per)fluoroalkylation of arenes and heteroarenes, i.e., they belong to the more general C—H functionalization type of reaction. However, it is also possible to prepare (per)fluoroalkylated arenes or heteroarenes starting from other substrates, in particular aryl or heteroaryl boronic acids. Thus, by merging photoredox catalysis with copper catalysis, the (per)fluoroalkylation of a wide range of boronic acids was efficiently carried out with CF_3I **2a**, C_4F_9I **2u**, or $C_{10}F_{21}I$ **2x**, as perfluoroalkyl radical sources[42] (Scheme 14.14).

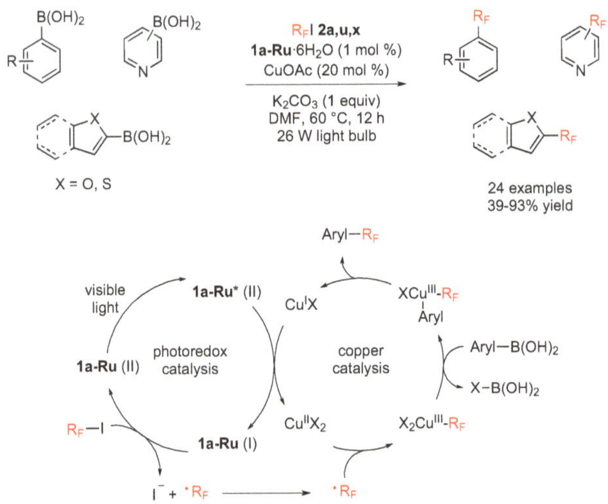

Scheme 14.14 Trifluoromethylation of boronic acids under photoredox and copper catalyses.

A plausible mechanism of this transformation is detailed in Scheme 14.14. After irradiation of **1a(Ru)** (II), the photoexcited catalyst **1a(Ru)*** (II) oxidizes Cu^I catalyst into Cu^{II}, generating a strong reductant **1a(Ru)** (I), which reduces the perfluoroalkyl iodide into the corresponding perfluoroalkyl radical $\bullet R_F$. This radical then reacts with Cu^{II} to form a $Cu^{III}(CF_3)$ intermediate. Subsequent transmetalation with the boronic acid substrate followed by reductive elimination finally affords the desired perfluoroalkylated product, while regenerating the Cu^I catalyst.

2.3 (Per)fluoroalkylation of Alkenes and Alkynes

Some visible light–driven processes for the trifluoromethylation of enols or arenes described in the preceding paragraphs are also compatible with alkenes as substrates.[24,30,32,35] Research works have been also specifically devoted to this topic. The use of terminal alkenes[43,44] or (poly)substituted styrenes[45,46] allowed the control of the addition of the radical (Scheme 14.15). With such substrates, the (per) fluoroalkylation reaction occurs via the more substituted (i.e., also the more stable) radical intermediate, and finally gives the desired (per)fluoroalkylated alkenes in good to excellent yields.

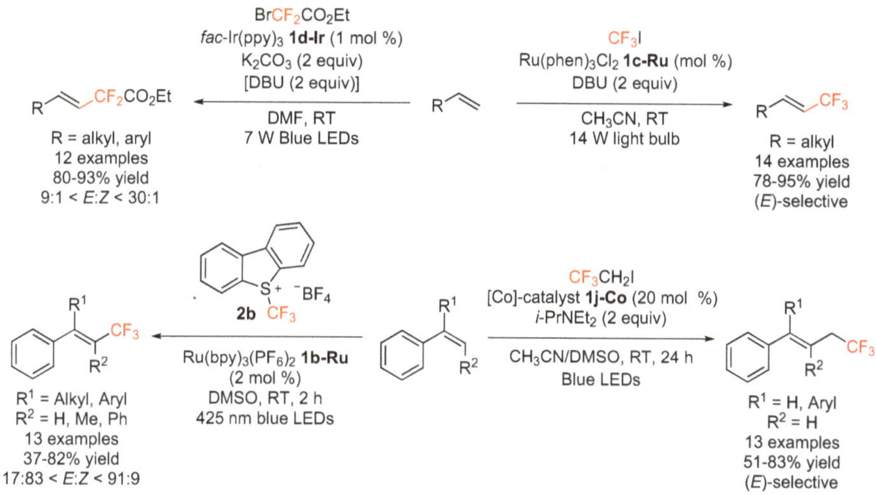

Scheme 14.15 Visible light–promoted (per)fluoroalkylation of alkenes and styrenes.

The regioselectivity for the addition of the (per)fluoroalkyl radical is total, whereas the stereoselectivity of the resulting (per)fluoroalkylated alkene may be an issue in particular with styrene derivatives. The access to the less thermodynamically stable (*Z*) isomer was more deeply studied by Qing et al. in 2014.[47] In this report, the authors exploited the ability of excited PCs to take part in SET processes as well as in triplet–triplet energy transfer (TTET) processes. Indeed, when ruthenium PC **1a(Ru)** was used in the optimized conditions, the (*E*) isomer was obtained. Inversely, when the reaction was carried out in the presence of PC *fac*-Ir(ppy)$_3$ **1d(Ir)** having a higher triplet energy, an additional TTET catalytic cycle took place, enabling the alkene isomerization and leading mostly to the less stable (*Z*) isomer in good yields (Scheme 14.16).

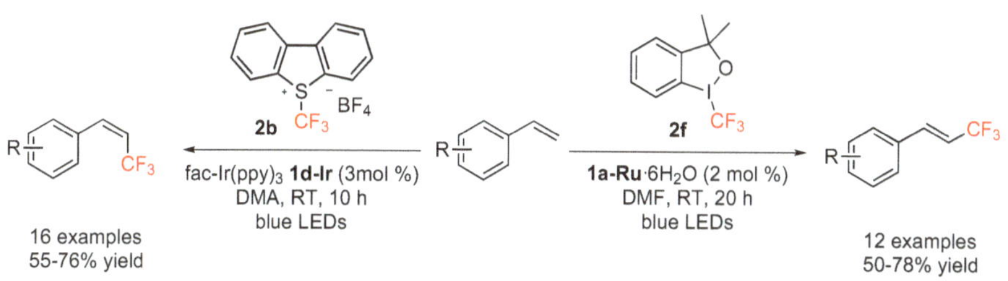

Scheme 14.16 Tunable stereoselective synthesis of trifluoromethylated alkenes.

In all the previous examples, among the possible side reactions, only traces of bis(trifluoromethylated) alkenes or allylic CF_3 alkenes were observed in some cases.

However, by smoothly tuning the reaction conditions with substrates such as α-alkyl styrenes, Koike and Akita demonstrated that such side products could become major products of trifluoromethylation reactions.[48]

In 2013, Gouverneur et al. reported the visible light–mediated trifluoromethylation of allylsilanes.[49] With such substrates, a complete selectivity for allyl trifluoromethylation was observed, without any formation of CF_3-substituted alkenes. All the CF_3 allylic products were obtained in 33–83% yield with general high (*E*)-selectivity. *Anti* α-substituted β-silyl-(*E*)-crotylsilanes were transformed into *syn*-(*E*) branched CF_3 allylic products, in accordance with an *anti*-S_E2' mode of addition of the CF_3 radical. Interestingly, enantioenriched allylsilanes (*R*,*E*) gave the major compounds (*S*,*E*) with a small erosion of enantiomeric ratio (Scheme 14.17).

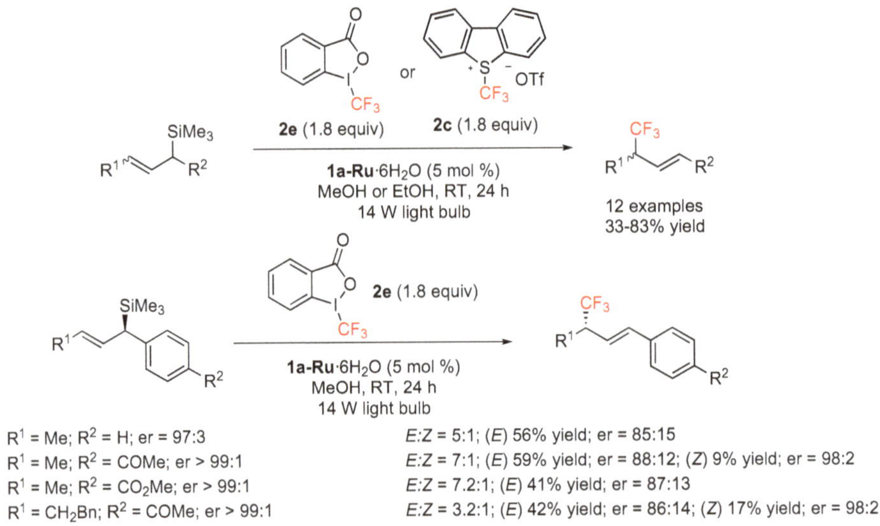

Scheme 14.17 Regio- and stereoselective trifluoromethylation of allylsilanes. *e.r.*, enantiomeric ratio.

The C–H trifluoromethylation reactions were also successfully performed with alkynes.[50] Phenylacetylene derivatives efficiently reacted with CF_3I **2a** and *fac*-Ir(ppy)$_3$ **1d(Ir)** to provide alkynyl-CF_3 compounds in 60–68% yield (Scheme 14.18). Importantly, inorganic bases, such as Cs_2CO_3 or KO*t*-Bu, should be used to avoid iodo- and hydrotrifluoromethylation (cf. vide infra). Aliphatic alkynes were not compatible with the reaction conditions.

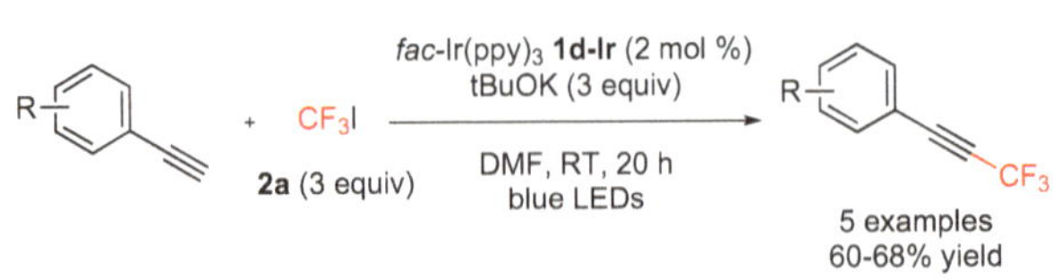

Scheme 14.18 Synthesis of alkynyl-CF_3 compounds.

As with aromatic substrates, deborylative trifluoromethylation of potassium vinyltrifluoroborates can also be an efficient way to prepare CF_3 alkenes. Thus, treatment of various β-substituted potassium vinyltrifluoroborates with Togni reagent **2e** in presence of a catalytic amount of PC **1b(Ru)** allowed for the synthesis of trifluoromethylated alkenes in good yields, with excellent *E*/*Z* selectivities (>94:6, except for β-alkyl-substituted substrates).[51] In a plausible mechanism, the CF_3 radical generated from **2e** adds to the substrate in a regioselective manner, giving a β-borato-stabilized radical intermediate **A**. Further oxidation of **A** by **1b(Ru) (III)** produces β-borato-stabilized carbocation **B**, regenerating ground-state catalyst **1b(Ru) (II)**. Final *trans*-selective Peterson elimination furnishes the desired (*E*)-trifluoromethylated alkene (Scheme 14.19).

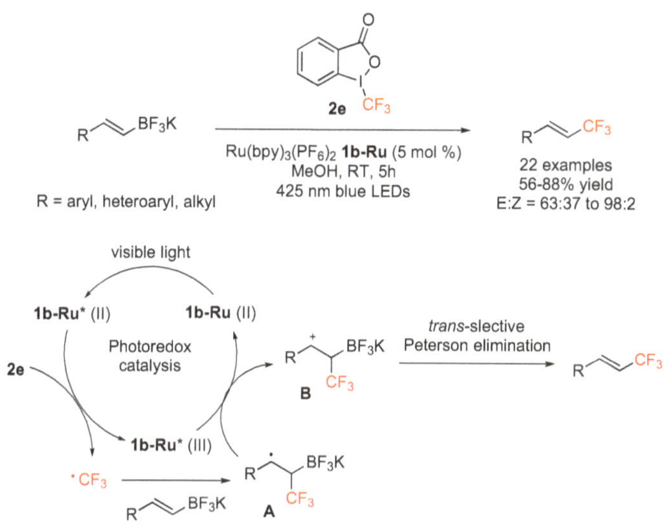

Scheme 14.19 Photoredox-induced trifluoromethylation of potassium vinyltrifluoroborates.

3. (Per)fluoroalkylation Using Photoredox-Induced Difunctionalization of Unsaturated Systems

Unless otherwise noted, difunctional (per)fluoroalkylation reactions described in this paragraph follow the following general scheme: (per)fluoroalkyl radical $\cdot R_F$ is generated either by an oxidative quenching pathway (Scheme 14.20A) or a reductive quenching pathway (Scheme 14.20B) of the PC. Subsequent addition of this radical to the unsaturated substrate affords a β-CF_3 carboradical, which can be further oxidized into the corresponding β-CF_3 carbocation (for example, by PC^+ in the oxidative quenching). Final trapping of this cation with the appropriate nucleophile gives the difunctionalized CF_3 adduct.

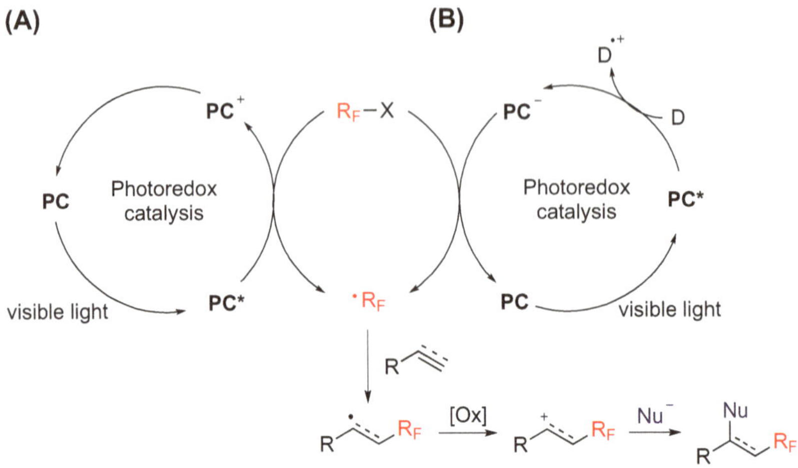

Scheme 14.20 General mechanism for (per)fluoroalkylation of alkenes and alkynes by difunctionalization. (A) Oxidative quenching pathway; (B) reductive quenching pathway.

3.1 Oxy(per)fluoroalkylation of Alkenes and Alkynes

One of the earliest examples of photoredox-catalyzed trifluoromethylation using difunctionalization of alkenes was developed by Koike and Akita in 2012.[52] By using Umemoto reagent **2b** as the CF_3 precursor, and water, alcohol, or carboxylic acid as the *O*-nucleophile, a wide range of styrene derivatives underwent oxytrifluoromethylation. Both electron-donating and electron-withdrawing groups are compatible with the reaction conditions, and α- or β-substituted styrenes are also well tolerated. In the latter case, oxytrifluoromethylated compounds are obtained with moderate diastereoselectivities (1:1 < d.r. < 10:1, Scheme 14.21 (diastereomeric ratio)).

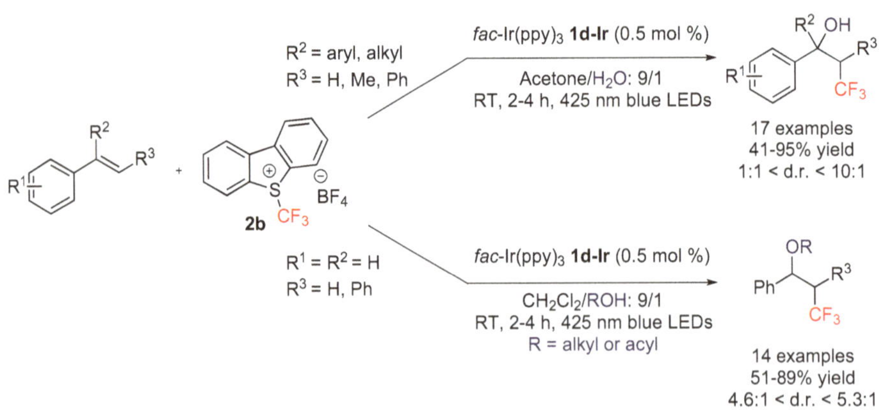

Scheme 14.21 Photoredox-induced oxytrifluoromethylation of styrene derivatives.

This three-component intermolecular process was later applied by Magnier and Masson to the oxytrifluoromethylation of enecarbamates.[53] In this report, Umemoto reagent **2b** was found to be unreactive and had to be replaced by Togni reagent **2e** as the $\cdot CF_3$

source to lead to the expected β-trifluoromethylated α-alkoxyamines in high yield (Scheme 14.22).

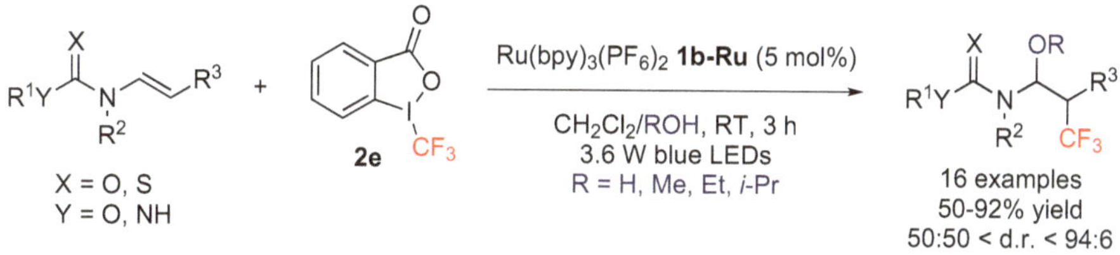

Scheme 14.22 Photoredox-induced oxytrifluoromethylation of enecarbamates.

Other *O*-nucleophiles are compatible with such intermolecular processes. For instance, DMSO was successfully used to access α-trifluoromethyl ketones from styrene derivatives.[54,55]

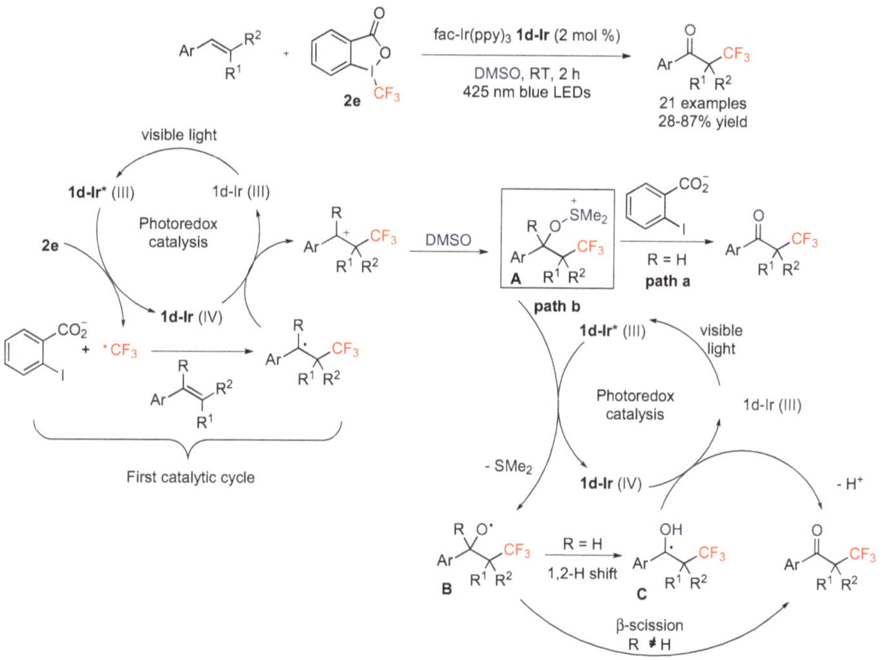

Scheme 14.23 Photoredox-induced synthesis of α-trifluoromethylated ketones.

In this transformation, PC *fac*-Ir(ppy)$_3$ **1d(Ir)** was found to take part in two different catalytic cycles. First, trifluoromethylation of styrenes via a standard oxidative quenching pathway (see Scheme 14.20A) could afford a key cationic intermediate **A**. This species can then react following two different pathways. In path A, a Kornblum-like oxidation takes place with *o*-iodobenzoate, the by-product of the SET reduction of Togni reagent **2e**, to give the desired compound. In path B, reduction of cation **A** promoted by **1d(Ir)*** leads to alkoxy radical **B**, which directly undergoes a 1,2-hydrogen atom shift to give α-oxo radical intermediate **C**. Further SET oxidation of **C** with **1d(Ir)** (IV) affords the final α-trifluoromethylated ketone. The evidence for the formation of intermediate **B** was made by using α-alkyl-substituted styrenes as substrates: in this case (R ≠ H), a β-scission of this tertiary alkoxy radical is observed, leading to the same α-CF_3 ketone (Scheme 14.23).

Similarly, molecular oxygen O_2 turned out to be effective for the hydroxytrifluoroethylation of styrenes under visible light photocatalysis.[56] Following a reductive quenching pathway (Scheme 14.20B), a wide range of γ-trifluoromethyl alcohols were synthesized in 28–81% yield using CF_3CH_2I **2ag** as the fluoroalkylating reagent under an air atmosphere. Water added in the reaction promoted the difunctionalization, but isotope labeling experiments showed that the oxygen atom present in the resulting hydroxytrifluoroethyl adduct came from O_2 and not from water (Scheme 14.24).

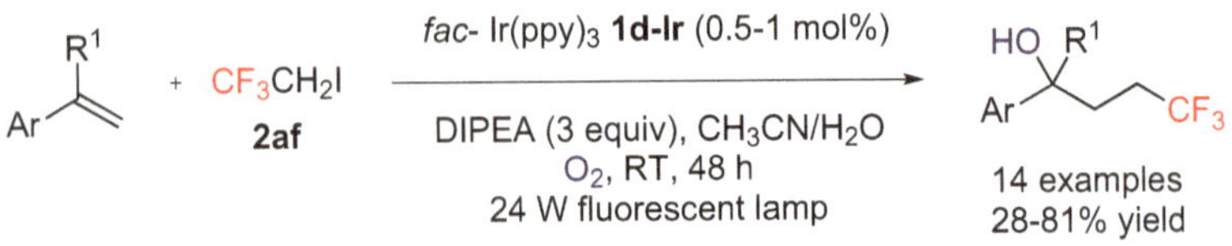

Scheme 14.24 Hydroxytrifluoroethylation of styrenes with molecular O_2.

Additionally, alkynes were also found to be suitable unsaturated systems to undergo photoredox-catalyzed intermolecular oxytrifluoromethylation reactions.[57] Thus, a variety of internal aryl alkynes readily reacted with Yagupol'skii reagent **2d** in the presence of a catalytic amount of $[Ir(ppy)_2(dtbbpy)](PF_6)$ **1e(Ir)**, leading to the expected tetrasubstituted trifluoromethylated alkenes in moderate to good yields (30–86%) and diastereoselectivities (61:39 < *E*/*Z* < 96:4). Notably, in the case of aryl alkynes bearing an electron-donating group, no desired trifluoromethylated alkenyl triflates were isolated presumably owing to their high propensity for hydrolysis. Alternatively, the use of an external nucleophile such as tetraalkylammonium tosylates delivered the corresponding trifluoromethylated alkenyl tosylates (Scheme 14.25).

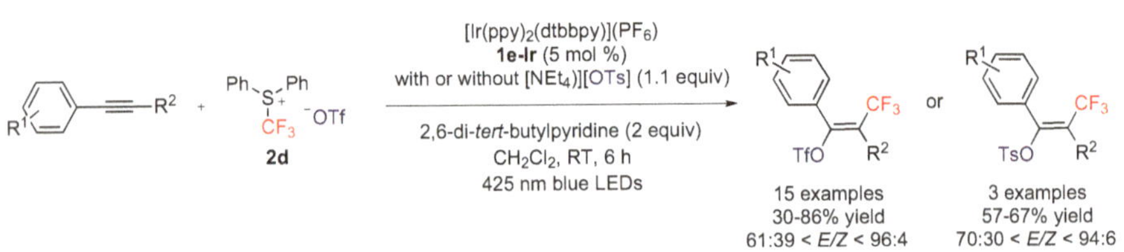

Scheme 14.25 Stereoselective conversion of alkynes into tetrasubstituted trifluoromethylated alkenes.

Along with these intermolecular oxytrifluoromethylation reactions, intramolecular procedures were successfully developed. The first example was reported by Cho et al. in 2013.[58] Following a reductive quenching pathway (Scheme 14.20B), a wide range of terminal allylic alcohols containing a variety of functional groups (hydroxy group, silyl ether, ester, carbonate, aryl halide) reacted smoothly with CF_3I **2a**, C_3F_7I **2s**, or C_4F_9I **2u** generating the desired perfluoroalkylated epoxides in 77–91% yield (Scheme 14.26).

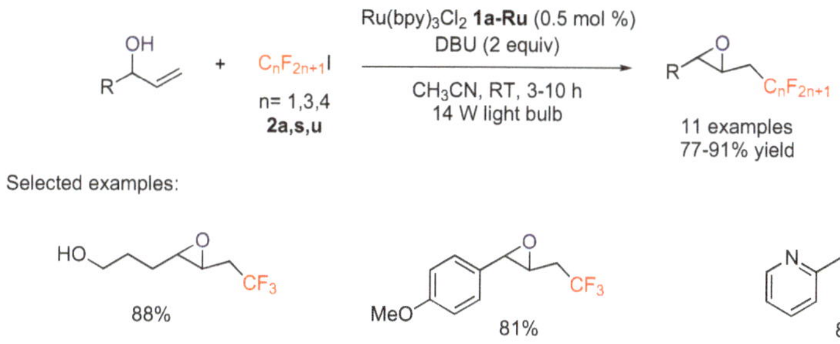

Scheme 14.26 Perfluoroalkylation of allylic alcohols into fluorinated epoxides.

This photoredox-catalyzed intramolecular strategy was later efficiently extended to other *O*-nucleophiles such as carboxylic acids,[59] amides,[60,61] or oximes,[62] giving access to CF_3-containing lactones, oxazolines and oxazines, or isoxazolines, respectively. Interestingly, in all cases, the trifluoromethylation reaction followed an oxidative quenching pathway (Scheme 14.20A) with $Ru(bpy)_3(PF_6)_2$ **1b(Ru)** as the PC and Umemoto reagent **2b** as the CF_3 source (Scheme 14.27).

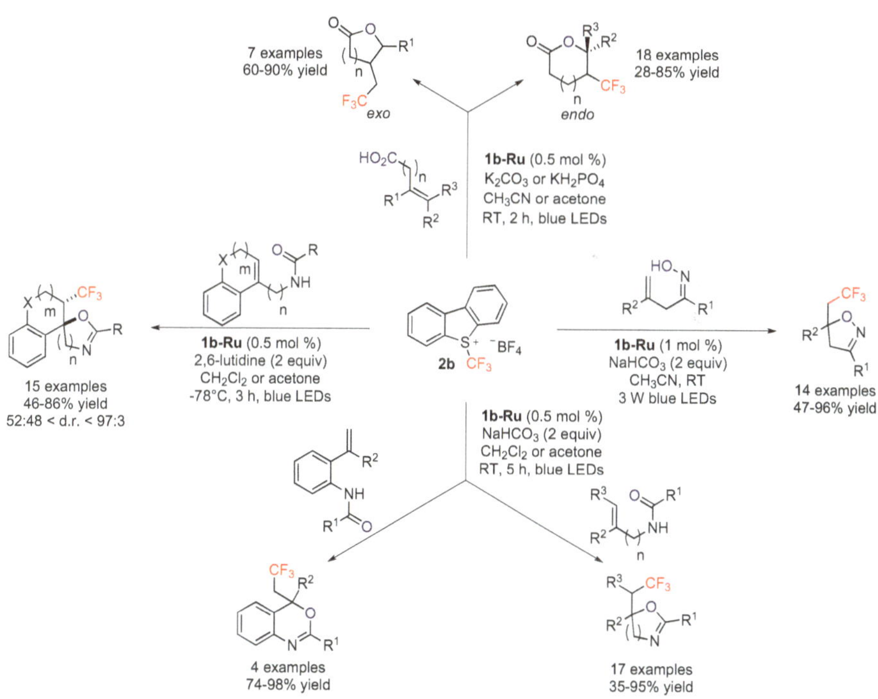

Scheme 14.27 Intramolecular oxytrifluoromethylation strategies under photoredox catalysis.

3.2 Amino- and Azido(per)fluoroalkylation of Alkenes

Studies on oxytrifluoromethylation reported by Masson et al.,[53] Cho et al.,[58] and Chen et al.[62] also deal with azido- or aminotrifluoromethylation of enecarbamates, allyl amines, and β,γ-unsaturated hydrazones, respectively, and will not be detailed in this paragraph.

The first example of photoredox-catalyzed aminotrifluoromethylation was developed by Akita et al. for the synthesis of β-trifluoromethyl amides.[63] This intermolecular process is compatible with styrene derivatives bearing various functional groups. Following an oxidative quenching cycle (Scheme 14.20A), the benzylic carbocation intermediate is trapped by the nitrile group via a Ritter-type reaction, furnishing the desired amidotrifluoromethylated products in good yield (Scheme 14.28).

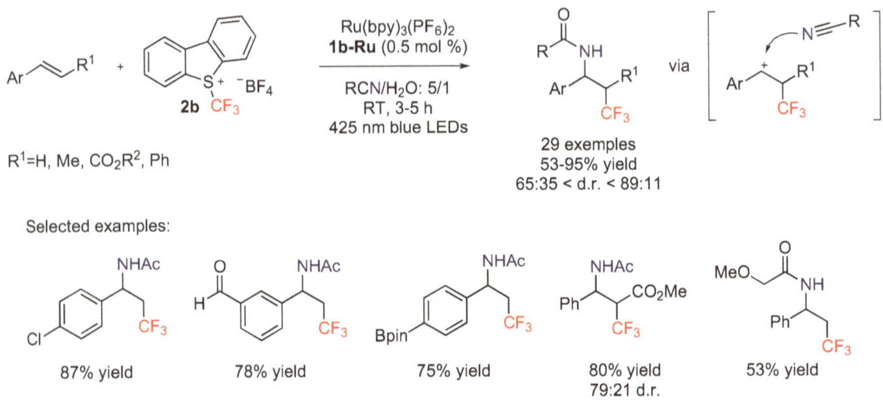

Scheme 14.28 Visible light–promoted amidotrifluoromethylation of styrenes.

Other *N*-nucleophiles were successfully used in similar three-component trifluoromethylation reactions. Azidotrimethylsilane as well as various anilines or amide derivatives were compatible with photoredox conditions and led to the synthesis of a wide range of β-trifluoromethylated amine derivatives in 39–81% yield (Scheme 14.29).[64]

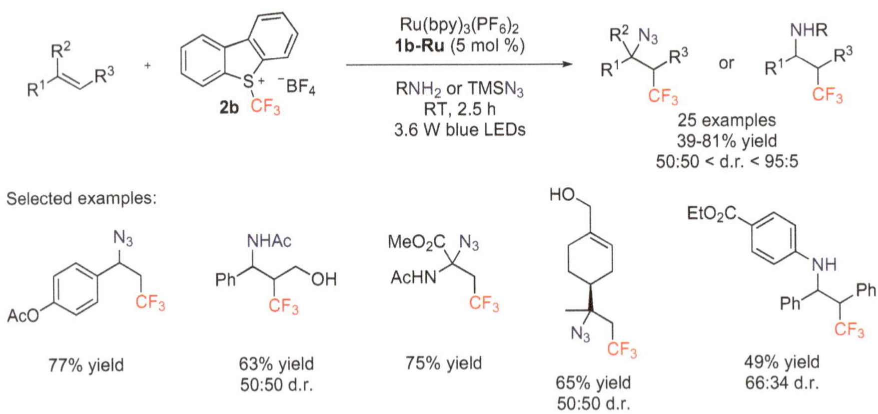

Scheme 14.29 Intermolecular azido- and aminotrifluoromethylation of alkenes.

Additionally, intramolecular aminodifluoromethylation was efficiently performed under photoredox catalysis by using HCF_2SO_2Cl **2p** as the fluoroalkylating agent and $Cu(dap)_2Cl$ **1g(Cu)** as the PC.[65] Silver carbonate was used to trap the chlorine anion formed during the reaction and to prevent the formation of halodifluorinated compounds (cf. vide infra). The reaction mechanism is proposed to follow an oxidative quenching pathway (Scheme 14.20A). Interestingly, the reaction could not be extended to the use of FCH_2SO_2Cl **2af** or CF_3SO_2Cl **2g**. It was indeed probably too difficult to reduce reagent **2af**, and the oxidation of the carboradical intermediate was presumably inhibited when the more electronegative CF_3 group was present in β-position (Scheme 14.30).

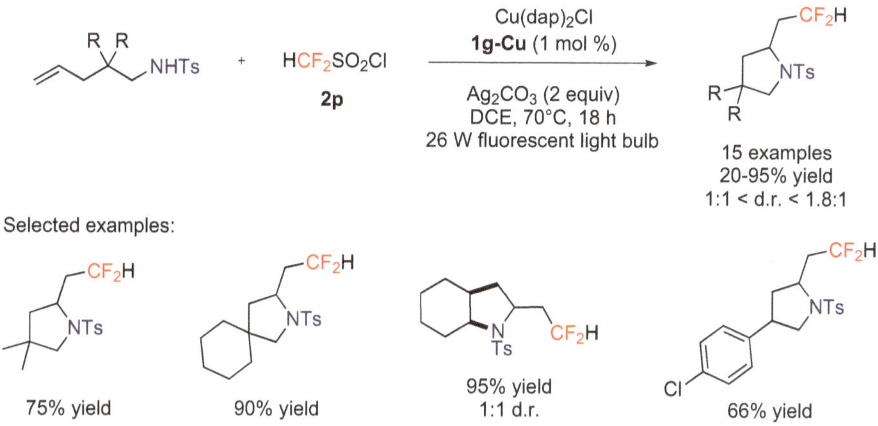

Scheme 14.30 Intramolecular aminodifluoromethylation of terminal alkenes.

3.3 Carbo(per)fluoroalkylation Reactions

3.3.1 Carbo(per)fluoroalkylation of Alkenes and Alkynes

The report of Masson et al. on oxy- and aminotrifluoromethylation of enecarbamates[53] was also applied to carbotrifluoromethylation with potassium cyanide as the nucleophile, generating the corresponding α-cyano β-trifluoromethyl carbamates, and will not be described in this paragraph.

The first example of photoredox-catalyzed carbotrifluoromethylation reaction was developed by Zhu in 2013.[66] Thus, treatment of various *N*-aryl acrylamides with Togni reagent **2e** under visible- light irradiation allowed for the synthesis of CF_3-containing oxindole derivatives bearing a quaternary carbon center (Scheme 14.31).

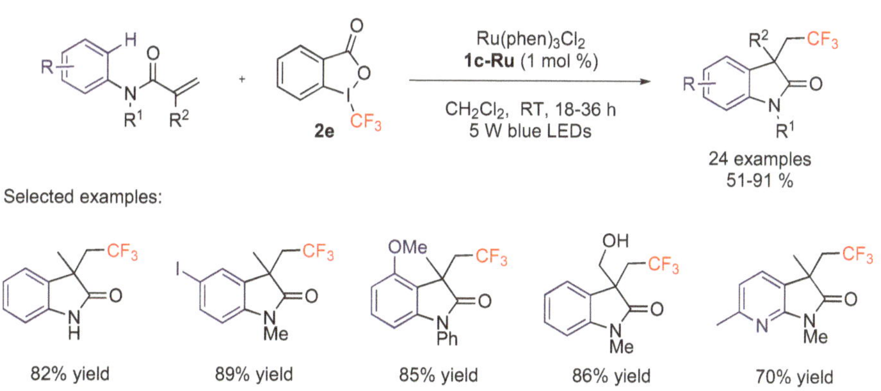

Scheme 14.31 Intramolecular aryltrifluoromethylation of *N*-aryl acrylamides.

To develop similar carbo(per)fluoroalkylation reactions, these *N*-aryl acrylamides were later broadly used, in particular for BiOBr **1i**-catalyzed aryltrifluoromethylation,[67] **1d(Ir)**-catalyzed aryl trifluoromethylation,[68] difluoromethylation,[69] trifluoroethylation,[70] and (per)fluoroalkylation[71] (Scheme 14.32).

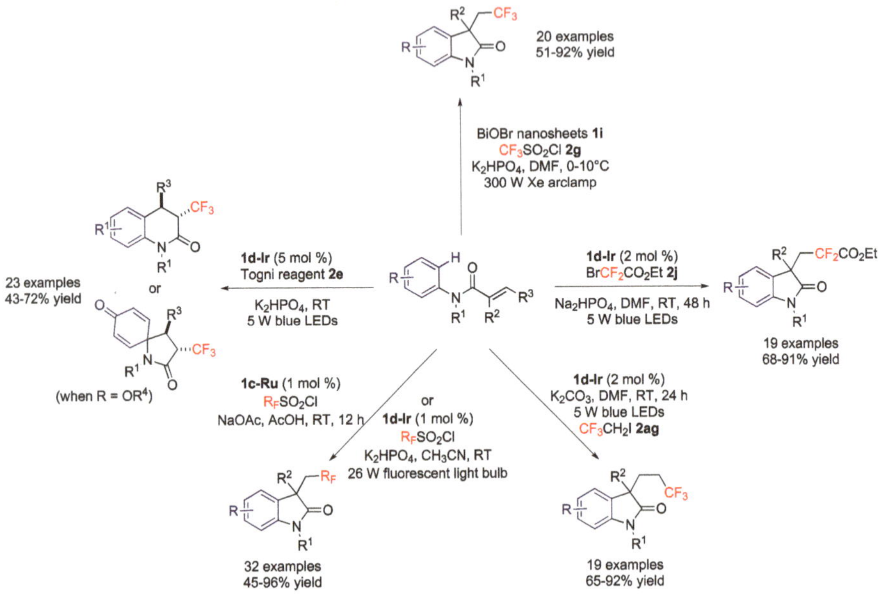

Scheme 14.32 Aryl(per)fluoroalkylation of *N*-aryl acrylamides.

Interestingly, different kinds of reactivities were observed with the modulation of the substituents on the nitrogen atom of the acrylamide scaffold.[72] Thus, *N*-benzamido acrylamides furnished the corresponding trifluoromethylated isoquinolinediones. *N*-sulfonyl *N*-alkyl acrylamides led to the synthesis of CF_3-containing oxindoles via a trifluoromethyl arylation/desulfonylation sequence. *N*-sulfonyl *N*-aryl acrylamides afforded linear CF_3 amides via a trifluoromethyl arylation/1,4-aryl shift/desulfonylation cascade reaction (Scheme 14.33).

Other unsaturated substrates are compatible with such carbo(per)fluoroalkylation reactions. Notably, Fu et al. successfully developed the aryldifluoroacetylation of alkynes with **2j** as the difluoroalkyl source.[73] Various phenyl alkynoate derivatives were transformed into the desired difluoromethylated coumarins in 52–87% yield (Scheme 14.34).

For aryl(per)fluoroalkylation reactions described earlier, during the oxidative quenching mechanism, the aryl moiety that traps the radical (or cationic) intermediate is already present in the unsaturated substrates. However, another intramolecular aryldifluoroalkylation reaction was designed by Zhu et al., the aryl moiety being present in this case in the difluoroalkyl source.[74] Indeed, by using α-bromo-α,α-difluoroacetophenone derivatives **2l**, a wide range of polycyclic lactones were efficiently generated from the corresponding butenolides under photoredox catalysis (Scheme 14.35).

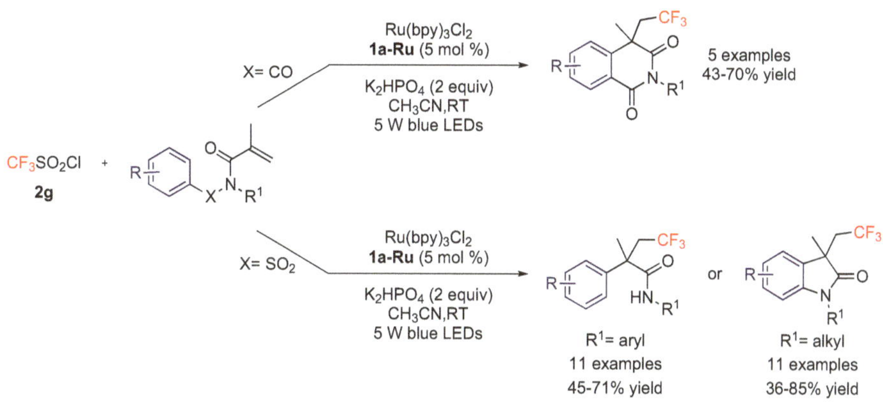

Scheme 14.33 Different reactivities for the photoredox-catalyzed trifluoromethylation of acrylamides.

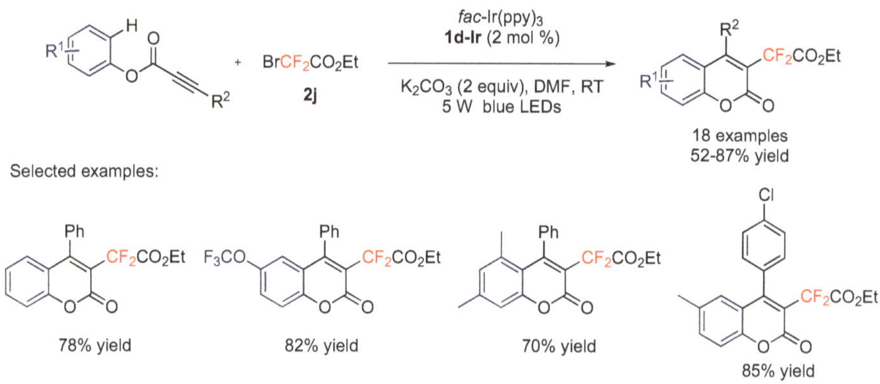

Scheme 14.34 Photoredox-catalyzed difluoroacetylation of alkynes.

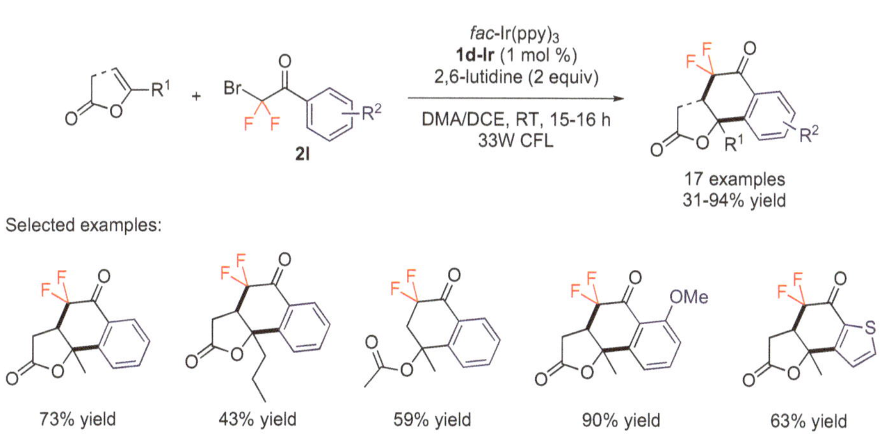

Scheme 14.35 Photoredox-catalyzed synthesis of fluoroinated polycyclic lactones. *CFL*, compact fluorescent lamp.

Besides all the examples of intramolecular aryl(per)fluoroalkylations described earlier, the more challenging intermolecular aryltrifluoromethylation was also successfully reported by Masson et al. in 2014.[75] Such three-component process was performed with Umemoto reagent **2b** as the CF_3 source and with a variety of electron-rich arenes and heteroarenes, generating the corresponding trifluoromethylated adducts in up to 99% yield (Scheme 14.36).

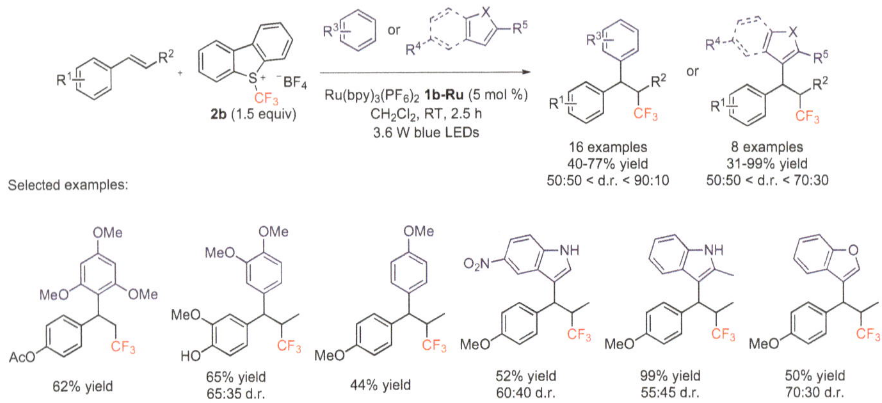

Scheme 14.36 Intermolecular aryl- and heteroaryltrifluoromethylation of styrenes.

In 2015, Glorius et al. reported another interesting example of carbotrifluoromethylation reaction, which involved a tandem trifluoromethylation/ring expansion via a semipinacol-type rearrangement.[76] As depicted in Scheme 14.37, a variety of cycloalkanol-substituted styrene derivatives reacted smoothly in the presence of PC **Ru(1b)**, Umemoto reagent **2c**, and 1.2 equivalents of Trimethylsilyl trifluoromethanesulfonate (TMSOTf) to prevent the formation of the undesired trifluoromethylated epoxide (see Scheme 14.26 for oxytrifluoromethylation of allylic alcohols).

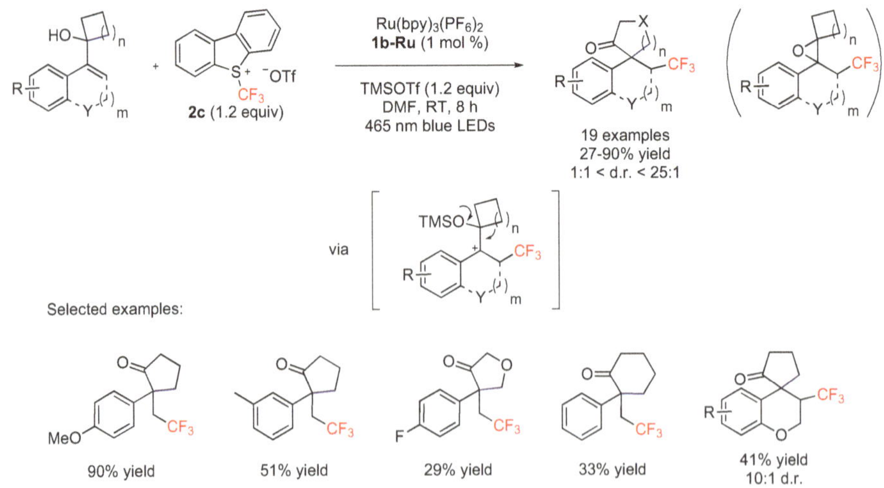

Scheme 14.37 Visible light–mediated trifluoromethylation/ring expansion via a semipinacol-type rearrangement.

Similar reactions involving a (per)fluoroalkylation/1,2-alkyl migration sequence were also developed with either Umemoto reagent **2b**[77] or $BrCF_2CO_2Et$ **2j**.[78] Another interesting study, reported by Zhu et al., implied the carbo(per)fluoroalkylation of α,α-diaryl allylic alcohols via a tandem (per)fluoroalkylation/1,2-aryl migration process.[79] Remarkably, in the case of unsymmetrical diaryl allylic alcohols ($Ar^1 \neq Ar^2$), the migration of the aryl moiety bearing an electron-deficient group was preferred, indicating that a neophyl rearrangement via a spiro[2.5]octadienyl radical is more likely than a semipinacol rearrangement (Scheme 14.38).

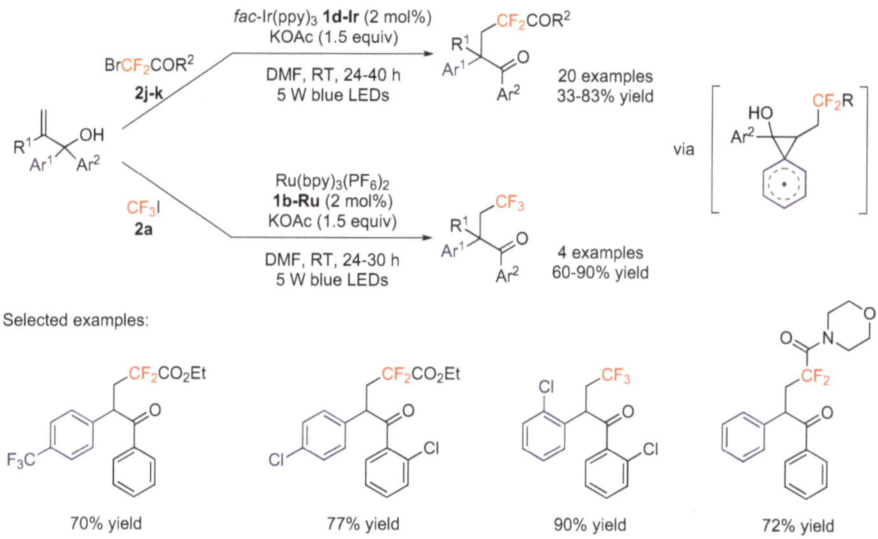

Scheme 14.38 Photoredox-catalyzed tandem (per)fluoroalkylation/1,2-aryl migration reaction.

3.3.2 Carbo(per)fluoroalkylation of Isocyanides

Besides carbo(per)fluoroalkylation reactions involving vicinal difunctionalization of alkenes or alkynes, another interesting carbo(per)fluoroalkylation implied the *gem*-difunctionalization of isocyanides. Thus, visible light irradiation of various biphenyl isocyanides with ethyl bromofluoroacetate **2ae** or ethyl bromodifluoroacetate **2j** in the presence of PC **Ir(1d)** led to the synthesis of mono- or difluoromethylated phenanthridine derivatives, respectively, in good to excellent yields (Scheme 14.39).[80]

Subsequent to this work, similar studies were reported with slight modifications on the isocyanide substrates, such as vinyl isocyanides, leading to trifluoromethylated isoquinolines,[81] or by changing the (per)fluoroalkyl source, which could be Umemoto reagent **2b**,[82] trifluoroethyl iodide **2ag**,[83] bromodifluoromethylphosphonate **2m**,[84,85] or HCF_2SO_2Cl **2p** and $BrCF_2R$ **2n** (R = Me, Ph)[86] (Scheme 14.40).

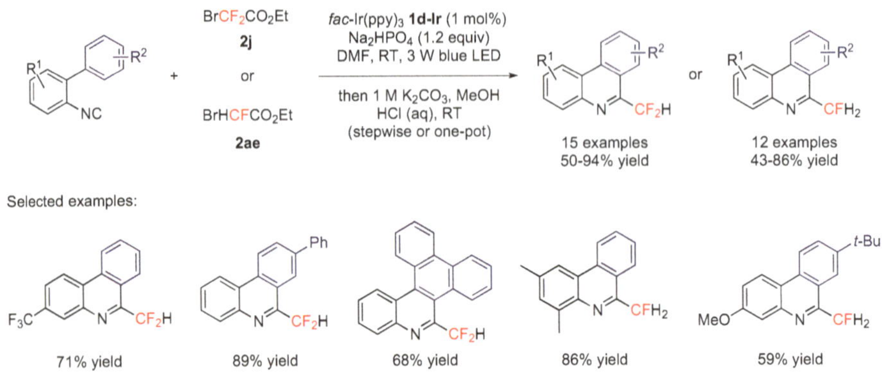

Scheme 14.39 Photoredox-catalyzed fluoroalkylation of isocyanides.

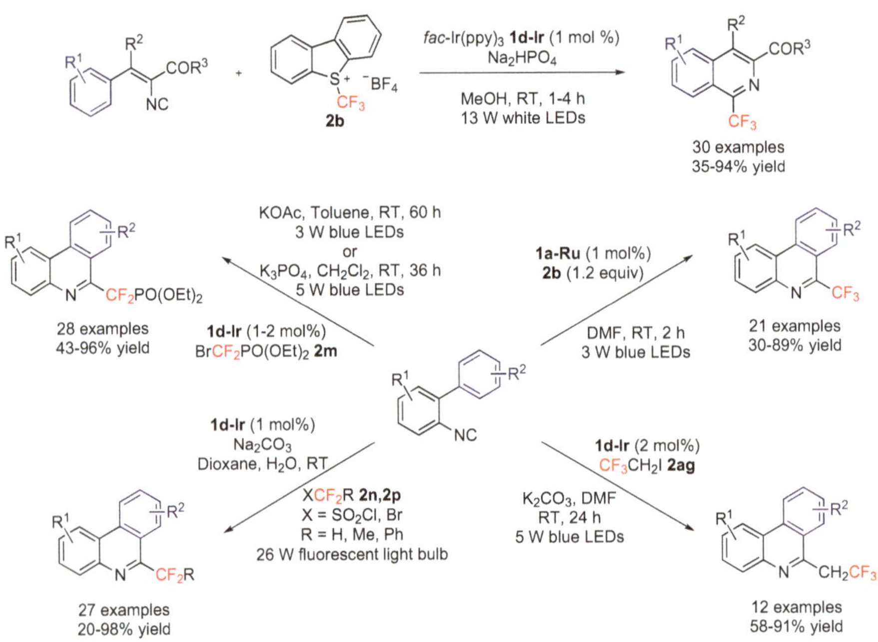

Scheme 14.40 Carbo(per)fluoroalkylation of isocyanides under photoredox catalysis.

3.4 Halo(per)fluoroalkylation of Alkenes and Alkynes

Seminal studies on halo(per)fluoroalkylation reactions were first reported by Stephenson et al. in 2011.[87,88] Interestingly, depending on the (per)fluoroalkylating source used for the reaction, two different sets of conditions were successfully developed to follow either a reductive or an oxidative quenching cycle. On the one hand, when CF_3I **2a**, CF_2Br_2 **2r**, $BrCF_2CO_2Et$ **2j**, and $BrHCFCO_2Et$ **2ae** were used, the reaction was performed in the presence of LiBr as Lewis acid to favor the direct reduction of the alkyl halide into the corresponding fluorinated radical (Scheme 14.41, oxidative quenching pathway). This carboradical can then add to the alkene substrate to form a second radical intermediate **A**, which can then be easily oxidized into the corresponding carbocation **B** by Ir(IV) PC species. Even though a large amount of experimental evidence provides support for such radical–polar cross-over mechanism, a radical chain transfer mechanism may also be operative, and its overall contribution most likely depends on the nature of the substrate and the conditions used. On the other hand, when perfluoroalkyl iodides **2a,u,w–y** were used, the use of sodium ascorbate as a reductant to follow a reductive quenching cycle was found to be more efficient. Once again, both radical–polar cross-over and radical propagation mechanisms are presumably competing as shown in Scheme 14.41 (reductive quenching cycle). For instance, in the case of alkynes, the radical propagation is more likely than the thermodynamically unfavored oxidation of the vinyl radical intermediate into a vinyl cation.

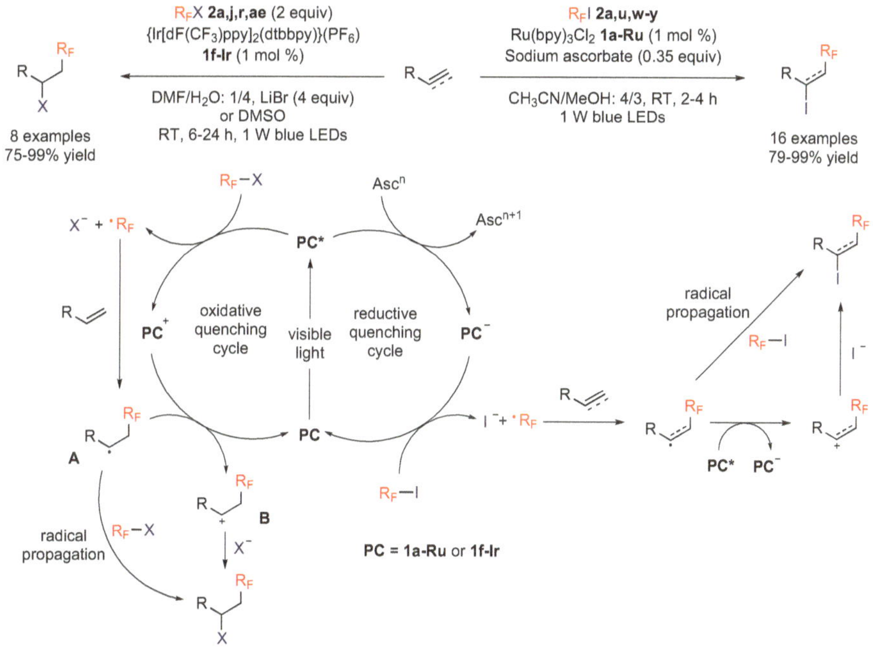

Scheme 14.41 Halo(per)fluoroalkylation of alkenes and alkynes under photoredox catalysis.

Evidence for a radical chain mechanism in the case of iodotrifluoromethylation of alkynes was also reported by Cho et al. in 2014.[50] This radical chain is initiated by $Ru(phen)_3Cl_2$ **1c(Ru)**, which is reductively quenched by tetramethylethylenediamine (TMEDA). The expected trifluoromethylated alkenyl iodides are obtained in good to excellent yields with high stereoselectivity (Scheme 14.42).

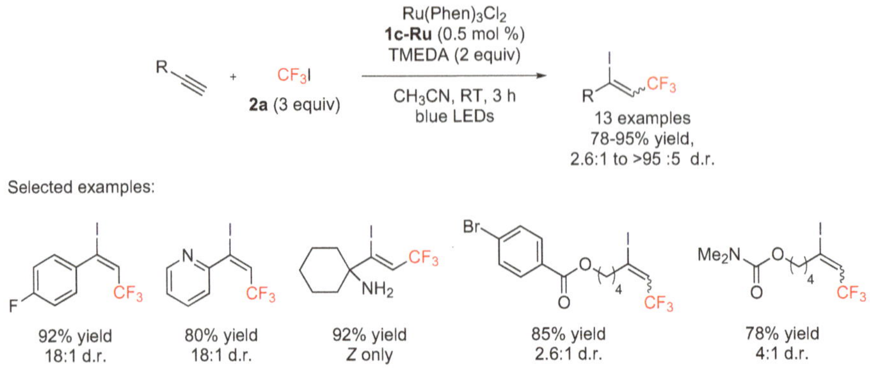

Scheme 14.42 Visible light–mediated synthesis of trifluoromethylated alkenyl iodides.

Apart from the aforementioned iodo- and bromo(per)fluoroalkylation, chloro(per) fluoroalkylation was also successfully developed by Han et al.[89] as well as Reiser et al. (cf. vide infra).[90] The reaction was performed by using CF_3SO_2Cl **2g** as the source of both CF_3 radical and chlorine nucleophile. Indeed, oxidative quenching of the excited **1c(Ru)*** by **2g** affords the CF_3 radical along with chloride anion after loss of the gaseous SO_2. A radical chain mechanism cannot be ruled out, but experimental evidence are in favor of a radical-polar cross-over mechanism via oxidation of the radical intermediate by **1c(Ru) (III)**. Notably, mono-, di-, and trisubstituted alkenes are compatible with the reaction conditions, leading to the corresponding trifluoromethylated adducts in 54–99% yield (Scheme 14.43).

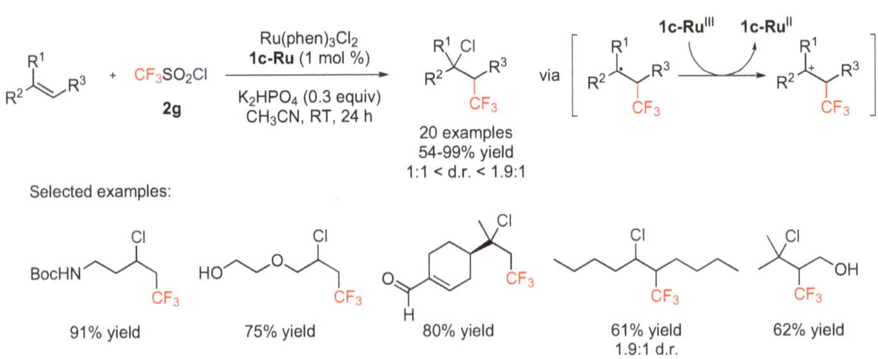

Scheme 14.43 Chlorotrifluoromethylation of alkenes under photoredox catalysis.

However, similar reaction conditions could not be generalized to electron-poor alkenes. Indeed, a chain propagation is unfavored because the radical intermediate is too electrophilic to abstract a halogen atom from CF_3SO_2Cl, and the radical–polar cross-over mechanism is also unlikely because this radical cannot be oxidized into the corresponding cation (Scheme 14.44). An elegant way to circumvent this issue was reported by Dolbier et al.,[91] which relied on the use of Cu(dap)$_2$Cl **1g(Cu)**. This PC has indeed good redox properties for the reduction of triflyl chloride **2g**, as well as good Cl transfer properties to propagate the atom transfer radical addition (ATRA) reaction. Moreover, by performing the reaction at higher temperatures, a wide range of other (per)fluoroalkylsulfonyl chlorides (**2p**, **2ab**, **2af,** and **2ah**) underwent reaction with α,β-unsaturated amides, esters, and ketones in good yields (Scheme 14.44).

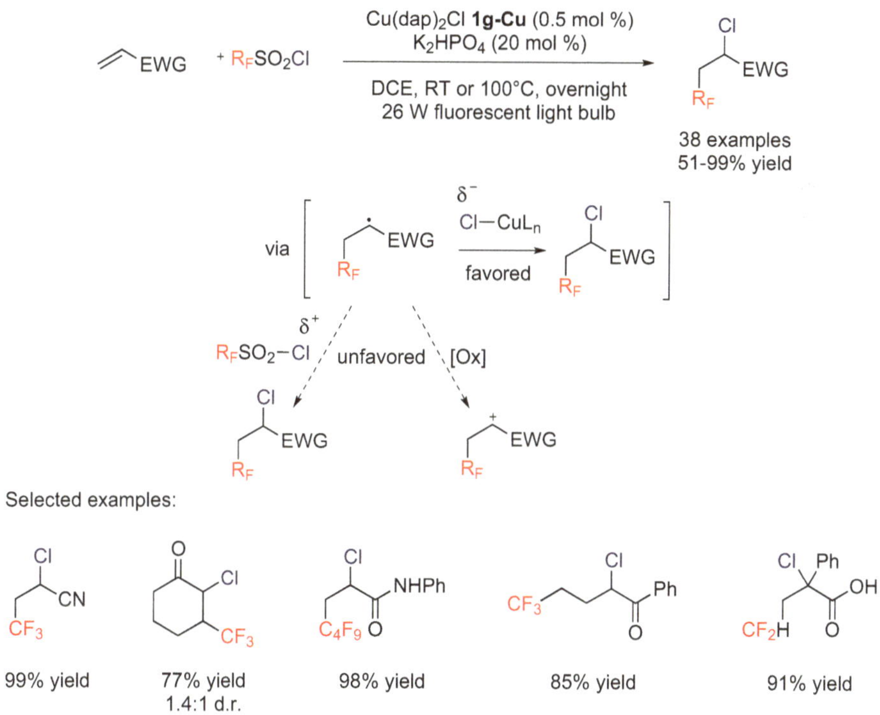

Scheme 14.44 Chlorotrifluoromethylation of electron-poor alkenes under photoredox catalysis. *EWG*, electron-withdrawing group.

In every example described earlier, the fluorine-containing reagent plays the role of both the fluoroalkyl radical source and the halogen source. However, halotrifluoromethylation of alkenes can also be efficiently performed in a three-component manner. Such intermolecular process between alkenes, Umemoto reagent **2b**, and TMSCl, CsBr, or CsI was reported by Masson et al. in 2015.[92] Under blue LED irradiation, a wide range of chloro-, bromo-, and iodotrifluoromethylated adducts were generated in moderate to good yields (Scheme 14.45).

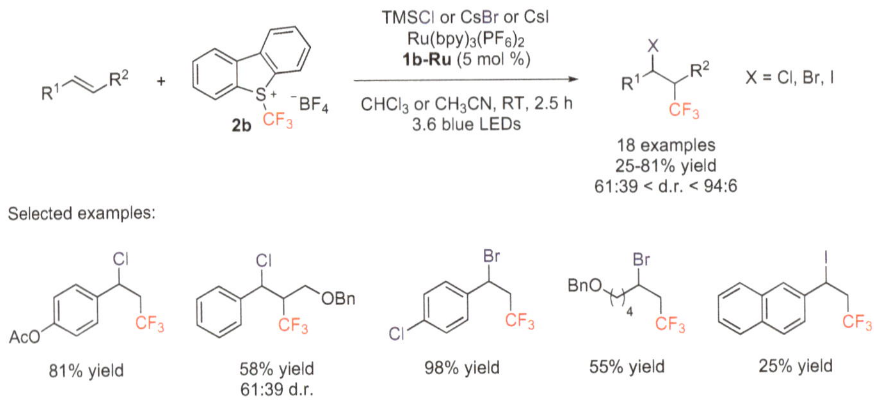

Scheme 14.45 Three-component chloro-, bromo-, and iodotrifluoromethylation of alkenes.

3.5 Trifluoromethylchlorosulfonylation of Alkenes

When the work of Han et al. on chlorotrifluoromethylation was tried by Reiser et al. with CF_3SO_2Cl **2g** and $Cu(dap)_2Cl$ **1g(Cu)**, the authors observed the formation of unexpected trifluoromethyl chlorosulfonylated compounds without extrusion of SO_2.[90] During this new trifluoromethylchlorosulfonyl process, copper PC **1g(Cu)** is assumed to weakly coordinate SO_2Cl, as shown in intermediate **A** in Scheme 14.46. The CF_3 radical is subsequently added to the alkene, giving rise to intermediate **B**, and the final transfer of SO_2Cl from copper leads to the desired ATRA adduct. Two main observations are in favor of such mechanistic hypothesis. First, when alkenes that contain donor atoms are used in the reaction conditions, only the chlorotrifluoromethylated adduct is observed, which suggests that the donor atom might displace SO_2Cl^- and facilitate SO_2 extrusion. Second, when a sterically more hindered alkene or PC is used, the formation of the chlorotrifluoromethylated adduct is once again favored, suggesting that the SO_2Cl group is probably transferred to the alkene by a Cu-SO_2Cl intermediate described in Scheme 14.46.

3.6 Hydro(per)fluoroalkylation of Alkenes and Alkynes

Unlike most of the difunctional (per)fluoroalkylation reactions in the previous paragraphs, hydro(per)fluoroalkylation reactions described in this section do not follow the general mechanism depicted in Scheme 14.20. In such reactions, the β-R_F carboradical intermediate **A** is not oxidized into the corresponding carbocation but abstracts a hydrogen atom from a suitable H atom donor to give rise to the desired hydro(per)fluoroalkylated adduct (Scheme 14.47).

The first example of hydrotrifluoromethylation was reported by Gouverneur et al. in 2013.[93] Thus, irradiation of various unactivated alkenes and alkynes in the presence of photocatalyst **1a(Ru)** and Umemoto reagent **2c** as the CF_3 source in methanol led to the corresponding trifluoromethylated compounds in moderate to good yields

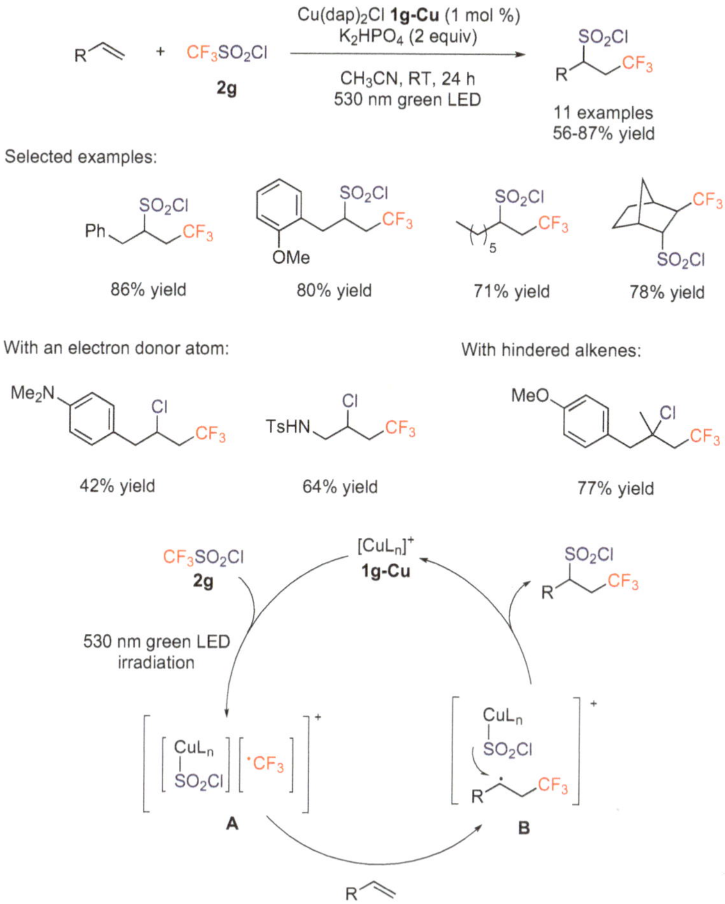

Scheme 14.46 Trifluoromethylchlorosulfonylation of alkenes under $Cu(dap)_2Cl$ photoredox catalysis.

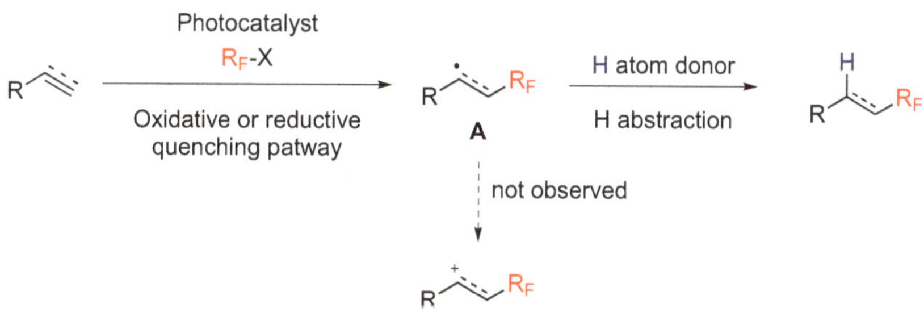

Scheme 14.47 General mechanism for hydro(per)fluoroalkylation of alkenes and alkynes.

(Scheme 14.48). Control experiments with deuterated methanol confirmed that the α-C−H bond of methanol serves as the hydrogen atom source in this process.

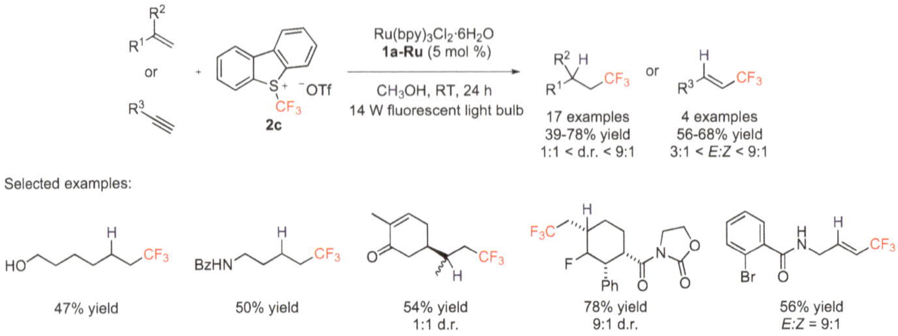

Scheme 14.48 Visible light−mediated hydrotrifluoromethylation of alkenes and alkynes.

This work was then generalized by Qing et al. for the hydrodifluoroalkylation of alkenes and alkynes using CF_2Br_2 **2r** as the fluoroalkyl reagent and eosin Y **1k** as the PC.[94] In this case, to avoid the formation of the ATRA product of bromodifluoroalkylation, methanol was replaced by THF. Once again, labeled experiments with deuterated THF unambiguously confirmed that the H abstraction occurs from a molecule of solvent (Scheme 14.49).

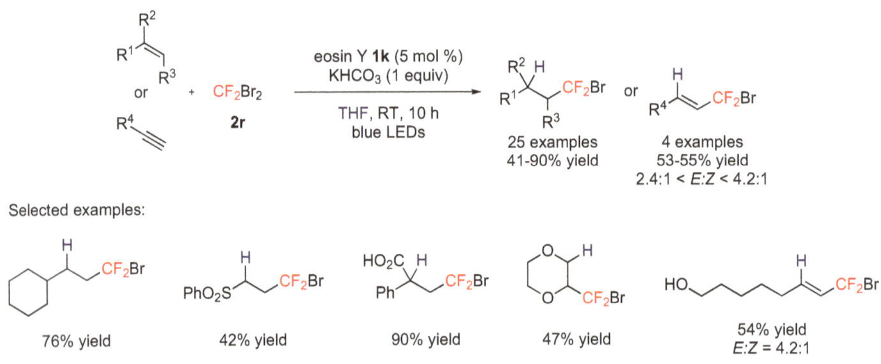

Scheme 14.49 Hydrodifluoroalkylation of alkenes and alkynes under photoredox catalysis.

In the last two examples, both reactions are presumably following an oxidative quenching mechanism, where Umemoto reagent **2c** and CF_2Br_2 **2r** are directly reduced by the excited state PC* of the PC. However, hydro(per)fluoroalkylation can also occur via a reductive quenching pathway, where an external tertiary amine is first oxidized by the excited photocatalyst PC*, giving rise to a strong reductant PC^- and the corresponding aminium radical cation. The β-R_F carboradical intermediate finally abstracts a hydrogen atom from this aminium radical cation to furnish the expected hydro(per)fluoroalkylated adduct (Scheme 14.50).

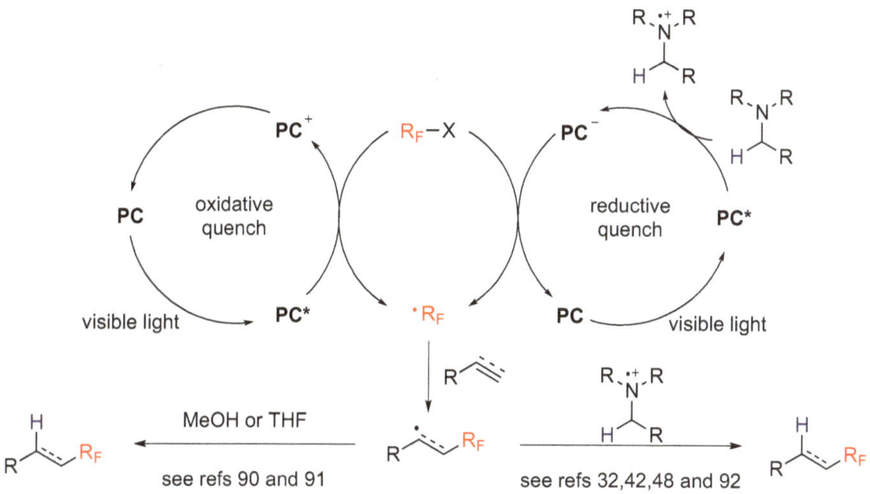

Scheme 14.50 Mechanisms for hydro(per)fluoroalkylation of alkenes and alkynes.

This reductive quenching strategy was successfully applied to hydro(per)fluoroalkylation of alkenes and alkynes by Cho et al.[95] and Scaiano et al.[34] by using either 1,8-diazabicyclo[5.4.0]undec-7-ene (DBU) or TMEDA as the H atom donor. This method was found to be quite general for a wide range of alkenes and terminal alkynes. Notably, in the case of propargylic alcohols, the hydrotrifluoromethylation step was followed by a migration of the generated allylic alcohol **A** into the corresponding ketone **B** (Scheme 14.51).

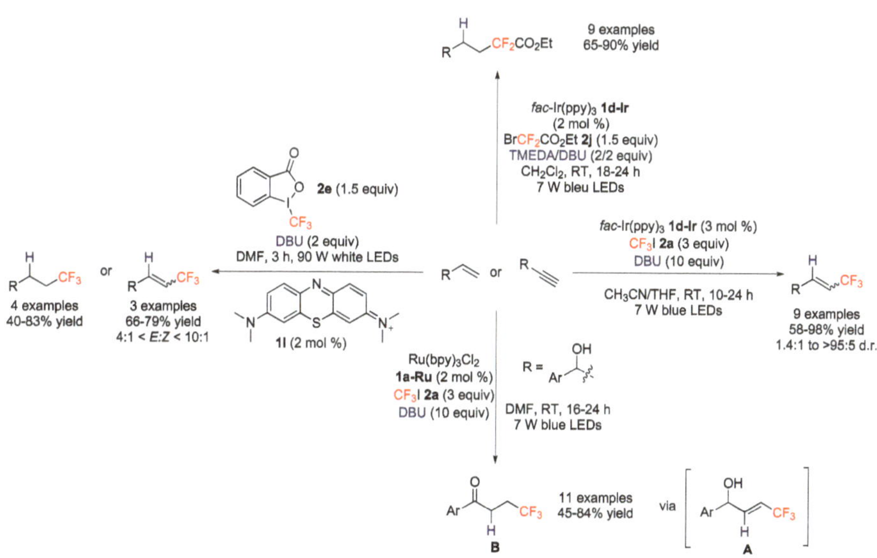

Scheme 14.51 Hydro(per)fluoroalkylation reactions with a tertiary amine as the H atom donor.

A last method of hydrotrifluoromethylation developed by Nicewicz et al. in 2013 employed Langlois reagent CF_3SO_2Na **2h** as the CF_3 source.[96] Importantly, this report is a rare example of photocatalyzed (per)fluoroalkylation in which the (per)fluoroalkyl radical is generated by an SET oxidation from the (per)fluoroalkyl source. To perform such oxidation, *N*-methyl-9-mesityl acridinium **1m** was used as a strong oxidizing organic photoredox catalyst. The reaction proceeded best with 2,2,2-trifluoroethanol as cosolvent, which is proposed to act as the H atom donor in this reaction, generating a trifluoromethyl ketyl radical **A**. In addition, substoichiometric quantities of thiol **B** greatly improved the yield of the reaction. Presumably, **B** could perform an H atom transfer with **A** to give rise to a thiyl radical **C**, which could be more easily reduced by acridine radical **D** (Scheme 14.52). A large variety of aliphatic alkenes as well as styrene derivatives were smoothly hydrotrifluoromethylated in 25–74% yield.

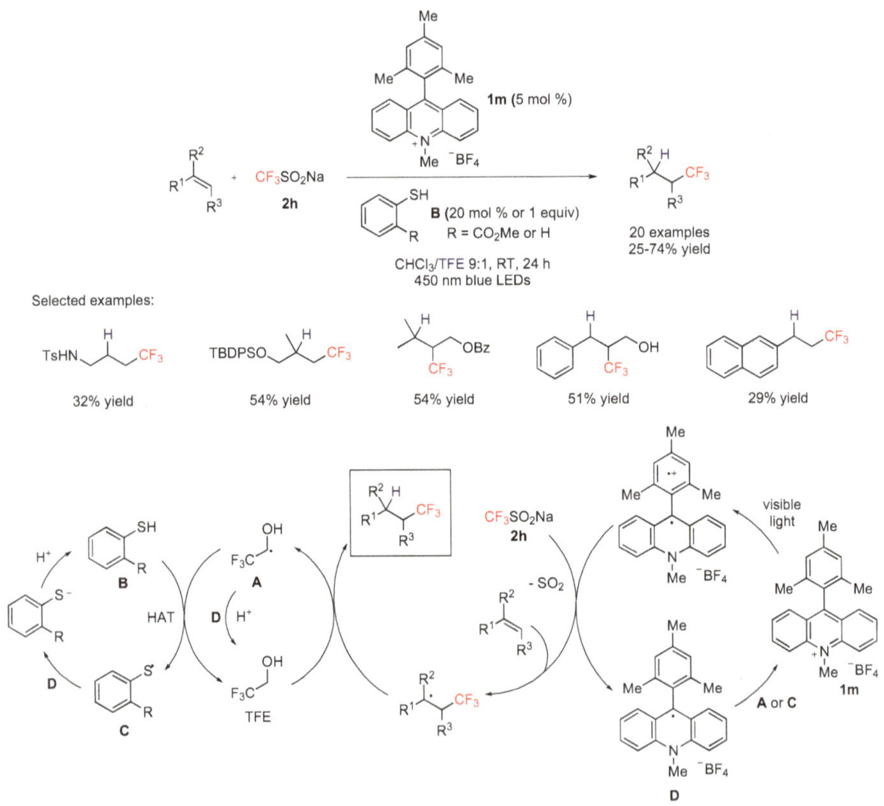

Scheme 14.52 Photoredox-catalyzed hydrotrifluoromethylation of alkenes with Langlois reagent. *TBDPS*, *tert*-butyldiphenylsilyl; *TFE*, 2,2,2-trifluoroethanol; *Ts*, *para*-toluenesulfonyl.

4. Conclusion

As discussed, visible light photoredox catalysis has quickly become a powerful and efficient tool in (per)fluoroalkylation reactions. Such process has proved to be an efficient and versatile approach and has allowed rapid access to a variety of intermediates that can serve as synthons in the preparation of pharmaceutically and agrochemically relevant molecules. Despite remarkable achievements in photoredox-catalyzed trifluoromethylation reagents, the search for inexpensive and more atom-economic trifluoromethylated reagents still continues. In addition, there are only very few examples demonstrating high diastereo-/enantioselectivity. The development of highly stereoselective methods is an aspect that requires further investigation. We are expecting breakthroughs in this direction in the near future.

References

1. Ischay, M. A.; Anzovino, M. E.; Du, J.; Yoon, T. P. Efficient Visible Light Photocatalysis of [2+2] Enone Cycloadditions. *J. Am. Chem. Soc.* **2008,** *130,* 12886–12887.
2. Nicewicz, D. A.; MacMillan, D. W. C. Merging Photoredox Catalysis with Organocatalysis: The Direct Asymmetric Alkylation of Aldehydes. *Science* **2008,** *322,* 77–80.
3. Rueda-Becerril, M.; Mahe, O.; Drouin, M.; Majewski, M. B.; West, J. G.; Wolf, M. O.; Sammis, G. M.; Paquin, J.-F. Direct C–F Bond Formation Using Photoredox Catalysis. *J. Am. Chem. Soc.* **2014,** *136,* 2637–2641.
4. Ventre, S.; Petronijevic, F. R.; MacMillan, D. W. C. Decarboxylative Fluorination of Aliphatic Carboxylic Acids via Photoredox Catalysis. *J. Am. Chem. Soc.* **2015,** *137,* 5654–5657.
5. Wu, X.; Meng, C.; Yuan, X.; Jia, X.; Qian, X.; Ye, J. Transition-metal-free Visible-light Photoredox Catalysis at Room-temperature for Decarboxylative Fluorination of Aliphatic Carboxylic Acids by Organic Dyes. *Chem. Comm.* **2015,** *51,* 11864–11867.
6. Xu, P.; Abdukader, A.; Hu, K.; Cheng, Y.; Zhu, C. Room Temperature Decarboxylative Trifluoromethylation of A,β-unsaturated Carboxylic Acids by Photoredox Catalysis. *Chem. Comm.* **2014,** *50,* 2308–2310.
7. Wilger, D. J.; Grandjean, J. -M. M.; Lammert, T. R.; Nicewicz, D. A.The Direct Anti-Markovnikov Addition of Mineral Acids to Styrenes. *Nat. Chem.* **2014,** *6,* 720–726.
8. Pan, Y.; Wang, S.; Kee, C. W.; Dubuisson, E.; Yang, Y.; Loh, K. P.; Tan, C. -H. Graphene Oxide and Rose Bengal: Oxidative C–h Functionalisation of Tertiary Amines Using Visible Light. *Green Chem.* **2011,** *13,* 3341–3344.
9. Fu, W.; Guo, W.; Zou, G.; Xu, C. Selective Trifluoromethylation and Alkynylation of Tetrahydroisoquinolines Using Visible Light Irradiation by Rose Bengal. *J. Fluor. Chem.* **2012,** *140,* 88–94.
10. a. Li, W.; Zhu, X.; Mao, H.; Tang, Z.; Cheng, Y.; Zhu, C. Visible-light-induced Direct $C(sp^3)$–h Difluoromethylation of Tetrahydroisoquinolines with the In Situ Generated Difluoroenolates. *Chem. Comm.* **2014,** *50,* 7521–7523.
    b. Li, W.; Zhu, X.; Duan, Y.; Zhang, M.; Zhu, C. Monofluoromethylation of Tetrahydroisoquinolines by Visible-light Induced Direct $C(sp^3)$–h Bond Activation. *Adv. Synthesis Catal.* **2015,** *357,* 1277–1282.

11. a. Nappi, M.; Bergonzini, G.; Melchiorre, P. Metal-free Photochemical Aromatic Perfluoroalkylation of α-cyano Arylacetates. *Angew. Chem. Int. Ed.* **2014,** *53,* 4921−4925.
b. Wozniak, L.; Murphy, J. J.; Melchiorre, P. Photo-organocatalytic Enantioselective Perfluoroalkylation of β-ketoesters. *J. Am. Chem. Soc.* **2015,** *137,* 5678−5681.
c. Filippini, G.; Nappi, M.; Melchiorre, P. Photochemical Direct Perfluoroalkylation of Phenols. *Tetrahedron* **2015,** *71,* 4535−4542.
12. Cheng, Y.; Yuan, X.; Ma, J.; Yu, S. Direct Aromatic C−H Trifluoromethylation via an Electron-donor−acceptor Complex. *Chem. Eur. J.* **2015,** *21,* 8355−8359.
13. Umemoto, T. Electrophilic Perfluoroalkylating Agents. *Chem. Rev.* **1996,** *96,* 1757−1778.
14. Yagupol'skii, L. M.; Kondratenko, N. Y.; Timofeeva, G. N. *Zhurnal Organicheskoi Khimii* **1984,** *20,* 115−118.
15. Eisenberger, P.; Gischig, S.; Togni, A. Novel 10-I-3 Hypervalent Iodine-based Compounds for Electrophilic Trifluoromethylation. *Chem. Eur. J.* **2006,** *12,* 2579−2586.
16. Koike, T.; Akita, M. Trifluoromethylation by Visible-light-driven Photoredox Catalysis. *Top. Catal.* **2014,** *57,* 967−974.
17. Egami, H.; Sodeoka, M. Trifluoromethylation of Alkenes with Concomitant Introduction of Additional Functional Groups. *Angew. Chem. Int. Ed.* **2014,** *53,* 8294−8308.
18. Cao, M. -Y.; Ren, X.; Lu, Z. Olefin Difunctionalizations via Visible Light Photocatalysis. *Tetrahedron Lett.* **2015,** *56,* 3732−3742.
19. Alonso, C.; Martínez de Marigorta, E.; Rubiales, G.; Palacios, F. Carbon Trifluoromethylation Reactions of Hydrocarbon Derivatives and Heteroarenes. *Chem. Rev.* **2015,** *115,* 1847−1935.
20. Barata-Vallejo, S.; Bonesi, S. M.; Postigo, A. Photocatalytic Fluoroalkylation Reactions of Organic Compounds. *Org. Biomol. Chem.* **2015,** *13,* 11153−11183.
21. Nagib, D. A.; Scott, M. E.; MacMillan, D. W. C. Enantioselective α-trifluoromethylation of Aldehydes via Photoredox Organocatalysis. *J. Am. Chem. Soc.* **2009,** *131,* 10875−10877.
22. Neumann, M.; Füldner, S.; König, B.; Zeitler, K. Metal-free, Cooperative Asymmetric Organophotoredox Catalysis with Visible Light. *Angew. Chem. Int. Ed.* **2011,** *50,* 951−954.
23. Pham, P. V.; Nagib, D. A.; MacMillan, D. W. C. Photoredox Catalysis: a Mild, Operationally Simple Approach to the Synthesis of α-trifluoromethyl Carbonyl Compounds. *Angew. Chem. Int. Ed.* **2011,** *50,* 6119−6122.
24. Sladojevich, F.; McNeill, E.; Börgel, J.; Zheng, S. -L.; Ritter, T. Condensed-phase, Halogen-bonded CF_3I and C_2F_5I Adducts for Perfluoroalkylation Reactions. *Angew. Chem. Int. Ed.* **2015,** *54,* 3712−3716.
25. Cantillo, D.; de Frutos, O.; Rincón, J. A.; Mateos, C.; Kappe, C. O. Continuous Flow α-trifluoromethylation of Ketones by Metal Free Visible Light Photoredox Catalysis. *Org. Lett.* **2014,** *16,* 896−899.
26. Jiang, H.; Cheng, Y.; Zhang, Y.; Yu, S. Sulfonation and Trifluoromethylation of Enol Acetates with Sulfonyl Chlorides Using Visible-light Photoredox Catalysis. *Eur. J. Org. Chem.* **2013,** 5485−5492.
27. Thomoson, C. S.; Dolbier, W. R., Jr. Carbomethoxydifluoromethylation of Enol Acetates with Methyl (Chlorosulfonyl)difluoroacetate Using Visible-light Photoredox Catalysis. Synthesis of 2,2-difluoro-γ-ketoesters. *J. Fluor. Chem.* **2015,** *178,* 327−331.
28. Jiang, H.; Huang, C.; Guo, J.; Zeng, C.; Zhang, Y.; Yu, S. Direct C−H Functionalization of Enamides and Enecarbamates by Using Visible-light Photoredox Catalysis. *Chem. Eur. J.* **2012,** *18,* 15158−15166.
29. Nagib, D. A.; Macmillan, D. W. C. Trifluoromethylation of Arenes and Heteroarenes by Means of Photoredox Catalysis. *Nature* **2011,** *480,* 224−228.

30. Iqbal, N.; Choi, S.; Ko, E.; Cho, E. J. Trifluoromethylation of Heterocycles via Visible Light Photoredox Catalysis. *Tetrahedron Lett.* **2012,** *53,* 2005–2008.
31. Choi, W. J.; Choi, S.; Ohkubo, K.; Fukuzumi, S.; Cho, E. J.; You, Y. Mechanisms and Applications of Cyclometalated Pt(II) Complexes in Photoredox Catalytic Trifluoromethylation. *Chem. Sci.* **2015,** *6,* 1454–1464.
32. a. Xie, J.; Yuan, X.; Abdukader, A.; Zhu, C.; Ma, J. Visible-light-promoted Radical C–H Trifluoromethylation of Free Anilines. *Org. Lett.* **2014,** *16,* 1768–1771.
b. Zhu, C.; Xie, J.; Xu, P. *Process for Preparation of* O-*(trifluoromethyl)aniline and Derivatives*. Chinese Patent CN 2013/10528325, February 5, 2014.
33. Cui, L.; Matusaki, Y.; Tada, N.; Miura, T.; Uno, B. Itoh a. Metal-free Direct C-H Perfluoroalkylation of Arenes and Heteroarenes Using a Photoredox Organocatalyst. *Adv. Synthesis Catal.* **2013,** *355,* 2203–2207.
34. Pitre, S. P.; McTiernan, C. D.; Ismaili, H.; Scaiano, J. C. Metal-free Photocatalytic Radical Trifluoromethylation Utilizing Methylene Blue and Visible Light Irradiation. *ACS Catal.* **2014,** *4,* 2530–2535.
35. Straathof, N. J. W.; Gemoets, H. P. L.; Wang, X.; Schouten, J. C.; Hessel, V.; Noël, T. Rapid Trifluoromethylation and Perfluoroalkylation of Five-membered Heterocycles by Photoredox Catalysis in Continuous Flow. *ChemSusChem* **2014,** *7,* 1612–1617.
36. Beatty, J. W.; Douglas, J. J.; Cole, K. P.; Stephenson, C. R. J. A Scalable and Operationally Simple Radical Trifluoromethylation. *Nat. Comm.* **2015,** *6,* 7919. http://dx.doi.org/10.1038/ncomms8919.
37. Jung, J.; Kim, E.; You, Y.; Cho, E. J. Visible Light-induced Aromatic Difluoroalkylation. *Adv. Synthesis Catal.* **2014,** *356,* 2741–2748.
38. Lin, Q.; Chu, L.; Qing, F. -L. Direct Introduction of Ethoxycarbonyldifluoromethyl-group to Heteroarenes with Ethyl Bromodifluoroacetate via Visible-light Photocatalysis. *Chin. J. Chem.* **2013,** *31,* 885–891.
39. Wang, L.; Wei, X. -J.; Jia, W. -L.; Zhong, J. -J.; Wu, L. -Z.; Liu, Q. Visible-light-driven Difluoroacetamidation of Unactive Arenes and Heteroarenes by Direct C–H Functionalization at Room Temperature. *Org. Lett.* **2014,** *16,* 5842–5845.
40. Su, Y. -M.; Hou, Y.; Yin, F.; Xu, Y. -M.; Li, Y.; Zheng, X.; Wang, X. -S. Visible Light-mediated C–H Difluoromethylation of Electron-rich Heteroarenes. *Org. Lett.* **2014,** *16,* 2958–2961.
41. Baar, M.; Blechert, S. Graphitic Carbon Nitride Polymer as a Recyclable Photoredox Catalyst for Fluoroalkylation of Arenes. *Chem. Eur. J.* **2015,** *21,* 526–530.
42. Ye, Y.; Sanford, M. S. Merging Visible-light Photocatalysis and Transition-metal Catalysis in the Copper-catalyzed Trifluoromethylation of Boronic Acids with CF_3I. *J. Am. Chem. Soc.* **2012,** *134,* 9034–9037.
43. Iqbal, N.; Choi, S.; Kim, E.; Cho, E. J. Trifluoromethylation of Alkenes by Visible Light Photoredox Catalysis. *J. Org. Chem.* **2012,** *77,* 11383–11387.
44. Yu, C.; Iqbal, N.; Park, S.; Cho, E. J. Selective Difluoroalkylation of Alkenes by Using Visible Light Photoredox Catalysis. *Chem. Comm.* **2014,** *50,* 12884–12887.
45. Tomita, R.; Yasu, Y.; Koike, T.; Akita, M. Direct C–H trifluoromethylation of di- and trisubstituted alkenes by photoredox catalysis. *Beilstein J. Org. Chem.* **2014,** *10,* 1099–1106.
46. Kreis, L. M.; Krautwald, S.; Pfeiffer, N.; Martin, R. E.; Carreira, E. M. Photocatalytic Synthesis of Allylic Trifluoromethyl Substituted Styrene Derivatives in Batch and Flow. *Org. Lett.* **2013,** *15,* 1634–1637.
47. Lin, Q. -Y.; Xu, X. -H.; Qing, F. -L. Chemo-, Regio-, and Stereoselective Trifluoromethylation of Styrenes via Visible Light-driven Single-electron Transfer (SET) and Triplet–triplet Energy Transfer (TTET) Processes. *J. Org. Chem.* **2014,** *79,* 10434–10446.

48. Asano, M.; Tomita, R.; Koike, T.; Akita, M. Trifluoromethylation of Aromatic Alkenes by Visible-light-driven Photoredox Catalysis: Direct Conversion of Alkenes to 3-trifluoromethyl-1-propenyl and 1,3-bis(trifluoromethyl)- 1-propenyl Derivatives. *J. Fluor. Chem.* **2015,** *179,* 83−88.
49. Mizuta, S.; Engle, K. M.; Verhoog, S.; Galicia-Lopez, O.; O'Duill, M.; Medebielle, M.; Wheelhouse, K.; Rassias, G.; Thompson, A.; Gouverneur, V. Trifluoromethylation of Allylsilanes under Photoredox Catalysis. *Org. Lett.* **2013,** *15,* 1250−1253.
50. Iqbal, N.; Jung, J.; Park, S.; Cho, E. J. Controlled Trifluoromethylation Reactions of Alkynes through Visible-light Photoredox Catalysis. *Angew.Chem. Int. Ed.* **2014,** *53,* 539−542.
51. Yasu, Y.; Koike, T.; Akita, M. Visible-light-induced Synthesis of a Variety of Trifluoromethylated Alkenes from Potassium Vinyltrifluoroborates by Photoredox Catalysis. *Chem. Comm.* **2013,** *49,* 2037.
52. Yasu, Y.; Koike, T.; Akita, M. Three-component Oxytrifluoromethylation of Alkenes: Highly Efficient and Regioselective Difunctionalization of C=C Bonds Mediated by Photoredox Catalysts. *Angew. Chem. Int. Ed.* **2012,** *51,* 9567−9571.
53. Carboni, A.; Dagousset, G.; Magnier, E.; Masson, G. Photoredox-induced Three-component Oxy-, Amino-, and Carbotrifluoromethylation of Enecarbamates. *Org. Lett.* **2014,** *16,* 1240−1243.
54. Tomita, R.; Yasu, Y.; Koike, T.; Akita, M. Combining Photoredox-catalyzed Trifluoromethylation and Oxidation with DMSO: Facile Synthesis of α-trifluoromethylated Ketones from Aromatic Alkenes. *Angew. Chem. Int. Ed.* **2014,** *53,* 7144−7148.
55. Li, L.; Chen, Q. -Y.; Guo, Y. Synthesis of α-CF_3 Ketones from Alkenes and Electrophilic Trifluoromethylating Reagents by Visible-light Driven Photoredox Catalysis. *J. Fluor. Chem.* **2014,** *167,* 79−83.
56. Li, L.; Huang, M.; Liu, C.; Xiao, J. -C.; Chen, Q. -Y.; Guo, Y.; Zhao, Z. -G. 2,2,2-Trifluoroethylation of Styrenes with Concomitant Introduction of a Hydroxyl Group from Molecular Oxygen by Photoredox Catalysis Activated by Visible Light. *Org. Lett.* **2015,** *17,* 4714−4717.
57. Tomita, R.; Koike, T.; Akita, M. Photoredox-catalyzed Stereoselective Conversion of Alkynes into Tetrasubstituted Trifluoromethylated Alkenes. *Angew. Chem. Int. Ed.* **2015,** *54,* 12923−12927.
58. Kim, E.; Choi, S.; Kim, H.; Cho, E. J. Generation of CF_3-containing Epoxides and Aziridines by Visible-light-driven Trifluoromethylation of Allylic Alcohols and Amines. *Chem. Eur. J.* **2013,** *19,* 6209−6212.
59. Yasu, Y.; Arai, Y.; Tomita, R.; Koike, T.; Akita, M. Highly Regio- and Diastereoselective Synthesis of CF_3-substituted Lactones via Photoredox-catalyzed Carbolactonization of Alkenoic Acids. *Org. Lett.* **2014,** *16,* 780−783.
60. Deng, Q. -H.; Chen, J. -R.; Wei, Q.; Zhao, Q. -Q.; Lu, L. -Q.; Xiao, W. -J. Visible-light-induced Photocatalytic Oxytrifluoromethylation of N-allylamides for the Synthesis of CF_3-containing Oxazolines and Benzoxazines. *Chem. Comm.* **2015,** *51,* 3537−3540.
61. Noto, N.; Miyazawa, K.; Koike, T.; Akita, M. *Anti*-diastereoselective Synthesis of CF_3-containing Spirooxazolines and Spirooxazines via Regiospecific Trifluoromethylative Spirocyclization by Photoredox Catalysis. *Org. Lett.* **2015,** *17,* 3710−3713.
62. Wei, Q.; Chen, J. -R.; Hu, X. -Q.; Yang, X. -C.; Lu, B.; Xiao, W. -J. Photocatalytic Radical Trifluoromethylation/cyclization Cascade: Synthesis of CF_3-containing Pyrazolines and Isoxazolines. *Org. Lett.* **2015,** *17,* 4464−4467.
63. Yasu, Y.; Koike, T.; Akita, M. Intermolecular Aminotrifluoromethylation of Alkenes by Visible-light-driven Photoredox Catalysis. *Org. Lett.* **2013,** *15,* 2136−2139.

64. Dagousset, G.; Carboni, A.; Magnier, E.; Masson, G. Photoredox-induced Three-component Azido- and Aminotrifluoromethylation of Alkenes. *Org. Lett.* **2014,** *16,* 4340–4343.
65. Zhang, Z.; Tang, X.; Thomoson, C. S.; Dolbier, W. R., Jr. Photoredox-catalyzed Intramolecular Aminodifluoromethylation of Unactivated Alkenes. *Org. Lett.* **2015,** *17,* 3528–3531.
66. Xu, P.; Xie, J.; Xue, Q.; Pan, C.; Cheng, Y.; Zhu, C. Visible-light-induced Trifluoromethylation of N-aryl Acrylamides: a Convenient and Effective Method to Synthesize CF_3-containing Oxindoles Bearing a Quaternary Carbon Center. *Chem. Eur. J.* **2013,** *19,* 14039–14042.
67. Liu, C.; Zhao, W.; Huang, Y.; Wang, H.; Zhang, B. Light-induced BiOBr Nanosheets Accelerated Highly Regioselective Intermolecular Trifluoromethylation/arylation of Alkenes to Synthesize CF_3-containing Aza-heterocycles. *Tetrahedron* **2015,** *71,* 4344–4351.
68. Gao, F.; Yang, C.; Gao, G. -L.; Zheng, L.; Xia, W. Visible-light Induced Trifluoromethylation of N-arylcinnamamides for the Synthesis of CF_3-containing 3,4-disubstituted Dihydroquinolinones and 1-azaspiro[4.5]decanes. *Org. Lett.* **2015,** *17,* 3478–3481.
69. Fu, W.; Zhu, M.; Zou, G.; Xu, C.; Wang, Z. Visible-light-mediated Radical Aryldifluoroacetylation of N-arylacrylamides to Give Difluoroacetylated Oxindoles. *Asian J. Org. Chem.* **2014,** *3,* 1273–1276.
70. Fu, W.; Zhu, M.; Zou, G.; Xu, C.; Wang, Z. Visible-light-mediated Trifluoroethylation of N-arylacrylamides with Trifluoroethyl Iodide: Synthesis of CF3-containing Oxindoles. *Synlett* **2014,** *25,* 2513–2517.
71. Tang, X. J.; Thomoson, C. S.; Dolbier, W. R., Jr. Photoredox-catalyzed Tandem Radical Cyclization of N-arylacrylamides: General Methods to Construct Fluorinated 3,3-disubstituted 2-oxindoles Using Fluoroalkylsulfonyl Chlorides. *Org. Lett.* **2014,** *16,* 4594–4597.
72. Zheng, L.; Yang, C.; Xu, Z.; Gao, F.; Xia, W. Difunctionalization of Alkenes via the Visible-light-induced Trifluoromethylarylation/1,4-aryl Shift/desulfonylation Cascade Reactions. *J. Org. Chem.* **2015,** *80,* 5730–5736.
73. Fu, W.; Zhu, M.; Zou, G.; Xu, C.; Wang, Z.; Ji, B. Visible-light-mediated Radical Aryldifluoroacetylation of Alkynes with Ethyl Bromodifluoroacetate for the Synthesis of 3-difluoroacetylated Coumarins. *J. Org. Chem.* **2015,** *80,* 4766–4770.
74. Qu, C.; Xu, P.; Ma, W.; Cheng, Y.; Zhu, C. A Novel Visible Light Mediated Radical Cyclization of Enol Lactones: a Concise Method for Fluorinated Polycyclic Lactone Scaffolds. *Chem. Comm.* **2015,** *51,* 13508–13510.
75. Carboni, A.; Dagousset, G.; Magnier, E.; Masson, G. One Pot and Selective Intermolecular Aryl- and Heteroaryl-trifluoromethylation of Alkenes by Photoredox Catalysis. *Chem. Comm.* **2014,** *50,* 14197–14200.
76. Sahoo, B.; Li, J. -L.; Glorius, F. Visible-light Photoredox-catalyzed Semipinacol-type Rearrangement: Trifluoromethylation/ring Expansion by a Radical–polar Mechanism. *Angew. Chem. Int. Ed.* **2015,** *54,* 11577–11580.
77. Woo, S. B.; Kim, D. Y. Visible-light-induced Photocatalytic Trifluoromethylation/1,2-carbon Migration Sequences for the Synthesis of CF_3-substituted Cyclic Ketones. *J. Fluor. Chem.* **2015,** *178,* 214–218.
78. Suh, C. W.; Kim, D. Y. Visible-light-mediated Photocatalytic Difluoroalkylation/1,2-carbon Migration Sequences: Synthesis of Difluoroalkyl-substituted Cyclic Ketones. *Tetrahedron Lett.* **2015,** *56,* 5661–5664.
79. Xu, P.; Hu, K.; Gu, Z.; Cheng, Y.; Zhu, C. Visible Light Promoted Carbodifluoroalkylation of Allylic Alcohols via Concomitant 1,2-aryl Migration. *Chem. Comm.* **2015,** *51,* 7222–7225.
80. Sun, X.; Yu, S. Visible-light-mediated Fluoroalkylation of Isocyanides with Ethyl Bromofluoroacetates: Unified Synthesis of Mono- and Difluoromethylated Phenanthridine Derivatives. *Org. Lett.* **2014,** *16,* 2938–2941.

81. Cheng, Y.; Yuan, X.; Jiang, H.; Wang, R.; Ma, J.; Zhang, Y.; Yu, S. Regiospecific Synthesis of 1-trifluoromethylisoquinolines Enabled by Photoredox Somophilic Vinyl Isocyanide Insertion. *Adv. Synthesis Catal.* **2014,** *356,* 2859−2866.
82. Wang, R.; Jiang, H.; Cheng, Y.; Kadi, A. A.; Fun, H. -K.; Zhang, Y.; Yu, S. Somophilic Isocyanide Insertion: Synthesis of 6-arylated and 6-trifluoromethylated Phenanthridines. *Synthesis* **2014,** *46,* 2711−2726.
83. Fu, W.; Zhu, M.; Xu, C.; Zou, G.; Wang, Z.; Ji, B. Visible-light-mediated Trifluoroethylation of 2-isocyanobiaryl with Trifluoroethyl Iodide: Synthesis of 6-trifluoroethyl-phenanthridines. *J. Fluor. Chem.* **2014,** *168,* 50−54.
84. Zhu, M.; Fu, W.; Zou, G.; Xu, C.; Wang, Z. Visible-light-mediated Radical Difluoromethylenephosphonation of 2-isocyanobiaryls with Bromodifluoromethylphosphonate for the Synthesis of 6-difluoromethylenephosphonyl-phenanthridines. *J. Fluor. Chem.* **2015,** *180,* 1−6.
85. Wang, S.; Jia, W. -L.; Wang, L.; Liu, Q. Preparation of 6-difluoromethylphosphonated Phenanthridines by Visible-light-driven Radical Cyclization of 2-isocyanobiphenyls. *Eur. J. Org. Chem.* **2015,** 6817−6821.
86. Zhang, Z.; Tang, X.; Dolbier, W. R., Jr. Photoredox-catalyzed Tandem Insertion/cyclization Reactions of Difluoromethyl and 1,1-difluoroalkyl Radicals with Biphenyl Isocyanides. *Org. Lett.* **2015,** *17,* 4401−4403.
87. Nguyen, J. D.; Tucker, J. W.; Konieczynska, M. D.; Stephenson, C. R. J. Intermolecular Atom Transfer Radical Addition to Olefins Mediated by Oxidative Quenching of Photoredox Catalysts. *J. Am. Chem. Soc.* **2011,** *133,* 4160−4163.
88. Wallentin, C. -J.; Nguyen, J. D.; Finkbeiner, P.; Stephenson, C. R. J. Visible Light-mediated Atom Transfer Radical Addition via Oxidative and Reductive Quenching of Photocatalysts. *J. Am. Chem. Soc.* **2012,** *134,* 8875−8884.
89. Oh, S. H.; Malpani, Y. R.; Ha, N.; Jung, Y. -S.; Han, S. B. Vicinal Difunctionalization of Alkenes: Chlorotrifluoromethylation with CF_3SO_2Cl by Photoredox Catalysis. *Org. Lett.* **2014,** *16,* 1310−1313.
90. Bagal, D. B.; Kachkovskyi, G.; Knorn, M.; Rawner, T.; Bhanage, B. M.; Reiser, O. Trifluoromethylchlorosulfonylation of Alkenes: Evidence for an Inner-sphere Mechanism by a Copper Phenanthroline Photoredox Catalyst. *Angew. Chem. Int. Ed.* **2015,** *54,* 6999−7002.
91. Tang, X. -J.; Dolbier, W. R., Jr. Efficient Cu-catalyzed Atom Transfer Radical Addition Reactions of Fluoroalkylsulfonyl Chlorides with Electron-deficient Alkenes Induced by Visible Light. *Angew. Chem. Int. Ed.* **2015,** *54,* 4246−4249.
92. Carboni, C.; Dagousset, G.; Magnier, E.; Masson, G. Three-component Photoredox-mediated Chloro-, Bromo-, or Iodotrifluoromethylation of Alkenes. *Synthesis* **2015,** *47,* 2439−2445.
93. Mizuta, S.; Verhoog, S.; Engle, K. M.; Khotavivattana, T.; O'Duill, M.; Wheelhouse, K.; Rassias, G.; Médebielle, M.; Gouverneur, V. Catalytic Hydrotrifluoromethylation of Unactivated Alkenes. *J. Am. Chem. Soc.* **2013,** *135,* 2505−2508.
94. Lin, Q.-Y.; Xu, X. -H.; Qing, F.-L. Visible Light-induced Selective Hydrobromodifluoromethylation of Alkenes with Dibromodifluoromethane. *Org. Biomol. Chem.* **2015,** *13,* 8740−8749.
95. Park, S.; Joo, J. M.; Cho, E. J. Synthesis of β-trifluoromethylated Ketones from Propargylic Alcohols by Visible Light Photoredox Catalysis. *Eur. J. Org. Chem.* **2015,** 4093−4097.
96. Wilger, D. J.; Gesmundo, N. J.; Nicewicz, D. A. Catalytic Hydrotrifluoromethylation of Styrenes and Unactivated Aliphatic Alkenes via an Organic Photoredox System. *Chem. Sci.* **2013,** *4,* 3160−3165.

Synthesis of Side Chain Fluorinated Amino Acids and Their Effects on the Properties of Peptides and Proteins

15

S. Vukelić, J. Moschner, S. Huhmann, R. Fernandes, A.A. Berger, B. Koksch
Freie Universität, Berlin, Germany

Chapter Outline

1. Introduction

Fluorine's unique stereoelectronic properties arises from an unprecedented combination of small size, very low polarizability, and the strongest inductive effect found among the chemical elements. Although the general effect of fluorine substitution is context dependent, its presence within a molecule often favorably and, in many cases predictably, changes the biophysical and chemical properties such as hydrophobicity, acidity/basicity, as well as the reactivity and conformation of a given target.[1–3] Moreover, perfluorinated molecules exhibit a tendency to segregate from both aqueous and

Modern Synthesis Processes and Reactivity of Fluorinated Compounds. http://dx.doi.org/10.1016/B978-0-12-803740-9.00015-9

hydrophobic phases, affording a third "fluorous" phase that has found broad applicability in the fields of catalysis and separation processes.[4] The aforementioned properties make fluorine substitution an interesting strategy for protein modification as well as de novo design of artificial peptides.[3,5]

Fluorinated variants of hydrophobic amino acids are promising in modifying the stability and folding of proteins whose interactions are mostly a consequence of complementary hydrophobic side chain packing.[6–8] They have also been shown to have an effect on the proteolytic stability of the peptides.[9] Here, the polarity of fluoroalkyl groups seems to play an important role in the interaction of the fluorine-carrying substrate with the enzyme. To determine the effect of fluorine on the interaction of polypeptides with native proteins, model systems with precisely defined interaction patterns have been established by our group. It is crucial that these model systems be stable, structurally homogeneous and easy to synthesize. Also, small model systems allow for easier interpretation of the experimental data. In this context, α-helical coiled-coil peptides have become important tools in recent years, representing a class of small ubiquitous biologically active molecules.[10] A second class of intensively investigated systems using these model substrates are β-sheets, which generally play a crucial role in folding as well as in the formation of amyloids.

This chapter will summarize the synthetic approaches toward fluorinated analogs of aminobutyric acid (Abu), which have enabled the rational design of fluorinated peptides. We will then discuss the properties of these peptides and what the crucial factors are for the formation of secondary structures in proteins (size, topology, hydrophobicity, secondary structure propensity, and electronic characteristics of the amino acid side chains). Finally, the impact of side chain fluorination on the amyloid formation and protein–protein interactions within enzyme active sites will be discussed.

2. Synthesis of Fluorinated Aliphatic α-Amino Acids

Numerous fluorinated amino acids have been described in the literature.[11,12] Aliphatic fluorinated amino acids are variants of natural α-L-amino acids where the side chain carries fluorine substituents (CF_x). In the course of our studies, we have incorporated side chain fluorinated amino acids into peptides to systematically characterize the impact of fluorination on their properties such as steric demand, hydrophobicity, lipophilicity, polarity, helix propensity, degradation profile, and proteolytic stability. In this context, we have especially focused on the fluorinated derivatives of Abu, also named ethylglycine, 4-monofluoroethylglycine (MfeGly), 4,4-difluoroethylglycine (DfeGly), 4,4,4-trifluoroethylglycine (TfeGly), and 4,4-difluoropropylglycine (DfpGly). We have extended our studies and included sterically more demanding and higher fluorinated amino acids such as 4,4,4-trifluorovaline (4^3-F_3Val), 4′,4′,4′-trifluoroisoleucine ($4'^3$-F_3Ile), 5,5,5-trifluoroisoleucine (5^3-F_3Ile), 5,5,5-trifluoroleucine (5^3-F_3Leu), and hexafluoroleucine (HfLeu). These amino acids have been introduced into peptides always in their enantiomerically pure form. The challenge is that each derivative poses a number of synthetic obstacles. Herein we will discuss the synthetic routes toward these compound, subdivided by the fluorination pattern.

2.1 Synthesis of Fluorinated Derivatives of Aminobutyric Acid—MfeGly, DfeGly, TfeGly, and DfpGly

The synthesis of MfeGly (**1**) applied in our laboratory, is established by Laue et al.'s group.[13] The entry uses the enantiopure auxiliary (1*R*,2*R*,5*R*)-(+)-2-hydroxy-3-pinanone, which is easily obtained from the inexpensive (−)-α-pinene.[14] The auxiliary is coupled to glycine *tert*-butyl ester to obtain the Schiff base **2**, which is in turn deprotonated and stereoselectively alkylated by the use of 1-bromo-2-fluoroethane (**3**) to yield the precursor **4**. Subsequent deprotection of the *tert*-butyl ester with citric acid and acidic hydrolysis of the imine of **4** gives pure (*S*)-MfeGly **1** (Scheme 15.1).

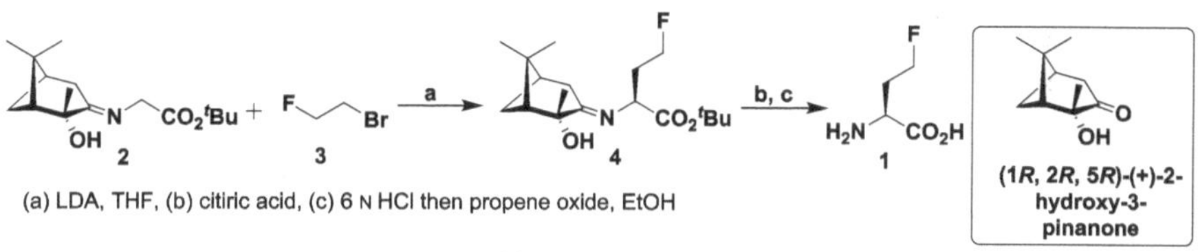

Scheme 15.1 Synthesis of MfeGly (**1**).[13]

DfeGly **5** is synthesized according to the procedure published by the Burger group.[15] Aspartic acid **6** is reacted with hexafluoroacetone to form the 2,2-bis(trifluoromethyl)-1,3-oxazilidin-5-one **7**, by simultaneously protecting the carboxylic group and the α-amino group, whereas the carboxylic group in the side chain remains unaffected. The latter is converted to the corresponding acid chloride by the use of thionyl chloride and hydrogenated in the presence of Rosemund catalyst to give the β-semialdehyde **8**. Fluorination of the aldehyde **8** is achieved with diethylaminosulfur trifluoride (DAST) under mild conditions. Hydrolysis of the resulting side chain−fluorinated 1,3-oxazolidin-5-one in a water/*iso*-propanol mixture gives the desired (*S*)- DfeGly (**5**, Scheme 15.2).

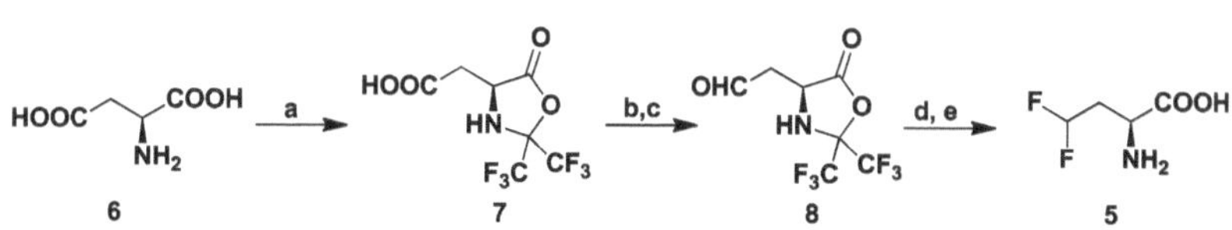

(a) Hexafluoroacetone, (b) $SOCl_2$, (c) H_2, Pd/$BaSO_4$, (d) DAST, (e) H_2O/*i*-PrOH

Scheme 15.2 Synthesis of DfeGly **5** according to Winkler and Burger.[15]

A convenient synthetic entry for TfeGly (**9**) is the procedure established by Schedel et al. that follows the synthetic strategy of DfeGly **5** described earlier.[16] Again, hexafluoroacetone is used as a protecting reagent to convert (−)-malic acid **10** into the

2,2-bis(trifluoromethyl)-1,3-dioxolan-4-one **11**. The key step of the synthesis is the transformation of the ω-carboxyl group of **11** to the trifluoromethyl group upon treatment with gaseous sulfur tetrafluoride in an autoclave to get **12**. After hydrolysis of **12** with methanol the corresponding α-hydroxymethyl ester **13** is generated and subjected to triflic anhydride to set stage for the nucleophilic attack of benzylamine to introduce the amine and provide **14** in a stereoselective manner. Acidic hydrolysis of the methyl ester, hydrogenation, and stirring in ethanol with propene oxide lead to the generation of enantiopure (*R*)- TfeGly (**9**, Scheme 15.3).

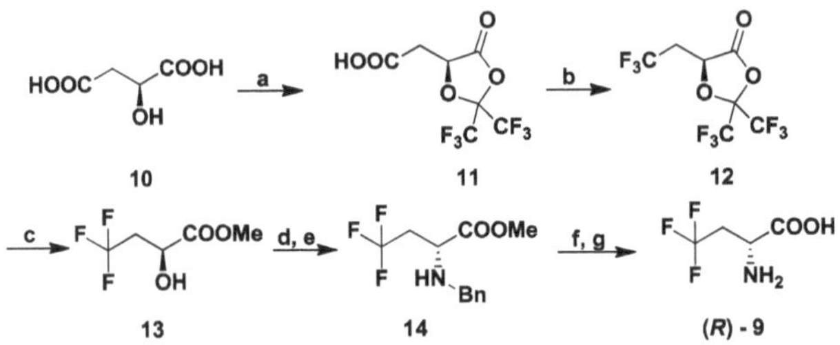

(a) Hexafluoroacetone, (b) SF_4, (c) MeOH, (d) Tf_2O, pyridine, (e) H_2NBn, (f) 6 N HCl, (g) 1. Pd/C, H_2, 2. EtOH, propene oxide

Scheme 15.3 Synthesis of (*R*)-TfeGly **9** according to Schedel et al.[16]

A semisynthetic entry toward (*S*)-TfeGly **9** involves the method presented by Tsushima et al.[17] The carbanion of diethyl acetamidomalonate (**15**) reacts with trifluoroethyl triflate to afford the desired diethyl-*N*-acetylamino-(2,2,2-trifluoroethyl)malonate **16**. Acidic hydrolysis of the ethyl esters and simultaneous decarboxylation of **16** gives (DL)-2-amino-4,4,4-trifluorobutanoic acid (**17**), which is in turn *N*-acylated and subjected to enzymatic kinetic resolution using *acylase* I to give (*S*)-TfeGly **9** (Scheme 15.4).

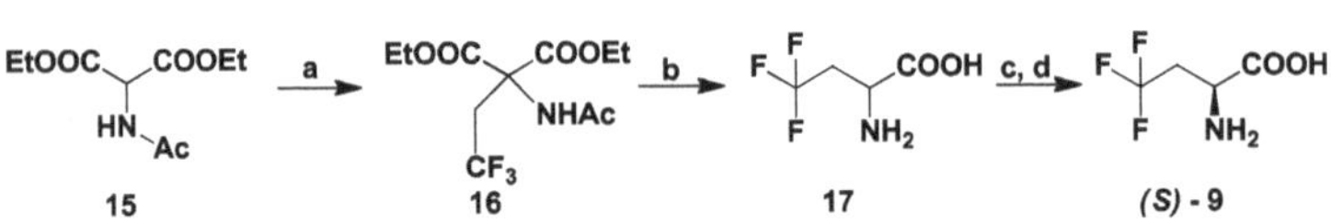

(a) ^{t}BuOK, $CF_3SO_3CH_2CF_3$, (b) HCl conc., (c) Ac_2O, NaOH aq., (d) *Acylase* I/AcOLi, pH 7

Scheme 15.4 Synthetic entry toward (*S*)-TfeGly **9** according to Tsushima et al.[17]

Two publications report the diastereoselective synthesis of TfeGly **9**. First, Soloshonok et al. have shown the deracemization of (DL)-2-amino-4,4,4-trifluorobutanoic acid (**17**) using a chiral nickel(II) complex.[18] Treatment of **17** with (*S*)-**18** in the presence of nickel(II)acetate results in the formation of the chiral complex **19**, which upon acidic workup and ion exchange chromatography gives (*R*)-TfeGly **9** and the recycled phenylmethylamine-derived ligand (*S*)-**18**. By using (*R*)-**18** and applying the same methodology to **17** the enantiomeric (*S*)-form of TfeGly **9** is obtained (Scheme 15.5).

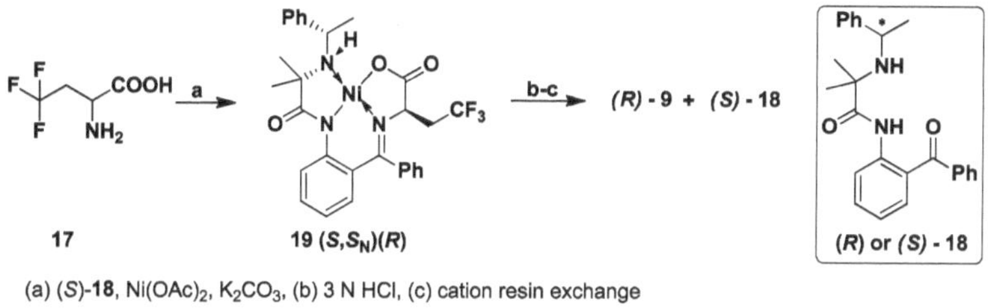

Scheme 15.5 Synthesis of both enantiomers of TfeGly **9** using a chiral ligand and a nickel complex according to Sorochinsky et al.[18]

Wang et al. have established a novel copper-catalyzed intermolecular trifluoromethylazidation of alkenes.[19] They access a variety of CF_3-containing organoazides from a wide range of olefins, including activated and unactivated alkenes. In this context the acrylamide **20**, containing the chiral (*S*)-camphorsultam (NR*), is subjected to standard reaction conditions using Togni II, as the trifluoromethyl cation donor, and tetrakis(acetonitrile)copper(I) hexafluorophosphate to generate the reactive intermediate, which readily reacts with trimethylsilyl azide to yield **21**. Formal basic cleavage of the chiral auxiliary and reduction of the azide would lead to (*S*)-TfeGly **9** (Scheme 15.6).

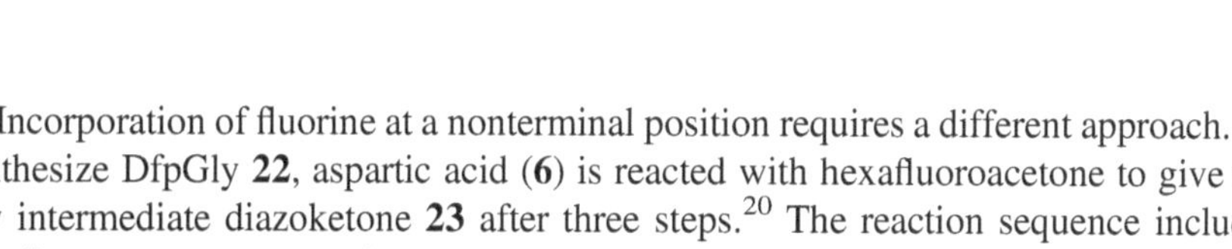

Scheme 15.6 Synthesis of TfeGly **9** according to Wang et al.[19]

Incorporation of fluorine at a nonterminal position requires a different approach. To synthesize DfpGly **22**, aspartic acid (**6**) is reacted with hexafluoroacetone to give the key intermediate diazoketone **23** after three steps.[20] The reaction sequence includes hexafluoroacetone protection/activation, conversion into the corresponding acid chloride by reaction with thionyl chloride, and finally, treatment with diazomethane to give compound **23**. Subjecting diazoketone **23** to concentrated HBr, bromoketone **24** is formed within minutes and subsequently converted into 5-bromo-4,4-difluoropentanoic acid derivative **25** by fluorodeoxygenation with DAST. Reductive debromination by using tributyltin hydride and azobisisobutyronitril gives **26**. Simultaneous deprotection of the carboxyl and the adjacent amino groups is achieved by treating compound **26** with a mixture of water and *iso*-propanol to give the desired DfpGly **22** (Scheme 15.7).

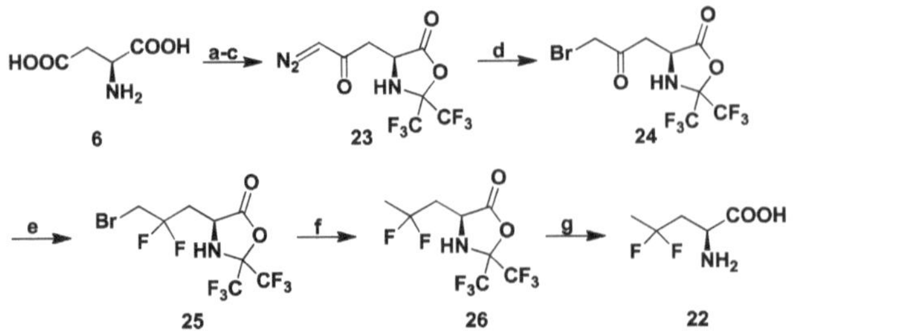

(a) Hexafluoroacetone, (b) $SOCl_2$, (c) CH_2N_2, (d) HBr, (e) DAST, (f) Bu_3SnH/AIBN, (g) H_2O/*i*-PrOH

Scheme 15.7 Synthesis of DfpGly **22** according to El-Kousy et al.[20]

Another method of synthesis of DfpGly **22** is published by Larrson et al. where chirality is introduced applying the Evans method.[21] The approach uses the carboxylic acid **27**, which is easily prepared by treatment of the γ-ketoester **28** with sulfur tetrafluoride to give **29** and saponification of the ethyl ester to yield 4,4-difluoropentanoic acid (**27**, Scheme 15.8).

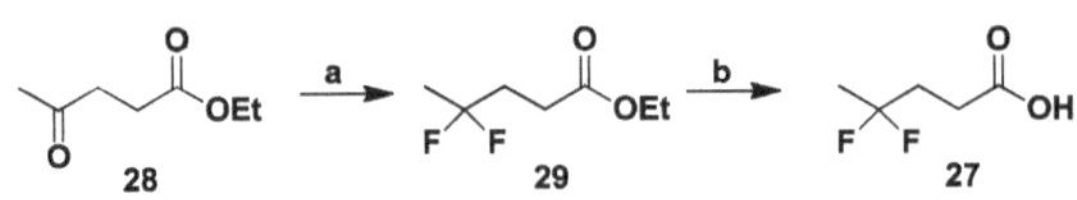

(a) SF_4/HF, (b) NaOH

Scheme 15.8 Synthesis of fluorinated carboxylic acid **29** according to Larrson et al.[21]

The in situ−generated mixed anhydride of **27** and pivaloyl chloride is reacted with a freshly prepared solution of (*R*)-4-benzyloxazolidin-2-one (**30**, Evans auxiliary) and *n*-butyllithium to provide **31**. Next, a stereoselective bromination is performed by generating the enolate of **31** using dibutylboron triflate, which in turn attacks an electrophilic bromine species (*N*-bromosuccinimide) to afford **32**. The latter is reacted with *N*,*N*,*N*′,*N*′-tetramethylguanidinium azide to yield the azide **33**. With the azide **33** in hand, the oxazolidin-2-one moiety is cleaved with lithium hydroperoxide and the azide moiety is hydrogenated on palladium to give DfeGly (**22**, Scheme 15.9).

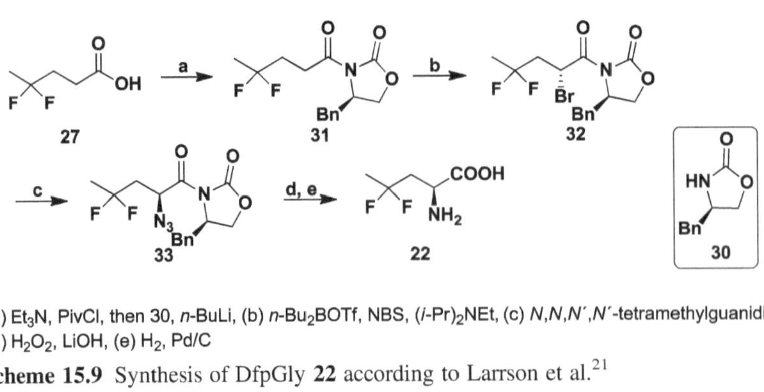

(a) Et_3N, PivCl, then 30, *n*-BuLi, (b) *n*-Bu_2BOTf, NBS, $(i\text{-}Pr)_2NEt$, (c) *N*,*N*,*N*′,*N*′-tetramethylguanidinium azide, (d) H_2O_2, LiOH, (e) H_2, Pd/C

Scheme 15.9 Synthesis of DfpGly **22** according to Larrson et al.[21]

2.2 Synthesis of Fluorinated Derivatives of Leucine, Isoleucine, and Valine

The first method to obtain all four optically pure diastereomers of 4^3-F_3Val (**34a–34d**) was disclosed in 2002 by Xing et al. using a derivatization/enzymatic resolution approach.[22] After derivatization of the *N*-Boc-protected trifluorovalines **35a–35d** via esterification with methyl iodide and reduction of the so-obtained methyl ester with sodium borohydride, separation of the generated *N*-Boc trifluorovalinols **36a–36d** into the two enantiomeric pairs is possible by column chromatography. Oxidation of the hydroxyl groups of **36a/36b** and **36c/36d** and removal of the *Boc*-protecting group followed by acylation of the free amino group affords the *N*-acetyl amino acids **37a/37b** and **37c/37d**, respectively. Subsequent enzymatic deacylation of the *N*-acetyl enantiomeric pairs of amino acids (**37a/37b** and **37c/37d**) using *porcine kidney acylase* I gives the (2*S*)-configured products (**34a** and **34c**), whereas the (2*R*)-configured 4^3-F_3Val (**34b** and **34d**) are obtained in an additional step using hydrochloric acid (Scheme 15.10).

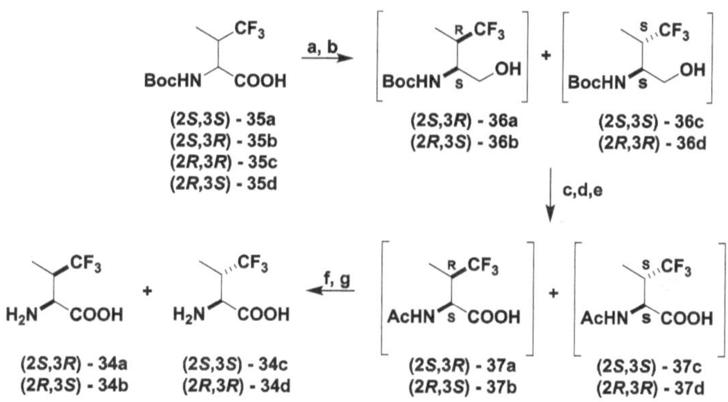

(a) $NaHCO_3$, CH_3I, (b) $NaBH_4$, (c) pyridinium dichromate, (d) 30% CF_3COOH, (e) NaOH/H_2O, Ac_2O, (f) *Porcine kidney acylase* I, pH=7.5, (g) 3 N HCl only for the (2*R*)-diastereomers **34b** and **34d**

Scheme 15.10 Resolution of all four diastereomers of 4^3-F_3Val **34a–34b** according to Xing et al.[22]

Additionally, the very same methodology is used to prepare all four diastereomers of 5^3-F_3Leu (**38a–38d**) by transformation of the *N*-Boc-protected trifluoroleucines (**39a–39d**) to the corresponding leucinols (**40a–40d**) and subsequent oxidation to generate the pairs **41a/41b** and **41c/41d**. Enzymatic resolution with *porcine kidney acylase* I gives the (2*S*)-configured products (**38b** and **38c**), with the (2*R*)-configured 5^3-F_3Leu (**38a** and **38d**) being generated in an additional step using hydrochloric acid (Scheme 15.11).

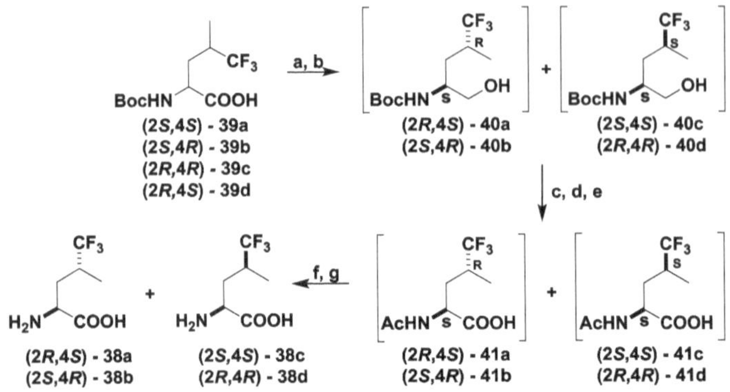

(a) $NaHCO_3$, CH_3I, (b) $NaBH_4$, (c) pyridinium dichromate, (d) 30% CF_3COOH, (e) $NaOH/H_2O$, Ac_2O, (f) *Porcine kidney acylase* I, pH=7.5, (g) 3 N HCl only for the (2*R*)-diastereomers **38a** and **38d**

Scheme 15.11 Resolution of all four diastereomers of 5^3-F_3Leu **38a**–**38d** according to Xing et al.[22]

A well-established method for stereoselective introduction of trifluoromethyl group uses Evans oxazolidin-2-ones.[23] Development of a reductive trifluoromethylation procedure of internal electron–deficient alkenes like **42** allows access to optically active β-trifluoromethylated oxazolidin-2-ones like **43**. The alkenes **43** themselves are accessible starting from (*S*)-4-benzyloxazolidin-2-one and coupling with crotonic acid and (*E*)-2-pentenoic acid (Scheme 15.12).[24]

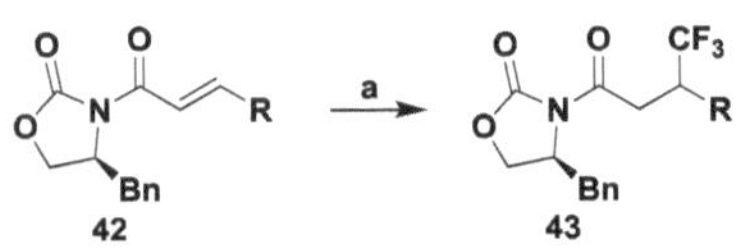

R = Me, Et
(a) CF_3I, Et_3B/O_2, $Yb(OTf)_3{\cdot}xH_2O$

Scheme 15.12 Reductive trifluoromethylation of electron-deficient alkenes according to Erdbrink et al.[24]

All four isomers of *N*-Boc-protected trifluorovaline **35a**–**35d** are synthesized employing standard conditions via the corresponding α-azides **44a**–**44d**, respectively. The *N*-acyloxazolidinones **45a** and **45b** are transformed into the α-azido derivative **44c** and **44d**, which upon reduction and auxiliary removal gives the *N*-Boc-4^3-F_3Val **35c** and **35d**. *N*-Boc-4^3-F_3-Val **35a** and **35b** can be prepared by α-bromination of **45a** and **45b** to give the bromines **46a** and **46b** followed by nucleophilic displacement of the bromine by an azide resulting in stereochemical inversion to give **44a** and **44b**. Reduction and auxiliary removal gives *N*-Boc-4^3-F_3-Val **35a** and **35b**. The described synthesis of *N*-Boc-4^3-F_3Val (**35a**–**35d**) also allowed the access to *N*-Boc-4-trifluoroisoleucine (4^3-F_3Ile) **47a** by starting with the oxazolidin-2-one **48** (Scheme 15.13).

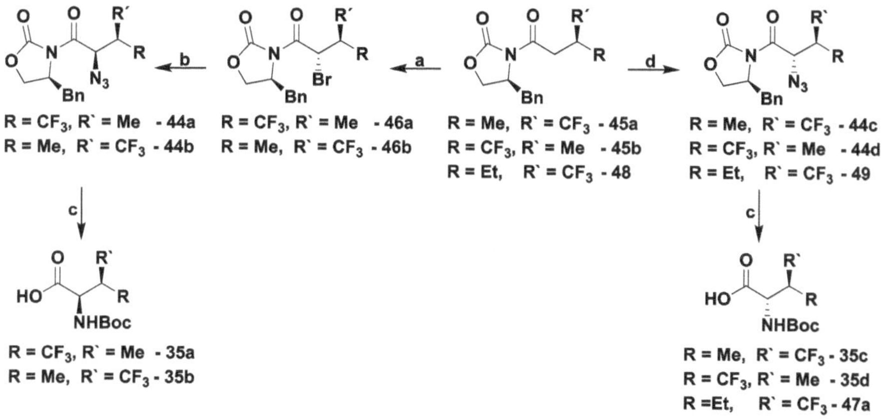

(a) 1. *i*-Pr_2NEt, Bu_2BOTf, 2. NBS, (b) *N,N,N′,N′*-tetramethylguanidinium azide, (c) 1. H_2, Pd/C, Boc_2O, 2. H_2O_2, LiOH, (d) 1. KHDMS, Trisyl-N_3 then HOAc

Scheme 15.13 Synthesis of *N*-Boc-4^3-F_3Val (**35a–35d**) and *N*-Boc-4^3-F_3Ile (**47a**) according to Erdbrink et al.[24]

Using the same approach, Erdbrink et al. report the synthesis of *N*-Boc–protected (2*S*,3*S*)-5,5,5-trifluoroisoleucine 5^3-F_3Ile **49a** and (2*R*,3*S*)-5,5,5-trifluoro-*allo*-isoleucine (Ile), 5^3-F_3-*allo*-Ile **49b**. The chiral acid **50** is coupled to (*S*)-4-benzyloxazolidin-2-one via the mixed anhydride to give **51**.[24] *N*-acyloxazolidinone **51** is transformed into the α-azido derivative **52**, which upon reduction and auxiliary removal gives the *N*-Boc-5^3-F_3Ile **49a** (*N*-Boc-5^3-F_3Ile). *N*-Boc-5^3-F_3-*allo*-Ile **49b** can be prepared by α-bromination of **51** to give the bromine **53** followed by nucleophilic displacement of the bromine by an azide resulting in a stereochemical inversion to give **54**. Reduction and auxiliary removal gives *N*-Boc-5^3-F_3-*allo*-Ile **49b** (Scheme 15.14).

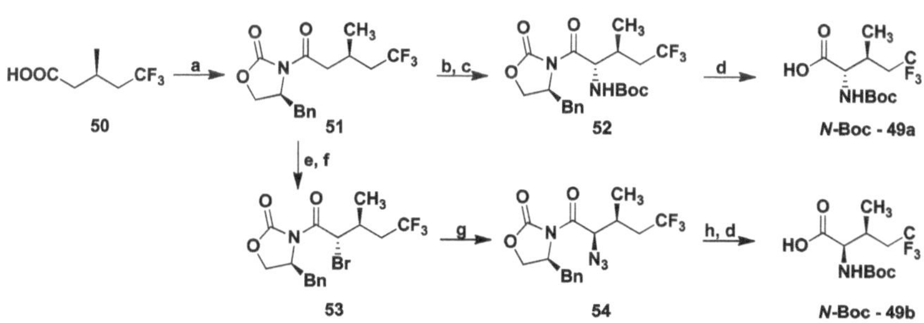

(a) NEt_3, PivCl then *n*-BuLi, (*S*)-4-benzyloxazolidin-2-one, (b) KHDMS, Trisyl-N_3 then HOAc, (c) H_2, Pd/C, Boc_2O, (d) H_2O_2, LiOH, (e) *i*-Pr_2NEt, Bu_2BOTf, (f) NBS, (g) *N,N,N′,N′*-tetramethylguanidinium azide, (h) H_2, Pd/C, Boc_2O

Scheme 15.14 Synthesis of *N*-Boc-5^3-F_3Ile **49a** and *N*-Boc-5^3-F_3-*allo*-Ile **49b** according to Erdbrink et al.[24]

Chen et al. have developed a diastereoselective allylation method for the synthesis of (2*R*,3*S*)-4^3-F_3Val **34b**.[25] Conversion of (*R*)-2,3-*O*-isopropylidene glyceraldehyde (**55a**) with benzylamine to the corresponding imine and indium-mediated allylation of the so-obtained (*R*)-*N*-benzyl-2,3-*O*-isopropylidene glyceraldimine (**56**) with (*E*)-4-bromo-1,1,1-trifluoro-2-butene affords the homoallylic amine **57** (Scheme 15.15).

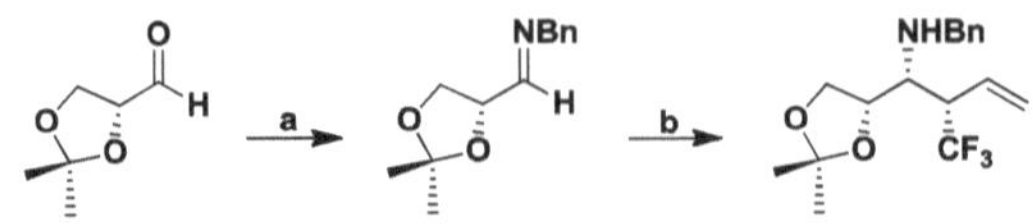

55a 56 57

(a) $BnNH_2$, $MgSO_4$, (b) F_3C Br, In

Scheme 15.15 Synthesis of the homoallylic precursor **57** according to Chen et al.[25]

As next, the homoallylic amine, **57** is exposed to triflouroacetic anhydride followed by ozonolysis of the double bond with reductive workup to obtain the alcohol **58**. With **58** in hands, a change of the protecting group from *N*-benzyl to *N*-Boc is performed by standard conditions and the hydroxyl function is reacted with *O*-phenyl chlorothionoformate, affording the thiocarbonyl compound **59**, to set stage for the upcoming Barton radical dehydroxylation. Exposing **59** to tributyltin hydride and acidic hydrolysis of the 1,3-dioxolane moiety gives the diol **60**, which is in turn subjected to a ruthenium-promoted periodate cleavage/oxidation and *N*-Boc deprotection to yield (2*R*,3*S*)-4^3-F_3Val **34b** (Scheme 15.16).

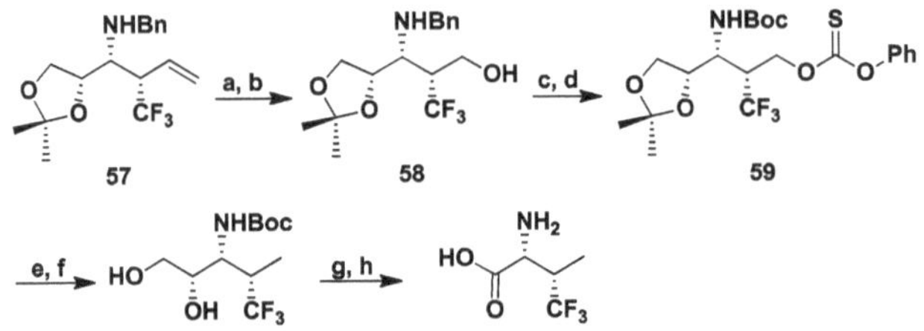

60 34b

(a) $(CF_3CO)_2O$, (b) O_3, $NaBH_4$, (c) Pd/C, H_2, Boc_2O, (d) PhOC(S)Cl, DMAP, (e) Bu_3SnH, (f) TsOH, (g) $RuCl_3/NaIO_4$, (h) TFA

Scheme 15.16 Synthetic access to (2*R*,3*S*)-4^3-F_3Val **34b** starting from **57** according to Chen et al.[25]

Chen et al. also report the synthesis of optically pure *N*-Cbz-(2*R*,3*S*)-$4'^3$-F_3Ile **61a** and *N*-Cbz-(2*S*,3*R*)-$4'^3$-F_3Ile **61b** starting from **57**. By concomitant reduction of the double bond and hydrogenation of the *N*-Bn group **62** is obtained. Cbz protection of the amine affords **63** and 1,3-dioxolane hydrolysis gives **64**, which is in turn subjected to oxidative periodate cleavage to get *N*-Cbz-(2*R*,3*S*)-$4'^3$-F_3Ile **61a** (Scheme 15.17).

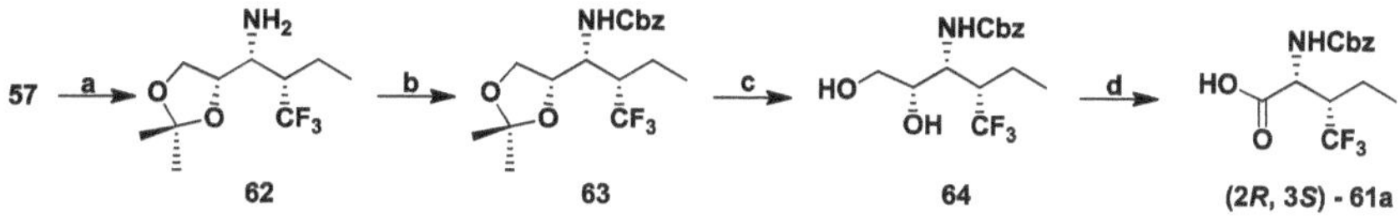

(a)Pd/C, H_2, (b) CbzCl, $NaHCO_3$, (c) TsOH, (d) $RuCl_3/NaIO_4$

Scheme 15.17 Synthesis of *N*-Cbz-(2*R*,3*S*)-$4'^{3}$-F_3Ile **61a** from intermediate **57** according to Chen et al.[25]

To synthesize the enantiomer *N*-Cbz-(2*S*,3*R*)-$4'^{3}$-F_3Ile **61b**, the same reaction sequence can be employed. Consequently, (*S*)-2,3-*O*-isopropylidene glyceraldehyde **55b** is used as the starting material to obtain *N*-Cbz-(2*S*,3*R*)-$4'^{3}$-F_3Ile **61b** (Scheme 15.18).[25]

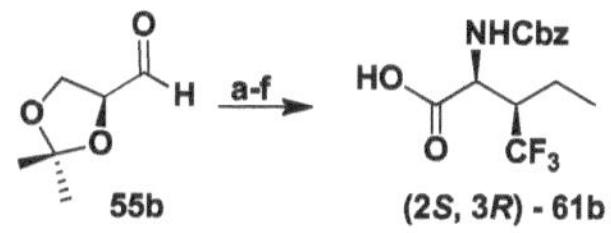

(a) $BnNH_2$, $MgSO_4$, (b) F_3C–CH=CH–CH_2Br, In, (c) Pd/C, H_2, (d) CbzCl, $NaHCO_3$, (e) TsOH, (f) $RuCl_3/NaIO_4$

Scheme 15.18 Synthesis of (2*S*,3*R*)-$4'^{3}$-F_3Ile **61b** according to Chen et al.[25]

In 2009, Pigza et al. have developed a stereoselective synthesis of (2*S*,3*S*)-4^{3}-F_3Val **34c** starting from 4,4,4-trifluoro-3-methylbutanoic acid **65**.[26] Coupling of the latter with (*S*)-camphorsultam and hydrogenation of the double bond using palladium gives **66**. Hydrolysis of the camphorsultam under basic conditions and immediate coupling with (*R*)-2-amino-2-phenylethanol gives the optically active amide **67**. The amide **67** undergoes diaminosulfur trifluoride–mediated cyclization without loss of the stereochemical integrity and gives oxazinone **68**, which is in turn reacted with SeO_2 to give in an oxidative rearrangement **69**. The so-obtained imine is face-selectively hydrogenated on platinoxide to give **70**. A deprotection sequence consisting of hydrogenation, step-wise acidification, and ion exchange chromatography gives the desired product (2*S*,3*S*)-4^{3}-F_3Val **34c** (Scheme 15.19).

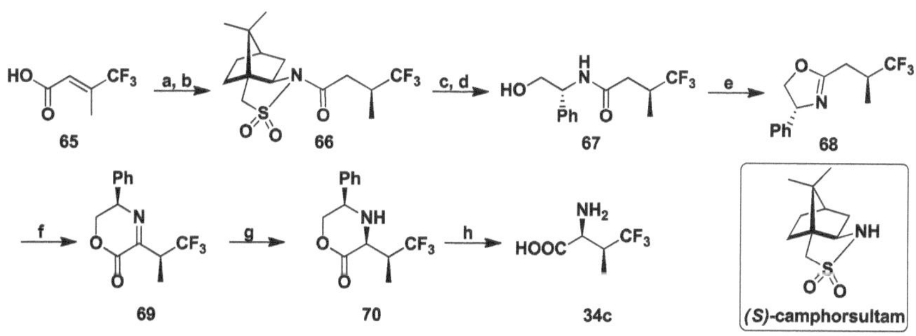

(a) $(COCl)_2$, DMF (cat.) then *(S)*-camphorsultam, NaH, (b) H_2, Pd/C, (c) LiOH, (d) 2-amino-2-phenylethanol, HATU, *i*-Pr_2NEt, (e) DAST, (f) SeO_2, (g) H_2, PtO_2, (h) 1. H_2, $Pd(OH)_2$, 2. 4 M HCl, $MeOH/H_2O$ 3. 6 M HCl 4. H^+ ion exchange

Scheme 15.19 Stereoselective synthesis of (2*S*,3*S*)-4^{3}-F_3Val **34c** according to Pigza et al.[26]

Benhaim et al. report a three-step enantioselective synthesis of the hydrochloric salt of (2*R*,3*R*)-4^3-F_3Val **34d** by stereoselective hydrogenation using the [((*R*)-trichickenfootphos)Rh(cod)]BF_4 catalyst (**71**) in the key step.[27] 1,1,1-Trifluoroacetone (**72**) is reacted with methyl cyanoacetate to provide the prochiral alkene **73**. Diastereoselective hydrogenation using **71** and subsequent acidic deprotection gives the hydrochloric salt of (2*R*,3*R*)-4^3-F_3Val **75** (Scheme 15.20).

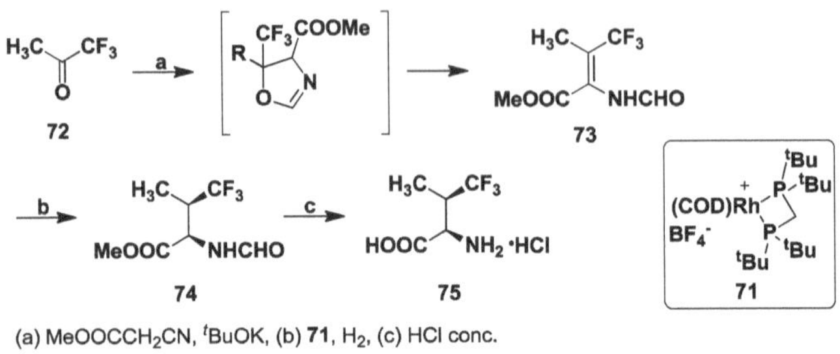

Scheme 15.20 Synthesis of 4^3-F_3Val hydrochloric salt **75** according to Benhaim et al.[27]

Budisa and coworkers have established a straightforward methodology for the synthesis of 2-amino-5,5,5-trifluoro-3-methylpentanoic acid **50** as a mixture of diastereoisomers.[28] Reaction between ethyl (triphenylphosphoranylidene)acetate and 4,4,4-trifluoro-2-butanone **76** yields the Wittig product **77**. The α,β-unsaturated ester **77** is hydrogenated to generate the corresponding saturated ester, which is hydrolyzed to obtain the free acid **78**. Subsequent addition of thionyl chloride and bromine led to the corresponding α-brominated acid **79** as a mixture of diastereoisomers. By nucleophilic substitution of α-bromine, **79** is transformed to the corresponding amino acid **50**. The latter is converted into enantiomerically pure samples of 5^3-F_3-Ile **50a** and 5^3-F_3-*allo*-Ile **50c** through enzymatic resolution of their *N*-acetyl-derivate **80**. A preference of *Aspergillus melleus acylase* I for one of the diastereomers leads to deacetylation of a single isomer. The so-obtained (2*S*,3*S*)-5^3-F_3-Ile **50a** and *N*-Ac-(2*S*,3*R*)-5^3-F_3-*allo*-Ile **80a** can be separated, and the latter is deacetylated with *porcine kidney acylase* I to give (2*S*,3*R*)-5^3-F_3-*allo*-Ile **50c** and **80b**, which is deacetylated and racemized to regenerate **50** (Scheme 15.21).

The approach Wang et al. uses for the synthesis of (2*S*,3*R*)-5^{3}-F_3Ile **50c** includes a diastereoselective methylation using the Evans oxazolidinone method and a chiral Strecker reaction via the Davis sulfonimine protocol.[29] Commercially available 4,4,4-trifluorobutyric acid **81** is converted into a mixed anhydride and then coupled with (*R*)-benzyl-2-oxazolidinone **30** to afford carboximide **82**. Asymmetric methylation gives **83**, and reductive cleavage of the auxiliary with lithium borohydride gives (*R*)-4,4,4-trifluoro-2-methyl-1-butanol (**84**). After oxidation to the corresponding aldehyde and imine formation with 4-methylbenzene-1-sulfonamide to get **85**, the so-obtained sulfinimine is subjected to an asymmetric Strecker reaction to yield **86**. Simultaneous hydrolysis of the sulfonamide and the nitrile gives (2*S*,3*R*)-5^3-F_3Ile **50c** (Scheme 15.22).

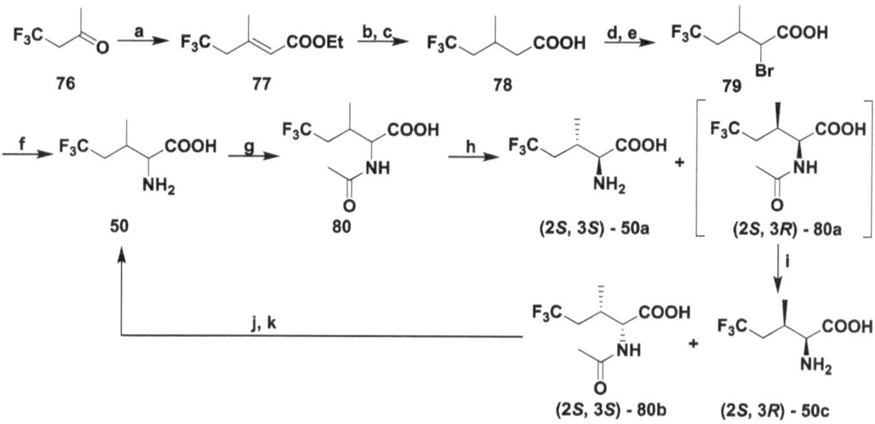

(a) $Ph_3PCHCOOEt$, (b) H_2, Pd/C, (c) 1 M NaOH then HCl conc., (d) $SOCl_2$, (e) Br_2 then H_2O, (f) NH_3 aq., (g) Ac_2O, NaOH then HCl conc., (h) *Aspergillus melleus acylase* I, pH=7.5, (i) *Porcine kidney acylase* I, pH= 7.5, (j) 3 M HCl, (k) AcOH, benzaldehyde

Scheme 15.21 Synthesis of **50** and enzymatic resolution of *N*-acetyl-5^3-F_3-Ile **80a** and *N*-acetyl-5^3-F_3-*allo*-Ile **80b** according to Biava and Budisa.[28]

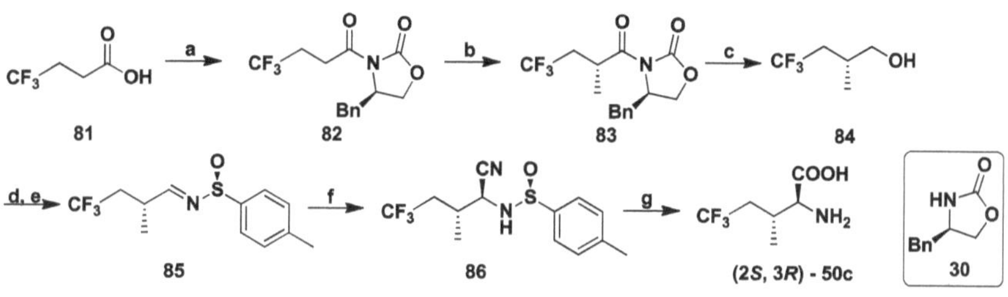

(a) $(CH_3)_3CCOCl$ then *n*-BuLi, **30** (b) NaHDMS, MeI, (c) $LiBH_4$, (d) Dess Martin Periodinane, (e) *p*-$MePhSONH_2$, $Ti(OEt)_4$, (f) Et(*i*-PrO)AlCN, (g) HCl conc.

Scheme 15.22 Synthesis of 5^3-F_3Ile **50c** according to Wang and Resnick.[29]

HfLeu hydrochloric salt **87** is obtained following Wittig reaction between hexafluoroacetone and ethyl bromopyruvate **88**, which gives the α-keto ester **89**.[30] Hydrogenation of the double bond of **89** affords the α-keto ester **90**, which is in turn enantioselectively reduced by either bakers' yeast or by Itsuno–Corey oxazaborolidine with catecholborane to yield the (*R*)-hydroxy ester **91**. The hydroxyl group of **91** is activated with triflic anhydride and reacted with (*R*)-1-phenylethylamine to give the separable diastereomers **92a** and **92b**. Deprotection and hydrolysis affords **87** (Scheme 15.23).

Xing et al. report the synthesis of (*S*)-HfLeu **95** using Garner aldehyde **96** as a chiral synthon (for its synthesis see Campbell et al.[30a]). In the key step, the aldehyde **96** is converted to the bis-trifluoromethyl olefin **97** by a Wittig reaction using tetrakis(tri fluoromethyl)-1,3-dithiethane and triphenylphosphine. The so-obtained olefin **97** is reduced by catalytic hydrogenation over Pd/C to give the suitably substituted oxazolidine **98**. Subjecting the oxazolidine to an acid-catalyzed ring opening gives the

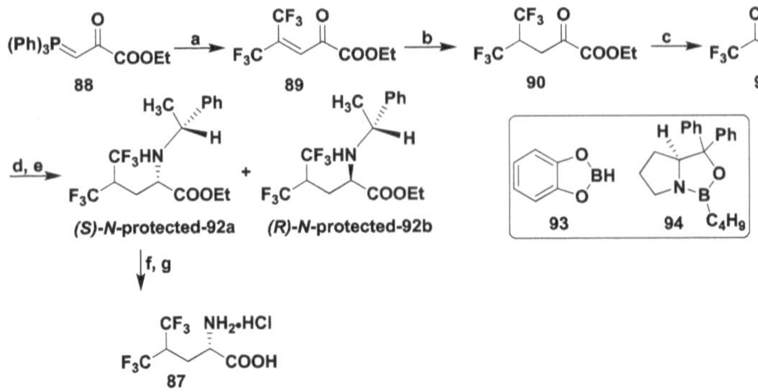

(a) Hexafluoroacetone, (b) H_2, Pd/C, (c) *baker's yeast*, sucrose or **93** and **94**, (d) Tf_2O, pyridine, (e) (*R*)-1-phenylethylamine, (f) H_2, Pt/C, (g) 6 N HCl,

Scheme 15.23 Synthesis of hexafluoroleucine hydrochloric salt **87**.[30]

alcohol **99**, which is oxidized to the carboxylic acid **100** using pyridinium dichromate. Finally, the *tert*-butoxycarbonyl group is removed using trifluoroacetic acid to yield HfLeu **95** (Scheme 15.24).[31]

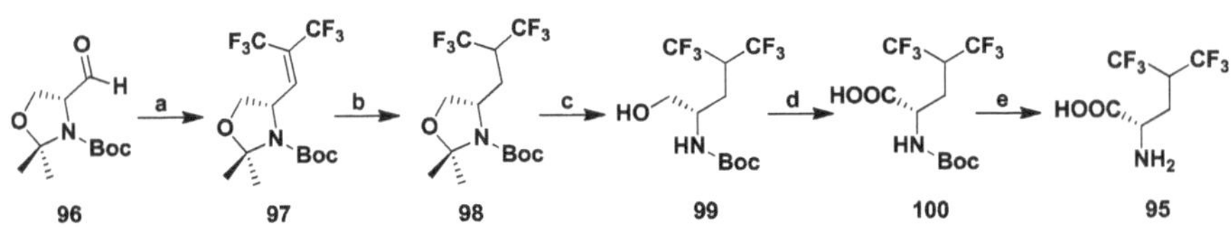

(a) PPh_3 ($((CF_3)_2C)_2S_2$, (b) H_2, 10% Pd/C, (c) *p*TsOH, (d) PDC, (e) 40% CF_3COOH, HCl

Scheme 15.24 Synthesis of HfLeu **95** from Garner aldehyde **96** according to Xing et al.[31]

N-Cbz-L-serine (**101**) is the starting material used by Anderson et al. in their HfLeu **95** synthesis.[32] **101** Is converted to the corresponding *tert*-butyl ester **102**, which is in turn reacted with methyltriphenylphosphonium iodide to obtain the iodoalanine compound **103**. The latter is transformed to the corresponding organozincate and reaction with in situ generated hexafluoroacetone gives the tertiary alcohol **104**. Radical-mediated deoxygenation of the tertiary alcohol **104** reveals the protected amino acid **105**, which after deprotection gives the desired HfLeu **95** (Scheme 15.25).

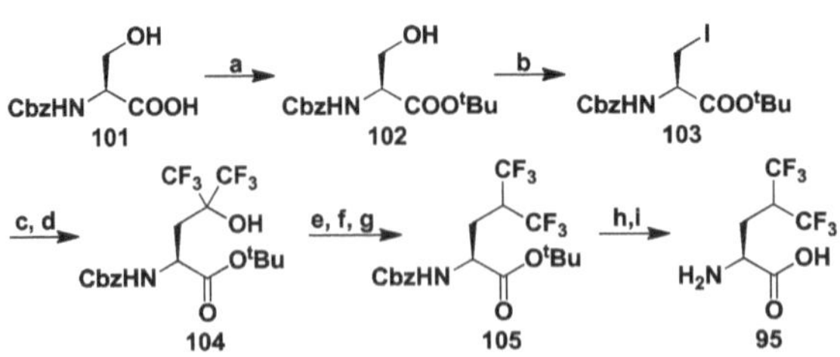

(a) K_2CO_3, $BnNEt_3Cl$, $^{t}BuBr$, (b) $(PhO)_3PMeI$, (c) Zn (dust), 1,2-dibromoethane, TMSCl, $CuBr\text{-}SMe_2$, (d) Hexafluoroacetone (gas), (e) ClCOCOPh, pyridine, (f) Bu_3SnH, AIBN, (g) KF-celite, (h) TFA, (i) H_2, Pd/C

Scheme 15.25 Synthesis of HfLeu **95** according to Anderson et al.[32]

The chemoenzymatic approach of Chiu et al. leads to the synthesis of *N*-Fmoc-HfLeu **106** starting from the already described ylide (see Scheme 15.23).[33] The ylide **88** is reacted with hexafluoroacetone to give the unsaturated pyruvate ester **89**. The latter is reduced using an aqueous titanium trichloride solution to afford **90**. Upon hydrolyzing compound **90**, reductive amination is performed with NADH mediated by *phenylalanine dehydrogenase* and *formate dehydrogenase* to yield HfLeu **95**. With **95** in hands a protection of the amino group using standard conditions is performed to get *N*-Fmoc-HfLeu **104** (Scheme 15.26).

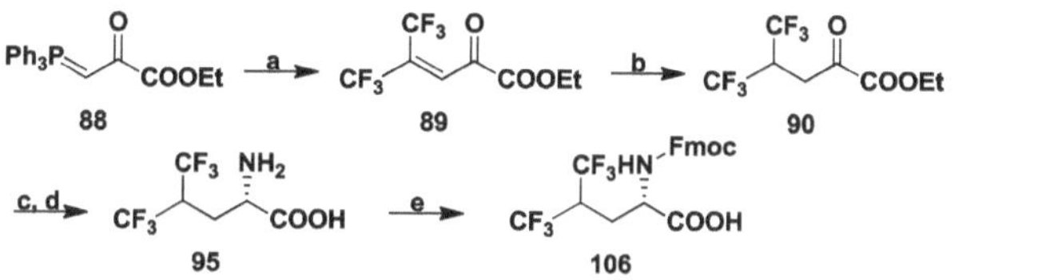

(a) Hexafluoroacetone, (b) $TiCl_3$ aq., (c) Na_2CO_3 aq., (d) *Phenylalanine dehydrogenase*, *Formate dehydrogenase*, NADH, $HCOONH_4$ aq., pH=8.5, (e) FmocOSu

Scheme 15.26 Synthesis of *N*-Fmoc-HfLeu **106** according to Chiu and Cheng.[33]

Fluorination of the side chain in aliphatic fluorinated amino acids can synthetically be achieved either by using a fluorinated building block or by direct fluorination. We have seen examples of both in the described synthesis of ω-fluorinated amino acids (MfeGly, DfeGly, TfeGly) and DfpGly. The stereocenter is either present in the starting material or it is introduced enzymatically or by using a chiral auxiliary. The trifluoromethyl group in 4^3-F_3Val is also commonly introduced by using fluorinated building blocks; however, we have presented an example of direct trifluoromethylation, where CF_3I is used as the CF_3 source. In general, the synthetic pathways to obtain 5^3-F_3Ile and $4'^3$-F_3Ile mostly rely on available fluorinated starting materials. Furthermore, there is an example of introducing chirality by enzymatic resolution, but the stereocenter is commonly incorporated using Evans oxazolidinone. Finally, the synthesis of HfLeu usually exploits hexafluoroacetone or similar starting materials $((CF_3)_2C)_2S_2$ containing the bistrifluoromethyl group. Chirality is routinely introduced enzymatically or through the use of a chiral starting material.

3. Fluorinated Amino Acids in Peptides

3.1 Hydrophobicity

The side chain hydrophobicity of an amino acid is an important factor for protein folding. The relationship between size and hydrophobicity of fluorinated amino acids when compared with that of the natural aliphatic hydrophobic amino acids has been investigated by plotting their side chain van der Waals volume versus their retention time from a reverse-phase-high-performance liquid chromatography (RP-HPLC) experiment.[6–8]In this experiment, the nonpolar character of a reverse-phase column

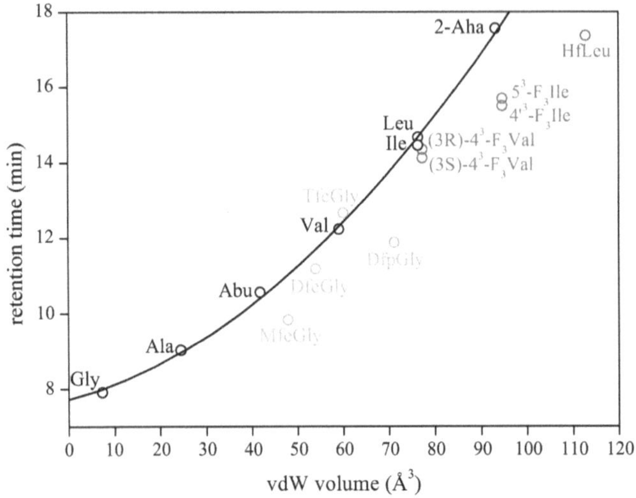

Figure 15.1 Retention times of Fmoc amino acids plotted against the van der Waals volume of their side chains.[6–8] The van der Waals volumes of the amino acid side chains were calculated according to Zhao et al.[36]

mimics the hydrophobic environment of a protein core.[34] For nonfluorinated amino acids, it was shown that the enlargement of the aliphatic side chain results in an increase in hydrophobicity and that the retention time increases nonlinearly with increasing side chain volume.[6,7] In contrast, the retention times of the investigated fluorinated amino acids do not fit into the correlation between side chain volume and retention time (Fig. 15.1).[6–8,24] A single fluorine substitution in the Abu side chain was shown to lead to a decrease in hydrophobicity; however, a significant increase of the van der Waals volume has been observed with each addition of a fluorine atom.[7] Despite being larger than Abu, MfeGly is less hydrophobic. The size of DfpGly is close to leucine (Leu) while being more polar. In fact, it is even more polar than valine (Val), which is much smaller. TfeGly is comparable to Val in both size and hydrophobicity. Encouraged by these findings, we expanded these initial studies by examining 2-aminoheptanoic acid (2-Aha) and the fluorinated amino acids $4'^3$-F_3Ile, 5^3-F_3Ile, (3*S*)-4^3-F_3Val, (3*R*)-4^3-F_3Val, and HfLeu. The two fluorinated stereoisomers of Ile ($4'^3$-F_3Ile and 5^3-F_3Ile) are similar in size but less hydrophobic than 2-Aha.[6,24] (3*S*)-4^3-F_3Val and (3*R*)-4^3-F_3Val are comparable to Leu and Ile with regard to both properties.[6,8] HfLeu is similar to 2-Aha in hydrophobicity, while possessing a much larger volume.[6] These findings, together with the fact that the hydration energy of HfLeu was shown to be 1.1 kcal/mol higher than that of Leu,[35] suggest that there are two factors determining the overall hydrophobicity of fluorinated amino acids.[7] First, the substitution of hydrogen by fluorine increases the solvent-accessible surface area and leads to an increase in hydration energy. Second, the C—F bond is more polarized than the C—H bond, and thus, electrostatic interactions of the fluorinated group with the solvent are energetically more favored. As a consequence, fluoroalkyl side chains possess two seemingly contrary physicochemical properties, hydrophobicity and polarity, and the combination of both leaves fluorinated amino acids less hydrophobic than their surface area would suggest.[6]

3.2 Helix Propensity

Besides its hydrophobicity the helix propensity of an amino acid is a crucial factor for the formation of helical assemblies. It can be determined by introducing the amino acid of interest in the guest position **X** of the monomeric α-helical polyalanine model peptide Ac-YGGKAAAAKA**X**AAKAAAAK-NH_2 (K-X).[37] The helix propensities of this particular amino acid can then be calculated from (circular dichroism) CD data by using the modified Lifson–Roig theory.[37b,38] Chiu et al. generally observed lower helix propensities for fluorinated amino acids compared with their hydrocarbon analogs.[33,37a] This trend was also shown for the helix propensities of the two different diastereomers of trifluorovaline, (3*S*)-4^3-F_3Val, (3*R*)-4^3-F_3Val, and trifluoroisoleucine, $4'^3$-F_3Ile[8,24] and 5^3-F_3Ile,[6] when compared with their natural analogs (Table 15.1).

The calculated helix propensities of Val and Ile are comparable with previously published data,[38b] but (3*S*)-4^3-F_3Val, (3*R*)-4^3-F_3Val, and $4'^3$-F_3Ile show a helix

Table 15.1 Mean Residue Ellipticity [θ] at 222 nm, Fraction Helix (f_{helix}), and Helix Propensity ω of K-X Peptides for Various Amino Acids

Amino Acid	$[\theta]_{222\ nm}$	f_{helix}	ω
Ala[8]	−19477 ± 478	0.56 ± 0.01	1.50 ± 0.17
Leu[8]	−17457 ± 465	0.50 ± 0.01	0.994 ± 0.093
HfLeu[33,37a]	−8720 ± 200	0.251 ± 0.006	0.123 ± 0.023
Abu[33,37a]	−18100 ± 200	0.522 ± 0.006	1.22 ± 0.14
MfeGly[8]	−16814 ± 379	0.48 ± 0.01	0.873 ± 0.068
DfeGly[8]	−13969 ± 569	0.40 ± 0.02	0.497 ± 0.060
TfeGly[8]	−7469 ± 475	0.22 ± 0.01	0.057 ± 0.022
Val[24]	−13054 ± 452	0.38 ± 0.01	0.41 ± 0.04
(3*S*)-4^3-F_3Val[24]	−2685 ± 526	0.08 ± 0.02	0
(3*R*)-4^3-F_3Val[24]	−3887 ± 547	0.11 ± 0.02	0
Ile[6]	−13813 ± 156	0.40 ± 0.01	0.52 ± 0.05
5^3-F_3Ile[6]	−10776 ± 216	0.31 ± 0.01	0.26 ± 0.03
$4'^3$-F_3Ile[24]	−3602 ± 130	0.10 ± 0.01	0

Ala, alanine.

propensity of zero. Already, the CD spectra of K-(3*S*)-4^3-F_3Val, K-(3*R*)-4^3-F_3Val, and K-$4'^3$-F_3Ile demonstrate a complete loss of helicity.[24] Remarkably, the CD spectrum of K-5^3-F_3Ile clearly indicates a helical structure as evidenced by the two distinct minima at 208 and 222 nm, and the helix propensity of 5^3-F_3Ile could be calculated as half of the value of Ile.[6] Generally, the decrease in helix propensity upon fluorination might be attributed to the easier burial of fluorocarbon side chains in the unfolded state of the model peptide[37a] and the exposure of these side chains in the helical state. In addition, associated to its branching on the β-carbon atom, the side chain atoms of Ile come in close proximity to the peptide backbone,[39] leading to reduced α-helicity when compared with its unbranched analog. This is underlined by the direct comparison of Ile to $4'^3$-F^3Ile, which carries the voluminous CF_3 group on its β-carbon atom. The close proximity of this sterically more demanding group to the peptide backbone seems to prevent the formation of α-helical structures, resulting in a very low helix propensity.[24] The two diastereomers of trifluorovaline ((3*S*)-4^3-F_3Val and (3*R*)-4^3-F_3Val)) exhibit no helix propensity, which is in agreement with the previously disclosed results.[24] These findings suggest that fluorinated amino acids should be better accommodated by β-sheet structures such as β-hairpins and amyloid-like fibrils.[37a] However, it remains to be elucidated to what extent helix propensity affects the protein stability at buried positions, e.g., within hydrophobic cores of proteins, since the reduced helix propensity may be connected to unfavorable solvent interactions at exposed positions of the applied monomeric model peptide.

3.3 *Fluorinated Amino Acids in Helical Model Systems*

Coiled-coil arrangements are often used as a protein model to study fluorinated amino acids in a native protein environment.[40] The primary structure of a coiled-coil arrangement is based on a (pseudo-) repetitive sequence of seven amino acids (abcdefg)*n*, the heptad repeat, in which hydrophobic side chains are generally located at the a- and d-positions on one side of the helix, whereas the other positions are hydrophilic (positions b, c, and f) or charged (positions e and g).[3,41] When the hydrophobic core of a coiled coil is globally substituted with fluorinated amino acids, usually a stabilization of the structure is observed.[42] In contrast, single amino acid substitutions often have a destabilizing effect on the coiled-coil stability.[43] Recent studies showed the influence of fluorinated analogs of Abu within two different hydrophobic microenvironments (positions a16 and d19) of the parallel, heterodimeric coiled-coil model system VPE/VPK (Fig. 15.2A and B).[44] Here, the incorporation of the fluorinated Abu analogs MfeGly, DfeGly and TfeGly led to a decrease in coiled-coil stability compared with the native system. The degree of destabilization depends on how efficiently the fluorinated residue packs against neighboring side chains.[44,45] Furthermore, at position a16, the stability is mainly determined by the side chain volume and hydrophobicity, whereas at position d19, the destabilizing impact of fluorine-induced polarity prevails.[44]

In addition, the substitution of L_{19} with HfLeu and of V_{16} with two different diastereomers of 4^3-F_3Val within the VPE/VPK model system was investigated.[44,46] The dimeric oligomerization state of the parent system remains unchanged when the

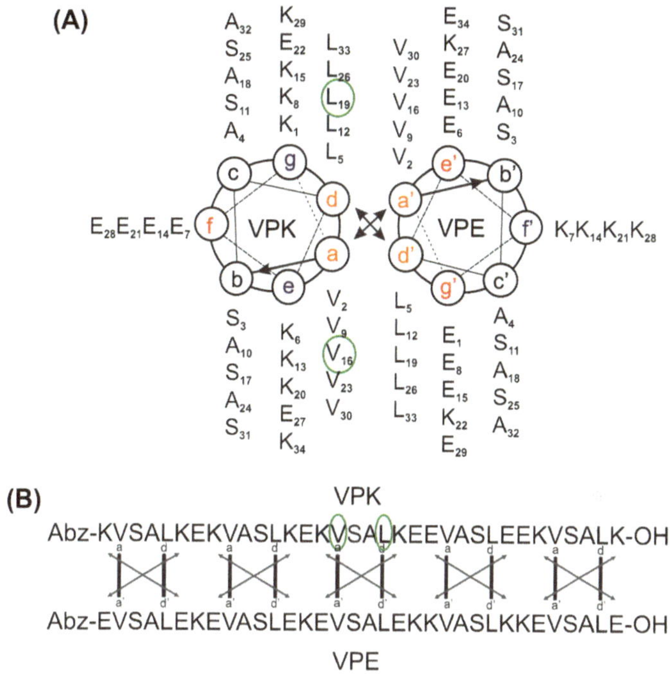

Figure 15.2 (A) Helical wheel representation of the parallel VPE/VPK heterodimer.[44] (B) Amino acid sequence of the VPE/VPK system. Substitution positions V_{16} and L_{19} are highlighted with a green circle.

fluorinated amino acids are incorporated (Table 15.2), and CD experiments reveal that all modified VPK analogs form α-helical bundles with VPE, as demonstrated by the two distinct characteristic minima at 208 and 222 nm (Fig. 15.3A). Thus, the substitution of V_{16} and L_{19} by their fluorinated analogs causes no significant structural perturbation.

In agreement with the previous findings for the Abu analogs,[44] the VPE/VPK-(3*S*)-4^3-F^3Val_{16} and VPE/VPK-(3*R*)-4^3-F_3Val_{16} show reduced or similar melting point than the parent bundle (67.5°C vs. 70.7°C and 70.8°C vs. 70.7°C, respectively; Fig. 15.3B). However, there is no great difference between (3*R*)-4^3-F_3Val and (3*S*)-4^3-F_3Val, and thus, the local orientation of the CF_3 group seems to have only a minor influence.[46] In contrast, for VPE/VPK-$HfLeu_{19}$, a higher thermal stability is observed (74.4°C vs. 70.7°C; Fig. 15.3B). In the parallel heterodimeric VPE/VPK model system the fluorinated residues directly interact with natural amino acids: VPK's position d19 with the central hydrophobic positions a′16, d′19, and a′23 of VPE, and VPK's position a16 with positions d′12, a′16, and d′19 of VPE.[45a,45b] Therefore, the enhancement in the thermal stability for the HfLeu-containing analogs is likely a result of greater hydrophobicity[6] and efficient packing[42d] with the hydrocarbon side chains in the hydrophobic core, whereas the fluorine content and packing behavior of 4^3-F_3Val appears to contribute less to hydrophobic core formation, due to the shorter side chain.[46]

Table 15.2 Theoretically and Experimentally Determined Molecular Weights of VPE/VPK Heteromers

Peptide	Theoretical Dimer Mass (Da)	SEC/SLS-Determined Mass (Da)
VPE/VPK	7576	7578 ± 16
VPE/VPK-HfLeu$_{19}$	7684	7685 ± 40
VPE/VPK-(3*R*)-4^3-F$_3$Val$_{16}$	7630	7674 ± 121
VPE/VPK-(3*S*)-4^3-F$_3$Val$_{16}$	7630	7634 ± 15
Sequences		
VPE	Abz-EVSALEKEVASLEKEVSALEKKVASLKKEVSALE-OH	
VPK	Abz-KVSALKEKVASLKEKVSALKEEVASLEEKVSALK-OH	
VPK-HfLeu$_{19}$	Abz-KVSALKEKVASLKEKVSA-HfLeu-KEEVASLEEKVSALK-OH	
VPK-(3*R*)-4^3-F$_3$Val$_{16}$	Abz-KVSALKEKVASLKEK-(3*R*)-4^3-F$_3$Val-SALKEEVASLEEKVSALK-OH	
VPK-(3*S*)-4^3-F$_3$Val$_{16}$	Abz-KVSALKEKVASLKEK-(3*S*)-4^3-F$_3$Val-SALKEEVASLEEKVSALK-OH	

Abz, *o*-aminobenzoic acid; *SEC*, size exclusion chromatography; *SLS*, static light scattering.
Standard deviations were determined from three independent measurements. Sequences of the peptides are also given.[46]

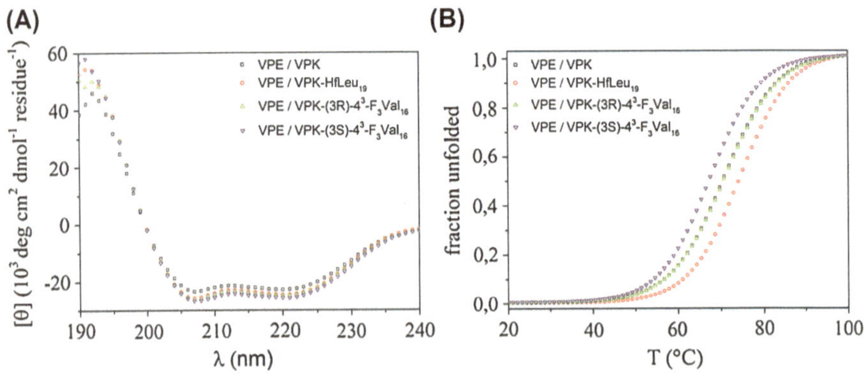

Figure 15.3 (A) CD spectra of VPK/VPE heteromers. Spectra were recorded at 20°C in 100 mM phosphate buffer at pH 7.4. The spectra were normalized and represent the mean of three independent measurements at an overall peptide concentration of 20 μM (10 μM in each monomer); (B) Fitted thermal denaturation curves of investigated equimolar mixtures of VPE/VPK. The spectra were recorded at pH 7.4 in 100 mM phosphate buffer with an overall peptide concentration of 20 μM (10 μM in each monomer).[46]

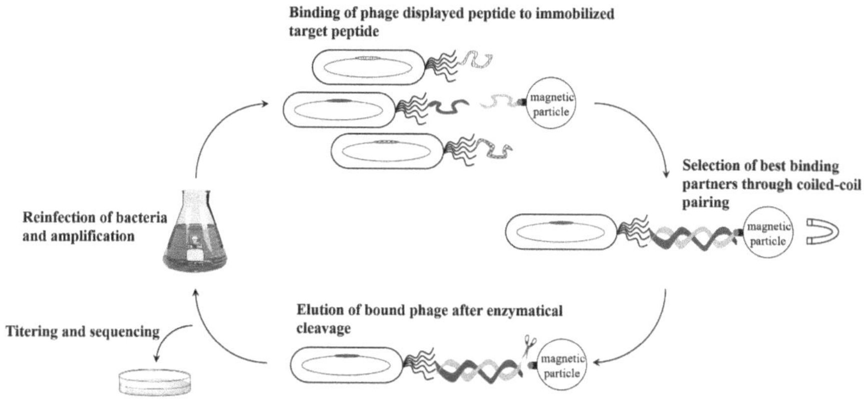

Figure 15.4 Selection of binding partners for a synthetic peptide out of a phage library via biopanning.

To identify the preferred binding partners interacting with DfeGly, TfeGly, or DfpGly at position a16 of the VPE/VPK heterodimer, a phage display was performed (Fig. 15.4).[45a] In position a16, the polarized β-methylene group points away from the hydrophobic core, and the fluorinated amino acids pair with amino acids similar to those of their natural control. In this context, we also reported the identification of suitable protein environments for MfeGly, DfeGly, TfeGly, and DfpGly in VPK's position d19, where polarized β-carbon atoms point directly into the hydrophobic core.[45b]

The phage display library was designed based on the randomization of the three central hydrophobic positions a′16, d′19, and a′23 of VPE that directly interact with position d19 of VPK. Screening of the phage library with fluorinated VPK variants revealed VPE analogs in which all three randomized positions were predominantly occupied by hydrophobic amino acids. At position d′19, Leu turned out to be the most conserved residue. It is also primarily selected in positions a′16 and a′23, but the β-branched Val and more frequently Ile were selected at these positions as well. The predominant selection of Leu and Ile in the randomized a-positions might be attributed to their higher hydrophobicity.[7] However, Leu seems to provide the best shielding against water for the fluorinated Abu analogs substituted in d-positions, since it has been primarily selected in all three randomized positions.[45b] The frequency of β-branched Ile in a'23 decreases with increasing side chain volume of the residues in VPK's d19, which might refer to the unfavorable packing behavior of Ile, since the bulkiness increases with the degree of Abu fluorination. Probably due to the more distinct constitutional differences compared with the other amino acids, DfpGly deviates from this trend. The combination of polarity, spatial properties, and packing characteristics influences the oligomerization state and, therefore, it was determined for all VPK-X_{19}/VPE-$L_{16}L_{23}$ by applying size exclusion chromatography (SEC) in combination with static light scattering (SLS). It was observed, that the dimeric oligomerization state of the VPE/VPK parent system is primarily retained (Table 15.3).[45b]

Table 15.3 Theoretically and Experimentally Determined Molecular Masses of VPK-X_{19}/VPE-$L_{16}L_{23}$ Heteromers

	Theoretical Dimer Mass (Da)	SEC/SLS-Determined Mass (Da)
VPK-Abu_{19}/VPE-$L_{16}L_{23}$	7581	7471 ± 571
VPK-$MfeGly_{19}$/VPE-$L_{16}L_{23}$	7599	7929 ± 279
VPK-$DfeGly_{19}$/VPE-$L_{16}L_{23}$	7617	7884 ± 90
VPK-$TfeGly_{19}$/VPE-$L_{16}L_{23}$	7635	7790 ± 114
VPK-$DfpGly_{19}$/VPE-$L_{16}L_{23}$	7631	7979 ± 67

SEC, size exclusion chromatography; *SLS*, static light scattering.
The standard deviations were determined from three independent measurements.[45b]

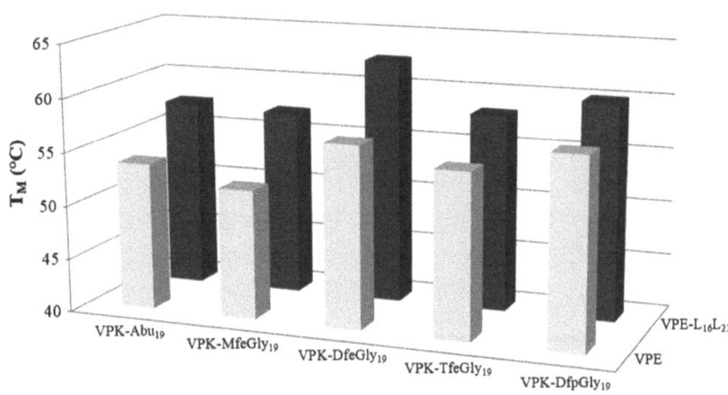

Figure 15.5 Melting points (T_M) of VPK-X_{19}/VPEL$_{16}$L$_{23}$ given in °C (defined as the temperature at which the unfolded fraction is 0.5). Melting points of VPK-X_{19}/$VPE_{wild\ type}$ are shown in the foreground, for comparison. The standard deviations determined from three independent measurements were 0.1°C.[45b]

For further characterization, the thermal stabilities of all VPK-X_{19}/VPE-$L_{16}L_{23}$ coiled-coil pairs were determined. All the 1:1 mixtures formed coiled coils with improved thermal stability, when compared with the 1:1 mixtures of the original VPE and the fluorinated VPK variants (Fig. 15.5).[45b]

3.4 Fluorinated Amino Acids in Amyloid Formation

Fluorinated amino acids are often studied within the context of helical folding, but barely in β-sheet and amyloid formation. To evaluate the impact of fluorine on amyloid formation, we substituted natural amino acids within a model peptide by amino acids with different fluorine content in their side chains. The amyloid-forming process was investigated by comparing the size, hydrophobicity, and secondary structure propensities (Fig. 15.6).[8] The design of the model peptide P1 is based on an engineered

(A)

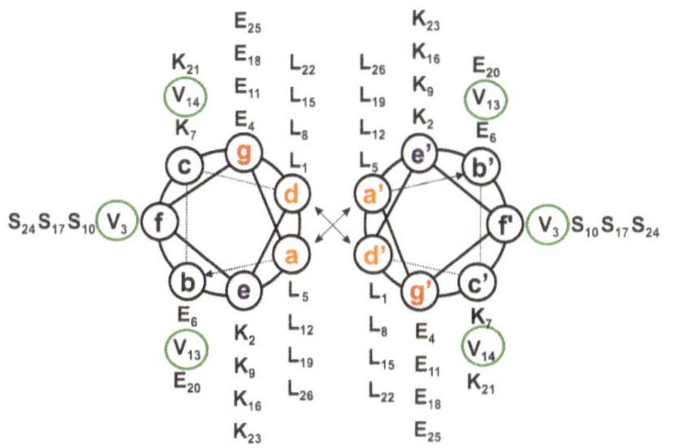

(B)

Peptide	Sequence
P1	LK-**Val**-ELEKLKSEL-**Val-Val**-LKSELEKLKSEL
P1_MfeGly 13	LKVELEKLKSEL-**MfeGly**-VLKSELEKLKSEL
P1_MfeGly 14	LKVELEKLKSELV-**MfeGly**-LKSELEKLKSEL
P1_DfeGly 13	LKVELEKLKSEL-**DfeGly**-VLKSELEKLKSEL
P1_DfeGly 14	LKVELEKLKSELV-**DfeGly**-LKSELEKLKSEL
P1_TfeGly 13	LKVELEKLKSEL-**TfeGly**-VLKSELEKLKSEL
P1_TfeGly 14	LKVELEKLKSELV-**TfeGly**-LKSELEKLKSEL
P1_TfeGly 3,13,14	LK-**TfeGly**-ELEKLKSEL-**TfeGly-TfeGly**-VLKSELEKLKSEL
P1_(3*S*)-4'-F_3Val 14	LKVELEKLKSELV-**(3*S*)-4'-F_3Val**-LKSELEKLKSEL
P1_(3*R*)-4'-F_3Val 14	LKVELEKLKSELV-**(3*R*)-4'-F_3Val**-LKSELEKLKSEL
P1_Leu 13	LKVELEKLKSEL-**Leu**-LLKSELEKLKSEL
P1_Leu 14	LKVELEKLKSELL-**Leu**-LKSELEKLKSEL
P1_Leu 3,13,13	LK-**Leu**-ELEKLKSEL-**Leu-Leu**-LLKSELEKLKSEL

Figure 15.6 (A) Helical wheel diagram of peptide P1. Residues that promote amyloid formation are highlighted in green. Single substitutions occurred at one of the positions V_3, V_{13}, and V_{24}. Leu and TfeGly were also used for triple substitution of V_3, V_{13}, and V_{24}. (B) Sequences of fluorinated and nonfluorinated P1 variants.[8]

coiled-coil folding motif that provides an α-helical starting structure but can fold into β-sheet-rich amyloids under controlled conditions (Fig. 15.6).[8,47]

P1 contains three β-sheet-preferring Val residues at positions 3, 13, and 14. These residues are initially exposed to the solvent when the peptide adopts the coiled-coil structure, and thus, induce a structural transition to a β-sheet-rich amyloid structure within 24 h under physiological pH.[48] Substitution with a set of fluorinated amino acids containing different fluorine stoichiometries was accomplished for Val residues in position 13 and 14. The latter were replaced by the nonfluorinated amino acid Leu, to create a second set of reference peptides. All members of this peptide library were investigated with regard to their amyloid formation potential and kinetics.[8] The amyloid formation process was followed with CD spectroscopy and thioflavin T fluorescence staining assays. The oligomerization state of the P1 peptide variants was determined with SEC in combination with SLS. P1 and the variants containing TfeGly, DfeGly, MfeGly, and Leu at either position 13 or 14, as well as the all-Leu variant initially form dimeric coiled coils. However, the determination of the initial oligomerization state of the peptide variants P1-TfeGly$_{3,13,14}$, P1 (*3R*) 4^3-F_3Val_{14}, and P1_(*3S*)-4^3-F_3Val_{14} was not possible because of a too fast amyloid formation. The rates of the fluorinated peptide variants were evaluated by a thioflavin T fluorescence staining assay and found to be strongly influenced by the nature of the incorporated amino acid. With a reduction of the fluorine content per side chain (TfeGly > DfeGly > MfeGly), a dramatic decrease in amyloid formation rates were observed. For a single TfeGly substitution the structural transition only needs few hours, whereas for the MfeGly peptides it takes several days. This effect was observed for both positions, 13 and 14, although the absolute folding rates show a dependency on the particular substituted position as well. A decrease of the fluorine content per side chain for the series TfeGly, DfeGly, and MfeGly led to a systematic decrease of hydrophobicity and side chain volume of the fluorinated amino acids.[7] DfeGly and MfeGly are less hydrophobic and smaller than the replaced canonical amino acid Val. However, the Leu-containing reference peptides (P1-Leu_{13} and P1-Leu_{14}), which are even more hydrophobic than P1, show reduced amyloid formation rates. The effects that were obtained for single TfeGly and Leu substitutions were even enhanced in the triply substituted variants. Thus, hydrophobicity and side chain volume are not the only driving forces for the structural rearrangement of the model peptide. The structural transition of all peptides was additionally investigated with CD spectroscopy. Except for the two 4^3-F_3Val-containing peptides, all P1 variants showed clear initial helical structures in the CD spectra with characteristic minima at 208 and 222 nm.

The intrinsic propensity of canonical amino acids to promote certain secondary structures is an important factor in peptide and protein folding. We studied the α-helix propensity of fluorinated amino acids applying the peptide model established by Cheng et al., and found that each addition of a fluorine atom in the Abu side chain additively reduced the α-helix propensities.[8] MfeGly has a lower propensity for an α-helical structure than Abu. However, the value is still twice as high as the one for Val. This means that MfeGly is much better accommodated in the initial α-helical structure than Val. The value of DfeGly is comparable to Val, and TfeGly has a very low, almost nonexistent, α-helix propensity.[8,24] Since Val is known to be a β-sheet-preferring amino acid,[39] it is not surprising that the fluorination of one of its methyl groups increases this effect even further. Both diastereomers of 4^3-F_3Val showed an α-helix propensity of zero.[24] Interestingly, a local influence of the orientation of the CF_3 group of both enantiomers of 4^3-F_3Val on amyloid formation was not observed.

Based on our results, it can be stated that the nature of the investigated fluorinated amino acids has a strong influence on the initially formed α-helical coiled coil of P1 and that stabilization or destabilization of the coiled coil occurs due to the different α-helix propensities of the incorporated amino acids. Furthermore, the rate of the structural rearrangement of P1 into β-sheet-rich amyloids strongly correlates with the stability of the initial coiled-coil structure (Fig. 15.7). Thus, secondary structure propensities play a key role in amyloid formation and should, besides other properties of fluorinated amino acids such as size, hydrophobicity, and induced polarity, generally be considered in the design of fluorinated peptides and proteins.[8]

3.5 Proteolytic Stability of Fluorinated Peptides

The main reason for reduced bioavailability of therapeutically active peptides is their proteolytic stability. Studies have shown that the effect that incorporation of

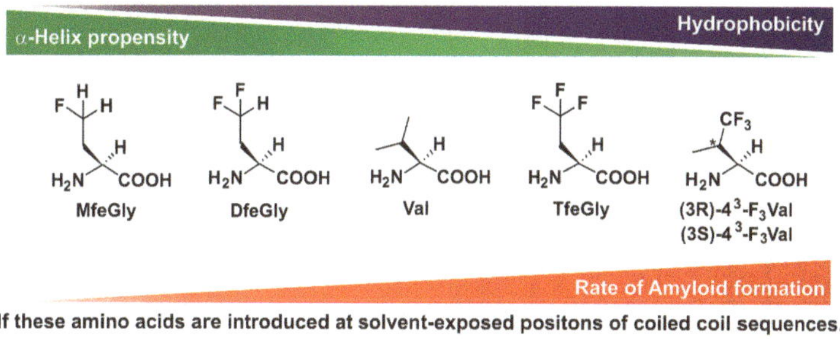

Figure 15.7 Summary of the generally observed trend in amyloid formation rates of fluorinated P1 variants as a result of the intrinsic properties of the fluorinated amino acids.[8]

fluorinated amino acids has on the stability and activity of the peptides is difficult to predict. To systematically elucidate the impact of side chain fluorination on the protease resistance, a library of model peptides was synthesized in which Abu, DfeGly, and TfeGly were incorporated at different positions and subjected to enzymatic degradation using the serine proteases α-chymotrypsin, elastase, pepsin, or human blood plasma. Time-dependent proteolytic stability of these peptides upon incubation with the different enzymes or human blood plasma was measured using RP-HPLC, and the degradation products were identified using (electrospray ionization time-of-flight) ESI-ToF-mass spectrometry.[9b] The model peptide FA was designed according to the specificities of α-chymotrypsin and pepsin with a central phenylalanine residue in the P1 position. All peptides carry 2-aminobenzoic acid as a fluorescent label for the high-performance liquid chromatography assay. Each alanine (Ala) residue in positions P2, P1′, or P2′ was singly substituted with the fluorinated amino acids TfeGly or DfeGly. For determining the actual impact of fluorination, Abu was also incorporated in these studies as control (Fig. 15.8). In addition, molecular modeling was used to identify potential noncovalent interactions between the enzymes and their substrates, to rationalize the experimentally determined proteolytic stabilities.[9b]

The results of the proteolytic degradation are shown in Fig. 15.9. When exposed to human blood plasma, in almost all cases, the nonnatural peptides degraded faster than the native control. P1′-TfeGly is the only variant that shows an improved stability toward proteolysis. The impact of fluorine was strictly position dependent since the P2′-TfeGly variant was the one that was degraded the fastest among all studied peptides. Depending on the position within the peptide substrate, fluorination can either induce a faster degradation (P2 variants) or improve stability toward proteolysis (P1′ variants). Immobilization of plasma proteins gave only one major cleavage product, which was identified to be the result of the action of carboxypeptidase N. However, in solution (nonimmobilized), this reaction gave four to six distinct cleavage products depending on the peptide sequence. Analysis of the latter strongly indicated the action of the leukocyte elastase. Thus, digestion of the fluorinated peptides by isolated elastase was carried out. The incubation of the peptides with elastase leads to cleavage products similar to those obtained from the nonimmobilized plasma reaction.[9a]

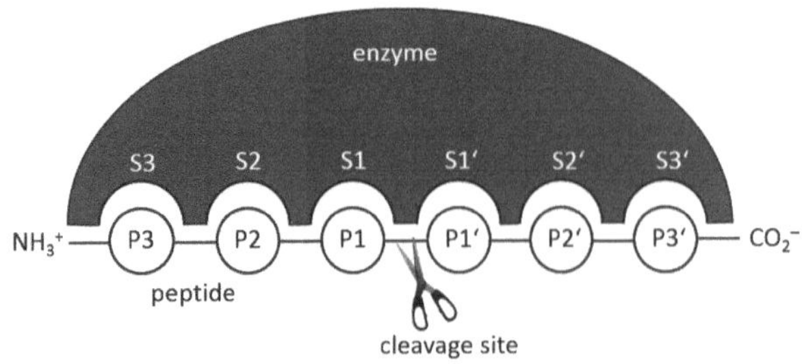

FA	Abz	Lys	Ala	Ala	Phe	Ala	Ala	Ala	Ala	Lys
P2-TfeGlyFA	Abz	Lys	Ala	**TfeGly**	Phe	Ala	Ala	Ala	Ala	Lys
P2-DfeGlyFA	Abz	Lys	Ala	**DfeGly**	Phe	Ala	Ala	Ala	Ala	Lys
P2-AbuFA	Abz	Lys	Ala	**Abu**	Phe	Ala	Ala	Ala	Ala	Lys
P1′-TfeGlyFA	Abz	Lys	Ala	Ala	Phe	**TfeGly**	Ala	Ala	Ala	Lys
P1′-DfeGlyFA	Abz	Lys	Ala	Ala	Phe	**DfeGly**	Ala	Ala	Ala	Lys
P1′-AbuFA	Abz	Lys	Ala	Ala	Phe	**Abu**	Ala	Ala	Ala	Lys
P2′-TfeGlyFA	Abz	Lys	Ala	Ala	Phe	Ala	**TfeGly**	Ala	Ala	Lys
P2′-DfeGlyFA	Abz	Lys	Ala	Ala	Phe	Ala	**DfeGly**	Ala	Ala	Lys
P2′-AbuFA	Abz	Lys	Ala	Ala	Phe	Ala	**Abu**	Ala	Ala	Lys

Figure 15.8 Schematic representation of the enzyme binding pocket named according to Schechter and Berger,[49] and table of the studied peptides.
Adapted from Asante, V.; Mortier, J.; Wolber, G.; Koksch, B. Amino Acids **2014,** 46 (12), 2733–2744.

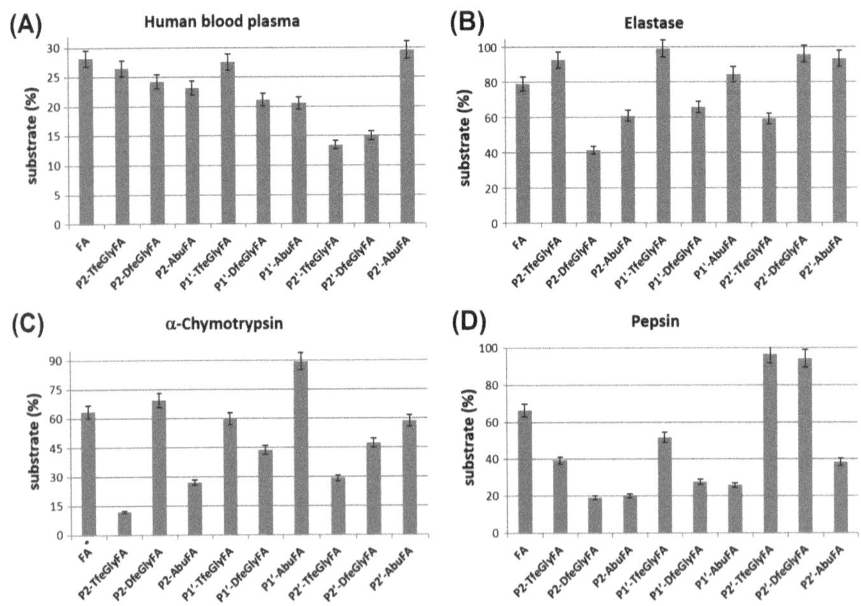

Figure 15.9 Percentage of peptide variants with substitution at positions P2′, P1, and P2 remaining after 120 min of hydrolysis by (A) human blood plasma (phosphate-buffered saline, pH 7.4) at 37°C, (B) elastase (100 mM Tris–HCl, pH 8.0) at 37°C, (C) α-chymotrypsin (10 mM phosphate buffer, pH 7.4) at 30°C, and (D) pepsin (10 mM acetate buffer, pH 4.0) at 30°C.[9a]

α-Chymotrypsin is a serine endopeptidase, which cleaves peptide bonds preferably on the C-terminal side of large hydrophobic residues, such as phenylalanine, tyrosine, and tryptophan.[50] For in-position P2-substituted peptides the following trend was observed from the most stable to the least stable: P2-DfeGlyFA > FA >> P2-AbuFA > P2-TfeGlyFA (see Fig. 15.9C). The S2 subsite of α-chymotrypsin forms a shallow hydrophobic groove and generally accommodates hydrophobic residues.[51] P2-TfeGlyFA seems to be well taken in by the active α-chymotrypsin since this peptide is the most rapidly degraded of all studied P2 variants. Molecular docking studies showed that the residue in the P2 position is likely to face the Ile99 side chain of the S2 pocket, creating an environment that is well suited to shelter the trifluoroalkyl side chain. This explains the lower proteolytic stability of this variant when compared with the FA peptide.[9b] In contrast, DfeGly in the P2 position significantly reduces the rate of hydrolysis, which might be attributed to the polarized γ-C-H methane moiety of the difluoromethyl group,[44,52] thus disfavoring the incorporation of the P2-DfeGlyFA variant within the S2 subsite of the protease. The measured trend in peptide stability in P1′ position−substituted peptides (from the most stable to the least stable) is P1′-AbuFA > FA > P1′-TfeGlyFA > P1′-DfeGlyFA (see Fig. 15.9C).[9b] The S1′ subsite of α-chymotrypsin generally harbors basic residues with long side chains, due to aspartates 35 and 64.[53] The slightly higher degradation of P1′-DfeGlyFA compared with P1′-TfeGlyFA might be explained by its ability to create a hydrogen bond between the difluoromethyl group of the substrate (H-bond donor) and the Cys42−58 disulfide bridge in the S1′ subsite.[9b] Observed proteolytic stability toward α-chymotrypsin of the P2′ residue decreased with the hydrophilicity[7,44,52] of the incorporated amino acid (see Fig. 15.9C): FA > P2′-AbuFA > P2′-DfeGlyFA > P2′-TfeGlyFA (from the most stable to the least stable).[9b] This might be explained with the hydrophobic character of the S2′ subsite of α-chymotrypsin, thus preferring hydrophobic residues.[54] Pepsin, an aspartic endopeptidase, is known to hydrolyze peptide bonds that connect bulky hydrophobic/aromatic residues, such as Phe−Trp, Phe−Tyr, and Phe−Phe.[50d,55,56] Peptides modified in P2 position are generally degraded faster, indicating a better geometric fit for pepsin compared with the FA peptide. The extent of proteolytic stability toward pepsin observed for these peptides is as follows (from the most stable to the least stable): FA > P2-TfeGlyFA > P2-DfeGlyFA = P2-AbuFA.[9b] The DfeGly side chain is in the vicinity of the Met_{289} side chain within the S2 pocket, and the modeled distance between the sulfur of Met_{289} and the DfeGly γ-carbon could indicate the existence of a weak hydrogen bond.[9b] Thus, the P2-DfeGlyFA peptide seems to be better accommodated within the enzyme resulting in the higher degradation compared with P2-TfeGlyFA. For the proteolytic stability of the P1′-substituted peptides, the following trend for was observed: FA > P1′-TfeGlyFA >> P1′-DfeGlyFA ≥ P1′-AbuFA (from the most stable to the least stable).[9b] Pepsin prefers to hydrolyze peptides that have hydrophobic residues in the P1′ position.[57] This specificity is due to the nature of the S1′ subsite constituents, namely, Ile_{213}, Ile_{300}, Leu_{291}, and Thr_{218}. However, the obtained results correspond rather to the steric bulk of the side chain of the non-proteinogenic amino acid than to their hydrophobicity. A strong increase in stability toward pepsin is observed when fluorinated amino acids occupy the P2′ position of the FA peptide. Although it is known that the S2′ subsite of pepsin preferentially binds hydrophobic amino acids, the stability toward proteolysis in P2′ position decreases with increasing fluorine content and hydrophobicity of the side chain and can be summarized from the most stable to the least stable as follows: P2′- TfeGlyFA > P2′-DfeGlyFA > FA > P2′-AbuFA (see Fig. 15.9D).[58] This is likely due to the larger solvent-accessible surface area of both TfeGly and DfeGly compared with that of Ala and Abu.[9b]

These studies showed that the influence of fluorination on the stability of the peptides is very difficult to predict, and that the incorporation of fluorine does not automatically improve the stability of peptides toward proteases.[9b] These results are more important in the field of medicinal chemistry, as this tendency cannot be reconciled with the generally improved metabolic stabilities observed for small drug molecules containing fluorine atoms. Hence, the impact of side chain fluorination on the proteolytic stability of peptides is a complex phenomenon that depends on the position of the substitution relative to the cleavage site, the type of enzyme, and the chemical nature and fluorination state of the side chain. Thus, systematic investigations will always be required to further elucidate the effect of fluorine, because there is still no general rule for predicting its impact.

3.6 Site-Specific Incorporation of Fluorinated Amino Acids Into Basic Pancreatic Trypsin Inhibitor

As a means to develop our understanding of the impact of fluorine substitution in a protein environment, the incorporation of fluorinated nonnatural amino acids should be expanded to larger globular proteins with diverse tertiary structures. Therefore, the incorporation of Abu and its fluorinated derivatives DfeGly and TfeGly into basic pancreatic trypsin inhibitor (BPTI) was recently carried out by our group. BPTI is a well-characterized small globular protein belonging to the serine protease inhibitor family that has been extensively used for studies of protein folding and protein–protein interactions. This protein is characterized by high thermal stability and resistance toward common protein denaturants.[59] Structurally, BPTI is composed of two β-strands forming a twisted antiparallel β-sheet, two α-helical regions, and three disulfide bonds. Crystal structures have shown that BPTI has a lysine residue (Lys_{15}) at the P1 position of the interacting loop that is fully solvent exposed in the free inhibitor and makes crucial contacts to the S1 binding pocket of the enzyme in the trypsin–BPTI complex.[60] Thus, to optimally determine the impact of fluorine substitution at a solvent-exposed position, with respect to protein folding and thermal stability, and to assess its influence on inhibitor activity in the context of a natural protease, Abu, DfeGly, and TfeGly (Fig. 15.7) were incorporated into position Lys_{15} within BPTI, and these variants were fully characterized.[61] The effect of single site-specific amino acid substitution, realized by means of solid-phase peptide synthesis and native chemical ligation, on the global structure of BPTI was determined by means of CD spectroscopy. The CD spectra show that structural perturbation caused by the nonnatural amino acids is minimal (Fig. 15.10).

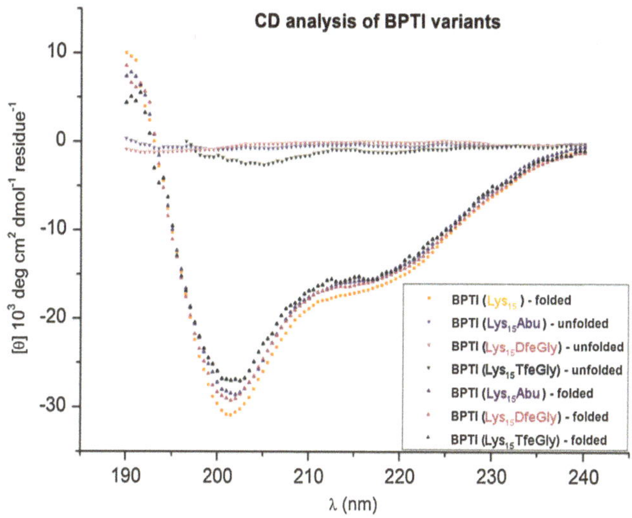

Figure 15.10 CD analysis of folded and unfolded mutant basic pancreatic trypsin inhibitor (BPTI) species compared with folded wild-type BPTI.

To study the influence of the incorporation of Abu, DfeGly, and TfeGly on the thermal stability of BPTI, denaturation studies were conducted. Due to the fact that the lysine side chain present in wild-type BPTI bears a formal positive charge, whereas the nonnatural Abu analogs are neutral, two distinct sets of conditions, 6 M GdmCl and of 8 M urea, were used (Table 15.4). The data show that the fluorination of Abu leads to restoration of the thermal stability of BPTI when urea is the denaturant, and even an enhancement in thermal stability when GdmCl is the denaturant. This points to the contribution of the positive charge of Lys_{15} to the thermal stability of the wild-type inhibitor.[62]

Table 15.4 Thermal Denaturation Parameters for BPTI Variants

	Tm/GdmCl[a]	ΔG^{ob}	Tm/urea[a]	ΔG^{ob}
Lys_{15}(wt)	66.5	12.0	75.8	14.6
Lys_{15}Abu	63.4	11.3	71.7	13.8
Lys_{15}DfeGly	68.8	12.4	74.8	14.3
Lys_{15}TfeGly	68.3	12.1	73.8	14.6

[a]Tm (°C) is defined as the temperature at which the fraction of folded species is 0.5. Errors are typically not higher than 0.5°C.
[b]ΔG° values were calculated for 1 mol at the standard state, errors are typically not higher than 0.2 kcal/mol.

The inhibitor activities of the nonnatural BPTI variants toward β-trypsin were compared with those of wild-type BPTI (Fig. 15.11). Standard inhibition assays[63] using the chromogenic substrate N_{α}-benzoyl-L-arginine 4-nitroanilide hydrochloride revealed that substitution of Lys with Abu at the P1 position dramatically reduces inhibitor activity. Remarkably, the introduction of two or three fluorine atoms into the Abu side chain (DfeGly or TfeGly) fully restores inhibitor activity to the wild-type level.

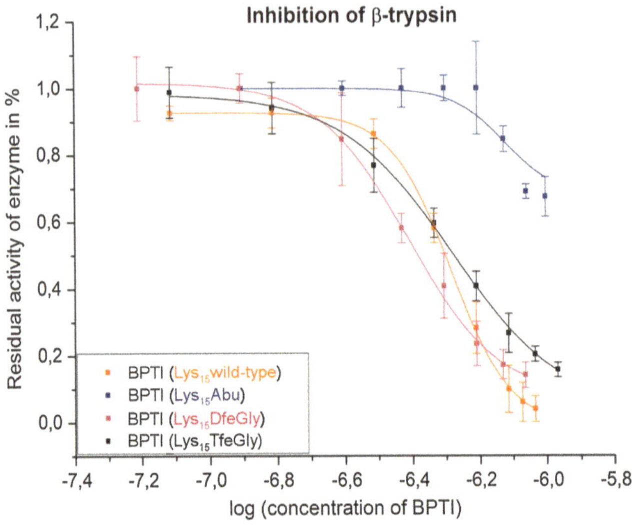

Figure 15.11 Schematic representation of inhibitor activity of synthetic BPTI variants toward β-trypsin.

To explore the molecular mechanisms by which fluorination of Abu is able to restore inhibitor activity to BPTI, high-resolution X-ray structures for all four β-trypsin-BPTI complexes were solved. As depicted in Fig. 15.12C, all four β-trypsin–BPTI complexes

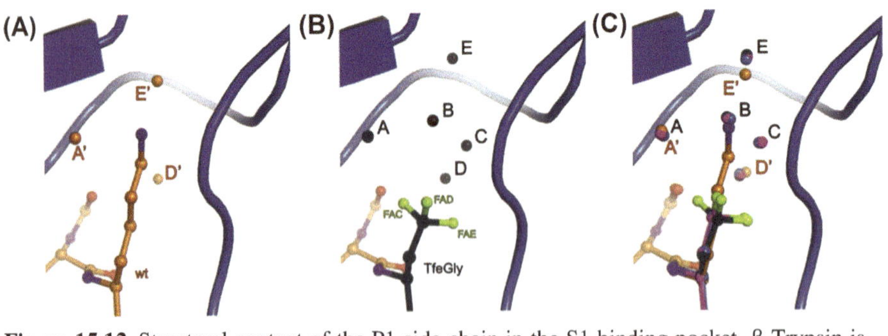

Figure 15.12 Structural context of the P1 side chain in the S1 binding pocket. β-Trypsin is shown in cartoon representation and the modified BPTI in stick representation. (A) Lys15 and structural water molecules A′, D′, and E′; (B) Lys_{15}TfeGly and structural water molecules A, B, C, D, and E; and (C) superimposition of P1 side chains and structural water molecules.

are highly similar in conformation and thus can be superimposed well, with the exception of the Lys15 side chain and the number of water molecules within the S_1 binding site. In the wild-type BPTI complex, three water molecules (Fig. 15.12A; referred to as A′, D′, and E′) are found within the S1 binding pocket, whereas in the case of the nonnatural BPTI species, there are five water molecules (Fig. 15.12B; referred to as

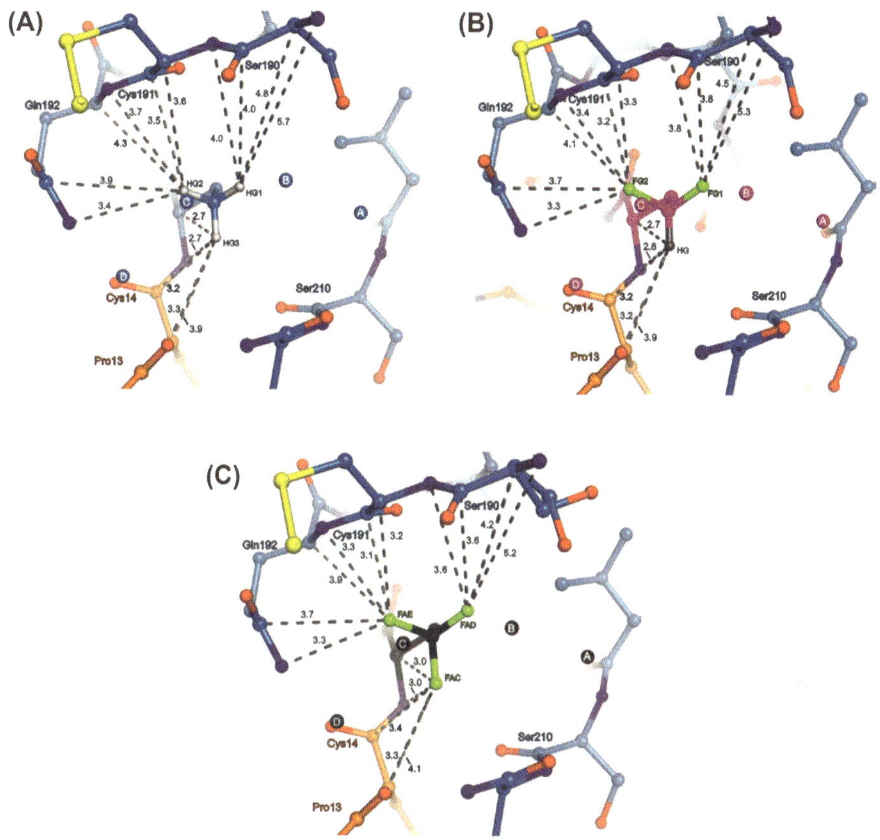

Figure 15.13 Distance analysis and top view of the unnatural side chain at position 15 of BPTI depicted in orange within the **S1** binding pocket of β-trypsin drawn in blue. Values indicate the distance in angstroms between the individual hydrogen or fluorine atoms of the terminal gC of the side chain and the 10 nearest enzyme atoms (clockwise, $Gln_{192}NE$, $Gln_{192}CD$, $Gln_{192}CA$, $Gln_{192}N$, $Cys_{191}C$, $Cys_{191}CA$, $Cys_{191}N$, $Ser_{190}C$, $Ser_{190}CA$, and $Ser_{190}N$), as well as the five nearest BPTI atoms (from top to bottom, Xaa15CA, Xaa15N, $Cys_{14}C$, $Cys_{14}CA$, and $Pro_{13}O$): (A) Abu structure with side chain and structural waters A–D in light blue and the three gC hydrogen substituents HG1 (ABA^{HG1}), HG2 (ABA^{HG2}), and HG3 (ABA^{HG3}) in light gray; (B) DfeGly structure with side chain and structural waters A–D in magenta, the two gC fluorine substituents FG1 (OBF^{FG1}) and FG2 (OBF^{FG2}) in green, and the single gC hydrogen substituent HG (OBF^{HG}) in gray; (C) TfeGly structure with side chain and structural waters A–D in black and the three gC fluorine substituents FAD ($3EG^{FAD}$), FAE ($3EG^{FAE}$), and FAC ($3EG^{FAC}$) in green. Nitrogen atoms are shown in dark blue and oxygen atoms in red.

A, B, C, D, E) present. Hence, in the latter case, the water molecules A, B, C, and D form a "hydration shell" about the side chains of the unnatural amino acids in the space vacated by the significantly longer side chain of the wild-type Lys_{15} side chain.

The distance analysis with respect to the P1 side chain substituents (hydrogens or fluorines) and the trypsin contacts revealed a trend common to both fluorinated Abu derivatives. Specifically, the fluorine atoms of the noncanonical side chains of DfeGly (FG1 and FG2) and TfeGly (FAD and FAE) that are oriented toward the enzyme backbone, namely, residues Ser_{190}, Cys_{191}, and Gln_{192}, and the side chain of the latter, are generally closer to the S1 contacts than are the two proximal hydrogens of the Abu side chain (HG1 and HG2). These particular distances (10 values ranging from 3.1 to 5.7 Å) are depicted in Fig. 15.13, and this trend appears to offer evidence of the type of "fluorophilic environment," described by Diederich, which has shown to result from multipolar interactions between C–F bonds and protein backbone fragments ($H{-}C_{\alpha}{-}C{=}O$) or main/side chain amides.

Second, the distance analysis with respect to the P1 side chain substituents (hydrogens or fluorines) and the structural water molecules revealed one significant difference between TfeGly and Abu/DfeGly (Fig. 15.14): water molecule C is closer to the fluorine atom $3EG^{FAC}$ in the TfeGly structure (3.4 Å) than it is to the hydrogen atom ABA^{HG3} in the Abu structure (3.6 Å) or the hydrogen atom OB^{FHG} in the DfeGly structure (3.7 Å). There could be three possible explanations for this observation. One explanation could be that the closer $3EG^{FAC}$–water C distance is a consequence of the slight shift in conformation within the side chain

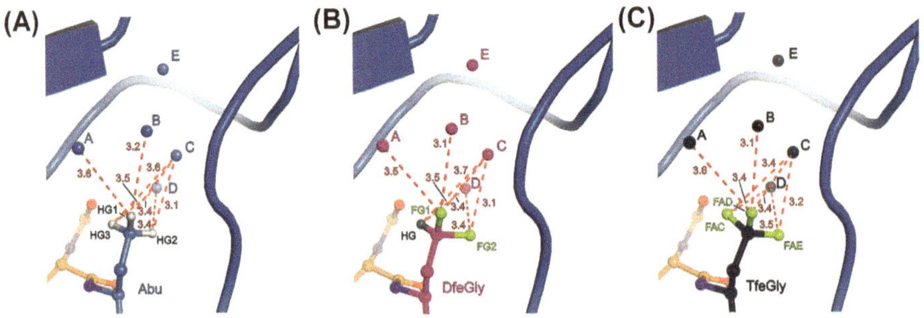

Figure 15.14 Distance analysis and side view of the unnatural side chain at position 15 of BPTI within the S1 binding pocket of β-trypsin. β-trypsin is shown in cartoon representation and the modified BPTI in stick representation. Values indicate the distance in angstroms between the individual hydrogen or fluorine atoms of the terminal gC of the side chain and the oxygen atom of the respective structural water molecule A–D: (A) Abu structure with side chain and structural waters A–D in light blue and the three gC hydrogen substituents HG1 (ABA^{HG1}), HG2 (ABA^{HG2}), and HG3 (ABA^{HG3}) in white; (B) DfeGly structure with side chain and structural waters A–D in magenta, the two gC fluorine substituents FG1 (OBF^{FG1}) and FG2 (OBF^{FG2}) in green, and the single gC hydrogen substituent HG (OBF^{HG}) in gray; (C) TfeGly structure with side chain and structural waters A–D in black and the three gC fluorine substituents FAD ($3EG^{FAD}$), FAE ($3EG^{FAE}$), and FAC ($3EG^{FAC}$) in green. Nitrogen atoms are shown in dark blue and oxygen atoms in red.

that occurs due to the electrostatic repulsion between $3EG^{FAC}$ and $3EG^{N}$ (3.0 Å apart). A second explanation could be based on the greater van der Waals radius of fluorine compared with that of hydrogen; however, this argument would not suffice, as no such phenomenon is seen between water molecule C and atoms OBF^{FG2} (3.1 Å) or $3EG^{FAE}$ (3.2 Å), compared to ABA^{HG2} (3.1 Å), that is, these distances do not decrease when hydrogen is replaced by fluorine. A third explanation could be that a favorable interaction, a weak OH···FC H-bond, exists between the TfeGly side chain and water C.

Taken together, the distance analysis described in the previous paragraphs suggests that the S1 subpocket of trypsin provides a "fluorophilic environment" for DfeGly and TfeGly, and that the five structural waters provide an extended water-mediated H-bond network that may include a weak OH···FC H-bond between water molecule C and fluorine atom $3EG^{FAC}$ of the TfeGly side chain. Furthermore, the average *B*-factors (Table 15.5) of the "new" water molecules B and C are

Table 15.5 Comparison of B-Factors Given in Square Angstroms of Amino Acid Side Chains and Water Molecules in the S1 Binding Pocket

β-Trypsin-BPTI Complexes				
B-Factor (Å^2)	**Weight**	**Lys15Abu**	**Lys15DfeGly**	**Lys15TfeGly**
Average B-factor of protein complex	20.7	21.3	19.6	21.2
All water molecules on average	31.0	33.8	31.2	32.0
BPTI Pro_{13} - Arg_{17}				
All atoms	13.0	13.2	12.4	13.4
Main chain	12.0	12.2	11.1	12.3
Side chain	14.0	14.6	13.9	14.6
BPTI P1 Residues				
All atoms	12.1	11.9	13.2	12.9
Main chain	11.6	11.5	11.0	11.8
Side chain	12.5	12.8	15.5	13.7
Water Molecules in S1 Pocket				
A/A′	–/13.5	21.8/–	17.5/-	18.4/–
B	–	27.7	20.0	20.5
C	–	31.2	22.2	20.7
D/D′	–/15.4	23.0/–	17.8/–	17.9/–
E/E′	–/14.1	20.7/–	19.1/–	19.4/–

significantly smaller in the Lys15DfeGly and Lys15TfeGly complexes (four values ranging from 20.0 to 22.2) than the corresponding values in the Lys15Abu structure (27.7 and 31.2, respectively). These results indicate that the fluorinated side chains more tightly hold waters B and C than does the hydrocarbon parent side chain of Abu, suggesting a "chemical complementation" mechanism in which fluorine atoms and water molecules team up to restore the inhibitor activity to the Abu mutant of BPTI.[64]

4. Summary and Prospects

Fluorination has long been recognized as a useful tool in medicinal chemistry, and this powerful approach is increasingly being applied to the unique problems posed by the field of peptide and protein engineering. In particular, amino acid building blocks that retain the connectivity found within the canonical aliphatic residues Val, Ile, and Leu, namely, monosubstitution of the alpha carbon by a side chain two to four carbon units in length, with β- or γ-branching, and that contain one to six fluorine atoms, have proven to be a particularly interesting family of structures. The chemical synthesis of these analogs is challenging due to the degree and diversity of functionality they possess, and the obvious stereochemical requirements. The synthetic approaches described herein for obtaining fluorinated analogs of Abu, Leu, Val, and Ile have enabled the advent of the rational design of fluorinated peptides and the exploration of their properties in the context of natural biomolecules.

Crucial factors for the formation of secondary structures in proteins include size, shape (topology), hydrophobicity, propensity (the tendency of an amino acid to adopt a certain secondary structure), and electronic characteristics of the amino acid side chains. Fluorinated amino acids tend to be larger than their hydrocarbon analogs and exhibit, except for MfeGly, greater relative hydrophobicities. In general, they also exhibit lower α-helical propensities than their natural analogs. In spite of the latter observation, we and others have found that the Abu-based building blocks are readily accommodated within the α-helical coiled-coil motif, common among native proteins, and that the impact they have on the properties of such assemblies is highly differentiated. For example, a single substitution of the VPE/VPK system with HfLeu leads to an increase in thermal stability, whereas replacement with 4^3-F_3Val does not. HfLeu preserves the topology of the Leu side chain and thus is well accommodated within the coiled-coil hydrophobic core, structurally virtually nonperturbative. In contrast, 4^3-F_3Val exhibits unfavorable steric clashes and electronic repulsion of the large electron-rich γ-CF_3 group with the helix backbone.

In other studies described in this chapter, the impact of side chain fluorination was determined in the context of amyloid formation and protein−protein interactions within enzyme active sites. In the former case, the rate of formation of amyloids was found to strongly correlate with the stability of the initial α-helical coiled coil of P1, which in turn is influenced by the α-helical propensities of the incorporated fluorinated amino acids. In the latter case, one set of studies used a small library of peptidic protease substrates to systematically probe how fluorination affects proteolytic stability. These results underlined the subtle interplay between (1) the position of the substitution relative to the cleavage site, (2) the particular enzyme, and (3) the degree of fluorination of the side chain. A separate set of investigations found that the replacement of Abu with DfeGly or TfeGly at position 15 of the 580amino-acid protease inhibitor BPTI results in an increase in thermal stability and inhibitor activity against β-trypsin. The high-resolution crystal structures of these complexes revealed that the fluorinated side chains are favorably accommodated by the "fluorophilic" environment of the S1 subsite of trypsin, and lead to significantly lower *B*-factors in the proximal structural waters. These results suggest "chemical complementation" in which fluorine restores inhibitor activity to the noncanonical hydrocarbon parent.

The future work in this area of research will rely on improved chemical access to more diverse fluorinated amino acid building blocks, and to more such systematic comprehensive studies of their effects in protein environments. This is because, at this time, no single general rule for the rational design of more proteolytically stable peptide therapeutics can be formulated. We have found that the incorporation of this type of "bio-orthogonal" functionality often requires subtle rearrangements of the neighboring natural sequence, to optimize their environment; phage display has been proved to be a powerful tool for identifying "fluorine-friendly" environments. If these environments can be identified, they might help to improve the predictability of fluorine's effects on protein folding, and thereby support the de novo design of protein folds with superior properties such as protease or thermal stability and biological activity.

Acknowledgments

The authors gratefully acknowledge funding from the DFG Research Training Group "Fluorine as key element" (DFG GRK 1582/2).

References

1. Horng, J. -C.; Raleigh, D. P. *J. Am. Chem. Soc.* **2003,** *125* (31), 9286–9287.
2. Smits, R.; Koksch, B. *Curr. Top. Med. Chem.* **2006,** *6* (14), 1483–1498.
3. Salwiczek, M.; Nyakatura, E. K.; Gerling, U. I. M.; Ye, S.; Koksch, B. *Chem. Soc. Rev.* **2012,** *41* (6), 2135–2171.
4. Budisa, N.; Pipitone, O.; Siwanowicz, I.; Rubini, M.; Pal, P. P.; Holak, T. A.; Gelmi, M. L. *Chem. Biodiversity* **2004,** *1* (10), 1465–1475.
5. Montclare, J. K.; Tirrell, D. A. *Angew. Chem. Int. Ed.* **2006,** *45* (27), 4518–4521.
6. Erdbrink, H.; Nyakatura, E. K.; Huhmann, S.; Gerling, U. I. M.; Lentz, D.; Koksch, B.; Czekelius, C. *Beilstein J. Org. Chem.* **2013,** *9,* 2009–2014.
7. Samsonov, S. A.; Salwiczek, M.; Anders, G.; Koksch, B.; Pisabarro, M. T. *J. Phys. Chem. B* **2009,** *113* (51), 16400–16408.

8. Gerling, U. I. M.; Salwiczek, M.; Cadicamo, C. D.; Erdbrink, H.; Czekelius, C.; Grage, S. L.; Wadhwani, P.; Ulrich, A. S.; Behrends, M.; Haufe, G.; et al. *Chem. Sci.* **2014,** *5* (2), 819–830.
9. a. Asante, V.; Mortier, J.; Schluter, H.; Koksch, B. *Biorg. Med. Chem.* **2013,** *21* (12), 3542–3546.
   b. Asante, V.; Mortier, J.; Wolber, G.; Koksch, B. *Amino Acids* **2014,** *46* (12), 2733–2744.
10. Mason, J. M.; Arndt, K. M. *ChemBioChem* **2004,** *5* (2), 170–176.
11. Qiu, X. -L.; Qing, F. -L. *Eur. J. Org. Chem.* **2011,** *2011* (18), 3261–3278.
12. Qiu, X. -L.; Meng, W. -D.; Qing, F. -L. *Tetrahedron* **2004,** *60* (32), 6711–6745.
13. Laue, K.; Kröger, S.; Wegelius, E.; Haufe, G. *Eur. J. Org. Chem.* **2000,** *2000* (22), 3737–3743.
14. a. Harry, S. *Chem. Ber.* **1960,** *93* (11), 2485–2490.
   b. Krishnamurthy, V.; Landi, J. J.; Roth, G. P. *Synth. Commun.* **1997,** *27* (5), 853–860.
15. Winkler, D.; Burger, K. *Synthesis* **1996,** *1996* (12), 1419–1421.
16. Schedel, H.; Dmowski, W.; Burger, K. *Synthesis* **2000,** *2000* (12), 1681–1688.
17. Tsushima, T.; Kawada, K.; Ishihara, S.; Uchida, N.; Shiratori, O.; Higaki, J.; Hirata, M. *Tetrahedron* **1988,** *44* (17), 5375–5387.
18. Sorochinsky, A. E.; Ueki, H.; Aceña, J. L.; Ellis, T. K.; Moriwaki, H.; Sato, T.; Soloshonok, V. A. *J. Fluorine Chem.* **2013,** *152,* 114–118.
19. Wang, F.; Qi, X.; Liang, Z.; Chen, P.; Liu, G. *Angew. Chem. Int. Ed.* **2014,** *53* (7), 1881–1886.
20. El-Kousy, S. M.; Burger, K. *Synthesis* **2004,** *2004* (11), 1821–1829.
21. Larsson, U.; Carlson, R.; Leroy, J. *Acta Chem. Scand.* **1993,** *47* (4), 380–390.
22. Xing, X.; Fichera, A.; Kumar, K. *J. Org. Chem.* **2002,** *67* (5), 1722–1725.
23. Evans, D. A.; Britton, T. C.; Ellman, J. A.; Dorow, R. L. *J. Am. Chem. Soc.* **1990,** *112* (10), 4011–4030.
24. Erdbrink, H.; Peuser, I.; Gerling, U. I. M.; Lentz, D.; Koksch, B.; Czekelius, C. *Org. Biomol. Chem.* **2012,** *10* (43), 8583–8586.
25. Chen, Q.; Qiu, X. -L.; Qing, F. -L. *J. Org. Chem.* **2006,** *71* (10), 3762–3767.
26. Pigza, J. A.; Quach, T.; Molinski, T. F. *J. Org. Chem.* **2009,** *74* (15), 5510–5515.
27. Benhaim, C.; Bouchard, L.; Pelletier, G.; Sellstedt, J.; Kristofova, L.; Daigneault, S. *Org. Lett.* **2010,** *12* (9), 2008–2011.
28. Biava, H.; Budisa, N. *J. Fluorine Chem.* **2013,** *156,* 372–377.
29. a. Wang, Z.; Resnick, L. *Tetrahedron* **2008,** *64* (27), 6440–6443.
   b. Davis, F. A.; Portonovo, P. S.; Reddy, R. E.; Chiu, Y. -h. *J. Org. Chem.* **1996,** *61* (2), 440–441.
   c. Davis, F. A.; Srirajan, V.; Fanelli, D. L.; Portonovo, P. *J. Org. Chem.* **2000,** *65* (22), 7663–7666.
   d. Portonovo, P.; Liang, B.; Joullié, M. M. *Tetrahedron: Asymmetry* **1999,** *10* (8), 1451–1455.
30. Zhang, C.; Ludin, C.; Eberle, M. K.; Stoeckli-Evans, H.; Keese, R. *Helv. Chim. Acta* **1998,** *81* (1), 174–181.
   a. Campbell, A. D.; Raynham, T. M.; Taylor, R. J. K. *Synthesis* **1998,** *12,* 1707–1709.
31. Xing, X.; Fichera, A.; Kumar, K. *Org. Lett.* **2001,** *3* (9), 1285–1286.
32. Anderson, J. T.; Toogood, P. L.; Marsh, E. N. G. *Org. Lett.* **2002,** *4* (24), 4281–4283.
33. Chiu, H. -P.; Cheng, R. P. *Org. Lett.* **2007,** *9* (26), 5517–5520.
34. Kovacs, J. M.; Mant, C. T.; Hodges, R. S. *Peptide Sci.* **2006,** *84* (3), 283–297.
35. Pendley, S. S.; Yu, Y. B.; Cheatham, T. E. *Proteins Struct. Funct. Bioinf.* **2009,** *74* (3), 612–629.

36. Zhao, Y. H.; Abraham, M. H.; Zissimos, A. M. *J. Org. Chem.* **2003,** *68* (19), 7368–7373.
37. a. Chiu, H. -P.; Suzuki, Y.; Gullickson, D.; Ahmad, R.; Kokona, B.; Fairman, R.; Cheng, R. P. *J. Am. Chem. Soc.* **2006,** *128* (49), 15556–15557.
    b. Chakrabartty, A.; Kortemme, T.; Baldwin, R. L. *Protein Sci.* **1994,** *3* (5), 843–852.
38. a. Doig, A. J.; Chakrabartty, A.; Klingler, T. M.; Baldwin, R. L. *Biochemistry* **1994,** *33* (11), 3396–3403.
    b. Andersen, N. H.; Tong, H. *Protein Sci.* **1997,** *6* (9), 1920–1936.
39. Minor, D. L.; Kim, P. S. *Nature* **1994,** *367* (6464), 660–663.
40. a. Hakelberg, M.; Koksch, B. *Chim. Oggi* **2007,** *25* (4).
    b. Scheike, J. A.; Baldauf, C.; Spengler, J.; Albericio, F.; Pisabarro, M. T.; Koksch, B. *Angew. Chem. Int. Ed.* **2007,** *46* (41), 7766–7769.
41. Woolfson, D. N. The Design of Coiled-coil Structures and Assemblies. In *Adv. Protein Chem,* Vol. 70; Academic Press, 2005; pp 79–112.
42. a. Tang, Y.; Ghirlanda, G.; Petka, W. A.; Nakajima, T.; DeGrado, W. F.; Tirrell, D. A. *Angew. Chem.* **2001,** *113* (8), 1542–1544.
    b. Tang, Y.; Ghirlanda, G.; Vaidehi, N.; Kua, J.; Mainz, D. T.; Goddard, W. A.; DeGrado, W. F.; Tirrell, D. A. *Biochemistry* **2001,** *40* (9), 2790–2796.
    c. Lee, H. -Y.; Lee, K.-H.; Al-Hashimi, H. M.; Marsh, E. N. G. *J. Am. Chem. Soc.* **2006,** *128* (1), 337–343.
    d. Buer, B. C.; Meagher, J. L.; Stuckey, J. A.; Marsh, E. N. G. *Proc. Natl. Acad. Sci. U.S.A.* **2012,** *109* (13), 4810–4815.
    e. Yoder, N. C.; Kumar, K. *Chem. Soc. Rev.* **2002,** *31* (6), 335–341.
43. a. Jäckel, C.; Seufert, W.; Thust, S.; Koksch, B. *ChemBioChem* **2004,** *5* (5), 717–720.
    b. Jäckel, C.; Salwiczek, M.; Koksch, B. *Angew. Chem. Int. Ed.* **2006,** *45* (25), 4198–4203.
44. Salwiczek, M.; Samsonov, S.; Vagt, T.; Nyakatura, E. K.; Fleige, E.; Numata, J.; Colfen, H.; Pisabarro, M. T.; Koksch, B. *Chem. – Eur. J.* **2009,** *15* (31), 7628–7636.
45. a. Vagt, T.; Nyakatura, E. K.; Salwiczek, M.; Jackel, C.; Koksch, B. *Org. Biomol. Chem.* **2010,** *8* (6), 1382–1386.
    b. Nyakatura, E. K.; Reimann, O.; Vagt, T.; Salwiczek, M.; Koksch, B. *RSC Adv.* **2013,** *3* (18), 6319–6322.
    c. Salwiczek, M.; Mikhailiuk, P.; Afonin, S.; Komarov, I.; Ulrich, A.; Koksch, B. *Amino Acids* **2010,** *39* (5), 1589–1593.
46. Huhmann, S.; Nyakatura, E. K.; Erdbrink, H.; Gerling, U. I. M.; Czekelius, C.; Koksch, B. *J. Fluorine Chem.* **2015,** *175,* 32–35.
47. a. Pagel, K.; Wagner, S. C.; Samedov, K.; von Berlepsch, H.; Böttcher, C.; Koksch, B. *J. Am. Chem. Soc.* **2006,** *128* (7), 2196–2197.
    b. Pagel, K.; Koksch, B. *Curr. Opin. Chem. Biol.* **2008,** *12* (6), 730–739.
    c. Pagel, K.; Vagt, T.; Koksch, B. *Org. Biomol. Chem.* **2005,** *3* (21), 3843–3850.
48. Brandenburg, E.; von Berlepsch, H.; Gerling, U. I. M.; Böttcher, C.; Koksch, B. *Chem. – Eur. J.* **2011,** *17* (38), 10651–10661.
49. Schechter, I.; Berger, A. *Biochem. Biophys. Res. Commun.* **1967,** *27* (2), 157–162.
50. a. Hedstrom, L. *Chem. Rev.* **2002,** *102* (12), 4501–4524.
    b. Polgar, L. *Cell. Mol. Life Sci.* **2005,** *62* (19–20), 2161–2172.
    c. Czapinska, H.; Otlewski, J. *Eur. J. Biochem.* **1999,** *260* (3), 571–595.
    d. Keil, B. *Specificity of Proteolysis;* Springer: Berlin Heidelberg, 1992.
51. Brady, K.; Abeles, R. H. *Biochemistry* **1990,** *29* (33), 7608–7617.
52. Salwiczek, M.; Vagt, T.; Koksch, B. Artificial Model System Designed to Investigate Fluorinated Amino Acids in Native Protein Environments. In *Fluorine in Bioorganic and Medicinal Chemistry;* Ojima, I., Ed.; Blackwell Publishers, 2009; pp 391–409.

53. a. Schellenberger, V.; Jakubke, H.-D. *Biochim. Biophys. Acta Protein Struct. Mol. Enzymol.* **1986,** *869* (1), 54–60.
b. Schellenberger, V.; Turck, C. W.; Rutter, W. J. *Biochemistry* **1994,** *33* (14), 4251–4257.
54. Antal, J.; Pal, G.; Asboth, B.; Buzas, Z.; Patthy, A.; Graf, L. *Anal. Biochem.* **2001,** *288* (2), 156–167.
55. a. Fruton, J. S. The Specificity and Mechanism of Pepsin Action. In *Adv. Enzymol. Relat. Areas Mol. Biol;* John Wiley & Sons, Inc., 2006; pp 401–443.
b. Powers, J. C.; Harley, A. D.; Myers, D. V. Subsite Specificity of Porcine Pepsin. In *Acid Proteases: Structure, Function, and Biology,* Vol. 92, Tang, J., Ed.; Springer: US, 1977; pp 141–157.
56. a. Cornish-Bowden, A. J.; Knowles, J. R. *Biochem. J.* **1969,** *113* (2), 353–362.
b. Suguna, K.; Padlan, E. A.; Smith, C. W.; Carlson, W. D.; Davies, D. R. *Proc. Natl. Acad. Sci. U.S.A.* **1987,** *84* (20), 7009–7013.
57. a. Kageyama, T. *Cell. Mol. Life Sci.* **2002,** *59* (2), 288–306.
b. Ben, M. D. *Chem. Rev.* **2002,** *102* (12), 4431–4458.
c. Dunn, B. M. *Chem. Rev.* **2002,** *102* (12), 4431–4458.
58. a. Dunn, B. M.; Jimenez, M.; Parten, B. F.; Valler, M. J.; Rolph, C. E.; Kay, J. *Biochem. J.* **1986,** *237* (3), 899–906.
b. Dunn, B. M.; Valler, M. J.; Rolph, C. E.; Foundling, S. I.; Jimenez, M.; Kay, J. *Biochim. Biophys. Acta Protein Struct. Mol. Enzymol.* **1987,** *913* (2), 122–130.
c. Dunn, B. M.; Hung, S.-H. *Biochim. Biophy. Acta Protein Struct. Mol. Enzymol.* **2000,** *1477* (1–2), 231–240.
59. Moses, E.; Hinz, H. -J. *J. Mol. Biol.* **1983,** *170* (3), 765–776.
60. a. Marquart, M.; Walter, J.; Deisenhofer, J.; Bode, W.; Huber, R. *Acta Crystallogr. Sect. B* **1983,** *39* (4), 480–490.
b. Hanson, W. M.; Domek, G. J.; Horvath, M. P.; Goldenberg, D. P. *J. Mol. Biol.* **2007,** *366* (1), 230–243.
61. Ye, S.; Loll, B.; Berger, A. A.; Mulow, U.; Alings, C.; Wahl, M. C.; Koksch, B. *Chem. Sci.* **2015,** *6* (9), 5246–5254.
62. Krowarsch, D.; Otlewski, J. *Protein Sci.* **2001,** *10* (4), 715–724.
63. Wenzel, H. R.; Tschesche, H. Reversible Inhibitors of Serine Proteinases: Naturally Ocurring Miniproteins, Semisynthetic Variants, Recombinant Homologs and Synthetic Peptides. In *Peptides: Synthesis, Structures, and Applications;* Gutter, B., Ed.; 1995; pp 321–362.
64. van Straaten, K. E.; Kuttiyatveetil, J. R. A.; Sevrain, C. M.; Villaume, S. A.; Jiménez-Barbero, J.; Linclau, B.; Vincent, S. P.; Sanders, D. A. R. *J. Am. Chem. Soc.* **2015,** *137* (3), 1230–1244.

Ionic Liquids and Polymers for Battery and Fuel Cells

16

C. Iojoiu[1,2], O. Danyliv[1,2], F. Alloin[1,2,3]
[1]University Grenoble Alpes, LEPMI, Grenoble, France; [2]CNRS, LEPMI, Grenoble, France;
[3]Réseau sur le Stockage Electrochimique de l'Energie (RS2E), Amiens Cedex, France

Chapter Outline

1. Introduction

Fossil fuel limitations, global warming, and air pollution make necessary the development of reliable, safe, cheap, and efficient electrochemical energy sources. Electrochemical energy devices currently extensively imply lithium batteries, fuel

Modern Synthesis Processes and Reactivity of Fluorinated Compounds. http://dx.doi.org/10.1016/B978-0-12-803740-9.00016-0

cells, photovoltaic cells, etc. Initially stimulated by the nomad electronic market needs, the trend has been tremendously accelerated by the electronic market of electric transportation (electric and hybrid vehicles) and of delocalized energy management needs (smart grids). This growing market will definitely explode in the future years following the partial electrification of automotive fleets. Therefore a perpetual effort in improving energy storage systems must be addressed to develop all these new technologies of mobility. In particular, research in the field of battery systems and fuel cell has intensified in the 2000s to meet consumers' requirements.

Rechargeable batteries are the oldest form of electricity storage and the most widely used. Among the different technologies, lithium-ion batteries arouse particular interest due to their high gravimetric and volumetric energy densities, when compared with other technologies of energy storage such as lead-acid (Pb), nickel–cadmium (Ni–Cd), or nickel-metal hydride (Ni–MH) batteries.[1] Thus, lithium batteries have progressively replaced Ni–Cd and Ni–MH systems, as they provide up to 250 Wh/kg, whereas other technologies provide only 100 Wh/kg.

In parallel to rechargeable batteries, proton exchange membrane fuel cell (PEMFC) appears as a promising alternative clean power for a wide variety of applications ranging from portable/micropower and transportation to large-scale stationary power systems for buildings and distributed generation.

The electrolyte component in rechargeable batteries and fuel cells is one of the main compounds as it conditions both the electrochemical performances and the safety of the system. The fluorine atom is omnipresent in the definition of a performant electrolyte of the systems; in this chapter, attention will be focused on fluorinated polymer and ionic liquid (IL) used as electrolyte component in the frame of lithium-ion battery and PEMFC.

2. Fluorinated Compounds in Lithium-Ion Battery

Lithium-ion battery electrolytes consist mainly of a molar solution of a superacid lithium salt, thus generally a fluorinated anion is used such as PF_6^-, BF_4^-, $(CF_3SO_2)_2N^-$ (TFSI) bis(trifluoromethane)sulfonimide lithium, and $(FSO_2)_2N^-$ (FSI (bis (fluorosulfonyl) imide)) in a solvating medium. The solvents used can be divided into several families (1) a mixture of two or three liquid solvents swelling a dense or porous polymer membrane, (2) a polymer that presents solvation ability and (3) an IL swelling a porous separator or a poly(IL).

2.1 Fluorinated Solvents for Liquid Electrolyte

Carbonate-based liquid electrolyte is commonly used in lithium-ion technology. Thus a mixture of cyclic carbonate such as ethylene carbonate (EC), which provides high dielectric constant, and linear carbonate such as dimethyl carbonate or diethyl carbonate (DEC), which improves ionic species mobility, is generally used. In addition to the electrochemical criteria (ionic conductivity, electrochemical stability), attention must be paid to other physicochemical properties such as toxicity, melting point, boiling point, flash point, and flammability.

The state of the art in commercially available lithium-ion batteries, is the use of lithium hexafluorophosphate ($LiPF_6$) as salt. In fact, $LiPF_6$-based electrolytes offer a good compromise of salt dissociation and ion mobility, thus, providing high ionic conductivity.[2] Furthermore, the presence of $LiPF_6$ suppresses the oxidation of the aluminum current collector at high cell potentials by forming a protective aluminum fluoride film on it.[3,4] However, the limited thermal and chemical stabilities restrict the use of $LiPF_6$ in large-scale applications: it is widely accepted that $LiPF_6$ decomposes into LiF and PF_5, of which the latter readily hydrolyzes to form HF and PF_3O,[5] which has a detrimental effect on battery performances.[6,7]

Fluorinated solvents were reported to be a good alternative to classical solvents as they present lower flammability, higher flash points, and frequently, an enhanced electrochemical stability toward oxidation compared with nonfluorinated solvents.[8,9] In the case of fluorinated carbonates, several authors[10–14] reported promising cyclability and coulombic efficiency, mostly ascribed to the ability of the fluorinated products to form a stable solid–electrolyte interface (SEI) at the negative electrode. Interestingly, the utilization of fluorinated solvents, such as methyldifluoroacetate,[15] fluorinated carbonate, and carbamate[16–18] was reported to reduce and even to suppress the aluminum corrosion in the presence of LiTFSI (lithium trifluoromethanesulfonyl imide) (Fig. 16.1).

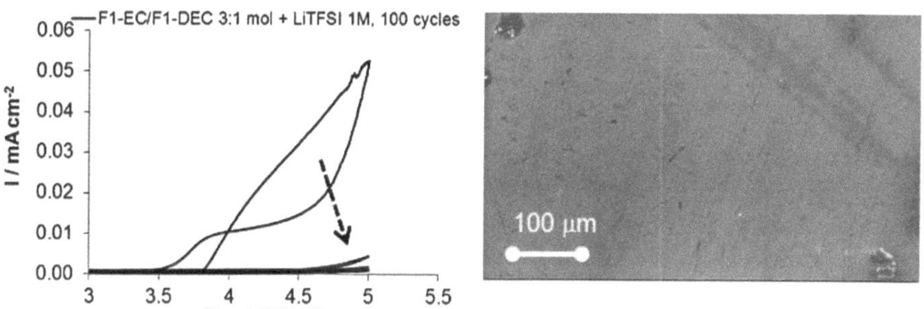

Figure 16.1 Investigation of the anodic aluminum dissolution in the presence of LiTFSI using fluorinated carbonate as electrolyte solvent and blank aluminum foil by means of cyclic voltammetry and subsequent scanning electron micrographic analysis (potential range: 3.3–5.1 V; sweep rate: 5 mV/s). *DEC*, diethyl carbonate; *EC*, ethyl carbonate.

Indeed LiTFSI is certainly one of the most promising candidates to replace $LiPF_6$ due to its high ionic conductivity,[19] high thermal and electrochemical stabilities, and substantially lower sensitivity toward moisture. Fluorinated and partially fluorinated carbonate mixtures provide a solution to overcome the detrimental anodic aluminum current collector dissolution observed with LiTFSI by forming a protective aluminum fluoride film on the aluminum surface.[16] Moreover, the investigation of these electrolytes in lithium-ion cells, composed of commercially available graphite as negative electrode and $Li[Ni_{1/3}Mn_{1/3}Co_{1/3}]O_2$ as the positive material, revealed very stable cycling performance and good rate capability.[16]

2.2 Fluorinated Ionic Liquid for Liquid Electrolytes

2.2.1 Electrolyte Properties

ILs have attracted growing attention as electrolytes for lithium-ion batteries, due to their high flash point (nonflammability), low vapor pressure, high thermal stability, and wide electrochemical stability window.[20–22] ILs offer great ability in the designing of cationic and anionic structures and their combinations. For lithium battery application, the desired properties, in addition to security aspect, are low viscosity to promote ionic species mobility, low melting point to exhibit high conductivity at low temperature, and chemical and electrochemical stabilities at the electrode/electrolyte interfaces.

The modulation of both the cation and anion was useful to decrease both the viscosity and the melting point. Imidazolium and pyrrolidinium-based ILs were thus the most studied, as these cations exhibit the lowest viscosities. The fluorinated anion also has a large impact on both the IL melting point and the viscosity. The first development of ILs, with low melting point, was associated with the used of highly fluorinated delocalized anion as bis(trifluoromethylsulfonyl)imide [$(CF_3SO_2)_2N$, TFSI, trifluoromethane- (CF_3SO_3) and nonafluorobutanesulfonate ($C_4F_9SO_3$) anions.[23] More recently, other anions were developed for both ILs and lithium salts as the requirements are similar, i.e., electrochemical stability and highly delocalized anion. The electrochemical fluorination process is a power tool to synthesize a large variety of anions, for example, the (bis(fluorosulfonyl)amide) FAP anion ($Li[(C_2F_5)_3PF_3]$) was synthesized by this route.[24] The substitution of fluorine atom by C_6F_5 increases significantly the stability of the salt in the presence of water, and this new anion can usefully replace PF_6^-. The FAP-based ILs was intensively studied.[25–31] Indeed, FAP-based ILs present interesting properties such as high stability in oxidation and large hydrophobicity, which can manage interface properties.[29] However the FAP-based ILs exhibit conductivity lower than ILS based on $TFSI^-$, which can be associated to the bulky character of the anion. In parallel, the works of Matsumoto et al.[32] reported the synthesis of new ILs with asymmetric 2,2,2-trifluoro-N-(trifluoromethylsulfonyl) acetamide anion [CF_3SO_2-N-$COCF_3$, (TSAC)]. TSAC-based ILs showed low melting point and viscosity; however, the conductivities obtained are similar to the ones exhibited by TFSI ILs.

The FSI anion, $(FSO_2)N^-$, is one of the most studied anion nowadays. The use of the FSI^- anion allows obtaining ILs with very low viscosity and high conductivity. The ethyl methyl imidazolium FSI (EMI-FSI) exhibits a conductivity of 17 mS/cm at 20°C,[33] higher than that of common liquid electrolyte EC-DEC $LiPF_6$ 1M. The viscosity of the FSI^--based IL was 60% lower than that of the IL with $TFSI^-$ and the same cation.[34] Another advantage of the FSI^- anion is the inhibition of Al current collector corrosion at high potential[35] and the formation of an SEI with adapted properties on lithium and graphite negative electrodes.[33,36] However, the use of the FSI^- anion presents some drawbacks, of which the main one is its weak thermal stability. Indeed, exothermic decomposition occurs at 183°C in dry FSI-based ILs. Moreover, in the presence of water, the degradation temperature decreases drastically, thus a very exothermic decomposition reaction (200 J/g) occurs at 120°C, which greatly attenuates the benefits of using ILs.[37]

For lithium-ion battery application, high lithium-ion conductivity is needed; for this a lithium salt generally with the same anion as that of the IL is added, at a concentration between 0.5 and 1 M. The lithium salt induces a neat decrease in ionic conductivity.[38,39] For example, the addition of 1 M LiTFSI decreases the conductivity by a factor of four in aliphatic quaternary ammonium.[38] The addition of lithium salt increases both the glass-transition temperature (T_g) and the viscosity. However, the viscosity increase cannot fully explain the huge conductivity decrease. To provide quantitative data on the effect of LiTFSI addition on electrolyte dissociation, diffusion coefficients of the different species (ammonium and lithium cations, anion) were measured by pulsed field gradient nuclear magnetic resonance.[38] However, experimentally, the diffusion coefficient of Li^+ is the lowest. Thus, the apparent charge carrier size of Li^+ is much higher than that of Li^+ and may attribute to the formation of ionic clusters, $[Li(TFSI)_{n+1}]^{n-}$.[40] Hence even if ILs exhibit high total ionic conductivity, low lithium-ion conductivity was obtained, much lower than that in common liquid electrolytes. In addition, correlation signals (^7Li-^1H nuclear overhauser effect spectroscopy (HOSEY) spectrum) are observed between the Li^+ cation and the *N*-methyl groups[38] (Fig. 16.2).

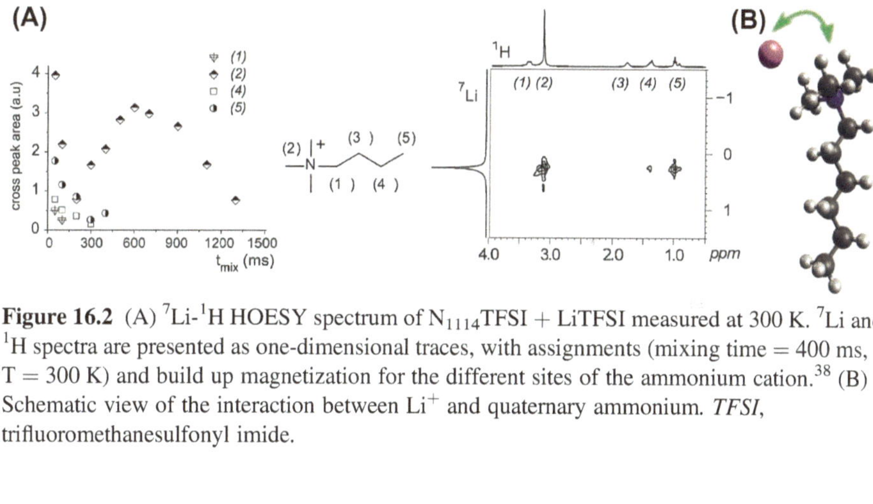

Figure 16.2 (A) ^7Li-^1H HOESY spectrum of $N_{1114}TFSI + LiTFSI$ measured at 300 K. ^7Li and ^1H spectra are presented as one-dimensional traces, with assignments (mixing time = 400 ms, T = 300 K) and build up magnetization for the different sites of the ammonium cation.[38] (B) Schematic view of the interaction between Li^+ and quaternary ammonium. *TFSI*, trifluoromethanesulfonyl imide.

Such pair correlation would be totally unexpected and establish obviously some nanoscale organization inside these electrolytes with segregation of the polar and apolar domains.

In addition to the use of fluorinated anion, fluorinated cation can be useful to modulate the IL properties. The properties of fluorinated ammonium ILs were not intensively studied; however, some specific properties were noticed.[41–43] For example, the solubility of carbon dioxide is significantly higher in the fluorine-substituted ILs.[41,42] For dye-sensitized solar cells, fluorinated imidazolium cation can permit high performances. Imidazolium iodide–based IL with a perfluoroalkyl polyethylene (PE) glycol tail exhibits high iodide diffusion, associated with the IL nanostructuration.[43] Fluorinated ammonium ILs can also be used as surfactant that promotes the formation and stabilization of perfluorocarbon dispersions in conventional ILs.[21] In the field of lithium battery, fluorination of the alkyl chain of aliphatic ammonium TFSI increases the stability toward oxidation because of the withdrawing effect of fluorine function.[38] However, fluorinated aliphatic ammonium TFSI ILs exhibit low conductivity values associated with high viscosity, although a weak increase of the dissociation was obtained. An improvement in the conductivity was obtained using fluorinated pyrrolidinium-based IL.[44] In addition, piperidium- and pyrrolidinium-based ILs with fluorinated group exhibit like aliphatic ILs high electrochemical stability in oxidation.

2.2.2 Ionic Liquid as Lithium-Ion Battery Electrolyte

Within the past few years, it has been shown that ILs + lithium salt can permit highly reversible plating/stripping of lithium.[45–48] For example, 1-butyl-1-methylpyrrolidinium TFSI with LiTFSI (0.5 M) has been shown to permit uniform deposition on the lithium metal electrode, thus reducing dendrite formation at rates up to 1 A/cm^2 without additional additives. The improved lithium plating behavior was associated with the formation of crystallinelike liquid layers at the interface lithium/electrolyte.[49] Several studies have also investigated the interaction of (room temperature ionic liquid) RTIL-based electrolyte solutions with lithium-intercalating positive electrode materials.[33,50–53] Layered oxides such as $LiCoO_2$ do not seem to be stable with respect to metal leaching,[50] whereas $LiFePO_4$ appears compatible with IL-based electrolytes.[33,50] Li/$LiCoO_2$ cells were cycled with the *N*-methyl-*N*-propylpyrrolidinium bis(fluorosulfonyl)imide as electrolyte with different lithium salt concentrations. An improvement of the capacity was observed with the salt concentration increase. When the salt concentration is 3.2 mol/kg, high capacity can be obtained even at high C-rate (4C, 5C). However, weak cyclability was reached, which may be linked to the $LiCoO_2$ electrode degradation in the presence of IL.[51,52] The use of functional IL, 1-allyl-1-methylpiperidinium TFSI, allows to electrochemically modify the surface of the $LiCoO_2$ electrode. This modified surface inhibited the decomposition of the electrolyte. A high coulombic efficiency, higher than 97%, is thus obtained.[53]

Studies performed on $LiFePO_4$-positive electrode[33,50] have shown that pyrrolidinium sulfonylimide–based ILs are suitable for use in a cell containing a lithium metal electrode and $LiFePO_4$ electrode. In particular, the combination of *N*-methyl-*N*-propylpyrrolidinium bis(fluorosulfonyl)imide + 0.5 mol/kg LiTFSI has been shown to work effectively as an electrolyte solution with metallic lithium. Excellent cyclability of the $LiFePO_4$ is obtained even at 4C discharge rate; the discharge capacity is still 110 mAh/g, i.e., 72% of the C/10 value at 50°C.[50]

Only a few studies have applied IL electrolyte to high-voltage electrode.[54–56] An optimized IL-based electrolyte (butylmethylpyrrolidium-TFSI, 0.4M LiTFSI, 0.6M $LiPF_6$) permits to exhibit good cyclability performances at 50°C, without Al current collector corrosion.[57] However, at room temperature, the performances are worse than the one obtained with carbonate-based electrolyte. The same conclusions can be obtained with regard to the working of lithium-rich layered oxide positive electrode. Pyrrolidinium-based TFSI IL electrolytes exhibit interesting performances at high temperature, 50°C, with lower capacity feeding than liquid electrolyte.[58]

Regarding the negative electrode, reversible capacity of a graphite electrode was obtained with pure EMI-FSI electrolyte during 30 cycles at 0.2C. The FSI anion is able to form stable SEI on graphite surface, which is not the case with TFSI anion.

In addition to intrinsic IL thermal stability, the presence of the charged electrode has a significant impact on the thermal run away in battery configuration. Using accelerating rate calorimetry, the reactivity between several ILs (with and without added $LiPF_6$) and charged electrode materials (Li_xSi, $Li_7Ti_4O_{12}$, $Li_{0.45}CoO_2$) is evaluated. In contrast to what is expected, not all ILs are safer than the commonly used electrolytes/solvents (EC + DEC with or without $LiPF_6$). EMI-FSI shows the worst safety properties, and the stability is lower than that obtained with conventional electrolyte. EMI-TFSI shows similar reactivity to carbonate-based electrolyte, whereas 1-butyl-2,3-dimethylimidazolium TFSI, methyl-N-propyl piperidinium $TFSI^-$, and trimethylbutylammonium TFSI are more safe than carbonate-based liquid electrolytes.[59]

2.3 Polymeric Ionic Liquid

The polymeric ionic liquids (PILs) (Fig. 16.3A) have created a keen interest. Indeed, PILs combine both the advanced properties of conventional ILs and the flexibility and mechanical properties of polymer. There has been a growing interest in PILs as solid polymer electrolytes for application in electrochemical devices, such as lithium-ion batteries,[60–62] solar cells,[63] and supercapacitors.[64] The use of PILs swelling by ILs is a good way to produce a flexible polymer electrolyte with good conductivity (especially at high temperature) and nonflammability. For obtaining high ionic conductivity, the anion should be small, be mobile with delocalized charge (fluorinated anion), and have low interaction with the main polymer chain. To optimize the conductivity, the main polymer chain must be flexible, the ionic group must be placed at the end of side chains, and the spacer (between the main polymer chain and the ionic function) should include oxyethylene unit to promote solvation properties. A review in 2005 was dedicated to the advance in PIL-based electrolytes.[65] The maximum conductivity reported for dried PIL is equal to 3×10^{-5} S/cm at 25°C.[65] An improvement in the conductivity is obtained with IL addition. In general, gel electrolytes (PIL + IL) present intermediate ionic conductivity between PIL and IL ones. The maximum conductivity reported for gels may reach 10^{-3}–10^{-2} S/cm at room temperature.

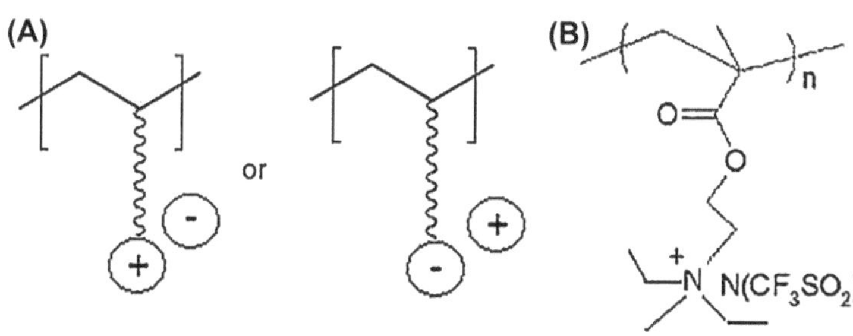

Figure 16.3 (A) Schematic representation of PILs, (B) poly(*N*, *N*-diethyl-*N*-(2-methacryloylethyl)-*N*-methylammonium TFSI).[66] *TFSI*, trifluoromethanesulfonyl imide.

Controlled molecular weight poly(*N*, *N*-diethyl- *N*-(2-methacryloylethyl)-*N*-methylammonium TFSI) can be prepared by control transfer radical polymerization (Fig. 16.3B).[66] Even if low-molecular-weight PILs permit to reduce T_g, low ionic conductivity is obtained. The addition of free IL is needed to improve the ionic conductivity. Cyclic test was performed using $Li_4Ti_5O_{12}/LiMn_2O_4$ as electrode; high capacities were obtained, but only at 40°C, at low C-rate, and with only 5% of PILs in the electrolyte.

2.4 Fluorinated Polymer as Separator

Liquid electrolytes swelling a polymer membrane allow a high conductivity level to be obtained at ambient temperature. However, a significant part of the internal resistance of the batteries originates from the electrolyte. An answer consists, therefore, in decreasing the electrolyte thickness used to separate the electrodes. The main parameter determining porous separator performances are the thickness; degree of porosity; pore size; wettability; electrolyte adsorption and retention; as well as chemical, electrochemical, dimensional, and thermal stabilities. Hence commercial porous separators dedicated to lithium batteries are manufactured from apolar polyolefins.[67] The fluorinated polymer that is most used in lithium battery application is polyvinylidene fluoride (PVdF) and its copolymers. These polymers present high chemical and electrochemical stabilities, high polarity, and high dielectric constant, which improve the polar electrolyte wettability and then the ionic conductivity when compared with polyolefin-based separators.[68]

Before introducing the properties of electrolytes based on the PVdF family, the main characteristics of PVdF and PVdF copolymers must be detailed. $(CH_2\text{-}CF_2)_n$ i.e., PVdF is a semicrystalline linear thermoplastic whose crystallinity varies between 35% and 70%, according to the method of synthesis and to the thermal history of the sample. This crystallinity makes it possible to have a good mechanical resistance even after swelling by the electrolyte. These properties are conditioned by the average molecular weights and the head-to-head content. For instance, PVdF samples with high enough molecular weight, i.e., $Mn > 100{,}000$ g/mol, and low head-to-head defects ($<7\%$) make it possible to limit the swelling ratio by the liquid electrolyte, thus preserving the mechanical integrity of the swollen membrane.[69] The thermal degradation of PVdF starts from 390°C, producing HF by-product.[70] PVdF is very resistant to the majority of inorganic acids, halogens, and oxidants, and this is observed even at high temperature; however, it is sensitive to bases that induce a dehydrofluoration and generate double bonds. Vinylidene fluoride can be polymerized with a variety of comonomers leading to statistical or random copolymers,[71] the main ones used as separator being poly(vinylidene fluoride-co-hexafluoropropene) (PVdF-HFP), poly(vinylidene fluoride-co-trifluoroethylene), and poly(vinylidene fluoride-co-chlorotrifluoroethylene.[72]

The main advantages of the fluorinated copolymer concern (1) its availability (2) its stability in oxidation, (3) its presence as binder in the electrode, and (4) its high wettability compared with the electrolyte. The main recent scientific investigations using PVdF-based electrolytes were dedicated to macroporous membrane instead of gelled ones, therefore this paragraph focuses on the new strategies involved in improving the PVdF-based porous separator properties. The porous PVdF membranes are generally obtained by phase inversion processes using solvent and nonsolvent[68,73–75] and electrospinning[76–82] (Fig. 16.4).

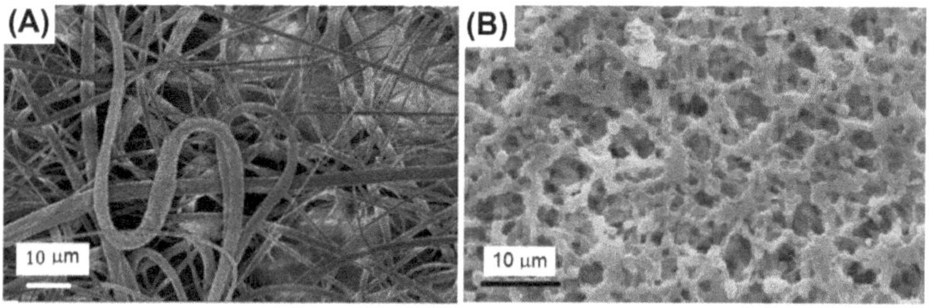

Figure 16.4 Scanning electron micrographic image of polyvinylidene fluoride membranes (A) elaborated by electrospinning and (B) phase inversion.

The obtained porosity ranges between 10% and 80%, and the pore size ranges from 0.5 to 16 µm.[72] The characteristics of the developed PVdF-based electrolytes (conductivity, porosity, electrolyte uptake) were reported in a recent review.[72] During the 2010s, electrospinning was largely developed to obtain PVdF-based porous membranes using several homopolymers, copolymers, polymer blends, composite membranes, and multilayers or core–shell structures. Electrospinning produces continuous fibers with diameter in the range of a nanometers to a few micrometers and builds highly porous structure with porosity near 80%. Ionic conductivity above 1 mS/cm and electrochemical stability corresponding to that of electrolytes were reported. The ionic conductivity of the PVdF-based membrane is directly associated with the total electrolyte uptake, related to the amount of porosity and the swelling property of the membrane. The effect of the fiber orientation was investigated[76]; the mechanical properties increase with fiber alignment. However, the orientation increases the membrane tortuosity and then decreases the ionic conductivity.

2.4.1 Blend Polymers

One strategy for both enhancing the ion mobility and the improving mechanical properties is the elaboration of polymer blends. The idea is to use one polymer with good mechanical property and the another one exhibiting good affinity for the electrolyte. The polymers more used with PVdF and its copolymers are the polyethylene oxide (PEO), polymethylmethacrylate (PMMA), and polyacrylonitrile (PAN). In these blends, PVdF assures the mechanical strength, whereas PEO, PMMA, or PAN improves the interface properties and the ionic conductivity.[72,83–87] To improve the compatibility between PEO and PVdF, fluorinated amphiphilic copolymer P(HFBMA-co-PEGMA), i.e., poly(hexafluorobutyl methacrylate)-co-poly(ethyleneglycol) methacrylate was synthesized and blended with PVdF.[88] The blend and PVdF membranes have the same morphology, with the amount of porosity and the size of the pore increasing with the addition of the amphiphilic copolymer. These properties improve the electrolyte uptake and then the conductivity. In the same vein, an original triethylene glycol methacrylate monomer was synthesized and copolymerized with VDF (1,1-Difluoroethylene) monomers. The strategy permits to control the introduced quantity of the monomer and then the properties of the obtained membranes. However, only low-molecular-weight copolymer was obtained inducing weak mechanical property.[89] Moreover, a too high affinity between the membrane and the electrolyte can induce creep and even dissolution of the membrane at high temperatures, which is detrimental for ionic conduction, thus the affinity level must be controlled and adjusted. Furthermore, the presence of ether function may induce a neat decrease of the oxidation stability of the electrolyte.

2.4.2 Composite Membranes

Composite membranes based on PVdF and charge were largely investigated. Both inorganic charges (TiO_2, SiO_2, Al_2O_3, montmorillonite)[90–93] and organic ones (cellulose whiskers, polyhedral oligomeric silsesquioxanes)[94,95] were evaluated. The addition of charge can have a significant impact on the separator properties (ionic conductivity, mechanical property), as addition of charge can modify the PVdF crystallinity, the porosity of the membrane, and the wettability of the composite separator. Therefore it is difficult to relate directly the composite membrane properties to the charge characteristics, i.e., surface area, dielectric constant, and Lewis acid–base interactions.

New strategies have been developed, in particular, multilayer and core–shell structures.

2.4.3 Multilayer Structure

To improve the thermal stability and electrolyte uptake without compromising mechanical strength, multilayer separators have been developed such as PVdF/PMMA/PVdF, PVdF/PAN/PVdF, and poly(phthalazinone ether sulfone ketone) (PPESK)/PVdF/PPESK.[96–99] PPESK due to its high T_g ranging from 263 to 305°C permits to improve the thermal dimensional stability without compromising the electrochemical performances (capacity, cyclability). Sandwiched polymer membrane PVdF/PMMA/PVdF presented interesting performances due to the good compatibility between the liquid electrolyte and the inner PMMA layer.[100] Applying a porous coating layer of PVdF or an inorganic fiber-based membrane on PE separator was investigated by several groups.[78,98,101] The addition of a porous PVdF layer on PE separator permits to enhance the ionic conductivity without decreasing the mechanical properties.[101] A multilayer approach was also chosen with an Al_2O_3 nanopowder layer and an electrospun PVdF nanofiber layer formed on the PE separator. This multilayer strategy is effective in improving the performance and safety; however, a large increase in separator thickness is obtained, and the adhesion between PE separator and Al_2O_3 layer must be improved.[78]

The elaboration of core–shell material may permit to propose a separator that exhibits a good compromise between high mechanical properties and excellent electrochemical stability and good ion mobility behavior. Core–shell structured polysulfonamide (PSA)@PVdF-HFP nonwoven separator was obtained by coaxial electrospinning technique[102] (Fig. 16.5).

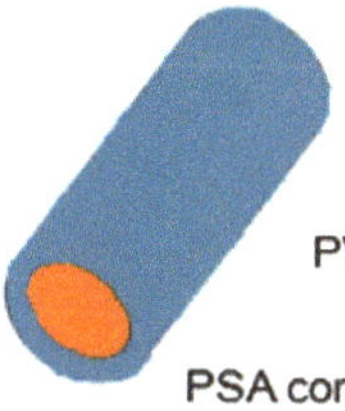

Figure 16.5 Schematic representation of core–shell structure with polysulfonamide (PSA) in the core and poly(vinylidene fluoride-co-hexafluoropropylene) (PVdF-HFP) shell.[102]

PSA was chosen for the core material due to its high mechanical property and thermal stability. PVdF-HFP was selected for the shell material because of its electrochemical stability and high swelling behavior in presence of liquid electrolyte. The electrolyte uptake at room temperature is very high, 300%, related to the high porosity of the membrane and good affinity between PVdF and electrolyte. This property induces low internal resistance, and the high ion mobility then permits to improve the performances of the battery at high C-rate.

3. Aromatic Ionomers Bearing Perfluorosulfonic Acid Side Chains

Proton exchange membranes (PEMs) in PEMFCs play a critical role in transporting hydrated protons. A successful PEM material must meet multiple stringent requirements, such as proton conductivity, good mechanical strength, high oxidative and chemical stability, low fuel and oxidant permeability, as well as easy fabrication and excellent water management under low relative humidity (RH) cycling.[103,104] Commercially available materials are perfluorosulfonic acid (PFSA) polymer membranes such as Nafion. These membranes consist of an extremely high hydrophobic perfluorinated backbone along with hydrophilic perfluorosulfonic acid side chains ($PFSA_{sc}$s). Their excellent conduction and overall fuel cell performance originate from two key properties: the highly phase-separated membrane morphology with a well-defined hydrophilic/hydrophobic interface and the acidity of the ionic function.[105,106] Although Nafion has been the most commonly used PEM during the recent decades, its properties need to be improved for broader fuel cell application. Considerable efforts have therefore been devoted to the synthesis of sulfonated aromatic polymers such as poly(arylene ether)s, poly(arylene ether sulfone)s, poly(arylene ether ketone)s, and poly(imide)s.[107–117] The sulfonated aromatic ionomers were selected due to their very good thermal, chemical, and electrochemical stability; lower cost[118]; and more environment-friendly synthesis process than that of PFSA ionomers. However, high proton conductivities are achieved within these types of PEM only at fully hydrated conditions, and the conductivities drop sharply when the RH is decreased.

The p*Ka* value of sulfonated polyether ketone, estimated to be approximately −1 when compared with −6 for Nafion, is one of the reasons for the lower conductivity of sulfonated aromatic ionomers, when compared with PFSA.[118] Additionally, it has been demonstrated that the microstructure of PEMs, is crucial for their performance in the operation of fuel cells.[109] Compared with PFSA, the sulfonated aromatic ionomers have poorer organization and phase separation and their ionic domains are narrower and less connected. Therefore it was considered that the key success of advanced PEM is to design membranes with long-range continuity of both the conducting and the mechanically robust domains.[119–124] Consequently, a variety of synthesis strategies aiming at increasing the acidity of the ionic function or/and at improving the membrane morphology have been explored to optimize the conductivity of aromatic ionomers. Regarding the search for increased acidity, aromatic ionomers, bearing aryl sulfonimide acid[125,126] or perfluoroalkylsulfonic acid side chains[127–135] instead of aryl sulfonic acid, were recently proposed. The poly(arylene ether)s containing pendant PFSA groups showed more phase-separated morphology and higher proton conductivity than the poly(arylene ether)s, having sulfonic acid groups grafted to the main chain. In this chapter an overview on the synthesis and characterization of aromatic ionomers bearing $PFSA_{sc}$ is presented, and the correlation between ionomer chemical structure, morphology, conductivity, and fuel cell performance is briefly discussed.

3.1 Ionomer Synthesis

The aromatic ionomers, bearing PFSAs, with different main chain and side chain structures were synthesized by three different methods, summarized in Table 16.1. They are: (1) postfunctionalization of an aromatic polymer, (2) postfunctionalization of a halogenated aromatic ionomer, and (3) bottom-up synthesis from ionic monomers. Regarding $PFSA_{sc}$, in the majority of syntheses 5-iodooctafluoro-3-oxapentanesulfonyl fluoride (CAS # 66137-74-4), which is commercially available, was used as a precursor.

3.1.1 Postfunctionalization of an Aromatic Polymer

This synthesis method is either postfunctionalization of a commercial poly(ether sulfone) (PES)[127,133,136,137] or PES is synthesized by copolymerization of commercial monomers, and then it is functionalized.[138–141] All the ionomers from this first synthesis method are statistical, except the ones of Iojoiu et al., who provided block copolymers by "one-pot two-reaction" method. Functionalization of the polymers with $(CF_2)_2O(CF_2)_2SO_3H$ was done by copper-mediated Ullmann-type coupling between the halogenated polymer (here, only brominated) and 5-iodooctafluoro-3-oxapentanesulfonate. This method usually includes two steps: bromination of the

Table 16.1 Summary on Existing Aromatic Ionomers, Bearing $PFSA_{sc}$

Synthesis Method	Sequence of Synthesis	Type of the Produced Ionomer	PFSA Type (R)[a]	Starting Materials for Synthesis	Ionomer Structure	Original Code of the Ionomer	References
Postfunctionalization of an aromatic polymer	1. Polycondensation of commercial monomers or the use of a ready polymer 2. Selective halogenation (**H**)/lithiation (**L**)/borylation (**B**) of the polymer 3. Grafting of PFSA through copper-mediated coupling (Ullmann-type reaction)	Poly(ether sulfone) *statistical copolymer*	$(CF_2)_2O(CF_2)_2SO_3H$ (**H**)	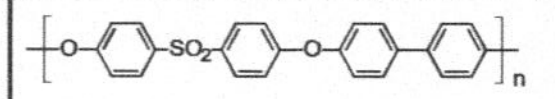	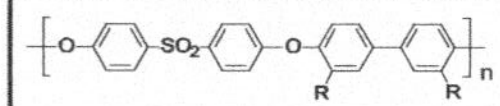	PES-PSA	127
				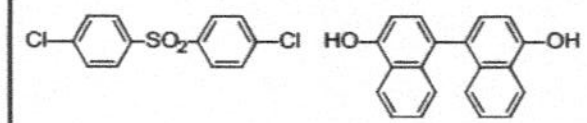	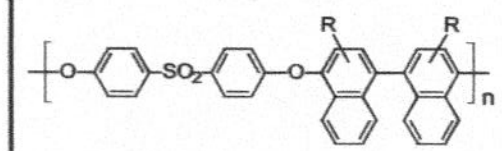	BNSH-PSA	138
				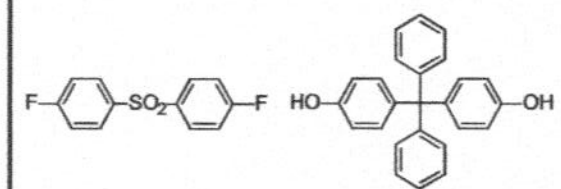	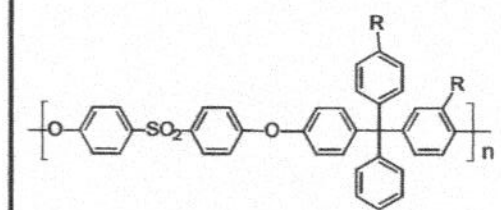	FSPE-1	139
		Poly(ether sulfone) *block copolymer*		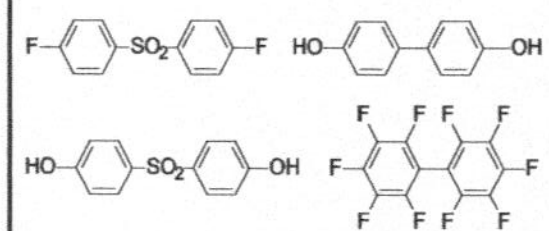	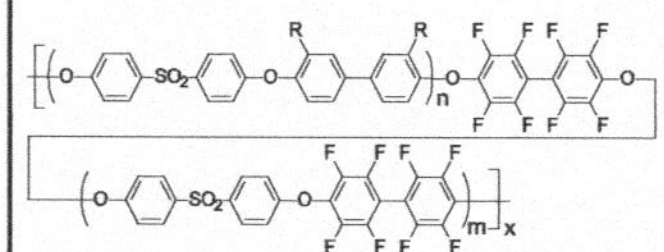	Ps-PES-FPES	140,141
		Poly(ether sulfone) *statistical copolymer*	$C(O)(CF_2)_3SO_3H$ (**L**)	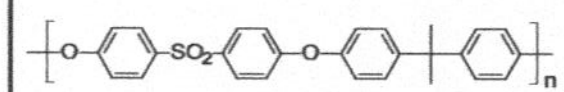	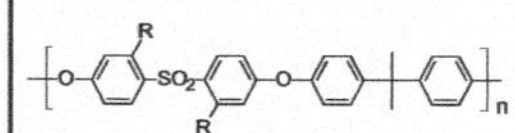	PSU-PSA	136
			$O(CF_2)_2SO_3HS(CF_2)_2SO_3HOCF_2SO_3HC(CF_3)(PhO(CF_2)_2SO_3H)_2$		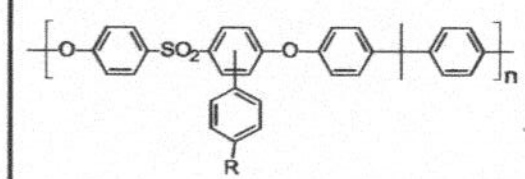	PSU-S1, $PSU\text{-}S_4$ $PSU\text{-}S_5$, $PSU\text{-}S_6$	133,137
Postfunctionalization of a halogenated aromatic polymer	1. Halogenation of a commercial monomer 2. Polycondensation of this monomer with other nonmodified monomers 3. Grafting of PFSA through copper-mediated coupling (Ullmann-type reaction)	Poly(ether) poly(ether sulfone ketone) *statistical polymer*	$(CF_2)_2O(CF_2)_2SO_3H$	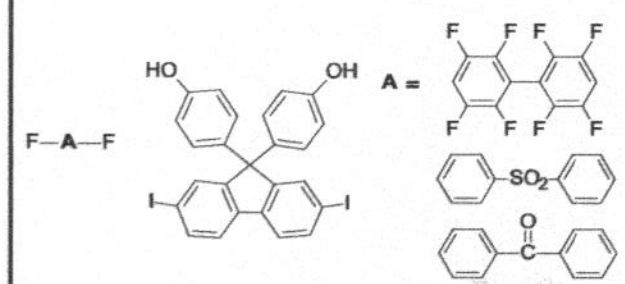	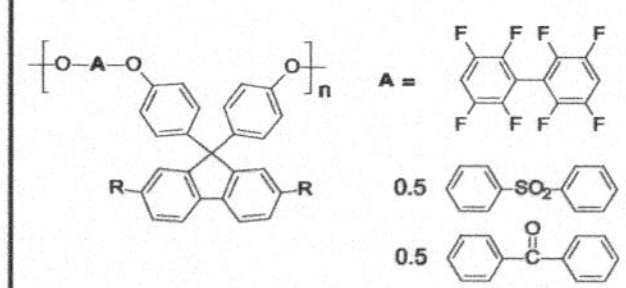	2a2b	131,132
						6a–c	135

Continued

Table 16.1 Summary on Existing Aromatic Ionomers, Bearing $PFSA_{sc}$—cont'd

Synthesis Method	Sequence of Synthesis	Type of the Produced Ionomer	PFSA Type (R)[a]	Starting Materials for Synthesis	Ionomer Structure	Original Code of the Ionomer	References
		Poly(ether sulfone ketone) *block copolymer*		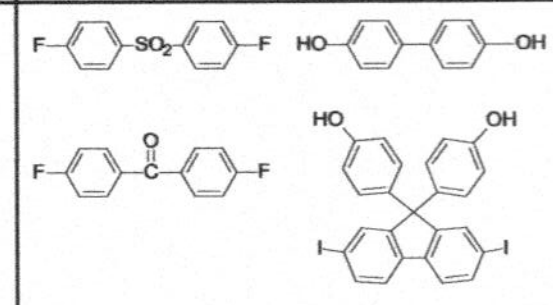	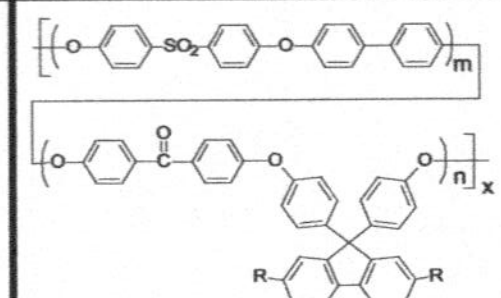		
		Poly(ether) *statistical polymer*		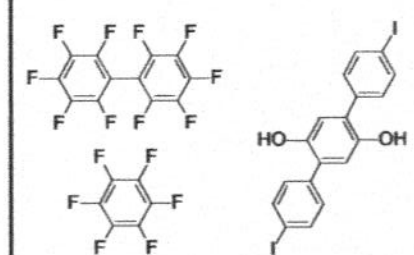	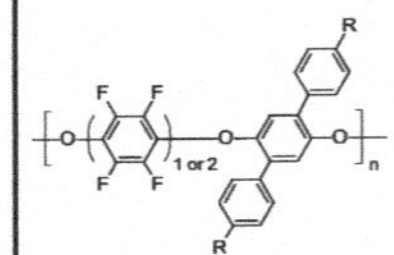	4a–d, 5a	129
Bottom-up synthesis	1. Preparation of a monomer, bearing PFSA function [by nucleophilic substitution (**S$_N$**), or by multistep addition (**A$_{MS}$**)] 2. Copolymerization of this monomer with other nonmodified monomers	Poly(ether sulfone) *statistical and block copolymer*	CF_2SO_3H (**A$_{MS}$**)	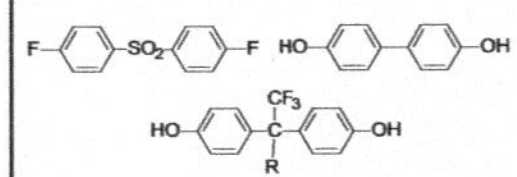	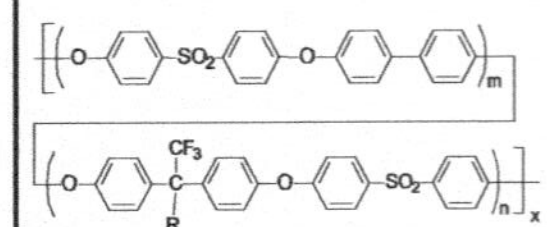	PAES (*statistical*) MB (*block-copolymers*)	128,142
		Poly(ether) Poly(phenylene) *statistical polymer*	$OCF_2CF_2SO_3H$ (**S$_N$**)	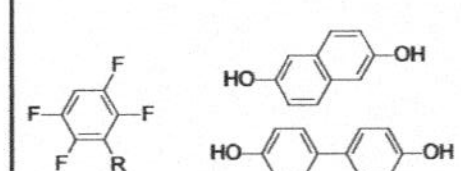	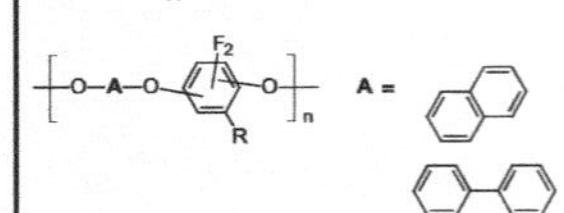	PAE-NA and PAE-BP	130
		Poly(imide)s *statistical polymer*		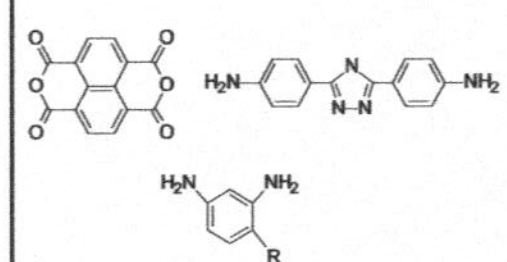	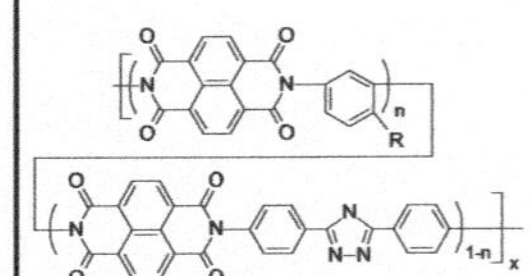	6	143

BNSH, Poly(ether sulfone) containing binaphthyl units; *FSPE*, perfluorosulfonated poly(arylene ether)s; *MB*, poly(arylene ether sulfone) multi-block copolymers bearing perfluoroalkylsulfonic acid moieties; *PAE-BP*, poly(arylene ether) containing biphenyl units; *PAE-NA*, poly(arylene ether) containing naphthyl units; *PES*, poly(ether sulfone); *PFSA*, perfluorosulfonic acid; *PFSA$_{sc}$*, perfluorosulfonic acid side chain; *PSA*, polysulfonamide.

[a]R is in acid form, when at the ionomer, and in salt form, when at starting materials.

polymer in a moderately polar solvent (dichloromethane or chloroform) at low temperature (from 0°C to room temperature) and the subsequent coupling of the brominated polymer with the 5-iodooctafluoro-3-oxapentanesulfonate in the presence of fine metallic copper powder in a highly polar solvent (dimethylsulfoxide (DMSO) or more rarely dimethylacetamide (DMAc)) at 120°C (sometimes the reaction is reported at 160°C[139]) during a wide time range (6–96 h). As a result of the electrophilic nature of bromination, *ortho*-positions to the ether bond become functionalized. Ghassemi et al.[136] performed modification of the commercial PES with $C(O)(CF_2)_3SO_3H$ group: they performed lithiation (with butyllithium), which occurs at the most acidic sites of the polymer, at the *ortho*-carbons to SO_2-bridge. Very recently another method of $PFSA_{sc}$ attachment to an aromatic backbone was reported by Bae et al.[133,137]—borylation and subsequent Suzuki coupling. The authors grafted $PFSA_{sc}$ with four different lengths.

It is important to underline that the postmodifications of the polymer has the disadvantage of control of functionalization of the active sites; e.g., not all the structural units contain two $PFSA_{sc}$ groups. The degree of functionalization varied from one research group to another: Ghassemi et al.[136] got approximately 55% of the maximal possible functionalization, Watanabe et al.[139] improved that result up till 68%, but Yoshimura's work,[127] which was one of the first researches on aromatic ionomers bearing $PFSA_{sc}$ groups, stated 79% functionalization. The highest degree of PFSA impregnation was reported by the three other groups,[133,137,138,140,141] which obtained >90%, even up till 100%. However, Ueda et al.[138] synthesized PES having a naphthalene moiety in a polymer backbone, therefore substitution to $PFSA_{sc}$ groups occurred at some additional sites of the binaphthyl part. Meanwhile, the success of $PFSA_{sc}$ introduction by Bae's group[133,137] might testify that borylation of PES is easier to conduct, rather than bromination, but to make such an assumption more work must be compared.

3.1.2 Postfunctionalization of a Halogenated Aromatic Polymer

The second method of polymer synthesis to obtain aromatic polymers with $PFSA_{sc}$ groups is similar to the first one; however, it is not the polymer itself, but one of the initial monomers that is halogenated. Then it undergoes polycondensation with a commercial monomer resulting in a halogenated polymer that is further functionalized with $PFSA_{sc}$. In this method, only $(CF_2)_2O(CF_2)_2SO_3H$ was investigated as a $PFSA_{sc}$. Miyatake et al.[131,132,135] paid much attention to the iodination of fluorene and to the synthesis of polymers thereof, having different aromatic backbone and monomer sequence (statistical and block copolymers). However, the most effective polymerization (resulting in a high-molecular-weight polymer) was reported with decafluorobiphenyl. At the same time this monomeric part increases the hydrophobicity of the main chain, which ought to favor the separation of hydrophilic and hydrophobic phases within the membrane. The researchers introduced two phenyl iodine groups to a dihydroxybenzene, and copolymerized with decafluorobiphenyl and hexafluorobenzene. After functionalization with $PFSA_{sc}$, they achieved ionomers of ionic exchange capacity (IEC) 1.8 and 1.6 meq/g, which was equal to 85% and 80% of iodine conversion, respectively.

3.1.2.1 Bottom-Up Synthesis

The last method is the bottom-up synthesis (polycondensation) of ionomers, starting from ionic and nonionic monomers (Table 16.1). The main challenge of this method lies in the difficulty of production of an ionic monomer of high purity. Storey et al.[128,142] produced an ionic monomer, with difluoromethyl spacer between the aromatic ring and sulfonic function, i.e., 2,2-bis(*p*-hydroxyphenyl)-pentafluorosulfonate. The group proposed the full study on both statistical and block copolymers. Longer $PFSA_{sc}$ lateral chains were introduced by Wang et al.[130] and Watanabe et al.[143] The synthesis procedures of the ionic monomers in both groups were common: aromatic phenolic precursors were subjected to nucleophilic substitution with 1,2-dibromotetrafluoroethylene with subsequent sulfination and oxidation.

3.2 Membrane Manufacturing

For comparing the ionomers one must consider identical conditions of membrane manufacturing and characterization, which is not the case for most of the studied materials. Yet, it was revealed that when casting a membrane the thermal treatments (casting temperature, annealing) changes drastically the morphology of the polymer chains and, as a consequence, its performance.[144] In addition, the solvent, in which the casting polymer solution is prepared is also a very important parameter influencing membrane morphology.[141]

The common feature for all the reports was that ionomers were cast in their salt forms (usually K^+ and Na^+), were solubilized in a highly polar aprotic solvent (mostly DMAc, more rarely DMSO[133,137] and (N-Methyl-2-pyrrolidone) NMP[136]), and then poured to a glass plate (an exception was the work of Storey et al.,[128,142] who used a Teflon supporting plate). Sometimes prior to casting on the plates, the ionomer solutions were subjected to filtering[131,132,135] or centrifugation.[140,141] The neutralized form of ionic polymers, especially with perfluorosulfonate lateral chains, is recommended to avoid backbone scission, which might occur in the presence of superacidic proton (of $PFSA_{SC}$), solvent, and water molecules at increased temperature during membrane elaboration.

The acidification procedure could also influence the membrane properties. Different conditions of membrane acidification have been applied: some researchers acidify the ionomers at room temperature, whereas the others while boiling in 0.5−2 M acids, varying also the nature of acidic solution. Hence, all this mismatching must be considered when comparing the performance between the proton-conducting membranes.

3.2.1 Morphology

As it was already mentioned, morphology is an important factor that influences PEM performance. The hydrophilic/hydrophobic phase separation with the formation of interconnected hydrophilic and hydrophobic domains, especially at low hydration levels, seems to be the key for high proton conductivity. The morphology can be evidenced by microscopy [transmission electron microscopy (TEM), scanning TEM, atomic force microscopy] or by small-angle scattering techniques (of X-rays—SAXS (small-angle x-ray scattering), or of neutrons—SANS (small-angle neutron scattering)).[133−137]

All authors working with aromatic ionomers, bearing perfluorosulfonic side chains, observed clear and homogeneous phase separation within the ionomers. The hydrophilic domains (clusters) measured approximately 2–3 nm in diameter,[130–133,137,139] which is lower than those of Nafion (4 nm). However, these clusters were interconnected: as an example, here Fig. 16.6 represents the TEM images of Nafion 112 and poly(sulfone) Udel, functionalized with *p*-phenoxytetrafluoroethylenesulfonic acid groups (PSU-S_1). Narrower hydrophilic phase within the aromatic ionomers was explained by the presence of a more rigid backbone structure, compared with the Teflon-like main chain in Nafion.[139] To overcome this problem it was proposed to pay more attention to the research on block copolymers. In this case, the high density of ionic function on the hydrophilic block might facilitate the aggregation and organization of ionic domains.

The advantage of block structures was indeed confirmed by several groups.[135,140,142] However, the size of the ionic domains and the membrane conductivities depend on the lengths of the blocks, the IEC, and the chemical structure of the ionomer backbone and the $PFSA_{sc}$. Thus, by increasing the lengths of blocks the connectivity of ionic domains is improved and an important increase in conductivity was observed (Fig. 16.7).[141]

As concerning the impact of the side chain, recently Bae et al.[137] observed significant differences between the morphologies of the PSU, functionalized with $PFSA_{sc}$ of different length and chemistry (Fig. 16.8). PSU-S_5 has the shortest $PFSA_{sc}$ lateral group ($PhOCF_2SO_3H$) and similar IEC (1.99 meq/g) and water uptake (29% at 30°C and 98%RH) as PSU-S_4 [containing the lateral chains of $PhS(CF_2)_2SO_3H$]; however, its structure creates ionic clusters of much smaller size and of poorer connectivity, compared with PSU-S_4. The same comparison could be made between PSU-S_5 and

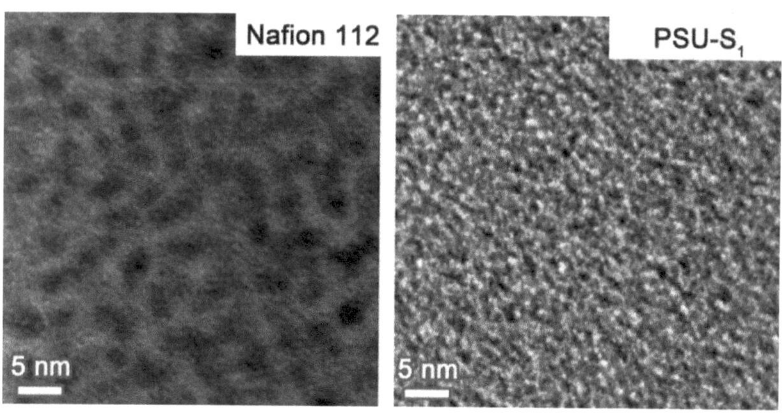

Figure 16.6 Transmission electron microscopic images of Nafion 112 and an aromatic ionomer, bearing $O(CF_2)_2SO_3H$ ionic lateral functions. *PSU*, *p*-phenoxytetrafluoroethylenesulfonic acid. Adapted with permission from Chang, Y.; Mohanty, A. D.; Smedley, S. B.; Abu-Hakmeh, K.; Lee, Y. H.; Morgan, J. E.; Hickner, M. A.; Jang, S. S.; Ryu, C. Y.; Bae, C. Effect of Superacidic Side chain Structures on High Conductivity Aromatic Polymer Fuel Cell Membrane. *Macromolecules* **2015,** *48*, 7117–7126. Copyright 2015 American Chemical Society.

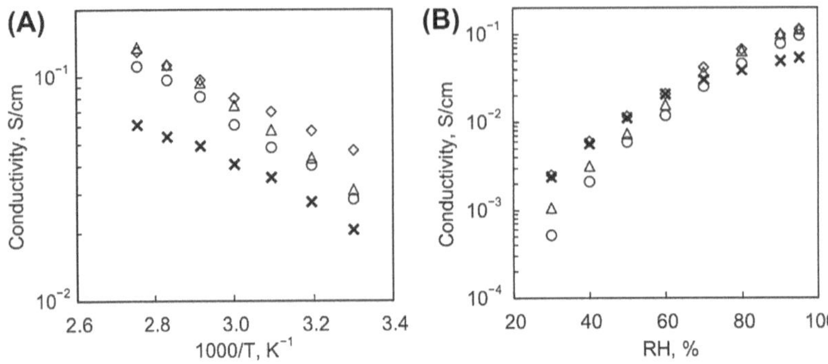

Figure 16.7 Dependence of conductivity (A) on temperature at 95%RH, and (B) on RH at 80°C for block copolymers of PES, bearing perfluorosulfonic acid pendant groups. Samples of equal lengths of hydrophilic and hydrophobic blocks in the main backbone, which are: (○) 5 kg/mol, (Δ) 10 kg/mol, and (◇) 15 kg/mol. Nafion 117 (×) is given as a reference. *PES*, poly(ether sulfone); *RH*, relative humidity.

Adapted with permission from Assumma, L.; Nguyen, H. D.; Iojoiu, C.; Lyonnard, S.; Mercier, R.; Espuche, E. Effects of Block Length and Membrane Processing Conditions on the Morphology and Properties of Perfluorosulfonated Poly(arylene ether sulfone) Multiblock Copolymer Membranes for PEMFC. *ACS Appl. Mater. Interfaces* **2015,** *7*, 13808−13820. Copyright 2015 American Chemical Society.

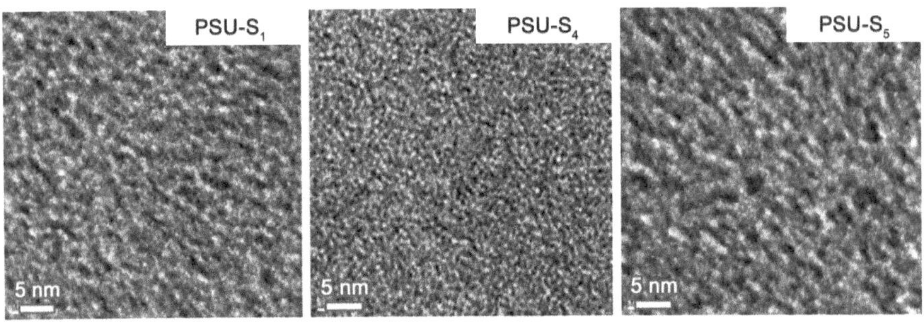

Figure 16.8 Transmission electron microscopic images of polyaromatics, bearing different perfluorosulfonic acid side chain groups: *p*-phenoxytetrafluoroethylenesulfonic acid (PSU)-S_1 has $O(CF_2)_2SO_3H$ lateral chains, PSU-S_4 has OCF_2SO_3H ionic groups, and PSU-S_5 contains $C(CF_3)(PhO(CF_2)_2SO_3H)_2$ superacid pendant chains.

Adapted with permission from Chang, Y.; Mohanty, A. D.; Smedley, S. B.; Abu-Hakmeh, K.; Lee, Y. H.; Morgan, J. E.; Hickner, M. A.; Jang, S. S.; Ryu, C. Y.; Bae, C. Effect of Superacidic Side chain Structures on High Conductivity Aromatic Polymer Fuel Cell Membrane. *Macromolecules* **2015,** *48*, 7117−7126. Copyright 2015 American Chemical Society.

PSU-S_1, containing $PhO(CF_2)_2SO_3H$ lateral chains and having even lower IEC (1.83 meq/g) and water uptake (24%), than the former sample. Therefore, it is concluded that the chemistry of the side chain does really influence the morphology.

Morphology at the nanoscale was estimated exclusively by groups of Yoshimura,[127] Bae,[133,137] and Iojoiu,[140,141] using SAXS and SANS techniques. The scattering profiles are described by the scattering intensity as a function of scattering vector q or angle 2θ. The ionic cluster dimensions are calculated from the value of q_{max}, which corresponds to the peak of the scattering profile, by the formula:

$$d = 2\pi/q_{max}$$

As it is evident from Refs. 127,133,137, statistical ionomers show one broad peak (ionomer peak) at q_{max} 0.17–0.25 $Å^{-1}$, depending on the lateral $PFSA_{sc}$ group. The ionomer peak is attributed to the hydrophilic/hydrophobic phase separation at the nanometric scale and is usually considered as a fingerprint of the ionomer membrane's nanostructure. The characteristic distance d, describing the current peak, is the mean separation distance between interconnected ionic domains[145,146] or elongated hydrophobic aggregates embedded in a continuous ionic phase.[147–149] One of the most interesting observations in phase separation, at both macro- and nanoscales, was in the block copolymers by Iojoiu et al.[140] First, the authors revealed two transition and two relaxation temperatures of the ionomers by thermal and thermomechanical analysis, which point out a phase separation between the hydrophilic and hydrophobic domains at the macroscopic scale. Moreover, by SANS two maxima in correlation vectors were observed (Fig. 16.9): the first at 0.15–0.20 $Å^{-1}$ gave the same

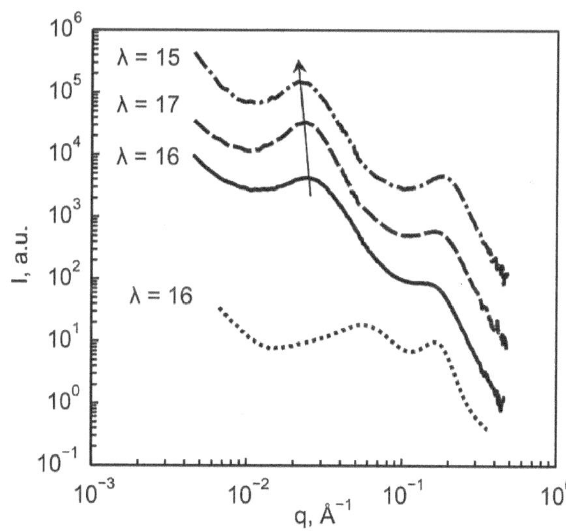

Figure 16.9 SANS spectrum of poly(ether sulfone) block copolymers, bearing $(CF_2)_2O(CF_2)_2SO_3H$ ionic groups, having hydrophobic blocks of the same length of 5000 kg/mol and different lengths of the backbones of the hydrophilic blocks of: (—), 5000 kg/mol; (---), 10,000 kg/mol; and (-·-·-), 15,000 kg/mol (······). Nafion 117 is given as a reference. Each sample contains approximately the same number of water molecules per ionic site (lambda values). Adapted with permission from Assumma, L.; Iojoiu, C.; Mercier, R.; Lyonnard, S.; Nguyen H.D., Planes E. Synthesis of Partially Fluorinated Poly(arylene ether sulfone) Multiblock Copolymers Bearing Perfluorosulfonic Functions. *J. Polym. Sci. Part A Polym. Chem.* **2015,** *53* 1941–1956. Copyright 2015 John Wiley and Sons.

information, as the other ionomers, described in the previous paragraph, that concerns the presence of the ionomer peak; the second peak at approximately 0.02 $Å^{-1}$ informed about the existence of the well-ordered scattering objects of bigger dimensions—blocks. The position of the two characteristic peaks is affected by the ionic block length in a nontrivial manner, as the two peaks are oppositely shifted upon block length increase (from 2500 to 7500 g/mol), whereas the hydrophobic block length is constant (5000 g/mol). It was assumed from these data that the block copolymers possess complex morphology with structural organization induced at different scales (typically at 3 and 30 nm).

To summarize, aromatic ionomers, bearing $PFSA_{SC}$ were characterized by a more developed morphology of interconnected ionic clusters, compared with directly sulfonated polyaromatics, or with statistical polymers. This could be related to the ability of $PFSA_{SC}$ to accommodate more water molecules even at low RH.[131,132] Even though the rigidity of the main chain is known to be an important obstacle toward ionomer organization similar to Nafion, the higher concentration of ionic groups into blocks was shown to partially overcome this problem.

3.3 Stability

3.3.1 Thermal Stability

The majority of ionomers, described earlier (Table 16.1) showed the degradation temperature at approximately 200°C under air. The degradation occurs at side chains, since mass spectroscopy, coupled with TGA (thermogravimetric analysis), evidenced the presence of SO, SO_2, CFO, and CF_2 species.[132,135] Stability of the aromatic ionomers bearing the PFSA group is lower than that of Nafion (c. 280°C). However, this lower degradation temperature for polyaromatic materials does not hinder their use in fuel cells, since the polymer materials are required to operate in the temperature range up till maximum 160°C.

3.3.2 Thermal Transitions

Depending on the ionomer structure the Tg of PFSA-modified polyaromatics is similar or higher than that of PFSA-functionalized perfluoroethylenes. As an example, Wang et al.[130] found T_g of 142°C for the homopoly(ether) PAE-NA (poly(arylene ether) containing naphthyl units), bearing naphthyl moiety in the main chain and 160°C for the homopoly(ether) with *p*-diphenyl species in the backbone. Both ionomers have an IEC of approximately 2.0 meq/g. In the case of block copolymerization, the overall transition temperature may be increased with the presence of a more robust hydrophobic block.[140] The ionomers showed two transitions: the one at lower temperature occurred at 155°C and was related to the movements of the ionic chains, and the other at 200−205°C characterized the transition of the hydrophobic block.

3.3.3 Mechanical Properties

Concerning the mechanical properties of PFSA-bearing ionomers, those with an aromatic backbone showed much higher storage and loss moduli, both at room temperature and at > 120°C. As examples, in dry conditions, statistical PES[138] was characterized by an E' of 200 MPa at 120°C, and block copolymer PES[140] showed values of E' of 1800 MPa. Even though up till 160°C the storage modulus decreased due to relaxation of the hydrophilic blocks, still it remained higher than that of Nafion (<10 MPa): $E' = 70$ MPa for the statistical PES and $E' = 1000$ MPa for the block copolymer PES. When comparing the block copolymers to the statistical copolymers, the former are characterized by 10 times higher mechanical strength.[140]

3.3.4 Oxidative Stability

Testing of ionomeric membranes for their oxidative stability has not been extensively performed until now. Thus, no standardized method for conducting this measurement was reported. Majority of researchers exposed the membrane to a 3% solution of H_2O_2 in the presence of 2–3 ppm of Fe^{2+} ions at 80°C or to a 30% solution of peroxide at room temperature. Undoubtedly, such conditions are very harsh and far more aggressive than those in a fuel cell. However, Nafion does not degrade in either of these experimental conditions for days. On the contrary, ionomers, having aromatic backbones, start the degradation at the backbone in the first hours. Such degradation is especially remarkable for poly(imide)s[143]: after 1 h in hot 3% solution of H_2O_2 24% loss in weight and 13% loss in molecular weight was reported. The poly(ether sulfone)s dissolved after 12 h in 30% solution of peroxide at room temperature.[138] Several authors[138,139,150] assumed that the degradation occurs by chain breaking; it was mainly supposed that the carbons at the *ortho*-positions to ether bonds are the most susceptible to the attack of radicals.

Poly(ether)s with partially fluorinated main chain[129] were proposed as the most stable ionomers of all the polyaromatics with $PFSA_{sc}$; they sustained oxidative test in hot peroxide solution for more than 100 h. In general, when testing the ionomers for their oxidative stability molecular weight and IEC must be taken to consideration, since the samples of low molecular weight and high IEC are more prompt to dissolve faster in the presence of hydroxyl radicals.

3.4 Conductivity and Fuel Cell Tests

The conductivities of membranes bearing PFSA side chains depend very much on the chemical structure and IEC, especially at low RH. Compared with directly sulfonated ionomers, despite the presence of $PFSA_{sc}$ on the backbone, which was expected to render better proton conductivity at low RH,[138] similar values (ranging between 10^{-5} and 10^{-4} S/cm at 20% RH, at 80°C for membranes with IEC 0.8–1.5 meq/g) were obtained. These results are represented in Fig. 16.10 by the lowest circle, named

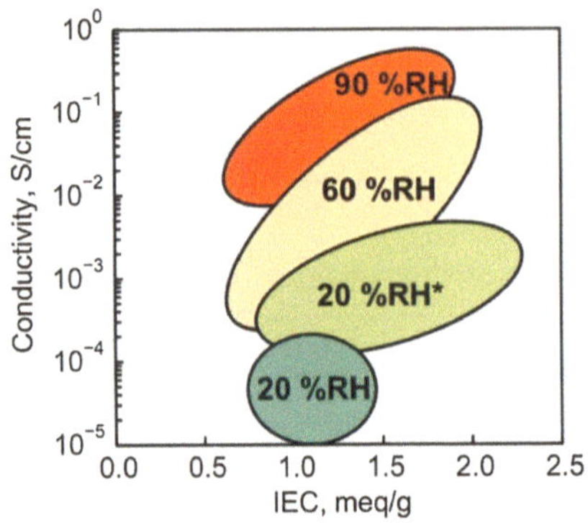

Figure 16.10 Dependence of conductivity on ion exchange capacity for aromatic ionomers, bearing PFSA side chains, at different values of relative humidity. *IEC*, ionic exchange capacity; *RH*, relative humidity.

20% RH. Nevertheless, the ionomers having fluorinated moieties in their backbone (as octafluorobiphenyl or tetrafluorophenyl) and/or bulky species like fluorene[129,132,135] and high concentration of PFSA chains (>2.0 meq/g)[133,137] show much better conductivities, ranging between 10^{-4} and 3×10^{-3} S/cm and closer to those of PFSA ionomers. They are marked with the ellipse, called 20% RH* in Fig. 16.10. Ionomers containing highly fluorinated side chains and backbone conserve their high conductivities at higher RH as well, even if they have lower IEC than those with a fully hydrogenated backbone.

Only four aromatic ionomers, bearing $PFSA_{sc}$ groups, were tested for their performance in a fuel cell. The ionomer chemical structure and IEC and the conditions of the fuel cell tests are gathered in Table 16.2. The evolution of voltage with current and power densities is presented in Fig. 16.11.

All the samples, except the one produced by Watanabe et al.,[132] were compared with Nafion at the same experimental conditions, as the testing of the polyaromatic materials. Their performance was similar to that of Nafion (not shown at the graph of Fig. 16.11, but the type of Nafion used is given in Table 16.2). However, to conclude on the best proton-conducting material among those proposed in Fig. 16.11, conditions of the fuel cell testing must be identical for all the samples.

As a conclusion on research on aromatic ionomers bearing $PFSA_{sc}$, we can say that these materials are very promising for application in fuel cells. The grafting of the $PFSA_{sc}$ on the aromatic backbone plays an important role in the morphology and functional properties of PEM. The PEM performances are improved significantly when in the ionomer (1) both the backbone and the side chain contain fluorinated moieties, (2) the side chain has a bulky structure, and (3) the backbone has multiblock structure. For the moment the majority of studies was performed on statistical structures, bearing different superacidic chains of various length and chemical structure. However, it is assumed that further investigation must be continued within the block structures, concentrating PFSA groups closer to each other at the aromatic backbone and creating highly connected hydrophilic channels.

Table 16.2 Generalized Information of Test Conditions and the Measured Samples in a Fuel Cell

References	Chemical Structure	IEC, meq/g	Thickness, μm	Conditions of the Fuel Cell Test		Ionomer of Comparison
				Temperature, °C	Humidity, %RH	
127	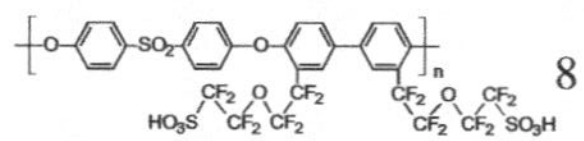 8	1.34	47	80	N/A	Nafion 112
132	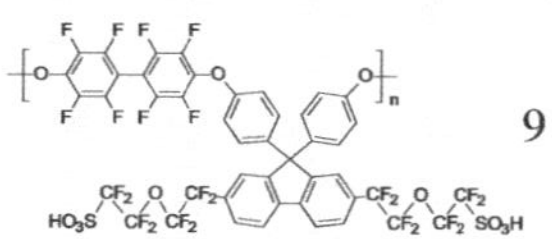 9	1.40	40	80	78	–
136	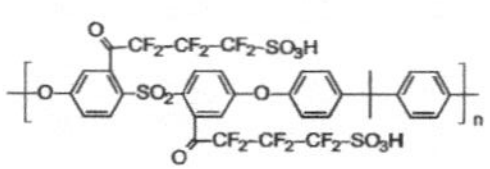	1.12	70	30	75	Nafion 112
130	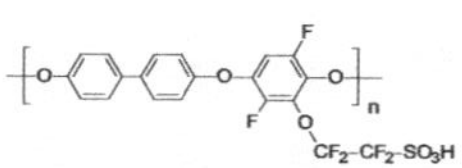	1.96	125	120	80	Nafion 115

IEC, ionic exchange capacity; *RH*, relative humidity.

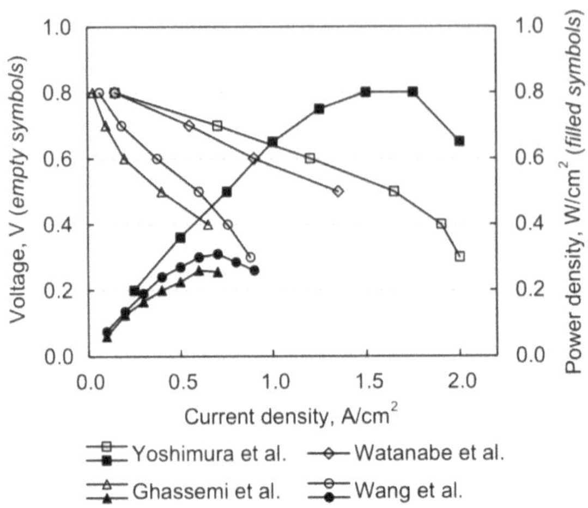

Figure 16.11 Results of fuel cell testing of aromatic materials, bearing perfluorosulfonic acid side chain. Conditions of the measurements are summarized in Table 16.2.

References

1. Tarascon, J. M.; Armand, M. Issues and Challenges Facing Rechargeable Lithium Batteries. *Nature* **2001,** *414,* 359–367.
2. Chiba, K.; Ueda, T.; Yamaguchi, Y.; Oki, Y.; Saiki, F.; Naoi, K. Electrolyte Systems for High Withstand Voltage and Durability II. Alkylated Cyclic Carbonates for Electric Double-Layer Capacitors. *J. Electrochem. Soc.* **2011,** *158,* A1320–A1327.
3. Myung, S. -T.; Hitoshi, Y.; Sun, Y. -K. Electrochemical Behavior and Passivation of Current Collectors in Lithium-Ion Batteries. *J. Mater. Chem.* **2011,** *21,* 9891–9911.
4. Zhang, X.; Winget, B.; Doeff, M.; Evans, J. W.; Devine, T. M. Corrosion of Aluminum Current Collectors in Lithium-Ion Batteries with Electrolytes Containing $LiPF_6$. *J. Electrochem. Soc.* **2005,** *152,* B448–B454.
5. Sloop, S. E.; Pugh, J. K.; Wang, S.; Kerr, J. B.; Kinoshita, K. Chemical Reactivity of PF_5 and $LiPF_6$ in Ethylene Carbonate/Dimethyl Carbonate Solutions. *Electrochem. Solid-State Lett.* **2001,** *4,* A42–A44.
6. Schaefer, J. L.; Lu, Y.; Moganty, S. S.; Agarwal, P.; Jayaprakash, N. A.; Archer, L. Electrolytes for High-Energy Lithium Batteries. *Appl. Nanosci.* **2012,** *2,* 91–109.
7. Kawamura, T.; Okada, S.; Yamaki, J. I. Decomposition Reaction of $LiPF_6$-Based Electrolytes for Lithium Ion Cells. *J. Power Sources* **2006,** *156,* 547–554.
8. Scheers, J.; Jonsson, E.; Jacobsson, P.; Johansson, P. Lithium Imides: Effect of –F, $-CF_3$, and –CN Substituents on Lithium Battery Salt Stability and Dissociation. *Electrochemistry* **2012,** *80,* 18–25.
9. Sasaki, Y. Physical and Electrochemical Properties and Application to Lithium Batteries of Fluorinated Organic Solvents. In *Fluorinated Materials for Energy Conversion;* Elsevier, 2005; pp 285–304.
10. Arai, J. No-flash-point Electrolytes Applied to Amorphous Carbon/$Li_{1+x}Mn_2O_4$ Cells for EV Use. *J. Power Sources* **2003,** *119-121,* 388–392.

11. Sasaki, Y.; Satake, H.; Tsukimori, N.; Nanbu, N.; Takehara, M.; Ue, M. Physical and Electrolytic Properties of Partially Fluorinated Methyl Carbonate and Its Application to Lithium Batteries. *Electrochemistry* **2010,** *78,* 467–470.
12. Achiha, T.; Nakajima, T.; Ohzawa, Y.; Koh, M.; Yamauchi, A.; Kagawa, M.; Aoyama, H. Electrochemical Behavior of Non-flammable Organo-Fluorine Compounds for Lithium Ion Batteries. *J. Electrochem. Soc.* **2009,** *156,* A483–A488.
13. McMillan, R.; Slegr, H.; Shu, Z. X.; Wang, W. Fluoroethylene Carbonate Electrolyte and Its Use in Lithium Ion Batteries with Graphite Anodes. *J. Power Sources* **1999,** *81-82,* 20–26.
14. Hu, L.; Zhang, Z.; Amine, K. Fluorinated Electrolytes for Li-Ion Battery: An FEC-based Electrolyte for High Voltage $LiNi_{0.5}Mn_{1.5}O_4$/Graphite Couple. *Electrochem. Commun.* **2013,** *35,* 76–79.
15. Kawamura, T.; Tanaka, T.; Egashira, M.; Watanabe, I.; Okada, S.; Yamaki, J. -I. Methyl Difluoroacetate Inhibits Corrosion of Aluminum Cathode Current Collector for Lithium Ion Cells. *Electrochem. Solid-State Lett.* **2005,** *8,* A459–A463.
16. Kalhoff, J.; Bresser, D.; Bolloli, M.; Alloin, F.; Sanchez, J. -Y.; Passerini, S. Enabling LiTFSI-Based Electrolytes for Safer Lithium-Ion Batteries by Using Linear Fluorinated Carbonates as (Co)solvent. *ChemSusChem* **2014,** *7,* 2939–2946.
17. Bolloli, M.; Kalhoff, J.; Alloin, F.; Bresser, D.; Le, M. L. P.; Langlois, B.; Passerini, S.; Sanchez, J. -Y. Fluorinated Carbamates as Suitable Solvents for LiTFSI-Based Lithium-Ion Electrolytes: Physico-Chemical Properties and Electrochemical Characterization. *J. Phys. Chem. B* **2015,** *119,* 22404–22414.
18. Bolloli, M.; Alloin, F.; Kalhoff, J.; Bresser, D.; Passerini, S.; Judeinstein, P.; Leprêtre, J. -C.; Sanchez, J. -Y. Effect of Carbonates Fluorination on the Properties of LiTFSI-Based Electrolytes for Li-Ion Batteries. *Electrochim. Acta* **2015,** *161,* 159–170.
19. Dudley, J. T.; Wilkinson, D. P.; Thomas, G.; LeVae, R.; Woo, S.; Blom, H.; Horvath, C.; Juzkow, M. W.; Denis, B.; Juric, P.; Aghakian, P.; Dahn, J. R. Conductivity of Electrolytes for Rechargeable Lithium Batteries. *J. Power Sources* **1991,** *35,* 59–82.
20. Wasserschied, P.; Welton, T. *Ionic Liquids in Synthesis;* Wiley-VCH: Weinhiem, 2003.
21. Ohno, H. *Electrochemical Aspects of Ionic Liquids;* John Wiley& Sons, Inc.: New Jersey, 2005.
22. Davis, J. -H. Task-Specific Ionic Liquids. *Chem. Lett.* **2004,** *33,* 1072–1077.
23. Bonhôte, P.; Dias, A. -P.; Papageorgiou, N.; Kalyanasundaram, K.; Grätzel, M.; Armand, M. Hydrophobic, Highly Conductive Ambient-Temperature Molten Salts. *Inorg. Chem.* **1996,** *35,* 1168–1178.
24. Ignat'ev, N. V.; Willner, H.; Sartori, P. Electrochemical Fluorination (Simons Process) – a Powerful Tool for the Preparation of New Conducting Salts, Ionic Liquids and Strong Bronsted Acids. *J. Fluor. Chem.* **2009,** *130,* 1183–1191.
25. Ignat'ev, N. V.; Welz-Biermann, U.; Kucheryna, A.; Bissky, G.; Willner, H. New Ionic Liquids with Tris(perfluoroalkyl)trifluorophosphate (FAP) Anions. *J. Fluor. Chem.* **2005,** *126,* 1150–1159.
26. Rogers, E. I.; Sljukic, B.; Hardacre, C.; Compton, R. G. Electrochemistry in Room Temperature Ionic Liquids: Potential Windows at Mercury Electrodes. *J. Chem. Eng. Data* **2009,** *54,* 2049–2053.
27. Barrosse-Antle, L. E.; Bond, A. M.; Compton, R. G.; Rogers, E. I.; Silvester, D. S. Voltammetry in Room Temperature Ionic Liquids: Comparisons and Contrasts with Conventional Electrochemical Solvents. *Chem. Asian J.* **2010,** *5,* 202–230.
28. Hayyan, M.; Mjalli, F. S.; Hashim, M. A.; Alnashef, I. M.; Mei, T. X. Investigating the Electrochemical Windows of Ionic Liquids. *J. Ind. Eng. Chem.* **2013,** *19,* 106–112.

29. Seki, S.; Serizawa, N.; Hayamizu, K.; Takei, K.; Miyashiro, H. Physicochemical and Electrochemical Properties of 1-Ethyl-3-methylimidazolium Tris(pentafluoroethyl) Trifluorophosphate and 1-Ethyl-3-methylimidazolium Tetracyanoborate. *J. Electrochem. Soc.* **2012,** *159,* A967–A971.
30. Ignat'ev, N. V.; Finze, M.; Sprenger, J. A. P.; Kerpen, C.; Bernhardt, E.; Willner, H. New Hydrophobic Ionic Liquids with Perfluoroalkyl Phosphate and Cyanofluoroborate Anions. *J. Fluor. Chem.* **2015,** *177,* 46–54.
31. Li, P.; Compton, R. G. Electrochemical High Concentration Oxygen Sensing Using a Phosphonium Cation Based Room Temperature Ionic Liquid: Analytical Studies. *Electroanalysis* **2015,** *27,* 1550–1555.
32. Matsumoto, H.; Sakaebe, H.; Tatsumi, K. Preparation of Room Temperature Ionic Liquids Based on Aliphatic Onium Cations and Asymmetric Amide Anions and Their Electrochemical Properties as a Lithium Battery Electrolyte. *J. Power Sources* **2005,** *146,* 45–50.
33. Guerfi, A.; Duchesne, S.; Kobayashi, Y.; Vijha, A.; Zaghib, K. $LiFePO_4$ and Graphite Electrodes with Ionic Liquids Based on Bis(fluorosulfonyl)imide $(FSI)^-$ for Li-Ion Batteries. *J. Power Sources* **2008,** *175,* 866–873.
34. Matsumoto, H.; Sakaebe, H.; Tatsumi, K.; Kikuta, M.; Ishiko, E.; Kono, M. Fast Cycling of Li/$LiCoO_2$ Cell with Low-Viscosity Ionic Liquids Based on Bis(fluorosulfonyl)imide [FSI]. *J. Power Sources* **2006,** *160,* 1308–1313.
35. Han, H. -B.; Zhou, S. -S.; Zhang, D. J.; Feng, S. W.; Li, L. F.; Liu, K.; Feng, W. F.; Nie, J.; Li, H.; Huang, X. J. Lithium Bis(fluorosulfonyl)imide (LiFSI) as Conducting Salt for Nonaqueous Liquid Electrolytes for Lithium-Ion Batteries: Physicochemical and Electrochemical Properties. *J. Power Sources* **2011,** *196,* 3623–3632.
36. Ishikawa, M. T.; Kikuta, M.; Ishiko, E.; Kono, M. Pure Ionic Liquid Electrolytes Compatible with a Graphitized Carbon Negative Electrode in Rechargeable Lithium-Ion Battery. *J. Power Sources* **2006,** *162,* 658–662.
37. Huang, J.; Hollenkamp, A. F. Thermal Behaviour of Ionic Liquids Containing the FSI Anion and the Li^+ Cation. *J. Phys. Chem. C* **2010,** *114,* 21840–21847.
38. Le, M. L. P.; Alloin, F.; Strobel, P.; Leprêtre, J. C.; Pérez del Valle, C.; Judeinstein, P. Structure-Properties Relationships of Lithium Electrolytes Based on Ionic Liquid. *J. Phys. Chem. Part B.* **2010,** *114,* 894–903.
39. Girard, G. M. A.; Hilder, M.; Zhu, H.; Nucciarone, D.; Whitbread, K.; Zavorine, S.; Moser, M.; Forsyth, M.; MacFarlane, D. R.; Howlett, P. C. Electrochemical and Physicochemical Properties of Small Phosphonium Cation Ionic Liquid Electrolytes with High Lithium Salt Content. *Phys. Chem. Chem. Phys.* **2015,** *17,* 8706–8713.
40. Monteiro, M. J.; Bazito, F. F. C.; Siqueira, L. J. A.; Ribeiro, M. C. C.; Torresi, R. M. Transport Coefficients, Raman Spectroscopy, and Computer Simulation of Lithium Salt Solutions in an Ionic Liquid. *J. Phys. Chem. B* **2008,** *112,* 2102–2109.
41. Almantariotis, D.; Gefflaut, T.; Padua, A. A. H.; Coxam, J. Y.; Costa Gomes, M. F. Effect of Fluorination and Size of the Alkyl Side-Chain on the Solubility of Carbon Dioxide in 1-Alkyl-3-methylimidazolium Bis(trifluoromethylsulfonyl)amide ionic liquids. *J. Phys. Chem. B* **2010,** *114,* 3608–3617.
42. Muldoon, M. J.; Aki, S. N. V. K.; Anderson, J. L.; Dixon, J. K.; Brennecke, J. F. Improving Carbon Dioxide Solubility in Ionic Liquids. *J. Phys. Chem. B* **2007,** *111,* 9001–9009.
43. Ganapatibhotla, L. V. N. R. L.; Zheng, W. J.; Roy, X. J. D.; McLaughlina, J. B.; Krishnan, S. Ionic Liquids with Fluorinated Block-oligomer Tails: Influence of Self-assembly on Transport Properties. *J. Mater. Chem.* **2011,** *21,* 19275–19285.

44. Le, M. L. P.; Alloin, F.; Strobel, P.; Leprêtre, J. C.; Cointeaux, L.; Pérez del Valle, C. Electrolyte Based on Fluorinated Cyclic Quaternary Ammonium Ionic Liquid. *Ionics* **2012,** *18,* 817–827.
45. Howlett, P. C.; Brack, N.; Hollenkamp, A. F.; Forsyth, M.; MacFarlane, D. R. Characterization of the Lithium Surface in *N*-methyl-N-alkylpyrrolidinium Bis(trifluoromethanesulfonyl)amide Room-Temperature Ionic Liquid Electrolytes. *J. Electrochem. Soc.* **2006,** *153,* A595–A606.
46. Liu, H.; Liu, Y.; Li, J. Ionic Liquid in Surface Electrochemistry. *Phys. Chem. Chem. Phys.* **2010,** *12,* 1685–1697.
47. Bhatt, A. I.; Best, A. S.; Huang, J.; Hollenkamp, A. F. Application of the *N*-propyl-*N*-methyl-pyrrolidinium Bis(Fluorosulfonyl) Imide RTIL Containing Lithium Bis(Fluorosulfonyl) Imide in Ionic Liquid Based Lithium Batteries. *J. Electrochem. Soc.* **2010,** *157,* A66–A74.
48. Lane, G. H.; Best, A. S.; MacFarlane, D. R.; Hollenkamp, A. F.; Forsyth, M. An Azo-Spiro Mixed Ionic Liquid Electrolyte for Lithium Metal-$LiFePO_4$ Batteries. *J. Electrochem. Soc.* **2010,** *157,* A876–A884.
49. Valencia, H.; Kohyama, M.; Tanaka, S.; Matsumoto, H. Ab Initio Study of EMIM-BF_4 Crystal Interaction with a Li (100) Surface as a Model for Ionic Liquid/Li Interfaces in Li-Ion Batteries. *J. Chem. Phys.* **2009,** *131,* 244705.
50. Lewandowski, A. P.; Hollenkamp, A. F.; Donne, S. W.; Best, A. S. Cycling and Rate Performance of Li–$LiFePO_4$ Cells in Mixed FSI–TFSI Room Temperature Ionic Liquids. *J. Power Sources* **2010,** *195,* 2029–2035.
51. Yoon, H.; Howlett, P. C.; Best, A. S.; Forsyth, M.; MacFarlane, D. R. Fast Charge/Discharge of Li Metal Batteries Using an Ionic Liquid Electrolyte. *J. Electrochem. Soc.* **2013,** A1629–A1637.
52. Snook, G. A.; Huynh, T. D.; Hollenkamp, A. F.; Best, A. S. Rapid SECM Probing of Dissolution of $LiCoO_2$ Battery Materials in an Ionic Liquid. *J. Electroanal. Chem.* **2012,** *687,* 30–34.
53. Mun, J.; Yim, T.; Park, J. H.; Ryu, J. H.; Lee, S. Y.; Kim, Y. G.; Oh, S. M. Allylic Ionic Liquid Electrolyte-Assisted Electrochemical Surface Passivation of $LiCoO_2$ for Advanced, Safe Lithium-Ion Batteries. *Sci. Rep.* **2014,** *4,* 5802.
54. Borgel, V.; Markevich, E.; Aurbach, D.; Semrau, G.; Schmidt, M. On the Application of Ionic Liquids for Rechargeable Li Batteries: High Voltage Systems. *J. Power Sources* **2009,** *189,* 331–336.
55. Mun, J.; Yim, T.; Park, K.; Ryu, J. H.; Kim, Y. G.; Oh, S. M. Surface Film Formation on $LiNi_{0.5}Mn_{1.5}O_4$ Electrode in an Ionic Liquid Solvent at Elevated Temperature. *J. Electrochem. Soc.* **2011,** *158,* A453–A457.
56. Wongittharom, N.; Lee, T. C.; Hung, I. M.; Lee, S. W.; Wang, Y. C.; Chang, J. K. Ionic Liquid Electrolytes for High-Voltage Rechargeable Li/$LiNi_{0.5}Mn_{1.5}O_4$ Cells. *J. Mater. Chem. A: Mater. Energy Sustain. A* **2014,** *2,* 3613–3620.
57. Swiderska-Mocek, A.; Naparstek, D. Compatibility of Polymer Electrolyte Based on N-propylpiperidinium Bis(trifluoromethanesulphonyl)imide Ionic Liquid with $LiMn_2O_4$ Cathode in Li-Ion Batteries. *Solid State Ionics* **2014,** *267,* 32–37.
58. Patra, J.; Dahiya, P. P.; Tseng, C. J.; Fang, J.; Lin, Y. W.; Basu, S.; Majumder, S. B.; Chang, J. K. Electrochemical Performance of $0.5Li_2MnO_3$–$0.5Li(Mn_{0.375}Ni_{0.375}Co_{0.25})O_2$ Composite Cathode in Pyrrolidinium-Based Ionic Liquid Electrolytes. *J. Power Sources* **2015,** *294,* 22–30.
59. Wang, Y.; Zaghib, K.; Guerfi, A.; Bazito, F. F. C.; Torresi, R. M.; Dahn, J. R. Ionic Liquid in Surface Electrochemistry Accelerating Rate Calorimetry Studies of the Reaction

Between Ionic Liquid and Charged Lithium Ion Battery Electrode Materials. *Electrochim. Acta* **2007,** *52,* 6346–6352.

60. Döbbelin, M.; Azcune, I.; Bedu, M.; Luzuriaga, A. R.; Genua, A.; Jovanovski, V.; Cabañero, G.; Odriozola, I. Synthesis of Pyrrolidinium-Based Poly(ionic liquid) Electrolytes with Poly(ethylene glycol) Side Chains. *Chem. Mater.* **2012,** *24,* 1583–1590.
61. Shaplov, A. S.; Vlasov, P. S.; Lozinskaya, E. I.; Ponkratov, D. O.; Malyshkina, I. A.; Vidal, F.; Okatova, O. V.; Pavlov, G. M.; Wandrey, C.; Bhide, A.; Schönhoff, M.; Vygodskii, Y. S. Polymeric Ionic Liquids: Comparison of Polycations and Polyanions. *Macromolecules* **2011,** *44,* 9792–9803.
62. Li, M. T.; Yang, L.; Fang, S. H.; Dong, S. M. Novel Polymeric Ionic Liquid Membranes as Solid Polymer Electrolytes with High Ionic Conductivity at Moderate Temperature. *J. Membr. Sci.* **2011,** *366,* 245–250.
63. Chen, X. J.; Li, Q.; Zhao, J.; Qiu, L. H.; Zhang, Y. G.; Sun, B. Q.; Yan, F. Ionic Liquid-Tethered Nanoparticle/poly(ionic liquid) Electrolytes for Quasi-Solid-State Dye-Sensitized Solar Cells. *J. Power Sources* **2012,** *207,* 216–221.
64. Kim, T. Y.; Lee, H. W.; Stoller, M.; Dreyer, D. R.; Bielawski, C. W.; Ruoff, R. S.; Suh, K. S. High-Performance Supercapacitors Based on Poly(ionic liquid)-modified Graphene Electrodes. *ACS Nano* **2011,** *5,* 436–442.
65. Shaplov, A. S.; Marcilla, R.; Mecerreyes, D. Recent Advances in Innovative Polymer Electrolytes Based on Poly(ionic liquid)s. *Electrochim. Acta* **2015,** *175,* 18–34.
66. Sato, T.; Morinaga, T.; Ishizuka, T. Ionic Liquids – New Aspects for the Future. In *Preparation, Physicochemical Properties and Battery. Applications of a Novel Poly(ionic liquid);* Kadokawa, J. I., Ed.; 2013.
67. Djian, D.; Alloin, F.; Martinet, S.; Lignier, H.; Sanchez, J.-Y. Lithium-Ion Batteries with High Charge Rate Capacity: Influence of the Porous Separator. *J. Power Sources* **2007,** *172,* 416–421.
68. Djian, D.; Alloin, F.; Martinet, S.; Lignier, H. Macroporous Poly(vinylidene fluoride) Membrane as a Separator for Lithium-Ion Batteries with High Charge Rate Capacity. *J. Power Sources* **2009,** *187,* 575–580.
69. Saunier, J.; Alloin, F.; Sanchez, J. Y.; Caillon, G. Thin and Flexible Lithium-Ion Batteries: Investigation of Polymer Electrolytes. *J. Power Sources* **2003,** *119-121,* 454–459.
70. Good, W. D.; Lacina, J. C.; De Prater, B. L.; McCullough, J. P. New Approach to the Combustion Calorimetry of Silicon and Organosilicon Compounds. Heats of Formation of Quartz, Fluorosilicic Acid, and Hexamethyldisiloxane. *J. Phys. Chem.* **1964,** *68,* 579–586.
71. Ameduri, B.; Boutevin, B. In *Well-Architectured Fluoropolymers: Synthesis, Properties and Applications;* Elsevier: Amsterdam, 2004.
72. Costa, C. M.; Silva, M. M.; Lanceros-Mendez, S. Battery Separators Based on Vinylidene Fluoride (VDF) Polymers and Copolymers for Lithium Ion Battery Applications. *RSC Adv.* **2013,** *3,* 11404–11417.
73. Boudin, F.; Andrieu, X.; Jehoulet, C.; Olsen, I. I. Microporous PVdF Gel for Lithium-Ion Batteries. *J. Power Sources* **1999,** *81-82,* 804–807.
74. Kataoka, H.; Saito, Y.; Sakai, T.; Quartarone, E.; Mustarelli, P. Conduction Mechanisms of PVDF-Type Gel Polymer Electrolytes of Lithium Prepared by a Phase Inversion Process. *J. Phys. Chem. B* **2000,** *104,* 11460–11464.
75. Sousa, R. E.; Kundu, M.; Gôren, A.; Silva, M. M.; Liu; Costa, C. M.; Lanceros-Mendez, S. Poly(vinylidene fluoride-co-chlorotrifluoroethylene) (PVDF-CTFE) Lithium-Ion Battery Separator Membranes Prepared by Phase Inversion. *RSC Adv.* **2015,** *5,* 90428–90436.

76. Costa, C. M.; Nunes-Pereira, J.; Sencadas, V.; Silva, M. M.; Lanceros-Mendez, S. Effect of Fiber Orientation in Gelled Poly(vinylidene fluoride) Electrospun Membranes for Li-Ion Battery Applications. *J. Mater. Sci.* **2013,** *48,* 6833–6840.
77. Zhang, F.; Ma, X.; Cao, C.; Li, J.; Zhu, Y. Poly(vinylidene Fluoride)/SiO_2 Composite Membranes Prepared by Electrospinning and Their Excellent Properties for Nonwoven Separators for Lithium-Ion Batteries. *J. Power Sources* **2014,** *251,* 423–431.
78. An, M. Y.; Kim, H. T.; Chang, D. R. Multilayered Separator Based on Porous Polyethylene Layer, Al_2O_3 Layer, and Electro-Spun PVdF Nanofiber Layer for Lithium Batteries. *J. Solid State Electrochem.* **2014,** *18,* 1807–1814.
79. Kim, J. R.; Choi, S. W.; Jo, S. M.; Lee, W. S.; Kim, B. C. Characterization and Properties of P(VdF-HFP)-Based Fibrous Polymer Electrolyte Membrane Prepared by Electrospinning. *J. Electrochem. Soc.* **2005,** *152,* A295–A300.
80. Gao, K.; Hu, X.; Dai, C.; Yi, T. Crystal Structures of Electrospun PVDF Membranes and Its Separator Application for Rechargeable Lithium Metal Cells. *Mater. Sci. Eng. B Solid-State Mater. Adv. Technol.* **2006,** *131,* 100–105.
81. Alcoutlabi, M.; Lee, H.; Watson, J. V.; Zhang, X. Preparation and Properties of Nanofiber-Coated Composite Membranes as Battery Separators via Electrospinning. *J. Mater. Sci.* **2013,** *48,* 2690–2700.
82. Croce, F.; Focarete, M. L.; Hassoun, J.; Meschini, I.; Scrosati, B. A Safe, High-Rate and High-Energy Polymer Lithium-Ion Battery Based on Gelled Membranes Prepared by Electrospinning. *Energy Environ. Sci.* **2011,** *4,* 921–927.
83. Deng, F.; Wang, X.; He, D.; Hu, J.; Gong, C.; Ye, Y. S.; Xie, X. Microporous Polymer Electrolyte Based on PVDF/PEO Star Polymer Blends for Lithium Ion Batteries. *J. Membr. Sci.* **2015,** *49,* 82–89.
84. Kim, D. W.; Sun, Y. K. Electrochemical Characterization of Gel Polymer Electrolytes Prepared with Porous Membranes. *J. Power Sources* **2001,** *102,* 41–45.
85. Liu, X.; Kusawake, H.; Kuwajima, S. Preparation of a PVdF-HFP/polyethylene Composite Gel Electrolyte with Shutdown Function for Lithium-Ion Secondary Battery. *J. Power Sources* **2001,** *97–98,* 661–663.
86. Rajendran, S.; Mahendran, O.; Kannan, R. Characterisation of [(1−x)PMMA−xPVdF] Polymer Blend Electrolyte with Li^+ Ion. *Fuel* **2002,** *81,* 1077–1081.
87. Seidel, S. M.; Jeschke, S.; Vettikuzha, P.; Wiemhoefer, H. D. PVDF-HFP/Ether-modified Polysiloxane Membranes Obtained via Airbrush Spraying as Active Separators for Application in Lithium Ion Batteries. *Chem. Commun.* **2015,** *51,* 12048–12051.
88. Zhang, H.; Ma, X.; Lin, C.; Zhu, B. Gel Polymer Electrolyte-Based on PVDF/Fluorinated Amphiphilic Copolymer Blends for High Performance Lithium-Ion Batteries. *RSC Adv.* **2014,** *4,* 33713–33719.
89. Alaaeddine, A.; Vergnaud, J.; Rolland, J.; Vlad, A.; Gohy, J. F.; Ameduri, B. Synthesis of an Original Fluorinated Triethylene Glycol Methacrylate Monomer and its Radical Copolymerisation with Vinylidene Fluoride. Its Application as a Gel Polymer Electrolyte for Li-Ion Batteries. *Polym. Chem.* **2015,** *6,* 6021–6028.
90. Miao, R.; Liu, B.; Zhu, Z.; Liu, Y.; Li, J.; Wang, X.; Li, Q. PVDF-HFP-based Porous Polymer Electrolyte Membranes for Lithium-Ion Batteries. *J. Power Sources* **2008,** *184,* 420–426.
91. Du Pasquier, A.; Warren, P. C.; Culver, D.; Gozdz, A. S.; Amatucci, G. G.; Tarascon, J. M. Plastic PVDF-HFP Electrolyte Laminates Prepared by a Phase-Inversion Process. *Solid State Ionics* **2000,** *135,* 249–257.

92. Wang, M.; Zhao, F.; Guo, Z.; Dong, S. Poly(vinylidene fluoride-hexafluoropropylene)/organo-montmorillonite Clays Nanocomposite Lithium Polymer Electrolytes. *Electrochim. Acta* **2004,** *49,* 3595–3602.
93. Takemura, D.; Aihara, S.; Hamano, K.; Kise, M.; Nishimura, T.; Urushibata, H.; Yoshiyasu, H. A Powder Particle Size Effect on Ceramic Powder Based Separator for Lithium Rechargeable Battery. *J. Power Sources* **2005,** *146,* 779–783.
94. Kelley, J.; Simonsen, J.; Ding, J. Poly(vinylidene fluoride-co-hexafluoropropylene) Nanocomposites Incorporating Cellulose Nanocrystals with Potential Applications in Lithium Ion Batteries. *J. Appl. Polym. Sci.* **2013,** 487–493.
95. Song, X.; Ding, W.; Cheng, B.; Xing, J. Electrospun Poly(vinylidene-fluoride)/POSS Nanofiber Membrane-Based Polymer Electrolytes for Lithium Ion Batteries. *Polym. Comp.* **2015;** http://dx.doi.org/10.1002/PC.
96. Lu, C.; Qi, W.; Li, L.; Xu, J.; Chen, P.; Xu, R.; Han, L.; Yu, Q. Electrochemical Performance and Thermal Property of Electrospun PPESK/PVDF/PPESK Composite Separator for Lithium-Ion Battery. *J. Appl. Electrochem.* **2013,** *43,* 711–720.
97. Raghavan, P.; Zhao, X. H.; Shin, C.; Baek, D. H.; Choi, J. W.; Manuel, J.; Heo, M. Y.; Ahn, J. H.; Nah, C. W. Preparation and Electrochemical Characterization of Polymer Electrolytes Based on Electrospun Poly(vinylidene Fluoride-co-hexafluoropropylene)/Polyacrylonitrile Blend/Composite Membranes for Lithium Batteries. *J. Power Sources* **2010,** *195,* 6088–6094.
98. Xiao, Q. Z.; Li, Z. H.; Gao, D. S.; Zhang, H. L. A Novel Sandwiched Membrane as Polymer Electrolyte for Application in Lithium-Ion Battery. *J. Membr. Sci.* **2009,** *326,* 260–264.
99. Kritzer, P. Nonwoven Support Material for Improved Separators in Li–Polymer Batteries. *J. Power Sources* **2006,** *161,* 1335–1340.
100. Zhang, H. P.; Zhang, P.; Li, Z. H.; Sun, M.; Wu, Y. P.; Wu, H. Q. A Novel Sandwiched Membrane as Polymer Electrolyte for Lithium Ion Battery. *Electrochem. Commun.* **2007,** *9,* 1700–1703.
101. Fang, L. F.; Shi, J. L.; Li, H.; Zhu, B. K.; Zhu, L. P. Construction of Porous PVDF Coating Layer and Electrochemical Performances of the Corresponding Modified Polyethylene Separators for Lithium Ion Batteries. *J. Appl. Polym. Sci.* **2014,** *131,* 41036/1–41036/9.
102. Zhou, X.; Yue, L.; Zhang, J.; Kong, Q.; Liu, Z.; Yao, J.; Cui, G. A Core-Shell Structured Polysulfonamide-Based Composite Nonwoven towards High Power Lithium Ion Battery Separator. *J. Electrochem. Soc.* **2013,** *160,* A1341–A1347.
103. Carmo, M.; Fritz, D. L.; Mergel, J.; Stolten, D. A Comprehensive Review on PEM Water Electrolysis. *J. Hydrogen Energy* **2013,** *38,* 4901–4934.
104. Kerres, J. A. Development of Ionomer Membranes for Fuel Cells. *J. Membr. Sci.* **2001,** *185,* 3–27.
105. Gebel, G. Structural Evolution of Water Swollen Perfluorosulfonated Ionomers from Dry Membrane to Solution. *Polymer* **2000,** *41,* 5829–5838.
106. Gebel, G.; Lyonnard, S.; Mendil-Jakani HMorin, A. The Kinetics of Water Sorption in Nafion Membranes: a Small-angle Neutron Scattering Study. *J. PhyS. Condens. Mater.* **2011,** *23,* 234107.
107. Wang, F.; Hickner, M.; Kim, Y. S.; Zawodzinski, T. A.; McGrath, J. E. Direct Polymerization of Sulfonated Poly(arylene ether Sulfone) Random (Statistical) Copolymers: Candidates for New Proton Exchange Membranes. *J. Membr. Sci.* **2002,** *197,* 231–242.
108. Iojoiu, C.; Sanchez, J. -Y. Polysulfone-Based Ionomers for Fuel Cell Applications. *High Perform. Polym.* **2009,** *21,* 673–692.

109. Chikashige, Y.; Chikyu, Y.; Miyatake, K.; Watanabe, M. Poly(arylene ether) Ionomers Containing Sulfofluorenyl Groups for Fuel Cell Applications. *Macromolecules* **2005,** *38,* 7121–7126.
110. Summer, M. J.; Harrison, W. L.; Weyers, R. M.; Kim, Y. S.; McGrath, J. E.; Riffle, J. S.; Brink, A.; Brink, M. H. Novel Proton Conducting Sulfonated Poly(arylene ether) Copolymers Containing Aromatic Nitrile. *J. Membr. Sci.* **2004,** *239,* 199–211.
111. Einsla, B. R.; Kim, Y. S.; Hickner, M. A.; Hong, Y. T.; Hill, M. L.; Pivovar, B. S.; McGrath, J. E. Sulfonated Naphthalene Dianhydride Based Polyimide Copolymers for Proton-Exchange-Membrane Fuel Cells: II. Membrane Properties and Fuel Cell Performance. *J. Membr. Sci.* **2005,** *255,* 141–148.
112. Schoenberger, F.; Chromik, A.; Kerres, J. Synthesis and Characterization of Novel (Sulfonated) Poly(arylene ether)s with Pendent Trifluoromethyl Groups. *Polymer* **2009,** *50,* 2010–2024.
113. Seyb, C.; Kerres, J. Novel Partially Fluorinated Sulfonated Poly(arylenethioether)s and Poly(aryleneether)s Prepared from Octafluorotoluene and Pentafluoropyridine and Their Blends with PBI-Celazol. *Eur. Polym. J.* **2013,** *49,* 518–531.
114. Katzfuß, A.; Krajinovic, K.; Chromik, A.; Kerres, J. Partially Fluorinated Sulfonated Poly(arylene sulfone)s Blended with Polybenzimidazole. *J. Polym. Sci. Part A Polym. Chem.* **2011,** *49,* 1919–1927.
115. Li, X. -F.; Paoloni, F. P. V.; Weiber, E. A.; Jiang, Z. -H.; Jannasch, P. Fully Aromatic Ionomers with Precisely Sequenced Sulfonated Moieties for Enhanced Proton Conductivity. *Macromolecules* **2012,** *45,* 1447–1459.
116. Takamuku, S.; Jannasch, P. Multiblock Copolymers Containing Highly Sulfonated Poly(arylene sulfone) Blocks for Proton Conducting Electrolyte Membranes. *Macromolecules* **2012,** *45,* 6538–6546.
117. Saito, J.; Tanaka, M.; Miyatake, K.; Watanabe, M. Proton Conductive Polyimide Ionomer Membranes: Effect of NH, OH, and COOH Groups. *J. Polym. Sci. Part A Polym. Chem.* **2010,** *48,* 2846–2854.
118. Kreuer, K. D. On the Development of Proton Conducting Polymer Membranes for Hydrogen and Methanol Fuel Cells. *J. Membr. Sci.* **2001,** *185,* 29–39.
119. Choi, J. -H.; Ye, Y.; Elabd, Y. A.; Winey, K. I. Network Structure and Strong Microphase Separation for High Ion Conductivity in Polymerized Ionic Liquid Block Copolymers. *Macromolecules* **2013,** *46,* 5290–5300.
120. Hoarfrost, M. L.; Segalman, R. A. Ionic Conductivity of Nanostructured Block Copolymer/Ionic Liquid Membranes. *Macromolecules* **2011,** *44,* 5281–5288.
121. Young, W. -S.; Epps, T. H. Ionic Conductivities of Block Copolymer Electrolytes with Various Conducting Pathways: Sample Preparation and Processing Considerations. *Macromolecules* **2012,** *45,* 4689–4697.
122. Weber, R. L.; Ye, Y.; Schmitt, A. L.; Banik, S. M.; Elabd, Y. A.; Mahanthappa, M. K. Effect of Nanoscale Morphology on the Conductivity of Polymerized Ionic Liquid Block Copolymers. *Macromolecules* **2011,** *44,* 5727–5735.
123. Virgili, J. M.; Hoarfrost, M. L.; Segalman, R. A. Morphologies of Block Copolymers Composed of Charged and Neutral Blocks. *Macromolecules* **2010,** *43,* 5417–5423.
124. Torquato, S. Optimal Design of Heterogeneous Materials. *Annu. Rev. Mater. Sci.* **2010,** *40,* 101–129.
125. Assumma, L.; Iojoiu, C.; Albayrakari, G.; Cointeaux, L.; Sanchez, J. -Y. Polyethersulfone Containing Sulfonimide Groups as Proton Exchange Membrane Fuel Cells. *Int. J. Hydrogen Energy* **2014,** *39,* 2740–2750.

126. Gi Cho, C.; Kim, Y. S.; Yu, X.; Hill, M.; McGrath, J. E. Synthesis and Characterization of Poly(arylene ether sulfone) Copolymers with Sulfonimide Side Groups for a Proton-Exchange Membrane. *J. Polym. Sci. Part A Polym. Chem.* **2006,** *44,* 6007–6014.
127. Yoshimura, K.; Iwasaki, K. Aromatic Polymer with Pendant Perfluoroalkyl Sulfonic Acid for Fuel Cell Applications. *Macromolecules* **2009,** *2009* (42), 9302–9306.
128. Li, H.; Jackson, A. B.; Kirk, N. J.; Mauritz, K. A.; Storey, R. F. Poly(arylene ether sulfone) Statistical Copolymers Bearing Perfluoroalkylsulfonic Acid Moieties. *Macromolecules* **2011,** *2011* (44), 694–702.
129. Nakabayashi, K.; Higashihara, T.; Ueda, M. Polymer Electrolyte Membranes Based on Poly(phenylene ether)s with Pendant Perfluoroalkyl Sulfonic Acids. *Macromolecules* **2011,** *44,* 1603–1609.
130. Xu, K.; Oh, H.; Hickner, M. A.; Wang, Q. Highly Conductive Aromatic Ionomers with Perfluorosulfonic Acid Side Chains for Elevated Temperature Fuel Cells. *Macromolecules* **2011,** *44,* 4605–4609.
131. Miyatake, K.; Shimura, T.; Mikami, T.; Watanabe, M. Aromatic Ionomers with Superacid Groups. *Chem. Commun.* **2009,** 6403–6405.
132. Mikami, T.; Miyatake, K.; Watanabe, M. Poly(arylene ether)s Containing Superacid Groups as Proton Exchange Membranes. *ACS Appl. Mater. Interfaces* **2010,** *2,* 1714–1721.
133. Chang, Y.; Brunello, G. F.; Fuller, J.; Disabb-Miller, M. L.; Hawley, M. E.; Kim, Y. S.; Hickner, M. A.; Soon Jang, S.; Bae, C. Acidity Effects in Poly(sulfone) – Based Proton Exchange Membranes. *Polym. Chem.* **2013,** *4,* 272–281.
134. Chang, Y.; Brunello, G. F.; Fuller, J.; Hawley, M.; Seung Kim, Y.; Disabb-Miller, M.; Hickner, M. A.; Soon Jang, S.; Bae, C. Aromatic Ionomers with Highly Acidic Sulfonate Groups: Acidity, Hydration, and Proton Conductivity. *Macromolecules* **2011,** *44,* 8458–8469.
135. Mikami, T.; Miyatake, K.; Watanabe, M. Synthesis and Properties of Multiblock Copoly(arylene ether)s Containing Superacid Groups for Fuel Cell Membranes. *J. Polym. Sci. Part A Polym. Chem.* **2011,** *49,* 452–464.
136. Ghassemi, H.; Schiraldi, D.; Zawodzinski, T.; Hamrock, S. Poly(arylene ether)s with Pendant Perfluoroalkyl Sulfonic Acid Groups as Proton-Exchange Membrane Materials. *Macromol. Chem. Phys.* **2011,** *212,* 673–678.
137. Chang, Y.; Mohanty, A. D.; Smedley, S. B.; Abu-Hakmeh, K.; Lee, Y. H.; Morgan, J. E.; Hickner, M. A.; Jang, S. S.; Ryu, C. Y.; Bae, C. Effect of Superacidic Side chain Structures on High Conductivity Aromatic Polymer Fuel Cell Membrane. *Macromolecules* **2015,** *48,* 7117–7126.
138. Nakagawa, T.; Nakabayashi, K.; Higashihara, T.; Ueda, M. Polymer Electrolyte Membrane Based on Poly(ether sulfone) Containing Binaphthyl Units with Pendant Perfluoroalkyl Sulfonic Acids. *J. Polym. Sci. Part A Polym. Chem.* **2011,** *49,* 2997–3003.
139. Shimura, T.; Watanabe, M.; Miyatake, K. Synthesis of Superacid-Modified Poly(arylene ether sulfone)s via Post-Bromination. *RSC Adv.* **2012,** *2,* 5199–5204.
140. Assumma, L.; Iojoiu, C.; Mercier, R.; Lyonnard, S.; Nguyen, H. D.; Planes, E. Synthesis of Partially Fluorinated Poly(arylene ether sulfone) Multiblock Copolymers Bearing Perfluorosulfonic Functions. *J. Polym. Sci. Part A Polym. Chem.* **2015,** *53,* 1941–1956.
141. Assumma, L.; Nguyen, H. D.; Iojoiu, C.; Lyonnard, S.; Mercier, R.; Espuche, E. Effects of Block Length and Membrane Processing Conditions on the Morphology and Properties of Perfluorosulfonated Poly(arylene ether sulfone) Multiblock Copolymer Membranes for PEMFC. *ACS Appl. Mater. Interfaces* **2015,** *7,* 13808–13820.

142. Li, H.; Kirk, N. J.; Mauritz, K. A.; Storey, R. F. Poly(arylene ether sulfone) Multi-block Copolymers Bearing Perfluoroalkylsulfonic Acid Groups. *Macromolecules* **2011,** *52,* 3550–3559.
143. Saito, J.; Tanaka, M.; Hirai, M.; Nanasawa, M.; Miyatake, K.; Watanabe, M. Polyimide Ionomer Containing Superacid Groups. *Polym. Adv. Technol.* **2011,** *22,* 1305–1310.
144. Chen, Y.; Guo, R.; Lee, C. H.; Lee, M.; McGrath, J. E. Partly Fluorinated Poly(arylene ether sulfone ketone) Hydrophilic–Hydrophobic Multiblock Copolymers for Fuel Cell Membranes. *Int. J. Hydrogen Energy* **2012,** *37,* 6132–6139.
145. Gierke, T. D.; Munn, G. E.; Wilson, F. C. The Morphology in Nafion Perfluorinated Membrane Products, as Determined by Wide- and Small-angle X-ray Studies. *J. Polym. Sci. Polym. Phys.* **1981,** *19,* 1687–1704.
146. Schmidt-Rohr, K.; Chen, Q. Parallel Cylindrical Water Nanochannels in Nafion Fuel-Cell Membranes. *Nat. Mater.* **2008,** *7,* 75–83.
147. Kreuer, K. D.; Portale, G. A. A Critical Revision of the Nano-Morphology of Proton Conducting Ionomers and Polyelectrolytes for Fuel Cell Application. *Adv. Funct. Mater.* **2013,** *23,* 5390–5397.
148. Rubatat, L.; Rollet, A. L.; Gebel, G.; Diat, O. Evidence of Elongated Polymeric Aggregates in Nafion, Macromolecules. *Macromolecules* **2002,** *35,* 4050–4055.
149. Rubatat, L.; Gebel, G.; Diat, O. Fibrillar Structure of Nafion: Matching Fourier and Real Space Studies of Corresponding Films and Solutions. *Macromolecules* **2004,** *37,* 7772–7783.
150. Hübner, G.; Roduner, E. EPR Investigation of HO/Radical Initiated Degradation Reactions of Sulfonated Aromatics as Model Compounds for Fuel Cell Proton Conducting Membranes. *J. Mater. Chem.* **1999,** *9,* 409–418.

Strategic Incorporation of Fluorine for Drug Discovery and Development

17

T. Honda, I. Ojima
Stony Brook University–State University of New York, Stony Brook, NY, United States

Chapter Outline

Modern Synthesis Processes and Reactivity of Fluorinated Compounds. http://dx.doi.org/10.1016/B978-0-12-803740-9.00017-2

1. Introduction

Special properties of fluorine have been successfully exploited for the development of new and effective biochemical tools as well as therapeutic agents. In fact, the extraordinary potential of fluorine-containing biologically relevant molecules has been widely explored in medicinal chemistry, chemical biology, pharmacology, drug discovery and development, as well as diagnostic and therapeutic applications in the past three decades. In particular, the fact that a large number of fluorine-containing agents have been approved by the US Food and Drug Administration (FDA) for medical use clearly demonstrates the importance of fluorine in drug discovery and development.[1–3] Thus, fluorine is now recognized as the second "favorite heteroatom" in the current drug design, after nitrogen.[4] The replacement of a C–H or C–O bond with a C–F bond in biologically active compounds often introduces beneficial properties such as higher metabolic stability, increased binding to target molecules, and increased membrane permeability.[5] In addition, positron emission tomographic (PET) imaging using ^{18}F-PET tracers enables rapid and accurate assessment of biodistribution and metabolic profiles of drug candidates in preclinical studies for drug addiction and neurological disorders in the brain. There has been a remarkable advance in the biomedical applications of ^{18}F-PET imaging as a rapid and reliable diagnostic tool, especially for Alzheimer disease (AD).

This chapter describes selected new fluorine-containing drugs, which have recently been approved by the FDA or evaluated in phase III clinical trials, highlighting the significance of the strategic incorporation of fluorine(s) in these drugs and the modern synthetic processes employed for the industrial production of these drugs. It is noteworthy that these selected drugs fulfill unmet medical needs in several challenging therapeutic areas, i.e., hypercholesterolemia, AD (diagnosis and treatment), non–small cell lung cancer (NSCLC), hepatitis C, diabetes, and diabetic neuropathy.

2. Case Study 1: Incorporation of Fluorine Increased the Binding Affinity of a Drug to Its Protein Target

2.1 *Anacetrapib: a Drug for the Treatment of Hypercholesterolemia*

Cardiovascular disease is the most common cause of morbidity and mortality in developed nations.[6] Atherogenic dyslipidemia, characterized by an abnormal circulating lipid profile, including low levels of high-density lipoprotein cholesterol (HDL-C), elevated levels of low-density lipoprotein cholesterol (LDL-C) and small-dense LDL-C, or elevated triglycerides (TG), is often found in patients who are obese and have type 2 diabetes or metabolic syndrome.[7] These individuals are at high risk for premature cardiovascular disease. Despite aggressive lowering of LDL-C with 3-hydroxy-3-methylglutaryl-coenzyme A (HMG-CoA) reductase inhibitors (statins),[8] two-thirds of statin-treated patients continued to develop cardiovascular disease. In addition, several epidemiological studies have established that a low level of HDL-C is a critical determinant of cardiovascular disease, independent of LDL-C levels.[9]

Cholesteryl ester transfer protein (CETP) is a plasma glycoprotein[10] that transfers cholesterol ester from HDL to LDL and very-low-density lipoprotein (VLDL), thereby decreasing antiatherogenic HDL and increasing proatherogenic LDL and VLDL, while in return, it transfers TG from VLDL (or LDL) to HDL. The resulting TG-enriched HDL are more readily hydrolyzed by hepatic lipase to produce smaller HDL particles, which are more effective in promoting reverse cholesterol transport.[11]

A compound that inhibits CETP may therefore be effective in increasing the levels of HDL-C and result in reducing the risk of atherosclerosis. Many pharmaceutical companies have tried to develop this class of inhibitors. As of 2015, however, these drugs have generally failed in clinical trials, either causing a marked increase in deaths (torcetrapib, Pfizer), or having no meaningful clinical improvement despite HDL increases (dalcetrapib, Roche/Japan Tobacco; evacetrapib, Eli Lilly). Nevertheless, anacetrapib developed by Merck has survived and is currently undergoing phase III clinical trials. Anacetrapib is expected to be the first CETP inhibitor that used to treat hypercholesterolemia and prevent cardiovascular disease.

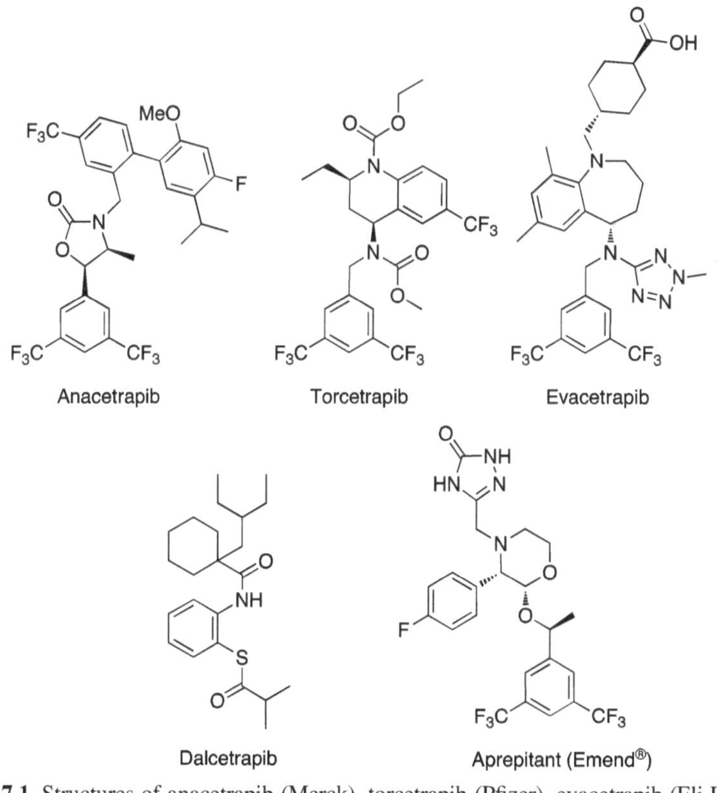

Figure 17.1 Structures of anacetrapib (Merck), torcetrapib (Pfizer), evacetrapib (Eli Lilly), dalcetrapib (Roche/Japan Tobacco), and aprepitant (Emend, Merck).

2.1.1 Significance of the Introduction of a 3,5-Di(trifluoromethyl)phenyl Group[12]

Anacetrapib and other two CETP inhibitors (Fig. 17.1) have a 3,5-di(trifluoromethyl) phenyl group. In the development of a series of neurokinin 1 (NK1) receptor antagonists[13] [for example, aprepitant is clinically used for the prevention of nausea and vomiting], it was found that this functional group dramatically increases binding affinity between ligands and NK1 receptor. Many NK1 antagonists have this functionality. On the basis of this information, to increase the affinity between CETP and ligands, this functionality was also introduced to a series of CETP inhibitors.

2.1.2 Synthesis of Anacetrapib[14]

Anacerapib was synthesized in seven steps from 2-fluoro-4-methoxyacetophenone (**1**) by the sequence shown in Fig. 17.2. (*Note*: the yields for the synthetic steps in Fig. 17.2 were, unfortunately, not disclosed in this patent.) 2-Fluoro-4-methoxyacetophenone (**1**) was treated with MeMgCl in the presence of $CsCl_3$ in tetrahydrofuran (THF) to give **2**. Hydrogenation of **2** with Pd/C under acidic conditions in EtOH afforded **3**. Bromination of **3** with *N*-bromosuccinimide (NBS) in CH_3CN produced **4**. Bromide **4** was treated with triisopropylborate and *n*-BuLi to give boronic acid **5**. Compound **7** was prepared by Suzuki coupling between **5** and **6** in the presence of bis(di-*t*-butylphosphino)ferrocene palladium dichloride in aqueous K_2CO_3 solution and THF. Chlorination of **7** with $SOCl_2$ in *N*,*N*-dimethylformamide (DMF) produced **8**. Anacetrapib was obtained by alkylation of **9** with **8** in the presence of NaHMDS in DMF, followed by crystallization from heptane.

Oxazolidinone **9** was prepared from benzoyloxy-L-alanine (**10**) as follows. Weinreb amide **11** was prepared by condensation of **10** with methoxymethylamine (Weinreb amine) hydrochloride in the presence of 1-hydroxybenzotriazole (HOBT) and 1-ethyl-3-(3-dimethylaminopropyl)carbodiimide (EDC) in THF. Ketone **13** was obtained by condensation between **11** and **12** with *i*-PrMgCl in THF. Reduction of the ketone **13** with $Al(Oi\text{-}Pr)_3$ in isopropyl alcohol (IPA) and toluene (Meerwein–Ponndorf–Verley reduction) gave **9**.

2.2 Idalopirdine: a Drug for the Treatment of Alzheimer Disease

AD is a progressive brain disorder in which the brain gradually degenerates. It frequently occurs in people older than 65 years. People with AD develop distressing changes in memory, thought, function, and behavior, which worsen over time. These changes increasingly affect patients' daily life and reduce their independence until ultimately these patients are entirely dependent on others.[15]

AD also has a huge impact on patients' caregivers. Most caregivers are close relatives who provide care at home—a demanding and exhausting role that represents a significant emotional and physical burden. Worldwide, it is estimated that about 36 million people have dementia. With the shift toward an increasingly elderly population, it is predicted that the number of people affected by dementia will almost double

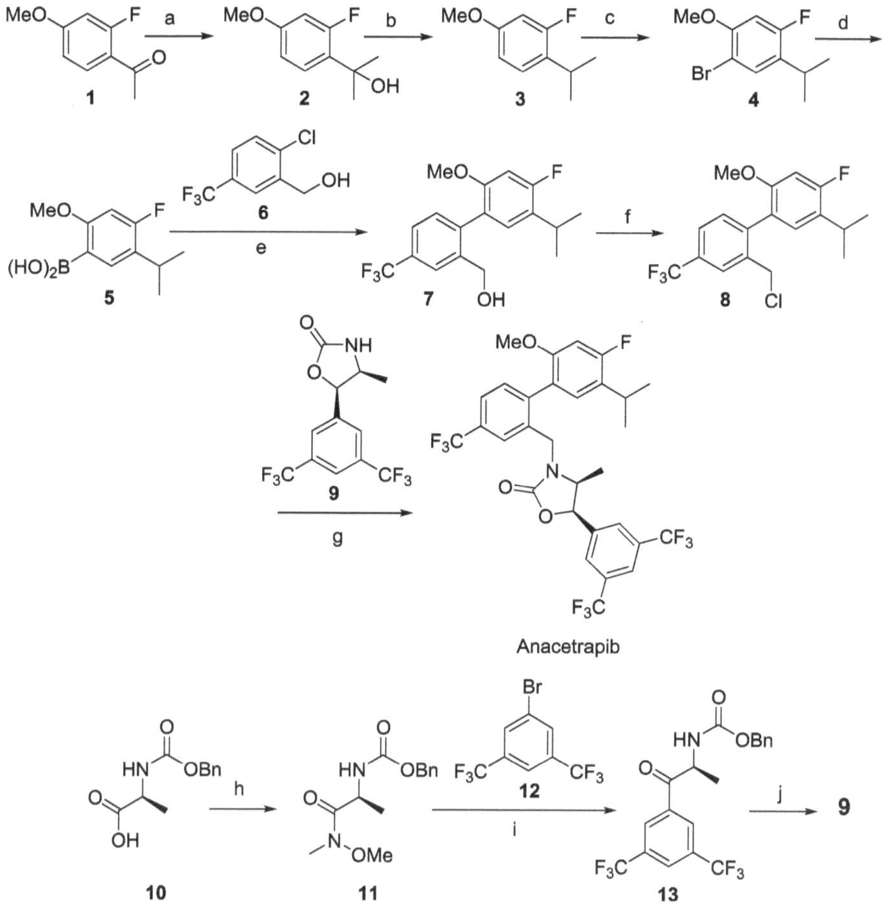

Figure 17.2 Synthesis of anacetrapib. Reagents: (a) MeMgCl, $CeCl_3$,THF; (b) H_2, Pd/C, HCl, EtOH; (c) NBS, CH_3CN; (d) $(i\text{-PrO})_3B$, *n*-BuLi; (e) Pd catalyst, aqueous K_2CO_3, THF; (f) $SOCl_2$, DMF; (g) NaHMDS, DMF; (h) $Me(MeO)NH_2Cl$, HOBt, EDC, THF; (i) *i*-PrMgCl, THF; (j) $Al(Oi\text{-Pr})_3$, IPA, toluene.

every 20 years, and by the year 2050, 115 million people will have the condition. AD is the most common cause of dementia, accounting for 50−80% of these patients. The worldwide costs of dementia (estimated at around US $600 billion in 2010) amount to more than 1% of the gross domestic product.[15,16]

In the past 20 years, the serotonin 6 (5-HT_6) receptor has received cumulative attention and become a promising target for improving cognition. The 5-HT_6 receptor subtype is expressed primarily in the brain, particularly in the cerebral cortex and hippocampus, where it has been proposed to play a role in cognitive impairments associated with schizophrenia and AD. The 5-HT_6 receptor antagonists are speculated to enhance cholinergic, glutamatergic, noradrenergic, and dopaminergic neurotransmission. Apart from some affinity for adrenergic receptors, a 5-HT_6 receptor antagonist, idalopirdine, has been reported to be highly selective over other G-protein coupled

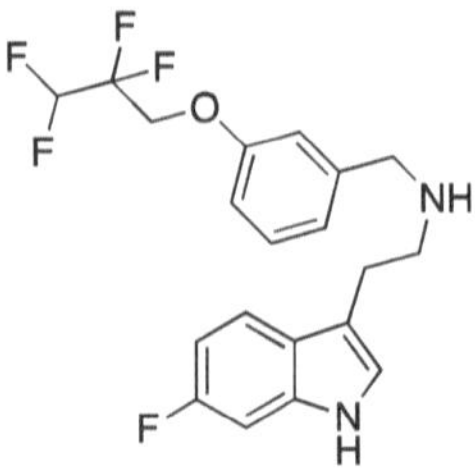

Figure 17.3 Structure of idalopirdine.

receptors (Fig. 17.3). The compound enters the brain and dose dependently reverses deficits in a rat model of cognitive impairment.[17,18]

Idalopirdine is currently under development as a symptomatic adjunct to cholinesterase inhibitor treatment in AD. Idalopirdine was originally discovered by Eli Lilly, which licensed it to a biotechnology company, Saegis, for clinical development in cognitive impairment in thinking disorders such as schizophrenia. In 2006, Saegis was merged by Lundbeck, which in October 2013 launched a global phase III clinical trials in AD in collaboration with Otsuka Pharmaceutical. This program consists of four trials planned to enroll a total of about 3000 patients.

2.2.1 Significance of Fluorine Introduction[19]

In a series of tetracyclic ergot analogs (compounds **14–17**, Fig. 17.4, Table 17.1), introduction of a fluorine atom at C3 position (compound **17**) drastically increased 5-HT_6 receptor–binding affinities when compared with hydrogen, chlorine, and a methyl group due to its high lipophilicity. For the same reasons, idalopirdine with five fluorine atoms has higher binding affinities than Ro 63-0563 and SB-271046, which are the initial selective 5-HT_6 receptor antagonists.

2.2.2 Synthesis of Idalopirdin[20]

Idalopirdin was synthesized in 11 steps from commercially available 6-fluoroindole by the sequence shown in Fig. 17.5. 6-Fluoroindole was converted to **18** by a Mannich reaction using diethoxymethane and dimethylamine (95% yield). Compound **18** with potassium cyanide in aqueous DMF under reflux conditions gave **19** in 85% yield. The nitrile **19** was converted to amine L-tartarate salt **20** by reduction in the presence of Raney Ni, followed by the treatment with L-tartaric acid in a mixture of ethyl acetate and methanol. After the salt **20** was converted to a free amine, the amine was condensed with benzaldehyde **21**, which was prepared from 3-hydroxybenzaldehyde and 2,2,3,3-tetrafluoropropyl *p*-toluenesulfonate, under basic conditions to afford an imine. The imine was reduced with $NaBH_4$, followed by treatment of hydrochloric acid to produce idalopirdine (80% yield from **20**).

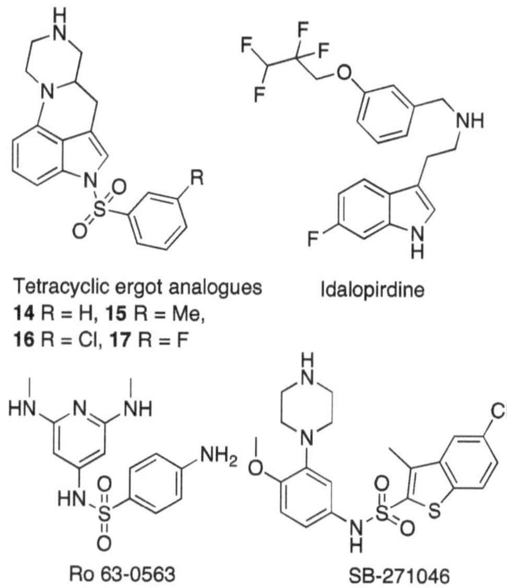

Figure 17.4 Serotonin 5-HT_6 receptor antagonists.

Table 17.1 **Binding Affinities of Serotonin 5-HT_6 Receptor Antagonists[a]**

Compound	K_i (nM)	$pK_i \pm$ SEM
14: R = H	9	8.05 ± 0.04
15: R = Me	24	7.62 ± 0.04
16: R = Cl	47	7.33 ± 0.10
17: R = F	0.75	9.12 ± 0.04
Idalopirdine	0.83	9.08 ± 0.06
Ro 63-0563	12	7.91 ± 0.02
SB-271046	3	8.52 ± 0.09

SEM, standard error of mean.
[a]Serotonin 5-HT_6 affinity: displacement of specific [^{3}H] lysergic acid diethylamide (LSD) binding to membranes of BHK-h5HT_6 cells (**14–17**,SB-271046 and idalopirdine); displacement of specific [^{3}H]LSD binding to membranes of HeLa cells (Ro 63-0563).

2.3 Rociletinib: a Drug for the Treatment of Non–Small Cell Lung Cancer

Despite years of research and prevention strategies, lung cancer remains the most common cancer worldwide, with approximately 1.35 million new cases annually. NSCLC accounts for almost 85% of all lung cancers.[21] In addition, lung cancer continues to be

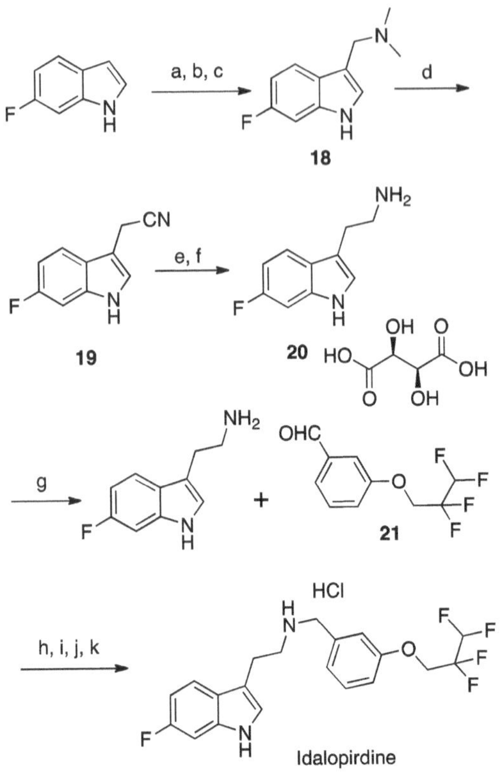

Figure 17.5 Synthesis of idalopirdine. Reagents: (a) $(EtO)_2CH_2$, H_2O, HCO_2H; (b) AcOH, H_2O, Me_2NH; (c) 3M aqueous NaOH; (d) KCN, aqueous DMF; (e) RaNi, H_2, MeOH, 25% aqueous NH_3; (f) L-tartaric acid, MeOH, EtOAc; (g) aqueous NaOH, toluene, THF; (h) *i*-PrOH; (i) $NaBH_4$, *i*-PrOH; (j) toluene, CH_3CN, aqueous HCl; (k) acetone, aqueous HCl.

the most common cause of cancer-related deaths worldwide, with a 5-year survival rate of less than 20% in patients in the United States.

NSCLC patients with activating epidermal growth factor receptor (EGFR) mutations initially respond to first-generation reversible EGFR tyrosine kinase inhibitors. However, clinical efficacy is limited by acquired resistance, frequently driven by the $EGFR^{T790M}$ mutation. Rociletinib (CO-1686)[22] is a novel, irreversible, and orally delivered kinase inhibitor that specifically targets the mutant forms of EGFR, including T790M, while exhibiting minimal activity toward the wild-type (WT) receptor (Fig. 17.6). Oral administration of rociletinib as single agent induces tumor regression in EGFR-mutated NSCLC tumor xenograft and transgenic models. Minimal activity of rociletinib against the WT EGFR receptor was observed. In NSCLC cells with acquired resistance to rociletinib in vitro, there was no evidence of additional mutations or amplification of the EGFR gene, but resistant cells exhibited signs of epithelial–mesenchymal transition and demonstrated increased sensitivity to AKT (protein kinase B) inhibitors. These results suggest that rociletinib may offer a novel therapeutic option for patients with mutant EGFR NSCLC.[22]

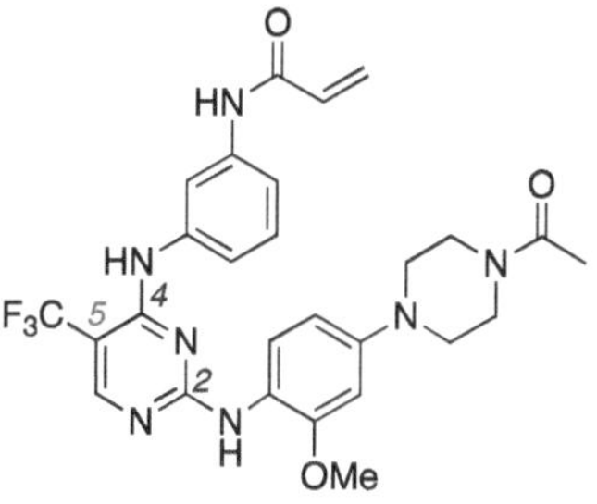

Figure 17.6 Structure of rociletinib.

Thus, Clovis Oncology Inc. is currently carrying out a number of clinical trials to assess rociletinib for patients with NSCLC. The phase II/III (TIGER 1) trial is comparing rociletinib with erlotinib as first-line therapy in treatment-naive patients with EGFR mutations who have not been screened for T790M status. A single-arm phase II (TIGER 2) trial is evaluating rociletinib as second-line therapy in patients with EGFR-mutated T790M-positive NSCLC. The randomized phase III (TIGER 3) trial is evaluating rociletinib versus chemotherapy in patients who have progressed on a first-line EGFR TKI and are also T790M positive.

2.3.1 Significance of Fluorine Introduction[22]

Using a structure-based approach, a potent 2,4-disubstituted pyrimidine molecule, rociletinib was designed as follows: (1) Rociletinib should covalently modify the conserved Cys797 in the ATP-binding pocket of the EGFR kinase domain and thus, the meta-acrylamide was introduced to point to Cys797 and to form covalent bond with $EGFR^{T790M}$ (see Fig. 17.7); (2) The aminopyrimidine was designed to bind to the hinge residue Met793 through hydrogen bonding (see Fig. 17.7, dotted line); (3) The CF_3 group was introduced at C5 to point to the gatekeeper residue Met790 because a CF_3 group has much more affinity on a methyl sulfide residue of methionine than on a CH_3 group; (4) Both C2 and C4 substitutions were designed to adapt a U-shaped binding mode; (5) The piperazine ring was introduced to be facing an open space in the solvent exposure area.

2.3.2 Synthesis of Rociletinib[22]

Rociletinib (CO-1686) was synthesized in four steps from 2,4-dichloro-5-trifluoromethylpyrimidine by the sequence shown in Fig. 17.8. Compound **22** was prepared in 39% yield by condensation between 2,4-dichloro-5-trifluoromethylpyrimidine and *N*-Boc-1,3-diaminobenzene in the presence of *N*,*N*-diisopropylethylamine (DIPEA) in *n* butanol. The Boc group of **22** was removed under acidic conditions to give **23** in quantitative yield. Compound **23** was treated with acryloyl chloride to afford **24** in 81% yield. Condensation between **24** and **25** under acidic conditions produced rociletinib in 46% yield.

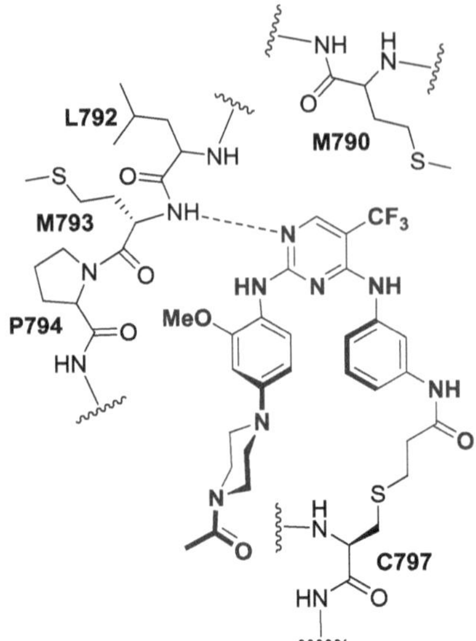

Figure 17.7 Schematic view of rociletinib binding to $EGFR^{T790M}$ based on molecular modeling.[22]

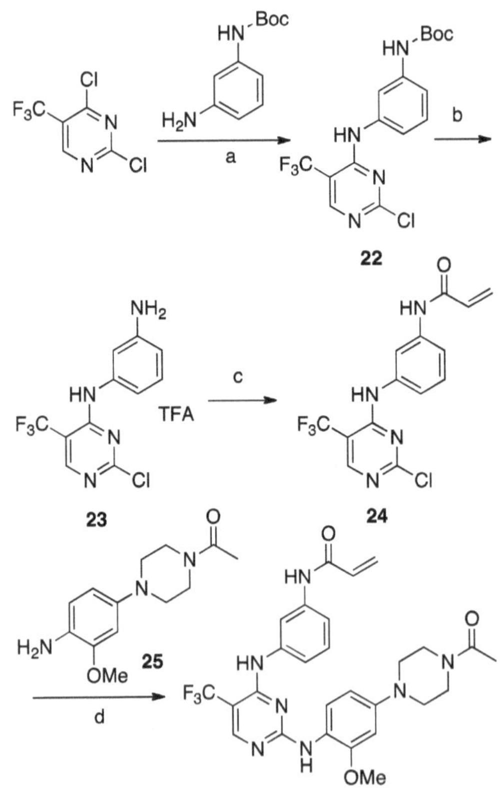

Figure 17.8 Synthesis of rociletinib. Reagents: (a) *n*-BuOH, DIPEA; (b) TFA; (c) acryloyl chloride; (d) *n*-BuOH, 1N aqueous HCl (0.8 equiv.).

3. Case Study 2: Incorporation of Fluorine Increased Potency In Vitro

3.1 *Ledipasvir: a Drug for the Treatment of Hepatitis C Virus Genotype 1 Infection*

Hepatitis C virus (HCV) was discovered in 1989 in the United States.[23] Over 90% of patients who had previously been diagnosed with non-A non-B hepatitis and more than half of those who had been previously diagnosed with alcohol liver disease were found to have liver disease caused by HCV. HCV infection presents a global heath problem with 130–170 million individuals who have detectable antibodies to HCV worldwide, corresponding to 3% of the world's population. Approximately 60–85% of such cases develop into chronic disease, leading to liver cirrhosis (5–25%), end-stage liver disease, and hepatocellular carcinoma (1–3%) over the course of the disease.[24] The current standard of care (SOC) for HCV-infected patients consists of regular injections of pegylated interferon (PEG) and oral administration of ribavirin (RBV), a nucleoside analog. However, SOC has been proved to be effective in producing a sustained virological response (SVR) in only 40–60% of patients treated, depending on the viral genotype and other predictors of host immune responsiveness. HCV genotype 1 (1b, 70% of the cases in Japan) is more resistant to SOC than genotype 2 (2a, 2b). Also, SOC is less effective against the cases with high viral loads than those with low viral loads. In addition, drug discontinuations may be high because of adverse side effects associated with the SOC treatment regimen.[25,26] Therefore, PEG-free therapy is necessary to serve a broader patient population, to improve outcomes, and to provide a safer, simpler regimen.

Ledipasvir (LDV)[27] is an HCV nonstructural protein (NS) 5A replication complex inhibitor, which has been developed by Gilead Sciences Inc. In June 2014, Gilead announced striking results from a phase III clinical trial in Japan evaluating the investigational once-daily fixed-dose combination of the NS5A inhibitor LDV and the nucleotide analog NS5B polymerase inhibitor sofosbuvir (SOF), with and without RBV, for the treatment of genotype 1 chronic HCV infection, which is the most common strain of HCV in Japan, accounting for approximately 70% of the more than 1 million people chronically infected with the disease. Among patients receiving 12 weeks of LDV/SOF without RBV, 100% ($n = 83/83$) of treatment-naive and 100% ($n = 88/88$) of treatment-experienced patients achieved a sustained virologic response 12 weeks after completing therapy (SVR12). Among patients receiving LDV/SOF plus RBV, 96% ($n = 80/83$) of treatment-naive and 100% of treatment-experienced patients ($n = 87/87$) achieved SVR12. Across all arms of the study, patients with cirrhosis achieved a 99% ($n = 75/76$) SVR12.

NS5A is a zinc-binding and proline-rich hydrophilic phosphoprotein that plays a key role in HCV RNA replication. Despite considerable amount of significant studies, the mechanistic role of NS5A in the HCV life cycle remains enigmatic.[28] NS5A has no known enzymatic activity and has no homologs in prokaryotes or eukaryotes. Nevertheless, the protein is critical for HCV viability; in clinical monotherapy studies, early NS5A inhibitors produce the most rapid viral load declines of any HCV antiviral class. It has been postulated that this rapid decline in HCV RNA is based on the inhibition of viral replication (as with NS3 and NS5B inhibitors) and additional inhibition of virion assembly or secretion from infected cells.[29]

Early NS5A replication complex inhibitors were identified through screening of compounds for their inhibitory activity against the genotype 1b replicon. Several series

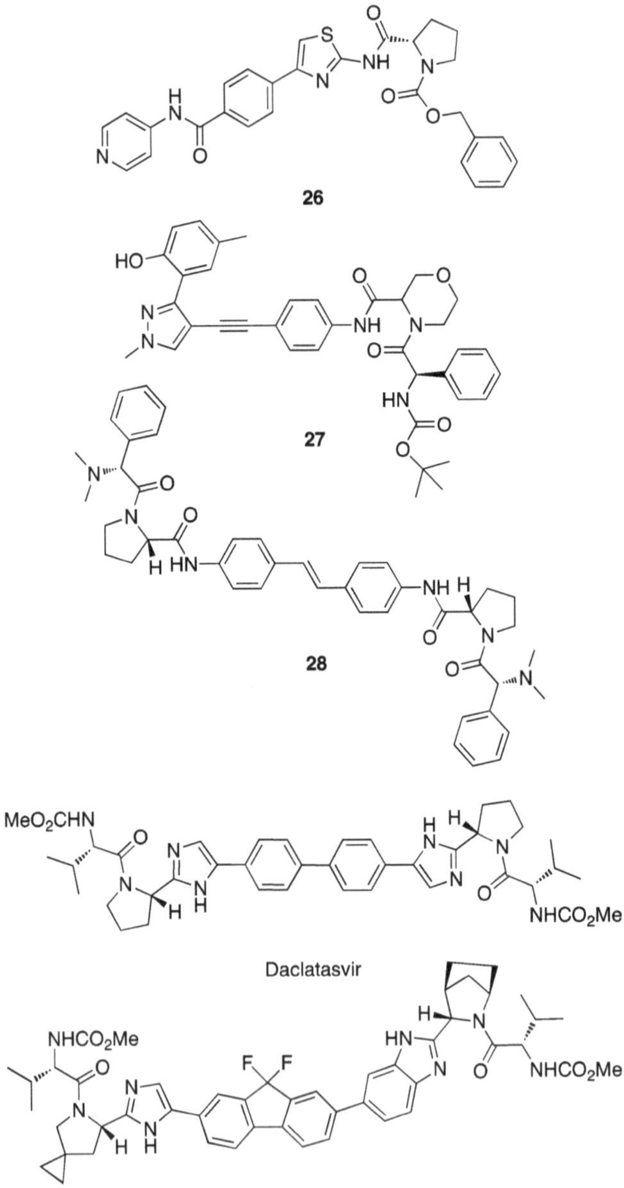

Figure 17.9 Structures of NS5A replication complex inhibitors.

of lipophilic proline or alanine-amide inhibitors of the genotype 1b replicon have been discovered (**26** and **27**; Fig. 17.9),[30] but these inhibitors typically have ~1000-fold weaker activity against the genotype 1a replicon.[31] Dimeric series provide potent genotype 1b-active stilbene diamide inhibitor **28**[32] and highly potent genotypes 1a- and 1b-active *bis*-imidazole biphenyl inhibitor, daclatasvir (BMS-790052), which achieved clinical proof of concept for the NS5A mechanism (Fig. 17.9).[33] NS5A has emerged as an important drug target for the treatment of HCV infection.[34]

Thus, to develop new NS5A inhibitors that have higher potency and a longer plasma half-life than previous inhibitors, a symmetric *bis*-benzimidazole series and an unsymmetric imidazole/benzimidazole series were investigated by the Gilead group. As a result, LDV was discovered. It is noteworthy that only LDV has difluoromethylene bridge between two phenyl groups (see Fig. 17.9).

3.1.1 Significance of Fluorine Incorporation[27]

According to the structure–activity relationship (SAR) study, the unsymmetric imidazole-benzimidazole series was better than the symmetric *bis*-benzimidazole and *bis*-imidazole series in both anti-HCV potency and plasma half-life (Fig. 17.10, Table 17.2). At almost the final stage of optimization, fluorine atoms were incorporated into the unsymmetric core. Both **31** and LDV clearly exhibited higher potency than **29** and **30** in both HCV (genotypes 1a and 1b) (Fig. 17.10, Table 17.2).

3.1.2 Synthesis of Ledipasvir[27]

LDV was synthesized in eight steps from the commercially available 2-bromofluorene **32** by the sequence shown in Fig. 17.11. Iodination of **32** in the presence of HIO_3 and H_2SO_4 in CCl_4 gave **33** in 90% yield.[35] Fluorene **33** was difluorinated with *N*-fluorobenzenesulfonimide in the presence of KHMDS to give **34**. The iodine-bearing carbon of **34** was selectively metalated by reaction with *i*-PrMgCl and converted to single chloroketone **35** by quenching the Grignard species with the Weinreb amide (2-chloro-*N*-methoxy-*N*-methylacetamide). The previously reported carboxylic acid **36**[36] was alkylated with **35**, and the resulting ketoester was condensed to imidazole **37** by heating with ammonium acetate. Suzuki coupling between **37** and boronate **38** gave **39**. Boc removal was followed by peptide coupling with *N*-methoxycarbonyl-L-valine (**40**) to afford LDV.

Boronate **38** was prepared as follows (see Fig. 17.12). Removal of the phenylethyl chiral auxiliary from **41**[37] by hydrogenolysis, followed by Boc protection and subsequent hydrolysis of the methyl ester gave [2.2.1]azabicyclic carboxylic acid **42**. [2.2.1]Azabicyclic carboxylic acid **42** was converted to benzimidazole **43** by condensation with 4-bromo-1,2-diaminobenzene in the presence of O-(7-aza-1H-benzotriazol-1-yl)-1,1,3,3-tetramethyluronium hexafluorophosphate (HATU) and 4-methylmorpholine in DMF, followed by dehydration at elevated temperature in EtOH. Boronate **38** was prepared by borylation of **43** with bis(pinacolate)diboron in the presence of $PdCl_2(dppf)_2$ and KOAc in 1,4-dioxane.

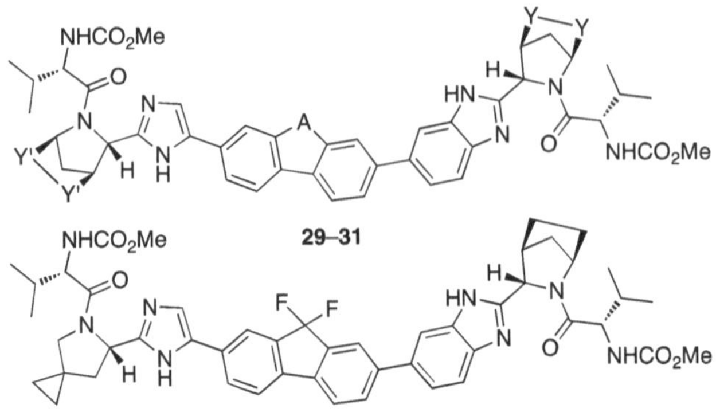

Figure 17.10 Unsymmetrical imidazole–benzimidazoles **29–31** and ledipasvir.

Table 17.2 **In Vitro Anti-HCV (Genotypes 1a and 1b) Potency of Unsymmetric Imidazole–Benzimidazoles 29–31 and Ledipasvir**

Compound	Y′	A	Y	EC_{50} (1a, nM)	EC_{50} (1b, nM)
29	–	–	CH_2	0.16	0.006
30	CH_2	–	–	0.66	0.021
31	–	CF_2	CH_2	0.056	0.004
Ledipasvir	–	CF_2	CH_2	0.031	0.004

HCV, hepatitis C virus.

4. Case Study 3: Incorporation of Fluorine Increased Both In Vitro and In Vivo Potency

4.1 Canagliflozin: a Drug for the Treatment of Diabetes

Diabetes mellitus is a worldwide public health concern, and projections for the future are alarming. According to the World Health Organization, diabetes currently affects over 170 million people worldwide, and this number will rise to more than 300 million by 2030.[38]

Plasma glucose is filtered in the glomerulus and then reabsorbed in the proximal tubules in the kidney. Sodium-glucose linked transporters (SGLTs) contribute to the renal glucose reabsorption. In the normoglycemic state, all filtered glucose is transported from the tubular lumen to the blood. However, under hyperglycemic conditions, the reabsorption process is saturated and urinary glucose excretion (UGE) increases linearly.[39] SGLT1 is distributed in the intestine, heart, and trachea besides the kidney, whereas SGLT2 is located solely in the kidney.[40] Also, an SGLT2 inhibitor enhanced UGE and consequently decreased blood glucose levels in diabetic animal

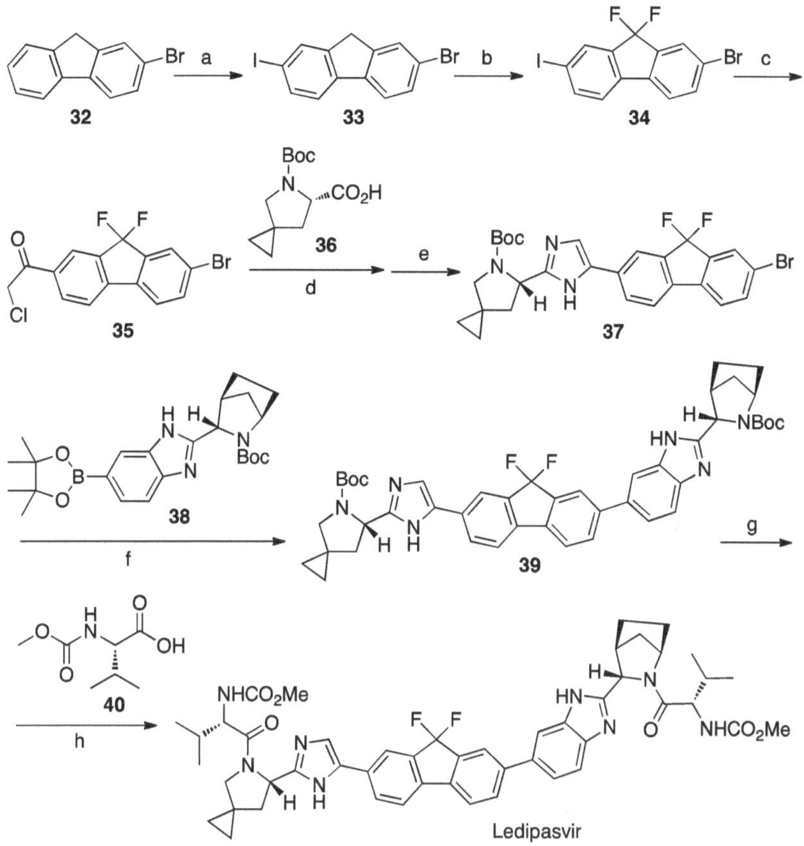

Figure 17.11 Synthesis of ledivasvir. Reagents: (a) I_2, HIO_3, H_2SO_4, CCl_4; (b) *N*-fluorobenzenesulfonimide, KHMDS, THF; (c) *i*-PrMgCl, 2-chloro-*N*-methoxy-*N*-methyl-acetamide, THF; (d) K_2CO_3, KI, acetone; (e) NH_4OAc, PhMe; (f) $Pd(OAc)_2$, PPh_3, $NaHCO_3$, aqueous DME; (g) HCl, dioxane, DCM; (h) HATU, *i*-Pr_2NEt, DMF.

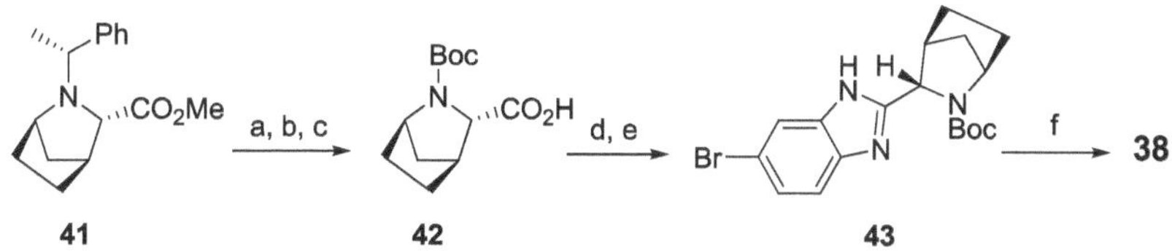

Figure 17.12 Synthesis of compound **38**, Reagents: (a) $Pd(OH_2)/C$, H_2, EtOH; (b) Boc_2O, *i*-Pr_2NEt, CH_2Cl_2; (c) LiOH, THF, aqueous MeOH; (d) 4-bromo-1,2-diaminobenzene, HATU, 4-methylmorpholine, DMF; (e) EtOH; (f) bis(pinacolate)diboron, $PdCl_2(dppf)_2$, KOAc, 1,4-dioxane.

models independent of insulin action.[41,42] Therefore, selective SGLT2 inhibitors would be desirable for antidiabetic agents.

Gliflozin drugs (see Fig. 17.13) are a class of pharmaceuticals that inhibit renal glucose reabsorption and therefore decrease blood glucose. Such biological potency is derived from SGLT2 inhibition. Among them, canagliflozin[43] is the first SGLT2 inhibitor approved by the FDA in 2013. Canagliflozin (Invokana) was developed by Mitsubishi–Tanabe and is marketed under license by Janssen, a division of Johnson & Johnson. This is the only gliflozin with fluorine atom so far. The 4-fluorophenyl group remarkably increased potency in a series of gliflozin analogs.

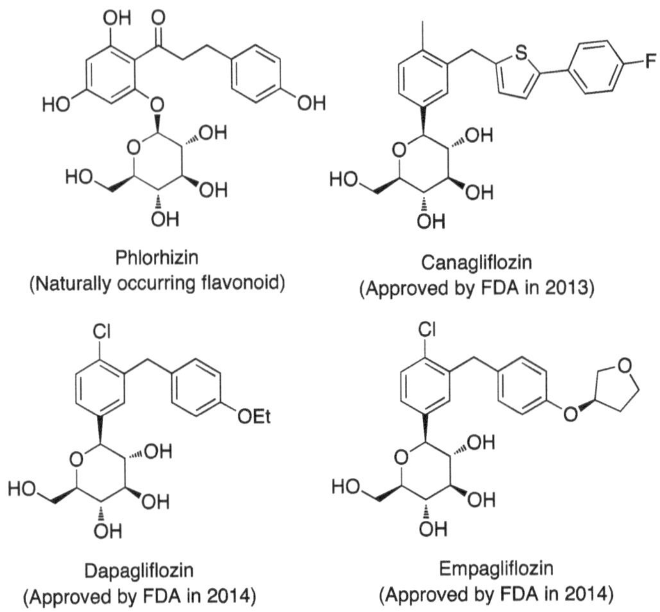

Figure 17.13 Structures of gliflozins. *FDA*, US Food and Drug Administration.

4.1.1 Significance of the Introduction of a 4-Fluorophenyl Group[43]

Effects of furan, thiophene, pyrazole, pyridine, and thiazole derivatives **44–49** were evaluated on human SGLT2 (hSGLT2) activity and on UGE in male Sprague–Dawley rats per 200 g of body weight over 24 h. The SAR of these compounds is shown in Table 17.3 and Fig. 17.14. Ethylthiophene derivative **45** possessed good hSGLT2 inhibitory potency compared with the corresponding furan and pyridine derivatives **44** and **48**. Introduction of a 4-fluorophenyl group to thiophene (compounds **46** and **49**) remarkably increased hSGLT2 inhibitory activity and UGE effect when compared with compounds without fluorine atom. Compound **46** (canagliflozin) was selected as a clinical candidate.

Table 17.3 SAR of the Representative *C*-Glucosides with Heteroaromatic Ring

Compound	R[1]	Het-R[2]	hSGLT2 IC_{50} (nM)	UGE[a] (mg/day)
44	H	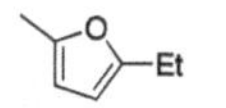	920	ND
45	H	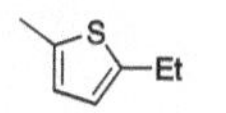	17	373
Canagliflozin **46**	Me	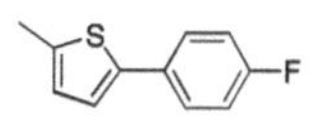	2.2	3696
47	Cl	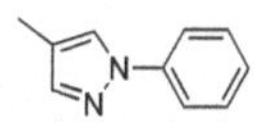	32	ND
48	Cl	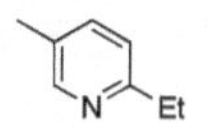	28	ND
49	Me	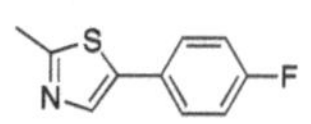	8.1	1277

hSGLT, human sodium-glucose linked transporter; *ND*, not determined; *SAR*, structure–activity relationship; *UGE*, urinary glucose excretion.
[a]Each compound was orally administered at a dose of 30 mg/kg to male Sprague–Dawley rats. UGE data over 24 h were normalized per 200 g body weight.

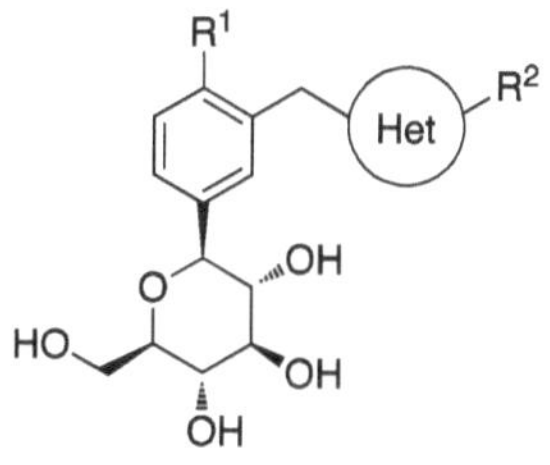

Figure 17.14 *C*-glucosides **44–49**.

4.1.2 Synthesis of Canagliflozin[43]

Canagliflozin was synthesized in five steps from 2-(4-fluorophenyl)thiophene (**50**)[44] by the sequence shown in Fig. 17.15. Thiophene aglycone **53** was prepared in 67% yield by the Friedel–Crafts acylation of **50** with benzoyl chloride **51** in CH_2Cl_2, followed by reduction of the resulting ketone with Et_3SiH and BF_3OEt_2 in CH_2Cl_2. Aglycon **53** was treated with *n*-BuLi in THF and toluene at −78°C to generate the corresponding aryllithium, followed by the addition of 2,3,4,6-tetra-*O*-trimethylsilyl-β-D-gluconolactone (**54**).[45] The resulting anomeric mixture of lactols was immediately converted into desilylated methyl ether **55** by the addition of methanesulfonic acid in methanol. Finally, canagliflozin was obtained by stereoselective reduction of **55** using a combination of Et_3SiH and BF_3OEt_2 in CH_2Cl_2.[46,47] The stereochemistry of canagliflozin was

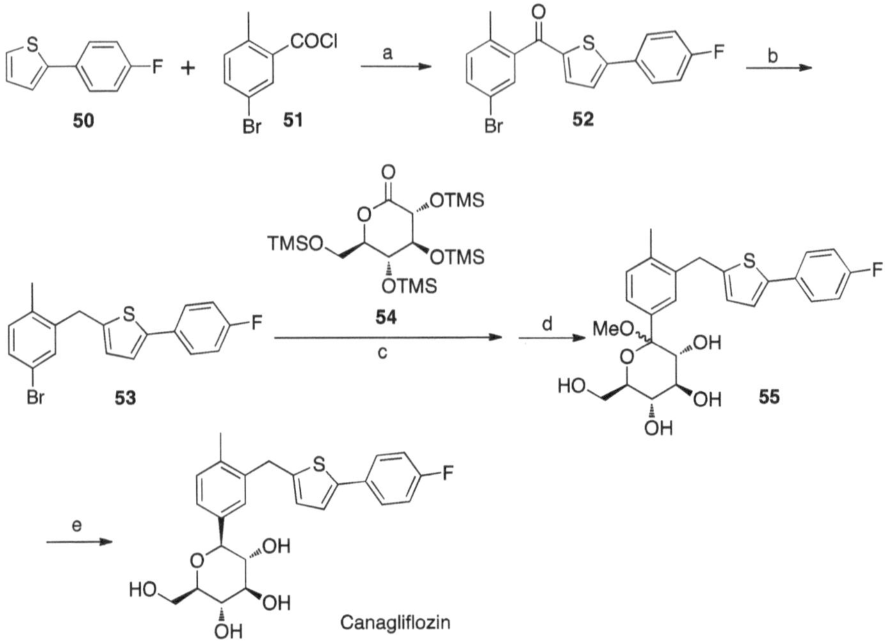

Figure 17.15 Synthesis of canagliflozin. Reagents: (a) $AlCl_3$, CH_2Cl_2; (b) Et_3SiH, BF_3OEt_2, CH_2Cl_2; (c) *n*-BuLi, THF, toluene; (d) $MeSO_3H$, MeOH; (e) Et_3SiH, BF_3OEt_2, CH_2Cl_2.

determined as β-configuration based on the coupling constant between anomeric C–H and adjacent C–H ($J \approx 9.5$ Hz) in the ^{1}H nuclear magnetic resonance spectrum.

4.2 *Ranirestat: a Drug for the Treatment of Diabetic Neuropathy*

Aldose reductase (AR) catalyzes the reduction of various aldehydes and carbonyls including monosaccharides with NADPH. This enzyme is located in the eye, kidney, and the myelin sheath, which are often involved in diabetic complications.[48]

Cellular glucose is mainly metabolized in vivo by the glycolytic pathway, but a small amount of nonphosphorylated glucose enters into the polyol pathway, an alternative route of glucose metabolism.[49] The rate-determining step of the polyol pathway is the reduction of glucose to sorbitol catalyzed by AR.[50] Under hyperglycemic conditions, the increased flux of glucose is metabolized in the polyol pathway,[50,51] inducing various metabolic imbalances such as the accumulation of sorbitol and excessive consumption of NADPH associated with the enhanced activity of AR.[52,53] Such metabolic disorders eventually cause the onset and progression of diabetic complications such as neuropathy and vascular disorders.[54] Thus, many AR inhibitors have been developed as drug candidates for the treatment of diabetic complications. However, most of them (for example, alrestatin and tolrestat) have almost failed. Currently, only epalrestat (Kinedax, Ono)[55] is clinically used for the treatment of diabetic neuropathy in Japan.

Ranirestat,[56] a new AR inhibitor having a 4-bromo-2-fluorobenzyl group, was discovered by Dainippon Sumitomo, and is currently undergoing phase III studies for the treatment of diabetic neurophathy (Fig. 17.16).

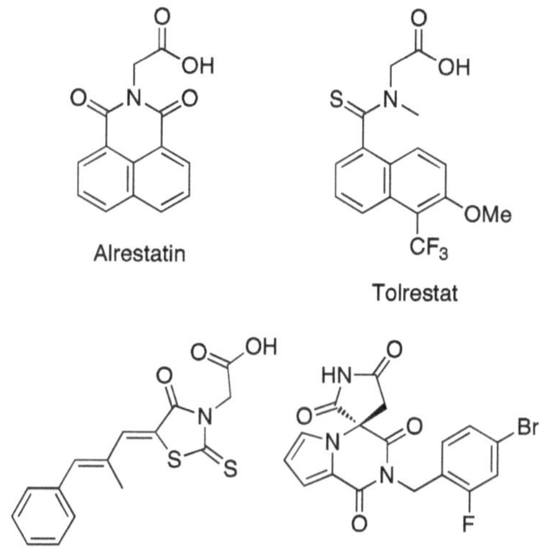

Figure 17.16 Structures of aldose reductase inhibitors.

4.2.1 Significance of the Introduction of a 4-Bromo-2-fluorobenzyl Group[56]

The biological potency in vitro and in vivo of compounds with a variety of substituents on the benzyl group at the N-2 position of the pyrrolopyrazine is summarized in Table 17.4 and Fig. 17.17. The AR inhibitory potencies were significantly influenced by substituents on the benzyl group at the N-2 position of the pyrrolopyrazine moiety. For instance, substitution by electron-donating groups such as methyl, methoxy, and amino groups led to significant decrease in the in vivo potency (**62**–**64**). On the other hand, substitution by electron-withdrawing groups such as halogens produced more potent compounds than the parent **56**, except for **58**, in both the AR inhibition and in vivo potencies. In particular, monosubstituted 4-bromobenzyl derivative **61** and disubstituted 4-chloro-2-fluoro- and 4-bromo-2-fluorobenzyl derivatives **65** and SX3030 showed excellent in vivo efficacy (ED_{50}: 0.3–0.7 mg/kg/day). Among these compounds, the best oral potency was observed for SX-3030. The 4-bromo-2-fluorobenzyl group increased not only the in vitro potency but also the in vivo efficacy. In the in vitro potency, this functionality increased interaction between AR and SX-3030, whereas in the in vivo efficacy, this group increased oral bioavailability because this group has higher lipophilicity than 4-bromo- and 4-chloro-2-fluorobenzyl groups (**61** and **65**).

Table 17.4 **In Vitro and In Vivo Activities of AR Inhibitors 56–65, (±)-SX3030, (−)-AS3201, and (+)-SX3202**

		AR Inhibition		
		In Vitro	In Vivo	
		Porcine Lens AR	% Inhibition[b]	ED_{50}[c]
Compound	R	IC_{50} (μM)[a]	30 mg/kg/day	mg/kg/day
56	H	0.099	69.1 ± 5.1	
57	2-F	0.061	98.0 ± 0.8	1.3
58	4-F	0.120	58.8 ± 3.3	
59	2-Br	0.052	82.6 ± 2.7	
60	3-Br	0.037	102.1 ± 0.5	
61	4-Br	0.047	108.1 ± 0.4	0.48
62	4-Me	0.052	65.8 ± 4.4	
63	4-OMe	0.046	46.7 ± 7.6	
64	4-NH_2	0.140	33.4 ± 3.3	
65	2-F, 4-Cl	0.041	109.5 ± 0.9	0.74
(±)-SX-3030	2-F, 4-Br	0.045	114.6 ± 0.2	0.30
(−)-AS-3201	2-F, 4-Br	0.015		0.18
(+)-SX-3202	2-F, 4-Br	0.19		>100

AR, aldose reductase.
[a]The concentration of test compounds required for 50% inhibition of aldose reductase in a partially purified porcine lens preparation.
[b]Percent inhibition of sorbitol accumulation in the sciatic nerve of streptozotocin-induced diabetic rat. Test compounds were orally given once a day for 5 days at the indicated doses. Values are mean ± standard error of mean; mean of four to six rats.
[c]The dose of test compounds required for 50% inhibition of sorbitol accumulation in the sciatic nerve of streptozotocin-induced diabetic rat.

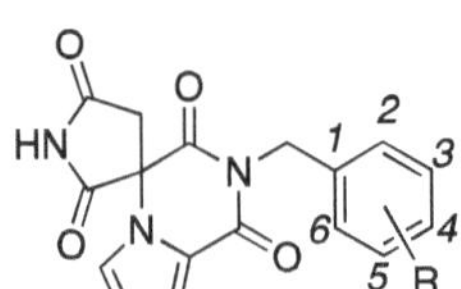

Figure 17.17 AR inhibitors **56–65**, (±)-SX3030, (−)-AS3201, and (+)-SX3202.

4.2.2 *Synthesis of Ranirestat*[56–58]

The first synthesis of ranirestat using a chiral resolution was achieved by the Dainippon group in 1998.[56] This synthesis gave enantiomers, i.e., (−)-AS-3201 (ranirestat) and (+)-SX-3202, which were separated by optical resolution, and their activity was assayed. Then, it was found that (−)-enantiomer is active, whereas (+)-enantiomer is inactive. Thus, the asymmetric synthesis of (−)-AS-3201 (ranirestat) became necessary. The first asymmetric synthesis of ranirestat was achieved by the Shibasaki group in 2007, using a lanthanide-catalyzed asymmetric conjugate addition of a succinimide derivative.[57] Another asymmetric synthesis of ranirestat was accomplished by the Trost group in 2010, using a Pd-catalyzed asymmetric allylic alkylation.[58]

4.2.2.1 First Synthesis of Ranirestat (Dainippon Synthesis)[56]

Ranirestat and its (+)-enantiomer were synthesized in seven steps from ethyl (benzyloxycarbonylamino)cyanoacetate (**66**)[59] by the sequence shown in Fig. 17.18. Alkylation of **66** with ethyl bromoacetate, followed by hydrolysis using hydrogen peroxide–sodium carbonate, brought about ring closure to give succinimide **68** (59% yield from **66**). Treatment of **68** with cinchonidine gave a cinchonidium salt, which was twice recrystallized from ethanol to give the diastereomerically enriched cinchonidium salt (>99.9% de). This salt was then treated with hydrochloric acid to afford the (−)-enantiomer **69** (>99.5% ee). On the other hand, recrystallization of the residue that was obtained from the acidified mother liquors gave (+)-enantiomer **70** (>99.4% ee). Each enantiomer **69** or **70** was then deprotected by hydrogenolysis with palladium on carbon, followed by treatment with 2,5-dimethoxytetrahydrofuran, to give pyrrole derivative **73** (69% yield from **69**) or **74**.·The treatment of **73** or **74** with excess trichloroacetyl chloride gave **75** (92% yield) or **76**. 2-Trichloroacetylpyrrole **75** or **76** was subjected to ring closure by condensation with 4-bromo-2-fluorobenzylamine in the presence of triethylamine in DMF to give the final product (−)-**77** (ranirestat) (>99.4% ee, 45% yield) or (+)-**78** (SX-3202) (>99.4% ee). The enantiomeric purities of enantiomers **69** and **70**, as well as **77** and **78** were determined by high-performance liquid chromatography (HPLC) analysis on a chiral column. The absolute configuration of (−)-enantiomer **77** was established as *R* by single-crystal X-ray analysis (Fig. 17.19).

4.2.2.2 Second and Enantioselective Synthesis of Ranirestat (Shibasaki Synthesis)[57]

The first enantioselective synthesis of ranirestat (**77**) was achieved in six steps from succinimide derivative **80** as shown in Fig. 17.20. Ten grams of **80** was subjected to catalytic asymmetric amination with (*E*)-di-*tert*-butyl diazene-1,2-dicarboxylate (**79**) in the presence of 2 mol% of La/(*R*)–**81** complex at 0°C to give 23.4 g of (−)-**82** in >99% yield with 92% ee. Boc groups were quantitatively removed by treatment with HCl in toluene at 0°C. The N–N bond cleavage of the hydrazine moiety proceeded smoothly with Raney nickel, and the following recrystallization gave **71** in optically pure form (66% yield). Compound **71** was subjected to modified Clauson–Kaas pyrrole synthesis to give **73** in 85% yield. Installation of a trichloroacetyl group and subsequent treatment with 4-bromo-2-fluorobenzylamine led to the formation of a pyrrolopyrazine core, affording ranirestat (**77**) (64% yield from **73**).

4.2.2.3 Third and Enantioselective Synthesis of Ranirestat (Trost Synthesis)[58]

The second enantioselective synthesis of ranirestat was accomplished in eight steps from commercially available 2-trichloroacetylpyrrole (**83**) by the sequence shown in Fig. 17.21. The key intermediate **87** in this synthesis was initiated by the *N*-alkylation

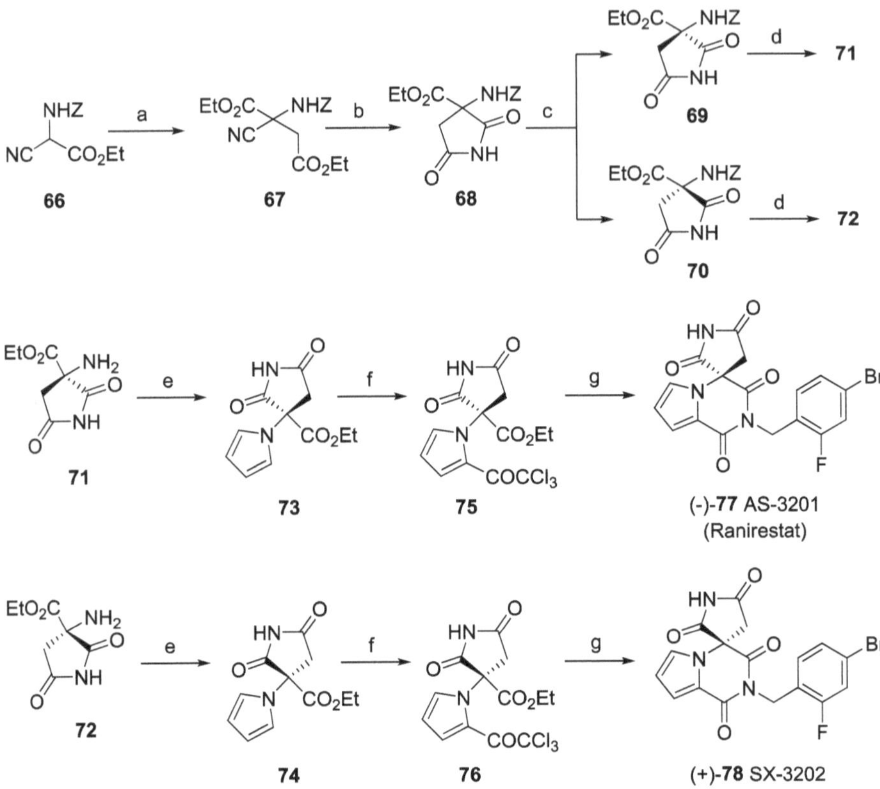

Z = benzyloxycarbonyl

Figure 17.18 First synthesis of ranirestat **77** and its enantiomer **78**. Reagents: (a) $BrCH_2CO_2Et$, K_2CO_3, Me_2CO; (b) H_2O_2-Na_2CO_3, aqueous Me_2CO; (c) cinconidine, EtOH; then HCl; (d) H_2, Pd/C, EtOH; (e) 2,5-dimethoxytetrahydrofuran, AcOH; (f) CCl_3COCl, EtOAc; (g) 4-bromo-2-fluorobenzylamine, Et_3N, DMF.

of **83** with methyl 2-bromoacetate, providing trichloroacetyl ketone **84** in 78% yield. Reaction of trichloroacetyl ketone **84** with 4-bromo-2-fluorobenzylamine in the presence of triethylamine in DMF facilitated the tandem condensation–cyclization sequence, forming the B-ring to provide imide **85** in 68% yield. An attempt to acylate **85** using methyl chloroformate only provided an *O*-acylated product. This problem was overcome by treatment of **85** with Mander's reagent **86**, which provided the desired *C*-acylated product **87** exclusively in nearly quantitative yield. It is noteworthy that all compounds **84**, **85**, and **87** were prepared on a multigram scale and did not require column chromatography for purification. Compound **87** was enantioselectively (84% ee) converted to olefin **90** in 69% yield in one pot by allylation with *tert*-butyl (2-triethoxysilylallyl)carbonate (**88**) in the presence of catalytic amounts of Pd(0) and chiral ligand (*R*,*R*)-**Ls** in hexafluorobenzene, followed by protodesilylation with AgF.

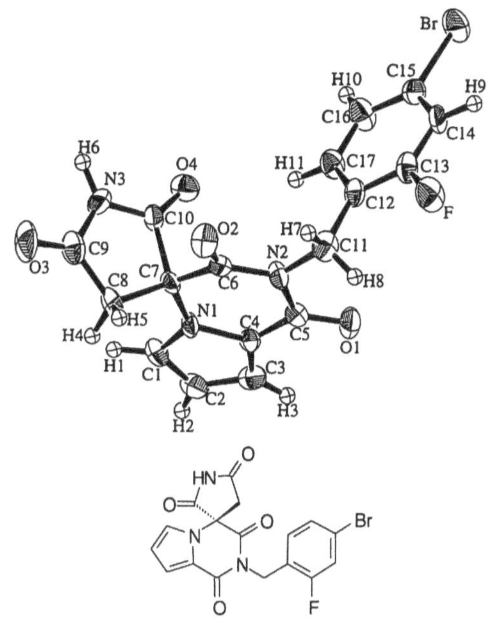

Figure 17.19 X-ray structure of ranirestat (**77**).
Copied from Negoro, T.; Murata, M.; Ueda, S.; Fujitani, B.; Ono, Y.; Kuromiya, A.; Komiya, M.; Suzuki, K.; Matsumoto, J. Novel, Highly Potent Aldose Reductase Inhibitors: (*R*)-(−)-2-(4-Bromo-2-fluorobenzyl)-1,2,3,4-tetrahydropyrrolo[1,2-*a*]pyrazine-4-spiro-3′-pyrrolidine-1,2′,3,5′-tetrone (AS-3201) and Its Congeners. *J. Med. Chem.* **1998,** *41*, 4118–4129 with permission.

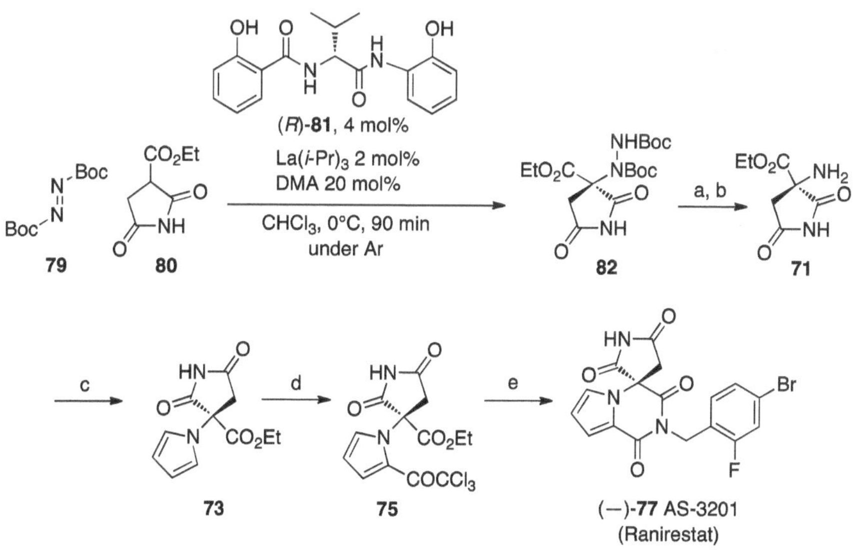

Figure 17.20 Enantioselective synthesis of ranirestat by the Shibasaki group. Reagents: (a) HCl gas, toluene; (b) Raney Ni, H_2, EtOH, rt, recrystallization; (c) 2,5-dihydroxytetrahydrofuran, aqueous THF; (d) CCl_3COCl; (e) 4-bromo-2-fluorobenzylamine, DMF.

Oxidative cleavage of the terminal methylene fragment of olefin **90** was achieved by the combination of $RuCl_3$, $NaIO_4$, and K_2CO_3 to give carboxylic acid **91**. Without further purification, the crude acid **91** was converted to the corresponding acyl chloride by treatment with thionyl chloride in THF, followed by treatment with ammonium hydroxide to afford amide **92** in 59% yield from olefin **90**. At this stage, amide **92** was conveniently cyclized to form the A-ring by treatment with LiHMDS in THF, providing ranirestat in near quantitative yield. The efficiency of the synthesis is demonstrated by requiring only eight steps, 14% overall yield, and the minimal use of column chromatography.

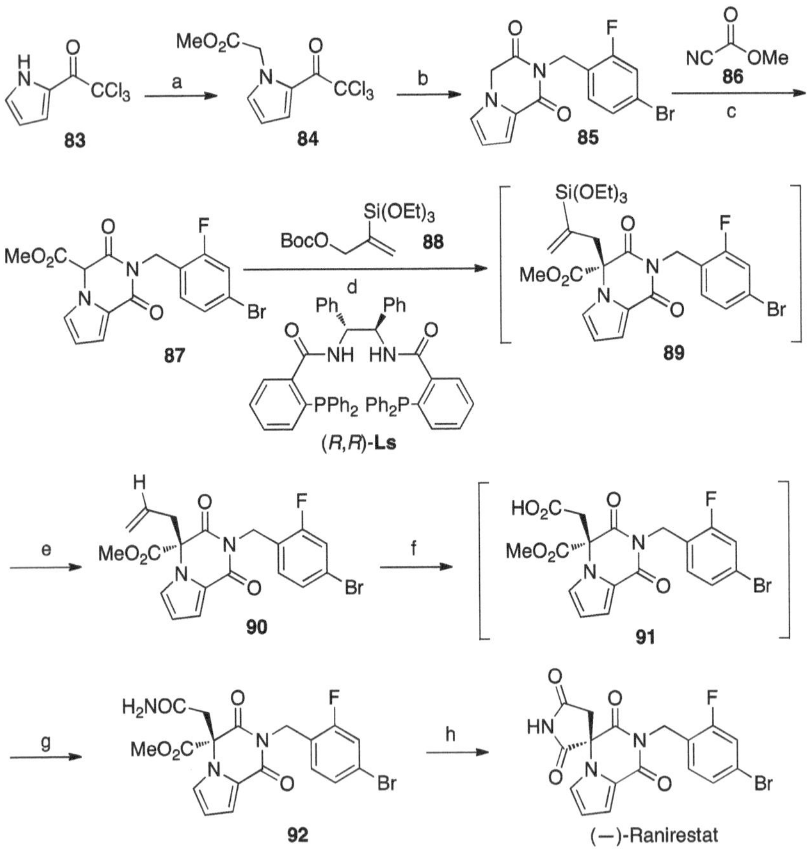

Figure 17.21 Enantioselective synthesis of ranirestat by the Trost group. Reagents: (a) methyl 2-bromoacetate, NaH, DMF, TBAI; (b) 4-bromo-2-fluorbenzylamine, NEt_3, DMF; (c) LiHMDS, *N,N'*-dimethylpropyleneurea (DMPU), THF; (d) $Pd_2(dba)_3CHCl_3$, C_6F_6; (e) AgF, THF, MeOH, dimethyl sulfoxide (DMSO), H_2O; (f) $RuCl_3$, $NaIO_4$, K_2CO_3, MeCN, CCl_4, H_2O; (g) $SOCl_2$, DMF, THF; then NH_4OH, EtOAc; (h) LiHMDS, THF.

5. Case Study 4: Use of a ^{18}F-Positron Emission Tomographic Tracer Replaced an Expensive ^{11}C-Positron Emission Tomographic Tracer for Practical Diagnosis of Alzheimer Disease

5.1 *^{18}F-Florbetaben: an Amyloid-β Positron Emission Tomographic Tracer for the Diagnosis of Alzheimer Disease*

Fluorine-18 (^{18}F) is a fluorine radioisotope, which is an important source of positrons. It has a mass of 18.0009380(6) u, and its half-life is 109.771(20) min. It decays by positron emission 97% of the time and electron capture 3% of the time. Both modes of decay yield stable oxygen-18. Because ^{18}F is easily introduced to compounds with hydroxyl groups and its half-life is relatively long, ^{18}F is often incorporated in diagnostic drugs for use in PET scans.

Amyloid-β (Aβ) plaques are a major pathological hallmark of AD. The noninvasive detection of Aβ plaques may increase the accuracy of clinical diagnosis as well as monitor therapeutic interventions. Although ^{11}C-Pittuburgh compound (^{11}C-PiB) (Fig. 17.22) is the most widely used Aβ PET radiotracer, due to the short half-life of ^{11}C (20 min), its application is expensive and limited to medical centers with an on-site cyclotron and ^{11}C radiochemistry expertise.[60,61] Therefore, ^{18}F-florbetaben has been developed as a novel ^{18}F-labeled Aβ PET tracer for practical diagnosis by Piramal Healthcare.

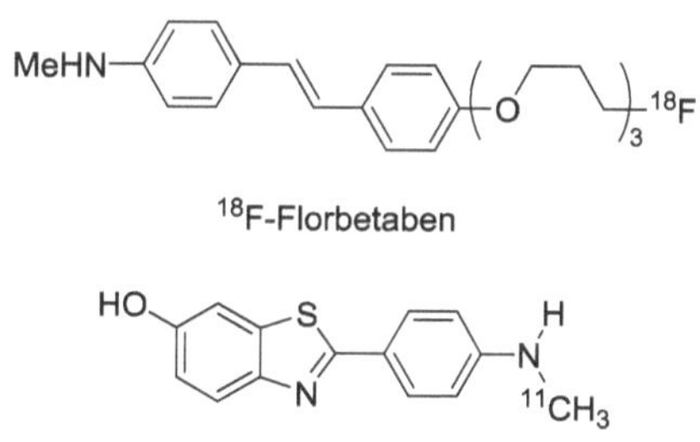

^{11}C-PiB

Figure 17.22 Structures of ^{18}F-florbetaben and ^{11}C-PiB.

^{18}F-florbetaben-PET can differentiate individuals diagnosed with AD from healthy elderly controls, as well as Parkinson's disease and other dementia patients. The ^{18}F-florbetaben-PET images overlaid with coregistered MRI images of AD patients and patients with Parkinson disease, dementia with Lewy bodies (DLB), mild cognitive impairment, frontotemporal lobe dementia (FTLD), and vascular dementia, exhibited clear differences between AD and other neurodisorders.[62]

^{18}F-Autoradiography in vitro demonstrated that ^{18}F-florbetaben exclusively bound to Aβ plaques in AD brain sections at low nanomolar concentrations, but did not bind to α-synuclein or *tau* aggregates in the brain sections of dementia with DLB and FTLD. Therefore, ^{18}F-florbetaben-PET has proved to be a highly selective radiotracer to assess Aβ plaque deposition in the brain.[63] ^{18}F-Florbetaben was approved as a novel ^{18}F-labeled Aβ PET tracer by the European Medicines Agency in February 2014 and by the US FDA in March 2014.

5.2 Automated Synthesis of ^{18}F-Florbetaben

Initially, ^{18}F-florbetaben was synthesized by the nucleophilic replacement of the mesylate group of Boc-stilbene-PEG$_3$-OMs **95** with activated ^{18}F-fluoride in the presence of Kryptofix 222 (K222) in DMSO, followed by removal of the Boc group under acidic conditions (Fig. 17.23, paths b, c, and d). The precursor **95** was prepared by Boc protection of the methylamino group of stilbene-PEG$_3$-OMs **94**, which was obtained by condensation between stilbene **93** and PEG$_3$-dimesylate in the presence of K_2CO_3 in acetone.[64]

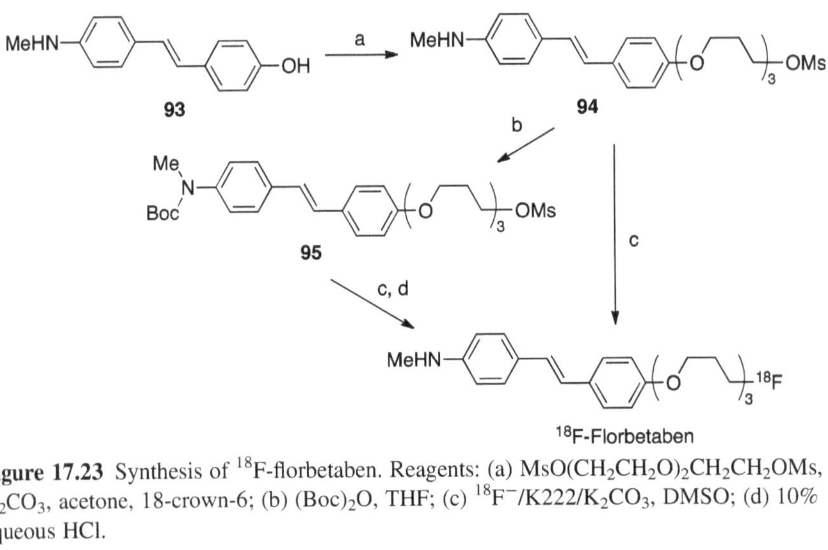

Figure 17.23 Synthesis of ^{18}F-florbetaben. Reagents: (a) $MsO(CH_2CH_2O)_2CH_2CH_2OMs$, K_2CO_3, acetone, 18-crown-6; (b) $(Boc)_2O$, THF; (c) $^{18}F^-/K222/K_2CO_3$, DMSO; (d) 10% aqueous HCl.

Then, more convenient one-step synthesis of ^{18}F-florbetaben was developed. This method was very easy to translate into automated synthesis of ^{18}F-florbetaben. This method gave ^{18}F-florbetaben in one step from mesylate **94** with activated ^{18}F-fluoride in the presence of K222 in DMSO (Fig. 17.23, path c).[65]

Furthermore, the automated synthesis of ^{18}F-florbetaben was performed on the PET-MF-2V-IT-1 synthesis module (PET Co. Ltd., China).[66] This synthesis module was originally designed for the routine production of a large amount of ^{18}F-fluorodeoxyglucose with two reaction vessels. The module automatically runs under the control of a computer-programmed synthesis protocol with a high level of automation and convenience. It is also a highly reliable system for the development of new ^{18}F-labeled tracers, and it can perform new automated radiosynthesis protocols after minimal modifications. In this work, the ^{18}F-florbetaben was obtained by a one-step ^{18}F-fluorination of mesylate **94** as the precursor (Fig. 17.23). The automated synthesis of ^{18}F-florbetaben was developed based on this module by connecting the reaction portion with built-in semipreparative HPLC and solid-phase extraction units (SPE units). Steps for the preparation of ^{18}F-florbetaben are summarized in Table 17.5. The first seven steps (A–G) are carried out in the ^{18}F-florbetaben synthesis unit, then the next two steps (H–I) are performed with the HPLC purification unit and the last four steps (J–M) are performed with the SPE unit. The reagents are loaded into vials 1–4, and the formulation solutions are loaded into the vials G3-2 and G4 of the synthesizer, as illustrated in Fig. 17.24.

Table 17.5 Summary of Operation Steps to Get Injectable Does of ^{18}F-Florbetaben

Fluorination of the Precursor Stilbene-PEG$_3$-OMs 94	
1.	$^{18}F^-$ trapped by a QMA cartridge
2.	Addition of the eluent to elute $^{18}F^-$ from a QMA cartridge
3.	Drying of $^{18}F^-$ with acetonitrile
4.	Addition of the precursor solution to the reaction vessel
5.	^{18}F-fluorination at 120°C for 6 min
6.	Addition of HPLC eluent to the reaction vessel
7.	Transfer of the solution to the HPLC loop
HPLC Purification	
8.	Elution using aqueous acetonitrile/ammonium formate (6/4) (0.1M); flow rate: 3 mL/min; detection: UV254 nm
9.	Collection of the radioactive peak of the ^{18}F-fluorination product (retention time, 19–22 min)
Solid-Phase Extraction with Cartridge	
10.	Dilute the fraction with water
11.	Transfer of the solution through a C18 Sep-Pak cartridge, and the product trapped on a C18 Sep-Pak cartridge
12.	Elute the final product with absolute ethanol
13.	Sterile filtration

HPLC, high-performance liquid chromatography; *QMA*, quaternary methylammonium; *UV*, ultraviolet.
Copied from Wang, H.; Guo, X.; Jiang, S.; Tang, G. Automated Synthesis of [^{18}F]Florbetaben as Alzheimer's Disease Imaging Agent Based on a Synthesis Module System. *Appl. Radiat. Isot.* **2013,** *71*, 41–46 with permission.

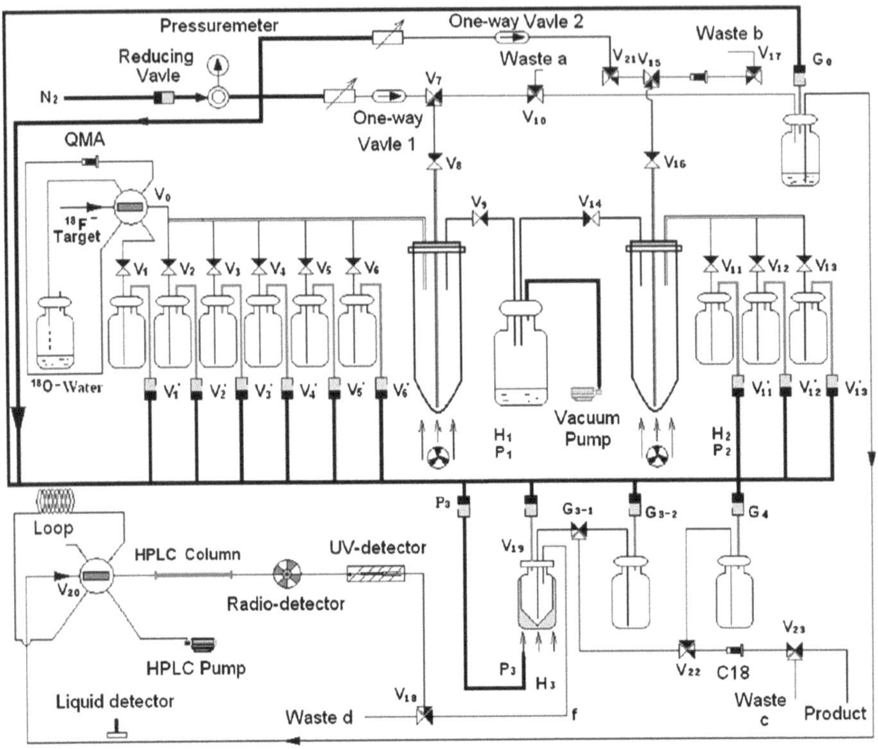

Figure 17.24 Illustration of the PET-MF-2V-IT-1 synthesis module (PET Co. Ltd., China). *HPLC*, high-performance liquid chromatography; *QMA*, quaternary methylammonium cartridge; *UV*, ultraviolet.
Copied from Wang, H.; Guo, X.; Jiang, S.; Tang, G. Automated Synthesis of [^{18}F]Florbetaben as Alzheimer's Disease Imaging Agent Based on a Synthesis Module System. *Appl. Radiat. Isot.* **2013,** *71*, 41–46 with permission.

6. Conclusions

This chapter presented four case studies on the remarkable effects of ^{19}F incorporations into therapeutic drugs, as well as ^{18}F incorporation into a diagnostic agent. In case study 1, the critical incorporation of fluorine to enhance the binding affinity of a drug to its protein target was highlighted using anacetrapib (for hypercholesterolemia treatment), idalopirdine (for AD treatment), and rociletinib (for NSCLC treatment) as examples. In case study 2, an essential boost to in vitro potency by incorporation of *gem*-fluorines was exemplified for the discovery and development of LDV (for hepatitis C treatment). In case study 3, crucial incorporation of fluorine to enhance in vitro potency and secure pharmacological property for in vivo efficacy was discussed using canagliflozin (for diabetes treatment) and ranirestat (for diabetic neuropathy treatment). In these case studies, the practical synthetic processes employed for these selected drugs were illustrated and discussed. For the synthesis of enantiomerically pure ranirestat, three methods of syntheses were presented with evolution of chemical technologies. In case study 4, the impressive development of a powerful PET tracer, ^{18}F-florbetaben, for the diagnosis of AD through the specific detection of Aβ in patient's brain was discussed, together with an efficient radiosynthesis, as well as an automated system for routine production in clinical research. These case studies and successful examples of fluorine-containing drugs with FDA approval or close to it clearly indicate the indisputable significance of the strategic incorporation of fluorine(s) into biologically relevant molecules for the advancement of biomedical sciences, and the importance of highly efficient synthetic methods enabling such incorporations.

References

1. Ojima, I., Ed. *Fluorine in Medicinal Chemistry and Chemical Biology;* Wiley-Blackwell: Chichester, 2009.
2. Begue, J.-P.; Bonnet-Delpon, D. Recent Advances (1995–2005) in Fluorinated Pharmaceuticals Based on Natural Products. *J. Fluor. Chem.* **2006,** *127,* 992–1012.
3. Wang, J.; Sánchez-Roselló, M.; Aceña, J. L.; Del Pozo, C.; Sorochinsky, A. E.; Fustero, S.; Soloshonok, V. A.; Liu, H. Fluorine in Pharmaceutical Industry: Fluorine-Containing Drugs Introduced to the Market in the Last Decade (2001–2011). *Chem. Rev.* **2014,** *114,* 2432–2506.
4. Cottet, F.; Marull, M.; Lefebvre, O.; Schlosser, M. Recommendable Routes to Trifluoromethyl-Substituted Pyridine- and Quinolonecarboxylic Acids. *Eur. J. Org. Chem.* **2003,** 1559–1568.
5. Yamazaki, T.; Taguchi, T.; Ojima, I. Unique Properties of Fluorine and Their Relevance to Medicinal Chemistry and Chemical Biology. In *Fluorine in Medicinal Chemistry and Chemical Biology;* Ojima, I., Ed.; Wiley-Blackwell: Chichester, 2009; pp 3–46.
6. Bays, H.; Stein, E. A. Pharmacotherapy for Dyslipidemia – Current Therapies and Future Agents. *Expert Opin. Pharmacother.* **2003,** *4,* 1901–1938.
7. Eckel, R. H.; Grundy, S. M.; Zimmet, P. Z. The Metabolic Syndrome. *Lancet* **2005,** *365,* 11415–11428.
8. Evans, M.; Roberts, A.; Davies, S.; Rees, A. Medical Lipid-Regulating Therapy: Current Evidence, Ongoing Trials and Future Developments. *Drugs* **2004,** *64,* 1181–1196.
9. Gordon, T.; Castelli, W. P.; Hjortland, M. C.; Kannel, W. B.; Dawber, T. R. High Density Lipoprotein as a Protective Factor Against Coronary Heart Disease. The Framingham Study. *Am. J. Med.* **1977,** *62,* 707–714.
10. Hesler, C. B.; Swenson, T. L.; Tall, A. R. Purification and Characterization of a Human Plasma Cholesteryl Ester Transfer Protein. *J. Biol. Chem.* **1987,** *262,* 2275–2282.
11. Sikorski, J. A. Oral Cholesteryl Ester Transfer Protein (CETP) Inhibitors: a Potential New Approach for Treating Coronary Artery Disease. *J. Med. Chem.* **2006,** *49,* 1–22.
12. Böhm, H.-J.; Banner, D.; Bendels, S.; Manfred Kansy, M.; Kuhn, B.; Müller, K.; Obst-Sander, U.; Stahl, M. Fluorine in Medicinal Chemistry. *Chem. Bio. Chem.* **2004,** *5,* 637–643.
13. Swain, C.; Rupniak, N. M. J. Progress in the Development of Neurokinin Antagonists. *Annu. Rep. Med. Chem.* **1999,** *34,* 51–60.

14. Miller, R. A.; Cote, A. S. *Process for Synthesizing a CETP Inhibitor,* PCT WO 2007/005572 A1, 2007.
15. Burns, A.; Iliffe, S. Altzheimer's Disease. *BMJ* **2009,** *338,* 467–471.
16. Burns, A.; Iliffe, S. Dementia. *BMJ* **2009,** *338,* 405–409.
17. Upton, N.; Chuang, T. T.; Hunter, A. J.; Virley, D. J. 5-HT6 Receptor Antagonists as Novel Cognitive Enhancing Agents for Alzheimer's Disease. *Neurotherapeutics* **2008,** *5,* 458–469.
18. Arnt, J.; Bang-Andersen, B.; Grayson, B.; Bymaster, F. P.; Cohen, M. P.; DeLapp, N. W.; Giethlen, B.; Kreilgaard, M.; McKinzie, D. L.; Neill, J. C.; Nelson, D. L.; Nielsen, S. M.; Poulsen, M. N.; Schaus, J. M.; Witten, L. M. Lu AE58054, a 5-HT_6 Antagonist, Reverses Cognitive Impairment Induced by Subchronic Phencyclidine in a Novel Object Recognition Test in Rats. *Int. J. Neuropsychopharmacol.* **2010,** *13,* 1021–1033.
19. Krogsgaard-Larsen, N.; Jensen, A. A.; Schrøder, T. J.; Christoffersen, C. T.; Jan Kehler, J. Novel Aza-Analogous Ergoline Derived Scaffolds as Potent Serotonin 5-HT_6 and Dopamine D_2 Receptor Ligands. *J. Med. Chem.* **2014,** *57,* 5823–5828.
20. Therkelsen, F.; Rock, M. H.; Treppendahl, S. *Processes for the Manufacture of a Pharmaceutically Active Agent,* Canadian Patent Application CA 2785440 A1, 2011.
21. Herbst, R. S.; Heymach, J. V.; Lippman, S. M. Lung Cancer. *N. Engl. J. Med.* **2008,** *359,* 1367–1380.
22. Walter, A. O.; Tjin Tham Sjin, R. T. T.; Haringsma, H. J.; Ohashi, K.; Sun, J.; Lee, K.; Dubrovskiy, A.; Labenski, M.; Zhu, Z.; Wang, Z.; Sheets, M.; St Martin, T.; Karp, R.; Van Kalken, D.; Chaturvedi, P.; Niu, D.; Nacht, M.; Petter, R. C.; Westlin, W.; Lin, K.; Jaw-Tsai, S.; Raponi, M.; Dyke, T. V.; Etter, J.; Weaver, Z.; Pao, W.; Singh, J.; Simmons, A. D.; Harding, T. C.; Allen, A. Discovery of a Mutant-Selective Covalent Inhibitor of EGFR that Overcomes T790M Mediated Resistance in NSCLC. *Cancer Discov.* **2013,** *3,* 1404–1415.
23. Choo, Q. L.; Kuo, G.; Weiner, A. J.; Overby, L. R.; Bradley, D. W.; Houghton, M. Isolation of a cDNA Clone Derived from a Blood-borne Non-A, Non-B Viral Hepatitis Genome. *Science* **1989,** *244,* 359–362.
24. Lavanchy, D. Evolving Epidemiology of Hepatitis C Virus. *Clin. Microbiol. Infect.* **2011,** *17,* 107–115.
25. Fried, M. W.; Shiffman, M. L.; Reddy, K. R.; Smith, C.; Marinos, G.; Goncales, F. L.; Haussinger, D.; Diango, M.; Carosi, G.; Dhumeaux, D.; Craxi, A.; Lin, A.; Hoffman, J.; Yu, J. Peginterferon Alfa-2a Plus Ribavirin for Chronic Hepatitis C Virus Infection. *N. Engl. J. Med.* **2002,** *347,* 975–982.
26. Zeuzem, S.; Berg, T.; Moeller, B.; Hinrichsen, H.; Mauss, S.; Wedemeyer, H.; Sarrazin, C.; Hueppe, D.; Zehnter, E.; Manns, M. P. Expert Opinion on the Treatment of Patients with Chronic Hepatitis C. *J. Viral Hepat.* **2009,** *16,* 75–90.
27. Link, J. O.; Taylor, J. G.; Xu, L.; Mitchell, M.; Guo, H.; Liu, H.; Kato, D.; Kirschberg, T.; Sun, J.; Squires, N.; Parrish, J.; Keller, T.; Yang, Z.-Y.; Yang, C.; Matles, M.; Wang, Y.; Wang, K.; Cheng, G.; Tian, Y.; Mogalian, E.; Mondou, E.; Cornpropst, M.; Perry, J.; Desai, M. C. Discovery of Ledipasvir (GS-5885): A Potent, Once-Daily Oral NS5A Inhibitor for the Treatment of Hepatitis C Virus Infection. *J. Med. Chem.* **2014,** *57,* 2033–2046.
28. Cordek, D. G.; Bechtel, J. T.; Maynard, A. T.; Kazmierski, W. M.; Cameron, C. E. Targeting the NS5A Protein of HCV: an Emerging Option. *Drugs Future* **2011,** *36,* 691–711.
29. Guedj, J.; Dahari, H.; Rong, L.; Sansone, N. D.; Nettles, R. E.; Cotler, S. J.; Layden, T. J.; Uprichard, S. L.; Perelson, A. S. Modeling Shows that the NS5A Inhibitor Daclatasvir has

Two Modes of Action and Yields a Shorter Estimate of the Hepatitis C Virus Half-life. *Proc. Natl. Acad. Sci. U.S.A.* **2013,** *110,* 3991–3996.
30. Romine, J. L.; Martin, S. W.; Snyder, L. B.; Serrano-Wu, M. H.; Deshpande, M. S.; Whitehouse, D.; Lemm, J. A.; O'Boyle, D. R., II; Gao, M.; Colonno, R. *Iminothiazolidinones as Inhibitors of HCV Replication,* February 19, 2004. Patent WO 2004/014852 A2.
31. Schmitz, U.; Tan, S. L. NS5A – from Obscurity to New Target for HCV Therapy. *Recent Pat. Anti-infect. Drug Discov.* **2008,** *3,* 77–92.
32. Serrano-Wu, M.; Belema, M.; Snyder, L. B.; St Laurent, D.; Kakaria, R.; Nguyen, V. N.; Qui, Y.; Yang, X.; Leet, J. E.; Gao, M.; O'Boyle, D. R.; Lemm, J. A.; Yang, F. *Inhibitors of HCV Replication,* December 14, 2006. Patent WO 2006/133326 A1.
33. a. Gao, M.; Nettles, R. E.; Belema, M.; Snyder, L. B.; Nguyen, V. N.; Fridell, R. A.; Serrano-Wu, M. H.; Langley, D. R.; Sun, J. H.; O'Boyle, D. R., 2nd; Lemm, J. A.; Wang, C.; Knipe, J. O.; Chien, C.; Colonno, R. J.; Grasela, D. M.; Meanwell, N. A.; Hamann, L. G. Chemical Genetics Strategy Identifies an HCV NS5A Inhibitor with a Potent Clinical Effect. *Nature* **2010,** *465,* 96–100.
b. St Laurent, D. R.; Serrano-Wu, M. H.; Belema, M.; Ding, M.; Fang, H.; Gao, M.; Goodrich, J. T.; Krause, R. G.; Lemm, J. A.; Liu, M.; Lopez, O. D.; Nguyen, V. N.; Nower, P. T.; O'Boyle, D. R., 2nd; Pearce, B. C.; Romine, J. L.; Valera, L.; Sun, J. H.; Wang, Y. K.; Yang, F.; Yang, X.; Meanwell, N. A.; Snyder, L. B. HCV NS5A Replication Complex Inhibitors. Part 4.1 Optimization for Genotype 1a Replicon Inhibitory Activity. *J. Med. Chem.* **2014,** *57,* 1976–1994.
34. Hamatake, R.; Maynard, A.; Kazmierski, W. M. HCV Inhibition Mediated through the Nonstructural Protein 5A (NS5A) Replication Complex. *Annu. Rep. Med. Chem.* **2012,** *47,* 331–345.
35. Anémian, R.; Mulatier, J.-C.; Andraud, C.; Stéphan, O.; Vial, J.-C. Monodisperse Fluorene Oligomers Exhibiting Strong Dipolar Coupling Interactions. *Chem. Commun.* **2002,** 1608–1609.
36. Tymtsunik, A. V.; Bilenko, V. A.; Ivon, Y. M.; Grygorenko, O. O.; Komarov, I. V. Synthesis of a Novel Boc-protected Cyclopropane-modified Proline Analogue. *Tetrahedron Lett.* **2012,** *53,* 3847–3849.
37. Stella, L.; Abraham, H.; Feneau-Dupont, J.; Tinant, B.; Declercq, J. P. Asymmetric Aza-Diels-Alder Reaction Using the Chiral 1-phenyl-ethyl Imine of Methyl Glyoxylate. *Tetrahedron Lett.* **1990,** *31,* 2603–2606.
38. Wild, S.; Roglic, G.; Green, A.; Sicree, R.; King, H. Global Prevalence of Diabetes. Estimates for the Year 2000 and Projections for 2030. *Diabetes Care* **2004,** *27,* 1047–1053.
39. Deetjen, P.; Baeyer, H. V.; Drexel, H. Renal Handling of D-glucose and Other Sugars. In *Textbook of Nephrology,* Vol. 1, .; 3rd ed.; Textbook of Nephrology, Vol 1, Massry, S. G., Glassock, R. J., Eds.; Williams &Wilkins Press: Baltimore, MD, 1995; pp 90–94.
40. Nishimura, M.; Naito, S. Tissue-Specific mRNA Expression Profiles of Human ATP-Binding Cassette and Solute Carrier Transporter Superfamilies. *Drug Metab. Pharmacokinet.* **2005,** *20,* 452–477.
41. Tsujihara, K.; Hongu, M.; Saito, K.; Kawanishi, H.; Kuriyama, K.; Matsumoto, M.; Oku, A.; Ueta, K.; Tsuda, M.; Saito, A. Na^{+}-Glucose Cotransporter (SGLT) Inhibitors as Antidiabetic Agents. 4. Synthesis and Pharmacological Properties of 4′-Dehydroxyphlorizin Derivatives Substituted on the B Ring. *J. Med. Chem.* **1999,** *42,* 5311–5324.
42. Oku, A.; Ueta, K.; Arakawa, K.; Ishihara, T.; Nawano, M.; Kuronuma, Y.; Matsumoto, M.; Saito, A.; Tsujihara, K.; Anai, M.; Asano, T.; Kanai, Y.; Endou, H. T-1095, an Inhibitor of

Renal Na^+-Glucose Cotransporters, may Provide a Novel Approach to Treating Diabetes. *Diabetes* **1999,** *48,* 1794–1800.

43. Nomura, S.; Sakamaki, S.; Hongu, M.; Kawanishi, E.; Koga, Y.; Sakamoto, T.; Yamamoto, Y.; Ueta, K.; Kimata, H.; Nakayama, K.; Tsuda-Tsukimoto, M. Discovery of Canagliflozin, a Novel *C*-glucoside with Thiophene Ring, as Sodium-dependent Glucose Cotransporter 2 Inhibitor for the Treatment of Type 2 Diabetes Mellitus. *J. Med. Chem.* **2010,** *53,* 6355–6360.
44. Srogl, J.; Allred, G. D.; Liebeskind, L. S. Sulfonium Salts. Participants par Excellence in Metal-Catalyzed Carbon–Carbon Bond-Forming Reactions. *J. Am. Chem. Soc.* **1997,** *119,* 12376–12377.
45. Horton, D.; Priebe, W. Synthetic Routes to Higher-Carbon Sugars. Reaction of Lactones with 2-Lithio-1,3-Dithiane. *Carbohydr. Res.* **1981,** *94,* 27–41.
46. Ellsworth, B. A.; Doyle, A. G.; Patel, M.; Caceres-Cortes, J.; Meng, W.; Deshpande, P. P.; Pullockaran, A.; Washburn, W. N. *C*-Arylglucoside Synthesis: Triisopropylsilane as a Selective Reagent for the Reduction of an Anomeric *C*-phenyl Ketal. *Tetrahedron Asymmetry* **2003,** *14,* 3243–3247.
47. Yang, G.; Franck, R. W.; Byun, H.-S.; Bittman, R.; Samadder, P.; Arthur, G. Convergent *C*-Glycolipid Synthesis via the Ramberg–Bäcklund Reaction: Active Antiproliferative Glycolipids. *Org. Lett.* **1999,** *1,* 2149–2151.
48. Petrash, J. M. All in the Family: Aldose Resuctase and Closely Related Aldoketo Reductases. *Cell. Mol. Life Sci.* **2004,** *61,* 737–749.
49. a. Yabe-Nishimura, C. Aldose Reductase in Glucose Toxicity: a Potential Target for the Prevention of Diabetic Complications. *Pharmacol. Rev.* **1998,** *50,* 21–33.
b. Oka, M.; Kato, N. Aldose Reductase Inhibitors. *J. Enzyme Inhib.* **2001,** *16,* 465–473.
50. Kinoshita, J. H.; Nishimura, C. The Involvement of Aldose Reductase in Diabetic Complications. *Diabetes Metab. Rev.* **1988,** *4,* 323–337.
51. González, R. G.; Barnett, P.; Aguayo, J.; Cheng, H. M.; Chylack, L. T., Jr. Direct Measurement of Polyol Pathway Activity in the Ocular Lens. *Diabetes* **1984,** *33,* 196–199.
52. a. Green, D. A.; Lattimer, S. A. Action of Sorbinil in Diabetic Peripheral Nerve: Relationship of Polyol (Sorbitol) Pathway Inhibition to a *myo*-inositol-mediated Defect in Sodium-Potassium ATPase Activity. *Diabetes* **1983,** *33,* 712–716.
b. Hamada, Y.; Kitoh, R.; Raskin, P. Crucial Role of Aldose Reductase Activity and Plasma Glucose Level in Sorbitol Accumulation in Erythrocytes from Diabetic Patients. *Diabetes* **1991,** *40,* 1233–1240.
53. Cameron, N. E.; Cotter, M. A.; Hohman, T. C. Interactions between Essential Fatty Acid, Prostanoid, Polyol Pathway and Nitric Oxide Mechanisms in the Neurovascular Deficit of Diabetic Rats. *Diabetologia* **1996,** *39,* 172–182.
54. Green, D. A.; Sima, A. A. F. Effects of Aldose Reductase Inhibitors on the Progression of Nerve Damage. *Diabetic Med.* **1993,** *10,* 31S–32S.
55. a. Kikkawa, R.; Hatanaka, I.; Yasuda, H.; Kobayashi, N.; Shigeta, Y.; Terashima, H.; Morimura, T.; Tsuboshima, M. Effects of a New Aldose Reductase Inhibitor, (*E*)-3-carboxymethyl-5-[(2*E*)-methyl-3-phenylpropenilidene]rhodanine (ONO-2235), on Peripheral Nerve Disorders in Streptozotocin-Diabetic Rats. *Diabetlogia* **1983,** *24,* 290–292.
b. Terashima, H.; Hama, K.; Yamamoto, R.; Tsuboshima, M.; Kikkawa, R.; Hatanaka, I.; Shigeta, Y. Effects of a New Aldose Reductase Inhibitor on Various Tissues in Vitro. *J. Pharmacol. Exp. Ther.* **1984,** *229,* 226–230.
56. Negoro, T.; Murata, M.; Ueda, S.; Fujitani, B.; Ono, Y.; Kuromiya, A.; Komiya, M.; Suzuki, K.; Matsumoto, J. Novel, Highly Potent Aldose Reductase Inhibitors: (*R*)-(−)-2-(4-

Bromo-2-fluorobenzyl)-1,2,3,4-tetrahydropyrrolo[1,2-*a*]pyrazine-4-spiro-3′-pyrrolidine-1,2′,3,5′-tetrone (AS-3201) and Its Congeners. *J. Med. Chem.* **1998,** *41,* 4118–4129.
57. Mashiko, T.; Hara, K.; Tanaka, D.; Fujiwara, Y.; Kumagai, N.; Shibasaki, N. En Route to an Efficient Catalytic Asymmetric Synthesis of AS-3201. *J. Am. Chem. Soc.* **2007,** *129,* 11342–11343.
58. Trost, B. M.; Osipov, M.; Dong, G. A Concise Enantioselective Synthesis of (−)-Ranirestat. *Org. Lett.* **2010,** *12,* 1276–1279.
59. Warren, C. B.; Minard, R. D.; Matthews, C. N. Synthesis of α-Cyanoglycine N-Carboxyanhydride and α-Cyanoglycine. *J. Org. Chem.* **1974,** *39,* 3375–3378.
60. Rowe, C. C.; Ng, S.; Ackermann, U.; Gong, S. J.; Pike, K.; Savage, G.; Cowie, T. F.; Dickinson, K. L.; Maruff, P.; Darby, D.; Smith, C.; Woodward, M.; Merory, J.; Tochon-Danguy, H.; O'Keefe, G.; Klunk, W. E.; Mathis, C. A.; Price, J. C.; Masters, C. L.; Villemagne, V. L. Imaging Beta-Amyloid Burden in Aging and Dementia. *Neurology* **2007,** *68,* 1718–1725.
61. Klunk, W. E.; Engler, H.; Nordberg, A.; Wang, Y.; Blomqvist, G.; Holt, D. P.; Bergström, M.; Savitcheva, I.; Huang, G. F.; Estrada, S.; Ausén, B.; Debnath, M. L.; Barletta, J.; Price, J. C.; Sandell, J.; Lopresti, B. J.; Wall, A.; Koivisto, P.; Antoni, G.; Mathis, C. A.; Långström, B. Imaging Brain Amyloid in Alzheimer's Disease with Pittsburgh Compound-B. *Ann. Neurol.* **2004,** *55,* 306–319.
62. Villemagne, V. L.; Ong, K.; Mulligan, R. S.; Holl, G.; Pejoska, S.; Jones, G.; O'Keefe, G.; Ackerman, U.; Tochon-Danguy, H.; Chan, J. G.; Reininger, C. B.; Fels, L.; Putz, B.; Rohde, B.; Masters, C. L.; Rowe, C. C. Amyloid Imaging with ^{18}F-Florbetaben in Alzheimer Disease and Other Dementias. *J. Nucl. Med.* **2011,** *52,* 1210–1217.
63. Fodero-Tavoletti, M. T.; Brockschnieder, D.; Villemagne, V. L.; Martin, L.; Connor, A. R.; Thiele, A.; Berndt, M.; McLean, C. A.; Krause, S.; Rowe, C. C.; Masters, C. L.; Dinkelborg, L.; Dyrks, T.; Cappai, R. In Vitro Characterization of [^{18}F]-Florbetaben, an Aβ Imaging Radiotracer. *Nuc. Med. Biol.* **2012,** *39,* 1042–1048.
64. Zhang, W.; Oya, S.; Kung, M.-P.; Hou, C.; Maier, D. L.; Kung, H. F. F-18 Polyethyleneglycol Stilbenes as PET Imaging Agents Targeting Aβ Aggregates in the Brain. *Nucl. Med. Biol.* **2005,** *32,* 799–809.
65. Wang, H.; Shi, H.; Yu, H.; Jiang, S.; Tang, G. Facile and Rapid One-Step Radiosynthesis of [^{18}F]BAY94-9172 with a New Precursor. *Nucl. Med. Biol.* **2011,** *38,* 121–127.
66. Wang, H.; Guo, X.; Jiang, S.; Tang, G. Automated Synthesis of [^{18}F]Florbetaben as Alzheimer's Disease Imaging Agent Based on a Synthesis Module System. *Appl. Radiat. Isot.* **2013,** *71,* 41–46.

Telomerization Reaction of 3,3,3-Trifluoropropene

18

G. Kostov[1], *B. Ameduri*[2]
[1]Bourgas As. Zlatarov University, Bourgas, Bulgaria; [2]Institut Charles Gerhardt, UMR5253 (CNRS) E.N.S. Chimie Montpellier, Montpellier Cedex, France

Chapter Outline

Modern Synthesis Processes and Reactivity of Fluorinated Compounds. http://dx.doi.org/10.1016/B978-0-12-803740-9.00018-4

1. Introduction

With the persistent need for increasingly better performing materials endowed with specific properties, macromolecular engineering has become a crucial tool to design well-architectured polymers. Telomers, alternating, telechelics, block and graft copolymers, and dendrimers are currently undergoing an increasing development in both their variety and number. Among them, fluoropolymers, which display outstanding properties, have already achieved distinction as high-added-value materials in many applications ranging from surfactants, optical and electroactive devices, high-performance elastomers, and coatings to gel electrolytes for lithium ion batteries and membranes for fuel cells. Indeed, interest in the relationship between the functional monomer's structures and the resulting polymer's properties is nowadays undergoing a fast growth, and the properties can be tuned for the most appropriate and relevant applications. This chapter contains three parts; it starts with a general introduction outlining the basic concepts, and then, the main section reports recent developments, and its subsections summarize the synthesis of well-defined fluorotelomers based on 3,3,3-trifluoropropene (TFP) from various chain transfer agents (CTAs), their properties, characterization, and finally, their applications.

In contrast to polymerization, telomerization reactions, pioneered by the DuPont Company in 1942,[1] usually yield low-molecular-weight polymers, named telomers (1), or even monoadducts with well-defined end groups. Such products 1 are usually produced from a telogen or transfer agent (X−Y), and one or more (n) molecules of a polymerizable derivative M (called taxogen or monomer) having ethylenic unsaturation, under radical initiation conditions (mainly reported in the chapter dealing with it since this strategy is easier to process, although ionic or organometallic[2] initiations are also possible), as follows:

$$\text{X—Y} + n\text{M} \xrightarrow{\text{free radicals}} \text{X—(M)}_n\text{—Y}$$

1

Radicals are usually released from various systems thermally, photochemically (even from sunlight), by visible light, and by irradiation (^{60}Co, electron beam, X-rays, ozone, or from metallic complexes, redox catalysts, or radical organic initiators). Advantageously, X−Y telogens bring functional X and Y end groups and must be cleavable by free radicals (generated according to the conditions of initiation) leading to an X• radical able to further react onto the less hindered site of the monomer. After the propagation of monomer, the final step consists of the transfer of the telogen onto the growing telomeric chain. This is the driving force of the telomerization.

The strength of the X−Y bond induces a more-or-less efficient transfer rate. The weaker the bond dissociation energy, the higher the transfer efficiency, and thus the lower the molecular weights of the resulting telomers.

Actually, telomers 1 are intermediate derivatives between organic compounds (e.g., n = 1) and macromolecular products (n = 30−100). Hence, in certain cases, end groups evidence a chemical importance that can induce useful opportunities for further functionalizations. All monomers and olefins used in radical polymerization can be involved in telomerization.

The initiation steps and mechanisms involved in telomerization have previously been reported,[1] as well as the various processes used, the nature of the CTAs, the monomers, and their respective reactivities. The general process of telomerization requires a source of radical, and in the case of a radical initiator, A_2, the overall mechanism is displayed in Scheme 18.1 (where A_2 represents an azo, peroxide, peroxicarbonate, or perester initiator that undergoes a homolytic cleavage under heating and generates two A° radicals[3]).

Initiation

$$A_2 \xrightarrow[\Delta]{k_i} 2\,A^{\bullet}$$

$$A^{\bullet} + XY \longrightarrow X^{\bullet} + AY$$

$$X^{\bullet} + F_2C{=}CRR' \longrightarrow X(C_2F_2RR')^{\bullet}$$

Propagation

$$XC_2F_2RR'^{\bullet} + F_2C{=}CRR' \xrightarrow{k_{p1}} X(C_2F_2RR')_2^{\bullet}$$

$$X(C_2F_2RR')_n^{\bullet} + F_2C{=}CRR' \xrightarrow{k_{pn}} X(C_2F_2RR')_{n+1}^{\bullet}$$

Termination

$$X(C_2F_2RR')_n^{\bullet} + X(C_2F_2RR')_p^{\bullet} \xrightarrow{k_{Te}} X(C_2F_2RR')_{n+p}X$$

Transfer

$$X(C_2F_2RR')_n^{\bullet} + XY \xrightarrow{k_{tr}} X(C_2F_2RR')_nY + X^{\bullet}$$

Scheme 18.1 Mechanism of telomerization (where XY, $F_2C{=}CRR'$, k_i, k_{p1}, k_{Te}, and k_{tr} represent the telogen, the monomer, as well as the rate constants of initiation, propagation, termination, and transfer, respectively).

The scope of telomerization was first outlined by Friedlina[4] in 1966, further developed by Starks in 1974,[5] and improved upon by Gordon and Loftus.[6] We reviewed such a reaction,[7,8] and the most recent comprehensive overview was published in a book chapter[9] in which the mechanisms and kinetics of radical and redox telomerizations were described, as well as the radical telomerization of the main commercially available fluorinated alkenes [tetrafluoroethylene (TEE), vinylidene fluoride (VDF), chlorotrifluoroethylene (CTFE), trifluoroethylene, hexafluoropropylene, etc.].

This chapter deals with the radical telomerization of TFP with various CTAs, their cotelomerization, and the applications therefrom.

2. Telomerization of 3,3,3-Trifluoropropene

2.1 Homotelomers of 3,3,3-Trifluoropropene

The fluoroalkene TFP, prepared and isolated for the first time by Haszeldine[10] by dehydroiodination of $F_3CCH_2CH_2I$, is the precursor of a commercially available fluorosilicone called Silastic, marketed by Dow Corning (Scheme 18.2). This interesting polymer preserves its performances (inertness, fair surface tension, chain mobility) at low and rather high temperatures. Interestingly, this alkene is quite reactive, in contrast to longer perfluorinated chain vinyl-type monomers (e.g., $C_nF_{2n+1}CH{=}CH_2$ with $n > 1$).

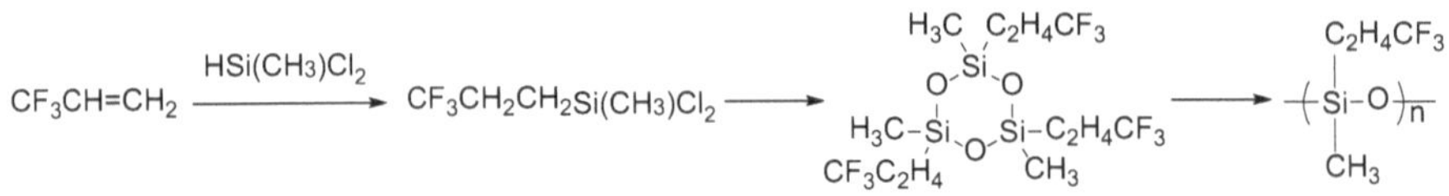

Scheme 18.2 Synthesis of poly(methyl-3,3,3-trifluoropropyl siloxane) from hydrosylilation of 3,3,3-trifluoropropene.

Nowadays, this fluoroolefin is commercialized by the Chemtura (formerly Great Lakes Juhua) companies. This monomer has been employed in telomerization by many authors who used a wide variety of telogens (Table 18.1) from different types of initiation: ultraviolet (UV) or $\gamma-$ radiations, redox, or in the presence of radical initiators.

In $CF_3CH{=}CH_2$, a double bond flanked by a CF_3 group is polarized in a way to make the end carbon atom more positive than the center carbon, i.e., $CF_3^{-\delta}CH{=}^{+\delta}CH_2$, and is sensitive to nucleophilic attack. For example, the presence of HCl (or HBr) led to $CF_3CH_2CH_2Cl$ (or $CF_3CH_2CH_2Br$) exclusively.[11]

In this field, the most pertinent studies were performed by Russian teams (mainly by Terent'ev and Vasil'eva in the 1980s) who chose esters,[12,13] alcohols,[14] or halogenated compounds as telogens.[15–17] In addition, Vasil'eva's team even proposed mechanisms of formation for unexpected telomers[15,16] and determined various chain transfer constants of telogens, showing that, in certain cases, the transfer to telogen is more efficient than the monomer propagation (or radical chain growth).[13] They also observed that $C_6H_5CH_2Cl$[15] and $C_6H_5CH_2Br$[13] undergo both C—H and C-halogen cleavages since their reaction products are $C_6H_5CH_2(TFP)_n$-X and $C_6H_5CHX(TFP)_nH$, where X = Cl, Br.

In the same way, bromoform, in the presence of iron complex or peroxide [such as dibenzoyl peroxide or di-*tert*-butylperoxide (DTBP)] also leads to two series of telomers arising from its C—H and C—Br cleavages.[18] Consequently, the formation of "reverse adduct" $CF_3CHBrCH_2CBr_2[CH_2CH(CF_3)]_2H$ is not unexpected. The Russian group also used tetrabromomethane as CTA.[19] In telomerization of TFP, CF_3I was shown to be an effective telogen from which both thermal and photochemical initiations were successfully achieved, the latter being more selective than the former.[20] Low et al.[21] reported that this reaction under UV radiation at various temperatures produced the monoadduct composed of a mixture of normal and reverse isomers. They observed that a higher proportion of reverse derivative was formed at higher temperatures. In addition, they noted the formation of higher adducts above 175°C.

Table 18.1 Telomerization of 3,3,3-Trifluoropropene with Various CTAs

Telogen or CTA	Initiation	Structure of Telomers	References
$H_3C{-}OCOC(CH_3)_2H$	Peroxides	$H_3C{-}OCOC(CH_3)_2[CH_2CH(CF_3)]_nH$	12,13
$(CH_3)_2CHOH$ alcohols	γ-rays (T < 90°C)	$(CH_3)_2(OH)[CH_2CH(CF_3)]_nH$ n = 1, 2	14
$C_6H_5CH_2{-}Cl$	$Fe(CO)_5$	$C_6H_5CHX[CH_2CH(CF_3)]_nY$ n = 1, 2 X = Cl, H Y = H, Cl	15
$C_6H_5CH_2{-}Br$	$Fe(CO)_5$	As the previous structure with X = Br, H and Y = H, Br	13
CCl_4	$Fe(CO)_5$	$CC_{13}[CH_2CH(CF_3)]_nCl$ (n = 1−3)	16
CCl_4	CuCl/LiCl	$CCl_3[CH_2CH(CF_3)]_nCl$ (n = 1−7)	17
$Br_2CH{-}Br$	$Fe(CO)_5$	$BrCH[CH_2CH(CF_3)]_nBr$ n = 1−3 $CF_3CBrHCH_2CBr_2[CH_2CH(CF_3)]_2H$	18
$Br_2CH{-}Br$	Peroxides	$XCBr_2[CH_2CH(CF_3)]_nY$ Y = X = H or Br n = 1, 2	18
$BrCH_2Br$	$Fe(CO)_5$	$BrCH[CH_2CH(CF_3)]_nBr$, n = 1, 2	18
CBr_4	DBP	$Br_3C[CH_2CH(CF_3)]_nBr$, n = 1−3	19
CF_3I	UV/5 d	$CF_3[CH_2CH(CF_3)]_nI$, n = 1, 2	20
CF_3I	225°C/36 h	$CF_3[CH_2CH(CF_3)]_nI$, n = 1−3	20
CF_3I	UV/various T	Normal and reverse monoadducts few amounts of n = 2	10,21
CF_3I	UV/T > 175°C	$CF_3[CH_2CH(CF_3)]_nI$, n = 1−8	10a
$C_6F_{13}I$	DTBP, Trig.101	$C_6F_{13}\text{-}[CH_2\text{-}CH(CF_3)]_n{-}I$	22
$(CF_3)_2CFI$	DTBP, Trig.101	$(CF_3)_2CF[CH_2{-}CH(CF_3)]_n{-}I$	22
$IC_6F_{12}I$	DTBP/140°C	$I(TFP)n(CF_2)_4(TFP)_nI$	23
$(CF_3)_2CF(VDF)_xI$	TBPPi/DTBP	$(CF_3)_2CF(VDF)_x[CH_2{-}CH(CF_3)]_n{-}I$	24
$Cl_3Si{-}H$	UV	$Cl_3Si[CH_2CH(CF_3)]_nH$ n = 1, 2	25
$(EtO)_2P(O)H$	DTBP/130°C	$(C_2H_5O)_2P(O)CH_2CH_2CF_3$ (39%)	26
$(EtO)_2P(O)H$	DTBP/140°C	$(C_2H_5O)_2P(O)[CH_2{-}CH(CF_3)]_n{-}H$	27
CH_3SSCH_3	UV	$CH_3SCH_2CH(CF_3)SCH_3$	28
HBr	UV	$CF_3CH_2CH_2Br$	20
Cyclopentadiene	180°C/72 h	Exo:endo = 62:38	29
2-Me-1,3-dioxolane	$Fe(CO)_5$	2,4-bis(TFP)-2-Me-dioxolane	30

CTA, chain transfer agents; *DBP*, dibenzoyl peroxide; *DTBP*, di-*tert*-butylperoxide; *TBPPi*, *tert*-butylperoxypivalate; *TFP*, 3,3,3-trifluoropropene; *Trig.101*, −2,5-*bis*(*tert*-butylperoxy)-2,5-dimethylhexane; *UV*, ultraviolet; *VDF*, vinylidene fluoride.

Among other transfer agents, silanes,[25] phosphonates,[26,27] disulfides[28] or hydrogen bromide,[20] cyclopentadiene,[29] and 2-methyl-1,3-dioxolane[30] were reported to react with TFP, some of them yielding the monoadduct mainly (Table 18.1); the details on most recent studies are given later.

2.1.1 *Telomerization of 3,3,3-Trifluoropropene With Perfluorinated Telogens*

The telomerization of TFP involving various initiating systems (thermal, photochemical, UV, radical, metal complexes) and iodoperfluoroalkanes, R_FI, as the telogens (CTAs) ($C_6F_{13}I$ and i-C_3F_7I preferably) was performed[22] and led to ω-iodofluorinated compounds (Scheme 18.3).

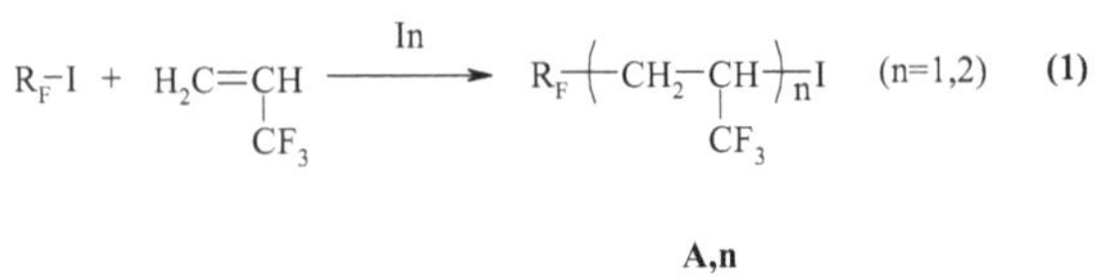

Scheme 18.3 Radical telomerization of 3,3,3-trifluoropropene (TFP) with 1-iodoperfluoroalkane.[22]

Furthermore, the photochemical and metal complexes reactions were carried out in batch conditions, whereas thermal and peroxide initiations were achieved under pressure, in autoclave at temperature ranging from 60 to 180°C depending on the initiators used: the most suitable initiators were DTBP and 2,5-*bis*(*tert*-butylperoxy)-2,5-dimethylhexane. The telogen $[T]_0$/monomer $[M]_0$ initial molar ratio (R_0) varied from 0.25 to 1.50, and the reaction time ranged between 4 and 22 h. The yield of mono- ($n = 1$) and diadducts ($n = 2$), normal and reverse addition ratios of monomer units as a function of reaction conditions, were determined. The yield (Y) of diadducts obtained was higher than 65% at $R_0 = 0.5$, whereas the monoadduct ($Y < 50\%$) was produced more selectively from an excess of CTA ($R_0 = 1.4$).

The structure of telomers and the mechanism of radical telomerization of TFP were evidenced by nuclear magnetic resonance (NMR) analysis.

Two possibilities can be suggested for the mechanism: the first one is the classical mechanism for telomerization in which the produced $R_FCH_2{-}CH(CF_3)^{\bullet}$ radical may undergo either telogen transfer or the propagation of TFP. In the latter case, the reverse adduct can be formed in the propagation step (Scheme 18.4). However, a stepwise mechanism may also be suggested in which the $R_FCH_2{-}CH(CF_3)I$ formed by R_FI transfer acts as a telogen in the further telomerization of TFP as shown in Scheme 18.4.[22]

Functionalization of such TFP telomers was achieved from their radical addition onto allyl acetate (Scheme 18.5). Different radical initiators such as peroxides (dibenzoyl peroxide, *tert*-amyl peroxypivalate, DTBP) or azobisisobutyronitrile (AIBN) can be used. Because of the high exothermicity of the reaction,[31,32] the initiator used was

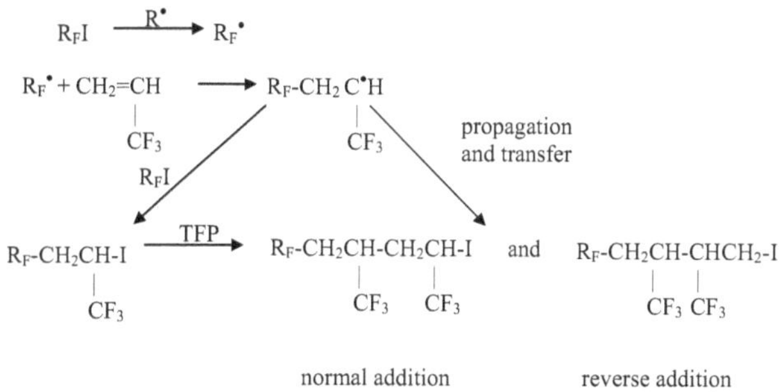

Scheme 18.4 Mechanism of the radical telomerization of 3,3,3-trifluoropropene (TFP) with 1-iodoperfluoroalkane (R_FI) via stepwise transfer or propagation and transfer for the formation of diadduct.[22]

of great importance (especially for an initiator that possesses a low decomposition temperature). It was previously observed that when the starting temperature is higher than 90°C (e.g., in the presence of dibenzoyl peroxide) the reaction temperature rose to 160−180°C and at such high temperatures, the iodoacetate obtained underwent a thermal rearrangement into $R_FCH_2CH(OAc)CH_2I$.[31,32]

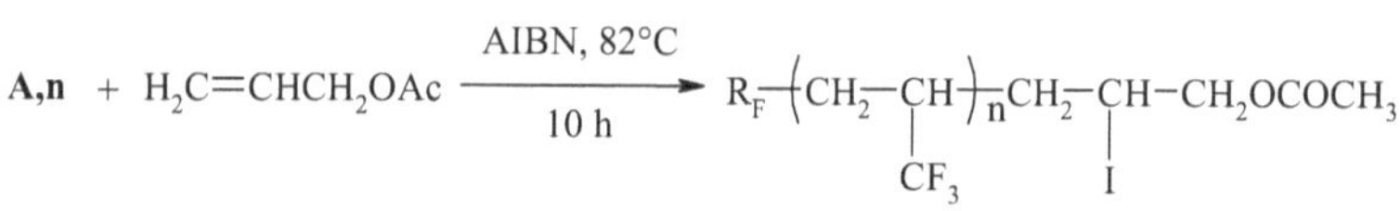

B,n

Scheme 18.5 Radical addition of iodinated TFP telomers onto allyl acetate.[33]

As described previously,[31,33] as well as in fluorinated bis(iodoacetate) series,[34] fluorinated allyl monomers **C,n** can be obtained by deiodoacetalization reaction from **B,n** (Scheme 18.6).

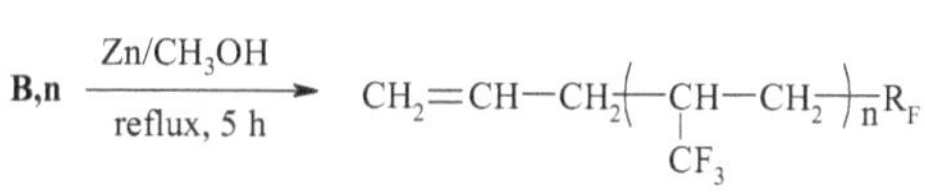

C,n

Scheme 18.6 Synthesis of fluorinated allylic monomers based on TFP from iodoacetate precursors.[33]

B,n conversion was quantitative, and the overall yield of the reaction product **C,n** was close to 75% after distillation. The functionalized TFP telomers obtained could be suitable intermediates to prepare new fluorinated derivatives applied as coating materials, surfactants, block copolymers, hybrid fluorosilicones, etc.

2.1.2 *Telomerization of 3,3,3-Trifluoropropene With Diethyl Hydrogenophosphonate*

The radical telomerization of TFP with diethyl phosphite (or diethyl hydrogenophosphonate, DEHP) was studied to synthesize fluorinated telomers bearing a phosphonate end group, as $H(TFP)_nP(O)(OEt)_2$ (Scheme 18.7).[26,27] These studies complete previous works involving similar phosphorous-containing chain transfer agents with tetrafluorethylene,[35] CTFE,[36] or VDF.[37] We revisited this reaction, initiated by DTBP at 140°C (Scheme 18.7).

$$HP(O)(OC_2H_5)_2 + n\ H_2C{=}CH(CF_3) \xrightarrow[140\ °C]{DTBP} H[CH(CF_3)CH_2]_n\text{-}P(O)(OC_2H_5)_2$$

Scheme 18.7 Radical telomerization of 3,3,3-trifluoropropene with diethyl phosphate (DTBP, di-*tert*-butylperoxide).

Acetonitrile was used as the most efficient solvent, and after reaction, the overall TFP conversion was c. 60%. It was also noted that the higher the initiator concentration, the higher the yield and the average degree of telomerization, $\overline{DP}_n$. The total product mixture, analyzed by gas chromatography, evidenced the formation of a classical telomeric distribution and identified the first seven adducts[27] (although the higher adducts were in negligible concentrations). The average cumulated degree of the telomerization in number, $\overline{DP}_{ncum}$, of these telomers was determined to be 2.1 or 2.6 according to the experimental conditions, corresponding to an average cumulated molecular weight, $\overline{M}_{ncum}$, of c. 330, assuming that each adduct has the formula $H(TFP)_nP(O)(OEt)_2$.

The first adducts were produced (well-evidenced by NMR analysis) and could not be isolated separately.

$CH_3\text{-}CH(CF_3)\text{-}P(O)(OC_2H_5)_2$ (1.1) (85 %)

$CF_3\text{-}CH_2\text{-}CH_2\text{-}P(O)(OC_2H_5)_2$ (1.2) (15 %)

This structural study brings up the evidence of the major attack of the phosphoryl radical onto the nonfluorinated carbon atom of TFP (Scheme 18.8) and confirmed previous works[36,37] on the radical telomerizations of CTFE and VDF with DEHP that also evidenced the selective addition of that phosphoryl radical onto CF_2 of CTFE[36] or the H_2C site of VDF.[37]

The structure of the diadduct was also characterized by NMR, in terms of the regioselectivity, and for the formula, $H(TFP)_2P(O)(OEt)_2$, four isomers of diadducts (2.1–2.4), resulting from TFP propagation, were produced as follows (Scheme 18.9):

$$CF_3CH_2CH_2{-}CH(CF_3)CH_2{-}P(O)(OCH_2CH_3)_2 \quad 75\% \tag{2.1}$$

5,5,5-trifluoro-2- trifluoromethyl-1-diethyl phosphonate pentane

$$CH_3CH(CF_3){-}CH(CF_3)CH_2{-}P(O)(OCH_2CH_3)_2 \quad 15\% \tag{2.2}$$

2,3-bis-trifluoromethyl-1-diethyl phosphonate butane

$$CH_3CH(CF_3)CH_2CH(CF_3)P(O)(OCH_2CH_3)_2 \quad \sim 5\% \tag{2.3}$$

1,3-bis-trifluoromethyl-1-diethyl phosphonate butane

$$CF_3CH_2CH_2CH_2CH(CF_3)P(O)(OCH_2CH_3)_2 \quad \sim 5\% \tag{2.4}$$

5,5,5-trifluoro-1-trifluoromethyl-1-diethyl phosphonate pentane as evidenced by ^1H, ^{19}F, and ^{31}P NMR spectroscopies.

Initiatior : tBuO-OtBu ⟶ 2tBuO•

$H\text{-}P(O)(OEt)_2 + tBuO^\bullet \longrightarrow {}^\bullet P(O)(OEt)_2 + tBuOH$

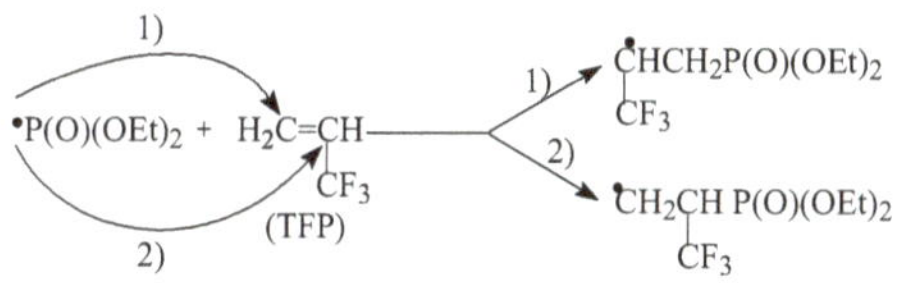

Propagation : $^\bullet TFP\text{-}(TFP)_xP(O)(OEt)_2 + TFP \xrightarrow{k_p} {}^\bullet TFP(TFP)_{x+1}P(O)(OEt)_2$

Transfer: $^\bullet TFP(TFP)_aP(O)(OEt)_2 + H\text{-}P(O)(OEt)_2 \xrightarrow{k_{tr}} H\text{-}(TFP)_{a+1}P(O)(OEt)_2 + {}^\bullet P(O)(OEt)_2$

Recombination : $^\bullet TFP(TFP)_{a-1}P(O)(OEt)_2 + {}^\bullet TFP(TFP)_{b-1}P(O)(OEt)_2 \longrightarrow (EtO)_2P(O)\text{-}(TFP)_{a+b}P(O)(OEt)_2$

Scheme 18.8 General scheme of radical telomerization of 3,3,3-trifluoropropene with diethyl hydrogenophosphonate.[27]

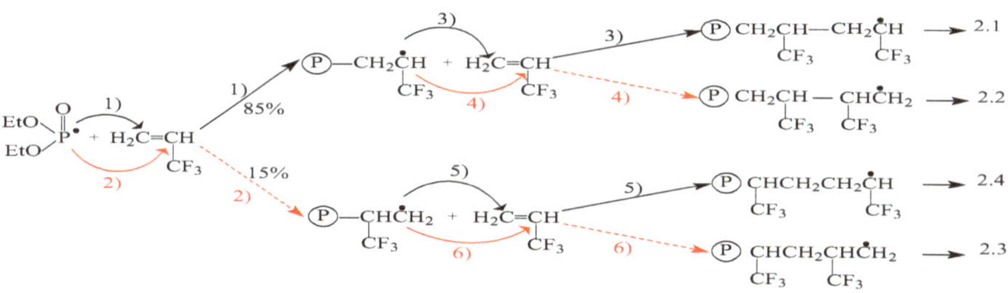

Scheme 18.9 Mechanism of telomerization of 3,3,3-trifluoropropene with diethyl hydrogen phosphonate as a chain transfer agent, where P stands for $P(O)(OEt)_2$.[27]

As for the triadducts produced from the telomerization of TFP, eight isomers can exist, and following are three of them (the other isomers changing only the position of the CF_3- groups):

$$CF_3CH_2CH_2{-}CH(CF_3)CH_2{-}CH(CF_3)CH_2{-}P(O)(OCH_2CH_3)_2 \quad (3.1)$$

7,7,7-trifluoro-2,4-bis(trifluoromethyl diethyl)-1-phosphonate heptane

$$CF_3CH_2{-}CH_2{-}CH_2CH(CF_3){-}CH(CF_3)CH_2{-}P(O)(OCH_2CH_3)_2 \quad (3.2)$$

7,7,7-trifluoro-2,3-bis(trifluoromethyl diethyl)-1-phosphonate heptane

$$CH_3CH(CF_3){-}CH(CF_3)CH_2{-}CH(CF_3)CH_2{-}P(O)(OCH_2CH_3)_2 \quad (3.3)$$

2,4,5-tris(trifluoromethyl)-1-diethyl phosphonate hexane.

The kinetic study of the telomerization reaction of TFP with diethyl hydrogen phosphonate enabled us to assess the transfer constants of first orders C_T^n.[27]

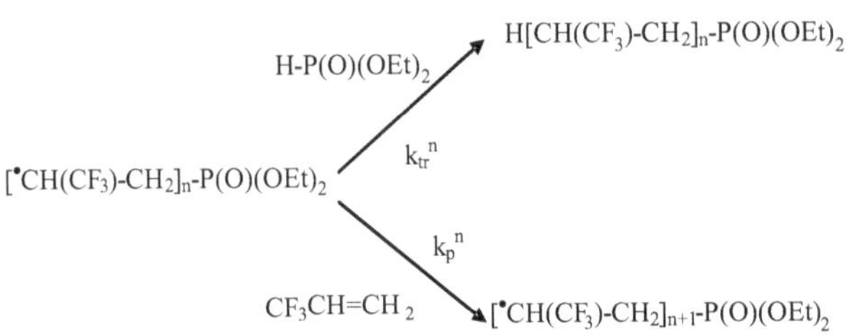

Scheme 18.10 Transfer and propagation steps for the 3,3,3-trifluoropropene telomerization with diethyl hydrogenophosphonate.

From David & Gosselain's theory,[38] plotting the molar fractions, F_n, of each telomer to the C_T^n values versus the initial $[\text{telogen}]_o/[\text{TFP}]_o$ ratio, R':

$$C_T^n = \frac{F_n}{R' \cdot \sum_{j=n+1}^{\infty} F_j}$$

where F_n, C_T^n, and R' stand for the molar fraction of nth telomer, the transfer constant of n order, and the initial $[\text{DEHP}]_o/[\text{TFP}]_o$ molar ratio, respectively.

Hence, the representation of the straight lines arising from $\frac{F_n}{\sum_{j=n+1}^{\infty} F_j}$ versus R_0 allowed us to determine the transfer constants C_T^n of the first adducts.

From these results, the C_T^{∞} value was deduced from the limit value of calculated C_T^n by extrapolation of C_T^n versus n: $C_{T\ (HP(O)(OEt)2/TFP)}^{\infty} = 0.75$ at 140°C.

This transfer constant, kTr/kp, lower than 1.0, is in favor of the propagation step rather than the transfer step (Scheme 18.10).[27]

It is also observed that the value of the transfer constant, C_T^{∞}, characteristic of the reactivity of the transfer of diethyl hydrogen phosphonate in the presence of TFP, is overall of the same order as those reported in the literature, obtained in the presence of TFE ($C_T^{\infty} = 0.2$ at 80°C[35]), CTFE ($C_T^{\infty} = 0.34$ at 140°C[36]), or VDF ($C_T^{\infty} = 0.38$ at 140°C[37]).

In addition, C_T^{∞} values less than 1 also indicate that the telomers produced are not able to act as potential CTAs as this was noted in the case where $R_F(CH_2CF_2)_nI$ and $R(CH_2CF_2)_nSC(S)Et$ for which C_T values were found to be about 7 at 74°C[39] and 49 at 73°C,[40] respectively. This also demonstrates that the C–P bond of $H(CF_2CH_2)_nP(O)(OR)_2$ telomers is too strong and cannot be cleaved to enable high-molecular-weight telomers.

2.1.3 Synthesis of Fluorinated Telechelic Diols Based on 3,3,3-Trifluropropene Telomers

Well-architectured polymers are synthesized by controlled radical polymerization (CRP),[9] polycondensation, or polyaddition, which need α,ω-difunctional precursors or telechelics (i.e., the functional groups are located at both extremities of the chains), to obtain high-molecular-weight materials with satisfactory properties. The synthesis of fluorinated telechelics is well-described in the literature[9] (e.g., they can be prepared from α,ω-diiodides[34,41]). Fluorinated α,ω-difunctionals bearing the fluorinated group in a dangling position from the polymeric backbone are used for their relevant surface properties. In contrast, those containing a fluorinated chain within the backbone display much better resistance against solvents, chemicals, UV radiation, and aging.[9]

Nevertheless, fluorodiols are scarce and few are commercially available. Original fluorinated telechelics containing TFP[23] were prepared from two strategies:

The first common step (Scheme 18.11, step 1), dealt with the radical telomerization of TFP with 1,6-diiodoperfluorohexane. As for the radical telomerization of TFP with 1-iodoperfluoroalkanes, the best initiator was DTBP, whereas initial reactant molar ratios were $[DTBP]_0/[TFP]_0 = 6$ mol% and $[IC_6F_{12}I]_0/[TFP]_0 = 0.35$. In the presence of 2.5-fold excess of TFP about $I{-}C_6F_{12}{-}I$ and given that the BDE of $CF_2{-}I$ is quite low (c. 200 kJ/mol),[42] the formation of diadduct $\mathbf{n_2}$ was favored over that of $\mathbf{n_1}$. In fact, the former contains two strong $CF_2{-}CH_2$ bonds (in the case of a tail addition), whereas the latter contains only one $CF_2{-}CH_2$ bond and a low energetic $CF_2{-}I$ bond. $\mathbf{n_3}$ was much more difficult to produce using that particular $[TFP]_o/[I{-}R_F{-}I]_o$ molar ratio, which did not allow a high propagation rate of TFP.

TFP telomers $\mathbf{n_2}$ were ethylene end-capped, initiated by *tert*-butylperoxypivalate (TBPPi), yielding **1** in 80% yield[23] (Scheme 18.11, step 2). The hydrolysis of such telomers **1** was carried out in step 3 in the presence of dimethylformamide (DMF)/H_2O medium and led to a mixture of diol **2** (35%) and alcohol formiate **3.** The use of sulfuric acid allowed converting alcohol formiate **3** into diol **2** in 70% yield. Fluorinated diol **2** was obtained in 25% overall yield in four steps. This fluorinated diol displayed a Tg of −56°C and fair to suitable thermal stability. Given that diols of type HO$-C_2H_4C_6F_{12}C_2H_4-$OH are crystalline,[41] the amorphous state of **2** is thus clearly linked to the insertion of TFP moieties (and especially CF_3 bulky side groups).

Another generation of telechelic diols based on TFP with a C_3 spacer was achieved from the radical addition of $\mathbf{n_2}$ diiodides onto allyl alcohol or allyl acetate, followed by chemical modifications,[23] as displayed in Scheme 18.12.

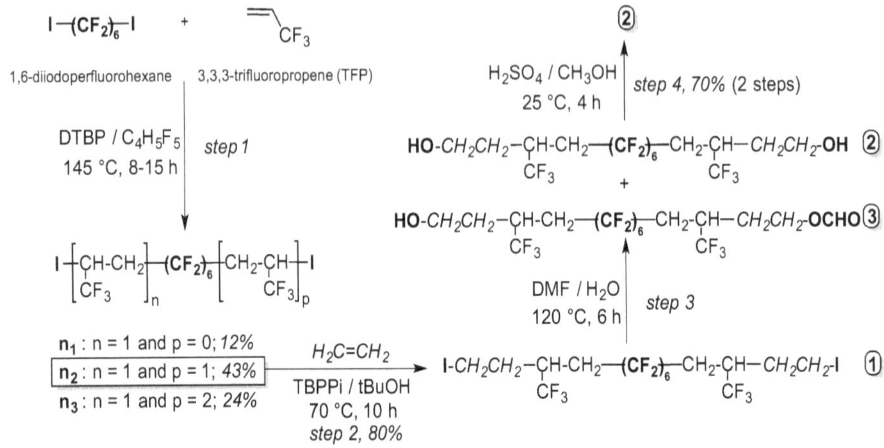

Scheme 18.11 Synthesis of fluorinated telechelic diols based on TFP by radical telomerization of TFP with 1,6-diiodoperfluorohexane followed by ethylenation. DTBP, *di-tert-butylperoxide*; *TBPPi*, *tert*-butylperoxypivalate.[23]

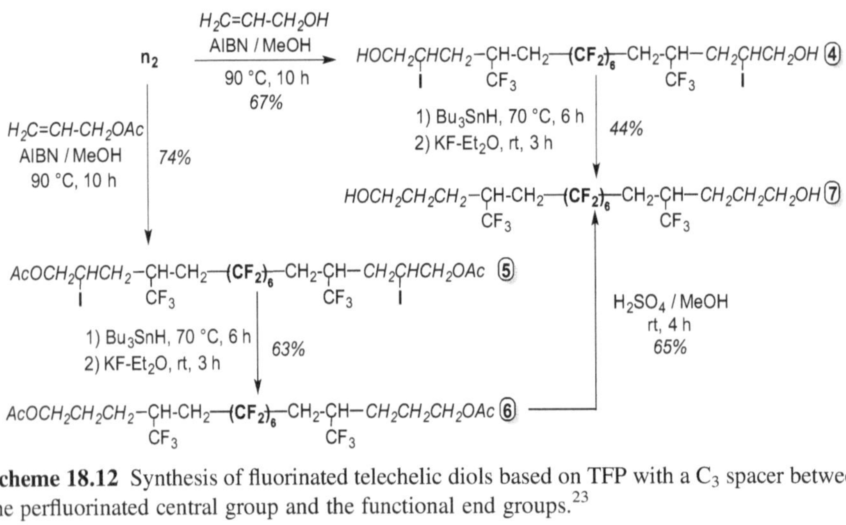

Scheme 18.12 Synthesis of fluorinated telechelic diols based on TFP with a C_3 spacer between the perfluorinated central group and the functional end groups.[23]

All C_2 and C_3 diols were characterized by infrared and NMR spectroscopies.[23]

2.2 Cotelomers of 3,3,3-Trifluoropropene

In contrast to the radical copolymerizations of TFP with either VDF[43,44] or with ethylene [via (phosphine sultonate) Pd-(Me)(DMSO) catalysis insertion-copolymerization[2]] and a more recent study on the terpolymerization of VDF with trifluoroethylene (Scheme 18.13) in which TFP was reported to be strongly incorporated in the polymer chain during the early moments of the radical terpolymerization, there are a few references on the cotelomers based on VDF and TFP.

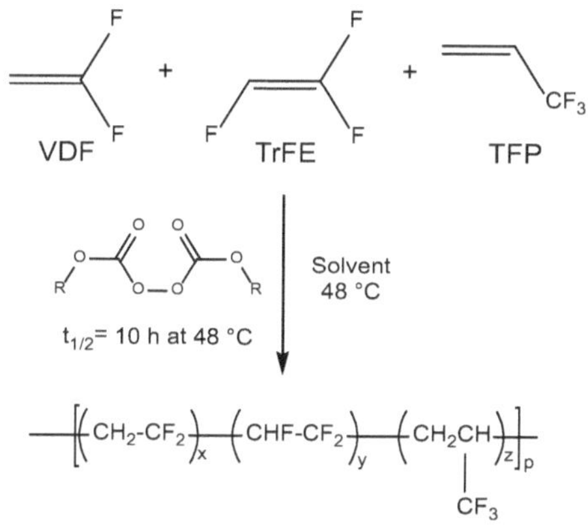

Scheme 18.13 Conventional radical solution terpolymerization of TFP, vinylidene fluoride (VDF), and trifluoroethylene (TrFE) [where R stands for ethyl or *tert*-butyl(cyclohexyl)].[45]

Hence, the objectives of this section concern the synthesis of TFP cotelomers especially those of the original poly(TFP-*co*-VDF) and poly(TFP-*co*-PFP) cotelomers as precursors of original surfactants with potential biodegradability and nonbioaccumulation.

2.2.1 Stepwise and Direct Radical Cotelomerization of 3,3,3-Trifluoropropene With Vinylidene Fluoride in the Presence of 1-Iodoperfluoroalkanes

The synthesis of original cotelomers based on TFP and VDF with the general formula $R_F{-}[(CH_2{-}CF_2)_n{-}(CH_2{-}CH(CF_3)_m)]_p{-}I$ [where n*p = 1−63, m*p = 2−640, and $R_F{=}(CF_3)_2CF$] was achieved by stepwise or direct cotelomerizations in the presence of 1-iodoperfluoroalkanes.[24] The radical cotelomerizations were initiated by thermal decomposition of different peroxide and persulfate initiators in bulk, in solution (in the presence of acetonitrile or 1,1,1,3,3-pentafluorobutane as the solvents), and in emulsion processes.

2.2.1.1 Stepwise Cotelomerization of 3,3,3-Trifluoropropene and Vinylidene Fluoride

TFP and VDF were successively cotelomerized to prepare poly(TFP-*co*-VDF) cotelomers, as depicted in Scheme 18.14. The first step consists of the telomerization of the monomer A (either TFP or VDF) in the presence of 2-iodoperfluoropropane ($i\text{-}C_3F_7I$) as the telogen.[24] The second step deals with the telomerization of monomer B (VDF or TFP, respectively) in the presence of the monoadduct produced from the first telomerization, to obtain the cotelomers. The reaction, carried out in bulk, produced the highest proportion of monoadduct (≥50%) and the lowest amount of diadduct with a small amount of unreacted (c. 20 %) telogen irrespective of the reaction conditions.

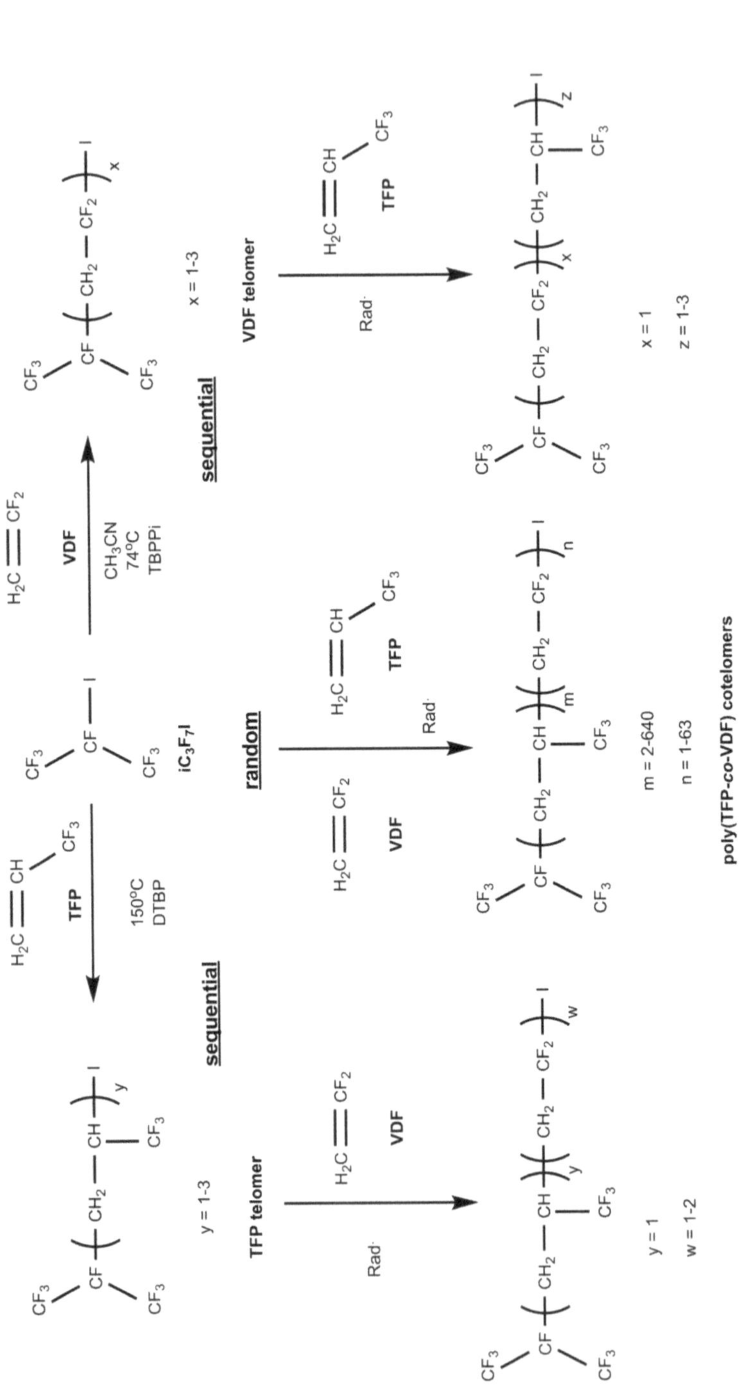

Scheme 18.14 Stepwise (co)telomerization of vinylidene fluoride (VDF) and 3,3,3-trifluoropropene (TFP) in the presence of 2-iodoperfluoropropane, and composition of the products (up). Direct (random) radical cotelomerization of TFP and VDF with 2-iodoperfluoropropane (middle and down).[24]

The molecular weights of these cotelomers were assessed by ^{19}F NMR spectroscopy, the isoperfluoropropyl end group acting as a label.

The monoadduct was isolated in 50−60% yield TFP and VDF. We reported that the telomerization of TFP in the presence of perfluorohexyl iodide ($C_6F_{13}I$) led to both normal and reverse adducts in contrast to that involving 2-iodoperfluoro-*iso*-propane (i-C_3F_7I), from which no reverse addition product was observed.[22]

In the case of the radical telomerization of TFP with R_F-VDF-I, better results were obtained when a mixture of two initiators were used (TBPPi and DTBP).[24] TFP conversion was fair to good (up to 60 wt%) with yet nonnegligible amount of triadducts or higher adducts. It seems that the reaction was favored at higher dilution where the reaction mainly led to the monoadduct in contrast to reactions with higher TFP concentrations. Scaling that reaction up in the same conditions led to better conversion (up to 81 wt%).[24]

As for the radical telomerization of VDF with R_F-TFP-I, VDF conversion was lower than that of TFP in the previous series. This tends to demonstrate that TFP is more reactive than VDF as recently confirmed in terpolymerization with TrFE[45]. The conversion was constant for all samples (31−50 wt.%). Predominantly, mono- and diadduct were obtained, indicating that no propagation occurred noticeably.

Indeed, the stepwise cotelomerization of VDF and TFP with R_F-VDF-I was not only more efficient than that of TFP and VDF but also led to polyadducts and evidenced obvious propagation. Surprisingly, it was observed that the cotelomerization was better achieved at low pressure, although a significant amount of unreacted telogen still remained in the mixture.

Since the yield was low, a further interest dealt with the study of the direct radical cotelomerization of both VDF and TFP with 2-iodoperfluoroisopropane.

As expected, the main difference with the previous subsection lies in the fact that cotelomers do not exhibit a block structure but a statistical one. From these strategies, the Chemtura Company patented the radical cotelomerization of VDF and TFP for further chain extension to yield amphiphilic block cooligomers as surfactants (Scheme 18.15 and Section 3.3).[46]

As expected, the emulsion cotelomerization led to slightly lower yields (by lack of optimization in the process) but to products of higher molecular weights (the reaction was supposed to take place mainly in the dispersed phase). Scale-up reaction led to comparable products with a substantially higher yield (80% vs. <60%). The higher the CTA concentration, the lower the molecular weight of the telomers. Furthermore, it was also noted that the number of TFP units in the cotelomer was unexpectedly higher than those of VDF units, although the percentage of VDF in the feed was generally higher. The kinetics of copolymerization led to the assessment of the reactivity ratios, $r_{VDF} = 0.28$ and $r_{TFP} = 2.35$ at 74°C.[24] However, cotelomers richer in VDF (≥50 mol%) could be obtained starting from VDF feed compositions higher than 70 mol%. Hence, a wide range of products was achieved by direct cotelomerization and the observed compounds had molecular weights ranging from 700 to ~66,000 g/mol.

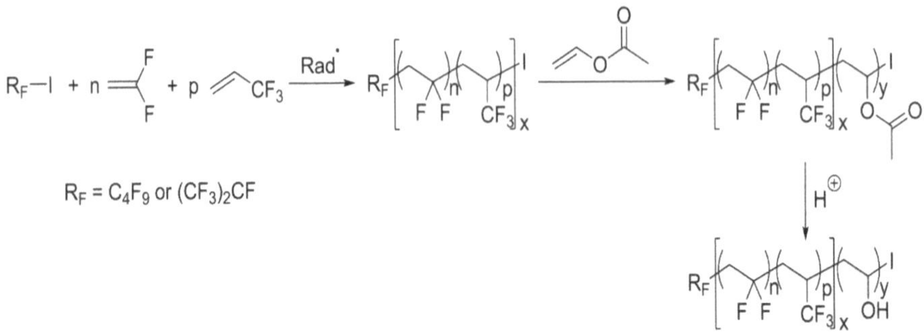

Scheme 18.15 Oligo(TFP-*co*-VDF)-*b*-oligo(vinyl acetate) block cotelomers by stepwise terpolymerization of TFP, VDF and vinyl acetate, followed by the hydrolysis of oligo(vinyl acetate) block to obtain amphiphilic surfactants.[54] *TFP*, 3,3,3-trifluoropropene; *VDF*, vinylidene fluoride.

2.2.2 Radical Cotelomerization of 3,3,3-Trifluoropropene and 1,1,3,3,3-Pentafluoropropylene

Only one article reports the radical cotelomerization of TFP with 1,1,3,3,3-pentafluoropropylene (PFP) in the presence of 1-iodoperfluoroalkane.[47] Compared with the aforementioned cotelomers, poly(TFP-*co*-PFP) cotelomers were obtained in lower yield, as expected, because of the quite low reactivity of PFP compared with that of TFP. In addition, as PFP does not homopropagate ($r_{PFP} = 0$), the cotelomer composition in the resulting poly(TFP-*co*-PFP) cotelomers had a PFP percentage much lower than that in the feed. It was also found that the nature of the fluorinated CTA (linear or branched) or the presence of solvent did not significantly influence the yield of the reaction.

First, $R_F^{\bullet}$ radical (arising from the CTA) selectively reacted onto the CH_2 site of TFP because of the less electrophilic character of TFP compared with that of PFP (Scheme 18.16), and this was evidenced by NMR spectroscopy. In addition, beside the monoadducts, two isomers were produced: $(CF_3)_2CFCH_2CH(CF_3)CF_2CH(CF_3)I$ (88 mol%) and $(CF_3)_2CFCH(CF_3)CF_2CH_2CH(CF_3)I$ (12 mol.%), although ^{19}F NMR enabled to determine the number of TFP units (1.3), in good agreement with the formation of mono- and diadducts.

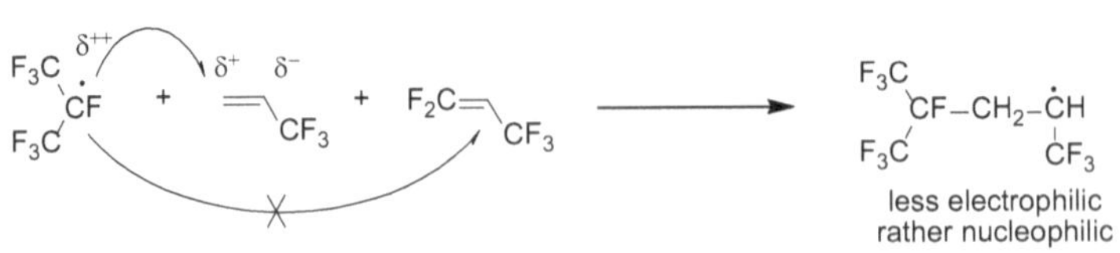

Scheme 18.16 Proposed mechanism for the radical cotelomerization of PFP with 3,3,3-trifluoropropene.[47]

3. Synthesis of Surfactants Based on 3,3,3-Trifluoropropene

3.1 Introduction and Current Issues on Fluorinated Surfactants

The most efficient surfactants are fluorinated amphiphilic ones, composed of a polar head and one or more perfluorinated chain(s) mainly C_7F_{15} or C_8F_{17},[48,49] such as perfluorooctanoic acid, $C_7F_{15}CO_2H$ (PFOA) or perfluorooctanesulfonic acid, $C_8F_{17}SO_3H$ (PFOS). These fluorosurfactants have been used in more than 200 applications including cosmetics, emulsifiers, waxes, adhesives, detergents, fire fighting foams, soil and stain repellents, coatings, electroplating, antifogging, and antistatic agents.[48,49] However, studies directed by the US Environmental Protection Agency[50] reported that such surfactants are toxic, persistent, and bioaccumulable[51] (arising from the highly stable perfluorinated chain; for example, PFOS' half-life is ~3.26 years in human blood, and these amphiphiles can be found at concentrations of 22.3 μg/L (for PFOS) and 6.8 μg/L (for PFOA) in the plasma of human beings[51]), and that decreasing their production by as much as 95% was scheduled at the end of 2010, followed by definitive stop of their production by 2015. Many analyses[49,51,52] around the planet have demonstrated that such derivatives are still present in many rivers, seas, animals, ice cap, and also in the liver of the polar bears.

The main companies (Arkema; Asahi Glass; Ciba, which was further bought by BASF; Clariant; Daikin; 3M/Dyneon; Dupont that recently became Chemours; and Solvay Specialty Polymers) that manufacture these derivatives and fluorinated polymers gathered efforts via the "Stewardship Program" consortium[53] that aimed at decreasing these productions and searching new alternatives to PFOA and PFOS. Indeed, these PFOA or PFOS have been synthesized from TFE telomers.[9] Thus, it was of interest to find out new nonbioaccumulable alternatives to PFOA.

Hence, several approaches have been claimed and also published by Chemtura Company[54,55] on the synthesis and characterizations of cationic, anionic, and nonionic surfactants.[33,56] Two main strategies were implemented: by chemical modifications and by controlled radical copolymerization of TFP and VDF.

3.2 Chemical Modifications of 3,3,3-Trifluoropropene Telomers

Original TFP telomers were achieved by simple and efficient modifications of fluorinated telomers obtained by iodine transfer polymerization.[33] The concept is illustrated using the telomerization of TFP and its modifications in two to four steps to obtain the various types of cationic surfactants, which bear weak points able to be metabolized or enzymatically degraded. In contrast to perfluorinated materials, hydrogenofluorinated compounds can be degraded by HF elimination, creating a double bond that can be oxidized and further cleaved. This degradation can be advantageously achieved by the methylene and methyne groups borne by each TFP unit. The overall pathway for the syntheses of these new surfactants is detailed in Scheme 18.17. From the radical telomerization of TFP, several surfactants can be obtained (cationic and nonionic) in satisfactory yields (50% overall yields from i-C_3F_7I).

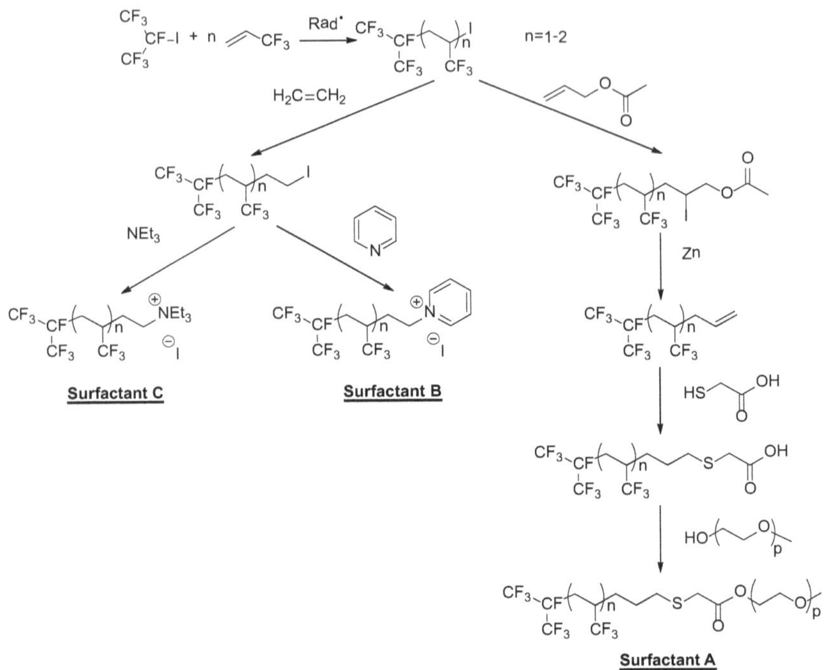

Scheme 18.17 Straightforward strategies for the preparation of 3,3,3-trifluoropropene (TFP)—based cationic and nonionic surfactants from the radical telomerization of TFP with 2-iodoperfluoroisopropane.[33]
Reproduced with permission from Royal Societey of Chemistry , 2012.

On the right side of Scheme 18.17, the allylic TFP-containing derivatives (see Section 2.1.1) were further reacted with thioglycolic acid to lead to carboxylic compounds (as an anionic surfactant, PFOA type). This reaction occurred either photochemically (81% yield) or in the presence of radical initiators such as TBPPi (73% yield). Finally, esterification with an ω-hydroxypoly(ethylene oxide) oligomer (PEO-OH) led to the formation of a new R_F—$(TFP)_x$—CH_2—CH_2—CH_2—S—CH_2—CO_2—PEO nonionic surfactant.

The left side of Scheme 18.17 displays a strategy to obtain cationic surfactants. First, ethylene end-capping of the TFP telomers, initiated by an organic peroxide, quantitatively led to R_F—$(TFP)_x$—CH_2—CH_2—I in 89% yield with a quantitative conversion of $(CF_3)_2CF(TFP)_nI$. The nucleophilic substitution of R_F—$(TFP)_x$—CH_2—CH_2—I by either triethylamine (NEt_3) or pyridine (C_5H_5N) at room temperature or at 40°C, respectively led to cationic surfactants bearing ammonium groups (Scheme 18.17) in satisfactory yield (>75%).

Advantageously, these surfactants display satisfactory overall inertness to acids and bases,[33] appropriate to be used in aggressive media (e.g., oil wells, acidic buffers for emulsion polymerization of fluorinated olefins), and especially i-$C_3F_7(TFP)(CH_2)_3$ $SCH_2CO_2(PEO)CH_3$ with good solubility in water and methanol but poor solubility in apolar solvents and acetone.

Although no attempts on the degradation of these surfactants were made (supposedly because of the presence of methylene and methyne groups in TFP units), their surface tensions were determined at different concentrations to access their critical micellar concentration (CMC) in water at 25°C, to be 0.06, 4.1, 3.2, and 3.0 g/L for surfactants **A**, **B**, **C** (Scheme 18.17), and PFOA, respectively.[33] Both cationic **B** and **C** surfactants behaved as PFOA does, whereas the nonionic surfactant (**A**) had a higher surface tension but a lower CMC value (0.06 g/L) than that of the commercially available surfactant.

3.3 Controlled Radical Copolymerization (or Reversible Deactivation Radical Copolymerization) of Vinylidene Fluoride With 3,3,3-Trifluoropropene

In contrast to surfactants **B** and **C** that contained a small polar end group, and as surfactant **A** (possessing an oligomeric hydrophilic chain), another route to get oligo(vinyl alcohol) polar block was reported by CRP (Schemes 18.15 and 18.18). Several techniques of controlled radical copolymerization (or reversible deactivation radical copolymerization) are known (all based on the fast exchange between active and dormant species) and are not reported in this chapter; hence, the readers will find useful information on CRP of fluoroalkenes in another book[9] and various articles.[57–59]

This technology offers a wide range of polymeric architectures and especially block copolymers suitable as relevant materials. For a more efficient copolymerization or cotelomerization of TFP, VDF inserted in the media was recently achieved successfully[54,55] by iodine transfer copolymerization[57–59] or by stepwise cotelomerization of both these coolefins to lead to poly(VDF-*co*-TFP) cotelomers.[54,55]

However, the insertion of a polar group [i.e., oligo(vinyl alcohol)] (Scheme 18.15) in the produced poly(VDF-*co*-TFP)I copolymer was achieved in low to fair yields, confirming the results of a previous study.[60]

Interestingly, macromolecular design via interchange of xanthates, MADIX, is a suitable technique for the controlled polymerization of vinyl acetate (VAc)[61] as that monomer is difficult to control, although iodine transfer polymerization was reported once by Iovu et al.[60], but has been attempted a few times on fluorinated olefins.[62–64]

3.3.1 Synthesis of Diblock Cooligomers by Copolymerization of 3,3,3-Trifluoropropene and Vinylidene Fluoride Copolymer Followed by Chain Extension of Vinyl Acetate

First, an original fluorinated xanthate was synthesized in two steps from $C_6F_{13}CH_2CH_2OH$ for further use as CTA to control the radical copolymerization of TFP and VDF via MADIX process (in 63% yields). Hence, a second block of poly(-VAc) was inserted in a controlled way via the poly(TFP-*co*-VDF) macroxanthate in 88% yields (Schemes 18.2). In the final step, the poly(VAc) block of the copolymer was chemically changed into poly(vinyl alcohol) via acid hydrolysis (hence avoiding dehydrofluorination of VDF units) to obtain poly(VDF-*co*-TFP)-*b*-poly(VAc)-Xa amphiphilic block cooligomers.

3.3.2 Synthesis of Diblock by First the Preparation of Poly(Vinyl Acetate) Homopolymer via MADIX Followed by a Chain Extension with a Poly(3,3,3-Trifluoropropene-co-Vinylidene Fluoride) Cooligomer

A similar strategy was reported by the first incorporation of VAc followed by the radical copolymerization of TFP with VDF leading to poly(VAc)-*b*-poly(VDF-*co*-TFP)-Xa diblock copolymer. First, we studied the CRP of such a monomer in the presence of fluorinated xanthate.[56] AIBN was chosen as the initiator in 1,1,1,3,3-pentafluorobutane, with an initial $[CTA]_o/[VAc]_o$ molar ratio of 0.05 to target an average degree of polymerization of c. 20. Size exclusion chromatography indicated an average molecular weight, M_n, of 1960 g/mol (that matched well with the targeted M_n) and a narrow molecular weight distribution (PDI = 1.45).[56] Hence, the presence of both the xanthate's C_6F_{13} end group and a fluorinated solvent did not disturb the CRP of VAc.

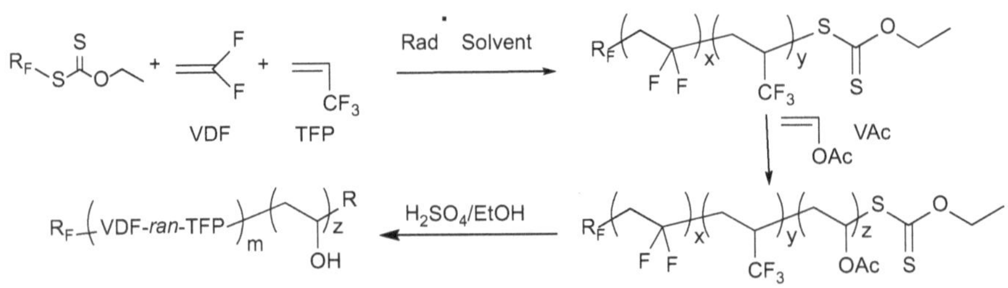

Scheme 18.18 Sequenced MADIX terpolymerization of TFP, VDF, and VAc controlled by a fluorinated xanthate for the synthesis of poly(VDF-*ran*-TFP)-*b*-oligo(VAc) block cooligomers further hydrolyzed into original surfactants.[56]

Then, chain extension with a fluorinated block for the synthesis of PVAc-*b*-poly(TFP-*co*-VDF) cooligomers was carried out in the presence of PVAc macroxanthate. As for the iodine transfer copolymerization of TFP and VDF in the presence of 1-iodoperfluoroalkanes,[24] a similar higher reactivity of TFP over VDF was noted, and this sequential copolymerization led to narrow polydispersity indices and almost the same experimental and targeted molecular weights.[56]

Among the different results above, it was more suitable to insert the poly(TFP-*co*-VDF) sequence first, followed by the insertion of the oligo(VAc) as the second block. This may be due to better stabilization of the radical formed by the macro-CTA. The secondary radical with an electron-withdrawing group is much more stable than a radical formed of the fluorinated block with VDF addition to the xanthate. On the other hand, using 2,5-bis(*tert*-butylperoxy)-2,5-dimethylhexane as an initiator led to the best results to insert an oligo(TFP-*co*-VDF) block.

3.3.3 Evaluation of the Surfactant Properties

The surfactant properties of i-$C_3F_7(VDF)_2(TFP)_3(VOH)_5$ were assessed by means of tensiometry and conductimetry and compared with those of ammonium perfluorooctanoate (APFO) (Fig. 18.1).[56]

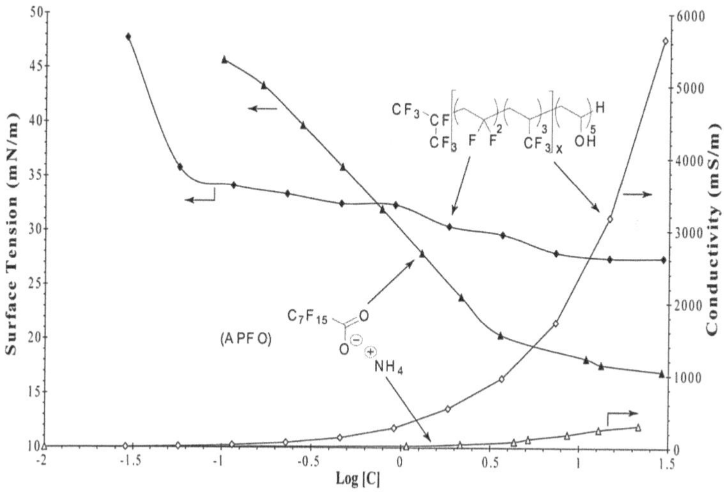

Figure 18.1 Surface tensions and conductimetry diamonds versus poly(TFP-*co*-VDF)-*b*-oligo(VOH) surfactant's concentration compared with those of ammonium perfluorooctanoate, APFO (triangles).
Reproduced with the permission of Kostov, G.; Boschet, F.; Buller, J.; Badache, L.; Brandstadter, S. M.; Ameduri, B. *Macromolecules* **2011,** 44, 1841–1855.

It can be seen that the poly(TFP-*co*-VDF)-*b*-oligo(vinyl alcohol) block oligomer displayed a reduction of both the surface tension (about 27.4 mN/m) and CMC.

Hence, novel surfactants with valuable surface properties have been designed by sequential controlled radical copolymerization, and, although degradation tests have not been attempted, it can be expected that methylene and methyne groups, regarded as "weak" points, should undergo some enzymatic or metabolic decompositions.

4. Conclusions

This chapter describes the synthesis of original fluorinated (co)telomers based on TFP from a wide range of CTAs and their functionalization. This olefin is interesting since it can also be used in copolymerization and because the resulting cotelomers and copolymers exhibit CF_3 side groups that should bring added value, especially for surface properties. Hence, many derivatives based on TFP have been prepared: allylic monomers, functional derivatives, telechelics, and block copolymers, in contrast to Silastic fluorosilicones that are the only commercially available products containing TFP units. Such new compounds obtained could be suitable intermediates to prepare new fluorinated products applied as coating materials, surfactants, monomers for polycondensation, hybrid fluorosilicones, etc. The radical telomerization of TFP with DEHP allowed to synthesize original TFP telomers bearing a phosphonate end group and then phosphonic acid, and its kinetics enabled to determine the infinite transfer constant of DEHP: 0.75 at 140°C. In addition, syntheses of original TFP-containing telechelic diols were reported with various C2 or C3 spacers between the central oligo(TFP) chain and the hydroxyl end functions, either via ethylene end-capping followed by a two-step process or by radical addition of $I(TFP)_nC_6F_{12}(TFP)_pI$ onto allyl alcohol or ally acetate followed by reduction of iodine atoms.

Then, the *stepwise cotelomerization* of TFP and VDF with 1-iodoperfluoroisopropane led to $(CF_3)_2CFCH_2CF_2CH_2CHICF_3$ and $(CF_3)_2CFCH_2CH(CF_3)CH_2CF_2I$ monoadducts in various yields. These reactions showed that TFP is more reactive than VDF. In contrast, the direct cotelomerization of VDF and TFP carried out in solution or in emulsion led to a wider range of statistic cotelomers, the molecular weights of which ranged from 700 to c. 66,000 g/mol.

Conversely, the cotelomerization of TFP with 2-hydroperfluoropropylene (PFP) led to low-yield cotelomers that showed the lower reactivity of PFP about the TFP comonomers, and low degree of telomerization was produced ($n = 1.5$).

Original non-ionic surfactants and cationic surfactants based on TFP that showed satisfactory properties as alternatives to perfluorooctanoic acid (PFOA) were synthesized from chemical modifications of TFP telomers. These original surfactants displayed satisfactory inertness to bases and acids, as well as good surface properties. They exhibit interesting CMC values comparable to that of PFOA (0.06, 4.1, and 3.2 versus 3.0 g/L).

Other *fluorinated surfactants* from block copolymers based on poly(TFP-*co*-VDF), and oligo(vinyl alcohol) sequences were synthesized by CRP in the presence of a fluorinated xanthate or by iodine transfer copolymerization. The radical copolymerization of these aforementioned fluorinated monomers via MADIX technology in the presence of poly(VAc)−xanthate also gave good yields (>60 wt%). Then, the polymerization of VAc, achieved from the fluorinated poly(TFP-*co*-VDF) macroxanthate, led to diblock cooligomers in good yields. Then, the surfactant was produced from acidic hydrolysis of the oligo(VAc) block. The final product displayed suitable solubility in water with comparable surface tension as that of APFO.

Hence, telomerization is a powerful technique to supply many well-defined compounds of various molecular weights, and from the suitable choice of CTA(s), appropriate function(s) can be inserted in the resulting TFP telomers. Nevertheless, the chain transfer constant requires to be assessed to target the molecular weights from Mayo's law.

However, to the best of our knowledge, neither graft copolymers nor dendrimers, stars, or multiarms containing TFP have been reported, and further research in these areas can be expected.

Acknowledgments

All the companies (Great Lakes now Chemtura, Solvay Fluor, Elf Atochem, Tosoh, and Akzo-Nobel) that have contributed to this work by providing us with free reactants are acknowledged. The authors would like to thank the Great Lakes/Chemtura Company for financial support, as well as Prof. Bernard Boutevin, Dr F. Boschet, and coworkers involved in the TFP project for their valuable inputs.

References

1. Hanford, W. E.; Joyce, R. M., Jr. *Halogenated Hydrocarbons*. US patent 1948/2 **1948,** *440*; p 800 (assigned to E. I. du Pont de Nemours & Co.).
2. Lanzinger, D.; Giuman, M. M.; Anselment, M. M. J.; Rieger, B. *ACS Macro Lett.* **2014,** *3,* 931–934.
3. Masson, J. Decomposition Rates of Organic Free Radical Initiators. In *Polymer Handbook;* Brandrup, J., Immergut, E. H., Eds.; John Wiley: New York, 1989; pp 1–44.
4. Freidlina, R. K.; Terent'ev, A. B.; Khorlina, M. Y.; Aminov, S. N. *Zh. Vses. Khim.* **1966,** *11,* 211–215.
5. Starks, C. *Free Radical Telomerisation;* Academic Press: New York, 1974.
6. Gordon, R.; Loftus, R. Telomerisation. In *Ency Poly Sci Tech;* Kirk, R. O., Othmer, D. F., Eds.; Wiley: New York, 1989; pp 533–572.
7. a. Boutevin, B.; Pietrasanta, Y. Telomerisation. In *Comprehensive Polymer Science;* Allen, G. B., Bevington, J. C., Eastmond, A. L., Russo, A., Eds.; Pergamon: Oxford, 1989; pp 185–198.
   b. Améduri, B.; Boutevin, B. Telomerisation. In *Encyclopedia Advanced Materials;* Bloor, D. B., Brook, R. J., Flemings, R. D., Mahajan, S., Eds.; Pergamon: Oxford, 1994; pp 2767–2774.
8. Ameduri, B.; Boutevin, B. *Top. Curr. Chem.* **1997,** *192,* 165–233.
9. a. Ameduri, B.; Boutevin, B. In *Well-Architectured Fluoropolymers: Synthesis, Properties, and Applications;* Elsevier: Amsterdam, 2004.
   b. Smith, D. W., Jr.; Iacono, S. T.; Iyer, S. S. In *Encyclopedia of Fluoropolymers;* Wiley: New York, 2014. ISBN-13: 978-0470079935.
   c. Ameduri, B.; Sawada, H.; Fluoropolymers: Synthesis, properties, processing and simulation; Royal Society of Chemistry; Fluoropolymers; ISBN: 9781782624158.
10. a. Haszeldine, R. N. *J. Chem. Soc. Abstr.* **1951,** 2495–2504.
    b. Maxson, M. T.; Norris, A. W.; Owen, M. J.. Chapter 20 In *Modern Fluoropolymers;* Scheirs, J., Ed.; John Wiley & Sons: Chichester, UK, 1997; pp 359–372.
    c. Owen, M. J. In *Handbook of Fluoropolymer Science and Technology;* Smith D. W. Jr., Iacono, S. T., Iyer, S. S., Eds.; John Wiley & Sons: Hoboken, NJ, 2014; pp 183–200. Chapter 9.
    d. Pasquet, C.; Longuet, C.; Hamdani-Devarennes, S.; Ameduri, B.; Ganachaud, F. In *Silicone Surface Science;* Owen, M. J., Dvornic, R., Eds.; Springer: Washington, USA, ISBN 978-94-007-3875-1; 2012; pp 115–178. Chapter 5.
11. Ikonnikov, N. S.; Lamova, N. I.; Terentev, A. B. *Izv. Akad. Nauk SSSR Ser. Khim.* **1988,** 117–121 (Chem. Abstr. 109:169831).
12. Haszeldine, R. N.; Steele, B. R. *J. Chem. Soc.* **1955,** 3005–3009.
13. Terent'ev, A. B.; Vasil'eva, T. T. *Ind. Chem. Library* **1995,** *7,* 180–201.

14. a. Zamyslov, R. A.; Shostenko, A. G.; Dobrov, I. V.; Myshkin, V. E. *Zh. Org. Khim.* **1980,** *16,* 897 (Chem. Abstr. 93:167229).
    b. Shostenko, A. G.; Dobrov, I. V.; Chertorizhskii, A. V. *Khim. Prom.* **1983,** 339 (Chem. Abstr. 99:123027).
    c. Zamyslov, R. A. *Zh. Vses. Khim. Obshch. Im. D. I. Mendeleeva* **1986,** *31,* 589 (Chem. Abstr. 106:175716).
    d. Zamyslov, R. A.; Shostenko, A. G.; Dobrov, I. V.; Tarasova, N. P. *Kinet. Katal.* **1987,** *28,* 977 (Chem. Abtsr. 108, 74741).
15. a. Vasil'eva, T. T.; Fokina, I. A.; Vitt, S. V. *Izv. Akad. Nauk SSSR Ser. Khim.* **1991,** 1384–1388 (Chem. Abstr. 116:40981).
    b. Gasanov, R. G.; Vasil'eva, T. T.; Gapusenko, S. I. *Kinet. Katal.* **1991,** *32,* 1466–1470 (Chem. Abstr. 116:105413).
16. Vasil'eva, T. T.; Fokina, I. A.; Vitt, S. V.; Dostovalova, V. I. *Izv. Akad. Nauk SSSR Ser. Khim.* **1990,** *8,* 1807–1811 (Chem. Abstr. 114:41974).
17. Keim, W.; Raffeis, G. H.; Kurth, D. *J. Fluor. Chem.* **1990,** *48,* 229–237.
18. Vasil'eva, T. T.; Kochetkova, V. A.; Dostovalova, V. I.; Nelyubin, B. V.; Freidlina, R. K. *Izv. Akad. Nauk SSSR Ser. Khim.* **1989,** 2558–2562 (Chem. Abstr.113:5666).
19. Vasil 'eva, T. T.; Kochetkova, V. A.; Nelyubin, B. V.; Dostovalova, V. I.; Freidlina, R. K. *Izv. Akad. Nauk SSSR Ser. Khim.* **1987,** 808–811.
20. Haszeldine, R. N. *J. Chem. Soc. Abstr.* **1952,** 2504–2513.
21. Low, H. C.; Tedder, J. M.; Walton, J. C. *J. Chem. Soc. Farad. Trans. 1* **1976,** *72,* 1300–1309.
22. Kostov, G. K.; Ameduri, B.; Brandstadter, S. M. *Coll. Czech. Chem. Commun.* **2008,** *73,* 1747–1763.
23. Kostov, G. K.; Lopez, G.; Brandstadter, S. M.; Edwards, B. E.; Ameduri, B. *Org. Lett.* **2014,** *16,* 3516–3519.
24. Kostov, G.; Boschet, F.; Brandstadter, S.; Ameduri, B. *J. Polym. Sci. Part A Polym. Chem.* **2009,** *47,* 3964–3981.
25. Haszeldine, R. N.; Newlands, M. J.; Plumb, J. B. *J. Chem. Soc.* **1965,** 2101.
26. Block, H. D. German Patent DE 1976/2,514,640, **1976** (assigned to Bayer).
27. Kostov, G.; Ameduri, B.; Brandstadter, S. M. *J. Fluor. Chem.* **2007,** *128,* 910–918. http://dx.doi.org/10.1016/j.jfluchem.2007.03.015.
28. Haran, G.; Sharp, D. W. A. *J. Chem. Soc. Perkin Trans. 1* **1972,** 34–38.
29. Feast, W. J.; Gimeno, M.; Koshravi, E. *Polymer* **2003,** *44,* 6111–6121.
30. Terent'ev, A. B.; Pastushenko, E. V.; Kruglov, D. E.; Rybininia, R. E. *Izv. Akad. Nauk SSSR Ser. Khim.* **1992,** 2768–2772 (Chem. Abstr. 120:106821).
31. a. Ameduri, B.; Boutevin, B.; Nouiri, M.; Talbi, M. *J. Fluor. Chem.* **1995,** *74,* 191–197.
    b. Ameduri, B.; Boutevin, B.; Kostov, G. K.; Petrova, P. *J. Fluor. Chem.* **1998,** *92,* 69–76.
32. Cirkva, V.; Ameduri, B.; Boutevin, B.; Kvichala, J.; Paleta, O. *J. Fluor. Chem.* **1995,** *74,* 97.
33. Boschet, F.; Kostov, G.; Ameduri, B.; Jackson, A. *Polym. Chem.* **2012,** *3,* 217–223.
34. a. Manseri, A.; Ameduri, B.; Boutevin, G.; Caporiccio, G.; Kotora, M.; Hajek, M. *J. Fluor. Chem.* **1995,** *73,* 151–158.
    b. Manseri, A.; Boulahia, D.; Ameduri, B.; Boutevin, B.; Caporiccio, G. *J. Fluor. Chem.* **1997,** *81,* 103–113.
35. Brace, N. O. *J. Org. Chem.* **1961,** *3,* 3197.
36. Gaboyard, M.; Hervaud, Y.; Boutevin, B. *J. Fluor. Chem.* **2001,** *107,* 5.

37. Duc, M.; Ameduri, B.; Boutevin, B. *J. Fluor. Chem.* **2001,** *112,* 3.
38. David, C.; Gosselain, P. A. *Tetrahedron* **1962,** *70,* 639.
39. a. Boyer, C.; Valade, D.; Sauguet, L.; Ameduri, B.; Boutevin, B. *Macromolecules* **2005,** *38,* 10353–10362.
    b. Valade, D.; Boyer, C.; Lacroix Demazes, P.; Ameduri, B.; Boutevin, B. *J. Polym. Sci Part A Polym. Chem.* **2006,** *44,* 5763–5777.
40. Guerre, M.; Rahaman, S.M.W.; Ameduri, B.; Poli, R.; Ladmiral, V. *Macromolecules* **2016,** 49, Article ASAP.
41. Lahiouel, D.; Ameduri, B.; Boutevin, B. *J. Fluor. Chem.* **2001,** *107,* 81.
42. Okafo, E. N.; Whittle, E. *Int. J. Chem. Kinet.* **1975,** *7,* 287–300.
43. Patil, Y.; Alaaeddine, A.; Ono, T.; Ameduri, B. *Macromolecules* **2013,** *46,* 3092.
44. Watanabe, J.; Fujiara, M. *Eur. Patent, 1990/0302784,* 1990 (assigned to Shin Etsu).
45. Soulestin, T.; Ladmiral, V.; Lannuzel, T.; Dos Santos, F. S.; Ameduri, B. *Macromolecules* **2015,** *48,* 7861–7871.
46. Kostov, G. K.; Ameduri, B.; Brandstader, S. M.; Edwards, B. E. US Patent 2007/937,408, (assigned to Chemtura).
47. Boschet, F.; Kostov, G. K.; Raynova, H.; Ameduri, B. *Eur. Polym. J.* **2015,** *73,* 487–499.
48. Kissa, E. *Fluorinated Surfactants: Synthesis, Properties and Applications;* CRC Press: New York, 2001.
49. a. Krafft, M.-P.; Riess, J.-G. *Chem. Rev.* **2009,** *109,* 1714–1792.
    b. Krafft, M.-P.; Riess, J.-G. *Curr. Opin. Coll. Interface Sci.* **2015,** *20,* 192–212.
    c. Krafft, M.-P.; Riess, J.-G. *Chemosphere* **2015,** *129,* 4–19.
50. U.S. Environmental Protection Agency. *Preliminary Risk Assessment: Perfluorooctanic Acid (PFOA) and Fluorinated Telomers,* April 2003.
51. a. Kostov, G.; Boschet, F.; Ameduri, B. *J. Fluor. Chem.* **2009,** *130,* 1192–1199.
    b. Kostov, G. K.; Ameduri, B.; Brandstadter, S. M. US 2007/60/992,845 (assigned to Chemtura).
    c. Zaggia, A.; Ameduri, B. *Curr. Opin. Coll. Interface Sci.* **2012,** *17,* 188–195.
52. a. Fromme, H.; Midasch, O.; Twardella, D.; Angerer, J.; Boehmer, S.; Liebl, B. *Int. Arch. Occup. Environ. Health* **2007,** *80,* 313–319.
    b. Kovarova, J.; Svobodova, Z. *Neuroendocrinol. Lett.* **2008,** *29,* 599–608.
    c. Ritter, S. K. *Chem. Eng. News* **2010,** *88,* 12–17.
53. PFOA stewardship Program. http://www.epa.gov/oppt/pfoa/pubs/pfoastewardship.htm, .
54. Ameduri, B.; Kostov, G. K.; Brandstader, S. M.; Edwards, B. E. PCT 2006/835645. US Patent 2008/0076892 (assigned to Great Lakes Chem. Co.).
55 a. Jackson, A.; Sharma, V.; Edwards, B. E.; Boggs, S.; Hedrick, V.; Brandstadter, S. M.; Chien, J.; Norman, E.; Kaufman, R.; Ameduri, B.; Kostov, G. K. US Patent 2007/027349 (assigned to Chemtura).
    b. Ameduri, B.; Brandstadter S. M.; Kostov, G. K. W.O. Patent 2008/019068 A2 (assigned to Great Lakes Chemical Corporation, USA).
56. Kostov, G.; Boschet, F.; Buller, J.; Badache, L.; Brandstadter, S. M.; Ameduri, B. *Macromolecules* **2011,** *44,* 1841–1855.
57. a. Tatemoto, M. US Patent 1979/4158678 (assigned to Daikin Kogyo Ltd).
    b. Tatemoto, M. *The First Regular Meeting of Soviet-Japanese Fluorine Chemists*. Tokyo, February 15–16, 1979.

c. Tatemoto, M.; Shimizu, T. Thermoplastic Elastomers. In *Modern Fluoropolymers;* Scheirs, J., Ed.; Wiley: New-York, 1997; pp 565–576. Chapter 30.

58. a. Ameduri, B. *Macromolecules* **2010,** *43,* 10163.
    b. Ameduri, B. *J. Taiwan. Inst. Chem. Eng.* **2014,** *45,* 3124–3133.
59. a. Asandei, A. D.; Adebolu, O. I.; Simpson, C. P. *J. Am. Chem. Soc.* **2012,** *134,* 6080–6083.
    b. Asandei, A. D.; Adebolu, O. I.; Simpson, C. P.; Kim, J.-S. *Angew. Chem. Int. Ed.* **2013,** *52,* 10027–10030.
    c. Simpson, C. P.; Adebolu, O. I.; Kim, J. -S.; Vasu, V.; Asandei, A. D. *Macromolecules* **2015,** *48,* 6404–6420.
    d. Cernoch, P.; Petrova, S.; Cernochova, Z.; Kim, J. -S.; Simpson, C. P.; Asandei, A. D. *Eur. Polym. J.* **2015,** *68,* 460–470.
60. Iovu, M. C.; Matyjaszewski, K. *Macromolecules* **2003,** *36,* 9346–9354.
61. a. Corpart, P.; Charmot, D.; Biadatti, T.; Zard, S. Z.; Michelet, D. WO 1998/58,974 (assigned to Rhodia).
    b. Destarac, M.; Charmot, D.; Franck, X.; Zard, S. Z. *Macromol. Rapid Commun.* **2000,** *21,* 1035–1039.
    c. Destarac, M. *Macromol. React. Eng.* **2010,** *4,* 165–179.
62. Girard, E.; Marty, J. D.; Ameduri, B.; Destarac, M. *ACS Macro Lett.* **2012,** *1,* 270–274.
63. a. Ameduri, B.; Patil, Y. WO 2013/160621 and US 2014/0154611 (assigned to Arkema France, CNRS and l'Ecole Nationale Supérieure de Chimie de Montpellier).
    b. Patil, Y.; Ameduri, B. *Polym. Chem.* **2013,** *4,* 2783–2799.
64. Guerre, M.; Campagne, B.; Gimello, O.; Parra, K.; Ameduri, B.; Ladmiral, V. *Macromolecules* **2015,** *48,* 7810–7822.

Graphical Abstract

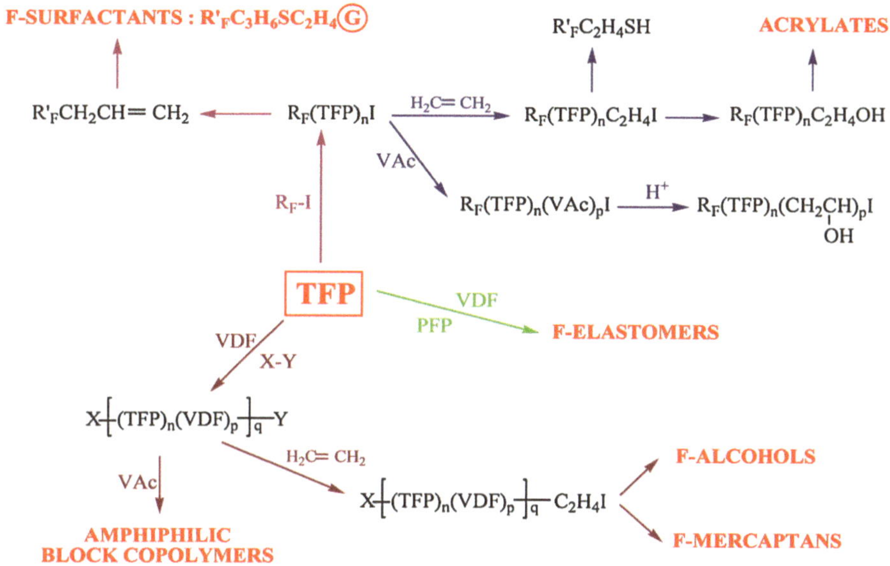

General scheme of varieties of compounds prepared from the telomerization of TFP.

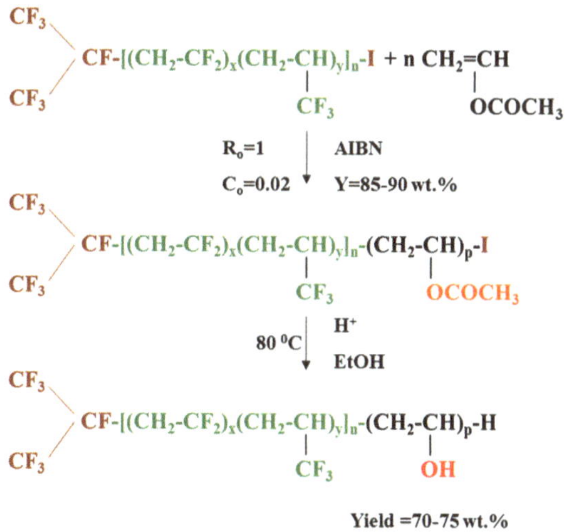

High Oxidation States in Transition Metal Fluorides

19

A. Higelin, S. Riedel
Freie Universität Berlin, Berlin, Germany

Chapter Outline

Modern Synthesis Processes and Reactivity of Fluorinated Compounds. http://dx.doi.org/10.1016/B978-0-12-803740-9.00019-6

1. Introduction

Fluorine has been essential in facilitating high and extremely high oxidation states in transition metal compounds. Only the extreme electronegativity of fluorine allows the oxidation and stabilization of many of these metals in their highest attainable oxidation states. Highly oxidized metal centers must be stabilized by negative ligands, whereby only fluorine and, to a lesser extent oxygen, resist oxidation by the metal. In oxidation states above +V, of especially the lighter transition metals, steric crowding of the fluoride ligands is often a limiting factor. In these cases, oxygen facilitates the formation of higher oxidation states, being able to carry twice the charge in about the same space as fluorine. Generally speaking, although both fluorine and oxygen are able to form highly oxidized transition metal compounds, fluorine stabilizes the most extreme oxidation states and oxygen stabilizes the highest oxidation states in the 3d elements. Mixed oxofluorides do also frequently occur in high-oxidation-state transition metal chemistry.[1] The highest oxidation state of osmium is, for example, stabilized by such a ligand system in OsO_3F_2 and OsO_2F_4. Beyond this also $Os^{VIII}O_4$ is well known, whereas $Os^{VIII}F_8$ is disfavored by the large coordination number, which results in steric repulsions between the fluoride ligands. Therefore the reported osmium octafluoride synthesis is most likely doubtful, as discussed later.

All these unusual compounds show unique properties and are potent reagents with in partly extreme reactivity. They allow synthetic access to spheres of chemical reactivity usually unattainable by ordinary chemical routes. Although these compounds are clearly fascinating in themselves, they were also fundamental in preparing, for example, the first noble gas compounds.[2]

Highly oxidized binary transition metal fluorides are mostly molecular compounds. As highly electron-deficient compounds, they easily coordinate fluoride anions to form complex anions, which are frequently more stable than their parent compounds when paired with the matching cation. Oligomers formed by aggregation of several monoanions and the neutral parent molecules are also a common feature. Complex cations, which are formed by abstracting a fluoride ion from the parent compound, are also known for several of the transition metals, but are usually unstable and highly reactive ($[CoF_4]^+$,'$[NiF_3]^+$' or $[ReF_6]^+$, see later). As a general rule, the oxidation potential of these compounds increases tremendously in acidic conditions upon fluoride abstraction.

The preparation of the herein discussed compounds is incredibly simple—on paper, yet highly challenging in the laboratory, demanding skilled and experienced experimenters, specialized equipment, and high safety precautions. Most compounds are prepared from the elements steered by stoichiometry and driven by high reaction enthalpies, high temperatures, and high pressures. These harsh reaction conditions of high temperature and high pressure combined with incredibly reactive compounds require robust reaction vessels made from Monel or stainless steel. Naturally, working with fluorine requires certain precautions and specialized equipment. Furthermore, high-pressure syntheses are conducted in autoclaves and often dedicated equipment. Few of these compounds are stable in air. Frequent by-products and decomposition products are anhydrous HF and again fluorine. Several of the herein discussed compounds are so reactive or unstable that they can only be observed in the gas phase or low-temperature matrices. These conditions again require specialized and dedicated equipment, albeit of different specification of ultralow temperatures and extremely high-quality vacuum.

Being highly oxidized, stable, storable, and manageable compounds, many transition metal fluorides are ideal oxidizers for a variety of specialized and tricky applications. The large range of available compounds provides a range of gradually increasing power to finely tune the oxidation strength to the respective task. Table 19.1 and Fig. 19.1 illustrate the increasing power to oxidize. This array of oxidation potential can also be used to rank the oxidation strength of a given compound experimentally by comparison.

These very strong oxidizers were ranked by experimental comparison, physical measurement of electron affinities, as well as theoretical studies through quantum chemical calculations. The well-known series of 5d transition metal hexafluorides displays an increase in oxidation strength of about 1 eV per step.[9] WF_6 does not even oxidize I_2, whereas ReF_6 and OsF_6 are able to oxidize NO and IrF_6 reacts with ONF and O_2NF. PtF_6 finally extracts electrons even from O_2 and Xe. It played a legendary role in the preparation of the first ever noble gas compound "$XePtF_6$," thereby fundamentally changing our understanding of the periodic table of the elements.[2] It fails to oxidize Kr, N_2, NF_3, and F_2O. The logical next compound in this line, AuF_6, has not yet been prepared or observed. It is, however, expected to be stable and would be the strongest known oxidizer once it is accessible. The tremendous stability of the corresponding anion $[AuF_6]^-$ is shown by the corresponding cations, which are usually known as strong oxidizing agents $[O_2]^+$,[21] $[KrF]^+$,[22] $[Xe_2F_3]^+$,[22] and $[NO]^+$.[22] This underlines the extreme electron affinity of the hypothetical gold hexafluoride. It is therefore a worthwhile and promising target molecule for future research and would once more widen the horizon of experimentally achievable oxidized compounds. All these experimentally known hexafluorides of the transition metals are very similar in their physical properties. The highly corrosive and hydrolysis-sensitive compounds of octahedral molecular shape have melting points in the range between 0°C and 70°C, and are of yellow, red, brown, or black color. They can be stored in steel cylinders and handled via the gas phase.

The 4d series at large shows the same trend in oxidative strength, with a strong increase along the series. RuF_6 and RhF_6, however, are reversed, albeit very close together in absolute strength. This twist in the oxidation strength is likely due to an ideal electronic configuration of t_{2g}^3 of the corresponding $[RuF_6]^-$ anion, which is very close to that of the strongest oxidizer of the hexafluoride PtF_6.[9,23] The 4d hexafluorides are considerably smaller than their heavier homologs, leading to instability

Table 19.1 Oxidation Potential of the Herein Discussed Highly Potent Transition Metal Fluoride Oxidizers

Oxidizer	Experimental Value (eV)	Theoretical Value (eV)
AuF_6		8.1[a], 8.20[b], 8.47[c], 9.56[d]
Ag^{III}/Ni^{IV}		
RuF_6	6.47[e], 7.5[f]	6.98[g], 7.01[h]
PtF_6	7.00[e]	6.36[i], 6.95r, 6.78[j], 7.09 [b], 7.4[a]
RhF_6		6.44[g], 6.60[g], 6.80[h]
IrF_6	6.50[e]	5.34[i], 5.99[b], 7.2[a]
OsF_6	5.93[e]	5.55[i], 5.92[b], 6.0[a]
TcF_6		5.64[g], 5.89[h]
ReF_6	>3.8[k]	4.50[i], 4.58[b], 4.8[a]
MoF_6	3.6[l], 3.8[f], 4.50[m], >5.1[n]	4.23[h], 4.11[g], 5.37[o]
WF_6	3.36[p], 3.50[q], >5.14[f]	3.16 [b], 3.34 [i], 3.85[r], 5.33 [o]

Experimental as well as theoretical values based on quantum chemical calculations are given.
[a]DV-Xa, Gutsev and Boldyrev.[3]
[b]Craciun et al.[4]
[c]CCSD(T), Riedel and Kaupp.[5]
[d]CISDþQ, Miyoshi and Sakai.[6]
[e]Selig et al.[7]
[f]Rudnyi et al.[8]
[g]MoF_6, TcF_6 computed at B3LYP level, RuF_6, RhF_6 at CCSD(T) level, Drews et al.[9]
[h]Values computed at CCSD(T) level with correction of CBS, ZPE, CV, rel, SO.[10]
[i]DFT/BP86, Macgregor and Moock.[11]
[j]B3LYP, Wesendrup and Schwerdtfeger.[12]
[k]Friedman et al.[13]
[l]Sidorov et al.[14]
[m]Mathur et al.[15]
[n]Compton et al.[16]
[o]Adiabatic EA computed at MP2 level, Sakai and Miyoshi.[17]
[p]Viggiano et al.[18]
[q]George and Beauchamp.[19]
[r]CISDþQ, Miyoshi et al.[20]

and higher reactivity of the fluorides. All attempts to prepare the missing PdF_6 have been unsuccessful so far. Even under cryogenic conditions in rare gas matrices no PdF_6 could be observed.[24]

In general, the relative oxidation strengths can be shown by these reactions. The element xenon can be oxidized by PtF_6, RuF_6, and RhF_6.[2,25] However, Xe is not oxidized by IrF_6, which is able to oxidize chlorine as shown by the $[Cl_4]^+[IrF_6]^-$ complex.[26] OsF_6 is not able to oxidize Cl_2, whereas MoF_6 oxidizes NO, which is not oxidized by WF_6.[23,27] This knowledge about the oxidizing strength can be extremely helpful, as was shown by the oxidation of, e.g., C_6F_6 using OsF_6.[28]

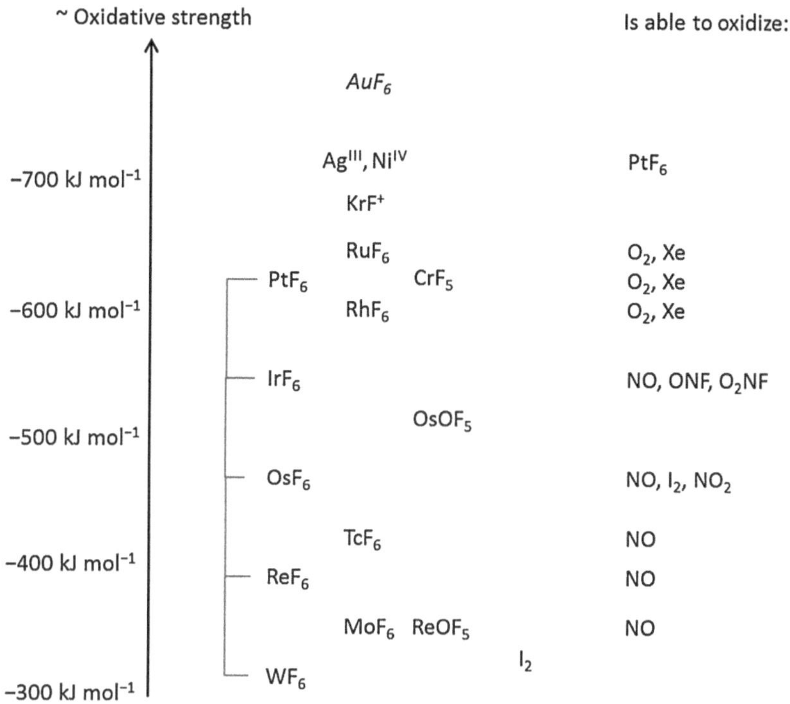

Figure 19.1 Oxidative ability and relative oxidation strength of the hexafluoride series as well as other highly potent transition metal fluorides. Other oxidizers are added for comparison.[9,31–34]

Beyond these hexafluoride one-electron oxidizers, there exists a second group of powerful oxidizers. The not entirely characterized amorphous Ag^{III} and Ni^{IV} species are the strongest known and synthetically manageable oxidizers at present. The active species are formed from anionic precursors like $[AgF_4]^-$ and $[NiF_6]^{2-}$ via activation by strong Lewis acids such as BF_3, AsF_5, or SbF_5 in anhydrous HF at low temperatures such as 208K. The active species are most likely cationic and polymeric but have so far evaded precise characterization. Ag^{III} and Ni^{IV} have an oxidizing potential of a very similar magnitude and have not been ranked in relation to each other so far. They are both able to oxidize O_2, Xe, PtF_6, and RuF_6, but fail to oxidize $[AuF_6]^-$ and are therefore stronger than $[KrF]^+$ as the $[KrF]^+[AuF_6]^-$ complex has been characterized experimentally.[29] They were, for example, used in the highly challenging oxidation of $[HCB_{11}F_5(CF_3)_6]^-$,[30] but have so far not found widespread application. This is not only due to the borderline chemical reactivity that is seldom challenged by only few chemists but also due to the intricate experimental procedure.

In summary, transition metal fluorides like AgF_3 or NiF_4 are the most powerful oxidizers known at present. The nine hexafluorides of the transition metal series represent a very useful tool box due to their tuneable oxidizer strength, which is shifted by approximately 1 eV per element along the period. Beyond transition metal oxidizers there is a further category of potent oxidizers called oxidative fluorinators. These compounds usually go along the main-group elements and often contain a rare gas atom such as $[KrF]^+$, KrF_2, XeF_6, or XeF_2.[35] A quantitative scale for the strength of such oxidative fluorinators was introduced by Christe et al.[36] However, most transition metals in their highest oxidation states can also be used as fluorinating agent or powerful oxidizers.

2. 3d Elements

2.1 Scandium

Scandium achieves its highest oxidation of +III as ScF_3, the only known fluoride of scandium. It is a high-melting white solid, but molecular ScF_3 of D_{3h} symmetry has been observed in low-temperature gas matrices.[37] It can be prepared by fluorination of the metal or the oxide with HF or other fluorination agents.[38] Ionic species such as $[ScF_2]^+$ and $[ScF_4]^-$ have been observed under cryogenic conditions.[39]

2.2 Titanium

TiF_4 can be prepared by direct fluorination from Ti, $TiCl_4$, or TiO_2.[40,41] The colorless hygroscopic compound is used as a fluorination agent and catalyst in organic synthesis. The ionic species $[TiF_6]^{2-}$ is stable in salts with K^+, Rb^+, or Cs^+. Higher anionic species such as $[Ti_2F_9]^-$, $[Ti_2F_{11}]^{3-}$, $[Ti_4F_{19}]^{3-}$, $[Ti_5F_{23}]^{3-}$, $[Ti_8F_{36}]^{4-}$, or $[Ti_8F_{33}]^-$ are also known.[41,42]

2.3 Vanadium

Direct fluorination of elemental vanadium followed by low-temperature distillation affords VF_5,[43] which can alternatively be prepared from V_2O_5 or VOF_3 by high-pressure fluorination.[44] It readily forms hexafluorovanadates $M[VF_6]$ (M^+ = alkali metal, Tl^+, Ag^+, NH_4^+) when offered a fluoride donor.[41]

2.4 Chromium

The oxidation state of +VI in chromium is achieved with oxygen, but not with fluorine, although CrF_6 has long been thought to exist.[45] There are several publications that discuss its possible formation.[46,47] However, in 2014, a detailed quantum chemical and matrix isolation investigation showed that the assigned CrF_6 band in the infrared (IR) spectra is due to a CrF_5 band, as it was suggested by Willner.[48] Its thermochemical evaluation at coupled-cluster level shows an exothermic decomposition channel for its concerted fluorine elimination as well as for the bimolecular reaction. Therefore the highest known binary chromium fluoride is the brick red CrF_5, which can be prepared from the elements in a Monel autoclave with an F_2 atmosphere of 3 bar and

200°C.[47,49] In the solid form, it is composed of *cis*-fluorine-bridged polymeric CrF_5 units.[49] It has an oxidation potential of similar magnitude as PtF_6 and is capable of oxidizing xenon and O_2, as discussed in a previous section.[34] The corresponding anion CrF_6^- is also known and has been reported as salts with the alkali metals.[50]

2.5 Manganese

As is the case for chromium, very high oxidation states of manganese can only—and even relatively easily—be achieved with oxygen. The highest oxidation state of +VII is present in the "household" chemical potassium permanganate. The contrary is true for fluorine; the highest known manganese fluoride is only MnF_4.[51]

MnF_4 is a synthetically very useful compound as it can be used for the storage of F_2, which it releases in an equilibrium at higher temperatures and low pressures.[52]

This $MnF_4/MnF_3 + F_2$ equilibrium can also be used for the purification of fluorine when small amounts of highly purified fluorine are needed, thereby avoiding risky techniques such as distilling to liquid O_2−cooled containers. Molecular MnF_4 has so far not undoubtedly experimentally been observed.[53] Even though it has been predicted to be thermochemically stable by coupled-cluster calculations, its preparation under cryogenic conditions in rare gas matrices has not been successful.[54] It is needless to mention that the preparation of MnF_5 has unsuccessfully been attempted even in cryogenic matrices.[55,56] The highest oxidation state for manganese in a fluorine compound has to date only been observed in the oxofluoride MnO_3F.[57]

2.6 Iron

The situation for iron is extremely similar to that of manganese. The highest possible oxidation state of +VI is readily achieved with oxygen in ferrate, but only FeF_3 is known in bulk quantities.[58,59] A quantum chemical and matrix isolation investigation has shown that FeF_4 can be prepared by laser ablation of elemental iron under excess elemental fluorine in rare gases such as neon and argon. The novel compound was characterized by IR matrix isolation spectroscopy at low-temperature matrices at 5K.[60] Based on Born−Haber cycles, its existence above the cryogenic temperatures is unlikely.

2.7 Cobalt

The highest oxidation state observed for cobalt is +V in the alkali salts of $[CoF_4]^+$.[38] The highest oxidation state has also been shown for fluoride, $[CoF_4]^+$, but only in the gas phase in mass spectrometric experiments.[61] The existence of CoF_4 is thereby also deduced, but has not unambiguously been characterized.[62] CoF_3 can be prepared from cobalt(II) oxide or halides by direct fluorination at 300°C[59,63] and is of industrial importance in such processes like the Flutec or Fowler process.[64] Both processes use cobalt(III) fluoride for fluorination. The hydrocarbon reacts in a fluidized bed reactor with CoF_3 forming the perfluorinated compound HF and cobalt difluoride, which in the next stage is refluorinated by elemental fluorine.

$$CoF_3 + R\text{-}H \rightarrow CoF_2 + R\text{-}F + HF$$

$$2CoF_2 + F_2 \rightarrow 2CoF_3$$

The anionic fluoride complexes of CoF_4 are, however, stable as the $M^+ = Cs^+$, Rb^+, or K^+ salts $M_2[CoF_6]$.[65,66]

2.8 Nickel

The highest oxidation state of nickel + IV is known in the highly stable complex anion $[NiF_6]^{2-}$. It can be prepared from NiF_2 with KrF_2 and XeF_6[67] or with F_2 under irradiation by ultraviolet (UV) light at room temperature.[68] The commercially available K_2NiF_6 can be stored under argon for an indefinite time. It is prepared by direct fluorination of a stoichiometric mixture of KCl and $NiCl_2$. When this bright magenta-colored salt (see Fig. 19.2) is dissolved in very cold (−65°C) anhydrous HF, it can be activated by strong Lewis acids such as AsF_5, SbF_5, or BF_3 to form formal "NiF_4."[56,67,69–71] As is the case with most other transition metal fluorides, abstraction of fluoride under acidic conditions greatly increases the oxidation potential. In a first step, neutral "NiF_4" most likely precipitates from the solution to then redissolve as a cationic species with an excess of Lewis acid.

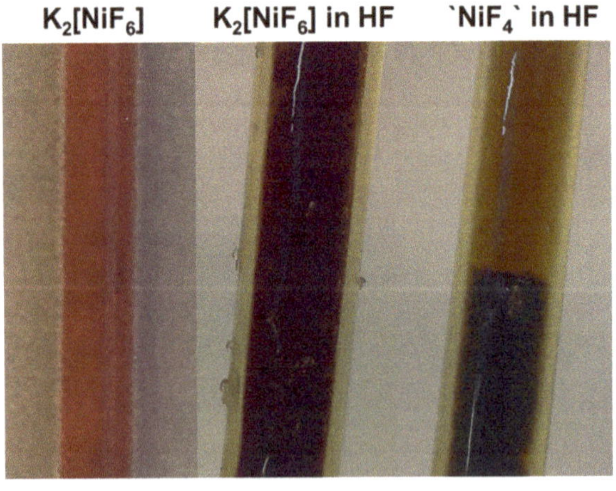

Figure 19.2 Photographs illustrating the activation of Ni(IV). Solid $K_2[NiF_6]$ is shown on the left, and the dark purple suspension of $K_2[NiF_6]$ in anhydrouss HF in the middle. After the addition of BF_3, a brown slurry, as shown on the right, is formed.

The subsequently observed thick brown slurry contains an extremely strong oxidizing species of not precisely determined composition. It is thought to contain polymeric and cationic Ni(IV) fluoride species, which is equivalent to the active species on the surface of the electrodes during electrochemical fluorination via the Simons process. Addition of fluoride bases, such as XeF_6, leads to the re-formation of $[NiF_6]^{2-}$. The potent oxidation agent "NiF_4" has been shown to oxidize $[RuF_6]^-$ and $[PtF_6]^-$, but unfortunately not $[AuF_6]^-$. This rates Ni(IV), along with Ag(III), as the to date strongest known oxidizers. The direct comparison with the oxidative strength of $[KrF]^+$ is not an easy one to make as Ni(IV) is an excellent one-electron oxidizer, whereas $[KrF]^+$ is not and thus behaves quite differently.[72] Ni(IV) has great potential for tricky synthetic challenges, but is complicated to handle. The Ni(IV) slurry starts to release fluorine when warmed above −65°C and fully decomposes within several hours at 18°C. Solid NiF_4 releases F_2 above −55°C to form different modifications of NiF_3, depending on the exact conditions. NiF_3 again releases fluorine above 20°C to form NiF_2.[67] NiF_3 can alternatively be prepared by comproportionation and is also a very strong oxidizer, able to oxidize xenon and to fluorinate organocompounds.[67]

2.9 Copper

Similar to several of the previous elements, copper shows its highest oxidation state of +IV with fluorine but only as anionic fluoride complex ($M_2[CuF_6]$ with $M^+ = Cs^+$ and Rb^+), not as binary fluoride.[66,73–75] The highest binary fluoride CuF_3 is a low-temperature compound, only observed after solvolysis of $[CuF_6]^{3-}$ HF at 195K.[56,70,76]

2.10 Zinc

Zinc forms due to its fully occupied d^{10} shell only zinc difluoride. It is shown by quantum chemical calculations that further oxidation of zinc to, e.g., ZnF_3 seems impossible.[77] A quantum chemical prediction[78] of the further oxidation by the hypothetical AuF_6 ligand to $Zn^{III}(AuF_6)_3$ was shown to be doubtful.[79]

3. 4d Metals

3.1 Yttrium

The highest possible oxidation state of yttrium + III is achieved in YF_3, which was detected in cryogenic matrices of Ar and N_2.[39,80] Later matrix isolation experiments have also shown the related species $[YF_2]^+$, $[YF_4]^-$, and Y_2F_6.[39]

3.2 Zirconium

Zirconium does of course also display its expected oxidation state of +IV in a bivalent fluorine compound. ZrF_4 can be prepared from elemental zirconium or zirconium-containing compounds by reaction with elemental fluorine.[81] It occurs in two interconvertible modifications and does form ionic complexes ($[ZrF_5]^-$, $[ZrF_6]^{2-}$, ...) and oligomers ($[Zr_2F_9]^-$, $[Zr_4F_{19}]^{3-}$, ...).[82,83]

3.3 Niobium

Direct fluorination of niobium metal results in the formation of NbF_5, displaying the highest thinkable oxidation state of +V.[84,85] Alternatively, oxides, chlorides, or other niobium-containing compounds can be used as precursors. The strong Lewis acid NbF_5 forms tetramers in the solid state, small oligomers ($n \leq 4$) in the gas phase,[86] and, like SbF_5, a great variety of anionic complexes, such as $M[NbF_6]$ (M = alkali metal, $[NH_4]^+$, Ag^+, Cu^+, Tl^+, $[NO]^+$, $[O_2]^+$, $[XeF]^+$), $M_2[NbF_7]$ (M = alkali), $M_3[NbF_8]$ (M = Li^+, Na^+, K^+, $[NH_4]^+$, Tl^+), exist.[41,87,88] It is a weaker Lewis acid than SbF_5, as $[NbF_4][SbF_6]$ is formed from NbF_5 and SbF_5.[89]

3.4 Molybdenum

MoF_6 can be prepared by direct fluorination of the metal or various molybdenum compounds.[90] The colorless solid is a stronger oxidizer than its homolog WF_6. WF_6 only oxidizes Tl to +I, whereas MoF_6 oxidizes it to +III.[91] It also forms a number of anionic fluoride complexes, as, for example, $M[MoF_7]$ and $M_2[MoF_8]$ ($M^+ = Li^+$, Na^+, K^+, Rb^+, Cs^+, $[NH_4]^+$ $[NO]^+$, $[NO_2]^+$).[92]

3.5 Technetium

TcF_7 has not been prepared, but is expected to be stable under cryogenic conditions.[93] However, the highest oxidation of +VII is not only observed in the binary oxide Tc_2O_7[94] but also in several oxofluorides ($[TcO_4]^-$, TcO_3F, TcO_2F_3, $TcOF_5$, $[TcO_2F_9]^+$, $[TcOF_4]^+$, $[TcO_2F_4]^-$).[93,95] The highest known binary fluoride is the hexafluoride TcF_6.[9,96] It can be prepared from $[NH_4][TcO_4]$ in fluorine atmosphere at 600°C in an autoclave.[9] It is interesting to note that this process involves reduction of the metal during the progress of fluorination. Alternatively, the elemental metal can be fluorinated at high pressure by an excess of fluorine at 400°C.[9]

3.6 Ruthenium

RuF_8 has previously been reported,[97] but its existence is doubted nowadays, and the highest oxidation state of +VIII is only believed to exist in the oxide RuO_4.[98] The highest oxidation state achieved with fluorine is only +VI in RuF_6 and $RuOF_4$. Dark brown RuF_6 can be prepared from powdery ruthenium metal in a stream of 1:7 F_2/Ar at 400–450°C.[9,99] RuF_6^- is oxidized by activated Ag(III) to RuF_6.[100] RuF_7 and RuF_8 were shown by quantum chemical calculation to be exothermic compounds.[101]

3.7 Rhodium

RhF_6 contains the highest known oxidation state for rhodium and is an extremely strong oxidizer, which can, for example, oxidize O_2.[102] It is noted that in contrast to the two previous elements, fluorine stabilizes higher oxidations states than oxygen in rhodium. Black crystalline RhF_6 can be prepared by continuous flow fluorination[99] as well as by oxidation of $KRhF_6$ with activated Ag(III).[100]

3.8 Palladium

The highest oxidation state is found in oxides and fluorides.[103] PdO_2 is known as well as PdF_4, and several $[PdX_6]^{2-}$ (X = F, Cl, OH) complexes exist. Earlier reports of PdF_6 in gas-phase experiments[104] were shown by quantum chemical calculations to be based on ill-assigned vibrational bands.[105] Similarly, reports of Pd(V) in $[PdF_6]^-$ salts must be doubted.[106]

A detailed quantum chemical and matrix isolation investigation of palladium fluorides has only confirmed that PdF_4 is the highest fluoride formed.[107]

3.9 Silver

The highest oxidation state known for silver is +III, but even Ag^{2+} is an extremely strong oxidizer.[108]

Intriguingly, Ag(II) is the strongest divalent oxidizer and indeed the only divalent metal among the very strong oxidizers.[109] AgF_2 is able to oxidize O_2 when sufficiently activated under acidic conditions. Under basic conditions, the reactivity is reversed, and $[O_2]^+$ oxidizes Ag(II) to Ag(III).

$$Ag^{2+} + O_2 + 2[AsF_6]^- \leftrightarrow {O_2}^+[AsF_6]^- + AgAsF_6$$

$$AgF_2 + [O_2]^+ + 2F^- \rightarrow [AgF_4]^- + O_2$$

In solid-state powder reactions under inert conditions, AgF_2 was shown to release O_2 from many binary and ternary oxides and Cl_2 from a variety of chlorides.[110]

The extremely high oxidation of +III is observed in AgF_3[33,69,111,112] and stabilized in the $[AgF_4]^-$ complex.[113,114] $KAgF_4$ is prepared by UV irradiation of a HF solution of KF and AgF_2, by the reaction between KF, AgF_2, and KrF_2 in anhydrous HF as a solvent or by fluorinating an equimolar mixture of KF and AgF_2 with 4 bar of F_2 for 4 days at 300°C.[115]

Similar to the nickel species described earlier, it can be activated by strong Lewis acids such as BF_3 at low temperatures in anhydrouss HF (*a*HF) to form an extremely reactive oxidizing Ag(III) species.[33,71,74] This HF-solvated AgF_3 species is one of the strongest known oxidants of similar magnitude as NiF_4, and just like the latter, it is not precisely characterized.

The activation of Ag(III) and subsequent formation of the highly oxidizing species is somewhat sensitive to the exact strength of the Lewis acid employed. AsF_5 cleaves fluoride from $[AgF_4]^-$ in *a*HF solution and is in excess able to remove a second fluoride to form cationic species that eventually decompose by releasing fluorine.

$$[AgF_4]^- + AsF_5 \rightarrow AgF_3 + [AsF_6]^-$$

$$AgF_3 + AsF_5 \rightarrow [AgF_2]^+ + [AsF_6]^-$$

$$[AgF_2]^+ + [AsF_6]^- \rightarrow AgFAsF_6 + 0.5F_2$$

BF_3 in contrast is only acidic enough to remove the first fluoride to produce neutral AgF_3. $[AgF_4]^-$ and PF_5 form equilibrium with AgF_3 and $[PF_6]^-$ (Fig. 19.3).

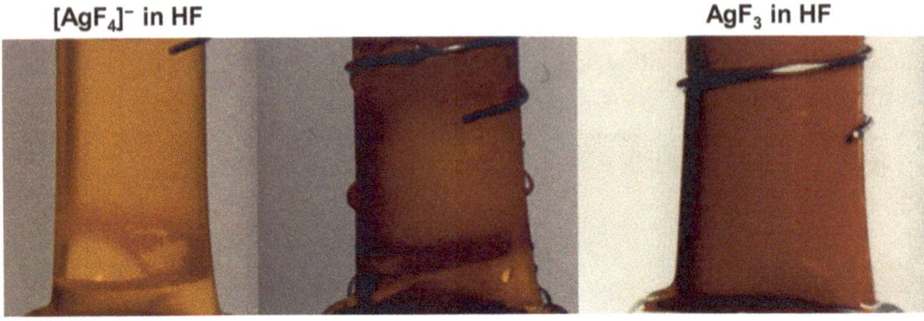

Figure 19.3 Photographs illustrating the precipitation of AgF_3 from a solution of $[AgF_4]^-$ in a HF.

Higher oxidation states have been discussed for silver ($[AgF_6]^-$ and $[AgF_6]^{2-}$) but are not well established and most likely not realistic.[75,111,113,116]

3.10 Cadmium

Cadmium forms, similar to its lighter group member zinc, only CdF_2. As it was reported by quantum chemical calculations, higher cadmium fluorides will most likely not exist due to its thermochemical instability.[77]

4. 5d Elements

4.1 Lanthanum

LaF_3 can be prepared in the form of a hydrate in water. Water-free LaF_3 can be prepared from La_2O_3 or $LaCl_3$ with aHF[41,117] or by thermal decomposition of their double salts with ammonia.[118]

4.2 Hafnium

White crystalline HfF_4 is prepared by fluorination of the elemental metal or hafnium compounds.[119] It occurs in monoclinic, tetragonal, and cubic modifications. Various fluoride complexes are also known, such as $[HfF_5]^-$, $[HfF_6]^{2-}$, $[HfH_7]^{3-}$.[83,120]

4.3 Tantalum

Tantalum pentafluoride was prepared by treating tantalum chlorides with *a*HF[121] or by the reaction of the metal with F_2 at 1000°C.[122] It forms tetramers in the solid state, while monomers, trimers, as well as TaF_4 and $[TaF_5]^+$ were detected in the gas phase and cryogenic matrices.[85,123] A great variety of anionic fluoride complexes are also known.[41,88,124]

4.4 Tungsten

The highest oxidation state of tungsten can be stabilized by fluorine as well as oxygen in WF_6[23,125] and WO_3, and several oxyfluorides also exist. Together with the hexafluorides of the following elements, it constitutes an interesting series of hexafluorides of increasing oxidative reactivity, as discussed previously.[9,23,125,126] WF_6 is prepared by fluorination of the metal or tungsten compounds.[127] It is a weaker oxidizer than the homologous MoF_6. Ionic fluoride complexes of the type $M[WF_7]$ and $M_2[WF_8]$, where $M = Li^+$, Na^+, K^+, Rb^+, Cs^+, $[NH_4]^+$, $[NO]^+$, $[NO_2]^+$, are known.[41,92,128]

4.5 Rhenium

Rhenium shows its highest possible oxidation state in the only known transition metal heptafluoride ReF_7.[56] The yellow solid is formed as the final and highest fluorination product at 400°C and a 3-bar fluorine atmosphere. With NOF and NO_2F, the fluoride complexes $NO[ReF_8]$ and $NO_2[ReF_8]$ are formed. The corresponding cation $[ReF_6]^+$ is stable with $[SbF_6]^-$, $[Sb_2F_{11}]^-$, and $[Au_2F_{11}]^-$. ReF_6 has to be noted as well, as it follows WF_6 in the aforementioned series of hexafluorides. Pure ReF_6 is obtained as a yellow liquid in a stoichiometric comproportionation of ReF_7 und Re powder in a Monel autoclave at 300°C.[129]

4.6 Osmium

OsO_4 shows the highest possible oxidation state of +VIII. However, with fluorine osmium merely reaches the oxidation state of +VI. OsF_6 is a member of the hexafluoride series; it is a yellow crystalline solid and can be prepared by either high-pressure fluorination in autoclaves or by reduced-pressure fluorination in a continuous flow apparatus.[9,99] OsF_7 is expected to be stable, but has not been observed so far. All experimental attempts have led to the formation of pure OsF_6.[130] This is most likely due to the exothermic bimolecular reaction two $2OsF_7 \rightarrow 2OsF_6 + F_2$ as predicted by quantum chemical calculations.[131] OsF_8 was reported, but later shown to be nonexistent.

4.7 Iridium

Analogous to osmium, iridium can be oxidized to + VIII by oxygen in IrO_4[132] and even beyond. It was shown by photodissociation spectroscopy that $[IrO_4]^+$ can be prepared in the gas phase by laser ablation of iridium under excess of oxygen.[133] This compound is so far the only one that can be formally assigned to the extreme oxidation state of IX.[134] However, the highest valent fluoride of iridium is iridium hexafluoride. IrF_6 is a yellow solid and can be prepared by either high-pressure fluorination in autoclaves or reduced pressure fluorination in a continuous flow apparatus.[9,99] IrF_7 and IrF_6^+ are expected to be stable and are valuable targets for matrix and gas-phase experiments.[135]

4.8 Platinum

PtF_6 is the final and most reactive member of the hexafluoride series. It is highly oxidizing and has played a prominent historical role in the first oxidation of O_2 and the preparation of the first true noble gas compound "$XePtF_6$." Red PtF_6 has been prepared by continuous flow fluorination of elemental platinum[99] or by electrically heating a thin platinum wire in a fluorine atmosphere at −196°C.[31] It decomposes at temperatures above 100°C to release F_2.

4.9 Gold

A number of $[AuF_6]^-$ salts represent gold in the very high oxidation state of +V, considering that gold is a noble metal. These salts are prepared by heating gold in an atmosphere of oxygen and fluorine or from AuF_3 with strong oxidizers such as O_2/F_2, NOCl, XeF_6, or KrF_2, resulting in reactive cations such as $[O_2]^+$, $[Xe_2F_{11}]^+$, $[KrF]^+$, and $[NO]^+$.[22,136] A number of AuF_6^- salts with divalent metal cations (Hg^{2+}, Cd^{2+}, and Mg^{2+}) have also been reported.[137] These salts decompose upon warming at different temperatures to yield the strong oxidizer AuF_5. It forms dimers in the solid, dimers and trimers in the gas phase, and is able to oxidize xenon.[138] AuF_7 was claimed but later shown to be nonexistent.[139,140] AuF_5 is such a strong Lewis acid that it could also complex fluorine as, for example, in $AuF_5{\cdot}F_2$, as it was predicted by quantum chemical calculations.[139]

4.10 Mercury

The most prominent oxidation state of mercury is +II as, for example, in HgF_2. However, it was already predicted in 1979 by Jørgensen[141] that a higher oxidation state of IV could be possible. Indeed, the stabilization of the HgF_4 species is due to the relativistic destabilization of HgF_2 compound. Based on the relativistic contraction of the 6s orbital and its corresponding reduction of the ionic contributions, a decrease in the binding energy of HgF_2 results. This relativistic destabilization favors the formation of the higher oxidation state in HgF_4. Several quantum chemical investigations exist that underline the stability of HgF_4 in the gas phase.[142] Based on these predictions, HgF_4 has been characterized under cryogenic conditions in neon matrices at 5K.[143]

5. Outlook

In summary, we know plenty of transition metal fluorides that are very potent and useful oxidizers. Especially the hexafluorides represent promising oxidation agents with an oxidizing strength tunable to a great variety of tasks. The most powerful of all oxidation agents, Ag(III) and Ni(IV), are also representatives of this part of the periodic table. Obviously, the investigation of these very powerful compounds is no easy task and needs some experience in handling of such material. As there are still species predicted by quantum chemistry, but so far experimentally unknown, the search for even more powerful oxidizers will most likely continue. Novel techniques like matrix isolation or photodissociation spectroscopy are therefore very important and useful methods to venture into this terra incognita.

Acknowledgments

The authors thank the Deutsche Forschungsgemeinschaft (GRK 1582/2 "Fluorine as a Key Element") for financial support. Prof. Boris Zemva is gratefully acknowledged for providing Figs. 19.2 and 19.3 of the solvated Ni(IV) and Ag(III) compounds.

References

1. Kiselev, Y. M.; Tretiyakov, Y. D. The Problem of Oxidation State Stabilisation and Some Regularities of a Periodic System of the Elements. *Russ. Chem. Rev.* **1999,** *68* (5), 365–379.
2. Bartlett, N. Xenon Hexafluoroplatinate(V), $Xe^+[PtF_6]$. *Proc. Chem. Soc.* **1962,** 218.
3. Gutsev, G. L.; Boldyrev, A. I. An Explanation of the High Electron Affinities of the 5d-metal Hexafluorides. *Chem. Phys. Lett.* **1983,** *101* (4-5), 441–445.
4. Craciun, R.; Picone, D.; Long, R. T.; Li, S.; Dixon, D. A.; Peterson, K. A.; Christe, K. O. Third Row Transition Metal Hexafluorides, Extraordinary Oxidizers, and Lewis Acids: Electron Affinities, Fluoride Affinities, and Heats of Formation of WF_6, ReF_6, OsF_6, IrF_6, PtF_6, and AuF_6. *Inorg. Chem.* **2010,** *49* (3), 1056–1070.
5. Riedel, S.; Kaupp, M. Has AuF_7 Been Made? *Inorg. Chem.* **2006,** *45* (3), 1228–1234.
6. Miyoshi, E.; Sakai, Y. Theoretical Study on Electronic Structures of Gold Fluoride (AuF_6) and Its Anions. *J. Chem. Phys.* **1988,** *89* (12), 7363–7366.
7. Selig, H.; Vaknin, D.; Davidov, D.; Yeshurun, Y. Graphite Intercalation Compounds with Rhenium Hexafluoride and Technetium Hexafluoride: Electronic and Magnetic Properties. *Synth. Met.* **1985,** *12* (1-2), 479–484.
8. Rudnyi, E. B.; Kaibicheva, E. A.; Sidorov, L. N. The Electron Affinity of Uranium Oxide (UO_3). *Rapid Commun. Mass Spectrom.* **1992,** *6* (5), 356–360.
9. Drews, T.; Supel, J.; Hagenbach, A.; Seppelt, K. Solid State Molecular Structures of Transition Metal Hexafluorides. *Inorg. Chem.* **2006,** *45* (9), 3782–3788.
10. Craciun, R.; Long, R. T.; Dixon, D. A.; Christe, K. O. Electron Affinities, Fluoride Affinities, and Heats of Formation of the Second Row Transition Metal Hexafluorides: MF_6 (M = Mo, Tc, Ru, Rh, Pd, Ag). *J. Phys. Chem. A* **2010,** *114* (28), 7571–7582.
11. Macgregor, S. A.; Moock, K. H. Stabilization of High Oxidation States in Transition Metals. 2. WCl_6 Oxidizes $[WF_6]^-$, but Would $PtCl_6$ Oxidize $[PtF_6]^-$? an Electrochemical and Computational Study of 5d Transition Metal Halides: $[MF_6]^z$ versus $[MCl_6]^z$ (M = Ta to Pt; $z = 0, 1-, 2-$). *Inorg. Chem.* **1998,** *37* (13), 3284–3292.
12. Wesendrup, R.; Schwerdtfeger, P. Structure and Electron Affinity of Platinum Fluorides. *Inorg. Chem.* **2001,** *40* (14), 3351–3354.
13. Friedman, J. F.; Stevens, A. E.; Miller, T. M.; Viggiano, A. A. *J. Chem. Phys.* **2006,** *124,* 224306.
14. Sidorov, L. N.; Borshchevskii, A. Y.; Rudnii, E. B.; Butskii, V. D. Electron Affinities of Higher Molybdenum Fluorides as Determined by the Effusion Technique. *Chem. Phys.* **1982,** *71* (1), 145–156.

15. Mathur, B. P.; Rothe, E. W.; Reck, G. P. Ionization Reactions of Metal Hexafluorides with Alkali Atoms and Dimers. *J. Chem. Phys.* **1977,** *67* (2), 377−381.
16. Compton, R. N.; Reinhardt, P. W.; Cooper, C. D. Collisional Ionization between Fast Alkali Atoms and Selected Hexafluoride Molecules. *J. Chem. Phys.* **1978,** *68* (5), 2023−2036.
17. Sakai, Y.; Miyoshi, E. Theoretical Study of the Electron Affinities of Metal Hexafluoride and Hexafluorometalate(1−) (MF_6 and MF_6^-) (M = Chromium, Molybdenum, and Tungsten) Using a Model Potential Method. *J. Chem. Phys.* **1987,** *87* (5), 2885−2892.
18. Viggiano, A. A.; Paulson, J. F.; Dale, F.; Henchman, M.; Adams, N. G.; Smith, D. Ion Chemistry and Electron Affinity of Tungsten Hexafluoride. *J. Phys. Chem.* **1985,** *89* (11), 2264−2267.
19. George, P. M.; Beauchamp, J. L. The Electron and Fluoride Affinities of Tungsten Hexafluoride by Ion Cyclotron Resonance Spectroscopy. *Chem. Phys.* **1979,** *36* (3), 345−351.
20. Miyoshi, E.; Sakai, Y.; Murakami, A.; Iwaki, H.; Terashima, H.; Shoda, T.; Kawaguchi, T. On the Electron Affinities of Chromium, Molybdenum, and Tungsten Hexafluorides (CrF_6, MoF_6, and WF_6). *J. Chem. Phys.* **1988,** *89* (7), 4193−4198.
21. Vasile, M. J.; Richardson, T. J.; Stevie, F. A.; Falconer, W. E. *J. Chem. Soc. Dalton Trans.* **1976,** *4,* 351.
22. Holloway, J. H.; Schrobilgen, G. J. Krypton Fluoride Chemistry. Route to Pentafluorogold, Fluorokrypton Hexafluoroaurate, Trifluorodixenon Hexafluoroaurate, and Nitrosonium Hexafluoroaurate. The Fluorokrypton(1+)-xenon Oxotetrafluoride System. *J. Chem. Soc. Chem. Commun.* **1975,** *15,* 623−624.
23. Molski, M. J.; Seppelt, K. The Transition Metal Hexafluorides. *Dalton Trans.* **2009,** *18,* 3379−3383.
24. Wilson, A. V.; Nguyen, T.; Brosi, F.; Wang, X.; Andrews, L.; Riedel, S.; Bridgeman, A. J.; Young, N. A. A Matrix Isolation and Computational Study of Molecular Palladium Fluorides: Does PdF_6 Exist? *Inorg. Chem.* **2016,** *55* (3), 1108−1123.
25. Chernick, C. L.; Claassen, H. H.; Fields, P. R.; Hyman, H. H.; Malm, J. G.; Manning, W. M.; Matheson, M. S.; Quarterman, L. A.; Schreiner, F.; Selig, H. H.; Sheft, I.; Siegel, S.; Sloth, E. N.; Stein, L.; Studier, M. H.; Weeds, J. L.; Zirin, M. H. Fluorine Compounds of Xenon and Radon. *Science* **1962,** *138,* 136−138.
26. Seidel, S.; Seppelt, K. The Cl_4^+ Ion. *Angew. Chem. Int. Ed.* **2000,** *39* (21), 3923−3925.
27. Gutsev, G. L.; Boldyrev, A. I. The Theoretical Investigation of the Electron Affinity of Chemical Compounds. *Adv. Chem. Phys.* **1985,** *61,* 169−221.
28. Shorafa, H.; Mollenhauer, D.; Paulus, B.; Seppelt, K. The Two Structures of the Hexafluorobenzene Radical Cation $C_6F_6^{+}$. *Angew. Chem. Int. Ed.* **2009,** *48* (32), 5845−5847.
29. Lehmann, J. F.; Schrobilgen, G. J. Structural and Vibrational Characterization of [KrF][AuF_6] and [O_2][AuF_6] Using Single Crystal X-ray Diffraction, Raman Spectroscopy and Electron Structure Calculations. *J. Fluor. Chem.* **2003,** *119* (2), 109−124.
30. Fete, M. G.; Havlas, Z.; Michl, J. *J. Am. Chem. Soc.* **2011,** *133,* 4123.
31. Bartlett, N. The Oxidizing Properties of the Third Transition Series Hexafluorides and Related Compounds. *Angew. Chem. Int. Ed.* **1968,** *7* (6), 433−439.
32. Bartlett, N.; Yeh, S.; Kourtakis, K.; Mallouk, T. The Fluorobasicities of Rhenium Fluoride (ReF_7) and Iodine Fluoride (IF_7) as Measured by the Enthalpy Change ΔH° ($EF_{7(g)}$ -> $EF_6^+ + F^-_{(g)}$. *J. Fluor. Chem.* **1984,** *26* (1), 97−116.
    a. Bartlett, N.; Beaton, S. P.; Jha, N. K. Oxidizing Trends in the Third-transition-series Hexafluorides. *Chem. Commun.* **1966**, (6), 168−197.

33. Lucier, G.; Shen, C.; Casteel, W. J., Jr.; Chacon, L.; Bartlett, N. Some Chemistry of High Oxidation State Transition Metal Fluorides in Anhydrous HF. *J. Fluor. Chem.* **1995,** *72* (2), 157–163.
34. Bougon, R.; Wilson, W. W.; Christe, K. O. Synthesis and Characterization of Tetrafluoroammonium Hexafluorochromate and Reaction Chemistry of Chromium Pentafluoride. *Inorg. Chem.* **1985,** *24* (14), 2286–2292.
35. Riedel, S. High-valent Fluorides and Fluoro-oxidizers. In *Comprehensive Inorganic Chemistry II,* Vol. 2, Reedijk, J., Poeppelmeier, K., Eds.; Elsevier: Oxford, 2013; pp 187–221.
36. Christe, K. O.; Dixon, D. A. A Quantitative Scale for the Oxidizing Strength of Oxidative Fluorinators. *J. Am. Chem. Soc.* **1992,** *114* (8), 2978–2985.
37. Hauge, R. H.; Hastie, J. W.; Margrave, J. L. Force Constants and Geometries of Matrix-isolated Rare-earth Trifluorides. *J. Less Common Met.* **1971,** *23* (4), 359–365.
38. Holleman, A. F.; Wiberg, E. *Lehrbuch der Anorganischen Chemie,* Vol. 102; Walter de Gruyter: Berlin, 2007; p 1209.
39. Wang, X.; Andrews, L. Infrared Spectra of MF_2, MF_2^+, MF_4^-, MF_3, and M_2F_6 Molecules (M = Sc, Y, La) in Solid Argon. *J. Phys. Chem. A* **2010,** *114* (6), 2293–2299.
40. Ehrlich, P.; Pietzka, G. Titanium Trifluoride. *Z. Anorg. Allg. Chem.* **1954,** *275,* 121–140.
a. Leng, B.; Moss, J. H. Reactions of Titanium Dioxide with Metal Fluorides. *J. Fluor. Chem.* **1975,** *5* (2), 93–98.
41. Naumann, D. *Spezielle Anorganische Chemie* In: *Fluor und Fluorverbindungen,* Vol. 2; Dr. Dietrich Steinkopff Verlag: Darmstadt, 1980; p 173.
42. Hofmann, B.; Hoppe, R. Knowledge of the Ammonium Heptafluorosilicate ($(NH_4)_3SiF_7$) Type Structure. New Metal Fluorides A_3MF_7 with M = Si, Ti, Cr, Mn, Ni, and A = Rb, Cs. *Z. Anorg. Allg. Chem.* **1979,** *458,* 151–162.
a. Dean, P. A. W. Fluorine-19 Nuclear Magnetic Resonance Study of Polymeric Fluoroanions of Tin(IV) and Titanium(IV). *Can. J. Chem.* **1973,** *51* (24), 4024–4030.
b. Dadachov, M. S.; Tang, L. Q.; Zou, X. D. Crystal structure of Dipiperazine Decafluorodititanate Dihydrate, $(C_4H_{12}N_2)^2[Ti_2F_{10}]Â\cdot 2H_2O$. *Z. Kristallogr. New Cryst. Struct.* **2000,** *215* (4), 605–606.
c. Tang, L. Q.; Dadachov, M. S.; Zou, X. D. Crystal structure of Dipyridine Oxonium Undecafluorodititanate Hydrate, $(C_5H_6N)^2(H_3O)[Ti_2F_{11}].H_2O$. *Z. Kristallogr. New Cryst. Struct.* **2001,** *216* (3), 387–388.
d. Decken, A.; Jenkins, H. D. B.; Knapp, C.; Nikiforov, G. B.; Passmore, J.; Rautiainen, J. M. The Autoionization of $[TiF_4]$ by Cation Complexation with [15]crown-5 to Give $[TiF_2([15]crown\text{-}5)][Ti_4F_{18}]$ Containing the Tetrahedral $[Ti_4F_{18}]^{2-}$ Ion. *Angew. Chem. Int. Ed.* **2005,** *44* (48), 7958–7961.
e. Mazej, Z.; Goreshnik, E. Poly[perfluorotitanate(IV)] Salts of $[H_3O]^+$, Cs^+, $[Me_4N]^+$, and $[Ph_4P]^+$ and about the Existence of an Isolated $[Ti_2F_9]^-$ Anion in the Solid State. *Inorg. Chem.* **2009,** *48* (14), 6918–6923.
f. Bialowons, H.; Mueller, B. G. $CsTi_8F_{33}$-after $(O_2)2Ti_7F_{30}$ the Last Step on the Way to TiF4. *Z. Anorg. Allg. Chem.* **1995,** *621* (7), 1223–1226.
g. Bialowons, H.; Mueller, M.; Mueller, B. G. Titanium Tetrafluoride – a Surprising Simple Column Structure. *Z. Anorg. Allg. Chem.* **1995,** *621* (7), 1227–1231.
h. Shylapnikov, I.; Goreshnik, E. A.; Mazej, Z. The Cubic $[Ti_8F_{36}]^{4-}$ Anion Found in the Crystal Structures of $K_4Ti_8F_{36}\cdot 8HF$ and $Rb_4Ti_8F_{36}\cdot 6HF$. *Chem. Commun.* **2013,** *49* (26), 2703–2705.
i. Shlypanikov, I.; Mercier, H. P. A.; Goreshnik, E. A.; Schrobilgen, G. J.; Mazej, Z. Crystal structures and Raman Spectra of Imidazolium Poly[perfluorotitanate(IV)] Salts Containing the $[TiF_6]^{2-}$, $([Ti_2F_9]^-$ and $[Ti_2F_{11}]^{3-}$ and the New $[Ti_4F_{20}]^{4-}$ and $[Ti_5F_{23}]^{3-}$ Anions. *Inorg. Chem.* **2013,** *52* (15), 8315–8326.

j. Mazej, Z.; Goreshnik, E. A. Synthesis and Characteristion of $[XeF_5]_3[Ti_4F_{19}]$ Containing a Discrete $[Ti_4F_{19}]^{3-}$ Anion. *Eur. J. Inorg. Chem.* **2009**, (29/30), 4503–4506.

43. Emeleus, H. J.; Gutmann, V. The Fluorides of Molybdenum and Vanadium. *J. Chem. Soc.* **1949,** 2979–2982.
44. Šmalc, A. Synthesis of VF_5 from V_2O_5 and Fluorine. *Monatsh. Chem.* **1967,** *98* (1), 163–164.
45. Glemser, O.; Roesky, H.; Hellberg, K. H. Preparation of Chromium Pentafluoride and Chromium Hexafluoride. *Angew. Chem. Int. Ed.* **1963,** *2* (5), 266–267.
a. Hellberg, K. H.; Mueller, A.; Glemser, O. The Infrared Spectra of Solid CrF_6, MoF_6, and OsF_6 at Low Temperatures. *Z. Naturforsch. B* **1966,** *21* (2), 118–121.
b. Hope, E. G.; Jones, P. J.; Levason, W.; Ogden, J. S.; Tajik, M. A New Synthesis and IR Characterization of Chromium Hexafluoride. *J. Chem. Soc. Chem. Commun.* **1984**, (20), 1355–1356.
c. Hope, E. G.; Jones, P. J.; Levason, W.; Ogden, J. S.; Tajik, M.; Turff, J. W. Spectroscopic Studies on the Higher Binary Fluorides of Chromium: CrF_4, CrF_5, and CrF_6, Both in the Solid State and Isolated in Inert Gas Matrixes. *J. Chem. Soc. Dalton Trans.* **1985**, (7), 1443–1449.
d. Hope, E. G.; Levason, W.; Ogden, J. S. Is Chromium Hexafluoride Octahedral? Experiment Still Suggests "yes!". *Inorg. Chem.* **1991,** *30* (26), 4873–4874.
46. Jacob, E.; Willner, H. Spectroscopic Studies of Higher Chromium Fluorides. Controversy over the Existence of Chromium Hexafluoride. *Chem. Ber.* **1990,** *123* (6), 1319–1321.
a. Jacobs, J.; Müller, H. S. P.; Willner, H.; Jacob, E.; Buerger, H. Vibrational and Electronic Spectra of Molecular Chromium Tetrafluoride, CrF_4, and Chromium Pentafluoride, CrF_5. Comments on the Existence of Chromium Hexafluoride, CrF_6. *Inorg. Chem.* **1992,** *31* (26), 5357–5363.
47. von Wartenberg, H. Higher Chromium Fluorides (CrF_4, CrF_3 and CrO_2F_2). *Z. Anorg. Allgem. Chem.* **1941,** *247,* 135–146.
48. Schloeder, T.; Brosi, F.; Freyh, B. J.; Vent-Schmidt, T.; Riedel, S. New Evidence in an Old Case: The Question of Chromium Hexafluoride Reinvestigated. *Inorg. Chem.* **2014,** *53* (11), 5820–5829.
49. Shorafa, H.; Seppelt, K. The Structures of CrF_5 and $CrF_5 \cdot SbF_5$. *Z. Anorg. Allgem. Chem.* **2008,** *35* (1).
50. Mazej, Z.; Goreshnik, E. Alkali Metal (Li^+-Cs^+) Salts with Hexafluorochromate(V), Hexafluorochromate(IV), Pentafluorochromate(IV), and Undecafluorodichromate(IV) Anions. *Eur. J. Inorg. Chem.* **2008,** *11,* 1795–1812.
51. Hoppe, R.; Dahne, W.; Klemm, W. Manganese Tetrafluoride. *Naturwissenschaften* **1961,** *48,* 429.
a. Mueller, B. G.; Serafin, M. The Crystal Structure of Manganese Tetrafluoride. *Z. Naturforsch. B Chem. Sci.* **1987,** *42* (9), 1102–1106.
b. Roesky, H. W.; Glemser, O.; Hellberg, K. H. Preparation of Metal Fluorides in the Fluidized Bed. *Chem. Ber.* **1965,** *98* (6), 2046–2048.
c. Lutar, K.; Jesih, A.; Žemva, B. Krypton Difluoride/manganese Tetrafluoride Adducts from Krypton Difluoride/manganese Difluoride Interaction in Hydrofluoric Acid as a Route to High Purity Manganese Tetrafluoride. *Polyhedron* **1988,** *7* (13), 1217–1219.
d. Mazej, Z. Room Temperature Syntheses of MnF_3, MnF_4 and Hexafluoromanganate(IV) Salts of Alkali Cations. *J. Fluor. Chem.* **2002,** *114* (1), 75–80.
52. Seseke-Koyro, U.; Garcia-Juan, P.; Palsherm, S.; Schulz, A. *Method for Preparing Manganese Tetrafluoride and Its Use,* 2009. 2008-EP67085, 2009074560, 20081209.
53. Osin, S. B.; Davlyatshin, D. I.; Shevelkov, V. F.; Mitkin, V. N. *Russ. J. Phys. Chem.* **1995,** *69,* 794–799.

a. Cesaro, S. N.; Rau, J. V.; Chilingarov, N. S.; Balducci, G.; Sidorov, L. N. *Inorg. Chem.* **2001,** *40,* 179–181.

54. Brosi, F.; Schloeder, T.; Schmidt, A.; Beckers, H.; Riedel, S. Are There Molecular Manganese Fluorides Beyond MnF3? – A Quantum-Chemical And Matrix-Isolation Investigation. *Dalton Trans.* **2016,** *45,* 5038–5044.
55. Nikitin, M. I.; Rakov, E. G. Manganese Pentafluoride: Arguments for and against. *Zh. Neorg. Khim.* **1998,** *43* (3), 375–379.
   a. Levason, W.; McAuliffe, C. A. Higher Oxidation State Chemistry of Manganese. *Coord. Chem. Rev.* **1972,** *7* (4), 353–384.
56. Žemva, B. High Oxidation State Transition Metal Fluorides. In *Advanced Inorganic Fluorides;* Nakajima, T., Žemva, B., Tressaud, A., Eds.; Elsevier: Lausanne, 2000; pp 79–115.
57. Engelbrecht, A.; Grosse, A. V. The Fluorides of Permanganic and Perrhenic Acids, MnO_3F and ReO_3F. *J. Am. Chem. Soc.* **1954,** *76* (8), 2042–2045.
58. Levason, W.; McAuliffe, C. A. Higher Oxidation State Chemistry of Iron, Cobalt, and Nickel. *Coord. Chem. Rev.* **1974,** *12* (2), 151–184.
59. Ketelaar, J. A. A. Structure of the Trifluorides of Aluminum, Iron, Cobalt, Rhodium and Palladium. *Nature* **1931,** *128,* 303.
60. Schlöder, T.; Vent-Schmidt, T.; Riedel, S. Beyond Iron Trifluoride: A Matrix-isolation and Quantum-chemical Investigation of FeF_4. *Angew. Chem. Int. Ed.* **2012,** *51* (48), 12063–12067.
61. Korobov, M. V.; Savinova, L. N.; Sidorov, L. N. Stabilities of Cobalt Tetrafluoride and Chromium Pentafluoride in the Gas Phase. *J. Chem. Thermodyn.* **1993,** *25* (9), 1161–1168.
62. Rau, J. V.; Nunziante Cesaro, S.; Chilingarov, N. S.; Balducci, G. Identification of Gaseous Cobalt Tetrafluoride: MS and FTIR Spectroscopic Studies. *Inorg. Chem.* **1999,** *38* (25), 5695–5697.
63. Rau, J. V.; Chilingarov, N. S.; Sidorov, L. N. Mass Spectrometric Determination of Cobalt Trifluoride Saturated Vapor Pressure. Enthalpy of Formation of Gaseous CoF_4 and CoF_4. *Rapid Commun. Mass Spectrom.* **1997,** *11* (18), 1977–1979.
64. Banks, R. E.; Sharp, D. W. A.; Tatlow, J. C. *Fluorine – the First Hundred Years;* Elsevier Sequoia: Lusanne, 1986.
65. Quail, J. W.; Rivett, G. A. Complex Fluorides of Tetravalent Cobalt. *Can. J. Chem.* **1972,** *50* (15), 2447–2450.
   a. Klemm, W.; Brandt, W.; Hoppe, R. Fluorocobaltates(III) and -(IV) and Fluoronickelates(III). *Z. Anorg. Allgem. Chem.* **1961,** *308,* 179–189.
66. Sorbe, P.; Grannec, J.; Portier, J.; Hagenmüller, P. Synthesis and Study of Fluoride Compounds of Tetravalent Copper. *Compt. Rend. C.* **1976,** *282* (15), 663–666.
67. Žemva, B.; Lutar, K.; Chacon, L.; Fele-Beuermann, M.; Allman, J.; Shen, C.; Bartlett, N. Thermodynamically Unstable Fluorides of Nickel: NiF_4 and NiF_3 Syntheses and Some Properties. *J. Am. Chem. Soc.* **1995,** *117* (40), 10025–10034.
68. Whalen, J. M.; Lucier, G. M.; Chacon, L.; Bartlett, N. The Room Temperature Conversion of Nickel Difluoride to Hexafluoronickelate(IV) Salts of Alkali Cations. *J. Fluor. Chem.* **1998,** *88* (2), 107–110.
69. Žemva, B.; Lutar, K.; Jesih, A.; Casteel, W. J.; Bartlett, N. A General Method for the Synthesis of Polymeric Binary Fluorides Exemplified by AgF_3, NiF_4, RuF_4, and OsF_4. *J. Chem. Soc. Chem. Commun.* **1989**, (6), 346–347.
70. Bartlett, N.; Lucier, G.; Shen, C.; Casteel, W. J., Jr.; Chacon, L.; Munzenberg, J.; Žemva, B. The Oxidizing Properties of Cationic High Oxidation State Transition-element Fluoro Species. *J. Fluor. Chem.* **1995,** *71* (2), 163–164.

71. Bartlett, N. Low Temperature Preparation and Uses of Potent Oxidizers. *J. Fluor. Chem.* **2006,** *127* (10), 1285–1288.
72. Schroer, T.; Christe, K. O. Novel Synthesis of ClF6+ and BrF6+ Saltsâ€. *Inorg. Chem.* **2001,** *40* (10), 2415–2419.
73. Levason, W.; Spicer, M. D. The Chemistry of Copper and Silver in Their Higher Oxidation States. *Coord. Chem. Rev.* **1987,** *76,* 45–120.
a. Popova, T. V.; Aksenova, N. V. Complexes of Copper in Unstable Oxidation States. *Russ. J. Coord. Chem.* **2003,** *29* (11), 743–765.
b. Burdett, J. K.; Sevov, S. Stability of the Oxidation States of Copper. *J. Am. Chem. Soc.* **1995,** *117* (51), 12788–12792.
c. Harnischmacher, W.; Hoppe, R. Tetravalent Copper. Cesium Hexafluorocuprate(IV). *Angew. Chem. Int. Ed.* **1973,** *12* (7), 582–583.
d. Kissel, D.; Hoppe, R. On Fluorides with Tetravalent Copper: Cesium Copper Fluoride ($Cs_2[CuF_6]$). *Z. Anorg. Allg. Chem.* **1988,** *559,* 40–48.
74. Kiselev, Y. M. Recent Advances in the Chemistry of Unstable Higher Oxidation States of Transition Metals. *Russ. J. Inorg. Chem.* **1999,** *44* (5), 652–658.
75. Hoppe, R. New Data on Metals with Oxygen or Fluorine as Ligands. *Angew. Chem. Int. Ed.* **1981,** *20* (1), 63–87.
76. Jesih, A.; Lutar, K.; Žemva, B. (personal communication). **1991,** (unpublished work).
77. Riedel, S.; Kaupp, M.; Pyykkö, P. Quantum Chemical Study of Trivalent Group-12 Compounds. *Inorg. Chem.* **2008,** *47* (8), 3379–3383.
78. Samanta, D.; Jena, P. Zn in the +III Oxidation State. *J. Am. Chem. Soc.* **2012,** *134* (20), 8400–8403.
79. Schlöder, T.; Kaupp, M.; Riedel, S. Can Zinc Really Exist in its Oxidation State +III? *J. Am. Chem. Soc.* **2012,** *134* (29), 11977–11979.
80. Hastie, J. W.; Hauge, R. H.; Margrave, J. L. Geometries and Entropies of Metal Trifluorides from Infrared Spectra. Scandium, Yttrium, Lanthanum, Cerium, Neodymium, Europium, and Gadolinium. *J. Less Common Met.* **1975,** *39* (2), 309–334.
a. Wesley, R. D.; DeKock, C. W. Infrared Spectra and Geometries of Matrix Isolated Yttrium Tri- and Difluorides. *J. Phys. Chem. A* **1973,** *77* (4), 466–468.
81. Chretien, A.; Gaudreau, B. Existence of Two Crystalline Varieties of Zirconium Tetrafluoride. *Compt. Rend.* **1958,** *246,* 2266–2268.
a. Chretien, A.; Gaudreau, B. Different Varieties of ZrF_4. *Compt. Rend.* **1959,** *248,* 2878–2879.
82. Skokan, E. V.; Sorokin, I. D.; Sidorov, L. N.; Nikitin, M. I. Mass Spectrometric Determination of the Enthalpies of Dissociation of Gaseous Complex Fluorides into Neutral and Charged Particles. VI. Enthalpy of Formation of Pentafluorozirconate (ZrF_5^-). *Int. J. Mass Spectrom. Ion Phys.* **1982,** *43* (4), 309–325.
a. Fischer, J.; Weiss, R. Crystal structure of K_2 $Cu(ZrF_6)_2 \cdot 6H_2O$. *Chem. Commun.* **1968**, (18), 1137.
b. Hoppe, R.; Dahne, W. Crystal structure of Li_2ZrF_6. *Naturwissenschaften* **1960,** *47,* 397.
c. Poulain, M.; Lucas, J. New Transition Metals Fluorozirconates. *C. R. Acad. Sci., Ser. C* **1970,** *271* (14), 822–824.
d. Hurst, H. J.; Taylor, J. C. Crystal structure of Ammonium Heptafluorozirconate and the Disorder of the Heptafluorozirconate Ion. *Acta Crystallogr. B* **1970,** *26* (Pt. 4), 417–421.
e. Misyul, S. V.; Mel'nikova, S. V.; Bovina, A. F.; Laptash, N. M. Optical and X-ray Diffraction Studies of the Symmetry of Distorted Phases of the $(NH_4)_3ZrF_7$ Crystal. *Phys. Solid State* **2008,** *50* (10), 1951–1956.
f. Fischer, J.; Elchinger, R.; Weiss, R. Stereochemistry of Zirconium and Copper in the

Hydrated Copper Fluorozirconates. IV. Complex Octafluorozirconate(4-) Ion in Dicopper Octafluorozirconate Dodecahydrate. *Acta Crystallogr. B* **1973,** *29* (Pt. 9), 1967–1971.
g. Glavič, P.; Bole, A.; Slivnik, J. Reactions of Zirconium, Hafnium, and Cerium Tetrafluoride with Hydrazine. *J. Inorg. Nucl. Chem.* **1975,** *37* (5), 1316–1318.
h. Brunton, G. Crystal structure of Rubidium Zirconium Fluoride $Rb_5Zr_4F_{21}$. *Acta Crystallogr. B* **1971,** *27* (10), 1944–1948.
i. Davidovich, R. L. Stereochemistry of Zirconium and Hafnium Fluoro Complexes. *Russ. J. Coord. Chem.* **1998,** *24* (11), 751–768.

83. Holloway, C. E.; Walker, I. M.; Melnik, M. Zirconium and Hafnium Coordination Compounds: Analysis and Classification of Crystallographic and Structural Data. *Rev. Inorg. Chem.* **1986,** *8,* 169–219.
84. Greenberg, E.; Natke, C. A.; Hubbard, W. N. Fluorine Bomb Calorimetry. X. The Enthalpies of Formation of Niobium and Tantalum Pentafluorides. *J. Phys. Chem.* **1965,** *69* (6), 2089–2093.
85. Edwards, A. J. The Structures of Niobium and Tantalum Pentafluorides. *J. Chem. Soc.* **October 1964,** 3714–3718.
86. Falconer, W. E.; Jones, G. R.; Sunder, W. A.; Vasile, M. J.; Muenter, A. A.; Dyke, T. R.; Klemperer, W. Gas-phase Structures and Mass Spectra of Binary Pentafluorides. *J. Fluor. Chem.* **1974,** *4* (2), 213–234.
a. Alexander, L. E. Polymer-monomer Equilibriums in the Pentafluorides of Niobium, Tantalum, and Antimony. Shape of Monomeric Antimony Pentafluoride. *Inorg. Nucl. Chem. Lett.* **1971,** *7* (10), 1053–1056.
b. Boghosian, S.; Pavlatou, E. A.; Papatheodorou, G. N. Establishing the Gas Phase Dimerization of Niobium(V) Fluoride and Tantalum(V) Fluoride by Quantitative Raman Spectroscopy. *Vib. Spectrosc.* **2005,** *37* (1), 133–139.
c. Konings, R. J. M. The Composition of Niobium Pentafluoride Vapor. *Struct. Chem.* **1994,** *5* (1), 9–13.
d. Girichev, G. V.; Petrova, V. N.; Petrov, V. M.; Krasnov, K. S.; Goncharuk, V. K. Electron Diffraction Study of the Structure of Trimeric Molecules of Molybdenum and Niobium Pentafluorides. *Izv. Vyssh. Uchebn. Zaved. Khim. Khim. Tekhnol.* **1981,** *24* (1), 131–132.
e. Blanchard, S. Infrared Spectrum of Niobium Pentafluoride. *J. Chim. Phys. Phys. Chim. Biol.* **1965,** *62* (7-8), 919–920.
f. Romanov, G. V.; Spiridonov, V. P. Electron Diffraction Investigation of Niobium and Tantalum Pentafluorides. *Vestn. Mosk. Univ. Khim.* **1968,** *23* (5), 7–11.
g. Vasile, M. J.; Jones, G. R.; Falconer, W. E. Molecular Beam-mass Spectrometric Study of Binary Pentafluorides. *J. Chem. Soc. D* **1971**, (21), 1355–1356.
h. Alexander, L. E.; Beattie, I. R.; Jones, P. J. Vibrational Spectra of Niobium and Tantalum Pentafluorides in the Gas Phase. Vapor Density of Niobium Pentafluoride. *J. Chem. Soc. Dalton Trans.* **1972**, (2), 210–212.
87. Žemva, B.; Slivnik, J. On the Syntheses of Xenon(VI) Fluoroniobates(V). *J. Fluor. Chem.* **1976,** *8* (4), 369–371.
a. Holloway, J. H.; Knowles, J. G. Reactions of Xenon Difluoride with the Pentafluorides of Tantalum, Niobium, Ruthenium, and Osmium. *J. Chem. Soc. A* **1969,** *5,* 756–761.
b. Bizot, D.; Malek-Zadeh, M. NbF_5-MF (M = Lithium, Sodium, Potassium) Systems. *Rev. Chim. Miner.* **1974,** *11* (6), 710–721.
c. Bizot, D.; Vestegaine, H. The Niobium(V) Fluoride-thallium(I) Fluoride System. *Inorg. Nucl. Chem. H. H. Hyman Mem.* **1976,** 67–68.
88. Agulyansky, A. *The Chemistry of Tantalum and Niobium Fluoride Compounds;* Elsevier: Amsterdam, 2004.

89. Edwards, A. J. Crystal structure of the Adduct of Niobium Pentafluoride and Antimony Pentafluoride. *J. Chem. Soc. D* **1970**, (13), 820.
a. Davies, T. K.; Moss, K. C. Nuclear Magnetic Resonance Studies on Fluorine-containing Compounds. I. Antimony Pentafluoride and its Reactions with Antimony Trifluoride, Arsenic Trifluoride, and Niobium Pentafluoride. *J. Chem. Soc. A* **1970**, (7), 1054–1058.
90. Cady, G. H.; Hargreaves, G. B. Vapor Pressures of Some Fluorides and Oxyfluorides of Molybdenum, Tungsten, Rhenium, and Osmium. *J. Chem. Soc* **1961,** 1568–1574.
a. Osin, S. B.; Davlyatshin, D. I.; Ogden, D. S. Infrared Spectra in Solid Argon Matrix of the Products of Reactions of Tungsten and Molybdenum with Fluorine. *Zh. Fiz. Khim.* **2001,** *75* (2), 294–302.
b. Siegel, S.; Northrop, D. A. X-ray Diffraction Studies of Some Transition Metal Hexafluorides. *Inorg. Chem.* **1966,** *5* (12), 2187–2188.
91. Prescott, A.; Sharp, D. W. A.; Winfield, J. M. Oxidation of Metals by Molybdenum and Tungsten Hexafluorides in Acetonitrile. *J. Chem. Soc. Dalton Trans.* **1975**, (10), 936–939.
a. Ajiboye, S. I.; Iqbal, J.; Peacock, R. D.; Prouff, N.; Taylor, G. J.; Winfield, J. M.; Liu, X. Redox Reactions between Molybdenum or Tungsten Hexafluorides and P, F or D Block Elements in Acetonitrile: Comparisons with Reactions Involving Nitrosonium Fluorometallates, the Effect of Fluoride Ligand Transfer and Redox Inhibition Due to Surface Oxide. *J. Fluor. Chem.* **1998,** *91* (2), 213–218.
92. Giese, S.; Seppelt, K. Structural Principles of Side-group Compounds with Coordination Number Seven as Illustrated by MoF_7-, WF_7-, and $ReOF_6$. *Angew. Chem. Int. Ed.* **1994,** *33* (4), 461–463.
a. Beuter, A.; Kuhlmann, W.; Sawodny, W. Vibrational Spectra of Hepta- and Octafluorocomplexes of Molybdenum(VI), Tungsten(VI), and Rhenium(VI). *J. Fluor. Chem.* **1975,** *6* (4), 367–378.
93. Riedel, S.; Renz, M.; Kaupp, M. High-Valent Technetium Fluorides. Does TcF_7 Exist? *Inorg. Chem.* **2007,** *46* (14), 5734–5738.
94. Boyd, G. E.; Cobble, J. W.; Nelson, C. M.; Smith, W. T., Jr. Chemistry of Technetium. I. Preparation of Technetium Heptoxide. *J. Am. Chem. Soc.* **1952,** *74,* 556–557.
a. Smith, W. T., Jr.; Cobble, J. W.; Boyd, G. E. Thermodynamic Properties of Technetium and Rhenium Compounds. I. Vapor Pressures of Technetium Heptoxide, Pertechnetic Acid, and Aqueous Solutions of Pertechnetic Acid. *J. Am. Chem. Soc.* **1953,** *75,* 5773–5776.
b. Krebs, B. Technetium(VII) Oxide: a Transition Metal Oxide with a Molecular Structure in the Solid State. *Angew. Chem. Int. Ed.* **1969,** *8* (5), 381–382.
95. Schwochau, K. *Technetium: Chemistry and Radiopharmaceutical Applications;* John Wiley & Sons, Ltd.: Chichester, UK, 2000; p 458.
96. Selig, H.; Chernick, C. L.; Malm, J. G. Preparation and Properties of Technetium Hexafluoride. *J. Inorg. Nucl. Chem.* **1961,** *19,* 377.
97. Courtois, C.; Kikindai, T. Study of the Products Formed during Fluorination of Ruthenium and its Dioxide and an Attempt at Characterizing Ruthenium Octafluoride. *Compt. Rend. C* **1976,** *283* (15), 679–682.
98. Sakurai, T.; Takahashi, A. Behavior of Ruthenium in Fluoride Volatility Processes. V. Conversions of $RuOF_4$, RuF_4, and RuF_5 into RuO_4. *J. Inorg. Nucl. Chem.* **1979,** *41* (5), 681–685.
a. Gulliver, D. J.; Levason, W. The Chemistry of Ruthenium, Osmium, Rhodium, Iridium, Palladium and Platinum in the Higher Oxidation States. *Coord. Chem. Rev.* **1982,** *46,* 1–127.
99. Holloway, J. H.; Hope, E. G.; Stanger, G.; Boyd, D. A. Preparation of High-oxidation-state Platinum Group Metal Fluorides in Flow Systems. Synthesis of the Metal Hexafluorides,

MF_6 (M = Ru, Rh, Os, Ir, Pt), Ruthenium Pentafluoride Tetramer, $(RuF_5)_4$, and the Novel Trimer, $(RuF_5)_3$. *J. Fluor. Chem.* **1992,** *56* (1), 77–84.
100. Botkovitz, P.; Lucier, G. M.; Rao, R. P.; Bartlett, N. *Acta Chim. Slov.* **1999,** *46,* 141–154.
101. Riedel, S.; Renz, M.; Kaupp, M. *Quantum-Chemical Investigation of RuF_8 and RuF_7* (Personal Communication); Universität Würzburg: Würzburg, 2006.
102. Chernick, C. L.; Claassen, H. H.; Weinstock, B. Rhodium Hexafluoride. *J. Am. Chem. Soc.* **1961,** *83,* 3165–3166.
103. Bartlett, N.; Rao, P. R. New Fluorides of Palladium: Palladium(II) Hexafluoropalladate (IV) and Related Compounds and Palladium Tetrafluoride. *Proc. Chem. Soc.* **December 1964,** 393–394.
a. Ivanov-Emin, B. N.; Venskovskii, N. U.; Lin'ko, I. V.; Zaitsev, B. E.; Borzova, L. D. Synthesis and Properties of Sodium Hexahydroxopalladate. *Koord. Khim.* **1980,** *6* (6), 928–931.
b. Mazej, Z.; Lutar, K. Metathetical Synthesis of $La_2(PdF_6)_3$ and the Vibrational Spectra of the PdF_6^{2-} Anion. *J. Fluor. Chem.* **2001,** *107* (1), 63–69.
104. Timakov, A. A.; Prusakov, V. N.; Drobyshevskii, Y. V. Formation of Palladium Hexafluoride. *Zh. Neorg. Khim.* **1982,** *27* (12), 3007–3009.
105. Aullon, G.; Alvarez, S. On the Existence of Molecular Palladium(VI) Compounds: Palladium Hexafluoride. *Inorg. Chem.* **2007,** *46* (7), 2700–2703.
106. Falconer, W. E.; DiSalvo, F. J.; Edwards, A. J.; Griffiths, J. E.; Sunder, W. A.; Vasile, M. J. Dioxygenyl Hexafluoropalladate(V) $O^{2+}PdF_6^-$: a Quinquevalent Compound of Palladium. *Inorg. Nucl. Chem. H. H. Hyman Mem.* **1976,** 59–60.
a. Sokolov, V. B.; Drobyshevskii, Y. V.; Prusakov, V. N.; Ryzhkov, A. V.; Khoroshev, S. S. Preparation of Higher Fluorides of Platinum Metals in Nonaqueous Solvents. *Dokl. Akad. Nauk SSSR* **1976,** *229* (3), 641-4 [Chem].
b. Lutar, K.; Borrmann, H.; Žemva, B. *Proceedings of 15th International Symposium on Fluorine Chemistry, Vancouver, Canada.* Vancouver, Canada, 1997.
c. Lucier, G. M.; Shen, C.; Elder, S. H.; Bartlett, N. Facile Routes to NiF_6^{2-}, AgF_4^-, AuF_6^-, and PtF_6^- Salts Using O_2^+ as a Source of O_2F in Anhydrous HF. *Inorg. Chem.* **1998,** *37* (15), 3829–3834.
107. Wilson, A. V.; Nguyen, T.; Brosi, F.; Wang, X.; Andrews, L.; Riedel, S.; Bridgeman, A. J.; Young, N. A. *Inorg. Chem.* **2016,** *55* (3), 1108–1123.
108. Grzybowska, D.; Malinowski, P.; Mazej, Z.; Grochala, W. Probing the Reactivity of the Potent AgF_2 Oxidizer. Part 1, Organic Compounds. *Collect. Czechoslov. Chem. Commun.* **2008,** *73* (12), 1729–1746.
a. Grochala, W.; Mazej, Z. Chemistry of Silver(II): a Cornucopia of Peculiarities. *Philos. Trans. R. Soc. A* **2015,** *373* (2037), 1–12.
109. Leszczynski, P. J.; Grochala, W. Strong Cationic Oxidizers: Thermal Decomposition, Electronic Structure and Magnetism of Their Compounds. *Acta Chim. Slov.* **2013,** *60,* 455–470.
110. Malinowski, P.; Mazej, Z.; Grochala, W. Probing the Reactivity of the Potent AgF_2 Oxidizer. *Z. Anorg. Allg. Chem.* **2008,** *634,* 2608–2616.
111. Žemva, B.; Lutar, K.; Jesih, A.; Casteel, W. J., Jr.; Wilkinson, A. P.; Cox, D. E.; Von Dreele, R. B.; Borrmann, H.; Bartlett, N. Silver Trifluoride: Preparation, Crystal Structure, Some Properties, and Comparison with AuF_3. *J. Am. Chem. Soc.* **1991,** *113* (11), 4192–4198.
112. Bougon, R.; Bui Huy, T.; Lance, M.; Abazli, H. Synthesis and Properties of Silver Trifluoride AgF_3. *Inorg. Chem.* **1984,** *23* (22), 3667–3668.
113. Hoppe, R. Tetrafluoroargentate(III). *Z. Anorg. Allgem. Chem.* **1957,** *292,* 28–33.

114. Hoppe, R.; Homann, R. Fluoro Complexes with Trivalent Silver and Gold. *Z. Anorg. Allg. Chem.* **1970,** *379* (2), 193–198.
115. Lucier, G. M.; Whalen, J. M.; Bartlett, N. High Yield Room Temperature Syntheses of $KAgF_4$ and AgF_3 and the Preparation and Unit Cell of $LiAgF_4$. *J. Fluor. Chem.* **1998,** *89* (1), 101–104.
a. Grochala, W.; Egdell, R. G.; Edwards, P. P.; Mazej, Z.; Zemva, B. On the Covalency of Silver-fluorine Bons in Compounds of Silver(I), Silver(II) and Silver(III). *ChemPhysChem* **2003,** *4,* 997–1001.
b. Kurzydlowski, D.; Mazej, Z.; Jaglicic, Z.; Filinchuk, Y.; Grochala, W. Structural Transition and Unusually Strong Antiferromagnetic Superexchange Coupling in Perovskite $KAgF_3$. *Chem. Commun.* **2013,** *49* (56), 6262–6264.
116. Popov, A. I.; Kiselev, Y. M.; Sukhoverkhov, V. F.; Spitsyn, V. I. Some Properties of the Highest Fluorides of Silver and Alkali Metals. *Dokl. Akad. Nauk SSSR* **1987,** *296* (3), 615–618.
a. Popov, A. I.; Kiselev, Y. M. Preparation and Study of Properties of Higher Silver Alkali Metal Fluorides. *Zh. Neorg. Khim.* **1988,** *33* (4), 965–970.
b. Sorbe, P.; Grannec, J.; Portier, J.; Hagenmüller, P. A New Silver Compound: Dicesium Hexafluoroargentate. *C. R. Hebd. Seances Acad. Sci. Ser. C* **1977,** *284* (5), 231–233.
c. Lutar, K.; Jesih, A.; Leban, I.; Žemva, B.; Bartlett, N. Crystal structures of Pentafluoroxenon(1+) Tetrafluorometallate(1−) (Metal = Silver, Gold) and Their Relevance to the Basicity and Oxidizability of Tetrafluorometallate(1−). *Inorg. Chem.* **1989,** *28* (18), 3467–3471.
d. Bartlett, N. *Personal Communication. Synthesis of $PtOF_3$ Was $PtOF_2$ Ed,* 2008.
e. Sorbe, P.; Grannec, J.; Portier, J.; Hagenmuller, P. Two New Silver Fluoride Compounds: Cesium Silver Fluoride (Cs_2AgF_6) and Cesium Gallium Silver Fluoride ($Cs_2Ga_{0.5}Ag_{0.5}F_6$). *J. Fluor. Chem.* **1978,** *11* (3–4), 243–250.
117. Holleman, A. F.; Wiberg, E.; 71–101th ed. *Lehrbuch der Anorganischen Chemie*, Vol. 101; Walter de Gruyter: Berlin, 1995, 1209 pp.
a. Staritzky, E.; Asprey, L. B. Lanthanum Trifluoride, LaF_3, Neodymium Trifluoride, NdF_3. *Anal. Chem.* **1957,** *29,* 856–857.
b. Mazej, Z.; Lutar, K.; Zemva, B. Novel Syntheses of Some Binary Fluorides : the Role of Anhydrous Hydrogen Fluoride. *Acta Chim. Slov.* **1999,** *46* (2), 229–238.
118. Batsanova, L. P.; Grigor'eva, G. N.; Batsanov, S. S. The Infrared Spectra of Fluorides of the Rare Earth Metals. *J. Struct. Chem.* **1963,** *4* (1), 33–37.
a. Batsanova, L. R.; Grigor'eva, G. N.; Batsanov, S. S. The Infrared Spectra of Fluorides of the Rare Earth Metals. *Zh. Strukt. Khim.* **1963,** *4,* 37–42.
119. v. Hevesy, G.; Dullenkopf, W. The Tetrafluorides of Zirconium and Hafnium. *Z. Anorg. Allg. Chem.* **1934,** *221,* 161–166.
a. Korenev, Y. M.; Sorokin, I. D.; Chirina, N. A.; Novoselova, A. V. Vapor Pressure of Hafnium Tetrafluoride. *Zh. Neorg. Khim.* **1972,** *17* (5), 1195–1198.
120. Shchepochkin, B. V.; Sazhin, N. P.; Yagodin, G. A. Behavior of Potassium Fluorohafnates when Heated. Trudy Instituta – Moskovskii Khimiko-tekhnologicheskii Institut Imeni D. I. *Mendeleeva* **1963**, (44), 35–36.
a. Sazhin, N. P.; Shchepochkin, B. V.; Yagodin, G. A. Interaction between Hafnium Tetrafluoride and Ammonium Fluoride in Aqueous Solution. *Izv. Akad. Nauk SSSR, Ser. Khim.* **1965**, (7), 1127–1130.
b. Sheiko, I.; Mal'tsev, V. T.; Bukhalova, G. A. Binary System NaF-hff4. *Dopovidi Akad. Nauk. Ukrains'koi RSR* **1966**, (5), 628–631.

c. Sheiko, I. N.; Bukhalova, G. A.; Mal'tsev, V. T. Ternary System NaF-kf-hff4. *Dopovidi Akad. Nauk. Ukrains'koi RSR* **1966**, (7), 917–919.
d. Nikitin, M. I.; Sorokin, I. D.; Skokan, E. V.; Sidorov, L. N. Negative Ions in the Saturated Vapor of Potassium Fluoride-hafnium Fluoride and Potassium Fluoride-beryllium Fluoride Systems. *Zh. Fiz. Khim.* **1980,** *54* (5), 1337–1338.
121. Ruff, O.; Zedner, J.; Schiller, E.; Heinzelmann, A. New Fluorides. *Ber. Dtsch. Chem. Ges.* **1910,** *42,* 492–497.
a. Ruff, O.; Schiller, E. Tantalum and Columbian Pentafluorides, and the Preparation of Pure Tantalic and Columbic Acids. *Z. Anorg. Chem.* **1912,** *72,* 329–358.
122. Osin, S. B.; Davlyatshin, D. I.; Ogden, J. S. The Infrared Spectra of Products of Tantalum and Ruthenium Reactions with Fluorine Isolated in Solid Argon. *Zh. Fiz. Khim.* **2001,** *75* (8), 1440–1447.
a. Spoliti, M.; Collini, D.; Nunziante Cesaro, S. Infrared Matrix Study on the Molecular Nature of Tantalum Pentafluoride in the Saturated Vapors. *Gazz. Chim. Ital.* **1973,** *103* (10–12), 1315–1316.
123. Gibson, J. K. Synthesis and Mass Spectrometry of Halide and Oxide Halide (Fluoride and Chloride) Vapor Species of Technetium, Tantalum and Tungsten. *J. Fluor. Chem.* **1991,** *55* (3), 299–311.
a. Zmbov, K. F.; Margrave, J. L. Mass Spectrometric Studies at High Temperatures. XXII. Stabilities of Tantalum Pentafluoride and Tantalum Oxytrifluoride. *J. Phys. Chem.* **1968,** *72* (4), 1099–1101.
124. Cady, G. H.; Aubert, J. Systems Xenon Hexafluoride-tantalum Pentafluoride, Xenon Hexafluoride-manganese Fluoride, Xenon Hexafluoride-uranium Hexafluoride. *Inorg. Chem.* **1970,** *9* (11), 2600–2602.
125. Quinones, G. S.; Haegele, G.; Seppelt, K. MoF_6 and WF_6: Nonrigid Molecules? *Chem. Eur. J.* **2004,** *10* (19), 4755–4762.
126. Seppelt, K. Nonoctahedral Structures. *Acc. Chem. Res.* **2003,** *36* (2), 147–153.
127. Priest, H. F.; Schumb, W. C. Tungsten Fluorides and Related Compounds. *J. Am. Chem. Soc.* **1948,** *70,* 3378–3379.
a. Ruff, O.; Eisner, F.; Heller, W. On the Preparation and Properties of Fluorides of Hezavalent Tungsten. *Z. Anorg. Chem.* **1907,** *52,* 256–269.
b. Ruff, O.; Ascher, E. Some Physical Constants of SiF_4, WF_6 and MoF_6. *Z. Anorg. Allg. Chem.* **1931,** *196,* 413–420.
128. Kaupp, M. "Non-VSEPR" Structures and Bonding in d^0 Systems. *Angew. Chem. Int. Ed.* **2001,** *40* (19), 3534–3565.
129. Malm, J. G.; Selig, H. Vapor Pressures and Other Properties of ReF_6 and ReF_7. *J. Inorg. Nucl. Chem.* **1961,** *20,* 189–197.
130. Shorafa, H.; Seppelt, K. Osmium(VII) Fluorine Compounds. *Inorg. Chem.* **2006,** *45* (19), 7929–7934.
131. Riedel, S.; Kaupp, M. Where Is the Limit of Highly Fluorinated High-Oxidation-State Osmium Species? *Inorg. Chem.* **2006,** *45,* 10497–10502.
132. Gong, Y.; Zhou, M.; Kaupp, M.; Riedel, S. Formation and Characterization of the Iridium Tetraoxide Molecule with Iridium in the Oxidation State +VIII. *Angew. Chem. Int. Ed.* **2009,** *48* (42), 7879–7883.
133. Wang, G.; Zhou, M.; Goettel, J. T.; Schrobilgen, G. J.; Su, J.; Li, J.; Schloeder, T. *Nature* **2014,** *514,* 475–477.
134. Pyykkö, P.; Xu, W. -H. On the Extreme Oxidation States of Iridium. *Chem. Eur. J.* **2015,** *21* (26), 9468.
135. Riedel, S.; Kaupp, M. Revising the Highest Oxidation States of the 5d Elements: the Case of Iridium(+VII). *Angew. Chem. Int. Ed.* **2006,** *45* (22), 3708–3711.

136. Leary, K.; Zalkin, A.; Bartlett, N. Crystal structure of Undecafluorodixenon(1+) Hexafluoroaurate. *J. Chem. Soc. Chem. Commun.* **1973**, (4), 131–132.
a. Bartlett, N.; Leary, K. *Rev. Chim. Min.* **1976,** *13,* 82.
137. Mazej, Z.; Goreshnik, E. A.; Tavcar, G. X-ray Single Crystal Structures of $Hg(AuF_6)_2$ and $AgFAuF_6$. *J. Fluor. Chem.* **2011,** *132* (10), 686–689.
a. Mazej, Z.; Goreshnik, E. A. X-ray Single Crystal Structures of $Cd(AuF_6)_2$, Mg(HF) AuF_4AuF_6 and $KAuF_6$ and Vibrational Spectra of $Cd(AuF_6)_2$ and $KAuF_6$. *Solid State Sci.* **2006,** *8,* 671–677.
b. Mazej, Z. Recent Achievements in the Synthesis and Characterization of Metal Hexafluorantimonates and Hexafluoroaurates. *J. Fluor. Chem.* **2004,** *125* (11), 1723–1733.
138. Brunvoll, J.; Ischenko, A. A.; Ivanov, A. A.; Romanov, G. V.; Sokolov, V. B.; Spiridonov, V. P.; Strand, T. G. Composition and Molecular Structure of Gaseous Gold Pentafluoride by Electron Diffraction. *Acta Chem. Scand. A* **1982,** *A36* (9), 705–709.
139. Himmel, D.; Riedel, S. After 20 Years, Theoretical Evidence that AuF_7 Is Actually $AuF_5 \cdot F_2$. *Inorg. Chem.* **2007,** *46* (13), 5338–5342.
140. Timakov, A. A.; Prusakov, V. N.; Drobyshevskii, Y. V. Gold Heptafluoride. *Dokl. Akad. Nauk SSSR* **1986,** *291* (1), 125–128.
141. Jørgensen, C. K. HgF_4. *J. Chim. Phys.* **1979,** *76,* 630.
a. Jørgensen, C. K. Conditions for Stability of Oxidation States Derived from Photoelectron Spectra and Inductive Quantum Chemistry. *Z. Anorg. Allg. Chem.* **1986,** *540–541,* 91–105.
142. Kaupp, M.; von Schnering, H. G. Molecular Mercury(IV) Fluoride (HgF_4): an Ab Initio Study. *Angew. Chem. Int. Ed.* **1993,** *32* (6), 861–863.
143. Wang, X.; Andrews, L.; Riedel, S.; Kaupp, M. Mercury Is a Transition Metal: The First Experimental Evidence for HgF_4. *Angew. Chem. Int. Ed.* **2007,** *46,* 8371–8375.

Photochemical Syntheses of Fluorides in Liquid Anhydrous Hydrogen Fluoride

20

Z. Mazej
Jožef Stefan Institute, Ljubljana, Slovenia

Chapter Outline

Modern Synthesis Processes and Reactivity of Fluorinated Compounds. http://dx.doi.org/10.1016/B978-0-12-803740-9.00020-2

1. Introduction

For the preparation of fluorides with transition or main group elements in the highest oxidation states,[1] quite drastic conditions such as high temperature and high pressure of fluorine are usually required. At ambient or room temperature (RT) they can be synthesized in liquid anhydrous hydrogen fluoride (aHF) by the use of strong oxidizers such as KrF_2[2] or by rarely used approaches involving PtF_6[3]; in situ−generated solvated O_2F[4]; aHF solutions of cationic Ag(II), Ag(III), or Ni(IV)[5,6,7]; etc. ultraviolet (UV)-irradiated F_2 has been widely used in the past for the preparation of various fluorine compounds in the gas phase.[8,9] In 1998, a photochemical approach using UV-irradiated F_2 in liquid aHF at ambient temperature was reported for the first time.[10] The reader is referred to an extensive reviews about the various methods of syntheses of fluorides reported by Grannec and Lozano[11] and Lutar et al.[9]

In this chapter, I present in more detail the photochemical approach of syntheses of various transition metal fluorocompounds in the highest oxidation states at ambient temperature. Additionally, some examples of the syntheses of main group fluorides are given. The primary intention of this contribution is to summarize results of photochemical preparations of various fluorides in liquid aHF at ambient temperature using elemental fluorine as the fluorinating agent. For comparison, some reactions showing how far oxidation with elemental fluorine at ambient temperature and using neutral or basic aHF as a solvent will go without using UV source are also given. The expression neutral aHF means that there are no fluoride ion donors, such as AF (A = Li, Cs), present in the starting reaction mixture. On the contrary, the term basic aHF refers to reactions in which such sources take part in chemical reactions. The highest oxidation states are more likely to be achieved in anionic parts of ternary fluorides rather than in binary fluorides.[12] On the other side, since the electronegativity of a given oxidation state must be higher in a cation than in neutral or anionic relatives, it follows that low oxidation states are more likely to be achieved if oxidations with F_2 are carried out in aHF solutions made acidic[13] with fluoride ion acceptors as Lewis fluoroacids (e.g., AsF_5, SbF_5).

The requirements of photochemical syntheses with elemental fluorine in aHF and the reaction mechanism are described in Sections 2 and 3. In Section 4, examples of F_2 fluorination in neutral and basic aHF with or without the presence of UV light, which I managed to collect from the available literature, are listed.

2. Requirements of Photochemical Syntheses With Elemental F_2 in Anhydrous HF

2.1 Caution

aHF, elemental fluorine, and some fluorides are highly reactive, toxic, and corrosive. Therefore working with them is demanding and potentially dangerous. They must be handled using appropriate apparatus and protective gear, and experiments should be performed in well-ventilated hoods. The exposure to UV sources should also be avoided because of possible damage to the skin and specially the eyes. One should never look directly at an unshielded operating lamp unless protecting glasses are worn. UV lamps can generate unsafe amounts of ozone. Before starting for the first time any photochemical experiment, it is recommended to consult a member of an experienced fluorine group.

2.2 Reaction Vessels and Ultraviolet Lamps

Since aHF etches glass and some reactions require the use of higher pressures of elemental fluorine (up to 6 bars), translucent fluorocarbon plastics are the only choice of materials for reaction vessels. Two such most common materials are FEP (hexafluoropropylene-tetrafluoroethylene copolymer) and PFA (tetrafluoroethylene-perfluoroether copolymer). Although, their chemical compositions are slightly different, their mechanical and chemical properties, for example, chemical inertness against aHF and remarkable kinetic stability to elemental fluorine and other strong oxidizers, are practically the same. Both materials are transparent to visible and near-UV light. The visible transparency is important because it is possible to visually follow what is happening during the reaction and the UV transparency because it provides effective photodissociation of elemental fluorine inside the reaction vessel by a UV source that is outside it. Reaction vessels are equipped with valves. Their bodies and needles are made from polytetrafluoroethylene (PTFE), which are fitted with metal collars with appropriately located metal treads. In Fig. 20.1 there is a typical reaction vessel equipped with valve (modified version of that shown in Ref. 14 used in our laboratory). Before using, they should be checked for the pressure that they could sustain. In some of our reactions, pressures of elemental fluorine and aHF inside reaction vessels at RT reach up to 6 bars.

Photochemical reactions can be performed using sunlight, although this is not a very common approach because of the variability of this source (intensity changes with time, its spectrum varies because of the impurities in atmosphere, etc.).[15] By far the most useful and widely used UV light sources are the mercury arc lamps made of two metal electrodes encased in glass or silica bulb and filled with mercury.[15,16] There are three basis types, i.e., low-, medium-, and high-pressure Hg arc

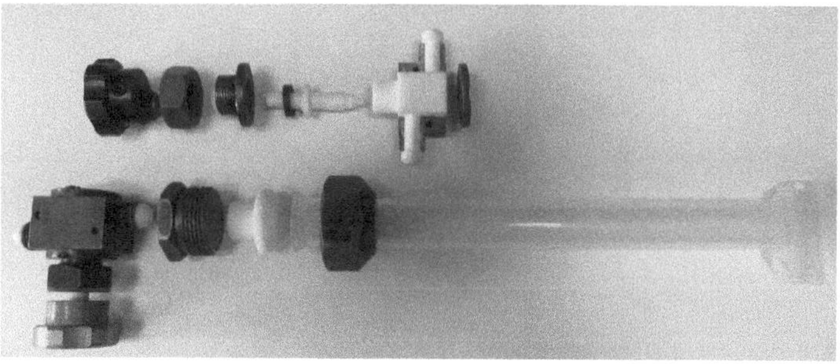

Figure 20.1 Disassembled FEP/PFA reaction vessel equipped with polytetrafluoroethylene valve used for reactions in liquid anhydrous HF.

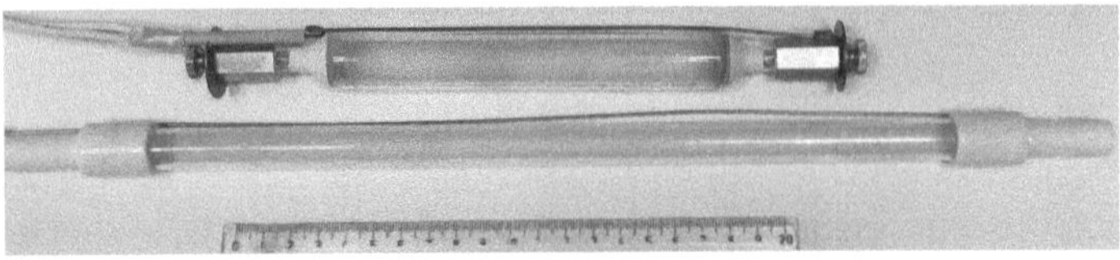

Figure 20.2 Two different models of medium-pressure Hg arc lamps (450 W, Ace Glass, USA).

lamps (Fig. 20.2). The low-pressure arc lamp gives practically a single line at 253.7 nm. The most useful ones are medium- and high-pressure Hg arc lamps, which emit many lines in the UV and visible regions (Fig. 20.3).

All Hg arc lamps have to be used with a proper power supply, and they should not be connected directly to the electricity grid. During their operation, Hg arc lamp

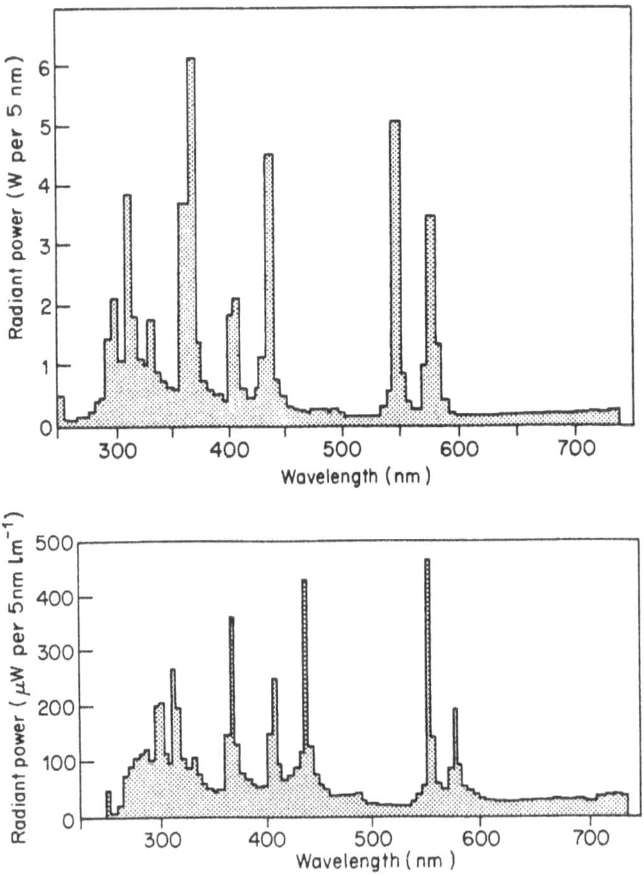

Figure 20.3 Emission spectra of medium-pressure (top) and high-pressure (bottom) Hg arc lamps.
Suppan, P. 1994. Chemistry and Light, The Royal Society of Chemistry, Cambridge - Reproduced by permission of The Royal Society of Chemistry; http://dx.doi.org/10.1039/9781847550439-00216.

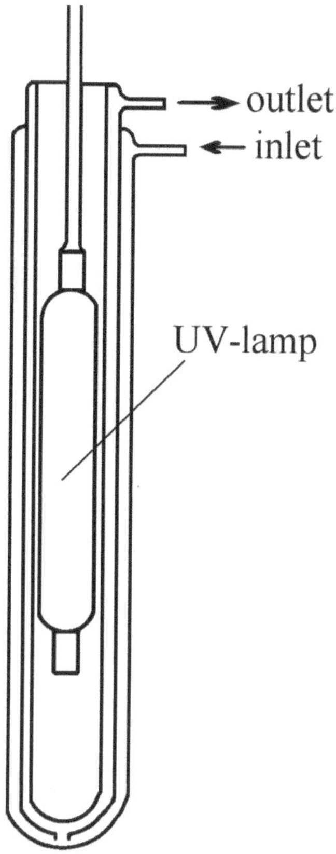

Figure 20.4 Tap-water-cooled immersion well.

surfaces get extremely hot and therefore they should be cooled down. The satisfactory cooling system for the lamp housing is shown in Fig. 20.4. The water-cooled immersion well is made of glass and cooled with tap water.

2.3 How to Perform Synthesis

In a typical reaction, solid starting materials are loaded into a reaction vessel in a dry box. aHF (4–8 mL) is condensed onto the solid at 77 K. With its long liquid range (melting point, −83.55°C; boiling point, 19.51°C), wide electrochemical window (useable potential range of about 4.5 V between anodic and cathodic generation of F_2 and H_2), excellent ionizing solvent (dielectric constant 83.6 at 0°C and 175 at −73°C), the widest range of acidity–basicity of any of superacids, and remarkably robust nature both oxidatively and reductively, the use of superacid aHF seems to be the best choice.[17,18] In referring to aHF we must have in mind the convention used by most chemists, certainly by fluorine chemists. Aqueous solutions of hydrogen fluoride are referred to as "hydrofluoric acid," whereas the term "anhydrous hydrogen fluoride (aHF)" refers to the pure compound.

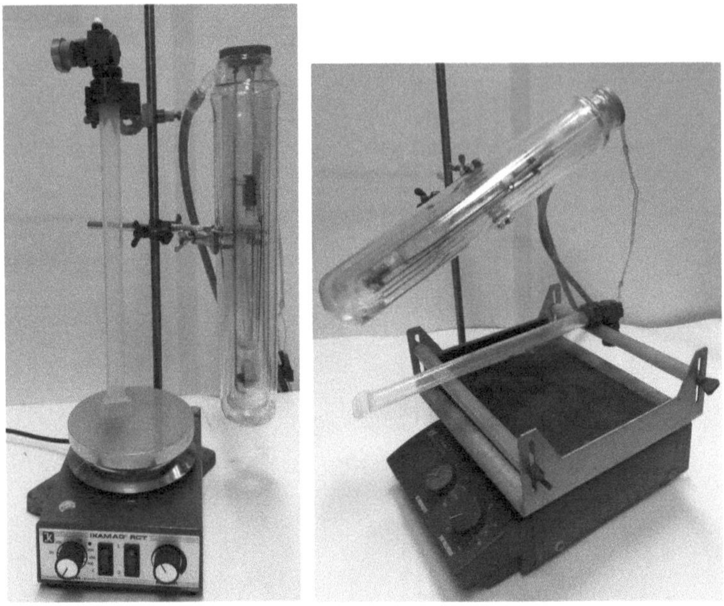

Figure 20.5 Experimental setup for photochemical syntheses using magnetic stirrer (left) or mechanical orbital shaker (right).

Concerning addition of fluorine, there are two possibilities. In the first, the reaction mixture is brought to ambient temperature and fluorine is slowly added to a preferred pressure. In the second method, fluorine is condensed onto frozen aHF/reactants mixture at 77 K from the known volume with known pressure. In both cases, it is better to start with lower pressure of fluorine and later increase it by step-by-step additions.

After the reaction vessel is filled with all reactants, it is placed near tap-water-cooled UV source. The lamp assembly should be operated in a hood or well-ventilated area, which is additionally protected against exposure to UV light. The reaction mixture should be well mixed by a magnetic stirrer (when PTFE-coated stirring bar is added inside the reaction vessel) or by mechanical orbital shaker on which the reaction vessel is mounted in horizontal position (Fig. 20.5). In the cases of prolonged reaction times (more than few days), the second method is more preferable, specially if the products are only sparingly soluble or insoluble in aHF. When PTFE-coated stirring bars were used in such cases, their abrasion was observed, i.e., their shape changed to egglike shape. Additionally, when the products were hydrolyzed in distilled water for chemical analysis, small PTFE particles were observed floating in the solutions.

3. Reaction Mechanism

The first thing that has to be stressed is that in the case of UV-photodissociated F_2 reactions, UV light is not a catalyst.[15] In photochemical reactions, UV light (and light generally) acts as a reactant because in those reactions light is absorbed and its energy is used to split F_2 molecule. It does not come out again, and it cannot be used after the reaction. The most important industrial process (at least in terms of quantity of produced chemicals) is photochemical chlorination of polymers.[15] By utilizing the UV energy by various approaches, the Cl_2 molecules are homolytically dissociated into Cl• atoms, which are in fact free radicals. They react with hydrocarbons through substitution and insertion processes. Photochemical reactions between UV-irradiated F_2 molecules and various inorganic starting materials in liquid aHF as a solvent have a similar beginning, i.e., photodisscosciation of diatomic molecules, but then the similarity ends. They are not chain reactions in which new F• radicals are formed as in the case of reactions of UV-irradiated Cl_2 with hydrocarbons. The explanation of light-promoted F_2 reactions was given in the first article about the use of photochemical reactions with elementary F_2 in liquid aHF.[10] The light absorption maximum of F_2 molecule is close to 300 nm with a broad tail into the visible part. This absorption is associated with the promotion of the electron from bonding to nonbonding molecular orbital. The consequence is dissociation of F_2 molecule to F• atoms (i.e., radicals). The important thing is the length of their effective lifetime. If F• radicals would recombine with each other very quickly, then there would be no time to react with reactants in aHF, and therefore photochemical reactions would make no sense. It was proposed that in the strongly basic aHF, the effective lifetime of F• atom is prolonged by the formation of species such as $F_2^{\bullet -}$ or its hydrogen-bonded solvates such as $[F_2^{\bullet}HF]^-$ or $[F_2^{\bullet}2HF]^-$.[10] Because of charge repulsion is less likely that such species quickly regenerate F^2 as in the case of uncharged $F^{\bullet}$. On the basis of comparison of electron affinities of $F_{(g)}^{\bullet}$ (3.399 eV) and $F_{2(g)}$ (3.10 eV), the $F_2^{\bullet -}$ must be less bound than the $F_{2(g)}$, i.e., $F_2^{\bullet -}$ or its HF solvates are better fluorine atom sources than F_2 itself. Additionally, charged $F_2^{\bullet -}$, $[F_2^{\bullet}HF]^-$, and $[F_2^{\bullet}2HF]^-$ species are also nucleophilic. This mechanism has been given for basic aHF in which well-soluble fluoride-ion donors (e.g., alkali metal fluorides) are present in the solutions increasing the concentrations of solvated F^- species. The extent of self dissociation of pure aHF is very small ($K_{ap} \sim 10^{-11}$).[17] When binary fluorides are synthesized in neutral aHF, there are no additional F^- donors in the form of alkali metal fluorides present in the reaction mixture. The concentrations of $F_2^{\bullet -}$, $[F_2^{\bullet}HF]^-$, and $[F_2^{\bullet}2HF]^-$ species must be much lower. However, the reactions proceed anyway, although in general, much longer times are requested than for the syntheses of ternary fluorides. The low or nonsolubility of starting binary fluorides and final products is also slowing down the reactions. The reactions are of course much faster when products are removed from the surface of the reactants by dissolving in aHF.

4. Examples of F_2 Fluorinations in Liquid Anhydrous HF at Ambient Temperature

The goal of this chapter is to summarize the photochemical syntheses of fluorides since the first published report in 1998.[10] For some elements how far oxidation with elemental fluorine at ambient temperature and in aHF as a solvent goes without using

UV source is also summarized. Many examples from our laboratory, which have not been published before, are reported. The given examples are limited mostly to reactions with elemental F_2, proceeding in aHF at ambient temperature (RT) with or without the use of UV light. Therefore, the syntheses of various binary and ternary fluorides using other strong oxidizers as KrF_2, KrF^+, and PtF_6 are not treated in this chapter. Similarly, with a few exceptions, the reactions that proceed with HF, F_2, or HF/F_2 mixtures at elevated temperatures are not included. The photochemical syntheses that take place in the presence of UV source at low or ambient temperatures without the presence of liquid aHF are also not part of this contribution (e.g., synthesis of XeF_2, KrF_2, O_2F_2, SF_6, ClF_5, NF_4^+ salts).

4.1 Fluorides of the Transition Metal Elements

4.1.1 Titanium, Zirconium, and Hafnium

The highest oxidation state +4 in MF_4 (M = Ti, Zr, Hf) tetrafluorides can be quite easily achieved, and their syntheses are well described in many general books of inorganic chemistry. Herein, I would like just to mention that the reaction between $TiCl_3$ and aHF at RT produces TiF_3, which can be further fluorinated with elemental F_2 in aHF at RT yielding pure TiF_4.[19]

In the available literature, I have not been able to find any data about fluorination of those metals or their compounds in lower oxidation states with F_2 in aHF at RT. There is only a report that Ti, Zr, and Hf react with gaseous aHF at 200–225°C yielding TiF_3, ZrF_4, and HfF_4, respectively.[20]

4.1.2 Vanadium, Niobium, and Tantalum

With V, Nb, and Ta the situation is very similar as in the case of Ti, Zr, and Hf with the main difference being that the highest oxidation state for V, Nb, and Ta is +5. The reports about preparation of V, Nb, and Ta compounds in 5+ oxidation state at RT with F_2 and aHF are scarce.

My attempt to prepare VF_5 at RT by reaction between VF_3 and F_2 in aHF resulted in impure VF_5 contaminated with VOF_3. Niobium powder reacts with gaseous aHF at 250°C and tantalum below 200°C.[21,20] For Nb, the volatile product is NbF_5 and the residue nonattacked metal, whereas for Ta, the product is nonattacked metal, TaF_5, and probably TaF_3.

There are no data regarding what happens if fluorine is added onto metal/liquid aHF mixtures, but the preparation of $LiMF_6$ (M = Nb, Ta) from the elements and LiF in aHF by F_2[22] shows that Nb and Ta can be easily oxidized at RT to their highest oxidation state +5 using F_2 in aHF made basic with alkali fluorides with no need for UV light.

4.1.3 Chromium, Molybdenum, and Tungsten

Attempts to fluorinate Cr powder (100 mesh) with ClF_3 in the presence of F_2 (2 bar) in aHF solution at RT for 16 h were unsuccessful.[23] This inertness should be of kinetic origin, wherein in liquid aHF, insoluble CrF_3 formed on the surface of metal can block further reaction. My attempt to prepare CrF_4 by F_2 fluorination of CrF_3 in aHF at RT was also unsuccessful.

The $ACrF_6$ (A = Li–Cs) compounds containing Cr(V) can be prepared by photochemical reactions with elemental F_2 on AF/CrF_3 mixtures in liquid aHF at RT. In the presence of UV source, all Cr^{3+} is oxidized by fluorine to Cr^{5+} yielding deep red clear solutions from which pure $ACrF_6$ compounds are isolated.[24] The K, Rb, and Cs salts could be recovered at RT by pumping away the excess fluorine and aHF solvent. When $NaCrF_6$ had been isolated at the same conditions, release of CrF_5/F_2 was observed, resulting in $NaCrF_6/Na_2CrF_6$. This could be avoided with the isolation at temperatures below 243 K. Dry $NaCrF_6$ is stable at RT and could be stored in a dry box. $LiCrF_6$ could be isolated in the same manner, but in contrast to $NaCrF_6$ it immediately starts to decompose at RT. All $[CrF_6]^-$ salts are deep red, moisture-sensitive crystalline solids. Single crystals of $ACrF_6$ (A = Na, K, Rb, Cs) were obtained upon crystallization from corresponding solutions, and their crystal structures were determined. $KCrF_6$, $RbCrF_6$, and $CsCrF_6$ crystallize in well-known $KOsF_6$ type structure, whereas $NaCrF_6$ exhibits polymorphism. The trigonal phase of $NaCrF_6$ is analogous to the well-known $LiSbF_6$ structure, whereas the crystal structure of orthorhombic modification appears to be a previously unknown structure.

As a fluorinating agent, ClF has been placed below elemental F_2 in reactivity. Since reactions of ClF with elemental tungsten and molybdenum proceed smoothly at ambient temperature,[25] it could be assumed that their fluorination with F_2 in liquid aHF at RT will also give WF_6 or MoF_6 in the highest known oxidation state +6.

4.1.4 Manganese and Rhenium

Elemental fluorine oxidizes MnF_2 in liquid aHF as a solvent at RT yielding pure MnF_3.[26] In the presence of UV source, MnF_2 is completely oxidized by photodissociated F_2 in aHF to MnF_4.[27] The MnF_4 obtained was poorly crystallized and, therefore it was not possible perform X-ray powder diffraction analysis. In its Raman spectrum, a vibrational band at 1834 cm^{-1} typical for the O_2^+ cation was observed showing that an O_2^+ salt was present in the final product. However, the O_2^+ cation is a very good Raman scatterer and its mass amount could be less than 1%. The presence of oxygen can be attributed to impure MnF_2 or to slow diffusion of O_2 or H_2O through the wall of FEP/PFA reaction vessel.

Elemental fluorine is also capable of oxidizing Mn^{II} to Mn^{IV} in aHF made basic with AF (A = alkali metal) at RT in aHF in the absence of UV source. The effectiveness of the oxidation for the same period of time depends on the basicity of aHF (starting amount of alkali metal fluoride). However, in the absence of a UV source, these reactions are incomplete and give mixtures of products, e.g., $AF \cdot nHF$, $AMnF_4$, A_2MnF_6, and MnF_3. Pure A_2MnF_6 (A = Li, K, Rb, Cs) phases can be obtained only by photochemical reactions.[27] An attempt to prepare $CsMn^VF_6$ by oxidation of MnF_2 with UV-irradiated F_2 in the presence of equimolar amount of CsF in aHF failed. The final product of this reaction was a mixture of MnF_4 and Cs_2MnF_6. Similarly, my attempt to prepare $CsMnF_6$ using KrF_2 instead of photodissociated F_2 in liquid aHF at RT was also unsuccessful.

I tried the reaction between rhenium metal (powder, 210 mg) and elemental F_2 (6 bar) in liquid aHF (6 mL) at RT. Since there was no visible reaction, the reaction vessel was exposed to UV light. After a few days, Re metal dissolved and transparent solution was obtained. The remaining F_2 was pumped away at 77 K. Raman spectrum of the solution and IR spectrum of the gaseous product corresponded to literature data of ReF_6. The solution was cooled down to −70°C, and volatiles were pumped away. First there was yellow solid visible in liquid aHF, but after 8 h of pumping, nothing was left in the reaction vessel. According to this result, only yellow ReF_6 has been formed with no indication of known ReF_7. The reactions with addition of CsF (molar ratio CsF:Re = 1:1); with or without the presence of UV light proceeded in the same manner, yielding only a mixture of ReF_6 and CsF·nHF.

4.1.5 Iron, Ruthenium, and Osmium

The fluorination of FeF_2 by elemental F_2 in liquid aHF at RT readily proceeds to FeF_3. All my attempts to prepare $Cs_2Fe^{IV}F_6$ by reaction between 2CsF, FeF_3, and UV-irradiated F_2 (8 bar in 35 mL FEP reaction vessel) in liquid aHF (8 mL) at RT or by reaction between 2CsF (0.5 mmol), FeF_3, (0.25 mmol) and KrF_2 (0.5 g) in liquid aHF (4 mL) at RT failed. Similarly, the attempt to synthesize $CsFe^{V}F_6$ by reaction of CsF (0.45 mmol) with FeF_3 (0.45 mmol) and KrF_2 (0.3 g) in aHF was also unsuccessful. In all cases, FeF_3 and CsF·nHF were recovered.

In 1996, two articles from two different groups working on fluorine were published, both reporting about the room temperature oxidative fluorination of platinum metals and gold by elemental fluorine in aHF.[28,29] After reaction between ammonium fluoride (NH_4F), Ru metal, and fluorine gas in aHF at RT, a cream-colored insoluble precipitate and a small quantity of unreacted metal powder were obtained. On drying, the precipitate turned sandy brown. According to IR data, the obtained product was Ru(V) compound $(NH_4)RuF_6$, which was readily reduced by moisture to $(NH_4)_2(RuF_6)$.[29] The F_2 with aHF made basic with KF dissolves Ru metal to give Ru(V) compound $KRuF_6$; however, when LiF is substituted for KF, the oxidation is slower and the ultimate product is Ru(IV) compound Li_2RuF_6. This difference is attributed to the low solubility of Li_2RuF_6 and the high solubility of K_2RuF_6 in the aHF solvent.[28]

Reaction between NH_4F, Os metal, and elemental F_2 in aHF at RT produced a yellow-green solution and some unreacted metal powder from which a white powder contaminated with the blue-gray metal was recovered after removal of the solvent. The IR spectrum could be assignable to Os(V) compound $(NH_4)OsF_6$.[29] When LiF was used instead of NH_4F, the $LiOsF_6$ with Os(V) was obtained.[28] It is unusual and unexplained why in the reaction done in the same manner, but using KF instead of LiF, there was no detectable formation of either K_2OsF_6 or $KOsF_6$.[28]

For Ru and Os metals, there are no reports regarding what happens when they, or their binary or ternary fluorides in oxidation state +5 (or lower), are treated with UV-irradiated elemental fluorine in neutral or basic liquid aHF. Otherwise, the highest oxidation state achieved in corresponding fluorides is +6 (RuF_6 and OsF_6).[1] Neither

Ru nor Os hexafluoride reacts with alkali metal fluorides, although $NO[OsF_7]$ was reported to be formed along with $NO[OsF_6]$ from NOF and OsF_6.[30]

4.1.6 Cobalt, Rhodium, and Iridium

Reaction between pink CoF_2 with excess of elemental fluorine (approximately 1.5–4) in liquid aHF at RT proceeds smoothly, and the course of reaction could be followed by the change in colors of the solids (starting compound and product are virtually insoluble in aHF), and by the consumption of fluorine followed by the drop in its pressure. The obtained product is pure light brown CoF_3.[26] No attempts have been made to prepare ternary Co(IV) compounds (e.g., Cs_2CoF_6) using photodissociated F_2, since it is known that $[CoF_6]^{2-}$ is unstable in liquid aHF at RT and it decomposes above 10°C with the formation of F_2 and some mixed Co(III)/Co(IV) solid product.[31] The rate of decomposition can be slowed by a factor of four by the addition of excess CsF to the solution.

When Rh metal was treated with elemental F_2 in the presence of NH_4F or LiF at RT in liquid aHF there was no indication of any complex rhodium fluoride having been formed. When the solvent was removed, only unreacted Rh metal and NH_4F or LiF, respectively, were recovered.[28,29] On the contrary, when a better fluoride ion donor such as CsF was used in our laboratory, $CsRhF_6$ was quantitatively obtained.

When I tried the reaction between Rh metal with UV-photolyzed F_2 in aHF at RT, the liquid phase turned yellow or sometimes yellow-brown. The large amount of insoluble dark residue was of different color than the starting Rh. After a few days, the reaction vessel was cooled down to −15°C and volatiles were pumped away for 3 days through the trap cooled in liquid nitrogen, wherein no coloration was observed. Finally, the sample was warmed to RT and pumped further. A small amount of yellow-red frozen solid was observed in the cold trap. Raman spectrum and X-ray powder diffraction analysis of the recovered dark red product (in which larger metallic pieces were still visible) were in agreement with the presence of starting Rh and RhF_5, whereas the Raman spectrum of the yellow-red compound caught in the trap corresponded to a mixture a RhF_5 and O_2RhF_6. The presence of the latter could be explained by the possible formation of RhF_6, which is able to oxidize traces of O_2 forming O_2RhF_6. It is evident that oxidation of Rh metal under the described conditions goes to Rh(V)/Rh(VI). However, direct fluorination of Rh metal with UV-irradiated F_2 in liquid aHF at RT is not a very suitable method for the preparation of pure RhF_5 or even less for the synthesis of RhF_6.

Fluorination of Ir metal by F_2 in the presence of NH_4F in liquid aHF at RT resulted in a gray, highly moisture-sensitive HF-insoluble precipitate.[29] Beside the characteristic vibrational bands for the $[NH_4]^+$ cations, an additional vibrational band at 633 cm^{-1} assignable to $[IrF_6]^-$ was observed in the IR spectrum of the isolated product showing that the latter is an Ir(V) compound NH_4IrF_6. Reactions of LiF or KF with Ir metal oxidized by elemental F_2 in liquid aHF at RT yield sparingly soluble yellow $LiIrF_6$ and $KIrF_6$, which could be crystallized from yellow (Li) and roseate (K) solutions.[28,32]

Similarly, as in the case of Ru and Os, there are no reports about reactions of Rh and Ir metals (or pentafluorides) by UV-irradiated F_2 in aHF to prepare binary or ternary

Rh(VI) and Ir(VI) fluorides. The highest known oxidation state achieved in corresponding binary fluorides is +6 (RhF_6 and IrF_6).[1] Neither of these react with alkali metal fluorides.

4.1.7 *Nickel, Palladium, and Platinum*

The report in 1998, about NiF_2 oxidizing at around 20°C to $[NiF_6]^{2-}$ by sunlight or UV-irradiated F_2 in liquid aHF containing dissolved alkali metal fluorides (LiF, NaF, KF, CsF), was the first one connected with the main topic of this chapter, i.e., photochemical syntheses of fluorides in liquid aHF at RT using elemental F_2.[10] The conversion to $[NiF_6]^{2-}$ was most effective with a saturated solution of KF in aHF and a large excess of alkali metal fluoride over NiF_2. With increasing concentrations of alkali metal solutions, fluorination becomes quicker and more efficient. This way, a dark purple-red Li_2NiF_6 was prepared for the first time. When syntheses were carried in sunlight, the particles of the aHF-insoluble pale yellow-green NiF_2 became black, before any coloring of the aHF solution occurred. This suggests that NiF_2 is first fluorinated to NiF_{2+x} ($x \leq 1$). This is the same phenomenon that I observed when NiF_2/F_2 reaction mixture in aHF was irradiated by UV light. However, the X-ray powder diffraction pattern of the isolated product corresponded to NiF_2 only. My attempt to prepare $CsNiF_5$ from stoichiometric amounts of CsF and NiF_2 by photochemical reaction, as described for Cs_2NiF_6, failed. The obtained solid was a mixture of Cs_2NiF_6 and $CsF \cdot nHF$.

Reaction between Pd powder (black or gray) in liquid aHF with excess of elemental fluorine proceeds smoothly at RT, and the course of reaction is followed by the consumption of fluorine, which is followed by the drop in its pressure. The product obtained is black, mixed oxidation state Pd_2F_6 ($Pd^{II}Pd^{IV}F_6$).[26] Even when PdO (1.6 mmol) and F_2 (13 mmol) in aHF were left to react for a longer time, the reaction was not completed. Only about 60 mol% of starting PdO was converted into Pd_2F_6 after 48 days.[33]

When I treated Pd metal/F_2 reaction mixture in liquid aHF with UV light, black Pd_2F_6, which was formed first, started to change its color to brick-red, whereas the liquid aHF stayed colorless. Raman spectrum of in aHF-insoluble solid showed, beside vibrational bands, which could be assigned to PdF_4, the presence of O_2^+ salt. After fluorine was pumped away, small amount of Lewis acid AsF_5 was condensed the inside reaction vessel. The liquid phase was separated from the insoluble one. The product recovered from decanted solution was colorless O_2AsF_6, and the insoluble residue was brick-red PdF_4.

In aHF made basic with AF (A = Li, K, Cs or $[NH_4]^+$), the fluorination of Pd metal with elemental F_2 in aHF at RT resulted in the formation of yellow-orange solutions from which yellow A_2PdF_6 and orange $(NH_4)_2PdF_6$ compounds were isolated.[28,29] The obtained Li_2PdF_6 is only slightly soluble in aHF and it is isostructural with tetragonal Li_2PtF_6. X-ray powder diffraction analysis of K_2PdF_6 showed the presence of both the known trigonal (K_2GeF_6-type structure) and hexagonal phases (K_2MnF_6-structure type), whereas that of Cs_2PdF_6 corresponded to cubic phase (K_2PtCl_6-type structure).

In the case of the product prepared by the reaction between Pd and excess of CsF in the presence of fluorine in aHF, it was noticed that the isolated solid turned from yellow to brown and finally to black, although the sample was kept all the time in the dry box.[34] In the Raman spectrum, besides vibrational bands of PdF_6^{2-} anion (561, 546, and 235 cm^{-1}), two additional bands at 410 and 510 cm^{-1} appeared. With time, their intensity increased and the intensity of the bands belonging to the PdF_6^{2-} anion decreased. Such change in color was also observed for $(NH_4)_2PdF_6$, wherein the isolated product turned from yellow to brown overnight in the dry box.[29] The most likely cause of the color change of the isolated Cs_2PdF_6 is the presence of CsF·nHF. Reactions between CsF and aHF yield CsF·nHF phases where n depends on the vacuum conditions and the temperature. When there is a mixture of Cs_2PdF_6 and CsF·nHF, the latter slowly releases aHF causing the decomposition of the former, similar to the case of yellow solid $La_2(PdF_6)_3$ (prepared by mixing of aHF solutions of $La(AsF_6)_3$ and Cs_2PdF_6 in molar ratio 2:3) solvolysis when it is exposed to liquid aHF.[34] When Cs_2PdF_6 was prepared from reactants in molar ratio CsF:Pd = 2:1 or by deficit of CsF, there was no sign of decomposition even after weeks. The color of the bright yellow solid did not change, and the Raman spectrum did not show any additional bands except those belonging to the PdF_6^{2-} anion.

The attempted fluorination by elemental F_2 (with UV irradiation) of K_2PdF_6 and Cs_2PdF_6 in aHF at RT to prepare Pd(V) compounds were unsuccessful. The constant product weights and the negligible fluorine consumption indicated that there was no oxidation.[35]

Platinum metal was fluorinated by elemental F_2 in liquid aHF made basic with NH_4F at RT to give lemon-yellow solution from which air-stable yellow crystalline $(NH_4)_2PtF_6$ containing Pt(IV) was isolated on removal of all volatiles in vacuum.[29] The obtained formulation was supported by ^{19}F nuclear magnetic resonance and IR spectroscopy. The reaction with LiF instead of NH_4F gives very soluble light yellow tetragonal form of Li_2PtF_6.[28,29] The X-ray powder diffraction analysis of K_2PtF_6 prepared in the same manner showed only the trigonal form (K_2GeF_6-type structure).

Li_2PtF_6 has low solubility in aHF. When it is partly dissolved at ~20°C in liquid aHF, it yields a faintly yellow solution. After fluorine is added and the reaction mixture is placed (with agitation) in sunlight, the color of solution gradually intensifies, yielding bright yellow solution. The latter is decanted, and after all volatiles are removed, the Pt(V) compound $LiPtF_6$ is obtained ($LiSbF_6$-type structure). The residue is unreacted Li_2PtF_6. Although the remaining valence electrons of $[PtF_6]^{2-}$ constitute a weakly antibonding set of t_{2g}^6 symmetry, which has highly favorable exchange energy, the F atoms (radicals from sunlight irradiated F_2) in aHF solution break this configuration and generate $[PtF_6]^-$.[35] Analogous reaction with $[PdF_6]^{2-}$ does not occur since the tighter binding of 4d electrons relative to 5d electrons causes Pd(V) to be inaccessible, at least by using UV-irradiated F_2.[35]

Only MF_6 (M = W, Mo and Re) seem to be able to increase their coordination numbers by adding F^-. It seems that higher members of transition metal hexafluorides (M = Ru, Rh, Os, Ir, Pt) do not undergo expansion of their coordination sphere[36] forming AMF_7 or A_2MF_8 (A = alkali metal) compounds.

4.1.8 Copper, Silver, and Gold

The triply charged $[CuF_6]^{3-}$ anion with Cu(III) undergoes solvolysis in aHF. CuF_3 is formed from K_3CuF_6 as a red solid at −78°C, and it loses F_2 at −40°C.[37] Since the

$[MF_6]^{2-}$ anions contain one negative charge less than the $[MF_6]^{3-}$ ones, they are less basic and consequently less likely to solvolyze in superacid aHF. However, even at low temperature, the Cu(IV) compound Cs_2CuF_6 reacts with aHF to form dark brown solid.[38] At ambient temperature the decomposition is completed and the evolution of F_2 occur. Because of this, it is not expected that oxidation of Cu(II) by UV-irradiated F_2 in liquid aHF at RT will result in Cu(III) or Cu(IV) compounds.

AgF_2 was reported to be prepared by passing fluorine diluted with nitrogen through a solution of AgF in aHF.[39] Much more conveniently, AgF_2 is synthesized by fluorination of $AgNO_3$ with F_2 in aHF at RT with occasional removal of volatiles and reintroduction of F_2. With this method, used in our laboratory, fine powdered AgF_2 can be prepared. Such AgF_2 is very reactive toward many organic and inorganic compounds.[40,41]

Neither AgF nor AgF_2 gave $KAgF_4$ when treated with F_2 in solution of KF in liquid HF at ~20°C.[28] When yellow Ag(I) compound $AgAuF_4$ with very low solubility in liquid aHF was pressurized with F_2 and left in the sunlight without agitation for two days, it became dark green. After the solution was decanted, away the X-ray powder diffraction pattern of the residue showed the patterns of starting Ag(I) compound $AgAuF_4$ and newly formed Ag(II) compound $AgFAuF_4$.[42] In this experiment it had been anticipated that some $AgFAuF_6$ with Au(V) would form. But, even in the strong sunlight, the F_2 concentration in the described experiment must have vastly exceeded that of F atoms. The Ag^+ cation is effectively attacked by the highly electrophilic F_2 species, which will not oxidize $[AuF_4]^-$, the F• radicals are required for that.

AgF_2 dissolves at ~20°C in aHF containing large excess of KF (~20 times more than the stoichiometric amount), with sunlight or UV-irradiated F_2 yielding bright yellow solution. Nearly quantitative conversion to Ag(III) compound $KAgF_4$ can be achieved with such concentrated KF solutions.[43] KHF_2 impurity can be removed as KPF_6 with PF_5 at 0°C. Preparation of $LiAgF_4$ is less efficient. My attempts to prepare pure $NaAgF_4$ and $CsAgF_4$ by reaction between AF (A = Na, Cs) and AgF_2 (molar ratio AF:AgF_2 = 1:1) resulted in $AAgF_4$ samples contaminated with Ag_3F_8 (i.e., mixed oxidation state fluoride $Ag(AgF_4)_2$).

Reactions between metallic Au and elemental F_2 in aHF at RT provide yellow Au(III) compounds, i.e., $AAuF_4$ (A = Li, K) salts.[28] Even prolongated fluorination failed to generate $[AuF_6]^-$ salts in liquid aHF at RT and fluorine pressures below 1 atm. When solutions of $[AuF_4]^-$ salts are further oxidized by photodissociated F_2 (using sunlight or UV light), the Au(V) compounds $AAuF_6$ (A = Li, K) are formed.[35] A similar reaction, which I tried between CsF, Au, and UV-irradiated F_2 in aHF at RT, gives light yellow $CsAuF_6$, sparingly soluble in aHF. O_2AuF_6 was also synthesized unintentionally in an attempt to prepare AuF_5 from AuF_3 and F_2 contaminated by O_2 in aHF employing photodissociation of F_2 by 20-day exposure to sunlight.[35] Otherwise, canary yellow O_2AuF_6 can be easily produced by treating of AuF_3 with O_2/F_2 in liquid aHF at RT in the presence of a UV source as done in our laboratory.

Beside M^IAuF_6 many $M^{II}(AuF_6)_2$ (M = Ni, Cu, Ag, Zn, Cd, Hg, Mg, Ca, Sr, Ba) compounds can be prepared from corresponding MF_2, AuF_3 and UV-photodissociated F_2 in liquid aHF at RT.[44] The only condition is that metallic cations in MF_2 are resistant to oxidation. In some cases the isolated products were contaminated with traces of

O_2AuF_6 due to the impure MF_2 or slow diffusion of O_2 through the wall of FEP/PFA reaction vessel. Mg, Ca, Sr, Ba, and Cd hexafluoridoaurates(V) are well soluble in aHF yielding yellow solutions. The Ni, Cu, and Zn salts are sparingly soluble, whereas the Hg and Ag salts are almost insoluble in aHF. Except $Ag(AuF_6)_2$, which is green, all other $M(AuF_6)_2$ compounds are yellow (from canary to dark yellow). Raman spectra and X-ray powder diffraction data indicate that Ni, Cu, Zn, and Mg salts are structurally related. Crystal structures of $Cd(AuF_6)_2$ and $Hg(AuF_6)_2$ were determined on single crystals.[45,46] Single crystals of $Mg(HF)AuF_4AuF_6$ were grown during the attempt to grow single crystals of $Mg(AuF_6)_2$.[45] $Mg(HF)AuF_4AuF_6$ is the only example of a mixed oxidation state Au(III)/Au(V) compound. An attempt to grow single crystal of $Ag(AuF_6)_2$ resulted in $AgFAuF_6$ instead.[46]

4.1.9 Zinc, Cadmium, and Mercury

It was reported that Zn and Cd react with gaseous aHF at elevated temperatures (160–220°C) yielding the corresponding difluorides.[20] Since there are many other much more convenient methods for their synthesis (e.g., preparations from carbonates, chlorides), their preparation by fluorination of the corresponding metals in liquid aHF is not the best choice. Additionally, reactions could be hindered by the passivation of metal surfaces by forming difluorides, which are only sparingly soluble in aHF.

4.1.10 Scandium, Yttrium, Lanthanum to Lutetium

Metallic Eu, Yb, and Sm react with aHF and dissolve with the evolution of H_2 and producing colorless (Eu, Yb) or yellow-orange (Sm) solutions containing the corresponding solvated Ln(II) cations and solvated fluoride anions.[47] Lanthanum and thulium metals showed slow reaction with aHF, but no evidence could be found for Ln(II) in solution. The respective LnF_3 and hydrogen were the only products observed. There was also a report about reactions of gaseous aHF with Sc, Y, La, Ce, Pr, Nd, and Sm at elevated temperatures (150–250°C), which yielded the corresponding trifluorides.[20]

When CeF_3 is oxidized at RT by F_2 in aHF, CeF_4 is obtained. The reactions of UV-photolyzed F_2 with III/IV mixed oxidation state Pr_6O_{11} or Tb_4O_7 in aHF at RT were reported to produce PrF_4 or TbF_4, respectively, in quantitative yield.[48] Although the results of elemental chemical analysis of so obtained PrF_4 were in agreement with theoretical values, the X-ray powder diffraction analysis done later indicated that sometimes there could be some PrF_3 still present in the final product. When TbF_3 or PrF_3 was used as the starting material, the final products were always LnF_3/LnF_4 mixtures.

4.2 Fluorides of the Main Group Elements

4.2.1 Dioxygenyl Salts

As already shown in the example of O_2AuF_6, various O_2^+ salts can be conveniently prepared by treating gaseous, liquid, or solid fluorides (e.g., AsF_5, BF_3, SbF_5, AuF_3, SbF_3) by UV-photolyzed O_2/F_2 mixtures in liquid aHF at RT. A typical example of the photochemical synthesis of O_2^+ salt in liquid aHF, is the preparation of O_2BF_4 synthesized by reaction between O_2, BF_3, and UV-irradiated F_2 (~1.2 mmol each) in 3–4 mL of liquid aHF.[49] After 12 h of intensive stirring, the volatiles were pumped away at −20°C in dynamic vacuum. Colorless precipitate was recovered from colorless transparent solution. Its Raman spectrum, recorded at RT [1860(100), 1048(5), 767(1009), 727(3), 354(90) cm^{-1}] was in agreement with the formation of O_2BF_4 compound.[49]

The published synthetic attempt (without liquid aHF) of synthesizing O_3^+ salts resulted in synthesis of O_2^+ salts.[50] Therefore it is not expected that the additional use of liquid aHF would change something concerning the preparation of O_3^+ salts.

4.2.2 Xenon

XeF_2 (Xe^{II}) was oxidized by elemental fluorine under UV light in aHF at RT to XeF_4 (Xe^{IV}). Its purity was confirmed by gas-phase IR spectroscopy and by Raman spectroscopy of the isolated solid.[51] When similar syntheses are done with the addition of fluorides (AuF_3, TiF_4, SbF_5, etc.), the oxidation goes further to Xe(VI) yielding XeF_5^+ or $Xe_2F_{11}^+$ salts depending on the starting XeF_6: fluoride ratios.[52] A typical example of such an approach is the preparation of $3XeF_6 \cdot 4TiF_4$ synthesized by the reaction between XeF_2, TiF_4, and UV-irradiated elemental fluorine in aHF at RT.[53] The advantage of this approach is that instead of XeF_6 the solid XeF_2 is used as a starting material and as precursor for the preparation of XeF_6 in situ. The XeF_2/TiF_4 mixtures could be prepared in the requested molar ratio and loaded into reaction vessels in a dry box, whereas the XeF_6 could be added onto TiF_4 only by condensation in a vacuum system. The XeF_2 was converted to XeF_6 by photochemical reaction with UV-irradiated elemental fluorine in liquid aHF at RT. Of course, using XeF_2 or XeF_6 is a matter of choice. For using XeF_2, elemental fluorine, and repeated UV photolysis instead of XeF_6 one pays a stiff price in terms of long reaction times. The crystal structure determination of $3XeF_6 \cdot 4TiF_4$ showed that it can be formulated as $[XeF_5]_3[Ti_4F_{19}]$ consisting from $[XeF_5]^+$ cations and discrete $[Ti_4F_{19}]^{3-}$ anions.

Later, I prepared many other compounds (e.g., $[XeF_5]_2[MnF_6]$, XeF_5MnF_5, $[XeF_5]_4[Mn_8F_{36}]$, $[XeF_5]_5[Ti_{10}F_{45}]$, $[XeF_5][Ti_3F_{13}]$, and $[Xe_2F_{11}]_2[PbF_6]$) and managed to grow corresponding single crystals. Their determined crystal structures still await to be published.

4.2.3 The Rest of the Main Group Elements

For the preparation of binary and ternary fluorine compounds of main group elements, various well-established routes exist that do not require the use of liquid aHF or UV-irradiated F_2. However, few examples can be found in the literature, and some experiments were done by myself, which I would like to mention in this paragraph. The alkali metals, barium, strontium, calcium, and thallium are soluble in aHF with evolution of H_2 and formation of corresponding binary fluorides.[54] Other metals do not dissolve in aHF, although those with standard electrode reductive potentials below

H_2 develop a protective coating of an aHF-insoluble metal fluoride on the surface, e.g., magnesium. At elevated temperatures (180−500°C), Group III (B, Al, Ga, In, Tl), Group IV (Si, Sn Ge, Pb), and Group V (P, As, Sb, Bi) elements react with gaseous aHF yielding binary fluorides.[20]

It was reported that TlF_3 was formed by passing of fluorine, diluted with N_2, through a solution of thallium(I) fluoride. Even though a large excess of F_2 was used, much of the TlF failed to react.[39] SnF_2 is readily soluble in aHF at 20°C. Such solution was found to react smoothly with F_2 ($p < 1$ bar) to yield in aHF-insoluble mixed oxidation state Sn_3F_8 rather expected SnF_4.[55] However, my experiments showed that with larger amounts of F_2 (i.e., higher pressure of F_2), SnF_2 can be quantitatively converted to the Sn(IV) compound SnF_4.

An interesting case represents the chemistry of bismuth. Fluorination of Bi metal with elemental F_2 in aHF at RT results in gray-white Bi(III) compound BiF_3.[26] When powdered Bi is completely covered by liquid aHF, the reaction proceeds smoothly because the released heat is efficiently removed by aHF. The problem arises when some particles of Bi metal are left on the reaction vessel walls above liquid aHF. When they come in contact with elemental fluorine, the reaction is very vigorous and releases a lot of heat in very short time in a single spot on the inner wall of the reaction vessel. As a consequence, the reaction vessel wall becomes very soft on a very small surface area and it cannot resist anymore to overpressure inside the reaction vessel resulting in an explosion.

As found later by me, further reaction of BiF_3 with UV-irradiated F_2 in aHF at RT resulted only in partial oxidation of Bi(III) to Bi(V) yielding an aHF-insoluble colorless product close to the composition $BiF_3 \cdot BiF_5$. It can be also be prepared by direct reaction between BiF_3 and BiF_5 (molar ratio 1:1) in aHF at RT. When BiF_3: BiF_5 starting molar ratios are 2:1 or 1:2, the BiF_3 or BiF_5, respectively, are detected by Raman spectroscopy and X-ray powder diffraction analysis in the recovered products. The Raman spectrum of photochemically prepared $BiF_3 \cdot BiF_5$ [585(100), 537(10), 496(8), 310(1), and 235(5) cm^{-1}] is significantly different from that of BiF_5 [595(100), 570(10), 255(10), 167(5) cm^{-1}][56] and is similar to that of the product obtained by reaction between BiF_5 and H_2O_2.[57] In some experiments I observed the presence of O_2BiF_6/$O_2Bi_2F_{11}$ (very intense vibrational bands at 1848 and/or 1852 cm^{-1} in Raman spectra) in the isolated $BiF_3 \cdot BiF_5$. This must be due to the contamination by oxygen. Without knowing the crystal structure of $BiF_3 \cdot BiF_5$ it is not possible to say if this compound could be formulated as ionic $[BiF_2][BiF_6]$ or more plausible as $BiF_3 \cdot BiF_5$ adduct. The studies of $SbF_3 \cdot SbF_5$ system showed that such systems can be very complex.[58]

5. Conclusions

The described fluorinations by elemental F_2 in aHF, and the photochemical syntheses of binary, ternary, and quaternary compounds using UV-photodissociated F_2 in liquid aHF, offer a milder approach for the formation of these compounds at ambient temperature. All reactions are carried out in reaction vessels made of fluorinated polymers, which are very resistant and inert materials, and therefore there is no danger of final product contamination by by-products formed between the reagents and reaction vessel material. This possibility should be taken into account when metal reaction vessels and very high temperatures are involved. Other advantages of such reactions in aHF are (1) courses of reactions can be visually followed (change in color, dissolution, precipitation, etc.), (2) fluorination can be easily controlled by step-by-step addition of fluorine, and (3) very homogeneous products with small uniform particle size are obtained. If there is a need, various very pure fluorides can be prepared in large quantities. This was, for example, necessary when the influence of cointercalated HF on the electrochemical behavior of highly fluorinated graphite was studied,[59] when reactions between C_{60} and d-block metal fluorides were investigated,[60] when gaseous products of thermal decomposition of some transition metal complex fluorides were studied by high-temperature mass spectroscopy,[61] etc.

One of the obstacles of photochemical reactions is moisture or oxygen contamination, which in many cases results in the formation of undesired by-products (i.e., O_2^+ salts),[62] especially in attempts to prepare binary fluorides. The presence of oxygen can be avoided or reduced by "freeze-pump-thaw-stirring" procedure. The reaction vessel with the sample is frozen at 77 K by liquid nitrogen. The volatiles are pumped away on a vacuum line. The valve is closed, the reaction vessel is warmed up to RT, and the reaction mixture is thoroughly stirred to force gases bubble out. Then the whole procedure is repeated by freezing the reaction vessel, etc.[15]

Acknowledgments

The author gratefully acknowledges the Ministry of Education, Science and Sport of the Republic of Slovenia for financial support of Research Program P1-0045 (Inorganic Chemistry and Technology). The author is also grateful to all his collaborators and coauthors of the articles referenced here and other joint contributions.

References

1. Riedel, S. High-valent Fluorides and Fluoro-Oxidizers. In *Comprehensive Inorganic Chemistry II;* Reedijk, J., Poeppelmeier, K., Eds., 2nd ed.; Elsevier: Amsterdam, 2013.
2. Lehmann, J. F.; Mercier, H. P. A.; Schrobilgen, G. J. The Chemistry of Krypton. *Coord. Chem. Rev.* **2002,** *233/234,* 1–39.
3. Christe, K. O.; Wilson, W. W.; Wilson, R. D. Coordinatively Saturated Fluoro Anions. Oxidative Reactions with KrF^+ Salts and PtF_6. *Inorg. Chem.* **1984,** *23,* 2058–2063.
4. Lucier, G. M.; Shen, C.; Elder, S. H.; Bartlett, N. Facile Routes to NiF_6^{2-}, AgF_4^-, AuF_6^-, and PtF_6^- Salts Using O_2^+ as a Source of O_2F in Anhydrous HF. *Inorg. Chem.* **1998,** *37,* 3829–3834.
5. Bartlett, N. Low Temperature Preparation and Uses of Potent Oxidizers. *J. Fluorine Chem.* **2006,** *127,* 1285–1288.

6. Goreshnik, E.; Mazej, Z. Hexafluoridoiodium(VII) Hexafluoroarsenate. *Acta Cryst.* **2006,** *C62,* i59–i60.
7. Schroer, T.; Christe, K. O. Novel Synthesis of ClF_6^+ and BrF_6^+ Salts. *Inorg. Chem.* **2001,** *40,* 2415–2419.
8. Christe, K. O.; Schack, C. J.; Wilson, R. D. Low-Temperature Ultraviolet Photolysis and Its Application to the Synthesis of Novel and Known NF_4^+ Salts. *Inorg. Chem.* **1976,** *15,* 1275–1282.
9. Lutar, K.; Borrmann, H.; Leblanc, M. Recent Developments in the Synthesis of Inorganic Fluorides. In *Advanced Inorganic Fluorides: Synthesis, Characterization and Applications;* Nakajima, T., Žemva, B., Tressaud, A., Eds.; Elsevier: Amsterdam, 2000; pp 5–49.
10. Whalen, J. M.; Lucier, G. M.; Bartlett, N. The Room Temperature Conversion of Nickel Difluoride to Hexafluoronickelate(IV) Salts of Alkali Cations. *J. Fluorine Chem.* **1998,** *88,* 107–110.
11. Grannec, J.; Lozano, L. Preparative Methods. In *Inorganic Solid Fluorides, Chemistry and Physics;* Hagenmuller, P., Ed.; Academic Press, Inc.: London, 1985; pp 18–76.
12. Bartlett, N. Introduction. In *Advanced Inorganic Fluorides: Synthesis, Characterization and Applications;* Nakajima, T., Žemva, B., Tressaud, A., Eds.; Elsevier: Amsterdam, 2000; pp 1–4.
13. Žemva, B. Protonic Superacid Anhydrous Hydrogen Fluoride as a Solvent in the Chemistry of High Oxidation States. *C.R. Acad. Sci. Paris* **1998,** *1,* 151–156.
14. O'Donnell, T. A. The Chemistry of Fluorine. In *Comprehensive Inorganic Chemistry;* Bailar J. C. Jr., Emeléus, H. J., Nyholm, R., Trotman-Dickenson, A. F., Eds.; Pergamon Press Ltd: Oxford, 1973; p 1018.
15. Suppan, P. *Chemistry and Light;* The Royal Society of Chemistry: Cambridge, 1994.
16. Calvert, J. G.; Pitts, J. N., Jr. *Photochemistry;* John Wiley & Sons, Inc.: New York, 1966.
17. O'Donnell, T. A. *Superacids and Acidic Melts as Inorganic Chemical Reaction Media;* VCH Publishers, Inc.: New York, 1993.
18. O'Donnell, T. A. On the Commonality of Speciation of Inorganic Solutes in Superacids, Strong Bases, Molten Salts, and Water. *Eur. J. Inorg. Chem.* **2001,** *2001,* 21–34.
19. Mazej, Z.; Goreshnik, E. Poly[perfluorotitanate(IV)] Salts of $[H_3O]^+$, Cs^+, $[Me_4N]^+$ and $[Ph_4P]^+$ and about the Existence of an Isolated $[Ti_2F_9]^-$ Anion in the Solid State. *Inorg. Chem.* **2009,** *48,* 6918–6923.
20. Muetterties, E. L.; Castle, J. E. Reactions of Hydrogen Fluoride with Metals and Metalloids. *J. Inorg. Nucl. Chem.* **1966,** *18,* 148–153.
21. Emeléus, H. J.; Gutmann, V. The Fluorides of Niobium, Tantalum, Tungsten, and Rhenium. *J. Chem. Soc.* **1950,** 2115–2118.
22. Graudejus, O.; Wilkinson, A. P.; Chacón, L.; Bartlett, N. M-F Interatomic Distances and Effective Volumes of Second and Third Transition Series MF_6^- and MF_6^{2-} Anions. *Inorg. Chem.* **2000,** *39,* 2794–2800.
23. Bougon, R.; Wilson, W. W.; Christe, K. O. Synthesis and Characterization of NF_4CrF_6 and Reaction Chemistry of CrF_5. *Inorg. Chem.* **1985,** *24,* 2286–2292.
24. Mazej, Z.; Goreshnik, E. Alkali Metal (Li^+-Cs^+) Salts with Hexafluorochromate(V), Hexafluorochromate(IV), Pentafluorochromate(IV), and Undecafluorodichromate(IV) Anions. *Eur. J. Inorg. Chem.* **2008,** *2008,* 1795–1812.
25. Pitts, J. J.; Jache, A. W. Reaction of Chlorine Monofluoride with Some Group IV Elements and Derivatives. *Inorg. Chem.* **1968,** *7,* 1661–1663.
26. Mazej, Z.; Lutar, K.; Žemva, B. Novel Syntheses of Some Binary Fluorides: the Role of Anhydrous Hydrogen Fluoride. *Acta Chim. Slov.* **1999,** *46,* 229–238.

27. Mazej, Z. Room Temperature Syntheses of MnF_3, MnF_4 and Hexafluoromanganate(IV) Salts of Alkali Cations. *J. Fluorine Chem.* **2002,** *114,* 75–80.
28. Lucier, G. M.; Elder, S. H.; Chacón, Bartlett, N. The Room Temperature Oxidation of Gold, Ruthenium, Osmium, Iridium, Platinum, or Palladium with Fluorine in Anhydrous Hydrogen Fluoride. *Eur. J. Solid State Inorg. Chem* **1996,** *33,* 809–820.
29. Holloway, J. H.; Hope, E. G.; Puxley, C. D. The Oxidative Fluorination of the Platinum Metals by Elemental Fluorine in Solution at Room Temperature. *Eur. J. Solid State Inorg. Chem.* **1996,** *33,* 821–829.
30. Bartlett, N.; Beaton, S. P.; Jha, N. K. Oxidizing Trends in the Third-Transition-Series Hexafluorides. *Chem. Commun. (London)* **1966,** 168–169.
31. Court, T. L.; Dove, M. F. A. Solution Chemistry of Cobalt in Liquid Hydrogen Fluoride. *J. Fluorine Chem.* **1975,** *6,* 491–498.
32. Fitz, H.; Müller, B. G.; Graudejus, O.; Bartlett, N. Single Crystal Investigations on $LiMF_6$ (M = Rh, Ir), Li_2RhF_6, and K_2IrF_6. *Z. Anorg. Allg. Chem.* **2002,** *628,* 133–137.
33. Mazej, Z.; Benkič, P.; Tressaud, A.; Žemva, B. Palladium Chemistry in Anhydrous HF/AsF_5 Superacid Medium. *Eur. J. Inorg. Chem.* **2004,** *2004,* 1827–1834.
34. Mazej, Z.; Lutar, K. Metathetical Synthesis of $La_2(PdF_6)_3$, and the Vibrational Spectra of the PdF_6^{2-} Anion. *J. Fluorine Chem.* **2001,** *107,* 63–69.
35. Graudejus, O.; Elder, S. H.; Lucier, G. M.; Shen, C.; Bartlett, N. Room Temperature Syntheses of AuF_6^- and PtF_6^- Salts, $Ag^+AuF_6^-$, $Ag^{2+}PtF_6^{2-}$, and $Ag^{2+}PdF_6^{2-}$, and an Estimate for $E(MF_6^-)$ (M = Pt, Pd). *Inorg. Chem.* **1999,** *38,* 2503–2509.
36. Seppelt, K. Molecular Hexafluorides. *Chem. Rev.* **2015,** *115,* 1296–1306.
37. Bartlett, N.; Lucier, G.; Shen, C.; Castell, W. J., Jr.; Chacón, L.; Munzenberg, J.; Žemva, B. The Oxidizing Properties of Cationic High Oxidation State Transition-Element Fluoro Species. *J. Fluorine Chem.* **1995,** *71,* 163–164.
38. Christe, K. O.; Wilson, W. W.; Wilson, R. D. Synthesis and Characterization of $(NF_4)_2MnF_6$. *Inorg. Chem.* **1980,** *19,* 3254–3256.
39. Jache, A. W.; Cady, G. H. Solubility of Fluorides in Liquid Hydrogen Fluoride. *J. Phys. Chem.* **1952,** *56,* 1106–1109.
40. Grzybowska, D.; Malinowski, P.; Mazej, Z.; Grochala, W. Probing the Reactivity of the Potent AgF_2 Oxidizer. Part 1: Organic Compounds. *Collect. Czech. Chem. Commun.* **2008,** *73,* 1729–1746.
41. Malinowski, P.; Mazej, Z.; Grochala, W. Probing the Reactivity of the Potent AgF_2 Oxidizer. Part 2: Inorganic Compounds. *Z. Anorg. Allg. Chem.* **2008,** *634,* 2608–2616.
42. Graudejus, O.; Wilkinson, A. P.; Bartlett, N. Structural Features of $Ag[AuF_4]$and $Ag[AuF_6]$ and the Structural Relastionship of $Ag[AgF_4]_2$ and $Au[AuF_4]_2$ to $Ag[AuF_4]_2$. *Inorg. Chem.* **2000,** *39,* 1545–1548.
43. Lucier, G. M.; Whalen, J. M.; Bartlett, N. High Yield Room Temperature Synthesis of $KAgF_4$ and AgF_3 and the Preparation and Unit Cell of $LiAgF_4$. *J. Fluorine Chem.* **1998,** *89,* 101–104.
44. Mazej, Z. Recent Achievements in the Syntheis and Characterization of Metal Hexafluoroantimonates and Hexfluoroaurates. *J. Fluorine Chem.* **2004,** *125,* 1723–1733.
45. Mazej, Z.; Goreshnik, E. X-ray Single Crystal Structures of $Cd(AuF_6)_2$, $Mg(HF)AuF_4AuF_6$ and $KAuF_6$ and Vibrational Spectra of $Cd(AuF_6)_2$ and $KAuF_6$. *Solid State Sci.* **2006,** *8,* 671–677.
46. Mazej, Z.; Goreshnik, E.; Tavčar, G. X-ray Single Crystal Structures of $Hg(AuF_6)_2$ and $AgFAuF_6$. *J. Fluorine Chem.* **2011,** *132,* 686–689.

47. Barraclough, C. G.; Cockman, R. W. Electron Spectra of the Cations of Europium(II), Samarium(II), and Ytterbium(II) in Anhydrous Hydrogen Fluoride. *Inorg. Chem.* **1991,** *30,* 340–343.
48. Mazej, Z. Room Temperature Syntheses Lanthanoid Tetrafluorides. *J. Fluorine Chem.* **2002,** *118,* 127–129.
49. Mazej, Z.; Goreshnik, E. Syntheses and Characterization of $ANi(AsF_6)_3$ ($A = H_3O^+$, O_2^+, NO^+, NH_4^+, K^+, Rb^+ and Cs^+) Compounds. *J. Fluoirne Chem.* **2009,** *130,* 399–405.
50. Liebmann, J.; Paige, H. L.; Passmore, J. Why Aren't There Any Trioxygenyl or Ozonium Salts? or, Why Does Our Reaction of O_3 and PtF_6 Give O_2PtF_6. *Struct. Chem.* **1998,** *9,* 315–317.
51. Tavčar, G.; Žemva, B. XeF_4 as Ligand for Metal Ion. *Angew. Chem. Int. Ed.* **2009,** *48,* 1432–1434.
52. Mazej, Z.; Goreshnik, E. Single-Crystal Structure Determination of NO_2SbF_6, XeF_5SbF_6 and $XeF_5Sb_2F_{11}$. *J. Fluorine Chem.* **2015,** *175,* 47–50.
53. Mazej, Z.; Goreshnik, E. Synthesis and Characterization of $[XeF_5]_3[Ti_4F_{19}]$ Containing a Discrete $[Ti_4F_{19}]^{3-}$ Anion. *Eur. J. Inorg. Chem.* **2009,** *2009,* 4503–4506.
54. Dove, M. F. A.; Clifford, A. F. Inorganic Chemistry in Liquid Hydrogen Fluoride. In *Chemistry in Anhydrous, Prototropic Inorganic Solvents, Volume II, Part 1;* Jander, G., Spandau, H., Addison, C. C., Eds.; Pergamon Press: Oxford, 1971; pp 154–155.
55. Dove, M. F. A.; King, R.; King, T. J. Preparation and X-ray Crystal Structure of Sn_3F_8. *J. Chem. Soc., Chem. Commun.* **1973,** 944–945.
56. Beattie, I. R.; Livingston, K. M. S.; Ozin, G. A.; Reynolds, D. J. Single-Crystal Raman Spectrum of Bismuth Pentafluoride and of Antimony Tetrachloride Fluoride and the Vibrational Spectrum of Vanadium, Niobium, Tantalum, and Antimony Pentafluorides. *J. Chem. Soc. (A)* **1969,** 958–965.
57. Christe, K. O.; Wilson, W. W.; Curtis, E. C. Novel Onium Salts. Synthesis and Characterization of the Peroxonium Cation H_2OOH^+. *Inorg. Chem.* **1979,** *18,* 2578–2586.
58. Fawcett, J.; Holloway, J. H.; Russell, D. R. Preparations and Crystal Structure of $3SbF_3 \cdot 4SbF_5$. *Can. J. Chem.* **1989,** *67,* 2041–2046.
59. Nakajima, T.; Gupta, V.; Ohzawa, Y.; Groult, H.; Mazej, Z.; Žemva, B. Influence of Cointercalated HF on the Electrochemical Behaviour of Highly Fluorinated Graphite. *J. Power Sources* **2004,** *137,* 80–87.
60. Shustova, N. B.; Mazej, Z.; Chen, Y.-S.; Popov, A. A.; Strauss, S. H.; Boltalina, O. V. Saturene Revealed: X-ray Crystal Structure of D_{5d}-$C_{60}F_{20}$ Formed in Reactions of C_{60} with A_xMF_y Fluorinating Agents (A = Alkali Metal; M = 3d Metal). *Angew. Chem. Int. Ed.* **2010,** *49,* 812–815.
61. Leskiv, M. S.; Abramov, S. V.; Oleinik, L. I.; Kepman, A. V.; Sukhoverkhov, V. F.; Mazej, Z.; Rau, D. V.; Chilingarov, N. S.; Sidorov, N. Vaporization Products of Transition-metal and Rare Earth Complex Fluorides Studied by High-temperature Mass Spectrometry. *Inorg. Mater.* **2005,** *41,* 1327–1333.
62. Mazej, Z.; Ponikvar-Svet, M.; Liebman, J. F.; Passmore, J.; Jenkins, H. D. B. Nitrosyl and Dioxygenyl Cations and Their Salts. Similar but Further Investigations Needed. *J. Fluorine Chem.* **2009,** *130,* 788–791.

Sol–Gel Synthesis of Metal Fluorides: Reactivity and Mechanisms

21

G. Scholz, E. Kemnitz
Humboldt University of Berlin, Berlin, Germany

Chapter Outline

1. Introduction

Metal fluorides are of interest for many different applications in metallurgy, as optical materials (e.g., solid-state lasers, luminophores, scintillators, or antireflective coatings), uranium isotope separation, sensing (e.g., fluoride-sensitive electrodes), and catalysis (e.g., heterogeneously catalyzed fluorination reactions). An outstanding bust of interest over the recent years arose from energy storage applications (see Refs. 1,2,36).

On the other hand, nanomaterials have gained enormous attraction over the past 20 years because properties of nanoscopic compounds usually differ drastically

Modern Synthesis Processes and Reactivity of Fluorinated Compounds. http://dx.doi.org/10.1016/B978-0-12-803740-9.00021-4

from those of classically prepared analogues. The high potential of nanomaterials for industrial applications is the major driving force for scientists to search for new synthesis routes toward these materials. Consequently, a variety of new synthesis techniques have been developed over the past decades that opens the door to the fascinating world of nanomaterials. It is worth mentioning that especially in this context the main focus was laid on metal oxides with their extremely wide area of applications. Among many new synthesis routes toward nanoscopic metal oxides, the sol–gel synthesis certainly is one of the most powerful approaches in terms of variety of synthesis techniques and technical applications. Although in fact very complex, this sol–gel synthesis is based on a very simplified description of the hydrolytic reaction of a metal alkoxide (often silicon alkoxide) with water.

However, motivated by new developments for applications of such nanoscopic materials, nonaqueous sol–gel synthesis approaches like atomic layer deposition and others have been developed recently, thus extending the synthesis access toward new nanoscopic materials.

Inspired by the great potential of industrial applications that other nanomaterials have already gained, there is an increasing interest in the development of nano-metal-fluoride-based new materials. This mainly emerged from their attractive perspectives in application areas like catalysis, optics, photonics, optical amplifiers, optoceramics, composite materials, biosensing, and biolabeling.

Consequently, chemists have been strongly engaged in developing synthesis routes that allow access to nano metal fluorides as is the case for nano metal oxides. Equally potent as the sol–gel synthesis of metal fluorides, the so-called fluorolytic sol–gel synthesis has been developed over the past years and might be considered as the most powerful synthesis approach to nano metal fluorides.

The intention of this review is to provide an overview on the mechanistic aspects of the nonaqueous fluorolytic sol–gel synthesis that yields for the first time direct access to novel nanoscale metal fluorides.

The growing interest in nanoscopic metal fluorides can be easily evidenced by checking Thomson Reuters' Web of Science[55] for the keyword "nano" in combination with "fluoride" and "materials" as displayed in Fig. 21.1. Starting on a low level at

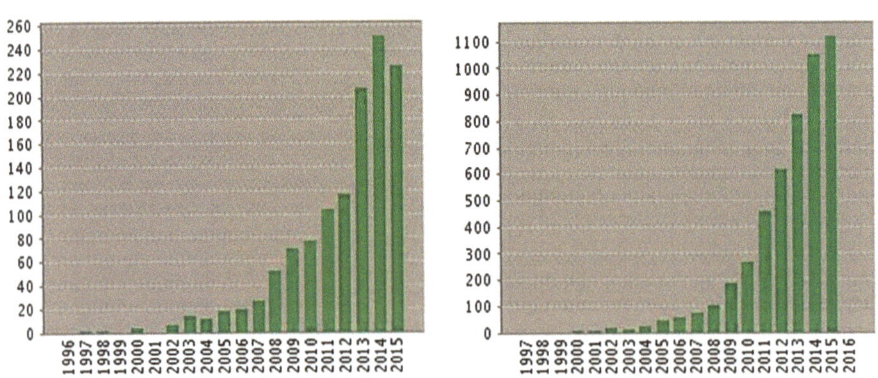

Figure 21.1 Published items in each year for the past 20 years.

around 2000 with just a few publications and even citations, in 2010, about 80 publications and c.270 citations can be counted, whereas in 2014, around 250 publications appeared and more than 1000 citations can be found.

It fits perfectly into this reflection that the first book emphasizing synthesis and applications of new nano metal fluorides appeared just in 2010.[55]

Based on the recently developed fluorolytic sol-gel synthesis, a substantial progress in nano metal fluoride chemistry has been achieved. Due to this progress, some few review articles were published already reflecting the great potential of nano metal fluorides for different applications.[18–20,40] Although a brief description of the chemistry and mechanistic aspects is already given in these reviews, this chapter will exclusively give a comprehensive overview and summary of the present "state of the art" regarding the chemical principles and essential mechanistic aspects of this new synthesis approach.

Consequently, the following sections will focus mainly on the chemistry of these metal fluoride systems. Moreover, special emphasis will be laid on the synthesis of optically transparent sols containing homodispersed nanoscopic metal fluoride particles in the solvent. Because several exciting applications of nanoscopic metal fluorides, briefly reflected in Chapter 4, are only possible based on such water-clear sols, the chemical routes leading to such materials will be discussed comprehensively in Chapter 2.

2. The Principle Chemistry of the Fluorolytic Sol–Gel Synthesis

2.1 The Chemical Pathway

The classical sol–gel synthesis of metal oxides, starting with the very early work by Stöber et al.[50] is based on the reaction of a metal alkoxide dissolved in a suitable organic solvent with water. In the first reaction step, the metal alkoxide group (M–OR) becomes hydrolyzed by water thus forming a metal hydroxide group (cf. Eq. 21.1).

$$M(OR)_n + nH_2O \rightarrow M(OH)_n + nROH \tag{21.1}$$

These M–OH groups formed by the hydrolysis reaction may, in the second step, undergo condensation reactions either with a further M–OH group forming water (Eq. 21.2) or with an unconverted M–OR group producing alcohol (Eq. 21.3).

$$-M-OH + HO-M- \rightarrow -M-O-M- + H_2O \tag{21.2}$$

$$-M-OH + RO-M- \rightarrow -M-O-M- + ROH \tag{21.3}$$

Although these three reactions are an oversimplification of the real reactions involved, the fundamental reaction path is correctly reflected. After separation of the solvent and thermal posttreatment, nanoscopic metal oxides will be obtained if the reaction is properly performed. Since the hydrolysis and the condensation reactions are very often incomplete, meaning still noticeable proportions of unconverted OR groups remain in the product, thermal posttreatment is ultimately demanded to finally obtain pure metal oxides. For more details, see representative reviews by Kistler,[21] Brinker and Scherer,[3] as well as Hüsing and Schubert.[14]

At first glance, the chemistry of *fluorolytic* sol–gel synthesis of metal fluorides parallels the chemical pathway of the classical sol–gel synthesis of metal oxides; only here, instead of water, anhydrous hydrogen fluoride is used as reactant.

$$M(OR)_{n(sov)} + nHF_{solv} \rightarrow MF_{n(solv)} + nROH_{(solv)} \quad (21.4)$$

Because of this formality we named this route "fluorolytic," thus pointing out its similarity to the classical sol–gel reaction, which in analogy might be better described as "hydrolytic" sol–gel reaction.

There are, however, significant differences. In case of the hydrolytic sol–gel reaction M–OH groups are formed, which then undergo further condensation reactions, thus forming M–O–M-bridges and finally resulting in metal oxide formation, whereas in the fluorolytic sol–gel reaction, direct formation of M–F bonds takes place, which cannot undergo further condensation reactions. However, as a result of the transformation of an M–OR into an M–F moiety, the ionic character of the bond drastically increases, and as a consequence, neither molecular units (e.g., molecules, clusters) nor terminal fluorine atoms are stable, thus resulting in ultimate bridging of fluorine atoms and the consequent formation of inorganic polymeric (dominantly ionic) compounds. Further consequences resulting from this distinguished difference will be discussed later.

The overall synthesis approach of this fluorolytic sol–gel route is presented by Scheme 21.1.

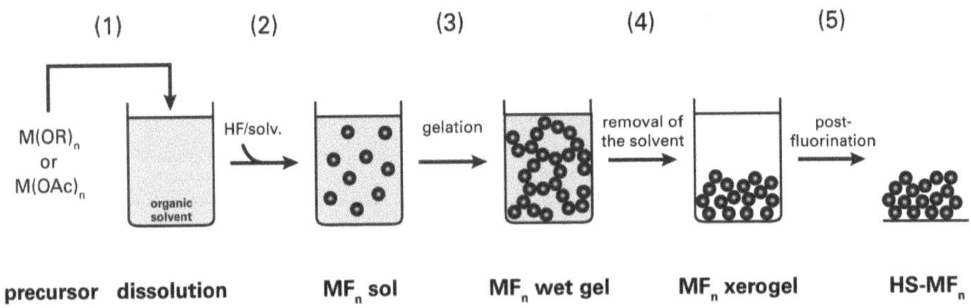

Scheme 21.1 Formation of a nanoscopic metal fluoride according to the *fluorolytic* sol–gel synthesis by reacting a metal alkoxide with anhydrous HF in a suitable organic solvent.

In general, a suitable metal precursor is dissolved in an organic solvent, preferentially alcohol (step 1 in Scheme 21.1). A solution of anhydrous hydrogen fluoride in, e.g., alcohol or ether, is added in stoichiometric amounts to the precursor solution, eventually resulting in a clear, translucent sol (step 2). Depending on the reaction conditions, the metal fluoride concentration, the type of precursor and solvent, and/or the water content, the sol may undergo further gelation (step 3). Since the metal fluoride xerogels obtained after solvent removal (step 4) may contain some remaining organic compounds, being either unconverted organic groups bound to the metal or trapped solvent molecules, organic-free, high surface (*HS*) metal fluoride xerogels can be obtained by thermal posttreatment (step 5) employing either gaseous HF or chlorofluorocarbons like CCl_2F_2 or $CHClF_2$ as fluorinating agents.[39]

The first publication on the fluorolytic sol–gel approach reported about the synthesis of nanoscopic (high surface) HS–AlF_3.[16] Although in fact valuable detailed mechanistic information was gained especially based on extensive magic angle spinning (MAS) nuclear magnetic resonance (NMR) investigations for the reaction of aluminum isopropoxide with anhydrous HF, which will be reflected in more detail in Chapter 3, the impact of different reaction parameters on sol formation is better understood for the formation of nanoscopic alkaline earth metal fluorides, especially for MgF_2. Hence the following chapters will focus mainly on the chemistry of these metal fluoride systems.

Several synthesis methods for water-clear sols of *nano*-MgF_2,[27,31–33,43,44] *nano*-CaF_2,[37,38] and *nano*-SrF_2[45] have been published already. These sols contain homodispersed nanoparticles (NPs) of diameter less than 20 nm (see Fig. 21 2).

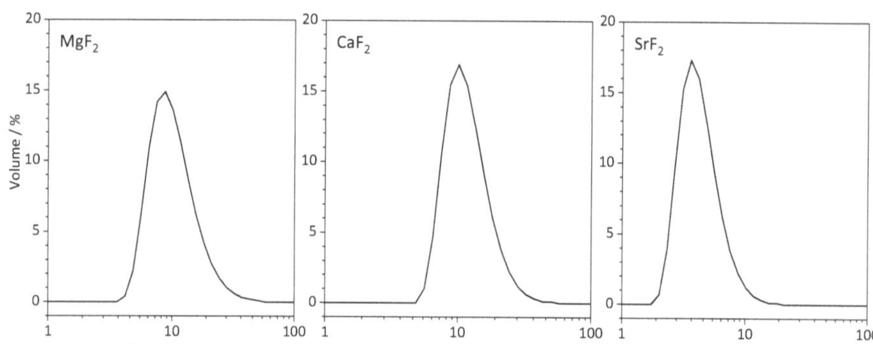

Figure 21.2 Hydrodynamic diameters of particles in sols of MgF_2, CaF_2, and SrF_2.

2.2 The Reaction Parameters

Metal fluoride NPs can be easily synthesized by the reaction of HF as fluorinating agent with several metal precursors. The classical synthesis of metal fluorides is usually performed in an aqueous system resulting in almost complete formation of either well-crystalline microscopic metal fluorides or thermodynamically favored metal fluoride hydrates. To obtain nanoscopic materials, crystal growth and hydration have to be prevented, meaning the reaction conditions have to be established in a way that fast oversaturation is achieved. By working in suitable organic solvents that usually show extremely low solubility for metal fluorides this precondition is fulfilled. Hence, nucleus formation is a very fast process, dominating crystal growth due to very limited availability of solved metal fluorides in the solvent. For this reason, working in organic solvents is strongly recommended, not only in case of metal fluoride NP but also with other materials like metal oxides. This way many different nano metal fluorides are accessible, the fundamental syntheses approaches being briefly reviewed in Refs. 6,19 Unfortunately, in most of these cases agglomerates of nano metal fluorides but not homidispersed NPs are obtained. However, for many applications, clear, optically transparent metal fluoride sols are desired, which requires performing the synthesis in a way that homidispersed MF_n NPs are formed, which, moreover, should be stable in the sol over a long time. Reflecting the chemical equation for the fluorolytic sol–gel synthesis (cf. Eq. 21.4) several reaction parameters have impact on the performance of the resulting metal fluoride NPs. These are:

1. the kind of metal precursor
2. the solvent
3. the concentration
4. the reaction temperature

2.2.1 Metal Precursors

Despite the tremendous negative impact of water on the reaction path, which will be discussed later, the solubility of the precursor compound in the solvent is another decisive factor for the successful synthesis of homodispersed metal fluoride sols.

2.2.1.1 Metal Acetates

Alkaline earth metal (M) acetates are in general suitable precursors for the synthesis of nanoscopic MF_2 materials. These acetates usually are traded in their hydrated form, $M(CH_3COO)_2 \cdot xH_2O$. Especially magnesium acetate tetrahydrate, $Mg(CH_3COO)_2 \cdot 4H_2O$, exhibits good solubility in alcohols, and therefore has been widely used for the synthesis of MgF_2 NPs.[30,35,52,53,56] However, several disadvantages, which have never been explicitly mentioned by other authors arise from this precursor.

A first drawback is that resolvation occurs already within 2 h when dissolving the precursor $Mg(OAc)_2 \cdot 4H_2O$ or pure $Mg(OAc)_2$ in EtOH resulting in the formation of solvated magnesium acetates $Mg(OAc)_2 \cdot (H_2O)_3 \cdot (EtOH)$ or $Mg_3(OAc)_6 \cdot 2EtOH$, respectively.[42] Unfortunately, these solvates are hardly soluble because of their oligomeric, one-dimensional (1D) or three-dimensional structures, and so the reaction with HF during the sol–gel synthesis process is limited. Formation of a protecting MgF_2 layer on acetate-solvated particles prevents full conversion due to limited access of HF through this dense product layer. Therefore, when using magnesium acetate as precursor, immediately after dissolution of the acetate the addition of HF has to be carried out to suppress undesired resolvation processes.

A second drawback originates generally from the introduction of any water, because fast gelation proceeds. In Fig. 21.3, a clear sol obtained from predried $Mg(OAc)_2$ is shown (left) compared with the same sol after addition of c.7 vol% water (right). Not surprisingly, we repeatedly found an almost steady increase of viscosity of metal acetate–based metal fluoride sols finally resulting in gelation after a few weeks.

Figure 21.3 Photographs of a MgF_2 sol (left) and MgF_2 gel with air bubbles (right) prepared from dried and tetrahydrated $Mg(CH_3COO)_2$, respectively.
Reproduced from Scheurell, K.; Noack, J.; König, R.; Hegmann, J.; Jahn, R.; Hofmann, T.; Löbmann, P.; Lintner, B.; Garcia-Juan, P.; Eicher, J.; Kemnitz, E. Optimization of a Sol-Gel Synthesis Route for the Preparation of MgF_2 Particles for a Large Scale Coating Process. Dalton Trans. **2015**, 44, 19501—19508 with permission of the Royal Society of Chemistry.

Hence, drying of the acetate precursor is an ultimate precondition to obtain clear transparent metal fluoride sols of stable viscosity at least for some weeks. Drying at ambient pressure does not give an appropriate precursor because basic metal acetates will form as has been proven for the magnesium acetate system. Here, a so far unknown basic magnesium acetate, $Mg_5(\mu_3\text{-}OH)_2(OAc)_8 \cdot 1.19H_2O$, is formed as minority phase, which almost prevents the formation of clear homodispersed sols due to its insolubility in water and alcohols, and thus, also limited reactivity against HF.[42] However, soluble $Mg(OAc)_2$ can be obtained by extracting the water under mild conditions (100°C, vacuum). Water-free $Mg(CH_3COO)_2$, free of basic magnesium acetate, obtained this way is suitable for sol—gel synthesis of transparent MgF_2 sols in ethanol, which have been thoroughly investigated for manufacturing antireflective coatings on glass.[44]

But there is another disadvantage with this precursor. The reaction of predried $Mg(OAc)_2$ with HF in ethanol leads to the formation of acetic acid, which can consecutively react with ethanol forming the corresponding ethyl acetate and equimolar amounts of water (Eqs. 21.5 and 21.6).

$$Mg(OOCH_3)_2 + 2HF \rightarrow MgF_2 + 2CH_3COOH \qquad (21.5)$$

$$CH_3COOH + C_2H_5OH \rightarrow CH_3COOC_2H_5 + H_2O \qquad (21.6)$$

As this reaction progresses, the water content in the system increases continuously and causes gelation and/or agglomeration within a few days as has been proven by long-term NMR investigations.[43] After 1 week, the ratio of acetic acid to ethyl acetate is 97:3, but after 4 weeks, it is already 84:16. The question arises, how much water can be tolerated before gelation occurs. Our investigations on sols obtained from water-free magnesium acetate as precursor showed that water content above c.7 vol% leads to immediate gelation (see Fig. 21.3); above 5 vol%, gelation starts after 2 days; above 2.5 vol%, gelation occurs after 2 weeks; and at c.0.5 vol%, gelation needs about 3 months (cf. also Fig. 21.4).

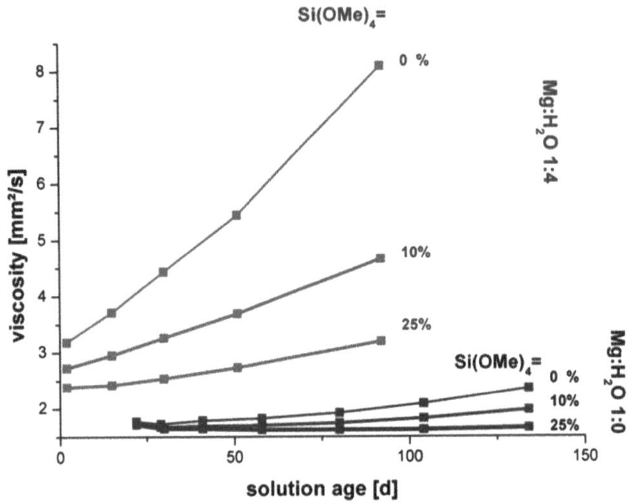

Figure 21.4 Viscosities of two MgF_2 sols depending on the water content and solution age. Reproduced from Scheurell, K.; Noack, J.; König, R.; Hegmann, J.; Jahn, R.; Hofmann, T.; Löbmann, P.; Lintner, B.; Garcia-Juan, P.; Eicher, J.; Kemnitz, E. Optimization of a Sol-Gel Synthesis Route for the Preparation of MgF_2 Particles for a Large Scale Coating Process. Dalton Trans. **2015**, 44, 19501−19508 with permission of the Royal Society of Chemistry.

Water content below 0.1 vol% can be tolerated, meaning the sols are stable against gelation over several months (>3). In spite of these facts, it is amazing that sol−gel syntheses for the preparation of clear and low-viscous MgF_2 sols starting from magnesium acetate tetrahydrate or syntheses in water-containing solutions with aqueous HF are described in the literature without commenting on the problems connected with the water content.[30,35,52]

2.2.1.2 Metal Alkoxides

The fluorolytic sol−gel synthesis of metal fluorides has been originally developed using metal alkoxides as precursor, which are dissolved in an appropriate alcohol or ether.[16,57] The advantage of alkoxides is that after reaction with HF (cf. Eq. 21.4) just the respective alcohol is formed, hence, esterification reactions and consequently water formation are excluded.

The precursor **magnesium methoxide**, easily obtained by the reaction of magnesium with dry methanol, is known to form tetrameric $[Mg_4(OCH_3)_8(CH_3OH)_8]$ units in solution and to crystallize as $Mg(OMe)_2 \cdot 3.55CH_3OH$.[59] Commercial magnesium methoxide is hardly available and, moreover, contains impurities that circumvent its application. Therefore, a nonaqueous fluorolytic sol–gel method has been developed by our group for the preparation of transparent MgF_2 sols.[58,28] This includes a two-step reaction as presented by Eqs. (12.7) and (21.8).

$$Mg + 2CH_3OH \rightarrow Mg(OCH_3)_2 + H_2 \quad (21.7)$$

$$Mg(OCH_3)_2 + 2HF \rightarrow MgF_2 + 2CH_3OH \quad (21.8)$$

The drawback of this reaction sequence is the evolution of H_2 (explosive), but the reaction runs very reproducibly. Hence a lot of mechanistic insight was gained for this reaction system.

When adding HF to the clear solution of $Mg(OCH_3)_2$ in methanol, a very turbid suspension is formed, which over a period of 2–4 weeks totally cleared off. Parallel to this, dynamic light scattering (DLS) indicates big particles of about 150 nm at the very beginning, which break down to about 10 nm after 1 month (cf. Fig. 21.5).

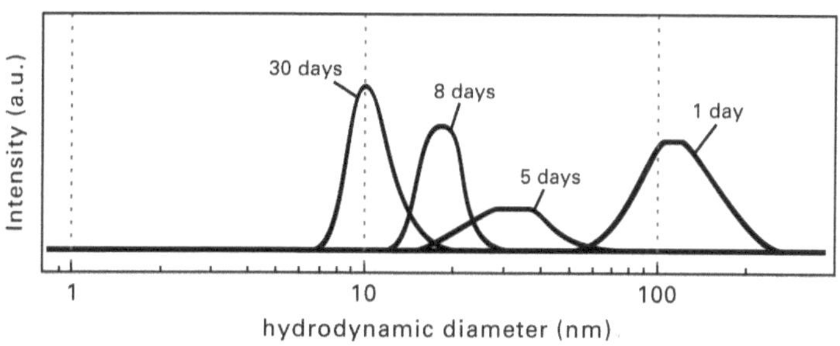

Figure 21.5 Particle size distribution (dynamic light scattering) depending on the sol age. Reproduced from Scheurell, K.; Noack, J.; König, R.; Hegmann, J.; Jahn, R.; Hofmann, T.; Löbmann, P.; Lintner, B.; Garcia-Juan, P.; Eicher, J.; Kemnitz, E. Optimization of a Sol-Gel Synthesis Route for the Preparation of MgF_2 Particles for a Large Scale Coating Process. Dalton Trans. **2015**, 44, 19501–19508 with permission of the Royal Society of Chemistry.

The conclusion is that primary nanoscaled MgF_2 NPs form agglomerates (secondary particles) at the very beginning of the reaction, which after a while break down to the homodispersed primary particles thus forming a transparent clear MgF_2–sol. In several cases, it was detected by ^{19}F NMR that still detectable traces of unreacted HF were immobilized at the surface of MgF_2 particles formed, which fully disappeared as the original turbid MgF_2 suspension cleared off. It was speculated that due to incomplete conversion Mg–OR groups and/or the immobilized surface -HF moieties might cause agglomeration. The addition of a stronger acid should suppress this effect. In fact, by addition of several "additives" like phosphonic acids and trifluoroacetic acid sols clear off within a few minutes instead of weeks. Fig. 21.6 shows examples of isolated MgF_2 NPs that are stabilized by the addition of c.2 mol% phenylphosphonic acid to the sol. In line with the DLS results of such MgF_2 sols, the particle size is below 5 nm.

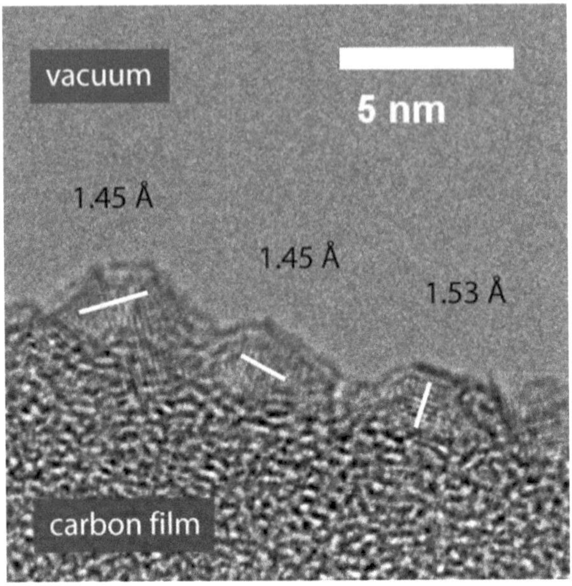

Figure 21.6 High-resolution transmission electron microscopic (TEM) image of magnesium fluoride particles stabilized by phenylphosphonic acid at the edge of a carbon film on a Holey carbon TEM grid.
Reproduced from Noack, J.; Emmerling, F.; Kirmse, H.; Kemnitz, E. Sols of Nanosized Magnesium Fluoride: Formation and Stabilisation of Nanoparticles. J. Mater. Chem. **2011**, 21, 15015–15021 with permission of the Royal Society of Chemistry.

To investigate structural changes with aging time, WAXS diffraction patterns were recorded, which give information about the crystallographic phases and crystallinity of the suspended particles. The use of an ultrasonic levitator as sample holder in combination with intense synchrotron X-ray radiation allowed to investigate the sols directly without altering the sample in the process of drying.[32] The results of the WAXS experiments are given in Fig. 21.7 for MgF_2 sols in a time range from 1 day to 6 months after HF addition.

Reflections of (111) and (211) interrecticular distances are the most distinct, whereas the (110) reflex does not show the expected intensity ratio to the others. For $2\theta = 64$ degrees peak is associated with the (130) lattice plane. With longer aging time, this reflection diminishes in intensity and two other reflections (002) and (301) appear. At the same time, the (110) reflection becomes more pronounced and almost reaches the expected intensity after 1 month of aging. It is noteworthy that the reflections from (111) and (211) lattice planes show almost no changes. If recrystallization would be a case, the crystallite size should increase, which—as a consequence—would lead to decreased reflex widths. In line with DLS results, which do not indicate any change in particle size, Ostwald ripening is obviously not the case for aging of nanoscopic magnesium fluoride particles derived from the fluorolytic sol–gel synthesis route. Instead, structural reorganization and healing of a highly distorted crystal

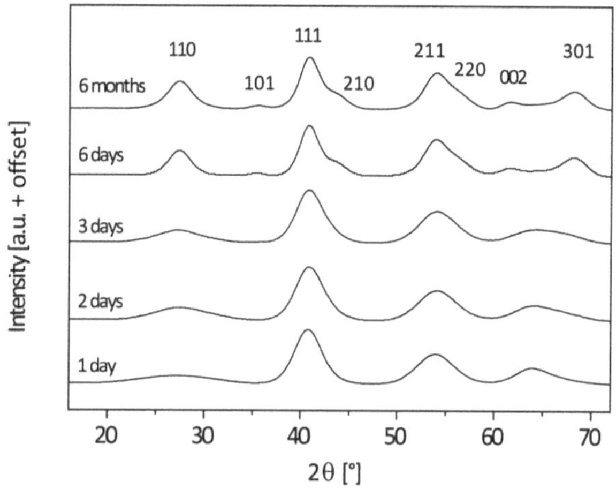

Figure 21.7 WAXS diffractograms of magnesium fluoride with different degrees of aging, measured for 0.25 M sols in an ultrasonic levitator at the μspot Beamline at BESSY II. Reproduced from Noack, J.; Emmerling, F.; Kirmse, H.; Kemnitz, E. Sols of Nanosized Magnesium Fluoride: Formation and Stabilisation of Nanoparticles. J. Mater. Chem. **2011**, 21, 15015–15021 with permission of the Royal Society of Chemistry

structure resulting in an almost perfect rutile structure is observed, which goes along with thermodynamic relaxation and decrease of surface energy. It is not clear whether or not this process might also have an impact on the observed deagglomeration of the magnesium fluoride NPs.

By following the intermediate compounds formed when HF is added stepwise to the methanolic $Mg(OCH_3)_2$ solution, further mechanistic insight was gained.[32] Wide angle X-ray powder scattering (WAXS) clearly evidenced the formation of two different intermediate phases in the course of reaction with a F:Mg ratio of 0.3 and 0.4. One (F:Mg = 0.4) is the crystalline hexanuclear dicubane $[Mg_6F_2(OCH_3)_{10}(CH_3OH)_{14}]$, which can even be isolated on a preparative scale.[59] The other one (F:Mg = 0.3) could be obtained neither in single crystalline form nor in suitable purity for Rietveld structure determination so far, but it is structurally very close to the one described before.

Unfortunately, commercially available $Mg(OCH_3)_2$ does not react properly, and therefore, it has to be prepared freshly by the reaction of Mg metal with water-free methanol. Large amounts of hydrogen gas are liberated during this reaction, and the toxicity of methanol is a further drawback. Attempts to transfer this synthesis to ethanol have failed so far due to the insolubility of the magnesium methoxide in ethanol.

2.2.1.3 Magnesium Ethoxide

Because of the aforementioned drawbacks with magnesium methoxide as precursor, magnesium ethoxide seems to be a highly recommended alternative. However, all attempts to start from commercially available magnesium ethoxide failed since it is insoluble in methanol and even in ethanol or other organic solvents, hence, no transparent sols can be obtained due to the deposition of a protective MgF_2 layer onto the suspended solid $Mg(OEt)_2$ precursor particles. However, it has been reported that stable solutions of $Mg(OEt)_2$ in ethanol can be obtained when CO_2 is passed through the suspension.[9] This approach has been adapted by us and hence, transparent MgF_2 sols can be obtained via this approach even starting with commercially available magnesium ethoxide.[26] The chemical action of CO_2 is that it reacts with suspended magnesium ethoxide, forming a soluble species of magnesium alkylcarbonate $Mg(ROCO_2)_2$ ($R = CH_3, C_2H_5$).

$$Mg\,(C_2H_5O)_2 + 2CO_2 \xrightarrow{EtOH} Mg(C_2H_5OCO_2)_2 \tag{21.9}$$

Surprisingly, partially hydrolyzed impurities in the commercial product also dissolve completely during the introduction of CO_2. Oxo bridges and hydroxyl groups can react with CO_2, more probably with intermediately formed monoethylcarbonic acid, to form carbonate CO_3^{2-} and hydrogen carbonate HCO_3^-. Obviously, partial hydrolysis of the reactant $Mg(OEt)_2$ does not disturb the formation of clear precursor solutions.

As a result of fluorination, the free acid $C_2H_5OCO_2H$ is formed. This acid is unstable and immediately decomposes into ethanol and CO_2. The same accounts for HCO_3^- and CO_3^{2-}. Upon protonation, carbonic acid is formed, which decomposes into H_2O and CO_2. A schematic overview of the reactions in this system is given in Fig. 21.8.

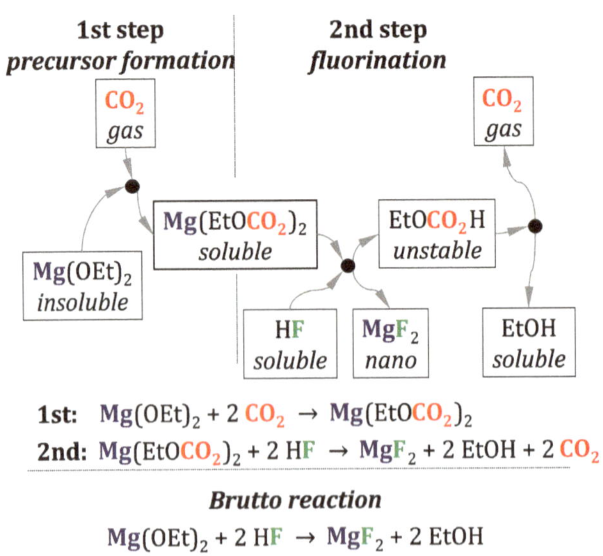

Figure 21.8 Schematic representation of the reactions occurring during synthesis using the CO_2 method ($Et = C_2H_5$). Both steps are carried out separately.
Reproduced from Krahl, T.; Broßke, D.; Scheurell, K.; Lintner, B.; Kemnitz, E. High Performance Antireflective MgF_2 Layers on Glass. J. Mater. Chem. C **2015**, accepted with permission of the Royal Society of Chemistry.

The role of CO_2 is formally that of a Lewis acid, which breaks the strong oxo bridges in solid $Mg(OEt)_2$. Since dissolution of magnesium ethoxide in ethanol by passing CO_2 through the solvent became possible, nanoscopic MgF_2 from this precursor can be prepared nowadays, although for a long time it was believed to be impossible. Based on this knowledge an even more elegant approach was derived because other acids, especially strong acids like HCl should have the same effect. Upon reaction of HCl with $Mg(OEt)_2$, soluble $MgCl_2$ is formed. When HF is present in the reaction mixture, $MgCl_2$ will react immediately with it to form nanoscale MgF_2, thus forming HCl again. This HCl can react with still unreacted $Mg(OEt)_2$, and hence, a catalytic cycle is established. Based on this reaction cycle a simple one-pot reaction can be performed just by adding $MgCl_2$ to $Mg(OEt)_2$ as precursor. $MgCl_2$ and $Mg(OEt)_2$ are provided in ethanol. The $MgCl_2$ quickly dissolves in alcohol, whereas $Mg(OEt)_2$ remains as a suspended solid. Upon fluorination, dissolved $MgCl_2$ reacts first with HF forming HCl beside MgF_2, and thus the catalytic cycle starts—very similar to the one in the "CO_2 route." A schematic overview of these reactions is given in Fig. 21.9.

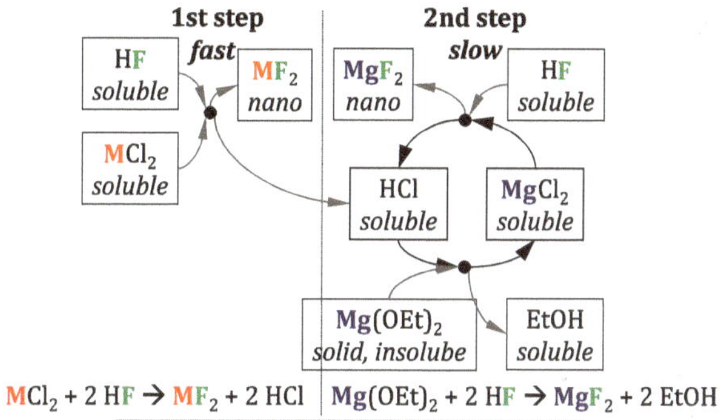

Brutto reaction

$$(1\text{-}x)\ Mg(OEt)_2 + x\ MCl_2 + 2\ HF$$
$$\downarrow$$
$$(1\text{-}x)\ MgF_2 + x\ MF_2 + 2x\ HCl + (2\text{-}2x)\ EtOH$$

Figure 21.9 Schematic representation of the reactions occurring during the one-pot synthesis using the chloride method (M = Mg; $x = 0.05\ldots0.30$; Et = C_2H_5).
Reproduced from Krahl, T.; Broßke, D.; Scheurell, K.; Lintner, B.; Kemnitz, E. High Performance Antireflective MgF_2 Layers on Glass. J. Mater. Chem. C **2015**, accepted with permission of the Royal Society of Chemistry.

Because of the addition of $MgCl_2$, these sols are acidic, because they always contain HCl depending on the amount of $MgCl_2$ added.

Other alkoxides like magnesium methoxyisopropoxide, $Mg[(OCH(CH_3)CH_2OCH_3)]_2$, and magnesium-*tert*-butoxide, $Mg[OC(CH_3)_3]_2$ were also investigated. After fluorination no differences in the bulk structure were detected, neither in X-ray diffraction (XRD) patterns nor in infrared spectra. The surface areas are different. A reduction in surface area and total pore volume of the respective nanoscopic MgF_2 powders was observed when the chain length of the alkoxy groups increases relative to that of $Mg(OCH_3)_2$, whereas the surface area of MgF_2 increased by about 50% when branched magnesium-*tert*-butoxide was used as precursor.

Basically, any other metal alkoxides can be used for the synthesis of nanoscaled MF_n. The solubility of the respective precursor in the solvent is one decisive factor, and another one is the tendency of the carboxylic acid formed during the reaction to undergo esterification reactions as outlined earlier. For instance, clear CaF_2 sols can be obtained by employing calcium lactate for this reaction.

$$Ca[OOCCH(OH)CH_3]_2 + 2HF \rightarrow CaF_2 + 2CH_3CH(OH)COOH \quad (21.10)$$

The lactic acid formed during this reaction has only a limited tendency for esterification; however, with 122°C its boiling point is comparatively high and it also acts as a complexing agent. Thus, it is not that easy to obtain organic-free nanoscopic CaF_2 afterward.

Inorganic precursors like water-free metal chlorides MCl_n might be used as well. In case of magnesium chloride[33] and calcium chloride,[37] the sols obtained from these precursors clear off immediately after HF addition to the alcoholic MCl_2 solution. This fast clearing off can be rationalized based on mechanistic considerations. Our theoretical calculations for the reaction of aluminum isopropoxide with HF in alcohol (cf. Eq. 21.11) show that protonation of an M–OR group is the first activation step.

$$Al(O^iPr)_3 + 3HF \rightarrow AlF_3 + 3^iPrOH \quad (21.11)$$

In a rather simplified presentation, a proton will first electrophilically attack an alkoxide oxygen atom and then a fluoride anion, F^-, will nucleophilically bind to the activated metal atom. By splitting off the respective alcohol, the first fluorination step is finished (Scheme 21.2A).

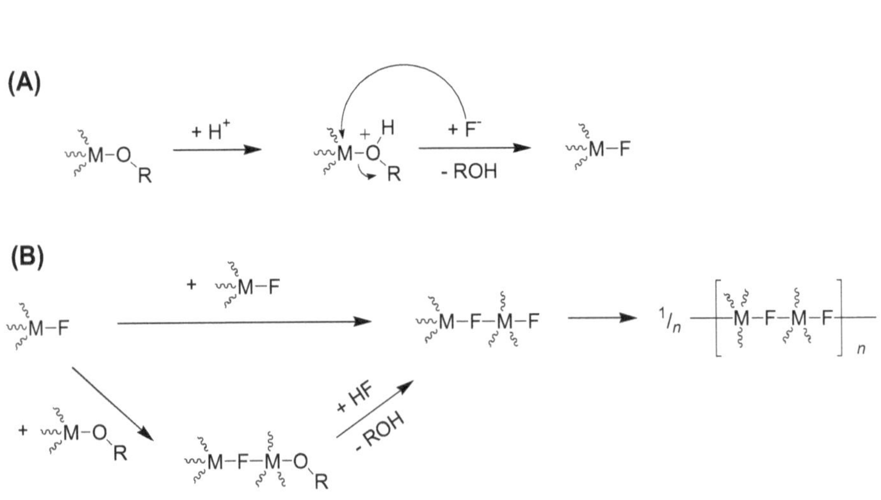

Scheme 21.2 The induction of the first fluorination step by electrophilic protonation of a metal alkoxide oxygen atom (A) and following fluorination (B).

Since fluoride ions almost immediately undergo bridging, oligomers of metal alkoxide fluorides are immediately formed. Bridging and further fluorination proceed parallel (Scheme 21.2B), thus resulting finally in the formation of three-dimensional polymeric metal fluorides. As a matter of fact, dissociation of any acid in a less polar organic solvent is drastically lower than in the usual solvent water. For example, whereas calcium carbonate undergoes immediate reaction when added to an aqueous HF solution, it is fully inert when added to an ethanolic HF solution. This clearly indicates that HF is almost molecularly dissolved but does not undergo detectable dissociation. If water is added stepwise to such an alcoholic HF solution, one can observe a reaction beginning when about 30 vol% water has been added to this solution. Since HCl is a significantly stronger acid than HF, the extent of proton dissociation, even in alcohol, should be larger. Consequently, the activation barrier for protonation will be drastically lower resulting in accelerated conversion rates as observed in all these cases.

Despite fast clearing off, these sols also exhibit long-term stability as long as water contamination is prevented, but unfortunately, they contain large amounts of hydrogen chloride, which causes corrosion issues. Other inorganic precursors like bromides, iodides, or even nitrates[11] might be used for this synthesis approach.

To sum up this section, many different precursors can be used for the synthesis of nanoscopic metal fluorides. Since the reaction is in general a metathesis reaction, the protonated form of the counteranion of the precursor will be formed in any case. Thus, the applications that are foreseen strongly define which of the different possible precursors might be chosen.

2.2.2 *Impact of Solvents*

As a rule, there are no big differences in the reaction path, the performance of the reaction, and the topological properties of the metal fluoride NPs depending on the organic solvent used for the synthesis. However, there might be significant differences in the concentrations of the MF_n sols depending on the viscosity, boiling point, and polarity of the solvents. First, the solubility of the metal precursor in the solvent is a very important issue, as has been extensively discussed earlier in the article.

The polarity of the solvent can also play an essential role that is closely connected with the reaction mechanism. As outlined previously (see Scheme 21.2), protonation of an M–OR-group is the first essential step of these reactions. The extent of dissociated protons (originated from HF) is strongly determined by the polarity of the solvent. Since common solvents like ethanol, isopropanol, butanol, etc., exhibit similar polarities, methanol is distinctly more polar. As a rule, faster formation of clear sols will be observed by working in methanolic solutions. An interesting observation in this context is the formation of MgF_2 sols from ethanolic $Mg(OC_2H_5)_2/MgCl_2$ suspensions. When working in ethanol as solvent, the fluorination of the precursor with *methanolic* HF will produce clear sols faster and more easily than the fluorination with *ethanolic* HF.

The most plausible explanation of this behavior is the different reaction rate of the initial nucleation in both systems. In methanol the extent of HF dissociation is higher so the reaction is faster than in ethanol. Moreover, the reaction between HF and the precursor is restricted mostly to the interface between the HF droplet and the dissolved precursor. When using *methanolic* HF, the concentration of methanol at this interface is high. The higher polarity of methanol causes a higher degree of HF dissociation, which results in extremely fast nuclei formation that consumes most of the dissolved Mg^{2+} ions, Consequently, nuclei growth is suppressed due to lack of dissolved precursor that can cause growth.

Therefore, due to the higher reactivity of HF in methanol, the initially formed MgF_2 nuclei do not contain any unconverted OR groups, and as a result, less agglomerated species are formed during the later nucleus growth. This results in a lower degree of agglomeration of the primary particles, and hence, the deagglomeration process is faster.

Another issue is the use of high-boiling solvents like ethylene glycol, propylene glycol, and glycerine. These solvents have significantly higher boiling points, which make separation and purification more difficult. However, probably mainly due to the higher viscosity of these solvents, significantly higher concentrations of clear transparent sols may be obtained. For example, in ethanol, clear sols of maximum 0.8 M can be synthesized; working in ethylene glycol may even allow for 2 M clear sols. However, for most of the desired applications the high boiling points of these solvents may make their use impossible.

2.2.3 Impact of Sol Concentration and Reaction Temperature

These two reaction parameters will just very briefly be touched upon since not too much impact on the reactions has been observed so far from these investigations. The most noticeable difference can be observed when clear transparent sols are desired. As a rule, the lower the concentration, the more easily homodispersed clear sols of metal fluoride particles in an organic solvent will be obtained. The upper concentration limit is mainly defined by the individual metal, by the kind of precursor (counter anion of the metal), the polarity, and/or viscosity of the solvent. These parameters have to be optimized to obtain transparent sols of appreciably high concentration with long life time stability. If transparent sols are not an issue, even higher concentrations can be prepared this way. The primary metal fluoride particles are mainly of the same size, the only difference is that larger agglomerates or aggregates of these primary NPs are formed. However, for applications, e.g., in heterogeneous catalysis or ceramic powders this does not matter.

Temperature is an even less influencing factor. The heats of conversion for these reactions are moderately low, that is, the temperature might increase during the reaction for up to 20° but no more. Since most of the syntheses are preferentially performed in low-boiling alcohols, these boiling points constitute the limiting factor for the synthesis. In some cases, running the synthesis at very low temperature might help to prevent agglomeration, obviously because diffusion-controlled material transport suppresses crystal growth. However, in some cases clear sols that are formed at low temperatures become turbid during heating up to room temperature obviously due to the enhanced mobility of the metal fluoride NPs resulting in the formation of agglomerates.

3. Mechanistic Aspects of Fluorolytic Sol–Gel Synthesis as Seen by Nuclear Magnetic Resonance Methods

Beginning with our first activities in this field in 2003[16] many educts, intermediates and products of the fluorolytic sol–gel process were widely studied exploring a large variety of spectroscopic, microscopic and diffraction methods. Among them the application of NMR methods—both in the liquid state and in the solid state—plays an outstanding role. Spin probes like ^{19}F, ^{27}Al, ^{1}H and also ^{13}C allow to determine local structures and to follow their changes in solutions, sols, wet gels, xerogels, and, in general, in nanostructured solid fluorides without the need of lattice periodicity. We gained most experience for the fluorolytic sol–gel synthesis of both nanostructured AlF_3 and alkaline earth metal fluorides MF_2 (M: Mg, Ca, Sr). Therefore, basic mechanistic aspects of this synthesis concept regarding the formation of nanostructured aluminum fluoride as well as nanostructured alkaline earth metal fluorides are briefly summarized in the following sections.

3.1 Sol–Gel Synthesis of Nanoscaled Aluminum Fluoride

The fluorolytic sol–gel synthesis of nanostructured AlF_3 is composed of many steps starting from aluminum alkoxide precursors $Al(OR)_3$ [e.g., aluminum isopropoxide, $Al(O^iPr)_3$]. The first steps involve their dissolution in organic or aqueous solvents, their fluorination with HF, and the subsequent formation of sols and wet gels. After drying solid aluminum alkoxide fluoride xerogels are formed, which serve as precursors for postfluorination reactions to obtain highly disordered AlF_3 with high surface area (HS–AlF_3). For all those steps, liquid- and solid-state NMR experiments were performed using ^{19}F, ^{27}Al, ^{1}H, and ^{13}C as spin probes. As already recapitulated in Refs. 17,19 local structures in these liquid and solid samples were addressed and compared and their changes were followed with progressive fluorination degree on the basis of mainly 1D NMR experiments. Based on these findings, a possible reaction pathway was derived, which is shown in Scheme 21.3 in a simplified manner (see also Refs. 17,19 for more details).

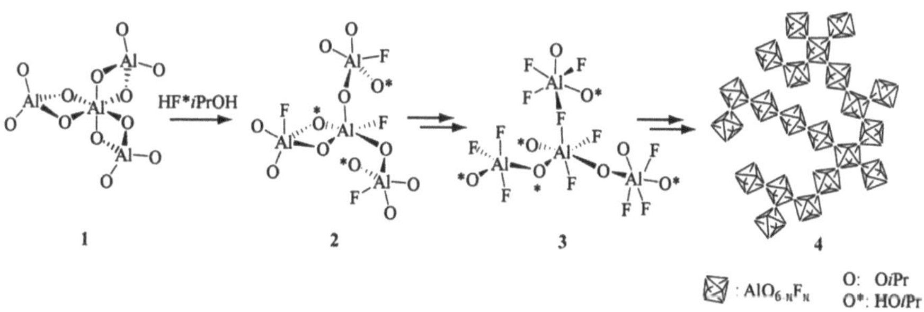

Scheme 21.3 Schematic and simplified representation of the formation of highly distorted nanoscopic AlF_3.
Reproduced from Kemnitz, E.; Noack, J. The Non-aqueous Fluorolytic Sol-Gel Synthesis of Nanoscaled Metal Fluorides. Dalton Trans. **2015**, 44, 19411–19431 with permission from the Royal Society of Chemistry.

The existence of a large variety of four-, five- and sixfold coordinated Al species like $AlF_x(OR)_{4-x}$, $AlF_x(OR)_{5-x}$, or $AlF_x(OR)_{6-x}$ was proven with 1D NMR and 1D MAS NMR studies supported by quantum chemical calculations. The presence of four- and fivefold coordinated Al species derived from 1D MAS NMR experiments is conceivable, but the resolution is not adequate to distinguish them more clearly. Although it was extremely difficult to grow single crystals of intermediate species, several of such intermediates were successfully crystallized from the sols with low fluorine supply and with the help of pyridine or dimethyl sulfoxide (DMSO) coordination to oligomeric aluminum isopropoxide fluorides; see Ref. 19 and references therein). Their single crystal structures support as well the existence of local partially fluorinated Al sites with coordination numbers between 4 and 6. Such disordered X-ray amorphous samples with intermediate fluorination degree [$AlF_x(OR)_{3-x}$, $0 < x < 3$] are finally key substances for clarifying local structures present in precursors of HS$-AlF_3$ and for the disclosure of the reaction mechanism.

To deepen and to confirm previous results, additional 2D MAS NMR experiments combined with high-field MAS NMR experiments were exemplarily performed for aluminum isopropoxide fluorides with the initial molar ratio Al:F = 1:1 [$AlF(O^iPr)_2$] and Al:F = 1:2 [$AlF_2(O^iPr)$].[23] The ^{27}Al 3-quantum magic-angle spinning nuclear magnetic resonance (3QMAS NMR) spectra of the two samples were taken at a magnetic field of 14.1 T for an enhanced spectral resolution. They are shown in Fig. 21.10 and exhibit several contributions for Al species in four-, five- and sixfold coordination. A total of eight different Al species were identified,[23] and their chemical shifts and quadrupolar parameters were determined. On the basis of these ^{27}Al 3QMAS NMR spectra not only reliable parameters but also knowledge about the shape and amount of contributing species were obtained.

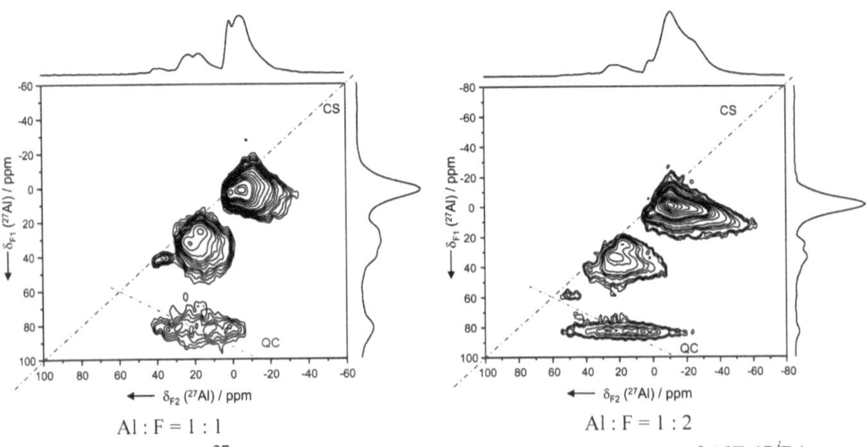

Figure 21.10 Sheared ^{27}Al 3QMAS nuclear magnetic resonance spectra of $AlF_x(O^iPr)_{3-x}$ samples resulting from different initial molar ratios recorded at $B_0 = 14.1$ T ($\nu_{rot} = 27.5$ kHz). Reproduced from König, R.; Scholz, G.; Pawlik, A.; Jäger, C.; van Rossum, B.; Kemnitz, E.; Identification of $AlF_x(OR)_y$ Species in Strongly Disordered Aluminium Isopropoxide Fluoride Solids: A Field-Dependent MAS NMR Study. J. Phys. Chem. C **113**, 15576–15585 with permission of the American Chemical Society.

Calculation of the strongly superimposed ^{27}Al MAS NMR spectra recorded at magnetic fields up to 21.1 T was now possible. These spectra along with their composition are shown for both samples in Fig. 21.11.

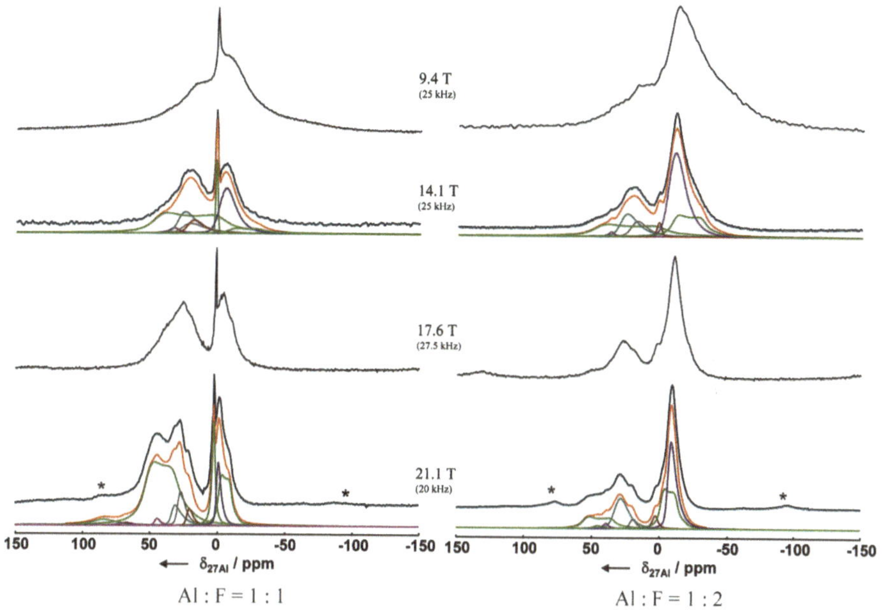

Figure 21.11 ^{27}Al MAS nuclear magnetic resonance spectra of samples $AlF(O^iPr)_2$ and AlF_2O^iPr obtained at different magnetic fields from 9.4 up to 21.1 T. Shown is the central region along with possible decomposition for two magnetic fields based on the data obtained by 3QMAS.
Reproduced from König, R.; Scholz, G.; Pawlik, A.; Jäger, C.; van Rossum, B.; Kemnitz, E.; Identification of $AlF_x(OR)_y$ Species in Strongly Disordered Aluminium Isopropoxide Fluoride Solids: A Field-Dependent MAS NMR Study. J. Phys. Chem. C **113**, 15576–15585 with permission of the American Chemical Society.

The existence of several four-, five-, and sixfold coordinated $AlF_x(OR)_{CN-x}$ species (where CN stands for coordination number) was unambiguously shown using ^{27}Al 3QMAS NMR experiments together with MAS NMR experiments at different magnetic fields up to ultrahigh fields (Figs. 21.10 and 21.11). With such a strong basis the sum of intensities of each Al species group with coordination numbers 4, 5, and 6 was calculated depending on the degree of fluorination. An insight into changes of the appropriate local structures and their amounts in aluminum alkoxide fluorides during the fluorolytic sol–gel process is now possible. The development of these approximated contributions is given in Fig. 21.12.

Surprisingly, this plot of the sum of intensities of each species group versus fluorination degree gives three almost linear correlations (Fig. 21.12). Differences between the two given approaches in Fig. 21.11, based on spectra taken at 21.1 and 14.1 T can be neglected. With increasing fluorination degree, the amount of

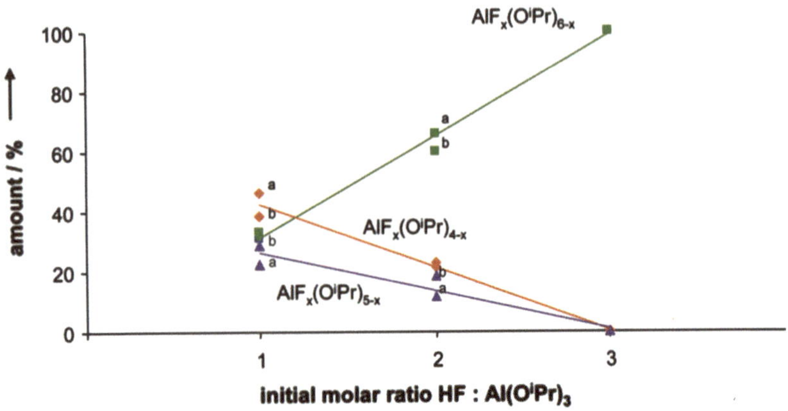

Figure 21.12 Development of the approximated contributions to the ^{27}Al MAS NMR spectra recorded at 21.1 T (A) and contributions based on the deconvolution of the spectra taken at 14.1 T (B) depending on the initial molar ratio F:Al.
Reproduced from König, R.; Scholz, G.; Pawlik, A.; Jäger, C.; van Rossum, B.; Kemnitz, E.; Identification of $AlF_x(OR)_y$ Species in Strongly Disordered Aluminium Isopropoxide Fluoride Solids: A Field-Dependent MAS NMR Study. J. Phys. Chem. C **113**, 15576−15585 with permission of the American Chemical Society.

four- and fivefold coordinated Al species decreases and the amount of sixfold coordinated Al species increases.

As for ^{27}Al, the interpretation of the ^{19}F MAS NMR spectra of the aluminum alkoxide fluorides is challenging as well. Their careful deconvolution results in up to 10 different species with chemical shifts ranging from −130.4 to −198.9 ppm.[23] A comparable intensity plot of different fluorine species depending on the fluorine supply allowed a rough assignment of aluminum sites with different coordination numbers to fluorine sites.[23] However, a clear assignment still remains difficult due to the fact that species superimpose at similar positions. In addition, the ^{19}F MAS NMR spectra of these highly disordered solids showed a noticeable narrow line at −163 ppm, which was supposed to belong to an ordered, possibly crystalline environment. Contributions for ordered local environments attributable in the ^{27}Al MAS NMR spectra cannot be unambiguously distinguished except for $Al(O^iPr)_3$ (Figs. 21.10 and 21.11). Here, the isolation of suitable single crystals with subsequent structure elucidation can provide additional opportunities to understand changes of local structures during sol−gel synthesis and can support deduced ideas of the mechanistic pathway. The growth of well-defined crystalline aluminum alkoxide fluorides, however, appears to be nontrivial. Previously isolated single crystals were stabilized by the incorporation of additional external donor molecules like DMSO or pyridine.[5,19,39] Later, three new crystalline aluminum isopropoxide fluorides were obtained from the sols with molar ratios of Al:F equal to 1:1 and 1:1.25:$Al_4F_4(\mu_4\text{-O})(\mu\text{-O}^iPr)_5[H(O^iPr)_2]$,$\cdot$ $Al_5F_5(\mu_5\text{-O})(\mu\text{-O}^iPr)_8$, and $Al_{16}F_{20}(\mu_4\text{-O})_4(\mu\text{-O}^iPr)_{20}\cdot 2(^iPrOH)$. Actually, no further external donor molecules are incorporated in these single crystals.[25] Their typical local tetranuclear Al unit, present in all three mentioned compounds, is

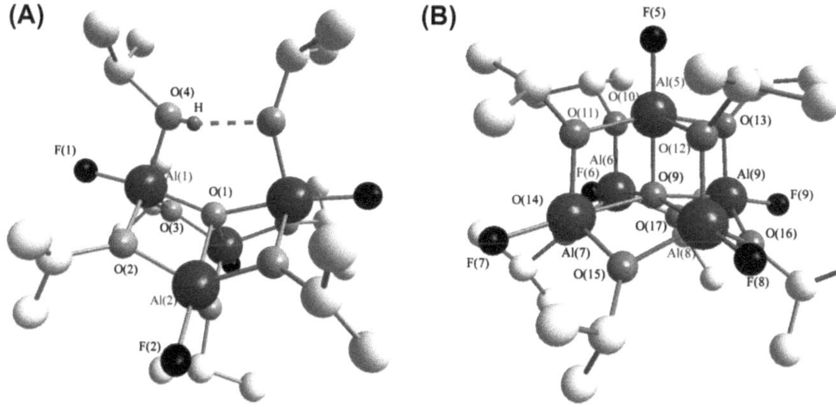

Figure 21.13 Molecular structures of the tetranuclear unit $Al_4F_4(\mu_4\text{-}O)(\mu\text{-}O^iPr)_5[H(O^iPr)_2]$ (A) and the pentanuclear unit $Al_5F_5(\mu_5\text{-}O)(\mu\text{-}O^iPr)_8$ (B) present in crystalline aluminum alkoxide fluorides without additional donor molecules.
Reproduced from König, R.; Scholz, G.; Veiczi, M.; Jäger, C.; Troyanov, S. I.; Kemnitz, E. New Crystalline Aluminium Alkoxide Fluorides: Evidence of the Mechanism of the Fluorolytic Sol-Gel Reaction. Dalton Trans. **2011**, 40, 8701–8710 with permission of the Royal Society of Chemistry.

shown in Fig. 21.13A. The second compound contains both the tetranuclear unit (Fig. 21.13A) and a pentanuclear Al unit as depicted in Fig. 21.13B.

Interestingly, the tetranuclear Al unit (Fig. 21.13A) consists of two different fivefold-coordinated $AlFO_4$ units, with fluorine exclusively in the terminal position. The aluminum atoms in the pentanuclear unit are five- and sixfold coordinated (Fig. 21.13B). Such mixed but defined F/O environments along with terminal F sites are very exceptional for crystalline Al–F–O compounds. The ^{27}Al MAS NMR spectra of the compound $Al_4F_4(\mu_4\text{-}O)(\mu\text{-}O^iPr)_5[H(O^iPr)_2]$ with the tetranuclear Al unit were recorded at two different external magnetic fields. They are given in Fig. 21.14. The observed line shapes are typical for Al units in well-defined and ordered environments.

The ^{27}Al isotropic chemical shifts of the two aluminum sites were determined as 35 ppm [$AlFO(OR)_3$] and 31 ppm [$AlFO(OR)_2(ROH)$], typical for fivefold-coordinated partially fluorine-coordinated Al sites, as also obtained for species in strongly disordered powders (see Figs. 21.10 and 21.11). Moreover, the ^{19}F MAS NMR spectrum of $Al_4F_4(\mu_4\text{-}O)(\mu\text{-}O^iPr)_5[H(O^iPr)_2]$ exhibits two lines at −164.7 and −165.7 ppm. This position agrees nicely with the narrow ^{19}F signal at −163 ppm observed in disordered aluminum alkoxide fluorides as discussed previously. Obviously, the incorporation of solvent molecules as well as the existence of terminal F sites in all fluorination states holds true for both the crystalline and the X-ray amorphous substances.[22–24] A comparison of the ^{19}F and ^{27}Al NMR shifts of $Al_4F_4(\mu_4\text{-}O)(\mu\text{-}O^iPr)_5[H(O^iPr)_2]$ with those of $AlF_x(OR)_{3-x}$-sols and xerogels is summarized in Table 21.1. An almost perfect correlation of "sol–gel/powder" with single

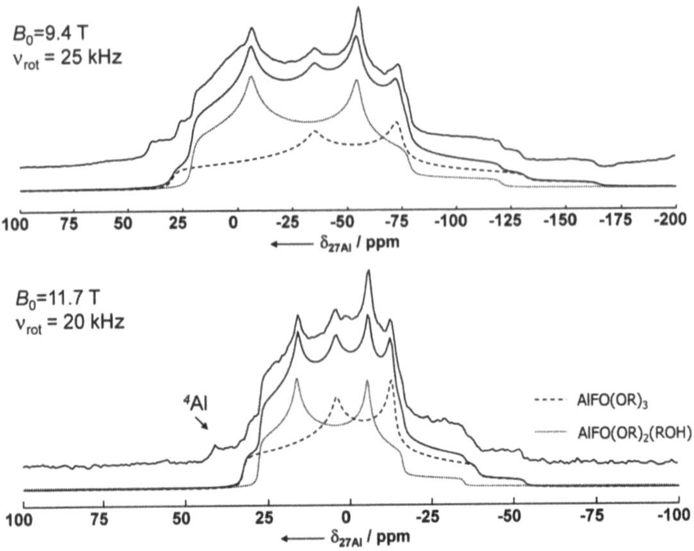

Figure 21.14 ^{27}Al MAS nuclear magnetic resonance spectra of the compound $Al_4F_4(\mu_4\text{-}O)(\mu\text{-}O^iPr)_5[H(O^iPr)_2]$ at 9.4 and 11.7 T (ν_{rot} = 20 kHz). Their decomposition is given as dashed/dotted lines within the figure.
Reproduced from König, R.; Scholz, G.; Veiczi, M.; Jäger, C.; Troyanov, S. I.; Kemnitz, E. New Crystalline Aluminium Alkoxide Fluorides: Evidence of the Mechanism of the Fluorolytic Sol-Gel Reaction. Dalton Trans. **2011**, 40, 8701–8710 with permission of the Royal Society of Chemistry.

Table 21.1 **Comparison of ^{19}F- and ^{27}Al Nuclear Magnetic Resonance Shifts of $Al_4F_4(\mu_4\text{-}O)(\mu\text{-}O^iPr)_5[H(O^iPr)_2]$ With Those of $AlF_x(OR)_{3\text{-}x}$ Sols and Xerogels**

Sample	^{27}Al: δ_{iso}/ppm	^{19}F: δ_{iso}/ppm
1: Solid State		
$Al_4F_4(\mu_4\text{-}O)(\mu\text{-}O^iPr)_5[H(O^iPr)_2]$ units: $AlF(OR)_3O$, $AlFO(OR)_2(ROH)$	31, 35	−165, −166
$AlF_x(OR)_{3-x}$ xerogels (Al: F = 4:1 to 2:1)	*58*, **38**, 2, −6	−156, **−162**, And broad contributions
$AlF_x(OR)_{3-x}$ xerogels (Al: F = 1:1)	*58*, **40**, **29**, 22, *3*, −4, −5	**−162**, **−172**, And broad contributions
1: Solution		
$Al_4F_4(\mu_4\text{-}O)(\mu\text{-}O^iPr)_5[H(O^iPr)_2]$ units: $AlF(OR)_3O$, $AlFO(OR)_2(ROH)$	37	−162, −164
$AlF_x(OR)_{3-x}$ xerogels (Al: F = 4:1−2:1)	*58*, **36**, 10, 2, −20 With different intensities	**−161**, **−163**, **−165** With different intensities

Bold, with **1** as the correlating species; *italics*, contributions of $Al(O^iPr)_3$.
Reproduced from König, R.; Scholz, G.; Veiczi, M.; Jäger, C.; Troyanov, S. I.; Kemnitz, E. New Crystalline Aluminium Alkoxide Fluorides: Evidence of the Mechanism of the Fluorolytic Sol-Gel Reaction. Dalton Trans. **2011**, 40, 8701–8710 with permission of the Royal Society of Chemistry.

crystal data can be drawn. The excellent agreement strongly corroborates the fundamental findings concerning the molecular picture drawn for the sol–gel fluorolysis of aluminum alkoxides.

In addition, several crystalline aluminum alkoxide fluorides are structurally closely related to fluorine-free, pure aluminum alkoxides, derivatives, or partially hydrolyzed aluminum alkoxides, as visible in Fig. 21.15.

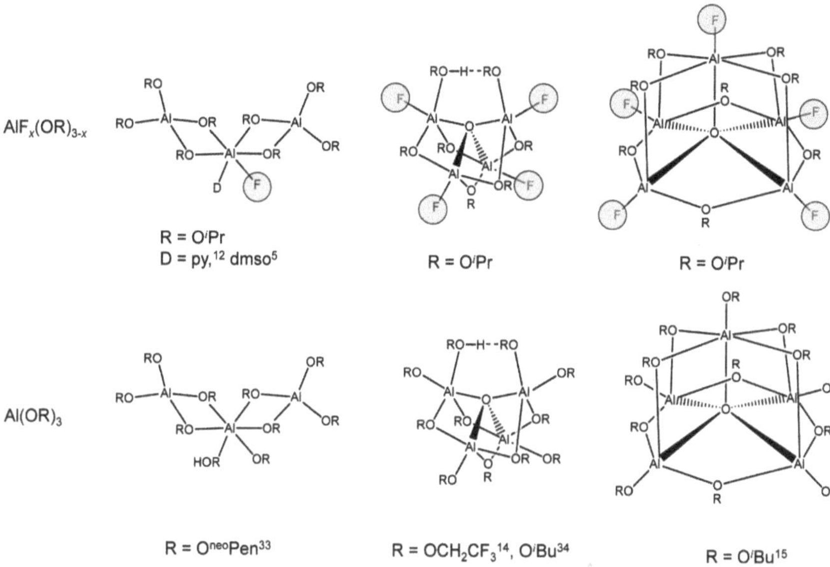

Figure 21.15 Molecular structures of some aluminum isopropoxide fluorides with low fluorination degree (top, Al:F up to 1:1, F atoms are highlighted) in comparison with related structures of aluminum alkoxides (bottom).
Reproduced from König, R.; Scholz, G.; Veiczi, M.; Jäger, C.; Troyanov, S. I.; Kemnitz, E. New Crystalline Aluminium Alkoxide Fluorides: Evidence of the Mechanism of the Fluorolytic Sol-Gel Reaction. Dalton Trans. **2011**, 40, 8701–8710 with permission of the Royal Society of Chemistry.

3.2 *Sol–Gel Synthesis of Nanoscaled Alkaline Earth Metal Fluorides MgF_2, CaF_2, and SrF_2*

Many studies regarding mechanistic aspects of the sol–gel synthesis were also performed for magnesium fluoride, and among the alkaline earth metal fluorides, it is the best system studied in most detail. Structural and topologic changes were mainly followed by small-angle x-ray scattering (SAXS) and wide-angle x-ray scattering (WAXS), along with transmission electron microscopic (TEM) and DLS measurements.[19,47,57,59] Compared with AlF_3, the accessibility of spin probes to follow local structural changes is quite different. Theoretically, the quadrupolar nucleus ^{25}Mg ($I = 5/2$) can be used as spin probe. However, there are many problems associated with this nucleus like its low natural abundance (10.1%), but especially its low gyromagnetic ratio. The latter is responsible for the low sensitivity at moderate magnetic fields and also contributes to the large second-order quadrupolar broadening of the central transition. All properties of the magnesium nucleus together considerably hamper the acquisition of high-quality NMR spectra,[29] which was also shown for nanoscopic hydroxide fluorides.[48] Therefore, basic studies about local structures are restricted to ^{19}F as spin probe both in liquids and in solids.

In contrast to the sol–gel synthesis of nanoscaled AlF_3, only one crystalline intermediate $Mg_6F_2(OCH_3)_{10}(CH_3OH)$ with a dicubane-like structure was successfully isolated at the preparation of nanoscaled MgF_2. With the help of ^{19}F as spin probe, it is now possible to follow the formation of local structures in magnesium methoxide fluoride sols, to address time-dependent changes during the fluorination process, and to follow the influence of acids on the formation of local structures of MgF_2 NPs. To do so it is quite important to make the right choice of the field sweep in the ^{19}F NMR spectra, otherwise broad lines belonging to sol particles are not visible.[15]

In this chapter, the ^{19}F NMR spectra of MgF_2 sols with the nominal composition Mg:F = 1:1 and 1:2 are implemented. All of them are prepared by fluorolytic sol–gel reaction of magnesium methoxide with HF in methanol. Fig. 21.16 shows the static ^{19}F NMR spectra of an opaque sol $MgF_{0.9}(OR)_{1.1}$ for different aging times. Immediately after the addition of HF two narrow signals with chemical shifts of −174 and −183 ppm can be observed. These signals can be attributed to $Mg_6F_2(OCH_3)_{10}$ (CH_3OH) (−174 ppm)[59] and unconsumed HF (−183 ppm). Both signals completely disappear after 7 days of aging. In addition, a very broad line with a spectral width of about 130 ppm is visible (Fig. 21.16). The time-dependent ^{19}F NMR spectra of the sols with higher fluorine supply (Mg:F = 1:2) are shown in Fig. 21.17A and B. Now the typical signal of ^{19}F in MgF_2 can be assigned at −198 ppm. In addition,

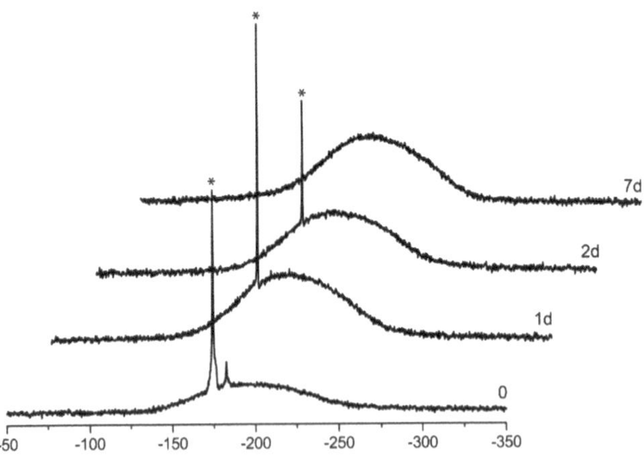

Figure 21.16 Static ^{19}F nuclear magnetic resonance spectra of the sol $MgF_{0.9}(OR)_{1.1}$. Signals with * have a chemical shift at −174 ppm.
Reproduced from Karg, M.; Scholz, G.; König, R.; Kemnitz, E. Mechanistic Insight Into Formation and Changes of Nanoparticles in MgF_2 Sols Evidenced by Liquid and Solid State NMR. Dalton Trans. **2012**, 41, 2360–2366 with permission of the Royal Society of Chemistry.

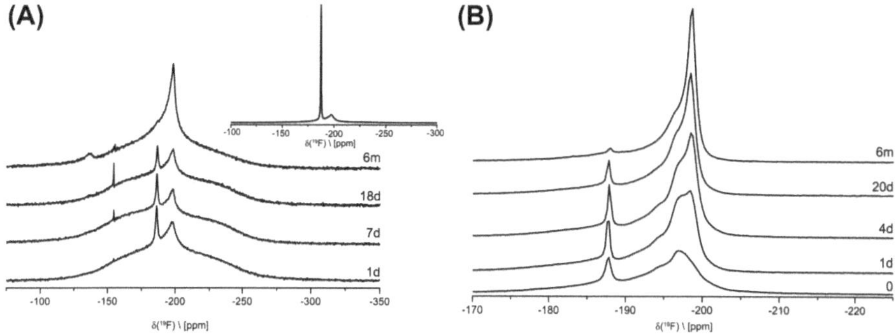

Figure 21.17 (A) Static ^{19}F nuclear magnetic resonance (NMR) spectra of the MgF_2 sol taken at different aging times. The inset shows the spectrum recorded immediately after the addition of HF; (B) Static ^{19}F NMR spectra of the MgF_2 sol with HNO_3 as additive, taken at different aging times.
Reproduced from Karg, M.; Scholz, G.; König, R.; Kemnitz, E. Mechanistic Insight Into Formation and Changes of Nanoparticles in MgF_2 Sols Evidenced by Liquid and Solid State NMR. Dalton Trans. **2012**, 41, 2360–2366 with permission of the Royal Society of Chemistry.

the narrow signal at −187 ppm belongs to HF species adsorbed on magnesium fluoride. With increasing aging time, the amplitude of the signal at −198 ppm increases relative to the adsorbed HF species. But as for the first examples (Fig. 21.16), a broad background signal exists.

The addition of nitric acid caused the sol to become clear and to attain low viscosity values within a few hours. In contrast to Figs. 21.16 and 12.17A the broad background signal almost disappeared. In addition to the static ^{19}F NMR spectra, MAS NMR spectra of all three sols were recorded with a rotation frequency of 13 kHz. They are shown in Fig. 21.18. All of them give the typical shape of solids with broad lines and spinning side bands. They are comparable to solid-state spectra of a magnesium fluoride powder (Fig. 21.18) with the typical signal at −198 ppm. The signal of the sol with lower fluorine supply $MgF_{0.9}(OR)_{1.1}$ (Fig. 21.18A) has a shoulder in its low-field region. This indicates, as expected, the presence of a higher portion of oxygen coordination in the second coordination sphere of fluorine [FMg_3].

Although ^{19}F static and ^{19}F MAS NMR spectra were performed on colloidal suspensions, we were able to identify five different F-containing species. They were observed at −198 ppm (MgF_2 NPs), at −187 ppm (unconsumed HF, adsorbed on MgF_2 particles), at −174 ppm [$Mg_6F_2(OCH_3)_{10}(CH_3OH)$, dicubane structure], and at −154 ppm [BF_4^-, with boron dissolved from the glass vessel], and a very broad signal was seen originating from larger particles and agglomerates. With substoichiometric fluorine supply, the dicubane structure is immediately formed and disappears with time. The typical narrow ^{19}F NMR signal of MgF_2 NPs was observed immediately after the addition of 2 eq. HF to the magnesium methoxide suspension. Therefore, the formation of MgF_2 NPs starts very rapidly but is then slowed down due to structural rearrangements (Fig. 21.17A). Adding nitric acid to the sol supports the rapid formation and stabilization of MgF_2 NPs and reduces the number of

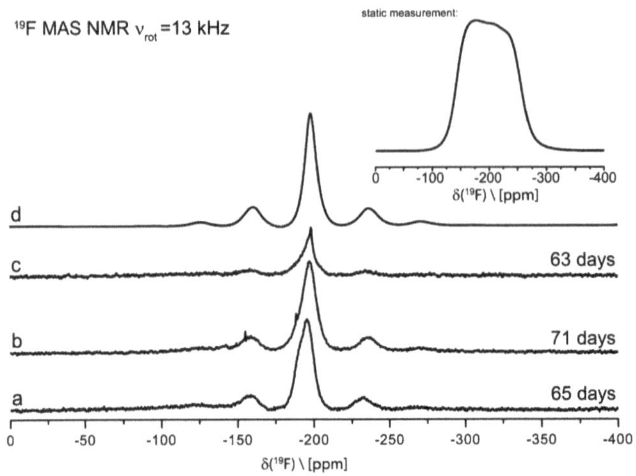

Figure 21.18 ^{19}F MAS nuclear magnetic resonance spectra (ν_{rot} = 13 kHz) of the three different sols (a) $MgF_{0.9}(OR)_{1.1}$, (b) MgF_2, and (c) MgF_2 + 1 mol% HNO_3, and (d) crystalline MgF_2 powder.
Reproduced from Karg, M.; Scholz, G.; König, R.; Kemnitz, E. Mechanistic Insight Into Formation and Changes of Nanoparticles in MgF_2 Sols Evidenced by Liquid and Solid State NMR. Dalton Trans. **2012**, 41, 2360–2366 with permission of the Royal Society of Chemistry.

agglomerates and larger particles (Fig. 21.17B). MAS of the sols allowed an unambiguous assignment of the broad ^{19}F lines to those typical for magnesium fluoride or magnesium fluoride xerogels (Fig. 21.18).

Such ^{19}F MAS NMR spectra of xerogels, obtained after drying of the respective sols with different Mg/F ratios, are shown in Figs. 21.19 and 21.20.

An Mg/F ratio of 10 leads to a narrow ^{19}F MAS NMR signal at −174 ppm, which experiences a broadening up to ~850 Hz with increasing fluorine supply (Fig. 21.19A–C). Simultaneously, new and broad signals occur with chemical shift positions at −186 and −193 ppm. These broad signals are more and more shifted into the high-field part of the spectrum with increasing fluorine content (Fig. 21.19D) and in the stoichiometric case (Mg/F = 0.5, Fig. 21.19E), the typical ^{19}F spectrum of amorphous MgF_2 is recorded. It is interesting to note that additional postfluorination of the $MgF_x(OCH_3)_{2-x}$ xerogels results in almost identical spectra (see Fig. 21.19F and G), independent of the fluorine supply during the sol–gel preparation. It means that for the Mg/F ratio of 2 (Fig. 21.19C) a complete change of the ^{19}F spectral shape takes place. Obviously, the understoichiometric fluorine supply in the anhydrous sol–gel route results unambiguously in the formation of the hexanuclear dicubane unit containing μ_4-fluorine atoms, indicated by the ^{19}F signal at −174 ppm. With increasing fluorine content the formation of the [FMg_4] dicubane environment is more and more suppressed and rutile-like local coordinations [FMg_3] with varying amounts of fluorine/oxygen in the second coordination sphere are formed. They are represented by broad signals between −185 and −200 ppm (Fig. 21.19). A proof for a higher oxygen content in these xerogels is supported by the change in the spectral

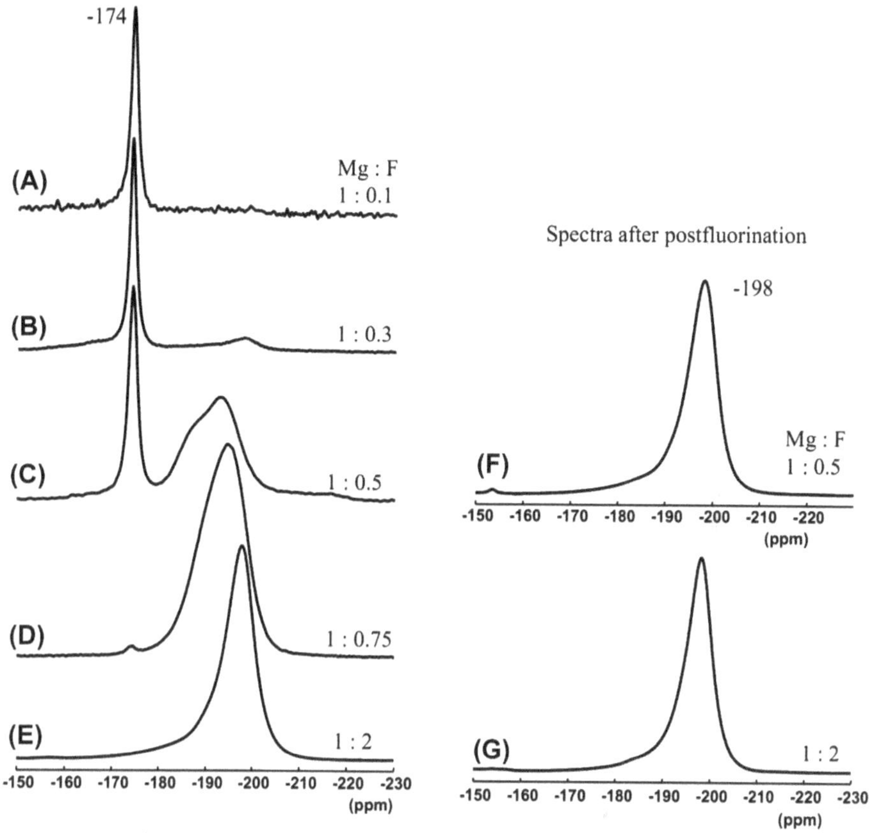

Figure 21.19 ^{19}F MAS nuclear magnetic resonance spectra (ν_{rot} = 28 kHz, central lines) of nanoscopic HS−MgF_2 samples prepared by the anhydrous sol−gel route applying different Mg:F molar ratios as given in the figure. [(A−E) xerogels, (F, G) samples after postfluorination].
Reproduced from Scholz, G.; Stosiek, C.; Noack, J.; Kemnitz, E. Local Fluorine Environments in Nanoscopic Magnesium Hydr(oxide) Fluorides Studied by ^{19}F MAS NMR. J. Fluor. Chem. **2011**, 132, 1079−1085 with permission of Elsevier.

shape and position after postfluorination processes, removing the remaining oxygen content, exemplarily shown for the Mg:F ratio of 1:0.5 (see Fig. 21.19F). For comparison, these changes are marginal if the Mg/F ratio is stoichiometric during the preparation (Fig. 21.19E and G).

Using aqueous HF for the sol−gel synthesis of MgF_2 samples, two effects can be expected to influence the shape of the ^{19}F signals: (1) the water content of the HF used, and (2) different Mg/F ratios. In both cases a competition between fluorolysis and hydrolysis of $Mg(OCH_3)_2$ has to be considered leading to magnesium hydroxide fluorides (xerogels) $MgF_x(OH)_{2-x}$ as reaction products. This interplay of fluorine content, fluorolysis, and hydrolysis is manifested especially in Fig. 21.20 using 48% $HF_{(aq)}$ for fluorination.

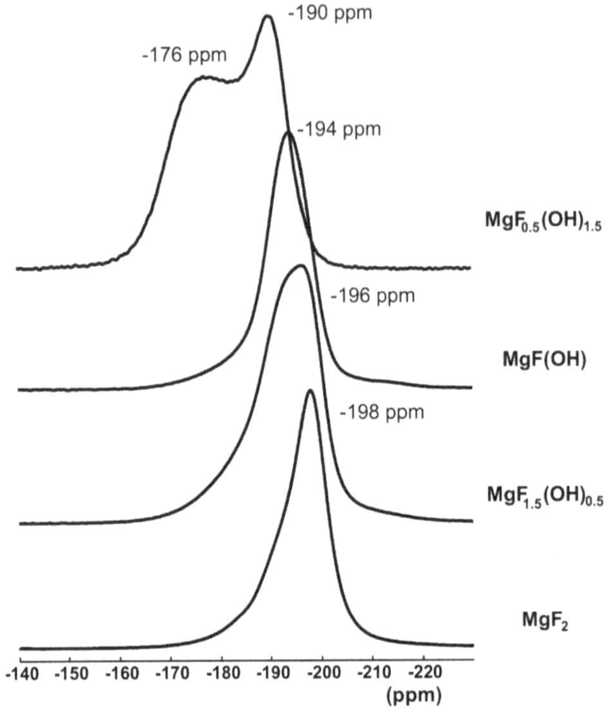

Figure 21.20 ^{19}F MAS nuclear magnetic resonance spectra ($\nu_{rot} = 25$ kHz, central lines) of $MgF_x(OH)_{2-x}$ samples prepared with 48% $HF_{(aq)}$ and different Mg:F molar ratios. Reproduced from Scholz, G.; Stosiek, C.; Noack, J.; Kemnitz, E. Local Fluorine Environments in Nanoscopic Magnesium Hydr(oxide) Fluorides Studied by ^{19}F MAS NMR. J. Fluor. Chem. **2011**, 132, 1079–1085 with permission of Elsevier.

Although the spectrum of $MgF_{0.5}(OH)_{1.5}$ is dominated by the broad signal at about −176 ppm, the higher F content induces not only a shift of the signal to lower δ-values, but also changes in the line shapes covering different species. For $MgF_{0.5}(OH)_{1.5}$ the dominating signal at −176 ppm foreshadows a related local structure as known for the dicubane structure ($[FMg_4]$) accompanied by distributions and a wide variety of mixed O/F coordinations in the second coordination sphere.

A comparison of all ^{19}F MAS NMR spectra shown in Figs. 21.18–21.20 shows that ^{19}F signals of $MgF_x(OH(R))_{2-x}$ samples ($0 < x \leq 2$) span a chemical shift range of about 50 ppm from −150 up to −200 ppm. Moreover, it is obvious that with under-stoichiometric fluorine content of the samples prepared via anhydrous or hydrous sol–gel routes, respectively, ^{19}F signals appear in the low-field region of the spectra or their intensities increase. This holds for the dried xerogels (cf. Figs. 21.19 and 21.20) and thermally treated samples.[46] These results are supported by findings obtained for mechanochemically synthesized samples.[46] Here incomplete mechanochemical fluorination reactions also lead to considerable ^{19}F signal intensity in the low-field part of the spectra between −150 and −185 ppm. On the basis of these experimental findings together with calculations exploring the superposition model to estimate ^{19}F chemical shifts,[4] chemical shift ranges for ^{19}F sites coordinated by Mg^{2+} cations were specified. A scheme of these ranges is given in Scheme 21.4.

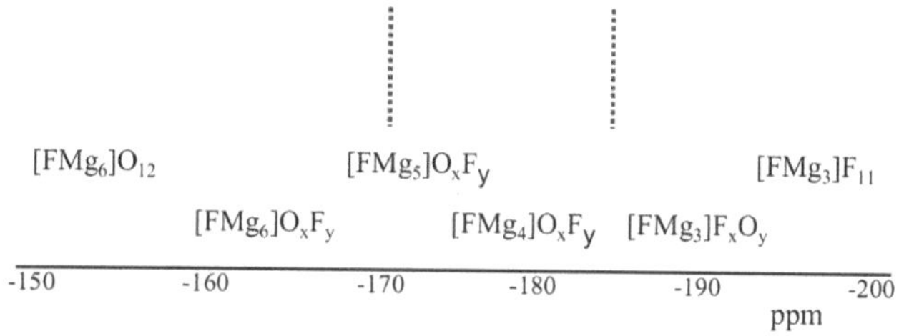

Scheme 21.4 Possible chemical shift ranges of ^{19}F sites coordinated by Mg^{2+} cations. Reproduced from Scholz, G.; Stosiek, C.; Noack, J.; Kemnitz, E. Local Fluorine Environments in Nanoscopic Magnesium Hydr(oxide) Fluorides Studied by ^{19}F MAS NMR. J. Fluor. Chem. **2011**, 132, 1079–1085 with permission of Elsevier.

The spanned ^{19}F chemical shift range can be roughly divided into three main areas: (1) −150 to −170 ppm, representing F sites with a high number of Mg^{2+} cations in the first coordination sphere of fluorine and mainly oxygen in the second coordination sphere; (2) −170 to −185 ppm, F species with a decreased number of Mg^{2+} cations in the first coordination sphere of fluorine, e.g., $[FMg_5]$ and $[FMg_4]$ environments, but still a high number of oxygen in the second coordination sphere; and (3) −185 to −200 ppm, fluorine sites with three Mg^{2+} ions in the nearest neighborhood, i.e., $[FMg_3]$ environments, similar to those in magnesium fluoride, but with slight variations of the fluorine content in the second coordination sphere.

Meanwhile calcium fluoride and strontium fluoride NPs were also successfully synthesized exploring the fluorolytic sol–gel route.[37,45] Bearing in mind the aim to prepare antireflective coatings with CaF_2 sols, different calcium precursors were studied. Finally, only the reaction of $CaCl_2$ with ethanolic HF resulted in transparent CaF_2 sols, which allowed to obtain antireflective layers with good optical and mechanical properties.[37] As for MgF_2, insights into the formation of local structures were possible with ^{19}F liquid and solid-state NMR spectroscopy. ^{19}F NMR spectra of sols with different fluorine supply are shown in Fig. 21.21.

The amount of HF to calcium chloride was varied between 0.1 and 2 to follow the appearance of new species formed in the course of the reaction (Fig. 21.21). The signal at −108 ppm, typical for CaF_2 NPs, appears immediately, even for the lowest fluorine portion. Only with a higher F/Ca ratio, especially visible for the ratio 2 given as insert, a narrow signal at about −180 ppm occurs. In analogy to the synthesis of MgF_2, this signal can be attributed to unconsumed HF, adsorbed on the surface of $CaF_x(OR(Cl))_{2-x}$ NPs. The latter form larger particles and agglomerates, which are responsible for the broad line visible with higher fluorine portion (Fig. 21.21). The small and narrow signal at about −93 ppm visible in Fig. 21.21 was originally attributed to calcium chlorofluoride species. However, separate solid-state chemical reactions between CaF_2 and $CaCl_2$ lead to crystalline CaClF, which finally gives a

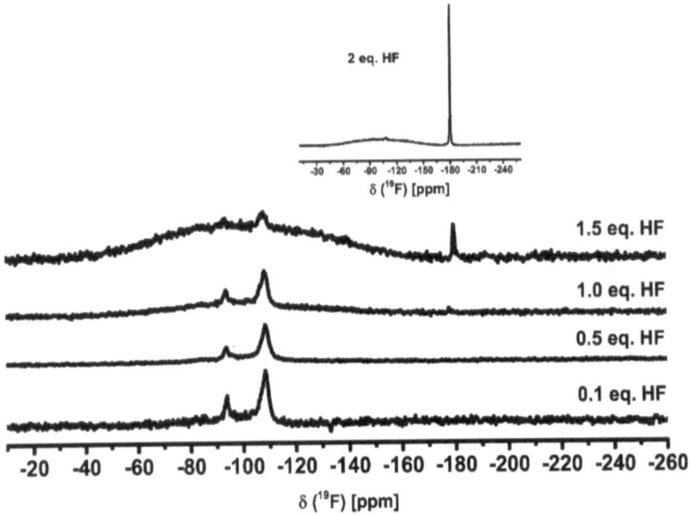

Figure 21.21 Static ^{19}F nuclear magnetic resonance spectra of CaF_2 sols prepared with different fluorine supply.
Reproduced from Rehmer, A.; Scheurell, K.; Kemnitz, E. Formation of Nanoscopic CaF_2 via a Fluorolytic Sol-Gel Process for Antireflective Coatings. J. Mater. Chem. C **2015**, 3, 1716–1723 with permission of the Royal Society of Chemistry.

^{19}F signal at −83 ppm. Also, thermal treatment of CaF_2 xerogels at different temperatures supported the assumption that a new intermediate fluorine-containing species is formed during the fluorolysis reaction of $CaCl_2$. Drying at lower temperature, here 40°C, shows the ^{19}F signals at −108 ppm (CaF_2) and −93 ppm (Fig. 21.22A). This signal disappears at higher temperatures, and the new signal at −83 ppm can be assigned to CaClF (Fig. 21.22B).

The new, so far unknown fluorine species that occurs at −93 ppm was assigned to an oxide fluoride species formed during the fluorination process. The latter was not only recorded for $CaCl_2$ as precursor compound but also for other calcium precursors like calcium bromide or calcium oxide.

The fluorolytic sol–gel synthesis of nanoscopic strontium fluoride was successfully performed using commercially available strontium acetate hemihydrate as precursor.[45] The latter was first dissolved in acetic acid, followed by adding ethanol to the solution. After that strontium acetate precipitates, and to the formed suspension a stoichiometric amount of HF was added. Methods like XRD, DLS, and TEM confirm the formation of SrF_2 NPs below 10 nm. The role of the organic solvent within the sol–gel process was followed by ^{1}H NMR measurements over a period of days and months. The spectra are given in Fig. 21.23. Immediately after fluorination, the ^{1}H spectrum is dominated by the signals of acetic acid and ethanol. The rapid proton exchanges allow observing only an average signal at about 7.53 ppm. With increasing time this signal shifts to a higher field, indicating the progressive esterification of acetic acid with ethanol to ethylacetate and the formation of water.

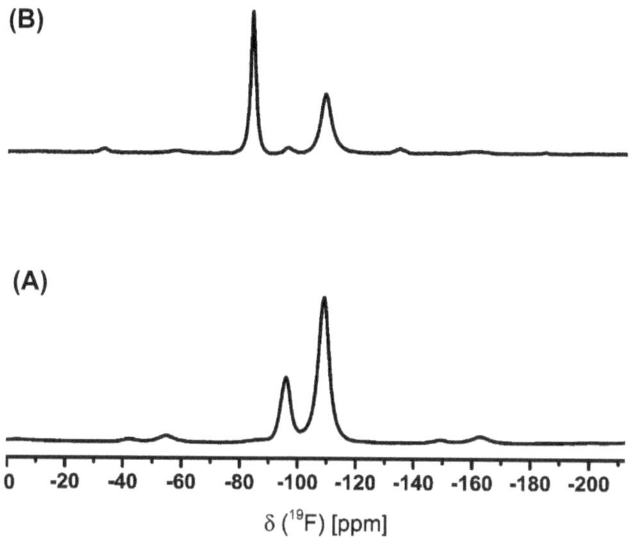

Figure 21.22 ^{19}F MAS nuclear magnetic resonance spectra ($\nu_{rot} = 20$ kHz) of a CaF_2 xerogel prepared from $CaCl_2$ with one equivalent of HF_{EtOH} (A) dried at 40°C and (B) dried at 600°C. Reproduced from Rehmer, A.; Scheurell, K.; Kemnitz, E. Formation of Nanoscopic CaF_2 via a Fluorolytic Sol-Gel Process for Antireflective Coatings. J. Mater. Chem. C **2015**, 3, 1716–1723 with permission of the Royal Society of Chemistry.

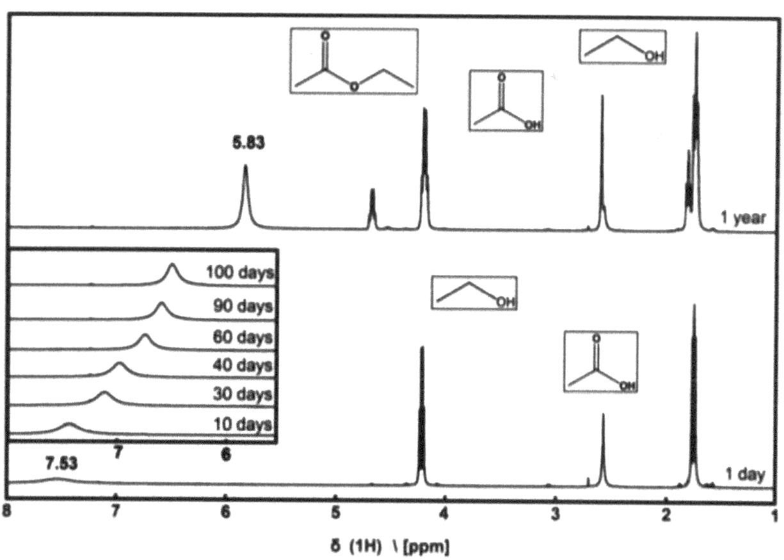

Figure 21.23 ^{1}H nuclear magnetic resonance spectra of a 0.2 M SrF_2 sol taken at different aging times.
Reproduced from Schmidt, L.; Emmerling, F.; Kirmse, H.; Kemnitz, E. Sol-Gel Synthesis and Characterisatopn of Nanoscopic Strontium Fluoride. RSC Adv. **2014**, 4, 32–38 with permission of the Royal Society of Chemistry.

To get more insight into the reaction progress, both X-ray powder diffractograms and ^{19}F MAS NMR spectra were recorded for SrF_2 xerogels with increasing fluorine supply, as it was also followed during the synthesis for other nanoscaled fluorides. The XRD powder patterns of this series are given in Fig. 21.24.

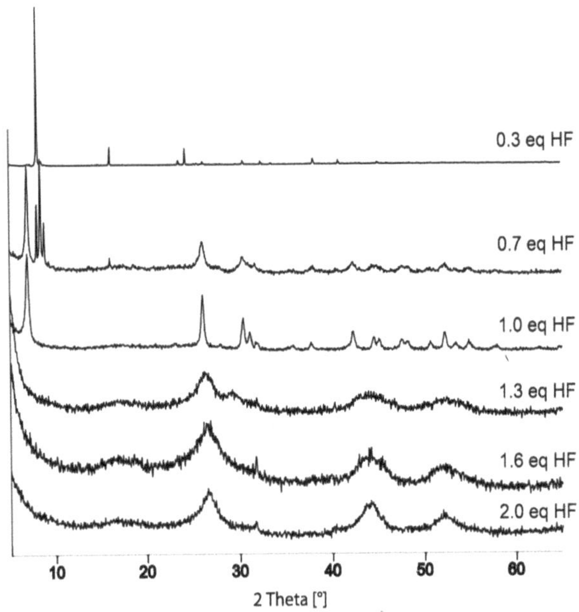

Figure 21.24 X-ray diffraction patterns of samples with varying F:Sr ratio. Reproduced from Schmidt, L.; Emmerling, F.; Kirmse, H.; Kemnitz, E. Sol-Gel Synthesis and Characterisatopn of Nanoscopic Strontium Fluoride. RSC Adv. **2014**, 4, 32–38 with permission of the Royal Society of Chemistry.

For F:Sr ratio of only 0.3 the narrow and intense reflections can be clearly assigned to the precursor $Sr(OAc)_2 \cdot 1/2H_2O$. These reflections disappear with higher fluorine supply. The formation of a new crystalline intermediate species is observed with a HF amount of 1.0 (Fig. 21.24). The formation of nanocrystalline SrF_2 is indicated with a F:Sr ratio of 1.3 and higher. This overall sequence is confirmed by solid-state NMR experiments. The ^{19}F MAS NMR spectra are summarized in Fig. 21.25.

The ^{19}F signal with a chemical shift of −79 ppm indicates the formation of a fluorine-containing crystalline species as the only fluorination product with F:Sr ratio below 1. With increasing HF amount, the intensity of this peak decreases, whereas two other broad and overlapping signals appear at −83 and −88 ppm. Following this sequence and comparing with crystalline SrF_2, the ^{19}F signal at −88 ppm can be unambiguously assigned to fluorine in the cubic structure of SrF_2. The asymmetry and broadening of this line toward the low field, is characteristic for fluorine with a varying chemical environment of oxygen and fluorine in the second coordination sphere.

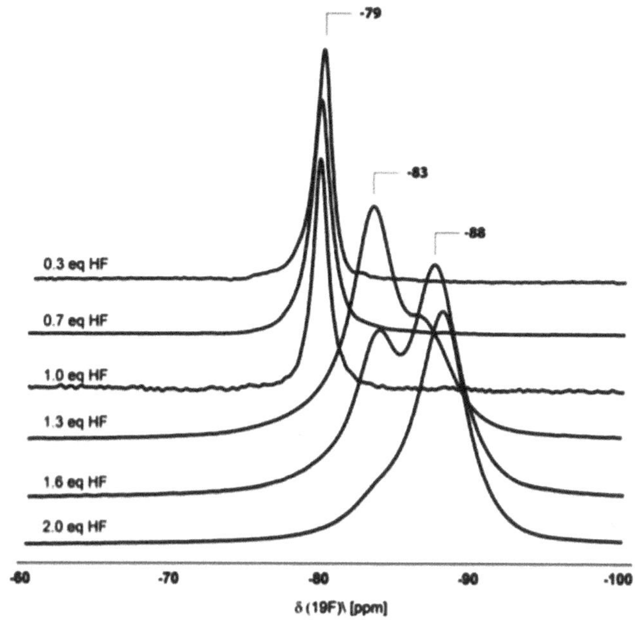

Figure 21.25 ^{19}F MAS nuclear magnetic resonance spectra of xerogels with different F:Sr ratio recorded at 25 kHz.
Reproduced from Schmidt, L.; Emmerling, F.; Kirmse, H.; Kemnitz, E. Sol-Gel Synthesis and Characterisatopn of Nanoscopic Strontium Fluoride. RSC Adv. **2014**, 4, 32–38 with permission of the Royal Society of Chemistry.

This situation is quite similar to the MgF_2 system, as shown in detail in Figs. 21.19 and 21.20 and Scheme 21.4, and also goes parallel to observations made for the formation of CaF_2 NPs via fluorolytic sol–gel synthesis.

4. Applications

There are several fields of applications arising from the availability of nanoscopic metal fluorides. One field with a tremendous potential is that of heterogeneous catalysis. However, it is not only the Lewis acidic character of nanoscopic metal fluorides but also extension of the synthesis approach that enables easy access to a wide variety of metal hydroxide fluorides and oxide fluorides, respectively, according to the following general equations:

$$M(OR)_n + (n-x)HF + xH_2O \rightarrow MF_{n-x}(OH)_x + nROH \quad (21.12)$$

$$MF_{n-x}(OH)_x \xrightarrow{T} MF_{n-x}O_{x/2} + x/2H_2O \quad (21.13)$$

These new catalyst systems and their wide application have been already comprehensively described in two recent reviews,[20,18] and hence, for further details, we refer to these reviews.

It is worth mentioning that even ternary or quaternary nanoscopic metal fluoride systems are accessible exhibiting interesting catalytic potential (see e.g., refs.11,12).

More interesting in this context are applications that are exclusively possible based on the accessibility of optically transparent metal fluoride sols. Because they are composed of homodispersed NPs, they allow applications that are not at all possible with classically prepared metal fluorides. Thus, this chapter will give a very brief overview about new fields of application based on materials available via the fluorolytic sol–gel synthesis route.

4.1 Antireflective Coatings

Thin films of organic and inorganic materials find wide industrial applications, e.g., in decorative and antireflective coatings of glass.[10] Metal fluorides, in particular AlF_3, MgF_2, and CaF_2, are inorganic materials with the lowest refractive indices. Sols obtained via sol–gel methods have been proved to be interesting for the processing of thin, defined films with high homogeneity.[13,27,33,41] Fig. 21.26 shows the antireflective behavior of MgF_2-coatings on glass. On the left side of the glass slide one can see the reflection of the image above, whereas the right side is highly transparent.

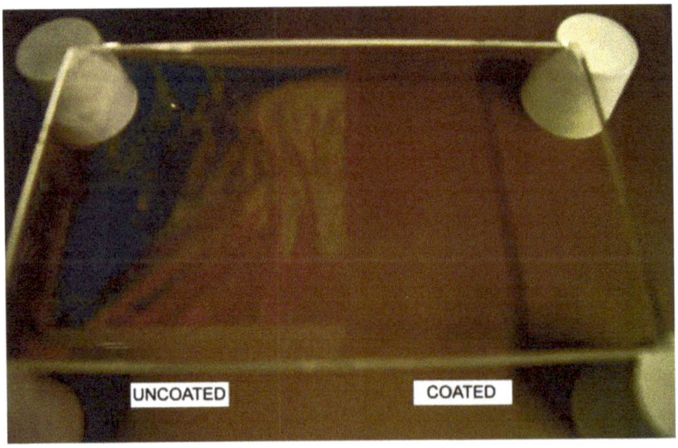

Figure 21.26 Glass substrate coated by MgF_2 (right), whereas the left side is untreated. An image is shown above the glass, and the mirror (left) or nonmirror (right) effect of each area while taking a picture from above can be seen.
Reproduced from Kemnitz, E.; Noack, J. The Non-aqueous Fluorolytic Sol-Gel Synthesis of Nanoscaled Metal Fluorides. Dalton Trans. **2015**, 44, 19411–19431 with permission of the Royal Society of Chemistry.

As a result of porosity inside the films (cf. Fig. 21.27) the refractive index of a sol–gel-prepared MgF_2 film on silicon wafers is significantly lower than that of bulk material ($n_{MgF_2\ layer} = 1.18$ to 1.30, $n_{bulkMgF_2} = 1.38$).

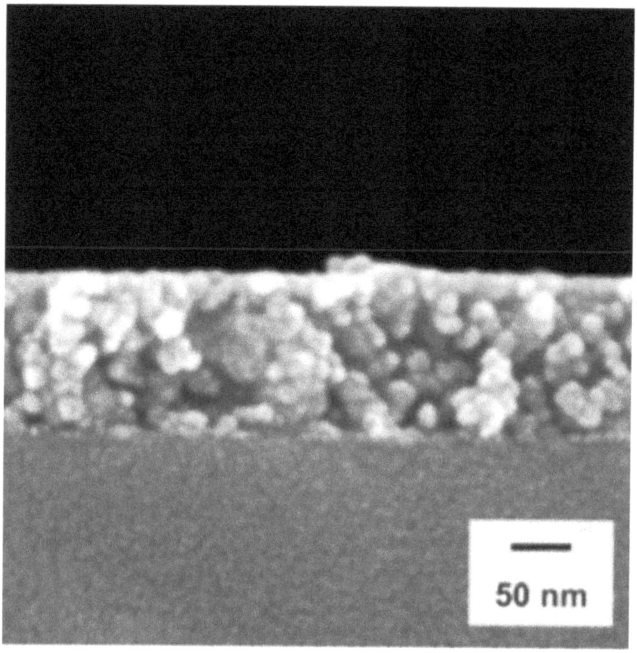

Figure 21.27 Cross-sectional scanning electron microscopic image of MgF_2 thin film annealed at 500°C. Reproduced from (MgF_2 anti-reflective coatings by sol–gel processing: film preparation and thermal densification.
Reproduced from Noack, J.; Scheurell, K.; Kemnitz, E.; Garcia-Juan, P.; Rau, H.; Lacroix, M.; Eicher, J.; Lintner, B.; Sontheimer, T.; Hofmann, T.; Hegmann, J.; Jahn, R.; Löbmann, P. MgF_2 Antireflective Coatings by Sol-Gel Processing: Film Preparation and Thermal Densification. J. Mater. Chem. **2012**, 22, 18535–18541 with permission of The Royal Society of Chemistry.

With a minimum reflectance of 0.2% at 600 nm, the films show excellent anti-reflective properties, which are, due to the porosity, much better than for bulk MgF_2. The films exhibit superior mechanical stability as proven by crockmeter tests applying steel wool pads.[26]

4.2 Up- and Downconversion Materials

In the past years, the wet chemical synthesis of rare-earth (RE)-doped calcium fluoride by precipitation methods has been reported by several groups.[7,51,54] The fluorolytic sol–gel method offers great advantages in contrast to classical precipitation routes since it can be carried out at room temperature in large batches using standard glass vessels and usually no surfactants are required. Recently, we reported the sol–gel synthesis of luminescent CaF_2 NP sols by the reaction of calcium lactate dissolved in methanol with anhydrous HF.[38] Doping of up to 40 mol% with RE metal cations can be achieved by addition of lanthanide acetate before fluorination. The reaction yields fully transparent sols of crystalline RE-doped CaF_2 particles with a size in

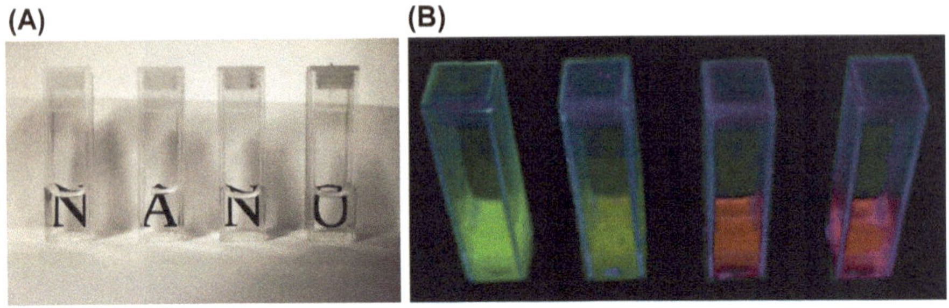

Figure 21.28 Transparent Eu^{3+}-and Tb^{3+}-doped CaF_2 sols (A) and the same sols excited with 366 nm (B). From left to right: CaF_2:Tb10, mixture of 52% of CaF_2:Tb10 + 48% of CaF_2: Eu10, CaF_2:Tb5.2, Eu4.8, and CaF_2:Eu10.
Reproduced from Ritter, B.; Krahl, T.; Rurack, K.; Kemnitz, E. Nanoscale CaF_2 Doped With Eu^{3+} and Tb^{3+} Through Fluorolytic Sol-Gel Synthesis. J. Mater. Chem. C **2014**, 2, 8607–8613 with permission of The Royal Society of Chemistry.

the lower nanometer range, as seen in Fig. 21.28A. Under illumination with 366 nm radiation, the Eu^{3+}- and Tb^{3+}-doped CaF_2 sols appear in the characteristic colors (Fig. 21.28B) of $^5D_0 \rightarrow {}^7F_J$ and $^5D_4 \rightarrow {}^7F_J$ transitions for Eu^{3+} and Tb^{3+}, respectively. Codoping of CaF_2 with Eu and Tb exhibits a different emission spectrum than the physical mixture of the pure sols which show the expected bright yellow emission. This suggests that there is an interaction between both rare earth ions in the codoped particles that has not been observed for the mixture of singly doped particles. Apparently, separate doping in two different particles prevents interaction between two different rare earth metals, and hence, energy transfer from one rare earth ion to another is not possible.[44]

4.3 Ceramics

Nanoscopic metal fluorides, especially MgF_2, have recently been reported for their superior properties as sintering additives in corundum ceramic manufacturing.[49] Without interfering with the technological process but just by addition of approximately 0.1% nanoscopic MgF_2 to the starting ceramic powder, optically transparent Al_2O_3 discs can be obtained with >70% optical transmittance (see Fig. 21.29). As can be seen from Fig. 21.17, the addition of nanoscopic MgF_2 resulted in highly transparent corundum ceramic materials. Remarkably enough, the Martens hardness, which for corundum ceramic is usually in the region of 2100, increases significantly up to 3200, thus corresponding to values that are characteristic for silicon carbide and/or boron nitride ceramics.

4.4 Inorganic–Organic Nanocomposites

Composite materials made by combination of inorganic NPs and organic polymers are of high interest for materials science. As in the case of other inorganic NP materials nanoscopic homodispersed metal fluorides are also now available for manufacturing

Figure 21.29 Picture taken wherein the left side of the camera lens (two-thirds of the image) is covered by the "new transparent ceramic" disc, whereas the right side shows the original photo. Reproduced from Kemnitz, E.; Noack, J. The Non-aqueous Fluorolytic Sol-Gel Synthesis of Nanoscaled Metal Fluorides. Dalton Trans. **2015**, 44, 19411–19431 with permission of the Royal Society of Chemistry.

similar composite materials. These investigations were driven by the fact that metal fluorides, and especially magnesium and aluminum fluoride, exhibit lower refractive indices than metal oxides and can, therefore, be utilized for the adjustment of the optical properties of polymers. Since the scattering of light at the filler particles strongly relates to their diameter, particles smaller than 20 nm have to be introduced into the

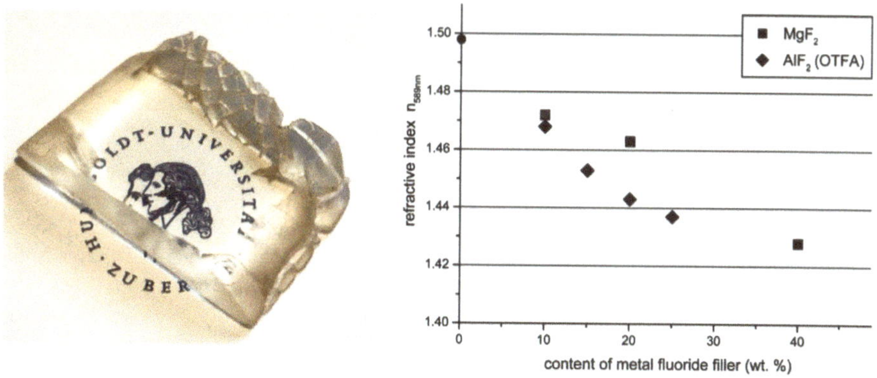

Figure 21.30 Photograph of a representative composite sample with 10 wt% MgF_2 in a polyHEMA matrix as bulk material (left) and refractive indices of thin film coatings based on magnesium and aluminum fluoride with different filler content in a matrix of poly-Butanediol Dimethacrylate (1,4-BDDMA) determined by ellipsometry (right).
Reproduced from Noack, J.; Fritz, C.; Flügel, C.; Hemmann, F.; Gläsel, H.J.; Kahle, O.; Dreyer, C.; Bauer, M.; Kemnitz, E. Metal Fluoride-Based Transparent Nanocomposites With Low Refractive Indices. Dalton Trans. **2013**, 42 (16), 5706–5710 with permission of the Royal Society of Chemistry.

polymer matrix and be effectively stabilized against agglomeration. If processed properly, optically transparent composites with a defined refractive index can be obtained, such as shown in Fig. 21.30 of a representative composite sample with 10 wt% MgF_2 in a polyHEMA (Hydroxyethylmethacrylate) matrix.[34] Fig. 21.30 illustrates the trends of decreasing refractive indices of the resulting composite films depending on the amount and type of filler material. The pure polymer has a refractive index of 1.498 at 589 nm wavelength, whereas the introduction of metal fluoride filler strongly reduces the overall refractive index of the composite. The filler material does not contribute to absorption or light scattering in the visible spectral range as determined by ellipsometry. Since aluminum fluoride has an even lower refractive index than magnesium fluoride, the effects of its incorporation are much more pronounced. Note that porosity, which essentially accounts for the very low refractive indices of thin films, cannot be considered in the case of composites. The resulting optical constants are described by the sum of the volume fractions of individual components.[34]

The introduction of metal fluoride NPs also positively influences the Martens hardness of the composites.[8,31,34] For instance, introduction of 15 wt% MgF_2 NPs into an HDDMA polymer causes the Martens hardness (determined by microindentation) to be more than doubled from 107 to 222 N/nm^2, and also, the storage modulus G' rises from 2.2 to 2.8 GPa.[31]

Acknowledgments

The authors gratefully acknowledge financial support from the German Federal Ministry of Economics and Technology (grant 0329800, synonym: TRex and grant 03ET1235C, synonym KrAnich) and the Deutsche Forschungsgemeinschaft (DFG) for the graduate school GRK 1582 "Fluorine as a Key Element."

References

1. Amatucci, G. G.; Pereira, N. Fluoride Based Electrode Materials for Advanced Energy Storage Devices. *J. Fluor. Chem.* **2007,** *128,* 243–262.
2. Amatucci, G. G.; Pereira, N.; Badway, F.; Sina, M.; Cosandey, F.; Ruotolo, M.; Cao, C. Formation of Lithium Fluoride/Metal Nanocomposites for Energy Storage Through Solid State Reduction of Metal Fluorides. *J. Fluor. Chem.* **2011,** *132,* 1086–1094.
3. Brinker, C. J.; Scherer, G. W. *Sol-Gel Science: The Physics and Chemistry of Sol-Gel Processing;* Academic Press: San Diego, 1990.
4. Bureau, B.; Silly, G.; Buzaré, J. Y.; Emery, J. Superposition Model for ^{19}F Isotropic Chemical Shift in Ionic Fluorides: From Basic Metal Fluorides to Transition Metal Fluoride Glasses. *Chem. Phys.* **1999,** *249,* 89–104.
5. Dimitrov, A.; Koch, J.; Troyanov, S. I.; Kemnitz, E. Aluminium Alkoxide Fluorides Involved in the Sol-Gel Synthesis of Nanoscopic AlF_3. *Eur. J. Inorg. Chem.* **2009,** 5299–5301.

6. Fedorov, P. P.; Luginina, A. A.; Kuznetsov, S. V.; Osiko, V. V. Nanofluorides. *J. Fluor. Chem.* **2011,** *132* (12), 1012–1039.
7. Feldmann, C.; Roming, M.; Trampert, K. Polyol-Mediated Synthesis of Nanoscale CaF_2 and CaF_2:Ce,Tb. *Small* **2006,** *2* (11), 1248–1250.
8. Fritz, C.; Scholz, G.; Feist, M.; Kemnitz, E. Preparation and Stabilization of Aluminium Trifluoroacetate Fluoride Sols for Optical Coatings. *Dalton Trans.* **2012,** *41* (37), 11351–11360.
9. German Patent. **1994**. EP 0 597 210 A1.
10. Glaubitt, W.; Löbmann, P. Antireflective Coatings Prepared by Sol-Gel Processing: Principles and Applications. *J. Eur. Ceram. Soc.* **2012,** *32* (11), 2995–2999.
11. Guo, Y.; Gaczynski, P.; Becker, K. -D.; Kemnitz, E. Sol-Gel Synthesis and Characterization of Nanoscopic FeF_3-MgF_2 Heterogeneous Catalysts with Bi-acidic Properties. *ChemCatChem* **2013,** *5,* 2223–2232.
12. Guo, Y.; Lippitz, A.; Saftien, P.; Unger, W. E. S.; Kemnitz, E. Tuning the Surface Properties of Novel Ternary Iron(III) Fluoride Based Catalysts Using the Template Effect of the Matrix. *Dalton Trans.* **2015,** *44,* 5076–5085.
13. Hegmann, J.; Löbmann, P. Sol-Gel Preparation of TiO_2 and MgF_2 Multilayers. *J. Sol Gel Sci. Technol.* **2013,** *67* (3), 436–441.
14. Hüsing, N.; Schubert, U. Aerogels-Airy Materials: Chemistry, Structure and Properties. *Angew. Chem. Int. Ed.* **1998,** *37,* 22–45.
15. Karg, M.; Scholz, G.; König, R.; Kemnitz, E. Mechanistic Insight Into Formation and Changes of Nanoparticles in MgF_2 Sols Evidenced by Liquid and Solid State NMR. *Dalton Trans.* **2012,** *41,* 2360–2366.
16. Kemnitz, E.; Groß, U.; Rüdiger, S.; Shekar, C. S. Amorphous Metal Fluorides With Extraordinary High Surface Areas. *Angew. Chem. Int. Ed.* **2003,** *42,* 4251–4254.
17. Kemnitz, E.; Scholz, G.; Rüdiger, S. Sol-Gel Synthesis of Nano-Scaled Metal Fluorides – Mechanism and Properties. In *Functionalized Inorganic Fluorides;* Tressaud, A., Ed.; J. Wiley and Sons, Ltd, 2010; pp 1–38.
18. Kemnitz, E.; Wuttke, S.; Coman, S. M. Tailor-Made MgF_2-Based Catalysts by Sol-Gel Synthesis. *Eur. J. Inorg. Chem.* **2011,** 4773–4794.
19. Kemnitz, E.; Noack, J. The Non-aqueous Fluorolytic Sol-Gel Synthesis of Nanoscaled Metal Fluorides. *Dalton Trans.* **2015,** *44,* 19411–19431.
20. Kemnitz, E. Nanoscale Metal Fluorides: A New Class of Heterogeneous Catalysts. *Catal. Sci. Technol.* **2015,** *5,* 786–806.
21. Kistler, S. S. Coherent Expanded Aerogels and Jellies. *Nature* **1931,** *127,* 741.
22. König, R.; Scholz, G.; Kemnitz, E. New Inserts and Low Temperature-Two Strategies to Overcome the Bottleneck in MAS NMR on Wet Gels. *Solid State Nucl. Magn. Reson.* **2007,** *32,* 78–88.
23. König, R.; Scholz, G.; Pawlik, A.; Jäger, C.; van Rossum, B.; Kemnitz, E. Identification of $AlF_x(OR)_y$ Species in Strongly Disordered Aluminium Isopropoxide Fluoride Solids: A Field-Dependent MAS NMR Study. *J. Phys. Chem. C* **2009,** *113,* 15576–15585.
24. König, R.; Scholz, G.; Kemnitz, E. Local Structural Changes in Aluminium Isopropoxide Fluoride Xerogels and Solids as a Consequence of the Progressive Fluorination Degree. *J. Phys. Chem. C* **2009,** *113,* 6426–6438.
25. König, R.; Scholz, G.; Veiczi, M.; Jäger, C.; Troyanov, S. I.; Kemnitz, E. New Crystalline Aluminium Alkoxide Fluorides: Evidence of the Mechanism of the Fluorolytic Sol-Gel Reaction. *Dalton Trans.* **2011,** *40,* 8701–8710.
26. Krahl, T.; Broßke, D.; Scheurell, K.; Lintner, B.; Kemnitz, E. High Performance Antireflective MgF_2 Layers on Glass. *J. Mater. Chem. C* **2016,** *4,* 1454–1466.

27. Krüger, H.; Kemnitz, E.; Hertwig, A.; Beck, U. Transparent MgF_2-Films by Sol-Gel Coating: Synthesis and Optical Properties. *Thin Solid Films* **2008,** *516,* 4175–4177.
28. Krüger, H.; Kemnitz, E.; Hertwig, A.; Beck, U. Moderate Temperature Sol-Gel Deposition of Magnesium Fluoride Films for Optical UV-Applications: A Study on Homogeneity Using Spectroscopic Ellipsometry. *Phys. Status Solidi A* **2008,** *205,* 821–824.
29. MacKenzie, K. J. D.; Smith, M. E. *Multinuclear Solid-State NMR of Inorganic Materials. Pergamon Materials Series,* Vol. 6; Pergamon Press, 2002.
30. Murata, T.; Ishizawa, H.; Motoyama, I.; Tanaka, A. Investigations of MgF_2 Optical Thin Films Prepared From Autoclaved. *J. Sol Gel Sci. Technol.* **2004,** *32,* 161–165.
31. Noack, J.; Schmidt, L.; Gläsel, H.-J.; Bauer, M.; Kemnitz, E. Inorganic-Organic Nanocomposites Based on Sol-Gel Derived Magnesium Fluoride. *Nanoscale* **2011,** *3,* 4774–4779.
32. Noack, J.; Emmerling, F.; Kirmse, H.; Kemnitz, E. Sols of Nanosized Magnesium Fluoride: Formation and Stabilisation of Nanoparticles. *J. Mater. Chem.* **2011,** *21,* 15015–15021.
33. Noack, J.; Scheurell, K.; Kemnitz, E.; Garcia-Juan, P.; Rau, H.; Lacroix, M.; Eicher, J.; Lintner, B.; Sontheimer, T.; Hofmann, T.; Hegmann, J.; Jahn, R.; Löbmann, P. MgF_2 Antireflective Coatings by Sol-Gel Processing: Film Preparation and Thermal Densification. *J. Mater. Chem.* **2012,** *22,* 18535–18541.
34. Noack, J.; Fritz, C.; Flügel, C.; Hemmann, F.; Gläsel, H. J.; Kahle, O.; Dreyer, C.; Bauer, M.; Kemnitz, E. Metal Fluoride-Based Transparent Nanocomposites With Low Refractive Indices. *Dalton Trans.* **2013,** *42* (16), 5706–5710.
35. Raut, H. K.; Dinachali, S. S.; Ansah- Antwi, K. K.; Ganesh, V. A.; Ramakrishna, S. Fabrication of Highly Uniform and Porous MgF Anti-reflective Coatings by Polymer-Based Sol-Gel Processing on Large-Area Glass Substrates. *Nanotechnology* **2013,** *24,* 505201.
36. Reddy, M. A.; Fichtner, M. Fluoride Cathodes for Secondary Batteries. In *Advanced Fluoride-Based Materials for Energy Conversion;* Nakajima, T., Groult, H., Eds.; Elsevier Inc., 2015; pp 51–76.
37. Rehmer, A.; Scheurell, K.; Kemnitz, E. Formation of Nanoscopic CaF_2 via a Fluorolytic Sol-Gel Process for Antireflective Coatings. *J. Mater. Chem. C* **2015,** *3,* 1716–1723.
38. Ritter, B.; Krahl, T.; Rurack, K.; Kemnitz, E. Nanoscale CaF_2 Doped With Eu^{3+} and Tb^{3+} Through Fluorolytic Sol-Gel Synthesis. *J. Mater. Chem. C* **2014,** *2,* 8607–8613.
39. Rüdiger, S.; Groß, U.; Feist, M.; Prescott, H.; Shekar, C. S.; Troyanov, S. I.; Kemnitz, E. Non-aqueous Synthesis of High Surface Area Aluminium Fluoride – A Mechanistic Investigation. *J. Mater. Chem.* **2005,** *15,* 588–597.
40. Rüdiger, S.; Kemnitz, E. The Fluorolytic Sol-Gel Route to Metal Fluorides – A Versatile Process Opening a Variety of Application Fields. *Dalton Trans.* **2008,** 1117–1127.
41. Rywak, A. A.; Burlitch, J. M. Sol-Gel Synthesis of Nanocrystalline Magnesium Fluoride: Its Use in the Preparation of MgF_2 Films and MgF_2-SiO_2 Composites. *Chem. Mater.* **1996,** *8* (1), 60–67.
42. Scheurell, K.; König, R.; Troyanov, S. I.; Kemnitz, E. Structural Chemistry of Magnesium Acetates. *Z. Anorg. Allg. Chem.* **2012,** *638,* 1265–1273.
43. Scheurell, K.; Noack, J.; König, R.; Hegmann, J.; Jahn, R.; Hofmann, T.; Löbmann, P.; Lintner, B.; Garcia-Juan, P.; Eicher, J.; Kemnitz, E. Optimization of a Sol-Gel Synthesis Route for the Preparation of MgF_2 Particles for a Large Scale Coating Process. *Dalton Trans.* **2015,** *44,* 19501–19508.
44. Scheurell, K.; Kemnitz, E.; Garcia-Juan, P.; Eicher, J.; Lintner, B.; Hegmann, J.; Jahn, R.; Hofmann, T.; Löbmann, P. Porous MgF_2 Antireflextive $\lambda/4$ Films Prepared by Sol-Gel Processing: Comparison of Synthesis Approaches. *J. Sol Gel Sci. Technol.* **2015,** *76,* 82–89.

45. Schmidt, L.; Emmerling, F.; Kirmse, H.; Kemnitz, E. Sol-Gel Synthesis and Characterisatopn of Nanoscopic Strontium Fluoride. *RSC Adv.* **2014,** *4,* 32—38.
46. Scholz, G.; Stosiek, C.; Noack, J.; Kemnitz, E. Local Fluorine Environments in Nanoscopic Magnesium Hydr(oxide) Fluorides Studied by ^{19}F MAS NMR. *J. Fluor. Chem.* **2011,** *132,* 1079—1085.
47. Scholz, G.; Stosiek, C.; Feist, M.; Kemnitz, E. Magnesium Hydroxide Fluorides-New Materials With Adjustable Composition and Properties. *Eur. J. Inorg. Chem.* **2012,** 2337—2340.
48. Scholz, G.; Heidemann, D.; Kemnitz, E. Local Structure of Nanoscopic Magnesium Hydroxide Fluorides Studies by Natural Abundance ^{25}Mg Solid State NMR Spectroscopy. *Z. Anorg. Allg. Chem.* **2013,** *639,* 694—701.
49. Stosiek, C.; Ludwig, H.; Reichel, U.; Scholz, G.; Kemnitz, E. Nanoscopic Metal Fluorides as Promising Sintering Aids for High-Performance Alumina Ceramics. *J. Ceram. Sci. Technol.* **2011,** *2* (1), 31—38.
50. Stöber, W.; Fink, A.; Bohn, E. Controlled Growth of Monodisperse Silica Spheres in the Micron Size Range. *J. Coll. Interface Sci.* **1968,** *26,* 62—69.
51. Sun, X. Y.; Gu, M.; Huang, S. M.; Jin, X. J.; Liu, X. L.; Liu, B.; Ni, C. Luminescence Behavior of Tb^{3+} Ions in Transparent Glass and Glass-Ceramics Containing CaF_2 Nanocrystals. *J. Lumin.* **2009,** *129* (8), 773—777.
52. US-Patent. **2011**. US20110122497A1.
53. US-Patent. **2014**. US20140147594A1.
54. Wang, J. S.; Miao, W. R.; Li, Y. X.; Yao, H. C.; Li, Z. J. Water-Soluble Ln^{3+}-Doped Calcium Fluoride Nanocrystals: Controlled Synthesis and Luminescence Properties. *Mater. Lett.* **2009,** *63* (21), 1794—1796.
55. Web of Science. **2015**. http://apps.webofknowledge.com/.
56. World-Patent. **2011**. WO2011116980A1.
57. Wuttke, S.; Scholz, G.; Rüdiger, S.; Kemnitz, E. Variation of Sol-Gel Synthesis Parameters and Their Consequence for the Surface Area and Structure of Magnesium Fluoride. *J. Mater. Chem.* **2007,** *17,* 4980—4988.
58. Wuttke, S.; Coman, S. M.; Scholz, G.; Kirmse, H.; Vimont, A.; Daturi, M.; Schröder, S. L. M.; Kemnitz, E. Novel Sol-Gel Synthesis of Acidic $MgF_{2-x}(OH)_x$ Materials. *Chem. Eur. J.* **2008,** *14,* 11488—11499.
59. Wuttke, S.; Lehmann, A.; Scholz, G.; Feist, M.; Dimitrov, A.; Troyanov, S. I.; Kemnitz, E. Investigation of the Fluorolysis of Magnesium Methoxide. *Dalton Trans.* **2009,** 4729—4734.

Solution-Based Synthesis of Nano-Sized TiO_2 Anatase in Fluorinating Media

22

M. Jiwei Ma, W. Li, D. Dambournet
Sorbonne Universities, CNRS, UMR 8234, Paris, France

Chapter Outline

1. Introduction

Titanium dioxide (TiO_2) is a multifunctional material of interest for a broad range of applications ranging from (photo)catalysis to energy storage.[1–6] The low toxicity and the abundance of titanium have favored the emergence of Ti-based compounds. Over the years, several approaches have been developed to modify/improve the properties of TiO_2. These included reducing the particle size to the nanoscale, controlling their morphology, and doping with heteroatoms to tune the electronic structure and structural features.[6]

Anatase, a natural mineral, is one of the polymorph of TiO_2. The name "anatase" is derived from the Greek word "ana," which means "elongated" and refers to the mineral crystal's shape (Fig. 22.1). Using Wulff construction and calculated surface energies, the equilibrium shape of a TiO_2 anatase crystal has been predicted to consist of a truncated octahedron, which agrees with experimental observations.[7] The crystal exposes only two types of surfaces with 96% of (101) and 4% of (001) surface. Calculations of

Modern Synthesis Processes and Reactivity of Fluorinated Compounds. http://dx.doi.org/10.1016/B978-0-12-803740-9.00022-6

Figure 22.1 Photograph of a crystal of anatase, St Christophe-en-Oisans, Bourg d'Oisans, Isère, Rhône-Alpes, France Bourg d'Oisans, France.

surface energies indicate 0.44 J/m^2 and 0.90 J/m^2 for (101) and (001) surface, respectively, highlighting the highest stability for the (101)-type surface. The difference of stability of the two types of surfaces has been explained in terms of the density and nature of undercoordinated Ti species. The stable (101) surface exhibits 50% titanium in a sixfold coordination mode and 50% titanium in a fivefold coordination mode, whereas the metastable (001) surface contains only fivefold coordinated Ti featuring enhanced interfacial properties. Such a difference in surface reactivity has led to extensive researches on the preparation of TiO_2 crystals with specific facets.[8]

The role of anions during solution-based synthesis of inorganic compounds is multiple. Depending on their complexing ability toward cations, anions can drive the nucleation/crystallization toward a specific crystal structure. They can also adsorb onto surfaces thus orienting, in a particular direction, the growth of particles.

Originally, fluorine was widely used due to its mineralizing properties particularly for the synthesis of zeolites.[9] Concerning the synthesis of TiO_2, it is used as a structure-directing agent toward the anatase phase at the expense of the thermodynamically stable rutile structure. Moreover, it was used to control the morphology of the anatase crystal. In this scope, a significant breakthrough was made in 2008 by Yang et al. who have stabilized micrometer-sized anatase single crystals with 47% of (001) surface using hydrofluoric acid as a morphology-controlling agent.[10] Since then, the understanding and development of new chemical methods allowing stabilizing a large content of reactive metastable surface structure has received impressive attention. Most of the syntheses that have been explored are based on the sol–gel chemistry. Strikingly, most of the articles focusing on the stabilization of reactive surfaces using fluorinated media poorly discussed the possible stabilization of fluorine within the lattice of anatase and its possible implications on the observed enhanced properties.[8]

In the first part of this chapter, we present a brief review of the different roles of fluorine in the solution-based synthesis of TiO_2 anatase. In the second part, we discuss the reactivity of titanium alkoxide precursors toward HF and expose the report on the stabilization of highly fluorinated anatase, demonstrating that the synthesis of TiO_2 anatase on fluorinated media does not lead to a pure oxide.

2. General Interests of Fluorinating Media

2.1 Phase Selectivity

The crystal structure formed by solution-based methods strongly relies on the presence of a complexing agent that can favor the stabilization of one particular structural arrangement. In that sense, it is well established that fluoride always leads to the stabilization of anatase-type structure at the expense of the thermodynamically stable phase.[11–19]

For instance, Yu et al. have demonstrated that the addition of NH_4F—water mixture in the solution of titanium isopropoxide suppresses the formation of brookite phase and favors the crystallization of TiO_2 anatase, highlighting the structure-directing effect of fluorine on the selective formation of anatase phase.[19] Dong et al. have reported a rutile to anatase phase transition by increasing the fluorine concentration in $TiCl_4$ solution.[20]

The ability of fluoride to bind with Ti^{4+} at the molecular level influences the geometry of polymeric building units, which eventually reflects the final arrangement of the structure. Thus, fluoride anion is a structure-directing agent favoring nonlinear arrangements as found in anatase and as opposed to linear arrangements displayed by the rutile structure. From the structural standpoint, the natural allotropic forms of titanium dioxide, i.e., anatase, rutile, and brookite, differ from the way the TiO_6 octahedral subunits are connected to each other (Fig. 22.2). The crystal parameters of TiO_2 anatase, rutile, and brookite are summarized in Table 22.1. TiO_2 anatase involves one cation site for Ti and one anion site for O with multiplicities of 4 and 8, respectively. The structure is built from distorted TiO_6 octahedra sharing four edges. Two different Ti—O distances can be distinguished. Larger distance (1.964 Å) originates from two Ti—O bonds along c-axis, whereas shorter distances (1.937 Å) come from four Ti—O bonds within the (ab) plane. Each oxygen atom is coordinated with three titanium atoms. The 3D anatase network is formed by the stacking of zigzag chains along the c-axis (Fig. 22.2B).

2.2 Surface Engineering

Another particular role of fluoride anions is their ability to direct crystal growth of TiO_2. Since the interaction between molecules/ions and surfaces of materials largely influences their physicochemical properties, tuning the nature of the facets exposed by crystals has emerged as a fine way to enhance the material's properties. For instance, TiO_2 with different exposed surfaces displays different photocatalytic and

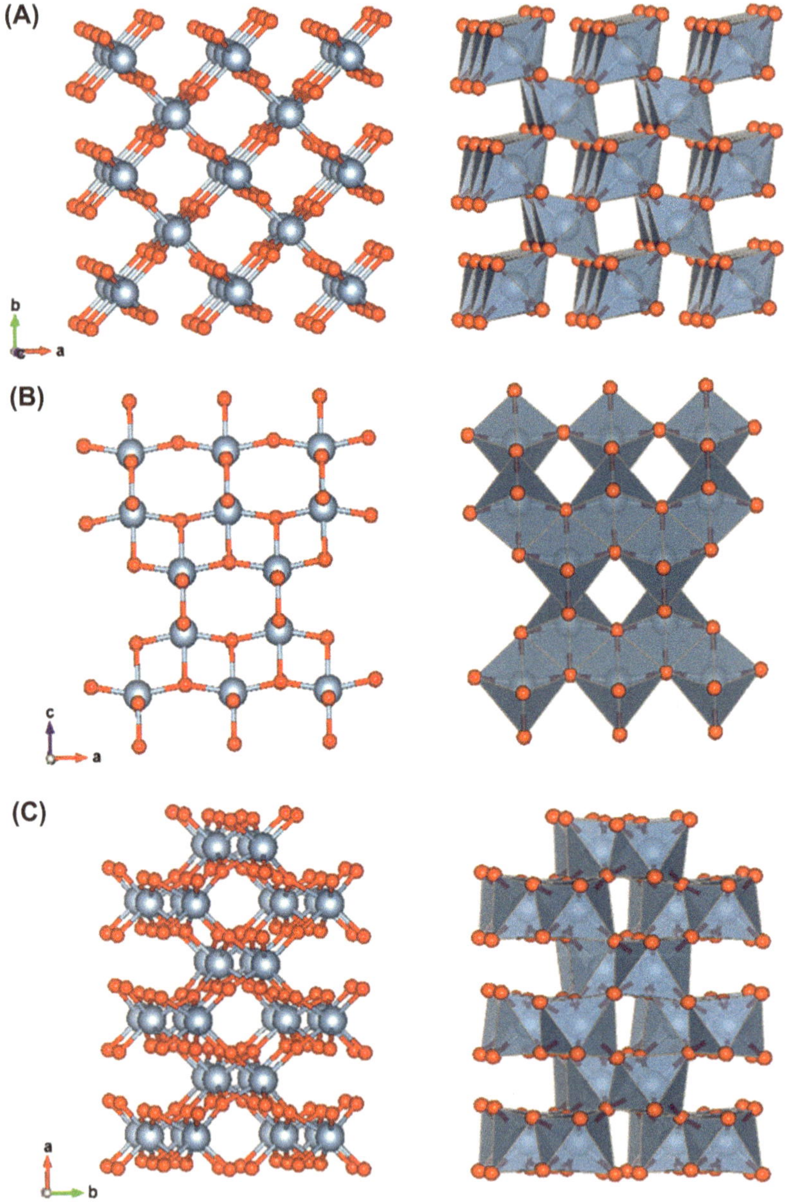

Figure 22.2 Structures of three natural TiO_2 phases rendered in ball-and-stick models and polyhedron models. (A) Rutile, (B) anatase, and (C) brookite. *Red balls*, O atoms; *blue balls*, Ti atoms.

Table 22.1 Structural Characteristics of Three Natural TiO_2 Crystalline Phases

Phase	Rutile	Anatase	Brookite
Crystal system	Tetragonal	Tetragonal	Orthorhombic
Space group	$P4_2/mnm$	$P4_1/amd$	*Pbca*
a (Å)	4.5941	3.7842	9.184
b (Å)	4.5941	3.7842	5.447
c (Å)	2.9589	9.5146	5.146
α (°)	90	90	90
β (°)	90	90	90
γ (°)	90	90	90
Density[a] (g/cm^3)	4.248	3.895	4.123
Polyhedron type	Octahedron	Octahedron	Octahedron
#Polyhedra per unit cell	2	4	8
#Edge sharing[b]	2	4	3

[a]Calculated.
[b]Per polyhedron.
Reprinted with permission from Zhang et al. *Chem. Rev.* **2014,** *114*, 9613. Copyright 2014 American Chemical Society.

electrochemical properties due to different surface adsorption, charge transfer, and ion transport.[8] The adsorption states of molecules/ions depend on the surface atomic coordination and arrangement of TiO_2. The role of different surfaces affecting the interaction has been demonstrated by studying water adsorption and lithium insertion on TiO_2 anatase (001) and (101) surfaces.[21–23] The anatase (001) surface permits dissociative and nondissociative adsorption of water, whereas the (101) surface only permits nondissociative adsorption of water.[21] A lithium-ion intercalation study on TiO_2 anatase with different percentages of (001)/(101) surfaces has demonstrated that Li-ion insertion is favorable with dominant (001) surfaces.[22] Therefore, developing TiO_2 with controlled exposed surfaces is highly desirable for various energy-related applications ranging from energy conversion to storage.

To investigate the effects of various adsorbate atoms as capping agents on the stabilization of {001} facets, Yang et al. have theoretically explored 12 nonmetallic adsorbate atoms on the surface energy of {101} and {001} facets by performing first-principles calculations.[10] They found that fluorine-adsorbed surfaces not only have the lowest energy for both {101} and {001} facets, but also that the (001) surfaces become more stable than the (101) ones (Fig. 22.3A). Chemisorption of fluorine onto the (001) surfaces prevents crystal growth along [001] direction and favors anisotropic growth along the horizontal plane. These results show that highly truncated octahedra dominated by fluorine-terminated {001} facets is the thermodynamically

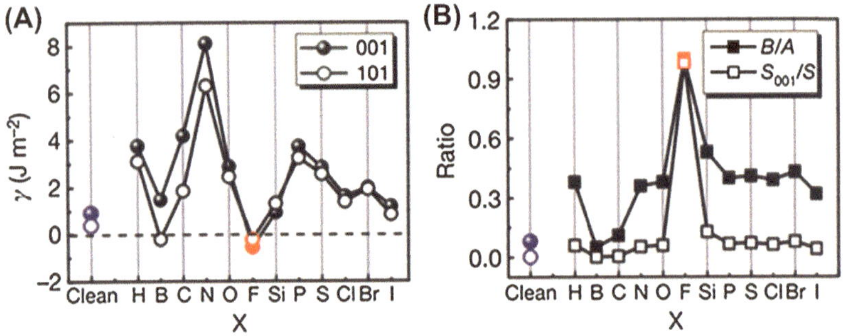

Figure 22.3 (A) Calculated energies of the (001) and (101) surfaces surrounded by X atoms. (B) Plots of the optimized value of B/A and percentage of (001) facets for anatase single crystals with various adsorbate atoms X.[10]
Reprinted with permission from Yang, H. G.; Sun, C. H.; Qiao, S. Z.; Zou, J.; Liu, G.; Smith, S. C.; Cheng, H. M.; Lu, G. Q. Anatase TiO_2 Single Crystals With a Large Percentage of Reactive Facets. *Nature* **2008,** *453* (7195), 638–641. Copyright 2014 Nature Publishing Group.

favorable morphology for TiO_2 anatase (Fig. 22.3B). The physical origin of why fluorine can lower the surface energy of (001) facets has been investigated by Ma and coworkers by using density functional theory (DFT) calculations.[24] They have demonstrated that the adsorption of fluorine on (101) surfaces significantly weakens the Ti–O bonds, whereas they are strengthened in (001) surfaces. Therefore, {001} facets become more stable than {101} facets when fluorine is bonded to the surface.

Mesoscopic properties and applications of (001) anatase surface were rarely reported until 2008 due to the synthetic difficulty in stabilizing anatase crystals with high proportion of {001} facets. The breakthrough made by Yang et al., who have synthesized TiO_2 anatase microcrystals with a large percentage of exposed {001} facets up to 47% in a fluorinated medium (Fig. 22.4),[10] has led to impressive research

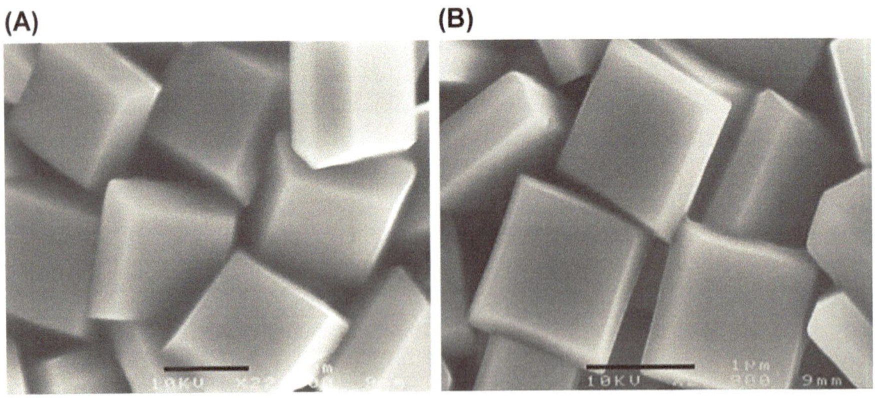

Figure 22.4 (A,B) Scanning electron microscopic images of TiO_2 anatase microcrystals using different concentration of fluorine.
Reprinted with permission from Yang, H. G.; Sun, C. H.; Qiao, S. Z.; Zou, J.; Liu, G.; Smith, S. C.; Cheng, H. M.; Lu, G. Q. Anatase TiO_2 Single Crystals With a Large Percentage of Reactive Facets. *Nature* **2008,** *453* (7195), 638–641. Copyright 2014 Nature Publishing Group.

activities with more than 1600 citations since 2008. In terms of synthesis methods, the wet chemistry methods, more particularly hydrothermal and solvothermal methods, are the dominant methods to synthesize TiO_2 with particle size ranging from nanoscale to microscale.[25] The capping agent plays a central role on controlling dominant {001} facets in TiO_2 by using wet chemistry. Indeed, fluorine is known to be the most effective capping agent to stabilize {001} facets based on the theoretical results.[10] In general, fluorine agent is believed to have dual roles: (1) to retard the hydrolysis rate of the titanium precursor and (2) and to reduce the surface energy and promote the growth of {001}-faceted surface.[10,26,27]

Many attempts have been made to achieve a higher percentage of exposed {001} facets than the initial percentage of 47% obtained by Yang et al. By using isopropanol as solvent together with diluted HF, Yang et al. have further improved the percentage to 64% exposed {001} facets using solvothermal synthesis.[27] Isopropanol not only provides the reaction medium but also acts as a protecting agent to strengthen the adsorption of fluorine onto (001) surfaces, thus favoring anisotropic growth of crystals. The synergistic effects of alcohol are also effective in the case of ethanol, tert-butanol, and benzyl alcohol.[28] Moreover, by hydrothermally treating tetrabutyl titanate, $(TiOBu)_4$, in 47 wt% hydrofluoric acid solution at 180°C, Han et al. obtained anatase nanosheets with average size of 130 nm and thickness of c. 8 nm, possessing an average percentage of {001} facets as high as 89%.[15] It seems that a relatively high concentration of HF and a low concentration of water are favorable for the stabilization of high percentage of {001} facets. This has been confirmed by the solvothermal synthesis of TiO_2 anatase with a thickness of 27.9 nm and a side length of 4140 nm using TiF_4 and 1-butanol mixture solution with additional HF at 210°C for 24 h, possessing 98.7% exposed {001} facets.[29]

A deeper understanding of the effect of stoichiometries of reagents and the reaction temperature on the proportion of the stabilized {001} facets has been conducted by Menzel et al.[14] Their study provides guidelines on controlling {001} facets in anatase crystals by adjusting synthesis parameters. Fig. 22.5 shows a ternary diagram of the dependence of TiO_2 nanoparticle morphology on the molar ratio of titanium source (titanium isopropoxide), fluoride source (aqueous HF, 50 wt%), and water. Syntheses have been performed hydrothermally at 180°C. They have shown that water and fluorine concentrations are competitive in determining the morphology of anatase nanoparticles. At high water concentration, hydroxyl groups that promote isotropic crystal growth by oxolation reaction will dominate the surface of anatase particles. Contrarily, high concentration of fluorine induces F-terminated {001} facets, thus favoring lateral growth. However, at very high fluorine concentration, the etching effect of HF led to reduced particle sizes and the formation of $TiOF_2$ or soluble titanium fluoride.

The etching properties of HF can also be used as to tailor the morphology of crystals. For instance, Wang et al. have provided an insight into the mechanistic role of the surface fluorination reaction on TiO_2 anatase and further highlight the etching properties of HF.[30] They have experimentally demonstrated the direct dependence of the etching of TiO_2 anatase {001} facets on the concentration of HF (HF is released under the hydrolysis of TiF_4) and proposed an etching mechanism based on DFT calculations

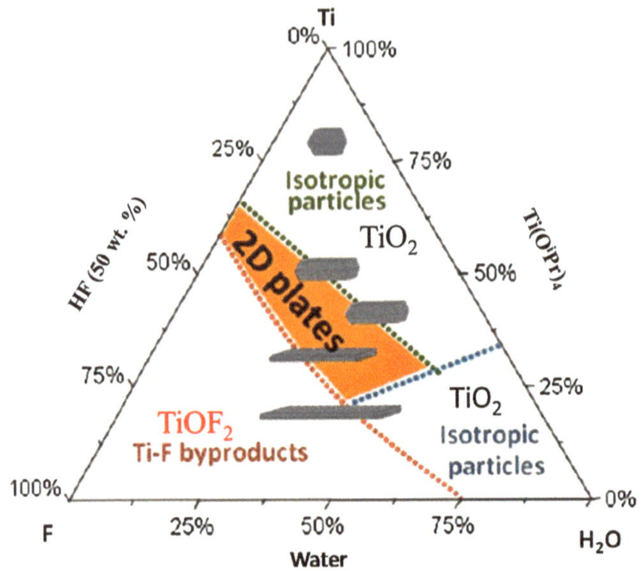

Figure 22.5 Ternary diagram illustrating the dependence of TiO_2 nanoparticle morphology on the molar ratio of titanium source (titanium isopropoxide), fluoride source (aqueous HF, 50 wt%), and water. Reaction temperature $T = 180°C$.[14] *2D*, two-dimensional.
Reprinted with permission from Menzel, R.; Duerrbeck, A.; Liberti, E.; Yau, H. C.; McComb, D.; Shaffer, M. S. P. Determining the Morphology and Photocatalytic Activity of Two-Dimensional Anatase Nanoplatelets Using Reagent Stoichiometry. *Chem. Mater.* **2013,** *25* (10), 2137–2145. Copyright 2013 American Chemical Society.

(Fig. 22.6). There are three key steps involved, which are: (1) the formation of a full HF-covered surface by the dissociative adsorption of HF, (2) the replacement of −OH surface with F on the completely fluorinated surface to form $-TiOF_2$ group, and (3) the dissolution of the $-TiOF_2$ group by further interaction with HF to create surface vacancies. Although the first two steps are thermodynamically favorable on both (101) and (001) surfaces, the third step is highly dependent on the crystallographic orientation. The calculated dissolution reaction energies are −0.41 and +0.04 eV for (101) and (001) surfaces, respectively, indicating that the proposed etching mechanism is energetically permitted solely for the (001) surface.

Reaction temperature is also an important parameter for synthesizing TiO_2 anatase crystals with a high percentage of (001) facets because it affects the chemical stability of Ti−F over Ti−OH bonds. High temperature favors nucleation and growth processes, leading to larger particles along with larger proportion of {001} facets. Nevertheless, particle growth via Ostwald ripening and oriented attachment mechanisms at high temperature competes with classic crystal growth, suggesting that the reaction temperature could not be too high. The hydrothermal stability of {001}-faceted TiO_2 anatase has been systematically studied by Yang et al.[31] They have proposed that under hydrothermal conditions, the hydrolysis of Ti−F groups on the surface of

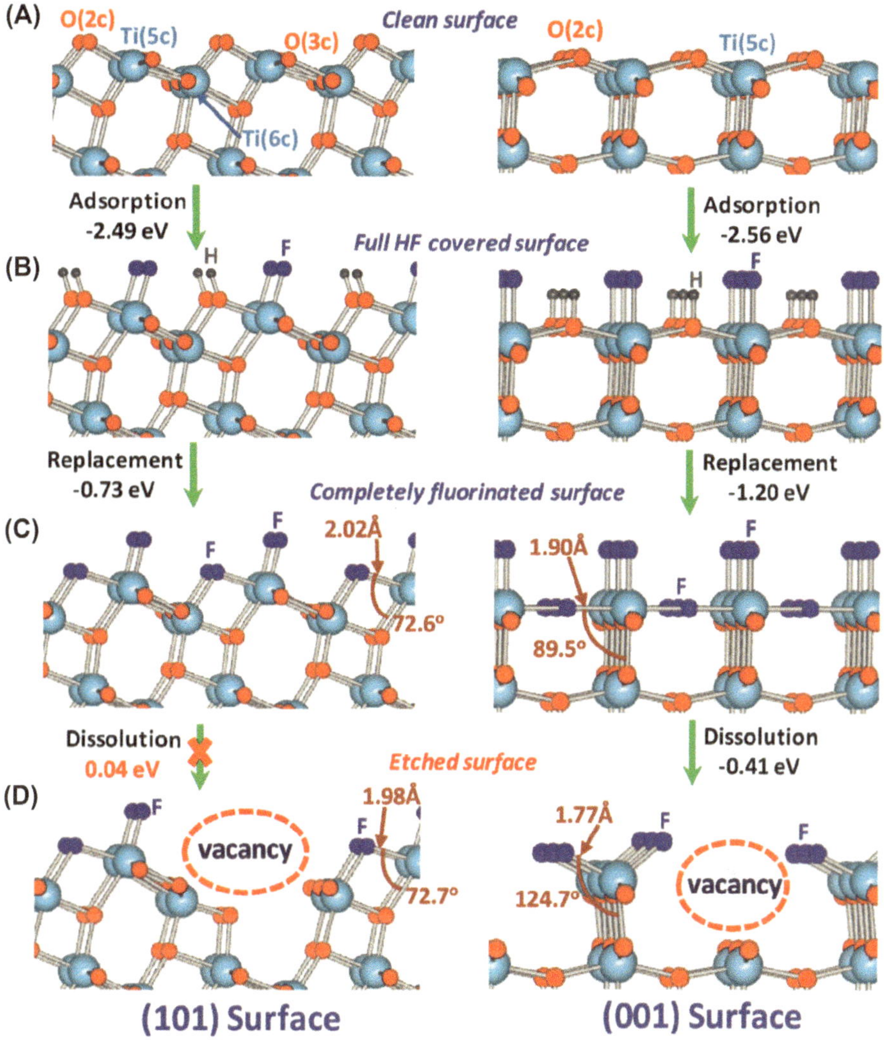

Figure 22.6 Density functional theory−calculated reaction energies and structures for different stages of HF interaction with single-crystal anatase TiO_2 (101) (left) and (001) (right) surfaces. (A) Clean surfaces, (B) full HF-covered surfaces, (C) completely fluorinated surfaces, and (D) etched surfaces. All are optimized structures.[30]

Reprinted with permission from Wang, Y.; Zhang, H.; Han, Y.; Liu, P.; Yao, X.; Zhao, H. A Selective Etching Phenomenon on {001} Faceted Anatase Titanium Dioxide Single Crystal Surfaces by Hydrofluoric Acid. *Chem. Commun.* **2011,** *47* (10), 2829−2831. Copyright 2011 Royal Society of Chemistry.

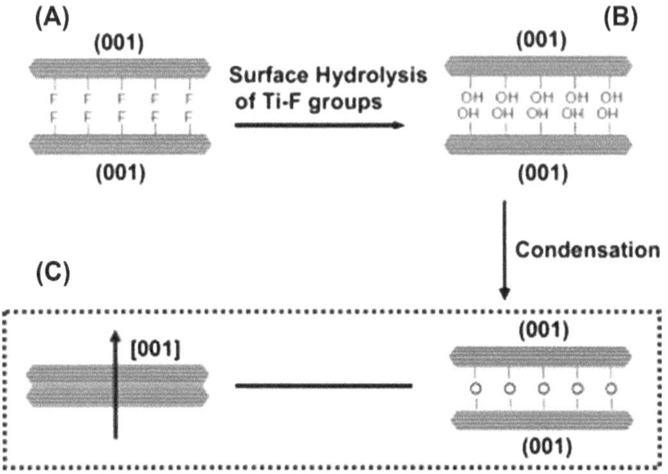

Figure 22.7 Proposed growth mechanism of stacked anatase TiO_2 nanosheets dominated by {001} facets at their interfacial regions. (A) Stacking of TiO_2 nanosheets with F-terminated surfaces, (B) hydrolysis of Ti—F groups on the surfaces and accordingly the formation of Ti—OH, and (C) formation of Ti—O—Ti linkages via condensation between Ti—OH on the (001) surfaces of the adjacent TiO_2 nanosheets and the resulting elimination of (001) surfaces.[31]

Reprinted with permission from Yang, X. H.; Li, Z.; Sun, C.; Yang, H. G.; Li, C. Hydrothermal Stability of {001} Faceted Anatase TiO_2. *Chem. Mater.* **2011,** *23* (15), 3486—3494. Copyright 2011 American Chemical Society.

TiO_2 nanosheets yields Ti—OH groups by replacing F atoms with hydroxyl groups. The subsequent (001) interfaces between adjacent TiO_2 nanosheets are eliminated through the condensation of Ti—OH groups yielding Ti—O—Ti linkages. A further surface fusion process is also involved in generating the well-faceted TiO_2 crystals. The aforementioned oriented attachment crystal growth mechanism is illustrated in Fig. 22.7. Additionally, they demonstrated that the reaction medium plays an important role in the morphological evolution of TiO_2 anatase nanosheets by controlling the hydrolysis of surface Ti—F groups. By replacing water with other solvents such as 1.5 M HCl solution, ethanol, propanol, and butanol, the surface hydrolysis of Ti—F groups can be effectively suppressed and TiO_2 anatase nanosheets can maintain their original morphology. This observation again proves the synergetic effect of solvents and HF in slowing the surface hydrolysis rate, thus preserving the as-grown {001} facets.

2.3 *Lattice Doping/Substitution*

Doping/substitution has been widely used as one of the most effective strategies to modify the structural features and properties of TiO_2.[6] Particularly in light-related applications, TiO_2 anatase suffers from two main drawbacks: little visible light adsorption due to its wide band gap of 3.2 eV and rapid recombination carriers of photoexcited electrons holes. These barriers can be mitigated by tuning the electronic structure of TiO_2 to extend its photosensitivity in the visible light wavelength range and enabling the promotion of charge separation of photoexcited electrons−holes. In this scope, fluorine atoms are shown to act as electron scavengers to limit the recombination of photoexcited electrons−holes.[14,19,32,33] Nevertheless, the mechanism by which fluorine atoms are incorporated into the bulk has been poorly discussed in the literature and mostly explained by the valence induction mechanism. In such a mechanism, the negative charge deficiency caused by the O^{2-}/F^- substitution is compensated by the reduction of Ti^{4+} to Ti^{3+} as expressed by the general chemical formula $Ti_x^{3+}Ti_{1-x}^{4+}O_{2-x}F_x$.[19,32,34] Czoska et al. have simulated this mechanism with different possible configurations in which different distances separating Ti^{3+} and F^- have been proposed (Fig. 22.8). Although the presence of fluorine has been probed

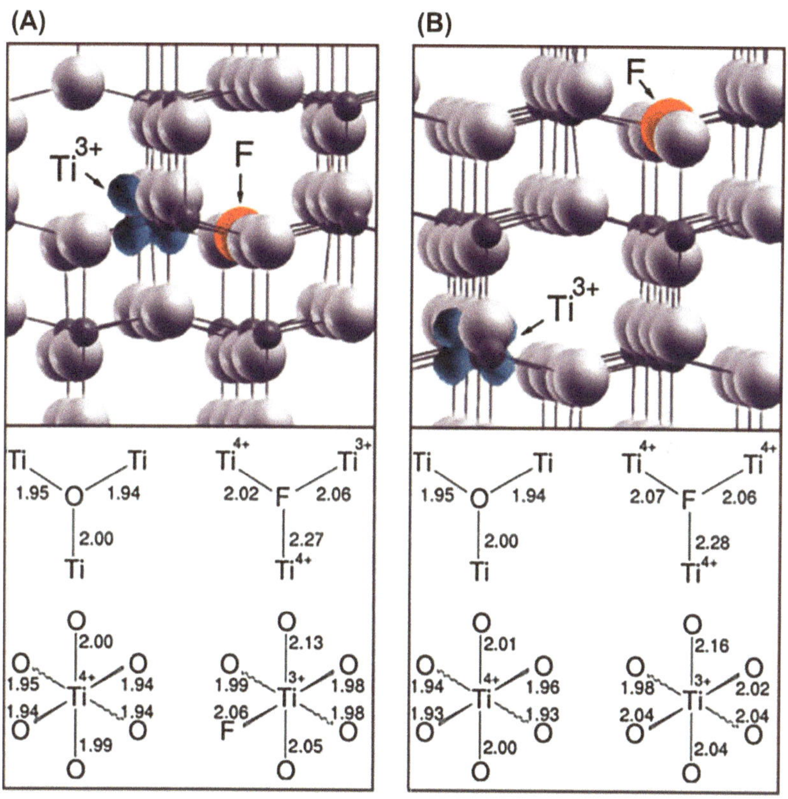

Figure 22.8 (Top) Ball-and-stick models for F-doped anatase TiO_2 reporting the spin density of the Ti^{3+} excess electron. (A) Ti^{3+} cation first neighbor to the F dopant; (B) Ti^{3+} cation at ~7 Å distance from the F dopant. (Bottom) Bond lengths in angstroms. The data refer to the H&HLYP calculations.[20]

by X-ray photoelectron spectroscopy (XPS) spectroscopy, they further confirmed the valence induction mechanism by electron paramagnetic resonance spectroscopy ruling out the presence of oxygen vacancies. For the synthesis of F bulk-doped TiO_2, the authors employed sol–gel-based chemistry with postthermal treatment performed at 500°C under air. Such a heating treatment likely induces the decomposition and release of fluorinated species, i.e., F_2 and/or HF. The departure of elemental fluorine could be the cause of the reduction of Ti^{4+} to Ti^{3+}.

3. Evidences for the Stabilization of Fluoride in Anatase Featuring Cationic Vacancies

3.1 General Considerations on Hydrolysis and Fluorolysis Reactions

The chemistry of the sol–gel synthesis relies on two main reactions including hydrolysis and condensation reactions of a molecular precursor, which is initially formed by dissolving a metal precursor in an appropriate solvent. Sol–gel synthesis performed in organic medium usually employs metal alkoxide as precursor, dissolved in its parent alcohol. The high reactivity of metal alkoxides toward nucleophilic reagents such as water can be utilized to form OH-reactive function according to Eq. (22.1).

$$-\text{M}-\text{OR} + \text{H}_2\text{O} \rightarrow -\text{M}-\text{OH} + \text{ROH} \tag{22.1}$$

Thereafter, reactive M–OH groups undergo condensation reactions such as olation (22.2), oxolation (22.3), and alcoxolation (22.4) reactions written as follows:

$$-\text{M}-\text{OH} + \text{H}_2\text{O}-\text{M}- \rightarrow -\text{M}-\text{OH}-\text{M}- + \text{H}_2\text{O} \tag{22.2}$$

$$-\text{M}-\text{OH} + \text{HO}-\text{M}- \rightarrow -\text{M}-\text{O}-\text{M}- + \text{ROH} \tag{22.3}$$

$$-\text{M}-\text{OR} + \text{HO}-\text{M}- \rightarrow -\text{M}-\text{O}-\text{M}- + \text{ROH} \tag{22.4}$$

These three reactions allow the formation of an infinite oxide network. In the presence of hydrofluoric acid, HF, the sol–gel process can deviate from the this conventional mechanism. A comparison between the nucleophilic characters of HF and H_2O indicates similar reactivity with a partial charge $\delta(\text{F}) = -0.42$ in HF and $\delta(\text{O}) = -0.4$ in H_2O.[25] Thus, similar to water, which can hydrolyze M–OR bonds, HF can react via a fluorolysis reaction as defined by Kemnitz's group[35] according to the Eq.(22.5).

$$\text{M(OR)}_n + x\cdot\text{HF} \rightarrow \text{M(OR)}_{n-x}\text{F}_x + x\cdot\text{ROH} \tag{22.5}$$

Hence, the use of aqueous HF can induce modifications of the metal precursor at the molecular level and thus its chemical reactivity,[36,37] eventually yielding new composition and structural features deviating from a well-ordered oxide network.

3.2 Reactivity of Titanium Alkoxide Toward HF Under Solvothermal Conditions

According to the aforementioned considerations, titanium alkoxide precursors, such as titanium tetraisopropoxide, can react through hydrolysis and fluorolysis reactions. Since the latter process involves the protonation of the oxygen atom of the alkoxide group, it is favored in acidic condition hydrofluoric acid a suitable fluorinating agent.[35]

To illustrate the impact of using a fluorinating medium in the sol–gel chemistry of titanium dioxide, the following part describes a synthesis protocol involving the reaction of titanium alkoxide with aqueous HF thermally activated under solvothermal conditions. This work refers to Ref. 38. The synthesis includes the following steps: (1) the dissolution of titanium alkoxide in its parent alcohol, (2) the addition of aqueous HF (with a fixed n_F/n_{Ti} molar ratio of 2) previously diluted in alcohol, and (3) the precipitation reaction induced by thermal activation in a stainless-steel Teflon-lined hydrothermal autoclave performed at 90°C for 12 h Fig. 22.9 presents the X-ray powder diffraction pattern of the as-synthesized compound.

The XRD pattern of the as-prepared compound can be fully indexed using tetragonal symmetry (space group: I41/amd), characteristic of the anatase network without the presence of any fluorinated phase. At this point, a conclusion drawn only on the indexation of the above XRD pattern would be biased as it suggests the stabilization of a pure titanium dioxide network. Moreover, the white color aspect of the recovered powder primary indicates the absence of reduced Ti^{3+} that based on literature data ruled out the valence induction mechanism. Thus, it can be suggested that the as-synthesized compound is free from fluorine atoms. However, the next part will demonstrate by establishing a new substitutional mechanism that deeper analyses enable to conclude about the stabilization of fluorine atoms within the oxide network.

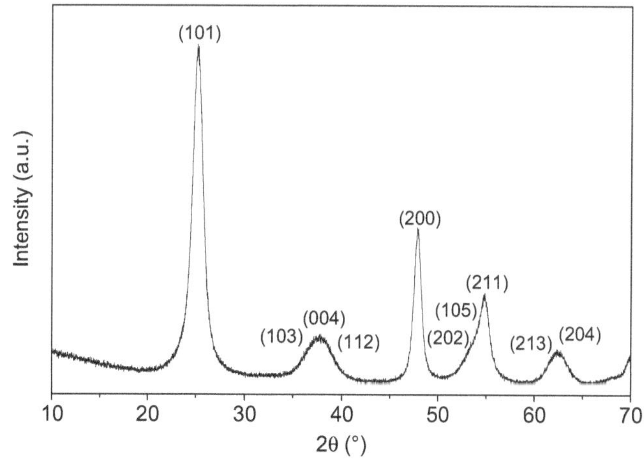

Figure 22.9 X-ray powder diffraction pattern of TiO_2 synthesized in a fluorinating medium.

3.3 Structural/Compositional Features and New Substitution-Mediated Mechanism

To probe the impact of using HF during the sol−gel synthesis of TiO_2, deeper analyses was required particularly on both the composition and structural features. High-energy synchrotron source has been used to obtain total scattering data containing all Bragg and diffuse scattering data. The Fourier transformation of the total structural function S(Q), obtained from a specialized scattering measurement, yields the pair distribution function (PDF), G(r), a histogram of all the atom−atom distances within a sample, independent of crystallinity. The atomic structure was investigated using a "real space Rietveld method."[39] According to X-ray diffraction analysis, the PDF has been refined using a single phase of anatase (Fig. 22.10).

The extracted crystal parameters are similar to those of pure anatase, except that the refinement of the Ti site occupancy yields c. 74(4)%, indicating the presence of a large content of cationic vacancies. Therefore, one can conclude that the synthesis does not lead to a pure titanium dioxide composition. The presence of cationic vacancies revealed by the PDF refinement indicates the formation of a complex structure that will be investigated in the following discussion.

First, the chemical composition was established using a combination of analytical and physical characterization tools such as high-energy synchrotron (pair distribution function), ^{19}F solid-state nuclear magnetic resonance (NMR) spectroscopy, XPS, and thermogravimetric analysis. The results converge toward the composition $Ti_{0.78}\square_{0.22}O_{1.12}F_{0.40}(OH)_{0.48}$, where $\square$ represents cationic vacancy. The cationic vacancies originate due to the partial substitution of O^{2-} by monovalent F^-/OH^- anions yielding negative charge deficiency counterbalanced by Ti vacancies. This substitution-mediated mechanism can be expressed by the general chemical formula $Ti_{1-x-y}\square_{x+y}O_{2-4(x+y)}F_{4x}(OH)_{4y}$.

Remarkably, the content of the cation vacancies can be synthetically controlled by tuning the temperature reaction, reaching an unprecedented variation of the chemical composition with respect to pure TiO_2 (Fig. 22.11).

Solid-state ^{19}F NMR reveals the occurrence of three different coordination environments of fluorine with respect to titanium and vacancies (Fig. 22.12). The occurrence of

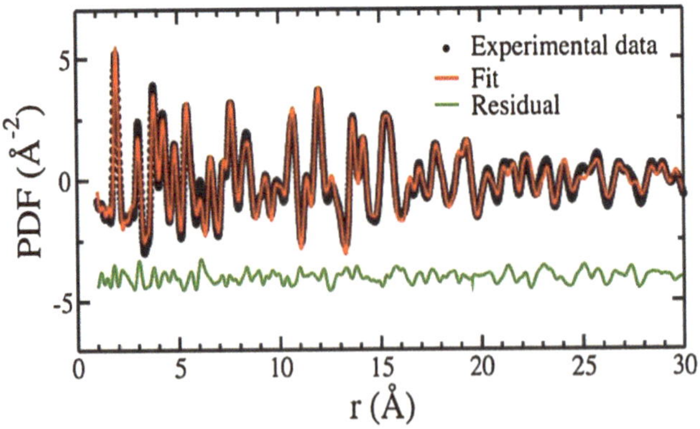

Figure 22.10 Refinement of the PDF of TiO_2 synthesized in a fluorinating medium.

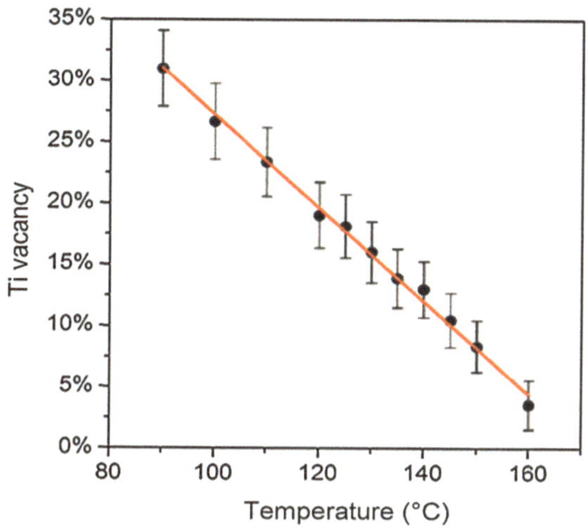

Figure 22.11 Correlation between the synthesis temperature and the content of Ti vacancy in $Ti_{1-x-y}\square_{x+y}O_{2-4(x+y)}F_{4x}(OH)_{4y}$ compounds. The Ti (4a) site occupancy was determined by fitting the PDF data.

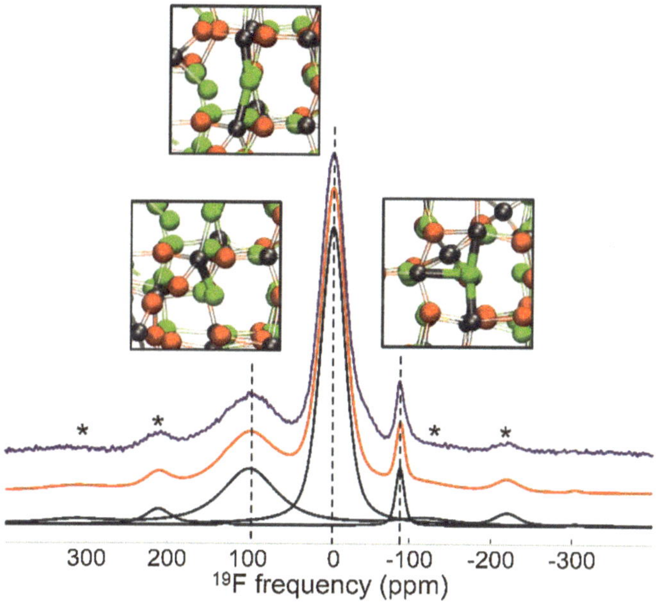

Figure 22.12 Experimental (in blue) ^{19}F solid-state magic angle spectroscopy (MAS) (34 kHz) NMR spectrum of the fluorinated anatase, fitted (in red) with three NMR lines (in black, indicated by *vertical dashed lines*), which are assigned to , and $Ti_1\square_2$−F, $Ti_2\square_1$−F, and Ti_3 −F (from left to right). The *asterisks* indicate the spinning sidebands. Corresponding structures extracted from a density functional theory−based molecular dynamics simulation are also shown (gray, Ti; green, F; red, O; on each snapshot the illustrated bonds are shown as *thick solid cylinders*, whereas the other bonds are shown as *thin transparent cylinders*. Two atoms are considered bonded when the distance between them is lower than 2.3 Å).

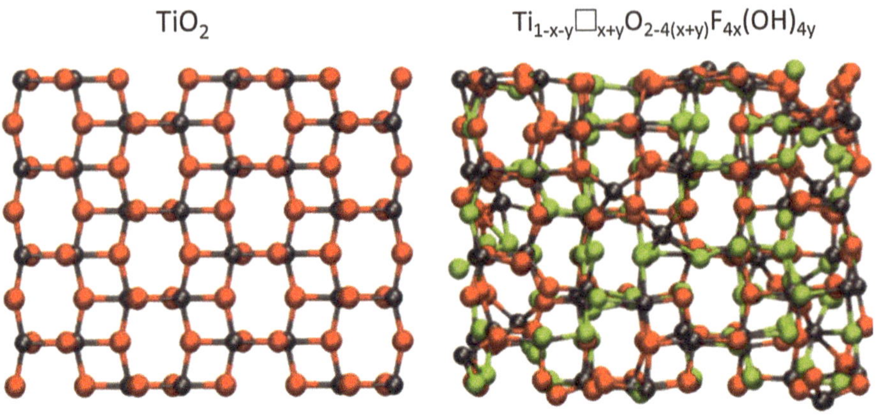

Figure 22.13 Structural representation of TiO_2 and $Ti_{1-x-y}\square_{x+y}O_{2-4(x+y)}F_{4x}(OH)_{4y}$ anatase obtained from density functional theory (DFT) (at 0K) and DFT-based molecular dynamics (at 300K), respectively. (gray, Ti; green, F/OH; red, O).

cation vacancies induces additional coordination modes for the anions. We thus considered the occurrence of three different environments such as $Ti_3{-}F$, $Ti_2\square_1{-}F$, and $Ti_1\square_2{-}F$ for fluoride. The relative intensity of the NMR resonances indicates preferential location of fluorine close to vacancies. This feature has been confirmed by DFT calculations showing that isolated or adjacent vacancies were more stabilized by fluoride than by oxide anions, which agrees with electrostatic considerations.

This part demonstrates that the synthesis of TiO_2 anatase in a fluorinating medium yields a drastic modification of the structure featuring cationic vacancies, three different anions, and up to three different coordination modes for fluorine atoms. The disorder induced by the modification of the chemical composition of TiO_2 can be visualized through structural representation derived from DFT calculation of TiO_2 and a fluorinated structure $Ti_{0.78}\square_2O_{1.12}F_{0.88}$ (Fig. 22.13). It can be seen that the anatase framework displays an impressive structural flexibility paving the way for the design of new functional materials. For example, band gap calculations performed on $Ti_{0.78}\square_2O_{1.12}F_{0.88}$ reveals that fluorination can lead to a band gap reduction of up to 0.8 eV as compared to pure TiO_2.[40] This result calls for deeper investigation on the synthesis and properties of the material.

Finally, this work provides important insight into materials' properties employing similar synthesis protocols as the rationalization of their properties should take into account the presence of defects that can alter their reactivity.

4. Conclusion

The use of fluorine in the synthesis of oxide particularly TiO_2 has proved to be a versatile approach to tune the physicochemical properties of the stabilized phase. Fluoride anions can play multiple roles such as structure-directing or morphology-controlling agents. The latter role has led to enthusiastic research activities motivated by enhanced properties related to various applications. Herein, we have posed the question about the possible stabilization of fluorine into the oxide lattice with an emphasis on the recent stabilization of highly fluorinated anatase featuring cationic vacancies (•), i.e., $Ti_{1-x-y}\square_{x+y}O_{2-4(x+y)}F_{4x}(OH)_{4y}$. A literature survey seems to suggest that most of the synthesis of TiO_2 employing sol−gel precursors likely contains fluorine associated with defects. Their presence is of great importance due to the impact on the physico-chemical properties of the materials, as recently exemplified in the field of Li-ion batteries.[38]

References

1. Fujishima, A.; Honda, K. Electrochemical Photolysis of Water at a Semiconductor Electrode. *Nature* **1972,** *238* (5358), 37−38.
2. O'Regan, B.; Gratzel, M. A Low-Cost, High-Efficiency Solar Cell Based on Dye-Sensitized Colloidal TiO_2 Films. *Nature* **1991,** *353* (6346), 737−740.
3. Gratzel, M. Photoelectrochemical Cells. *Nature* **2001,** *414* (6861), 338−344.
4. Kavan, L.; Grätzel, M.; Gilbert, S. E.; Klemenz, C.; Scheel, H. J. Electrochemical and Photoelectrochemical Investigation of Single-Crystal Anatase. *J. Am. Chem. Soc.* **1996,** *118* (28), 6716−6723.
5. Khan, S. U. M.; Al-Shahry, M.; Ingler, W. B. Efficient Photochemical Water Splitting by a Chemically Modified n-TiO_2. *Science* **2002,** *297* (5590), 2243−2245.
6. Chen, X.; Mao, S. S. Titanium Dioxide Nanomaterials: Synthesis, Properties, Modifications, and Applications. *Chem. Rev.* **2007,** *107* (7), 2891−2959.
7. Lazzeri, M.; Vittadini, A.; Selloni, A. Structure and Energetics of Stoichiometric TiO_2 Anatase Surfaces. *Phys. Rev. B* **2001,** *63* (15), 155409.
8. Liu, G.; Yang, H. G.; Pan, J.; Yang, Y. Q.; Lu, G. Q.; Cheng, H. -M. Titanium Dioxide Crystals With Tailored Facets. *Chem. Rev.* **2014,** *114* (19), 9559−9612.
9. Tressaud, A. *Functionalized Inorganic Fluorides: Synthesis, Characterization and Properties of Nanostructured Solids;* John Wiley & Sons, Ltd, 2010.
10. Yang, H. G.; Sun, C. H.; Qiao, S. Z.; Zou, J.; Liu, G.; Smith, S. C.; Cheng, H. M.; Lu, G. Q. Anatase TiO_2 Single Crystals With a Large Percentage of Reactive Facets. *Nature* **2008,** *453* (7195), 638−641.
11. Liu, N.; Zhao, Y.; Wang, X.; Peng, H.; Li, G. Facile Synthesis and Enhanced Photocatalytic Properties of Truncated Bipyramid-Shaped Anatase TiO_2 Nanocrystals. *Mater. Lett.* **2013,** *102−103,* 53−55.
12. Li, H.; Zeng, Y.; Huang, T.; Piao, L.; Yan, Z.; Liu, M. Hierarchical TiO_2 Nanospheres with Dominant {001} Facets: Facile Synthesis, Growth Mechanism, and Photocatalytic Activity. *Chem. Eur. J.* **2012,** *18* (24), 7525−7532.
13. Liu, G.; Sun, C.; Yang, H. G.; Smith, S. C.; Wang, L.; Lu, G. Q.; Cheng, H. -M. Nanosized Anatase TiO_2 Single Crystals for Enhanced Photocatalytic Activity. *Chem. Commun.* **2010,** *46* (5), 755−757.
14. Menzel, R.; Duerrbeck, A.; Liberti, E.; Yau, H. C.; McComb, D.; Shaffer, M. S. P. Determining the Morphology and Photocatalytic Activity of Two-Dimensional Anatase Nanoplatelets Using Reagent Stoichiometry. *Chem. Mater.* **2013,** *25* (10), 2137−2145.

15. Han, X.; Kuang, Q.; Jin, M.; Xie, Z.; Zheng, L. Synthesis of Titania Nanosheets with a High Percentage of Exposed (001) Facets and Related Photocatalytic Properties. *J. Am. Chem. Soc.* **2009,** *131* (9), 3152–3153.
16. Dai, Y.; Cobley, C. M.; Zeng, J.; Sun, Y.; Xia, Y. Synthesis of Anatase TiO_2 Nanocrystals with Exposed {001} Facets. *Nano Lett.* **2009,** *9* (6), 2455–2459.
17. Ma, X. Y.; Chen, Z. G.; Hartono, S. B.; Jiang, H. B.; Zou, J.; Qiao, S. Z.; Yang, H. G. Fabrication of Uniform Anatase TiO_2 Particles Exposed by {001} Facets. *Chem. Commun.* **2010,** *46* (35), 6608–6610.
18. Yu, J.; Wang, W.; Cheng, B.; Su, B. -L. Enhancement of Photocatalytic Activity of Mesporous TiO_2 Powders by Hydrothermal Surface Fluorination Treatment. *J. Phys. Chem. C* **2009,** *113* (16), 6743–6750.
19. Yu, J. C.; Yu, J. G.; Ho, W. K.; Jiang, Z. T.; Zhang, L. Z. Effects of F^- Doping on the Photocatalytic Activity and Microstructures of Nanocrystalline TiO_2 Powders. *Chem. Mater.* **2002,** *14* (9), 3808–3816.
20. Dong, Y.; Kapilashrami, M.; Zhang, Y.; Guo, J. Morphology Change and Band Gap Narrowing of Hierarchical TiO_2 Nanostructures Induced by Fluorine Doping. *CrystEngComm* **2013,** *15* (48), 10657–10664.
21. Vittadini, A.; Selloni, A.; Rotzinger, F. P.; Grätzel, M. Structure and Energetics of Water Adsorbed at TiO_2 Anatase (101) and (001). *Surfaces Phys. Rev. Lett.* **1998,** *81* (14), 2954–2957.
22. Hengerer, R.; Kavan, L.; Krtil, P.; Grätzel, M. Orientation Dependence of Charge-Transfer Processes on TiO_2 (Anatase) Single Crystals. *J. Electrochem. Soc.* **2000,** *147* (4), 1467–1472.
23. Sun, C. H.; Yang, X. H.; Chen, J. S.; Li, Z.; Lou, X. W.; Li, C.; Smith, S. C.; Lu, G. Q.; Yang, H. G. Higher Charge/Discharge Rates of Lithium-Ions Across Engineered TiO_2 Surfaces Leads to Enhanced Battery Performance. *Chem. Commun.* **2010,** *46* (33), 6129–6131.
24. Ma, X.; Dai, Y.; Wei, W.; Huang, B.; Whangbo, M. -H. Insights into How Fluorine-Adsorption and n-Type Doping Affect the Relative Stability of the (001) and (101) Surfaces of TiO_2: Enhancing the Exposure of More Active but Thermodynamically Less Stable (001). *J. Phys. Chem. Lett.* **2015,** *6,* 1876–1882.
25. Livage, J.; Henry, M.; Sanchez, C. Sol-Gel Chemistry of Transition Metal Oxides. *Prog. Solid State Chem.* **1988,** *18* (4), 259–341.
26. Liu, M.; Piao, L.; Zhao, L.; Ju, S.; Yan, Z.; He, T.; Zhou, C.; Wang, W. Anatase TiO_2 Single Crystals with Exposed {001} and {110} Facets: Facile Synthesis and Enhanced Photocatalysis. *Chem. Commun.* **2010,** *46* (10), 1664–1666.
27. Yang, H. G.; Liu, G.; Qiao, S. Z.; Sun, C. H.; Jin, Y. G.; Smith, S. C.; Zou, J.; Cheng, H. M.; Lu, G. Q. Solvothermal Synthesis and Photoreactivity of Anatase TiO_2 Nanosheets With Dominant {001} Facets. *J. Am. Chem. Soc.* **2009,** *131* (11), 4078–4083.
28. Zhu, J.; Wang, S.; Bian, Z.; Xie, S.; Cai, C.; Wang, J.; Yang, H.; Li, H. Solvothermally Controllable Synthesis of Anatase TiO_2 Nanocrystals With Dominant {001} Facets and Enhanced Photocatalytic Activity. *CrystEngComm* **2010,** *12* (7), 2219–2224.
29. Wen, C. Z.; Zhou, J. Z.; Jiang, H. B.; Hu, Q. H.; Qiao, S. Z.; Yang, H. G. Synthesis of Micro-Sized Titanium Dioxide Nanosheets Wholly Exposed With High-Energy {001} and {100} Facets. *Chem. Commun.* **2011,** *47* (15), 4400–4402.
30. Wang, Y.; Zhang, H.; Han, Y.; Liu, P.; Yao, X.; Zhao, H. A Selective Etching Phenomenon on {001} Faceted Anatase Titanium Dioxide Single Crystal Surfaces by Hydrofluoric Acid. *Chem. Commun.* **2011,** *47* (10), 2829–2831.

31. Yang, X. H.; Li, Z.; Sun, C.; Yang, H. G.; Li, C. Hydrothermal Stability of {001} Faceted Anatase TiO_2. *Chem. Mater.* **2011,** *23* (15), 3486–3494.
32. Czoska, A. M.; Livraghi, S.; Chiesa, M.; Giamello, E.; Agnoli, S.; Granozzi, G.; Finazzi, E.; Di Valentin, C.; Pacchioni, G. The Nature of Defects in Fluorine-Doped TiO_2. *J. Phys. Chem. C* **2008,** *112* (24), 8951–8956.
33. Calatayud, D. G.; Jardiel, T.; Peiteado, M.; Illas, F.; Giamello, E.; Palomares, F. J.; Fernández-Hevia, D.; Caballero, A. C. Synthesis and Characterization of Blue Faceted Anatase Nanoparticles Through Extensive Fluorine Lattice Doping. *J. Phys. Chem. C* **2015,** *119* (36), 21243–21250.
34. Chiesa, M.; Paganini, M. C.; Livraghi, S.; Giamello, E. Charge Trapping in TiO_2 Polymorphs as Seen by Electron Paramagnetic Resonance Spectroscopy. *Phys. Chem. Chem. Phys.* **2013,** *15* (24), 9435–9447.
35. Rudiger, S.; Kemnitz, E. The Fluorolytic Sol-Gel Route to Metal Fluorides-a Versatile Process Opening a Variety of Application Fields. *Dalton Trans.* **2008**, (9), 1117–1127.
36. Sanchez, C.; Livage, J.; Henry, M.; Babonneau, F. Glasses and Glass Ceramics from Gels Chemical Modification of Alkoxide Precursors. *J. Non-Cryst. Solids* **1988,** *100* (1), 65–76.
37. Jolivet, J. -P. *Metal Oxide Chemistry and Synthesis: From Solution to Solid State;* Wiley, 2000.
38. Li, W.; Corradini, D.; Body, M.; Legein, C.; Salanne, M.; Ma, J.; Chapman, K. W.; Chupas, P. J.; Rollet, A. -L.; Julien, C.; Zhagib, K.; Duttine, M.; Demourgues, A.; Groult, H.; Dambournet, D. High Substitution Rate in TiO_2 Anatase Nanoparticles With Cationic Vacancies for Fast Lithium Storage. *Chem. Mater.* **2015,** *27* (14), 5014–5019.
39. Billinge, S. J. L.; Levin, I. The Problem With Determining Atomic Structure at the Nanoscale. *Science* **2007,** *316* (5824), 561–565.
40. Corradini, D.; Dambournet, D.; Salanne, M. Tuning the Electronic Structure of Anatase Through Fluorination. *Sci. Rep.* **2015,** *5,* 11553.

Ionic Liquid Materials Based on Fluoroanions

23

K. Matsumoto, R. Hagiwara
Kyoto University, Sakyo-ku, Kyoto, Japan

Chapter Outline

1. Introduction

Ionic liquids are now widely used in the fields of chemistry and electrochemistry. Since various kinds of ionic liquids are commercially available today, their applications spread into a number of scientific fields.[1–3] Low vapor pressure and flammability are unique properties of ionic liquids, which clearly differentiate them from conventional volatile organic compounds. The wide temperature range and high chemical and electrochemical stabilities are also important characteristics of most ionic liquids. Incorporation of fluorine-containing anions played an important role in the syntheses and applications of ionic liquids. In 1992, two air-stable ionic liquids with BF_4^-[4] and $SO_3CF_3^-$[5] were reported, followed by ionic liquids based on other fluorine-containing anions, which are common today, including PF_6^-[6] and $N(SO_2CF_3)_2^-$[7]. Their higher stability against moisture, compared with previously known haloaluminate-based ionic liquids,[8–10] enabled the handling in the air and led to rapid spread of ionic liquids as electrolytes and reaction media.

Fluoride ion, F^-, is a strong Lewis base and forms fluorocomplex anions with Lewis acidic fluorides. The reaction of F^- with HF is a special case, which gives a series of fluorohydrogenate anions formulated as $(FH)_nF^-$ (Eq. (23.1)):

$$F^- + n\text{HF} \rightarrow (\text{FH})_n\text{F}^- \tag{23.1}$$

Modern Synthesis Processes and Reactivity of Fluorinated Compounds. http://dx.doi.org/10.1016/B978-0-12-803740-9.00023-8

Although fluorohydrogenate anions have been known to form stable salts with various cations including alkali metal and various organic cations,[11–15] room temperature ionic liquids with negligible HF dissociation pressure was first reported in 1999.[16] Although a series of crystallographic examples are known for $(FH)_nF^-$ with different HF compositions ($n = 1–7$),[17–27] important species in ionic liquids are FHF^-, $(FH)_2F^-$, and $(FH)_3F^-$ (see Fig. 23.1).

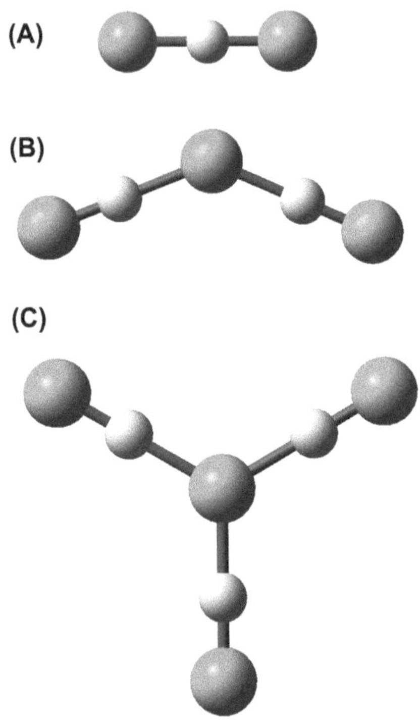

Figure 23.1 Molecular structures of (A) FHF^-, (B) $(FH)_2F^-$, and (C) $(FH)_3F^-$.

Other weakly coordinating fluorocomplex anions such as BF_4^- and PF_6^- are also combined with organic cations and form ionic liquids.[4,6,28–31] These ionic liquids are usually prepared by an ion exchange method, but alternative routes for their synthesis are required to obtain highly pure ionic liquids. Oxofluorocomplex anions, which have lower symmetry than homoleptic fluorocomplex anions, have different interaction with the cation and are also attractive as counteranion in ionic liquids.

Since quite a few review articles are available on the general and specific topics related to ionic liquids,[32–38] including the ones with fluorine-containing anions,[39–41] the present chapter focuses on topics of ionic liquids with fluorocomplex anions. The former half describes the synthesis and properties of fluorohydrogenate ionic liquids. In the latter half, their reactions with Lewis acidic fluorides as well as structures and properties of the obtained salts are discussed. Some properties of oxofluorocomplex anions are also described.

2. Syntheses and Structures of Ionic Liquids Based on Fluorohydrogenate Anions

Fluorohydrogenate ionic liquids of organic cations are synthesized by the reactions of the corresponding chlorides with anhydrous HF [see Eq. (23.2) for synthesis of a $C_2C_1im^+$ salt ($C_2C_1im^+$ = 1-ethyl-3-methylimidazolium cation) as an example].[42]

$$[C_2C_1im]Cl + (n+1)HF \rightarrow [C_2C_1im]\left[(FH)_nF\right] + HCl\uparrow \tag{23.2}$$

Bromide and iodide salts can be used as starting materials, but chloride salts are ideal in terms of removal of hydrogen halide. Such organic halide salts are prepared by quaternarization of amine with alkyl chloride[7]; e.g., the reaction of 1-methylimidazole with ethyl chloride gives $[C_2C_1im]Cl$. In the laboratory scale, perfluoroplastic reactors such as tetrafluoroethylene−perfluoroalkylvinylether copolymer can be used for this reaction. Since the reaction is exothermic and gives HCl gas, special attention should be paid to the sudden increase of temperature and pressure. Temperature control with an appropriate coolant is highly recommended. Addition of excess anhydrous HF and elimination of volatiles should be repeated for complete removal of HCl. The resulting products contain chloride impurities below the detection limit of the silver nitrate test. Water content of the fluorohydrogenate ionic liquids is typically below 100 ppm when rigorously dried chlorides and HF are used as starting materials.

A variety of organic cations shown in Fig. 23.2 form room temperature ionic liquids with $(FH)_nF^-$.[16,42–51] Although the HF content n in $(FH)_nF^-$ varies depending on the temperature of the final evacuation, weight change and elemental analysis confirmed that it was mostly in the vicinity of 2.3 at room temperature regardless of the organic cations. Some exceptional cases were observed for the cations with noncyclic frame or long alkyl chains, which lead to $n < 2.3$. The n value can be reversibly adjusted by addition (or elimination) of HF to (or from) fluorohydrogenate ionic liquids.[52,53] Mixing of two $(FH)_nF^-$ salts with different n values is another method to adjust the n value in fluorohydrogenate ionic liquids; $[C_2C_1im][(FH)_{1.5}F]$ is obtained by simply mixing $[C_2C_1im][FHF]$ and $[C_2C_1im][(FH)_2F]$ due to the immediate HF exchange between the anions. This method is convenient when a fluorohydrogenate ionic liquid with a specific n value is required. Although increase of temperature results in HF dissociation, irreversible decomposition occurs above 474 K. Fig. 23.3 shows the phase diagram of the $[C_2C_1im][FHF]$-HF system based on differential scanning calorimetry (DSC) analysis of the salt with different n values.[52] The DSC analysis suggested three compounds, $[C_2C_1im][FHF]$, $[C_2C_1im][(FH)_2F]$, and $[C_2C_1im][(FH)_3F]$ in this system, where $[C_2C_1im][FHF]$ was crystallographically confirmed.[54] The liquidus line drops from $n = 1$ to $n = 1.8$ and remained nearly constant in the region $n > 1.8$, suggesting a wide temperature range of $[C_2C_1im][(FH)_nF]$ ($n > 1.8$). Isolation of the compounds with $n < 1.0$ was difficult due to the extreme basicity of F^- in such salts, which accelerates self-decomposition.

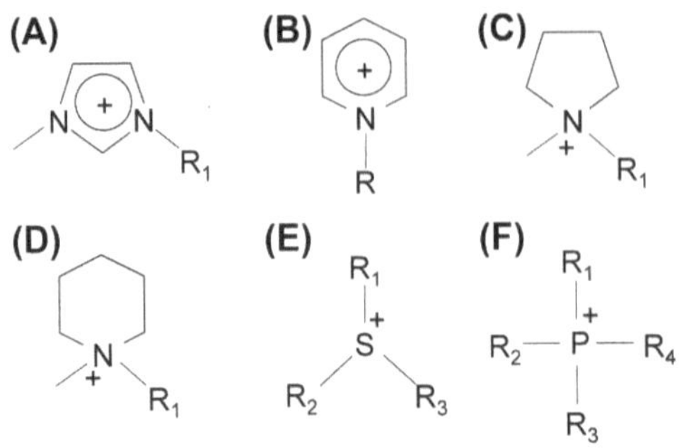

Figure 23.2 Structures of organic cations used for fluorohydrogenate ionic liquids. (A) 1-Alkyl-3-methylimidazolium ($C_nC_1im^+$), (B) *N*-alkylpyridinium (C_npy^+), (C) *N*-alkyl-*N*-methylpyrrolidinium ($C_nC_1pyrr^+$), (D) *N*-alkyl-*N*-methylpiperidinium ($C_nC_1pip^+$), (E) trialkylsulfonium $\left(S_{nml}{}^+\right)$, and (F) tetraalkylphosphonium $\left(P_{nmlk}{}^+\right)$. The carbon number in the cation is denoted as n in C_n or as n, m, l, and k in $P_{nmlk}{}^+$ and $S_{nml}{}^+$.

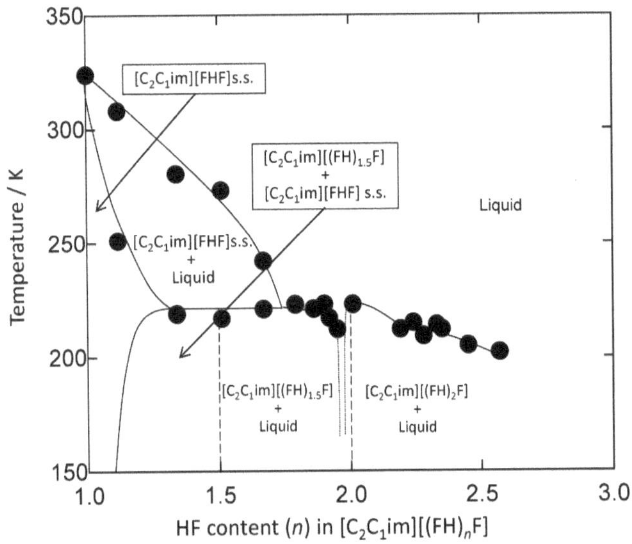

Figure 23.3 Phase diagram of the $[C_2C_1im][FHF]$-HF system based on differential scanning calorimetric analysis.[52]

Infrared spectroscopy is used to identify $(FH)_nF^-$ in fluorohydrogenate ionic liquids. Vibrational bands of $(FH)_nF^-$ were experimentally studied by infrared spectroscopy in some simple salts[55] and theoretically by quantum mechanical calculations.[56–58] Fig. 23.4 shows the infrared spectra of $[C_2C_1im][(FH)_nF]$ ($n = 1.3$, 1.5, 1.6, 1.8, and 2.3), which gradually change according to the anionic species. The FHF^- anion shows characteristic infrared bands around $\sim 1250\ cm^{-1}$ (Π_u mode) and $\sim 1500\ cm^{-1}$ (Σ_u^+ mode). The $(FH)_2F^-$ (C_{2v}) and $(FH)_3F^-$ (D_{3h}) anions show infrared bands at similar wavenumbers but can be distinguished by careful investigation [~500, 1000, 1800, and 2500 cm^{-1} for $(FH)_2F^-$ and ~1000 and 3000 cm^{-1} for $(FH)_3F^-$ (D_{3h})]. The nonvolatile fluorohydrogenate ionic liquids with $n = 2.3$ contains $(FH)_2F^-$ and $(FH)_3F^-$ (D_{3h}) in 7:3 ratio according to infrared spectroscopy. The $(FH)_3F^-$ (C_{2v}) anion, which can be formulated as $(FH)(FHF)(HF)^-$ and is the structural isomer of $(FH)_3F^-$ (D_{3h}),[24] was not observed in the infrared spectroscopic study of $[C_2C_1im][(FH)_{2.3}F]$. Nuclear magnetic resonance (NMR) spectroscopy revealed that the HF unit of $(FH)_2F^-$ and $(FH)_3F^-$ in $[C_2C_1im][(FH)_{2.3}F]$ is exchanged between the anions and that the H and F atoms in $(FH)_nF^-$ are magnetically equivalent with NMR time scale.[42]

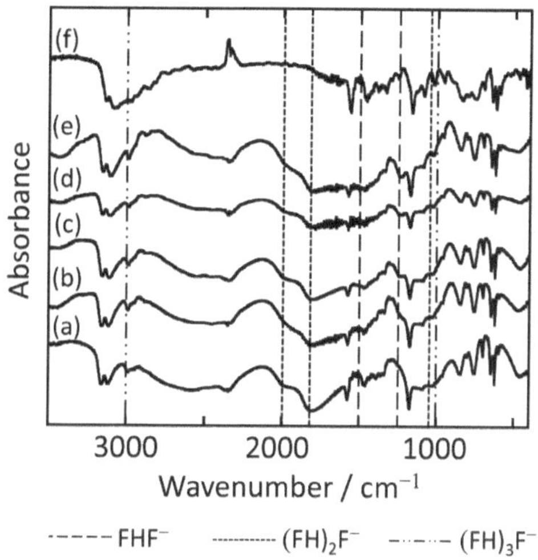

Figure 23.4 Infrared spectra of $[C_2C_1im][(FH)_nF]$; (a) $[C_2C_1im][(FH)_{2.3}F]$, (b) $[C_2C_1im][(FH)_{1.8}F]$, (c) $[C_2C_1im][(FH)_{1.6}F]$, (d) $[C_2C_1im][(FH)_{1.5}F]$, (e) $[C_2C_1im][(FH)_{1.3}F]$, and (f) $[C_2C_1im]Cl$.

Crystallographic works on $[C_2C_1im][FHF]$, $[C_1C_1im][(FH)_2F]$, and $[C_1C_1im][(FH)_3F]$ salts ($C_1C_1im^+$ = 1,3-dimethylimidazolium) confirmed the corresponding $(FH)_nF^-$ anions ($n = 1$, 2, and 3) in the solid state.[54,58] In the crystal structure of $[C_2C_1im][FHF]$, the imidazolium ring of $C_2C_1im^+$ and the entire FHF^- are located in the same layer, and the $C_2C_1im^+$ cations form one-dimensional (1D) columns, binding FHF^- between two ethyl groups. This configuration of ions in each layer is related to those in $[C_2C_1im][NO_2]$[4] and $[C_2C_1im][OCN]$.[59] The crystal structures of $[C_1C_1im][(FH)_2F]$ and $[C_1C_1im][(FH)_3F]$ contain 1D columns of the cations and anions.

3. Properties of Ionic Liquids Based on Fluorohydrogenate Anions

Table 23.1 lists the physical properties of selected fluorohydrogenate ionic liquids. In general, ionic conductivity of a fluorohydrogenate ionic liquid is significantly higher than those of other ionic liquids with the same cation. For example, 100 mS/cm of nonvolatile $[C_2C_1im][(FH)_{2.3}F]$ ionic liquid is considerably higher than 8.8 mS/cm of $[C_2C_1im][N(SO_2CF_3)_2]$,[7] 16.5 mS/cm of $[C_2C_1im][N(SO_2F)_2]$,[60] 13 mS/cm of $[C_2C_1im][BF_4]$,[61] and 9.2 mS/cm of $[C_2C_1im][SO_3CF_3]$.[5] The highest ionic conductivity of 131 mS/cm was observed for $[S_{111}][(FH)_{1.9}F]$ (S_{111}^+ = trimethylsulfonium) owing to the low viscosity and small size of S_{111}^+.[49] Such high ionic conductivity does not arise from a special conduction mechanism such as the Grotthuss mechanism but simply from low viscosity, which is confirmed by the Walden plot. The Walden plot is depicted with the horizontal axis of logarithm of η^{-1} and vertical axis of logarithm of λ (λ and η denote molar conductivity and viscosity, respectively).[62,63] According to Walden's rule, the product of λ and η is constant and the plot gives a straight line. This relation is applied to ionic liquids,[64–66] and fluorohydrogenate ionic liquids are not exceptional as shown in Fig. 23.5. Ionic conductivity of $[C_2C_1im]$ $[(FH)_nF]$ is a function of n; the ionic conductivity monotonously increases with increasing n in the range as shown in Fig. 23 6.

Table 23.1 Physical Properties of Some Organic Salts of Fluorohydrogenate Anions[a]

	T_m, T_g, T_{cl} (K)	ρ (g/cm^3)	η (mPa s)	σ (mS/cm)
$[C_1C_1im][(FH)_{2.3}F]$[b]	272	1.17	5.1	110
$[C_2C_1im][(FH)_{2.3}F]$[b]	208 (148g)	1.13	4.9	100
$[C_{allyl}C_1im][(FH)_{2.3}F]$[c]	–	1.11	5.5	90
$[C_3C_1im][(FH)_{2.3}F]$[b]	(152g)	1.11	7.0	61
$[C_4C_1im][(FH)_{2.3}F]$[b]	(154g)	1.08	19.6	33
$[C_5C_1im][(FH)_{2.3}F]$[b]	(158g)	1.05	26.7	27
$[C_6C_1im][(FH)_{2.3}F]$[b]	(157g)	1.00	25.8	16
$[C_8C_1im][(FH)_{2.1}F]$[d]	248 (288cl)	–	–	–
$[C_{10}C_1im][(FH)_{2.0}F]$[d]	278 (314cl)	–	–	–
$[C_{12}C_1im][(FH)_{2.0}F]$[d]	295 (324cl)	–	–	7.4 (parallel to the layer), 0.62 (perpendicular to the layer)
$[C_{14}C_1im][(FH)_{2.0}F]$[d]	307 (376cl)	–	–	–
$[C_{16}C_1im][(FH)_{2.0}F]$[d]	327 (424cl)	–	–	–
$[C_{18}C_1im][(FH)_{2.0}F]$[d]	340 (463cl)	–	–	–
$[C_2py][(FH)_{2.3}F]$[c]	(165g)	1.13	10.1	71
$[C_{all}py][(FH)_{2.3}F]$[c]	203	1.12	5.3	82

Table 23.1 Physical Properties of Some Organic Salts of Fluorohydrogenate Anions[a]—cont'd

	T_m, T_g, T_{cl} (K)	ρ (g/cm^3)	η (mPa s)	σ (mS/cm)
$[C_3py][(FH)_{2.3}F]$[c]	–	1.11	12.3	48
$[C_4py][(FH)_{2.3}F]$[c]	–	1.09	29.7	25
$[C_2C_1pyrr][(FH)_{2.3}F]$[e]	–	1.07	9.9	74.6
$[C_{allyl}C_1pyrr][(FH)_{2.3}F]$[c]	–	1.05	8.5	78
$[C_3C_1pyrr][(FH)_{2.3}F]$[e]	–	1.05	11.2	58.1
$[C_4C_1pyrr][(FH)_{2.3}F]$[e]	–	1.04	14.5	35.9
$[C_6C_1pyrr][(FH)_{2.3}F]$[f]		0.993	18.0	23.7
$[C_{1O1}C_1pyrr][(FH)_{2.2}F]$[g]	<153	1.13	7.9	74.7
$[C_{all}C_1mor][(FH)_{2.3}F]$[c]	–	1.16	34.6	35
$[C_2C_1pip][(FH)_{2.3}F]$[e]	217,237	1.07	24.2	37.2
$[C_{all}C_1pip][(FH)_{2.3}F]$[c]	–	1.06	16.7	63
$[C_3C_1pip][(FH)_{2.3}F]$[e]	(164g)	1.06	33.0	23.9
$[C_4C_1pip][(FH)_{2.3}F]$[e]	(162g)	1.04	37.1	12.3
$[S_{111}][(FH)_{1.9}F]$[h]	242	1.18	7.8	131
$[S_{112}][(FH)_{1.9}F]$[h]	227	1.14	8.2	111
$[S_{122}][(FH)_{1.9}F]$[h]	–	1.11	8.9	91
$[S_{222}][(FH)_{1.9}F]$[h]	217	1.09	8.3	83
$[P_{4441}][(FH)_{2.3}F]$[f]	249 (169g)	0.969	36	6.0
$[P_{4444}][(FH)_{2.3}F]$[f]	239, 255	0.954	47	3.7
$[P_{4448}][(FH)_{2.3}F]$[f]	(174g)	0.934	74	1.5
$[P_{222(1O1)}][(FH)_{2.3}F]$[i]	206, 239	1.07	44	11
$[P_{2225}][(FH)_{2.3}F]$[i]	218 (166g)	0.999	28	12.4
$[P_{2228}][(FH)_{2.3}F]$[i]	227, 246	0.968	47	6.1

[a]The symbols, ρ, η, and σ denote the density, viscosity, and ionic conductivity, respectively, at 298 K. See Fig. 23.2 for the abbreviated name of each cation. The subscript in the cation name means the number of carbon atom in the alkyl chain (e.g., C_2C_1im denotes 1-ethyl-3-methylimidazolium cation). Allyl and methoxymethyl groups are abbreviated as "all" and "1O1". T_m, T_g, and T_{cl} denote melting, glass transition, and clearing temperatures, respectively. T_g and T_{cl} are listed with "g" and "cl," respectively.
[b]Ref. 44.
[c]Ref. 51.
[d]Ref. 82.
[e]Ref. 47.
[f]Ref. 45.
[g]Ref. 48.
[h]Ref. 49.
[i]Ref. 43.

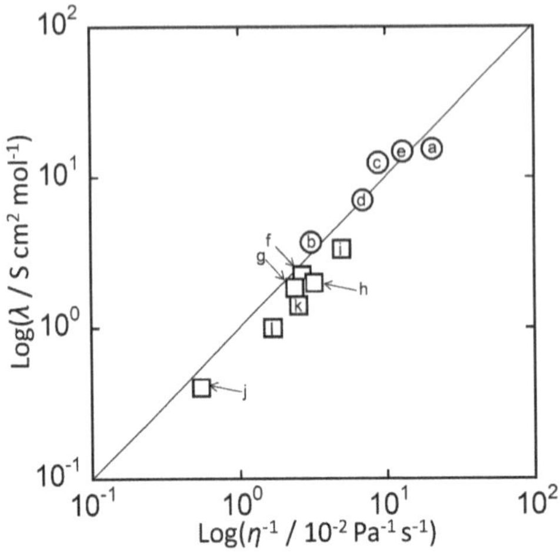

Figure 23.5 Walden plot [molar ionic conductivity (λ) against reciprocal viscosity (η)] for selected ionic liquids ((a) $[C_2C_1im][(FH)_{2.3}F]$, (b) $[C_4C_1im][(FH)_{2.3}F]$, (c) $[C_2C_1pyrr][(FH)_{2.3}F]$, (d)$[C_4C_1pyrr][(FH)_{2.3}F]$, (e) $[S_{111}][(FH)_{1.9}F]$, (f) $[C_2C_1im][BF_4]$, (g) $[C_2C_1im][SO_3CF_3]$, (h) $[C_2C_1im][N(SO_2CF_3)_2]$, (i) $[C_2C_1im][N(SO_2F)_2]$, (j) $[C_4C_1im][PF_6]$, (k) $[C_4C_1im][N(SO_2CF_3)_2]$, (l) $[C_4C_1pyrr][N(SO_2CF_3)_2]$).[7,44,47,49,60,61,67] Abbreviations of the cations are shown in Fig. 23.2. *Circles* and *squares* mean fluorohydrogenate and nonfluorohydrogenate ionic liquids, respectively.

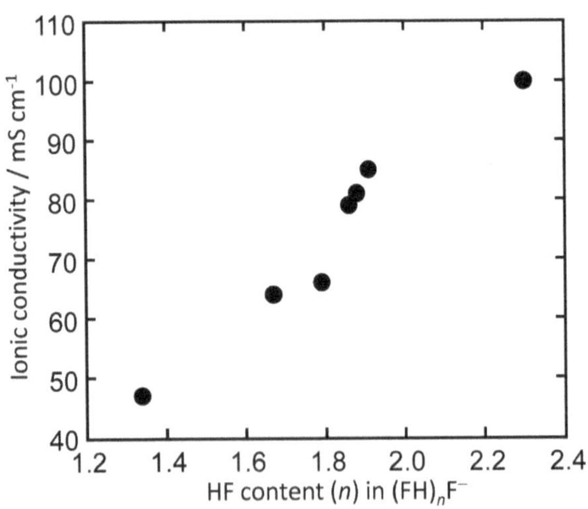

Figure 23.6 The HF composition dependence of ionic conductivity for $[C_2C_1im][(FH)_nF]$ at 298 K.[52]

Pulsed field gradient spin echo (PGSE) NMR is a powerful tool to determine self-diffusion coefficients of ionic species in ionic liquids.[61,68–70] According to the PGSE-NMR measurements, the self-diffusion coefficient of $(FH)_nF^-$ in fluorohydrogenate ionic liquids is larger than that of an anion in other ionic liquids, but the self-diffusion coefficient of the cation is also larger than that of the same cation in other ionic liquids with different anions[71] (e.g., 4.1×10^{-6} cm^2/s for the anion and 3.0×10^{-6} cm^2/s for the cation in $[C_2C_1im][(FH)_{2.3}F]$[71]; 0.42×10^{-6} cm^2/s for the anion and 0.50×10^{-6} cm^2/s for the cation in $[C_2C_1im][BF_4]$[61]), which denies the special conduction mechanism such as ion hopping.

Interaction energies between ions were evaluated for fluorohydrogenate ionic liquids based on the method proposed by Tsuzuki et al.[72] This method assesses ion–ion interaction energies by quantum mechanical calculations using the configuration of ions in crystal structures. Crystal structures of $[C_2C_1im][FHF]$, $[C_1C_1im][(FH)_2F]$, and $[C_1C_1im][(FH)_2F]$ were used for this calculation.[54,58] The plots of the interaction energies against the distance between the cation and anion exhibit a linear relationship in any case, suggesting that the electrostatic interaction is simply dominant in this system as in the cases of other imidazolium salts and that dynamic properties are responsible for the low viscosity of fluorohydrogenate ionic liquids.[58] The exchanging HF units in $(FH)_nF^-$ may work as a dielectric spacer and facilitates the diffusion of an ion (see Fig. 23.7).

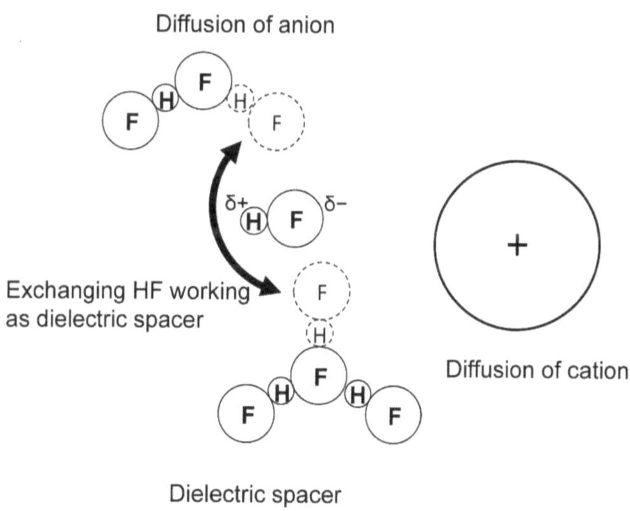

Figure 23.7 Model for the ion conduction mechanism of fluorohydrogenate ILs, where exchanging HF units work as a dielectric spacer and buffer the coulombic interaction between cation and anion.[58]

Fig. 23.8 shows cyclic voltammograms of a glass-like carbon electrode in (a) $[C_2C_1im][BF_4]$, (b) $[C_2C_1im][(FH)_{2.3}F]$, and (c) $[C_2C_1pyrr][(FH)_{2.3}F]$.[36,39,59] The electrochemical window of $[C_2C_1im][BF_4]$ is ~4.5 V, whereas that of $[C_2C_1im]$

$[(FH)_{2.3}F]$ is ~3 V, ranging from −2 to 1 V vs. Fc^+/Fc. Although replacement of $C_2C_1im^+$ with $C_2C_1pyrr^+$ improves the electrochemical stability of the ionic liquid, especially to the anodic direction, intrinsic stability is not significantly improved. This was confirmed by capacitor tests using activated carbon electrodes, which resulted in the decomposition of the fluorohydrogenate ionic liquids at much lower voltages.[48,73,74]

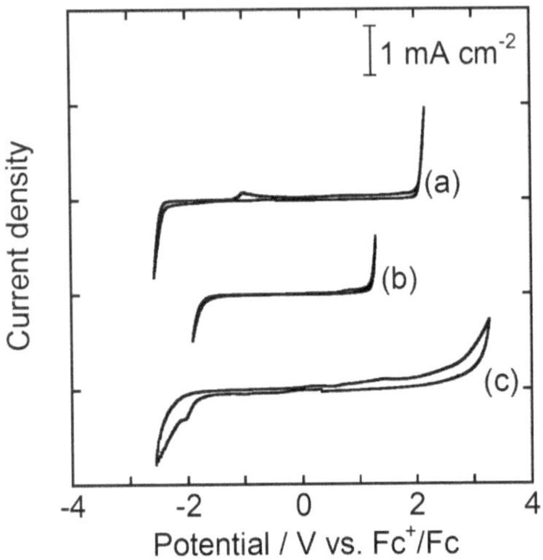

Figure 23.8 Cyclic voltammograms of a glass-like carbon electrode in (a) $[C_2C_1im][BF_4]$, (b) $[C_2C_1im][(FH)_{2.3}F]$, and (c) $[C_2C_1pyrr][(FH)_{2.3}F]$.[44,47]

Fluorohydrogenate salts with characteristic mesophase structures are derived from the ionic liquids by incorporating functional organic cations. Ionic liquid crystal is liquid crystalline material solely composed of ionic species[75,76] and attracts attention as anisotropic ion conductors owing to their anisotropic structures.[77–80] Fluorohydrogenates with some cyclic ammonium cations such as imidazolium and pyrrolidinium frames with long alkyl side chains exhibit ionic liquid crystal phases in limited temperature ranges owing to the alkyl−alkyl interactions between the long alkyl chains. According to the polarized optical microscopic and X-ray diffraction analysis, fluorohydrogenate ionic liquid crystals have smectic A_2-type bilayer structures regardless of the alkyl side chain length (carbon number of greater than 10 for 1-alkyl-3-methylimidazolium and 14 for *N*-alkyl-*N*-methylpyrrolidinium), as shown in Fig. 23.9.[81,82] Fluorohydrogenate ionic liquid crystals show anisotropy in ionic conductivities, that is, ionic conductivity parallel ($\sigma_{//}$) to the smectic layer is higher than that perpendicular ($\sigma\perp$) to it.[82] The ratio of $\sigma_{//}/\sigma\perp$ depends on the ionic species and is ~10 at 298 K for $[C_{12}C_1im][(FH)_2F]$ ($\sigma_{//} = 7.37$ and $\sigma\perp = 0.62$ mS/cm). For 1-alkyl-3-methylimidazolium fluorohydrogenate $[C_xC_1im][(FH)_2F]$, the $\sigma_{//}$ is nearly constant against alkyl chain length, whereas $\sigma\perp$ decreases with an increase in the length of the alkyl side chain. Effects of the n value in $(FH)_nF^-$ on the structures and properties of $[C_{12}C_1im][(FH)_nF]$ salts were investigated by several analytical methods.[83]

The $[C_{12}C_1im][(FH)_nF]$ salt is a mixed crystal system that has a boundary around $n = 1.9$. The temperature range of the smectic mesophase decreases with increasing n, whereas the interlayer distance decreases with an increase in n or in temperature. These materials are considered to be anion conductors because of the bulky cation.

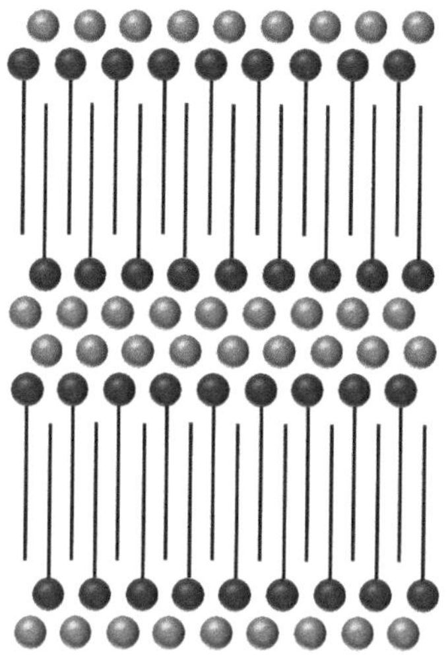

Figure 23.9 A smectic A_2 bilayer structure observed for $[C_xC_2im][(FH)_nF]$.[82,83]

Another interesting mesophase observed for fluorohydrogenate salts is plastic crystal. Ionic plastic crystals, which are characterized by highly disordered constituent ions, are interesting candidates as solid-state ion conductor[84–87] for electrochemical devices including lithium ion batteries.[88,89] The pyrrolidinium-based salts, *N,N*-dimethylpyrrolidinium and *N*-ethyl-*N*-methylpyrrolidinium fluorohydrogenates ($[C_1C_1pyrr][(FH)_2F]$ and $[C_2C_1pyrr][(FH)_2F]$) form ionic plastic crystal around room temperature (258–325 and 236–303 K, respectively).[90] They exhibit small entropy changes of melting (4.1 J/K mol for $[C_1C_1pyrr][(FH)_2F]$ and 2.0 J/K mol for $[C_2C_1pyrr][(FH)_2F]$) owing to their disordered structures. X-ray diffraction analysis revealed that these salts have a rock-salt-type structure ($a = 9.90$ Å for $[C_1C_1pyrr][(FH)_2F]$ and $a = 10.18$ Å for $[C_2C_1pyrr][(FH)_2F]$). The slightly larger a lattice parameter of $[C_2C_1pyrr][(FH)_2F]$ results from the larger size of $C_2C_1pyrr^+$. The constituent ions with nearly three-dimensional rotation satisfy high site symmetry in this cell. Another structural type, the inverse NiAs-type structure, is observed for $[P_{2222}][(FH)_2F]$ (P_{2222}^+ = tetraethylphosphonium) ($a = 8.02$ Å and $c = 11.84$ Å).[43] The HF content n in $[C_2C_1pyrr][(FH)_nF]$ (1.0–2.3) and $[C_1C_1pyrr][(FH)_nF]$ (1.0–2.0) significantly affects their phase behavior. Both the systems have ionic plastic phases with a rock-salt-type structure and the anions with different n randomly occupy the anionic sites in the lattice over a wide range.[53] Although ionic plastic crystal phase was observed at n near 1.0 and 2.0 for $[C_2C_1pyrr][(FH)_nF]$, $[C_1C_1pyrr][(FH)_nF]$ exhibits it over the range of investigation. Ionic conductivities of these ionic plastic crystal phases are quite high and range from 10^0 to 10^1 mS/cm (e.g., 10.3 mS/cm at 298 K for $[C_1C_1pyrr][(FH)_2F]$ and 14.4 mS/cm at 288 K for $[C_2C_1pyrr][(FH)_2F]$). According to the PGSE NMR spectroscopy, these salts are $(FH)_nF^-$-conducting materials with self-diffusion coefficients of $\sim 10^{-7}$ cm^2/s (see Fig. 23.10). The electrochemical window of $[C_1C_1pyrr][(FH)_2F]$ is nearly 4 V on a glass-like carbon, although it shows considerable pseudocapacitance on activated carbon even at 2.5-V charging.[91]

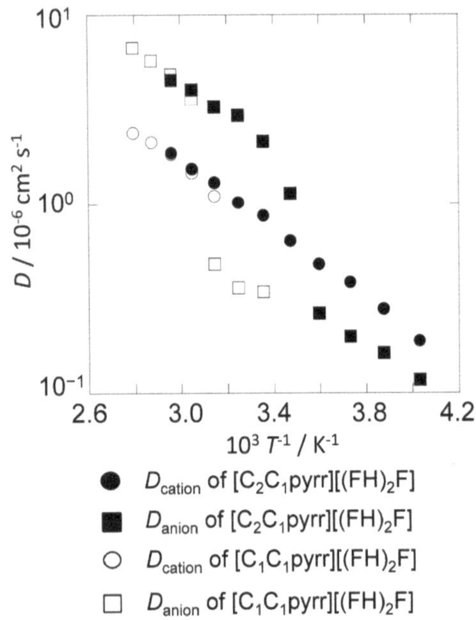

Figure 23.10 Arrhenius plots of self-diffusion coefficients for ions in plastic crystal phases of $[C_1C_1pyrr][(FH)_2F]$ and $[C_2C_1pyrr][(FH)_2F]$.[90]

4. Syntheses of Ionic Liquids Based on Fluorocomplex and Oxofluorocomplex Anions

Ionic liquids based on fluorocomplex anions are usually prepared by the ion exchange method [Eq. (23.3) for the synthesis of $[C_2C_1im][BF_4]$ as an example)]:

$$[C_2C_1im]X + M[BF_4] \rightarrow [C_2C_1im][BF_4] + MX\downarrow \tag{23.3}$$

$$\left(M^+ = \text{alkali metal, } NH_4{}^+, \text{or } Ag^+;\ X^- = \text{halide anion}\right)$$

This method is simple but suffers from contamination by the by-product MX when the resulting ionic liquid is hydrophilic. Since naked fluoride ion is not stable against organic cations commonly used for ionic liquids, another fluoride ion source weakly bound by Lewis acid is required for direct synthesis with Lewis acidic fluorides. Fluorohydrogenate ionic liquids can provide such a situation based on Eq. (23.4) (example for $[C_2C_1im][BF_4]$):

$$[C_2C_1im][(FH)_nF] + BF_3 \rightarrow [C_2C_1im][BF_4] + nHF \tag{23.4}$$

This reaction has an advantage to minimize impurities in the product because fluorohydrogenate ionic liquids with low halide impurities are readily available and the by-product HF is easily removed by evacuation at elevated temperatures. The Lewis acid–base reaction is highly exothermic and the reactor should be occasionally cooled by refrigerant. Reaction of $[C_2C_1im][(FH)_{2.3}F]$ with large excess of gaseous BF_3, PF_5, AsF_5, and WF_6 (WF_6 is condensed as liquid phase) followed by removal of volatiles gave $C_2C_1im^+$ salts of the corresponding fluorocomplex anions.[92] In the cases of solid fluorides or oxyfluorides (NbF_5, TaF_5, UF_5, VOF_3, WOF_4, and $MoOF_4$), a stoichiometric reaction gives the corresponding $C_2C_1im^+$ salts of fluoro- or oxofluorocomplex anions.[92–96] Our trial of the reaction with SbF_5, BiF_5, and VF_5 resulted in decomposition of the products due to the strong oxidizing power of these binary fluorides. In such a case we have to select an alternative route given by Eq. (23.3) in spite of the inconvenience of the formation of metal halide by-products. The metathesis of $[C_2C_1im]Cl$ and $KSbF_6$ gives an ionic liquid SbF_6^-. This synthetic route has been extended to the salts of other organic cations such as *N*-alkyl-*N*-methylpyrrolidinium and *N*-butylpyridinium.[93,94,97]

Oxofluorocomplex anions are interesting candidates as counteranion of ionic liquids since they are usually asymmetric and have a dipole moment compared with symmetric fluoroanions such as BF_4^- and PF_6^- (see Fig. 23.11). Because PO_2F is not available as a stable compound, ionic liquids based on $PO_2F_2^-$ are synthesized by ion exchange reaction between $K[PO_2F_2]$ and a chloride salt of the corresponding organic cation.[98] The pure starting material, $K[PO_2F_2]$, is prepared by the stoichiometric reaction of $[K][PO_3]$ and $[K][PF_6]$ in a Pt crucible under a dry Ar atmosphere at 320°C (see Eq. (23.5)):

$$2[K][PO_3] + [K][PF_6] \rightarrow 3K[PO_2F_2] \tag{23.5}$$

The crude ionic liquid obtained by the ion exchange reaction is purified through activated carbon column, which is necessary to reduce halide impurities. Another oxofluorocomplex anion, SO_3F^-, also gives stable ionic liquids. In this case, HSO_3F is a stable liquid and works as a good anion source. Although the ion exchange method was proposed for the synthesis of SO_3F^- ionic liquids,[100] the reaction between a chloride salt of the corresponding organic cation (e.g., $[C_2C_1im]Cl$) and HSO_3F is preferable to decrease impurities.[99] It should be noted that the 1:2 and 1:3 stoichiometric reactions of $[C_2C_1im]Cl$ and HSO_3F give the salts of polynuclear anions, $[C_2C_1im][(FO_2SOH)OSO_2F]$ and $[C_2C_1im][(FO_2SOH)_2O_2SOF]$, respectively. They are liquids at room temperature without HSO_3F dissociation pressure

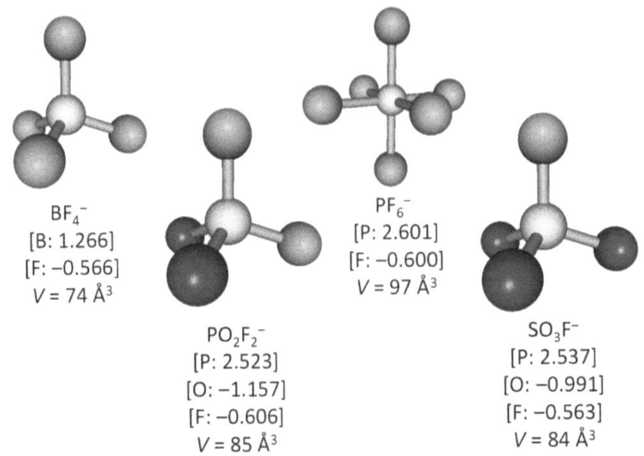

Figure 23.11 Molecular geometries, volumes, and natural bond orbital charges calculated at PBE1PBE/aug-cc-pVTZ for BF_4^-, PF_6^-, $PO_2F_2^-$, and SO_3F^-.[98,99]

at room temperature [see Eq. (23.6) for the reaction of $[C_2C_1im]Cl$ and HSO_3F at different ratios].

$$[C_2C_1im]Cl + nHSO_3F \rightarrow [C_2C_1im]\left[(FO_2SOH)_nO_nSO_{3-n}F\right] + HCl \qquad (23.6)$$

5. Structures and Properties of Ionic Liquids Based on Fluorocomplex Anions

Table 23.2 summarizes physical properties of some salts based on fluorocomplex anions. Crystal structures of $[C_2C_1im][AF_6]$ (A = P, As, Sb, Nb, and Ta) revealed that these five salts can be classified into two different structural types in the solid state: $[C_2C_1im][PF_6]$-type structure (PF_6, AsF_6, and SbF_6 salts) and $[C_2C_1im][NbF_6]$-type structure (NbF_6 and TaF_6 salts) (see Fig. 23.12).[6,92,101] In the former structure, cations and anions alternately stack to form a 1D pillar. Strong interactions other than coulombic interactions such as hydrogen bonding do not exist between the cation and the anion, and no cation−cation interactions are observed. On the other hand, the latter structure has an anion−anion stacking resulting from the packing dominated by the hydrogen bond or van der Waals cation−cation interaction. The tetrafluoroborate salt, $[C_2C_1im][BF_4]$, exhibits a unique structure wherein $C_2C_1im^+$ cations form 1D pillars where one imidazolium ring faces the adjacent imidazolium ring linked by H(methylene)...π electron interaction (see Fig. 23.12). The BF_4 anion also forms 1D pillars along the same direction with the nearest F...F contact distance of 3.368 Å.

Table 23.2 Physical Properties of Some Organic Salts of Fluorocomplex or Oxofluorocomplex Anions[a]

	T_m (K)	ρ (g/cm^3)	η (mPa s)	σ (mS/cm)
$[C_2C_1im][BF_4]$[b]	288	1.27	34	13.0
$[C_2C_1im][PF_6]$[c]	333	–	–	–
$[C_2C_1im][AsF_6]$[d]	326	1.78	–	–
$[C_2C_1im][SbF_6]$[d]	283	1.85	67	6.2
$[C_2C_1im][NbF_6]$[d]	272	1.67	49	8.5
$[C_2C_1im][TaF_6]$[d]	275	2.17	51	7.1
$[C_2C_1im][WF_7]$[d]	258	2.27	171	3.2
$[C_2C_1im][VOF_4]$[e]	348	–	–	–
$[C_2C_1im][WOF_5]$[f]	253	2.25	105	3.0
$[C_2C_1im][MoOF_5]$[e]	(190g)[l]	1.76	86	5.1
$[C_2C_1im][UF_6]$[g]	284	2.43	59	7.9
$[C_2C_1pyrr][BF_4]$[h]	567	–	–	–
$[C_2C_1pyrr][PF_6]$[h]	534	–	–	–
$[C_2C_1pyrr][SbF_6]$[h]	dec[m]	–	–	–
$[C_2C_1pyrr][NbF_6]$[h]	dec[m]	–	–	–
$[C_2C_1pyrr][TaF_6]$[h]	dec[m]	–	–	–
$[C_2C_1pyrr][WF_7]$[h]	dec[m]	–	–	–
$[C_4C_1pyrr][BF_4]$[i]	411	–	–	–
$[C_4C_1pyrr][PF_6]$[j]	343	–	–	–
$[C_4C_1pyrr][SbF_6]$[h]	326	–	–	–
$[C_4C_1pyrr][NbF_6]$[h]	320	–	164 (328 K)	3.0 (328 K)
$[C_4C_1pyrr][TaF_6]$[h]	323	–	165 (328 K)	3.0 (328 K)
$[C_4C_1pyrr][WF_7]$[h]	322	–	–	–
$[C_6C_1pyrr][BF_4]$[h]	345	–	–	–
$[C_6C_1pyrr][PF_6]$[h]	467	–	–	–
$[C_6C_1pyrr][SbF_6]$[h]	453	–	–	–
$[C_6C_1pyrr][NbF_6]$[h]	439	–	–	–
$[C_6C_1pyrr][TaF_6]$[h]	448	–	–	–
$[C_6C_1pyrr][WF_7]$[h]	456	–	–	–
$[C_4py][BF_4]$[k]	272	1.220	103	1.9
$[C_4py][PF_6]$[h]	351	–	–	–

Continued

Table 23.2 Physical Properties of Some Organic Salts of Fluorocomplex or Oxofluorocomplex Anions[a]—cont'd

	T_m (K)	ρ (g/cm^3)	η (mPa s)	σ (mS/cm)
$[C_4py][SbF_6]$[h]	351	–	–	–
$[C_4py][NbF_6]$[h]	357	–	–	–
$[C_4py][TaF_6]$[h]	351	–	–	–
$[C_4py][WF_7]$[h]	348	–	–	–

[a]See Table 23.1 for abbreviations.
[b]Ref. 61.
[c]Ref. 6.
[d]Ref. 92.
[e]Ref. 94.
[f]Ref. 95.
[g]Ref. 93.
[h]Ref. 97.
[i]Ref. 121.
[j]Ref. 122.
[k]Ref. 61.
[l]Glass transition
[m]Decomposition

The highest ionic conductivity in this series is often observed for BF_4^- ionic liquids (e.g., 13.6 mS/cm for $[C_2C_1im][BF_4]$, 6.2 mS/cm for $[C_2C_1im][SbF_6]$, 8.5 mS/cm for $[C_2C_1im][NbF_6]$, 7.1 mS/cm $[C_2C_1im][TaF_6]$, and 3.2 mS/cm $[C_2C_1im][WF_7]$ at 298 K).[61,92] Electrochemical window of the ionic liquids of fluorocomplex anions depends on the stability of the anion against reduction, as shown in Fig. 23.13. The $C_2C_1im^+$ ionic liquids with the reductively stable BF_4^- and TaF_6^- exhibit a wide electrochemical window of ~4.5 V, whereas those with SbF_6^-, NbF_6^-, and WF_7^- have narrower cathodic limit, resulting in electrochemical windows of ~3.5, 3.5, and 2.5 V, respectively. The $C_2C_1im^+$ ionic liquids of $PO_2F_2^-$ and SO_3F^- exhibit relatively high ionic conductivities of 12 and 10.8 mS/cm at 298 K, which is derived from the relatively smaller sizes of these anions compared to the hexafluorocomplex and bis(trifluorosulfonyl)amide anions. Their electrochemical windows are ~4.3 V and are slightly narrower than that of $[C_2C_1im][BF_4]$ due to the lower anodic stability of the oxofluorocomplex anions. The anionic structure with a dipole moment may contribute to a slightly higher capacitance on activated carbon electrodes.[102]

Donor ability of ionic liquid is important when interactions between chemical species and anion are discussed. Fig. 23.14 shows the trend of donor properties of typical fluorine-containing anions for ionic liquids estimated by a solvatochromic method with a copper salt, $[Cu(acac)(tmen)][BPh_4]$ (tmen = *N,N,N′,N′*-tetramethylethylenediamine, acac = acetylacetonate, and BPh_4^- = tetraphenylborate).[103,104] In this method, the position of the maximum wavelength (λ_{max}) depends on the donor property of the solvent or the ionic species in a solvent. There is a trend for the donor ability of fluorine-containing anions, $PF_6^- < BF_4^- < N(SO_2CF_3)_2^- < SO_3CF_3^- < SO_3F^- < PO_2F_2^- < CO_2CF_3^-$, which is related to the atomic charges on the F or O atoms in these anions.[67,98,99,105]

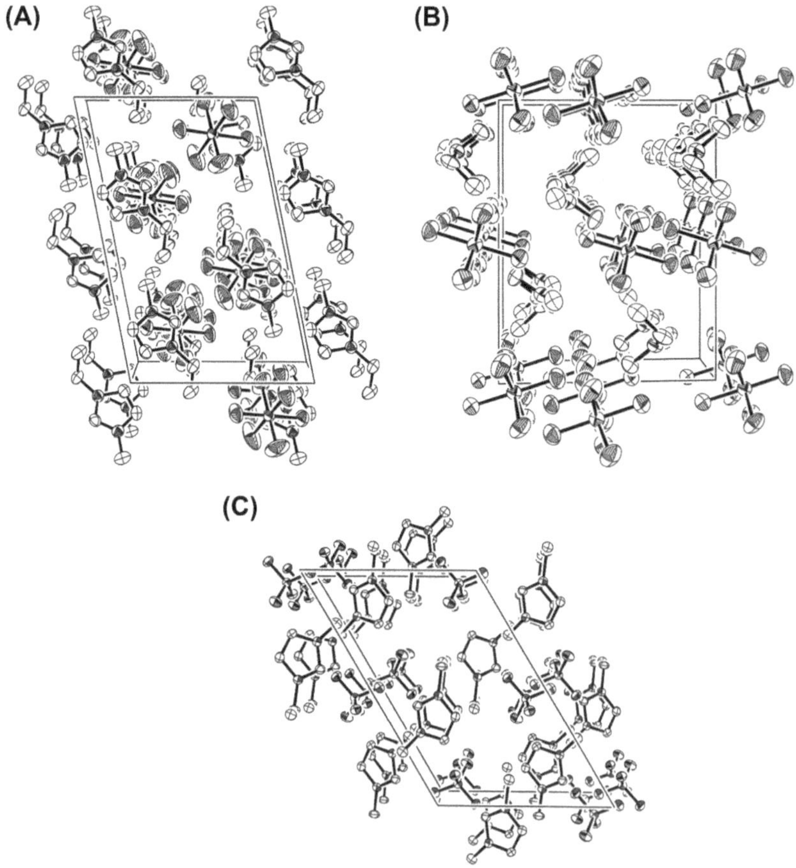

Figure 23.12 Crystal packing of (A) $[C_2C_1im][SbF_6]$, (B) $[C_2C_1im][NbF_6]$, and (C) $[C_2C_1im][BF_4]$ determined by single-crystal X-ray diffraction.[92,101]

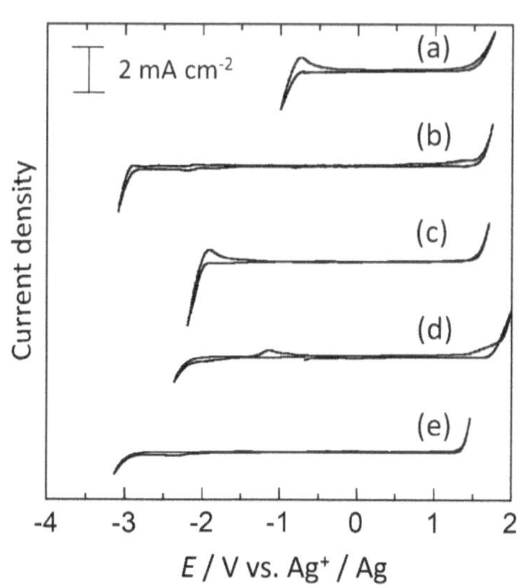

Figure 23.13 Cyclic voltammograms of a glass-like carbon electrode in (a) $[C_2C_1im][WF_7]$ and a Pt electrode in (b) $[C_2C_1im][TaF_6]$, (c) $[C_2C_1im][NbF_6]$, (d) $[C_2C_1im][SbF_6]$, and (e) $[C_2C_1im][BF_4]$. Scan rate: 30 mV/s.

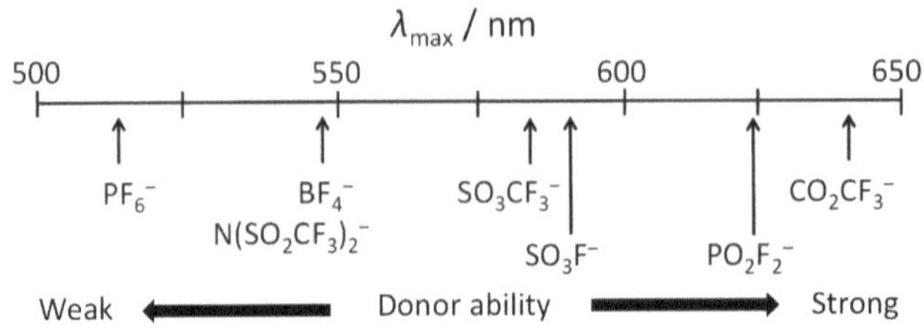

Figure 23.14 The trend of donor ability for fluorine-containing anions in ionic liquids estimated by a solvatochromic method with a $[Cu(acac)(tmen)][BPh_4]$ dye.[67,98,99,105]

As in the cases of the fluorohydrogenate salts mentioned previously, an increase in the carbon number of alkyl side chain in organic cations leads to the formation of PF_6^-- and BF_4^--based ionic liquid crystals.[29,30] The 1-alkyl-3-methylimidazolium salts, $[C_xC_1im][BF_4]$ and $[C_xC_1im][PF_6]$, exhibit liquid crystalline phase with the alkyl side chain longer than dodecyl and tetradecyl groups, respectively. As the crystal phase forms a layered structure with polar and apolar regions, such ionic liquid crystals have smectic-type structures. Similar behavior was observed for other hexafluorocomplex salts with long alkyl chains ($[C_xC_1im][AF_6]$ (A = As, Sb, Nb, and Ta).[106]

6. Conclusions

This chapter summarized the syntheses, structures, and properties of ionic liquids based on fluorocomplex and oxofluorocomplex anions. The high ionic conductivity of fluorohydrogenate ionic liquids is attractive as electrolytes in electrochemical devices. Applications to fuel cells and electrochemical capacitors showed interesting performance due to the unique properties of these ionic liquids.[48,73,107–111] Utilization of fluorohydrogenate ionic liquids as fluorinating reagents was also explored for some organic syntheses.[112–114] Further applications will be explored in future studies. Synthesis of ionic liquids based on fluorocomplex anions via fluorohydrogenate ionic liquids is an attractive way both in laboratory- and large-scale production. Specific applications for such ionic liquids will be investigated in future studies. There are more ionic liquids with fluorine-containing anions, which were not mentioned in detail here, such as $N(SO_2CF_3)_2^-$,[7,115] $N(SO_2F)_2^-$,[60,116] $SO_3CF_3^-$,[5,7] and $R_fBF_3^-$.[117–119] They are also quite appealing in a wide range of applications including electrolytes for electrochemical devices.[2,85,120]

References

1. Endres, F.; Abbott, A. P.; MacFarlane, D. R. *Electrodeposition from Ionic Liquids;* Wiley-VCH: Weinheim, Germany, 2008.
2. Ohno, H. *Electrochemical Aspects of Ionic Liquids,* 2nd ed.; John Wiley & Sons Inc: Hoboken, New Jersey, 2011.

3. Wasserscheid, P.; Welton, T. *Ionic Liquids in Synthesis;* Wiley-VCH: Weinheim, Germany, 2008.
4. Wilkes, J. S.; Zaworotko, M. J. Air and Water Stable 1-Ethyl-3-methylimidazolium Based Ionic Liquids. *J. Chem. Soc. Chem. Commun.* **1992,** 965–967.
5. Cooper, E. I.; Sullivan, E. J. M. In *Proceedings of the 8th International Symposium on Molten Salts*; Vol. 92–16; The Electrochemical Society: Pennington, NJ, 1992, pp 386–396. Proceeding.
6. Fuller, J.; Carlin, R. T.; Delong, H. C.; Haworth, D. Structure of 1-Ethyl-3-methylimidazolium Hexafluorophosphate – Model for Room-Temperature Molten-salts. *J. Chem. Soc. Chem. Commun.* **1994,** 299–300.
7. Bonhote, P.; Dias, A. P.; Papageorgiou, N.; Kalyanasundaram, K.; Gratzel, M. Hydrophobic, Highly Conductive Ambient-temperature Molten Salts. *Inorg. Chem.* **1996,** *35,* 1168–1178.
8. Fannin, A. A.; King, L. A.; Levisky, J. A.; Wilkes, J. S. Properties of 1,3-Dialkylimidazolium Chloride Aluminum-Chloride Ionic Liquids. 1. Ion Interactions by Nuclear Magnetic-Resonance Spectroscopy. *J. Phys. Chem.* **1984,** *88,* 2609–2614.
9. Hussey, C. L.; Scheffler, T. B.; Wilkes, J. S.; Fannin, A. A. Chloroaluminate Equilibria in the Aluminum Chloride- 1-methyl-3-ethylimidazolium Chloride Ionic Liquid. *J. Electrochem. Soc.* **1986,** *133,* 1389–1391.
10. Wilkes, J. S.; Levisky, J. A.; Wilson, R. A.; Hussey, C. L. Dialkylimidazolium Chloroaluminate Melts – A New Class of Room-Temperature Ionic Liquids for Electrochemistry, Spectroscopy, and Synthesis. *Inorg. Chem.* **1982,** *21,* 1263–1264.
11. Cady, G. H. Freezing Points and Vapor Pressures of the System Potassium Fluoride-Hydrogen Fluoride. *J. Am. Chem. Soc.* **1934,** *56,* 1431–1434.
12. Olah, G. A.; Shih, J. G.; Prakash, G. K. S. Fluorine-Containing Reagents in Organic-Synthesis. *J. Fluor. Chem.* **1986,** *33,* 377–396.
13. Wiechert, D.; Mootz, D.; Franz, R.; Siegemund, G. Amine-poly(hydrogen Fluoride) Solid Complexes: New Studies of Formation, Crystal Structures, and $H_{n-1}F_n^-$ Ion Diversity. *Chem. Eur. J.* **1998,** *4,* 1043–1047.
14. Winsor, R. V.; Cady, G. H. The System Cesium Fluoride Hydrogen Fluoride. *J. Am. Chem. Soc.* **1948,** *70,* 1500–1502.
15. Yoneda, N. The Combination of Hydrogen-Fluoride with Organic-Bases as Fluorination Agents. *Tetrahedron* **1991,** *47,* 5329–5365.
16. Hagiwara, R.; Hirashige, T.; Tsuda, T.; Ito, Y. Acidic 1-Ethyl-3-methylimidazolium Fluoride: a New Room Temperature Ionic Liquid. *J. Fluor. Chem.* **1999,** *99,* 1–3.
17. Beno, M. A.; Blackman, G. S.; Williams, J. M.; Bechgaard, K. Synthetic Metals Based on Tetramethyltetraselenafulvalene (TMTSF) – Synthesis, Structure (T = 298 and 125 K), and Novel Properties of $(TMTSF)_2H_2F_3$. *Inorg. Chem.* **1982,** *21,* 3860–3862.
18. Boenigk, D.; Mootz, D. Fluorides and Fluoro Acids. 17. The System Pyridine Hydrogen-Fluoride at Low-Temperatures – Formation and Crystal-Structures of Solid Complexes with Very Strong NHF and FHF Hydrogen-Bonding. *J. Am. Chem. Soc.* **1988,** *110,* 2135–2139.
19. Bozorth, R. M. The Crystal Structure of Potassium Hydrogen Fluoride. *J. Am. Chem. Soc.* **1923,** *45,* 2128–2132.
20. Forrester, J. D.; Templeton, D. H.; Zalkin, A.; Senko, M. E. Crystal Structure of KH_2F_3 and Geometry of $H_2F_3^-$ Ion. *Acta Cryst.* **1963,** *16,* 58–62.
21. Kruh, R.; Fuwa, K.; McEver, T. E. The Crystal Structures of Alkali Metal Bifluorides. *J. Am. Chem. Soc.* **1956,** *78,* 4256–4258.

22. Mcdonald, T. R. Electron Density in Hydrogen Atom of Bifluoride Ion – A Comparison of Theory and Experiment. *Acta Cryst. A* **1969,** *25,* 374–375.
23. Mcgaw, B. L.; Ibers, J. A. Nature of Hydrogen Bond in Sodium Acid Fluoride. *J. Chem. Phys.* **1963,** *39,* 2677–2684.
24. Mootz, D.; Boenigk, D. Fluorides and Fluoro Acids.12. Complex-Anion Homology and Isomerism in the Crystal-Structures of 2 Potassium Poly(hydrogen Fluorides), KF•2.5HF and KF•3HF. *J. Am. Chem. Soc.* **1986,** *108,* 6634–6636.
25. Mootz, D.; Poll, W. Fluorides and Fluoro Acids.6. Compound Formation and Crystal-Structures in the System Ammonia Hydrogen-Fluoride. *Z. Naturforsch. B* **1984,** *39,* 290–297.
26. Mootz, D.; Poll, W. Fluorides and Fluoro Acids. 9. Compound Formation in the System NOF-HF and Crystal-Structures of NOF•3HF and NOF•4HF. *Z. Naturforsch. B* **1984,** *39,* 1300–1305.
27. Ostlund, N. S.; Ballenger, L. W. Crystal Geometry of Bifluoride Anion. *J. Am. Chem. Soc.* **1975,** *97,* 1237–1238.
28. Chauvin, Y.; Mussmann, L.; Olivier, H. A Novel Class of Versatile Solvents for Two-Phase Catalysis: Hydrogenation, Isomerization, and Hydroformylation of Alkenes Catalyzed by Rhodium Complexes in Liquid 1,3-Dialkylimidazolium Salts. *Angew. Chem. Int. Ed.* **1995,** *34,* 2698–2700.
29. Gordon, C. M.; Holbrey, J. D.; Kennedy, A. R.; Seddon, K. R. Ionic Liquid Crystals: Hexafluorophosphate Salts. *J. Mater. Chem.* **1998,** *8,* 2627–2636.
30. Holbrey, J. D.; Seddon, K. R. The Phase Behaviour of 1-Alkyl-3-methylimidazolium Tetrafluoroborates; Ionic Liquids and Ionic Liquid Crystals. *J. Chem. Soc. Dalton Trans.* **1999,** 2133–2139.
31. Reichert, W. M.; Holbrey, J. D.; Swatloski, R. P.; Gutowski, K. E.; Visser, A. E.; Nieuwenhuyzen, M.; Seddon, K. R.; Rogers, R. D. Solid-State Analysis of Low-Melting 1,3-Dialkylimidazolium Hexafluorophosphate Salts (Ionic Liquids) by Combined X-ray Crystallographic and Computational Analyses. *Cryst. Growth Des.* **2007,** *7,* 1106–1114.
32. Angell, C. A.; Ansari, Y.; Zhao, Z. F. Ionic Liquids: Past, Present and Future. *Faraday Discuss.* **2012,** *154,* 9–27.
33. Hagiwara, R.; Lee, J. S. Ionic Liquids for Electrochemical Devices. *Electrochemistry* **2007,** *75,* 23–34.
34. Hallett, J. P.; Welton, T. Room-Temperature Ionic Liquids: Solvents for Synthesis and Catalysis. 2. *Chem. Rev.* **2011,** *111,* 3508–3576.
35. MacFarlane, D. R.; Tachikawa, N.; Forsyth, M.; Pringle, J. M.; Howlett, P. C.; Elliott, G. D.; Davis, J. H.; Watanabe, M.; Simon, P.; Angell, C. A. Energy Applications of Ionic Liquids. *Energy Environ. Sci.* **2014,** *7,* 232–250.
36. Torimoto, T.; Tsuda, T.; Okazaki, K.; Kuwabata, S. New Frontiers in Materials Science Opened by Ionic Liquids. *Adv. Mater.* **2010,** *22,* 1196–1221.
37. van Rantwijk, F.; Sheldon, R. A. Biocatalysis in Ionic Liquids. *Chem. Rev.* **2007,** *107,* 2757–2785.
38. Welton, T. Room-Temperature Ionic Liquids. Solvents for Synthesis and Catalysis. *Chem. Rev.* **1999,** *99,* 2071–2083.
39. Hagiwara, R.; Ito, Y. Room Temperature Ionic Liquids of Alkylimidazolium Cations and Fluoroanions. *J. Fluor. Chem.* **2000,** *105,* 221–227.
40. Matsumoto, K.; Hagiwara, R. Structural Characteristics of Alkylimidazolium-Based Salts Containing Fluoroanions. *J. Fluor. Chem.* **2007,** *128,* 317–331.
41. Xue, H.; Verma, R.; Shreeve, J. M. Review of Ionic Liquids with Fluorine-Containing Anions. *J. Fluor. Chem.* **2006,** *127,* 159–176.

42. Hagiwara, R.; Hirashige, T.; Tsuda, T.; Ito, Y. A Highly Conductive Room Temperature Molten Fluoride: EMIF•2.3HF. *J. Electrochem. Soc.* **2002,** *149,* D1–D6.
43. Enomoto, T.; Kanematsu, S.; Tsunashima, K.; Matsumoto, K.; Hagiwara, R. Physicochemical Properties and Plastic Crystal Structures of Phosphonium Fluorohydrogenate Salts. *Phys. Chem. Chem. Phys.* **2011,** *13,* 12536–12544.
44. Hagiwara, R.; Matsumoto, K.; Nakamori, Y.; Tsuda, T.; Ito, Y.; Matsumoto, H.; Momota, K. Physicochemical Properties of 1,3-Dialkylimidazolium Fluorohydrogenate Room-Temperature Molten Salts. *J. Electrochem. Soc.* **2003,** *150,* D195–D199.
45. Kanematsu, S.; Matsumoto, K.; Hagiwara, R. Electrochemically Stable Fluorohydrogenate Ionic Liquids Based on Quaternary Phosphonium Cations. *Electrochem. Commun.* **2009,** *11,* 1312–1315.
46. Matsumoto, K.; Hagiwara, R. Physical and Electrochemical Properties of a Room Temperature Molten Salt: 1-Ethyl-2,3-dimethylimidazolium Fluorohydrogenate. *Electrochemistry* **2005,** *73,* 730–732.
47. Matsumoto, K.; Hagiwara, R.; Ito, Y. Room-Temperature Ionic Liquids with High Conductivities and Wide Electrochemical Windows – *N*-alkyl-*N*-methylpyrrolidinium and *N*-alkyl-*N*-methylpiperidinium Fluorohydrogenates. *Electrochem. Solid-State Lett.* **2004,** *7,* E41–E44.
48. Senda, A.; Matsumoto, K.; Nohira, T.; Hagiwara, R. Effects of the Cationic Structures of Fluorohydrogenate Ionic Liquid Electrolytes on the Electric Double Layer Capacitance. *J. Power Sources* **2010,** *195,* 4414–4417.
49. Taniki, R.; Matsumoto, K.; Hagiwara, R. Trialkylsulfonium Fluorohydrogenate Giving the Highest Conductivity in Room Temperature Ionic Liquids. *Electrochem. Solid-State Lett.* **2012,** *15,* F13–F15.
50. Taniki, R.; Matsumoto, K.; Hagiwara, R. Synthesis and Characterization of Fluorohydrogenate Ionic Liquids Based on Azoniaspiro-Type Cations. *Chem. Lett.* **2013,** *42,* 1469–1471.
51. Yamagata, M.; Konno, S.; Matsumoto, K.; Hagiwara, R. Room-Temperature Fluorohydrogenate Ionic Liquids of Alkylpyridinium Cations and Allylated Quaternary Cyclic Ammonium Cations. *Electrochem. Solid-State Lett.* **2009,** *12,* F9–F12.
52. Hagiwara, R.; Nakamori, Y.; Matsumoto, K.; Ito, Y. The Effect of the Anion Fraction on the Physicochemical Properties of EMIm$(HF)_nF$ (n =1.0–2.6). *J. Phys. Chem. B* **2005,** *109,* 5445–5449.
53. Taniki, R.; Matsumoto, K.; Hagiwara, R. Effects of HF Content in the $(FH)(_n)F$- Anion on the Formation of Ionic Plastic Crystal Phases of *N*-ethyl-*N*-methylpyrrolidinium and *N,N*-dimethylpyrrolidinium Fluorohydrogenate Salts. *Phys. Chem. Chem. Phys.* **2014,** *16,* 1522–1528.
54. Matsumoto, K.; Tsuda, T.; Hagiwara, R.; Ito, Y.; Tamada, O. Structural Characteristics of 1-Ethyl-3-methylimidazolium Bifluoride: HF-Deficient Form of a Highly Conductive Room Temperature Molten Salt. *Solid State Sci.* **2002,** *4,* 23–26.
55. Gennick, I.; Harmon, K. M.; Potvin, M. M. Hydrogen-Bonding. 8. Preparation, Properties, and Low-Temperature IR Structural-Analysis of Ammonium and Alkylammonium Trihydrogen Tetrafluorides and Tetramethylammonium Dihydrogen Trifluoride. *Inorg. Chem.* **1977,** *16,* 2033–2040.
56. Chandler, W. D.; Johnson, K. E.; Campbell, J. L. E. Thermodynamic Calculations for Reactions Involving Hydrogen Halide Polymers, Ions, and Lewis-Acid Adducts. 1. Polyfluorohydrogenate(1–) Anions ($H_nF_{n+1}^-$), Polyfluorohydrogen(I) Cations ($H_{n+1}F_n^+$), and Hydrogen-fluoride Polymers ($(HF)_n$). *Inorg. Chem.* **1995,** *34,* 4943–4949.

57. Clark, J. H.; Emsley, J.; Overill, R. E. Strong Hydrogen-Bonding in the Polyfluorides – Ab Initio Calculations and a New Method of Preparing Potassium Hydrogendifluoride. *J. Chem. Soc. Dalton Trans.* **1981,** 1219–1222.
58. Enomoto, T.; Nakamori, Y.; Matsumoto, K.; Hagiwara, R. Ion–Ion Interactions and Conduction Mechanism of Highly Conductive Fluorohydrogenate Ionic Liquids. *J. Phys. Chem. C* **2011,** *115,* 4324–4332.
59. Janikowski, J.; Forsyth, C.; MacFarlane, D. R.; Pringle, J. M. Novel Ionic Liquids and Plastic Crystals Utilizing the Cyanate Anion. *J. Mater. Chem.* **2011,** *21,* 19219–19225.
60. Ishikawa, M.; Sugimoto, T.; Kikuta, M.; Ishiko, E.; Kono, M. Pure Ionic Liquid Electrolytes Compatible with a Graphitized Carbon Negative Electrode in Rechargeable Lithium-Ion Batteries. *J. Power Sources* **2006,** *162,* 658–662.
61. Noda, A.; Hayamizu, K.; Watanabe, M. Pulsed-Gradient Spin-Echo ^{1}H and ^{19}F NMR Ionic Diffusion Coefficient, Viscosity, and Ionic Conductivity of Non-chloroaluminate Room-Temperature Ionic Liquids. *J. Phys. Chem. B* **2001,** *105,* 4603–4610.
62. Izutsu, K. *Electrochemistry in Nonaqueous Solutions,* 2nd Revised and Enlarged ed.; Wiley-VCH, 2009.
63. Walden, P. Organic Solutions- and Ionisation Means. III. Chapter: Internal Friction and Its Connection with Conductivity. *Z. Phys. Chem.* **1906,** *55,* 207–249.
64. MacFarlane, D. R.; Forsyth, M.; Izgorodina, E. I.; Abbott, A. P.; Annat, G.; Fraser, K. On the Concept of Ionicity in Ionic Liquids. *Phys. Chem. Chem. Phys.* **2009,** *11,* 4962–4967.
65. Xu, W.; Cooper, E. I.; Angell, C. A. Ionic Liquids: Ion Mobilities, Glass Temperatures, and Fragilities. *J. Phys. Chem. B* **2003,** *107,* 6170–6178.
66. Yoshida, Y.; Muroi, K.; Otsuka, A.; Saito, G.; Takahashi, M.; Yoko, T. 1-Ethyl-3-methylimidazolium Based Ionic Liquids Containing Cyano Groups: Synthesis, Characterization, and Crystal Structure. *Inorg. Chem.* **2004,** *43,* 1458–1462.
67. Tokuda, H.; Tsuzuki, S.; Susan, M. A. B. H.; Hayamizu, K.; Watanabe, M. How Ionic Are Room-Temperature Ionic Liquids? An Indicator of the Physicochemical Properties. *J. Phys. Chem. B* **2006,** *110,* 19593–19600.
68. Tokuda, H.; Hayamizu, K.; Ishii, K.; Abu Bin Hasan Susan, M.; Watanabe, M. Physicochemical Properties and Structures of Room Temperature Ionic Liquids. 1. Variation of Anionic Species. *J. Phys. Chem. B* **2004,** *108,* 16593–16600.
69. Tokuda, H.; Hayamizu, K.; Ishii, K.; Susan, M.; Watanabe, M. Physicochemical Properties and Structures of Room Temperature Ionic Liquids. 2. Variation of Alkyl Chain Length in Imidazolium Cation. *J. Phys. Chem. B* **2005,** *109,* 6103–6110.
70. Tokuda, H.; Ishii, K.; Susan, M.; Tsuzuki, S.; Hayamizu, K.; Watanabe, M. Physicochemical Properties and Structures of Room-Temperature Ionic Liquids. 3. Variation of Cationic Structures. *J. Phys. Chem. B* **2006,** *110,* 2833–2839.
71. Saito, Y.; Hirai, K.; Matsumoto, K.; Hagiwara, R.; Minamizaki, Y. Ionization State and Ion Migration Mechanism of Room Temperature Molten Dialkylimidazolium Fluorohydrogenates. *J. Phys. Chem. B* **2005,** *109,* 2942–2948.
72. Tsuzuki, S.; Tokuda, H.; Mikami, M. Theoretical Analysis of the Hydrogen Bond of Imidazolium C_2-H with Anions. *Phys. Chem. Chem. Phys.* **2007,** *9,* 4780–4784.
73. Taniki, R.; Matsumoto, K.; Nohira, T.; Hagiwara, R. Evaluation of Double-Layer and Redox Capacitances of Activated Carbon Electrodes in *N*-ethyl-*N*-methylpyrrolidinium Fluorohydrogenate Ionic Liquid. *J. Electrochem. Soc.* **2013,** *160,* A734–A738.
74. Ue, M.; Takeda, M.; Toriumi, A.; Kominato, A.; Hagiwara, R.; Ito, Y. Application of Low-Viscosity Ionic Liquid to the Electrolyte of Double-Layer Capacitors. *J. Electrochem. Soc.* **2003,** *150,* A499–A502.

75. Axenov, K. V.; Laschat, S. Thermotropic Ionic Liquid Crystals. *Materials* **2011,** *4,* 206–259.
76. Binnemans, K. Ionic Liquid Crystals. *Chem. Rev.* **2005,** *105,* 4148–4204.
77. Abdallah, D. J.; Robertson, A.; Hsu, H. F.; Weiss, R. G. Smectic Liquid-crystalline Phases of Quaternary Group VA (Especially Phosphonium) Salts with Three Equivalent Long N-alkyl Chains. How Do Layered Assemblies Form in Liquid-Crystalline and Crystalline Phases? *J. Am. Chem. Soc.* **2000,** *122,* 3053–3062.
78. Yamanaka, N.; Kawano, R.; Kubo, W.; Kitamura, T.; Wada, Y.; Watanabe, M.; Yanagida, S. Ionic Liquid Crystal as a Hole Transport Layer of Dye-Sensitized Solar Cells. *Chem. Commun.* **2005**.
79. Yoshio, M.; Mukai, T.; Kanie, K.; Yoshizawa, M.; Ohno, H.; Kato, T. Layered Ionic Liquids: Anisotropic Ion Conduction in New Self-organized Liquid-Crystalline Materials. *Adv. Mater.* **2002,** *14,* 351–354.
80. Yoshio, M.; Mukai, T.; Ohno, H.; Kato, T. One-dimensional Ion Transport in Self-organized Columnar Ionic Liquids. *J. Am. Chem. Soc.* **2004,** *126,* 994–995.
81. Xu, F.; Matsubara, S.; Matsumoto, K.; Hagiwara, R. Effects of Alkyl Chain Length on Properties of N-alkyl-N-methylpyrrolidinium Fluorohydrogenate Ionic Liquid Crystals. *J. Fluor. Chem.* **2012,** *135,* 344–349.
82. Xu, F.; Matsumoto, K.; Hagiwara, R. Effects of Alkyl Chain Length on Properties of 1-Alkyl-3-methylimidazolium Fluorohydrogenate Ionic Liquid Crystals. *Chem. Eur. J.* **2010,** *16,* 12970–12976.
83. Xu, F.; Matsumoto, K.; Hagiwara, R. Phase Behavior of 1-Dodecyl-3-methylimidazolium Fluorohydrogenate Salts ($C_{12}Mim(FH)_nF$, $n = 1.0–2.3$) and Their Anisotropic Ionic Conductivity as Ionic Liquid Crystal Electrolytes. *J. Phys. Chem. B* **2012,** *116,* 10106–10112.
84. Jin, L. Y.; Nairn, K. M.; Forsyth, C. M.; Seeber, A. J.; MacFarlane, D. R.; Howlett, P. C.; Forsyth, M.; Pringle, J. M. Structure and Transport Properties of a Plastic Crystal Ion Conductor: Diethyl(methyl)(isobutyl)phosphonium Hexafluorophosphate. *J. Am. Chem. Soc.* **2012,** *134,* 9688–9697.
85. Macfarlane, D. R.; Forsyth, M.; Howlett, P. C.; Pringle, J. M.; Sun, J.; Annat, G.; Neil, W.; Izgorodina, E. I. Ionic Liquids in Electrochemical Devices and Processes: Managing Interfacial Electrochemistry. *Acc. Chem. Res.* **2007,** *40,* 1165–1173.
86. Pringle, J. M. Recent Progress in the Development and Use of Organic Ionic Plastic Crystal Electrolytes. *Phys. Chem. Chem. Phys.* **2013,** *15,* 1339–1351.
87. Pringle, J. M.; Howlett, P. C.; MacFarlane, D. R.; Forsyth, M. Organic Ionic Plastic Crystals: Recent Advances. *J. Mater. Chem.* **2010,** *20,* 2056–2062.
88. Huang, J. H.; Forsyth, M.; MacFarlane, D. R. Solid State Lithium Ion Conduction in Pyrrolidinium Imide-Lithium Imide Salt Mixtures. *Solid State Ionics* **2000,** *136,* 447–452.
89. MacFarlane, D. R.; Huang, J. H.; Forsyth, M. Lithium-Doped Plastic Crystal Electrolytes Exhibiting Fast Ion Conduction for Secondary Batteries. *Nature* **1999,** *402,* 792–794.
90. Taniki, R.; Matsumoto, K.; Hagiwara, R.; Hachiya, K.; Morinaga, T.; Sato, T. Highly Conductive Plastic Crystals Based on Fluorohydrogenate Anions. *J. Phys. Chem. B* **2013,** *117,* 955–960.
91. Taniki, R.; Matsumoto, K.; Nohira, T.; Hagiwara, R. All Solid-State Electrochemical Capacitors Using N,N-dimethylpyrrolidinium Fluorohydrogenate as Ionic Plastic Crystal Electrolyte. *J. Power Sources* **2014,** *245,* 758–763.
92. Matsumoto, K.; Hagiwara, R.; Yoshida, R.; Ito, Y.; Mazej, Z.; Benkič, P.; Žemva, B.; Tamada, O.; Yoshino, H.; Matsubara, S. Syntheses, Structures and Properties of 1-Ethyl-3-methylimidazolium Salts of Fluorocomplex Anions. *Dalton Trans.* **2004,** 144–149.

93. Kanatani, T.; Matsumoto, K.; Hagiwara, R. Syntheses and Physicochemical Properties of New Ionic Liquids Based on the Hexafluorouranate Anion. *Chem. Lett.* **2009,** *38,* 714–715.
94. Kanatani, T.; Matsumoto, K.; Hagiwara, R. Syntheses and Physicochemical Properties of Low-Melting Salts Based on VOF_4^- and $MoOF_5^-$, and the Molecular Geometries of the Dimeric $(VOF_4^-)_2$ and $Mo_2O_4F_6^{2-}$ Anions. *Eur. J. Inorg. Chem.* **2010,** 1049–1055.
95. Matsumoto, K.; Hagiwara, R. A New Room Temperature Ionic Liquid of Oxyfluorometallate Anion: 1-Ethyl-3-methylimidazolium Oxypentafluorotungstate (EMIm-wof$_5$). *J. Fluor. Chem.* **2005,** *126,* 1095–1100.
96. Matsumoto, K.; Hagiwara, R.; Ito, Y. Room Temperature Molten Fluorometallates: 1-Ethyl-3-methylimidazolium Hexafluoroniobate(V) and Hexafluorotantalate(V). *J. Fluor. Chem.* **2002,** *115,* 133–135.
97. Kanatani, T.; Ueno, R.; Matsumoto, K.; Nohira, T.; Hagiwara, R. Thermal properties of N-alkyl-N-methylpyrrolidinium and N-butylpyridinium Fluorometallates and Physicochemical Properties of Their Melts. *J. Fluor. Chem.* **2009,** *130,* 979–984.
98. Matsumoto, K.; Hagiwara, R. A New Series of Ionic Liquids Based on the Difluorophosphate Anion. *Inorg. Chem.* **2009,** *48,* 7350–7358.
99. Enomoto, T.; Matsumoto, K.; Hagiwara, R. Properties of Fluorosulfate-Based Ionic Liquids and Geometries of $(FO_2SOH)OSO_2F^-$ and $(FO_2SOH)_2O_2SOF^-$. *Dalton Trans.* **2011,** *40,* 12491–12499.
100. Laali, K. K.; Gettwert, V. J. Electrophilic Nitration of Aromatics in Ionic Liquid Solvents. *J. Org. Chem.* **2001,** *66,* 35–40.
101. Matsumoto, K.; Hagiwara, R.; Mazej, Z.; Benkič, P.; Žemva, B. Crystal structures of Frozen Room Temperature Ionic Liquids, 1-Ethyl-3-methylimidazolium Tetrafluoroborate ($EMImBF_4$), Hexafluoroniobate ($EMImNbF_6$) and Hexafluorotantalate ($EMImTaF_6$), Determined by Low-Temperature X-ray Diffraction. *Solid State Sci.* **2006,** *8,* 1250–1257.
102. Matsumoto, K.; Hagiwara, R. Electrochemical Properties of the Ionic Liquid 1-Ethyl-3-methylimidazolium Difluorophosphate as an Electrolyte for Electric Double-Layer Capacitors. *J. Electrochem. Soc.* **2010,** *157,* A578–A581.
103. Fukuda, Y.; Sone, K. *Bull. Chem. Soc. Jpn.* **1972,** *45,* 465–469.
104. Soukup, R. W.; Sone, K. *Bull. Chem. Soc. Jpn.* **1987,** *60,* 2286–2288.
105. Muldoon, M. J.; Gordon, C. M.; Dunkin, I. R. *J. Chem. Soc. Perkin Trans.* **2001,** *2,* 433–435.
106. Xu, F.; Matsumoto, K.; Hagiwara, R. Effects of Alkyl Chain Length and Anion Size on Thermal and Structural Properties for 1-Alkyl-3-methylimidazolium Hexafluorocomplex Salts (CxMImAF6, x = 14, 16 and 18; A = P, As, Sb, Nb and Ta). *Dalton Trans.* **2012,** *41,* 3494–3502.
107. Hagiwara, R.; Nohira, T.; Matsumoto, K.; Tamba, Y. A Fluorohydrogenate Ionic Liquid Fuel Cell Operating without Humidification. *Electrochem. Solid-State Lett.* **2005,** *8,* A231–A233.
108. Kiatkittikul, P.; Nohira, T.; Hagiwara, R. Nonhumidified Fuel Cell Using *N*-ethyl-*N*-methylpyrrolidinium Fluorohydrogenate Ionic Liquid-Polymer Composite Membranes. *J. Power Sources* **2012,** *220,* 10–14.
109. Kiatkittikul, P.; Nohira, T.; Hagiwara, R. Advantages of a Polyimide Membrane Support in Nonhumidified Fluorohydrogenate-Polymer Composite Membrane Fuel Cells. *Fuel Cells* **2015,** *15,* 604–609.
110. Kiatkittikul, P.; Yamaguchi, J.; Taniki, R.; Matsumoto, K.; Nohira, T.; Hagiwara, R. Influence of Cationic Structures on Oxygen Reduction Reaction at Pt Electrode in Fluorohydrogenate Ionic Liquids. *J. Power Sources* **2014,** *266,* 193–197.

111. Tani, Y.; Nohira, T.; Enomoto, T.; Matsumoto, K.; Hagiwara, R. Solubility and Diffusion Coefficient of Oxygen in 1-Ethyl-1-methylpyrrolidinium Fluorohydrogenate Room Temperature Ionic Liquid at 298–373 K. *Electrochim. Acta* **2011,** *56,* 3852–3856.
112. Yoshino, H.; Matsubara, S.; Oshima, K.; Matsumoto, K.; Hagiwara, R.; Ito, Y. Halofluorination of Alkenes with Ionic Liquid EMIMF$(HF)_{2.3}$. *J. Fluor. Chem.* **2004,** *125,* 455–458.
113. Yoshino, H.; Matsumoto, K.; Hagiwara, R.; Ito, Y.; Oshima, K.; Matsubara, S. Fluorination with Ionic Liquid EMIMF$(HF)_{2.3}$ as Mild HF Source. *J. Fluor. Chem.* **2006,** *127,* 29–35.
114. Yoshino, H.; Nomura, K.; Matsubara, S.; Oshima, K.; Matsumoto, K.; Hagiwara, R.; Ito, Y. A Mild Ring Opening Fluorination of Epoxide with Ionic Liquid 1-Ethyl-3-methylimidazorium Oligo Hydrogenfluoride (EMIMF$(HF)_{2.3}$). *J. Fluor. Chem.* **2004,** *125,* 1127–1129.
115. MacFarlane, D. R.; Meakin, P.; Sun, J.; Amini, N.; Forsyth, M. Pyrrolidinium Imides: A New Family of Molten Salts and Conductive Plastic Crystal Phases. *J. Phys. Chem. B* **1999,** *103,* 4164–4170.
116. Matsumoto, H.; Sakaebe, H.; Tatsumi, K.; Kikuta, M.; Ishiko, E.; Kono, M. Fast cycling of Li/$LiCoO_2$ Cell with Low-Viscosity Ionic Liquids Based on bis(fluorosulfonyl)imide $[FSI]^-$. *J. Power Sources* **2006,** *160,* 1308–1313.
117. Zhou, Z. B.; Matsumoto, H.; Tatsumi, K. Low-Melting, Low-Viscous, Hydrophobic Ionic Liquids: 1-Alkyl(alkyl Ether)-3-methylimidazolium Perfluoroalkyltrifluoroborate. *Chem. Eur. J.* **2004,** *10,* 6581–6591.
118. Zhou, Z. B.; Matsumoto, H.; Tatsumi, K. Low-Melting, Low-Viscous, Hydrophobic Ionic Liquids: Aliphatic Quaternary Ammonium Salts with Perfluoroalkyltrifluoroborates. *Chem. Eur. J.* **2005,** *11,* 752–766.
119. Zhou, Z. B.; Matsumoto, H.; Tatsumi, K. Cyclic Quaternary Ammonium Ionic Liquids with Perfluoroalkyltrifluoroborates: Synthesis, Characterization, and Properties. *Chem. Eur. J.* **2006,** *12,* 2196–2212.
120. Lewandowski, A.; Swiderska-Mocek, A. Ionic Liquids as Electrolytes for Li-Ion Batteries-an Overview of Electrochemical Studies. *J. Power Sources* **2009,** *194,* 601–609.
121. Forsyth, S.; Golding, J.; MacFarlane, D. R.; Forsyth, M. *N*-methyl-*N*-alkylpyrrolidinium Tetrafluoroborate Salts: Ionic Solvents and Solid Electrolytes. *Electrochim. Acta* **2001,** *46,* 1753–1757.
122. Golding, J.; Hamid, N.; MacFarlane, D. R.; Forsyth, M.; Forsyth, C.; Collins, C.; Huang, J. N-methyl-N-alkylpyrrolidinium Hexafluorophosphate Salts: Novel Molten Salts and Plastic Crystal Phases. *Chem. Mater.* **2001,** *13,* 558–564.

Reactivity of Surface Fluorinated TiO_2 and TiAl Particles

24

J.-H. Kim, S. Yonezawa, M. Takashima
University of Fukui, Fukui, Japan

Chapter Outline

1. Surface Fluorination of TiO_2 Particles

1.1 Introduction

Nanosized TiO_2 is one of the most promising photocatalysts currently available. To achieve high activities in solution-phase catalysis, it is important to facilitate good dispersion of the catalyst.[1–3] However, small particles tend to aggregate, resulting in low or complete absence of photocatalytic activity. Many studies have focused on the dispersion stability of TiO_2 particles in water.[4–6] Surface modification of TiO_2 nanoparticles is an effective method to minimize the agglomeration of TiO_2 particles.[7–9] Silane alkoxides with organic functional groups have also been widely used for this purpose. It is important to consider the high cost of surfactants required and the residues generated in the particles. Al_2O_3 and TiO_2 metal oxide particles tend to disperse in a highly polar solvent; by contrast, they tend to flocculate in a low-dielectric solvent because the Hamaker constant between the particles is high in such a solvent.[10,11] TiO_2 particles generally disperse well in water but not in other solutions such as acetone and ethanol.

Modern Synthesis Processes and Reactivity of Fluorinated Compounds. http://dx.doi.org/10.1016/B978-0-12-803740-9.00024-X

Many researchers have reported that the photocatalytic activity of TiO_2 can be improved using various fluorinating agents other than F_2 gas.[12–14] Park and Choi[15] reported that surface fluorination results in enhanced photocatalytic degradation of certain substrates because of the formation of electron-trapping sites. Li et al.[16–18] attributed the enhanced photocatalytic activity of F-doped powders mainly to the creation of surface oxygen vacancies and the beneficial effects of F doping. Yu et al.[19] also reported the incorporation of fluoride ions into a TiO_2 lattice using an NH_4F source. However, the surface fluorination of TiO_2 could have either positive or negative effect depending on the fluorine content in TiO_2 particles. The relationship between the photocatalytic activity and fluorine content in TiO_2 using F_2 gas has not been reported.

Our previous work proved that surface fluorination using F_2 gas is beneficial to the dispersion stability of TiO_2 particles in water.[20] F-TiO_2 particles with good dispersion stability were synthesized at a temperature less than 100°C using fluorine gas. The idea was to stabilize the TiO_2 particles by enabling their surfaces to carry an electric charge to create electrostatic repulsive forces that repel each other.

Here, we report the effects of surface fluorination on the dispersion stability of TiO_2 in various organic solvents and also investigate the relationship between fluorine contents in TiO_2 and the photocatalytic activity of TiO_2.

1.2 Experimental Details

TiO_2 particles (ST-21, anatase; 98% purity) were obtained from Ishihara Sangyou Kaisha, Ltd. Fluorine gas (99.5% purity) was supplied by Daikin Industries Ltd. Details of the fluorination apparatus have been given in our previous article.[21,22] Fluorinated TiO_2 (F-TiO_2) particles were prepared by direct fluorination using F_2 gas under various reaction conditions. Reaction temperature, fluorine pressure, and reaction time were set at 25°C–200°C, 1–50 kPa, and 1 h, respectively.

The structural and electronic properties of the samples were investigated using X-ray powder diffraction (XRD, XD-6100) and X-ray photoelectron spectroscopy (XPS, XPS-9010). The surface morphology of various samples was observed using scanning electron microscopy (SEM, s-2400; Hitachi Ltd.). The BET surface area was determined using a Micromeritics ASAP 2000 nitrogen adsorption apparatus.

Particle size distribution and zeta potential profiles were measured using a zeta potential/particle size measurement device (Otsuka Electronics Co., Ltd, ELSZ-2). A solid sample was suspended in distilled water, ethanol (99.5%, Kanto Chemical Co., Inc.), and acetone (99.0%, Kanto Chemical Co., Inc.). The pH of the suspension was adjusted using a 1 M NaOH or HCl solution. The dispersion stability of samples in various solvents was determined by a sedimentation experiment. A typical procedure was used to prepare the suspension; first, 15 mg of TiO_2 was mixed with 15 mL solvents and sonicated for 1 h.

The ultraviolet (UV)–visible (Vis) absorption spectra of samples were recorded on a Hitachi U-3900H spectrophotometer with an integrating sphere assembly. The photocatalytic activity of samples was evaluated using the photocatalytic decomposition of methylene blue (MB, $C_{16}H_{18}C1N_3S$).[23] Samples (5 mg) were dispersed in an MB

(1×10^{-5} mol/L) aqueous solution (50 mL). Three milliliters of test liquid was taken from this solution and fed into a quartz cell. The test solution was irradiated at 365 nm by a UV lamp (Spectroline Spectronics, 4 W), and the absorbance at 665 nm, which is the maximum absorption wavelength of MB, was measured using a UV–Vis spectrometer (U-3900H). The decomposition rate of MB-containing samples was evaluated from the absorbance obtained with irradiation times.

1.3 Characterization of Fluorinated TiO_2 Samples

Sample names, reaction conditions, total surface area (BET), and fluorine contents (x) in $TiO_{2-x}F_{2x}$ are summarized in Table 24.1, in which, fluorinated TiO_2 (F-TiO_2) samples prepared at various reaction conditions with F_2 gas were named as the F1-TiO_2, F2-TiO_2, F3-TiO_2, and F4-TiO_2, respectively. The fluorine contents in the surface region of F-TiO_2 particles were evaluated from the XPS data. Fluorination temperature and F_2 pressure were increased to enhance the fluorine content (x) in $TiO_{2-x}F_{2x}$. At a temperature higher than 200°C, x became 1.10, which suggests the formation of $TiOF_2$. The relative surface area of the samples was determined by the BET method. The BET surface areas of fluorinated TiO_2 samples were in the range of 56.4–61.3 m^2/g, which are obviously higher than that of untreated TiO_2 (54.6 m^2/g).

The effects of reaction temperature and F_2 pressure on the TiO_2 crystal structure are shown in Fig. 24.1A. Only a single phase of anatase TiO_2 was observed when fluorination was performed at room temperature (25°C). Fluorination did not cause any shift in the peak position of the TiO_2 phase. This is easily understood because the ionic radius of the fluorine atom (0.133 nm) is nearly the same as that of the replaced oxygen atom (0.132 nm).[16] However, as shown in Fig. 24.1B, the peak intensity ratio of (101)/(200) shown in Fig. 24.1A decreased when fluorine pressure increased to 5.5 kPa. This indicates that the crystallinity of TiO_2 gradually decreased by the fluorine substitution. After the fluorination temperature was increased from 25°C to 200°C, some peaks (●) assigned to $TiOF_2$ appeared for F4-TiO_2 ($TiO_{2-x}F_{2x}$, x = 1.10).

Table 24.1 Reaction Conditions of TiO_2 Particles Treated With F_2 Gas and the Fluorine Contents (x) in $TiO_{2-x}F_{2x}$

Sample Name	Temperature (°C)	F_2 Pressure (kPa)	Time (h)	Total Surface Area (BET, m^2/g)	x in $TiO_{2-x}F_{2x}$[a]
TiO_2	–	–	–	54.6	0.00
F1-TiO_2	25	1.3	1	56.4	0.13
F2-TiO_2	25	6.7	1	61.3	0.18
F3-TiO_2	25	50.5	1	57.2	0.44
F4-TiO_2	200	50.5	1	55.9	1.10

[a]Fluorine contents (x) in $TiO_{2-x}F_{2x}$ were evaluated from XPS results shown in Fig. 24.3.

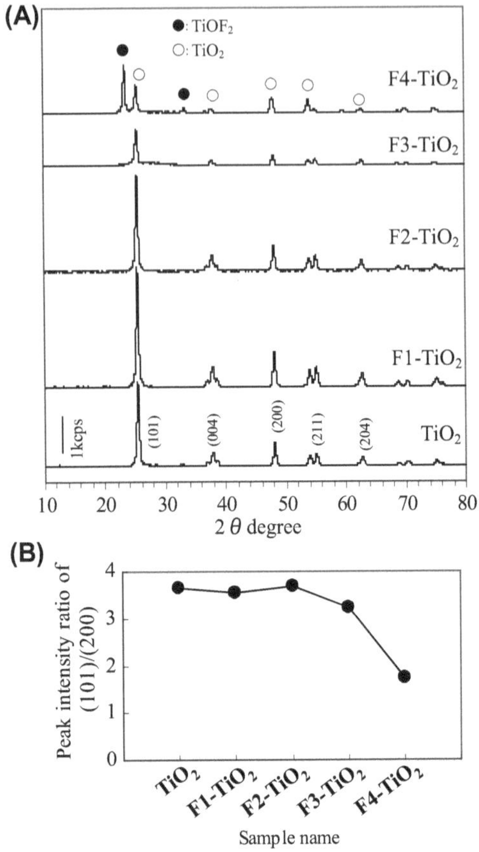

Figure 24.1 X-ray powder diffraction patterns (A) and peak intensity ratio (B) of (101)/(200) in XRD patterns of untreated TiO_2 and fluorinated TiO_2 (F-TiO_2) particles.

SEM images of untreated and fluorinated TiO_2 samples are presented in Fig. 24.2. As seen from Fig. 24.2A and C, obvious agglomerates composed of TiO_2 particles with an approximate size of 20 nm were found in all samples. The fluorination effect on the surface morphology of TiO_2 particles could not be detected in the SEM images. However, XPS data confirm the existence of fluorinated surface layer of TiO_2 particles, as shown in Fig. 24.3.

Fig. 24.3 shows F 1s (A and B), Ti 2p (C), and O 1s (D) spectra of untreated and fluorinated TiO_2 samples. All binding energies were calibrated to the C 1s peak at 284.8 eV of carbon. An F 1s peak located at the binding energy of 684.3 eV was observed in all F-TiO_2 samples, as shown in Fig. 24.3A. However, the peak at 684.3 eV disappeared after Ar^+ ion etching, as shown in Fig. 24.3B. Because the fluorine at 684.3 eV is assigned to fluorine atoms chemically adsorbed on TiO_2 surface, they can be easily eliminated by low-energy argon etching (300 V, 5 s). Moreover, an asymmetrical F 1s peak was observed for F3-TiO_2 and F4-TiO_2 samples, where

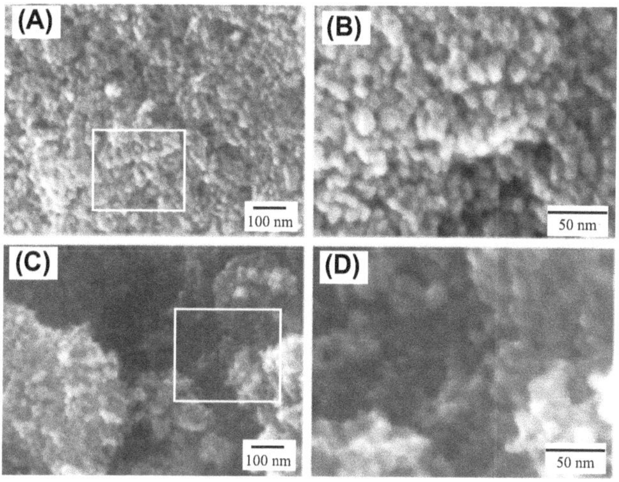

Figure 24.2 Scanning electron microscopic images of untreated TiO_2 (A, B) and fluorinated TiO_2 (F4-TiO_2) (C, D) particles.

a tailing peak could be found. This means that various chemical forms of F atoms might exist in the samples. Therefore, the F 1s peaks of F3-TiO_2 and F4-TiO_2 samples were deconvoluted into two separated peaks with Gaussian distributions, as shown in Fig. 24.3A. The peak located at 685.2 eV was attributed to the F atom in $TiOF_2$.[17] This is easily understood in case of F4-TiO_2 because an obvious $TiOF_2$ phase appeared in the XRD patterns, as shown in Fig. 24.1. The peak located at 687.2 eV may be attributed to substituted fluorine atoms in $TiO_{2-x}F_{2x}$.[24] In the case of Ti 2p (C), the Ti (IV) $2p_{1/2}$ and Ti (IV) $2p_{3/2}$ spin-orbital splitting photoelectrons of the original TiO_2 were located at binding energies of 464.9 eV and 458.9 eV, respectively. However, in the XPS spectrum after fluorination at 25°C, the peak of Ti $2p_{3/2}$ shifted to an energy lower than that of the original TiO_2. This result showed that the valence state of Ti^{4+} (458.9 eV) gradually changed to Ti^{3+} (457.0 eV) with increasing F_2 pressure. In the case of F3-TiO_2, a Ti $2p_{3/2}$ peak due to Ti^{2+} (455.2 eV) is especially evident in Fig. 24.3C. Furthermore, when the reaction temperature was increased from 25°C to 200°C, the Ti $2p_{3/2}$ peak of F4-TiO_2 shifted to high energy with the creation of Ti–F bond (461.2 eV) of $TiOF_2$. Fig. 24.3D shows the peaks of various samples found at 530.6 eV. The peak intensity decreased with an increase in F_2 pressure. In case of F4-TiO_2, similar to the F 1s and Ti 2p peaks, the O 1s peak shifted to higher energy. This result seems to be related to the formation of $TiOF_2$, as indicated in Fig. 24.1.

1.4 Dispersion Stability of Fluorinated TiO_2 Samples

Fig. 24.4 shows the suspension of samples dispersed in water (A) and ethanol (B) as polar protic solvents and in acetone (C) as a polar aprotic solvent over retention times. In the untreated TiO_2 suspension (a), TiO_2 particles re-formed into large agglomerates within 4 h in all solvents. Especially in ethanol (24.3) and acetone (20.7), which have

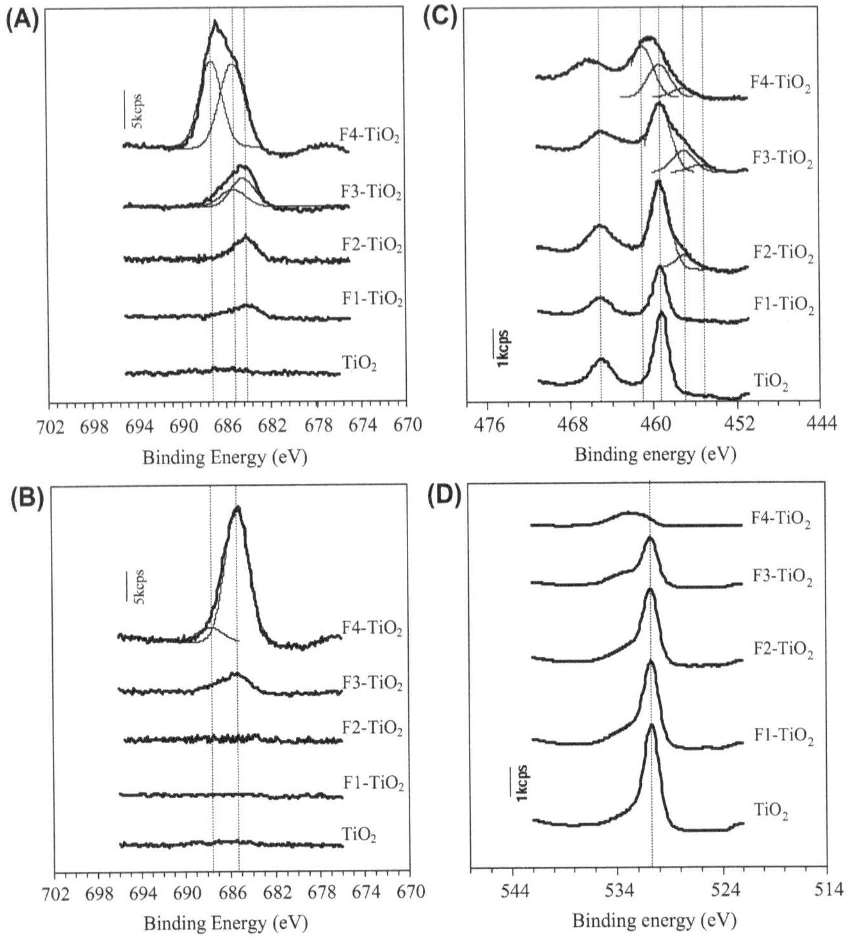

Figure 24.3 X-ray photoelectron spectroscopic spectra of F 1s (A), F 1s after Ar^+ ion etching (B), Ti 2p (C), and O 1s (D) for untreated TiO_2 and fluorinated TiO_2 (F1-TiO_2 ~ F4-TiO_2) particles.

lower dielectric constant values than water (78.3),[10] decreased turbidity in the untreated TiO_2 suspension (a) was observed within 1 h. The dielectric constant of a substance is closely related to the dipole moment. Also, the polarity of a substance with a high dielectric constant is considered to be large. The dispersion stability of fluorinated TiO_2 could be sustained in all solvents for 24 h. The stability of colloidal suspensions is primarily governed by interparticle (or surface) forces, especially by the repulsive electrostatic interaction of these charges. Because fluorine at the TiO_2 surface has high electronegativity and high acidity, it changes the OH to O^- by releasing H^+ and/or the fluorine (F) changes into the fluoride ion (F^-) by taking the electron (e^-) from the O^-. That is, the surface of F-TiO_2 particles becomes negatively charged and enhances the repulsive interaction between F-TiO_2 particles, as shown later in

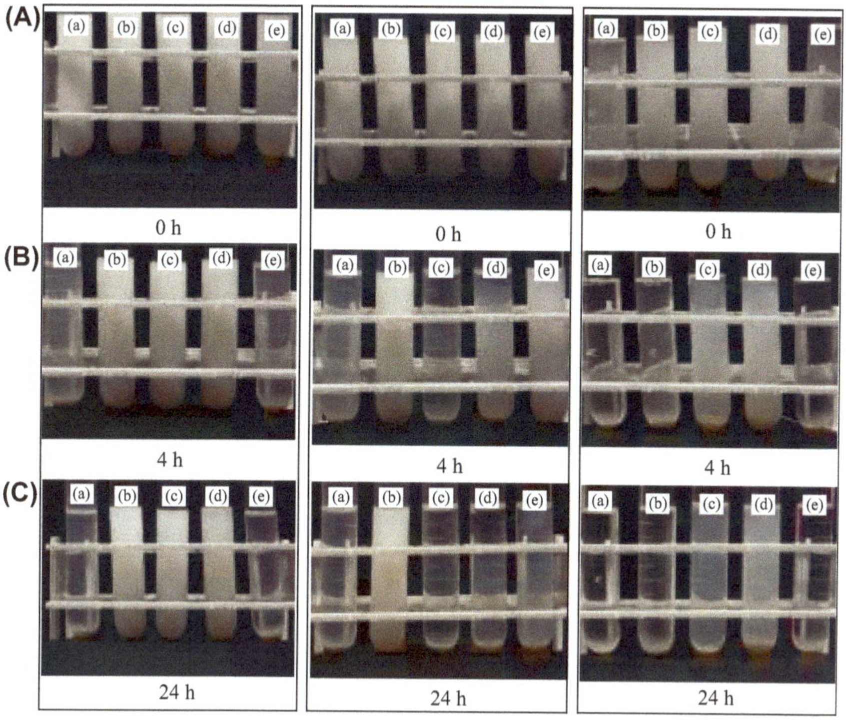

Figure 24.4 Suspension of the dispersed samples in water (A), ethanol (B), and acetone (C) with retention times [(a) TiO_2, (b) F1-TiO_2, (c) F2-TiO_2, (d) F3-TiO_2, (e) F4-TiO_2].

Fig. 24.6. Consequently, surface fluorination can alter the particle–particle interactions. However, the formation of $TiOF_2$ (e) in TiO_2 particles could adversely affect the dispersion stability in all solvents, as shown in Fig. 24.4.

Fig. 24.5 depicts the effects of surface fluorination on the particle size (A) and the zeta potential (B) of TiO_2 particles in water (■), ethanol (◆), and acetone (▲) at a constant pH of 6.5. The average size and zeta potentials of TiO_2 particles fluorinated at 25°C in all solvents were approximately 11 times smaller and 1.5 times larger, respectively, than those (2.5×10^3 nm and -19 mV) of untreated TiO_2 particles. Therefore, it can be said that the fluorinated particles can be stabilized against agglomeration by electrostatic forces because the charges increased by surface fluorination are able to create electrostatic repulsion between the particles, as shown in Fig. 24.6.[25] However, the dispersion stability of TiO_2 samples (e) fluorinated at 200°C decreased owing to the formation of $TiOF_2$ film at the particle surface. To improve the dispersion stability of TiO_2 particles, it is important to control the surface fluorination to maintain the state of fluorine adsorbed or partly bonded on the TiO_2 surface without $TiOF_2$ formation because the high electronegativity and high acidity of fluorine increase the repulsive interactions between TiO_2 particles, which is related to dispersion stability in solvents.

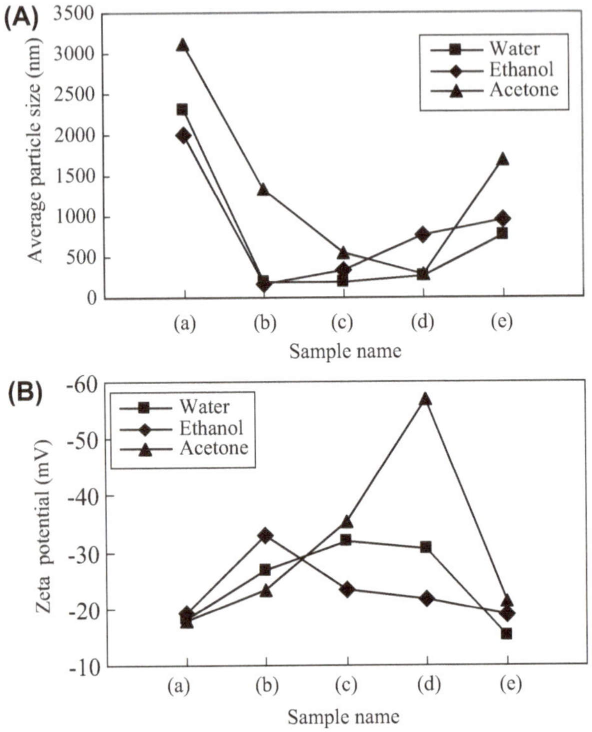

Figure 24.5 Average particle size (A) and zeta potential (B) of untreated and fluorinated TiO_2 particles in various solvents (at pH 6.5) [(a) TiO_2, (b) F1-TiO_2, (c) F2-TiO_2, (d) F3-TiO_2, (e) F4-TiO_2].

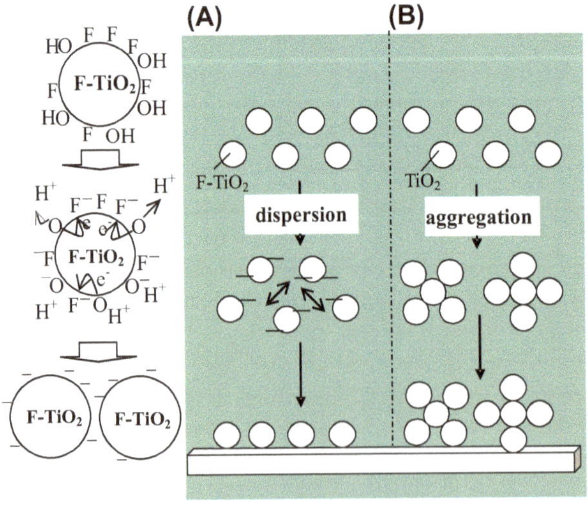

Figure 24.6 Schematic layout of the processes during sedimentation of (A) fluorinated TiO_2 and (B) untreated TiO_2 suspensions.

1.5 Photocatalytic Activity of Fluorinated TiO_2 Samples

Surface fluorination obviously affects the UV−Vis absorption characteristics of TiO_2, as shown in Fig. 24.7. The absorption spectra of the fluorinated TiO_2 samples showed a stronger absorption in the UV−Vis range and a red shift in the band gap transition than untreated TiO_2. This is probably because surface fluorination can expand the wavelength response range of TiO_2 to the visible region. For example, the wavelength range of F3-TiO_2 samples containing Ti^{3+} and Ti^{2+} valences, as shown in Fig. 24.3C, could expand to approximately the 500-nm range owing to a downshift of the conduction band edge. However, in the case of F4-TiO_2 samples, the wavelength range became narrower than that of F3-TiO_2, as indicated in Fig. 24.7. Also, the decrease in surface area (Table 24.1) caused by the $TiOF_2$ formation may seriously affect the wavelength response range.

To compare the photocatalytic activity of untreated and fluorinated TiO_2, the MB degradation reaction was performed. The results are shown in Fig. 24.8. The MB degradation ratio (73%) with fluorinated TiO_2 was much higher than that (18%) with untreated TiO_2 (○) at 4 h. As suggested by Yang,[26] a large number of holes created in fluorinated TiO_2 (F-TiO_2) yields many hydroxyl radicals:

$$h_{vb}^{+} + H_2O_{ad} \rightarrow H_2O^{\bullet +} \text{ (on F-TiO}_2\text{)} \tag{24.1}$$

$$H_2O^{\bullet +} \rightarrow H^{+} + OH_{free}^{\bullet} \text{ (on F-TiO}_2\text{)} \tag{24.2}$$

$$h_{vb}^{+} + \text{Ti-OH}_{surf} \rightarrow \text{Ti-OH}_{surf}^{\bullet} \text{ (on untreated TiO}_2\text{)} \tag{24.3}$$

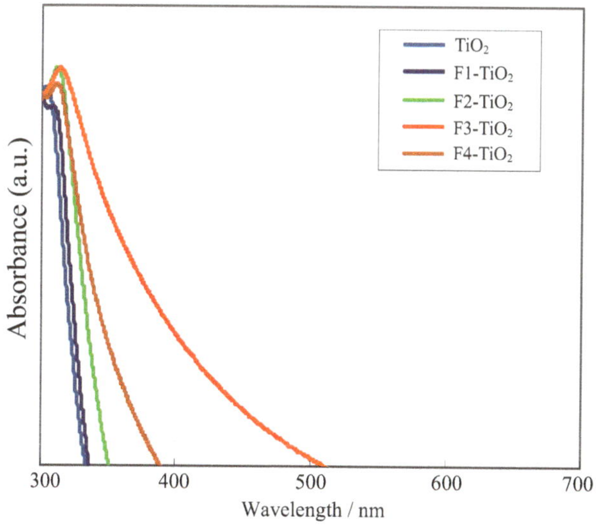

Figure 24.7 Ultraviolet−visible absorption spectra of various samples.

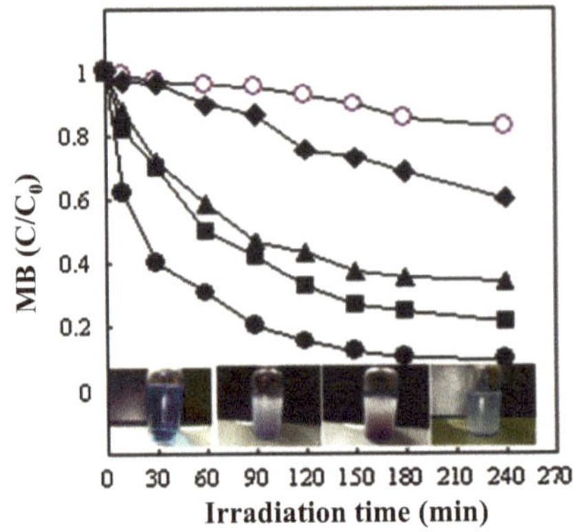

Figure 24.8 Photocatalytic degradation of methylene blue (MB) with various samples under ultraviolet irradiation [○, TiO_2; ◆, F1-TiO_2; ▲, F2-TiO_2; ●, F3-TiO_2; ■, F4-TiO_2].

The preferential formation of free OH radicals on F-TiO_2 should subsequently oxidize the MB molecules, and the adsorbed amount of MB near the surface will increase simultaneously. Consequently, the photocatalytic activity is enhanced by surface fluorination. In particular, the photocatalytic activity of the F3-TiO_2 sample (●) was superior to that of other samples because the Ti^{3+} and Ti^{2+} conduction band edges in F3-TiO_2 can trap the photogenerated electrons and lead to the reduction of the recombination rate between excited electrons and holes.

2. Surface Fluorination of TiAl Alloy Particles

2.1 *Introduction*

Titanium aluminides based on γ-TiAl are highly regarded for high-temperature applications in automotive, aerospace, and power generation industries.[27–31] Attractive properties such as low density, high stiffness, high yield strength, and good creep resistance up to moderately elevated temperatures are interesting for their use as structural materials at temperatures of 700–900°C. These benefits of TiAl alloys make them able to replace, to some degree, the heavy steels and nickel-based alloys that are used at present. However, some drawbacks remain to be overcome, such as limited creep strength and insufficient oxidation resistance at temperatures higher than 800 °C. Efforts for the improvement of high-temperature oxidation resistance of TiAl include surface modification and alloying design.[32–35] Many reports of the relevant literature indicate that improvement of oxidation resistance for TiAl favors surface modifications, rather than alloying design.[36,37]

Industrial-scale processing routes for TiAl have included ingot casting, powder processing, and ingot forging and sheet production by hot rolling.[31,38] Powder metallurgy (PM) processing and compaction offer an alternative for prematerial production through hot isostatic pressing (HIP) of TiAl-based alloy powder into billets that can be rolled into large sheets.[39] However, oxide layers such as Al_2O_3 or TiO_2 on TiAl prevent the sintering of TiAl particles in PM processing. The addition of AlF_3 reportedly improves the sintering properties of Al alloy in PM processing.[40–43] The AlF_3 reacts with the oxide layer formed on Al alloy in the sintering process. It produces the activation area on the Al alloy powder surface. Consequently, sintering between activated Al alloy powders proceeded by the addition of AlF_3.

In this study, we strove to prepare TiAl powders using PM processing. Furthermore, the surface fluorination effects on oxidation resistance and on TiAl powders were investigated.

2.2 Experimental Details

Fig. 24.9 portrays a total flow diagram illustrating the preparation of TiAl powders. Ti powders (>150 μm, 99.9%) and Al powders (>150 μm, 99.9%) in ratios of 50:50 were mixed using a pot mill. The mixture was pressed and converted to a molded rod (250 mm length × 50 mmϕ) under 400 MPa for 3 min using cold isostatic pressing. To synthesize TiAl alloy, the prepared molded rod was reacted in vacuum using self-propagating high-temperature synthesis.[44,45] TiAl powders (300 μm) were prepared from the TiAl rod using the plasma rotating electrode process. The prepared

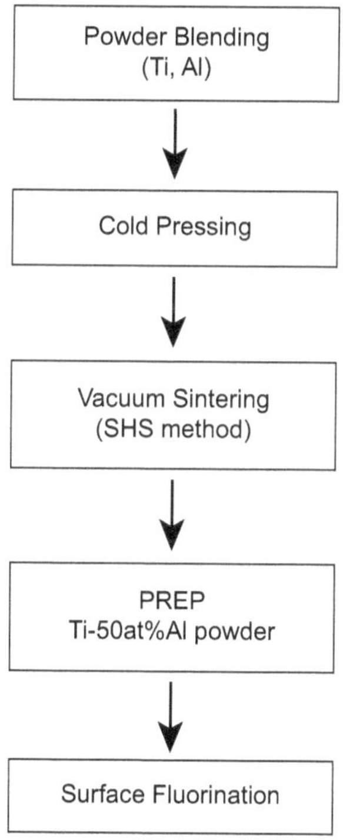

Figure 24.9 Flow diagram showing TiAl powder preparation. *PREP*, plasma rotating electrode process; *SHS*, self-propagating high-temperature synthesis.

TiAl powders were analyzed quantitatively using SEM/EDX (energy dispersive x-ray spectroscopy) (Ultra-Plus (Carl Zeiss Inc.) with a detector (XFlash 5030; Bruker Analytik GmbH). The TiAl particle cross-section was analyzed using the ZAF method (automatic PB-ZAF; Bruker). The weight ratio of Ti:Al was 31.47:68.53 (wt.%), which is 55.10:44.90 in atomic percent (at%). Fluorinated TiAl alloy particles were prepared by direct fluorination using F_2 gas under various reaction conditions. The reaction temperature, fluorine pressure, and reaction time were set, respectively, at 25°C–200°C, 101 kPa, and 3 h. Details of the fluorination apparatus have been explained in our previous articles.[21,22] The structural and electronic properties of the samples were investigated using XRD (XRD–6100; Shimadzu Corp.) and XPS (XPS–9010; JEOL). The surface morphology of various samples was observed using an SEM (s-2400; Hitachi Ltd.). The thermal behavior of TiAl samples was investigated using (thermal gravity/differential thermal analysis) TG/DTA (Seiko Instruments Inc., TG/DTA 6300) at a heating rate of 10°C/min up to 1000°C in air. To prepare the specimen in terms of mechanical strength, the sample powder was compacted into a rod (55 mm length × 45 mmϕ) and pressed at 1000°C and 180 MPa using HIP. The tensile test was conducted at 600°C in Ar gas.

2.3 Characterization of Fluorinated TiAl Samples

Sample names and reaction conditions are presented in Table 24.2. The SEM images of untreated and fluorinated TiAl samples are presented in Fig. 24.10. No change was found in the shape or the morphology of TiAl particles with different degrees of fluorination. The contrast of the grain and grain boundary on the particle surface in Fig. 24.10A was observed more clearly than that for the other samples, as presented in Fig. 24.10B–F. During the SEM observations, the samples after fluorination tend to charge up. Therefore, the electric conductivity of samples might be decreased by fluorination. The decrease in this electric conductivity might underlie the difference in the contrast in the SEM images before and after fluorination. Fluorination of the

Table 24.2 Reaction Conditions of TiAl Particles Fluorinated With F_2 Gas

Sample Name	Temperature (°C)	Time (h)	F_2 Pressure (kPa)
Untreated	–	–	–
F-RT	25	3	101
F-100	100	3	101
F-125	125	3	101
F-150	150	3	101
F-200	200	3	101

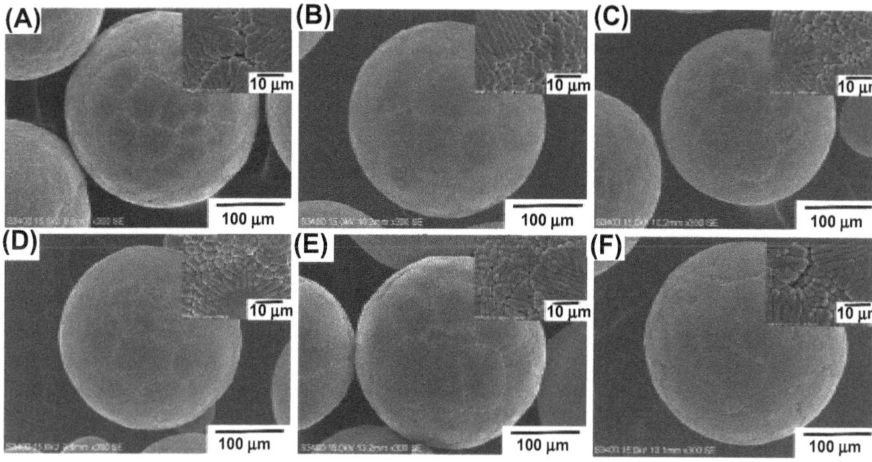

Figure 24.10 Scanning electron microscopic images of various TiAl samples [(A) untreated, (B) F-RT, (C) F-100, (D) F-125, (E) F-150, and (F) F-200].

TiAl alloy sample did not cause the change in the morphology, but rather in the electronic state of the sample surface.

Fig. 24.11 depicts the XRD profiles of various TiAl samples. The prepared TiAl powders exhibited two phases, viz., γ-TiAl (○) and α_2–Ti_3Al (△). Considering the atomic ratio (55.10:44.90) of Ti:Al in TiAl particles calculated with SEM/EDX analysis as explained in the experimental section, γ-TiAl and α_2-Ti_3Al exist at an

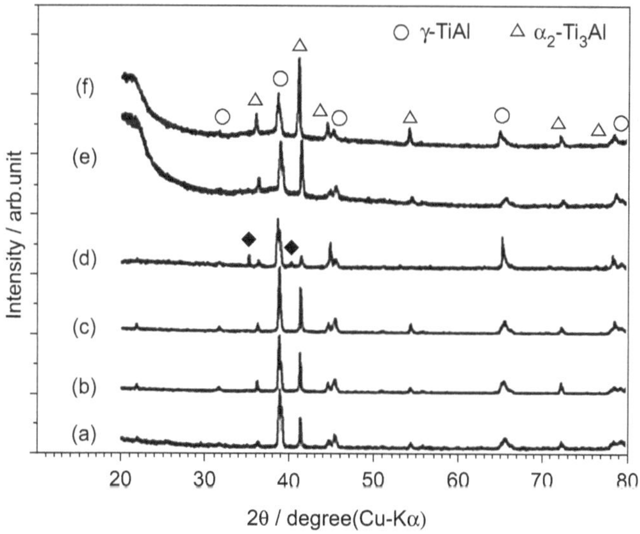

Figure 24.11 X-ray powder diffraction profiles of various TiAl samples [(a) untreated, (b) F-RT, (c) F-100, (d) F-125, (e) F-150, and (f) F-200].

8:1 molar ratio in the particle. Below the fluorination temperature of 125°C, no extra peak was observed in the profiles. New peaks (◆) were detected near 35 degree and 40 degree, which might be regarded as Ti-Al oxyfluorides in the case of the F-125 sample. Furthermore, the peak intensities attributable to γ-TiAl and α_2-Ti_3Al decrease simultaneously. Especially, the peak intensity for α_2-Ti_3Al decreased considerably. Above 125°C fluorination temperature, the peaks of Ti-Al oxyfluorides observed in Fig. 24.11D disappeared. The peaks corresponding to γ-TiAl and α_2-Ti_3Al were clearly detected. The signal/noise ratio of the XRD spectra became smaller, which might mean that some compounds having low crystallinity might be produced and the inner material part located below the oxide surface layer must be exposed. Compounds having low crystallinity must contain some Ti and Al containing fluorides and their decomposition products by reaction with the moisture in air. The oxyfluorides produced at 125°C are apparently converted into fluorides by further reaction at higher temperatures.

Fig. 24.12 depicts the XPS spectra of (A) Ti 2p, (B) Al 2p, (C) O 1s, and (d) F1s electrons for untreated and fluorinated TiAl samples. All binding energies were calibrated to the C 1s peak at 284.8 eV of carbon. For (A) Ti 2p, the Ti (IV) $2p_{1/2}$, and

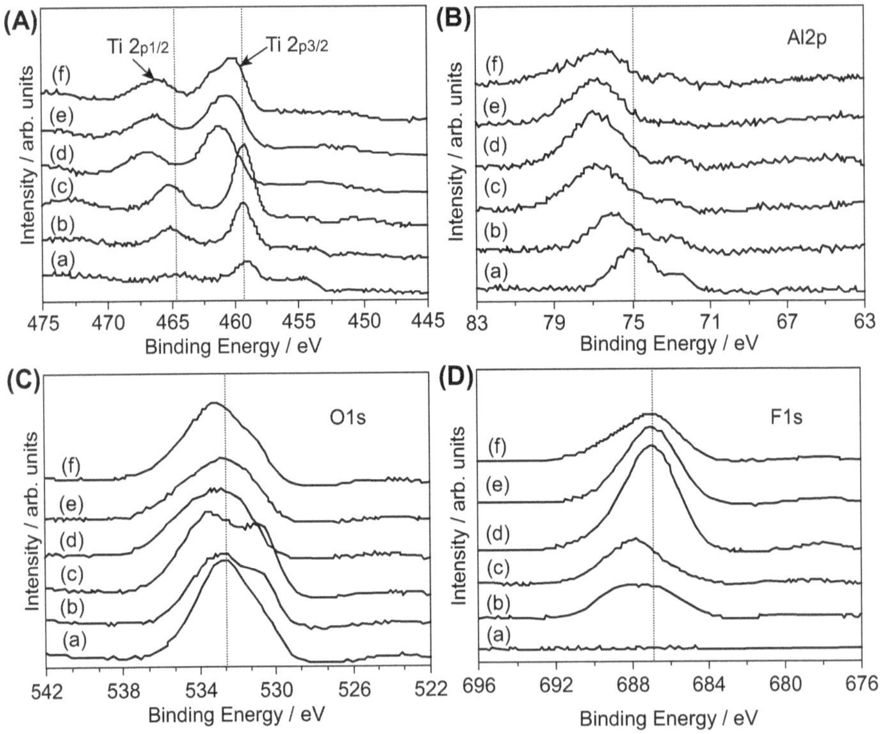

Figure 24.12 X-ray photoelectron spectroscopic spectra of (A) Ti 2p, (B) Al 2p, (C) O 1s, and (D) F 1s for untreated and fluorinated TiAl particles [(a) untreated, (b) F-RT, (c) F-100, (d) F-125, (e) F-150, and (f) F-200].

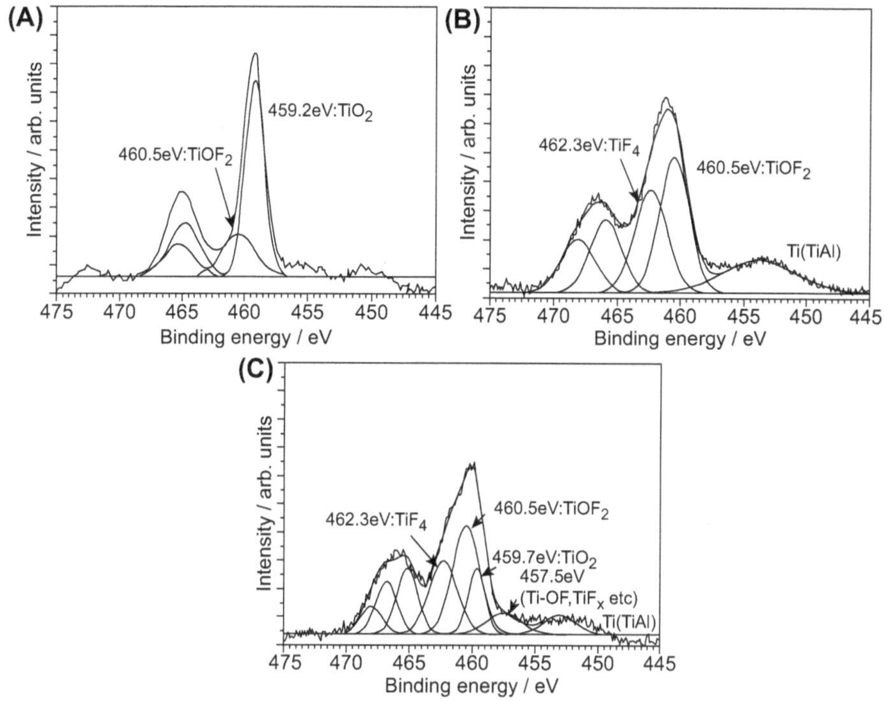

Figure 24.13 Peak separation of Ti 2p in Fig. 4 [(A) F-100, (B) F-125, and (C) F-200].

Ti (IV) $2p_{3/2}$ spin−orbital splitting photoelectrons of the untreated TiO_2 (a) were located, respectively, at binding energies of 464.9 and 459.2 eV. However, in the XPS spectrum after fluorination, especially at temperatures higher than 125°C, the peak of Ti $2p_{3/2}$ shifted to energy higher than that of the untreated material. This result demonstrated the creation of the Ti−F bond of $TiOF_2$ or TiF_4. Regarding the XPS results of Ti 2p, a detailed explanation is presented in Fig. 24.13. Fig. 24.13A shows that the Ti 2p peak of the F-100 sample was deconvoluted into two separate peaks with Gaussian distributions. The peaks located at 459.2 and 460.5 eV were attributed, respectively, to the Ti atom in TiO_2 and $TiOF_2$. Results show that $TiOF_2$ is prepared by fluorination of TiO_2 on the surface. Furthermore, when the reaction temperature was increased to 125°C, TiO_2 and $TiOF_2$ were converted into $TiOF_2$ and TiF_4, with fluorination progressing as presented in Fig. 24.13B. At temperatures higher than 200°C, however, TiO_2 and Ti(III)OF appeared with the decrease of TiF_4, as presented in Fig. 24.13C. The reason might be the hydrolysis and sublimation of titanium fluorides such as TiF_4 with moisture in air. For Al 2p (Fig. 24.12B), the peak (74.9 eV) of Al_2O_3 shifted to higher energy with the increase of fluorination temperature. At temperatures higher than 125°C, the AlF_3 peak was detected mainly near 77 eV. Similar to the result of Ti 2p, the peak intensity of AlF_3 was decreased by hydrolysis with moisture in air. Fig. 24.12C presents the O 1s peaks of various samples. The peak assigned to oxyfluorides was detected at 531 eV in all fluorinated samples (b)−(f). In addition,

Table 24.3 **Elemental Contents in TiAl Samples Evaluated From XPS Results (Fig. 24.12)**

	Element (Atom%)			
Sample	**Ti**	**Al**	**O**	**F**
Untreated	8.69	27.99	63.32	–
F-RT	11.80	16.19	44.07	27.64
F-100C	7.90	21.12	30.69	40.30
F-125	7.12	15.95	20.29	56.64
F-150	6.28	17.37	27.21	49.14
F-200	6.68	16.93	46.35	30.04

F-RT; *XPS*, X-ray photoelectron spectroscopy.

the oxygen contents in the surface region of fluorinated TiAl samples were decreased, as presented in Table 24.3. In the F-200 sample, the oxygen contents were increased markedly by the creation of oxyfluorides and/or oxides. In opposition to O 1s, the fluorine contents increased with the increase of reaction temperature up to 125°C. At temperatures higher than 125°C, the fluorine peak (687 eV) assigned to fluorides decreased as shown in Table 24.3.

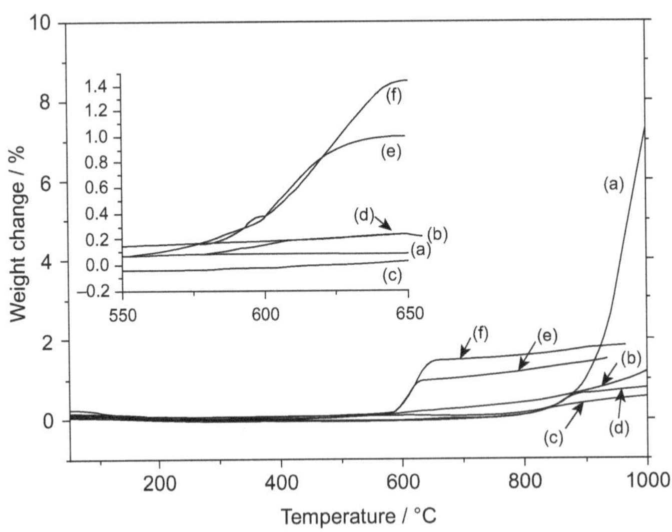

Figure 24.14 TG curves of various TiAl samples [(a) untreated, (b) F-RT, (c) F-100, (d) F-125, (e) F-150, and (f) F-200].

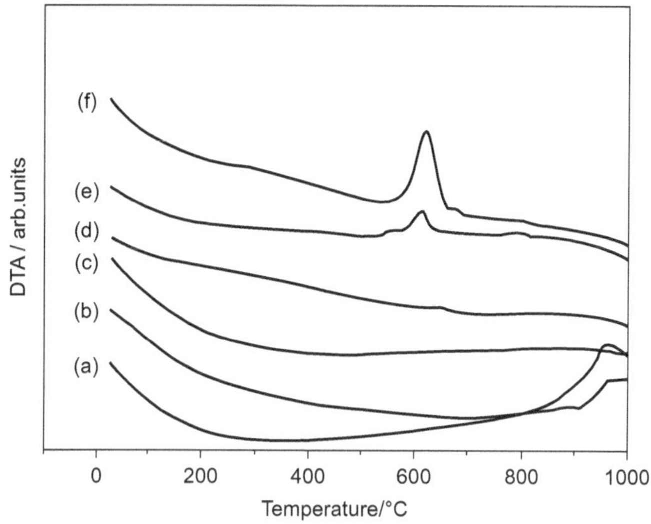

Figure 24.15 DTA curves of various TiAl samples [(a) Ti-Al, (b) F-RT, (c) F-100, (d) F-125, (e) F-150, and (f) F-200].

2.4 Oxidation Resistance of TiAl Samples

To investigate the resistance to oxidation of TiAl alloy, TG/DTA analyses of untreated and fluorinated samples were conducted as depicted in Figs. 24.14 and 24.15. Fig. 24.14 shows that the weight change of untreated sample (a) increased drastically at temperatures higher than 800°C. This might be due to the formation of oxidation layer by the reaction between metals and oxygen in air. However, the fluorinated TiAl samples can restrain the weight increase even at temperatures higher than 800°C. Especially, the weight change (<0.8%) of samples (c) and (d) was much less than that (7.2%) of untreated materials until 1000 °C. It might be considered that the substantial oxyfluorides on the surface play a role in protecting the metal oxidation. However, in cases of samples (e) and (f), the fluorides formed at temperatures higher than 150°C began to oxidize near 600°C. The weight change reached about 2% at 1000°C. The exothermic peak of fluorides caused by oxidation was found at 600°C in the DTA results, as indicated in Fig. 24.15. Based on these results, only oxyfluoride layers on the TiAl surface are effective for resistance to TiAl alloy oxidation.

Fig. 24.16 portrays the SEM images of cross-sections of compacts sintered from untreated (A) and F-100 (B) samples. At a glance, the sintering proceeds particularly well in the case of the fluorinated sample. The spherical shape of each TiAl particle is retained in cases of untreated samples, although the original shape of the TiAl particle did not remain in the case of the F-100 sample. In addition, no void was attributable to the interstitial space of the particle in the case of the F-100 sample. Surface fluorination can give new potential to TiAl particles as a material for sintering processes. The tensile strength of sintered bodies was measured. Sintered bodies made of untreated and

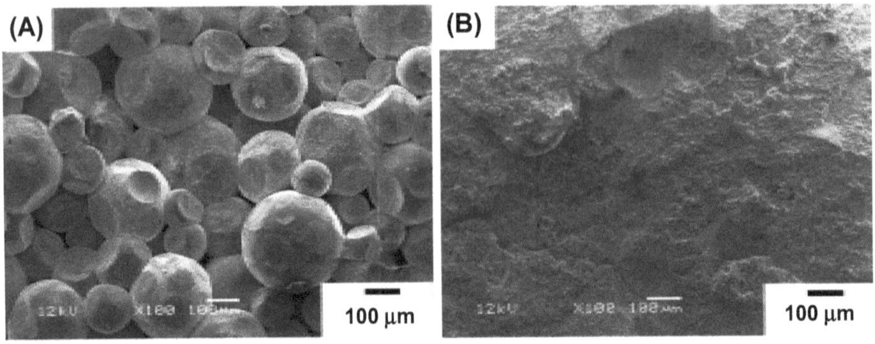

Figure 24.16 Scanning electron microscopic images of the cross-section of sintered compact [(A) untreated and (B) F-100].

fluorinated (F-100) Ti-Al powder have, respectively, tensile strength of 314.6 and 407.2 MPa. The F-100 sample exhibited 1.3 times higher tensile strength than an untreated TiAl sample. Considering the TG data, the sintering must proceed at temperatures higher than 900°C. This fact greatly influenced the sintering process. The voidless sintered body can be prepared using the surface-fluorinated Ti-Al powder sample. The fluorination, however, must be controlled precisely to avoid overfluorination.

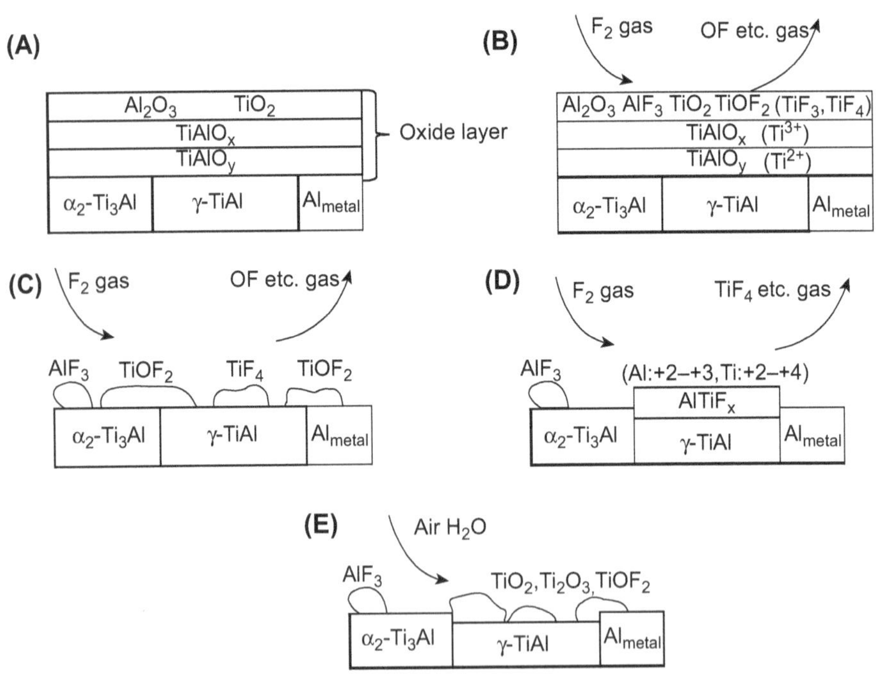

Figure 24.17 Schematic image of surface state on TiAl alloy at each fluorination temperature [(A) 25°C, (B)100°C, (C) 125°C, (D) ≥150, and (E) hydrolysis of (D)].

Only the oxide layer existing on the surface of the samples must be fluorinated to improve the sintering property of the sample.

Schematic images of surface states on TiAl alloy at each fluorination temperature estimated from these results are presented in Fig. 24.17. With increasing fluorination temperature, the oxide on TiAl is converted stepwise to fluoride by way of oxyfluoride. At temperatures lower than 125°C, the oxyfluoride layer mainly existed. When the fluorination temperature increases to temperatures higher than 150°C, the fluoride became the main layer on TiAl. Furthermore, Fig. 24.17 (E) shows that the fluorides formed on TiAl change easily to oxides and/or oxyfluorides through hydrolysis with moisture in air.

3. Conclusions

We have reported the effects of surface fluorination on the dispersion stability and photocatalytic activity of TiO_2 particles and on the oxidation resistance of TiAl alloy particles. In case of TiO_2 fluorination, fluorinated TiO_2 (F-TiO_2) was successfully prepared by direct fluorination using F_2 gas. The fluorine contents (x) in $TiO_{2-x}F_{2x}$ primarily depended on the reaction temperature and fluorine pressure. At room temperature (25°C) and fluorine pressure lower than 6.7 kPa, the fluorine contents (x) were controlled bellow 0.18, and the fluorine was chemically adsorbed on the TiO_2 surface. At 200°C, the fluorine content (x) in $TiO_{2-x}F_{2x}$ increased to 1.10 and the existence of $TiOF_2$ was confirmed. The chemically adsorbed fluorine on the TiO_2 surface might play a positive role toward both the dispersion stability and photocatalysis. Especially for the photocatalytic activity of TiO_2, F3-TiO_2 with Ti^{3+} and Ti^{2+} valences could expand to approximately 500 nm of the UV−Vis range, and the photocatalytic activity of F3-TiO_2 was also 4 times higher than the MB degradation ratio (18%) of untreated TiO_2. However, the dispersion stability and photocatalysis of F4-TiO_2 prepared at 200°C was negatively affected by the formation of $TiOF_2$. Therefore, considering the dispersion stability and photocatalytic activity of TiO_2, it is essential to control the fluorine content (x) in the range $0 < x < 0.5$ in $TiO_{2-x}F_{2x}$ to optimize the beneficial effects of surface fluorination.

In case of TiAl fluorination, the surface components and TiAl alloy structure are reaction-temperature dependent. At 25°C−125°C (fluorination temperature), the TiAl particle surface layer changed from oxide to oxyfluoride. Furthermore, when fluorination was conducted at temperatures higher than 150°C, the oxyfluorides were converted into fluorides on the TiAl particle surface. The TG/DTA results show that the weight increase (<0.8% at 1000°C) of oxyfluorides on TiAl was distinctly less than that (7.2%) of untreated TiO_2. However, the weight increase of fluorides on TiAl was about 2.0% because of the oxides created from hydrolysis with moisture in air. Consequently, control of F/O contents such as oxyfluorides on the surface is effective to produce higher resistance against TiAl particle oxidation.

References

1. Parmar, K. P. S.; Ramasamy, E.; Lee, J. W.; Lee, J. S. *Scr. Mater.* **2010,** *62,* 223–227.
2. Fujihira, M.; Satoh, Y.; Osa, T. *Nature* **1981,** *293,* 206.
3. Hotsenpiller, P. A. M.; Bolt, J. D.; Farneth, W. E.; Lowekamp, J. B.; Rohrer, G. S. *J. Phys. Chem. B* **1988,** *102,* 3216–3226.
4. French, R. A.; Jacobson, A. R.; Kim, B.; Isley, S. L.; Lee Penn, R.; Baveye, P. C. *Environ. Sci. Technol.* **2009,** *43,* 1354–1359.
5. Li, C. C.; Chang, S. J.; Tai, M. Y.; Am, J. *Ceram. Soc.* **2010,** *93,* 4008–4010.
6. Li, G.; Lv, L.; Fan, H.; Ma, J.; Li, Y.; Wan, Y.; Zhao, X. S. *J. Colloid Interface Sci.* **2010,** *348,* 342–347.
7. Yaremko, Z. M.; Tkachenko, N. H.; Bellmann, C. *J. Colloid Interface Sci.* **2006,** *296,* 565–571.
8. Hoogeveen, N. G.; Stuart, M. A. C.; Fleer, G. J. *J. Colloid Interface Sci.* **1996,** *182,* 133–145.
9. Iijima, M.; Kobayakawa, M.; Kamiya, H. *J. Colloid Interface Sci.* **2009,** *337,* 61–65.
10. Shibata, J.; Fuji, K.; Horai, K.; Yamamoto, H. *Kagaku Kogaku Ronbunshu* **2001,** *27,* 497–501.
11. Shibata, J.; Fuji, K.; Horai, K.; Yamamoto, H. *Kagaku Kogaku Ronbunshu* **2002,** *28,* 641–646.
12. Vijayabalan, A.; Selvam, K.; Velmurugan, R.; Swaminathan, M. *J. Hazard. Mater.* **2009,** *172,* 914–921.
13. Hattori, A.; Yamamoto, M.; Tada, H.; Ito, S. *Chem. Lett.* **1998,** *27,* 707–708.
14. Hattori, A.; Shimoda, K.; Tada, H.; Ito, S. *Langmuir* **1999,** *15,* 5422–5425.
15. Park, H.; Choi, W. *J. Phys. Chem. B* **2004,** *108,* 4086–4093.
16. Li, D.; Haneda, H.; Hishita, S.; Ohashi, N.; Labhsetwar, N. K. *J. Fluorine Chem.* **2005,** *126,* 69–77.
17. Li, D.; Haneda, H.; Labhsetwar, N. K.; Hishita, S.; Ohashi, N. *Chem. Phys. Lett.* **2005,** *401,* 579–584.
18. Li, D.; Ohashi, N.; Hishita, S.; Kolodiazhnyi, T.; Haneda, H. *J. Solid State Chem.* **2005,** *178,* 3293–3302.
19. Yu, J.; Yu, J. C.; Leung, M. K. -P. *J. Catal.* **2003,** *217,* 69–78.
20. Kim, J. H.; Sato, H.; Kubo, T.; Yonezawa, S.; Takashima, M. *Chem. Lett.* **2011,** *40,* 230–232.
21. Takashima, M.; Nosaka, Y.; Unishi, T. *Eur. J. Solid State Inorg. Chem.* **1992,** *29,* 691–703.
22. Kim, J. H.; Umeda, H.; Ohe, M.; Yonezawa, S.; Takashima, M. *Chem. Lett.* **2011,** *40,* 36–61.
23. Ohno, T.; Tsubota, T.; Nishijima, K.; Miyamoto, Z. *Chem. Lett.* **2004,** *33,* 750–751.
24. Yu, J. C.; Yu, J.; Ho, W.; Jiang, Z.; Zhang, L. *Chem. Mater.* **2002,** *14,* 3808–3816.
25. Liao, D. L.; Wu, G. S.; Liao, B. Q. *Colloids Surf. A* **2009,** *348,* 270–275.
26. Yang, S. Y.; Chen, Y. Y.; Zheng, J. G.; Cui, Y. J. *J. Environ. Sci.* **2007,** *19,* 86–89.
27. Zhang, W. J.; Reddy, B. V.; Deevi, S. C. *Scr. Mater.* **2001,** *45,* 645–651.
28. Yamaguchi, M.; Inui, H.; Ito, K. *Acta Mater.* **2000,** *48,* 307–322.
29. Tetsui, T.; Ono, S. *Intermetallics* **1997,** *7,* 689–697.
30. Noda, T. *Intermetallics* **1998,** *6,* 709–713.
31. Clemens, H.; Kestler, H. *Adv. Eng. Mater.* **2000,** *2,* 551–570.
32. Yang, M. R.; Wu, S. K. *Acta Mater.* **2002,** *50,* 691–701.
33. Hsu, I. C.; Wu, S. K.; Lin, R. Y. *Mater. Chem. Phys.* **1997,** *49,* 184–190.

34. Kim, J. P.; Jung, H. G.; Kim, K. Y. *Surf. Coat. Technol.* **1999,** *112,* 91–97.
35. Tang, Z.; Niewolak, L.; Schemet, V.; Singheiser, L.; Quadarkker, W. J.; Wang, F. *Mater. Sci. Eng. A* **2002,** *328,* 297–301.
36. Lee, J. K.; Oh, M. H.; Wee, D. M. *Intermetallics* **2002,** *10,* 347–352.
37. Zhou, C.; Yang, Y.; Gong, S.; Xu, H. *Mater. Sci. Eng. A* **2001,** *307,* 182–187.
38. Clemens, H. *Z. Metallkd.* **1995,** *86,* 814–822.
39. Appel, F.; Brossman, U.; Christoph, U.; Eggert, S.; Janschek, P.; Lorenz, U. *Adv. Eng. Mater.* **2000,** *2,* 699–720.
40. Masuoka, S.; Arami, Y.; Nagai, S. *J. Jpn. Soc. Powder Powder Metall.* **2002,** *49,* 1003–1008.
41. Ohtsuki, M.; Kohni, T. *Jpn. Soc. Powder Powder Metall.* **1989,** *36,* 723–726.
42. Masuoka, S.; Arami, Y. *J. Jpn. Soc. Powder Powder Metall.* **2007,** *55,* 64–66.
43. Inaba, T.; Nagai, S.; Iwatsu, O.; Nakanishi, N.; Shigematsu, T. *J. Jpn. Soc. Powder Powder Metall.* **1989,** *36,* 149–152.
44. Aminikia, B. *Power Technol.* **2012,** *232,* 78–86.
45. Guojian, J.; Jiayue, X.; Hanrui, Z.; Wenlan, L. *Mater. Lett.* **2011,** *65,* 2969–2971.

Chemical and Electrochemical Stability of Copper in Molten KF-2HF

25

D. Binet[1], *I. Crassous*[2], *A. Mantoux*[3], *C. Belhomme*[2], *A. Colisson*[2], *B. Morel*[2], *F. Lantelme*[1], *D. Devilliers*[1], *C.M. Julien*[1], *H. Groult*[1]
[1]Sorbonne Universités, UPMC, CNRS UMR 8234, Paris, France; [2]Areva, Pierrelatte Cedex, France; [3]Université Joseph Fourier, Saint Martin d'Hères, France

Chapter Outline

1. Introduction

Corrosion is considered as one of the most important problems our society has to face since it affects various key sectors (industry, gas and liquid transportation pipelines, energy…) and it is estimated that hundreds of billions of dollars are lost every year to protect or to replace the corroded devices. Corrosion mainly concerns chemical or electrochemical reaction of metals with their environment, for instance, the oxidation of metals in air or in liquids giving rise to metal oxide in the presence of an oxidant such as oxygen or metal fluoride in a fluorinated environment.[1–5] In electrochemical processes, corrosion is usually considered as an undesirable reaction and has a negative effect on the electrochemical process by reducing the yield of the required electrochemical reaction, by degrading the useful properties of materials, etc. In addition, corrosion

Modern Synthesis Processes and Reactivity of Fluorinated Compounds. http://dx.doi.org/10.1016/B978-0-12-803740-9.00025-1

induces undesirable and expensive handling. Thus, due to difficulties in handling, transporting, and maintaining a high purity level, the main part of F_2 production is carried out on site mainly to prepare UF_6 by direct fluorination of UF_4 with F_2. About 60% of fluorine gas produced in the world serves to prepare UF_6 containing between 3% and 5% of ^{235}U, mainly produced to be used in nuclear plants.[1] The enrichment process is being done since several decades (UF_6 being a volatile compound of uranium) not only because of UF_6's relative chemical stability and low sublimation temperature but also because fluorine has only a single stable naturally occurring isotope (^{19}F). Therefore, isotopomeres of UF_6 differ in their molecular weight only because of U.

Let us consider now the preparation process of F_2 gas, which consists in electrolysis of anhydrous hydrogen fluoride from molten KF-2HF at around 95°C. In this process, fluorine and the hydrogen gas evolve at the anode and the cathode, respectively, according to the following two half-cell reactions.

At the anode:

$$HF_2^- \rightarrow \frac{1}{2}F_2 + HF + e^- \qquad (25.1)$$

At the cathode:

$$2HF + e^- \rightarrow \frac{1}{2}H_2 + HF_2^- \qquad (25.2)$$

This gives rise to the global reaction:

$$HF_{(liq.)} \rightarrow 1/2F_{2(g)} + 1/2\ H_{2(g)} \qquad (25.3)$$

The fluorine evolution process is characterized by high current efficiency approaching 0.95 but poor energy efficiency (about 0.3).[1] As a consequence, large quantity of heat must be removed from the cell, and cooling coils are used to maintain a constant temperature of about 95°C. Concerning the applied potential reaction in electrolyzers, and even if the thermodynamic potential of HF decomposition is c. 2.9 V,[6,7] an anode–cathode applied voltage of about 10 V is usually necessary for obtaining a current density of 10 A/dm^2 in industrial cells. For such a cell, the total voltage is composed of five contributions[8]: the reversible decomposition voltage (≈2.9 V), the ohmic drop in the electrolyte (≈3 V), the ohmic drop in the electrodes (≈0.5 V), the cathode overvoltage (≈0.2 V), and finally the anode overvoltage (≈2.5 V). Concerning the anode reaction, inhibition of the fluorine evolution reaction is observed due to the electrogeneration of C-F compounds characterized by low surface energy, which repels the electrolyte from the electrode. The contact angle is about 120–160 degree[9–11]: fluorine bubbles are lenticular and are strongly attached to the carbon anode's surface. This induces a significant decrease in the electroactive surface area of the electrode. However, from a thermodynamic point of view, direct contact between carbon and F_2 gas should give rise to CF_4. Thus, the passivation of the carbon anode material allows protecting carbon and generating fluorine gas. The high anode overvoltage ($\approx$2.5 V) is commonly ascribed to the formation of a solid carbon–fluorine compound on carbon anodes during fluorine production.[6,7,10–27]

The main cathodic reaction consists in hydrogen evolution and induces electrolyte movements.[28–32] Nevertheless, metal ions (mainly copper ions) issued from corrosion are present in the electrolyte and give rise to a rough surface film composed of copper-based fluorides. The dendritic growth of such Cu-based electrodeposited films can cause severe damages including short circuits in the electrolyzers. Thus, even if the cathode overvoltage is about 12 times lower than the anode overvoltage, the reactions occurring at the cathode have a strong influence on the lifetime of the electrolyzer, as it will be studied in this chapter.

A typical industrial electrolyzer contains molten KF-2HF in which about 30 plate carbon anodes are screwed to a copper anodic feeder, called the busbar. The cells are also equipped with iron or Monel skirts to separate the gases to avoid explosive recombination. The presence of the molten fluoride electrolyte and evolved F_2 gas causes severe corrosion, which leads to contamination of the electrolyte with metallic ions (mainly copper, iron, and nickel). This reaction interferes with the main electrolysis reaction (Eq. (25.3)) and decreases the current efficiency of the industrial process. Moreover, copper ions migrate through the cathodic compartment and are reduced at the cathode concomitantly to H_2 evolution, forming a thick copper layer. Consequently, it is important to know the corrosion rate of metals in KF-2HF and to identify the mechanism involved during corrosion since the lifetime of a fluorine electrolyzer strictly depends on the corrosion of metallic pieces.

In this chapter, it will be shown that a simple corrosion process is not sufficient to explain the high degradation/corrosion of metallic pieces during electrolysis. Indeed, the corrosion mechanism involves a corrosion–erosion process, the erosion phenomenon being mainly related to the impact of F_2 bubbles onto the surface of the metallic pieces, including the busbar. Thus, the impact of F_2 bubbles on the corrosion rates has also been discussed. Such a phenomenon cannot be neglected since the lifetime of the electrolyzers depends directly on the stability of metallic pieces.

First, we will study the behavior of copper immersed in KF-2HF with and without polarization, and then with and without impact of bubbles on the surface. The aim is to determine the corrosion parameters of copper in molten KF-2HF such as current of corrosion and potential of corrosion. At each step of the fluorine generation process, the stability of material to corrosion is one of the key parameters since metallic pieces are in contact with aggressive fluorinated compounds such as F_2 and HF. Thus, the corrosion rates of all the metallic materials present in the electrochemical cell were carefully studied, to propose improvements to minimize this detrimental phenomenon.

2. Experimental Section

2.1 Electrochemical Experiments

All experiments were performed in industrial-grade KF-xHF melts with HF ratio (denoted THF hereafter) in the range 38–43 w% and a temperature θ in the range

85–105°C. The electrochemical cell was fitted with a copper plate (purity >99.5%) as the working electrode, a mild steel counterelectrode, and a Cu/CuF_2 reference electrode. A carbon anode is used for fluorine production during the erosion experiments. Data were collected by a VMP-2 potentiometer from Biologic Instruments.

2.2 Corrosion Rate Evaluation

Depending on the voltage applied to the material, the corrosion rates are estimated by either the Tafel plots or the residual current. For a sample under anodic polarization, the constant residual current is denoted I_{res}. The corrosion rate, v_{corr}, expressed in centimeters per second, is thus calculated using the following relation (Eq. (25.4)):

$$v_{corr} = \frac{M \times i_{res} \times t}{\rho \times n \times F \times S} \tag{25.4}$$

where M is the molecular weight (g/mol); i_{res}, residual current (A); t, time (s), ρ, density (g/cm^3); n, number of exchanged electrons; F, Faraday constant (96,500 C/mol); and S, geometric surface area (cm^2).

We have developed an original approach to evaluate the influence of the most important parameters in terms of corrosion. Indeed, it is well known that whatever be the metallic materials chosen, their corrosion rate (denoted v_{corr}) in molten KF-xHF depends on two parameters: the temperature θ (expressed in °C) and the HF ratio, THF (expressed in w% HF) in the KF-xHF mixture.[33] Whatever be the study of the interactions between these two parameters, our approach could allow a better understanding of the evolution of corrosion rate. An interesting tool for this is a multivariate analysis applied to a detailed experiment design. The two variables are the temperature and the HF ratio. The temperature values are selected in the range [85–105°C] and the HF ratio is chosen in the range [38–43 w%]. Weight change measurements also support the electrochemically derived results. Our analysis permits the definition of a multiple regression given below (Eq. (25.5)) where the dependency is expressed according to the mean values θ_0 and THF_0 of the experimental data.

$$\begin{aligned} v_{corr} = {} & a_0 + a_1(\theta - \theta_0) + a_2(THF - THF_0) + a_{12}(\theta - \theta_0)(THF - THF_0) \\ & + a_{11}(\theta - \theta_0)^2 + a_{22}(THF - THF_0)^2 \end{aligned} \tag{25.5}$$

where a_0 is a constant term, and the principal effects of the variables are denoted a_1 and a_2. A classical study allows measuring these parameters. The interaction effect between the two variables is denoted a_{12}. At least, the quadratic effects are denoted a_{11} and a_{22}.

This method is very effective and allows realizing maps where the corrosion rate is linked to the bath temperature and HF percentage. Thanks to the multiple regression (Eq. (25.5)), the plots that are drawn on graphs correspond to the whole couple (temperature vs. HF percentage) for a given corrosion rate.

2.3 Physicochemical Characterization

Several techniques were used to obtain useful information on the morphology and on the surface composition of our samples. Scanning electronic microscopic (SEM) analysis was performed using SEM-FEG ZEISS Ultra 55, whereas Electron Probe Microanalysis (EPMA) was performed using CAMECO SX-100. Moreover, XRD (X ray diffraction) measurements were made using RIGAKU Ultima-IV with the CuKα emission line.

The copper content in the molten salt was estimated by atomic absorption measurements after decantation of about 1 day, although the main drawback of this technique is that it cannot bring information on the speciation of copper in the electrolyte. It was also possible to increase the amount of copper in the electrolyte by anodic dissolution of a copper rod. The dissolved quantity was thus calculated from the corresponding electrode weight loss.

3. Results

3.1 Oxidation, Reduction, and Precipitation of Copper Compounds in Molten Salt KF-2HF

The behavior of copper in KF-2HF was briefly investigated in a very large potential window to determine the electroactivity domain. The standard potential of copper is between that of hydrogen evolution and fluorine. A typical I—E (current vs. potential) curve recorded in KF-2HF with a copper electrode is given in Fig. 25.1A.

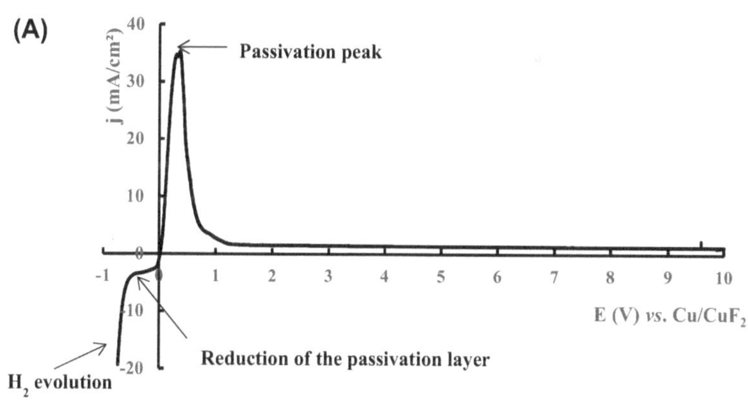

Figure 25.1A I-E curve obtained with a copper electrode in KF-2HF (40.5 % - 95°C). Scan rate: 100 mV/s).

After the occurrence of a passivating peak at 0.30 V vs. Cu/CuF_2, the current dropped down rapidly and reached a constant value (about 150 mA). It contributes to the growth of the passivation and protective layer on the surface of copper. One must notice that even for high oxidation potentials (>10 V), the production of F_2 gas does not occur on the copper anode. During the reverse scan, the reduction of the passivating layer takes place just before the evolution of H_2, which starts at around −0.50 V vs. Cu/CuF_2.

The oxidation peak at around 0.3 V is related to the formation of a passivating CuF_2 surface layer[33–41] according to the following half reaction (Eq. (25.6)):

$$Cu_{(s)} + 2HF_2^- \rightarrow CuF_{2(s)} + 2HF + 2e^- \quad (25.6)$$

This half reaction coupled with the reduction of HF to H_2 (half reaction (Eq. (25.7))) gives rise to the global reaction (Eq. (25.8)):

$$4HF + 2e^- \rightarrow 2HF_2^- + H_{2(g)} \quad (25.7)$$

$$Cu_{(s)} + 2HF \rightarrow CuF_{2(s)} + H_{2(g)} \quad (25.8)$$

For a better analysis of the involved mechanism, a second set of experiments was done in a narrow potential range from −0.60 V to +0.70 V vs. Cu/CuF_2 with several scan rates. Typical cyclic voltammograms recorded at 1 and 10 mV/s are given in Fig. 25.1B. The potential started from the initial potential [open circuit voltage (OCV) at about 0.06 V] and was increased up to 0.7 V vs. Cu/CuF_2. Then, the potential was scanned back until −0.6 V vs. Cu/CuF_2 where H_2 was evolved.

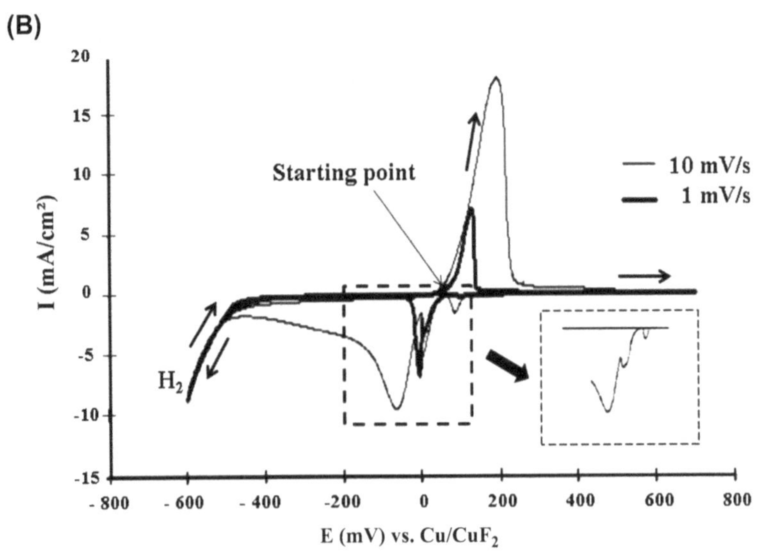

Figure 25.1B Cyclic voltammetry on a copper electrode in KF-2HF (40.5 w% - 90°C).

As observed in Fig. 25.1B, the shape of the oxidation peak is characteristic of the deposition of an insoluble film. This passivation peak corresponds to the formation of a CuF_2 passive layer[33,36,38] on the copper electrode according to Eq. (25.6). Its intensity depends roughly on the square root of the scan rate. The peak potential also depends on the scan rate; it appears between 0.10 and 0.20 V vs. Cu/CuF_2.

Moreover, in addition to the hydrogen evolution reaction beginning at about −0.5 V vs. Cu/CuF_2, several reduction peaks were observed between 0 and −0.15 V vs. Cu/CuF_2 and attributed to the reduction of the passivation layer according to a two-step mechanism:

$$CuF_{2(s)} + HF + e^- \Leftrightarrow CuF_{(s)} + HF_2^- \quad (25.9)$$

and

$$CuF_{(s)} + HF + e^- \rightarrow Cu + HF_2^- \quad (25.10)$$

Previous studies reported in the literature[36,37,39] concerning the corrosion of metals in KF-2HF have also revealed the existence of the (+I) and (+II) oxidation states for copper, giving rise to CuF and CuF_2. For instance, Tasaka et al.[37] have studied the theoretical electrode potentials deduced from thermodynamic approaches and found values of +0.458 V and +0.307 V vs. HF/H_2 for the CuF_2/Cu and CuF/Cu redox couples, respectively. However, the observed potentials were nonuniform on the copper electrode surface and the coexistence of CuF and CuF_2 on the surface could be considered in a first approximation, the latter being influenced by the diffusion of CuF as suggested by Conway et al.[36] It must be noticed that CuF is thermodynamically unstable in KF-2HF below 250°C, and a disproportionation reaction of CuF could take place to form metallic copper and CuF_2 according to reaction (25.11):

$$2CuF \rightarrow Cu + CuF_2 \quad (25.11)$$

However, one cannot exclude the fact that in some parts of the surface electrode, Cu(I) is still present. Other electrochemical techniques such as impedance spectroscopy can be used to provide information, but the impedance diagrams are often difficult to be fitted. Lamirault et al.[33,39] have proposed that the corrosion mechanism of copper in KF-2HF at 85°C is composed of five steps. The first step involves the formation of $Cu^{(+II)}{}_{ads}$ species, which can induce the precipitation of K_xCuF_{2+x} at the electrode surface. The authors have also considered that the partial desorption of $Cu^{(+II)}$ could lead to Cu^{2+} ions, which diffuse into the pores and cause the precipitation of the same copper fluoride. Their conclusions seem to be in good agreement with that of ours, but their experiments were not carried out at the same temperature and it could affect the adsorption or precipitation phenomenon. The precipitation of $KCuF_3$ on the electrode surface has also been mentioned by Beccu et al. with Cu^{2+} species and HF_2^-.[42]

Several observations support the hypothesis of the K_xCuF_{2+x} (with x = 1 or 2) precipitation process.[40,42,43] Indeed, as mentioned in the introduction, copper ions can be reduced at the cathode. The liquid movements, the liquid composition (local HF starvation near the electrode[34,44,45]), the liquid temperature just near the deposit surface, and the local current density distribution have a real impact on the morphology of the electrodeposits. For example, very sharp dendritic deposits can be obtained on the cathode of industrial cells, as illustrated in Fig. 25.2. EDX analysis reveals that bright areas are mainly composed of metallic copper and gray areas of KF-HF. Fig. 25.3 represents a diffractogram of this type of deposit and shows also that it contains KF-HF and metallic copper. So, the XRD and SEM (Scanning Electron Microscopy) analysis show the large amount of KF-HF merged with copper in this type of deposit.

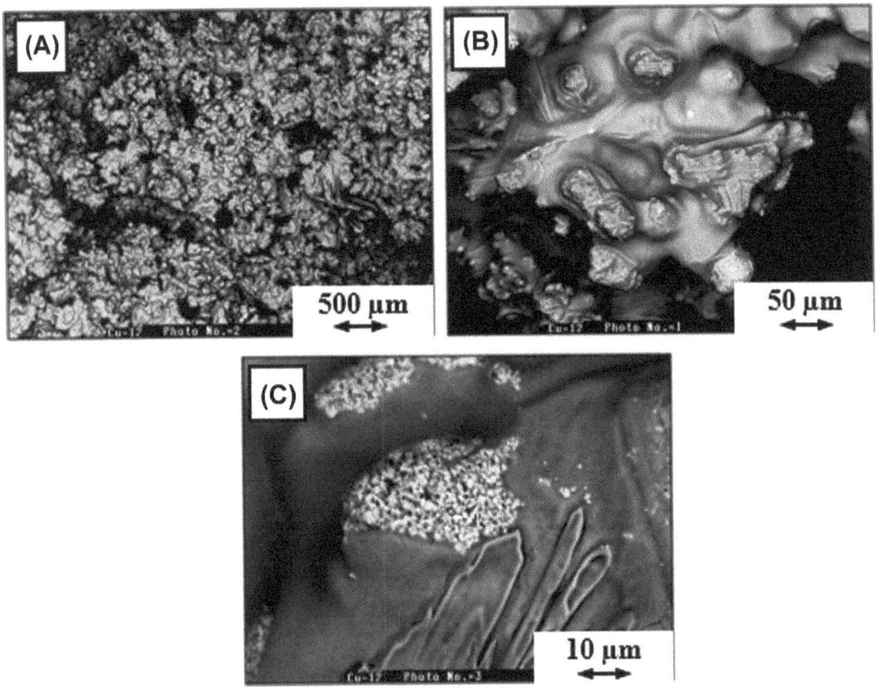

Figure 25.2 SEM pictures of industrial cathodic deposits.

3.2 *Behavior of Metallic Copper in Molten KF-2HF at Open Circuit Voltage*

Let us consider now the behavior of copper at OCV in KF-2HF.

First, a copper plate was put in a Teflon crucible containing KF-2HF and placed in an oven heated at 95°C, i.e., the temperature of the industrial cells. The weight of the copper plate was measured vs. time. Finally, the relative weight loss as a function of the immersion time is given in Fig. 25.4. The relative weight loss varies linearly with the immersion time, except at the beginning of the experiment. This result shows the good resistance to the corrosion of copper immersed in KF-2HF with a relative weight loss of only 0.2% after 100 days. It must be noticed that the HF content in the crucible decreased slightly vs. time due to evaporation and could give rise to erroneous results. Therefore, the corrosion current and corrosion

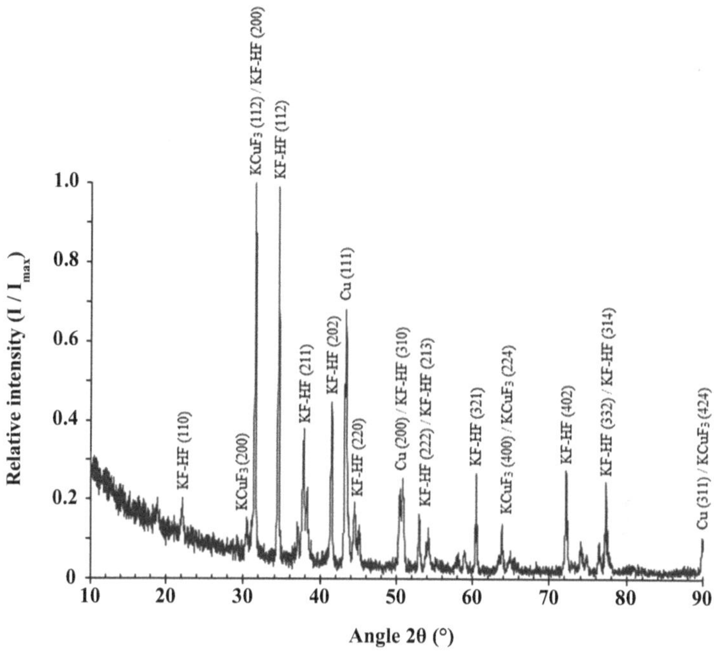

Figure 25.3 X-ray pattern obtained with industrial cathodic deposit ($\lambda_{CuK\alpha} = 0.15418$ nm).

potential were also determined using the Tafel method in a laboratory-scale cell containing a water-free electrolyte, purified by means of a careful preelectrolysis. In the present study, the values of corrosion rates (expressed in millimeters per year) were determined for several HF ratios and temperatures. A typical Tafel curve is given in Fig. 25.5.

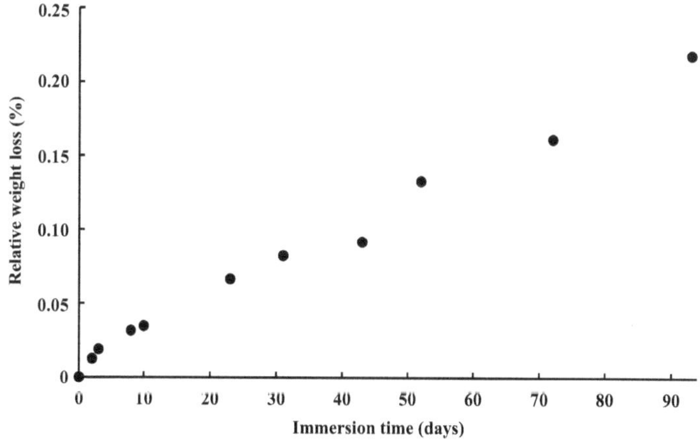

Figure 25.4 Variation of the relative weight of copper as a function of the time immersion in KF-2HF (40-41%m and 95°C).

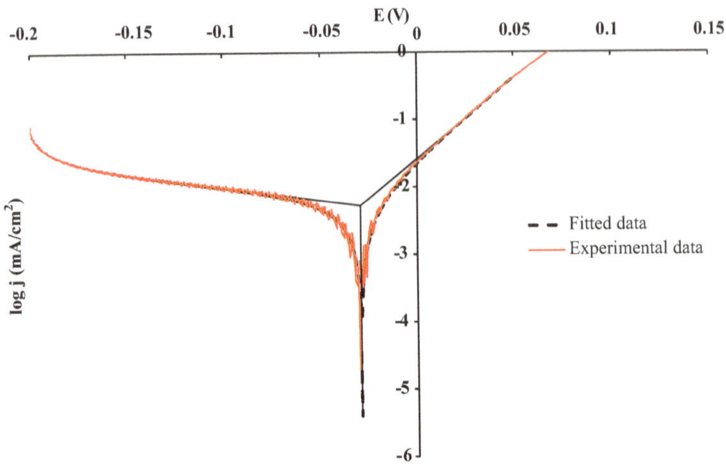

Figure 25.5 Extrapolation of the Tafel's plots for copper (40%m of HF and 95°C).

The experiments were performed in large domains of HF ratio (THF from 38 to 43 w%) and temperature (θ from 85 to 105°C), giving rise to about 30 different tested experimental conditions. From these experimental data, an analytical expression of the corrosion rate of copper was obtained using the Lumiere software. The corrosion rate is expressed as follows:

$$v_{corr} = 0.0746 + 0.0025(\theta - 95.3) - 0.0036(THF - 40.77) - 0.0043(THF - 40.77)^2$$

In Fig. 25.6, the seven curves correspond to seven different values of corrosion rate, from 0.04 mm/year to 0.10 mm/year. Five sets of experimental data (THF, θ) are reported in this figure. We can notice that the corrosion rate v_{corr} has weak values in the whole range of experimental parameters THF and θ.

In our experimental range (θ from 85 to 105°C and THF from 38 to 43 w%), the largest corrosion rate of copper was found to be close to 0.1 mm/year Fig. 25.6 shows that the highest corrosion rate was obtained for HF ratio close to 40.5 w% for a given temperature. The evolution and shape of the calculated curves could be linked to the formation of the same passivation layer on the whole experimental range.[39] One also may notice that very small values of corrosion rate were obtained in the usual industrial operating conditions (<0.1 mm/year). It indicates good stability of copper in KF-xHF melts at OCV. Finally, these low values are in good agreement with those deduced previously by weight measurements.

Physicochemical characterizations were then performed to examine the modifications that could occur at the surface and in the bulk of a copper sample dipped in KF-2HF at OCV. First, cross-sectional images were recorded by SEM analysis realized with the copper plate after 94 days of immersion in KF-2HF at 95°C and without any polarization (Fig. 25.7). To study the possible evolution of the interface without any degradation

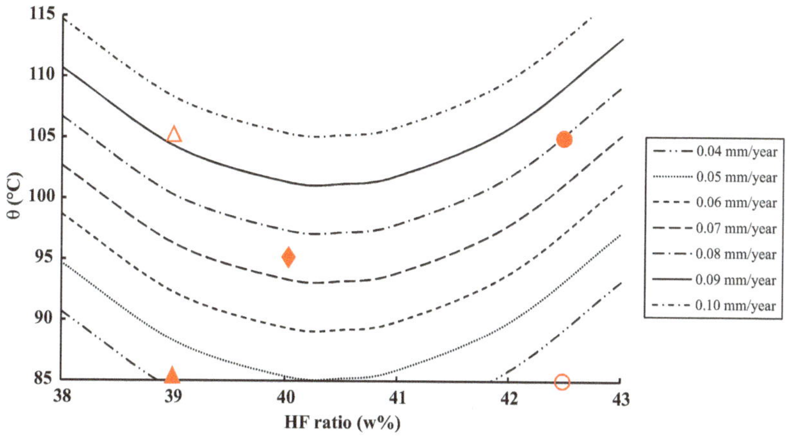

▲ : THF = 39.0 w%, θ = 85°C - v_{corr}: 0.04 mm/year / v_{corr} estimated : 0.04 ± 0.01 mm/year.

△: THF = 39.0 w%, θ = 105°C – v_{corr}: 0.09 mm/year / v_{corr} estimated : 0.09 ± 0.01 mm/year.

◆ : THF = 40.0 w%, θ = 95°C – v_{corr}: 0.07 mm/year / v_{corr} estimated : 0.07 ± 0.01 mm/year.

○: THF = 42.4 w%, θ = 85°C – v_{corr}: 0.03 mm/year / v_{corr} estimated : 0.03 ± 0.01 mm/year.

● : THF = 42.4 w%, θ = 105°C - v_{corr}: 0.08 mm/year / v_{corr} estimated : 0.08 ± 0.01 mm/year.

Figure 25.6 Corrosion rate of copper at open circuit voltage in KF-xHF melts for different values of the temperature θ, and the HF ratio, THF.

of the sample during polishing and washing, the sample was embedded in an epoxy resin and polished. These SEM images were compared with others, obtained with a new copper sample, only polished. For the latter, a nonadherent layer composed of copper oxide was revealed by SEM (Fig. 25.7A and B). The images obtained with the sample in contact with KF-2HF did not reveal the presence of passivating layer on the surface of the metal. Nevertheless, it has been evidenced that the immersion of Cu in KF-2HF induces its degradation due to the presence of many surface defects (Fig. 25.7D).

Considering that rather no passivation layer was detected by SEM during cross-sectional examinations, the surface of the copper sample was also examined carefully.

These analyses have shown that the copper surface is strongly modified by simple contact with KF-2HF. As seen in Fig. 25.8, the surface becomes highly porous. EDX analyses gave the following chemical composition of the surface layer: 93.0 at% of Cu, 1.4 at% of F, and 5.6 at% of O. The presence of oxygen can be attributed to the contamination of the sample during its contact with the atmosphere. The low fluorine amount proves the presence of a very thin fluorinated surface layer on the copper sample, even if it was impossible to detect it by SEM. Nevertheless, due to the very small amount of F within that film, its nature cannot be specified any more.

To summarize the first part, we have shown that when a piece of copper is put without any polarization in a crucible containing KF-2HF, the corrosion of the metal is very limited. Although it was not detected by SEM, the presence of a thin fluorinated surface layer was evidenced by EDX. Finally, the maximal corrosion rate of copper was estimated

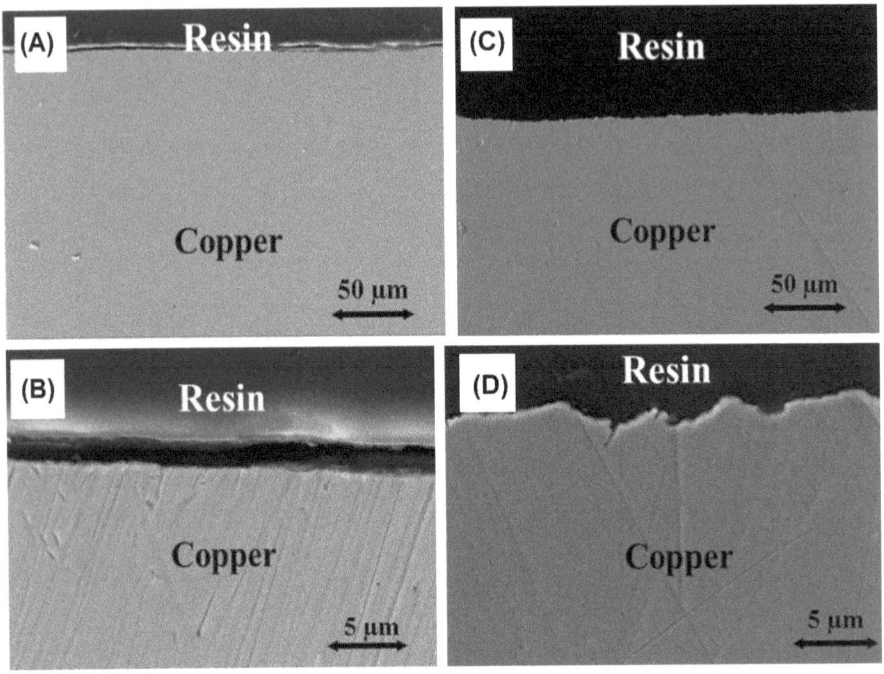

Figure 25.7 SEM cross sections images of copper before (images A (x1000)) and B (x10000)) and after (images C (x1000) and D (x10000)) immersion in KF-2HF) during 94 days.

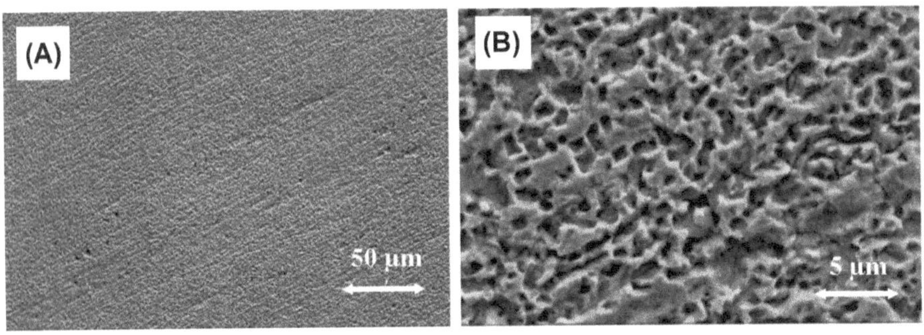

Figure 25.8 (A, B) Scanning electron microscopic images of the copper surface after 94 days in KF-2HF at 95°C and at open circuit voltage.

to be 0.1 mm/year. The ionic copper released in molten salt could be reduced at the cathodes or precipitated. Let us now study the stability of copper polarized at 6 V in KF-2HF, since the copper busbars are polarized at this potential during fluorine production.

3.3 Behavior of Copper Polarized at 6 V in Molten KF-2HF

A similar approach was conducted to study the corrosion of Cu dipped in KF-xHF and polarized at 6 V (vs. Cu/CuF_2). As it was mentioned previously, a large range of molten salt temperature and compositions was tested (θ from 85 to 105°C and THF from 38 to 43 w%). The measurements were performed twice to give a set of 34 accurate experimental data. Multiple linear regression was performed on these measurements to obtain an analytical expression of the corrosion rate of copper under an anodic polarization of 6 V:

$$v_{corr} = 3.25 + 0.18(\theta - 95.92) + 1.90(THF - 40.52) + 0.099(\theta - 95.92)(THF - 40.52) + 0.402(THF - 40.52)^2$$

As discussed previously, the curves presented in Fig. 25.9 correspond to a given corrosion rate.

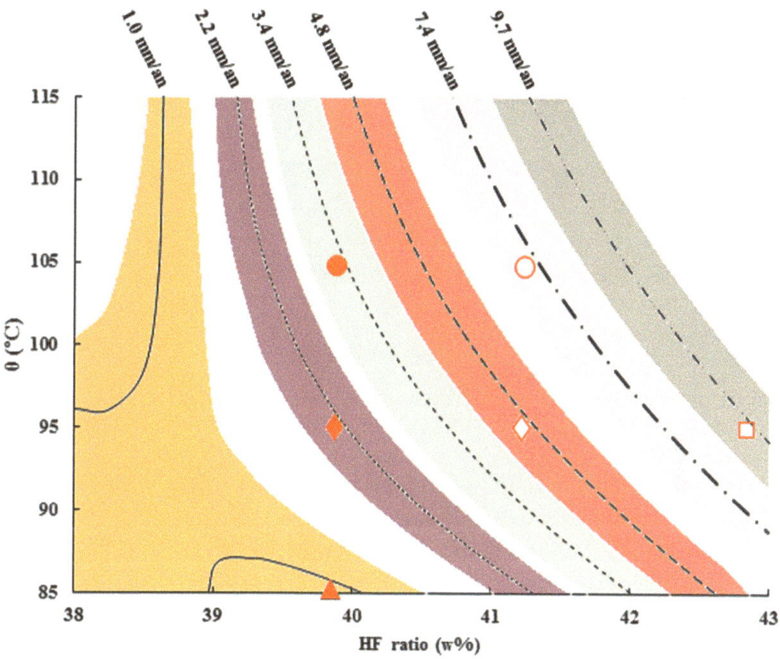

▲ : THF = 39.9 w%, θ = 85°C - v_{corr}: 1.2 mm/year / v_{corr} estimated : 1.0 ± 0.3 mm/year.

♦ : THF = 39.9 w%, θ = 95°C - v_{corr}: 2.0 mm/year / v_{corr} estimated : 2.2 ± 0.4 mm/year.

● : THF = 39.9 w%, θ = 105°C - v_{corr}: 3.2 mm/year / v_{corr} estimated : 3.4 ± 0.6 mm/year.

◊ : THF = 41.3 w%, θ = 95°C - v_{corr}: 5.3 mm/year / v_{corr} estimated : 4.8 ± 0.8 mm/year.

○ : THF = 41.3 w%, θ = 105°C - v_{corr}: 7.9 mm/year / v_{corr} estimated : 7.4 ± 1.0 mm/year.

□ : THF = 42.9 w%, θ = 95°C - v_{corr}: 9.9 mm/year / v_{corr} estimated : 9.7 ± 1.2 mm/year.

Figure 25.9 Corrosion rate of copper polarized at 6 V vs. Cu/CuF_2 in KF-xHF melts for different values of the temperature, θ, and the HF ratio, THF.

As shown in Fig. 25.9, the corrosion rate of copper in KF-xHF under polarization increases with an increase in the temperature or/and the HF ratio. Whatever be the values of θ and THF, the values of v_{corr} were of the same order of magnitude as those obtained by Germanaz et al.[33] The positive value obtained for the quadratic effect of THF ($a_{22} = +0.402$) is also an indication that the higher the HF amount, the higher is the effect of this parameter.

Previous studies have shown that copper under an anodic polarization was covered by a protective layer.[42,46,47] CuF_2 and $KCuF_3$ species have been observed in molten KF-xHF yet.[39,40,42,47] Thus, we performed physicochemical analyses with a copper plate polarized at 6 V in KF-2HF (THF = 40.8 w% and θ = 95°C) for 10 days. As previously described, each sample was embedded in an epoxy resin. Cross-sectional images of passivated Cu (Fig. 25.10) were compared with those obtained with raw Cu (Fig. 25.7A and B) and with Cu in contact with KF-2HF at OCV (Fig. 25.7C and D).

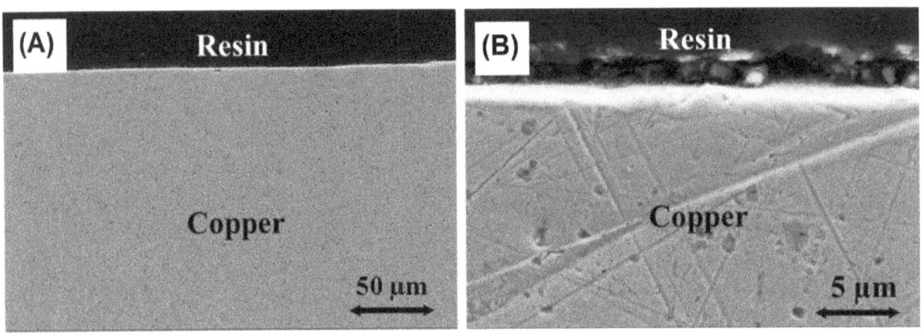

Figure 25.10 (A, B) Scanning electron microscopic cross-sections of a copper electrode after a 10-day polarization at 6 V vs. Cu/CuF_2 in KF-2HF (THF = 40.8 w% and θ = 95°C).

For a better understanding of the evolution of the corrosion rates with increasing HF ratio, three passivation layers were prepared at θ = 95°C for three different values of THF (38.4, 40.5 and 42.0 w%). These three samples were also polarized for 10 days at 6 V. Their surface composition was determined by EPMA. These experiments revealed the existence of two different kinds of copper fluorides: the first one with a very low content of potassium is equivalent to CuF_2 (green spots in Fig. 25.11) whereas the second one, the potassium-rich one, could correspond to $KCuF_3$ (red spots in Fig. 25.11). An average composition could be calculated for each analysis. In the top row, the green spots are related to a compound containing only Cu and F, with an atomic ratio close to 2:1. This result may indicate that a thin layer of CuF_2 is obtained. In the bottom row, the red spots are related to another compound with a quite different composition: the atomic ratios of K, Cu, and F being close to 1:1:3.

According to the images collected in Fig. 25.11, the most uniform coverage of the surface by the fluorinated compounds was observed with the sample prepared with the lowest HF ratio. Indeed, as soon as the THF increases, the surface becomes less homogeneous in terms of composition. For THF = 38.4 w%, the film deposited onto the surface is composed mainly of CuF_2 (in green) and only small spots of $KCuF_3$ (in red) were detected. These experiments have also shown that the higher the HF amount,

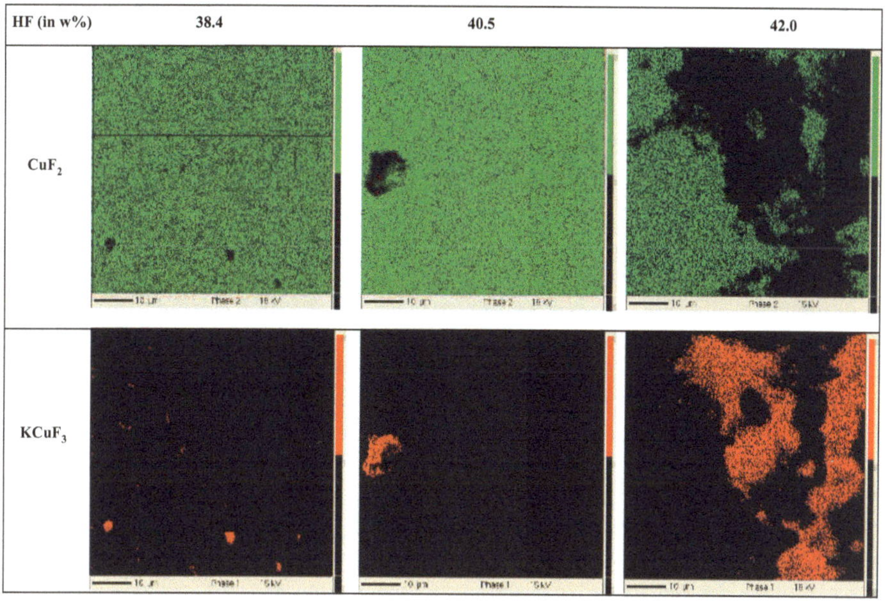

Figure 25.11 Surface composition of copper samples polarized at 6 V during 10 days at 38.4, 40.5 and 42.0 w.% HF.

the more important is the $KCuF_3$ coverage. This could be attributed to the increase in the corrosion rate of copper, which induced a larger concentration of copper near the electrode surface. This local saturation leads to the precipitation of $KCuF_3$ on the electrode surface according to previous studies.[39,40,42]

To conclude, during the fluorine production process, the carbon anodes that are polarized at about 6 V, are screwed in a copper busbar. Thus, it was useful to study the behavior of copper polarized at 6 V for various operating conditions (temperature, HF ratio) in an industrial KF-xHF melt. When a copper plate is in contact with KF-xHF at OCV, first traces of corrosion have been evidenced. The tests performed on copper samples polarized at 6 V have shown that the corrosion rates of copper increase strongly with an increase in the temperature or/and the amount of HF. Analysis by EPMA highlighted that the copper surface is covered by CuF_2 or $KCuF_3$ species. The amount of the latter becomes more important for the highest values of the HF ratio, for which a concomitant increase in the corrosion rate of copper is observed. $KCuF_3$ is probably formed by the precipitation of copper present in the electrolyte and resulting from the corrosion of the copper-based parts of the industrial cells, i.e., mainly the busbars.

3.4 Evolving Fluorine Influence During 6 V Anodic Polarization of Copper in KF-2HF

As discussed earlier, the choice of materials used in electrolyzers plays an important role in minimizing the damages caused by erosion−corrosion. Rough surfaces are generally undesirable since they facilitate this phenomenon. This erosion–corrosion phenomenon leads to the dissolution of the passivating layer, which protects the copper samples. A two-phase mechanism has been usually involved,[48] taking into account the porosity of the passivating layer. Briefly, the first phase is relative to the production of soluble copper species at the passivating Cu-based layer/electrolyte interface. It occurs with three parallel steps: (1) the oxidation of copper at the interface metal/fluoride, (2) the diffusion of soluble species from the metal to the electrolyte through the porous surface layer, and (3) the dissolution of the Cu-F-based film at the interface fluoride/electrolyte. In the second phase, transfer of ions in KF-2HF occurs.

In industrial fluorine electrolyzers, the corrosion rate of copper is larger than that expected from the results presented in the previous sections. Indeed, additional contributions must be taken into account: (1) the impact on the surface of the flow of F_2 bubbles produced by electrolysis and (2) the movement of electrolyte. Therefore, the degradation of copper should not only be due to corrosion under anodic polarization but also be due to a mechanical effect. To confirm this hypothesis, two tests were made on a copper surface at laboratory-scale electrolyzer containing KF-2HF at 95°C. For the first test, a copper rod was screwed onto a carbon log. The copper rod was polarized at 6 V and subject to incipient fluorine produced on carbon (Fig. 25.12A). For the second test, the copper sample at OCV was subject to fluorine bubbling (Fig. 25.12B). On both samples, the effect of the fluorine flow was visible.

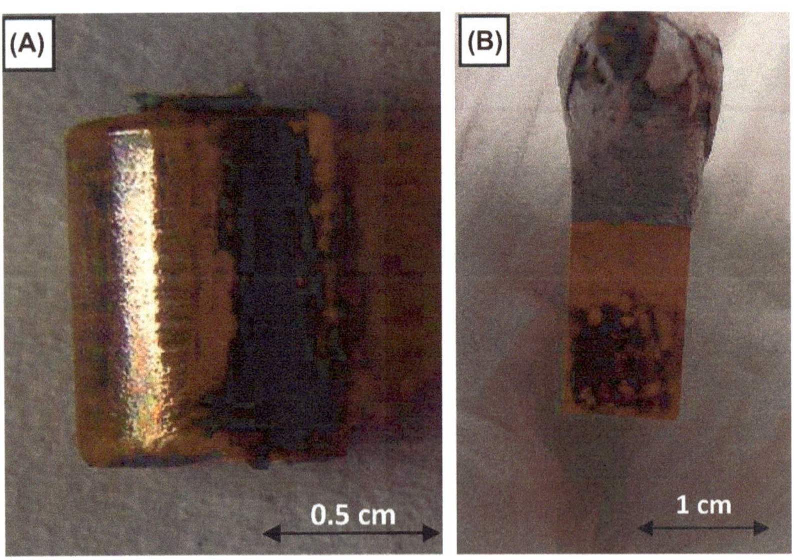

Figure 25.12 Pieces of copper corroded by an erosion-corrosion phenomena (A) 6 V polarized copper in touch with incipient fluorine (B) OCV copper in touch with fluorine.

These experiences have shown that copper was not only corroded but also undergoes erosion by fluorine flow in addition to the electrochemical corrosion occurring under polarization at 6 V. In Fig. 25.12A, the F_2 bubble pathway is visible, whereas in

Table 25.1 Relative weight loss of copper sample polarized at 6 V vs. Cu/CuF_2 with or without a fluorine bubble flow

	Without Bubbling	F_2 Bubbling
Mass loss (g)	0.139 ± 0.001	0.397 ± 0.001
Relative mass loss (%)	−0.8%	−2.2%

Fig. 25.12B, there is a difference between the top (no contact with fluorine) and the bottom of the sample (in contact with fluorine). The erosion−corrosion phenomenon could also be explained by mechanical strengths exerted by molten salt movements inducing particle impacts on the copper surface.[49]

To quantify the erosion phenomena, two identical copper samples were polarized at 6 V vs. Cu/CuF_2 in similar experimental conditions (KF-2HF melt at 95°C). One sample was subject to fluorine bubbling, whereas the other was not.

The mass loss of the samples was recorded after a 15-day experiment and was reported in Table 25.1.

A significant difference of the mass loss is obtained with a factor close to 3. It means that copper is highly sensitive to mechanical effects, inducing enhancement of its degradation. This phenomenon may be explained by the low hardness of this material. Indeed, copper is considered as a soft material (Eq. (25.7)).[50,51]

We can imagine that the mechanical effect of corrosion−erosion leads to the breakdown of the thin CuF_2 passivating layer present on the copper surface. Thus, the bare metal is exposed not only to corrosion by the electrolyte but also to oxidation by elemental fluorine. It results in higher values of the corrosion rate of copper busbars.

4. Conclusions

All these experiments have shown that copper is usually covered by a thin passivating copper fluoride layer that protects it from corrosion. A corrosion mechanism has been proposed in the literature to explain the presence of two different copper fluorides (CuF_2 and $KCuF_3$) when the metal is polarized at 6 V. The precipitation of Cu^{2+} species with H_2F^- from the molten salt is at the origin of $KCuF_3$ formation on the electrode surface.

We have also shown that copper is not only corroded by the molten salt since a simultaneous phenomenon of erosion and corrosion occurs. Erosion is explained by the fluorine flow at the copper surface, which induces not only the impacts of the fluorine bubbles but also movements of the bath. So, mechanical effects take place on the copper surface. In industrial electrolyzers, this effect could be larger thanks to the additional impact of particles contained in the electrolyte. These two phenomena cause the breakdown of the passivation layer and expose the bare copper surface to accelerated corrosion.

Acknowledgments

This study was financially supported both by CNRS and AREVA. The authors express their gratitude to J. Gomes, S. Leclerc, Dr. A. G. Porras Gutierrez (Laboratoire PHENIX, UPMC,

France), and E. Durand (ICMCB, France) for their helpful assistance in experimental works and to Dr. A. Tressaud (ICMCB, France) for fruitful advices and discussions. The authors express their gratitude to S. Borensztajn (IPGP, France) and F. Pillier (UPMC France) for SEM experiments and to M. Lahaye (University of Bordeaux) for EPMA experiments.

References

1. Groult, H.; Lantelme, F.; Devilliers, D.; Belhomme, C.; Morel, B.; Nicolas, F.; Caire, J. P. Le fluor, élément clé pour l'énergie nucléaire. *L'actualité Chim.* **2006,** *301−302,* 40−46.
2. Delpech, S.; Cabet, C.; Slim, C.; Picard, G. Molten Fluorides for Nuclear Applications. *Mater. Today* **2010,** *13* (12), 34−41.
3. Groult, H.; Lantelme, F.; Salanne, M.; Simon, C.; Belhomme, C.; Morel, B.; Nicolas, F. Role of Elemental Fluorine in Nuclear Field. *J. Fluor. Chem.* **2007,** *128,* 285−295.
4. Brown, P. E.; Crabtree, J. M.; Duncan, J. F. The Kinetics of the Reaction of Elementary Fluorine with Copper Metal. *J. Inorg. Nucl. Chem.* **1955,** *1,* 202−212.
5. Kreysa, G.; Schütze, M. *Corrosion Handbook,* Vol. 5; Wiley: New York, 2005.
6. Devilliers, D.; Lantelme, F.; Chemla, M. Study of the Decomposition Voltage of HF on Carbon Electrodes in a 2HF-KF Melt. *J. Chim. Phys.* **1979,** *76,* 428−432.
7. Brown, O. R. The equilibrium e.m.f. of fluorine cells. *Electrochim. Acta* **1980,** *25,* 367−368.
8. Groult, H. Electrochemistry of Fluorine Production. *J. Fluor. Chem.* **2003,** *119,* 173−189.
9. Watanabe, N.; Nakajima, T.; Touhara, H. *Anode Effect in Molten Fluorides* In: *Graphite Fluorides,* Vol. 8; Elsevier: Amsterdam, 1988; pp 1−22. Chapter 1.
10. Crassous, I. *Contribution à la mise au point d'un nouvel électrolyseur pour la production de fluor en milieu sel fondu KF-2HF* (Thesis); P&M Curie University: Paris, 2009.
11. Groult, H.; Simon, C.; Mantoux, A.; Lantelme, F.; Turq, P. Experimental and Theoretical Aspects of the Fluorine Evolution Reaction on Carbon Anode in Molten KF-2HF. In *Fluorinated Materials for Energy Conversion;* Nakajima, T., Groult, H., Eds.; Elsevier, 2005; pp 1−29. Chapter 1.
12. Rudge, A. J. Production of Elemental Fluorine by Electrolysis. In *Industrial Electrochemical Process;* Kuhn, A. T., Ed.; Elsevier: Amsterdam, 1971; pp 1−69. Chapter 1.
13. Groult, H.; Devilliers, D.; Vogler, M. Trends in the Fluorine Preparation Process. *Curr. Top. Electrochem.* **1997,** *4,* 23−40. Research trends, Poojappura, Trivandrum, India.
14. Devilliers, D.; Chemla, M. Carbon Anode Reaction in Fluorine Production. In *Fluorine-Carbon and Fluoride−Carbon Materials;* Nakajima, T., Ed.; Marcel Dekker: New York, 1995; pp 283−331. Chapter 8.
15. Nakajima, T. Synthesis, Structure and Physicochemical Properties of Fluorine-graphite Intercalation Compouds. In *Fluorine-Carbon and Fluoride-Carbon Materials;* Nakajima, T., Ed.; Marcel Dekker: New York, 1995; pp 1−31. Chapter 1.
16. Imoto, H.; Nakajima, T.; Watanabe, N. A Study on Anode Effect in KF-2HF System. *Bull. Chem. Soc. Jpn.* **1975,** *48,* 1633−1634.
17. Bai, L.; Conway, B. E. Electrochemistry of Anodic F_2 Evolution at Carbon Electrodes: Bubble Adherence Effects in the Kinetics at Rotating Cone Electrodes. *J. Appl. Electrochem.* **1988,** *18,* 839−848.
18. Groult, H.; Devilliers, D.; Lantelme, F.; Caire, J. P.; Combel, M.; Nicolas, F. Origin of the Anode Overvoltage Observed during Fluorine Evolution in KF-2HF. *J. Electrochem. Soc.* **2002,** *149,* E485−E492.

19. Groult, H.; Lantelme, F. Study of the Mass Transfer at Fluorine-Evolution Carbon Electrodes. *J. Electrochem. Soc.* **2001,** *148,* E13–E18.
20. Lantelme, F.; Groult, H.; Belhomme, C.; Morel, B.; Nicolas, F. Interfacial Tension and Fluorine Evolution on a Carbon Anode. *J. Fluor. Chem.* **2006,** *127,* 704–707.
21. Groult, H.; Devilliers, D.; Vogler, M.; Hinnen, C.; Marcus, P.; Nicolas, F. Electrochemical Behaviour and Surface Analysis of Crude and Modified Carbon Electrodes for Fluorine Production. *Electrochim. Acta* **1993,** *38,* 2413–2421.
22. Groult, H.; Devilliers, D.; Durand-Vidal, S.; Nicolas, F.; Combel, M. Electronic Properties of Passivating Compounds: Application to the Fluorine Evolution Reaction. *Electrochim. Acta* **1999,** *44,* 2793–2803.
23. Groult, H.; Lantelme, F. Interfacial Properties and Gas Bubble Formation during the Electrolytic Preparation of Fluorine. *J. Electrochem. Soc.* **2004,** *151,* D121–D126.
24. Crassous, I.; Groult, H.; Lantelme, F.; Devilliers, D.; Tressaud, A.; Labrugère, C.; Dubois, M.; Belhomme, C.; Colisson, A.; Morel, B. Study of the Fluorination of Carbon Anode in Molten KF-2HF by XPS and NMR Investigations. *J. Fluor. Chem.* **2009,** *130,* 1080–1085.
25. Devilliers, D. *Carbon Anodes for Fluorine Production* (Thesis); P&M Curie University: Paris, 1984.
26. Bai, L.; Conway, B. E. Electrochemistry of Anodic Fluorine Gas Evolution at Carbon Electrodes: Part II Studies of the 'CF' Film by the Current-Interruption, AC Impedance, ESCA and Auger Techniques. *J. Appl. Electrochem.* **1990,** *20,* 916–924.
27. Cadman, P.; Scott, J. D.; Thomas, J. M. The Fluorination of the Surfaces of Elemental Carbon,. *Carbon* **1977,** *15,* 75–86.
28. Roustan, H. *Modélisation numérique des transferts couples au sein d'un électrolyseur industriel* (Thesis); INPG Grenoble, 1998.
29. Roustan, H.; Caire, J. P.; Nicolas, F.; Pham, P. Modelling Coupled Transfer in an Industrial Fluorine Electrolyser. *J. Appl. Electrochem.* **1998,** *28,* 237–243.
30. Espinasse, G.; Peyrard, M.; Nicolas, F.; Caire, J. P. Effect of Hydrodynamics on Faradaic Current Efficiency in a Fluorine Electrolyser. *J. Appl. Electrochem.* **2007,** *37,* 77–85.
31. Pretorius, R. *A Transient Computational Fluid Dynamic Study of a Laboratory-scale Fluorine Electrolysis Cell* (Thesis); University of Pretoria: South Africa, 2011.
32. Pretorius, R.; Crouse, P. L.; Hattingh, C. J. A Multiphysics Simulation of a Fluorine Electrolysis Cell. *S. Afr. J. Sci.* **2015,** *111,* 7–8.
33. Germanaz, P.; Nicolas, F.; Rouquette, S.; Lamirault, S.; Ferry, D.; Picard, G. Etude électrochimique de la corrosion du cuivre et du magnésium dans les mélanges fondus HF-KF: influence de leurs teneurs en HF. *Mater. Sci. Eng.* **1989,** *A120,* 329–339.
34. Qian, S. Y.; Conway, B. E. On the Origins of Cathode Hyperpolarization Effects in Electrolytic Fluorine Production from KF-2HF Melts. *J. Appl. Electrochem.* **1994,** *24,* 195–200.
35. Dumont, H.; Qian, S. Y.; Conway, B. E. Passivation and Depassivation of Monel Electrodes in KF-2HF Melts in Comparison with the Behaviour of Nickel. *J. Appl. Electrochem.* **1997,** *27,* 1245–1253.
36. Dumont, H.; Qian, S. Y.; Conway, B. E. Dissolution and Passivation Processes in the Corrosion of Copper and Nickel in KF-2HF at 85°C. *J. Appl. Electrochem.* **1997,** *27,* 267–279.
37. Mimoto, A.; Miyazaki, T.; Yamashita, J.; Nagamine, S.; Inaba, M.; Tasaka, A. Measurement and Thermodynamic Analysis of M/MF_n (M = Cu and Fe) Electrode Potentials in a Few Fluoride Melts Containing HF. *J. Electrochem. Soc.* **2006,** *153,* D149–D154.
38. Burrows, B.; Jasinski, R. The Cu/CuF_2 Couple in Anhydrous Hydrogen Fluoride. *J. Electrochem. Soc.* **1968,** *115,* 348–354.

39. Lamirault, S. *Comportement du cuivre dans les mélanges HF-MF (M = K ou NH_4) fondus utilisés pour l'obtention électrolytique du fluor* (Thesis); P&M Curie University: Paris, 1987.
40. Binet, D. *Développement de matériaux innovants pour la préparation électrolytique du fluor en milieu fondu KF-2HF* (Thesis); P&M Curie University: Paris, 2015.
41. Filliaudeau, D.; Picard, G. Investigation of the Electrochemical Oxidation of Copper in Molten NH_4HF_2-HF. *Electrochim. Acta* **1993,** *38* (14), 1951–1955.
42. Beccu, K. D.; Giess, H.; Porta, A. An Inorganic Coating Process Providing Copper with an Electrically Insulating Layer by Anodisation in Non-aqueous Fluoride Electrolytes. *Electrodep. Surf. Treat.* **1972,** *1,* 253–265.
43. Fourie, E. *Establishing a Process to Reduce, Recycle and Reuse the Waste Electrolyte from Fluorine Generation* (Thesis); University of the Witwatersrand: Johannesburg, South Africa, 2004.
44. Conway, B. E.; Qian, S. Y. Unusual Kinetic Behaviour in Cathodic H_2 Evolution from KF-2HF Melts at Mild-Steel and Alloy Electrodes. *Chem. Ind.* **2004,** *58,* 259–270.
45. Vukasin, J.; Crassous, I.; Morel, B.; Sanchez-Marcano, J.; Namy, P. Transient Model of a Fluorine Electrolysis Cell. In *COMSOL Conference, Grenooble;* 2015.
46. Germanaz, P.; Lamirault, S.; Picard, G. Procédé d'anodisation passivante du cuivre en milieu fluorures fondus. Application à la protection de pièces en cuivre des électrolyseurs fluor. *Eur. Pat.* **1989**. EP 0 321 536 A1.
47. Hackerman, N.; Snavely, E. D.; Fiel, L. D. The Anodic Polarization Behaviour of Metals in Hydrogen Fluoride. *Corros. Sci.* **1967,** *7,* 39–50.
48. Landolt, D. Corrosion et chimie de surface des métaux. In *Traité des matériaux*; Vol. 12; Presses polytechniques et universitaires romandes, 2003, .
49. Postlethwaite, J.; Nesic, S. Erosion-Corrosion in Single and Multiphase Flow. In *Uhlig's Corrosion Handbook;* Revie, W., Ed.; Electrochem. Soc. Series: New York, 2000; pp 249–271. Chapter 15.
50. Béranger, G.; Mazille, H. *Corrosion des métaux et alliages;* Lavoisier: Paris, 2002.
51. Gale, W. F.; Totemeier, T. C. *Smithells Metals Reference Book,* 8th ed.; Elsevier: Amsterdam, 2004. Chapter 22.

Index

'*Note:* Page numbers followed by "f" indicate figures, "t" indicate tables.'

D

E

I

O

P

Q

R

FSC
www.fsc.org
MIX
Papier | Fördert
gute Waldnutzung
FSC® C083411

Kolibri 360 GmbH
Ferdinand-Jühlke-Straße 7
99095 Erfurt, Deutschland
produktsicherheit@kolibri360.de